Volume 3

Encyclopedia of Energy Engineering *and* Technology

Encyclopedias from Taylor & Francis Group

Agriculture Titles

Dekker Agropedia Collection (Eight Volume Set)
ISBN: 0-8247-2194-2   13-Digit ISBN: 978-0-8247-2194-7

Encyclopedia of Agricultural, Food, and Biological Engineering
Edited by Dennis R. Heldman
ISBN: 0-8247-0938-1   13-Digit ISBN: 978-0-8247-0938-9

Encyclopedia of Animal Science
Edited by Wilson G. Pond and Alan Bell
ISBN: 0-8247-5496-4   13-Digit ISBN: 978-0-8247-5496-9

Encyclopedia of Pest Management
Edited by David Pimentel
ISBN: 0-8247-0632-3   13-Digit ISBN: 978-0-8247-0632-6

Encyclopedia of Pest Management, Volume II
Edited by David Pimentel
ISBN: 1-4200-5361-2   13-Digit ISBN: 978-1-4200-5361-6

Encyclopedia of Plant and Crop Science
Edited by Robert M. Goodman
ISBN: 0-8247-0944-6   13-Digit ISBN: 978-0-8247-0944-0

Encyclopedia of Soil Science, Second Edition (Two Volume Set)
Edited by Rattan Lal
ISBN: 0-8493-3830-1   13-Digit ISBN: 978-0-8493-3830-4

Encyclopedia of Water Science
Edited by B.A. Stewart and Terry Howell
ISBN: 0-8247-0948-9   13-Digit ISBN: 978-0-8247-0948-8

Chemistry Titles

Encyclopedia of Chromatography, Second Edition (Two Volume Set)
Edited by Jack Cazes
ISBN: 0-8247-2785-1   13-Digit ISBN: 978-0-8247-2785-7

Encyclopedia of Surface and Colloid Science, Second Edition (Eight Volume Set)
Edited by P. Somasundaran
ISBN: 0-8493-9615-8   13-Digit ISBN: 978-0-8493-9615-1

Encyclopedia of Supramolecular Chemistry (Two Volume Set)
Edited by Jerry L. Atwood and Jonathan W. Steed
ISBN: 0-8247-5056-X   13-Digit ISBN: 978-0-8247-5056-5

Engineering Titles

Encyclopedia of Chemical Processing (Five Volume Set)
Edited by Sunggyu Lee
ISBN: 0-8247-5563-4   13-Digit ISBN: 978-0-8247-5563-8

Encyclopedia of Corrosion Technology, Second Edition
Edited by Philip A. Schweitzer, P.E.
ISBN: 0-8247-4878-6   13-Digit ISBN: 978-0-8247-4878-4

Dekker Encyclopedia of Nanoscience and Nanotechnology (Five Volume Set)
Edited by James A. Schwarz, Cristian I. Contescu, and Karol Putyera
ISBN: 0-8247-5055-1   13-Digit ISBN: 978-0-8247-5055-8

Encyclopedia of Optical Engineering (Three Volume Set)
Edited by Ronald G. Driggers
ISBN: 0-8247-0940-3   13-Digit ISBN: 978-0-8247-0940-2

Business Titles

Encyclopedia of Library and Information Science, Second Edition (Four Volume Set)
Edited by Miriam Drake
ISBN: 0-8247-2075-X   13-Digit ISBN: 978-0-8247-2075-9

Encyclopedia of Public Administration and Public Policy (Two Volume Set)
Edited by Jack Rabin
ISBN: 0-8247-4748-8   13-Digit ISBN: 978-0-8247-4748-0

Coming Soon

Encyclopedia of Public Administration and Public Policy, Second Edition
Edited by Jack Rabin
ISBN: 1-4200-5275-6   13-Digit ISBN: 978-1-4200-5275-6

Encyclopedia of Water Science, Second Edition (Two Volume Set)
Edited by Stanley W. Trimble
ISBN: 0-8493-9627-1   13-Digit ISBN: 978-0-8493-9627-4

Encyclopedia of Wireless and Mobile Communications (Three Volume Set)
Edited by Borko Furht
ISBN: 1-4200-4326-9   13-Digit ISBN: 978-1-4200-4326-6

These titles are available both in print and online. To order, visit:
www.crcpress.com
Telephone: 1-800-272-7737
Fax: 1-800-374-3401
E-Mail: orders@crcpress.com

Volume 3

Photovoltaic Systems / 1147
Physics of Energy / 1160
Pricing Programs: Time-of-Use and Real Time / 1175
Psychrometrics / 1184
Public Policy for Improving Energy Sector Performance / 1193
Public Utility Regulatory Policies Act (PURPA) of 1978 / 1201
Pumped Storage Hydroelectricity / 1207
Pumps and Fans / 1213
Radiant Barriers / 1227
Reciprocating Engines: Diesel and Gas / 1233
Regulation: Price Cap and Revenue Cap / 1245
Regulation: Rate of Return / 1252
Renewable and Decentralized Energy Options: State Promotion in the U.S. / 1258
Renewable Energy / 1265
Residential Buildings: Heating Loads / 1272
Run-Around Heat Recovery Systems / 1278
Savings and Optimization: Case Studies / 1283
Savings and Optimization: Chemical Process Industry / 1302
Six Sigma Methods: Measurement and Verification / 1310
Solar Heating and Air Conditioning: Case Study / 1317
Solar Thermal Technologies / 1321
Solar Water Heating: Domestic and Industrial Applications / 1331
Solid Waste to Energy by Advanced Thermal Technologies (SWEATT) / 1340
Space Heating / 1357
Steam and Hot Water System Optimization: Case Study / 1366
Steam Turbines / 1380
Sustainability Policies: Sunbelt Cities / 1389
Sustainable Building Simulation / 1396
Sustainable Development / 1406
Thermal Energy Storage / 1412

Volume 3 (Continued)

Thermodynamics / 1422
Tradable Certificates for Energy Savings / 1433
Transportation Systems: Hydrogen-Fueled / 1441
Transportation: Location Efficiency / 1449
Tribal Land and Energy Efficiency / 1456
Underfloor Air Distribution (UFAD) / 1463
Utilities and Energy Suppliers: Bill Analysis / 1471
Utilities and Energy Suppliers: Business Partnership Management / 1484
Utilities and Energy Suppliers: Planning and Portfolio Management / 1491
Utilities and Energy Suppliers: Rate Structures / 1497
Walls and Windows / 1513
Waste Fuels / 1523
Waste Heat Recovery / 1536
Waste Heat Recovery Applications: Absorption Heat Pumps / 1541
Water and Wastewater Plants: Energy Use / 1548
Water and Wastewater Utilities / 1556
Water Source Heat Pump for Modular Classrooms / 1564
Water-Augmented Gas Turbine Power Cycles / 1574
Water-Using Equipment: Commercial and Industrial / 1587
Water-Using Equipment: Domestic / 1596
Wind Power / 1607
Window Energy / 1616
Window Films: Savings from IPMVP Options C and D / 1626
Window Films: Solar-Control and Insulating / 1630
Window Films: Spectrally Selective versus Conventional Applied / 1637
Windows: Shading Devices / 1641
Wireless Applications: Energy Information and Control / 1649
Wireless Applications: Mobile Thermostat Climate Control and Energy Conservation / 1660

Volume 3

Encyclopedia of Energy Engineering *and* Technology

Edited by

Barney L. Capehart
University of Florida
Gainesville, USA

CRC Press is an imprint of the
Taylor & Francis Group, an informa business

CRC Press
Taylor & Francis Group
6000 Broken Sound Parkway NW, Suite 300
Boca Raton, FL 33487-2742

© 2007 by Taylor & Francis Group, LLC, except as noted on the opening page of the entry. All rights reserved.
CRC Press is an imprint of Taylor & Francis Group, an Informa business

No claim to original U.S. Government works
Printed in the United States of America on acid-free paper
10 9 8 7 6 5 4 3 2 1

International Standard Book Number-13: 978-0-8493-3653-9 (Hardcover)

This book contains information obtained from authentic and highly regarded sources. Reprinted material is quoted with permission, and sources are indicated. A wide variety of references are listed. Reasonable efforts have been made to publish reliable data and information, but the author and the publisher cannot assume responsibility for the validity of all materials or for the consequences of their use.

No part of this book may be reprinted, reproduced, transmitted, or utilized in any form by any electronic, mechanical, or other means, now known or hereafter invented, including photocopying, microfilming, and recording, or in any information storage or retrieval system, without written permission from the publishers.

For permission to photocopy or use material electronically from this work, please access www.copyright.com (http://www.copyright.com/) or contact the Copyright Clearance Center, Inc. (CCC) 222 Rosewood Drive, Danvers, MA 01923, 978-750-8400. CCC is a not-for-profit organization that provides licenses and registration for a variety of users. For organizations that have been granted a photocopy license by the CCC, a separate system of payment has been arranged.

Trademark Notice: Product or corporate names may be trademarks or registered trademarks, and are used only for identification and explanation without intent to infringe.

Library of Congress Cataloging-in-Publication Data

Encyclopedia of energy engineering and technology / edited by Barney L. Capehart.
    p. cm.
    Includes bibliographical references and index.
    ISBN-13: 978-0-8493-3653-9 (alk. paper : set)
    ISBN-10: 0-8493-3653-8 (alk. paper : set)
    ISBN-13: 978-0-8493-5039-9 (alk. paper : v. 1)
    ISBN-10: 0-8493-5039-5 (alk. paper : v. 1)
    [etc.]
    1. Power resources--Encyclopedias. 2. Power (Mechanics)--Encyclopedias. I. Capehart, B. L. (Barney L.)

TJ163.2.E385 2007
621.04203--dc22
    2007007878

Visit the Taylor & Francis Web site at
http://www.taylorandfrancis.com

and the CRC Press Web site at
http://www.crcpress.com

*In memory of my mother and father
and
for my grandchildren Hannah and Easton.
May their energy future be efficient and sustainable.*

Editorial Advisory Board

Barney L. Capehart, Editor

*Department of Industrial and Systems Engineering,
University of Florida, Gainesville, Florida, U.S.A.*

Mr. Larry B. Barrett
President, Barrett Consulting Associates, Colorado Springs, Colorado, U.S.A.

Dr. Sanford V. Berg
Former Director, Public Utility Research Center, University of Florida, Gainesville, Florida, U.S.A.

Dr. David L. Block, Director Emeritus
Florida Solar Energy Center, Cocoa, Florida, U.S.A.

Dr. Marilyn A. Brown
Director, Energy Efficiency, Reliability, and Security Program, Oak Ridge National Laboratory, Oak Ridge, Tennessee, U.S.A.

Dr. R. Neal Elliott, P.E.
Director of Industrial Energy Programs, American Council for an Energy Efficient Economy, Washington, D.C., U.S.A.

Mr. Clark W. Gellings
Vice President of Innovation, Electric Power Research Institute (EPRI), Palo Alto, California, U.S.A.

Dr. David Goldstein
Head of Energy Efficiency Section, Natural Resources Defense Council, San Francisco, California, U.S.A.

Dr. Alex E. S. Green
Professor Emeritus, College of Engineering, University of Florida, Gainesville, Florida, U.S.A.

Dr. David L. Greene
Corporate Fellow, Oak Ridge National Laboratory, Knoxville, Tennessee, U.S.A.

Dr. Jay Hakes
Director, Jimmy Carter Presidential Library, and Former Director, US Energy Information Agency, Atlanta, Georgia, U.S.A.

Dr. Richard F. Hirsh
Professor of History of Technology, Virginia Polytechnic University, Blacksburg, Virginia, U.S.A.

Mr. Ronald E. Jarnagin
Research Scientist, Pacific Northwest National Laboratory, Richland, Washington, U.S.A.

Dr. Xianguo Li, P.Eng.
Editor, International Journal of Green Energy, and Professor, Department of Mechanical Engineering, University of Waterloo, Waterloo, Ontario, Canada

Mr. Steve Nadel
Executive Director, American Council for an Energy Efficient Environment (ACEEE), Washington, D.C., U.S.A.

Mr. Dan Parker
Principal Research Scientist, Florida Solar Energy Center, Cocoa, Florida, U.S.A.

Mr. Steven A. Parker
Senior Engineer and Director of New Technology Applications, Pacific Northwest National Laboratory, Richland, Washington, U.S.A.

Mr. F. William Payne
Editor Emeritus, Strategic Planning for Energy and the Environment, Marble Hill, Georgia, U.S.A.

Mr. Neil Petchers
President, NORESCO, Westborough, Massachusetts, U.S.A.

Dr. Stephen A. Roosa, CEM, CIAQP, CMVP, CBEP, CDSM, MBA
Performance Engineer, Energy Systems Group, Inc., Louisville, Kentucky, U.S.A.

Dr. Wayne H. Smith
Interim Dean for Research and Interim Director, Florida Agricultural Experiment Station, University of Florida, Gainesville, Florida, U.S.A.

Prof. James S. Tulenko
Laboratory for Development of Advanced Nuclear Fuels and Materials, University of Florida, Gainesville, Florida, U.S.A.

Dr. Dan Turner
Director, Energy Systems Laboratory, Texas A&M University, College Station, Texas, U.S.A.

Dr. Wayne C. Turner
Editor, Energy Management Handbook, and Regents Professor, Industrial Engineering and Management, Oklahoma State University, Stillwater, Oklahoma, U.S.A.

Dr. Jerry Ventre
Former Research Scientist, Florida Solar Energy Center, Oviedo, Florida, U.S.A.

Dr. Richard Wakefield
Vice President, Transmission & Regulatory Services, KEMA, Inc., Fairfax, Virginia, U.S.A.

Contributors

Bill Allemon / *North American Energy Efficiency, Ford Land, Dearborn, Michigan, U.S.A.*

Paul J. Allen / *Reedy Creek Energy Services, Walt Disney World Co., Lake Buena Vista, Florida, U.S.A.*

Fatouh Al-Ragom / *Building and Energy Technologies Department, Kuwait Institute for Scientific Research, Safat, Kuwait*

Kalyan Annamalai / *Department of Mechanical Engineering, College Station, Texas, U.S.A.*

John Archibald / *American Solar Inc., Annandale, Virginia, U.S.A.*

Senthil Arumugam / *Enerquip, Inc., Medford, Wisconsin, U.S.A.*

M. Asif / *School of the Built and Natural Environment, Glasgow Caledonian University, Glasgow, Scotland, U.K.*

Essam Omar Assem / *Arab Fund for Economic and Social Development, Arab Organizations Headquarters Building, Shuwaikh, Kuwait*

U. Atikol / *Energy Research Center, Eastern Mediterranean University, Magusa, Northern Cyprus*

Rick Avery / *Bay Controls, LLC, Maumee, Ohio, U.S.A.*

Lu Aye / *International Technologies Centre (IDTC), Department of Civil and Environmental Engineering, The University of Melbourne, Victoria, Australia*

M. Babcock / *EMO Energy Solutions, LLC, Vienna, Virginia, U.S.A.*

Christopher G. J. Baker / *Chemical Engineering Department, Kuwait University, Safat, Kuwait*

K. Baker / *EMO Energy Solutions, LLC, Vienna, Virginia, U.S.A.*

Peter Barhydt / *Arctic Combustion Ltd., Mississauga, Ontario, Canada*

Larry B. Barrett / *Barrett Consulting Associates, Colorado Springs, Colorado, U.S.A.*

Rosemarie Bartlett / *Pacific Northwest National Laboratory, Richland, Washington, U.S.A.*

Fred Bauman / *Center for the Built Environment, University of California, Berkeley, California, U.S.A.*

David Beal / *Florida Solar Energy Center, Cocoa, Florida, U.S.A.*

Sanford V. Berg / *Director of Water Studies, Public Utility Research Center, University of Florida, Gainesville, Florida, U.S.A.*

Paolo Bertoldi / *European Commission, Directorate General JRC, Ispra (VA), Italy*

Asfaw Beyene / *Department of Mechanical Engineering, San Diego State University, San Diego, California, U.S.A.*

Ujjwal Bhattacharjee / *Energy Engineering Program, University of Massachusetts—Lowell, Lowell, Massachusetts, U.S.A.*

David Bisbee / *Customer Advanced Technologies Program, Sacramento Municipal Utility District (SMUD), Sacramento, California, U.S.A.*

John O. Blackburn / *Department of Economics, Duke University, Durham, North Carolina, U.S.A.*

Richard F. Bonskowski / *Office of Coal, Nuclear, Electric and Alternate Fuels, Energy Information Administration, U.S. Department of Energy, Washington, D.C., U.S.A.*

Robert G. Botelho / *Energy Management International, Seekonk, Massachussetts, U.S.A.*

Mark T. Brown / *Department of Environmental Engineering Sciences, University of Florida, Gainesville, Florida, U.S.A.*

Michael L. Brown / *Energy and Environmental Management Center, Georgia Institute of Technology, Savannah, Georgia, U.S.A.*

Alexander L. Burd / *Advanced Research Technology, Suffield, Connecticut, U.S.A.*

Galina S. Burd / *Advanced Research Technology, Suffield, Connecticut, U.S.A.*

James Call / *James Call Engineering, PLLC, Larchmont, New York, U.S.A.*

Norm Campbell / *Energy Systems Group, Newburgh, Indiana, U.S.A.*

Barney L. Capehart / *Department of Industrial and Systems Engineering, University of Florida College of Engineering, Gainesville, Florida, U.S.A.*

Lynne C. Capehart / *Consultant and Technical Writing Specialist, Gainesville, Florida, U.S.A.*

Cristián Cárdenas-Lailhacar / *Department of Industrial and Systems Engineering, University of Florida, Gainesville, Florida, U.S.A.*

Jack Casazza / *American Education Institute, Springfield, Virginia, U.S.A.*

Yunus A. Cengel / *Department of Mechanical Engineering, University of Nevada, Reno, Nevada, U.S.A.*

Guangnan Chen / *Faculty of Engineering and Surveying, University of Southern Queensland, Toowoomba, Queensland, Australia*

David E. Claridge / *Department of Mechanical Engineering, and Associate Director, Energy Systems Laboratory, Texas A & M University, College Station, Texas, U.S.A.*

James A. Clark / *Clark Energy, Inc., Broomall, Pennsylvania, U.S.A.*

Gregory Cmar / *Interval Data Systems, Inc., Watertown, Massachusetts, U.S.A.*

Kevin C. Coates / *Transportation & Energy Policy, Coates Consult, Bethesda, Maryland, U.S.A.*

Bruce K. Colburn / *EPS Capital Corp., Villanova, Pennsylvania, U.S.A.*

Stephany L. Cull / *RetroCom Energy Strategies, Inc., Elk Grove, California, U.S.A.*

Rick Cumbo / *Operations and Maintenance Manager, Armstrong Service, Inc., Orlando, Florida, U.S.A.*

Leonard A. Damiano / *EBTRON, Inc., Loris, South Carolina, U.S.A.*

James W. Dean / *Office of Strategic Analysis and Governmental Affairs, Florida Public Service Commission, Tallahassee, Florida, U.S.A.*

Steve DeBusk / *Energy Solutions Manager, CPFilms Inc., Martinsville, Virginia, U.S.A.*

Harvey E. Diamond / *Energy Management International, Conroe, Texas, U.S.A.*

Craig DiLouie / *Lighting Controls Association, National Electrical Manufacturers Association, Rosslyn, Virginia, U.S.A.*

Ibrahim Dincer / *Faculty of Engineering and Applied Science, University of Ontario Institute of Technology (UOIT), Oshawa, Ontario, Canada*

Michael R. Dorrington / *Cameron's Compression Systems Division, Houston, Texas, U.S.A.*

Steve Doty / *Colorado Springs Utilities, Colorado Springs, Colorado, U.S.A.*

David S. Dougan / *EBTRON, Inc., Loris, South Carolina, U.S.A.*

John Duffy / *Department of Mechanical Engineering, University of Massachussetts—Lowell, Lowell, Massachussetts, U.S.A.*

Philip Fairey / *Florida Solar Energy Center, Cocoa, Florida, U.S.A.*

Carla Fair-Wright / *Maintenance Technology Services, Cameron's Compression Systems Division, Houston, Texas, U.S.A.*

Ahmad Faruqui / *The Brattle Group, San Francisco, California, U.S.A.*

J. Michael Fernandes / *Mobile Platforms Group, Intel Corporation, Santa Clara, California, U.S.A.*

Steven Ferrey / *Suffolk University Law School, Boston, Massachusetts, U.S.A.*

Andrew R. Forrest / *Government Communications Systems Division, Harris Corporation, Melbourne, Florida, U.S.A.*

R. Scot Foss / *IR Air Solutions, Davidson, North Carolina, U.S.A.*

Jackie L. Francke / *Geotechnika, Inc., Longmont, Colorado, U.S.A.*

Fred Freme / *Office of Coal, Nuclear, Electric and Alternate Fuels, Energy Information Administration, U.S. Department of Energy, Washington, D.C., U.S.A.*

Dwight K. French / *Energy Consumption Division, Energy Information Administration, U.S. Department of Energy, Washington, D.C., U.S.A.*

Kevin Fuller / *Interval Data Systems, Inc., Watertown, Massachusetts, U.S.A.*

Kamiel S. Gabriel / *University of Ontario Institute of Technology, Oshawa, Ontario, Canada*

Clark W. Gellings / *Electric Power Research Institute (EPRI), Palo Alto, California, U.S.A.*

Geoffrey J. Gilg / *Pepco Energy Services, Inc., Arlington, Virginia, U.S.A.*

Bill Gnerre / *Interval Data Systems, Inc., Watertown, Massachusetts, U.S.A.*

Fredric S. Goldner / *Energy Management and Research Associates, East Meadow, New York, U.S.A.*

Dan Golomb / *Department of Environmental, Earth and Atmospheric Sciences, University of Massachusetts—Lowell, Lowell, Massachusetts, U.S.A.*

John Van Gorp / *Power Monitoring and Control, Schneider Electric, Saanichton, British Columbia, Canada*

Alex E. S. Green / *Green Liquids and Gas Technologies, Gainesville, Florida, U.S.A.*

David C. Green / *Green Management Services, Inc., Fort Myers, Florida, U.S.A.*

L. J. Grobler / *School for Mechanical Engineering, North-West University, Potchefstroom, South Africa*

H. M. Güven / *Energy Research Center, Eastern Mediterranean University, Magusa, Northern Cyprus*

Mark A. Halverson / *Pacific Northwest National Laboratory, West Halifax, Vermont, U.S.A.*

Shirley J. Hansen / *Hansen Associates, Inc., Gig Harbor, Washington, U.S.A.*

Susanna S. Hanson / *Global Applied Systems, Trane Commercial Systems, La Crosse, Wisconsin, U.S.A.*

Glenn C. Haynes / *RLW Analytics, Middletown, Connecticut, U.S.A.*

Warren M. Heffington / *Industrial Assessment Center, Department of Mechanical Engineering, Texas A&M University, College Station, Texas, U.S.A.*

Arif Hepbasli / *Department of Mechanical Engineering, Faculty of Engineering, Ege University, Izmir, Turkey*

Keith E. Herold / *Department of Mechanical Engineering, University of Maryland, College Park, Maryland, U.S.A.*

James G. Hewlett / *Energy Information Administration, U.S. Department of Energy, Washington, D.C., U.S.A.*

Richard F. Hirsh / *Department of History, Virginia Tech, Blacksburg, Virginia, U.S.A.*

Kevin Hoag / *Engine Research Center, University of Wisconsin-Madison, Madison, Wisconsin, U.S.A.*

Nathan E. Hultman / *Science, Technology, and International Affairs, Georgetown University, Washington, D.C., U.S.A.*

W. David Hunt / *Pacific Northwest National Laboratory, Washington, D.C., U.S.A.*

H. A. Ingley III / *Department of Mechanical Engineering, University of Florida, Gainesville, Florida, U.S.A.*

Nevena H. Iordanova / *Senior Utility Systems Engineer, Armstrong Service, Inc., Orlando, Florida, U.S.A.*

Charles P. Ivey / *Aiken Global Environmental/FT Benning DPW, Fort Benning, Georgia, U.S.A.*

Mark A. Jamison / *Public Utility Research Center, University of Florida, Gainesville, Florida, U.S.A.*

Somchai Jiajitsawat / *Department of Mechanical Engineering, University of Massachussetts—Lowell, Lowell, Massachusetts, U.S.A.*

Gholamreza Karimi / *Department of Mechanical Engineering, University of Waterloo, Waterloo, Ontario, Canada*

Janey Kaster / *Yamas Controls, Inc., South San Francisco, California, U.S.A.*

Sila Kiliccote / *Commercial Building Systems Group, Lawrence Berkeley National Laboratory, Berkeley, California, U.S.A.*

Birol I. Kilkis / *Green Energy Systems, International LLC, Vienna, Virginia, U.S.A.*

Ronald L. Klaus / *VAST Power Systems, Elkhart, Indiana, U.S.A.*

Milivoje M. Kostic / *Department of Mechanical Engineering, Northern Illinois University, DeKalb, Illinois, U.S.A.*

Athula Kulatunga / *Department of Electrical and Computer Engineering Technology, Purdue University, West Lafayette, Indiana, U.S.A.*

James W. Leach / *Industrial Assessment Center, Mechanical and Aerospace Engineering, North Carolina State University, Raleigh, North Carolina, U.S.A.*

Larry Leetzow / *World Institute of Lighting and Development Corp./Magnaray International Division, Sarasota, Florida, U.S.A.*

Jim Lewis / *Obvius LLC, Hillsboro, Oregon, U.S.A.*

Xianguo Li / *Department of Mechanical Engineering, University of Waterloo, Waterloo, Ontario, Canada*

Todd A. Litman / *Victoria Transport Policy Institute, Victoria, British Columbia, Canada*

Mingsheng Liu / Architectural Engineering Program, Director, Energy Systems Laboratory, Peter Kiewit Institute, University of Nebraska—Lincoln, Omaha, Nebraska, U.S.A.

Robert Bruce Lung / Resource Dynamics Corporation, Vienna, Virginia, U.S.A.

Alfred J. Lutz / AJL Resources, LLC, Philadelphia, Pennsylvania, U.S.A.

David MacPhaul / CH2M HILL, Gainesville, Florida, U.S.A.

Richard J. Marceau / University of Ontario Institute of Technology, Oshawa, Ontario, Canada

Rudolf Marloth / Department of Mechanical Engineering, Loyola Marymount University, Los Angeles, California, U.S.A.

Sandra B. McCardell / Current-C Energy Systems, Inc., Mills, Wyoming, U.S.A.

William Ross McCluney / Florida Solar Energy Center, University of Central Florida, Cocoa, Florida, U.S.A.

Donald T. McGillis / Pointe Claire, Québec, Canada

Aimee McKane / Lawrence Berkeley National Laboratory, Washington, D.C., U.S.A.

D. Paul Mehta / Department of Mechanical Engineering, Bradley University, Peoria, Illinois, U.S.A.

Zohrab Melikyan / HVAC Department, Armenia State University of Architecture and Construction, Yerevan, Armenia

Ingrid Melody / Florida Solar Energy Center, Cocoa, Florida, U.S.A.

Mark Menezes / Emerson Process Management (Rosemount), Mississauga, Ontario, Canada

Adnan Midilli / Department of Mechanical Engineering, Faculty of Engineering, Nigde University, Nigde, Turkey

Ruth Mossad / Faculty of Engineering and Surveying, University of Southern Queensland, Toowoomba, Queensland, Australia

Naoya Motegi / Commercial Building Systems Group, Lawrence Berkeley National Laboratory, Berkeley, California, U.S.A.

Alex V. Moultanovsky / Automotive Climate Control, Inc., Elkhart, Indiana, U.S.A.

Martin A. Mozzo / M and A Associates, Inc., Robbinsville, New Jersey, U.S.A.

T. Muneer / School of Engineering, Napier University, Edinburgh, Scotland, U.K.

Greg F. Naterer / University of Ontario Institute of Technology, Oshawa, Ontario, Canada

Robert R. Nordhaus / Van Ness Feldman, PC, Washington, D.C., U.S.A.

Michael M. Ohadi / The Petroleum Institute, Abu Dhabi, United Arab Emirates

K. E. Ohrn / Cypress Digital Ltd., Vancouver, British Columbia, Canada

Svetlana J. Olbina / Rinker School of Building Construction, University of Florida, Gainesville, Florida, U.S.A.

Eric Oliver / EMO Energy Solutions, LLC, Vienna, Virginia, U.S.A.

Mitchell Olszewski / Oak Ridge National Laboratory, Oak Ridge, Tennessee, U.S.A.

Bohdan W. Oppenheim / U.S. Department of Energy Industrial Assessment Center, Loyola Marymount University, Los Angeles, U.S.A.

Leyla Ozgener / Department of Mechanical Engineering, Celal Bayar University, Manisa, Turkey

Graham Parker / Battelle Pacific Northwest National Laboratory, U.S. Department of Energy, Richland, Washington, U.S.A.

Steven A. Parker / Pacific Northwest National Laboratory, Richland, Washington, U.S.A.

Ken E. Patterson II / Advanced Energy Innovations—Bringing Innovation to Industry, Menifee, California, U.S.A.

Klaus-Dieter E. Pawlik / Accenture, St Petersburg, Florida, U.S.A.

David E. Perkins / Active Power, Inc., Austin, Texas, U.S.A.

Jeffery P. Perl / Chicago Chem Consultants Corp., Chicago, Illinois, U.S.A.

Neil Petchers / NORESCO Energy Consulting, Shelton, Connecticut, U.S.A.

Eric Peterschmidt / Honeywell Building Solutions, Honeywell, Golden Valley, Minnesota, U.S.A.

Robert W. Peters / Department of Civil, Construction & Environmental Engineering, University of Alabama at Birmingham, Birmingham, Alabama, U.S.A.

Mark A. Peterson / *Sustainable Success LLC, Clementon, New Jersey, U.S.A.*

Mary Ann Piette / *Commercial Building Systems Group, Lawrence Berkeley National Laboratory, Berkeley, California, U.S.A.*

Wendell A. Porter / *University of Florida, Gainesville, Florida, U.S.A.*

Soyuz Priyadarsan / *Texas A&M University, College Station, Texas, U.S.A.*

Sam Prudhomme / *Bay Controls, LLC, Maumee, Ohio, U.S.A.*

Jianwei Qi / *Department of Mechanical Engineering, University of Maryland, College Park, Maryland, U.S.A.*

Ashok D. Rao / *Advanced Power and Energy Program, University of California, Irvine, California, U.S.A.*

Ab Ream / *U.S. Department of Energy, Washington, D.C., U.S.A.*

Rolf D. Reitz / *Engine Research Center, University of Wisconsin-Madison, Madison, Wisconsin, U.S.A.*

Rich Remke / *Commercial Systems and Services, Carrier Corporation, Syracuse, New York, U.S.A.*

Silvia Rezessy / *Energy Efficiency Advisory, REEEP International Secretariat, Vienna International Centre, Austria, Environmental Sciences and Policy Department, Central European University, Nador, Hungary*

James P. Riordan / *Energy Department, DMJM + Harris / AECOM, New York, U.S.A.*

Vernon P. Roan / *University of Florida, Gainesville, Florida, U.S.A.*

Stephen A. Roosa / *Energy Systems Group, Inc., Louisville, Kentucky, U.S.A.*

Michael Ropp / *South Dakota State University, Brookings, South Dakota, U.S.A.*

Marc A. Rosen / *Faculty of Engineering and Applied Science, University of Ontario Institute of Technology, Oshawa, Ontario, Canada*

Christopher Russell / *Energy Pathfinder Management Consulting, LLC, Baltimore, Maryland, U.S.A.*

Hemmat Safwat / *Project Development Power and Desalination Plants, Consolidated Contractors International Company, Athens, Greece*

Abdou-R. Sana / *Montreal, Québec, Canada*

Diane Schaub / *Industrial and Systems Engineering, University of Florida, Gainesville, Florida, U.S.A.*

Diana L. Shankle / *Battelle Pacific Northwest National Laboratory, U.S. Department of Energy, Richland, Washington, U.S.A.*

S. A. Sherif / *Department of Mechanical and Aerospace Engineering, Wayne K. and Lyla L. Masur HVAC Laboratory, University of Florida, Gainesville, Florida, U.S.A.*

Brian Silvetti / *CALMAC Manufacturing Corporation, Fair Lawn, New Jersey, U.S.A.*

Carey J. Simonson / *Department of Mechanical Engineering, University of Saskatchewan, Saskatoon, Saskatchewan, Canada*

Amy Solana / *Pacific Northwest National Laboratory, Portland, Oregon, U.S.A.*

Paul M. Sotkiewicz / *Public Utility Research Center, University of Florida, Warrington College of Business, Gainesville, Florida, U.S.A.*

Amanda Staudt / *Board on Atmospheric Sciences and Climate, The National Academies, Washington, D.C., U.S.A.*

Nick Stecky / *NJS Associates, Denville, New Jersey, U.S.A.*

Kate McMordie Stoughton / *Battelle Pacific Northwest National Laboratory, U.S. Department of Energy, Richland, Washington, U.S.A.*

Therese Stovall / *Oak Ridge National Laboratory, Oak Ridge, Tennessee, U.S.A.*

Gregory P. Sullivan / *Battelle Pacific Northwest National Laboratory, U.S. Department of Energy, Richland, Washington, U.S.A.*

John M. Sweeten / *Texas Agricultural Experiment Station, Amarillo, Texas, U.S.A.*

Thomas F. Taranto / *ConservAIR Technologies, LLP, Baldwinsville, New York, U.S.A.*

Michael Taylor / *Honeywell Building Solutions, Honeywell, St. Louis Park, Minnesota, U.S.A.*

Albert Thumann / *Association of Energy Engineers, Atlanta, Georgia, U.S.A.*

Robert J. Tidona / *RMT, Incorporated, Plymouth Meeting, Pennsylvania, U.S.A.*

Jill S. Tietjen / *Technically Speaking, Inc., Greenwood Village, Colorado, U.S.A.*

Lorrie B. Tietze / *Interface Consulting, LLC, Castle Rock, Colorado, U.S.A.*

Greg Tinkler / *RLB Consulting Engineers, Houston, Texas, U.S.A.*

Alberto Traverso / *Dipartimento di Macchine Sistemi Energetici e Trasporti, Thermochemical Power Group, Università di Genova, Genova, Italy*

Douglas E. Tripp / *Canadian Institute for Energy Training, Rockwood, Ontario, Canada*

James S. Tulenko / *Laboratory for Development of Advanced Nuclear Fuels and Materials, University of Florida, Gainesville, Florida, U.S.A.*

W. D. Turner / *Director, Energy Systems Laboratory, Texas A & M University, College Station, Texas, U.S.A.*

Wayne C. Turner / *Industrial Engineering and Management, Oklahoma State University, Stillwater, Oklahoma, U.S.A.*

S. Kay Tuttle / *Process Heating, Boilers and Foodservices, Duke Energy, Charlotte, North Carolina, U.S.A.*

Robert E. Uhrig / *University of Tennessee, Knoxville, Tennessee, U.S.A.*

Sergio Ulgiati / *Department of Sciences for the Environment, Parthenope, University of Naples, Napoli, Italy*

Francisco L. Valentine / *Premier Energy Services, LLC, Columbia, Maryland, U.S.A.*

Ing. Jesús Mario Vignolo / *Instituto Ingeniería Eléctrica, Universidad de la República, Montevideo, Uruguay*

Paul L. Villeneuve / *University of Maine, Orono, Maine, U.S.*

T. G. Vorster / *School for Mechanical Engineering, North-West University, Potchefstroom, South Africa*

James P. Waltz / *Energy Resource Associates, Inc., Livermore, California, U.S.A.*

Devra Bachrach Wang / *Natural Resources Defense Council, San Francisco, California, U.S.A.*

John W. Wang / *Department of Mechanical Engineering, University of Massachussetts—Lowell, Lowell, Massachussetts, U.S.A.*

David S. Watson / *Commercial Building Systems Group, Lawrence Berkeley National Laboratory, Berkeley, California, U.S.A.*

William D. Watson / *Office of Coal, Nuclear, Electric and Alternate Fuels, Energy Information Administration, U.S. Department of Energy, Washington, D.C., U.S.A.*

Marty Watts / *V-Kool, Inc., Houston, Texas, U.S.A.*

Tom Webster / *Center for the Built Environment, University of California, Berkeley, California, U.S.A.*

Michael K. West / *Building Systems Scientists, Advantek Consulting, Inc., Melbourne, Florida, U.S.A.*

Robert E. Wilson / *ConservAIR Technologies Co., LLP, Kenosha, Wisconsin, U.S.A.*

Michael Burke Wood / *Ministry of Energy, Al Dasmah, Kuwait*

Kurt E. Yeager / *Galvin Electricity Initiative, Palo Alto, California, U.S.A.*

Contents

Topical Table of Contents	xxi
Foreword	xxix
Preface	xxxi
Common Energy Abbreviations	xxxiii
Thermal Metric and Other Conversion Factors	xxxvii
About the Editor	xxxix

Volume 1

Accounting: Facility Energy Use / *Douglas E. Tripp*	1
Air Emissions Reductions from Energy Efficiency / *Bruce K. Colburn*	9
Air Quality: Indoor Environment and Energy Efficiency / *Shirley J. Hansen*	18
Aircraft Energy Use / *K. E. Ohrn*	24
Alternative Energy Technologies: Price Effects / *Michael M. Ohadi and Jianwei Qi*	31
ANSI/ASHRAE Standard 62.1-2004 / *Leonard A. Damiano and David S. Dougan*	50
Auditing: Facility Energy Use / *Warren M. Heffington*	63
Auditing: Improved Accuracy / *Barney L. Capehart and Lynne C. Capehart*	69
Auditing: User-Friendly Reports / *Lynne C. Capehart*	76
Benefit Cost Analysis / *Fatouh Al-Ragom*	81
Biomass / *Alberto Traverso*	86
Boilers and Boiler Control Systems / *Eric Peterschmidt and Michael Taylor*	93
Building Automation Systems (BAS): Direct Digital Control / *Paul J. Allen and Rich Remke*	104
Building Geometry: Energy Use Effect / *Geoffrey J. Gilg and Francisco L. Valentine*	111
Building System Simulation / *Essam Omar Assem*	116
Carbon Sequestration / *Nathan E. Hultman*	125
Career Advancement and Assessment in Energy Engineering / *Albert Thumann*	131
Climate Policy: International / *Nathan E. Hultman*	137
Coal Production in the U.S. / *Richard F. Bonskowski, Fred Freme, and William D. Watson*	143
Coal Supply in the U.S. / *Jill S. Tietjen*	156
Coal-to-Liquid Fuels / *Graham Parker*	163
Cold Air Retrofit: Case Study / *James P. Waltz*	172
Combined Heat and Power (CHP): Integration with Industrial Processes / *James A. Clark*	175
Commissioning: Existing Buildings / *David E. Claridge, Mingsheng Liu, and W. D. Turner*	179
Commissioning: New Buildings / *Janey Kaster*	188
Commissioning: Retrocommissioning / *Stephany L. Cull*	200
Compressed Air Control Systems / *Bill Allemon, Rick Avery, and Sam Prudhomme*	207
Compressed Air Energy Storage (CAES) / *David E. Perkins*	214
Compressed Air Leak Detection and Repair / *Robert E. Wilson*	219

Compressed Air Storage and Distribution / *Thomas F. Taranto*	226
Compressed Air Systems / *Diane Schaub*	236
Compressed Air Systems: Optimization / *R. Scot Foss*	241
Cooling Towers / *Ruth Mossad*	246
Data Collection: Preparing Energy Managers and Technicians / *Athula Kulatunga*	255
Daylighting / *William Ross McCluney*	264
Demand Response: Commercial Building Strategies / *David S. Watson, Sila Kiliccote, Naoya Motegi, and Mary Ann Piette*	270
Demand Response: Load Response Resources and Programs / *Larry B. Barrett*	279
Demand-Side Management Programs / *Clark W. Gellings*	286
Desiccant Dehumidification: Case Study / *Michael K. West and Glenn C. Haynes*	292
Distributed Generation / *Paul M. Sotkiewicz and Ing. Jesús Mario Vignolo*	296
Distributed Generation: Combined Heat and Power / *Barney L. Capehart, D. Paul Mehta, and Wayne C. Turner*	303
District Cooling Systems / *Susanna S. Hanson*	309
District Energy Systems / *Ibrahim Dincer and Arif Hepbasli*	316
Drying Operations: Agricultural and Forestry Products / *Guangnan Chen*	332
Drying Operations: Industrial / *Christopher G. J. Baker*	338
Electric Motors / *H. A. Ingley III*	349
Electric Power Transmission Systems / *Jack Casazza*	356
Electric Power Transmission Systems: Asymmetric Operation / *Richard J. Marceau, Abdou-R. Sana, and Donald T. McGillis*	364
Electric Supply System: Generation / *Jill S. Tietjen*	374
Electricity Deregulation for Customers / *Norm Campbell*	379
Electricity Enterprise: U.S., Past and Present / *Kurt E. Yeager*	387
Electricity Enterprise: U.S., Prospects / *Kurt E. Yeager*	399
Electronic Control Systems: Basic / *Eric Peterschmidt and Michael Taylor*	407
Emergy Accounting / *Mark T. Brown and Sergio Ulgiati*	420
Emissions Trading / *Paul M. Sotkiewicz*	430
Energy Codes and Standards: Facilities / *Rosemarie Bartlett, Mark A. Halverson, and Diana L. Shankle*	438
Energy Conservation / *Ibrahim Dincer and Adnan Midilli*	444
Energy Conservation: Industrial Processes / *Harvey E. Diamond*	458
Energy Conservation: Lean Manufacturing / *Bohdan W. Oppenheim*	467
Energy Conversion: Principles for Coal, Animal Waste, and Biomass Fuels / *Kalyan Annamalai, Soyuz Priyadarsan, Senthil Arumugam, and John M. Sweeten*	476
Energy Efficiency: Developing Countries / *U. Atikol and H. M. Güven*	498
Energy Efficiency: Information Sources for New and Emerging Technologies / *Steven A. Parker*	507
Energy Efficiency: Low Cost Improvements / *James Call*	517
Energy Efficiency: Strategic Facility Guidelines / *Steve Doty*	524
Energy Information Systems / *Paul J. Allen and David C. Green*	535
Energy Management: Organizational Aptitude Self-Test / *Christopher Russell*	543
Energy Master Planning / *Fredric S. Goldner*	549
Energy Project Management / *Lorrie B. Tietze and Sandra B. McCardell*	556
Energy Service Companies: Europe / *Silvia Rezessy and Paolo Bertoldi*	566

Volume 2

Energy Star® Portfolio Manager and Building Labeling Program / *Bill Allemon*	573
Energy Use: U.S. Overview / *Dwight K. French*	580
Energy Use: U.S. Transportation / *Kevin C. Coates*	588
Energy: Global and Historical Background / *Milivoje M. Kostic*	601
Enterprise Energy Management Systems / *Gregory Cmar, Bill Gnerre, and Kevin Fuller*	616
Environmental Policy / *Sanford V. Berg*	625
Evaporative Cooling / *David Bisbee*	633
Exergy: Analysis / *Marc A. Rosen*	645
Exergy: Environmental Impact Assessment Applications / *Marc A. Rosen*	655
Facility Air Leakage / *Wendell A. Porter*	662
Facility Energy Efficiency and Controls: Automobile Technology Applications / *Barney L. Capehart and Lynne C. Capehart*	671
Facility Energy Use: Analysis / *Klaus-Dieter E. Pawlik, Lynne C. Capehart, and Barney L. Capehart*	680
Facility Energy Use: Benchmarking / *John Van Gorp*	690
Facility Power Distribution Systems / *Paul L. Villeneuve*	699
Federal Energy Management Program (FEMP): Operations and Maintenance Best Practices Guide (O&M BPG) / *Gregory P. Sullivan, W. David Hunt, and Ab Ream*	707
Fossil Fuel Combustion: Air Pollution and Global Warming / *Dan Golomb*	715
Fuel Cells: Intermediate and High Temperature / *Xianguo Li and Gholamreza Karimi*	726
Fuel Cells: Low Temperature / *Xianguo Li*	733
Geothermal Energy Resources / *Ibrahim Dincer, Arif Hepbasli, and Leyla Ozgener*	744
Geothermal Heat Pump Systems / *Greg Tinkler*	753
Global Climate Change / *Amanda Staudt and Nathan E. Hultman*	760
Green Energy / *Ibrahim Dincer and Adnan Midilli*	771
Greenhouse Gas Emissions: Gasoline, Hybrid-Electric, and Hydrogen-Fueled Vehicles / *Robert E. Uhrig*	787
Heat and Energy Wheels / *Carey J. Simonson*	794
Heat Exchangers and Heat Pipes / *Greg F. Naterer*	801
Heat Pipe Application / *Somchai Jiajitsawat, John Duffy, and John W. Wang*	807
Heat Pumps / *Lu Aye*	814
Heat Transfer / *Yunus A. Cengel*	822
High Intensity Discharge (HID) Electronic Lighting / *Ken E. Patterson II*	830
HVAC Systems: Humid Climates / *David MacPhaul*	839
Hybrid-Electric Vehicles: Plug-In Configuration / *Robert E. Uhrig and Vernon P. Roan*	847
Independent Power Producers / *Hemmat Safwat*	854
Industrial Classification and Energy Efficiency / *Asfaw Beyene*	862
Industrial Energy Management: Global Trends / *Cristián Cárdenas-Lailhacar*	872
Industrial Motor System Optimization Projects in the U.S. / *Robert Bruce Lung, Aimee McKane, and Mitchell Olszewski*	883
Insulation: Facilities / *Wendell A. Porter*	890
Integrated Gasification Combined Cycle (IGCC): Coal- and Biomass-Based / *Ashok D. Rao*	906
IntelliGrid℠ / *Clark W. Gellings and Kurt E. Yeager*	914
International Performance Measurement and Verification Protocol (IPMVP) / *James P. Waltz*	919
Investment Analysis Techniques / *James W. Dean*	930

LEED-CI and LEED-CS: Leadership in Energy and Environmental Design for Commercial Interiors and Core and Shell / *Nick Stecky*	937
LEED-NC: Leadership in Energy and Environmental Design for New Construction / *Stephen A. Roosa*	947
Life Cycle Costing: Electric Power Projects / *Ujjwal Bhattacharjee*	953
Life Cycle Costing: Energy Projects / *Sandra B. McCardell*	967
Lighting Controls / *Craig DiLouie*	977
Lighting Design and Retrofits / *Larry Leetzow*	993
Liquefied Natural Gas (LNG) / *Charles P. Ivey*	1001
Living Standards and Culture: Energy Impact / *Marc A. Rosen*	1009
Maglev (Magnetic Levitation) / *Kevin C. Coates*	1018
Management Systems for Energy / *Michael L. Brown*	1027
Manufacturing Industry: Activity-Based Costing / *J. Michael Fernandes, Barney L. Capehart, and Lynne C. Capehart*	1042
Measurement and Verification / *Stephen A. Roosa*	1050
Measurements in Energy Management: Best Practices and Software Tools / *Peter Barhydt and Mark Menezes*	1056
Mobile HVAC Systems: Fundamentals, Design, and Innovations / *Alex V. Moultanovsky*	1061
Mobile HVAC Systems: Physics and Configuration / *Alex V. Moultanovsky*	1070
National Energy Act of 1978 / *Robert R. Nordhaus*	1082
Natural Energy versus Additional Energy / *Marc A. Rosen*	1088
Net Metering / *Steven Ferrey*	1096
Nuclear Energy: Economics / *James G. Hewlett*	1101
Nuclear Energy: Fuel Cycles / *James S. Tulenko*	1111
Nuclear Energy: Power Plants / *Michael Burke Wood*	1115
Nuclear Energy: Technology / *Michael Burke Wood*	1125
Performance Contracting / *Shirley J. Hansen*	1134
Performance Indicators: Industrial Energy / *Harvey E. Diamond and Robert G. Botelho*	1140

Volume 3

Photovoltaic Systems / *Michael Ropp*	1147
Physics of Energy / *Milivoje M. Kostic*	1160
Pricing Programs: Time-of-Use and Real Time / *Ahmad Faruqui*	1175
Psychrometrics / *S. A. Sherif*	1184
Public Policy for Improving Energy Sector Performance / *Sanford V. Berg*	1193
Public Utility Regulatory Policies Act (PURPA) of 1978 / *Richard F. Hirsh*	1201
Pumped Storage Hydroelectricity / *Jill S. Tietjen*	1207
Pumps and Fans / *L. J. Grobler and T. G. Vorster*	1213
Radiant Barriers / *Ingrid Melody, Philip Fairey, and David Beal*	1227
Reciprocating Engines: Diesel and Gas / *Rolf D. Reitz and Kevin Hoag*	1233
Regulation: Price Cap and Revenue Cap / *Mark A. Jamison*	1245
Regulation: Rate of Return / *Mark A. Jamison*	1252
Renewable and Decentralized Energy Options: State Promotion in the U.S. / *Steven Ferrey*	1258
Renewable Energy / *John O. Blackburn*	1265
Residential Buildings: Heating Loads / *Zohrab Melikyan*	1272
Run-Around Heat Recovery Systems / *Kamiel S. Gabriel*	1278
Savings and Optimization: Case Studies / *Robert W. Peters and Jeffery P. Perl*	1283

Savings and Optimization: Chemical Process Industry / *Jeffery P. Perl and Robert W. Peters* 1302

Six Sigma Methods: Measurement and Verification / *Carla Fair-Wright and Michael R. Dorrington* 1310

Solar Heating and Air Conditioning: Case Study / *Eric Oliver, K. Baker, M. Babcock, and John Archibald* 1317

Solar Thermal Technologies / *M. Asif and T. Muneer* 1321

Solar Water Heating: Domestic and Industrial Applications / *M. Asif and T. Muneer* 1331

Solid Waste to Energy by Advanced Thermal Technologies (SWEATT) / *Alex E. S. Green* 1340

Space Heating / *James P. Riordan* 1357

Steam and Hot Water System Optimization: Case Study / *Nevena Iordanova and Rick Cumbo* 1366

Steam Turbines / *Michael Burke Wood* 1380

Sustainability Policies: Sunbelt Cities / *Stephen A. Roosa* 1389

Sustainable Building Simulation / *Birol I. Kilkis* 1396

Sustainable Development / *Mark A. Peterson* 1406

Thermal Energy Storage / *Brian Silvetti* 1412

Thermodynamics / *Ronald L. Klaus* 1422

Tradable Certificates for Energy Savings / *Silvia Rezessy and Paolo Bertoldi* 1433

Transportation Systems: Hydrogen-Fueled / *Robert E. Uhrig* 1441

Transportation: Location Efficiency / *Todd A. Litman* 1449

Tribal Land and Energy Efficiency / *Jackie L. Francke and Sandra B. McCardell* 1456

Underfloor Air Distribution (UFAD) / *Tom Webster and Fred Bauman* 1463

Utilities and Energy Suppliers: Bill Analysis / *Neil Petchers* 1471

Utilities and Energy Suppliers: Business Partnership Management / *S. Kay Tuttle* 1484

Utilities and Energy Suppliers: Planning and Portfolio Management / *Devra Bachrach Wang* 1491

Utilities and Energy Suppliers: Rate Structures / *Neil Petchers* 1497

Walls and Windows / *Therese Stovall* 1513

Waste Fuels / *Robert J. Tidona* 1523

Waste Heat Recovery / *Martin A. Mozzo* 1536

Waste Heat Recovery Applications: Absorption Heat Pumps / *Keith E. Herold* 1541

Water and Wastewater Plants: Energy Use / *Alfred J. Lutz* 1548

Water and Wastewater Utilities / *Rudolf Marloth* 1556

Water Source Heat Pump for Modular Classrooms / *Andrew R. Forrest and James W. Leach* 1564

Water-Augmented Gas Turbine Power Cycles / *Ronald L. Klaus* 1574

Water-Using Equipment: Commercial and Industrial / *Kate McMordie Stoughton and Amy Solana* 1587

Water-Using Equipment: Domestic / *Amy Solana and Kate McMordie Stoughton* 1596

Wind Power / *K. E. Ohrn* 1607

Window Energy / *William Ross McCluney* 1616

Window Films: Savings from IPMVP Options C and D / *Steve DeBusk* 1626

Window Films: Solar-Control and Insulating / *Steve DeBusk* 1630

Window Films: Spectrally Selective versus Conventional Applied / *Marty Watts* 1637

Windows: Shading Devices / *Svetlana J. Olbina* 1641

Wireless Applications: Energy Information and Control / *Jim Lewis* 1649

Wireless Applications: Mobile Thermostat Climate Control and Energy Conservation / *Alexander L. Burd and Galina S. Burd* 1660

Topical Table of Contents

I. Energy, Energy Sources, and Energy Use

Energy: Historical and Technical Background

Energy Conservation / *Ibrahim Dincer and Adnan Midilli*	444
Energy Efficiency: Developing Countries / *U. Atikol and H. M. Güven*	498
Energy Use: U.S. Overview / *Dwight K. French*	580
Energy Use: U.S. Transportation / *Kevin C. Coates*	588
Energy: Global and Historical Background / *Milivoje M. Kostic*	601
Living Standards and Culture: Energy Impact / *Marc A. Rosen*	1009
National Energy Act of 1978 / *Robert R. Nordhaus*	1082
Natural Energy versus Additional Energy / *Marc A. Rosen*	1088
Physics of Energy / *Milivoje M. Kostic*	1160
Public Policy for Improving Energy Sector Performance / *Sanford V. Berg*	1193
Public Utility Regulatory Policies Act (PURPA) of 1978 / *Richard Hirsh*	1201

Fossil Fuels

Coal Production in the U.S. / *Richard F. Bonskowski, Fred Freme, and William D. Watson*	143

Nuclear Energy

Nuclear Energy: Fuel Cycles / *James S. Tulenko*	1111
Nuclear Energy: Power Plants / *Michael Burke Wood*	1115
Nuclear Energy: Technology / *Michael Burke Wood*	1125

Renewable Energy

Alternative Energy Technologies: Price Effects / *Michael M. Ohadi and Jianwei Qi*	31
Biomass / *Alberto Traverso*	86
Geothermal Energy Resources / *Ibrahim Dincer, Arif Hepbasli, and Leyla Ozgener*	744
Green Energy / *Ibrahim Dincer and Adnan Midilli*	771
Photovoltaic Systems / *Michael Ropp*	1147
Renewable and Decentralized Energy Options: State Promotion in the U.S. / *Steven Ferrey*	1258
Renewable Energy / *John O. Blackburn*	1265
Solar Thermal Technologies / *M. Asif and T. Muneer*	1321
Solid Waste to Energy by Advanced Thermal Technologies (SWEATT) / *Alex E. S. Green*	1340
Wind Power / *K. E. Ohrn*	1607

Energy Storage and Derived Energy
Derived Energy

Coal-to-Liquid Fuels / *Graham Parker*	163
Transportation Systems: Hydrogen-Fueled / *Robert E. Uhrig*	1441

I. Energy, Energy Sources, and Energy Use (*cont'd.*)

Derived Energy (cont'd.)

Energy Storage

Compressed Air Energy Storage (CAES) / *David E. Perkins*	214
Pumped Storage Hydroelectricity / *Jill S. Tietjen*	1207

Fuel Cells

Fuel Cells: Intermediate and High Temperature / *Xianguo Li and Gholamreza Karimi*	733
Fuel Cells: Low Temperature / *Xianguo Li*	744

II. Principles of Energy Analysis and Systems/Economic Analysis of Energy Systems

Energy Systems Analysis

Accounting: Facility Energy Use / *Douglas E. Tripp*	1
Auditing: Facility Energy Use / *Warren M. Heffington*	63
Building System Simulation / *Essam Omar Assem*	116
Emergy Accounting / *Mark T. Brown and Sergio Ulgiati*	420
Exergy: Analysis / *Marc A. Rosen*	645
Facility Energy Use: Analysis / *Klaus-Dieter E. Pawlik, Lynne C. Capehart, and Barney L. Capehart*	680

Financial and Economic Analysis Principles

Benefit Cost Analysis / *Fatouh Al-Ragom*	81
Investment Analysis Techniques / *James W. Dean*	930
Life Cycle Costing: Electric Power Projects / *Ujjwal Bhattacharjee*	953
Life Cycle Costing: Energy Projects / *Sandra B. McCardell*	967

Principles of Electric Energy Systems

Electric Power Transmission Systems: Asymmetric Operation / *Richard J. Marceau, Abdou-R. Sana, and Donald T. McGillis*	364
Facility Power Distribution Systems / *Paul L. Villeneuve*	699

Principles of Thermal Energy Systems

Energy Conversion: Principles for Coal, Animal Waste, and Biomass Fuels / *Kalyan Annamalai, Soyuz Priyadarsan, Senthil Arumugam, and John M. Sweeten*	476
Heat Transfer / *Yunus A. Cengel*	822
Psychrometrics / *S. A. Sherif*	1184
Thermodynamics / *Ronald L. Klaus*	1422

III. Utilities, Suppliers of Energy, and Utility Regulation

Electric Supply System

Electric Power Transmission Systems / *Jack Casazza*	356
Electric Supply System: Generation / *Jill S. Tietjen*	374
Independent Power Producers / *Hemmat Safwat*	854
Integrated Gasification Combined Cycle (IGCC): Coal- and Biomass-Based / *Ashok D. Rao*	906
IntelliGrid^{SM} / *Clark W. Gellings and Kurt E. Yeager*	914
Nuclear Energy: Economics / *James G. Hewlett*	1101

Fuel Supply System

Coal Supply in the U.S. / *Jill S. Tietjen*	156
Liquefied Natural Gas (LNG) / *Charles P. Ivey*	1001

Utilities: Overview

District Energy Systems / *Ibrahim Dincer and Arif Hepbasli*	316
Electricity Enterprise: U.S., Past and Present / *Kurt E. Yeager*	387
Electricity Enterprise: U.S., Prospects / *Kurt E. Yeager*	399
Utilities and Energy Suppliers: Bill Analysis / *Neil Petchers*	1471
Utilities and Energy Suppliers: Rate Structures / *Neil Petchers*	1497
Water and Wastewater Utilities / *Rudolf Marloth*	1556

Utility Regulatory Issues

Demand Response: Load Response Resources and Programs / *Larry B. Barrett*	279
Demand-Side Management Programs / *Clark W. Gellings*	286
Electricity Deregulation for Customers / *Norm Campbell*	379
Net Metering / *Steven Ferrey*	1096
Pricing Programs: Time-of-Use and Real Time / *Ahmad Faruqui*	1175
Regulation: Price Cap and Revenue Cap / *Mark A. Jamison*	1245
Regulation: Rate of Return / *Mark A. Jamison*	1252
Utilities and Energy Suppliers: Planning and Portfolio Management / *Devra Bachrach Wang*	1491

IV. Facilities and Users of Energy

Building Envelope

Facility Air Leakage / *Wendell A. Porter*	662
Insulation: Facilities / *Wendell A. Porter*	890
Radiant Barriers / *Ingrid Melody, Philip Fairey, and David Beal*	1227
Walls and Windows / *Therese Stovall*	1513
Window Energy / *William Ross McCluney*	1616
Window Films: Savings from IPMVP Options C and D / *Steve DeBusk*	1626

IV. Facilities and Users of Energy (*cont'd.*)

Building Envelope (cont'd.)

Window Films: Solar-Control and Insulating / *Steve DeBusk*	1630
Window Films: Spectrally Selective versus Conventional Applied / *Marty Watts*	1637
Windows: Shading Devices / *Svetlana J. Olbina*	1641

Compressed Air Systems

Compressed Air Control Systems / *Bill Allemon, Rick Avery, and Sam Prudhomme*	207
Compressed Air Leak Detection and Repair / *Robert E. Wilson*	219
Compressed Air Storage and Distribution / *Thomas F. Taranto*	226
Compressed Air Systems / *Diane Schaub*	236
Compressed Air Systems: Optimization / *R. Scot Foss*	241

Facility Controls and Information Systems

Building Automation Systems (BAS): Direct Digital Control / *Paul J. Allen and Rich Remke*	104
Electronic Control Systems: Basic / *Eric Peterschmidt and Michael Taylor*	407
Energy Information Systems / *Paul J. Allen and David C. Green*	535
Enterprise Energy Management Systems / *Gregory Cmar*	616

Heat Recovery

Heat and Energy Wheels / *Carey J. Simonson*	794
Heat Exchangers and Heat Pipes / *Greg F. Naterer*	801
Heat Pipe Application / *Somchai Jiajitsawat, John Duffy, and John W. Wang*	807
Run-Around Heat Recovery Systems / *Kamiel S. Gabriel*	1278
Waste Heat Recovery / *Martin A. Mozzo Jr.*	1536
Waste Heat Recovery Applications: Absorption Heat Pumps / *Keith E. Herold*	1548

HVAC Systems

Cold Air Retrofit: Case Study / *James P. Waltz*	172
Cooling Towers / *Ruth Mossad*	246
Desiccant Dehumidification: Case Study / *Michael K. West and Glenn C. Haynes*	292
District Cooling Systems / *Susanna S. Hanson*	309
Evaporative Cooling / *David Bisbee*	633
Geothermal Heat Pump Systems / *Greg Tinkler*	753
Heat Pumps / *Lu Aye*	814
HVAC Systems: Humid Climates / *David MacPhaul*	839
Pumps and Fans / *L. J. Grobler and T. G. Vorster*	1213
Thermal Energy Storage / *Brian Silvetti*	1412
Underfloor Air Distribution (UFAD) / *Tom Webster and Fred Bauman*	1463

Industrial Facilities

- Combined Heat and Power (CHP): Integration with Industrial Processes / *James A. Clark* 175
- Drying Operations: Agricultural and Forestry Products / *Guangnan Chen* 332
- Drying Operations: Industrial / *Christopher G.J. Baker* 338
- Energy Conservation: Industrial Processes / *Harvey E. Diamond* 458
- Energy Conservation: Lean Manufacturing / *Bohdan W. Oppenheim* 467
- Industrial Classification and Energy Efficiency / *Asfaw Beyene* 862
- Industrial Energy Management: Global Trends / *Cristián Cárdenas-Lailhacar* 872
- Industrial Motor System Optimization Projects in the U.S. / *Robert Bruce Lung, Aimee McKane, and Mitchell Olszewski* 883
- Performance Indicators: Industrial Energy / *Harvey E. Diamond and Robert G. Botelho* 1140
- Savings and Optimization: Case Studies / *Robert W. Peters and Jeffery P. Perl* 1283
- Savings and Optimization: Chemical Process Industry / *Jeffery P. Perl and Robert W. Peters* 1302
- Waste Fuels / *Robert J. Tidona* 1523

Lighting and Motors

- Daylighting / *William Ross McCluney* 264
- Electric Motors / *H. A. Ingley III* 349
- High Intensity Discharge (HID) Electronic Lighting / *Ken E. Patterson II* 830
- Lighting Controls / *Craig DiLouie* 977
- Lighting Design and Retrofits / *Larry Leetzow* 993

On Site Electric Generation

- Distributed Generation / *Paul M. Sotkiewicz and Ing. Jesús Mario Vignolo* 296
- Distributed Generation: Combined Heat and Power / *Barney L. Capehart and D. Paul Mehta, and Wayne C. Turner* 303
- Reciprocating Engines: Diesel and Gas / *Rolf D. Reitz and Kevin Hoag* 1233
- Water-Augmented Gas Turbine Power Cycles / *Ronald L. Klaus* 1574

Space Heating, Water Heating, and Water Use

- Residential Buildings: Heating Loads / *Zohrab Melikyan* 1272
- Solar Heating and Air Conditioning: Case Study / *Eric Oliver, K. Baker, M. Babcock, and John Archibald* 1317
- Solar Water Heating: Domestic and Industrial Applications / *M. Asif and T. Muneer* 1331
- Space Heating / *James P. Riordan* 1357
- Water Source Heat Pump for Modular Classrooms / *Andrew R. Forrest and James W. Leach* .. 1564
- Water-Using Equipment: Commercial and Industrial / *Kate McMordie Stoughton and Amy Solana* 1587
- Water-Using Equipment: Domestic / *Amy Solana and Kate McMordie Stoughton* 1596

Steam Boilers and Steam Systems

- Boilers and Boiler Control Systems / *Eric Peterschmidt and Michael Taylor* 93
- Steam and Hot Water System Optimization: Case Study / *Nevena Iordanova and Rick Cumbo* 1366
- Steam Turbines / *Michael Burke Wood* 1380

V. Energy Management

Commissioning

Commissioning: Existing Buildings / *David E. Claridge, Mingsheng Liu, and W. D. Turner* . . .	179
Commissioning: New Buildings / *Janey Kaster*	188
Commissioning: Retrocommissioning / *Stephany L. Cull*	200

Energy Auditing and Benchmarking

Auditing: Improved Accuracy / *Barney L. Capehart and Lynne C. Capehart*	69
Auditing: User-Friendly Reports / *Lynne C. Capehart*.	76
Building Geometry: Energy Use Effect / *Geoffrey J. Gilg and Francisco L. Valentine*	111
Facility Energy Use: Benchmarking / *John Van Gorp*.	690
Federal Energy Management Program (FEMP): Operations and Maintenance Best Practices Guide (O&M BPG) / *Gregory P. Sullivan, W. David Hunt, and Ab Ream* .	707
Manufacturing Industry: Activity-Based Costing / *J. Michael Fernandes, Barney L. Capehart, and Lynne C. Capehart*	1042

Energy Codes and Standards

Air Quality: Indoor Environment and Energy Efficiency / *Shirley J. Hansen*	18
ANSI/ASHRAE Standard 62.1 - 2004 / *Leonard A. Damiano and David Dougan*	50
Energy Codes and Standards: Facilities / *Rosemarie Bartlett, Mark A. Halverson, and Diana L. Shankle*	438

Energy Management Programs for Facilities

Career Advancement and Assessment in Energy Engineering / *Albert Thumann*	131
Data Collection: Preparing Energy Managers and Technicians / *Athula Kulatunga*	255
Energy Efficiency: Information Sources for New and Emerging Technologies / *Steven A. Parker*	507
Energy Management: Organizational Aptitude Self-Test / *Christopher Russell*.	543
Energy Master Planning / *Fredric S. Goldner*	549
Management Systems for Energy / *Michael L. Brown*	1027

Energy Savings Projects

Demand Response: Commercial Building Strategies / *David S. Watson, Sila Kiliccote, Naoya Motegi, and Mary Ann Piette*	270
Energy Efficiency: Low Cost Improvements / *James Call*.	517
Energy Efficiency: Strategic Facility Guidelines / *Steve Doty*	524
Energy Project Management / *Lorrie B. Tietze and Sandra B. McCardell*	556
Facility Energy Efficiency and Controls: Automobile Technology Applications / *Barney L. Capehart and Lynne C. Capehart*.	671
Tribal Land and Energy Efficiency / *Jackie L. Francke and Sandra B. McCardell*	1456
Utilities and Energy Suppliers: Business Partnership Management / *S. Kay Tuttle*.	1484
Wireless Applications: Mobile Thermostat Climate Control and Energy Conservation / *Alexander L. Burd and Galina S. Burd*.	1660

Measurement and Verification of Energy Savings

International Performance Measurement and Verification Protocol (IPMVP) / *James P. Waltz*	919
Measurement and Verification / *Stephen A. Roosa*	1050
Measurements in Energy Management: Best Practices and Software Tools / *Peter Barhydt and Mark Menezes*	1056
Six Sigma Methods: Measurement and Verification / *Carla Fair-Wright and Michael R. Dorrington*	1310
Wireless Applications: Energy Information and Control / *Jim Lewis*	1649

Performance Contracting

Energy Service Companies: Europe / *Silvia Rezessy and Paolo Bertoldi*	566
Performance Contracting / *Shirley J. Hansen*	1134

VI. Current Energy and Environmental Issues

Environmental Regulation: Overview

Air Emissions Reductions from Energy Efficiency / *Bruce K. Colburn*	9
Carbon Sequestration / *Nathan E. Hultman*	125
Climate Policy: International / *Nathan E. Hultman*	137
Emissions Trading / *Paul M. Sotkiewicz*	430
Environmental Policy / *Sanford V. Berg*	625
Exergy: Environmental Impact Assessment Applications / *Marc A. Rosen*	655
Fossil Fuel Combustion: Air Pollution and Global Warming / *Dan Golomb*	715
Global Climate Change / *Amanda Staudt and Nathan E. Hultman*	760
Tradable Certificates for Energy Savings / *Silvia Rezessy and Paolo Bertoldi*	1433

Sustainable Buildings

Energy Star® Portfolio Manager and Building Labeling Program / *Bill Allemon*	573
LEED-CI and LEED-CS: Leadership in Energy and Environmental Design for Commercial Interiors and Core and Shell / *Nick Stecky*	937
LEED-NC: Leadership in Energy and Environmental Design for New Construction / *Stephen A. Roosa*	947
Sustainability Policies: Sunbelt Cities / *Stephen A. Roosa*	1389
Sustainable Building Simulation / *Birol I. Kilkis*	1396
Sustainable Development / *Mark A. Peterson*	1406

VII. Transportation and Other Energy Uses

Other Energy Uses

Water and Wastewater Plants: Energy Use / *Alfred J. Lutz*	1548

VII. Transportation and Other Energy Uses (*cont'd*.)

Transportation

Aircraft Energy Use / *K. E. Ohrn*	24
Greenhouse Gas Emissions: Gasoline, Hybrid-Electric, and Hydrogen-Fueled Vehicles / *Robert E. Uhrig*	787
Hybrid-Electric Vehicles: Plug-In Configuration / *Robert E. Uhrig and Vernon P. Roan*	847
Maglev (Magnetic Levitation) / *Kevin C. Coates*	1018
Mobile HVAC Systems: Fundamentals, Design, and Innovations / *Alex V. Moultanovsky*	1061
Mobile HVAC Systems: Physics and Configuration / *Alex V. Moultanovsky*	1070
Transportation: Location Efficiency / *Todd A. Litman*	1449

Foreword

The Association of Energy Engineers (AEE) is proud to be a sponsor of the *Encyclopedia of Energy Engineering and Technology, Three-Volume Set*, edited by Dr. Barney L. Capehart. In 2007 AEE is celebrating its 30th anniversary and it is a fitting tribute that the *Encyclopedia of Energy Engineering and Technology* is published at this time. AEE defined the energy engineering profession and this comprehensive work details the core elements for success in this field.

Dr. Capehart has performed a monumental task of facilitating over 300 researchers and practitioners who have contributed to this three-volume set. These distinguished authorities share a wealth of knowledge on approximately 190 topics. Dr. Capehart, through his training and publications, has significantly impacted the energy engineering profession and has helped make it what it is today.

Global climate change concerns and unstable energy prices have raised the importance of energy engineering. This encyclopedia will be a valuable tool in assisting energy engineers to reach their potential. The Association of Energy Engineers (AEE) and our network of 8,000 members in 77 countries would like to thank Dr. Barney Capehart and the numerous volunteers who have made this work possible.

Albert Thumann, P.E., CEM
Executive Director
The Association of Energy Engineers (AEE)

The Association of Energy Engineers

> The Association of Energy Engineers' mission is to promote the scientific and educational interests of those engaged in the energy industry and to foster action for sustainable development.

The Association of Energy Engineers (AEE) is proud to sponsor the Encyclopedia of Energy Engineering and Technology. The Encyclopedia of Engineering and Technology is an important contribution to the field of Energy Engineering and will be invaluable to the industry.

AEE provides a gateway for information on the dynamic field of energy efficiency, renewable energy and global warming. Celebrating its 30th year, AEE is in the forefront of energy engineering technology transfer. With a full array of information and outreach programs from technical seminars, conferences, books and journals to critical buyer-seller networking trade shows.

AEE presents three important industry events every year. These events take place across the U.S. to allow energy professionals from all regions to attend. The following events bring attendees from across the U.S. and around the world together to network and to discuss the most up-to-date technologies and innovations affecting the industry:

Globalcon
www.GLOBALCONevent.com

West Coast Energy Management Congress (EMC)
www.energyevent.com

World Energy Engineering Congress (WEEC)
www.energycongress.com

The Association also offers a variety of information resource tools. As a growing membership organization, the overall strength of AEE is highlighted by a strong membership base of over 8,000 professionals and recognized certification programs, including Certified Energy Managers (CEMs), Lighting Professionals, Indoor Air Quality Professionals, and Cogeneration Professionals. The Association's network of 67 local and regional chapters has powerful grassroots agendas and members. Further, AEE's roster of corporate members is a veritable "Who's Who" of the commercial, industrial, institutional, governmental, energy services, and utility sectors.

The Association of Energy Engineers is pleased that an Encyclopedia of this magnitude has been created for use by industry professionals and advocates.

For more details:
www.aeecenter.org

ASSOCIATION OF ENERGY ENGINEERS

Phone: 770-447-5083
Fax: 770-446-3969

4025 Pleasantdale Rd, Suite 420
Atlanta, GA 30340

www.aeecenter.org

Preface

Energy engineers and technologists have made efficient and cost effective devices for many years which provide the energy services society wants and expects. From air conditioners to waste fuels, energy engineers and technologists continue to make our lives comfortable and affordable using limited resources in efficient and renewable ways.

Over 300 researchers and practitioners, through 190 entries, provide ready access to the basic principles and applications of energy engineering, as well as advanced applications in the technologies of energy production and use. The global supply of energy is increasingly being stressed to provide for an expanding world population. Energy efficiency, energy conservation through energy management, and use of renewable energy sources are three of the major strategies that in the future will help provide the energy and energy services for the world's population and the world's economy.

This unique reference contains state-of-the-art progress on the most important topics in this field of energy engineering and technology. All entries in the encyclopedia have been written by experts in their specialties, and have been reviewed by subject matter authorities. This distinguished group of experts share a wealth of knowledge on topics such as:

- Energy, energy supplies and energy use
- Renewable and alternative energy sources
- Technical, economic and financial analysis of energy systems
- Energy uses in buildings and industry
- Energy efficiency and energy conservation opportunities and projects
- Commissioning, benchmarking, performance contracting, and measurement and verification
- Environmental regulation and public policy for energy supply and use
- Global climate change and carbon control
- Sustainable buildings and green development
- Hybrid electric and hydrogen fueled vehicles and maglev transportation

The *Encyclopedia of Energy Engineering and Technology, Three-Volume Set*, is a key reference work for professionals in academia, business, industry and government, as well as students at all levels. It should be regularly consulted for basic and advanced information to guide students, scholars, practitioners, the public, and policy makers. Contributions address a wide spectrum of theoretical and applied topics, concepts, methodologies, strategies, and possible solutions.

The On-Line Edition is a dynamic resource that will grow as time and knowledge progress. Suggestions for additional content are welcomed by the editor, and new authors should contact me at the e-mail address listed below.

As editor, I would like to thank the people who worked hard to initiate this encyclopedia, and make the project a success. Thanks go to Russell Dekker at Marcel Dekker, and Al Thumann, the Executive Director of the Association of Energy Engineers (AEE) for getting the project going. I appreciate their confidence in my ability to accomplish this immense project. Part of the purpose of this project is to help provide professional and educational support for new people coming into our area of energy engineering. A profession can only succeed and grow if new people have a resource to learn about a new area, and find this area interesting and exciting for their careers.

Directors Oona Schmidt and Claire Miller were both excellent in helping to organize and specify the work that had to be done to get the encyclopedia started and on track for completion. Editorial Assistants Andrea Cunningham, Lousia Lam, and Marisa Hoheb provided daily help to me, and kept all of the records and contacts with the authors. Their help was invaluable.

Preparation of this modern compendium on energy engineering and technology has only been possible through the commitment and hard work of hundreds of energy engineers from around the globe. I want to thank all of the authors for their outstanding efforts to identify major topics of interest for this project, and to write interesting and educational articles based on their areas of expertise. Many of the authors also served a dual function of both writing their own articles, and reviewing the submissions of other authors. Another important group of people were those on the Editorial Board who helped submit topics, organizational ideas, and lists of potential authors for the encyclopedia. This Board was a great help in getting the actual writing of articles started, as well as many of the Editorial Board members also contributed articles themselves.

I would also like to thank my wife Lynne for her continuing support of all of my projects over the years. And finally, I would like to dedicate this encyclopedia to my grandchildren Hannah and Easton. They are a part of the future that I hope will be using the efficient and sustainable energy resources presented in this encyclopedia.

Barney L. Capehart, Editor
Capehart@ise.ufl.edu

Common Energy Abbreviations

The following abbreviations were provided by the Energy Information Administration's National Energy Information Center. The Energy Information Administration is a statistical agency of the U.S. Department of Energy.

AC	alternating current
AFUDC	allowance for funds used during construction
AFV	alternative-fuel vehicle
AGA	American Gas Association
ANSI	American National Standards Institute
API	American Petroleum Institute
ASTM	American Society for Testing and Materials
bbl	barrel(s)
bbl/d	barrel(s) per day
bcf	billion cubic feet
BLS	Bureau of Labor Statistics within the U.S. Department of Labor
BOE	barrels of oil equivalent (used internationally)
Btu	British thermal unit(s)
BWR	boiling-water reactor
C/gal	cents per gallon
CAFE	corporate average fuel economy
CARB	California Air Resources Board
CDD	cooling degree-days
CERCLA	Comprehensive Environmental Response, Compensation, and Liability Act
CF	cubic foot
CFC	chlorofluorocarbon
CFS	cubic feet per second
CH_4	Methane
CHP	combined heat and power
CNG	compressed natural gas
cnt	cent
CO	carbon monoxide
CO_2	carbon dioxide
CPI	consumer price index
CWIP	construction work in progress
DC	direct current
DOE	Department of Energy
DRB	demonstrated reserve base
DSM	demand-side management
E85	A fuel containing a mixture of 85 percent ethanol and 15 percent gasoline
E95	A fuel containing a mixture of 95 percent ethanol and 5 percent gasoline
EAR	estimated additional resources
EIA	Energy Information Administration
EIS	Environmental Impact Statement
EOR	enhanced oil recovery
EPA	Environmental Protection Agency
EPACT	Energy Policy Act
EU	European Union
EWG	exempt wholesale generator
FASB	Financial Accounting Standards Board
FBR	fast breeder reactor

FERC	Federal Energy Regulatory Commission
FGD	flue-gas desulfurization
F.O.B	free on board
FPC	Federal Power Commission
FRS	Financial Reporting System
gal	gallon
GDP	gross domestic product
GNP	gross national product
GVW	gross vehicle weight
GW	gigawatt
GWe	gigawatt-electric
GWh	gigawatthour
GWP	global warming potential
HCFC	hydrochlorofluorocarbon
HDD	heating degree-days
HFC	hydrofluorocarbon
HID	high-intensity discharge
HTGR	high temperature gas-cooled reactor
HVAC	heating, ventilation, and air-conditioning
IEA	International Energy Agency
IOU	investor-owned utility
IPP	independent power producer
ISO	independent system operator
kVa	kilovolt-Ampere
kW	kilowatt
kWe	kilowatt-electric
kWh	kilowatthour
lb	pound
LDC	local distribution company
LEVP	Low Emissions Vehicle Program
LHV	lower heating value
LIHEAP	Low-Income Home Energy Assistance Program
LNG	liquefied natural gas
LPG	liquefied petroleum gases
LRG	liquefied refinery gases
LWR	light water reactor
M	thousand
Mcf	one thousand cubic feet
MECS	Manufacturing Energy Consumption Survey
MM	million (10^6)
MMbbl/d	one million (10^6) barrels of oil per day
MMBtu	one million (10^6) British thermal units
MMcf	one million (10^6) cubic feet
MMgal/d	one million (10^6) gallons per day
MMst	one million (10^6) short tons
MPG	miles per gallon
MSA	metropolitan statistical area
MSHA	Mine Safety and Health Administration
MSW	municipal solid waste
MTBE	methyl tertiary butyl ether
MW	megawatt
MWe	megawatt electric
MWh	megawatthour
N_2O	Nitrous oxide
NAAQS	National Ambient Air Quality Standards
NAICS	North American Industry Classification System

NARUC	National Association of Regulatory Utility Commissioners
NERC	North American Electric Reliability Council
NGL	natural gas liquids
NGPA	Natural Gas Policy Act of 1978
NGPL	natural gas plant liquids
NGV	natural gas vehicle
NOAA	National Oceanic and Atmospheric Administration
NOPR	Notice of Proposed Rulemaking
NO$_x$	nitrogen oxides
NRECA	National Rural Electric Cooperative Association
NUG	nonutility generator
NURE	National Uranium Resource Evaluation
NYMEX	New York Mercantile Exchange
O$_3$	Ozone
O&M	operation and maintenance
OCS	Outer Continental Shelf
OECD	Organization for Economic Cooperation and Development
OEM	original equipment manufacturers
OPEC	Organization of Petroleum Exporting Countries
OPRG	oxygenated fuels program reformulated gasoline
OTEC	ocean thermal energy conversion
PADD	Petroleum Administration for Defense Districts
PBR	pebble-bed reactor
PBR	performance-based rates
PCB	polychlorinated biphenyl
PFCs	perfluorocarbons
PGA	purchased gas adjustment
PPI	Producer Price Index
PUD	public utility district
PUHCA	Public Utility Holding Company Act of 1935
PURPA	Public Utility Regulatory Policies Act of 1978
PV	photovoltaic
PVC	photovoltaic cell; polyvinyl chloride
PWR	pressurized-water reactor
QF	qualifying facility
QUAD	quadrillion Btu: 10^{15} Btu
RAC	refiners' acquisition cost
RAR	reasonable assured resources
RBOB	Reformulated Gasoline Blendstock for Oxygenate Blending
RDF	refuse-derived fuel
REA	Rural Electrification Administration
RECS	Residential Energy Consumption Survey
RFG	reformulated gasoline
RSE	relative standard error
RVP	Reid vapor pressure
SEER	seasonal energy efficiency ratio
SF$_6$	sulfur hexafluoride
SI	International System of Units (Système international d'unités)
SIC	Standard Industrial Classification
SNG	synthetic natural gas
SO$_2$	sulfur dioxide
SPP	small power producer
SPR	Strategic Petroleum Reserve
SR	speculative resources
T	trillion 10^{12}
TVA	Tennessee Valley Authority

TW	Terawatt
U₃O₈	Uranium oxide
UF₆	uranium hexaflouride
ULCC	ultra large crude carrier
UMTRA	Uranium Mill Tailings Radiation Control Act of 1978
USACE	U.S. Army Corps of Engineers (sometimes shortened to USCE in EIA tables)
USBR	United States Bureau of Reclamation
V	Volt
VAWT	vertical-axis wind turbine
VIN	vehicle identification number
VLCC	very large crude carrier
VMT	vehicle miles traveled
VOC	volatile organic compound
W	watt
WACOG	weighted average cost of gas
Wh	watt hour
WTI	West Texas Intermediate

Thermal Metric and Other Conversion Factors

The following tables appeared in the January 2007 issue of the Energy Information Administration's *Monthly Energy Review*. The Energy Information Administration is a statistical agency of the U.S. Department of Energy.

Table 1 Metric Conversion Factors
These metric conversion factors can be used to calculate the metric-unit equivalents of values expressed in U.S. Customary units. For example, 500 short tons are the equivalent of 453.6 metric tons (500 short tons × 0.9071847 metric tons/short ton = 453.6 metric tons).

Type of Unit	U.S. Unit		Equivalent in Metric Units	
Mass	1 short ton (2,000 lb)	=	0.907 184 7	metric tons (t)
	1 long ton	=	1.016 047	metric tons (t)
	1 pound (lb)	=	0.453 592 37[a]	kilograms (kg)
	1 pound uranium oxide (lb U^3O^8)	=	0.384 647[b]	kilograms uranium (kgU)
	1 ounce, avoirdupois (avdp oz)	=	28.349 52	grams (g)
Volume	1 barrel of oil (bbl)	=	0.158 987 3	cubic meters (m^3)
	1 cubic yard (yd^3)	=	0.764 555	cubic meters (m^3)
	1 cubic foot (ft^3)	=	0.028 316 85	cubic meters (m^3)
	1 U.S. gallon (gal)	=	3.785 412	liters (L)
	1 ounce, fluid (fl oz)	=	29.573 53	milliliters (mL)
	1 cubic inch (in^3)	=	16.387 06	milliliters (mL)
Length	1 mile (mi)	=	1.609 344[a]	kilometers (km)
	1 yard (yd)		0.914 4[a]	meters (m)
	1 foot (ft)	=	0.304 8[a]	meters (m)
	1 inch (in)	=	2.54[a]	centimeters (cm)
Area	1 acre	=	0.404 69	hectares (ha)
	1 square mile (mi^2)	=	2.589 988	square kilometers (km^2)
	1 square yard (yd^2)	=	0.836 127 4	square meters (m^2)
	1 square foot (ft^2)	=	0.092 903 04[a]	square meters (m^2)
	1 square inch (in^2)	=	6.451 6[a]	square centimeters (cm^2)
Energy	1 British thermal unit (Btu)[c]	=	1,055.055 852 62[a]	joules (J)
	1 calorie (cal)	=	4.186 8[a]	joules (J)
	1 kilowatthour (kWh)	=	3.6[a]	megajoules (MJ)
Temperature[d]	32 degrees Fahrenheit (°F)	=	0[a]	degrees Celsius (°C)
	212 degrees Fahrenheit (°F)	=	100[a]	degrees Celsius (°C)

[a]Exact conversion.
[b]Calculated by the Energy Information Administration.
[c]The Btu used in this table is the International Table Btu adopted by the Fifth International Conference on Properties of Steam, London, 1956.
[d]To convert degrees Fahrenheit (°F) to degrees Celsius (°C) exactly, subtract 32, then multiply by 5/9.

Notes: • Spaces have been inserted after every third digit to the right of the decimal for ease of reading. • Most metric units belong to the International System of Units (SI), and the liter, hectare, and metric ton are accepted for use with the SI units. For more information about the SI units, see http://physics.nist.gov/cuu/Units/index.html.
Web Page: http://www.eia.doe.gov/emeu/mer/append_b.html.
Sources: • General Services Administration, Federal Standard 376B, *Preferred Metric Units for General Use by the Federal Government* (Washington, D.C., January 1993), pp. 9-11, 13, and 16. • U.S. Department of Commerce, National Institute of Standards and Technology, Special Publications 330, 811, and 814. • American National Standards Institute/Institute of Electrical and Electronic Engineers, ANSI/IEEE Std 268-1992, pp. 28 and 29.

Table 2 Metric Prefixes
The names of multiples and subdivisions of any unit may be derived by combining the name of the unit with prefixes as below.

Unit Multiple	Prefix	Symbol	Unit Subdivision	Prefix	Symbol
10^1	deka	da	10^{-1}	deci	D
10^2	hecto	h	10^{-2}	centi	c
10^3	kilo	k	10^{-3}	milli	m
10^6	mega	M	10^{-6}	micro	µ
10^9	giga	G	10^{-9}	nano	n
10^{12}	tera	T	10^{-12}	pico	p
10^{15}	peta	P	10^{-15}	femto	f
10^{18}	exa	E	10^{-18}	atto	a
10^{21}	zetta	Z	10^{-21}	zepto	z
10^{24}	yotta	Y	10^{-24}	yocto	y

Web Page: http://www.eia.doe.gov/emeu/mer/append_b.html.
Source: U.S. Department of Commerce, National Institute of Standards and Technology, *The International System of Units (SI)*, NIST Special Publication 330, 1991 Edition (Washington, D.C., August 1991), p. 10.

Table 3 Other Physical Conversion Factors
The factors below can be used to calculate equivalents in various physical units commonly used in energy analyses. For example, 10 barrels are the equivalent of 420 U.S. gallons (10 barrels x 42 gallons/barrel = 420 gallons).

Energy Source	Original Unit		Equivalent in Final Units	
Petroleum	1 barrel (bbl)	=	42[a]	U.S. gallons (gal)
Coal	1 short ton	=	2,000[a]	pounds (lb)
	1 long ton	=	2,240[a]	pounds (lb)
	1 metric ton (t)	=	1,000[a]	kilograms (kg)
Wood	1 cord (cd)	=	1.25[b]	shorts tons
	1 cord (cd)	=	128[a]	cubic feet (ft^3)

[a]Exact conversion.
[b]Calculated by the Energy Information Administration.
Web Page: http://www.eia.doe.gov/emeu/mer/append_b.html.
Source: U.S. Department of Commerce, National Institute of Standards and Technology, *Specifications, Tolerances, and Other Technical Requirements for Weighing and Measuring Devices*, NIST Handbook 44, 1994 Edition (Washington, D.C., October 1993), pp. B-10, C-17 and C-21.

About the Editor

Dr. Barney L. Capehart is a Professor Emeritus of Industrial and Systems Engineering at the University of Florida, Gainesville, Florida. He has BS and Master's degrees in Electrical Engineering and a Ph.D. in Systems Engineering. He taught at the University of Florida from 1968 until 2000, where he taught a wide variety of courses in energy systems analysis, energy efficiency and computer simulation. For the last 25–30 years, energy systems analysis has been his main area of research and publication. He is the co-author or chapter contributor on six books on energy topics, the author of over 50 energy research articles in scholarly journals, and has been involved in 11–12 funded energy research projects totaling over $2,000,000.

He has performed energy efficiency and utility research projects for the Florida Governor's Energy Office, the Florida Public Service Commission, and several utilities in the state of Florida. In addition, he has served as an Expert Witness in the development of the Florida Energy Efficiency and Conservation Act electric growth goals, the Florida Rules for Payments to Cogenerators, and the passage of the Florida Appliance Efficiency Standards Act. He is one of the state's leading experts on electric utility Demand Side Management programs for reducing customer costs and for increasing the efficiency of customer end-use.

He also has broad experience in the commercial/industrial sector, with his management of the University of Florida Industrial Assessment Center from 1990 until 1999. This center performs audits for small and medium sized manufacturing plants, and is funded by the U.S. Department of Energy. He has personally conducted over 100 audits of industrial facilities, and has helped students conduct audits of hundreds of office buildings, small businesses, government facilities, and apartment complexes. He regularly taught a University of Florida course on Energy Management to about fifty engineering students each year. He currently teaches Energy Management seminars around the country and around the world for the Association of Energy Engineers (AEE). In addition, he has been a prolific author of newspaper articles and has appeared on numerous radio and TV talk shows to discuss his work on energy efficiency and reducing people's electric bills. He is also a popular speaker at civic, professional, educational, and government meetings to talk about the benefits of energy efficiency.

Dr. Capehart is a Fellow of the Institute of Electrical and Electronic Engineers, the American Association for the Advancement of Science, and the Institute of Industrial Engineers. He is also a Member of the Hall of Fame of the AEE. He is listed in Who's Who in the World and Who's Who in America. In 1988 he was awarded the Palladium Medal by the American Association of Engineering Societies for his work on energy systems analysis and appliance efficiency standards. He has conducted many energy research projects, and has published over fifty journal articles. He was the Editor of the *International Journal of Energy Systems* from 1985–1988.

He is the editor of *Information Technology for Energy Managers—Understanding Web Based Energy Information and Control Systems*, Fairmont Press, 2004; senior co-editor of *Web Based Energy Information and Control Systems*, Fairmont Press, 2005; senior co-editor of *Web Based Enterprise Energy Management and BAS Systems*, Fairmont Press, 2007; senior co-author of the fifth edition of the *Guide to Energy Management*, Fairmont Press, 2006; and author of the chapter on *Energy Management* in the *Handbook of Industrial Engineering, Second Edition*, by Gavriel Salvendy. He also wrote the chapter on *Energy Auditing* for the *Energy Management Handbook, Sixth Edition* by Wayne C. Turner.

Encyclopedia of
Energy Engineering and Technology
First Edition

Volume 3
Pages 1147 through 1667
Phot–Wire

Photovoltaic Systems

Michael Ropp
South Dakota State University, Brookings, South Dakota, U.S.A.

Abstract

Photovoltaics (PV) is the direct conversion of sunlight to electricity. This entry contains a brief review of the history of PV; a survey of solar resource determination options; a summary of first, second and third-generation PV devices and concentrating PV devices; overviews of the power electronics (charge controllers and inverters) and batteries associated with PV systems; a discussion of PV systems and their design; and a brief examination of PV applications, with comments on future prospects.

HISTORY

Photovoltaics (PV) ("photo" meaning "light" and "voltaic" referring to electricity) is the direct conversion of sunlight into an electrical potential (a photovoltage) that can be used to provide electric power. The PV effect itself was discovered in 1839 by a French physicist, Edmund Becquerel, who observed that a photocurrent would flow between two electrodes in a solution when the apparatus was exposed to light.[1] Later, the effect was noticed in selenium by William Adams and Richard Day,[2] and the first solid-state solar cells were made from selenium by Charles Fritts and Werner Siemens.[3] However, many investigators were skeptical about these devices because the quantum physics required to explain the observed effect were not known yet. It wasn't until Max Planck's proposal of the quantum nature of light in 1900[4] that the theoretical foundations for understanding PV were established.

Selenium-based PV remained impractical because of its high cost and low efficiency (about 0.5%).[5] Then, in 1954, Calvin Fuller and Gerald Pearson were working on new silicon rectifier diode technology at Bell Laboratories, and during one experiment they found that their device produced a significant photocurrent when strongly illuminated.[5] At that same time, Daryl Chapin was working on selenium solar cells. When Pearson alerted Chapin to his silicon discovery, Chapin immediately abandoned his selenium work and switched to silicon, and after significant effort but a relatively short time, the result was the achievement of 6% conversion efficiency.[6] However, the energy cost for PV, which is the critical figure of merit for PV systems (usually expressed in $/kWh), was nearly a thousand times that of competing alternatives at that time.[7] Although technically successful, PV was still too expensive to be useful.

Fortunately, another PV application was identified—power for satellites.[8] Its reliability, relatively high specific power (power per unit weight), and lack of a need for refueling in orbit made PV a natural choice for spacecraft power. Vanguard I, in 1958, was the first PV-powered satellite.[9] The satellite power market supported the fledgling PV industry, which in turn aggressively worked to improve device performance. Efficiencies climbed to more than 10%[10] and energy costs declined. During the 1960s, a deep physical understanding of the operation of silicon solar cells was obtained by the PV research community,[11] a legacy that continues to pay dividends today. Still, PV's energy cost remained high, and terrestrial PV use was confined to small off-grid applications where other forms of energy supply or a grid connection were more costly than PV. Ironically, one of the biggest customers for the early terrestrial PV industry was the oil industry,[12] which used PV extensively in signaling systems for offshore oil drilling rigs.

Throughout the tumultuous 1970s and 1980s, the cost of PV-produced energy continued to fall, largely because of the growth of accumulated industrial and field experience, the "economies of scale" afforded by larger-scale production,[13] and the niche markets in which PV was cost-competitive continued to grow.

Then, in the 1990s, concerns mounted over the environmental impacts of the combustion of fossil fuels. This concern is perhaps best expressed by the term "the 3-E trilemma," explained by Yoshihiro Hamakawa as follows: we require the Energy needed to sustain Economic development without damaging the Environment.[14] An increasing awareness of the reality of the 3-E trilemma has spurred rising levels of interest in many forms of sustainable energy production, including PV. A variety of government programs worldwide, most notably in Germany and Japan, have spurred explosive growth in

Keywords: Photovoltaics; Solar electricity; Solar cells; Silicon; Batteries; Inverters; Charge controllers; System design.

the PV industry,[15] which in turn has led to higher levels of research support for PV and its associated technologies.

THE SOLAR RESOURCE

Photovoltaic systems use sunlight as their "fuel," and thus in PV system design it is crucial to know how much sunlight one might expect at the PV array site and how that sunlight is distributed day to day, season to season, and year to year. The terminology of solar energy can be a bit confusing because several commonly used but similar-sounding terms have critically different meanings.[16] The solar energy available on a site is quantified using the irradiation, which is an energy flux usually given in kilowatt hour per square meter per day and is also sometimes called insolation when referring specifically to sunlight. The solar power available on a site is given by irradiance, which is a power flux usually given in units of kilowatt hour per square meter. For PV system design, both of these quantities are important. The spectral properties of sunlight are also important to PV device operation, but at the system level this consideration is secondary.

The best way to gain an idea of what irradiation and irradiance can be expected on a site is to turn to a historical database of meteorological measurements taken at or near the site. Several such databases are available worldwide, and an ever-increasing number of solar databases are available online (although it should be noted that in many cases sunlight is not actually measured but rather modeled based on other meteorological measurements and observations). A few examples are:

- The Renewable Resource Data Center (http://rredc.nrel.gov), maintained by the National Renewable Energy Laboratory.
- Data from the Remote Automated Weather Station (RAWS) network are available from the Western Regional Climate Center (http://www.raws.dri.edu/index.html).
- The European Solar Radiation Atlas (see http://www.ensmp.fr/Fr/Services/PressesENSMP/Collections/ScTerEnv/Livres/atlas_tome1.htm).

If one requires solar resource estimates for sites far from any measuring station, there are procedures for computing such estimates based on a few simple inputs.[17,18] As one might expect, the accuracy of the computed estimates is not as good as estimates based on lengthy periods of historical data, but because of the high level of inaccuracy inherent in any long-term meteorological prediction, these computations coupled with some conservative design decisions are often quite sufficient for PV system design.

FIRST-GENERATION PV: CRYSTALLINE SI

Any PV cell has to accomplish two fundamental tasks: (1) it must absorb the energy of light and use it to generate mobile positive and negative electrical charges and (2) it must separate those mobile positive and negative electrical charges to produce a potential difference between them. There are many structures that can do this, with one of the most common being leaves—the process just described is a significant part of photosynthesis.[19]

Most man-made solar cells today are made from silicon (Si). Silicon is a semiconductor, which simply means that its electrical conductivity falls between that of metals and insulators. Silicon is in column IV of the periodic table, meaning that each silicon atom has four chemically-active electrons called valence electrons.

The detailed physics of solar cell operation are well described in many texts.[20–22] In the dark, each silicon atom's four valence electrons are tightly bound to their respective atoms so that they cannot move easily through the crystal. When photons of light strike the silicon, the photons' energy can be absorbed by the valence electrons, which are released from their bonds and become mobile charges. Also, each mobile electron leaves behind a "spot" where an electron could be, but isn't. These "spots" act for all practical purposes as if they were positive charge carriers or positive particles, and these positive charge carriers are called holes. Therefore, silicon can absorb light to create mobile electron-hole pairs (EHPs).

To separate the mobile charges, the silicon is used to make a structure called a diode. One key property of semiconductors is that their conductivities can be controlled through the addition of certain impurities, called dopants. In silicon, there are two main categories of dopants: elements from column V of the periodic table that have one extra electron and elements from column III that have one missing electron, or equivalently an extra hole. When added to the silicon crystal, the extra electron of the column V dopant becomes mobile in the crystal, and thus silicon doped with column V atoms will have an excess of mobile electrons. Similarly, the extra hole carried by the column III atoms is mobile in silicon, and a column III-doped silicon crystal will have an excess of mobile holes. We refer to the column V-doped silicon as *n*-type because its conductivity is controlled by the negative electrons, and column III-doped silicon is analogously called *p*-type.

The diode structure is shown in Fig. 1. It is essentially just a "sandwich" of an *n*-type and a *p*-type region. The interface between these two regions is called the *p–n* junction. Because of *p–n* junction physics, the junction acts as a one-way electrical valve; electrons can flow from the *p*-side to the *n*-side, but not the other way. Similarly, holes can flow from the *n*-side to the *p*-side but not back. Thus, when a photon is absorbed and generates an EHP, only one of the two charges can cross the junction, after which it can't get back. This separates the charges,

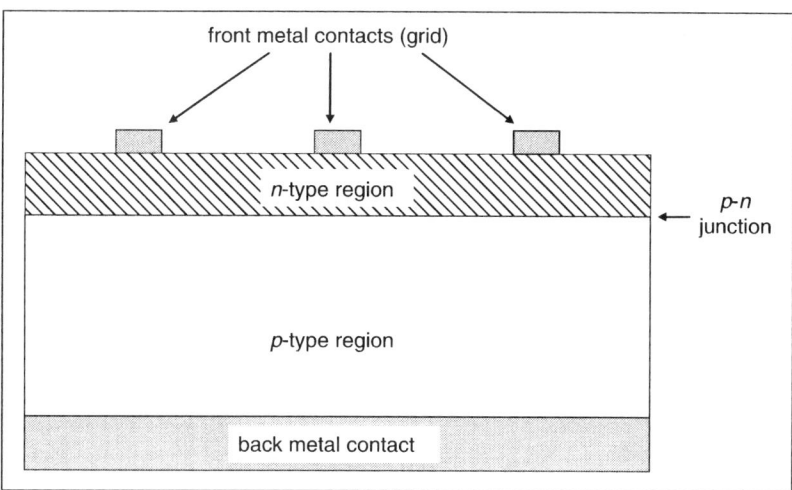

Fig. 1 Structure of a basic photovoltaic (PV) cell.

and a photovoltage builds up as one end of the device accumulates an excess of negative charge and the other end sees a buildup of positive charge. There is thus a positive and a negative terminal of the solar cell, just like a battery, and if electrical contact is made to the two ends of the device, the photovoltage will cause a current to flow. To turn the diode into a solar cell, adding electrical contacts is all that is required, remembering that the front contact must allow the light to enter the silicon. Usually a grid pattern is used for the front contact, as shown in Fig. 1.

Because this was the first "practical" PV technology, solar cells of the type described above are sometimes called first-generation PV devices. The structure in Fig. 1 is an extremely basic one. A wide variety of efficiency-enhancing features can be (and usually are) added to this structure.[23–25]

The efficiency of the above-described simple solar cell depends on many factors. One of the most important factors is the quality of the silicon material.[26] If the silicon is highly pure and has no defects in its crystal structure, then its electronic properties are favorable for high solar cell efficiency. Such material is referred to as single-crystal or crystalline silicon (c-Si). As might be expected, c-Si material leads to relatively high efficiencies but it is relatively expensive. Solar cells can also be made from lower-quality, less expensive materials such as multicrystalline silicon (mc-Si). In mc-Si, the silicon is composed of crystalline grains stuck together by regions of noncrystalline silicon. The grains have dimensions on the order of centimeters. Multicrystalline silicon is less costly than c-Si but has inferior electrical properties, and thus the efficiencies of solar cells made on mc-Si tend to be lower than those on c-Si.[27] These two types of silicon are shown in Fig. 2 below.

SECOND-GENERATION PV: THIN FILMS

First-generation PV devices suffer from a number of fundamental limitations. One of the most persistent is the aforementioned cost of the silicon material itself,[28] which combines with other factors to keep the energy cost of first-generation PV relatively high.

One fairly obvious way to reduce the cost of PV might be to reduce the amount of material required. Wafers of c-Si or mc-Si for PV cells are usually on the order of 150–300 μm (a micron is one one-millionth of a meter) in thickness. If it were possible to make solar cells much thinner but maintain high efficiency, then the material costs should be reduced.[29] This is the aim of so-called thin-film PV, or second-generation PV.

The basic structure of second-generation devices is usually essentially the same as that of first-generation devices, with a few modifications.[30] The main difference between first- and second-generation devices is that second-generation devices are not made from bulk crystalline material, but instead are fabricated on thin films of semiconductor formed using any of several deposition techniques, such as chemical vapor deposition (CVD).[31,32]

Silicon can be used to make thin-film devices. Using a variety of relatively inexpensive deposition techniques such as CVD, it is possible to create layers of silicon that are very thin—on the order of 10 μm—and make PV devices from these films. The resulting silicon usually has almost no crystal structure at all; it is neither crystalline nor multicrystalline, but amorphous silicon (a-Si). Amorphous silicon has very low electrical quality, which makes achieving high solar cell efficiencies difficult, although novel device structures and processing techniques have led to a-Si PV devices with efficiencies near 10%.[33] Unfortunately, a-Si-based devices have another drawback—they tend to degrade under illumination due to the Staebler–Wronski effect (S–W effect).[34] The physical mechanism behind the S–W effect is not yet fully understood, but it is well known that a-Si-based devices loses efficiency under sunlight exposure. Although thin film silicon devices still hold great promise because of their low-cost potential, considerable research

 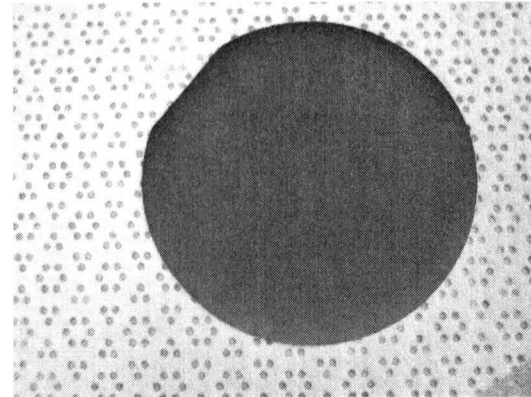

Fig. 2 Multicrystalline (left) and single-crystal (right) silicon wafers. Left photo courtesy of the National Renewable Energy Laboratory's Pictographic Information eXchange (NREL/PiX) (credit: Sandia National Laboratories); right photo by Jeff Moore.

remains to be done to improve their efficiencies and stability.

Silicon is not the only material that can be used to make second-generation PV devices. Successful PV devices have been made using a system containing cadmium telluride and cadmium sulfide layers (CdTe/CdS).[33] The chalcopyrite family is another family of alloys that has properties that could be very favorable for high-efficiency PV device fabrication.[35] This family includes copper indium diselenide (CIS) and a variety of quaternary alloys, such as copper indium gallium selenide (CIGS). Even in a-Si devices, some of the silicon layers are usually alloyed with germanium to produce SiGe.[36]

It is also possible to make second-generation PV devices using organic materials.[37] The advantages of using organic materials are that (i) the processing is relatively simple and inexpensive and (ii) the materials are cheap and readily available. Thus, in theory, these devices could have lower energy costs than PV cells based on inorganic semiconductors. That promise is yet to be realized, however, because the efficiencies of these devices remain low—in the range of 1%–5%.[38,39] Although organic devices must still produce mobile charge pairs and somehow separate the charges, the mechanisms by which they do this[37] are quite different than those of semiconductor-based devices, and are more akin to the process of photosynthesis.

THIRD-GENERATION PV: USING THE FULL SPECTRUM

Although there are a number of efficiency-limiting mechanisms in solar cells, both first- and second-generation PV devices suffer from two particular fundamental limitations.[40,41] First, there is a certain minimum photon energy, called the band gap energy and denoted E_g, below which the device cannot capture photons and convert them to electricity. Second, even if a photon does have energy greater than the band gap energy, E_g (actually, a bit less than E_g)[41] is the portion of the energy that the solar cell can convert to electrical energy. Those two loss mechanisms mean that first- or second-generation PV cells cannot use all of the energy of the solar spectrum.

It would be highly desirable to be able to circumvent this problem of inefficient use of the solar spectrum, and researchers worldwide are currently investigating approaches for doing that. The resulting full-spectrum PV devices are called third-generation PV devices.

One of the earliest and simplest approaches to improving the use of the solar spectrum is to use a multijunction PV cell or tandem PV cell[42] like that shown in Fig. 3. The multijunction device is basically a stack of cells designed such that each cell in the stack uses a different portion of the spectrum. Fig. 3 shows a three-cell stack. The materials of which the cells are made have different band gap energies, arranged from highest to lowest as shown. Cell 1, with the highest E_g (E_{g1}), captures the high-energy photons and can use most of their energy because of the cell's high E_g. The lower energy photons pass through it into Cell 2, which has an intermediate E_g (E_{g2}). Cell 2 absorbs midenergy photons. The photons with energy less than E_{g2} pass through Cell 2 and strike Cell 3, which has a low band gap energy (E_{g3}) and absorbs the low energy photons. The most efficient PV devices made to date (over 32%) are of this type,[43] and research is ongoing to continue to improve their performance and cost effectiveness. (It should be noted that almost all multijunction devices are also thin-film devices, which blurs the line between second- and third-generation PV.)

Other novel approaches are being enabled by advancements in our understanding of extremely small structures, with dimensions on the order of a few nanometers (so-called "nanostructures"). Nanostructures can behave very differently than larger-scale structures in some

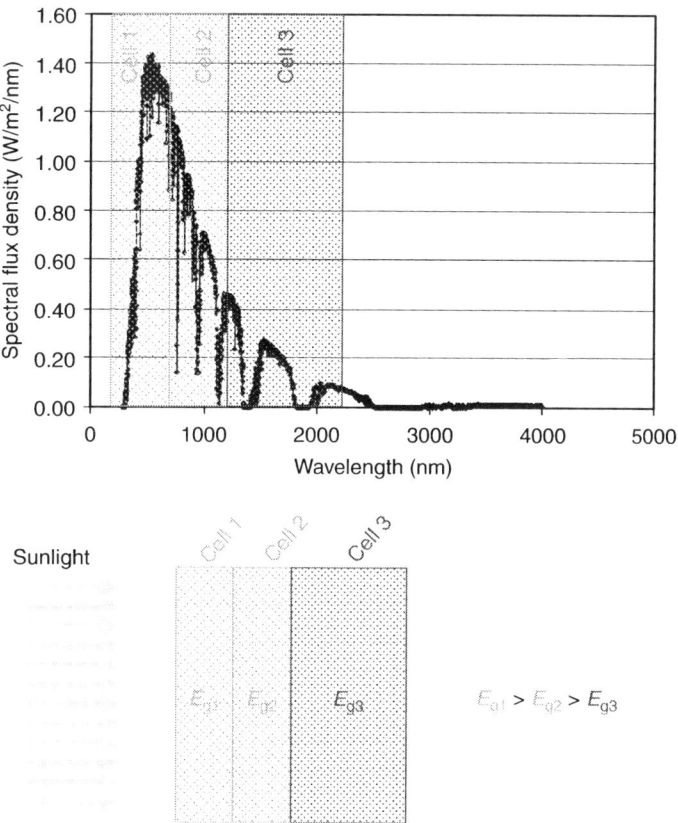

Fig. 3 Schematic representation of a multijunction photovoltaic (PV) cell.

respects, because at those small dimensions quantum physical effects begin to dominate the structure's behavior. One aspect of nanostructure behavior that is particularly important for PV is that the optical properties of nanostructures begin to change and become dependent on the size of the nanostructure. This potentially means that the optical properties can be controlled by simply controlling the size of the nanostructure. With such controllability, it may be possible to create devices that in effect split higher energy photons into multiple lower energy photons, or sum low energy photons into a single higher-energy photon. A device incorporating this feature would shift the energy of the solar spectrum into ranges that are used with maximum efficiency, thereby leading to significantly higher overall device efficiencies. Examples of such devices include:

- Intermediate band devices,[44,45]
- Hot-carrier cells,[46,47]
- Impact ionization devices,[48,49] or
- Devices utilizing spectral upconversion or spectral downconversion.[50]

Many researchers worldwide are pursuing third-generation devices, and initial results indicate that third-generation PV devices with efficiencies much higher than those of first-generation devices may be available in the future.

CONCENTRATION (FOCUSED SUNLIGHT)

One approach to reducing the cost of PV is to concentrate the sunlight. In this way, the total required area of expensive solar cells can be reduced, being replaced by large-area but less-expensive lenses or mirrors.[51] Also, most PV cells become more efficient under concentrated sunlight.[52] Concentrators have been shown to be capable of reducing PV energy costs.[53] Most concentrators require fairly accurate solar tracking, although one type, the luminescent concentrator, does not.[54] Generally, concentrators are feasible only in larger-scale PV power plants,[51] but significant research efforts are currently underway to continue driving down the energy cost from concentrating PV systems.

FROM PV CELLS TO PV POWER PLANTS

Consider the single silicon solar cell in Fig. 4A below. As a general rule of thumb, this cell typically produces on the order of 0.5 W at a voltage of about 0.5 V, depending on

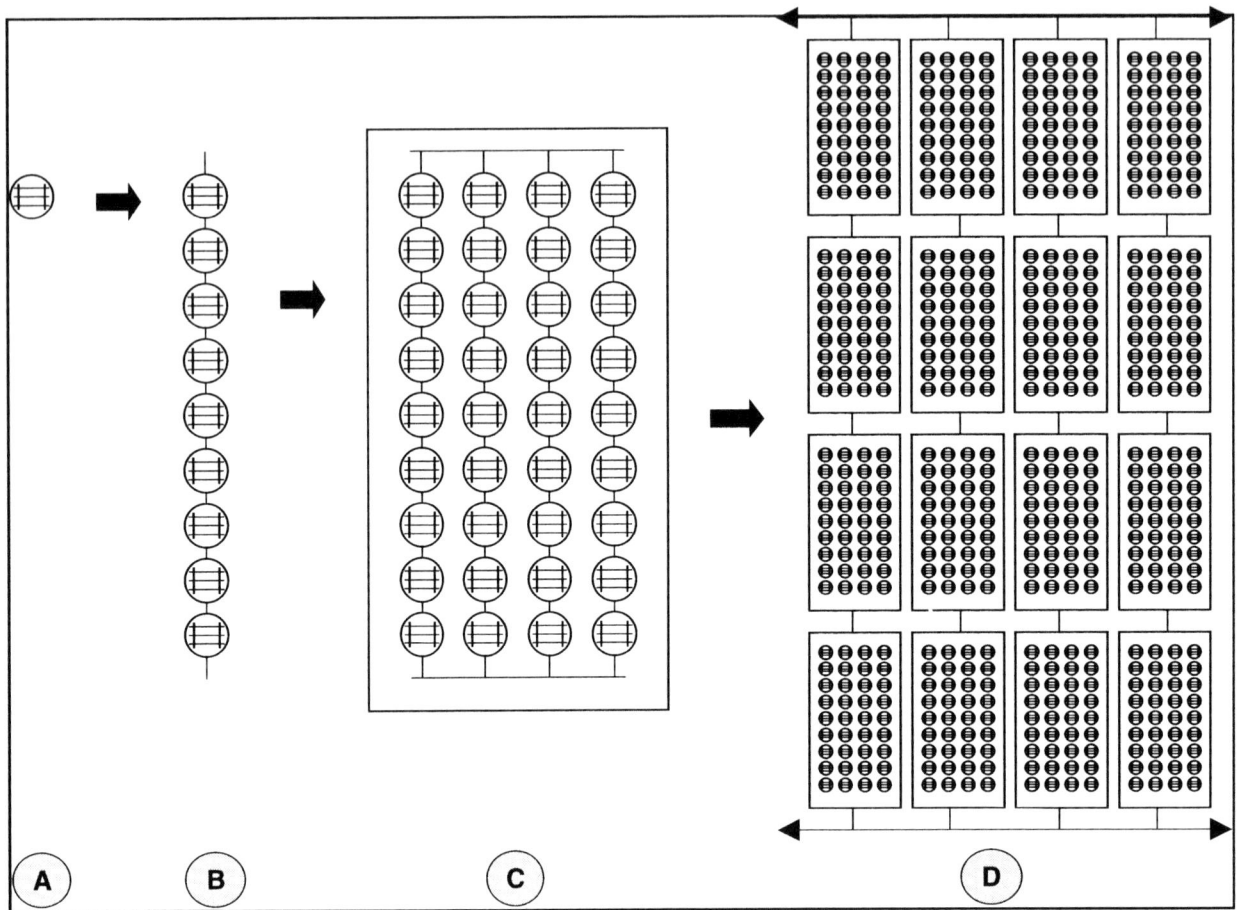

Fig. 4 Individual photovoltaic (PV) cells (A) are combined in series strings (B) to obtain higher voltage. Series strings are encapsulated together to form a module (C). Modules are connected in series strings, and module series strings are connected in parallel to form a PV array (D) that produces the desired voltage and power.

the amount of sunlight and the cell temperature. This is inadequate for most applications. To obtain higher voltages, solar cells are typically connected in series to form series strings, as shown in Fig. 4B. In silicon PV, series strings commonly contain around 36 cells in series.[55] Series strings can be connected in parallel to increase power output. A group of one or more series strings is then encapsulated in a module. Fig. 4C shows a schematic of a module and Fig. 5 is a photograph of a module at work in a water pumping application. The module typically has a glass front sheet, a plastic backsheet, an aluminum frame, and a terminal box that provides a means for electrically connecting to the PV cells. Modules are available in a wide array of sizes, from less than 5 W to over 300 W. Finally, to meet the voltage and power needs of a specific application, modules are connected in series and parallel as shown in Fig. 4D, forming a PV array.

The number of PV modules required in a series string or array depends on a number of factors, the most important of which is the energy requirement of the electrical load. Larger loads require more PV modules, but one must also remember that sunlight is a statistical parameter. Photovoltaic systems are usually designed to provide a specific level of availability (percent of time that the load is being powered—in other words, the percent of time during which there is not a power failure).[56,57] Cloudiness affects availability, as does the seasonal distribution of the load; if the load is larger in the wintertime when there is less sunlight, availability suffers. A high availability requirement will also increase the needed number of modules. Several references contain details on how to size a PV system for a given application.[58–62]

Photovoltaic arrays only provide electricity when the sun shines, so to provide for the needs of most loads a complete PV system must also contain either energy storage or backup generation. There are three basic PV system configurations, shown in Figs. 6–8 below. Fig. 6 shows a very general configuration of a stand-alone PV system in which 100% of the load energy must be supplied by the PV array, and the array must be designed according to load energy and availability requirements. The stand-alone system usually includes batteries to provide continuous power to the load, but not always; for example,

Fig. 5 Photovoltaic (PV) module in the field (this one is in a livestock watering application near Laramie Peak in Wyoming). Photograph by Susan Ropp.

water pumping applications, such as the one shown in Fig. 5, usually do not use batteries because in those cases the water can be stored.

Fig. 6 shows two different loads—one that requires DC power and another requiring AC power. The DC load often is powered directly from the battery bank, but if it requires a highly regulated voltage or a voltage different than the battery bank voltage there may be a DC–DC converter, as shown. Because PV arrays produce DC power, AC loads necessitate the inclusion of a DC–AC converter or inverter. In a system of this type, the inverter's output is AC voltage.

Stand-alone PV systems can be cost effective if the load is relatively small and if the load is sufficiently far from the grid that a grid extension, transformer, and service drop become prohibitively expensive. However, if either the load energy demand or required availability is too large, the large PV array needed may make the cost of a stand-alone system prohibitive. One solution to this problem is shown in Fig. 7, which shows a hybrid PV system configuration. The hybrid system includes an additional generator of some type, so that the PV array does not have to supply 100% of the load's energy needs. The system in Fig. 7 includes an engine-generator set (E-G). Gasoline, diesel or liquified propane gas (LPG)-powered generators are all used in hybrid PV systems, and in some cases wind turbines or small hydropower units can also be used. In the hybrid system, the PV array can be much smaller because it does not have to meet the entire load energy or availability requirement. Instead, the entire hybrid system can be subjected to an optimization process to determine the PV array size and generator run strategy that leads to the system with the lowest energy cost. This is a nontrivial process, and several software packages are available to assist with it.[62] Sometimes a simpler process can be used, in which the PV array is designed to minimize generator run time within some specified cost constraint.

In most locations in the industrialized world, the energy cost from PV is higher than that from the utility grid. However, this is not always the case. Some locations have exceptionally high grid-based energy costs; in others, special policies or incentives have been put into place to encourage the use of renewable energy. In these cases, one might find the system configuration shown in Fig. 8, a grid-connected or utility-interactive photovoltaic (UIPV) system. In a sense, the UIPV system is a special case of the hybrid system shown in Fig. 7. Utility-interactive photovoltaic systems use the utility grid as their backup generator and usually do not incorporate local energy storage (batteries). The lack of batteries greatly simplifies the system, as seen in Fig. 8. The point labeled "PCC" is the point of common coupling between the PV system and the grid.

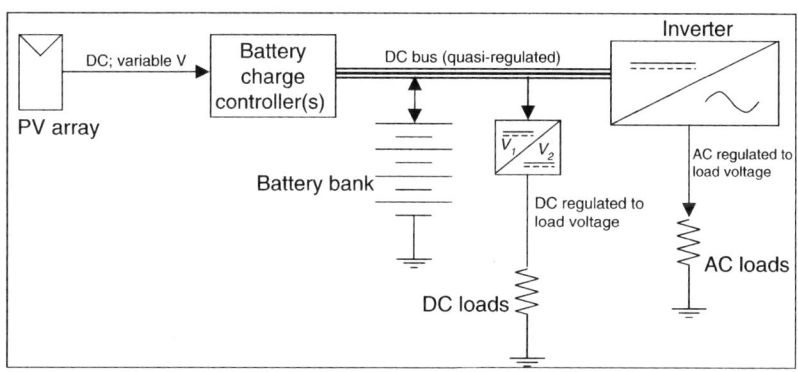

Fig. 6 Generalized configuration of a stand-alone photovoltaic (PV) system.

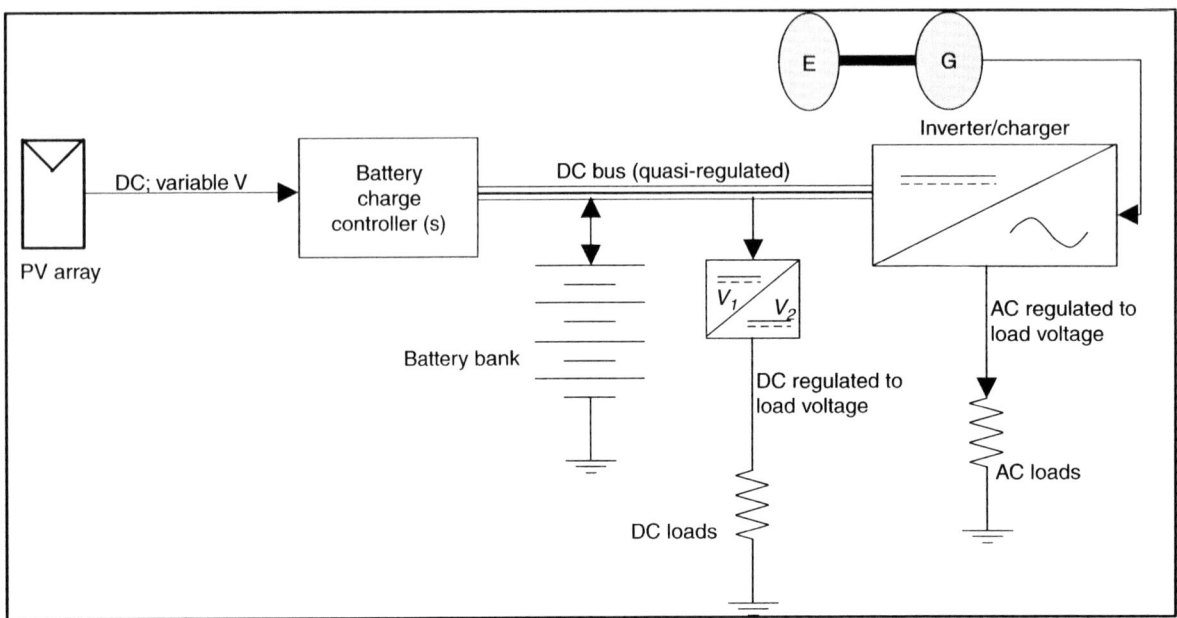

Fig. 7 Generalized configuration of a hybrid photovoltaic (PV) system, in this case using an engine-generator set (E-G) as the redundant generator.

PV POWER ELECTRONICS

As is clear in Figs. 6–8, power electronic converters, or so-called switch-mode power converters, are critically important in PV systems. The battery charge controllers and inverters fall into this category.[63–65]

Charge controllers are typically DC–DC converters (their inputs and outputs are both DC, but at different current/voltage levels—see Figs. 6 and 7). Inverters, by definition, have DC inputs but AC outputs (see Figs. 6–8). As shown in Fig. 7, some inverters are bidirectional in that power can flow from DC to AC (inverter mode) or from AC to DC. This latter mode is used to import power from the hybrid generator to recharge the batteries, and thus it is called the charger mode.

Photovoltaic cells have one operating point (that is, one voltage and current) at which they are maximally efficient. Thus, one job of PV power electronics is to attempt to keep the PV array operating at that highest-efficiency point, or maximum power point, at all times. Specialized control algorithms called maximum power point tracking (MPPT) algorithms accomplish this task. The MPPT algorithm would be in the charge controllers in Figs. 6 and 7 and in the inverter in Fig. 8.

The charge controller must also manage the charging and discharging of the batteries. As will be discussed below, batteries must be charged and discharged carefully to maximize their lifetimes and minimize operational problems. Charge controllers must balance the battery's needs with the requirement for MPPT. Notice that MPPT is not always possible—if the batteries are fully charged and the electrical load is being satisfied, then even if the PV array is capable of producing more energy, that extra energy has nowhere to go, and the PV array must be operated at a point other than its maximum power point.

The inverter's primary job is to supply AC power. In Figs. 6 and 7, the inverter produces an AC output voltage. The wave shape of the output voltage may be a sinewave, as in utility systems, but it may also be a square wave or a so-called modified sine wave (see Fig. 9). Square wave and modified sine wave inverters are generally more efficient

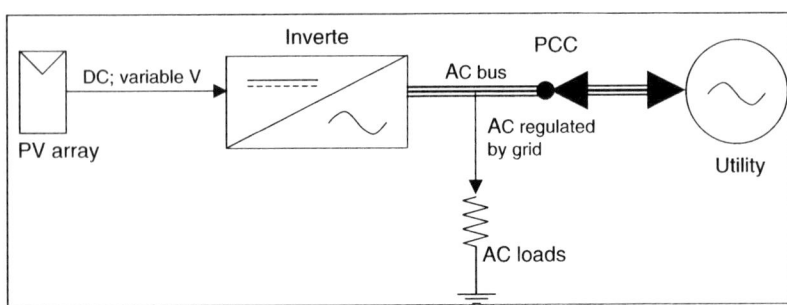

Fig. 8 Generalized configuration of a grid-connected photovoltaic (PV) system.

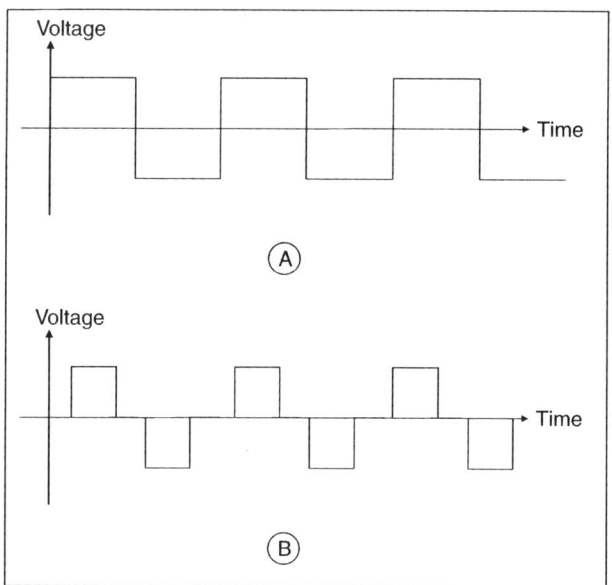

Fig. 9 Two common nonsinusoidal inverter output waveforms. (A) square wave. (B) so—called "modified sine wave."

and less expensive than sinewave inverters, and they are used whenever the load can tolerate the nonsinusoidal waveform. In Fig. 8, the inverter's output voltage is fixed by the utility grid, and thus the inverter's output in that case is an AC current (not voltage) and must be sinusoidal.

ENERGY STORAGE FOR PV

Photovoltaic arrays only convert energy when light strikes them, so it is necessary to use energy storage to provide power to loads at night and during extended cloudy periods. (Note that PV arrays do work when there are clouds; they simply don't produce as much energy on cloudy days.)

The most common type of storage in PV systems is lead-acid batteries.[66–68] There are several types of lead-acid battery used in PV. Flooded batteries use a liquid electrolyte. They are the longest lasting (if maintained properly), but water must be added to them periodically. It is possible to use special valves to reduce the need for addition of water, and batteries so equipped are called valve-regulated lead-acid (VRLA) batteries. These require less maintenance but are not as long-lived. One can also use different types of electrolyte, such as the gel used in gel-cell batteries. Gel cells require essentially no maintenance but are shorter-lived than VRLA.

Lead-acid battery technology is mature, but it is a bit challenging to use.[68,69] If they are discharged too deeply, they permanently lose storage capacity. If they are overcharged, some of the electrolyte can be vaporized and lost. The charging and discharging rates must also be carefully controlled to avoid permanent loss of capacity.[60] If any of these requirements are not met, the battery lifetime will be shortened.[68,69] Also, when they are discharged they can freeze at a relatively high temperature, so they must be protected against this.[69] Still, they are widely available and they are the least expensive type of battery,[68] so they continue to be the most common type used in PV.

Other types of batteries can be used. For example, pocket-plate nickel–cadmium batteries have some properties that are better than lead-acid for PV applications,[68,69] but Ni-Cads are becoming less available because of concerns over the toxicity of cadmium. Nickel metal hydride (NiMH) batteries retain some of the advantages of Ni-Cads without using cadmium, and thus NiMH may be a promising candidate in some applications. Both Ni-Cads and NiMH batteries are significantly more expensive than lead-acid at this time.

Eventually, hydrogen may prove to be the ultimate storage medium.[70] Photovoltaic could be used to electrolyze water into its hydrogen and oxygen constituents, which could be converted back to electricity by a fuel cell or combustion turbine or used to power vehicles. Although very promising because of its sustainability, its potential to displace oil use, and its reliance only on very

Fig. 10 A photovoltaic (PV)-powered rail yard switch near Alliance, Nebraska (Photo by Susan Ropp).

Fig. 11 A photovoltaic (PV)-powered instrumentation package near Gordon, Nebraska (Photo by Susan Ropp).

common resources, hydrogen storage is too expensive to be feasible at present.

PV APPLICATIONS

As mentioned before, the key figure of merit for PV energy is an economic one—the energy cost, usually in dollar per kilowatt hour. It is common practice to compute the life-cycle cost or levelized energy cost of the PV system, which is the energy cost considering all PV-related costs over the entire lifetime of the system.[71]

Even with current technology, there are many applications today in which PV is cost-competitive.[72–74] In particular, small off-grid loads are good candidates for PV. These include autonomous instrumentation systems, highway and marine navigation and signaling systems, remote communications repeaters, and livestock water pumping. A few examples are shown in Figs. 5, 10–13. In addition, in the developing world, almost two billion people live without electricity and are effectively "off grid." Photovoltaic can be cost effectively used to provide them with basic lighting, refrigeration, water pumping, and communications.

However, the present energy cost of dispatchable PV (that is, PV + batteries) is still on the order of five times higher than that of electricity from the grid. This situation is changing, though; the cost of externalities and other

Fig. 12 Traffic signaling applications: a temporary stoplight in a construction zone in Brookings, South Dakota (Photo by Michael Ropp).

Fig. 13 Traffic signaling applications: a sign warning of an upcoming intersection near Punkin Center, Colorado (Photo by Susan Ropp).

costs will continue to add to the cost of fossil-fuel power while technology and economies of scale continue to drive PV costs down. The range of applications in which PV is cost-competitive continues to expand, and perhaps one day dispatchable PV will reach a cost at which it provides a significant portion of the energy we get from the grid. Energy storage remains a critical need, but new advancements in PV device technology give good grounds for optimism that at least that part of the system costs will someday reach these goals.

REFERENCES

1. Zweibel, K. *Harnessing Solar Power: The Photovoltaics Challenge*; Plenum Press: New York, 1990; 2.
2. Perlin, J. The Story of Solar Cells. In *Organic Photovoltaics: Mechanisms, Materials and Devices*; Sun, S.-S., Sariciftci, N., Eds., Taylor and Francis: Boca Raton, FL, 2005; 3.
3. Komp, R. *Practical Photovoltaics: Electricity from Solar Cells*; Aatec Publications: Ann Arbor, MI, 1995; xii–xiii.
4. Messenger, R.; Ventre, J. *Photovoltaic Systems Engineering*, 2nd Ed.; CRC Press: Boca Raton, FL, 2004; 15.
5. Perlin, J. The story of solar cells. In *Organic Photovoltaics: Mechanisms, Materials and Devices*; Sun, S.-S., Sariciftci, N., Eds., Taylor and Francis: Boca Raton, FL, 2005; 4.
6. Zweibel, K. *Harnessing Solar Power: The Photovoltaics Challenge*; Plenum Press: New York, 1990; 4.
7. Zweibel, K. *Harnessing Solar Power: The Photovoltaics Challenge*; Plenum Press: New York, 1990; 5.
8. Zweibel, K. *Harnessing Solar Power: The Photovoltaics Challenge*; Plenum Press: New York, 1990; 6.
9. Perlin, J. The story of solar cells. In *Organic Photovoltaics: Mechanisms, Materials and Devices*; Sun, S.-S., Sariciftci, N., Eds., Taylor and Francis: Boca Raton, FL, 2005; 7.
10. Green, M. *Silicon Solar Cells*; Centre for Photovoltaic Devices and Systems: Sydney, Australia, 1995; 1.
11. Zweibel, K. *Harnessing Solar Power: The Photovoltaics Challenge*; Plenum Press: New York, 1990; 6.
12. Perlin, J. The story of solar cells. In *Organic Photovoltaics: Mechanisms, Materials and Devices*; Sun, S.-S., Sariciftci, N., Eds., Taylor and Francis: Boca Raton, FL, 2005; 9.
13. Green, M. *Silicon Solar Cells*; Centre for Photovoltaic Devices and Systems: Sydney, Australia, 1995; 2.
14. Hamakawa, Y. Background and motivation for thin-film solar cell development. *Thin-Film Solar Cells: Next Generation Photovoltaics and Its Applications*, Springer: Heidelberg, 2004; 3–4.
15. Maycock, P. World photovoltaic markets. In *Practical Handbook of Photovoltaics*; Markvart, T., Castañer, L., Eds., Elsevier: Oxford, England, 2003; 887–912.
16. Duffie, J.; Beckman, W. *Solar Engineering of Thermal Processes*; Wiley: New York, 1991; 10–11.
17. Duffie, J.; Beckman, W. *Solar Engineering of Thermal Processes*; Wiley: New York, 1991; 3–141.
18. Page, J. The role of solar radiation climatology in the design of photovoltaic systems. In *Practical Handbook of Photovoltaics*; Markvart, T., Castañer, L., Eds., Elsevier: Oxford, England, 2003; 5–66.
19. Blankenship, R. Natural organic photosynthetic solar energy transduction. In *Organic Photovoltaics: Mechanisms, Materials and Devices*; Sun, S.-S., Sariciftci, N., Eds., Taylor and Francis: Boca Raton, FL, 2005; 37–48.
20. Nelson, J. *The Physics of Solar Cells*; Imperial College Press: London, 2003; [Chapters 1–7].
21. Gray, J. The physics of the solar cell. In *Handbook of Photovoltaic Science and Engineering*; Luque, A., Hegedus, S., Eds., Wiley: Chichester, England, 2003; [Chapter 3].
22. Hovel, H. Solar cells In *Semiconductors and Semimetals*, Vol. 11; Academic Press: Orlando, FL, 1975; [Chapters 1–4].
23. Nelson, J. *The Physics of Solar Cells*; Imperial College Press: London, 2003; 188–198.
24. Green, M. *Silicon Solar Cells*; Centre for Photovoltaic Devices and Systems: Sydney, Australia, 1995; 201–282.
25. Messenger, R.; Ventre, J. *Photovoltaic Systems Engineering*, 2nd Ed.; CRC Press: Boca Raton, FL, 2004; 373–386.
26. Tobías, I.; del Cañizo, C.; Alonso, J. Crystalline silicon solar cells and modules. In *Handbook of Photovoltaic Science and Engineering*; Luque, A., Hegedus, S., Eds., Wiley: Chichester, England, 2003; 257.
27. Möller, H. *Semiconductors for Solar Cells*; Artech House: Norwood, MA, 1993; 4.
28. Hegedus, S.; Luque, A. Status, trends, challenges and the bright future of solar electricity from photovoltaics. In

Handbook of Photovoltaic Science and Engineering; Luque, A., Hegedus, S., Eds., Wiley: Chichester, England, 2003; 25.
29. Möller, H. *Semiconductors for Solar Cells*; Artech House: Norwood, MA, 1993; 65.
30. Nelson, J. *The Physics of Solar Cells*; Imperial College Press: London, 2003; 221–222.
31. Möller, H. *Semiconductors for Solar Cells*; Artech House: Norwood, MA, 1993; 5.
32. Saito, K.; Nishimoto, T.; Hayashi, R.; Fukae, K.; Ogawa, K. Production of a-Si:H/a-SiGe:H/a-SiGe:H stacked solar cells and their applications. *Thin-Film Solar Cells: Next Generation Photovoltaics and Its Applications*, Springer: Heidelberg, 2004; 132.
33. Deb, S. Recent advances and future opportunities for thin-film solar cells. *Thin-Film Solar Cells: Next Generation Photovoltaics and Its Applications*, Springer: Heidelberg, 2004; 19.
34. Nomoto, K.; Tomita, T. Development of amorphous-silicon single-junction solar cells and their application systems. *Thin-Film Solar Cells: Next Generation Photovoltaics and Its Applications*, Springer: Heidelberg, 2004; 113.
35. Schock, H.-W. Properties of chalcopyrite-based materials and film deposition for thin-film solar cells. *Thin-Film Solar Cells: Next Generation Photovoltaics and Its Applications*, Springer: Heidelberg, 2004; 163–178.
36. Saito, K.; Nishimoto, T.; Hayashi, R.; Fukae, K.; Ogawa, K. Production of a-Si:H/a-SiGe:H/a-SiGe:H stacked solar cells and their applications. *Thin-Film Solar Cells: Next Generation Photovoltaics and Its Applications*, Springer: Heidelberg, 2004; 121–137.
37. Lane, P.; Kafafi, Z. Solid-state organic photovoltaics: a review of molecular and polymeric devices. In *Organic Photovoltaics: Mechanisms, Materials and Devices*; Sun, S.-S., Sariciftci, N., Eds., Taylor and Francis: Boca Raton, FL, 2005; 49–97.
38. Lane, P.; Kafafi, Z. Solid-state organic photovoltaics: A review of molecular and polymeric devices. In *Organic Photovoltaics: Mechanisms, Materials and Devices*; Sun, S.-S., Sariciftci, N., Eds., Taylor and Francis: Boca Raton, FL, 2005; 50; see also page 97.
39. Deb, S. Recent advances and future opportunities for thn-film solar cells. *Thin-Film Solar Cells: Next Generation Photovoltaics and Its Applications*, Springer: Heidelberg, 2004; 37–38.
40. Nelson, J. *The Physics of Solar Cells*; Imperial College Press: London, 2003; 289–290.
41. Green, M. *Third-Generation Photovoltaics: Advanced Solar Energy Conversion*; Springer: Heidelberg, 2003; 35.
42. Green, M. *Third-Generation Photovoltaics: Advanced Solar Energy Conversion*; Springer: Heidelberg, 2003; 59–67.
43. Olson, J.; Friedman, D.; Kurtz, S. High-efficiency III–V multijunction solar cells. In *Handbook of Photovoltaic Science and Engineering*; Luque, A., Hegedus, S., Eds., Wiley: Chichester, England, 2003; 362.
44. Green, M. *Third-Generation Photovoltaics: Advanced Solar Energy Conversion*; Springer: Heidelberg, 2003; 95–107.
45. Nelson, J. *The Physics of Solar Cells*; Imperial College Press: London, 2003; 302–309.
46. Green, M. *Third-Generation Photovoltaics: Advanced Solar Energy Conversion*; Springer: Heidelberg, 2003; 69–79.
47. Nelson, J. *The Physics of Solar Cells*; Imperial College Press: London, 2003; 309–318.
48. Green, M. *Third-Generation Photovoltaics: Advanced Solar Energy Conversion*; Springer: Heidelberg, 2003; 81–93.
49. Nelson, J. *The Physics of Solar Cells*; Imperial College Press: London, 2003; 318–323.
50. Green, M. *Third-Generation Photovoltaics: Advanced Solar Energy Conversion*; Springer: Heidelberg, 2003; 107.
51. Swanson, R. Photovoltaic concentrators. In *Handbook of Photovoltaic Science and Engineering*; Luque, A., Hegedus, S., Eds., Wiley: Chichester, England, 2003; 449.
52. Verlinden, P. High-efficiency concentrator silicon solar cells. In *Practical Handbook of Photovoltaics*; Markvart, T., Castañer, L., Eds., Elsevier: Oxford, England, 2003; 436.
53. Jones, J. Time to concentrate. Renew. Energy **2005**, *September–October*, 78–84.
54. Slooff, L.H.; Kinderman, R.; Burgers, A.R.; et al. *The Luminescent Concentrator: A Bright Idea for Spectrum Conversion?*, Proceedings of the 20th EPVSEC, Barcelona, Spain, June 6–10 2005.
55. Messenger, R.; Ventre, J. *Photovoltaic Systems Engineering*, 2nd Ed.; CRC Press: Boca Raton, FL, 2004; 53–54.
56. Messenger, R.; Ventre, J. *Photovoltaic Systems Engineering*, 2nd Ed.; CRC Press: Boca Raton, FL, 2004; 72.
57. Lorenzo, E. Energy collected and delivered by PV modules. In *Handbook of Photovoltaic Science and Engineering*; Luque, A., Hegedus, S., Eds., Wiley: Chichester, England, 2003; 956–962.
58. Lasnier, F.; Ang, T. *Photovoltaic Engineering Handbook*; IOP Publishing Ltd.: Bristol, England, 1990; 339–370.
59. Messenger, R.; Ventre, J. *Photovoltaic Systems Engineering*, 2nd Ed.; CRC Press: Boca Raton, FL, 2004; [Chapters 4, 7, and 8].
60. Köthe, H. Solar electric power supply with batteries.In *Battery Technology Handbook* 2nd Ed.; Kiehne, H. Ed.; Marcel Dekker: New York, 2003; [Chapter 11].
61. Markvart, T. *Solar Electricity*, 2nd Ed.; Wiley: Chichester, England, 2000; [Chapter 4].
62. Silvestre, S. Review of system design and sizing tools. In *Practical Handbook of Photovoltaics*; Markvart, T., Castañer, L., Eds., Elsevier: Oxford, England, 2003; 543–561.
63. Ross, J. System electronics. In *Practical Handbook of Photovoltaics*; Markvart, T., Castañer, L., Eds., Elsevier: Oxford, England, 2003; 565–585.
64. Schmid, J.; Schmidt, H. Power conditioning for photovoltaic power systems. In *Handbook of Photovoltaic Science and Engineering*; Luque, A., Hegedus, S., Eds., Wiley: Chichester, England, 2003; 863–902.
65. Lasnier, F.; Ang, T. *Photovoltaic Engineering Handbook*; IOP Publishing Ltd.: Bristol, England, 1990; 139–162.
66. Sauer, D. Electrochemical storage for photovoltaics. In *Handbook of Photovoltaic Science and Engineering*; Luque, A., Hegedus, S., Eds., Wiley: Chichester, England, 2003; 799–862.
67. Vincent, C.; Scrosati, B. *Modern Batteries: An Introduction to Electrochemical Power Sources*, 2nd Ed.; Wiley: New York, 1997; 142–161.

68. Spiers, D. Batteries in PV systems. In *Practical Handbook of Photovoltaics*; Markvart, T., Castañer, L., Eds., Elsevier: Oxford, England, 2003; 589–631.
69. Lasnier, F.; Ang, T. *Photovoltaic Engineering Handbook*; IOP Publishing Ltd.: Bristol, England, 1990; 114–133.
70. Markvart, T. *Solar Electricity*, 2nd Ed.; Wiley: Chichester, England, 2000; 260–266.
71. Messenger, R.; Ventre, J. *Photovoltaic Systems Engineering*, 2nd Ed.; CRC Press: Boca Raton, FL, 2004; 145.
72. Wolfe, P. Solar-powered products. In *Practical Handbook of Photovoltaics*; Markvart, T., Castañer, L., Eds., Elsevier: Oxford, England, 2003; 771–789.
73. Preiser, K. Photovoltaic systems. In *Handbook of Photovoltaic Science and Engineering*; Luque, A., Hegedus, S., Eds., Wiley: Chichester, England, 2003; 753–785.
74. Lasnier, F.; Ang, T. *Photovoltaic Engineering Handbook*; IOP Publishing Ltd.: Bristol, England, 1990; 371–500.

Physics of Energy

Milivoje M. Kostic
Department of Mechanical Engineering, Northern Illinois University, DeKalb, Illinois, U.S.A.

Abstract

The concept and definition of energy are elaborated, as well as different forms and classification of energy are presented. Energy is a fundamental concept indivisible from matter and space, and energy exchanges or transfers are associated with all processes (or changes), thus indivisible from time. Actually, energy is "the building block" and fundamental property of matter and space, thus fundamental property of existence. Any and every material system in nature possesses energy. The structure of any matter and field is "energetic," meaning active, i.e., photon waves are traveling in space, electrons are orbiting around an atom nucleus or flowing through a conductor, atoms and molecules are in constant rotation, vibration or random thermal motion, etc. When energy is exchanged or transferred from one system to another it is manifested as work or heat. In addition, the First Law of energy conservation and the Second Law of energy degradation and entropy generation are presented along with relevant concluding remarks. In summary, energy is providing existence, and if exchanged, it has ability to perform change.

INTRODUCTION: FROM WORK TO HEAT TO GENERAL ENERGY CONCEPT

Energy is a fundamental concept indivisible from matter and space, and energy exchanges or transfers are associated with all processes (or changes), thus indivisible from time. Actually, energy is "the building block" and fundamental property of matter and space, thus a fundamental property of existence. Energy transfer is needed to produce a process to change other properties. Also, among all properties, energy is the only one which is directly related to mass and vice versa: $E = mc^2$ (known in some literature as *mass energy*, the c is the speed of light in a vacuum), thus the mass and energy are interrelated.[1,2]

Energy moves cars and trains, and boats and planes. It bakes foods and keeps them frozen for storage. Energy lights our homes and plays our music. Energy makes our bodies alive and grow, and allows our minds to think. Through centuries, people have learned how to use energy in different forms in order to do work more easily and live more comfortably. No wonder that energy is often defined as *ability to perform work*, i.e., as a potential for energy transfer in a specific direction (displacement in force direction) thus achieving a purposeful process, as opposed to dissipative (less-purposeful) energy transfer in form of heat. The above definition could be generalized as: *energy is providing existence, and if exchanged, it has the ability to perform change.*[3]

Any and every material system in nature possesses energy. The structure of any matter and field is "energetic," meaning active, i.e., photon waves are traveling in space, electrons are orbiting around an atom nucleus or flowing through a conductor, atoms and molecules are in constant rotation, vibration or random thermal motion, etc., see Table 1 and Fig. 1. Thus energy is a property of a material system (further on simply referred to as *system*), and together with other properties defines the system equilibrium state or existence in space and time.

Energy in transfer (E_{transfer}) is manifested as work (W) or heat (Q) when it is exchanged or transferred from one system to another, as explained next (see Fig. 2).

Work

Work is a mode of energy transfer from one acting body (or system) to another resisting body (or system), with an acting force (or its component) in the direction of motion, along a path or displacement. A body that is acting (forcing) in motion-direction in time, is doing work on another body which is resisting the motion (displacement rate) by an equal resistance force, including inertial force, in opposite direction of action. The acting body (energy source) is imparting (transferring away) its energy to the resisting body (energy sink), and the amount of energy transfer is the work done by the acting onto the resisting body, equal to the product of the force component in the motion direction multiplied with the corresponding displacement, or vice versa (force multiplied by displacement component in the force direction), see Fig. 3. If the force ($\vec{F}$) and displacement vectors ($d\vec{s} = d\vec{r}$) are not constant, then integration of differential work transfers from initial (1) to final state (2), defined by the corresponding position vectors $\vec{r}$, will be necessary, see Fig. 4.

Keywords: Energy; Entropy; Heat; Heat engine; Power; Thermal energy; Total internal energy; Work.

Table 1 Material system structure and related forces and energies

Particles	Forces	Energies
Atom nucleus	Strong and weak	Nuclear
Electron shell, or electron flow	Electromagnetic	Electrical, magnetic, electromagnetic
Atoms/molecules	Inter-atomic/molecular	Chemical
Molecules	Inertial due to random collision and, potential inter-molecular	Sensible thermal
Molecules	Potential inter-molecular	Latent thermal
Molecules	Potential inter-molecular	Mechanical elastic
System mass	Inertial and gravitational	Mechanical kinetic and gravitational potential

The work is a directional energy transfer. However, it is a scalar quantity like energy, and is distinctive from another energy transfer in form of *heat*, which is due to random motion (chaotic or random in all directions) and collisions of system molecules and their structural components.

Work transfer cannot occur without existence of the resisting body or system, nor without finite displacement in the force direction. This may not always be obvious. For example, if we are holding a heavy weight or pushing hard against a stationary wall, there will be no work done against the weight or wall (neglecting their small deformations). However, we will be doing work internally due to contraction and expansion of our muscles (thus force with displacement), and that way converting (spending) a lot of chemical energy via muscle work, then dissipating it into thermal energy and heat transfer (sweating and getting tired).

Heat

Heat is another mode of energy transfer from one system at higher temperature to another at lower temperature due to their temperature difference. Fire was civilization's first great invention, long before people could read and write, and wood was the main heat source for cooking and heating for ages. However, true physical understanding of the nature of heat was discovered rather recently, in the middle of the nineteenth century, thanks to the development of the kinetic theory of gasses. Thermal energy and heat are defined as the energy associated with the random motion of atoms and molecules. The prior concept of heat was based on the caloric theory proposed by the French chemist Antoine Lavoisier in 1789. The caloric theory defines heat as a massless, colorless, odorless, and tasteless, fluid-like substance called the *caloric* that can be transferred or "poured" from one body into another. When caloric was added to a body, its temperature increased and vice versa. The caloric theory came under attack soon after its introduction. It maintained that heat is a substance that could not be created or destroyed. Yet it was known that heat can be generated indefinitely by rubbing hands together or from mechanical energy during friction, like mixing or similar. Finally, careful experiments by James P. Joule published in 1843, quantified correlation between mechanical work and heat, and thus put the caloric theory to rest by convincing the skeptics that heat was not the caloric substance after all. Although the caloric theory was totally abandoned in the middle of the nineteenth century, it contributed greatly to the development of thermodynamics and heat transfer.

Heat may be transferred by three distinctive mechanisms: conduction, convection, and thermal radiation, see Fig. 5. Heat conduction is the transfer of thermal energy due to interaction between the more energetic particles of a substance, like atoms and molecules (thus at higher temperature), to the adjacent less energetic ones (thus at lower temperature). Heat convection is the transfer of thermal energy between a solid surface and the adjacent moving fluid, and it involves the combined effects of conduction and fluid motion. Thermal radiation is the transfer of thermal energy due to the emission of electromagnetic waves (or photons) which are products of random interactions between energetic particles of a substance (thus related to the temperature). During those interactions the electron energy level is changed, thus causing emission of photons, i.e., electromagnetic thermal radiation.

The Joule's experiments of establishing equivalency between work and heat paved the way of establishing the concept of internal thermal energy, to generalize the concept of energy, and to formulate the general law of energy conservation. The total internal energy includes all other possible but mechanical energy types or forms, including chemical and nuclear energy. This allowed extension of the well-established law of mechanical energy conservation to the general law of energy conservation, known as the First Law of Thermodynamics, which includes all possible energy forms that a system could possess, and heat and all types of work as all possible energy-transfers between the systems. The Law of Energy Conservation will be elaborated later.

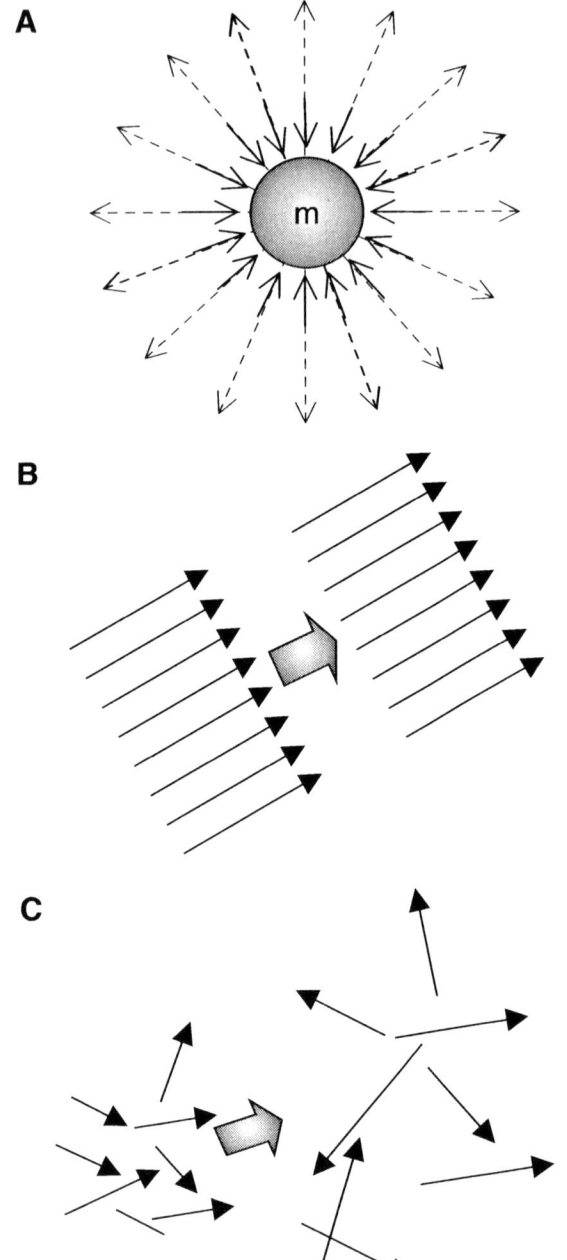

Fig. 1 Different types of energy: (A) Potential gravitational and electromagnetic radiation; (B) Organized energy as work transfer; (C) Disorganized thermal energy as heat transfer.

Energy

Energy is fundamental property of a physical system and refers to its potential to maintain a system identity or structure and to influence changes with other systems (via forced interaction) by imparting work (forced directional displacement) or heat (forced chaotic displacement/motion of a system molecular or related structures). Energy exists in many forms: electromagnetic (including light), electrical, magnetic, nuclear, chemical, thermal, and mechanical (including kinetic,

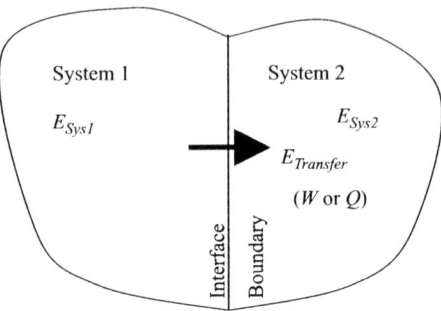

Fig. 2 Energy as material system property and energy transfer from one system to another.

elastic, gravitational, and sound); where, for example, electro-mechanical energy may be kinetic or potential, while thermal energy represents overall potential and chaotic motion energy of molecules and/or related micro structure.[3,4]

Energy, Work, and Heat Units

Energy, Work, and Heat Units. Energy is manifested via work and heat transfer, with corresponding Force × Length dimension for work (N m, kg_fm, and lb_fft, in SI, metric and English system of units, respectively); and the caloric units, in kilocalorie (kcal) or British-thermal-unit (Btu), the last two defined as heat needed to increase a unit mass of water (at specified pressure and temperature) for one degree of temperature in their respective units. Therefore, the water specific heat is 1 kcal/(kg °C) = 1 Btu/(lb °F) by definition, in metric and English system of units, respectively. It was demonstrated by Joule that 4187 N m of work, when dissipated in heat, is equivalent to 1 kcal. In his honor, 1 N m of work is named after him as 1 Joule, or 1 J, the SI energy unit, also equal to electrical work of 1 W s = 1 V A s. The SI unit for power, or work rate, is watt, i.e., 1 J/s = 1 W, and also corresponding units in other system of units, like Btu/h, etc. The Horse Power is defined as 1 HP = 550 lb_fft/s = 745.7 W. Other common units for energy, work and heat are given in Table 2.

$$W = \vec{F} \cdot \vec{d} = \underbrace{(F \cdot \cos\alpha)}_{F_d = \text{force in d-direction}} \cdot d = F \cdot \underbrace{(d \cdot \cos\alpha)}_{d_F = \text{displacement in F-direction}}$$

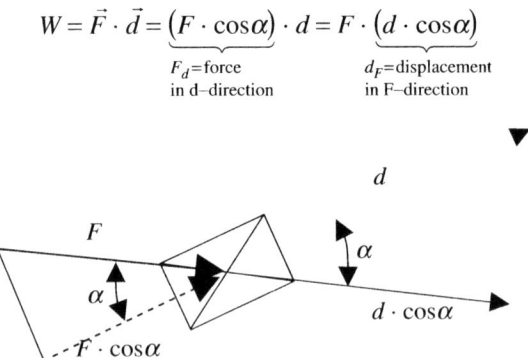

Fig. 3 Work, force, and displacement.

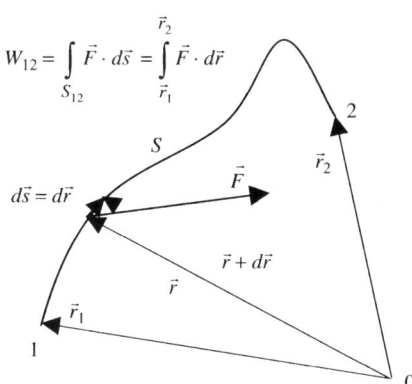

Fig. 4 Work along an arbitrary path.

ENERGY FORMS AND CLASSIFICATIONS: ENERGY-TRANSFER VS ENERGY-PROPERTY

Any and all changes or processes (happening in space and time) are caused by energy exchanges or transfers from one substance (*system* or *subsystem*) to another, see Fig. 2. A part of a system may be considered as a subsystem if energy transfer within a system is taking place, and inversely, a group of interacting systems may be considered as a larger isolated system, if they do not interact with the rest of the surroundings. Energy transfer may be in organized form (different types of work transfer due to different force actions) or in chaotic disorganized form (heat transfer due to temperature difference). Energy transfer into a system builds up energy–potential or generalized-force (called simply potential for short, like pressure, temperature, voltage, etc.) over energy—displacement (like volume, entropy, charge, etc.). Conversely, if energy is transferred from a system, its energy potential is decreased. That is why net-energy is transferred from higher to lower energy potential only, due to virtually infinite probability of equi-partition of energy over system micro-structure, causing system equilibrium, otherwise virtually impossible singularity of energy potential at infinite magnitude would result.[4]

There are many forms and classifications of energy, see Table 3, all of which could be classified as *microscopic* (or

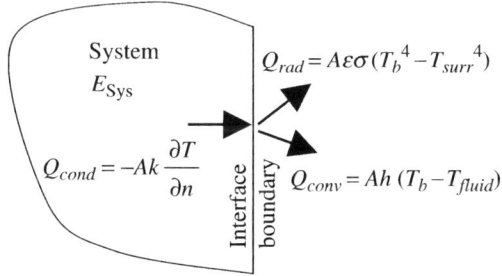

Fig. 5 Heat as energy-transfer by conduction, convection, and radiation is due to a difference in temperature.

internal within a system microstructure) and/or *macroscopic* (or external as related to the system mass as a whole with reference to other systems). Furthermore, energy may be "quasi-potential" (associated with a system equilibrium state and structure, i.e., system property) or "quasi-kinetic" (energy in-transfer from one system or one structure to another, in form of *work* or *heat*).

Every material system state is an equilibrium potential "held" by forces (space force-fields), i.e., the forces "define" the potential and state—action and reaction; otherwise a system will undergo dynamic change (in time), or quasi-kinetic energy exchange towards another stable equilibrium. Atoms (mass) are "held" by atomic and electromagnetic forces in small scale and by gravity in large scale, see Fig. 1A, otherwise mass would disintegrate ("evaporate" or radiate into energy) like partly in nuclear reactions—*nuclear* energy or electromagnetic radiation. Molecules are "held" by electro-chemical bonding (valence) forces (chemical reactions—*chemical* energy). Liquids are "held" by latent intermolecular forces (gas condensation, when kinetic energy is reduced by cooling—*latent thermal* energy). Solids are "held" by "firm" intermolecular forces (freezing/solidification when energy is further reduced by cooling—*latent thermal* energy again). *Sensible thermal* energy represents intermolecular potential energy and energy of random molecular motion, and is related to temperature of a system. "Holding" potential forces may be "broken" by energy transfer (e.g., radiation, heating, high-energy particles interactions, etc.). States and potentials are often "hooked" (i.e., stable) and thus need to be "unhooked" (or to "be broken") to overcome existing "threshold" or equilibrium, like in igniting combustion, starting nuclear reaction, etc.

As stated above, energy transfer can be directional (purposeful or organized) and chaotic (dissipative or disorganized). For example, mass-in-motion, *mechanical kinetic* energy, and electricity-in-motion, *electrical kinetic* energy, are organized kinetic energies (Fig. 1B), while *thermal* energy is disorganized chaotic energy of motion of molecules and atoms (Fig. 1C). System energy may be defined with reference to position in a vector-force field, like *elastic potential* (stress) energy, *gravitational potential* energy, or *electromagnetic* field energy. There are many different energy forms and types (see Table 3). We are usually not interested in (absolute) energy level, but in the change of energy (during a process) from an initial state (i) to a final state (f), and thus zero reference values for different energy forms are irrelevant, and often taken arbitrarily for convenience. The followings are basic correlations for energy changes of several typical energy forms, often encountered in practice: motion kinetic energy ($E_K = KE$) as a function of system velocity (V); potential elastic-deformational, e.g., pressure elastic or spring elastic energy ($E_{Pdeff} = E_{Pp} = E_{Ps}$) as a function of spring deformation displacement (x); gravitational potential energy ($E_{Pg} = PE_g$) as a function of gravitational

Table 2 Typical energy units with conversion factors

Energy units	J	kWh	Btu
1 Joule (J)	1	2.78×10^{-7}	9.49×10^{-4}
1 Kilowatt-hour (kWh)	3.6×10^{6}	1	3.412×10^{3}
1 Kilocalorie (kcal = Cal = 1000 cal)	4187	1.19×10^{-3}	3.968
1 British thermal unit (Btu)	1055	2.93×10^{-4}	1
1 Pound-force foot (lb$_f$·ft)	1.36	3.78×10^{-7}	1.29×10^{-3}
1 Electron volt (eV)	1.6×10^{-19}	4.45×10^{-26}	1.52×10^{-22}
1 Horse Power × second (HP sec)	745.7	2.071×10^{-4}	0.707

elevation (z); and sensible thermal energy ($E_U = U$) as a function of system temperature (T):

$$\Delta E_k = \frac{1}{2} m \left(V_f^2 - V_i^2 \right); \quad \Delta E_{Ps} = \frac{1}{2} k \left(x_f^2 - x_i^2 \right)$$

$$\Delta E_{Pg} = mg(z_f - z_i); \quad \Delta E_U = m c_v (T_f - T_i) \quad (1)$$

If the reference energy values are taken to be zero when the above initial (i) variables are zero, then the above equations will represent the energy values for the final values (f) of the corresponding variables. If the corresponding parameters, spring constant k, gravity g, or constant-volume specific heat c_v, are not constant, then integration of differential energy changes from initial to final state will be necessary.

Energy transfer via work W (net-out), and heat transfer Q (net-in), may be expressed for reversible processes as product of related energy–potentials (pressure P, or temperature T) and corresponding energy–displacements (change of volume V and entropy S, respectively), i.e.:

$$W_{12} = \vec{F} \cdot \vec{d}$$
$$= (P \cdot A\vec{n}) \cdot \underbrace{\vec{d}}_{\Delta V} = P \cdot \Delta V_{12} |_{P \neq \text{const}} = \int_{V_1}^{V_2} P dV \quad (2)$$

$$Q_{12} = T \Delta S_{12} |_{T \neq \text{const}} = \int_{S_1}^{S_2} T dS \quad (3)$$

Note, in Eq. 2, that force cannot act at a point but is distributed as pressure (P) over some area A (with orthogonal unit vector $\vec{n}$), which when displaced will cause the volume change ΔV. Also note that it is custom in some references to denote heat transfer "in" and work transfer "out" as positive (as they appear in a heat engine). In general, "in" (means "net-in") is negative "out" (means "net-out") and vice versa.

In general, energy transfer between systems is taking place at the system boundary interface and is equal to the product of energy–potential or generalized-force and the corresponding generalized-displacement[5]:

$$\delta E_{\text{Transfer}} = \delta Q_{\text{netIN}} - \left[\sum \delta W_{\text{netOUT}} \right]$$
$$= \delta Q_{\text{netIN}} + \left[\sum \delta W_{\text{netIN}} \right]$$
$$= TdS + \left[\underbrace{-PdV}_{\text{COMPR.}} + \underbrace{\sigma dA}_{\text{STRECHING}} \right.$$
$$+ \underbrace{\tau d(As)}_{\text{SHEARING}} + \underbrace{Vdq}_{\text{CHARGING}} + \underbrace{\vec{E} d(V\vec{P})}_{\text{POLARIZATION}}$$
$$\left. + \underbrace{\mu_o \vec{H} d(V\vec{M})}_{\text{MAGNETIZATION}} + \underbrace{\cdots}_{\text{ETC}} \right] \quad (4)$$

Where the quantities after the last equal sign are: temperature and entropy; pressure and volume; surface tension and area; tangential-stress and area with tangential-displacement, voltage and electrical charge; electric field strength and volume-electric dipole moment per unit volume product; and permeability of free space, magnetic field strength and volume-magnetic dipole moment per unit volume product; respectively.

The total system energy stored within the system, as energy property, is:

$$E_{\text{Sys}} = \underbrace{E_K + E_{Pg} + E_{Pdeff}}_{E_{\text{Mechanical}}}$$
$$+ \underbrace{\underbrace{E_{\text{Uth}}}_{\text{Thermal}} + E_{\text{Ch}} + E_{\text{Nucl}} + E_{\text{El}} + E_{\text{Magn}} + \underbrace{\cdots}_{\text{Etc.}}}_{\text{Internal(total)}} \quad (5)$$

Where the quantities after the equal sign are: kinetic, potential-gravitational, potential-elastic-deformational, thermal, chemical, nuclear, electric, magnetic energies, etc.

THE FIRST LAW OF ENERGY CONSERVATION: WORK–HEAT–ENERGY PRINCIPLE

Newton formulated the general theory of motion of objects due to applied forces (1687). This provided for

Table 3 Energy forms and classifications

Scale		Energy form (Energy storage)		Energy process	Potential (state or field)	Type — Kinetic (motion)		Transfer (release) — Kinetic (motion)	
Macro/external (mass-based)	Micro/internal (structure-based)					Directional[a]	Chaotic dissipative	Work[b] directional[a]	Heat dissipative
		Mechanical							
X		Kinetic	$mV^2/2$	Acceleration		X		X	
X		Gravitational[c]	mgz	Elevation	X			X	
	X	Elastic	$kx^2/2$ or $PV = mP/\rho$	Deformation	X			X	
		Thermal	U_{th}						
	X	Sensible	$U_{th} = mc_{avg}T$	Heating			X		X
	X	Latent	$U_{th} = H_{latent}$	Melting, Evaporation	X				X
	X	Chemical	U_{ch}	Chemical Reaction	X				X
	X	Nuclear	U_{nucl}	Nuclear Reaction	X				X
		Electrical	E_{el}						
	X	Electro-kinetic	$V(It)$ or $LI^2/2$	Electro-current flow		X		X	
	X	Electrostatic	$(It)^2/(2C)$	Electro-charging	X			X	
		Magnetic	E_{magn}	Magnetization	X			X	
	X[d]	Electromagnetic	E_{el_mag}	Radiation		X[d]		X	

[a]Electro-mechanical kinetic energy type (directional/organized, the highest energy quality) is preferable since it may be converted to any other energy form/type with high efficiency.
[b]All processes (involving energy transfer) are to some degree irreversible (i.e., dissipative or chaotic/disorganized).
[c]Due to mass position in a gravitational field.
[d]Electromagnetic form of energy is the smallest known scale of energy.

concepts of mechanical work, kinetic and potential energies, and development of solid-body mechanics. Furthermore, in absence of non-mechanical energy interactions, excluding friction and other dissipation effects, it is straightforward to derive (and thus prove) energy conservation, i.e.:

$$\underbrace{\int_{s_1}^{s_2} F_s ds}_{W_{Fs}} = \int_{s_1}^{s_2} \left(m \underbrace{\frac{dV_s}{dt}}_{a_s} \right) ds = \int_{s_1}^{s_2} \left(m \frac{dV_s}{ds} \underbrace{\left\{ \frac{ds}{dt} \right\}}_{V_s} \right) ds$$

$$= \int_{V_1}^{V_2} mV_s dV_s = \underbrace{\frac{1}{2} m \left(V_2^2 - V_1^2 \right)}_{KE_2 - KE_1} \quad (6)$$

$$(E_{Transfer} = W_{Fs}) = (KE_2 - KE_1 = E_2 - E_1)$$

The above correlation is known as the *work–energy principle*. The work–energy principle could be easily expended to include work of gravity force and gravitational potential energy as well as elastic spring force and potential elastic spring energy.[3]

During a free gravity fall (or free bounce) without air friction, for example, the potential energy is being converted to kinetic energy of the falling body (or vice versa for free bounce), and at any time the total mechanical energy (sum of kinetic and potential mechanical energies) is conserved, i.e., stays the same, see Fig. 6. The mechanical energy is also conserved if a mass freely vibrates on an ideally elastic spring, or if a pendulum oscillates around its pivot, both in absence of dissipative effects, like friction or non-elastic deformation. In general, for work of conservative forces only, the mechanical energy, E_{mech}, for N isolated systems, is conserved since

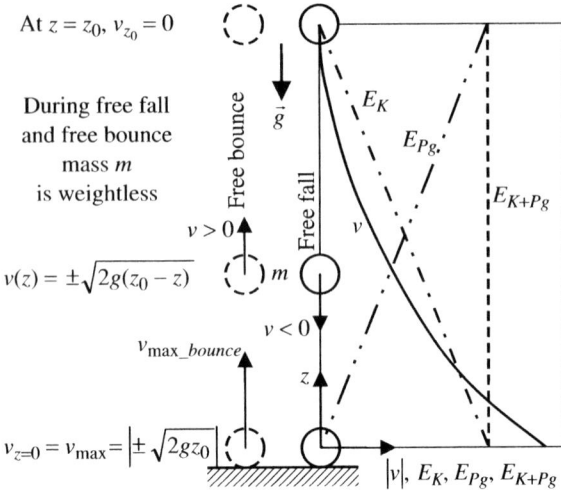

Fig. 6 Energy and work due to conservative gravity force.

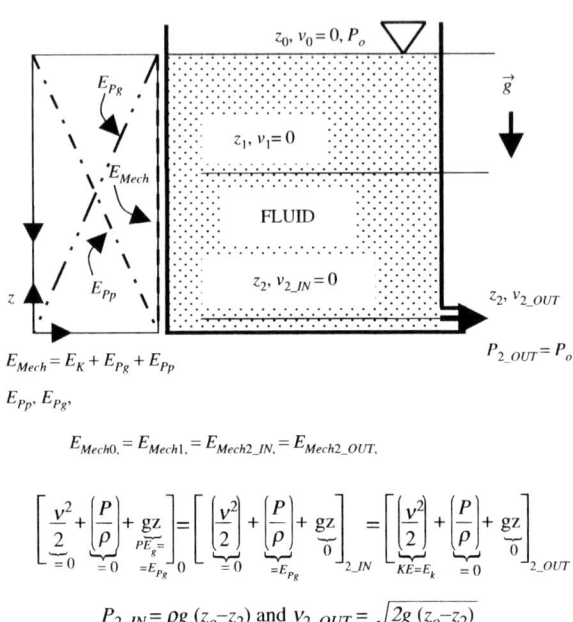

Fig. 7 Conservation of fluid mechanical energy: Bernoulli equation, hydrostatic equation, and Torricelli's orifice velocity.

there is no dissipative conversion into thermal energy and thus no heat transfer, i.e.:

$$E_{\text{mech}} = E_k + E_{Pg} + E_{Ps}$$
$$= \sum_{i=1}^{N} \left(\frac{1}{2} mV^2 + mgz + \frac{1}{2} kx^2 \right)_i$$
$$= \text{const} \quad (7)$$

The mechanical work–energy concept could also be expended to fluid motion by inclusion of elastic-pressure force and potential elastic-pressure energy (referred to in some references as flow work; however, note that elastic-pressure energy is a system property while the flow work is related energy transfer), see the Bernoulli equation below. For flowing or stationary fluid without frictional effects, the mechanical energy is conserved, including fluid elastic—pressure energy, $PV = mP/\rho$ (where V is volume, whereas v is used for velocity here), as expressed by the Bernoulli or hydrostatic equations below, see also Fig. 7.[3]

$$\frac{E_{\text{mech}}}{m} = \frac{1}{m}(E_K + E_{Ps} + E_{Pg})$$
$$= \underbrace{\frac{v^2}{2} + \frac{P}{\rho} + gz}_{\text{Bernoulli equation}} \bigg|_{v=0}$$
$$= \underbrace{\frac{P}{\rho} + gz}_{\text{Hydrostatic equation } (v=0)} = \text{const.} \quad (8)$$

Work against inertial and/or conservative forces (also known as internal, or volumetric, or space potential field), is path-independent and during such a process the mechanical energy is conserved. However, work of non-conservative, dissipative forces is process path-dependent and part of the mechanical energy is converted (dissipated) to thermal energy, see Fig. 8A.

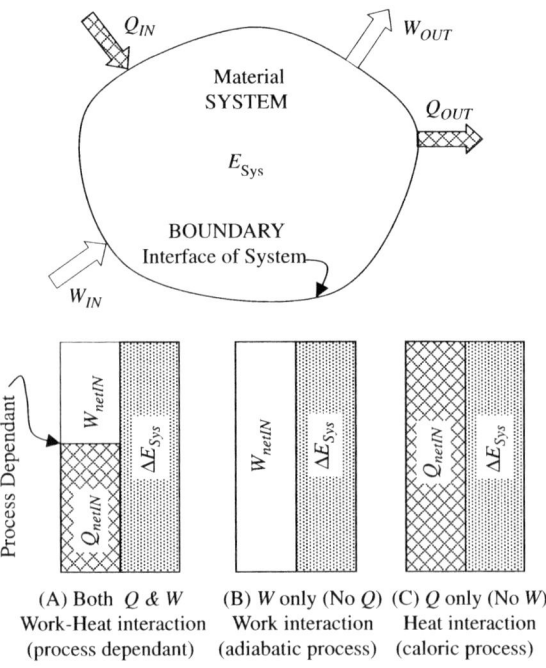

Fig. 8 System energy and energy boundary interactions (transfers) for (A) arbitrary, (B) adiabatic, and (C) caloric processes.

Physics of Energy

When work of non-conservative forces W_{nc}, is exchanged between N isolated systems, from an initial (i) to final state (f), then the total mechanical energy of all systems is reduced by that work amount, i.e.:

$$W_{nc,i \to f} = \left(\sum_{j=1}^{N} E_{\text{mech},j}\right)_i - \left(\sum_{j=1}^{N} E_{\text{mech},j}\right)_f \quad (9)$$

Regardless of the traveled path (or displacement), the work against conservative forces (like gravity or elastic spring in above cases) in absence of any dissipative forces, will depend on the final and initial position (or state) only. However, the work of non-conservative, dissipative forces (W_{nc}) depends on the traveled path since the energy is dissipated during the force displacement, and mechanical energy will not be conserved, but in part converted (via dissipation and heat transfer) into thermal energy, see Eq. 9. This should not be misunderstood with the total energy conservation, which is always the case, and it must include both work and heat transfer, see below.

As already stated, there are many different types of work transfer into (or out of) a system which will change the corresponding energy-form stored in (or released, discharged out of) the system. In addition to work, energy may be transferred as heat caused by temperature difference and associated with change of the thermal energy of a system. Furthermore, different forms of stored energy are often coupled so that one type of energy transfer may change more than one form of stored energy, particularly due to inevitable dissipative conversion of work to heat, and in turn to internal thermal energy. Conversely, heat and thermal energy may be converted into other energy forms. In the absence of nuclear reaction (no conversion of mass into energy, $E=mc^2$), mass and energy are conserved separately for an isolated system, a group of isolated systems, or for the Universe. Since the material system structure is of particulate form, then systems' interactions (collisions at different scale-sizes) will exchange energy during the forced displacement—and similarly to the mechanical energy conservation—the totality of all forms of energy will be conserved,[8] see Fig. 8, which could be expressed as:

$$\underbrace{\sum_{\text{All } i's} E_{i,\text{Trans.}}}_{\text{BOUNDARY}} = \underbrace{\Delta E_{\text{Change}}}_{\text{SYSTEM}} \quad \text{or}$$

$$\underbrace{\sum_{\text{All } j's} W_{j,\text{netIN}} + \sum_{\text{All } k's} Q_{k,\text{netIN}}}_{\text{BOUNDARY}} = \underbrace{\Delta E_{\text{netIncrease}}}_{\text{SYSTEM}} = \Delta E_{\text{Sys}}$$

(10)

Energy interactions or transfers across a system boundary, in the form of work, $W_{\text{netIN}} = \sum W_{\text{IN}} - \sum W_{\text{OUT}} = -(\sum W_{\text{OUT}} - \sum W_{\text{IN}}) = -W_{\text{netOUT}}$, and heat, $Q_{\text{netIN}} = \sum Q_{\text{IN}} - \sum Q_{\text{OUT}}$, will change the system total energy, $\Delta E_{\text{Sys}} = E_{\text{Sys},2} - E_{\text{Sys},1}$.

The boundary energy transfers are process (or process path) dependent for the same ΔE_{Sys} change, except for special cases for adiabatic processes with work interaction only (no heat transfer), or for caloric processes with heat interaction only (no work transfer), see Fig. 8A to C. For the latter caloric processes without work interactions (no volumetric expansion or other mechanical energy changes), the internal thermal energy is conserved by being transferred from one system to another via heat transfer only, known as *caloric*. This demonstrates the value of the caloric theory of heat that was established by Lavoisier and Laplace (1789), the great minds of the 18th century. Ironically, the caloric theory was creatively used by Sadi Carnot to develop the concept of reversible cycles for conversion of caloric heat to mechanical work as it "flows" from high to low temperature reservoirs (1824) that later helped in dismantling the caloric theory. The caloric theory was discredited by establishing the *heat equivalent of work*, e.g., *mechanical equivalent of heat* by Mayer (1842) and experimentally confirmed by Joule (1843), which paved the way for establishing the First Law of Energy Conservation and new science of Thermodynamics (Clausius, Rankine, and Kelvin, 1850 and later). Prejudging the caloric theory now as a "failure" is unfair and unjustified since it made great contributions in calorimetry and heat transfer, and it is valid for caloric processes (without work interactions). The coupling of work–heat interactions and conversion between thermal and mechanical energy are outside of the caloric theory domain and are further developed within *the First* and *the Second Laws* of Thermodynamics.

The First Law of energy conservation for the control-volume (CV, with boundary surface BS) flow process, see Fig. 9, is:

$$\underbrace{\frac{d}{dt} E_{CV}}_{\text{RATE OF ENERGY CHANGE IN CV}}$$

$$= \underbrace{\sum_{BS} \dot{Q}_{\text{netIN},i}}_{\text{BS TRANSFER RATE OF HEAT}}$$

$$- \underbrace{\sum_{BS} \dot{W}_{\text{netOUT},i}}_{\text{BS TRANSFER RATE OF WORK}}$$

$$+ \underbrace{\sum_{IN} \dot{m}_j (e + Pv)_j}_{\text{ENERGY TRANSPORT RATE WITH MASS IN}}$$

$$- \underbrace{\sum_{OUT} \dot{m}_k (e + Pv)_k}_{\text{ENERGY TRANSPORT RATE WITH MASS OUT}} \quad (11)$$

The First Law of energy conservation equation for a differential volume per unit of volume around a point

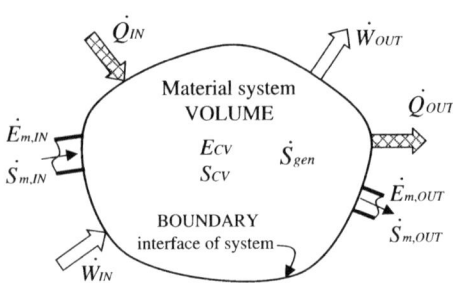

Fig. 9 Control-volume (CV) energy and entropy, and energy and entropy flows through the boundary interface of the control volume.

(x,y,z) in a flowing fluid is[6]:

$$\underbrace{\rho\frac{De}{Dt}}_{\text{energy change in time}}$$

$$= \underbrace{-\vec{V}\cdot(\nabla P)}_{\text{work rate of normal pressure stresses}}$$

$$+ \underbrace{\nabla\cdot\left(\vec{V}\cdot\tau_{ij}\right)}_{\text{work rate of shearing stresses}} + \underbrace{\nabla\cdot(k\nabla T)}_{\text{heat rate via thermal conduction}} \quad (12)$$

Where $e = \hat{u} + \frac{\vec{v}^2}{2} + gz$ NOTE: distinguish specific thermal energy $\hat{u}$, from velocity component u below.

Eq. 12 after substitution, $\nabla\cdot\left(\vec{V}\cdot\tau_{ij}\right) = \vec{V}\cdot\left(\nabla\cdot\tau_{ij}\right) + \Phi$, and using the momentum equation, reduces to:

$$\rho\frac{D\hat{u}}{Dt} = -p(\nabla\cdot\vec{V}) + \Phi_\kappa + \Phi + \nabla\cdot(k\nabla T) \quad (13)$$

Where, $\Phi_\kappa = \kappa(\nabla\cdot\vec{V})^2$ is the bulk-viscosity dissipation, and Φ is the shear-viscosity dissipation function, which is the rate of mechanical work conversion to internal thermal energy for a differential volume per unit of volume, with $[W/m^3]$ unit, is given for Newtonian fluid as:

$$\Phi = \frac{d\dot{W}_\Phi}{dV} = \left[\left(\frac{\partial u}{\partial x}\right)^2 + \left(\frac{\partial v}{\partial y}\right)^2 + \left(\frac{\partial w}{\partial z}\right)^2\right]$$

$$+ \mu\left[\left(\frac{\partial v}{\partial x} + \frac{\partial u}{\partial y}\right)^2 + \left(\frac{\partial w}{\partial y} + \frac{\partial v}{\partial z}\right)^2\right.$$

$$\left. + \left(\frac{\partial u}{\partial z} + \frac{\partial w}{\partial x}\right)^2\right]$$

$$- \frac{2}{3}\mu(\nabla\cdot\vec{V})^2 \quad (14)$$

The power or work rate of viscous dissipation (irreversible conversion of mechanical to thermal energy) in a control volume V is:

$$\dot{W}_\Phi = \int_V \Phi\, dV \quad (15)$$

NOTE: distinguish volume V, form velocity $\vec{V}$

THE SECOND LAW OF ENERGY DEGRADATION: ENTROPY AND EXERGY

Every organized kinetic energy will, in part or in whole (and ultimately in whole), disorganize/dissipate within the microstructure of a system (in time over its mass and space) into disorganized thermal energy. Entropy, as an energy–displacement system property, represents the measure of energy disorganization, or random energy redistribution to smaller-scale structure and space, per absolute temperature level. Contrary to energy and mass, which are conserved in the universe, entropy is continuously generated (increased) due to continuous redistribution and disorganization of energy in transfer and thus degradation of the quality of energy ("spreading" of energy towards and over lower potentials in time until equi-partitioned over system structure and space). Often, we want to extract energy from one system in order to purposefully change another system, thus to transfer energy in organized form (as work, thus the ultimate energy quality). No wonder that energy is defined as *ability to perform work*, and a special quantity *exergy* is defined as the maximum possible work that may be obtained from a system by bringing it to the equilibrium in a process with reference surroundings (called *dead state*). The maximum possible work will be obtained if we prevent energy disorganization, thus with limiting *reversible* processes where the existing non-equilibrium is conserved within interacting systems. Since the energy is conserved during any process, only in ideal reversible processes entropy (measure of energy disorganization or degradation) and exergy (maximum possible work with regard to the reference surroundings) will be conserved, while in real irreversible processes, the entropy will be generated and exergy will be partly (or even fully) destroyed. Therefore, heat transfer and thermal energy are universal manifestations of all natural and artificial (man-made) processes, where all organized potential and/or quasi-kinetic energies are disorganized or dissipated in the form of thermal energy, in irreversible and spontaneous processes.

Reversibility and Irreversibility: Energy Transfer and Disorganization, and Entropy Generation

Energy transfer (when energy moves from one system or subsystem to another) through a system boundary and in time, is of kinetic nature, and may be directionally organized as work or directionally chaotic and disorganized as heat. However, the net-energy transfer is in one direction only, from higher to lower energy–potential, and the process cannot be reversed. Thus *all processes are irreversible* in the direction of decreasing energy–potential (like pressure and temperature) and increasing energy–displacement (like volume and entropy) as a consequence of energy and mass conservation in the universe. This implies that the universe (as isolated and unrestricted system) is expending in space with entropy generation (or increase) as a measure of continuous energy degradation, i.e., energy redistribution and disorganization. It is possible in the limit to have an energy transfer process with infinitesimal potential difference (still from higher to infinitesimally lower potential, P). Then, if infinitesimal change of potential difference direction is reversed ($P + \mathrm{d}P \to P - \mathrm{d}P$, with infinitesimally small external energy, since $\mathrm{d}P \to 0$), the process will be reversed too, which is characterized with infinitesimal entropy generation, thus in the limit, without energy degradation (no further energy disorganization) and no entropy generation—thus achieving a limiting, ideal *reversible process*. Such processes at infinitesimal potential differences, called quasiequilibrium processes, maintain the system equilibrium at any instant but with incremental changes in time. Only quasiequilibrium processes are reversible and vice versa. In effect, the quasiequilibrium reversible processes are infinitely slow processes at infinitesimally small potential differences, but they could be reversed to any previous state, and forwarded to any future equilibrium state, without any "permanent" change to the rest of the surroundings. Therefore, the changes are "fully" reversible, and along with their rate of change and time, completely irrelevant, as if nothing is effectively changing—the time is irrelevant as if it does not exist since it could be reversed (no permanent change and relativity of time). Since the real time cannot be reversed, it is a measure of permanent changes, i.e., irreversibility, which is in turn measured by entropy generation. In this regard the time and irreversible entropy generation are related.[2]

Entropy is also a system property, which together with energy defines its equilibrium state, and actually represents the system energy–displacement or random energy disorganization (dissipation) per absolute temperature level over its mass and space it occupies. Therefore, the entropy as property of a system, for a given system state, is the same regardless whether it is reached by reversible heat transfer, Eq. 16, or irreversible heat or irreversible work transfer (adiabatic or caloric processes on Fig. 8B and C). For example, an ideal gas system entropy increase will be the same during a reversible isothermal heat transfer and reversible expansion to a lower pressure (heat-in equal to expansion work-out), as during an irreversible adiabatic unrestricted expansion (no heat transfer and no expansion work) to the same pressure and volume, as illustrated in Fig. 10A and B, respectively.

If heat or work at higher potential (temperature or pressure) than necessary, is transferred to a system, the energy at excess potential will dissipate spontaneously to a lower potential (if left alone) before a new equilibrium state is reached, with entropy generation, i.e., increase of entropy (energy degradation per absolute temperature level). A system will "accept" energy if it is transferred at minimum necessary (infinitesimally higher) or higher potential with regard to the system energy–potential. Furthermore, the higher potential energy will dissipate and entropy increase will be the same as with minimum necessary potential, like in reversible heating process, i.e.:

$$\mathrm{d}S = \frac{\delta Q}{T} \quad \text{or} \quad S = \int \frac{\delta Q}{T} + S_{\mathrm{ref}} \qquad (16)$$

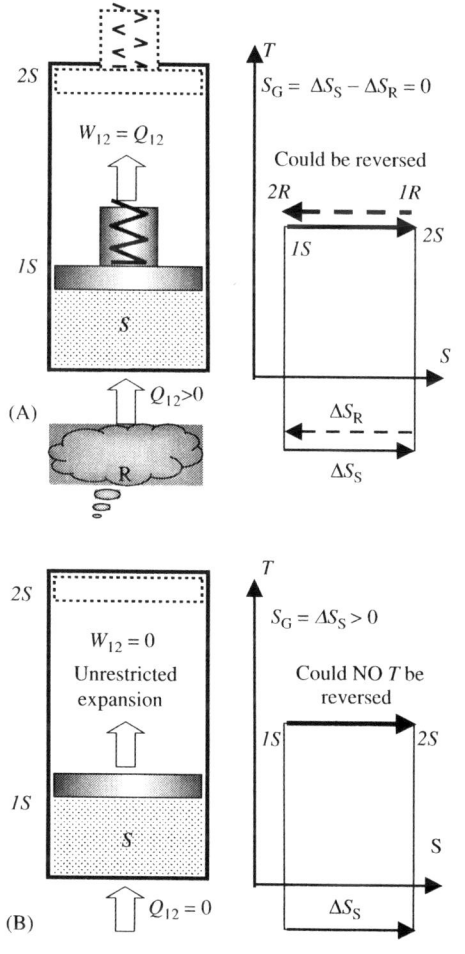

Fig. 10 (A) Isothermal reversible heat transfer and restricted reversible expansion; (B) Adiabatic unrestricted irreversible expansion of the same initial system to the same final state (Thus the same system entropy change).

However, the source entropy will decrease to a smaller extent over higher potential, thus resulting in overall entropy generation for the two (or all) interacting systems, which may be considered as a combined isolated system (no energy exchange with the rest of the surroundings). The same is true for energy exchange between different system parts (could be considered as subsystems) at different energy potentials (non uniform, not at equilibrium at a given time). Energy at higher potential (say close to boundary within a system) will dissipate ("mix") to parts at lower energy potential with larger entropy increase than decrease at higher potential, resulting in internal irreversibility and entropy generation, i.e., energy "expansion" over more mass and/or space with lower potential. Entropy is not displacement for heat only as often stated, but also displacement for any energy dissipation (energy disorganization) and the measure of irreversibility. Examples are unrestricted or throttling expansion with no heat exchange but entropy generation. Therefore, entropy generation is fundamental measure of irreversibility or "permanent change."

Even though directionally organized energy transfer as work, does not possess or generate any entropy (no energy disorganization, Fig. 1B), it is possible to obtain work from the equal amount of disorganized thermal energy or heat if such process is reversible. There are two typical reversible processes where disorganized heat or thermal energy could be entirely transferred into organized work, and vice versa. Namely, they are[2]:

1. reversible expansion at constant thermal energy, e.g., isothermal ideal-gas expansion ($\delta W = \delta Q$), Fig. 10A, and
2. reversible adiabatic expansion ($\delta W = -dU$).

During a reversible isothermal heat transfer and expansion of an ideal-gas system (S), for example, Fig. 10A, the heat transferred from a thermal reservoir (R) will reduce its entropy for (ΔS_R magnitude), wile ideal gas expansion in space (larger volume and lower pressure) will further disorganize its internal thermal energy and increase the gas entropy for (ΔS_S), while in the process an organized expansion work, equivalent to the heat transferred, will be obtained ($W_{12} = Q_{12}$). The process could be reversed, and thus it is reversible process with zero total entropy generation ($S_G = \Delta S_S - \Delta S_R = 0$). On the other hand, if the same initial system (ideal gas) is expanded without any restriction (Fig. 10B, thus zero expansion work) to the same final state, but without heat transfer, the system internal energy will remain the same but more disorganized over the larger volume, resulting in the same entropy increase as during the reversible isothermal heating and restricted expansion. However, this process can not be reversed, since no work was obtained to compress back the system, and indeed the system entropy increase represents the total entropy generation ($S_G = \Delta S_S > 0$). Similarly, during reversible adiabatic expansion, the system internal thermal energy will be reduced and transferred in organized expansion work with no change of system entropy (isentropic process), since the reduction of disorganized internal energy and potential reduction of entropy will be compensated with equal increase of disorder and entropy in expanded volume. The process could be reversed back-and-forth (like elastic compression–expansion oscillations of a system) without energy degradation and entropy change, thus an isentropic processes. In reversible processes energy is exchanged at minimum-needed, not higher than needed potential, and isolated systems do not undergo any energy–potential related degradation/disorganization, and with total conservation of entropy. The total non-equilibrium is conserved by reversible energy transfer within interacting systems, i.e., during reversible processes.

There are two classical statements of the Second Law (both negative, about impossibility), see Fig. 11.[7,8] One is the *Kelvin–Plank statement* which expresses the never violated fact that it is impossible to obtain work in a continuous cyclic process (*perpetual mobile*) from a single thermal reservoir (100% efficiency impossible, since it is not possible to spontaneously create non-equilibrium

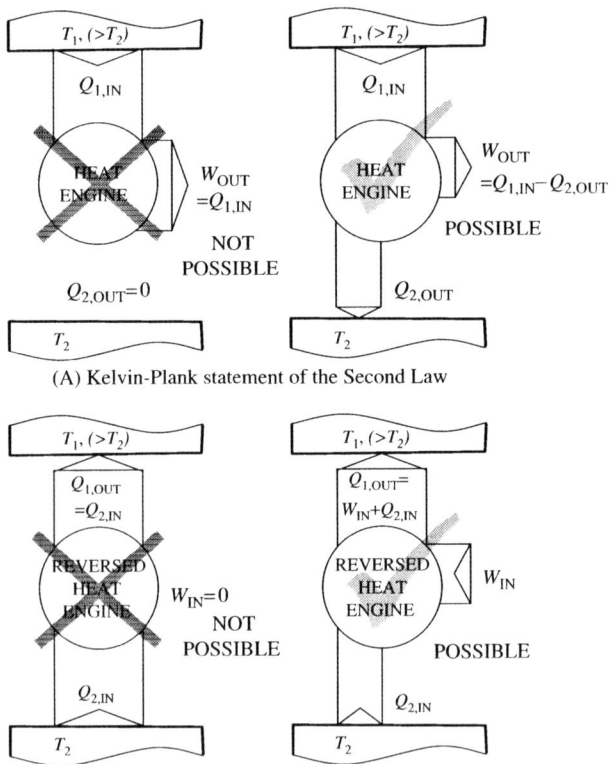

Fig. 11 The Second Law: (A) Kelvin–Plank, and (B) Clausius statements.

within the single reservoir equilibrium), and the second *Clausius statement* refers to the direction of heat transfer, expressing the never violated fact that it is impossible for heat transfer to take place by itself (without any work input) from a lower to higher temperature thermal reservoir (it is impossible to spontaneously create non-equilibrium). Actually the two statements imply each other and thus are the same, as well as they imply that all reversible cycles between the two temperature reservoirs (or all reversible processes between the two states) are the most and equally efficient with regards to extracting the maximum work out of a system, or requiring the minimum possible work into the system (thus conserving the existing non-equilibrium). A heat-engine cycle, energy conversion efficiency is defined as:

$$\eta_{cycle} = \frac{W_{cycle,netOUT}}{Q_{IN}} = \frac{Q_{IN} - Q_{OUT}}{Q_{IN}}$$
$$= 1 - \frac{Q_{OUT}}{Q_{IN}}\bigg|_{rev} \quad (17)$$
$$= 1 - \frac{T_{OUT}}{T_{IN}} = \eta_{rev} = \eta_{max}$$

The reversible cycle efficiency between the two thermal reservoirs does not depend on the cycling system, but only on the ratio of the absolute temperatures of the two reservoirs ($T = T_{IN}$ and $T_{ref} = T_{OUT}$), known as *Carnot* cycle efficiency (last part in the Eq. 17, see also below). The latter defines the *thermodynamic temperature* with the following correlation:

$$\frac{T}{T_{ref}} = \frac{Q}{Q_{ref}}\bigg|_{rev\ cycle} \quad (18)$$

The Second Law efficiency is defined by comparing the real irreversible processes or cycles with the corresponding ideal reversible processes or cycles:

$$\underbrace{\eta_{II,OUT} = \frac{W_{OUT}}{W_{OUT,rev}}}_{\text{Energy Production Process}} \quad \text{or} \quad \underbrace{\eta_{II,IN} = \frac{W_{IN,rev}}{W_{IN}}}_{\text{Energy Consumption Process}} \quad (19)$$

The irreversibility (I) is due to the entropy generation (S_{gen}) and represents the lost work potential or lost exergy ($E_{X,loss}$) with regard to reference system (surroundings at T_o absolute temperature), as expressed by the following correlation:

$$I = E_{X,loss} = T_o S_{gen} \quad (20)$$

The entropy balance equation for the control-volume flow process, complementing the related energy balance Eq. 11, see Fig. 9, is[5–7]:

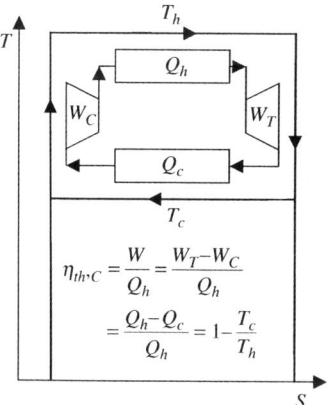

Fig. 12 Heat engine ideal Carnot cycle.

$$\underbrace{\frac{d}{dt} S_{CV}}_{\substack{\text{RATE OF ENTROPY}\\ \text{CHANGE IN CV}}} = \underbrace{\sum_{BS} \frac{\dot{Q}_i}{T_i}}_{\substack{\text{BS TRANSFER}\\ \text{RATE WITH } Q}} + \underbrace{\sum_{IN} \dot{m}_j s_j}_{\substack{\text{TRANSPORT RATE}\\ \text{WITH MASS IN}}}$$
$$- \underbrace{\sum_{OUT} \dot{m}_k s_k}_{\substack{\text{TRANSPORT RATE}\\ \text{WITH MASS OUT}}} + \underbrace{\dot{S}_{gen}}_{\substack{\text{IRREVERSIBLE}\\ \text{GENERATION RATE}}} \quad (21)$$

Heat Engines

Heat Engines are devices undergoing thermo-mechanical cycles (transforming thermal into mechanical energy), similar to one on Fig. 12, with mechanical expansion and compression net-work ($W = Q_h - Q_c$), obtained as the difference between the heat transferred to the engine from high temperature heat reservoir (at T_h) and rejected to a low (cold) temperature heat reservoir (at T_c), thus converting part of the thermal energy into mechanical work. In a close-system cycle the net-work-out is due to net-work of thermal-expansion and thermal-compression. Therefore, heat engine cycle cannot be accomplished without two thermal reservoirs at different temperatures, one at higher temperature to accomplish thermal expansion and work out, and another at lower temperature to accomplish thermal compression to initial volume and thus complete the cycle. The combustion process itself is an irreversible one, where chemical energy (electro-chemical energy binding atoms in reactants' molecules) is chaotically released during combustion, i.e., converted into random thermal energy of products' molecules, and cannot be fully converted into directional work energy. The *Second Law of Thermodynamics* limits the maximum amount of work that could be obtained from thermal energy between any two thermal reservoirs at different temperatures, hot T_h, and cold T_c, by using the ideal, reversible *Carnot cycle*, see Fig. 12, with thermal efficiency given by Eq. 22. As an example, consider $T_c = 293$ K and $T_h = 2273$ K:

$$\eta_{th,C\_ad} = \frac{W}{Q_h} = \frac{Q_h - Q_c}{Q_h}$$

$$= 1 - \frac{T_c}{T_h}\bigg|_{T_h = T_{ad} = 2273 \text{ K}, T_c = 293 \text{ K}}$$

$$= 1 - \frac{293}{2273} = 87.1\% \qquad (22)$$

where, $W = W_T - W_C$, is the net-work of expansion, usually turbine (W_T), and compression (W_C). The maximum efficiency is achieved if heat is supplied at the highest possible temperature T_h, and released at the lowest possible temperature T_c. However, both temperatures are limited by the fact that fuel combustion is performed using oxygen with ambient air, resulting in maximum so called adiabatic, stoichiometric combustion temperature T_{ad}, which is for most fuels about 2000°C, or $T_{ad} = 2273$ K. A part of the heat supplied at hot temperature T_h, must be released to the surroundings at cold temperature about $T_c = 20°C = 293$ K, which results in a Carnot efficiency of 87.1%, see Eq. 22 and Fig. 12. However, the fuel heating value energy $Q_{HV} = Q_{ad\_var}$, is not all available at the adiabatic temperature of the products, but is distributed over their variable temperature range from initial surrounding temperature before combustion T_c, to final adiabatic temperature T_{ad}. For such a variable heat reservoir, a large number (infinite in the limit) of ideal Carnot engines operating at different temperatures (with $dW = dQ$), must be employed to achieve a reversible cycle, resulting in the variable-temperature Carnot cycle with the maximum possible combustion-products-to-work conversion efficiency[1]:

$$\eta_{th,C \text{ var max}}$$

$$= \left(1 - \frac{\ln(T_{ad}/T_c)}{(T_{ad}/T_c) - 1}\right)\bigg|_{T_{ad} = 2273 \text{ K}, T_c = 293 \text{ K}}$$

$$= 69.7\% \qquad (23)$$

The above Eq. 23 is valid if the cyclic medium has constant specific heat, otherwise integration will be required. Due to engine material property limitations and other unavoidable irreversibilities, it is impossible to reach the ideal Carnot efficiency. Different actual heat engines undergo similar but different cycles, depending on the system design. For example, internal combustion engines undergo the *Otto* cycle with gasoline fuel and the *Diesel* cycle with diesel fuel, while the steam and gas turbine power plants undergo the *Rankine* and *Brayton* cycles, respectively. However, with improvements in material properties, effective component cooling, and combining gas and steam turbine systems, the efficiencies above 50% are being achieved, which is a substantial improvement over the usual 30%–35% efficiency. The ideal Carnot cycle is an important reference to guide researchers and engineers to better understand limits and possibilities for new concepts and performance improvements of real heat engines.

CONCLUDING REMARKS: ENERGY PROVIDES EXISTENCE AND IS CAUSE FOR CHANGE

Energy is fundamental property of a physical system and refers to its potential to maintain a system identity or structure and to influence changes with other systems (via forced interaction) by imparting work (forced directional displacement) or heat (forced chaotic displacement/motion of a system molecular or related structures). Energy exists in many forms: electromagnetic (including light), electrical, magnetic, nuclear, chemical, thermal, and mechanical (including kinetic, elastic, gravitational, and sound); where, for example, electro-mechanical energy may be kinetic or potential, while thermal energy represents overall potential and chaotic motion energy of molecules and/or related micro structure.

The philosophical and practical aspects of energy and entropy, including reversibility and irreversibility could be summarized as follows:

- Energy is a fundamental concept indivisible from matter and space, and energy exchanges or transfers are associated with all processes (or changes), thus indivisible from time.
- Energy is "the building block" and fundamental property of matter and space, and thus a fundamental property of existence. For a given matter (system structure) and space (volume) energy defines the system equilibrium state, and vice versa.
- For a given system state (structure and phase) addition of energy will spontaneously tend to randomly redistribute (disorganize, degrade) over the system smaller microstructure and space it occupies, called thermal energy, equalizing the thermal energy–potential (temperature) and increasing the energy–displacement (entropy), and vice versa.
- Entropy may be transferred from system to system by reversible heat transfer and also generated due to irreversibility of heat and work transfer.

- Energy and mass are conserved within interacting systems (all of which may be considered as a combined isolated system not interacting with its surrounding systems), and energy transfer (in time) is irreversible (in one direction) from higher to lower energy–potential only, which then results in continuous generation (increase) of energy–displacement, called entropy generation, which is fundamental measure of irreversibility, or permanent changes.
- Reversible energy transfer is only possible as a limiting case of irreversible energy transfer at infinitesimally small energy–potential differences, thus in quasiequilibrium processes, with conservation of entropy. Since such changes are reversible, they are not permanent and, along with time, irrelevant.

In summary, energy is providing existence, and if exchanged, it has ability to perform change.

Glossary

Energy: It is fundamental property of a physical system and refers to its potential to maintain a system identity or structure and to influence changes with other systems (via forced-displacement interactions) by imparting work (forced directional displacement) or heat (forced chaotic displacement/motion of a system molecular or related structures). Energy exists in many forms: electromagnetic (including light), electrical, magnetic, nuclear, chemical, thermal, and mechanical (including kinetic, elastic, gravitational, and sound).

Energy Conservation: It may refer to the fundamental law of nature that energy and mass are conserved, i.e., cannot be created or destroyed but only transferred from one form or one system to another. Another meaning of energy conservation is improvement of efficiency of energy processes so that they could be accomplished with minimal use of energy sources and minimal impact on the environment.

Energy Conversion: A process of transformation of one form of energy to another, like conversion of chemical to thermal energy during combustion of fuels, or thermal to mechanical energy using the heat engines, etc.

Energy Efficiency: Ratio between useful (or minimally necessary) energy to complete a process and actual energy used to accomplish that process. Efficiency may also be defined as the ratio between energy used in an ideal energy-consuming process vs energy used in the corresponding real process, or vice versa for an energy-producing process. Energy, as per the conservation law, cannot be lost (destroyed), but part of energy input which is not converted into *useful energy* is customarily referred to as *energy loss*.

Entropy: It is an integral measure of thermal energy redistribution (due to heat transfer or irreversible heat generation) within a system mass and/or space (during system expansion), per absolute temperature level. Entropy is increasing from orderly crystalline structure at zero absolute temperature (zero reference) during reversible heating and entropy generation during irreversible energy conversion, i.e., energy degradation or random equi-partition within system material structure and space.

Exergy: It is the maximum system work potential if it is reversibly brought to the equilibrium with reference surroundings, i.e., exergy is a measure of a system non-equilibrium with regard to the reference system.

Heat: It is inevitable (spontaneous) energy transfer due to temperature differences (from higher to lower level), to a larger or smaller degree without control (dissipative) via chaotic (in all directions, non purposeful) displacement/motion of system molecules and related microstructure, as opposed to controlled (purposeful and directional) energy transfer referred to as *work* (see below).

Heat Engine: It is a device undergoing thermo-mechanical cycle that partially converts thermal energy into mechanical work and is limited by the *Carnot cycle* efficiency. The cycle mechanical expansion and compression net-work is obtained due to difference between heat transferred to the engine from a high temperature heat reservoir and rejected to a low temperature reservoir, thus converting part of thermal energy into mechanical work.

Mechanical Energy: It is defined as the energy associated with ordered motion of moving objects at large scale (kinetic) and ordered elastic potential energy within the material structure (potential elastic), as well as potential energy in gravitational field (potential gravitational).

Power: It is the energy rate per unit of time and is related to work or heat transfer processes (different work power or heating power).

System: (also *Particle* or *Body* or *Object*) refers to any, arbitrary chosen but fixed physical or material system in space (from a single particle to system of particles), which is subject to observation and analysis. System occupies so called system volume within its own enclosure interface or system boundary, and thus separates itself from its surroundings, i.e., other surrounding systems.

Temperature: It is a measure of the average quasi-translational kinetic energy associated with the disordered microscopic motion of atoms and molecules relevant for inter-particulate collision and heat transfer, thus temperature is related to relevant particle kinetic energy and not to the particle density in space.

Thermal Energy: It is defined as the energy associated with the random, disordered motion of molecules and potential energy due to intermolecular forces (also associated with phase change), as opposed to the macroscopic ordered energy associated with moving objects at large scale.

Total Internal Energy: It is defined as the energy associated with the random, disordered motion of molecules and intermolecular potential energy (thermal), "binding" potential energy associated with chemical molecular structure (chemical) and atomic nuclear structure (nuclear), as well as with other structural potentials in force fields (electrical, magnetic, etc.). It refers to the "invisible" microscopic energy on the subatomic, atomic and molecular scale.

Work: It is a type of controlled energy transfer when one system is exerting force in a specific direction and thus making a purposeful change (forced displacement) of the other systems. It is inevitably (spontaneously) accompanied, to a larger or smaller degree (negligible in ideal processes), with dissipative (without control) energy transfer referred to as *heat* (see above).

REFERENCES

1. Kostic, M. Work, power, and energy, In *Encyclopedia of Energy*, Cleveland, C.J., Ed.; Elsevier Inc.: Oxford, U.K., 2004; Vol. 6, 527–538.
2. Kostic, M. Irreversibility and reversible heat transfer: the quest and nature of energy and entropy, IMECE2004. In *ASME Proceedings*, ASME, New York, 2004.
3. Kostic, M. *Treatise with Reasoning Proof of the First Law of Energy Conservation*, Copyrighted Manuscript; Northern Illinois University: DeKalb, IL, 2006.
4. Kostic, M. *Treatise with Reasoning Proof of the Second Law of Energy Degradation*, Copyrighted Manuscript; Northern Illinois University: DeKalb, IL, 2006.
5. Moran, A.J.; Shapiro, H.N. *Fundamentals of Engineering Thermodynamics*, 4th Ed. Wiley: New York, 2000.
6. Bejan, A. *Advanced Engineering Thermodynamics*; Wiley: New York, 1988.
7. Cengel, Y.A.; Boles, M.A. *Thermodynamics, An Engineering Approach*, 5th Ed.; WCB McGraw-Hill: Boston, MA, 2006.
8. Zemansky, M.W.; Dittman, R.H. *Heat and Thermodynamics*; McGraw-Hill: New York, 1982.

Pricing Programs: Time-of-Use and Real Time

Ahmad Faruqui
The Brattle Group, San Francisco, California, U.S.A.

Abstract

This article surveys numerous pricing designs for improving economic efficiency in all market segments. Electricity is a very capital-intensive industry characterized by a significant peak load problem. Expensive generating plants have to be installed to meet peak loads that are only encountered for a few hundred hours a year. This raises the cost of electricity to all consumers. Average cost pricing, the staple of the industry in which rates do not vary by time of use, compounds the problem by creating cross-subsidies. Customers with flatter load shapes subsidize those with peakier load shapes.

The problem can be alleviated by modifying electricity pricing practices to allow time-variation in costs. This would provide customers an incentive to lower peak usage, either by curtailing or shifting their activities. In addition, it would eliminate unfair and economically unjustified cross subsidies. But the potential benefits of time-varying pricing have yet to be fully realized. Many barriers stand in the way of reform, including economic, technological and political. Of all these barriers, the most formidable ones are the political ones. They have to be resolved by modifying the legal and regulatory framework through which electricity pricing is determined.

INTRODUCTION

Time-of-use (TOU) pricing and real time pricing (RTP) programs are designed to lower system costs for utilities and bring down customer bills by raising prices during expensive hours and lowering them during inexpensive hours. They differ in, that the former fixes the price and time periods in advance while the latter fixes neither the price nor the time period in advance. Thus, TOU rates can be considered static while RTP rates can be considered dynamic, even though before feature time-varying prices. Other rate designs bridge the gap between these two rate designs, as shown below.

Time-of-use pricing (TOU). This rate design features prices that vary by time period, and are higher in peak periods and lower in off-peak periods. The simplest rate involves just two pricing seasons, with prices being higher during the peaking season. A time-of-day rate is slightly more complex and involves two pricing periods within a day, a peak period and an off-peak period. More complex rates have one or more shoulder periods and seasonal variation.

Critical peak pricing (CPP). This rate design layers a very high price during a few critical hours of the year. It can also be combined with a TOU rate. Typically, a CPP rate is only used on 12–15 days a year. These days are called the day before or the day of the critical peak price.

Extreme day pricing (EDP). This rate design is similar to CPP, except that the higher price is in effect for all 24 h for a maximum number of critical days, the timing of which is unknown until a day ahead.

Extreme day CPP (ED-CPP). This rate design is a variation of CPP in which the critical peak price applies to the critical peak hours on extreme days but there is no TOU pricing on other days.

Real time pricing (RTP). This rate design features prices that vary hourly or sub-hourly all year long, for some or all of a customer's load. Customers are notified of the rates on a day-ahead or hour-ahead basis.

Each of these rates exposes customers to varying amounts of price variance. Customers can lower their expected (average) price by taking more risks. For example, RTP rates are riskiest from the customer's viewpoint since they face wholesale prices that vary in real time, but they will most likely be associated with the lowest average price. Critical peak pricing rates carry less pricing uncertainty for customers, since customers know the prices ahead of time and the time for which these prices will be in effect is limited. However, the average price is likely to be higher than that for RTP rates. At the other end of the spectrum are rates that do not vary over the hours of the day and only vary seasonally. They provide the highest rate predictability to customers but are also likely to carry the highest average price.

TIME-OF-USE PRICING

Time-of-use pricing is commonplace in developed economies at all stages of market restructuring. Electricite de France (EDF) operates the most successful example of

Keywords: Electricity pricing; Rate design; Economic efficiency; Demand response.

TOU pricing. Currently, a third of its population of 30 million customers is estimated to be on TOU pricing. This pricing design was first introduced for residential customers in 1965 on a voluntary basis, having been first applied in the country to large industrial customers as the Green Tariff in 1956. The French model served for many years as a benchmark for many countries in Latin America. For example, in Brazil, it was introduced as the "Horo-sazonal" tariff, which divides the day into peak and off peak periods and the year into dry and wet seasons. The idea was to continue all the way to the residential customer (yellow tariff), but it never came to fruition.

Time-of-use rates have been mandatory in California for all customers above 500 kW since 1978, as a statewide policy response to the energy crisis of 1973. These rates are mandatory in several U.S. states but the size threshold varies by state.

Residential TOU rates are offered on a voluntary opt-in basis by utilities in all types of climates within the U.S., including Pepco in the Washington, DC area and the Salt River Project in the Phoenix area. The simplest variation involves two time periods. An example is the residential rate design offered by Pacific Gas & Electric Company (PG&E) in central and northern California. During the summer months, from noon to six P.M. on weekdays, electricity costs three times as much as during all other hours of the week. During the winter months, the price differential is smaller.

Another example is the project that was implemented by Puget Sound Energy (PSE) in the suburbs of Seattle. In May 2001, as a response to the power crisis in the Western states, PSE designed and implemented a TOU rate for its residential and small commercial customers. It involved four pricing periods. The morning and evening periods were the most expensive periods, followed by the mid-day period and the economy period. Unlike most TOU rates, which feature significant differentials between peak and off-peak prices, PSE's TOU rate featured very modest price differentials between the peak and off-peak periods, reflecting the hydro-based system in the Northwest.

The peak price was about 15% higher than the average price customers had faced prior to being moved to the TOU rate and the off-peak price was about 15% lower. To keep the rate simple, there was no seasonal variation in prices.

Puget Sound Energy placed about 300,000 customers on the rate, but they could opt-out to the standard rate if they so desired. There was no additional charge to participate in the rate. During the first year of the program, less than half of one percent elected to opt-out of the rate. Customer satisfaction with the rate was high. In focus groups, customers identified several benefits of the TOU rate besides bill savings, including greater control over their energy use; choice about which rate to be on; social responsibility; and energy security. PSE also provided a website to customers where they could review their load shapes for the past seven days.

Puget Sound Energy had a rate case settlement in June 2002. Under the terms of the settlement, the program became an opt-in program for new customers. The peak/off-peak rate differential of the TOU rate was reduced from 14 to 12 mils/kWh (A mil is a thousandth of a dollar). A monthly fee of $1 a month, about 80% of the estimated variable cost of providing TOU meter reading, was levied on participating customers. Finally, each quarter PSE would notify customers of their savings (or losses) on the program, and it would switch all customers to the lower-cost rate (flat or TOU) in August 2003.

In October 2002, PSE sent customers their first quarterly report. For 94% of the customers, this report showed that they were paying an extra 80 cents/month by participating in the TOU pilot, comprised of the difference between 20 cents of power cost savings and a dollar of incremental meter reading costs. This was marked in contrast to the first year of the program when, prior to charging customers any part of the TOU meter reading costs, over 55% of residential customers experienced bill savings by being on the TOU rate.

Even though the report was for a single quarter, 10% of the participating customers chose to opt-out of the program between July 1 and October 31. At the same time, 1.8% of new customers opted into the program.

Media coverage was very negative and featured interviews with customers claiming that they had shifted almost half of their load from peak to off-peak periods, only to find out that they had lost money. PSE pulled the plug on a program that had become the most visible national symbol of a utility's commitment to time-varying pricing, and agreed to refund the increased amounts to participating customers.

Lessons Learned From the PSE TOU Rate

Five lessons can be drawn from PSE's TOU program.

- Customers do shift loads in response to a TOU price signal, even if the price signal is quite modest. According to an independent analysis, customers consistently lowered peak period usage by 5% per month over a 15-month period.
- It is important to manage customer expectations about bill savings.
- Customers should be educated on the magnitude of bill savings they can expect from specific load shifting activities.
- It is desirable to conduct a pilot program involving a few thousand customers before offering a rate to hundreds of thousands of customers.
- Finally, and most importantly, any program should make a majority of the customers better off, or it should not be offered.

Developing a TOU Rate

It is fairly straightforward to develop a TOU rate design. The following sidebar shows the steps involved in developing a "revenue-neutral" TOU rate. Such a rate would leave the average customer's bill unchanged if that customer chose to make no adjustments in their pattern of usage. Of course, a customer who uses less power in the peak period than the average customer would be made better off (compared to his or her situation on the standard rate) by the rate even without responding to the rate and a customer who uses proportionately more power in the peak period than the average customer would be made worse off by the rate if he or she did not respond to the rate.

The sidebar brings out the type of information that is needed to develop a TOU rate.

CRITICAL PEAK PRICING

Under this rate design, customers are on TOU prices for most hours of the year but additionally face a much higher price during a small number of critical hours when system reliability is threatened or very high prices are encountered in wholesale markets because of extreme weather conditions and similar factors. In 1993, EDF (France) introduced a new rate design, tempo, and now has over 120,000 residential customers on it. The program features two daily pricing periods and three types of days. The year is divided into three types of days, named after the colors of the French flag. The blue days are the most numerous (300) and least expensive; the white days are the next most numerous (43) and mid-range in price; and the red days are the least numerous (22) and the most expensive. The ratio of prices between the most expensive time period (red peak hours) and the least expensive time period (blue off-peak hours) is about 15–1, reflecting the corresponding ratio in marginal costs.

The tempo rate does not offer a fixed calendar of days, but customers can learn what color will take effect the next day by checking a variety of different sources:

- Consulting the Tempo Internet website: www.tempo.tm.fr
- Subscribing to an email service that alerts them of the colors to come
- Using Minitel (a data terminal particular to France, sometimes called a primitive form of Internet)
- Using a vocal system over the telephone
- Checking an electrical device (*Compteur Electronique*) provided by EDF that can be plugged into any electrical socket.

The tempo rate was preceded by a pilot program, in which prices were quite a bit higher than those that were ultimately implemented. The rates associated with the tempo program and with EDF's standard TOU rate are shown in Fig. 1.

Critical Peak Pricing With Enabling Technologies

Recently, a number of utilities have experimented with dynamic pricing options, sometimes in conjunction with enabling technologies that automate customer response during high priced periods. As seen below, dynamic pricing, especially when combined with enabling technologies, can produce much larger reductions in peak demand than traditional TOU or non-technology enabled CPP rates.

Two utilities, GPU in Pennsylvania and American Electric Power in Ohio, conducted small-scale pilot programs in the 1980s using a two-way communication and control technology called TransText. The TransText device allows for the creation of a fourth critical price period in which the retail price of electricity rises to a much higher level (e.g., 50 cents/kWh in the GPU pilot). The number of hours during which this price can be charged is small (e.g., 100–200 h) and the customer knows what the critical price will be ahead of time, but does not know when the price may be called.

Sidebar 1 Developing a TOU rate involves several steps

Existing flat rate	
Per-customer revenue requirement	$100
Per-customer monthly usage	1000 kWh
Average price	$0.10/kWh
Revenue neutral TOU rate	
Estimate peak usage	200 kWh
Estimate off-peak usage	800 kWh
Set peak price = peak marginal cost	$0.20/kWh
Set off-peak price = off-peak marginal cost	$0.075 kWh
Given class revenue requirement	$100
Given monthly usage	1000 kWh
TOU rate with load shifting	
Estimate price elasticity	−0.2
Estimate new peak usage	160 kWh[a]
Estimate new off-peak usage	840 kWh
Estimate new monthly usage	1000 kWh
Estimate new per-customer monthly bill	$95
Estimate bill savings = per-customer revenue loss	$100 − $95 = $5

[a]These changes in usage for the peak and off-peak period are estimated by using the percent changes in peak and off-peak prices and the estimated price elasticity of demand.

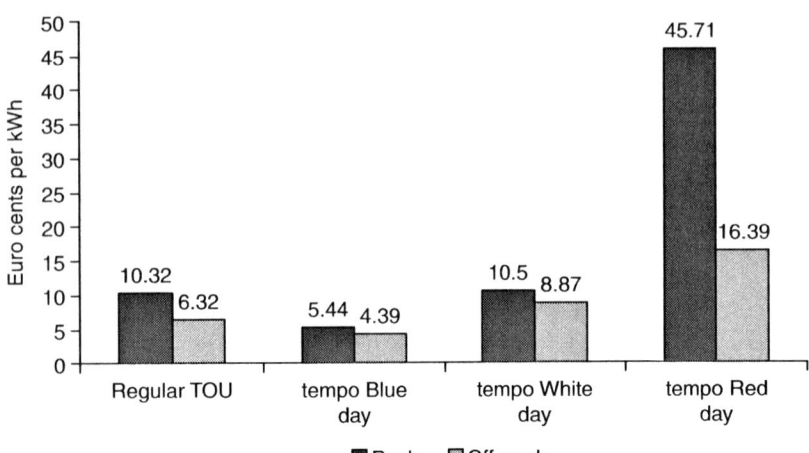

Fig. 1 EDF's *tempo* and standard TOU rates.

The TransText device incorporates an advanced communication feature that lets customers know that a critical period is approaching and it can be programmed so that the customer's thermostat is automatically adjusted when prices exceed a certain level. Using this technology, American Electric Power found significant load shifting, with estimated peak demand reductions of 2–3 kW per customer during on-peak periods and of 3.5–6.6 kW during critical peak periods. These critical peak reductions represented a drop of nearly 60% of a typical customer's peak load during the winter period.

The GPU experiment produced similar results, showing elasticities of substitution that ranged from -0.31 to -0.40, significantly higher than the elasticities associated with traditional TOU rates, which have averaged -0.14 in a range of studies. These elasticities were estimated by comparing customer loads on days when control was being exercised with days when control was not being exercised.

Another example is provided by Gulf Power Company's *Good Cents Select* program in Florida. Like the GPU experiment, the Gulf Power program uses dynamic pricing to obtain additional benefits beyond traditional TOU pricing. Under this voluntary program, residential consumers face a three-part TOU rate for 99% of all hours in the year, where the peak period price of $0.093/kWh is roughly 60% higher than the standard (flat) tariff price and approximately twice the intermediate (shoulder) price. For the remaining 1% of the hours, Gulf Power has the option of charging a critical period price equal to $0.29/kWh, more than three times the value of the peak-period price. The timing of this much higher price is uncertain and it is called during the day when critical conditions are encountered. In conjunction with this rate, participating customers are provided with a programmable/controllable thermostat that automatically adjusts their heating and cooling loads and up to three additional control points in the home such as water heating and pool pumps. The devices can be programmed to modify usage when prices exceed a certain level.

Gulf Power is seeing results similar to those of the GPU experiment. Peak-period reductions in energy use over a 2-year period have equaled roughly 22% compared with a control group, while reductions during critical-peak periods have equaled almost 42%. Diversified coincident peak demand reductions have exceeded more than 2 kW per customer. This voluntary program has been in place for less than a year, and Gulf Power has already signed up more than 3000 high use customers. It hopes to attract 40,000 customers over the next 10 years, representing about 10% of the residential population. Participating customers pay roughly $5/month to help offset the additional cost of the communication and control equipment. In a recent survey, the program received a 96% satisfaction rating.

The Gulf Power program is targeted at high use customers, just like the EDF program. Customer savings are large enough to offset the program costs. Both rates have significant peak to off-peak differentials as well. Because of these two factors, the programs have been successful. The PSE program failed in part because it had weak peak to off-peak differential and in part because it did not target the large customers.

California's Pricing Experiment

The state of California conducted a statewide pricing pilot (SPP) during the 2003–2005 timeframe to test customer response to a variety of pricing options, including TOU rates and CPP rates. In California, standard residential tariffs involve an "inverted tier" design in which the price of power rises with electricity usage. The typical residential customer pays an average price of about 13 cents/kWh. Within the SPP, customers on TOU and CPP rates pay a higher price during the five-hour peak period that lasts from 2 P.M. to 7 P.M. on weekdays and a lower price during the off-peak period, which applies during all other hours.

Pricing Programs: Time-of-Use and Real Time

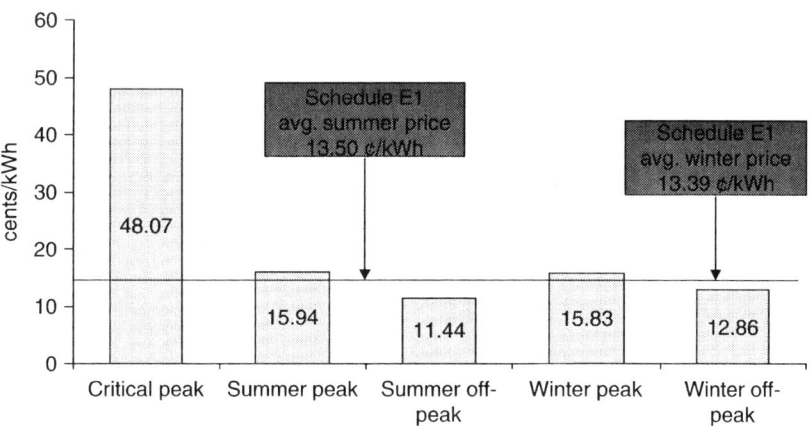

Fig. 2 Critical-peak pricing (CPP) tariff.

Each TOU and CPP rate involves two sets of peak/off-peak prices, to allow for precise estimation of the elasticities of demand. On average, customers on TOU rates are given a discount of 23% during the off-peak hours and are charged a price of around 10 cents. They are charged a price of 22 cents during the peak hours, which is 69% higher than their standard rate. Thus, with TOU rates, customers are given a strong incentive to curtail peak usage and to shift usage to off-peak periods. However, the incentive is much greater on selected days for customers on CPP rates, who are charged, on average, a price of 64 cents during the peak hours on 12 summer days, i.e., prices are nearly five times higher than the standard price. On the peak hours of other days and the off-peak hours of all days they face prices that are slightly lower than the prices faced by TOU customers during these periods. Fig. 2 shows the CPP tariffs that were used in the California experiment.

Analysis of data from the California experiment indicates that CPP rate customers face "rifle shot" price signals that can be very effective at reducing peak demand, thus dampening wholesale prices and obviating the need for building costly power plants that would run for only a few hundred hours a year. Customers are likely to respond to higher peak prices by reducing peak usage, e.g., by reducing air conditioning usage, and perhaps by shifting some peak period usage associated with laundry, dishwashing and cooking activities to lower cost off-peak periods. They may also be raising off-peak use in response to lower off-peak rates by raising air conditioning usage, increasing lighting levels, and so on. Finally, since prices have changed in the peak and off-peak periods, the average price for electricity over the day may have changed for some customers as well. This would trigger additional changes in usage.

Fig. 3 shows the changes in customer load shapes caused by the CPP tariff in customers who were located in the San Diego Gas and Electric service area. The black line shows the usage of the control group of customers. The gray line shows the usage of customers who were equipped with a smart thermostat that received a communication signal from the utility during critical hours, which raised the set point of the thermostat. Their tariff was unchanged from that of the control group. The difference between the two lines is noticeable and suggests that remotely controlling the thermostat lowers peak usage. The white line shows the

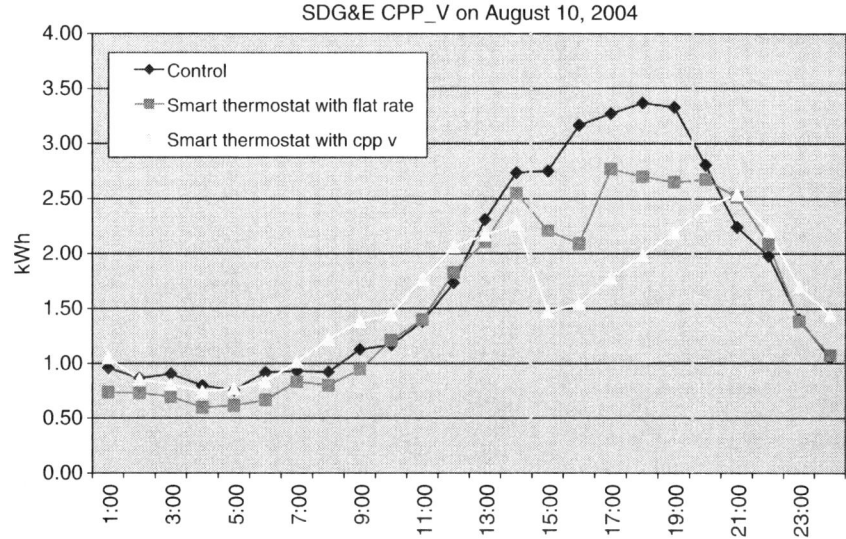

Fig. 3 Changes in customer load shapes.

usage of customers who were equipped with a smart thermostat and who also were placed on the CPP tariff. They show a greater drop than customers who had the smart thermostat but who were not placed on the CPP tariff.

REAL-TIME PRICING FOR RESIDENTIAL CUSTOMERS

The Chicago Community Energy Cooperative (Co-op) has implemented a market-based RTP pricing plan for residential customers, in conjunction with the local electric utility, CommEd. The utility provides the rate and the metering/billing system while the Co-op provides customer notification (via a web site, e-mail and telephone), education, and energy management tools (Fig. 4).

The pilot program is intended to model the bundled rate/market rate differential in the post-2006 market environment when the rate freeze is lifted. It involves RTP prices on a day-ahead basis for the generation portion of the rate. The prices are capped at 50 cents/kWh. The project is designed to estimate the magnitude of customer response to hourly energy pricing and understand the drivers of responsiveness. This is a 3-year experimental program that commenced in January 2003.

In the first year of the program, 750 customers were enrolled. Of these, 100 are in a control group. The summer of 2003 was mild in terms of both temperatures and prices. For example, the number of days with a maximum temperature higher than 90°C was 10 versus a historical average of 18. The maximum price was 12.39 cents/kWh, versus a price of 38.11 cents/kWh during the crisis years of 2000–2002.

Analysis of customer loads during the first year indicates that participants responded to the higher prices they faced during the peak periods. A price elasticity of −0.042 was estimated over the full range of prices. Over half of all participants showed significant response to high price notifications. Aggregate demand reduction was as high as 25% during the notification period. Over 80% of the participants reported modifying their air conditioning usage, and over 70% reported modifying their clothes-washing patterns.

Multifamily households as a group were more price-responsive than single-family households. Households with window air conditioners maintained their price responsiveness better across multiple high-priced hours than single-family households, who started out strong but whose responsiveness tended to taper off during the high priced periods.

Customer satisfaction was very high with the program. The program was "quick and easy" for 82% of the participants and "time consuming and difficult" for 1%. Participants saved on average $12/month or 20% of their monthly bill.

The project has shown that residential customers are a viable market for RTP. They represent a key target market, since residential load is a major contributor to system peak. And giving residential customers a choice of pricing options may be the only way to give them a meaningful choice in restructured power markets.

REAL-TIME PRICING FOR COMMERCIAL AND INDUSTRIAL CUSTOMERS

Utilities in the Southeastern U.S. have implemented RTP rates for about 2000 customers on a day-ahead or hour-ahead basis. These companies include Georgia Power,

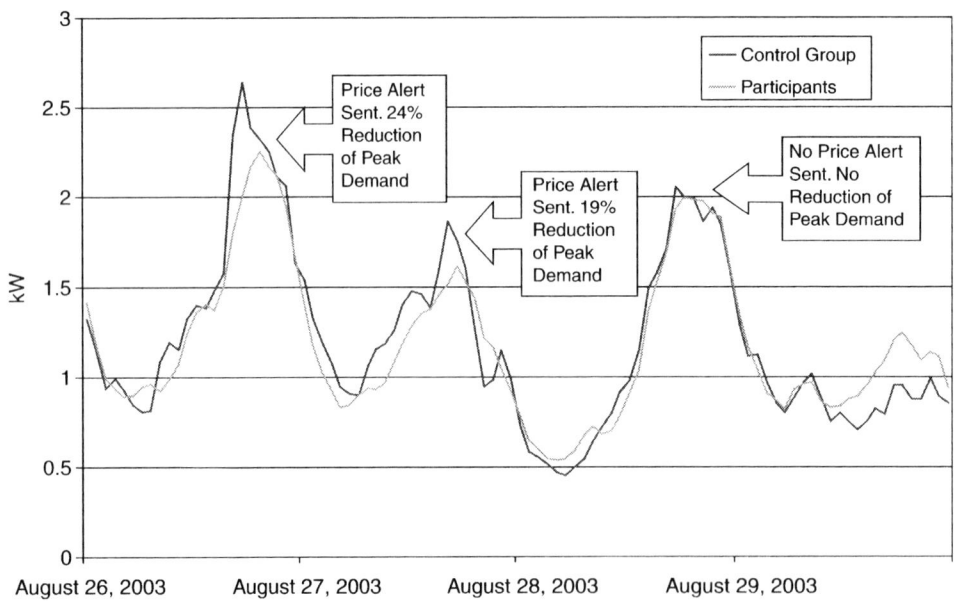

Fig. 4 Impact of real time pricing on Chicago Residences.

Duke Power and the Tennessee Valley Authority. The Georgia Power program is discussed in detail below.

Before describing the Georgia Power program, we note that RTP rates were probably first used by ESKOM, the state-owned utility in South Africa, for its largest customers, including the fabled gold mines. ESKOM has 1400 MW of load on day-ahead RTP. These customers drop their load by 350–400 MW for up to three consecutive hours when faced with high prices. While RTP is set up on a day-ahead basis, customer response is not used to optimize the dispatch of the power system. Electricity prices are based on the Pool Output Price, and do not change in response to changes in customer demand that may be induced by RTP. The utility is not aggressively marketing the program for this reason. It hopes that once a competitive energy market has been created, with a functioning Power Exchange, RTP will then be able to play its proper role in system operations.

If RTP had been implemented in California during the summer of 2000, much of the power crisis that developed in May 2000 would have abated within a month, rather than persisting for a year. If only a small proportion of the total customers had bought power on RTP, statewide peak demand would have dropped by 2.5%, or 1250 MW. During the peak hours, this would have lowered wholesale market prices by 20%. The state's power costs for the summer would have dropped by 6% (Faruqui, Ahmad, Hung-po Chao, Vic Niemeyer, Jeremy Platt and Karl Stahlkopf, "Getting out of the Dark," *Regulation*, Fall 2001, pp. 58–62).

RTP at Georgia Power

Georgia Power runs the worlds largest and possibly the most successful RTP program. The company estimates that during emergency conditions, its customers drop demand by 17%, freeing up 800 MW of capacity. A load drop of this magnitude eliminates the need for several expensive power plants that would otherwise be needed for meeting the peak load.

Program Background

Georgia law permits customers with 900 kW or more of connected load to put their load out to bid, and be served by any supplier in the state. In the late 1980s, Georgia Power was competing for these customers with almost 100 rural cooperatives and municipal utilities. In part to increase its competitiveness, Georgia Power began looking into RTP. In 1992, it began a 2-year controlled pilot, with the goals of increasing competitiveness; improving customer satisfaction by giving customers more control over their bills; and curtailing load when needed to balance supply and demand.

Georgia Power was one of the first utilities in the country to develop a two-part RTP tariff, following the lead of Niagara Mohawk in New York that had launched an Hourly Pricing Program in the late 1980s. The utility chose a two-part rather than a one-part rate for several reasons. First, the two-part rate allows the hourly price to more closely reflect the utility's true marginal cost. Second, the two-part rate best represents the "market price." Georgia Power believed a two-part rate would give it an opportunity to work with customers on price protection products. A discussion of price protection products is provided below. In addition, the utility was concerned about revenue stability; with a one-part rate, it would lose some of the contribution to fixed costs when customers curtailed in high priced hours. Georgia Power has expanded its RTP offerings since the 1992 pilot, but the basics of the program and tariff have remained relatively unchanged for almost a decade.

Rate Structure

Customers are billed for "baseline" use at their standard rate and pay (or receive credits) for energy used above (or below) the baseline each hour at the hourly price. The hourly price is composed of a measure of marginal energy costs, line losses, a "risk recovery factor" for forecasting risk (a fixed adder), and—near peaks—marginal transmission costs and outage cost estimates. Marginal transmission costs are triggered by load and temperature. Outage cost estimates are based on loss of load probabilities, as well as customer surveys on the costs of having an outage.

Georgia Power offers a "day-ahead" program, where customers are notified of price schedules by 4 P.M. the day before they go into effect, and an "hour-ahead" program, where customers are given an hour's notice on price. Currently, interruptible customers are served on the hour-ahead program. For these customers, their customer baseline (CBL) drops to their firm contract level during periods of interruption. Customers who do not interrupt to their firm levels pay interruption penalties plus the hourly prices. The utility has filed a tariff with the Public Service Commission that would allow interruptible customers on the day-ahead rate as well. The other difference between the day and hour-ahead rates is that the risk-recovery factor for the day-ahead rate is greater than that for the hour-ahead rate (4 mils/kWh versus 3 mils/kWh), since the utility bears a greater forecast risk.

Setting the Customer Baseline

When Georgia Power began its RTP program, it based a customer's baseline usage, or CBL, on an 8760-point hourly load profile. However, customers often found this CBL confusing, and therefore frustrating. In response to these customers, Georgia Power now offers 360-point CBLs (with 24 average hourly weekday loads per month and six average four-hour weekend day loads, for a total of 30 CBL points per month), and two-point CBLs.

The two-point CBLs simply average usage levels during the peak and off-peak periods.

The majority of customers (basically, the high-load-factor customers) now select the two-point CBL. If the two-point CBL does not seem appropriate based on a customer's usage profile, Georgia Power will usually use a 360-point CBL. Only a very few "unique loads" use the 8760-point CBL today (Our source noted that customers who can "really respond a lot" are typically on the higher point CBLs).

Price Protection Products

Georgia Power offers customers a variety of products that allow customers to influence their exposure to RTP price risk. One product, the adjustable CBL, allows customers to temporarily adjust their CBLs. For example, if customers want to lower their exposure to price volatility, they would increase CBLs. (Customers wanting to raise their CBLs must be on the RTP rate for a year, so that Georgia Power can determine how high the CBL can be raised.) Customers wanting to expose more loads to real-time prices—presumably because they believe it will be a cool summer—can lower their CBLs. Of the roughly 1650 customers on RTP, 600 currently have adjustable CBLs. About 60% of the incremental energy sold on the RTP rate (i.e., usage above baseline) is now protected by this product.

Georgia Power also offers a variety of financial products to limit customers' exposure to RTP price volatility. These products include price caps, contracts for differences, collars, index swaps, and index caps (Georgia Power's price-cap product guarantees that average RTP prices over a specific time period will not go above the cap. Its contract for differences gives a fixed price guarantee on the average RTP price. The collar has a cap and floor on the average RTP price over a specific time period. The index swap is a financial agreement that ties the RTP price to a commodity price index. If the commodity price index increases, so does the RTP price. If it decreases, so does the RTP price. The index cap is a financial agreement that ties an RTP price cap to a commodity price index. As the commodity price increases or decreases, so does the price cap). Georgia Power has sold these Price Protection Products, or PPPs, for 3 years. It currently has 250 contracts with about 90 customers. (Customers have multiple contracts to cover different time periods.) Georgia Power believes that offering these products has not probably increased the number of customers on the RTP program, but it has increased customer satisfaction. The utility has examined whether offering the PPPs has dampened price responsiveness, and has found no evidence of this.

LESSONS LEARNED

Georgia Power's experience highlights a number of lessons that have also been seen at other utilities. First, RTP can deliver substantial peak savingsssas, despite the fact that many customers are not very responsive to price. When the hourly price reached $6.40/kWh, Georgia Power saw 850 MW of load reduction (out of 1500–2000 MW of incremental or above-baseline load) from its RTP customers. Georgia Power also believes that customers have responded to the availability of low off-peak prices by expanding their facilities and business operations in Georgia. In other words, the rate has served to bring economic growth to the state and been a form of strategic electrification while also being a form of load management.

The utility's experience also supports the finding that customers join RTP programs to have access to lower cost power. When hourly prices went up in response to changing market conditions, customers sought price relief, and were granted it by the Georgia Public Services Commission.

Georgia Power has also found that a small percentage of customers are willing to pay for limited protection against price volatility. In response to customer requests, they developed and now sell a variety of risk-management products.

Georgia Power has also found that manufacturers with highly energy-intensive processes, such as chemical and pulp and paper companies are generally the most price responsive customers. It is also learnt that some commercial customers would respond to price. Office buildings, universities, grocery stores, and even a hospital (that changes chiller use based on hourly prices) are all responsive to real-time pricing.

Georgia Power states that the major lesson it has learnt is that education is the key to a successful RTP program. Georgia Power now holds annual, statewide meetings with all its customers to keep them informed about the RTP program. The utility believes its education program has paid off in customer satisfaction.

CONCLUSION

Electricity is a very capital-intensive industry characterized by a significant peak load problem. Expensive generating plants have to be installed to meet peak loads that are only encountered for a few hundred hours a year. This raises the cost of electricity to all consumers. Average cost pricing, the staple of the industry in which rates do not vary by time of use, compounds the problem by creating cross-subsidies. Customers with flatter load shapes subsidize those with peakier load shapes.

The problem can be alleviated by modifying electricity pricing practices to allow time-variation in costs. This would provide customers an incentive to lower peak usage, either by curtailing or shifting their activities. In addition, it would eliminate unfair and economically unjustified cross subsidies. As surveyed in this article, there are numerous pricing designs for improving economic efficiency in all market segments. But the potential benefits of time-varying pricing have yet to be fully realized. Many barriers stand in the way of reform, including economic, technological and political. Of all these barriers, the most formidable ones are the political ones. They have to be resolved by modifying the legal and regulatory framework through which electricity pricing is determined.

ACKNOWLEDGMENTS

Principal, The Brattle Group. This paper has benefited from conversations over the years with several colleagues. Without implicating them for any errors, I would like to specifically acknowledge the contributions of Andrew Bell, Steve Braithwait, Kelly Eakin, Robert Earle, Stephen S. George, Richard Goldberg, Phil Hanser, Roger Levy, Larry Ruff and Lisa Wood.

BIBLIOGRAPHY

1. Braithwait, S.; Faruqui, A. The choice not to buy: Energy savings and policy alternatives for demand response. Public Utilities Fortnightly **2001**, *March 15*.
2. Earle, R.; Faruqui, A. Toward a new paradigm for valuing demand response. The Electricity Journal **2006**, *19* (4), 21–31.
3. Faruqui, A.; Eakin, K.; Eds. *Electricity Pricing in Transition*, Kluwer Academic Publishing: Amsterdam, U.K., 2002.
4. Faruqui, A.; Eakin, K.; Eds. *Pricing in Competitive Electricity Markets*, Kluwer Academic Publishing: Amsterdam, U.K., 2000.
5. Faruqui, A.; George, S.S. Pushing the envelope on rate design. The Electricity Journal **2006**, *19* (2), 33–42.
6. Faruqui, A.; George, S.S. Quantifying customer response to dynamic pricing. The Electricity Journal **2005**, *18* (4), 53–63.
7. Faruqui, A.; George, S.S. Dynamic pricing for the mass market: California experiment. Public Utilities Fortnightly **2003**, *141* (13), 33–35.
8. Faruqui, A. Toward post-modern pricing. The Electricity Journal **2003**, *16* (6) [Guest Editorial].
9. Faruqui, A.; George, S.S. Demise of PSE's TOU program imparts lessons. Electric Light and Power **2003**, *January*.
10. Faruqui, A.; George, S.S. The value of dynamic pricing. The Electricity Journal **2002**, *15* (6).
11. Fox-Penner, P.; Broehm, R. Price-responsive electric demand: A national necessity, not an option. In *Electricity Pricing in Transition*; Faruqui, A., Eakin, B.K., Eds., Kluwer Academic Publishers: New York, 2002; [Chapter 10].
12. Fox-Penner, P.; Hanser, P.Q.; Wharton, J.B. Real-Time Pricing: Restructuring's Big Bang? Public Utilities Fortnightly March 1997.
13. Hanser, P.Q.; Caves, D.W.; Herriges, J.; Windle, R.J. Load impact of interruptible and curtailable rate programs. IEEE Transactions on Power Systems **1988**, *3* (4).
14. Wood, L. The new vanilla: Why making time-of-use the default rate for residential customers makes sense. Energy Customer Management **2002**, *5* (1).
15. Wood, L. *Effective Demand Response Programs for Mass Market Customers*. Presented at NYSERDA Time Sensitive Electricity Pricing Workshop, Albany, NY, October 3, 2002.
16. Wood, L. *The Customer's Critical Role in Demand Response*, Presented at U.S. Association for Energy Economics' (National Capital Area Chapter) New Developments in Electric Market Restructuring Conference, Washington, DC, May 21, 2002.

Psychrometrics

S. A. Sherif
Department of Mechanical and Aerospace Engineering, Wayne K. and Lyla L. Masur HVAC Laboratory, University of Florida, Gainesville, Florida, U.S.A.

Abstract
This entry presents the basics of psychrometric theory. This includes a brief discussion of moist air properties and psychrometric processes as well as provides examples of how the theory may be used in the HVAC design process for different summer and winter systems.

INTRODUCTION

Psychrometrics is the science of moist air properties and processes. It is used to illustrate and analyze air conditioning cycles, translating the knowledge of the building heating or cooling loads (which are in kW or tons) into volume flow rates (in m^3/s or cfm) for the air to be circulated into the duct system. The approximate composition of dry air by volume is as follows: nitrogen = 79.08%; oxygen = 20.95%; argon = 0.93%; carbon dioxide = 0.03%; other gases = 0.01%. Water vapor is lighter than dry air. The amount of water vapor that the air can carry increases with its temperature. Any amount of moisture that is present beyond what the air can carry at the prevailing temperature can only exist in the liquid phase as suspended liquid droplets (if the air temperature is above the freezing point of water) or in the solid state as suspended ice crystals (if the temperature is below the freezing point).

The most exact formulations of thermodynamic properties of moist air in the temperature range of $-100°C–200°C$ are based on the study performed by Hyland and Wexler.[4,5] More recent studies by Sauer et al.[8] and Nelson et al.[7] provided psychrometric data for moist air in the temperature range 200°C–320°C and humidity ratios from 0 to 1 kg/kg$_{air}$ at pressures of 0.07706 MPa (corresponding to an altitude of 2250 m), 0.101325, 0.2, 1.0, and 5 MPa. Both studies developed the psychrometric data using the most current values of the virial coefficients, enthalpy, and entropy of both air and water vapor. Other psychrometric data were generated by Stewart et al.,[9] who created psychrometric charts in SI units at low pressures. More psychrometric charts and tables are also available in the American Society of Heating, Refrigerating and Air-Conditioning Engineers, Inc. (ASHRAE) brochure on psychrometry.[1] Most recently, Mago and Sherif[6] presented new psychrometric charts and property formulations of supersaturated air in the temperature range from -40 to $+40°C$.

The most commonly used psychrometric quantities include the dry- and wet-bulb temperatures, dew point, humidity ratio, absolute humidity, specific humidity, relative humidity, and degree of saturation. These will be briefly defined and discussed.

Dry-Bulb Temperature

This is the temperature measured by a dry-bulb thermometer. There are several temperature scales commonly used in measuring the temperature. In the inch–pound (I–P) system of units, at standard atmosphere, the Fahrenheit scale has a water freezing point of 32°F and a boiling point of 212°F. In the International System (SI) of units, the Celsius scale has a water freezing point of 0°C and a boiling point of 100°C.

Wet-Bulb Temperature and Thermodynamic Wet-Bulb Temperature

The wet-bulb temperature is the temperature measured with a wet-bulb thermometer after the reading has stabilized in the air stream. Because of the evaporative cooling effect, the temperature measured with a wet-bulb thermometer is lower than the dry-bulb temperature, except when the air is saturated. In that case, the wet-bulb and dry-bulb temperatures are the same. The thermodynamic wet-bulb temperature, on the other hand, is the saturation temperature of moist air at the end of an ideal adiabatic saturation process. The latter process is defined as one of saturating an air stream by passing it over a water surface of infinite length in a well-insulated chamber.

Dew-Point Temperature

This is the temperature at which moisture will begin to condense out of the air.

Keywords: Psychrometrics; Moist air; Air conditioning; Psychrometric processes.

Psychrometrics

Saturation Pressure

Saturation pressure is needed to determine a number of moist air properties. Between the triple-point and critical-point temperatures of water, two states (saturated liquid and saturated vapor) may coexist in equilibrium. The saturation pressure over ice for the temperature range $-100°C \leq T \leq 0°C$ is given by[2]

$$\log_e(P_s) = \frac{a_0}{T + 273.15} + a_1 + a_2(T + 273.15) + a_3(T + 273.15)^2 + a_4(T + 273.15)^3 + a_5(T + 273.15)^4 + a_6\log_e(T + 273.15)$$

where $a_0 = -0.56745359 \times 10^4$; $a_1 = 6.3925247$; $a_2 = -0.9677843 \times 10^{-2}$; $a_3 = 0.62215701 \times 10^{-6}$; $a_4 = 0.20747825 \times 10^{-8}$; $a_5 = -0.9484024 \times 10^{-12}$; $a_6 = 4.1635019$.

The saturation pressure over liquid water for the temperature range of $0°C \leq T \leq 200°C$ can be written as[2]

$$\log_e(P_s) = \frac{b_0}{T + 273.15} + b_1 + b_2(T + 273.15) + b_3(T + 273.15)^2 + b_4(T + 273.15)^3 + b_5\log_e(T + 273.15)$$

where $b_0 = -0.58002206 \times 10^4$; $b_1 = 1.3914993$; $b_2 = -4.8640239 \times 10^{-2}$; $b_3 = 0.41764768 \times 10^{-4}$; $b_4 = -0.14452093 \times 10^{-7}$; $b_5 = 6.5459673$.

In both of the above equations, P_s is in Pa and T is in °C.

Humidity Ratio, Specific Humidity, Absolute Humidity, and Relative Humidity

The humidity ratio is the ratio of the mass of water vapor to the mass of dry air contained in the mixture of moist air. This is expressed as follows:

$$W_m = \frac{m_w}{m_a} = 0.62198 \frac{\varepsilon(P_s\phi)}{P - \varepsilon(P_s\phi)}$$

where m_w and m_a are the mass of water vapor and dry air, respectively, P_s is the saturation pressure of the water vapor, P is the total pressure, ϕ is the relative humidity of the air, and ε is an enhancement factor. The enhancement factor (ε) is a correction parameter that takes into account the effect of dissolved gases and pressure on the properties of the condensed phase as well as the effect of intermolecular forces on the properties of moisture itself.[2] The enhancement factor can be expressed in terms of virial coefficients, but the equation is rather complicated. An equally accurate but simpler expression can be determined from a polynomial equation which has been least-square curve-fitted to the data, as shown below.[6]

$$\varepsilon = 1.00391 - 7.82205 \times 10^{-6}T + 6.94682 \times 10^{-7}T^2 + 3.04059 \times 10^{-9}T^3 - 2.7852 \times 10^{-11}T^4 - 5.656 \times 10^{-13}T^5$$

where T as given above is the dry-bulb temperature in °C.

The specific humidity is the ratio of the mass of water vapor to the total mass of the moist air sample. This is expressed as follows:

$$\gamma = \frac{m_w}{(m_w + m_a)}$$

The absolute humidity is the ratio of the mass of water vapor to the total volume of the sample. This is expressed as follows:

$$d_v = \frac{m_w}{V}$$

The relative humidity is the ratio of the mole fraction of water vapor in a moist air sample to the mole fraction of water vapor in a saturated moist air sample at the same temperature and pressure. According to ASHRAE,[2] this is expressed as follows:

$$\phi = \left(\frac{x_w}{x_{ws}}\right)_{T,P}$$

where the subscript s in the denominator refers to the saturation condition.

Degree of Saturation

This is the ratio of the humidity ratio of moist air to the humidity ratio of saturated moist air at the same temperature and pressure. This is expressed as follows:

$$\mu_d = \frac{W_m}{W_s}$$

where W_m is the humidity ratio of moist air while W_s is the humidity ratio of saturated air at the same temperature and pressure. The degree of saturation may have a value between 0 and 1 for subsaturated air. Combining the definition of the degree of saturation and those of the humidity parameters above, the following equation can be reached:

$$\mu_d = \frac{\phi}{1 + (1 - \phi)W_s/0.62198}$$

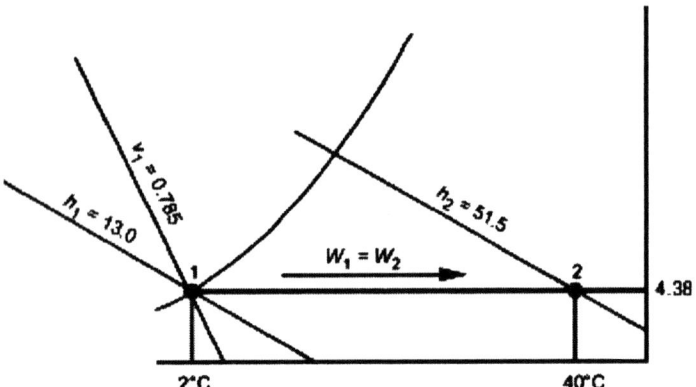

Fig. 1 Psychrometric depiction of sensible heating (1–2) and cooling (2–1).

PSYCHROMETRIC PROCESSES

Sensible Heating or Cooling

This is the process of heating or cooling the air without changing its moisture content. It is represented by lines of constant humidity ratio on the psychrometric chart (see Fig. 1). Sensible heating is accomplished when the air passes over a heating coil. Sensible cooling is accomplished when the air passes over a cooling coil whose surface temperature is above the dew point temperature of the air.

Humidification (with Heating or Cooling)

This is the process of introducing moisture into the air stream. In the winter season, humidification is frequently required because the cold outside air, infiltrating into a heated space or being intentionally brought in to satisfy the space ventilation requirement, is too dry. In the summer season, on the other hand, humidification is usually done as part of an evaporative cooling system. Humidification is achieved in a variety of ways, which range from using spray washers to passing the air over a pool of water to injecting steam. The process is represented on the psychrometric chart as a line of constant wet-bulb temperature when the water being sprayed is not externally heated or cooled. When there is external heating or cooling, the process is represented by a line to the right or to the left of the wet-bulb temperature line, respectively. Depending on the magnitude of water heating, the humidification-process line can be oriented in such a way that results in an increase in the dry-bulb temperature of the air stream at the exit of the dehumidifying device. In the extreme case of spraying cold water in the air stream such that the water temperature is less than the dew point temperature of water vapor in air, the spraying process will actually result in air dehumidification.

Cooling and Dehumidification

This process is used in air conditioning systems operating in hot and humid climates. It is accomplished by using a cooling coil whose surface temperature is below the dew point temperature of water vapor in air. On the psychrometric chart, the process is represented by a line in which both the dry-bulb temperature and the humidity ratio decrease (see Fig. 2). Because of the fact that not all of the air molecules going through the cooling coil make physical contact with the coil surface, the air condition at the exit of the coil is usually not saturated (but close to the saturation curve). This fact is reflected by the use of the so-called coil bypass factor, which is defined as the ratio of the temperature difference between the leaving coil air condition and the coil apparatus dew point and that between the entering air condition and coil apparatus dew point (see Fig. 3). This is expressed as follows:

$$B_F = \frac{T_{cc} - T_{cadp}}{T_m - T_{cadp}}$$

Coil bypass factors therefore range from 0 to 1. Fig. 3 shows the apparatus dew point and the coil bypass factor.

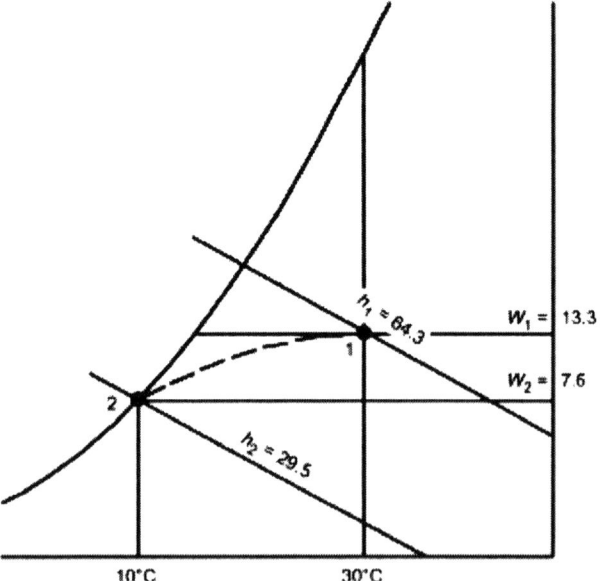

Fig. 2 Psychrometric depiction of the cooling and dehumidifying process (1–2).

Psychrometrics

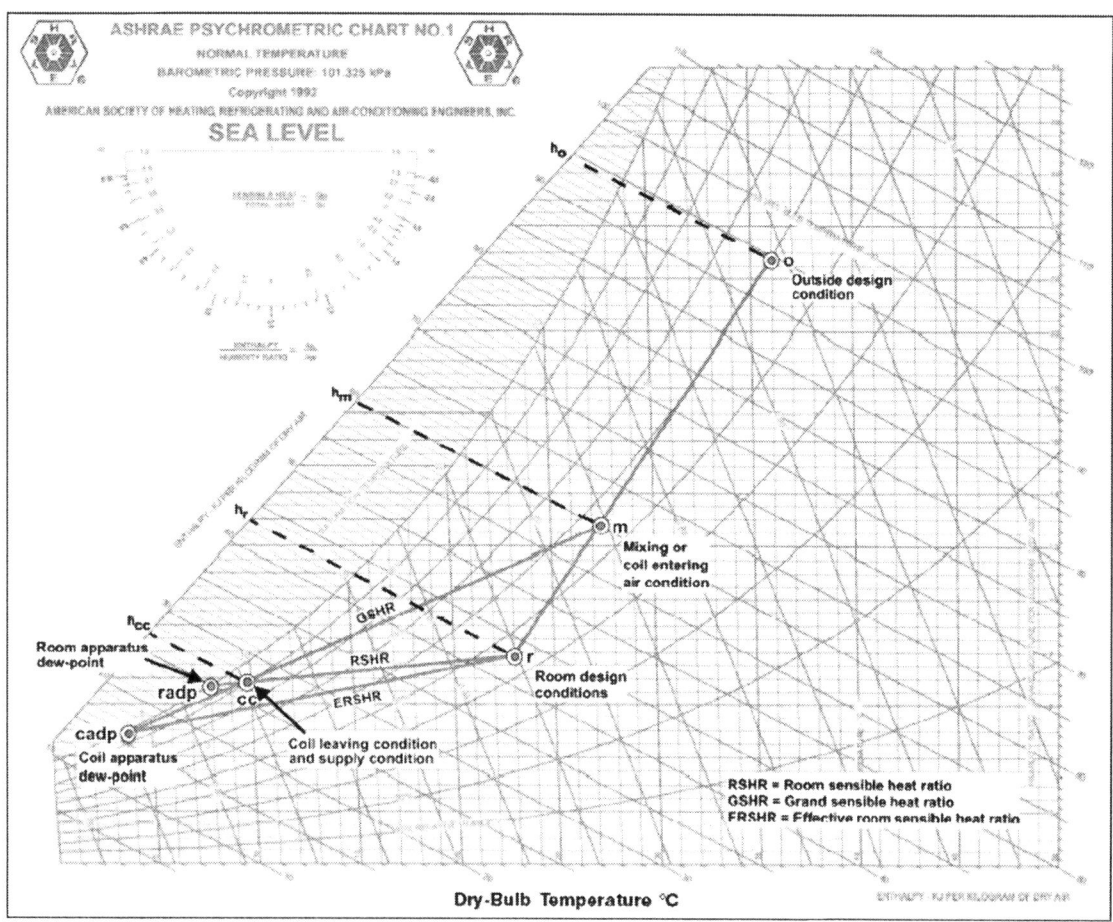

Fig. 3 Psychrometric depiction showing coil bypass factor and sensible heat ratios.

The figure also shows the lines representing the room sensible heat ratio (RSHR), the coil (or grand) sensible heat ratio (GSHR), and the effective room sensible heat ratio (ERSHR). These are defined according to the following equations:

$$\text{RSHR} = \frac{\dot{Q}_{R,S}}{\dot{Q}_{R,S} + \dot{Q}_{R,L}}$$

$$\text{GSHR} = \frac{\dot{Q}_{G,S}}{\dot{Q}_{G,S} + \dot{Q}_{G,L}}$$

$$\text{ERSHR} = \frac{\dot{Q}_{R,S} + B_F \dot{Q}_{O,S}}{[\dot{Q}_{R,S} + B_F \dot{Q}_{O,S}] + [\dot{Q}_{R,L} + B_F \dot{Q}_{O,L}]}$$

where the Q_S represent the loads. Subscripts R, O, and G represent room, outside, and grand, respectively, while S and L represent sensible and latent (loads), respectively. The effective sensible heat ratio is interwoven with both the coil apparatus dew point temperature and the coil bypass factor, with the sole intention of simplifying psychrometric calculations. The mass flow rate using the effective RSHR can be computed from the following equation:

$$\dot{m}_s = \frac{(\dot{Q}_{R,S} + B_F \dot{Q}_{O,S})}{\rho C_p (T_r - T_{\text{cadp}})(1 - B_F)}$$

where T_r is the indoor (room) design temperature and T_{cadp} is the coil apparatus dew point temperature. The mass flow rate term represents the quantity of air per unit time (kg/s or lb/hr) supplied to the conditioned space. The above equation is not exact because the product of the specific heat (C_p) and the temperature difference was used in lieu of the enthalpy difference. However, the specific heat of air does not change much with temperature in the air conditioning temperature range, and hence the error introduced by the above approximation is negligible.

Heating and Dehumidification

This is also referred to as desiccant (or chemical) dehumidification, which takes place when air is exposed to either solid or liquid desiccant materials. The mechanism of dehumidification in this case is either absorption (when physical or chemical changes occur) or adsorption (when there are no physical or chemical changes). During the sorption process, heat is released. This heat is the sum of the latent heat of condensation of the absorbed water vapor into liquid plus the heat of wetting. The latter quantity refers to either wetting of the surface of the solid desiccant by the water molecules or the heat of solution in the case of liquid desiccant. Dehumidification by solid desiccants is represented on the psychrometric chart by a process of increasing dry-bulb temperature and a decreasing humidity ratio. Dehumidification by liquid desiccants is also represented by a similar line, but when internal cooling is used in the apparatus, the process air line can also go from warm and moist to cool and dry on the chart.

Adiabatic Mixing of Air Streams

This usually refers to either adiabatic mixing of two or more air streams or to bypass mixing. In the former process, two or more streams are mixed together adiabatically forming a uniform mixture "m" in a mixing chamber. In this case, mass, energy, and water vapor balances yield the following equations:

$$\dot{m}_1 + \dot{m}_2 + \cdots + \dot{m}_j = \dot{m}_m$$

$$\dot{m}_1 h_1 + \dot{m}_2 h_2 + \cdots + \dot{m}_j h_j = \dot{m}_m h_m$$

$$\dot{m}_1 W_1 + \dot{m}_2 W_2 + \cdots + \dot{m}_j W_j = \dot{m}_m W_m$$

Bypass mixing, on the other hand, happens usually in air handling units where the air flow is divided into upper hot deck and lower cold deck streams. In a summer air conditioning operation, the upper hot deck acts as a bypass air stream in order to moderate the temperature of the otherwise overcooled air leaving the cooling and dehumidifying coil. In the winter season, on the other hand, the lower cold deck acts as the bypass stream by mixing with the warm air after it has passed over the heating coil (Fig. 4).

AIR CONDITIONING CYCLES AND SYSTEMS

An air conditioning cycle is a combination of several air conditioning processes connected together. Different systems are characterized by the type of air conditioning cycle they use. The main function that psychrometric analysis of an air conditioning system achieves is the determination of the volume flow rates of air to be pushed into the ducting system and the sizing of the major system components.

There are generally four extreme climatic conditions that an air conditioning system may face. In summer operation, for example, the dry-bulb temperature of the outdoor air is always high, but the humidity ratio may either be high or low. In hot and humid climates (such as Miami, Florida), the air conditioning system is typically composed

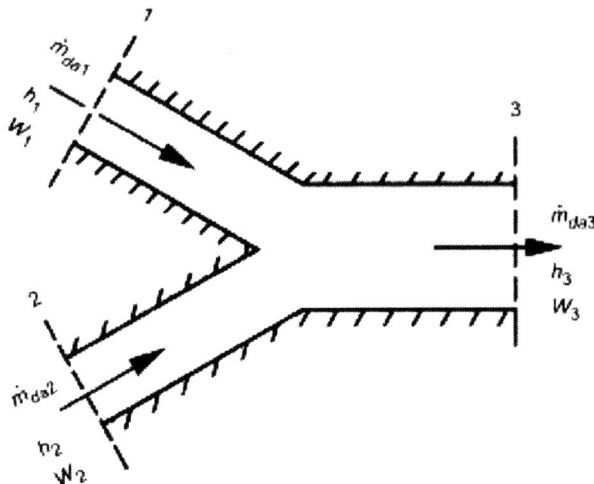

Fig. 4 Adiabatic mixing of air streams.

of a cooling coil whose surface temperature is below the dew point temperature (see Fig. 5). That way, both cooling and dehumidification can be achieved by the system. In hot and dry climates (such as Phoenix, Arizona), on the other hand, evaporative coolers are typically used.

Winter conditions may have similar extremes, with the dry-bulb and dew point temperatures being both low. In extremely cold conditions (such as Minneapolis, Minnesota), the environment is typically very dry. In this case, the air conditioning system is typically composed of a heating coil and a humidifying device, with the former being located upstream of the latter. The humidifying device may be a spray washer composed of a spray chamber in which a number of spray nozzles and risers are installed. Spray washers are typically used in industrial applications where the device performs the dual function of air humidification

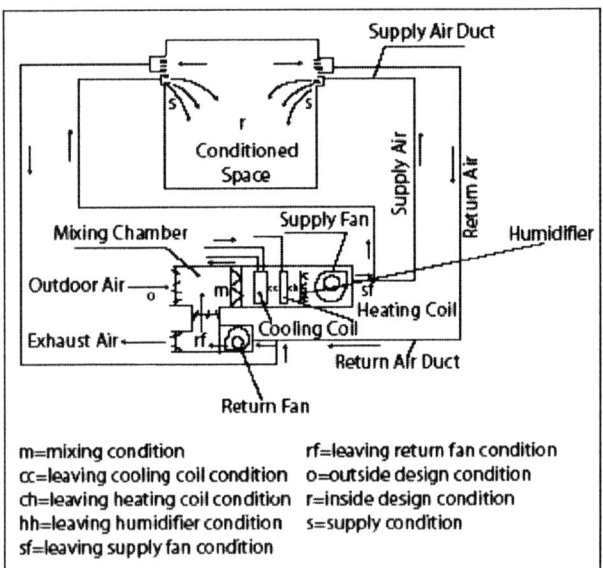

Fig. 5 Basic air-conditioning system.

and cleaning. The washer may have one or more banks of spray nozzles that have the capacity of injecting one to two gpm of atomized water per nozzle into the air stream. Adequate atomization of the water can be achieved by operating at pumping pressures ranging from 20 to 40 psi.[3] Washers typically employ baffles at the inlet air section in order to distribute the air uniformly throughout the chamber. At the exit section, on the other hand, moisture eliminators are employed to prevent carryover water from exiting the chamber into the conditioned space.

In cases where the outdoor air is cool but humid (such as Seattle, Washington), there may not be a need to humidify the air, and the air conditioning system is typically composed of a heating coil only. However, because of the fact that the humidity ratio at low temperatures is low even though the relative humidity may be high, some humidification may be required when the dew point is low during the winter season.

In addition to the four extreme climatic conditions described above, the air conditioning system may also have to operate in the fall and spring seasons. This constitutes a mixed-mode operation that usually requires switching back and forth between heating and cooling based on the value of the outdoor temperature. However, because of the potential energy waste associated with this mode of operation, the system is usually set to operate only if the outdoor temperature goes outside a prespecified wide band in order to minimize the frequency of cycling between the heating and cooling modes.

In order to provide the reader with a flavor of the psychrometric analyses that need to be performed on an air conditioning system, only three of the above systems (hot and humid, hot and dry, and cold and dry) will be described in more detail. The cool-and-humid mode of operation is identical to the cold-and-dry mode except in the absence of the humidifying device.

Summer Hot-and-Humid Mode

Fig. 5 shows a basic air conditioning system capable of both summer and winter mode operations. Fig. 6 shows the psychrometric representation of a hot-and-humid summer mode operation applicable to Miami, Florida. In this mode, outdoor air at State "o" is adiabatically mixed with

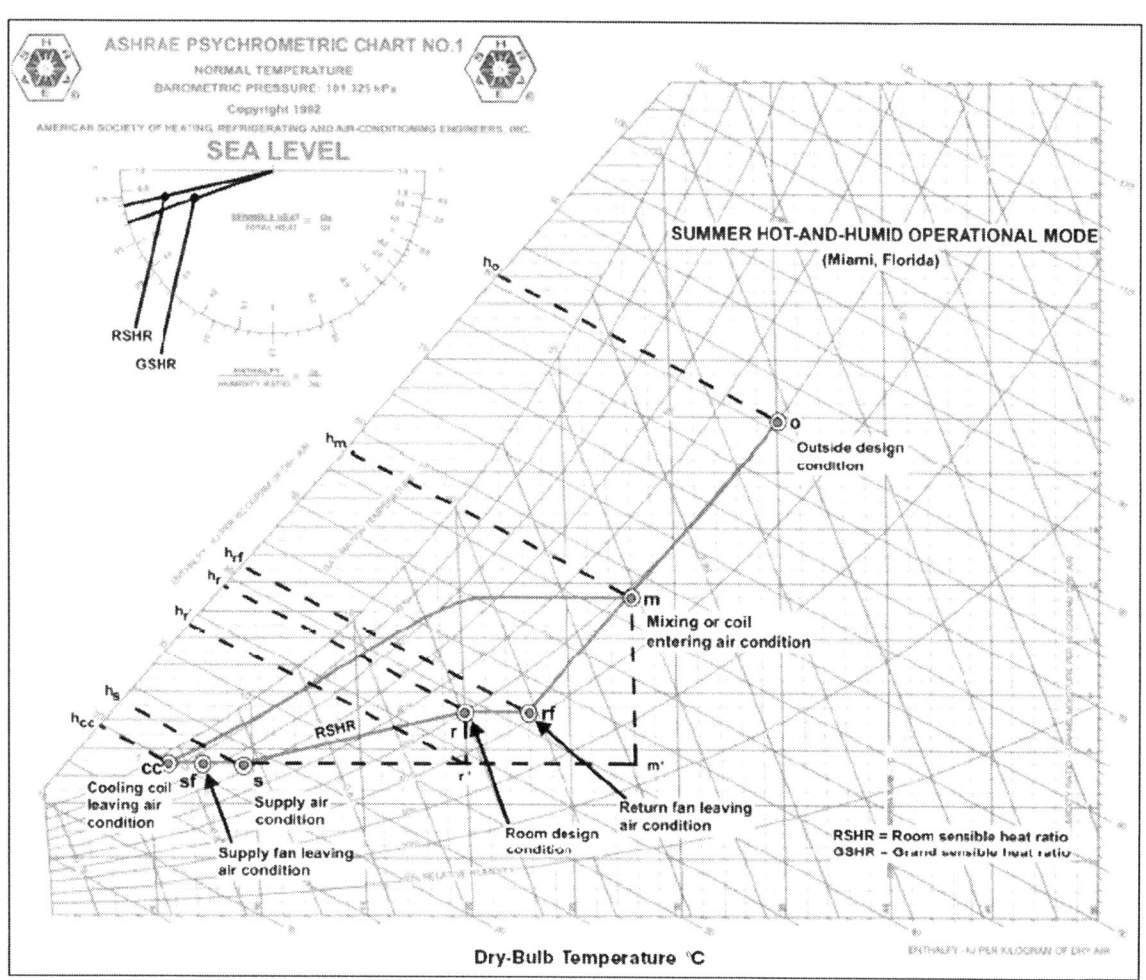

Fig. 6 Psychrometric depiction of hot, humid summer air conditioning system.

recirculated air from the room after passing through the ceiling plenum, the return duct, and the return fan (State "rf") to form the mixed air condition "m," which is also the entering state to the cooling coil. Air is then cooled and dehumidified until it exits the coil at State "cc." After that, air is reheated, due to passing over the supply fan and through the supply ducts, to State "s," the supply condition. The space sensible, latent, and total loads can, respectively, be expressed as follows:

$$\dot{Q}_{R,S} = \dot{m}_s(h_{r'} - h_s)$$

$$\dot{Q}_{R,L} = \dot{m}_s(h_r - h_{r'})$$

$$\dot{Q}_{R,T} = \dot{m}_s(h_r - h_s)$$

where, again, the Q_S represent the load quantities as before and the enthalpies have subscripts corresponding to the respective points identified on the psychrometric chart. Any one of the above equations can be used to compute the supply-mass flow rate. The supply-volume flow rate, on the other hand, is computed with the knowledge of the specific volume at the supply condition v_s. This is expressed as follows:

$$\dot{V}_s = \dot{m}_s v_s$$

The space (room) sensible heat ratio (RSHR) is graphically represented by line "rs" on the psychrometric chart. As for the cooling coil (or grand) loads, the following equations apply:

$$\dot{Q}_{G,S} = \dot{m}_s(h_{m'} - h_{cc})$$

$$\dot{Q}_{G,L} = \dot{m}_s(h_m - h_{m'})$$

$$\dot{Q}_{G,T} = \dot{m}_s(h_m - h_{cc})$$

Again, the subscripts of the enthalpy quantities correspond to state points identified on the chart. The above equations can be used to size the cooling coil. The GSHR is graphically represented by line "m cc." The relative humidity of the air exiting the cooling coil is a function of the coil design, fin spacing, coil surface area, and coil face velocity, among other factors. For a coil with 10 or more fins per inch and four rows of coil, the relative humidity of the leaving air is approximately 93%. For six- and eight-row coils with fin spacing of ten or more per inch, the exiting relative humidities are 96 and 98%, respectively.[10]

Summer Hot-and-Dry Mode

Psychrometric depiction of the processes involved in this mode of operation is shown in Fig. 7. As indicated earlier, the air conditioning equipment in this case is simply composed of a humidifying device where atomized water is introduced into the air stream. Outdoor air at State "o" is typically mixed with return air from the conditioned space after passing through the return ducts and over the return fan (State "rf") to form the mixed state "m." The evaporative cooling process that results from the spray equipment usually follows a constant wet-bulb temperature line, with the air exiting at State "hh." After that, the air passes over the supply fan and through the supply ducts until it enters the conditioned space at State "s." The humidifying effectiveness, η_H, can be expressed either in terms of a temperature deficit ratio or a humidity ratio deficit ratio as follows:

$$\eta_H = \frac{T_m - T_{hh}}{T_m - T_{sat}}$$

$$\eta_H = \frac{W_{hh} - W_m}{W_{sat} - W_m}$$

Special care has to be exercised in analyzing room and grand loads in this mode as the sensible and latent components have opposite signs. For example, the evaporative cooling process employed results in a decrease in the air temperature (sensible cooling), whereas it increases the humidity ratio (latent heating). The process results in a very small change in the air enthalpy (almost zero), as the adiabatic saturation process through the humidifying device follows a constant wet-bulb temperature line.

Winter Mode

There are two possible scenarios in a winter system—one involves the supply of heated air and the other involves the supply of unheated air to the conditioned space. The latter mode is applicable in cases when the outdoor temperature is not too cold that the fan and duct heating may be enough to provide for comfort level temperatures. Because of the similarities between these scenarios, this text will only consider the former scenario for further analysis. As mentioned earlier, for the cold-and-dry mode, there is a need to humidify the air stream. Fig. 8 shows the basic cycle represented on the psychrometric chart for this operational mode. Outside air at State "o" is mixed with recirculated air from the room after passing over the return fan (State "rf"), producing the mixed state "m." Air at this state passes over a heating coil and leaves at State "ch." After the sensible heating process "m ch," air enters the humidifying device where both the dry-bulb temperature and humidity ratio of the air stream are increased, producing state "hh" at the exit of the device. In the humidifying device, either steam or atomized hot water is injected. Air then passes over the supply fan and through the supply ducts to the conditioned space at the supply condition "s."

The amount of heating required by the heating coil typically needs to be coordinated with the heating performed by the humidifying device due to steam or hot water injection. The optimum amount of heating performed by both devices may be hard to pinpoint in an

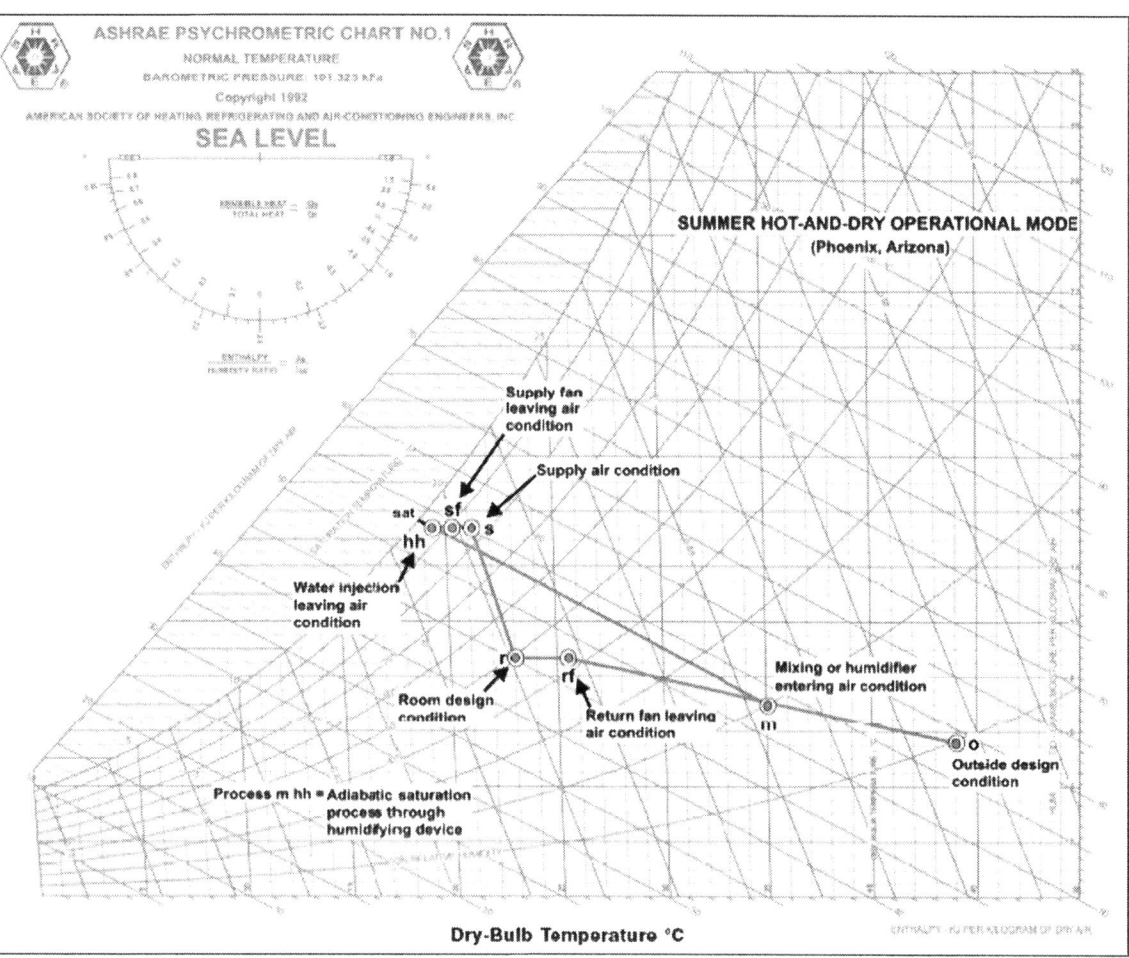

Fig. 7 Psychrometric depiction of a hot, dry summer air conditioning system.

exact way, but a careful choice of the state of air leaving the heating coil may go a long way towards minimizing unnecessary use of heating energy.

The sensible heat ratio for the humidifying device is represented by line "ch hh." The GSHR of the overall system is represented by line "m hh."

In cases when the outside air is humid enough (such as in the Pacific Northwest of the continental United States in winter months), injecting steam or atomized hot water may not be necessary and sensible heating is the only process that may be required.

CONCLUSIONS

This entry presented an overview of psychrometric processes and systems as applied to different operational modes. Proper execution of this stage is crucial in accurately computing the volume flow rates of air through the air conditioning ducts. This phase, in an air conditioning design process, follows the load calculation phase. While the latter phase produces quantities that represent the sensible and latent loads imposed on the conditioned space, the former phase (psychrometric analysis) is capable of incorporating the effects of introducing fresh outside air into the space for ventilation purposes. Cooling or heating equipment sizing has to take into account not only the load imposed on the conditioned space, but also the outside load. By specifying the amount of outside air to be introduced for a specified set of outdoor and indoor conditions, psychrometric analysis enables the HVAC system designer to compute the load imposed on the conditioning equipment (grand load). The analysis is inherently capable of distinguishing between the sensible and latent load quantities of outside and conditioned space (room) air, thus providing a truly insightful picture of how the moisture present should be handled. Psychrometric analysis also enables the designer to account for other smaller loads that may be imposed on the system, such as ducts and fans, in equipment sizing. By identifying the different state points of the air as it passes through the duct system and over the supply and return fans, the volume flow rates of air computed by the analysis become necessarily inclusive of the effects of the ducts and fans in equipment sizing.

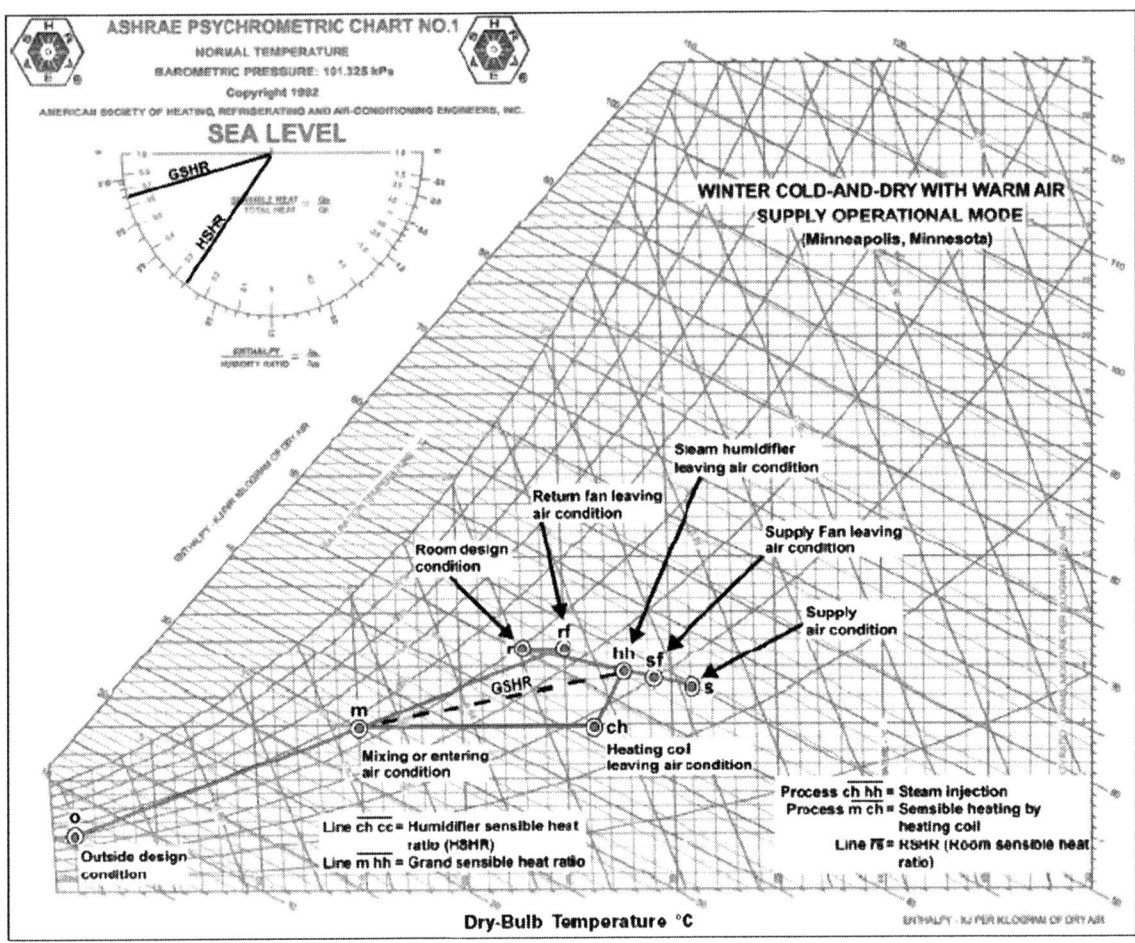

Fig. 8 Psychrometric depiction of a cold, dry winter air conditioning system.

REFERENCES

1. ASHRAE. *Psychrometrics: Theory and Practice*; The American Society of Heating, Refrigerating and Air-Conditioning Engineers, Inc.: Atlanta, GA, 1996.
2. ASHRAE. *2005 ASHRAE Handbook—Fundamentals*; The American Society of Heating, Refrigerating and Air-Conditioning Engineers, Inc.: Atlanta, GA, 2005.
3. Clifford, G. *Modern Heating, Ventilating, and Air Conditioning*; Prentice Hall: Englewood Cliffs, NJ, 1990.
4. Hyland, R.W.; Wexler, A. Formulations for the thermodynamic properties of the saturated phases of H$_2$O from 173.15 to 473.15 K. ASHRAE Trans. **1983**, *89* (Part 2A), 500–519.
5. Hyland, R.W.; Wexler, A. Formulations of the thermodynamic properties of dry air from 173.15 to 473.15 K, and of saturated moist air from 173.15 to 372.15 K, at pressures to 5 MPa. ASHRAE Trans. **1983**, *89* (Part 2A), 520–535.
6. Mago, P.J.; Sherif, S.A. Psychrometric charts and property formulations of supersaturated air. HVAC&R Res. **2005**, *11* (1), 147–163.
7. Nelson, H.F.; Sauer, H.J.; Huang, X. High temperature properties of moist air. ASHRAE Trans. **2001**, *107* (Part 2), 780–791.
8. Sauer, H.J.; Nelson, H.F.; Huang, X. The search for high temperature experimental psychrometric data. ASHRAE Trans. **2001**, *107* (Part 2), 768–779.
9. Stewart, R.B.; Jacobsen, R.T.; Becker, J.H. Formulations for thermodynamic properties of moist air at low pressures as used for construction of new ASHRAE SI unit psychrometric charts. ASHRAE Trans. **1983**, *89* (2A), 536–548.
10. Wang, S.K. *Handbook of Air Conditioning and Refrigeration*; McGraw-Hill Book Company: New York, 1993.

Public Policy for Improving Energy Sector Performance

Sanford V. Berg
Director of Water Studies, Public Utility Research Center, University of Florida, Gainesville, Florida, U.S.A.

Abstract

Public policy evolves in response to a number of forces, including public perceptions and special interest campaigns. Policy influences the basic conditions in the energy industry, in market structures, in corporate behavior, and in energy sector performance. Government intervention is often justified on the basis of market imperfections (market power and information problems), market failures (pollution), and income distributional concerns. It affects the performance of the energy sector through regulation, antitrust, environmental mandates, and other mechanisms.

INTRODUCTION

Public policy establishes the legal constraints facing decision-makers and determines the jurisdictional responsibilities of different levels of government. The basic rationale is that market imperfections (market power and information gaps) and market failures (such as pollution) require some form of government intervention. Energy policies tend to address three broad areas: market structure, corporate behavior, and sector performance. While antitrust regulation addresses mergers and anticompetitive behavior in the economy, energy sector economic regulation tends to address those elements of the market structure (including the supply chain) that are viewed as natural monopolies. Government intervenes when economic or social problems catch the attention of those involved in political processes. Politicians then craft legislation and create administrative agencies to implement the laws. The courts rule on the legality of different arrangements when laws come into conflict or are challenged in terms of constitutionality. In addition to antitrust agencies and energy sector regulators, environmental agencies have also been created to address the ecological and health impacts of energy production and distribution. Thus, pressure for changes in public policy builds as energy sector performance falls short of expectations. The relative roles of markets and government shift when citizens lack confidence in current institutional arrangements and when new social concerns arise.

Keywords: Market power; Environmental impacts; Regulation; Incentives; Restructuring.

CREATION OF LAWS AND RULES

The standard justification for government intervention in the economy is to "improve performance." Of course, different dimensions of performance will tend to be given different weights, according to how citizens view current social and economic outcomes. Energy policy is particularly sensitive to public opinion because the sector significantly affects citizens; in addition, energy suppliers are often politically powerful. The range of concerns means that political coalitions (based on regional alliances or ideological predispositions) form around issues and support policy initiatives that meet their concerns. Some groups are concerned with social justice (or fairness), particularly regarding the effect of energy prices on low-income citizens. Others worry about environmental impacts associated with energy and seek investments in research and development (R&D) and conservation to reduce those impacts. Current suppliers tend to focus on financial success and seek programs that reduce risks and increase returns. Potential entrants seek a level playing field so that incumbents do not benefit from artificial advantages. Customers understand that if firms are not efficient, prices will be higher than they need to be. In addition, particular customer groups are interested in whether they are subsidizing other groups, so the price structures also matter.[1]

All of these interest groups communicate their concerns to politicians. In addition, the media (newspapers, television, and, increasingly, the Internet) shape citizen attitudes by providing information about the implications of various policy options. Policy options include changes in sector taxes, subsidies, entry conditions, environmental mandates, and other regulatory rules. Because managers of private and publicly owned enterprises view government as constraining decision-making, they will track potential policy changes and lobby against actions that seem to harm the interests of those enterprises. Another term for self-interested policy promotion to obtain favorable treatment from governmental authorities is "political rent-seeking."[2]

The policy-making process involves convincing key stakeholders that the change is (or is not) needed. During the public debate, different groups provide information to policy-makers on alternative approaches to resolving issues. Public policies often concentrate benefits while dispersing costs over a number of groups. For example, large corn farms benefit from subsidies for ethanol. For groups with high per capita benefits, political lobbying will be intense. This pattern means that public intervention sometimes results in greater aggregate costs than social benefits, as when special interests gain protection from developments that threaten their positions of privilege. In other instances, new laws create win–win outcomes that can enhance energy sector performance.

Although public policy is partly based on legislation, it is also developed by government ministries, departments, and agencies established by the law and responsible for creating rules based on the policy. For example, energy sector regulatory commissions have the task of implementing the law by developing procedures that meet the objectives established by political authorities. Each nation has a unique history and legal structure, so public policies have tended to differ across countries.[3] Under federal systems such as in the United States, India, Brazil, and Russia, state and national authorities generally have a division of labor that gives states jurisdiction over local energy activity (such as electricity distribution). However, the division of responsibility in the area of energy depends on constitutional authority and on the interaction between public expectations and sector performance over time. Jurisdictional disputes are likely to arise when the status quo is challenged by one group or another.[4]

RATIONALE FOR INTERVENTION

Competitive markets often promote efficient outcomes, but problems occasionally arise when the market does not fully account for all benefits and costs or when it produces an unreasonably inequitable outcome. Government intervention is often justified on the basis of these market imperfections, market failures, and income distributional concerns.[5]

Market imperfections include market power and information asymmetries. Market power arises if there is a single supplier or if suppliers collude to earn financial returns far greater than required to attract investors, and consumers face high prices. Two major policies address market power: sector regulators deal with natural monopoly and antitrust authorities attempt to limit the artificial attainment (and exercise) of market power. Of course, the scale economies associated with natural monopoly can change. When technological change alters the production economies on which previous market structures had been predicated, public policy may change in response to competitive possibilities. The transition to new market structures may require that regulators develop rules for access to essential facilities (as when generators need access to the transmission grid).

Another market imperfection involves information problems. For example, do consumers have full information about the life-cycle costs of energy-using appliances? What time horizons do home buyers use when considering appropriate levels of insulation? Product labeling requirements and energy efficiency mandates are based on an assumed lack of consumer information.

Market failures arise when there are environmental impacts. In such cases, those who produce and consume the product are not the only ones affected by the transaction. The energy production chain (exploration, extraction, transport, refining/generation, transmission, distribution, and retailing) involves a number of sensitive areas. Thus, air pollution, water pollution, and land use impacts (plant siting and the placement of high voltage lines) have come under the purview of energy regulation. (See the entry on *Environmental Regulation*.) Environmental issues can also be examined in the context of common property resources—who "owns" the clean air? Another potential market failure arises in the context of fuel diversity: do competitive generation markets yield the optimal portfolio of fuel types (including renewable resources) so end users do not bear the excessive risk of price increases under particular fuel price scenarios?

Income distributional concerns, such as universal access to electricity, involve low-income or high-cost (rural) consumers. Nations generally devise subsidies of one form or another to promote network expansion into high-cost geographic areas. In addition, costs may be covered by public funds or cross-subsidies to promote usage by those otherwise unable to afford what is generally viewed as a necessity. Of course, not all subsidies are well-targeted, and once a group is receiving subsidies, governments find it difficult to cut them off even when circumstances change.[6] The ethical objective of social justice is difficult to define and to quantify.

All three rationales for intervention relate to sector performance that does not match citizen expectations. Besides performance, citizens also care about the process used to develop and implement policy. Was the process transparent and did it involve citizen input (participation)? Procedural fairness affects citizen perceptions regarding the legitimacy of public policy. The ultimate form of intervention (and potential inefficiencies) becomes a matter of the relative political power of different stakeholders. However, the key point is that public policy is not shaped in a vacuum. There is a rationale for intervention even when the specifics of the adopted policy include elements that favor particular groups or create inefficiencies.[7]

FACTORS AFFECTING POLICIES

The figure is a flow diagram that shows how public policy responds to sector performance.[8] The behavior of firms (in terms of pricing, cost-cutting, provision of services, and network expansion) follows from the market structure and rules affecting behavior. Ultimately, public perceptions are influenced by sector performance, including the adoption of innovations, network reliability, environmental impacts, and other outcomes. Indicators of sector performance include returns that are commensurate with risks, production efficiency, introduction of new technologies, safety in the production process, and sound environmental stewardship. Thus, public policy changes arise from weak sector performance relative to what is deemed possible. Perceptions are affected by information presented in the media and by opinion leaders (Fig. 1).

The diagram shows how social objectives and priorities influence the creation and operation of regulatory commissions, antitrust agencies, product information regimes, and environmental agencies. The law generally specifies the policy objectives and the approaches to be utilized by the agencies implementing public policy. However, regulatory commissions and other government agencies have a significant impact on the ultimate rules that are adopted, with court challenges serving as another influence on policy implementation. The boxes in the figure represent fourteen factors that are part of the policy environment, which can be understood as follows:

1. Objectives and Priorities: The broad economic and social objectives of citizens include freedom, equality, justice, high living standards, and technological advancement.[5] In the context of energy, political leaders attempt to discern (and shape) what citizens want from infrastructure sectors. Social values may reflect a consensus or be deeply divisive and lead to dramatic shifts in public policy. Events such as an oil crisis, a serious accident (like Three Mile Island), or a natural disaster can also trigger changes in public priorities and a willingness to move from the status quo.

2. Institutional Conditions: These represent the starting point for policy development. Mediated by the press and political leaders, they are the context in which agencies are created and "reformed."[9] Institutional conditions can be characterized by the extent of consensus between the legislative and executive branches of government, the judicial capabilities (and consistency in legal decisions), agency administrative capacity (whether highly professional or heavily politicized), informal norms, and formal rules. Informal norms are the customs that mold day-to-day behavior. When corruption is accepted behavior, firms and customers are subject to arbitrary treatment, which increases risk and raises costs. Formal rules are embedded in the legal framework in which organizations operate.

3. International Experience: History provides evidence regarding which institutions and policies have failed and which have been successful. Experience from other countries (or states within a nation) represent "natural experiments" regarding impacts of alternative energy policies and provide lessons for policy-makers. Thus, infrastructure performance across jurisdictions and over time yields a yardstick for determining whether performance is subnormal. Dubash[6] outlines how different national policies have attempted to address a broad range of social and economic objectives.

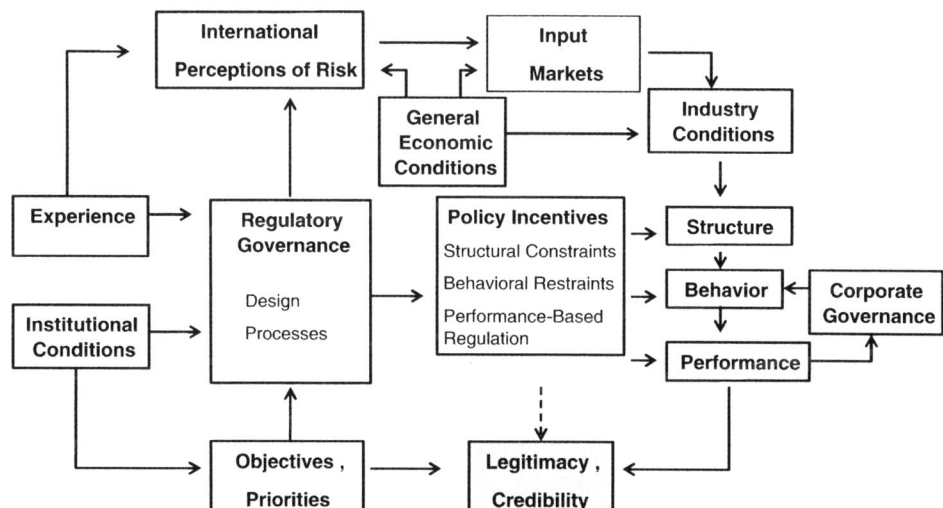

Fig. 1 Determinants of public policy and sector performance.

4. International Risk Perceptions: International perceptions regarding national security, energy independence, trends in the exchange rate, and concerns over cross-national environmental impacts affect investor attitudes and thus the cost of capital (reflected in required interest rates on bonds and expected returns on equity investments) but tend to be beyond the control of sector regulators or antitrust authorities. An exception concerns the predictability of government rules—an area where public policy does affect international perceptions. If the rules of the game are continually changing, investors are less willing to commit substantial funds to projects whose cash flows depend on a stable regulatory environment.[10]

5. General Economic Conditions: Macroeconomic factors are also outside the purview of those enforcing rules in the sector. Trends in the exchange rate will be affected by monetary and fiscal policy. Clearly, income growth has important consequences for energy demand and for the profitability of investment in the sector.

6. Input Markets: The presence or absence of key inputs determines earnings from investment. For example, operational effectiveness depends on the skills and incentives of managers and engineers. In addition, domestic and international capital markets affect the supply and demand of capital, thus the cost of capital. Similarly, the training of skilled engineers and thoughtful managers affects the availability of these key inputs. The availability of natural resources is another aspect of input markets affecting performance. Dependence on imported high-priced energy affects the costs that are ultimately borne by consumers. After considering demand growth and production technologies, these inputs create the basic conditions in the energy sector.

7. Basic Industry Conditions: Basic conditions include production technologies, input prices (fuel, capacity, materials, and labor), demand patterns (and growth), and ownership patterns. When the scale economies for generation are large relative to the market size, a natural monopoly may exist because having a single supplier can be least-cost. Such a situation leads to a market structure with a single (often vertically integrated) electric utility. Thus, a regulatory commission might be created to constrain the exercise of market power by the single supplier. If demand grows substantially or small-scale production becomes economical (as with combined cycle gas turbines or distributed generation), generation can become potentially competitive and lead to pressures by potential suppliers to lower entry barriers.[11]

Common property resources (clean air or water) create situations that invite government intervention. To some extent, technical standards (such as those developed under the auspices of the Institute of Electrical and Electronics Engineers) or energy efficiency standards mandated by law represent part of the information structure constraining utility decisions. Basic industry conditions change with national expenditures on energy R&D as new production technologies are developed and introduced. R&D might be performed in industry collaborative ventures (such as the Electric Power Research Institute in the United States), universities, and product suppliers. R&D funding comes from a variety of public and private sources. For nations with domestic natural gas production, regulation of gas transmission and distribution pipelines has often been added to the responsibilities of the entity having electricity oversight. Private and public ownership is another factor affecting downstream activities. Changes in basic conditions affect sector regulation in other ways. For example, the development of nuclear power precipitated the creation of specialized agencies responsible for setting standards for safety and public health aspects of the technology (including storage and disposal). National agencies like the Atomic Energy Commission (now the Nuclear Regulatory Commission) in the United States illustrate this trend. In addition, international agencies (International Atomic Energy Agency) have been created to address transnational issues and to monitor developments.

8. Regulatory Governance: Stepping back from how basic conditions affect the structure, behavior, and performance of an industry, we need to recognize the important role of the features characterizing the agency responsible for implementing public policy. Regulatory governance includes the legal mandate given to a government agency, the resources available for policy implementation, the organizational design of the agency, and the processes adopted by the agency, all of which affect regulatory activities. In addition, if there is no clarity in terms of which agency is responsible for implementing particular policies, public policy is likely to be compromised by intragovernmental rivalries. Resulting policy incentives affect the structure of the energy sector, the behavior of energy suppliers, and the performance of the industry. Baldwin and Cave[2] survey the key features of regulatory systems. Incentives can include structural constraints,

behavioral restraints, and performance-based initiatives (discussed below).

9. Policy Incentives: These can include taxes and subsidies that discourage and encourage a variety of activities. Regulatory incentives include constraints on market structure, restraints on corporate behavior, and performance-based rewards and penalties. The significant policy issues involve designing incentives that promote cost-cutting, service quality, and network expansion. In some cases, changes in public policy can significantly lower the cash flows that can be obtained from productive assets. For example, if allowing additional entry into the production of electricity means that "old" plants are operated for fewer hours per year, the net cash flows associated with those plants decline. Analysts can debate whether (and when) regulatory policy changes could have been anticipated and factored into investment decisions. On the practical side, if a restructuring initiative is adopted (to unbundle what was traditionally a vertically integrated industry), policy-makers generally try to address the issue of how to deal with the lost economic values stemming from the policy change. For example, some U.S. states have imposed competitive transition charges so consumers bear some of the burden of moving to a new market structure and a new regulatory framework (sometimes labeled "stranded costs").

10. Market Structure: For a homogenous product like electricity, market structure can be characterized in terms of entry conditions, the degree of vertical integration, and other factors. Government policies greatly influence the number and size distribution of suppliers through merger policy and the creation of franchise territories. Up until the last few decades, the electricity sector generally involved local monopolies, where entry was restricted by law. In the United States, these firms were often privately owned but regulated; in many nations, the governance system for publicly owned utilities was not very transparent, with policy-making, policy implementation, and operations often conducted within a government ministry. Policy-makers in many nations are currently exploring the extent to which competition can substitute for regulation. Of course, effective competition requires a sufficient number of firms that are operating independently from one another so that they become "price takers" (and lack market power). There is substantial evidence that competition can substitute for regulation when economies of scale and multiproduct economies make entry feasible. Thus, policies affecting market structure, vertical integration, and corporate ownership shape the evolution of the energy sector.

11. Corporate Behavior: Public policy also creates incentives involving behavioral restraints. These incentives are related to price, quality-of-service requirements, and mandates for system expansion. Antitrust laws attempt to address price-fixing, undue price discrimination, market foreclosure, and other issues. Sector regulators use cost of service (or rate-of-return regulation), price and revenue caps, and other mechanisms for constraining prices. For example, electricity regulators often include some fuel cost adjustment in the price mechanism to pass unavoidable cost increases (and decreases) through to consumers. The customer's bill is changed when the actual cost of fuel at the supplier's generating stations varies from a previously specified unit cost. Of course, such clauses insulate suppliers from input price fluctuations, reducing their incentive to seek low prices or to reduce the risk through hedging instruments. Other automatic adjustment clauses fund conservation programs or environmental programs (including "green" energy). Every regulatory rule has some impact on corporate behavior.

12. Sector Performance: Another set of policies affect sector performance. In the case of energy, regulators may mandate reliability requirements, network expansion targets, or profit-sharing plans. In the case of sharing plans, instead of establishing strict rate-of-return limitations (with associated weak incentives for cost containment), regulators may split the savings from performance improvements and allow investors to benefit from above-normal productivity gains. Meeting targets is often encouraged through performance-based ratemaking (PBR), which fits into a broad category of rate-setting mechanisms that link rewards to desired results or targets by setting rates (or rate components) for a given time according to external indices rather than a utility's actual cost of service. This type of regulation gives utilities better cost-reduction incentives than cost-of-service regulation. In developing performance standards for a PBR plan, a regulatory agency attempts to understand the utility's historic performance in order to develop an appropriate baseline for yardstick comparisons. The regulator collects information on service quality, determines the areas where cost savings may be realized and quality may be improved, and then develops measures for benchmarking performance. Brennan, Palmer,

and Martinez[4] survey the evolution of performance in the electricity industry and the issues confronting policy-makers.

13. Corporate Governance: Traditionally, activities internal to a firm are not micromanaged by regulators, but public perceptions about potential problems have changed in recent years. Corporate governance involves the allocation of decision rights within the firm (hierarchical vs team decision-making), the design of pay plans that compensate people for high levels of effort and performance, the development of performance evaluation mechanisms that monitor the effectiveness of internal reward systems, and dissemination of material about corporate activities. Policy may require that suppliers introduce reporting procedures that limit how insiders might adversely affect investor interests and sector performance. This factor has led to substantial revisions in reporting requirements in a number of nations (including the United States).

14. Legitimacy and Credibility: Ultimately, citizen acceptance of public policies, industrial outcomes, and the political processes that lead to those policies depends on how well industry performance matches shared national objectives. If there is no social consensus (or shared political vision), the legitimacy of the system is continually called into question. The lack of consensus could stem from disagreement regarding facts (whether global warming is occurring and whether industrial development causes it) or lack of consensus regarding values (whether creation of industrial jobs for low-income citizens is more important than devoting resources to improving environmental amenities). In either case, controversy is endemic in energy policy, where different stakeholders seek to influence policy outcomes. Another element affected by sector performance is the credibility of the system to private and public investors. Private investment is a voluntary activity and investors will shy away from projects, companies, or nations whose returns are low relative to the risks. Public investment (through legislative budget outlays or through cash flows to municipal utilities) is also affected by utility performance. If there is evidence of corruption or weak performance, pressures for change build up.

Thus, the figure depicts the circular dynamics of the larger decision environment in which government policy-makers, regulators, energy firms, and public and private sector investors operate and interact. Causation is seldom unidirectional, but changes in basic conditions generally result in altered market structures, corporate behavior, and sector performance. Similarly, changes in public perceptions can lead to significant public policy initiatives.

INFORMATION AND POLICY IMPLEMENTATION

Public policy can be designed to slow down the effects of changes in basic industry conditions or to speed up the system's response. Regulators have only limited information about a firm's commercial activities and its opportunities for cost containment. Regulatory institutions and incentives are partly designed in recognition of this information problem. Also, as noted, technological change and new national priorities can require modifications in regulatory rules and incentives. A nation's institutional endowment (including judicial, legislative, and executive institutions) partly determines the appropriate balance between rigid regulatory rules and flexibility to respond to changing basic conditions.

Regulatory Processes

The creation of a (relatively) independent regulatory commission is one mechanism for insulating those implementing public policy from daily political pressures, and in doing so, balancing the interests of customers, investors, suppliers, and current political authorities. In the context of regulation, a formal hearing procedure provides an avenue for stakeholder participation. A judicial or adjudicatory role is certainly appropriate when the agency needs to determine compliance with specific rules or evaluate behavior. Enforcement proceedings and complaint cases are examples for which this process is appropriate. However, hearings are basically "trial type" adjudicatory processes with a major defect—they tend to emphasize process instead of substance. In such a fact-specific setting, the stability and certainty of the process can cause all parties to overlook the substance of the outcome.

Some jurisdictions have emphasized informal processes for resolving issues. For example, as the issues become more complex, alternative dispute resolution procedures can be effective in identifying win–win options. An "all-party settlement" process represents an alternative way to gain stakeholder buy-ins. In this setting, a negotiated outcome reached by stakeholders is adopted if all parties support the outcome, the public interest is adequately represented, no part of the agreement abrogates any other rule or prior decision, and a factual record is developed to enable implementation.

Reversals or inconsistencies can undermine a regulator's credibility and hence effectiveness. Legislative policy intentions are usually general enough that

Table 1 Decisions and outcomes subject to regulation

Decision	Regulatory oversight
Profit level	"Excessive" returns addressed
Price structure	Rate design and prices to customer groups considered
Product mix	Accounting separations if entry into complementary services
Production process	Environmental mandates and R&D incentives established
Product quality	Reliability standards promulgated
Place	Facilities siting subject to review
Purchasing	Cost containment monitored
Promotion	Advertising and consumer information examined
Planning	Forecasts and plans placed into public record
Partnerships	Mergers and strategic alliances reviewed
Portfolio	Mix of debt and equity (corporate leverage) monitored
Public offerings	Public ownership vs private participation

regulators will need to balance competing goals and values in specific situations. As long as the public record provides clear evidence regarding the facts associated with the issues at hand, the judgment calls of regulators will at least be based on information that others can examine and verify.[7] Without such openness in the process, stakeholders supporting alternative policies will have a case for questioning regulatory decisions. Effective agencies establish procedures that promote accountability, credibility, and legitimacy. Transparency is one mechanism for addressing all three of these aspects of the regulatory process.

Regulatory Activities and Impacts

Effective regulation of prices involves a variety of factors, including the following:

1. Licensing specifies operating standards that affect costs/tariffs.
2. Performance standards for quality and reliability have cost and tariff implications.
3. Monitoring data on costs, revenues, and performance is essential for tariff determination.
4. Tariff setting determines revenue sufficiency for operating costs, capital costs (returns, asset values, deferred loans), etc.
5. Uniform accounting systems provide comparable cost data (generation, transmission, distribution) for tariffs.
6. Arbitration among firms and consumers is needed to settle disputes about tariffs access.
7. Management audits promote cost and tariff reduction via yardstick comparisons.
8. Human resource policies have operating costs and tariff implications.
9. Reports on costs and tariffs emphasize current and future performance and efficiency, both for individual firms and the sector as a whole.

These factors reflect how the implementation of public policy affects corporate decisions. The table lists some of the corporate decisions that might be constrained by public policy. Each of the listed items can be subject to regulatory rules or to statutory limitations and underscore the wide range of issues that arise in the context of energy industries (Table 1).

CONCLUSION

Public policy evolves in response to a number of forces, including public perceptions and special interest campaigns. Policy influences the basic conditions in the energy industry, market structures, corporate behavior, and energy sector performance. Additional factors that influence public policy and the types of agencies developed to implement policy include institutional features of the nation, lessons learned from others, economic growth, and the priorities placed on different performance outcomes. To the extent that actual industry performance falls short of what citizens view as possible or desirable, pressures build for changes in public policy (often labeled "policy reform"). Typical policy objectives include (but are not limited to) the following:

- Constraining the exercise of monopoly power by incumbent suppliers
- Stopping subsidies or creating tax incentives for the sector and discouraging or encouraging particular activities (which affects government deficits)
- Optimizing the structure of the sector (encouraging competition where feasible)

- Providing incentives for operating efficiency and improving reliability
- Promoting least-cost system expansion (incentives for new investment)
- Stimulating innovation and energy conservation

Even when subsidy programs are continued (like those benefiting particular fuels), supporters describe them in terms of how the programs might contribute to meeting other broad energy objectives. Achieving desired outcomes requires some prioritization because policy-makers must make political and economic trade-offs on a continual basis. Here, the focus has been on electricity and the relative roles of markets and government, but similar issues arise in other segments of the energy industry.

REFERENCES

1. Newbery, D.M. *Privatization, Restructuring, and Regulation of Network Industries*; MIT Press: Cambridge, MA, 1999.
2. Baldwin, R.; Cave, M. *Understanding Regulation: Theory, Strategy, and Practice*; Oxford University Press: Oxford, 1999.
3. Armstrong, M.; Cowan, S.; Vickers, J. *Regulatory Reform: Economic Analysis and British Experience*; MIT Press: Cambridge, MA, 1999.
4. Brennan, T.; Palmer, K.L.; Martinez, S.A. *Alternating Currents: Electricity Markets and Public Policy*; Resources for the Future: Washington DC, 2002; xi-210.
5. Greer, D.E. *Business, Government, and Society*; Macmillan: New York, 1993; xviii-606.
6. Dubash, N.K. *Power Politics: Equity and Environment in Electricity Reform*; World Resources Institute: Washington, DC, 2002; xv-175.
7. Viscusi, W.K.; Vernon, J.M.; Harrington, J.E. Jr *Economics of Regulation and Antitrust*; MIT Press: Cambridge, MA, 2000; xxvi-864.
8. Berg, S.V. Sustainable regulatory systems: laws, resources, and values. Utilities Policy **2000**, *9*, 159–170.
9. Levy, B.; Spiller, P.T. The institutional foundations of regulatory commitment: a comparative analysis of telecommunications regulation. J. Law, Econ. Organ. **1994**, *10* (2), 201–246.
10. Kessides, I.N. *Reforming Infrastructure: Privatization, Regulation, and Competition*; World Bank and Oxford University Press: Oxford, 2004; xv-306.
11. Hunt, S. *Making Competition Work in Electricity*; Wiley: New York, 2002.

Public Utility Regulatory Policies Act (PURPA) of 1978

Richard F. Hirsh
Department of History, Virginia Tech, Blacksburg, Virginia, U.S.A.

Abstract
The Public Utility Regulatory Policies Act (PURPA) of 1978 contributed to the restructuring the American electric utility system. It did so by providing incentives for nonutility companies to produce power using energy-efficient generating equipment and renewable-energy resources. By making electricity as cheaply as existing utilities in many cases, the new power companies injected competition into a formerly said, monopolistic industry. The law's unintentional consequences motivated policy makers to consider the value of introducing other free-market principles, some of which became elements of the Energy Policy Act of 1992.

INTRODUCTION

As one part of President Jimmy Carter's comprehensive energy strategy, the Public Utility Regulatory Policies Act (PURPA) of 1978 sought to encourage the efficient use of electricity and stimulate the production of power in nontraditional ways. For the first time, PURPA gave special privileges to nonutility companies that produced electricity using fossil fuels less wastefully than utilities and to firms that employed renewable resources. In the process, the law removed barriers that had previously hindered entry into the generation sector of the power business.

More important, some of the new nonutility producers sold electricity at prices that compared favorably to those offered by regulated power companies. By doing so, they challenged the status of utility companies as natural monopolies—companies that deserved special privileges because they could deliver services more cheaply and efficiently only if they remained in a noncompetitive business environment. The law's implementation further helped spur the introduction of free-market principles, and it forced some policy makers to reconsider the merits of traditional regulation. It, therefore, played a major role in efforts that led to deregulation and restructuring of the American electric utility system.

ORIGIN OF LAW DURING THE ENERGY CRISIS

Even though the so-called energy crisis had occurred more than 2 years earlier, newly inaugurated President Jimmy Carter felt much needed to be done to fix America's energy infrastructure.[1] He recalled how Arab members of the Organization of Petroleum Exporting Countries had imposed an embargo on the shipment of oil to the United States and other countries that supported Israel in its October 1973 war. After the embargo, oil flowed again, but at greatly increased prices that wreaked havoc on the economy and everyday life. Automobile drivers, who had grown accustomed to cheap and easily available gasoline found themselves on long lines for expensive and sometimes rationed fuel. Fortunately for these consumers, the lines disappeared fairly quickly, though gasoline did not return to preembargo prices. The next 2 years saw relatively stable prices for all energy supplies, and many Americans became complacent about energy policy. But President Carter worried about the increasing amount of imported petroleum being used by the United States—rising from 36% in 1973 to 42% in 1976. As his first major initiative after taking office in January 1977, Carter promoted an energy policy that would make the country more energy efficient and less dependent on foreign fuel.

Hoping to rally Americans to fight a "moral equivalent of war," Carter sought to win support for his National Energy Plan. The plan offered tax credits to people who employed energy-efficiency measures in homes and businesses, realizing that conservation was "the quickest, cheapest, most practical source of energy."[2] It also included a set of taxes on those who wasted energy, such as owners of gas-guzzling cars that failed to meet fuel-economy standards. Another proposed tax would have added 5 cents to the price of a gallon of gasoline each year that total gasoline consumption exceeded government-established goals. Additionally, Carter sought to remove price controls on oil and natural gas, which would have allowed the fuels to become more expensive and which would have encouraged users to be more efficient due to economic self interest. He also proposed new measures to spur nuclear power and the use of domestic coal rather than imported oil.

Keywords: Energy legislation; Deregulation; Restructuring; Wind turbines; Photovoltaics; Renewable energy; Cogeneration; Qualifying facilities.

Though legislators may have lauded the president's goals, they proved unwilling to accept all the president's measures, especially those that would have called for potential sacrifice and hardship for their constituents. Instead, they took more than a year to ponder the president's plan and passed a watered-down version of it in the form of five specific laws. At a signing ceremony in November 1978, Carter admitted that the laws did not accomplish everything he had hoped. Nevertheless, they constituted a first step toward a more sound energy policy.

FOCUS OF LAW

Of the five weakened measures, the PURPA had already been viewed as the least consequential. Consisting of 61 pages of text and 78 sections, the law focused on encouraging greater efficiency in the use of electricity.[3] It did so by calling attention to the way electric utility companies charged customers for electricity. Consequently, industry insiders earlier referred to it as the "rate reform" bill. Typically, customers (especially residential customers) paid higher prices per unit of electricity (a kilowatt-hour, or kWh) for the first increment of energy they used each month. This greater cost generally included a large amount of the fixed customer and capital costs associated with the production and distribution of electricity. Subsequent increments cost less per kilowatt-hour, representing only the cost of energy needed to generate the power. This "declining block" rate structure had been used for decades and was popular because it often encouraged customers to use more power, since later increments became relatively cheap. During an age of declining costs for the power industry—an age that lasted from the beginning of the twentieth century until around 1970—the rate structure made perfect sense. But with energy prices increasing, PURPA required utilities and state regulators to reconsider (but not necessarily replace) these rate structures and design new ones that encouraged wise use of electricity. Lobbyists for the utility industry viewed the law as tame, since it did not mandate too much change from the status quo.

SECTION 210

With PURPA's focus on rate reform and with concerns over more severe provisions of other laws that came out of the National Energy Plan (such as the requirement to shift from oil and natural gas to coal for producing power), utility executives and lobbyists paid little attention to the apparently inconsequential Section 210 of PURPA. Dealing with production of power from small-scale generators, the section reflected President Carter's hopes to produce more electricity through unconventional means, thus reducing the amount of oil, coal, or natural gas needed by traditional utilities to make electricity. As one promising approach, a cogeneration facility looked like a tiny utility power plant, but with a twist. It heated water using fossil fuels, and the steam went through a turbine that turned a generator, which then produced power. Instead of venting the waste heat into the environment, like utilities did, a cogeneration plant employed the heat for industrial processes (such as for manufacturing paper or chemicals) or for space conditioning. It, therefore, achieved a higher efficiency of converting raw fuel into economically valuable products. (Basically, cogeneration won double duty from fuel.) An old process, being common in the United States early in the twentieth century, cogeneration lost favor, as utilities found they could exploit economies of scale and produce cheap, central station power for industrial customers. With the new need to improve energy efficiency in light of the energy crisis, President Carter wanted to see a resurgence of cogeneration technology.

Public Utility Regulatory Policies Act's Section 210 provided the means by which cogeneration would become more practical. For example, it eliminated the major impediment for cogenerating firms by requiring utility companies to purchase excess power from the nonutilities. Implemented by state regulatory authorities, this requirement gave industrial companies an incentive to install equipment that produced process steam and electricity. The firms would use some of the power themselves, but they could also sell surplus electricity (produced more efficiently than utilities) to power companies, which would then distribute it over their grid to customers. Previously, if nonutility companies wanted to sell excess electricity, utilities had no obligation to purchase it, and they often offered low rates to sellers. Now, PURPA required utilities to buy the electricity at a rate that equaled the utilities' cost of producing the power themselves.

This section of PURPA also contained similar incentives for companies and individuals that wanted to produce power from renewable sources of energy. These sources typically used the movement of water in rivers, the flow of air, and the effect of sunlight to produce electricity in ways that did not require the combustion of fossil fuels. Research and development on water turbines, wind turbines, and solar-electricity conversion technologies had been going on slowly for decades, and the 1973 energy crisis provided a spur. But provisions in PURPA gave these efforts more momentum because they created a guaranteed market for the electricity produced by these nontraditional generating technologies.

IMPLEMENTATION OF THE LAW

Implementation of PURPA became a job first of the Federal Energy Regulatory Commission (FERC). Under the terms of the law, FERC needed to hold hearings and codify parts of the law for the country's regulatory commissions to employ when dealing with the new, privileged nonutility generators, known as qualifying facilities (QFs). In most cases, FERC chose to interpret elements of "Section 210" in ways that benefited these companies. It relaxed administrative procedures so the nontraditional power firms would not be subject to the same government oversight as regulated utilities. Federal Energy Regulatory Commission also interpreted the law liberally so cogenerators and renewable energy generators earned relatively high rates for the power they produced. Overall, FERC sought to encourage these new firms with as many incentives as possible.[4]

After completion of this review process—as well as two U.S. Supreme Court decisions that maintained FERC's interpretations—PURPA found its way to the states, where public utility commissions defined how the law would be put into practice. In most cases, implementation of PURPA did not elicit much excitement. However, in California, the law became one tool in the state's efforts to obtain environmentally sound electricity supplies. Moreover, energy efficiency and renewable energy fit nicely into the political agenda of Governor Jerry Brown, who took office in 1975, and to several of the activist members of the California Public Utilities Commission (CPUC) appointed by the unconventional chief executive.

The CPUC sought to make it easy for QFs to deal with utilities by designing "standard offer" contracts that contained the relevant terms and conditions for the producers and buyers of power. To provide flexibility to the QFs, the CPUC designed several standard offers, with the interim Standard Offer 4 becoming popular after its issuance in 1983. This contract provided for per-kilowatt hour payments from utilities that rose over the first 10 years of a 15–30 year contract period. The rising prices were designed to keep track of the cost of energy, which analysts believed would continue increasing during the 1980s. Moreover, the long-term contracts offered certainty to the fledgling nonutility companies and allowed them to obtain financing more easily from investors. Because of these generous provisions—especially the escalating prices that contrasted the declining price of energy in 1984 and later—scores of nonutility generating companies signed up for the interim contracts, promising to build as much as 15,000 MW of capacity. Though the companies suffered no penalties if they failed to live up to their promises, regulators realized that they could see a glut of unused capacity if they allowed more companies to accept this offer. Consequently, the CPUC suspended it in 1985 and developed other standard offer contracts for nonutility developers. Nevertheless, those firms that had obtained the Standard Offer 4 contract could still take advantage of its lucrative terms.[5]

SPUR FOR TECHNOLOGICAL INNOVATION

Combined with tax incentives offered to small power producers by federal and state governments, PURPA accomplished its goal of spurring innovation in several technologies. Cogeneration technology had roots that went back to the early 1900s, but the law encouraged manufacturing companies to make improvements in it to exceed requirements to obtain at least 42.5%–45% thermal efficiencies. (Utility power plants typically converted only about 35% of raw fuel into electricity.) More important, perhaps, manufacturers exploited mass production techniques for several of the cogeneration plants' components, allowing construction of a plant for between $800 and $1200/kW in 1986. This capital cost compared favorably to utility costs for traditional fossil fuel and nuclear plants, which cost between about $900 and $5000/kW in 1984. And despite the lack of economies of scale, something that utilities prized in their huge plants, the cogenerating companies produced two salable products—electricity and heat—which enabled them to earn money while simultaneously meeting PURPA's goal of producing electricity in a more efficient fashion. Not surprisingly, cogeneration took off, making up 56% of all nonutility produced power in California in 1990 and 59% of nonutility power capacity nationwide in 1991. By 1992, almost 40,700 MW of the country's capacity came from cogeneration sources, up from 10,500 MW in 1979. In short, PURPA stimulated a new segment of the electric power business, as companies, such as Cogentrix, AES Corporation, and affiliates of Bechtel Power, General Electric, Mobil Oil, and Destech Energy, exploited the law's provisions.

Public Utility Regulatory Policies Act also stimulated the use of gas turbines used for making electricity. This technology had a long heritage, used for turning generators and producing electricity during peak usage times. But the small turbines (usually rated from 10 to 25 MW) had a reputation for poor reliability and low fuel efficiency. During the 1980s, however, research on gas turbines expanded, largely because of funding from the U.S. Department of Defense, which wanted smaller and more energy-efficient gas turbines for use in military aircraft. Manufacturers of these aeronautical turbines, such as General Electric, shrewdly realized that the technology could easily be transferred for stationary use in power plants, and they produced improved turbines for this application. To employ the turbines in the manner sanctioned by PURPA, manufacturers created systems that recovered energy produced from the gas engine's exhaust and reused it to heat water in a conventional steam turbine generator. These "combined cycle" units reached thermal efficiencies of >40% in the 1980s. Small and easy to install, these units also gained popularity in the mid-to-late 1980s, when the price of natural gas declined. At the end of 1992, nonutility companies employed gas turbines to provide almost 39% of their capacity. The cost of their delivered electricity ranged from about 3.2 to 5.5 cents/kWh, comparing favorably to the average cost of power produced by regulated power companies.

Beyond these more traditional technologies, wind turbine technology developed rapidly as a result of PURPA. Though used for pumping water and for producing small amounts of electricity on farms for decades, wind turbine technology had previously not been exploited to produce enough power for distribution by electric utilities. After the energy crisis, the federal government spent heavily to produce large wind turbines for utility use—some as large as 3.2 MW each—but the hardware failed. Much more successfully, entrepreneurs working under the impetus of PURPA fashioned smaller turbines (between 0.05 and 0.5 MW) to use in clusters, with the aggregated electricity sold to power companies. Some components of the small turbines could be mass produced or adapted from other applications, thus allowing the wind companies to pare costs. And because of the beneficial terms of California's standard offer contracts, many wind firms took root in the Golden State, even though wind resources may have been better in other parts of the country. Wind powered generation increased dramatically in California, rising from 50 million kWh produced in 1983 to 3268 million kWh generated in 1994.[6] Research and development in the technology continued, such that the technology in 2005 produces power at costs comparable to fossil fuel (including natural gas) generators, and more cheaply than any other nonhydro renewable resource.[7]

Technologies that employed the sun to make electricity also benefited from PURPA (and other laws, especially in California). Entrepreneurs working on solar photovoltaic cells, for example, designed and tested novel designs and could obtain some income because of PURPA's guarantee of a market for the cells' power. Largely because of the research, the cost of solar cell electricity declined from about 90 cents/kWh in 1980 to about 20 cents/kWh in 1995. Though still not commercially viable without additional subsidies, except in niche applications, photovoltaic technologies nevertheless improved dramatically. Meanwhile, another solar technology almost reached commercial success as a result of PURPA. Explicitly taking advantage of the law's provisions (and, again, other incentives offered in California) Luz International, a small entrepreneurial firm, built a system that employed parabolic mirrors to focus sunlight onto tubes containing oil. The heated oil transferred energy to water, which turned into vapor and powered a turbine-generator to produce electricity. Between 1984 and 1990, the company built several iterations of the system and produced power that dropped from 25 to 8 cents/kWh during that period. Overall, PURPA stimulated the development of novel power generation technologies, a fact noted as early as 1985 by analysts at the Congressional Office of Technology Assessment, which viewed the law as the motivator of several "first generation commercial applications" of technologies that may have great value in the future.[8]

PURPA AND DEREGULATION: UNINTENDED CONSEQUENCES OF THE LAW

Beyond its stimulation of research and development on small-scale technologies, PURPA unintentionally began the process of deregulating the American electric utility industry.[9] It did so largely in two ways. First, as noted earlier, PURPA eliminated the barrier to entry in the generation sector of the utility business. Under the regulatory framework that had existed since the beginning of the twentieth century, utility companies enjoyed status as regional monopolies that permitted them to generate, transmit, and distribute electricity without competition. But PURPA gave a special class of unregulated companies the right to produce power, effectively enabling them to compete with the formerly protected firms. In short, PURPA invalidated the notion that only integrated monopolies could effectively operate in the utility business.

Second, and perhaps more important, the success of some PURPA QFs in making a profit by selling power to utilities suggested that the regulated power companies no longer had a legitimate claim as natural monopolies. Going back to the late nineteenth century, the notion of natural monopoly helped justify the existence of regulation of utility companies in the first place. The academic and political principle held that in some industries (such as the railroad, streetcar, water delivery, and electric power industries), companies needed to invest heavily in technology and would only do so if they obtained guaranteed markets. Competition would also be detrimental to the public good because of the need for duplicate equipment, which raised costs, to serve the same customers. (Before regulation, streets of some cities were cluttered with wires from different companies, each seeking to serve the same lucrative customers.) Moreover, when several firms competed for a fixed number of customers, none could exploit the largest and best generation technologies that emerged in the early twentieth century. Steam-turbine generators, in particular, showed huge economies of scale. If two or more companies contested the same customer base, none could employ the biggest generating unit. But if only one firm served all customers, it could buy the largest equipment and reduce its unit cost for the power. Swaying politicians, this logic encouraged the creation of regulatory bodies that oversaw the natural monopoly utility companies, ensuring (in principle), that the benefits of the utilities' unusual status would flow to customers in the form of lower rates and good service.

The Public Utility Regulatory Policies Act challenged this rationale for regulation and, therefore, started discussion of further deregulation of the utility industry. It did so by spurring creation of nonutility companies that produced power at competitive prices. Even though the QFs did not exploit economies of scale like utilities did, they produced power with higher thermal efficiencies (in the case of cogenerators), and they sold the heat byproduct to gain economic advantages over utilities. They, therefore, often could beat utility costs by a margin of 5%–15% in the mid-1990s. This fact was noted by some politicians and regulators, who argued that the financial success of some QFs indicated that utilities no longer constituted natural monopolies. After all, natural monopolies supposedly produced power more efficiently and more cheaply than could competitors. But some PURPA QFs demonstrated that they could produce lower priced electricity than could the protected utilities. Following this logic further, the questioning of the natural monopoly rationale led to challenges to the notion of regulation. Congressmen, federal regulators, and some utility leaders themselves started asking why regulation should exist when utility monopolies no longer could be justified as "natural" anymore.

This questioning of regulation in the utility sector did not occur in a vacuum. Throughout the 1970s and 1980s, economists and politicians in the United States and elsewhere began challenging the value of regulation of various industries as the best way to provide services to customers. Starting with President Carter's efforts, the airline industry began the process of deregulation. Similar moves to introduce market forces occurred in the natural gas, petroleum, railroad freight transportation, and financial services industries, only to be followed (during President Reagan's tenure in office) by deregulation of the telecommunications industry. Similar deregulation of various industries occurred in England and several other countries, yielding (in many cases) declining prices, improved services, and technological innovation in industries that had previously appeared stagnant.

With the questioning of natural monopoly status in the power industry, stimulated partly by the experience of PURPA, and the apparent success of deregulation in other businesses, many people called for restructuring of the electricity infrastructure. President George H.W. Bush heeded these calls and proposed, after the Gulf War of 1991, to employ market forces to increase domestic fuel production and to improve the efficiency of energy use. One provision of the Energy Policy Act of 1992 gave states the right to allow their transmission network to serve as a common carrier so any electricity producer could sell power to any customer. Essentially, the law enabled states to begin competition on the retail level. Taking advantage of the legislation, several states in the late 1990s established competitive retail frameworks. By September 2001, 23 states (and the District of Columbia) had passed restructuring laws, while other states used the regulatory process to reduce their oversight and introduce more market forces into the industry.[10]

CONCLUSION

While PURPA contributed to the deregulation of the power industry, that same deregulation also minimized the importance of PURPA. Though the law remains on the books, despite the best efforts of repeal proponents,[11] who think the law has led to overpriced and unneeded power, terms of the Energy Policy Acts of 1992 and 2005 make some elements of PURPA less significant. For example, the 1992 law created a class of independent power producers known as exempt wholesale generators. While these generators do not receive guaranteed payments from utilities, as do QFs, they also are not bound by PURPA's requirements for overall energy efficiency. Entrepreneurial companies have taken advantage of this feature of the new law and have built scores of plants that sell power to an open market of customers. They have eclipsed QFs as the primary nonutility providers of power to the nation's electricity grid.[12]

Nevertheless, the significance of PURPA should not be overlooked just because the law no longer has as much practical meaning as it did in the 1980s. Intended largely to reform the way utility companies charged customers for electricity, the statute unexpectedly had at least two major consequences. First, a benign-looking part of the law provided the financial incentive that motivated entrepreneurs to perform research and development activities on small-scale renewable energy technologies, such as wind turbines and solar energy systems. Combined with tax breaks offered by some states, these technologies saw rapid improvement and the ability to produce electricity at greatly reduced costs. Second, the law helped undermine the rationale for utilities' status as regulated natural monopolies. It did so by encouraging firms to improve cogeneration and gas-turbine technologies and to make them commercially feasible in small sizes. Owners of the PURPA-induced qualifying facilities employed these technologies to break down the barriers to entry in the utility business by allowing them to compete with utility companies in the generation sector. In the process, they challenged the notion that utilities deserved recognition as special noncompetitive enterprises and that they required oversight by government regulatory bodies. Within a political environment that valued the deregulation of industries, PURPA's questioning of regulation motivated, to a large extent, the restructuring of the utility industry. That restructuring process continues into the early part of the twenty-first century.

REFERENCES

1. This article draws heavily from Hirsh, R.F. *Power Loss: The Origins of Deregulation and Restructuring in the American Electric Utility System*; MIT Press: Cambridge, MA, 1999: Hirsh, R.F. PURPA: *The spur to competition and utility restructuring Elec. J.* **1999**, *12* (7), 60–72.
2. Carter, J. President's proposed energy policy. Vital Speeches of the Day **1977**, *43* (14), 418–420.
3. Public Utility Regulatory Policies Act, Public Law 95–617, signed 9 November 1978.
4. Federal Energy Regulatory Commission Small power production and cogeneration facilities; regulations implementing Section 210 of the public utility regulatory policies act of 1978. Fed. Regist. **1980**, *45* (25), 12–215: FERC Power production and cogeneration facilities—qualifying status. Fed. Regist. **1980**, March (20), 17–959.
5. Ahern, W.R. Implementing avoided cost pricing for alternative electricity generators in California. In *New Regulatory and Management Strategies in a Changing Market Environment*; Trebing, H.M., Mann, P.C., Eds., MSU Public Utilities Papers: East Lansing, MI, 1987; 408.
6. California Energy Commission, California Statistical Abstract 1996 http://www.dof.ca.gov/html/fs_data/STAT-ABS/sec_J.htm, obtained 18 July 1997.
7. Roth, I.F.; Ambs, L.L. Incorporating externalities into a full cost approach to electric power generation life-cycle costing. Energy **2004**, *29*, 2125–2144.
8. The previous paragraphs rely on information in Hirsh, *Power Loss*, 101–117. Also see Guey-Lee, L. Renewable electricity purchases: history and recent developments. In *Renewable Energy 1998: Issues and Trends*, U.S. Department of Energy, Energy Information Administration, at http://www.eia.doe.gov/cneaf/solar.renewables/rea_issues/renewelec_art.pdf, obtained 12 April 2005.
9. Strictly speaking, one should really be speak about "restructuring", rather than "deregulating", the electric utility system. Deregulation generally means the end of all government oversight and the existence of a totally free market. While the generation sector of the utility business became somewhat competitive under PURPA, it retained a large amount of regulation that oversaw the interactions between utilities and qualifying facilities. More accurately, restructuring describes events that maintain a government presence in all utility sectors to ensure that (in theory) the new wholesale and retail markets operate efficiently and in the public interest.
10. U.S. Department of Energy, Energy Information Administration, at http://www.eia.doe.gov/cneaf/electricity/chg_str/regmap.html, obtained 15 October 2001.
11. Salpukas, A. Utilities See Costly Time Warp In '78 Law. In *New York Times*, Vol. 5 1994. Section D; 1. See efforts of the PURPA Reform Group, an ad hoc organization made up of utility companies that objected to allegedly higher costs caused by implementation of the law. Adelberg, A. Way off on PURPA. Elec. J. **1999**, *12* (10), 2–3. Mr Adelberg was co-chair of the PURPA Reform Group.
12. U.S. Department of Energy, Energy Information Administration *Electric Power Annual 2003*. See 2, 8, 13 [OE/EIA-0348(2003)].

Pumped Storage Hydroelectricity

Jill S. Tietjen
Technically Speaking, Inc., Greenwood Village, Colorado, U.S.A.

Abstract

Pumped storage hydro is a form of hydroelectric power generation for electric utilities that incorporates an energy storage feature. The fuel, water, moves between two reservoirs—an upper and a lower—with a significant vertical distance between them. Water is stored in the upper reservoir until such time as the utility determines that it is economic to use the water to produce electricity for the system, usually to keep coal-fired and nuclear power plants operating at economic levels during low load periods, such as at night and on weekends. Pumped storage is the most widespread energy storage system in use on power networks and is used for energy management, frequency control, and provision of reserves. Efficiency of any specific pumped storage facility, which is primarily dependent on the height between the upper and lower reservoirs, ranges from 70 to 85%.

INTRODUCTION

Pumped storage hydro is a form of hydroelectric power generation for electric utilities that incorporates an energy storage feature.[1] The fuel, water, moves between two reservoirs—an upper and a lower—with a significant vertical separation (see Fig. 1). Water is stored in the upper reservoir until such time as the utility determines that it is economic to use the water to produce electricity for the system. The water in the upper reservoir is stored gravitational energy.[2] When the water is released, the force of that water spins the blades of a turbine that connect to a generator, which produces electricity.[3]

Water is generally released from the upper reservoir to produce electricity during the daylight or on-peak hours. After passing through the turbines, the water is discharged into the lower reservoir. At night and on weekends, or during light-load or off-peak hours, water is pumped from the lower reservoir into the upper reservoir by the turbines, which have been reversed to work as electric motor-driven pumps.

In the upper reservoir, the water is essentially stored energy. Water can be stored for a long or short time in the upper reservoir, depending on the needs of the utility. A vertical separation of at least 100 m (328 ft) is necessary to make a pumped hydro facility economic.[4] The height difference between the upper and lower reservoirs is called the head. The amount of potential energy in the water is directly proportional to the head, so the greater the height, the more energy that can be stored.[1,5]

Pumped storage hydro was first used in Italy and Switzerland in the 1890s. In 1929, the first major pumped storage hydroelectric plant in the United States, Rocky River, was built in New Milford, Connecticut. By 1933, reversible pump-turbines with motor generators had become available. The turbines could operate as both turbine-generators and in reverse, as electric motor-driven pumps.[6,7]

About 3% of total global generation capacity—more than 90 GW—is pumped storage capacity. In 2000, the United States had 19,500 MW of pumped storage facilities in operation.

The largest pumped storage facility in the United States is in Bath County, Virginia, which has a capacity of 2,100 MW. Pumped storage plants are characterized by long construction times and high capital expenditures.[6,8,9]

Pumped storage is the most widespread energy storage system in use on power networks and is used for energy management, frequency control, and provision of reserves. Efficiency of any specific pumped storage facility, which is primarily dependent on the height between the upper and lower reservoirs, ranges from 70 to 85%.

Environmental issues associated with the construction of pumped storage plants often parallel those for conventional hydroelectric plants and range from water-resource and biological effects to potential damage to archaeological, cultural, and historical sites.[6]

HYDROELECTRIC GENERATION

In conventional hydroelectric generation, hydraulic turbines rotate due to the force of moving water (its kinetic energy) as it flows from a higher to a lower elevation. This water can be flowing naturally in streams or rivers, or it can be contained in manmade facilities such as canals, reservoirs, and pipelines. Dams raise the water level of a stream or river to a height sufficient to create an adequate head (height differential) for electricity generation.[10]

Keywords: Energy storage; Hydroelectric; Reservoir; Pumped storage; Reversible pump-turbine; Francis turbine; Turbine-generator; Head; Flow.

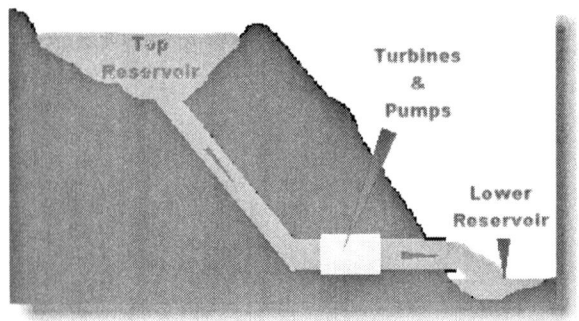

Fig. 1 Pumped storage hydro configuration.

If the dam stops the flow of the river, water pools behind the dam to form a reservoir or artificial lake. As hydroelectric generation is needed by the electric utility, the water is released to flow through the dam and powerhouse. In other cases, the dam is simply built across the river, and the water moves through the power plant or powerhouse inside the dam on its way downstream.[11]

In either case, as the water actually moves through the dam, the water pushes against the blades of a turbine, causing the blades to turn (see Fig. 2). The turbine converts the energy in the form of falling water to rotating shaft power to turn a generator, which produces electricity.[11,12]

The selection of the best turbine for any specific hydroelectric site is primarily dependent on the head (the vertical distance through which the water falls) and the water flow (measured as volume per unit of time) available. Generally, a high-head plant needs less water flow than a low-head plant to produce the same amount of electricity.[12,13]

The power available in a stream of water is

$$P = \eta \rho g h V$$

where: P, power (J/s or watts); η, turbine efficiency; ρ, density of water (kg/m^3); g, acceleration of gravity (9.81 m/s^2); h, head (m; the difference in height between the inlet and outlet water surfaces); V, flow rate (m^3/s).[14]

For SI units, this equation can be roughly approximated as

$$\text{POWER(kW)} = 5.9 \times \text{FLOW} \times \text{HEAD}$$

where FLOW is measured in cubic meters per second and HEAD is measured in meters.[15] In general terms, for IP units, 1 gal/s falling 100 ft can generate 1 kW of electrical power.[16]

Hydroelectric power is generally found in mountainous areas where there are lakes and reservoirs and along rivers. Hydroelectric power currently provides about 10% of all the electricity produced in the United States. Hydroelectricity provides about one-fifth of the world's electricity. Worldwide capacity is 650,000 MW.[11,16]

Hydroelectric power is characterized as a renewable resource. The fuel, water, is replenished by rain and snowfall. With an existing plant, it is the cheapest way today to generate electricity. Producing electricity from hydropower is so economical because when the dam is built and the equipment has been installed, the flowing water has no cost. In addition, the dams are very robust structures, and the equipment is relatively simple mechanically. Hydro plants are dependable and long lived, and their maintenance costs are low compared with those of most other forms of electricity generation, including fossil-fired and nuclear generation. Many hydroelectric plants do not even require staff, further contributing to their low ongoing maintenance costs.[13]

Pumped storage hydro generation is a specific kind of hydroelectric generation requiring an upper and lower reservoir, and special equipment that can both generate power as water flows downhill and then reverse and serve as a pump to move the water back uphill.

FACILITY DESCRIPTION

A pumped storage hydro power plant typically has three major components: the upper reservoir, the lower reservoir, and the pumping/generating facilities. The pumped storage facility represented in Fig. 3 is Duke Power Company's Jocassee plant in South Carolina.

The water from the upper reservoir used for electricity generation is allowed into the intake shaft through the opening of the headgates. Water moves through the high-pressure shaft and steel-lined power tunnel until it reaches the turbines in the powerhouse.[1] The water turns the turbines, which then drive the generators to produce electricity. Then the water moves through the tailrace tunnel until it is discharged into the lower reservoir, which in Fig. 3 is Lake Jocassee. Water discharge capacity in the upper reservoir can require several hours to several days.[6]

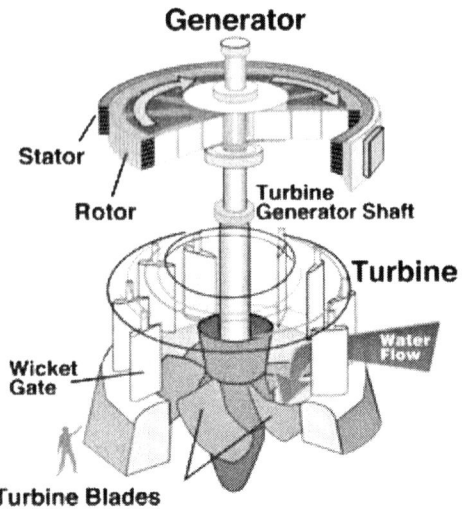

Fig. 2 Hydroelectric turbine-generator.

Pumped storage facilities can be categorized as "pure" or "combined." Pure pumped storage plants, also referred to as modular pumped storage or MPS, continually shift water between an upper and a lower reservoir. Combined pumped storage plants also generate their own electricity like conventional hydroelectric plants through natural stream flow. Modular pumped storage systems tend to be much smaller than conventional hydroelectric power plants. They use closed water systems that are artificially created instead of natural waterways or watersheds. The water for MPS usually is put into the system only when it begins operation, either from groundwater or possibly from municipal wastewater.[8,10]

When water is to be pumped back into the upper reservoir of a pumped storage plant, it flows from the lower reservoir into the powerhouse, where reversible pump-turbines (usually, a Francis turbine design) pump it back into the upper reservoir. The hydraulic units are designed to operate as pumps when rotating in one direction and as turbines when rotating in the opposite direction. Similarly, motors that drive the pumps can be reversed to act as generators. Fig. 4 contains a schematic diagram showing the directions of water flows and rotation of the pump-turbine for each mode of operation.[8,9]

Many turbines used at pumped storage plants are Francis turbines, named for American engineer James B. Francis (1815–1892), who worked to enhance turbine design. Francis turbines are capable of applications of 2–800 m (approximately 6.5–2,600 ft). The turbines are individually designed for each site to operate as efficiently as possible and to match each site's flow conditions. These turbines may be designed for a wide range of heads and water flows. Francis turbines are very expensive to design, manufacture, and install, but they operate for decades.[14,17–19]

The efficiency of pumped storage plants generally ranges from 70 to 85%. This means that 70%–85% of the electrical energy used to pump the water into the upper reservoir is actually generated when water flows back down through the turbines to the lower reservoir. The losses of energy occur due to evaporation from the exposed water surface (in both the upper and lower reservoirs) and to the mechanical efficiency losses during conversion from flowing water to electricity. In hot climates or windy areas, the evaporation losses will be higher. Similarly, mechanical losses are higher with older equipment.[8]

HOW PUMPED STORAGE IS USED BY UTILITIES

Pumped storage hydro serves an important function in the portfolio of resources available to utilities. They provide the most significant and most cost-effective means of storage of energy on a scale useful for a utility. It is not usually feasible and/or economic for utilities to turn down or reduce load on large nuclear and coal-fired (so-called thermal) units during low load periods—generally, at night. Using electricity from these thermal units to provide the power needed to pump water into the upper reservoir allows the utility to have the water to use during higher load periods the next day to generate electricity through the turbines of the pumped storage power plant and to keep their coal and nuclear units operating at more efficient levels.[5]

Pumped storage hydro is economical because it flattens the variations in load on the power grid, permitting thermal plants (coal-fired and nuclear) that provide electricity to continue operating at their most efficient capacity while reducing the need to build special power plants that will run only at peak demand, using more costly generation methods.[8]

Fig. 5 shows a typical pattern for pumped storage usage by a utility. Here, power is generated by the pumped storage facility during the higher load hours of 7:00 A.M. to 10:00 P.M., and pumping (which is usage of electricity from the grid) occurs in the off-peak hours.

In addition to energy management, pumped storage systems are important components in controlling electrical network frequency and in providing reserve generation. Thermal plants are much less able to respond to sudden

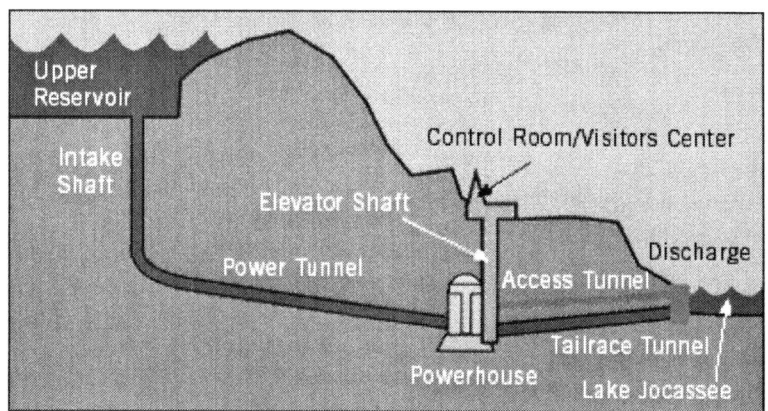

Fig. 3 Pumped storage power plant.

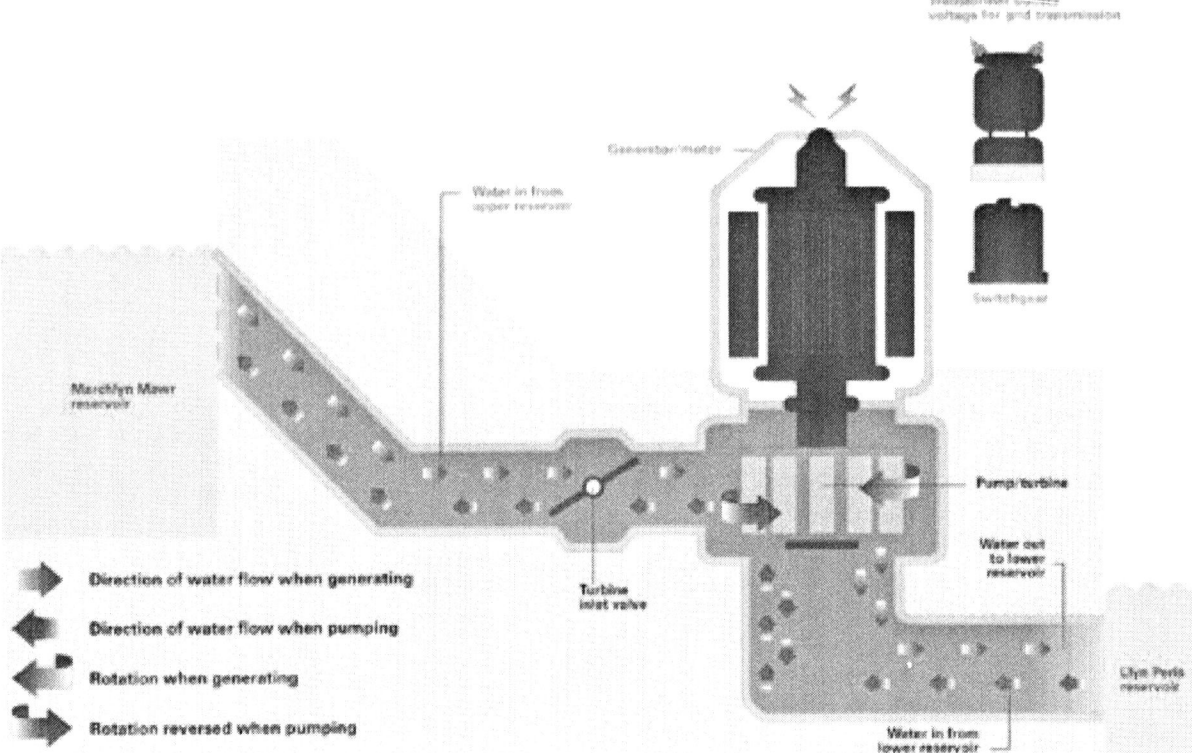

Fig. 4 Schematic of pumped storage hydro operation.

changes in electrical demand, which cause frequency and voltage instability. Pumped storage plants can respond to those changes in seconds. This is particularly desirable in the case of a unit's becoming unavailable or forced out of service or on utility systems with high amounts of intermittent resources, such as solar and wind. The Dinorwig pumped storage facility in north Wales, United Kingdom, for example, can go from 0 MW to full capacity of 1,320 MW in 12 s and usually can stay at this level until other generating units on the utility's system can be brought online.[1,4,8]

COST AND ECONOMICS

The cost effectiveness of a pumped storage hydroelectric facility depends on the topography of an area and on the types and sizes of power plants in the electric utility's system. The capital cost of the facility will be dependent on the height of the head and the geography in the area. Adequate resources need to be available consistently to provide the pumping energy required economically.

Higher head ranges, varying between 300 and 600 m (approximately 1,000–2,000 ft), and relatively steep topography are generally desirable for a cost-effective facility to be designed, built, and operated. Two main parameters affect the costs of pumped storage facilities: the ratio of waterway length to head (l/h) and the overall head of the project. A low l/h ratio will result in shorter water passages and will reduce the need for surge tanks to control transient flow conditions. Higher head projects require smaller volumes of water to provide the same level of energy storage and smaller waterway passages for the same level of power generation. Most desirable pumped storage sites have heads in excess of 300 m (approximately 1,000 ft) with l/h ratio of 6 or less.[20]

New pumped storage projects have not been built in the United States for quite a number of years. In 2006, the Lake Elsinore Advanced Pumped Storage Project is being proposed in California and is under review by the Federal Energy Regulatory Commission. Two separate alternatives have been proposed, including pumped storage and pumped storage plus a conventional hydroelectric facility, at costs ranging from $720 million to $1.3 billion, including the associated required transmission for 500–1,275 MW.[21]

The cost effectiveness of the pumped storage power plant during daily operation for an electric utility depends on the cost of the power used to pump the water back uphill, the efficiency of the unit, and the cost of the power at the margin during the utility's on-peak period. Consider the example of the following hypothetical utility (Utility A), whose level of so-called production costs—costs

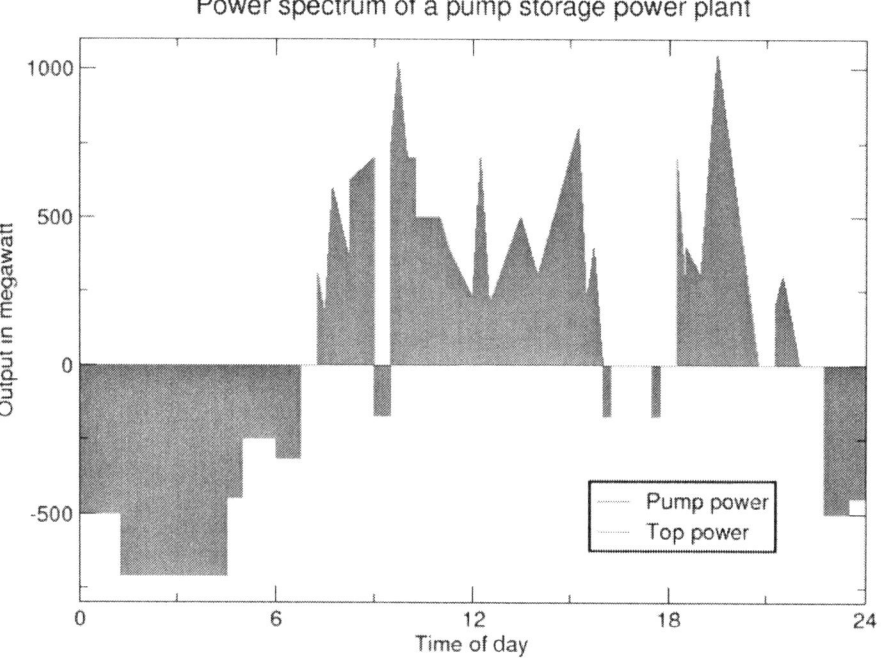

Fig. 5 Power spectrum of a pumped storage power plant.

including only fuel and operating and maintenance (O&M) costs—are as shown.

Utility A has a peak load of 10,000 MW. Utility A has three 750 MW nuclear power units that generate electricity at a production cost of 1.7 cents/kWh. Additional generation resources include a variety of coal-fired generating units totaling 7,000 MW that have an average production cost of 2.5 cents/kWh. Utility A also has 1,250 MW of natural gas-fired combustion turbines that produce power at an average price of 5.5 cents/kWh. There is also 1,000 MW of conventional hydroelectric capacity on Utility A's system.[22]

The following table summarizes the resources and the costs for Utility A. Utility A also has a 750 MW pumped storage facility that is 70% efficient.

Because the cost of using the pumping energy is significantly lower than the production cost of using natural gas, Utility A can cost-effectively use the nuclear and coal resources at night to pump water back uphill and then use generation from the pumped hydro facility during the day instead of running its natural gas-fired combustion turbine units.

ENVIRONMENTAL ISSUES

Issues related to permitting of conventional hydroelectric and on-stream pumped storage hydroelectric facilities include

- Water-resource impacts: stream flows, reservoir surface area, groundwater recharge, water temperature, turbidity, and oxygen content
- Biological impacts: displacement of terrestrial habitat, alteration of fish migration patterns, and other impacts due to changes in water quality and quantity
- Potential damage to archaeological, cultural, or historic sites
- Visual-quality changes

Generation type	MW	Production cost (cents/kWh)	Cost of energy from pumped hydro using this capacity to pump
Conventional hydroelectric	1,000	0	Already used to serve load, not available to use for pumping
Nuclear	2,250	1.7	2.21
Coal	7,000	2.5	3.25
Natural gas	1,250	5.5	Too expensive to use this fuel to pump

- Loss of scenic or wilderness resources
- Increased risks of landslides and erosion
- Gain in recreational resources

These concerns are much less significant for MPS plants, because MPS operate in a closed loop and are not associated with natural waterways and watersheds. Usually, MPS are specifically not located near existing rivers, lakes, streams, and other sensitive environmental areas to avoid the regulatory lag time and complexity associated with combined pumped storage hydroelectric facilities.[23]

CONCLUSION

Pumped storage hydroelectric facilities offer many benefits to utilities, including energy management, frequency control, and the provision of reserves. Utilizing water as a fuel, these facilities now provide about 3% of the world's generating capacity. These plants are the most economical in large utility systems, where the pumping energy can come from large coal-fired and nuclear units, and can keep those units from having to reduce load at night and on weekends.

REFERENCES

1. Andy Darvill, Energy Resources. Pumped Storage Reservoirs—storing energy to cope with big demands. Available at: http://home.clara.net/darvill/altenerg/pumped.htm (accessed March 2006).
2. U.S. Department of Energy—Energy Efficiency and Renewable Energy Distributed Energy Program. Pumped Hydro. Available at: www.eere.energy.gov/de/pumped_hydro.html (accessed January 2006).
3. Tennessee Valley Authority, Hydroelectric Power. Available at: www.tva.gov/power/hydro.htm (accessed March 2006).
4. University of Alaska, Fairbanks. Available at: www.uaf.edu/energyin/webpage/webpagewithframes/pumpedhydro.text.htm (accessed January 2006).
5. Hydroelectricity. From Wikipedia. Available at: http://en.wikipedia.org/wiki/Hydroelectricity (accessed March 2006).
6. Electricity Storage Association. Technologies & Applications, Technologies, Pumped Hydro Storage. Available at: www.electricitystorage.org/tech/technologies_technologies_pumpedhydro.htm (accessed March 2006).
7. U.S. Bureau of Reclamation. The History of Hydropower Development in the United States. Available at: www.usbr.gov/power/edu/history.html (accessed April 2006).
8. Pumped-Storage Hydroelectricity. From wikipedia. Available at: http://en.wikipedia.org/wiki/Pumped-storage_hydroelectricity (accessed March 2006).
9. Encyclopedia Brittanica. Waterpower: Pumped Hydro Storage. Available at: www.brittanica.com/ebi/article-210089 (accessed March 2006).
10. California Energy Commission. Hydroelectric Power in California. Available at: www.energy.ca.gov/electricity/hydro.html (accessed April 2006).
11. California Energy Commission. Energy Story: Chapter 12—Hydro Power. Available at: www.energyquest.ca.gov/story/chapter12.html (accessed April 2006).
12. Micro Hydropower Basics: Turbines. Available at: www.microhydropower.net/turbines.html (accessed April 2006).
13. Anderson High School NEED Team, Learn About Energy.org. Part 11, Hydropower. Available at: www.learnaboutenergy.org/focus/part11.htm (accessed April 2006).
14. Water Turbine. From wikipedia. Available at: http://en.wikipedia.org/wiki/Water_turbine (accessed April 2006).
15. ICLEI, Stuart Baird. Hydro-Electric Power. Available at: www.ecology.com/archived-links/hydroelectric-energy (accessed April 2006).
16. Union of Concerned Scientists. How Hydroelectric Energy Works. Available at: www.ucsusa.org/clean_energy/renewable_energy_basics/how-hydroelectric-energy-works_html (accessed April 2006).
17. James B. Francis. From wikipedia. Available at: http://en.wikipedia.org/wiki/James_B_Francis (accessed April 2006).
18. Francis turbine. From wikipedia Available at: http://en.wikipedia.org/wiki/Francis_turbine (accessed March 2006).
19. The Worlds of David Darling, The Encyclopedia of Alternative Energy and Sustainable Living. Francis turbine. Available at: www.daviddarling.info/encyclopedia/F/AE_Francis_turbine.html (accessed April 2006).
20. World Bank. Pumped-Storage Projects. Available at: www.worldbank.org/html/fpd/em/hydro/psp.stm (accessed March 2006).
21. Jose Carvajal. High Hydro Plant Price Tag Not Seen as a Dealbreaker. Available at: www.nctimes.com/articles/2006/03/09/news/californian/21_21_563_8_06.txt (accessed April 2006).
22. Nuclear Energy Institute. Comparative Measures of Power Plant Efficiency. Available at: www.nei.org/index.asp?catnum=2?catid=262 (accessed April 2006).
23. Claifornia Energy Commission. Hydroelectric Power in California. Available at: www.energy.ca.gov/electricity/hydro.html (accessed April 2006).

Pumps and Fans

L. J. Grobler
T. G. Vorster
School for Mechanical Engineering, North-West University, Potchefstroom, South Africa

Abstract
Pumps and fans are used in the commercial and industrial sectors to transfer fluids and gases and to consume a significant amount of energy. This entry focuses on energy management and cost saving opportunities that exist with pumping and fan systems. It gives an overview on the types of pumps and fans, their operating characteristics, theory, and power requirements, and highlights typical energy management opportunities through maintenance and retrofitting of pumping and fan systems.

INTRODUCTION

Pumps and fans are two of the most common pieces of equipment in use today in commercial and industrial applications. As such, they make up a large part of the energy used in these applications and therefore offer large scope for energy reduction measures.

Fans and pumps react within the same manner; therefore, the governing laws are similar for both fans and pumps. The difference between fans and pumps is in the fluid being moved; in the case of a fan, it is some type of gas, while the pump moves some type of liquid. A second, but less clear, distinction is that pumps are usually coupled directly to the drive motor while the fan uses a system of pulleys.

The aim of this article is to provide the reader with the basics of pump and fan system design, while highlighting the importance of proper design in energy usage reduction. The article also aims to highlight other energy reduction methods in use or which are currently being developed.

PUMPING SYSTEMS

The Internet Free Dictionary describes a pump as "a machine or device for raising, compressing, or transferring fluids."

This is achieved through the addition of energy to the fluid by various methods. These methods and the theory governing pumps and the types of pumps currently in use are discussed below.

Keywords: Pumping systems; Fan systems; Types of fans and pumps; Operating characteristics; Theory; Power requirements; Energy management opportunities; Retrofit opportunities.

Types of Pumps

There are two main categories of pumps that are identified by their basic principle of operation, namely positive displacement pumps and centrifugal pumps. Under these two basic types, there various subtypes. These are shown in Fig. 1.

The Centrifugal Pump

The term Centrifugal pump is a general description for machines that displace fluid through the addition of momentum provided by a spinning impeller. These pumps are commonly used in a wide range of industrial and commercial applications.

The basic centrifugal pump, known as a radial flow pump, as shown in Fig. 2, has rotational vanes that move the liquid away from the center of the impeller into the scrolled casing. A part of the kinetic energy imparted to the fluid is converted into pressure, which forces the liquid out of the discharge.

The other type of centrifugal pump is the axial flow centrifugal pump, which imparts energy to a fluid through a lifting motion by a set of propeller-shaped vanes. This motion results in an axial discharge.

A mixed flow centrifugal pump is a pump that uses a combination of radial and axial action.

The Positive Displacement Pump

The positive displacement pump works by trapping liquid in a cavity and then displacing it by some or other method to the pump discharge. These types of pumps provide an almost constant flow rate for a particular pump speed, which is independent of pressure differences and liquid characteristics. A gear type positive displacement pumps is shown in Fig. 3.

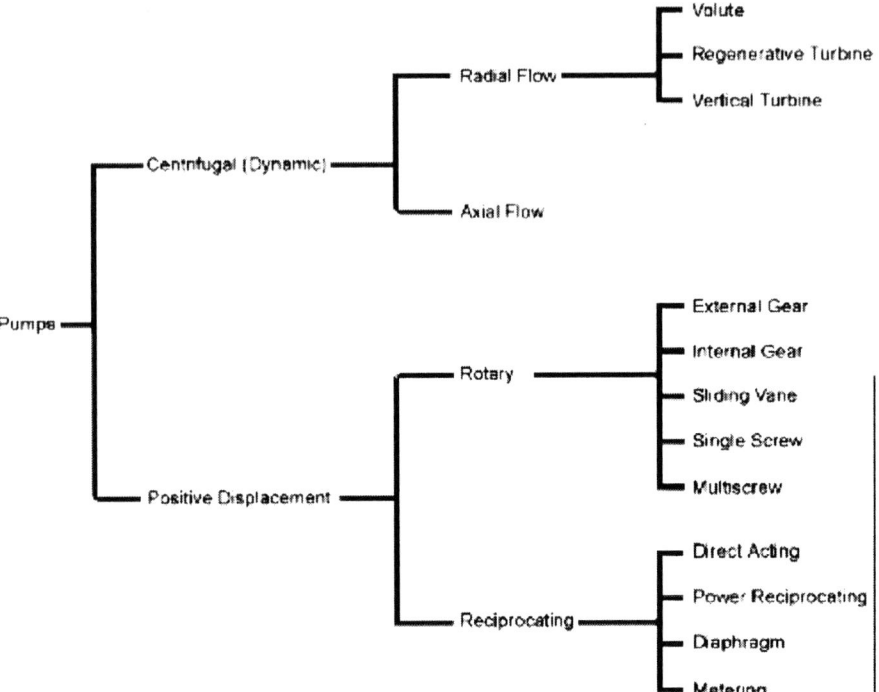

Fig. 1 Classification of pump types.
Source: From ENEWS (see Ref. 1).

Pressure reducing devices are normally installed at the outlet of positive displacement pumps to limit the total head to that of the rated pump output head. Data for these types of pumps are usually found in the form of a table listing the flow rate, total head, and power. As positive displacement pumps are usually used to pump liquids other than water, the assistance of an expert may be needed to select a pump for a particular application.

Pump Theory

Pump Operating Characteristics

Pump performance is defined by the flow rate and head. Head is the pressure required to support a certain height of fluid. Head is expressed in "meters of water."

To explain this, visualize a $1 \times 1 \times 1 \, m^3$ vessel containing water. The weight of the water contained in the vessel is 1000 kg. The pressure exerted on the bottom of the vessel is 1000 kg/m² which equates to 10 kPa. This is the equivalent pressure exerted by a 1 m column of water. This pressure is dependent on the specific gravity of the fluid, so if the vessel was to be filled with another liquid, the head would still be 1 m, but the equivalent pressure would change.

The relationship between equivalent pressure and head can now be equated as:

$$H = \frac{P}{9.81 \times SG} \tag{1}$$

where H, head of fluid (m); P, equivalent pressure (kPa); and SG, specific gravity of fluid (kg/m³).

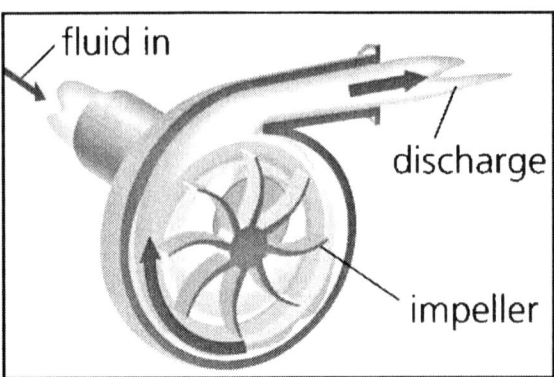

Fig. 2 A basic centrifugal pump.
Source: From The Free Dictionary.

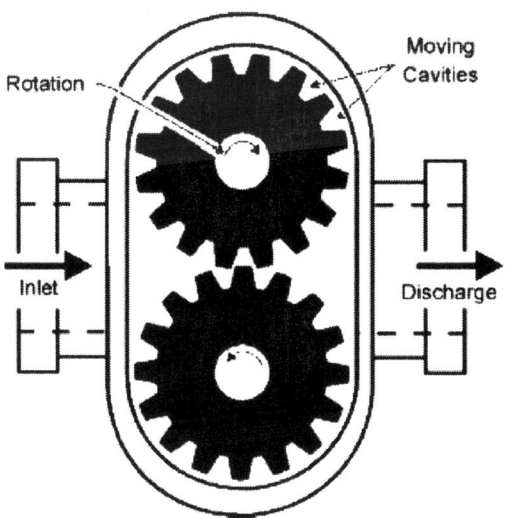

Fig. 3 A basic positive displacement pump.
Source: From ENEWS (see Ref. 1).

For water, this roughly equates to 1 m = 9.81 kPa.

The required head differential across a pump to move a liquid is called the total dynamic head. The output head or dynamic head of a pump must be equal or greater than the sum of the static head and friction head.

Friction Head. The friction and inertia losses, due to the resistance to the flow of a fluid through a pipe, is known as the friction head. These losses depend on the diameter, material, and length of pipe, as well as the type and amount of fittings used in the piping system. Friction head is proportional to the square of the flow rate, thus friction losses are zero if there is no flow.

Velocity Head. Velocity head is the amount of kinetic energy contained within a fluid for a specific flow rate. The relationship between the velocity head and fluid velocity is given in Eq. 2:

$$P_v = \frac{v^2}{2g} \qquad (2)$$

where P_v, velocity head (m); v, fluid velocity (m/s); and g, gravity (9.81 m/s^2).

Because of velocity head is usually small for typical systems, it is incorporated in the friction head, but for gravity and low-pressure applications, it should always be considered.

Total Static Head. The head across a pump when there is no flow is called the total static head. Total static head is the sum of the elevation head and system pressure differential. Elevation head is the difference in height or elevation between the pump and the discharge of the system. The system pressure differential is the difference between the static suction head (the static head at the pump intake) and the static discharge head (static pressure at pump outlet).

Centrifugal Pump Theory

Centrifugal pump performance is defined by the impeller diameter, pump speed, flow rate, head, power, and fluid characteristics. Manufacturers provide data in graph form for each pump type, pump size, and pump. These graphs are usually based on water as the pumped fluid. A typical example of such a graph is shown below in Fig. 4.

The set of graphs in Fig. 4 show the head, power, and efficiency of the pump for a certain flow rate. Each of the numbered curves represents a different pump speed. Some pump curves incorporate different impeller diameters into the curves instead of pump speed. Most manufacturers recommend that only the middle section of the curves be used, as the outside sections may prove unstable.

The variables determining centrifugal pump performance are related by the Affinity Laws, which are used to develop the pump curves. These laws are given in Eqs. 3–9.

D and δ constant:

$$Q_1 = Q_2 \left(\frac{N_1}{N_2}\right) \qquad (3)$$

$$P_1 = P_2 \left(\frac{N_1}{N_2}\right)^2 \qquad (4)$$

$$P_1 = P_2 \left(\frac{N_1}{N_2}\right)^2 \qquad (5)$$

N and δ constant:

$$Q_1 = Q_2 \left(\frac{D_1}{D_2}\right) \qquad (6)$$

$$P_1 = P_2 \left(\frac{D_1}{D_2}\right)^2 \qquad (7)$$

$$W_1 = W_2 \left(\frac{D_1}{D_2}\right)^3 \qquad (8)$$

N and D constant:

$$W_1 = W_2 \left(\frac{\delta_1}{\delta_2}\right) \qquad (9)$$

where D, impeller diameter (m); Q, flow rate (m^3/h); P, total head; W, power (kW); N, pump speed (RPM); and δ, fluid density (kg/m^3).

If conditions in a system change, but the performance curves remain unchanged, the laws can be used to determine the new performance of the system. They may, however, not be used where changes in the layout of the piping causes changes in the system performance curve. These laws may also not be used to analyze systems with a significant static head component, unless the change to the system is less than 10%.

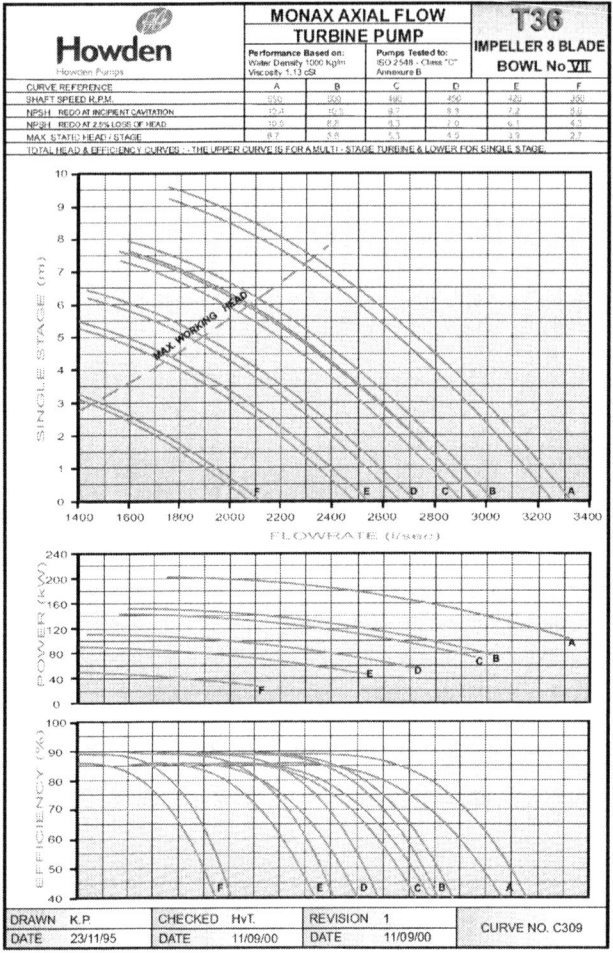

Fig. 4 Common pump curve.
Source: From Howden Pumps SA.

Positive Displacement Pump Theory

For positive displacement pumps, the power requirement of the pump is affected by flow rate, pressure, and fluid characteristics. The relationships used to determine performance for changes in total head, pump speed, or fluid density are given below in Eqs. 10–13.

$$Q_1 = Q_2 \left(\frac{N_1}{N_2} \right) \quad (10)$$

$$W_1 = W_2 \left(\frac{P_1}{P_2} \right) \quad (11)$$

$$W_1 = W_2 \left(\frac{Q_1}{Q_2} \right) \quad (12)$$

$$W_1 = W_2 \left(\frac{\delta_1}{\delta_2} \right) \quad (13)$$

where Q, flow rate (m³/h); P, total head (m); W, power (kW); N, pump speed (RPM); and δ, fluid density (kg/m³).

Pump Power Requirements

If one is to assume that a pump is 100% efficient, the theoretical power required for a specific flow rate against a given total head is determined by the relationship in Eq. 14.

$$W_t = \frac{QP_t}{367} \quad (14)$$

where Q, flow rate (m³/h); P_t, total head (m); and W_t, theoretical power (kW).

The relation shown above can be expanded to incorporate the pump efficiency, belt drive efficiency, and motor efficiency. The incorporation of these variables results in Eq. 15.

$$W_{\text{overall}} = \frac{QP_t}{\eta_p \eta_d \eta_m 367} \quad (15)$$

Q, flow rate (m³/h); P_t, total head (m); W_{overall}, theoretical power (kW); η_p, pump efficiency; η_d, belt drive efficiency; and η_m, motor efficiency.

The overall efficiency of the pump is given by the ratio of output energy to input energy in Eq. 16. This is the same the efficiency as given for the pump by the manufacturer.

$$\eta = \frac{\text{power output}}{\text{power input}} = \frac{\text{theoretical power}}{\text{actual power}} \quad (16)$$

Multiple Pumps Systems

Pumps can be installed in either parallel or series. For a series configuration, the combined pump curve is given by the sum of the heads of the individual pumps at equal flows. For pumps in parallel, the sum of the flows is taken at points of equal head. Typical curves for pumps in series and parallel are shown in Fig. 5.

Determining of Power Usage

One of the most frequent requirements of energy conservation work on pumps is to estimate the change in the power required for a change of head or flow rate. In ideal situations, the performance curves of the pump are available. By superimposing the new system curve on the pump performance curve, the changes in head, flow rate, and power can be read directly from the pump curve.

In cases where the performance curves for pumps are not available, the calculations to determine the system curve can become very laborious. In such a situation, use can be made of the affinity laws. It must be kept in mind, though, that the answers resulting from the use of the affinity laws may not coincide with the actual system performance curve. The error, however, is small enough if the changes in pump settings are small.

Pump Seals

One of the factors that has a large effect on pump efficiency is the type of pump seal used and the quality of maintenance. The seal uses up a portion of the shaft power

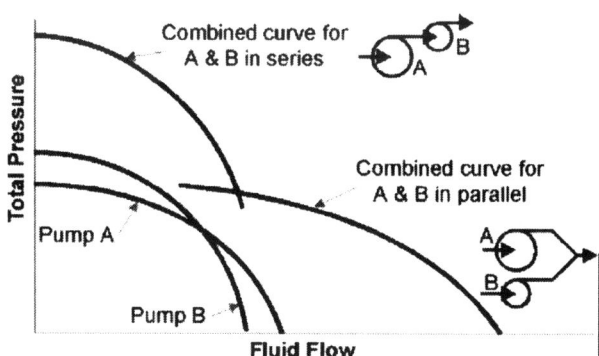

Fig. 5 Combined pump curves for series and parallel configurations.
Source: From ENEWS (see Ref. 1).

provided by the motor and the leakage caused by faulty seals represents a portion of the pumped fluid. The two most common types of pump seals are packing glands and mechanical seals. An example of each is in Fig. 6.

Packing Glands. Packing gland seals are made up of a set of rings, made of low friction material, compressed to achieve a close contact between the shaft and pump casing. The packing gland is cooled by forced lubrication. Common lubrication methods used are controlled leakage of pumped liquid, forced flushing by a separate liquid, and controlled force feeding of oil or grease. For any type of packing gland, the power used, the liquid lost, and life of the packing gland depends on the skilled adjustment of the pressure exerted on the packing rings by the retaining rings. A rule of thumb to keep in mind is that the power usage of packing glands is about six times greater than that of mechanical seals.

Mechanical Seals. Mechanical seals are made up of a set of rigid, spring-loaded rings, which slide against a finely finished mating surface. The rings are manufactured of low friction material, which is either self-lubricating or lubricated through a slight leakage of pumped liquid.

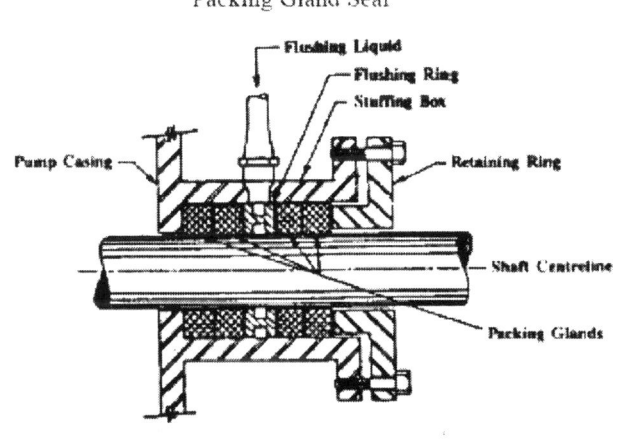
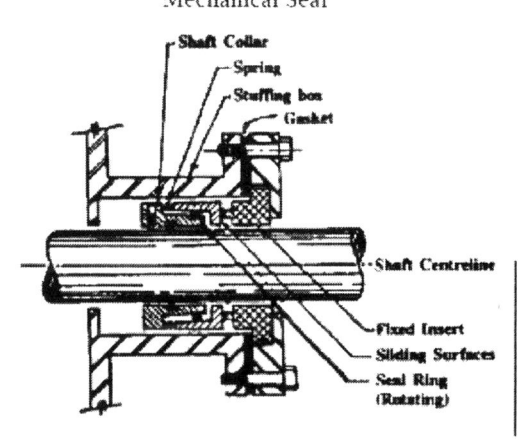

Fig. 6 Typical pump seals.
Source: From ENEWS (see Ref. 1).

Although the initial cost of mechanical seals is high, the lower power usage results in significant cost savings. For pumps that have packing glands, special types of mechanical seals are available as a retrofit option.

Energy Management Opportunities

Maintenance

Maintenance is one of the often-overlooked aspects of energy management. An effective maintenance program can result in viable savings within a very short time span. Some of the most important issues are given and discussed, cross-referencing the readings with the manufacturer's specifications.

Check Packing Glands. Packing glands should be checked regularly to ensure the glands are properly tightened and adjusted. For certain types of packing material, liquid temperature, shaft speed, optimal tightness, and allowable leakage may be established by monitoring the rate of leakage from the packing.

Check Critical Tolerances. Pump efficiency is affected by the leakage past the impeller from the intake (suction side) to the outlet (discharge side). On some pumps, use is made of replaceable wear rings, with small clearances between the moving surfaces, to minimize leakage and improve maintenance. The critical tolerances of mechanical seals stop leakage and air from being drawn into the fluid flow. The critical tolerances can be affected by the erosion of the impeller and wear rings if liquids are pumped that contain abrasive particles. So, to maintain high pump efficiency, the critical clearances and surfaces need to be checked periodically.

Check and Adjust Drives. If properly designed and maintained, drives such as belts or flexible couplings, can provide many years of hassle-free service. To ensure this, the following actions need to be carried regularly.

- Maintain alignment of pulleys and couplings
- Check tension of belts
- Lubricate bearings
- Replace or repair damaged belts, pulleys, clutches, drive keys, and couplings

The proper tension for various types of belts can usually be found in handbooks or catalogues available from component manufacturers.

Clean Pump Impeller. Pumps need to be cleaned regularly to maintain the efficiency of the pump. This is especially true of pumps that move "dirty" liquids. A build-up of dirt on the impeller and pump casing results in static pressure losses, which in turn reduces the efficiency of the pump.

Pump Scheduling. By switching off pumps that are not required to maintain the required fluid flow of the system, significant savings in energy and maintenance costs can be obtained. For example, hot water circulating pumps can be switched off in the summer, or process cooling water pumps can be switched of when the system in not working.

Establish Maintenance Program. Using the manufacturer's recommendations as guidelines, a maintenance plan should be set up. This maintenance plan should be designed for the use and type of pump. The program should typically include the following:

- Daily: Monitor pump sound, bearing temperature, leakage, and gauge readings.
- Semi-annually: Check and lubricate components in need of lubrication; check free movement of the stuffing box glands, and check the pump and drive alignment.
- Annually: Clean, inspect, and lubricate bearings and their seals, examine the packing and the shaft sleeve, recalibrate all associated instrumentation, and check performance against the design ratings.
- Replace worn components when test indicate a drop in performance.
- Adjust impeller clearance when test shows a loss of performance power requirement.

Retrofit Opportunities

In certain cases, opportunities arise which, if properly investigated and capitalized on through the use of a well-designed retrofit option, can result in viable savings of energy and costs. Some of the available retrofit options are discussed below.

Variable Speed Controller. Variable speed drives can be used to control the speed of the pump and also the flow rate and head. As such, the use of variable speed drives offers a potentially large saving in energy and cost for pumping systems.

If a throttle valve is removed from a pumping system, the system head curve will usually be lowered. This is due to elimination of the loss caused by the throttle valve, which can be as high as 25% of the total friction loss of the system. The reduction in the system head creates a potential for savings in energy if a new optimally sized pump is used. Although it is not always possible to replace the pump every time there is a change in the system curve, the installation of a VSD makes practical sense. A VSD-driven pump provides an infinite family of system curves depending on the speed.

In most cases, the high initial capital expenditure associated with a VSD can be justified; this is especially true of cases where the required system output varies over a wide range. If during the initial design of a new pumping system, the control valves and pump bypass piping is eliminated in favor of a VSD, the cost of the VSD will be defrayed. On existing systems, the cost of the retrofit will, in most cases, be paid back in less than two years. The

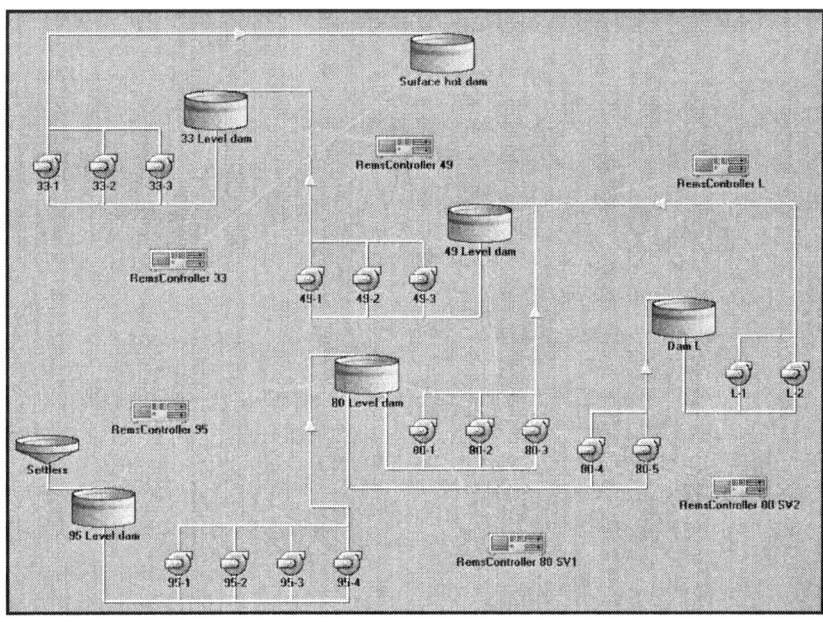

Fig. 7 Schematic layout of a typical intelligent pump scheduling system.
Source: From HVAC International.

savings and efficiency of the retrofit, however, is dependent on the type of VSD used.

Other advantages of the use of VSDs are:

- Improved process control
- Reduced wear on the pump and motor
- Reduced noise
- Reduced maintenance

For VSD applications, there is no single solution. Despite this, not all pump applications are good candidates for a VSD retrofit. Systems that typically fall in this category are:

- Applications with a low duty cycle
- Pumps that operate at a practically constant pressure
- Systems requiring a constant flow at a constant head

Intelligent Pump Scheduling. With this method, an automated scheduling system automatically schedules pumping activities out of peak periods. Extensive use is made of this method in South African mines on cold water drainage pump systems.

In this application, the system schedules the pumps to fill underground dams during the off-peak periods. This allows the pumps to be switched off during the peak period, resulting in cost savings. An example of such a control system is shown schematically in Fig. 7.

Although this method results in cost savings, the energy use remains the same. This is due to the fact that the energy use is just shifted from the peak to the off-peak periods.

Multiple Pumps. Savings can be had from using two smaller pumps in parallel instead of one large pump for systems where a wide range of flow rates is required at a reasonably constant pressure. For such a system, both pumps can operate simultaneously during peak demand periods, while only one would satisfy the need in lower demand periods. Energy savings result from avoiding the need to throttle the much larger pump during lower demand periods and from running the smaller pumps closer or on the optimum efficiency point.

Another option is to use one smaller pump in parallel to one or more larger pumps. This is a useful option where the capacity demand varies considerably. In such cases, the small pump can match the very lowest demand, while the larger pumps operate as normal.

The last option is to use one normally-driven pump in parallel with a VSD-driven pump. In such a setup, the VSD can be controlled until the desired capacity is matched. For very low demand, the normal pump can be switched off, while the VSD-driven pump is used to carry the load.

Power Recovery Turbines. In processes where high pressures are required, some of the residual energy can be used to drive a Hydraulic Power Recovery Turbine (HPRT) instead of dissipating it through a throttling valve. A good example is chilled water being fed down a mineshaft. The water is at a high pressure when it reaches the bottom, and instead of throttling the flow to avoid exceeding maximum pressure limits, an HPRT can be used to recover the latent energy contained in the water. This energy can be used to drive a generator, which in turn generates electricity to power either one of the pumps or some other application.

Another useful fact to remember is that practically all centrifugal pumps will perform as turbines when run in reverse.

Other methods that are not discussed above include:

- The replacement of outsized equipment
- Replacement of oversized motors

FAN SYSTEMS

A fan is defined by the Internet Free Dictionary as "a device for creating a current of air or a breeze, especially a machine using an electric motor to rotate thin, rigid vanes in order to move air, as for cooling."

Fans are one of the most misused, abused, and faultily applied types of equipment. The result of this misuse is high energy costs, which offers a large scope for the implementation of energy management methods, and accordingly also for sizeable savings.

Types of Fans

As with pumps, there are two basic types of fans—centrifugal and axial. These types are discussed in detail in the following paragraphs.

Centrifugal Fans

Centrifugal fans create a fluid flow through a centrifugal force, which is produced by moving a fluid between rotating impeller blades, and by the inertia generated by the velocity of the fluid leaving the impeller. Centrifugal fans are further differentiated by the shape of the fan blades. The four most common shapes are:

- Forward Curved
- Backward Inclined
- Airfoil
- Radial Blade

An example of each of the blade types, along with some typical performance curves, is shown in Fig. 8.

Fans with forward curved blades are used in low-pressure systems and run at a slower speed than backwardly inclined types. For this type of fan, a decrease in system static pressure at a constant operating speed results in a considerable increase in horsepower. This increase in horsepower could possibly lead to motor overload.

Backwardly inclined blades are used in high-pressure duct system applications. This type of blade is also non-overloading under normal conditions. This type of fan will have fewer blades on the wheel and these blades may be flat or slightly curved. This is in contrast with the forward curved fan on which the blades are more closely spaced and never flat. This type of fan also has a higher efficiency over the design range and operates at considerably higher speeds than the forward curved fan.

Airfoil fans are a special type of backward curved airfoil blade. They are used on all types of systems, from low to high pressure and run at very high speeds. They are very efficient, but this efficiency comes at a very high cost.

The last type is the radial blade. This type of fan is used for material handling and has self-cleaning blades. Radial bladed fans are the least effective type.

Axial Fans

Air is moved by an axial fan through the change in velocity of the air passing over the impeller. No energy is added to the fluid flow by the centrifugal forces. As with centrifugal fans, axial fans are distinguished by the shape of the fan blades. The most common types are:

- Propeller Fans
- Tubeaxial Fans
- Vaneaxial Fans

These types are shown in Fig. 9 along with common performance curves for some axial type fans.

The propeller fan usually is either belt or driven, and has either flat or slightly curved blades. This type of fan is used in applications associated with very low pressures and are seldom connected to a duct system. This type of fan is mostly used to move large volumes of air through large openings, which causes a small pressure loss.

Tubeaxial fans are capable of higher-pressure differentials and are more efficient than propeller fans. This type of fan typically has four to eight curved or airfoil type blades. The housings are made of cylindrical tubes, so that the clearance between the housing and blade tips is minimal. This type of construction results in the higher efficiency. The advantages of using tubeaxial fans are:

- Ease of installation
- Reasonable cost
- Low maintenance

Vaneaxial fans have blades with an airfoil design and can be operated against pressure. This type of fan is usually directly driven and has a higher efficiency than propeller type fans. For an increase in speed and/or system resistance, this type of fan will show a sharp increase in horsepower. This can result in motor overload.

Fan Measurements

Flow Measurement

The total pressure exerted by an air stream in a duct is the sum of the static pressure and the velocity pressure. A common device used to measure airflow is the Pitot tube shown in Fig. 10.

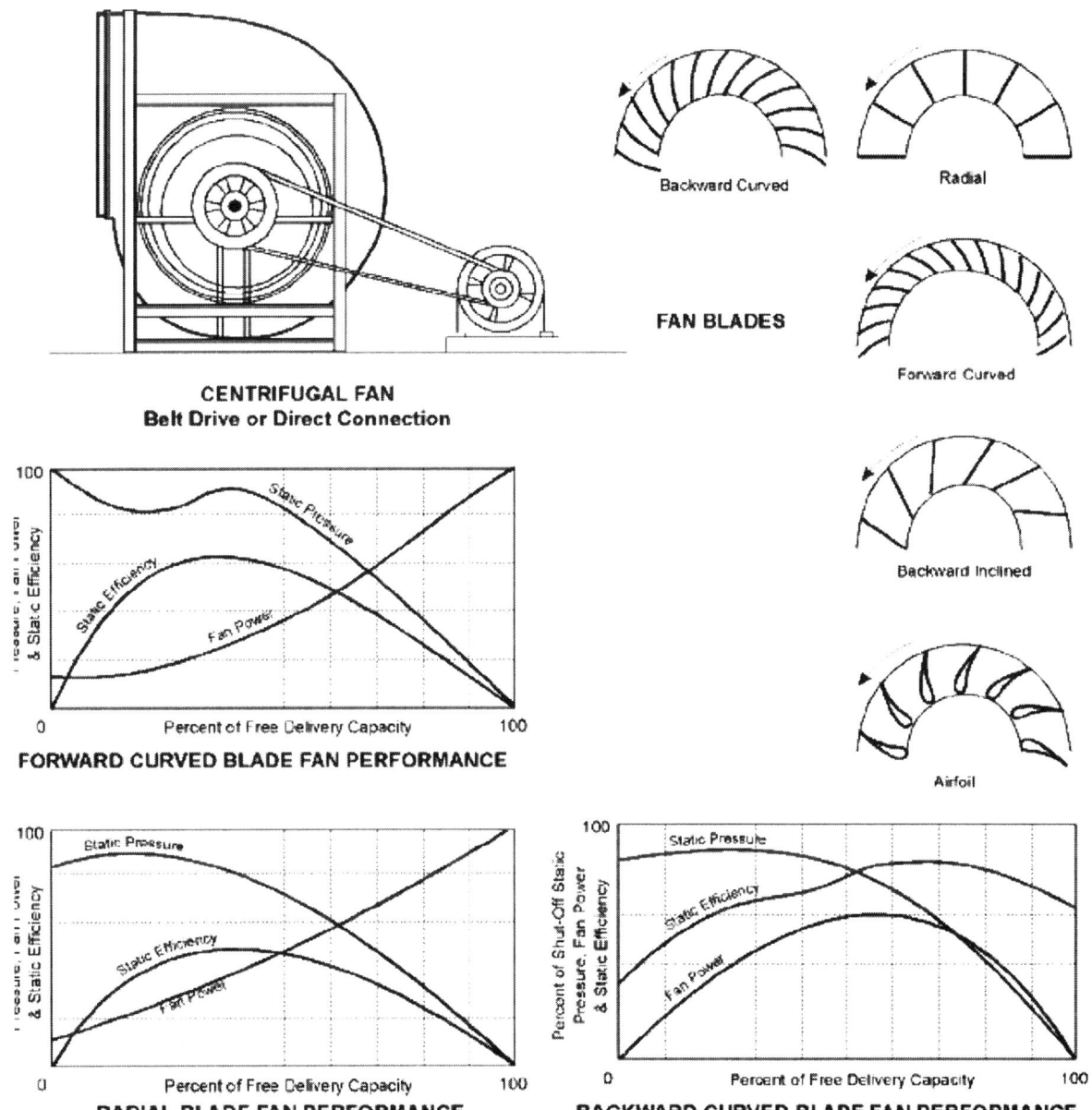

Fig. 8 Types of centrifugal fan blades.
Source: From ENEWS (see Ref. 1).

Measurements should be taken with the Pitot tube in accordance with the traverse detail shown in Fig. 10. Velocity pressure should be calculated for each of the traverse positions and then averaged. By using these measurements, the velocity of the flow in a duct can be calculated using Eq. 17.

$$\text{Velocity} = 0.764 \left(\frac{TP_v}{B} \right)^{0.5} \quad (17)$$

where velocity, average air velocity (m/s); T, temperature (°K); P_v, velocity pressure (Pa); B, barometric pressure (kPa absolute); and 0.764, equation constant and unit conversion.

This equation can be simplified to take into account the standard conditions of 20°C and 101.325 kPa. This simplification results in Eq. 18.

$$\text{Velocity} = 1.3 P_V^{0.5} \quad (18)$$

The volume airflow rate can now be calculated, using Eq. 19.

$$f_a = \text{Velocity } A_d 1000 \quad (19)$$

where f_a, volume air flow rate (l/s); A_d, area of duct (m²); and 1000, conversion of units.

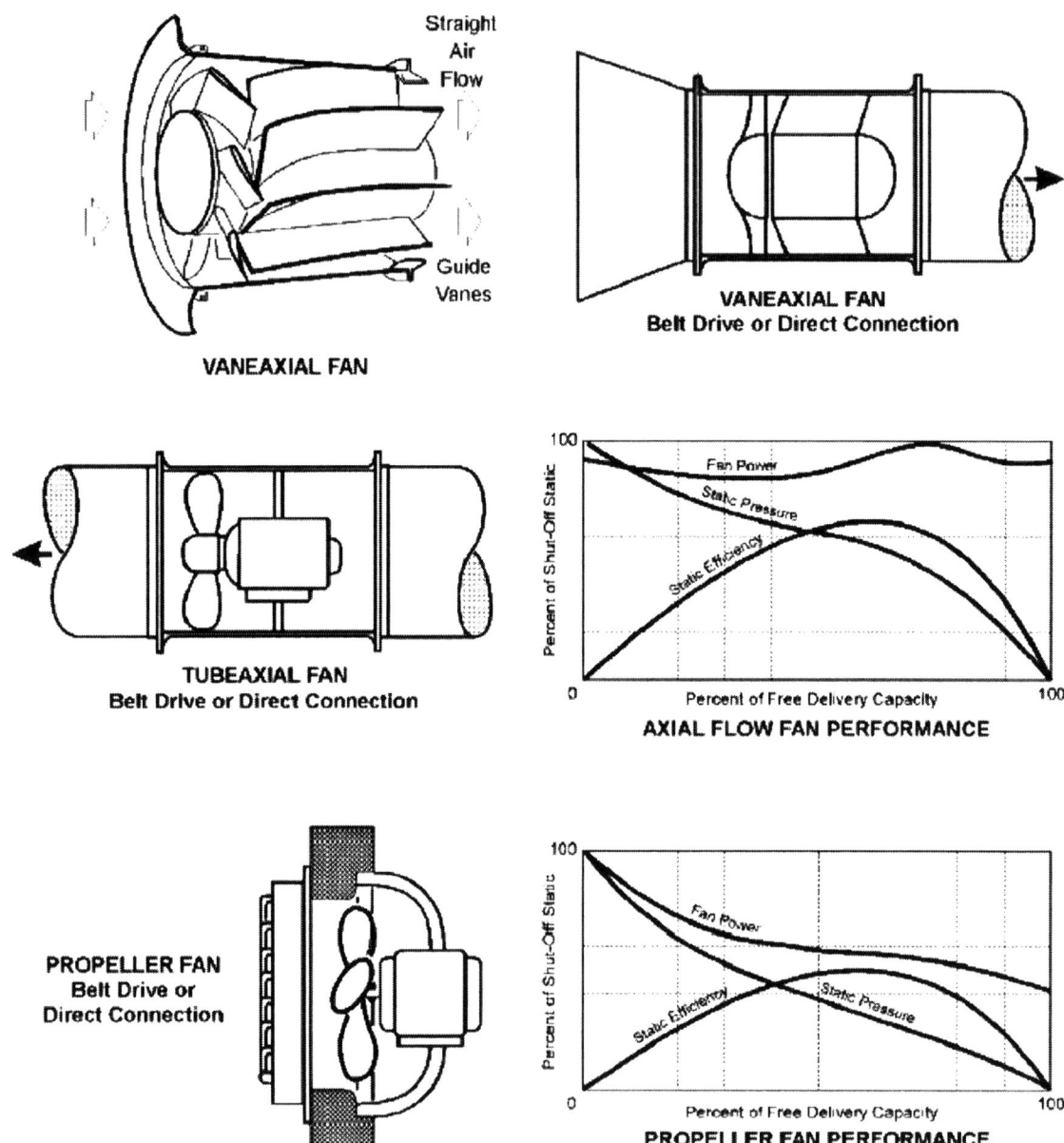

Fig. 9 Types of axial fan blades.
Source: From ENEWS (see Ref. 1).

Performance Measurement

Fan performance can be easily calculated using the measurements taken at the inlet and outlet through the method described in the previous section. There are several factors, however, which can affect the accuracy of these measurements. These factors are:

- Air flow not at right angles with the measurement plane
- Non-uniform velocity distribution
- Irregular cross sectional shape of the duct or passageway
- Air leaking between the measurement plane and the fan

For more precise measurements, refer to the Field Performance Measurements, Publication 203 by the Air movement and Control Association Inc. (AMCA).

Using the measured data, the total differential pressure across the fan can be calculated using Eq. 20.

$$DP_t = Ps_o + Pv_o - Ps_i - Pv_i \qquad (20)$$

where DP_t, total differential pressure (Pa); Ps_o, static pressure at outlet (Pa); Pv_o, velocity pressure at outlet (Pa); Ps_i, static pressure at inlet (Pa); and Pv_i, static pressure at inlet (Pa).

The total fan static differential pressure can also be calculated using Eq. 21

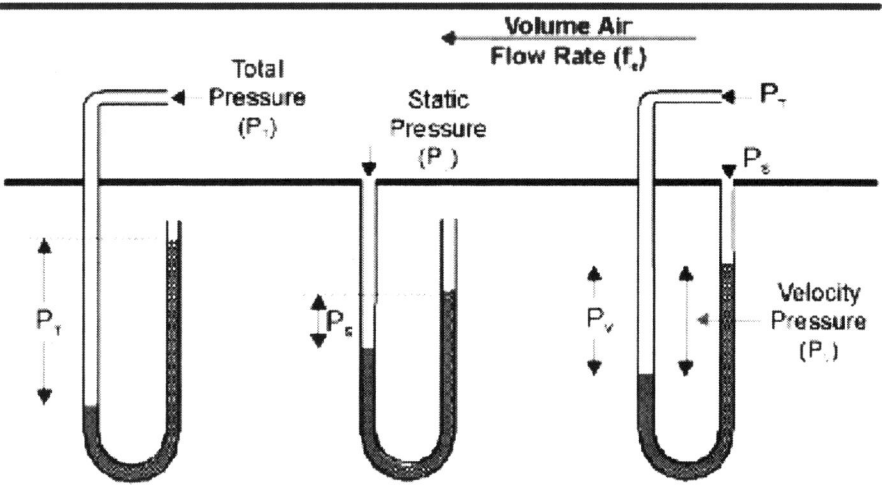

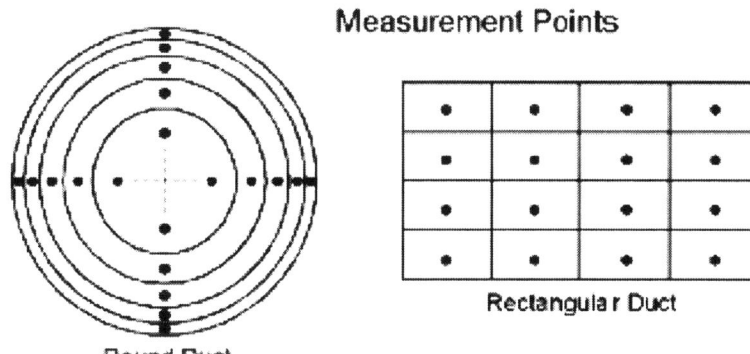

Fig. 10 Pitot tube measurements. Source: From ENEWS (see Ref. 1).

$$DP_s = Ps_o - Ps_i - Pv_i \quad (21)$$

where DP_s, total fan static differential pressure (Pa).

Although the effects of inlet and outlet conditions have not been included in the equations, the equations still provide a reasonable basis for the further calculation of static efficiency and power.

Fan Performance

Power Requirements

By measuring the quantity of air delivered, and the pressure against which it is delivered, it is possible to calculate the work done by a fan. If one is to assume that the fan is 100% efficient, the required theoretical power needed to move a volume of air against a total pressure can be calculated using Eq. 22.

$$W_t = \frac{QP_t}{1000} \quad (22)$$

where W_t, theoretical power to move air at 100% efficiency (kW); Q, static pressure at inlet (Pa); and P_t, static pressure at inlet (Pa).

The overall fan efficiency relates the theoretical output power of the fan against the input electrical power. This relation includes fan losses, belt drive losses, and motor losses.

Performance Curves

As with pumps, information on the performance of commercial fans is available from the manufacturers in the form of performance curves. An example of a performance curve is shown in Fig. 11.

A fan operates along a specific performance curve for a specific speed. Thus, at a fan speed of N_1, the fan will operate along the N_1 performance curve as shown in Fig. 11. The actual operating point on the curve is dependent upon the system's operating characteristics.

For any fan system, pressure increases with an increase in airflow. The pressure requirement of a system over a range of flows can be determined, and using this data a "system performance curve" can subsequently be developed (shown as SC_1 in Fig. 11). If the system curve is plotted on the fan curve, the point where the two curves intersect is the actual operating point of the fan (shown as point "A" on Fig. 11).

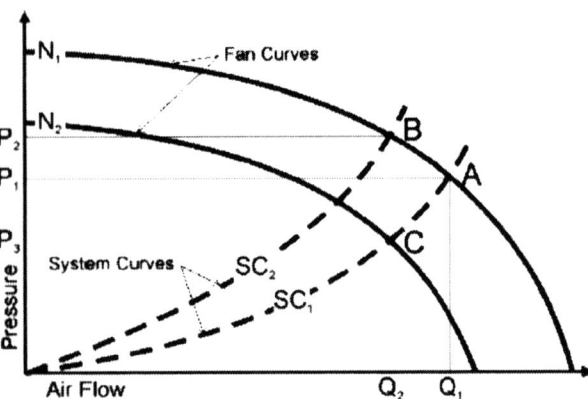

Fig. 11 Fan performance curve.
Source: From ENEWS (see Ref. 1).

Two methods can be used to reduce airflow. One method is to restrict the flow by partially closing a damper. This causes a new system performance curve (SC_2 on Fig. 11) to develop where the required pressure is greater for the new given airflow.

The second method is to reduce the speed of the fan (N_2 on Fig. 11) while keeping the damper fully open. The fan would now operate at point "C" to provide the same airflow, but at a lower pressure. The reduction of fan speed is a more efficient way to decrease airflow, since it requires less power and thus less energy is used.

Impact of Air Density

The performance data provided by manufacturers is based, unless otherwise specified, on dry air at a standard atmospheric pressure (101.325 kPa), temperature (20°C), and density (1.2 kg/m³).

In most applications, moist air at temperatures and pressures other than the standard conditions are processed. Because of this, the air density needs to be corrected to obtain the actual fan performance. By measuring the wet bulb, relative humidity, or dew point temperature, the moisture content of an air stream can be determined. Wet bulb or dry bulb temperatures are mostly determined at the fan inlet using a sling thermometer. When the air stream temperature is above 82°C, a more reliable determination of moisture content is made from the dew-point temperature. If the dry bulb temperature is between 5 and 38°C, density may be determined by using a psychrometric chart.

Fan Laws

Fan laws are very similar to the laws that are used to predict pump performance. The laws are made up of a set of relations concerning speed, power, and pressure. A change in RPM on any fan causes a predictable change in pressure and power. The basic fan laws are shown in Eqs. 23–25.

$$\frac{Q_1}{Q_2} = \frac{N_1}{N_2} \qquad (23)$$

$$\frac{P_1}{P_2} = \left(\frac{N_1}{N_2}\right)^2 \qquad (24)$$

$$\frac{kW_1}{kW_2} = \left(\frac{N_1}{N_2}\right)^3 \qquad (25)$$

where W, power (W); N, speed (RPM); P, pressure (Pa).

Energy Management

Air handling systems are usually balanced after installation, which requires the services of an experienced specialist. For this article, it is assumed that the air handling system under consideration has been balanced after installation. However, during the operating lifetime, the components comprising the air handling system wear down.

Dirt accumulated on fan blades can cause the blades to become unbalanced, which in turn causes vibrations and noise. Faulty bearings can also cause vibrations and damage to the fan blades. Fouled filters reduce airflow, and thus reduce motor load, and can create a false sense of savings. From this, it is apparent that the first step, as with pumps, is to bring air handing systems up to standard and to implement a proper maintenance schedule.

Maintenance

Some common maintenance actions are listed and discussed in the following paragraphs.

Check and adjust belt drives regularly. Misaligned drive sheaves damage belts and can increase the power requirement of the fan. In Fig. 12, drive losses for properly aligned drives are shown for different speed ranges. Losses due to misaligned drives fall outside the limits set in Fig. 12.

Lubricate fan components according to the manufacturer's instructions. Fan components, such as couplings, shaft bearings, adjustment linkage, and adjustable supports, need lubrication with a proper lubricant at intervals recommended by the manufacturer.

Clean fan components regularly. All fans should be cleaned regularly to maintain efficiency. This is especially true of fans handling dirty air. Dirt that has accumulated on the fan blades and housing causes higher static pressure losses, and these losses in turn reduce efficiency.

Correct excess noise and vibration. Fan noise and vibration can be caused by a number of factors:

- Out of balance fan wheel
- Bad bearings
- Insufficient insulation

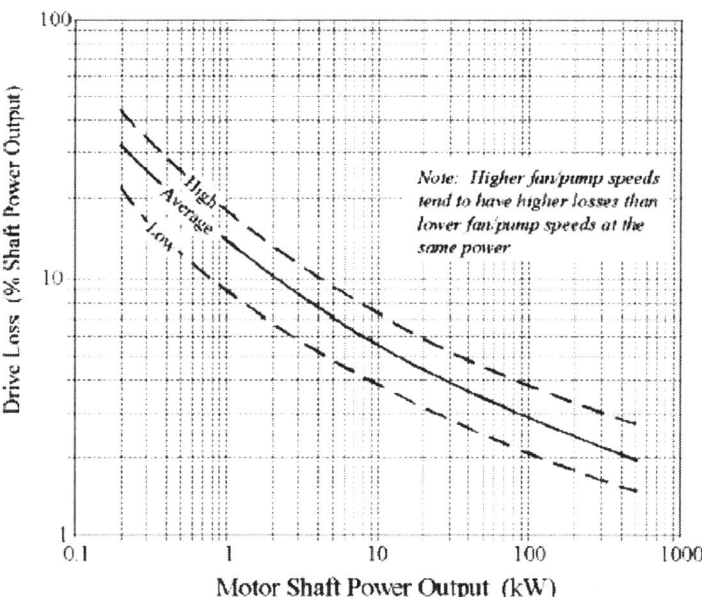

Fig. 12 Drive losses.
Source: From ENEWS (see Ref. 1).

- Misalignment of shaft seals
- Corrosion between shaft and bearing

Correct leaks. Energy savings can be attained by fixing air leaks from loose connections, improperly sized dampers and shaft openings, and unsealed expansion connections.

Replace loaded air filters. One common cause of poor system performance is loaded air filters. Manufacturers provide data on the point at which a filter is considered fully loaded. This data is rated in terms of pressure drop at various velocities. When it is found that the pressure drop over a filter reaches a point where the filter becomes loaded, the filter should be replaced. Furthermore, if a system with loaded filters is balanced, and the filters subsequently replaced, the airflow might become excessive, which would result in higher system energy.

Implement a maintenance program. Implement a maintenance program similar to that described for pumping systems above.

Low Cost Options

These are energy management options that are relatively low cost and can be implemented easily. Some low cost options worthy of evaluation are:

- Improve fan inlet and outlet duct connections to reduce losses at the inlet and outlet
- Shut off fans that are not needed
- Reduce fan speed to provide the optimum amount of airflow when the dampers are open in the maximum position for balanced air distribution

Retrofit Options

The implementation cost of any retrofit option is significantly higher than low cost options. Below are some typical energy management options that are available under the Retrofit option:

- Use a variable speed motor to allow the fan to adjust for changes in system requirements.
- Replace oversized motors.
- Replace outdated equipment with newer, optimal sized units.
- Install digital control system.

Flow Control

Most air handling systems are designed for 100% airflow. However, in some cases, they can operate at lower flow rates. What options are available to a manager to achieve a lower flow rate? The following options are available.

- Use dampers or inlet vanes to control the flow.
- Install a variable speed drive.

Fig. 13 shows the typical power required at a specific flow rate for different control methods. As most fans are usually oversized, the curves are based on the power requirement of a fan which is 10% oversized. Most fans are oversized because of:

- Leaks that develop and filters clogging
- Conservative design engineers
- The fan's inability to meet design specifications

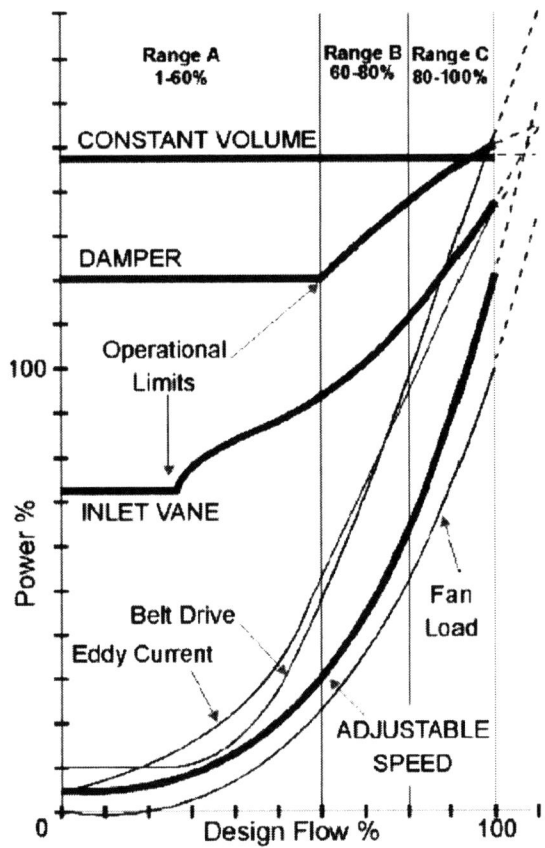

Fig. 13 Typical power to flow curves for different control methods.
Source: From ENEWS (see Ref. 1).

Furthermore, the curves are also based on AC motors running at full load and 90% efficiency, and with a typical efficiency drop for reduced loads.

There are, however, some limitations associated with each of these methods. Dampers have a limited turn down and may be become noisy. Inlet vanes, even if wide open, restrict flow. Although variable speed drives reduce noise and save energy by adjusting the speed of the fan to the exact needs of the system, they are very expensive, and the potential savings may not justify the initial capital expenditure.

CONCLUSION

This article focused on energy management opportunities that exist within pumping and fan systems. It provided an overview of the types of pumps and fans, their operating characteristics, theory, and power requirements, and provided typical cost and energy saving opportunities.

REFERENCE

1. SADC Industrial Energy Management Project. Core Training Manual. Zimbabwe.

BIBLIOGRAPHY

1. Reeves, G.; Gillespie, K.L.; Cowan, J.D.; et al. *ASHRAE Guideline 14-2002—Measurement of Energy and Demand Savings*; American Society of Heating, Refrigeration and Air-Conditioning Engineers: 1791 Tullie Circle, NE - Atlanta, GA, 2002.
2. Cooper, P. Pump Machinery. In Cooper, P., Ed.; American Society of Mechanical Engineers: United Engineering Centre, 345 East 47 Street, New York, 1989.
3. Warring, R.H. *Pumping Manual 7th Ed.*; Gulf Publishing Company: Houston, TX, 1984.
4. Karnassik, I.J.; Messina, J.D.; Cooper, P.; et al. *Pump Handbook; 3rd Ed.*; McGraw Hill Book Company: Singapore, 2001.
5. Hydraulic Institute. *Energy Reduction in Pumps and Pumping Systems*; Hydraulic Institute: U.S.A., 1997.

Radiant Barriers

Ingrid Melody
Philip Fairey
David Beal
Florida Solar Energy Center, Cocoa, Florida, U.S.A.

Abstract
Attic radiant barriers made of aluminum foil are becoming a popular way for homeowners to save energy and money in Southern states. They are increasing in popularity for two reasons. First, tests by the Florida Solar Energy Center (FSEC) and other groups show that they work. Second, manufacturers are improving the quality of radiant barrier materials.

To most homeowners, attic radiant barriers are a new energy conservation concept; many of them have questions about how radiant barriers work and how to use them. Herein are answers to some of the most commonly asked questions. Also included are recommended ways to install radiant barriers in existing attics and new homes.

RADIANT BARRIERS

A radiant barrier is a layer of aluminum foil placed in an airspace to block radiant heat transfer between a heat-radiating surface (such as a hot roof) and a heat-absorbing surface (such as conventional attic insulation). Fig. 1 illustrates the locations in which a radiant barrier may be installed in an attic. Only locations 1 and 2 are recommended for sheet radiant barriers because dust will accumulate on the radiant barrier if it is installed at location 3.

BENEFITS

In hot climates, benefits of attic radiant barriers include both dollar savings and increased comfort.

Without a radiant barrier, your roof radiates solar-generated heat to the insulation below it. The insulation absorbs the heat and gradually transfers it to the material it touches, principally, the ceiling. This heat transfer makes your air conditioner run longer and consume more electricity.

An aluminum foil radiant barrier blocks 95% of the heat radiated down by the roof so it can't reach the insulation.

In summer, when your roof gets very hot, a radiant barrier cuts air-conditioning costs by blocking a sizable portion of the downward heat gain into the building.

In the warm spring and fall, radiant barriers may save even more energy and cooling dollars by increasing your personal comfort. During these milder seasons, outdoor air temperatures are comfortable much of the time. Yet solar energy still heats up your roof, insulation, attic air, and ceiling to temperatures that can make you uncomfortably warm. An attic radiant barrier stops almost all of this downward heat transfer so that you can stay comfortable without air conditioning during mild weather.

You may also find that radiant barriers can expand the use of space in your home. For instance, uninsulated, unconditioned spaces such as garages, porches, and workrooms can be more comfortable with radiant barriers. In addition, because radiant barriers keep attics cooler, the space is more usable for storage.

One final benefit: a cooler attic transfers less heat into air conditioner ducts, so the cooling system operates more efficiently.

"BLOCKING" HEAT TRANSFER

Aluminum foil, the operative material in attic radiant barriers, has two physical properties of interest here. First, it reflects thermal radiation very well. Second, it emits (gives off) very little heat. In other words, aluminum is a good heat reflector and a bad heat radiator.

Your grandmother probably made use of these properties through "kitchen physics." She covered the Thanksgiving turkey with a loose "tent" of aluminum foil before she put it in the oven. The foil reflected the oven's thermal radiation, so the meat cooked as evenly on top as on the bottom. She removed the foil briefly to let the skin brown, but when she took the bird from the oven, she "tented" it with foil again. Since aluminum doesn't emit

---

* © 2005 by Florida Solar Energy Center. Reprinted with permission from Florida Solar Energy Center.

Keywords: Radiant barriers; Hot roofs; Attic insulation; Infrared heat; Low emissivity; Radiant heat transfer; Aluminum foil barriers.

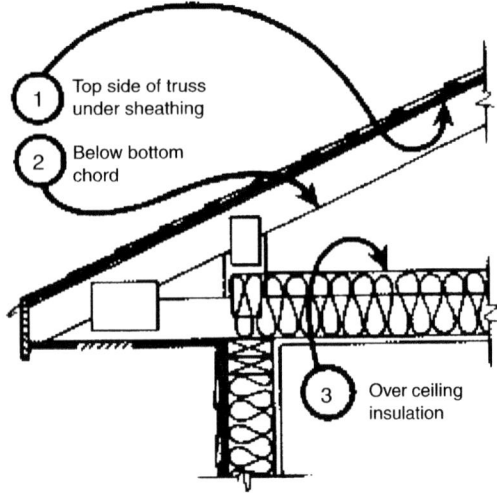

Fig. 1 Typical attic section with three possible locations for radiant barrier.

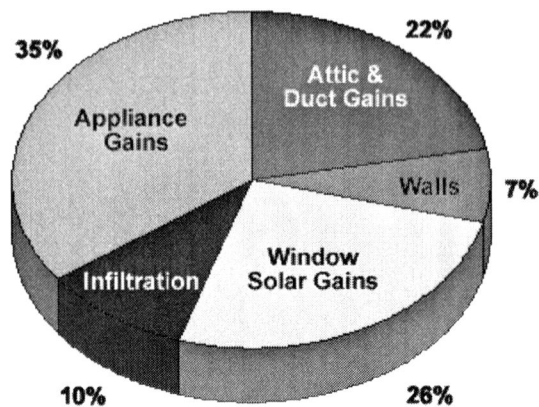

Fig. 2 Breakdown of cooling loads in a typical 1500-ft^2 central Florida home.

much heat, the turkey stayed hot until the rest of the meal was ready. To understand the concept of not emitting heat, let's use an analogy of a light bulb. When you turn on a light bulb, it emits light. If you were to paint the light bulb black, when it was turned on, it would not emit light. A radiant barrier has a similar effect on infrared heat. Your roof surface heats up in the sun and will emit infrared heat. When this infrared heat heats the radiant barrier it will not emit, or reradiate, the heat into your attic.

Cooking a turkey and painting light bulbs are simple analogies, but the same principles of physics apply to an attic radiant barrier. Aluminum foil across the attic airspace reflects heat radiated by the roof. Even if the radiant barrier material has only one aluminum foil side and that side faces down, it still stops downward heat transfer because the foil's low emissivity will not allow it to radiate the roof's heat to the insulation below it.

SAVINGS

Since everyone's home and lifestyle are different, we can't precisely calculate your personal savings from attic radiant barriers. However, it's reasonable to expect that an attic radiant barrier can save 8–12% of your annual cooling costs in the Southeast.

Savings from an attic radiant barrier depend on the amount of heat the roof and attic contribute to your home's cooling load. ("Cooling load" is the total amount of heat your air conditioner must remove to maintain comfortable indoor temperatures.) In general, the more energy efficient the rest of your home is, the larger the percentage of energy you save from an attic radiant barrier because the roof and attic make up a larger portion of the cooling load.

Fig. 2 shows a breakdown of cooling loads in a typical 1500-ft^2 Central Florida home. The attic (including heat gains to the duct system) accounts for 22% of the total cooling load. In this house, an attic radiant barrier could save 8–12% on the annual air-conditioning costs.

Although not as significant, heating savings may also accrue from the use of radiant barriers.

Results from a recent comprehensive field monitoring study conducted for Florida Power Corporation (FPC) by FSEC on the performance of attic radiant barrier systems in Central Florida homes may be viewed at http://www.fsec.ucf.edu/bldg/pubs/rbs/.

Claims of Greater Savings

As in most cases, claims for radiant barriers that *sound* too good to be true *are* too good to be true. If your roof accounts for less than 20% of your cooling load, then an attic radiant barrier can't possibly save more than 20% on your bills.

Claims of greater savings may simply be the results of misunderstanding. For instance, FSEC has measured and reported that radiant barriers can reduce heat gain through R-19 insulated ceilings by over 40%. If the ceiling portion of the total cooling load is 20%, that's a reduction of 40% of 20%, which amounts to 8% savings on the total cooling load.

If all the facts and figures tend to confuse you, just remember that an attic radiant barrier can save about 8–12% on your air-conditioning costs in the Southeast. Any Sunbelt homeowner knows that an 8–12% saving on air-conditioning bills can be significant.

AVAILABLE MATERIALS

There are many types of radiant barrier materials on the market, and more are being developed as radiant barriers become more widely used. Five generic types are most common:

- Single-sided foil (one foil side) with another material backing such as craft paper or polypropylene. Some products are further strengthened by fiber webbing sandwiched between foil and backing. The strength of the backing material is important since unreinforced foil tears very easily.
- Foil-faced roof sheathing materials that come from the manufacturer with a foil facing adhered to one side of the sheathing.
- Double-sided foil with reinforcement between the foil layers. Reinforcement may be cardboard, craft paper, mylar or fiber webbing.
- Foil-faced insulation. The insulating material may be polyisocyanurate, polyethylene "air-bubble" packing or other materials that impede heat conduction.
- Multilayered foil systems. When fully extended and installed so that the foil layers do not touch, these products also form insulating airspaces.
- Radiant barrier "chips" are also manufactured and sold. This product is slightly different than a conventional sheet-type radiant barrier in that the "chips," which are blown onto the floor of the attic—typically to a depth of 3 or more inches, act as a multilayer product with many "trapped" air pockets. These air pockets cause this product to function somewhat like traditional, fibrous insulation products. Even though this product may collect dust on its uppermost layer, the remaining layers and air spaces work to significantly reduce heat transfer through the ceiling assembly.

Some of these products may have R-values, which may be claimed only if the product was tested according to Federal Trade Commission regulations for insulation.

Although it is not by definition a radiant barrier, there is a low-emissivity paint available that can be applied directly to the underside of the roof decking.

Best Materials

While the Florida Solar Energy Center strongly recommends radiant barrier systems in attics, it doesn't endorse any particular brand of radiant barrier material. However, we suggest that you look for a few common-sense characteristics:

- Emissivity (the lower the better).
- Fire rating (as required by building codes).
- Ease of handling.
- Strength of reinforcement.
- Width appropriate for installation.
- Low cost.

COSTS

Costs for an attic radiant barrier depend on several factors, including the following:

- Whether purchase includes installation (which increases cost).
- Amount purchased (greater quantities can cost less per square foot).
- Manufacturing method and type of reinforcement.
- Presence of other insulation materials.
- Marketing methods.
- Aspects of supply and demand.

One other condition greatly affects the cost of a radiant barrier system to the homeowner—the individual's knowledge and willingness to do some comparison shopping. A few phone calls and a little research can save you money on most purchases. Radiant barriers are no exception.

Informal surveys show a wide range of material costs ($0.07–$1.00/ft^2) and installation costs ($0.10–$1.00/ft^2). The increases costs appear to be due more to marketing practices than to any inherent difference in thermal performance.

In some cases, radiant barriers are included in a package of energy-saving features sold to homeowners. When considering a "package deal," you may want to ask for an itemized list that includes material and installation costs for all measures included. Then shop around to see what each item would cost if you purchased them individually. You may see considerable savings. In addition, you may decide that you want to install the items yourself, including the radiant barrier.

FOIL SIDE

In attics, single-sided radiant barrier material should be installed with the foil side facing down. This may run counter to our intuitive feel for "how things work," but it does work, and work well.

To understand how it works, remember the two properties of aluminum foil from our Thanksgiving turkey and light bulb analogies; foil reflects radiant energy very well but does not radiate heat well. It does not emit heat to the cooler surfaces around it.

If you install a single-sided radiant barrier with the foil side facing up, the aluminum will (for a time) reflect the thermal energy radiated by the hot roof.

If you install a single-sided radiant barrier with the foil side facing down, the aluminum simply will not radiate the heat it gains from the roof to the cooler insulation it faces.

At first, a single-sided radiant barrier will work equally well with the foil facing up or down. However, over time, dust may accumulate on the surface of foil facing up. The dust will reduce the radiant barrier effect by allowing the foil to absorb rather than reflect thermal radiation. However, a radiant barrier with the foil side facing down will not collect dust on the foil and will continue to stop radiant heat transfer from the hot roof to the insulation over the life of the installation.

Even if you use a double-sided radiant barrier material, it is best to install it at the rafter level so that the bottom side faces the attic airspace and will not collect dust.

INSTALLATION

The most effective way to install a radiant barrier in an existing attic is simply to staple the foil material to the underside of the top chord of the roof trusses or to the underside of the roof decking.

It is not very easy to work in any attic, even one with a steep pitch. Always keep in mind that a misstep could be disastrous, since most attic "floors" are not floors at all, but rather 2×4s holding ceiling drywall topped by conventional insulation. You should consider safety first (see installation safety tips listed below).

Take care to avoid compressing existing insulation in the attic.

Tools and materials needed to install a radiant barrier include the following:

- Enough radiant barrier material to cover the underside of the roof.
- Measuring tape and flashlight.
- Heavy-duty scissors or utility knife.
- Staple guns and heavy-duty staples.
- Two movable support surfaces such as 3×2-ft sheets of 1 in. plywood or 3 ft lengths of 1×12 board.

Perhaps your most important aid will be a partner. Working in pairs in the attic makes the work go faster. Even more important, it adds to safety.

Begin by measuring the length of the attic roof from peak to soffit. Then, return to a stable, ground-or floor-level surface to measure and cut the radiant barrier material to size. The material usually comes in rolls of 50 to several hundred feet; it's easiest to cut and reroll all the lengths you'll need before returning to the attic.

At one end of the attic, place the plywood or 1×12 as a stable surface across two of the attic truss members. Try to minimize compression of existing insulation. Provide one surface at the peak and one at the soffit end so that both installers can work together.

Safety reminder: Be extremely careful at the sides of your support surface. If you step on an edge, the surface can tilt and drop you through the ceiling drywall below.

With your partner, unroll one length of the radiant barrier material from soffit to peak. Leaving 1 or 2 in. of free space at the roof peak, staple one corner of the material to the underside of the top chord of the first roof truss. Continue stapling the edge of the radiant barrier material down the truss at 6–12-in. intervals, stopping 2–3 in. from the ceiling insulation. Next, staple the other edge of the material to the underside of the adjacent roof truss. Continue the process at adjoining trusses until the underside of the roof is no longer visible except for a 1- or 2-in. strip at the roof peak.

As an alternative, you may staple the radiant barrier material to the underside of the roof decking, adjacent to the top chord of the truss. The weight of the material will allow it to drape naturally between trusses.

Safety Tips for Installation

- If you use a ladder for access to the attic, make sure it is stable and tall enough for easy entry and exit.
- Work in the attic only when temperatures are reasonable. Attic daytime temperatures can rise far above 100°F during much of the year in the Sunbelt. Install your radiant barrier system early in the morning, or wait until cool weather sets in.
- Work with a partner. Not only does it make the job go faster, it also means that you'll have aid should a problem occur.
- Watch where you walk and use a movable support surface. Step only on the attic trusses or rafters and your working surface. Never step on the attic insulation or the ceiling drywall below it.
- Step and stand only on the center of your movable working surface. Don't step on the edge; it can cause the surface to tip.
- Watch your head. In most attics, roofing nails penetrate through the underside of the roof. If you bump your head, it can cause a serious cut or puncture. If your skin is punctured by a nail, an up-to-date tetanus vaccination is a must. Avoid potential problems by wearing a hard hat.
- Be especially careful around electrical wiring, particularly around junction boxes and older wiring. Never staple through or over electrical wiring.
- Make sure that the attic space is well ventilated and well lighted. Bring in fans and extra work lights if necessary.
- If your attic has blown-in insulation, direct fans upward, away from the insulation material.
- Avoid exposure to mineral fiber insulation. Wear goggles, long pants, a long-sleeved shirt, and a particle mask or kerchief over your nose and mouth. Wear gloves if you are particularly sensitive to fiberglass.
- Wear a tool belt or utility apron to carry staples, staple gun, scissors, measuring tape, etc.
- Take frequent breaks, and pace yourself. It's better to get the job done over a longer period than to risk an accident due to fatigue or to end up with a poor-quality installation.

Airtightness

You're installing a barrier against radiated not convected heat, so you need not cut off air motion. In fact, ventilation from soffit to peak improves radiant barrier system performance.

Small tears and holes will not significantly lessen the performance of the radiant barrier, so don't worry if you must cut and patch around obstructions such as vent stacks and truss supports.

Placement

It's not recommended to place the material directly on top of insulation. In this type of installation, dust will accumulate on the foil surface facing the roof. In time, the dust will negate the radiant barrier effect. In addition, problems could develop with moisture condensation. An exception to this is the "chip" and multi layer products on the market. Their multiple layers give them resistance to the effects of duct build up, making them suitable for floor installations.

Installation in New Construction

There are two widely accepted methods of installing attic radiant barrier systems in new construction:

- Attach the radiant barrier to the roof decking before it is installed on the trusses, or
- Attach the radiant barrier to the roof deck or truss chords after the roof decking is installed, but prior to the installation of the ceiling drywall.
- Perhaps the easiest way to install a radiant barrier in a new house is to use one of the foil-faced roof sheathing materials on the market. The price increase for these products should be minimal, and the installation does not differ from regular decking, except for a little additional care in the handling of the product.

Note: When installing a single-sided radiant barrier material, remember to face the foil side down toward the attic floor/insulation.

FOIL-FACED BATT INSULATION

While some conventional batt-type insulations have an aluminum foil backing, it's probably not a good idea to simply flip your insulation over to use it as a radiant barrier. Not only will you encounter the eventual dust problem, you may also encounter a fire hazard in the glue that bonds the foil to the batt.

At least one batt insulation manufacturer has introduced a product with a foil face that is bonded to the insulation with a fire-retardant glue. This product meets fire codes, but it still has the potential for dust problems.

HEAT BUILDUP

The Florida Solar Energy Center has measured the temperatures of roof shingles above attic radiant barriers on hot, sunny summer days. Depending on the color of the shingles, their peak temperatures are only 2–5°F higher than the temperature of shingles under the same conditions without a radiant barrier.

Roofing materials are manufactured to withstand the high temperatures to which they are frequently exposed. A 2–5°F increase in peak temperatures that normally reach 160–190°F should have no adverse affect.

SHINGLE WARRANTIES

Shingle warranties should not be subject to cancellation by the manufacturer on the basis of radiant barrier installation. However, it may be wise to review the warranty to be sure that work of this nature will not void it. You may want to inquire directly of the manufacturer. Any changes in warranty should be substantiated in writing.

Reshingling

Remember, to be a radiant barrier, the aluminum foil must be installed facing *an airspace*. If there is no airspace, the foil acts as a conductor and quickly passes heat by conduction from a hot surface to a cooler one.

If your re-roofing job requires decking replacement, however, it is a good time to consider a radiant barrier, especially the foil-faced roof sheathing materials on the market.

DECREASING HEAT GAIN

While a radiant barrier is one effective way of reducing heat gain through attics, it's not the only one. Other options include:

- Continuous peak and soffit or gable vents (which also improve radiant barrier system performance).
- Light-colored shingles or white or light metal or tile roof coverings.
- Additional conventional insulation.

If you shop carefully, you will probably find that attic radiant barriers are one of the least costly and yet most effective of the attic conservation measures for Southern climates.

PAYBACK

Two things affect the performance of a radiant barrier system—the level of insulation in the attic, and the geographic location of the home. A radiant barrier system reduces the heat flow into the house from the attic by approximately 40%. Attic insulation levels have a large effect on the amount of heat flow that is reduced, in other words, if you have little or no insulation in your attic, a 40% reduction is very significant, but if your attic is insulated to R-30 or better, there is very little heat flow to reduce. The more of your energy bill that is concerned with heating, the less desirable having a radiant barrier becomes. When you are using your heater, any heat gain from the attic is desirable. There may be a reduction of heat loss through the roof during winter nights, but the climates where testing has been performed do not lend themselves to demonstrating this, as there is very often a brief or non-existent winter. A very helpful website for guidance to the cost effectiveness of installing a radiant barrier is found on Oak Ridge National Laboratory's web site, http://www.ornl.gov/sci/roofs+walls/radiant/rb_02.html.

In Florida, computer studies conducted in the development of the Florida Model Energy Code indicate that a typical attic radiant barrier installed in a Florida home will offer a 6–7 yr simple payback and a 15–19% return on investment.

CONCLUSIONS

Attic radiant barriers are an inexpensive but effective way for Sunbelt homeowners to save energy and money. While they are not a new concept, radiant barriers have only recently been proved effective for energy conservation.

Manufacturers are continuing to improve radiant barrier materials, which are becoming widely available throughout the southern states.

A radiant barrier may be installed in an existing attic or during construction of a new home. Both are relatively easy procedures.

ACKNOWLEDGMENT

Publication of this document was supported in part by Florida Power & Light Company.

BIBLIOGRAPHY

1. Parker, D.S.; Sonne, J.K.; Sherwin, J.R. Flexible Roofing Facility: 2002 Summer Test Results, Prepared for: U.S. Department of Energy Building Technologies Program, July 2003.
2. Parker, D.S.; Sherwin, J.R.; Anello, M.T. *FPC Residential Monitoring Project*: New Technology Development—Radiant Barrier Pilot Project, Contract Report FSEC-CR-1231-01, Florida Solar Energy Center, Cocoa, Florida, January 2001.
3. Parker, D.; Vieira, R. *Priorities for Energy Efficiency in New Residential Construction in Florida*; Florida Solar Energy Center: Cocoa, Florida, 2000; February.
4. Parker, D.; Sherwin, J. Comparative Summer Attic Thermal Performance of Six Roof Contructions, The 1998 ASHRAE Annual Meeting, Toronto, Canada, June 20–24, 1998.
5. Parker, D.; Sherwin, J.; Sonne, J.; Barkaszi, S.; Floyd, D.; Withers, C. Measured Energy Savings of a Comprehensive Retrofit in an Existing Florida Residence, For the Florida Energy Office, December 1997.
6. Fairey, P. Designing and Installing Radiant Barrier Systems, FSEC-DN-7, Florida Solar Energy Center, Cape Canaveral, FL, 1984.
7. Fairey, P. Effects of Infrared Radiation Barriers on the Effective Thermal Resistance of Building Envelopes, Proceedings of the ASHRAE/DOE Conference on Thermal Performance of the Exterior Envelopes of Buildings II, Las Vegas, NV, December 1982.
8. Fairey, P. The Measured Side-by-Side Performance of Attic Radiant Barrier Systems in Hot-Humid Climates, Proceedings of the 19th International Thermal Conductivity Conference, Cookeville, TN, October 1985.
9. Fairey, P. Radiant Energy Transfer and Radiant Barrier Systems in Buildings, FSEC-DN-6, Florida Solar Energy Center, Cape Canaveral, FL, 1984.
10. Joy, F.A. Improving attic space insulating values. ASHAE Trans. **1958**, *64*.
11. Levins, W.P.; Karnitz, M.A. Cooling-Energy Measurements of Unoccupied Single-Family Houses with Attics Containing Radiant Barriers, Oak Ridge National Laboratory, Contract Report, DE-ACO5-840R21400, July 1986.
12. Radiant Barriers: How They Work and How to Install Them, videotape, FSEC Producer, Cape Canaveral, FL, 1986.
13. Van Stratten, J.F. *Thermal Performance of Buildings*; Elsevier Publishing: New York, 1967.

Reciprocating Engines: Diesel and Gas

Rolf D. Reitz

Kevin Hoag
Engine Research Center, University of Wisconsin-Madison, Madison, Wisconsin, U.S.A.

Abstract
This entry first identifies the energy conversion principles that govern reciprocating piston devices. Next, the hardware and operating cycles used in practical reciprocating piston internal combustion engines are described. Finally, current issues, trends, and analysis approaches used in developing combustion and air handling systems are discussed.

INTRODUCTION

The goal of any engine is to convert the energy contained in a fuel into useful work as efficiently and cost-effectively as possible. Engines can be classified into various families based on how this conversion is accomplished. The internal combustion engine is one such family; in this family, a combustion reaction between a fuel and oxygen occurs within a mechanical device designed to harness the energy and produce work. This discussion addresses one mechanical configuration in the family of internal combustion engines. Other mechanical configurations of practical internal combustion engines are the gas turbine and rotary (Wankel) engine.

ENERGY CONVERSION AND RECIPROCATING PISTON DEVICES

Consider a sealed chamber in which a combustion reaction has just occurred, as shown in Fig. 1A. When a fuel reacts with oxygen, the chemical energy contained in the fuel is converted to sensible energy, which is physically sensed as a temperature increase in an open flame, and if the chamber is confined as shown here, an increase in temperature and pressure. The state within the chamber after combustion is indicated on coordinates of pressure and specific volume in the figure.

One can now begin considering various approaches for energy conversion. The possibilities include:

- Heat transfer from the chamber supplying energy to another device
- Mass transfer from the chamber and flow through a turbine
- Work transfer across a moveable chamber wall

Each possible approach follows some process line, as shown in Fig. 1B, which will ultimately bring the combustion gas to equilibrium with the outside environment. At equilibrium, the gas will have the same pressure, temperature, and chemical make-up as its surroundings. Unfortunately, any practical engine will release the gas to the environment before equilibrium is completely attained.[1] In other words, the temperature, pressure, or mixture concentration will still differ from the environment, but it becomes impractical to recover any further work.

If the curves shown were extended sufficiently to the right, each of these processes would have a different temperature when reaching atmospheric pressure. If work is being produced throughout the chosen process, intuition suggests that the process with the lowest temperature when atmospheric pressure is reached will be most efficient. The ideal process would simultaneously reach both pressure and temperature equilibrium with the environment. However, according to the Second Law of Thermodynamics, such a process does not exist.[1] The very best process—the highest efficiency process—is the one in which constant entropy is maintained. The temperature of the gas when atmospheric pressure is reached is still above atmospheric temperature, but all observations have demonstrated that a more efficient process cannot be achieved.

The reciprocating piston engine is one in which work transfer occurs across a moving combustion chamber wall (piston), as depicted in Fig. 1C. The pressure in the chamber forces the piston to the right, and work is performed on the piston. An important attraction of this moving piston device is that it has been found to follow a process line very closely resembling the isentropic process.

A very important practical concern arises: expansion to atmospheric pressure requires moving far to the right on the pressure versus specific volume coordinates. Because there is no change in the gas mass, the specific volume scale can be replaced with a volume scale, and it must be

Keywords: Thermodynamics; Piston engines; Energy conversion; Internal combustion engines; Spark ignition; Compression ignition.

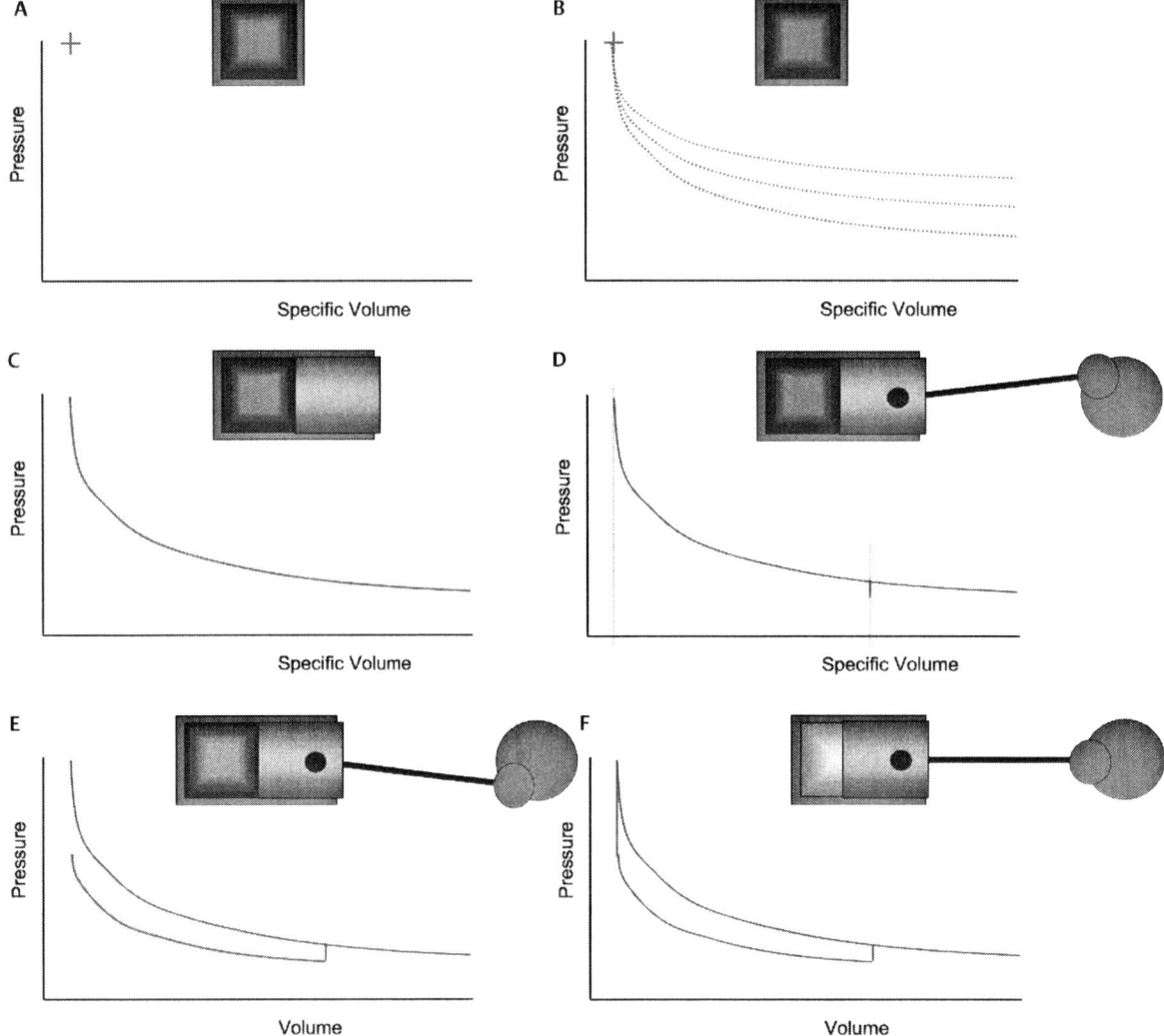

Fig. 1 Thermodynamic idealizations of reciprocating piston engine processes.

concluded that expansion to atmospheric pressure requires a very long chamber.

Fortunately, the isentropic process is quite non-linear. The pressure drops rapidly and the majority of the work extraction occurs early in the process. For practicality in regard to engine packaging, piston movement is stopped at some chosen maximum volume, and the remaining gas is expelled from the chamber.

In the arrangement shown in Fig. 1D, the piston is linked through a connecting rod to a crankshaft. This arrangement results in a defined maximum volume and the opportunity to create a repeated series of processes in which the piston is cycled between this maximum and some minimum volume. Thus, with suitable arrangements for bringing in a new mixture of fuel and oxygen and initiating combustion, work can be repeatedly produced. This arrangement also provides for convenient work extraction through the spinning crankshaft.

Fig. 1E shows a compression process. The linkage arrangement returns the piston to minimum chamber volume with a new charge in the cylinder. Work is required to compress the new charge, but here, too, the moving piston results in compression in a nearly 100% isentropic efficiency process.[1]

The cycle concludes as shown in Fig. 1F—with another combustion reaction from which the processes are repeated. This has been a conceptual discussion that demonstrates process idealizations. Much of engine development involves controlling the actual, practical processes to accomplish an engine operating cycle, and minimizing the deviations of the actual processes from the ideals. Practical processes for transferring fresh fuel and oxygen into the chamber, initiating and controlling the combustion process, and transferring the reacted gases out of the chamber are fundamental to the work of engine development.

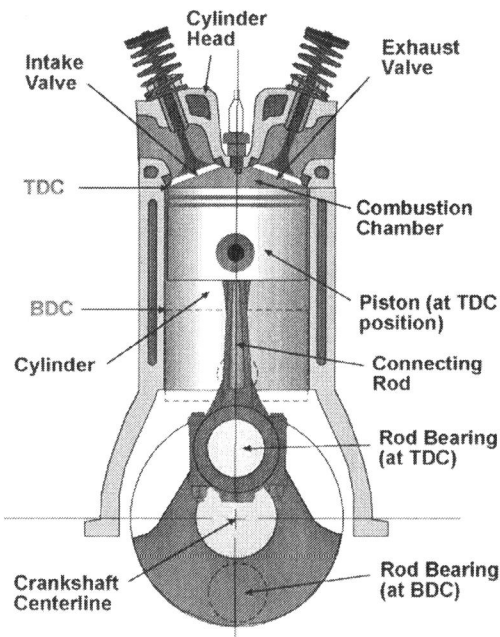

Fig. 2 Key components of reciprocating piston engines.

PRACTICAL ENGINE HARDWARE AND OPERATING CYCLES

The latter discussion identified the rationale behind the reciprocating piston engine. While many other mechanical configurations have been proposed, patented, and demonstrated over the years, few have enjoyed commercial success. Such success results from the ability to address both efficiency and cost-effectiveness.

The primary components of the reciprocating engine are shown in Fig. 2. The moving piston controls the volume of the combustion chamber between a minimum at top dead center (TDC) and a maximum at bottom dead center (BDC). The ratio between the volume at BDC and TDC is referred to as the compression ratio. The change in volume is the displacement of the cylinder, which, when multiplied by the number of cylinders, is the engine displacement.

The cylinder is sealed by the cylinder head opposite the moving piston. In most engines, the intake and exhaust valves are located in the cylinder head, as shown. The piston is linked to the crankshaft through a connecting rod. As the crankshaft spins about its centerline (the main bearing bore in the cylinder block), the offset of the rod bearing from the main bearing determines piston travel. As the crankshaft rotates a half revolution from the position shown in the figure, the piston moves from its TDC to its BDC position. The distance the piston travels is the engine's stroke, equal to twice the offset between the main bearing and rod bearing centerlines of the crankshaft. The cylinder diameter is termed its "bore"; the combination of bore and stroke determines the cylinder displacement.

Actual engine operation, based on the idealized operating cycle, can be further explained with reference to the pressure-volume diagrams of Fig. 3. The diagram in Fig. 3A is comprised of a four-stroke engine, while the Fig. 3B diagram depicts a two-stroke engine. The four-stroke engine requires four strokes of the piston—or two revolutions of the crankshaft—for completion of each cycle. The additional revolution is used to exhaust spent combustion products on the upward stroke of the piston, and draw in fresh reactants on the downward stroke. These additional strokes result in the relatively low pressure "pumping" loop at the base of Fig. 3A. The four strokes can be summarized as follows:

- Intake—the piston moves from TDC to BDC with the intake valve open, drawing fresh reactants into the cylinder.
- Compression—the valves are closed and the piston moves from BDC to TDC, compressing the reactants. The combustion reaction is initiated as the piston approaches TDC, increasing the pressure and temperature in the sealed chamber.
- Expansion—the high pressure forces the piston from TDC to BDC, transferring work to the crankshaft.
- Exhaust—the exhaust valve opens and the piston moves from BDC to TDC, forcing the spent exhaust products out of the chamber.

In the two-stroke engine, the operating cycle is completed in a single crankshaft revolution. Gas exchange is completed as the piston approaches BDC. Intake and exhaust passages are simultaneously opened, and a pressurized intake system is used to force the exhaust products out and fill the cylinder with a fresh charge in this scavenging process. The scavenging process results in the small process loop at the right of Fig. 3B (near BDC volume).

Today, considerable effort is devoted to gas exchange or "breathing" processes in engine development. Design goals include minimizing flow restrictions, maximizing spent products removal, and maximizing cylinder-filling with fresh reactants. Note that conventional spark-ignited gasoline engines operate with a premixed, nearly homogeneous fuel–air mixture. Load is controlled by throttling because of the relatively narrow ignitability limits of gasoline-air-residual mixtures, and this incurs substantial pumping losses at part load.

Another important aspect of engine development pertains to the combustion process. The idealization of Fig. 1F assumes an instantaneous combustion reaction, while the piston is at TDC. While this would result in maximum efficiency, real combustion processes are rapid, but not instantaneous. This fact, and the need to manage the peak pressure seen by the engine's structure, result in

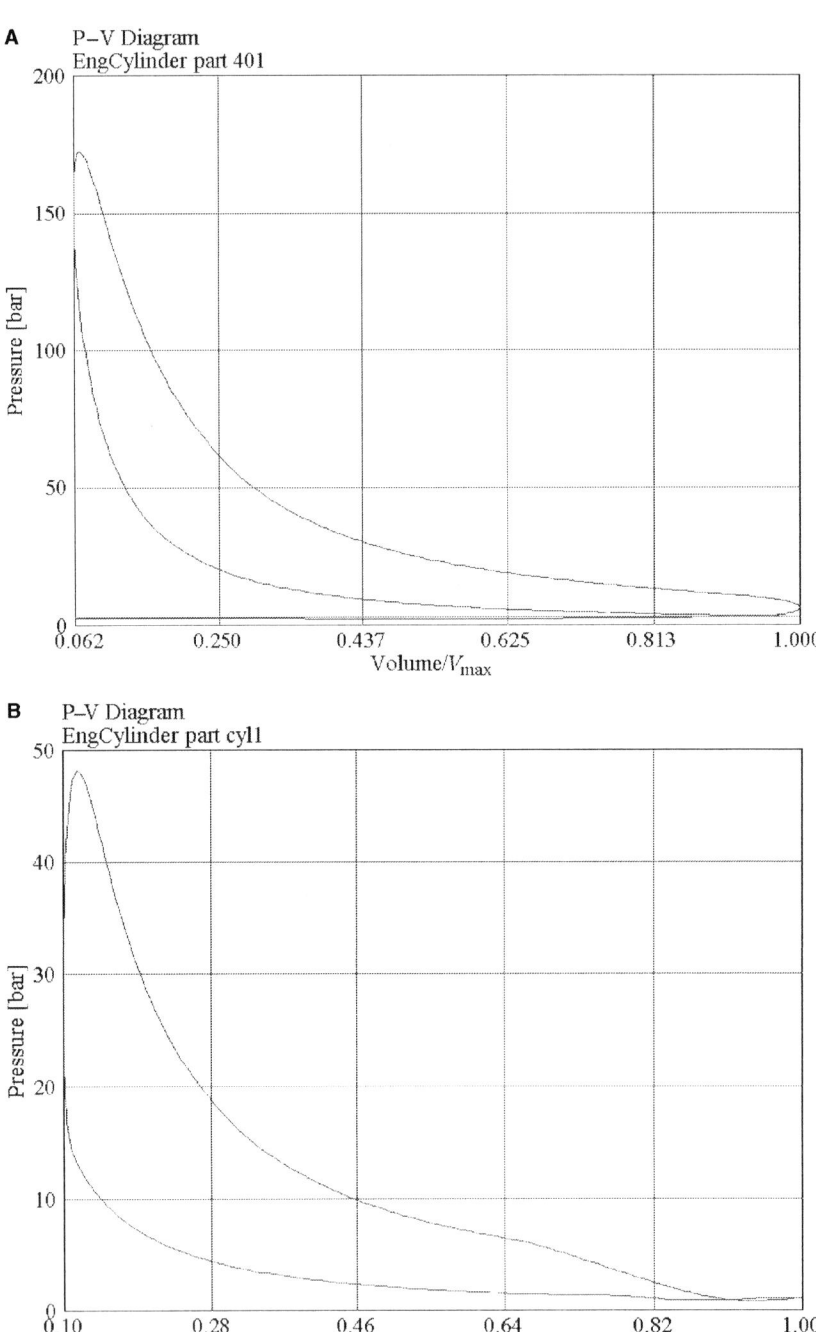

Fig. 3 (A) Four-stroke diesel pressure-volume diagram at full load. (B) Two-stroke spark-ignition pressure-volume diagram at full load.

combustion processes that are initiated slightly before TDC as the piston nears the end of the compression stroke. Combustion continues as the piston moves down during the expansion stroke.

Two approaches are prevalent in production engines. In a conventional spark-ignition engine, a mixture of air (the oxygen carrier) and fuel are drawn into the combustion chamber during the intake process. The mixture is compressed, and combustion is then initiated using a high-energy electrical spark. In the compression-ignition (diesel) engine, air alone is drawn into the cylinder and compressed. Fuel is injected directly into the cylinder near the end of the compression process. The fuel used in the compression-ignition engine is intended to easily spontaneously ignite when exposed to the temperature and pressure of the compressed air. While the diesel engine is often portrayed as having a slower combustion process (constant pressure instead of constant volume in the idealization of Fig. 1F), the goal of rapid combustion near TDC for maximum efficiency applies to both diesel and spark-ignition engines.

In order to increase the specific power output of an engine (power output per unit of displacement), some form of pre-compression (supercharging) is often considered. This is rapidly becoming standard practice in diesel engines, and is often seen in high-performance spark-ignition engines.

A crankshaft-driven compressor may be added to elevate the pressure of the air (or mixture) prior to drawing it into the cylinder. This allows more mixture to be burned in a given cylinder volume. Recognizing that the exhaust gases leaving the engine still contain a significant quantity of energy, an alternative is to use a portion of this energy to drive the compressor. This configuration is the turbocharged engine.

When air is compressed, its temperature increases. Its density can be further increased (and still more air forced into the cylinder) if it is cooled after compression. The charge air cooler, variously termed intercooler (cooling between stages of compression) and aftercooler (cooling after compression) may be used with either the turbocharger or crankshaft-driven supercharger.

ENGINE ANALYSIS

The remaining sections of this entry address analysis and development of practical engines, emphasizing engine breathing and combustion process optimization to meet efficiency and exhaust emission control needs. After a century of development it might be believed that the internal combustion engine would have little potential for further improvement. Nevertheless, new emissions standards have been promulgated that require the consideration of new technologies.[2] In response, engines continue to show substantial improvements in efficiency, power, and reduced emissions. Some of these advances have been due to the use of detailed analysis tools. Recent advances in computer speed and model development make the use of modeling increasingly attractive for engine design. Analysis tools include zero-dimensional thermodynamic models to analyze the in-cylinder combustion heat release, one-dimensional flow models to design the air handling and fuel injection systems, and multi-dimensional models that are useful in optimizing in-cylinder combustion and engine coolant systems.[3]

Gas Exchange Process

The gas exchange process controls the power achievable from the engine.[1] A conventional premixed-charge gasoline (i.e., spark-ignition) engine's intake system consists of the air filter, carburetor, and throttle plate or port fuel injector, intake manifold, intake port, and intake valves. Supercharging can be used to increase air mass into the cylinder in both gasoline and diesel engines. Spatial and temporal variations in flow and pressure throughout intake and exhaust manifolds must be considered to maximize cylinder filling and scavenging.

Intake system pressure drop occurs due to quasi-steady effects (flow resistances), and unsteady effects (wave action in the runners). Wave propagation is exploited in engine tuning and can be modeled effectively using commercial one-dimensional fluid flow software. Modeling is also needed to ensure equal airflows to each cylinder in a multi-cylinder engine. Engine breathing is affected by the intake and exhaust valve lifts and open areas (see Fig. 4), where much of the loss occurs, and valve overlap, which can cause exhaust gases to flow back into the intake system or intake gases to enter the exhaust (depending on the intake/exhaust pressure ratio), especially at low engine speeds.[4] At high engine speed, valve overlap can improve breathing since the inertia of the flowing gases can cause inflow into the cylinder even during the compression stroke. With Variable Valve Actuation (VVA) technologies, valve timing can be used to control the effective compression ratio (through early or late intake valve closure), or with exhaust gas re-induction (re-breathing) to control in-cylinder temperatures. Residual gas left from the previous cycle affects the engine combustion processes through its influence on charge mass, temperature, and dilution.

Accurate descriptions of valve flow losses require consideration of multidimensional flow separation

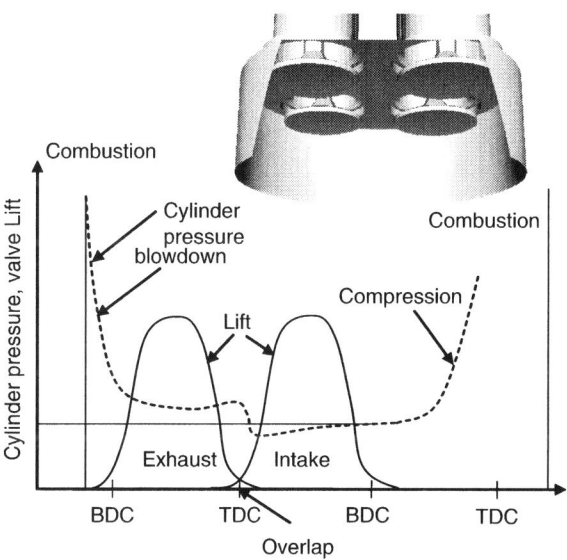

Fig. 4 Intake and exhaust valve arrangements for diesel engine.

Table 1 Heavy-duty diesel engine specifications

Engine details	Caterpillar 3401	Cummins N14
Cylinder bore (mm)	137.2	138.3
Stroke (mm)	165.1	152.4
Compression ratio	16.1	15.1
Displacement /cyl (l)	2.44	2.48
Fuel injection system		
Injectors	6 or 8 holes, 0.26 mm diameter	
Caterpillar port/valve data		
Exhaust valve opening	$-217°$ ATDC	
Intake valve opening	$220°$ ATDC	
Intake valve closing	$-143°$ ATDC	
Intake valve diameter	45 mm	
Exhaust valve diameter	41.8 mm	
Maximum valve lift	11.0 mm	
Intake port diameter	40.38 mm	
Exhaust port diameter	37.21 mm	
Port lengths	130 mm	

phenomena and their effect on the conditions within the cylinder at intake valve closure.[5] A typical valve timing diagram and predictions of the in-cylinder flow during the gas exchange process for a heavy-duty diesel engine (see Table 1) are shown in Figs. 4 and 5, respectively.[6] Fig. 5 shows velocity vectors and residual gas mass distributions in the engine just as the intake valves are about to close. The highest mixing of incoming fresh charge and combustion products occurs where the intake flow velocities are the largest due to high flow turbulence.

DIESEL ENGINES

Improvements in diesel engine technology in the last decade have been due to factors such as the introduction of high-pressure electronic fuel injection systems and multistage turbocharging, guided by insights obtained from computer modeling. Simulating the diesel engine represents one of the most complex and comprehensive modeling problems conceivable. The Navier–Stokes equations must be solved with consideration of when moving boundaries, turbulence, heat transfer (including radiation), sprays, combustion and emissions, and commercial and open source computer codes[7] are available.

Computer models are not entirely predictive due to the wide range of length and time scales needed to describe engine fluid mechanics. For example, a modern truck diesel engine operates with injection pressures as high as 200 MPa, with injector nozzle holes less than 200 μm in diameter. Thus, the fuel jet enters the high-temperature combustion chamber gas close to sonic velocity (~ 600 m/s) and breaks up into droplets with diameters less than 10 μm in microseconds. To begin to resolve these processes in a combustion chamber of bore 100 mm, more than $(10^4)^3 = 10^{12}$ grid points would be required. Current, practical computer storage capabilities are limited to about 10^5 or 10^6 grid points. Accordingly, sub-models are used to describe unresolved processes,[8] as summarized in Table 2.

The combustion chamber surface temperature is needed to specify wall boundary conditions (see Fig. 6[19]). In the model, heat flux from the gas-side is used for the metal component heat-conduction prediction. Localized high temperature regions are predicted near the piston bowl rim in agreement with engine experimental data.

Following fuel injection and atomization, injected fuel droplets vaporize and mix with the compressed air. Auto-ignition takes place and the burning rapidly consumes the premixed mixture. Thereafter, the burning that occurs while the injection continues occurs via mixing-controlled or diffusion-type combustion. Fig. 7A shows predicted spray plumes and soot formation regions for a Caterpillar diesel engine (see Table 1). Good agreement between measured and predicted in-cylinder pressures and soot and NO_X emissions are shown in Figs. 7B and 8, respectively.[5,12]

Fig. 8 also shows that staged or multiple injections are effective for emissions reduction, and the mechanism of emission reduction is summarized in Fig. 9.[20] Soot accumulates at the tip of the spray jet (see also Fig. 7A), as also confirmed experimentally in optical engines.[21] With single injection, the high-momentum injected fuel penetrates to the fuel-rich, relatively low-temperature jet tip and continuously replenishes it, producing soot. With a split-injection, the soot cloud of the first plume is not replenished with fresh fuel, but continues to oxidize.

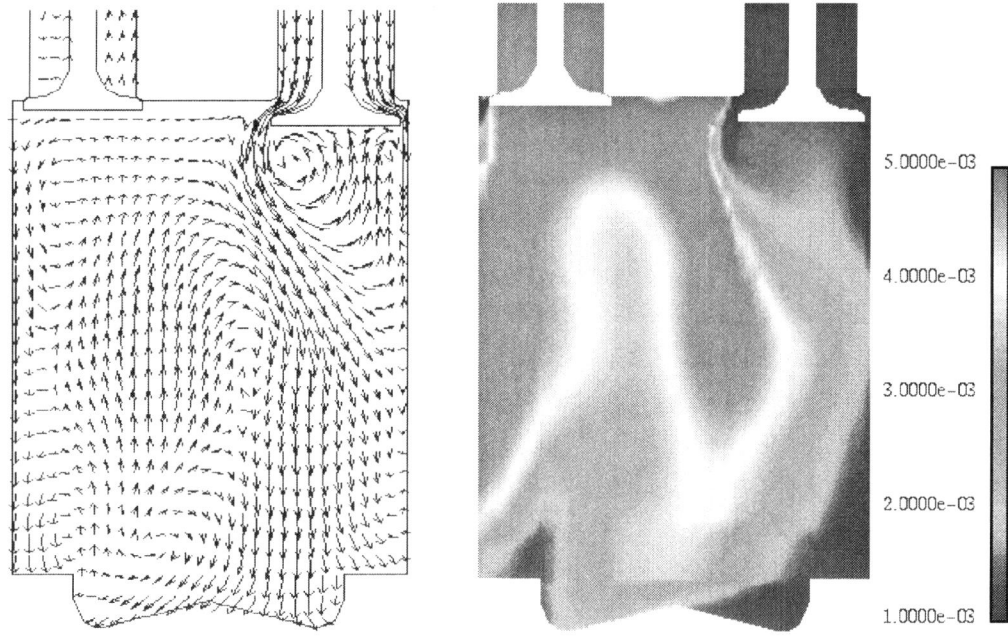

Fig. 5 Computed flow velocities (left) and residual gas distribution (right) during gas exchange in plane of valves (144 degrees ATDC—1600 rev/min)
Source: Reprinted with permission from SAE Paper 962052 © 1996 SAE International.

GASOLINE ENGINES

Gaseous pollutant emissions (NO_X, CO, and HC) from gasoline engines have been controlled effectively using the three-way catalyst. However, in order to permit both reductions of NO_X and the oxidation of hydrocarbons and CO, combustion must occur at the chemically balanced (stoichiometric) air–fuel ratio within very narrow limits. Issues such as cold-start emissions, fuel economy, and engine responsiveness have been the main drivers for further advancements. The fuel system of spark-ignition engines has evolved from carburetion to throttle-body injection, followed by Port Fuel Injection (PFI), including sequential-fire injection.

During cold start, a film of liquid fuel can form in the intake valve area of the port and some portion of it is drawn into the cylinder during induction, as shown in Fig. 10A. The fuel delivered to each cylinder can also differ from that metered by the injector, causing a fuel delivery delay and an associated metering error. Thus, extra fuel may be needed for cold starting, causing increased unburned hydrocarbons emissions.

With Gasoline Direct Injection (GDI), which is also called Direct Injection Spark Ignition (DISI), fuel is injected directly into the combustion chamber (Fig. 10B). This leads to higher fuel efficiency, better drivability, and better cold-start performance.[25] However, GDI engines suffer from higher hydrocarbon emissions at light-load, hampering introduction in some markets.

There are three GDI operation régimes.[25] The simplest use early injection to create a homogeneous stoichiometric mixture, and throttling for load control. Cooling due to vaporization of injected fuel allows for the use of higher compression ratios and lower octane fuels without knock (uncontrolled fast combustion). The next régime uses a lean homogeneous mixture with reduced throttling for greater load control. In practice, lean homogeneous operation is less desirable than heavy

Table 2 Example physical submodels used in diesel engine computer modeling

Physical process	Submodel
Turbulence	RNG k-ε [9]
Heat transfer	Log-law [9]
Nozzle flow	Cavitation [10]
Atomization	Jet instability [11]
Drop breakup/coalesence	Drop instability [7,12]
Drop distortion	Drop drag [12]
Hydrocarbon ignition	Reduced [13]/detailed chemistry [14]
Diesel combustion	Turbulent/chemistry timescale [15,16]
NOx emissions	Zeldo'vich/detailed [15]
Soot emissions	Formation/Oxidation [16–18]

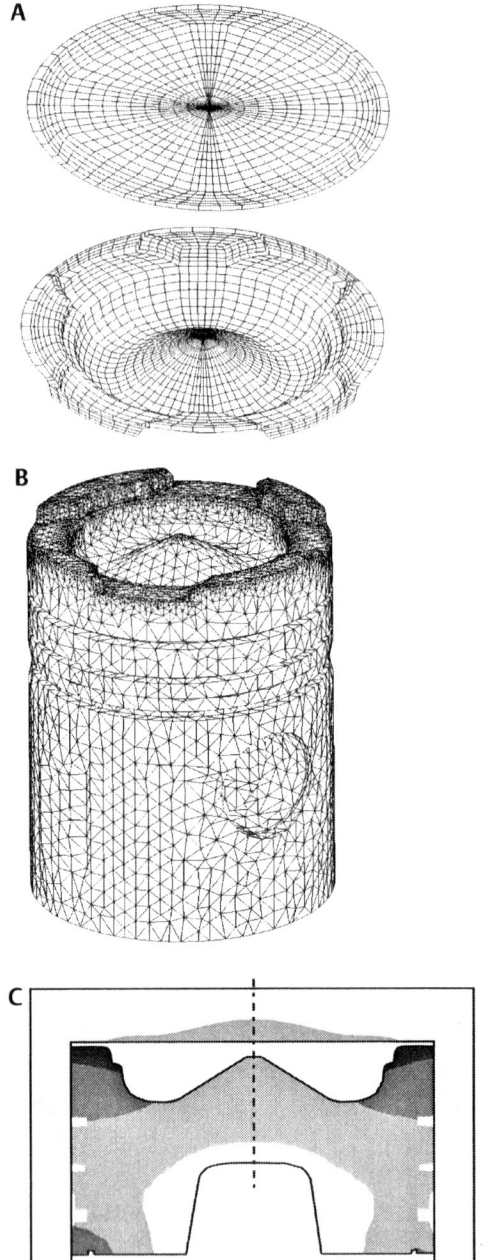

Fig. 6 Cummins diesel engine (A) in-cylinder CFD surface mesh, (B) finite element piston mesh, (C) predicted temperature distribution on head, liner, and piston.
Source: Reprinted with permission from SAE Paper 2003-01-0560 © 2003 SAE International. (see Ref. 19).

stratified mixture at part load and late-injection. This strategy should smoothly transition to homogeneous charge operation by injecting a larger mass of fuel early in the cycle. Engines utilizing all three GDI modes require complex control systems.

Accurate fuel delivery makes it possible to operate with more dilute mixtures and less cycle-to-cycle variation. Visualization experiments have provided substantial insights into mixture preparation and the causes of cyclic variability in ignition and combustion.[25,26] An inherent disadvantage of lean and stratified GDI engines is that with light-load operation, a three-way catalyst cannot be used effectively for NO_X control due to excess air in the exhaust stream. In addition, since the location of the ignition source is fixed, the mixture cloud must be controlled both temporally and spatially for optimal spark ignition, requiring precise matching of fuel injection and in-cylinder flow fields. CFD modeling has been useful for design,[24] and spark-ignition kernel growth and flame propagation models based on front tracking techniques, or flamelet models have been developed for the simulations.[27]

Two mixing concepts have been proposed: the "wide spacing" concept,[28] where the injector is side-mounted and fuel is injected toward the piston surface and deflected toward the spark plug (see Fig. 10C).[24] Spray momentum is independent of engine speed, and is the driving force for air–fuel mixing; adequate mixing can thus be realized over a wide range of engine speeds. The "close spacing" or "spray-guided" concept, with central injector placement (see Fig. 10B), is often seen in homogenous charge GDI engines,[23] but is also under development as a means to increase the speed-load range over which highly stratified operation can be maintained.[29] To improve the transition between low load and high load operation, optimally timed split injections during intake and compression have been proposed.[30]

HCCI ENGINES

There is great current interest in Homogeneous Charge, Compression Ignition (HCCI) engines for both light-duty automotive (gasoline and diesel) and heavy-duty diesel engines due to its potential for very low emissions and high fuel efficiency. In these engines, a mixture of fuel, air, and recycled combustion products is compressed until it auto-ignites.[31] Unlike diesel engines, combustion is not controlled by the fuel injection, and unlike spark-ignited engines, no discernable flame front is evident. At light load, low temperature combustion leads to more than 90% lower NO_X emissions compared to conventional diesel engines.[32] Low soot emissions are also generally found. Unburned hydrocarbon emissions can increase with fuel deposition or lean operation (at low loads), but can be

dilution with Exhaust Gas Recirculation (EGR) to reduce throttling since overall lean operation precludes the use of a three-way catalyst and requires more elaborate lean-NO_X after-treatment approaches, such as storage catalysts.

The greatest benefit of GDI is operation in the third regime, which is GDI sans throttling with an overall lean

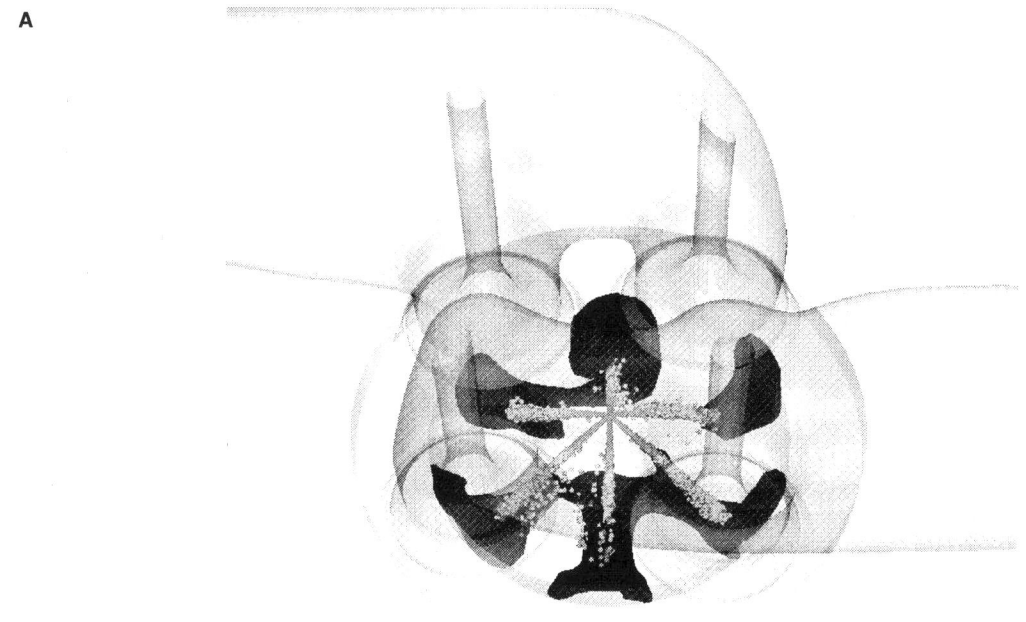
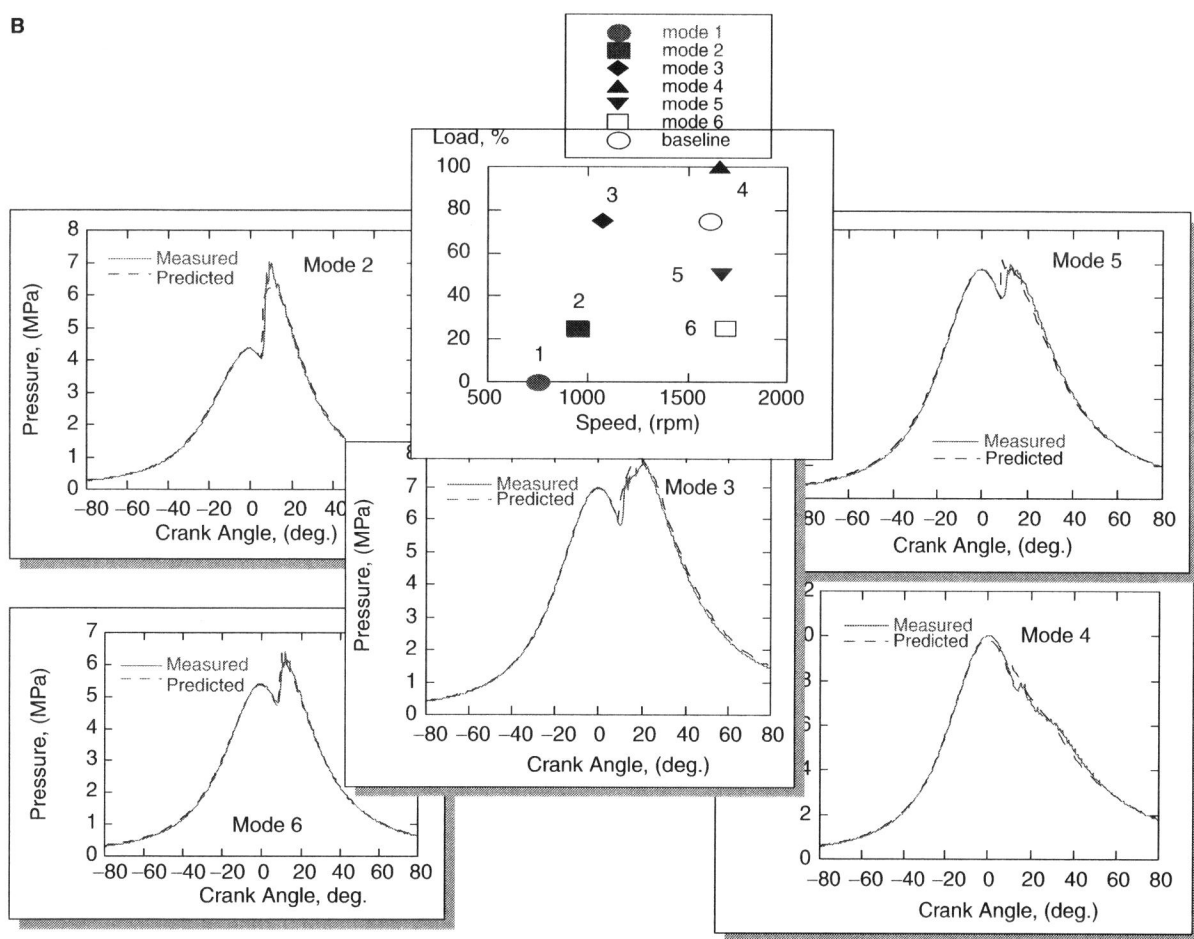

Fig. 7 (A) Fuel sprays with soot iso-surfaces (dark region at spray tips) during combustion, (B) measured and predicted cylinder pressure at steady-state test points for a heavy-duty diesel engine.
Source: Reprinted with permission from SAE Paper 2000-01-1178 © 2000 SAE International.

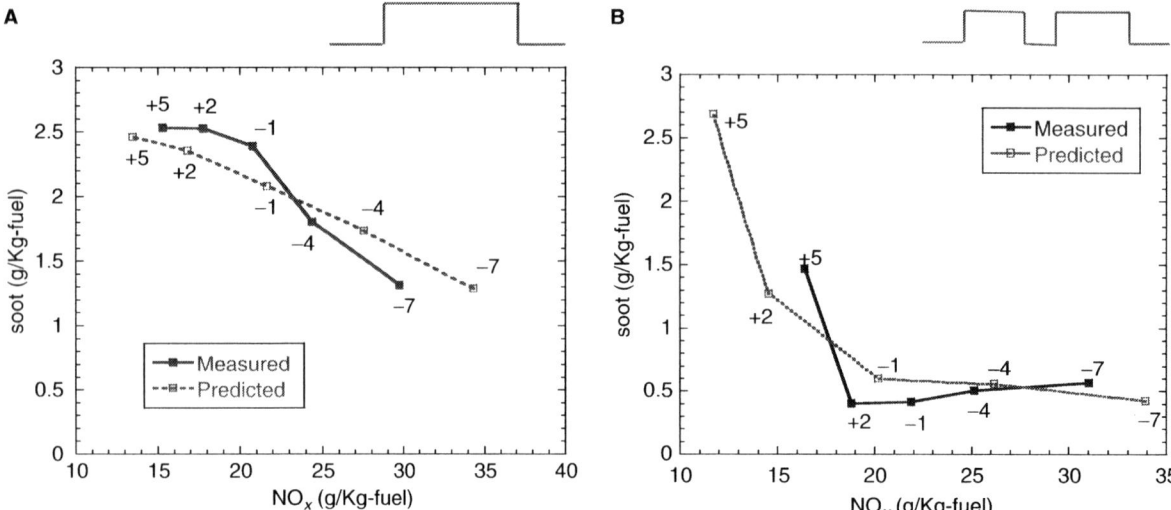

Fig. 8 Measured and predicted soot-NOx tradeoff for a heavy-duty diesel engine. (A) single and (B) double injection (1600 rev/min, 75% load). Numbers indicate start-of-injection timing in crank-angle degrees, where negative numbers are before TDC in the compression stroke and positive numbers are after TDC.
Source: Reprinted with permission from SAE Paper 2000-01-1178 © 2000 SAE International (see Ref. 5).

reduced with available exhaust gas oxidation catalysts. Fuel impingement can be reduced with low pressure, low penetration sprays. The requirement to use lean air–fuel mixtures limits the achievable power, favoring the use of high boost.

Challenges to HCCI operation include the fact that there is no direct method for controlling the start-of-combustion timing, leading to poor fuel efficiency. Ignition is controlled by chemical kinetics, and influenced by the fuel composition, mixture stoichiometry, and the thermodynamic state of the mixture. There are no external controls, such as the fuel injection or spark timings used in diesel or SI engines, respectively. However, charge stratification using multiple injection strategies has been explored for combustion phasing and heat release rate control.[33] Due to the rapid auto-ignition heat release, true HCCI combustion approaches the idealization described in the first section. This rapid heat release can cause engine noise and potentially damage the engine, and this places constraints on the maximum load or mixture strengths that can be used. Diesel ignition delay can be extended to allow time for mixing by using very early injection or late injection, as in Nissan's Modulated Kinetics (MK) concept.[34]

CONCLUSIONS

This entry provides a review of diesel and gasoline engine operating principles, and an introduction to current research issues and literature. Developments in engine hardware, coupled with advanced analysis techniques, such as computer modeling, allow efficient exploration of the large number of possible design configurations. As an

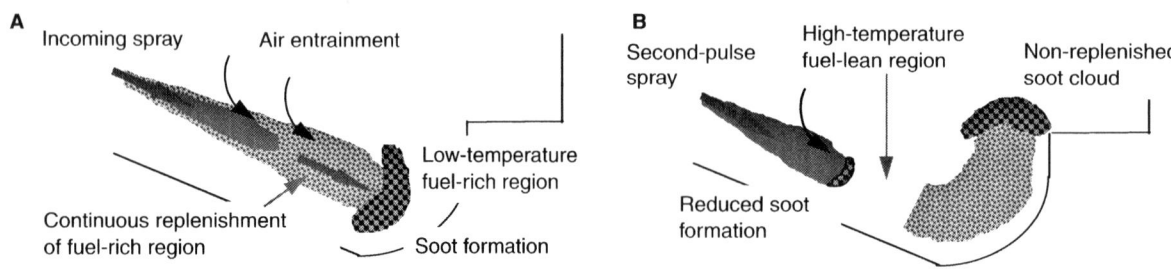

Fig. 9 Mechanism of emission reduction with multiple injections.
Source: Reprinted with permission from SAE Paper 960633 © 1996 SAE International (see Ref. 20). (A) single injection, (B) split injection.

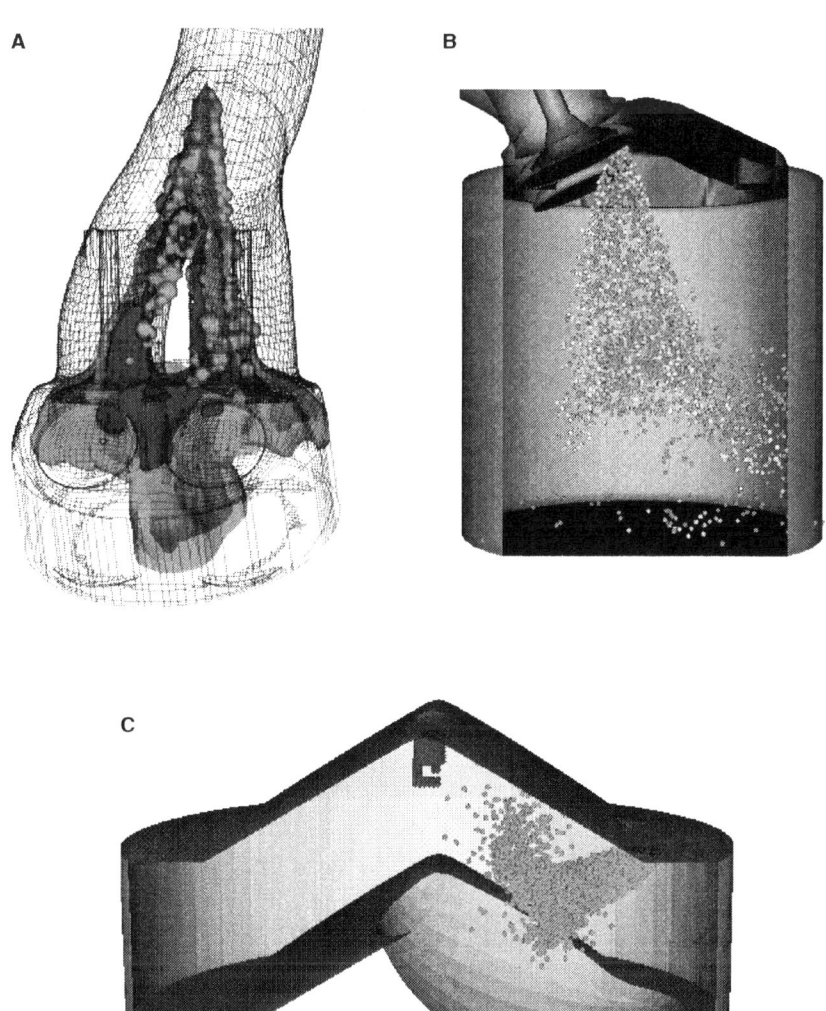

Fig. 10 Gasoline engine fuel–air mixing (A) Port Fuel Injection (PFI).
Source: Reprinted with permission from SAE Papers 2001-01-1231 © 2001(see Ref. 22), (B) spray-guided direct injection.
Source: Reprinted with permission from SAE Papers 970625 © 1997 SAE International (see Ref. 23), (C) wall-guided direct injection.

example of current technology, the combustion chamber geometry can be optimized using computer modeling for low emission, high efficiency engine operation.[35]

ACKNOWLEDGMENTS

Contributions of faculty, students, and staff at the Engine Research Center are greatly appreciated.

REFERENCES

1. Heywood, J.B. *Internal Combustion Engine Fundamentals*; McGraw-Hill Inc.: New York, 1988.
2. Anon, Emission Standards and Supplemental Requirements for 2007 and Later Model Year Diesel Heavy-Duty Engines and Vehicles, 40 CF—Chapter I—Part 86, 40CFR68.007-11, US EPA, March, 2002.
3. Reitz, R.D. Computational fluid dynamics modeling of diesel engine spray combustion and emissions. In *Diesel Engine Reference Book*, 2nd Ed.; Challen, B., Baranescu, R., Eds.; Butterworth-Heinemann: Woburn, MA, 1998.
4. Zhu, Y.; Reitz, R.D. A 1-D Gas Dynamics Code for Subsonic and Supersonic Flows applied to predict EGR levels in a heavy-duty diesel engine. Int. J. Vehicle Design **1999**, *22*, 227–252.
5. Yi, Y.; Hessel, R.; Zhu, G.; and Reitz, R.; The Influence of Physical Input Parameter Uncertainties on Multidimensional Model Predictions of Diesel Engine Performance and Emissions, SAE Paper 2000-01-1178, SAE Transactions, Engines, Vol. 109, Section 3, pp. 1298-1316, 2000.
6. Senecal, P.K.; Xin, J.; Reitz, R.D.; Predictions of Residual Gas Fraction in IC Engines, SAE Paper 962052, SAE Transactions, Vol. 105, Section 3, J. Engines, pp. 2243-2254, 1996.

7. Amsden, A.A.; KIVA-3V: A Block-Structured KIVA Program for Engines with Vertical or Canted Valves, Los Alamos National Laboratory Report, LA-13313-MS (1997).
8. Reitz, R.D.; Rutland, C.J. Development and testing of diesel engine CFD models. Progr. Energy Combust. Sci. **1995**, *21*, 173–196.
9. Han, Z.; Reitz, R.D. Turbulence modeling of internal combustion engines using RNG k-e models. Combust. Sci. Tech. **1995**, *106*, 267–276.
10. Von Kuensberg, S.C.; Kong, S.-C.; Reitz, R.D.; Modeling the Effects of Injector Nozzle Geometry on Diesel Sprays, SAE Paper 1999-01-0912, 1999.
11. Reitz, R.D. Modeling atomization processes in high-pressure vaporizing sprays. Atom. Spray Tech. **1988**, *3*, 309–337.
12. Xin, J.; Ricart, L.; Reitz, R.D. Computer modeling of diesel spray atomization and combustion. Combust. Sci. Tech. **1998**, *137*, 171–183.
13. Halstead, M.; Kirsh, L.; Quinn, C. Multidimensional autoignition of hydrocarbon fuels at high temperatures and pressures—fitting of a mathematical model. Combust. Flame **1977**, *30*, 45–60.
14. Westbrook, C.K.; Warnatz, J.; Pitz, W.J.; A Detailed Chemical Kinetic Reaction Mechanism for the Oxidation of *Iso-Octane* and *N*-Heptane Over an Extended Temperature Range and its Application to Analysis of Engine Knock, Proceedings of 22nd Symposium (Int.) on Combust, pp. 893–901, Combustion Institute, Pittsburg, PA, 1988.
15. Kong, S.-C.; Patel, A.; Yin, Q.; Klingbeil, A.; Reitz, R.D.; Numerical Modeling of Diesel Engine Combustion and Emissions Under HCCI-Like Conditions with High EGR Levels, SAE Paper 2003-01-1087, SAE Transactions, Volume 112, Section 3, J. Engines, pp. 1500–1510, 2003.
16. Kong, S.-C.; Sun, Y.; Reitz, R.D.; Modeling Diesel Spray Flame Lift-Off, Sooting Tendency and NO_x Emissions Using Detailed Chemistry with Phenomenological Soot Model, Proceedings of ASME Spring Technical Conference, (ICES05), McCormick Place Convention Center, Chicago, April 5–7, 2005.
17. Hiroyasu, H.; Nishida, K.; Simplified Three-Dimensional Modeling of Mixture Formation and Combustion in a D.I. Diesel Engine, SAE Paper 890269, 1989.
18. Nagel, J.; Strickland-Constable, R.F.; Oxidation of Carbon Between 1000-2000 C, Proceedings of the Fifth Carbon Conference; Pergammon Press: Elmsford, NY, Vol. 1, 1962; pp. 154–172.
19. Wiedenhoefer, J.F.; Reitz, R.D.; Multidimensional Modeling of the Effects of Radiation and Soot Deposition in Heavy-Duty Diesel Engines, SAE Paper 2003-01-0560, SAE Transactions, Volume 112, Section 3, J. Engines, pp. 784–804, 2003.
20. Han, Z.; Uludogan, A.; Hampson, G.; Reitz, R.D.; Mechanisms of Soot and NO_x Emission Reduction Using Multiple-Injection in a Diesel Engine, SAE Paper 960633, SAE Transactions, Vol. 105, Section 3, J. Engines, pp. 837–852, 1996.
21. Dec, J.E.; A Conceptual Model of DI Diesel Combustion Based on Laser-Sheet Imaging, SAE Paper 970873, 1997.
22. Zhu, G.-S.; Reitz, R.D.; Xin, J.; Takabayashi, T.; Characteristics of Vaporizing Continuous Multi-Component Fuel Sprays in a Port Fuel Injection Gasoline Engine, SAE Paper 2001-01-1231, SAE Transactions, Vol. 110, J. Engines, Section 3, pp. 1368–1384, 2001.
23. Han, Z.; Reitz, R.D.; Yang, J.; Anderson, R.W.; Effects of Injection Timing on Air-Fuel Mixing in a Direct-Injection Spark-Ignition Engine, SAE Paper 970625, SAE Transactions, Vol. 106, Section 3, J. Engines, pp. 848–860, 1997.
24. Fan, L.; Reitz, R.D. Spray and combustion modeling in gasoline direct-injection engines. Atom. Sprays **2000**, *10* (3–5), 219–250.
25. Zhao, F.Q.; Lai, M.C.; Harrington, D.L.; A Review of Mixture Preparation and Combustion Strategies for Spark Ignited Direct Injection Gasoline Engines, SAE Paper 970627.
26. Fansler, T.D.; Drake, M.C.; Stojkovic, B.; Rosalik, M.E. Local Fuel Concentration, Ignition and Combustion in a Stratified-charge spark-ignited direct-injection engine: Spectroscopic, imaging and pressure-based measurements. Intl. J. Engine Res. **2003**, *4*, 61–86.
27. Tan, Z.; Reitz, R.D.; Ignition and Combustion Modeling in Spark-Ignition Engines Using a Level Set Method, SAE Paper 2003-01-0722, SAE Transactions, Volume 112, Section 3, J. Engines, pp. 1028–1040, 2003.
28. Iwamoto, Y.; Noma, K.; Nakayama, O.; Yamauchi, H.; Ando, H.; Development of Gasoline Direct Injection Engines, SAE Paper 960627.
29. Drake, M.C.; Fansler, T.D.; Lippert, A.M. Stratified-charge combustion: Modeling and imaging of a spray-guided direct injection spark-ignition engine. Proc. Combust. Inst. **2005**, *30*, 2683–2691.
30. Stiesch, G.; Tan, Z.; Merker, G.P.; Reitz, R.D.; Modeling the Effect of Split Injections on Direct Injection Spark Ignition (DISI) Engine Performance, SAE Paper 2001-01-0965, 2001.
31. Najt, P.; Foster, D.; Compression-Ignited Homogeneous Charge Combustion, SAE Technical Paper, No. 830246, 1983.
32. Stanglmaier, R.H.; Roberts, C.E.; Homogeneous Charge Compression Ignition (HCCI): Benefits, Compromises, and Future Engine Applications, SAE Paper 1999-01-3682.
33. Marriott, C.M.; Reitz, R.D.; Experimental Investigation of Direct Injection-Gasoline for Premixed Compression-Ignited Combustion Phasing Control, SAE Paper 2002-01-0418, SAE Transactions, Volume 111, Section 3, J. Engines, pp. 863–874, 2002.
34. Kimura, S.; Aoki, O.; Kitahara, Y.; Aiyoshizawa, E.; Ultra-Clean Combustion Technology Combining a Low-Temperature and Premixed Combustion Concept for Meeting Future Emission Standards, SAE Paper 2001-01-0200, 2001.
35. Wickman, D.D.; Yun, H.; Reitz, R.D.; Optimized Split-Spray Piston Geometry for HSDI Diesel Engine Combustion, SAE Paper 2003-01-0348, 2003.

Regulation: Price Cap and Revenue Cap

Mark A. Jamison
Public Utility Research Center, University of Florida, Gainesville, Florida, U.S.A.

Abstract

Price cap regulation allows the operator to change its price level according to an index that is typically comprised of an inflation measure, I, and a "productivity offset," which is more commonly called the X-factor. Typically with price cap regulation, the regulator groups services into price or service baskets and establishes an I–X index, called a price cap index, for each basket. Establishing price baskets allows the operator to change prices within the basket as the operator sees fit as long as the average percentage change in prices for the services in the basket does not exceed the price cap index for the basket. Revenue cap regulation is similar to price cap regulation in that the regulator establishes an I–X index, which in this case is called a revenue cap index, for service baskets and allows the operator to change prices within the basket so long as the percentage change in revenue does not exceed the revenue cap index. Revenue cap regulation is more appropriate than price cap regulation when costs do not vary appreciably with units of sales.

INTRODUCTION

Price cap and revenue cap regulation are forms of incentive regulation, which is the use of rewards and penalties to induce the utility company to achieve desired goals and in which the operator is afforded some discretion in achieving goals.[1,2] With price cap regulation, the company's average price increase is restricted by a price index that generally includes an inflation measure (such as the U.S. Gross Domestic Product Implicit Price Deflator) and an offset that generally reflects expected changes in the company's productivity. (Revenue cap regulation is the same as price cap regulation except that the company's revenue is restricted by the inflation-productivity index. In this chapter, I simplify my discussion by focusing on price cap regulation.) With *pure price caps*, the regulator never directly observes the operator's profits. This form of price caps is rare and indeed may never be practiced except in instances where the regulator is prohibited by law from observing costs and adjusting prices. Most price cap regimes base prices on past costs or expected costs, and prohibit the regulator from adjusting prices according to new information for a set period—typically, 4–6 year.

Price caps were first developed in the United Kingdom in the 1980s to be the regulatory framework for the country's newly privatized utilities. The basic idea behind the country's price cap regulation was that the regulator would be at an information disadvantage relative to the utilities in terms of knowing how efficiently the utilities could operate. By adopting price cap regulation and allowing utilities to keep for a period of time profits they received by improving efficiency, the government believed that the companies would reveal their efficiency capabilities. In turn, this would allow the regulator eventually to set regulated prices that reflected the companies' true abilities. Price cap regulation did not work out entirely as planned, so adjustments have been made to the point that the United Kingdom's price cap regulation looks a lot like U.S. rate of return regulation. (Excellent summaries of the U.K. experience can be found in several studies.[3-5] A critical difference between U.S.-style rate of return regulation and U.K.-style price cap regulation is that the U.K. regimes have fixed time periods between price reviews, whereas under rate of return regulation, price reviews are triggered by high or low earnings [relative to the cost of capital].)

There are three important elements of an incentive regulation plan: (1) providing the reward/penalty structure; (2) allowing the company an opportunity to choose its goals; and (3) allowing the operator latitude in how it will achieve its goals. An example of a reward/penalty structure would be allowing the company to retain higher (lower) profits if it increases (decreases) its operating efficiency. Allowing the company a role in choosing its goals is referred to as "a menu of options," whereby the regulator matches greater potential rewards with more ambitious goals. The company may be allowed to choose, for example, between a goal of decreasing costs by 5% and keeping 50% of the profits it receives above its cost of capital, and a goal of decreasing costs by 10% and keeping 100% of the profits it receives above its cost of capital. (A company's cost of capital is the interest that the company pays on its debt plus the return that it must provide shareholders to ensure that they continue to invest in the company.) If the company chose the goal of decreasing costs by 10%, the operator would have the latitude to do this by, for example, negotiating lower input

Keywords: Price caps; Incentives; Revenue caps; Information asymmetry.

prices from suppliers, decreasing overhead, improving network reliability, obtaining lower-cost capital, or using some combination of methods.

The benefits of price cap regulation include providing companies incentives to improve efficiency, dampening the effects of cost information asymmetries between companies and regulators, and decreasing the incentives to overinvest in capital and cross-subsidize relative to rate of return regulation. In some instances, however, service quality and infrastructure development have suffered under price cap regulation. Furthermore, it is difficult for regulators to keep commitments that allow companies to retain profits above their cost of capital.

The remainder of this chapter is organized as follows. The next section describes the theory underlying price cap regulation. "The Basic Price Restriction" describes establishing the price index. "Service Baskets" discusses how regulators structure price baskets. "Case Studies in Price Caps" summarizes some cases followed by the "Conclusion."

UNDERLYING THEORY

Regulators and other policymakers have certain energy goals for their countries, including near-universal availability of service, affordable prices, and quality service. Achieving these goals requires that utilities incur costs and exert effort. The difficult question for regulators is how much cost and effort will be required. Utilities generally know more about the answers to these questions than regulators do. A company generally knows more than its regulator about how much it would cost to provide a certain level and quality of network expansion, for example. This is because the regulator cannot directly observe the operator's innate abilities and its degree of effort.

These problems are called information asymmetry or principal-agent problems. An information asymmetry arises from the company's having information—namely, about the utility's innate ability to achieve performance goals at a specific cost and the amount of effort the employees exert—that the regulator does not have. The name "principal-agent" arises from the nature of the relationship; the regulator (the principal) has goals that she wants the operator (the agent) to achieve. The company may agree with some of the principal's goals, but companies generally have other interests, such as maximizing profits for their shareholders and limiting the amount of effort exerted. To solve these problems, the regulator offers the operator financial rewards for controlling costs and/or exerting effort.

THE BASIC PRICE RESTRICTION

With price cap regulation, prices are initially set to allow the company to receive its cost of capital. Thereafter, prices are allowed to rise on average at the rate of inflation, less an offset, namely

$$\%\Delta p \leq I - X,$$

where $\%\Delta p$ is the average percentage change in prices allowed in a year, I is the inflation index, and X is the offset. The key issues are: What is the "offset"? What is the measure of inflation? And what does it mean that prices are allowed to rise on average?[6]

The underlying logic of the price cap restriction is that it emulates the competitive market. In a competitive market, prices reflect the costs of production. Prices rise when production costs unavoidably rise. Prices decline with productivity increases. As a result, in a competitive economy, the economy-wide inflation rate reflects unavoidable increases in production costs and accounts for productivity gains. If the regulated company is just like the average firm in the economy, its prices should rise at the general rate of inflation.[7]

Therefore, the X-factor should represent the difference between the regulated firm and the average firm in the economy. There are two key differences to consider—namely, the regulated company's ability to improve productivity and changes in its input costs. If the regulated company can improve its productivity more than the average firm in the economy, or if the regulated company's input prices increase less than input prices for the average firm, this would imply $X > 0$. The opposite situations would imply $X < 0$. If the regulated firm is just like the average firm, this would imply $X = 0$. Consider a situation in which the average firm in the economy improves its productivity by 3% per year, and its input prices increase 1% per year. Further assume that the regulated firm can improve its productivity by 5% per year and that its input prices actually decrease 2% per year. The appropriate X-factor would be $X = (5 - 3) - (-2 - 1) = 5$.

There are two basic approaches for establishing an X-factor—namely, the historical approach and the forecast approach. The historical approach compares estimates of the total factor productivity (TFP) for the average firm in the economy with estimates for the regulated company. The X-factor is set equal to the difference between the TFP estimates after adjusting for differences in input prices. A modification to this approach adds a stretch factor, S, that accounts for the effects of historic regulation and/or anticipated changes in industry conditions. Examples of explicit stretch factors include 0.5% for AT&T by the Federal Communications Commission and 1% for local-exchange telephone companies in Canada.

The forecast approach is a three-step process. The first step is to determine the rate base for year t, where t is the first year of the new pricing regime, according to the formula

$$B_{t-1} = B_0 + \sum_{i=1}^{t-1}(\text{Capex}_i - d_i),$$

where $t=0$ is the initial rate base of the company, for example, at the time of privatization; $Capex_i$ is the additional investment in rate base in year i; and d_i is the depreciation expense in year i. The next step is to project cash outflows (Capex), operating expenses (Opex), nonoperating expenses (Nopex), and unit sales for each year of the new pricing regime. The last step is to estimate the X-factor that will equate the present value of the cash flows of the company with the change in shareholder value, using the formula

$$\sum_{j=t}^{n+t} \frac{P_j Q_j - \text{Opex}_j - \text{Capex}_j - T_j \pm \text{Tr}}{(1 + \text{WACC})^j}$$
$$= B_{t-1} - \frac{B_{n+t}}{(1 + \text{WACC})^n},$$

where $P_j Q_j$ is the projected revenue for year j, $Opex_j$ is the forecasted operating expenses for year j, $Capex_j$ is the projected capital expenditures for year j, T_j is the projected taxes for year j, B_j is the rate base at end of year j, Tr represents cash transfers between the government and other entities (not counted in revenue, operating expenses, or capital expenditures), WACC is the weighted average cost of capital, and n is the length of time a price cap plan is in effect.[8] The WACC is the return the company is allowed to receive on its assets, and includes both the cost of debt the company uses to finance its rate base and the cost of equity. The cost of debt is simply the weighted average of the interest rates that the company pays on its long-term corporate bonds. The cost of equity is the return that shareholders need to ensure that they continue to finance the company.[9]

The United Kingdom used this approach in setting prices for Hydro Electric (HE) in 1995.[10] Table 1 shows the Monopoly and Merger Commission's (MMC) present value calculation for HE's price control for the period 1995–1996 to 1999–2000. The first three lines contain its allowances for operating costs, network capital expenditure, and nonoperational capital expenditure. These cash flows were discounted at 7% (the MMC's assumption about the cost of capital), which came to £457.9 million. Then the MMC added the present value of the opening less closing asset values of the distribution business, which represented another £128.2 million, giving a total of £586.1 million.

Asset values were calculated by taking an opening balance in 1990–1991 and rolling this forward by adding net distribution network capital expenditure. This was defined as network capital expenditure less depreciation. By the end of 1994–1995, this gave a total of £563 million and £610 million by the end of 1999–2000. The latter figure had a present value in 1995–1996 of £434.8 million.

The opening balance of £523.4 million in 1990–1991 was consistent with the figure used by the government in setting the original price control and the initial market value of HE. Table 2 shows the roll forward of the opening balance to £563 million at the start of the price control period in 1995–1996.

The total of £586.1 million in Table 1 represented the present value of the revenue that the MMC considered HE would need to raise to cover its allowable cash outflows and earn a 7% return on its asset value. The MMC calculated that the continuation of the existing price control would raise revenue with a present value of £462.1 million, which fell short of this amount. In the case of HE's distribution business, however, there was an additional source of revenue: the hydro benefit, which could be transferred from the generation business in accordance with HE's license. Taking this into account, the MMC decided that an appropriate relationship would be established and maintained if HE's price control required it to reduce prices by 0.3% in 1995–1996, followed by reductions of 2% per year for the next 4 years. Table 3 shows the MMC's projections of distribution business revenue. The present value of revenue and hydro

Table 1 MMC's calculation of HE's distribution business costs (1994/1995 prices)

	1995/1996	1996/1997	1997/1998	1998/1999	1999/2000	Total
Operating costs	60.7	59.5	58.3	57.1	56.0	
Network capital expenditure	43.5	43.2	43.8	44.1	44.6	
Non-operational expenditure	6.7	5.6	5.3	5.6	5.0	
Total	110.9	108.3	107.4	106.8	105.6	
PV of costs at 7%	107.2	97.8	90.7	84.3	77.9	457.9
PV of asset values a 7%	563.0				−434.8	128.2
						586.1

Table 2 MMC's calculation of HE's distribution asset base (1994/1995 prices)

	1990/1991	1991/1992	1992/1993	1993/1994	1994/1995
Opening value	523.4	534.6	534.4	536.1	545.1
Depreciation	27.2	27.9	28.7	29.7	31.0
Network capital expenditure	38.4	27.7	30.4	38.7	48.9
Closing value	534.6	534.4	536.1	545.1	563.0

benefit is £586.1 million, which is equal to the present value of costs and return on assets shown in Table 1.

The inflation index in the basic price restriction is generally one that is a good approximation of the previous year's inflation, reflects general price movements in the economy, is not focused on a particular segment of the economy, and is reliable and available in a timely manner. The regulator compares this price index with the average price change proposed by the company to determine whether the proposed price change is acceptable. The average price change is the weighted average change in prices, where a price's weight is the proportion of the company's revenue that the price generates.

Assume that a company has two services: service 1 and service 2. Service 1 provides 60% of the company's revenue, and service 2 provides 40%. The company proposes to increase the price of service 1 by 10% and the price of service 2 by 5%. The resulting average price change is: $(0.6 \times 10\% + 0.4 \times 5\%) \times 100 = 8\%$. If the basic restriction (inflation$-X$) is 8% or larger, the regulator approves the pricing proposal.

Extraordinary events may affect the utility disproportionately compared with the average firm in the economy. In these instances, regulators consider applying to the basic price cap formula an adjustment called an exogenous factor. Exogenous factors, also called Z-factors, reflect the effects of rare, one-time events whose occurrence and impacts are beyond the control of the regulated company and that affect the company differently from the average firm in the economy. An example might be a special tax placed on electric utilities. These exogenous factors increase or decrease the price index, depending on how the extraordinary event affected the utility.

SERVICE BASKETS

A service basket is a group of services placed under a common inflation$-X$ restriction. Services that the regulator wants to protect from price increases or decreases relative to certain other services are placed in a separate basket. If the regulator does not want the company to change urban prices relative to rural prices, for example, the regulator might place urban prices in one basket and rural prices in another.

The company is allowed to change the relative price levels of the services within a basket, subject to two possible restrictions. The first type of restriction is a limit on individual prices. Regulators may apply such a restriction by placing an absolute restriction on the price (e.g., the price per kWh for residential electricity cannot exceed 5 cents) or a percentage restriction (e.g., the price per kWh for residential electricity cannot increase more than 10% per year).

Regulators may also apply caps to subsets of services within a basket. The regulator may apply a restriction of inflation-5 to all services, for example, and a subrestriction of inflation-3 to residential services. In this case, the company's average overall price would need to decrease 5% in real terms, and residential prices would have to decrease 3% in real terms.

Table 3 MMC's projections of HE's distribution business revenue (1994/1995 prices)

	1995/1996	1996/1997	1997/1998	1998/1999	1999/2000	Total
Regulated revenue	105.2	104.6	103.8	102.9	102.1	
Unregulated revenue	5.5	5.3	5.1	5.0	4.8	
Hydro benefit	29.2	29.2	29.2	29.2	29.2	
Total	139.9	139.0	138.1	137.2	136.2	
PV of revenue at 7%	135.2	125.6	116.6	108.2	100.4	586.1

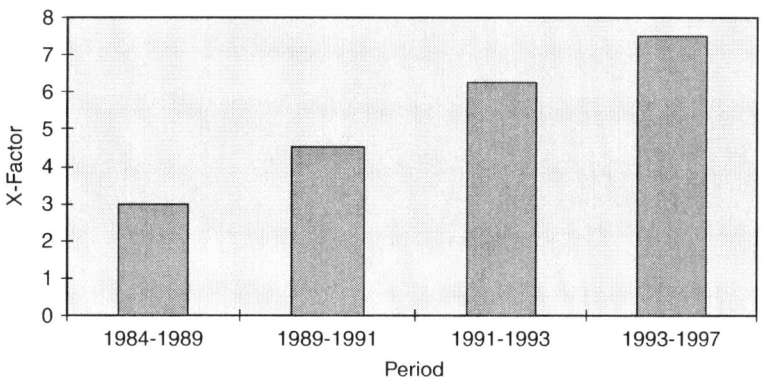

Fig. 1 Oftel's X-factors for BT.
Source: From Oxford University Press (see Ref. 11).

CASE STUDIES IN PRICE CAPS

Most applications of price cap regulation have been in telecommunications. Berg and Foreman[11] provide one of the earliest studies of the effects of price cap regulation, focusing on the U.K. regulation of British Telecom (BT), and the Federal Communications Commission's price cap regulation of AT&T and the Bell operating companies. The United Kingdom implemented price regulation for BT in 1984. There were four basic reasons why the United Kingdom adopted price regulation for BT:

- Price regulation would provide BT incentives to decrease costs.
- Because BT had been a government-owned service provider, information necessary for rate of return regulation was not available.
- The United Kingdom wanted to minimize the amount of adversarial litigation that had characterized U.S. rate of return regulation.
- The United Kingdom believed that regulation would serve primarily as a brief transitional mechanism to full competition.

The chart in Fig. 1 shows how the U.K. regulator changed the X-factor for BT over time. This was the general trend except for the 1997–2001 pricing decision. This growth in X may have related to the regulator's concomitant expansion of services covered under price caps (see Table 4), but it may also have reflected the regulator's increasing knowledge of how BT could improve its operating efficiency. Table 4 shows changes in services or elements subject to price control. Each price review has resulted in increasing numbers of services being subject to the price cap constraint. The chart in Fig. 2 shows the percentage of BT's turnover that is under price control for each period. This percentage grew from 48% during the first period to 71% during the 1993–1997 period.

Berg and Foreman[11] conducted their review using traditional rate evaluation criteria of simplicity and public acceptability, freedom from controversy, revenue sufficiency, revenue stability, price stability, fairness in apportionment of total costs, avoidance of undue rate discrimination, and encouragement of efficiency. They concluded the following:

- *Simplicity and public acceptability*. It is unlikely that price caps resulted in simplicity and administrative savings. Design of price caps required attention to service baskets and price bands, floors, and ceilings. The desire to increase public and other stakeholder acceptability created the need for additional control

Table 4 Changes in services subject to price cap

Period	Exchange line rentals	Domestic calls	Operator assisted calls	International calls	Connection charges
1984–1989	x	x			
1989–1991	x	x	x		
1991–1993	x	x	x	x	
1993–1997	x	x	x	x	x

Source: From Oxford University Press (see Ref. 11).

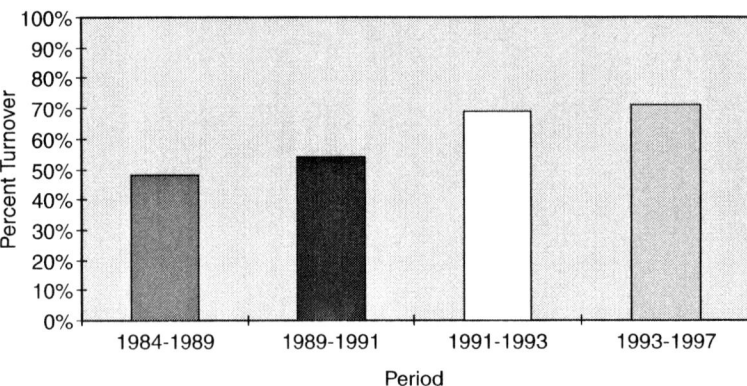

Fig. 2 Percent BT's turnover under price control.

features. Each feature has provided an opportunity for increased debate and litigation.

- *Freedom from controversy.* All the terms of price caps were controversial, including service quality, how to handle "excessive" returns, and public perceptions of the legitimacy of the regulation. Earnings-sharing assessments in the United States were sensitive to the same arbitrary cost allocations as rate of return regulation. Bell operating companies were given optional regulatory contracts and generally chose the lower productivity factors, even though these included higher earnings-sharing requirements.
- *Revenue sufficiency.* Competition complicated this objective. This would have been true regardless of the method of regulation. Weisman[12] concluded that regulators have less interest in revenue sufficiency when the price cap deal is struck.
- *Revenue stability.* This objective became one of net revenue stability (i.e., net income) under price cap regulation. As a result, using price caps increases the importance of making cross-subsidies explicit.
- *Price stability.* The price cap formula explicitly improves price predictability and stability relative to other prices in the economy by aligning price changes with changes in general inflation indices.
- *Fairness in apportionment of total costs.* Initial prices were part of a political compromise, so it was not immediately clear that price regulation results were different from rate of return regulation results with respect to this view of fairness.
- *Avoidance of undue rate discrimination.* Price caps use ceilings and floors to contain price discrimination. The U.K. regulator made three general changes to the price cap regime over time relative to this issue: (1) increased the X-factor, perhaps in response to high earnings by BT; (2) added special constraints to some prices, such as residential exchange line rental; and (3) added additional services and baskets (such as the median residential bill) over time.
- *Encouragement of efficiency.* British Telecom was allowed significant opportunity for rate rebalancing. Attenborough[13] found a total welfare gain of £2 billion per year in 1990–1991 prices, and 30% of this gain was from more efficient rate design. Price regulation allowed companies to improve economic efficiency by aligning prices with marginal costs, but competitive pressure, political constraints, and non-efficiency-related regulatory objectives may prevent this from happening in other situations.

CONCLUSION

Incentives and opportunities to improve efficiency are generally greater under price cap regulation than under rate of return regulation. This does not mean, however, that price cap regulation is the right form of regulation in all situations. Compared with rate of return regulation, price cap regulation decreases regulators' concern for revenue adequacy because they have less direct control over revenue. Also, regulators may come under pressure from consumer groups to break their commitment to allow higher earnings if the regulated company improves efficiency: consumers may view the higher profits as evidence that the regulator is not tough enough on the utility or is not knowledgeable. This challenge to regulatory legitimacy has led some regulators to roll back profits that they once said companies could keep.

When choosing a regulatory scheme, regulators should weigh these potential problems and benefits of price cap regulation against the corresponding costs and benefits of rate of return regulation. They may find that neither form of regulation is adequate by itself, and may adopt a hybrid system that applies different aspects of different forms of regulation to craft a regulatory scheme that makes sense for the regulator's institutional, political, and economic situation.[6]

REFERENCES

1. Lewis, T.R.; Garmon, C. *Fundamentals of Incentive Regulation*, 12th PURC/World Bank International Training Program on Utility Regulation and Strategy, Gainesville, FL, June 10–21, 2002.
2. Berg, S.V. *Introduction to the Fundamentals of Incentive Regulation*, 12th PURC/World Bank International Training Program on Utility Regulation and Strategy, Gainesville, FL, June 10–21, 2002.
3. Baldwin, R.; Cave, M. *Understanding Regulation: Theory, Strategy, and Practice*; Oxford University Press: Oxford, U.K., 1999.
4. Newbery, D.M. *Privatization, Restructuring, and Regulation of Network Utilities*; MIT Press: Cambridge, MA, 1999.
5. Lee, H. Price Cap: The U.K.'s Efforts to Regulate Regional Distribution Companies, Kennedy School of Government Case Program CR14-01-1619.0, Harvard University: Cambridge, MA, 2001.
6. Sappington, D.E.M. Price regulation. In *Handbook of Telecommunications Economics*, Cave, M.E. et al. Eds., North-Holland, Amsterdam: 2002; Vol. 1, 227–293.
7. Bernstein, J.I.; Sappington, D.E.M. How to determine the X in RPI-X regulation: A user's guide. Telecommunications Policy **2000**, *24* (1), 63–68.
8. Green, R.; Pardina, M.R. *Resetting Price Controls for Privatized Utilities: A Manual for Regulators*; World Bank: Washington, DC, 1999.
9. Bonbright, J.C.; Danielsen, A.L.; Kamerschen, D.R. *Principles of Public Utility Rates*; Public Utilities Reports, Inc.: Arlington, VA, 1988.
10. Office of Electricity Regulator. Transmission Price Control Review of the National Grid Company. Appendix D. Available at: http://cbdd.wsu.edu/kewlcontent/cdoutput/TR506/page29.htm (accessed on August 24, 2005).
11. Berg, S.V.; Foreman, R.D. Price cap policies in the transition from monopoly to competitive markets. Ind. Corp. Change **1995**, *4* (4), 671–681.
12. Weisman, D.L. Why less may be more under price-cap regulation. J. Regul. Econ. **1994**, *6*, 339–361.
13. Attenborough, N.; Foster, R.; Sandbach, J. *Economic Effects of Telephony Price Changes in the U.K.*, NERA Topics Paper No. 8; NERA Economic Consulting: London, U.K., 1992.

Regulation: Rate of Return

Mark A. Jamison
Public Utility Research Center, University of Florida, Gainesville, Florida, U.S.A.

Abstract
Rate of return regulation adjusts overall price levels according to the operator's accounting costs and cost of capital. In most cases, the regulator reviews the operator's overall price level in response to a claim by the operator that the rate of return that it is receiving is less than its cost of capital, or in response to a suspicion of the regulator or claim by a consumer group that the actual rate of return is greater than the cost of capital. Critical issues for the regulator include how to value the base, whether to add investments to the rate base as they are made or when the facilities go into service, the amount of depreciation, and whether expenditures have been prudently made and whether they relate to items that are used and useful for providing the utility service.

INTRODUCTION

This article describes rate of return regulation, which regulators often use to determine fair and reasonable prices for electricity sold by utility companies. Prices under rate of return regulation are considered fair and reasonable because they give the company an opportunity to recover the costs it has appropriately incurred in providing electricity service, and customers are protected from paying prices that would provide the company monopoly profits.[1] Rate of return regulation is sometimes criticized for not providing companies incentives to operate efficiently.

In performing this form of regulation, the regulator determines the appropriate amount for the company's rate base, cost of capital, operating expenses, and depreciation (Rate base is the gross value of the company's assets, minus accumulated depreciation. Cost of capital is also called the allowed rate of return and is the interest that the company pays on its debt, plus the return it must provide to shareholders to ensure that they continue to invest in the company). Based on these amounts, the regulator determines the amount of revenue the company needs to cover its operating expenses, depreciation, and cost of capital.

The emphasis on cost recovery in rate of return regulation is the source of the concern that companies may not operate efficiently.[2] If the regulator allows a rate of return that is higher than what the company actually needs to ensure that shareholders continue to provide capital for investment, the company could increase its returns to shareholders by making unnecessary investments (if the regulator does not catch the company doing so). This is called the Averch–Johnson effect.[3] Rate of return regulation, however, is also generally viewed as having the advantage of restricting opportunities for regulators to lower companies' prices arbitrarily.

Keywords: Rate of return; Asset; Rate base; Earnings; Cost; Revenue; Prudent; Used and useful.

BASICS FOR ASSESSING RATE OF RETURN

Assessing the company's rate of return involves evaluating the effects of price levels on earnings so that investors have an opportunity to receive a fair rate of return on their investments. There are five traditional criteria for determining whether a rate of return is appropriate.[1]

The first criterion is whether the rate of return is adequate to attract capital. One of the primary goals of utility regulation is to ensure that a sufficient level of service is available for customers. Service cannot be provided on an ongoing basis without continual reinvestment; therefore, capital attraction is a primary criterion for evaluating the rate of return.

A second criterion is the implementation of efficient management practices.

The third criterion measures efficient consumer-rationing of services. To encourage efficient consumption of services, prices should reflect marginal costs. In some situations, efficient prices may be different from those that attract capital, so there will be a need to balance these two criteria.

A fourth criterion is rate stability and predictability, which assists customers who value being able to plan their utility expenses.

The last criterion is fairness to investors. This may sound redundant with the capital-attraction criterion, but it is different. Consider a situation in which asset prices have been declining. A regulator may be tempted to adopt prices that are sufficient to attract investment for the new lower-priced assets but that are insufficient to recover the past costs of the historical assets. Such a decision would not satisfy the fairness criterion.

HOW RATE OF RETURN REGULATION WORKS

Rate of Return Regulation Basic Formula

Rate of return regulation combines a company's costs and allowed rate of return to develop a revenue requirement.

This revenue requirement then becomes the target revenue for setting prices. The basic formula for determining a revenue requirement is

$$R \equiv B \cdot r + E + d + T$$

where: R, revenue requirement; B, rate base, which is the amount of capital or assets the utility dedicates to providing its regulated services; r, allowed rate of return, which is the cost the utility incurs to finance its rate base, including both debt and equity; E, operating expenses, which are the costs of items such as supplies, labor (not used for plant construction), and items for resale that are consumed by the business in a short period (less than one year); d, annual depreciation expense, which is the annual accounting charge for wear, tear, and obsolescence of plant; T, all taxes not counted as operating expenses and not directly charged to customers.

Assume that the regulator determines that the company has a net asset base of $30 million, an after-tax cost of capital of 12%, a tax rate of 25%, operating expenses of $1 million, and depreciation expenses of $1,50,000. Further assume that the cost of capital is comprised of 50% debt and 50% equity; the cost of debt is 10%, and the cost of equity is 14%. (See the section on estimating the cost of capital later in this article.) To calculate the revenue requirement, we first need to determine the after-tax profit that the company would be allowed to receive when the new prices are in place:

Rate base (B)		$30,000,000
Cost of equity	×	14%
		$4,200,000
Percent of B financed by equity	×	50%
After-tax profit		$2,100,000

The company would pay a 25% tax on this profit—namely,

After-tax profit		$2,100,000
1−Income tax rate	÷	75%
Pre-tax profit		$2,800,000
After-tax profit	−	$2,100,000
Income taxes (T)		$700,000

We can now calculate the revenue requirement as:

Rate base (B)		$30,000,000
Allowed rate of return (r)	×	12%
Allowed after tax return (B × r)		$3,600,000
Expenses (E)		1,000,000
Depreciation expenses (d)		1,500,000
Taxes (T)		700,000
Revenue requirement (R)		$6,800,000

Advantages and Disadvantages of Rate of Return Regulation

There are several advantages to using rate of return regulation. The first is that it is sustainable if there is no competition, because prices can be adjusted to the company's changing conditions. It can also provide comfort to investors because rate of return regulation constrains the regulator's discretion in setting prices. This lowers investor risk, which lowers the cost of capital. Furthermore, company profits can be kept within acceptable levels from the perspectives of both investors and customers. Unless the regulator chronically underestimates the cost of capital (and courts do not reverse the regulator in this regard), investors can be confident that they have a fair opportunity to receive the profits they expect and, thus, are willing to make investments. Customers can observe that the regulator is limiting company profits to the cost of capital.

There are also several disadvantages to using rate of return regulation. First, it provides only weak incentives for companies to operate efficiently.[2] This weakness takes two forms. The first form is the Averch–Johnson effect. The second is that managers have less incentive to operate efficiently because regulators are unable to observe perfectly managers' efforts toward efficiency or the managers' innate abilities to be efficient. Another disadvantage is that rate cases have to occur frequently during times of high inflation (in the absence of a periodic adjustment for inflation between rate cases), and rate cases may be costly to perform. Last, rate of return regulation provides a mechanism for companies to shift costs from competitive markets to noncompetitive markets.

REVENUE IMPUTATION

In some situations, the company may receive revenue that should be attributed to the regulated operations but is recorded in the company's nonregulated operations accounts. In such cases, regulators can impute this revenue to the regulated operations. In the United Kingdom, for example, an electric distribution company receiving financial benefits from the government because the company used hydro power did not reflect this benefit in its regulated accounting books. Consequently, the regulator adjusted the revenue requirement to reflect this revenue.

The calculation for revenue imputation is fairly simple. First, the regulator calculates the revenue requirement. Then the regulator subtracts from the revenue requirement the revenue imputation amount. The adjusted revenue requirement is the amount of revenue the company is allowed to recover through prices charged for regulated services. Consider our previous example in which the revenue requirement was $5.275 million. If the regulator determined that there should be $2,00,000 in imputed revenue, the company would be required to charge prices designed to provide $5.075 million in revenue.

HOW RATE BASE IS DETERMINED

The Objective

When determining rate base, the objective is to identify the amount of capital the company uses and needs to use to provide regulated services. This capital includes the plant or facilities in service that the regulator determines to be prudent, as well as the working capital. The basic decisions include determining how to value the plant that is in service, for what time period the rate base is measured, and what plant is included. Each of these decisions is discussed below.

Methods for Valuing Plant in Service

There are three basic methods for valuing the plant that the company uses to provide its services.[1,4] The first of these is called fair value or economic valuation. There are two methods for determining fair value. The first is a nonmarket approach that bases fair value on the company's financial data, such as the discounted value of its cash flow. The second method uses the market valuation of the company. The fair-value approach can be problematic because it involves circular reasoning—namely, that the profitability of the company affects the asset value, which in turn affects profitability.

The most common method for valuing utility plant in the United States is the original-cost approach, or historical approach. In this approach, assets are valued at what the company originally paid to place it in service. If an electric company spent $5,00,000 for distribution wire and poles in 1999, and spent an additional $1,50,000 to construct the distribution facilities, that portion of the distribution network would be valued at $6,50,000, less depreciation, until it was retired from service.

The original-cost approach has the advantage of being objective because the values are tied to the financial records of the company, and provide a continual matching between the money the shareholders provide for investment and the cash flow that is provided back to investors. Disadvantages include difficulty of implementation where accounting and property records are poor, understating the economic value of assets during times of inflation, and providing misleading economic signals to markets as to the real economic costs of the electricity service.

The final approach to valuing assets is the replacement-cost approach, also called current-cost accounting, in which assets are valued based on what it would cost to replace them today. Under one version of this approach, the physical facilities in place are repriced at today's cost. If prices for wires, poles, and labor for the electric company in the previous example were increasing at a rate of 10% per year, the assets would be valued at $7,86,500, less depreciation, after two years. Replacement costs are generally determined by applying an inflation factor. They may also be valued by finding replacement prices in the marketplace. Because finding such prices is difficult, however, the inflation method is the most commonly used approach. Under an alternative version, sometimes called the virtual-company approach, the system is redesigned on paper to incorporate new technology not available when the original investment was made, and the components of this new system are valued at current prices.

Replacement-cost valuation has the advantage of being able to overcome some deficiencies of poor accounting records, although previous values are needed for applying the inflation approach. It also values assets near their economic value, which sends efficient price signals to customers and suppliers. The disadvantages include its subjectivity because of the difficulty of obtaining objective prices or inflation indices for old equipment, requiring exact inventories if current prices are used to estimate replacement costs, and returning to investors an amount of cash that is different from what they provided to the company. It is important to note that if rate base is defined in terms of current value, the cost of capital used to compute the revenue requirement should be adjusted to exclude inflation.

Some regulators use a combination of the historical and current-cost approaches. These regulators use historical costs for determining the total amount of revenue that the company is allowed to collect and current costs for designing prices. This combination gives the best of both worlds: investors receive back from customers exactly the cash they provided to the company, plus a return on their investment, and prices can send efficient economic signals to customers and managers.

Choosing a Test Period

A test period, which can be either a historical period or a future period, is the period of time chosen for quantifying the amount of plant that is being used to provide the utility service, expenses that are being incurred, and billing units that are being used to develop prices that will produce the allowed revenue requirement.[4] It is normally at least one year and may be several years. In choosing a test period, regulators generally attempt to choose a period that is representative of the time over which the prices will actually be charged. They also seek a period sufficiently long that it represents normal operations. In cases where the test period contains some unusual activities or events, such as an extraordinary amount of maintenance, regulators will generally normalize the costs or revenue that they believe may have been affected.

With historical periods, plant in service is measured for a recent period that is believed to be representative of the company's typical operations, for which the necessary accounting records are available, and for which all major adjustments to the accounting records have been concluded. This has an advantage of being objective and transparent, but the period for estimating plant in service is different from the period for which prices will actually be in effect.

When using a future period for the test period, plant in service is estimated based on projected changes. This has the advantage of being able to match the period for which plant in service is estimated with the period for which prices will actually be in effect. It has the disadvantage, however, of being subjective and may provide a greater incentive for companies to operate to meet the regulator's expectation rather than to be efficient. Alternatively, the company may project major additions to plant and then fail to make the investment and, in effect, receive a return and depreciation element on nonexistent plant.

Determining the Amount of Plant in Service During the Test Period

When the test period is chosen, the amount of plant in service may be valued in one of three ways: (1) average monthly balances, (2) end-of-period balance, or (3) average beginning-of-year and end-of-year balances. Average monthly balance simply estimates the arithmetic average of plant in service at the end (or beginning) of each month. End-of-period balance uses plant in service at the end of the year. Average beginning-of-year and end-of-year balances simply estimate the arithmetic average of the first month's and last month's plant in service.

Prudence Concept and Used and Useful Concept

In some jurisdictions, a utility plant must be considered both prudent and used and useful before being allowed into the rate base.[1,4] Prudent means the investment is reasonable based on cost-minimizing criteria. There are two perspectives. In one view, the investment is considered to be prudent if it was prudent at the time the decision was made. This requires accurately assessing what information management had available and used to make its decision. In the second perspective, the investment is prudent if management acted to minimize costs by fully considering changing conditions that would affect the investment. This requires assessing what management should have known and should have considered in making its decision.

Used and useful means that the plant is actually being used to provide service and that it is contributing to the provision of the service. If a company has excessive numbers of distribution lines carrying electricity to a neighborhood, for example, the regulator might not include some of the investment in the rate base, because even though all the lines are used, many are not needed, so they are not really useful.

Adjustments for Construction

Companies generally construct facilities over time. This means that investments are made prior to the plant's actually being used and useful. There are two ways to reflect this in the rate base.[1] One method, called construction work in progress (CWIP), includes the investment in the rate base as the investment is made. This is problematic because it violates the used-and-useful standard and causes current customers to pay for the plant that will be used by future customers. On the other hand, it provides cash flow for the construction project. The second method capitalizes the financing of construction projects and is called allowance for funds used during construction (AFUDC). AFUDC adds the cost of money used to finance the project during construction to the rate base once the plant is used and useful. Then the AFUDC is depreciated along with the plant. AFUDC does not provide cash flow to fund construction. The cash flow comes later and in some instances creates a cash surplus. This cash is returned to shareholders, held for future use, or invested outside the utility business. AFUDC makes current services more affordable to customers in the short run by capitalizing outlays and deferring returns until construction is complete. AFUDC can create "rate shock," however, when the full cost of the new plant plus accumulated financing costs enters rate base in one year.

Working Capital

Working capital is the average amount of capital, in excess of net plant, that is necessary for business operations.[4] Examples include inventories, petty cash, prepayments, minimum bank balances, and cash working capital. Cash working capital is the average amount of money that investors supply to bridge the gap between the time when expenses are incurred and the time when revenue is received. In some cases, customers prepay for services. This prepayment can be shown as an offset to cash working capital. When investors have provided working capital, it should be included in the rate base.

Accumulated Depreciation

Rate base excludes accumulated depreciation. In other words, the B in the formula includes the value of the plant less the amount by which it has been depreciated.

COST OF CAPITAL

The return the company is allowed to receive on its rate base is called the allowed rate of return or the cost of capital, and includes both the cost of debt that the company uses to finance its rate base and the cost of equity. The cost of debt is simply the weighted average of the interest rates that the company pays on its long-term corporate bonds. The cost of equity is the returns that shareholders need to ensure that they continue to finance the company. Regulators combine the cost of debt and cost of equity to form what is called the weighted average cost of capital (WACC).[1]

There are several ways to estimate the cost of equity, but the most popular is the capital asset pricing model (CAPM). Capital asset pricing model includes two basic components: the risk-free cost of capital and the risk premium. The risk-free cost of capital is generally considered to be the interest that the U.S. government pays on long-term bonds. The repayment of these bonds is generally considered to be secure, so the interest rate reflects only investors' time value of money. The risk premium is the amount of return that investors require because the actual earnings of the company are uncertain. This risk premium for a company is estimated by analyzing the degree to which the variation in the return on the stock follows the variation in the averaged returns on all the stocks in the market. When the costs of debt and equity are determined, regulators combine them into the WACC, using the company's capital structure as the weights.

OPERATING EXPENSES

Operating expenses[4] include costs of items such as supplies, labor (not used for plant construction), and items for resale that are consumed by the business in a short period of time (e.g., less than one year). Standards for accepting expenses include arm's-length bargaining, whether the expense is a legitimate expense for providing the utility services, whether the utility company has been inefficient or imprudent in incurring the expenses, and whether the expenses are representative. Arm's-length bargaining means that the company management, when deciding whether to incur an operating expense, has looked out for the financial interests of the company as though the company were in a competitive marketplace and had no financial interest in the expense payee, except as a purchaser of the payee's service or product.

Expenses are considered to be representative if they are being incurred at normal levels. Exceptional expenses may be disallowed if they do not represent the normal operations of the company. This applies primarily if the prices based on this revenue requirement will apply to multiple years. If this is the case, including expenses that are rarely incurred would cause these expenses to be recovered several times. Sometimes these expenses are normalized—that is, spread over multiple years. Also, expenses may be adjusted for known changes, such as pending wage increases or imminent decreases in numbers of employees. Expenses included in the revenue requirement are generally referred to as being "above the line." Expenses disallowed by the regulator are generally referred to as being "below the line."

DEPRECIATION

Depreciation is generally viewed as an annual accounting charge for wear, tear, and obsolescence. In regulation, depreciation is viewed as capital recovery—that is, the spreading of the plant investment over time to be recovered in revenue requirement.[4]

The determinants of depreciation are the useful life of the plant, salvage value, and depreciation method.[4] The useful life of the plant refers to the number of years over which the assets are depreciated. There are several ways to determine useful life. For many years, the physical life of the plant was used; this was usually longer than the economic viability of the plant, however. During times of technical and economic stability, actual experience with the length of time this type of plant is in use is appropriate for determining useful life. In most situations, though, historical experience does not match present or future needs because the stability preconditions are not met. Recently, the economic life of the plant, which is the projected length of time in which it will be economical to use the plant rather than replace it, has come to be favored. This approach has the down side of requiring accurate operating forecasts.

Salvage value refers to the market value of the plant at the time it will be removed from service, minus any removal and decommissioning costs. If this is a positive number, which means that the company can obtain some positive economic value for the asset when it is no longer used or useful, the salvage value is subtracted from the cost of the asset before depreciation is calculated. Assume that a building costs $20 million to build and that the company can sell it for scrap for $1 million after its useful life. The net cost to the shareholders for providing capital for the building is $19 million ($20 million minus $1 million); this is the amount that is depreciated. If, on the other hand, the building has contained hazardous waste, and it costs $5 million to clean up the site after the building is no longer in use and the materials are worthless, the amount to be depreciated is $25 million ($20 million plus $5 million).

Several methods are available for spreading the cost of the asset over time, but the most popular method is straight-line depreciation. With straight-line depreciation, the cost of the asset is spread uniformly over its useful life. Assume that the building will be used to store hazardous waste and will be used for 25 years. The annual depreciation using the straight-line method would be $1 million per year ($25 million divided by 25 years).

Straight-line depreciation is often criticized for not reflecting the rate at which the plant actually decreases in value. Generally, plant value decreases rapidly its first few years of service and more slowly in later years. Because straight-line depreciation assumes a constant rate of decrease, it understates actual depreciation in early years and overstates actual depreciation in later years. Slow depreciation rates create problems for companies whose markets are transitioning to competition or are experiencing increasing rates of technological change. If the regulatory depreciation rates are slower than economic depreciation, the company's book value of its plant may be greater than the economic value of its plant.

TAXES

Governments generally require utility companies to pay taxes, including income or profit taxes, franchise fees, property taxes, and excise taxes. Some of these taxes are passed directly to customers (it is common to see franchise fees and excise taxes listed on utility bills). Income taxes, however, are not passed on directly to customers, so it is necessary to include these taxes in the calculation of the revenue requirement. This is done by tax-effecting the revenue requirement. Consider a situation in which a regulator determines that a company's current revenue is $1 million less than the amount of revenue that would be needed to provide an adequate after-tax rate of return. Assume that the income tax rate is 20%. This means that the regulator would need to allow the operator prices that would increase before-tax revenue by $1.25 million ($1 million divided by 0.8, which is one minus the income tax rate).

The income taxes paid by the company may not match the income taxes that should be included in the rate base. This is because the timing in the accounting methods for tax purposes may be different from those used for regulatory purposes. Some countries may allow accelerated depreciation for tax purposes, for example, to encourage business investment. The regulator may prefer to use straight-line depreciation, however. The effect of using straight-line depreciation is that customers are in effect prepaying taxes and providing capital to the utility (one regulatory treatment, called "flow-through," eliminates this effect and removes the benefit to the utility of the favorable tax policy). This customer-provided capital typically receives one of two treatments in revenue requirements. Under the first treatment, the capital is deducted from the rate base. Under the second treatment, the capital is treated as cost-free capital.[5]

CONCLUSION

This chapter describes how regulators use rate of return regulation to control power companies' overall rate levels. Setting the overall rate level is just the first step in a two-step process for setting prices. The second step is rate design, which refers to the price structure or relationship among the individual prices.

In practice, regulators combine elements of rate of return regulation with other regulatory tools, such as benchmarking and price or revenue caps. These combinations allow regulators to customize the amounts of certainty and efficiency incentives that they believe are appropriate for their situation.

REFERENCES

1. Bonbright, J.C.; Danielsen, A.L.; Kamerschen, D.R. *Principles of Public Utility Rates*; Public Utilities Reports, Inc.: Arlington, VA, 1988.
2. Sappington, D.E.M.; Weisman, D.L. *Designing Incentive Regulation for the Telecommunications Industry*; MIT Press: Cambridge, MA, 1996.
3. Averch, H.; Johnson, L.L. Behavior of the firm under regulatory constraint. Am. Econ. Rev. **1962**, *52* (4), 1052–1069.
4. Phillips, C.F. *The Regulation of Public Utilities: Theory and Practice*; Public Utilities Reports, Inc.: Arlington, VA, 1993.
5. National Association of Regulatory Commissioners *Utility Regulatory Policy in the United States and Canada: Compilation 1995–1996*; The National Association of Regulatory Utility Commissioners: Washington, DC, 1996.

Renewable and Decentralized Energy Options: State Promotion in the U.S.

Steven Ferrey
Suffolk University Law School, Boston, Massachusetts, U.S.A.

Abstract

Renewable energy has emerged from the shadows of a novelty electric energy technology. In an era of fossil fuel volatility, renewable power which is immune from fossil fuel availability, pricing and delivery, makes power systems more resilient. Along with cogeneration, renewable energy can increase electric power efficiency, and renewable power can foster environmental benefits. Twenty-four of the 50 states promote these technologies with system benefit charges that underwrite renewable energy trust funds, and/or with renewable portfolio standards. These programs either subsidize or require a certain fraction of renewable electric power in the states' generation mix. Each of these states defines eligible renewable power differently, although all make wind power and solar power eligible. With implementation of these programs, the states have taken the lead on promotion of renewable energy.

MINIMIZING RISKS OF ENERGY SUPPLY DISRUPTION

The electric energy crisis in California during 2000–2001 demonstrated better than any study the unique and critical role of electricity in American society. In the course of a few months, the price of electricity quadrupled, skyrocketing the cost of living, and causing business dislocation. Nonetheless, blackouts rolled across the Golden State.

A shortage of electricity has dire social and political consequences—a blackout has been equated to a natural disaster. Allowing rolling blackouts is a tremendously inefficient way to balance supply and demand differences. Not every consumer attaches the same value to electricity at a given hour. For some industries, even a short blackout can ruin millions of dollars of production; for others, it is a minor inconvenience.

Since the attacks of September 11, 2001, the vulnerability of electric systems to systematic planned attacks has come under more analysis. The United States contains one quarter of all the electric generation capacity in the world. Because electricity cannot be easily stored or rerouted, supply must instantaneously match demand. System interruptions in the United States usually result from transmission and distribution difficulties within one-half mile of a customer. Where an electric system is centralized and integrated, a disruption from attack at a given point can temporarily destroy large parts of the integrated network.

Analysts argue that a distributed energy system, including increased use of cogeneration, is much less subject to disruption—whether it is from weather, terrorism, or other factors—than the centralized generation and distribution system employed in the United States. A distributed generation system typically is an electric generator placed on the consumer's side of the electric meter. It may be owned by the customer, the utility in rare instances, or a third party. By being so placed, it can either supply the host consumer, feed some or all power to the grid, or be wired to supply power to the host and selected abutters on dedicated distribution lines. When coupled with use of the thermal by-product of the generation process, distributed generation is known as cogeneration. A cogeneration system produces both electric energy and useful thermal energy. It thus uses some of the approximately 45%–70% of that portion of the output of a conventional electric power plant that is thermal energy and wasted as thermal pollution to the environment. Cogeneration is defined by federal law as a system that utilizes a minimum of 5% of the energy output as useful thermal energy while taking the remainder of the energy output as electric power. The robustness of a distributed, on-site, cogeneration-based system, likely fueled by natural gas, results from

- Reliance on a larger number of small generators; no one of which is critical to huge amounts of supply.
- Less reliance on a vulnerable centralized transmission and distribution grid.
- Reliance on the movement of natural gas in a more protected underground fashion to the electric generation source near the load center; rather than reliance on above-ground, more vulnerable electric transmission infrastructure. Gas can be stored in pipelines,

Keywords: System benefit charges; Renewable trust fund; Renewable Portfolio Standards; Green electricity; Renewable; Decentralized; Cogeneration.

while electricity cannot be stored in transmission lines, especially where they are knocked out.

EFFICIENCY AND CONTROL

From an efficiency point of view, there are significant reasons to promote decentralized on-site electricity supply. Decentralized electric production can transform electric production efficiency from approximately 33% for central station conventional utility supply to something approaching 80% for decentralized cogeneration. These decentralized electric supply technologies, in addition to greater potential efficiency and in certain circumstances, environmental benefits, tend to encourage the deployment of renewable energy sources and applications.

Because renewable energy sources are not under the control of any nation or cartel but are distributed across the earth, they are not subject to embargo or manipulation. Because decentralized renewable energy sources are developed in relatively small modules, this promotes reliability and resiliency of the system. Because decentralized energy resources are built close to their points of use, they are not as dependent on long transmission and distribution networks and are less vulnerable to supply disruption from an overloaded system line, storm, or intentional disruption.

Unlike finite fossil fuels, solar energy represents a constantly replenished flow, rather than an existing stock that is diminished by its use. By contrast, burning a barrel of oil or a cubic meter of natural gas diminishes permanently that quantity of fossil fuels for the next day and for future generations. While many nations—particularly developing nations—have no significant reserves of oil, coal, or natural gas, every nation has solar energy in some form—sunlight, wind, or ocean wave power.

GREATER EFFICIENCY OF COGENRATION AND DISTRIBUTED ENERGY OPTIONS

Both conventional electric generation technologies and industrial process heat applications are inefficient. Conventional electric generating technologies typically exhaust as much as two-thirds of the heat energy produced to power electric generators. Many industries use process steam in applications that run below 400°F; however, the combustion of fossil fuels required to produce that heat results in temperatures of more than 3000°F, much of which is wasted. The next major leap in efficiency must come from recovering and reusing waste heat. Machines that recover all waste heat and produce electricity have the capability to achieve efficiencies from 50 to 90%, much better than the typical 30+% of the existing central station utility fossil steam system. Thus, cogeneration facilities operate at overall thermal efficiencies as great as 250%–300% higher than conventional electric generating technologies. The very best cogeneration technologies are more than twice as efficient as new coal-fired power plants. This increase in efficiency results in savings of up to 31% in the fuel input needed to generate a unit of usable energy output by various cogeneration technologies when compared to conventional electricity generation technologies.

Distributed generation and cogeneration systems, because they are smaller, tend to be less efficient at electric production than large central station power generation facilities. It has been estimated that distributed generation would be 23% more expensive to implement in Florida (where cooling requirements dominate) and 27% more expensive to implement in New York state (where heating requirements dominate) than a new centralized system.

However, this comparison looks only at electric production. If one assumes that waste heat from distributed cogeneration can be employed productively, the economics change: a distributed cogeneration model realizes cost savings of 30 and 21% in New York and Florida, respectively. Interestingly, as the cogeneration units get smaller, total system savings increase. This is due to the improved cost profiles of new gas turbines and internal combustion engines, and the ability of smaller units to meet variable loads more efficiently. Also, small units can be sited where waste heat can be most productively used. Thus, the cogeneration value of distributed generation turns the economics from negative to positive because of the greater overall efficiency of energy production and use.

The heat recovered from a total cogenerating energy system can be used for direct application heat, for industrial process heat, or for preheating the combustion air for a utility boiler. This means that more useful energy can be produced while generating fewer environmental pollutants and emissions. It also means that less transmission capability would be required if there is development of dispersed electric and total energy systems, located close to load centers. Not only will additional transmission capacity not be required in certain areas, but capacity on existing transmission grids will be less burdened. One way to view this phenomenon is that if natural gas cogeneration or total energy systems replace centrally dispatched electricity, energy will be moved more in its primary form by natural gas pipelines and less in its derived form as electricity.

ENVIRONMENTAL BENEFITS

Conventional production of electricity by electric utilities in the United States is responsible for substantial shares of criteria pollutant emissions, including 68% of SO_2

emissions, 33% of NO_x emissions, and 33% of CO_2 emissions. Environmental costs associated with power plants occur at each of three stages of the energy process:

- Front-end costs at the point of extraction and processing of energy sources. These include the costs of drilling, mining, or otherwise extracting raw fuel sources; the processing, enrichment or concentration on these fuel sources; the manufacture of equipment to effectively utilize these fuel sources; and transportation costs for fuel and equipment.
- Direct costs associated with the use of energy sources. These include the emission of a variety of pollutants, health impacts from these emissions, impacts on the natural environment of such emissions, and human occupational exposure or illness at the power plant work site. The primary effects on human populations are the increased risk of mortality and morbidity, including chronic illness and increased risk of chronic disease.
- Back-end residual costs. These include waste disposal costs for residual elements of fuel and the eventual costs of decommissioning energy producing facilities.

The primary impacts on human health from direct production of electric energy are from emissions of the criteria pollutants carbon dioxide, sulfur dioxide (SO_2), NO_x, ozone, and particulates; and from acid deposition. Carbon dioxide is caused principally by the burning of fossil fuels, and it is a principal greenhouse gas responsible for global warming. Sulfur exerts a significant impact on human health directly, it is a precursor of aerosols that result in acid deposition, and it is transformed into sulfates, which pose independent problems. NO_x is formed by the conversion of chemically bound nitrogen in the fuel or from thermal fixation of atmospheric nitrogen in the combustion air. Ozone causes damage to human health, agriculture, and plant life. Particulates include solid particles and liquid matter which range in size from 1 micron to more than 100 microns in diameter. They are responsible for major health impairment, impairment of visibility by causing haze, and the creation of sulfated from SO_2 emissions. Acid deposition causes damage to forests, wildlife, water quality, and aquatic species. Conventional power facilities exert environmental impacts on health and the environment in the form of water pollution and impairment of land uses. The choice of fuels as well as the technology used for converting those fuels to electricity has profound implications for attaining CO_2 reduction targets to limit possible effects of global warming.

Cogeneration facilities should cause fewer environmental impacts than equivalent megawatts of conventional power production because cogeneration facilities simultaneously produce electricity and thermal energy by the same process, thereby recapturing and utilizing energy that would otherwise be wasted. This substitution of an integrated cogeneration technology for conventional separate electricity and thermal energy production technologies should save 15%–25% of the energy input otherwise consumed by separate energy production configurations, according to government studies.

Typical air emissions of technologies, without added emission controls, are displayed in Table 1.

STATE OPTIONS FOR PROMOTION OF RENEWABLE TECHNOLOGIES

There are several recognized techniques capable of deployment to promote renewable energy and demand-side management (DSM) investments. Each attempts to require or subsidize certain preferred technologies that otherwise might be less demanded by the market.

System Benefits Charge/Renewable Trust Funds

The system benefits charge is a tax or surcharge mechanism for collecting funds from electric consumers,

Table 1 Air quality impacts of cogeneration

Technological characteristics	Direct physical effects	Impact on air quality: positive, negative or mixed
Increased efficiency	Reduction in total emissions per unit of energy produced	Positive
Smaller scale of QF deployment	Change in emissions levels	Mixed or negative
	Change in level of environmental control, usually less	Negative
	Lesser emissions stack height	Mixed or negative
Change in energy production technology	Change in emissions and type of pollutants	Mixed
Change of fuels	Change in emissions and type of pollutants	Mixed or positive
Change of location of electricity generation	Change in location of emissions, density, and distribution	Mixed

Table 2 Funding levels and program duration

State	Approximate annual funding (millions of dollars)	Per capita annual funding	Per MWh funding	Funding duration
California	$135	$4.0	$0.58	1998–2011
Connecticut	$15–$30	$4.4	$0.50	2000–indefinite
Delaware	$1 (maximum)	$1.3	$0.09	10/1999–indefinite
Illinois	$5	$0.4	$0.04	1998–2007
Massachusetts	$30–$20	$4.7	$0.59	1998–indefinite
Montana	$2	$2.2	$0.20	1999–7/2003
New Jersey	$30	$3.6	$0.43	2001–2008
New Mexico	$4	$2.2	$0.22	2007–indefinite
New York	$6–$14	$0.7	$0.11	7/1998–6/2006
Ohio	$15–$5 (portion of)	$1.3	$0.09	2001–2010
Oregon	$8.6	$2.5	$0.17	10/2001–9/2010
Pennsylvania	$10.8 (portion of)	$0.9	$0.08	1999–indefinite
Rhode Island	$2	$1.9	$0.28	1997–2002
Wisconsin	$1–$4.8	$0.9	$0.07	4/1999–indefinite

Source: From U.S. Department of Energy (see Ref. 1).

the proceeds of which may then support a range of activities. In order to support DSM or renewable resources, funds are collected through a nonbypassable system benefits charge to users of electric distribution services. The money raised from the system benefits charge is then used to "buy down" the cost of power produced from sustainable technologies on both the supply and demand side, so that they can compete with more conventional technologies. A system benefits charge will raise the following issues: the level of the charge, the allocation to classes of customers, the rate design, the programs to be implemented, and the ongoing process for oversight and management of the fund.

Between 1998 and 2012, approximately $3.5 billion will be collected by 14 states with existing renewable energy funds. More than half the amount collected, at least $135 million per year, comes from just California. The funding levels range from $0.07/MWh in Wisconsin up to almost $0.6/MWh in Massachusetts. Most only provide assistance to new projects, and not existing renewable projects.

The form of administration of renewable trust funds varies. Many states administer them through a state agency, while others use a quasi-public business development organization. Some funds are managed by independent third-party organizations, and some are managed by existing utilities, while two states allow large customers to self-direct the funds. For distribution, some states utilize an investment model, making loans and equity investments. Other states provide financial incentives for production or grants to stimulate supply-side development. Some other states use research and development grants, technical assistance, education, and demonstration projects.

As Table 2 indicates, the funding level is in the range of $175–$250 million annually for the cumulative impact of the fourteen state system benefit charge programs. While many of these programs are set up to run indefinitely, others have set life spans. The level of per capita funding ranges between $0.90 and $4.40 annually for renewable energy. Expressed another way, for each megawatt-hour sold in the state, the level of subsidy ranges from $0.07 to $0.59.

Renewable trust funds are likely to be less efficient than portfolio standards in promoting the burgeoning renewable power industry. Portfolio standards set a requirement and challenge market participants to satisfy it in the most efficient manner possible. By contrast, trust funds create a discretionary gift program. This process will cause renewable projects to conform themselves to funding criteria rather than to take the initiative to operate most efficiently. Political manipulation of trust fund cash flows also is possible.

Renewable Resource Portfolio Requirements

A resource portfolio requirement requires certain electricity sellers or buyers maintain a predetermined percentage of designated clean resources in their wholesale supply mix. A number of variations of resource portfolios are possible, including a renewable resource portfolio requirement, a DSM portfolio requirement, and a fossil plant efficiency portfolio requirement.

While Massachusetts and Connecticut are the first two states in the initial wave or retail deregulation to adopt both a system benefit charge to fund renewable technologies and a resource portfolio standard mandating renewable wholesale power sources, 22 states and the District of Columbia have adopted the renewable portfolio standard. The key to making the portfolio requirements work is to establish trading schemes for "portfolio obligations." A trading scheme would allow distribution companies, energy service companies (ESCOs), or generation companies that are particularly effective at developing low-cost DSM programs or renewable resources to sell portfolio obligations to other distribution and generation companies that are less effective in developing these resources. (ESCOs are companies that achieve conservation savings on customers' premises, and often split such savings with the customer or charge a fee for service.)

Portfolio standards are flexible, in that certain technologies can be included in the renewables definition or certain subgroups of technologies can be targeted for inclusion at distinct levels. The standard allows market competition to decide how best to achieve these standards. The standards become self-enforcing as a condition of retail sale licensure.

The advantage of a portfolio standard is that it does not subsidize any particular technology or locus of that technology. Resource portfolio requirements can be applied under any wholesale or retail competition, without placing any entities at a disadvantage.

A disadvantage to implementing portfolio resource requirements is that the appropriate portfolio target level must be decided on, rather than relying on general financial incentives towards continuous improvements. The primary disadvantage for DSM standards is the logistical challenge of monitoring tradable obligations for DSM, which requires continuing regulatory oversight for measuring and monitoring DSM savings. A renewable standard will also involve subsidiary issues, such as how to define renewable resources and whether standards should be based on available capacity or actual generation; though from an environmental perspective, basing them on actual kWh of generation makes the most logical sense.

"Green" Electricity Pricing

Certain commodities with an "environmental" identification sell for more than generic commodities. This is true for organic foods, "pure" spring bottled water, and products made from recycled parts, for example. The average price of bottled water is 1000–4000 times the cost of tap water in the U.S. Bottled water often is tap water in a bottle. This indicates that certain brand identities or green products command a substantial premium in the market of commodities.

Table 3 Portfolio standards and trust funds in deregulated states

State name	Renewable energy trust fund	Portfolio standards
Arizona		x
California	x	x
Colorado		x
Connecticut	x	x
Delaware	x	x
District of Columbia	x	x
Hawaii		x
Illinois	x	x
Iowa		x
Maine	x	x
Maryland		x
Massachusetts	x	x
Minnesota	x	x
Montana	x	x
Nevada		x
New Jersey	x	x
New Mexico		x
New York	x	x
Ohio	x	
Oregon	x	
Pennsylvania	x	x
Rhode Island	x	x
Texas		x
Vermont		x
Wisconsin	x	x

The operational premise of so-called "green pricing" is that there are certain customers who want to use electricity produced by "clean" or renewable technology and are willing to undertake the expense in order to procure it. Electricity sellers can then use funds raised from such environmentally oriented customers to acquire less-polluting resources that would not otherwise be developed because of market costs or market barriers. One advantage of green pricing is that providers can protect customers from fuel price fluctuations, as most renewable resource prices do not experience the volatility of oil and gas-fired electricity, which fluctuate with fossil fuel commodity prices.

Renewable power that qualifies as "green" energy can be marketed and sold in different modes:

- Actual kWh sales, in which electricity and its green attributes are bundled together as a single product in a single transaction; there, the purchaser switches from

conventional power to a bilateral contract with the green energy supplier.
- Sale only of the green attributes without purchasing the actual kilowatt hours; the retail customer does not need to switch from buying conventional power but only purchases the renewable certificate.

The drawback of green power marketing is that it relies on individual consumer decisions to create a public good. The environmental benefits of green power consumption are not internalized to the consumer who elects to pay a premium for green power, but rather are shared by all in the region. This allows free riders to benefit without having to pay or contribute. This raises equity and efficiency issues. Therefore, green power marketing suffers from individual consumer motivation and equity impediments that are not raised by a portfolio standard, which imposes requirements on energy producers rather than consumers.

In the deregulated Pennsylvania market, over 500,000 residential customers switched to competitive suppliers. Of those, 15%–20% chose green electricity products, paying a premium of 0.8–1.5 ¢/kWh over standard electricity pricing. In California, almost 100% of customers who switched retail suppliers prior to the California restructuring debacle consumed a green power

Table 4 "Renewable" resources as defined in state statutes

State	Solar	Wind	Fuel cell	Methane/ landfill	Biomass	Trash-to-energy
California	x	x		x	x	x
Connecticut	x	x	x	x	x	x
Illinois	x	x			x	x
Maine	x	x	x		x	x
Massachusetts	x	x	x	x	x	x
Nevada	x	x	x			
New Jersey	x	x	x	x	x	x
New Mexico	x	x	x	x	x	x
New York	x	x		x		x
Oregon	x	x		x		x
Pennsylvania	x	x		x	x	x
Rhode Island	x	x		x	x	x
Texas	x	x		x	x	x
Wisconsin	x	x	x		x	x

State	Hydro	Tidal	Geothermal	Photovoltaic	Dedicated crops
California	x		x	x	
Connecticut	x			x	
Illinois	x			x	x
Maine	x	x	x	x	
Massachusetts	x	x		x	x
Nevada			x	x	
New Jersey	x	x	x	x	
New Mexico	x	x	x	x	
New York	x	x	x	x	
Oregon	x	x	x	x	x
Pennsylvania	x		x	x	x
Rhode Island	x			x	
Texas	x	x	x	x	
Wisconsin	x	x	x		x

Note: "Photovoltaic" is likely included within "solar" in some states; "methane" and/or "trash-to-energy" may be included within a broad definition of "biomass."

product, paying a premium of 1.1–2.5 ¢/kWh over a standard electricity price. In California, this resulted because of a state renewable energy credit of 1.5 ¢/kWh to green power suppliers, allowing them to undercut the default service price. When the California market collapsed, almost all of these green consumers were returned to standard utility service. By 2002, "green-e" certified products were available in Connecticut, New Jersey, Pennsylvania, and Texas, offered by four suppliers. Where utilities sell green power products, participation rates are typically 3%–4%, with customers paying premiums ranging from 1 to 7.6 ¢/kWh.

While public opinion surveys reveal that 50%–95% of Americans say they are willing to pay more for renewable power, in fact, only approximately 1% of residential consumers who have an ability to select and purchase renewable energy power have done so. On the other hand, bottled water reached 8% market penetration, socially responsible investing a 13% market penetration, and recycling a 25% market penetration over an extended period time of 1–2 decades from the original product launch. Customer participation rates in green power programs, as of 2001, were less than 1% in more than half of programs available.

The ability of green pricing mechanisms to promote cleaner resources will depend on how willing customers are to pay for them, as well as the success of the electricity sellers' marketing and promotion campaigns. Any rate impacts associated with environmental improvement are assigned to those customers who choose to accept them. As a result, green pricing does not cause any market distortions and is, in theory, reflective of true customer preferences. Green pricing effectively assigns an important social policy decision to individual customers, who may act on their own interest rather than in society's best interest.

STATE PROGRAMS FOR RENEWABLE ENERGY

As of 2004, 22 states and the District of Columbia had enacted either a law or an administrative order to create open power markets, thus allowing consumers to choose their electricity supplier. States deregulating their retail electric sectors have implemented renewable portfolio standards and/or trust funds. Twenty four states have elected one or both of these options (see Table 3).

The renewable resource measures that states have incorporated into electricity restructuring and deregulation statutes vary. Some renewable energy measures create portfolio standards; others create trust funds to invest in the development and utilization of renewable resources. Some adopt both concurrently. How each defines an eligible renewable resource varies significantly.

Table 3 illustrates how states have deployed these two options. Each defines what an eligible renewable resource is differently. The diverse pattern of "renewable" resources included under state definitions is set forth in Table 4.

REFERENCES

1. Bolinger, M.; et al. Clean Energy Funds, LBNL-47705 (2001) at ix.

Renewable Energy

John O. Blackburn
Emeritus of Economics, Department of Economics, Duke University, Durham, North Carolina, U.S.A.

Abstract

Renewable energy sources in the United States include solar heat and electricity and indirect solar sources like wind, falling water, biomass, wave, ocean current, and ocean temperature differences, as well as geothermal and tidal possibilities. The first two sources, solar radiation and wind, are very large. Solar radiation exceeds total U.S. energy use by a factor of 440, while the wind energy potential is three times the annual electricity use. Hydroelectric, biomass, and geothermal sources are commercially developed, as well, and can supply half the nation's energy. The largest sources are also intermittent in nature, thus raising issues of complementarity and storage in high-renewable energy scenarios.

U.S. energy use, now in the 100-quad range, can be greatly reduced by efficiency gains. Energy end-uses in the forms of heat, liquid fuels, and electricity are examined. Reduced energy demands can be met largely or wholly via renewable sources, thus preserving relatively high living standards in a sustainable manner, given time, commitment, and further development of two or three key technologies.

INTRODUCTION

Renewable energy is a category of energy supply that is receiving increasing attention in the United States. The term refers, of course, to energy sources that can be fully utilized without diminishing future supplies. The obvious contrast is with fossil fuels which, once used, are no longer available.

Renewable energy is certainly nothing new. Mankind lived on it almost exclusively until the industrial revolution in the 1700s. Falling water and wind provided mechanical energy for tasks like pumping water or grinding grain, while ships were propelled by wind. Wood fires provided heat for dwellings or for tasks like smelting metals.

This article examines the various renewable energy resources in the United States and the status of conversion technologies. Because the largest two sources are also intermittent, integration into reliable supply systems is also examined. This turns out to be a less formidable task than it first appears to be.

Present energy end uses are examined in order to see how renewable sources might be deployed in a highly renewable energy future.

RENEWABLE ENERGY FORMS

Renewable energy comes in many forms, though most sources come directly or indirectly from solar radiation. Solar energy may be used directly as a source of heat as, for example, in heating water. It may also produce electricity, either directly in photovoltaic cells or by using concentrating collectors to produce steam for turbines. Wind energy, which is derived from solar energy, is now used almost exclusively to produce electricity, as is falling water. Winds over ocean waters produce waves, which are beginning to be harnessed as a source of electricity. Ocean currents or temperature differences between surface and deep waters also offer opportunities to extract useful energy.

Trees, crops, and other plant matter (referred to as "biomass" in the renewable energy literature) likewise represent stored solar energy. Half the world still uses wood as a fuel, while forest product industries in the industrial world use wood waste to produce process heat, steam, and electricity. Energy crops, trees, or even grasses could be grown. Fuel alcohols are now produced mostly from sugarcane and corn, and they are beginning to be produced from cheaper, more abundant cellulosic biomass materials, such as wheat straw or other waste products. Other biomass sources include animal manures and sewage, which can produce methane through anaerobic digestion. There are a few renewable sources of energy that do not derive from solar sources.

The moon's gravitational pull on the earth creates tidal movements of ocean waters. In some locations, the rise and fall of water levels is sufficiently great to generate electricity.

Heat generated from nuclear processes deep in the earth sometimes comes close enough to the surface to be captured as useful energy. When underground water is present as well, steam or hot water may be available for electricity generation.

The principal difference between age-old uses of renewable energy and the present possibilities is the

Keywords: Solar energy; Wind energy; Renewable resources; Biomass energy; Renewable energy future; Intermittent energy sources.

availability of modern technologies with conversion efficiencies much higher than those previously available. Most importantly, all forms of renewable energy can be converted into electricity—the most versatile form of energy for human purposes.

POTENTIAL RENEWABLE RESOURCES IN THE UNITED STATES

Solar radiation falling on the lower forty-eight states amounts to an estimated 44,000 quads (quadrillion Btu) annually—440 times the annual energy use for the nation.[1] Few other resources are found in such abundance. Windpower, the next most abundant renewable energy source, has an estimated electricity-generating potential of 10,000 billion kilowatt-hours in the United States, almost three times present annual electricity consumption.[2] This estimate is for on-shore wind turbines only; the additional offshore potential is large, but has not been as carefully estimated. Moreover, recent studies find the U.S. potential to be underestimated in quantity and in the number of areas with economic wind-generating potential.[3]

Hydroelectric facilities already generate 7%–8% of the nation's electricity, with another 3%–4% that might still be developed.[4] Geothermal resources that are suitable for electricity generation are found only in the western third of the United States, but they amount to 25,000 megawatts—enough to provide 25%–30% of that area's present electricity use. Geothermal heat is available to more of the nation, with a much larger potential.[5] The wave-generating potential has not yet been fully assessed in the United States. It appears to be substantial along the Pacific coast and the North Atlantic coast. Tidal power possibilities in the United States are limited to Alaska, Maine, and Washington.

As for biomass sources, U.S. forests and farms now produce materials with an energy content of some 14 quads. With energy crops, this could increase fourfold, according to estimates of the National Renewable Energy Laboratory.[6] Other recent estimates give a near-term potential of 19–20 quads over and above materials taken for food, feed, and forest products, given relatively minor modifications in farm and forestry practices.[7]

There is clearly no shortage of renewable energy. It comes in many forms and, with respect to solar radiation and windpower, in great abundance.

In order to be useful in human activities, renewable energy must be captured and converted into forms which are consistent with energy demands. Time gaps between availability and need must also be bridged. For these tasks, equipment that is reliable, durable, and affordable must be available.

CONVERSION EFFICIENCIES

Conversion efficiencies for various forms of renewable energy vary considerably. The growth of trees and other plant life generally convert only 0.5%–1% of solar radiation into stored biomass energy, which can be used commercially. The fastest growing species in areas with long growing seasons (sugarcane, for example) convert 2%–3% of solar energy into recoverable energy.[8]

Other renewable technologies, fortunately, have much higher conversion ratios. Solar panels for heating water convert 40%–55% of the solar energy that falls on their surfaces to useful energy.[9] Crystalline silicon photovoltaic cells now on the market convert 11%–14% of solar radiation into electricity, lab models reach 33%, and nanotechnologies hold some promise of yet-higher figures.[10] It is these conversion efficiencies that facilitate the use of large quantities of renewable energy collected in a relatively small portion of a nation's total area.

THE STATE OF TECHNOLOGIES: COSTS

Technologies Now Fully Developed and Cost Competitive

Hydroelectricity has been competitive since the earliest days of grid electricity. Expansion in the United States is constrained by environmental considerations, though output can be expanded by repowering existing facilities or installing generators at small existing dams. Hydro capacity becomes extremely valuable in grids with high components of intermittent power from wind and solar sources because it enables them to operate over a considerable range without further backup.

Wind energy is competitive with conventional fuels in areas with good wind resources and its costs continue to fall.

Geothermal Electricity

This technology is now well established in suitable sites around the world. Initial capital costs are high, both for drilling into steam or hot water sources and for generating equipment. California has 2200 megawatts in operation, and new facilities are planned or under construction in several Western states.[11]

Biomass

About three quads are already used annually for energy in the United States. The forest products industries use 1.8 quads of wood and wood wastes for process heat and cogenerated electricity. Firewood heats three million homes and supplies some occasional heat to another 20 million. These uses are fully competitive with fossil fuels, as is methane collected from landfills.[12] Other biomass possibilities are listed in the following section.

Ocean Sources

A small number of tidal plants have been installed around the world, but none have been installed as of yet in the United States. Other technologies for using ocean energy, except for waves, are not under active development at this time.[13]

Technologies that are Well Developed but not Yet Fully Cost Competitive

Solar Water Heating

This is a relatively simple technology consisting basically of water pipes in a box covered with glass and painted black. Today's versions are, to be sure, much more sophisticated, with freeze-protection features to make the devices useable nationwide. Making costs competitive with present energy sources will be possible upon mass production and mass installation of the equipment, a feat long since accomplished in Israel.

Photovoltaic Electricity

Fifty years of research and development efforts have brought higher efficiencies and greatly reduced costs. Costs are still too high, by a factor of two or three, for photovoltaic electricity to be fully competitive on customer facilities in the United States. They are already competitive for users remote from a power grid or for utilities at peak summer hours.

Cost reductions in photovoltaic systems manufacturing are proceeding more slowly than anticipated in the 1980s and 1990s, while conventional generation, with which these systems compete, has shown declining inflation-adjusted cost, as well. Authoritative analyses in the 1980s projected cost reductions of photovoltaic modules and other system costs to about one dollar per peak watt (two dollars in 2005 prices), which implied Photovoltaic (PV) electricity costs of six to seven cents per kilowatt-hour.[14] Module prices have indeed declined to about $3.50 for volume purchases in 2005 (in current $).[15] Costs are not likely to fall significantly until some alternatives to the (now) standard silicon wafer are developed for volume production. Nonetheless, manufacturing costs have fallen below three dollars for modules.[16] With rapid market and manufacturing development, Japan can produce photovoltaic electricity at about 15 cents per kwh, competitive with (expensive) Japanese-delivered grid electricity.[17]

Photovoltaic equipment benefits from its potential location at the point of consumption, thus avoiding most of the costs of transmission and distribution. PV electricity, therefore, can be priced higher than electricity from central generating plants. Because there is already enough rooftop space to accommodate nearly any conceivable volume of photovoltaic power, site costs are avoided as well.

Other Biomass Sources

Subsidized fuel ethanol is made mostly from corn, but more abundant cellulose sources are beginning to be utilized. Scale economies and cost decreases with production experience should make this technology competitive in the next few years. Iogen, the Canadian producer of ethanol from wheat straw, estimates a price of $1.30 per gallon from a proposed plant in Idaho.[18] Economically feasible processes and equipment to produce methane from animal wastes are already available, but the technology is still not widely used.

Technologies for Which Further Development is Required

Solar Industrial Process Heat

Development of equipment for these tasks received some attention in the late 1970s, but was abandoned after that as fuel prices declined. Industrial process heat requirements and insolation overlap considerably in the United States, so their potential remains to be exploited.[19] Further development awaits higher fossil fuel costs and some renewed research attention.

Wave Energy

Several small demonstration facilities now operate. The first full-scale commercial generation facilities are about to be installed in Portugal by British and Portuguese firms using British designs.[20]

Ocean Currents and Thermal Differences between Surface and Deep Ocean Waters

There is very little activity at present.

In summary, hydroelectric, wind, and geothermal electricity are fully cost-competitive with fossil and nuclear-generated electricity in areas with good resources. These areas are, respectively, three-quarters of the states for wind, all but three or four states for hydroelectricity, and the western third of the nation for geothermal. Biomass-based heat and electricity are likewise well-established sources and they are available in every state. Widespread market penetration of solar hot water systems awaits economies of mass production and mass installation. Biomass-based liquid fuel production is set to expand rapidly. Sales of photovoltaic equipment are rising rapidly, but still with subsidies and incentives. Cost reduction proceeds apace at about 4%–5% per year, while several promising new and potentially much cheaper versions are moving into production.

Solar industrial process heat is receiving little attention, though it probably will as oil and gas prices trend upward in coming decades.

DEALING WITH INTERMITTENT SOURCES

Solar radiation and wind are the most abundant renewable energy sources, but they are also intermittent in nature. This feature raises issues of complementarity and energy storage. Utilities think of generating capacity in terms of baseload, intermediate, and peaking capacities. Because intermittent sources do not fit into any of these categories, they are likely to be relegated to marginal roles.

Inquiries that pose the question of usefulness in another way have produced much more favorable conclusions. Utilities are already accustomed to demand patterns that are subject to seasonal, daily, weather-related, and random fluctuations. Their supply sources, in contrast, can be turned on or off as they wish. The more fruitful approach in considering intermittent sources is to regard them as "negative demand"—the side which is already subject to variations. Models that start with actual hour-by-hour demand loads through the year and then subtract out wind or solar contributions, hour-by-hour, consistently show sizeable potential contributions from intermittent renewables without further backup. Needless to say, the presence of hydroelectric facilities or pumped storage units supports the ability to accommodate intermittent sources.

A simulation of the British electricity system showed that windpower, though not steadily available and not correlated with demand patterns, could still meet 25%–45% of system demand without additional backup, provided that the conventional components could be reconfigured to accommodate wind patterns.[21] A U.S. simulation found that in a system with demand patterns similar to those of most utilities, a grid with a substantial hydroelectric component could accommodate 50% of its outputs from intermittent renewable sources. (i.e., wind and solar) and at costs no higher than those for conventional generation.[22] These are certainly counter-intuitive results, and they would need to be supplemented by modeling studies of other utilities. The findings, though, are consistent with experiences in Denmark and North Germany. Denmark now obtains 21% of its electricity from wind turbines, without evident problems of integrating this output into the national grid. German reports of strains on the grid seem to have more to do with transmission bottlenecks than generation sources.

Solar electricity has the advantage, for most utilities, of coinciding in availability with summer peaks in air-conditioning use. A recent New Jersey study found that photovoltaic generation would not only supply peak power (and thereby displace fuel generation) but 40%–70% of PV capacity would displace the capital costs of conventional generation, fractions which rose quickly in the presence of grid storage capacity.[23]

These results are so much at variance with traditional thinking in utility managements that time, experience, and external pressures will be required to bring solar and wind technologies into widespread use, even when they are cost-competitive. Therein lays the case for incentives and regulatory requirements, such as the renewable portfolio standards now found in 18 states. Expanding markets enable the solar and wind equipment manufacturers to scale up and further reduce costs, while the utilities gain actual operating experience with intermittent sources.

MOVING TOWARD AN ALL-RENEWABLE ENERGY FUTURE

A consideration of energy futures in which renewable sources contribute most or all energy needs begins with an assessment of the quantity of energy that is likely to be needed, and its appropriate forms. As a starting point, one examines present energy sources and end uses in the economy, and then allows for increasing efficiencies. One must show that renewable resources are sufficient in quantity, quality and cost to provide for each of the end uses.

A word of caution is in order. It is much too easy to focus on energy supply, while neglecting the rich possibilities for improving energy efficiency in all end uses. One thus can miss the most economical opportunities for meeting energy demands because efficiency measures usually are less expensive than any new supplies, renewable or conventional. The reader is referred to the many articles in this encyclopedia that deal with the significant potential of increasing the efficiency with which energy is used.

Present U.S. Energy use

U.S. primary energy use is now running at about 100 quads per year and growing at about the same rate as population, unlike the growth pattern prior to 1973. Data for 2003 are shown in Table 1.

As the table indicates, primary energy use in the United States in 2003 was 98.7 quads. About 40% of this primary energy was used to generate electricity, and two-thirds (26.7 quads) of that was lost as waste heat in generation and transmission. End-use energy was thus 72 quads. Transportation and industry each take about one third of

Table 1 Fuel and energy use by sector, United States, 2003 (Quadrillion Btu's)

Sector	Fuel	Electricity	Total end energy use	Electricity losses	Total primary energy	Fuel end uses in sector
Residential	7.2	4.4	11.5	8.7	21.3	Space heat, Water heat
Commercial	4.3	4.1	8.4	9.2	17.6	Space heat, Water heat
Transportation	26.8	—[a]	26.8	0.1	26.9	Liquid fuel
Industry						
Manufacturing						
Agriculture, Mining	15.8	3.4	19.2	7.7	26.9	Process heat
Construction	3.0	—[a]	3.0		3.0	
Feed stocks	3.0	—	3.0		3.0	
Total industry	21.8	3.4	25.2	7.7	32.9	
Total all sectors	60.1	11.9	72.0	26.7	98.7	

[a] Less than 0.1 quad.
Source: Adapted from Energy Information Administration (see Ref. 25).

end-use energy, with the rest going to the residential and commercial sectors. In U.S. energy statistics, electricity losses are allocated to each of the sectors in order to see the primary energy demand occasioned by the activities of each sector.

Table 1 also lists the kinds of energy end-uses provided by fuels because these would have to be supplied from renewable sources in the absence of fuels. (Sector end-uses for electricity are not shown, since electricity is readily produced from renewable sources.) In the residential sector, for example, fuels are used mostly to heat space and water or, put differently, to provide low temperature (under 100 degrees centigrade) heat. A much smaller use of fuel is for cooking, which utilizes temperatures in the 100–250 degree centigrade range. Fuel use in the commercial sector is applied for similar purposes. Eighty percent of industrial energy use is in manufacturing, while 20% goes to mining, agriculture, and construction or is used for fuel-based feedstocks. Manufacturing end-use energy is 19 quads of which three to four quads is for refining petroleum.

Energy Efficiency Gains

All of these uses can be reduced with gains in energy efficiency, and some of them can be reduced significantly. Indeed, if the best practices already existing in building design and construction, lights, appliances, heating and cooling systems, vehicle mileage, and industrial processes were uniformly applied in the whole economy, end-use energy demands use would be 50%–70% lower, or in the 22–36 quad range—not the 72 quads shown in Table 1.[24] These potential reductions in energy use make renewable energy futures much more readily conceivable for high-income, high-output nations. Any energy supply system in a world of diminishing oil and gas output would be hard pressed to meet the energy demands of our present inefficient economy.

The Adequacy of Renewable Resources

The questions for renewable energies then become:

1. In the United States, can renewable energy sources provide 22–36 quads, of which six to nine or so would meet efficiency-reduced electricity demands, three to five quads would be used for heat in buildings, seven to ten quads or so would power the transportation sector, and six to ten quads would be composed of fuels for industry?
2. If that can be shown, then can these sources provide for any continuing growth in the economy?
3. What technological developments are needed for this potential to be realized?

The adequacy of renewable energy sources, considered in gross, is easy to establish. Wind and solar resources alone provide these amounts of energy many times over. In addition, the "nonintermittent" renewable sources (biomass, hydroelectric, and geothermal resources) sum to another 17 quads at a minimum.

As to end-uses, six to nine quads of electricity fall well under available sources, with storage capability being required only in areas lacking hydroelectric resources.

Space and water heat can be provided in part from direct solar sources, with the rest coming from biomass or perhaps electricity.

For those who envision a hydrogen-fueled transportation system, intermittent sources of renewable electricity are tailor-made. Hydrogen can be produced by electrolysis whenever solar or wind sources exceed grid demands. Liquid fuels from biomass are likely to be available, as well, especially with emerging technologies to produce liquid fuels from cellulosic materials.

Industrial process heat would come primarily from biomass sources, with some assistance from solar equipment or perhaps hydrogen produced from renewable electricity.

As to growth over time in energy demands over time, one can observe that economic growth in the United States, at least in per capita terms, is directed toward sectors that generally do not have high energy contents (services vs basic commodities). If the United States is ever to reach a measure of sustainability with respect to its physical resource demands, its population must level off, as must that of any other nation seeking sustainability.

There are two major technological developments that remain to be achieved in order to make an all-renewable future attainable. The first is cost reductions in the manufacture and installation of photovoltaic systems and the second is the rapid commercialization and cost reduction in processes which make fuel ethanol from cellulosic materials, rather than foods. These developments are already well underway.

Space does not permit an extended analysis of these matters; this brief discussion is intended to establish, in general terms, that a largely or wholly renewable energy future is possible, given several decades to adjust. This view is not now widely accepted, yet it is certainly the most hopeful one for mankind in an age of diminishing fossil fuel resources.

REFERENCES

1. Kendall, H.W., Nadis, S.J. Eds.; Energy strategies: toward a solar future. In *A Report of the Union of Concerned Scientists*. Ballenger: Cambridge, MA, 1979.
2. Grubb, M.J.; Meyer, N.I. Wind energy: resources, systems, and regional strategies. In *Renewable Energy: Sources for Fuel and Electricity*; Johansson, T.B., Kelly, H., Reddy, A.K.N. et al., Eds.; Island Press: Washington, DC, 1992.
3. Archer, C.L.; Jacobson, M.Z. Spatial and temporal distributions of U.S. winds and windpower at 80 m derived from measurements. J. Geophys. Res. Atmos. **2003**, *108* (D9), 4289.
4. National Hydropower Association, Hydrofacts. http://www.hydro.org/hydrofacts/forecast/asp (viewed May 24, 2005).
5. Wright, P.M. Status and potential of the U. S. geothermal industry. AAPG Bull. **1993**, *77* (8).
6. National Renewable Energy Laboratory, Biomass Energy, Potential, http://www.nrel.gov/documents/biomass_energy.html (viewed May 21, 2005).
7. Perlack, R.D.; Wright, L.L.; Turhollow, A.; Graham, R.L. *Biomass as Feedstock for a Bioenergy and Bioproducts Industry: The Technical Feasibility of a Billion-Ton Annual Supply*; Oak Ridge National Laboratory: Oak Ridge, TN, April, 2005.
8. Hall, D.O.; Rosillo-Calle, F.; Williams, R.H.; Woods, J. Biomass for energy: supply prospects. In *Renewable Energy: Sources for Fuels and Electricity*; Johansson, T.B., Kelly, H., Reddy, A.K.N., Williams, R.H., Eds.; Island Press: Washington, DC, 1992.
9. Florida Solar Energy Center. http://www.fsec.ucf.edu/solar/testcertr/collectr/tpndhw.htm (viewed May 23, 2005).
10. http://www.oja-services.nl/iea-pvps/pv/materials.htm (viewed May 24, 2005).
11. Renewable Energy Policy Project http://www.repp.org./geothermal_brief_power_technologyandgeneration.html (viewed May 24, 2005).
12. http://www.nrel.gov/documents/biomass_energy.html (viewed May 21, 2005).
13. http://www.poemsinc.org/FAQtidal.html (viewed May 25, 2005).
14. Maycock, P.D. *Photovoltaics: Sunlight to Electricity in One Step*; Brick House Publishing Co.: Andover, MA, 1981.
15. http://www.solarbuzz.com Price Survey, May 2005. (viewed May 20, 2005).
16. http://www.nrel.gov/docs/fy04osti/32578.pdf (viewed May 21, 2005).
17. Sawin, J. L. *Mainstreaming renewable energy in the 21st century*. Worldwatch Paper 169, Worldwatch Institute: Washington, DC, 2004. The author gives a figure of 11 cents per kwh in Japan, but this calculation does not include amortization of the equipment or annual maintenance costs.
18. Governors seek wider ethanol use. Wall Street Journal, April 12, 2005.
19. *A New Prosperity; Building a Sustainable Energy Future. The SERI Solar/Conservation Study*. Brick House Publishing Andover, MA, 1981.
20. http://www.renewableenergyaccess.com (viewed May 20, 2005).
21. Grubb, M.J.; Meyer, N.I. Wind energy: resources, systems, and regional strategies. In *Renewable Energy: Sources for Fuels and Electricity*; Johanssen, T.B., Kelly, H., Reddy, A.K.N., Williams, R.H. Eds.; Island Press: Washington, DC, 1992.

22. Kelly, H.; Weinberg, C. Utility strategies for using renewables. In *Renewable Energy: Sources for Fuels and Electricity*; Johanssen, T.B., Kelly, H., Reddy, A.K.N., Williams, R.H. Eds.; Island Press: Washington, DC, 1992.
23. Perez, R. Determination of Photovoltaic Effective Capacity for New Jersey. http://cleanpower.com/research/capacityvaluation/ELCC_New_Jersey.pdf (viewed May 24, 2005).
24. Blackburn, J.O. Possible efficiency gains identified for the U.S. as of the mid-1980's are examined. *The Renewable Energy Alternative*; Duke University Press: Durham, NC, 1987.
25. Energy Information Administration, Monthly Energy Review, March 2005. Additional detail for industry from International Energy Agency, IEA Energy Statistics, 2003, Energy Balance Table for the United States.

Residential Buildings: Heating Loads

Zohrab Melikyan
HVAC Department, Armenia State University of Architecture and Construction, Yerevan, Armenia

Abstract

In designing energy-saving heating systems, it is important to have the exact values of heating loads and seasonal heating demands of buildings. The existing methods for determining these values are not exact enough, because they do not take into account important factors such as the impact of solar radiation on south walls' and roofs' surfaces, the significant difference between day and night outside temperatures, and these temperatures' duration during the heating season. Apart from the stated disadvantage, the existing methods do not take into account the fact that each building has its own duration-of-heating period.

The mentioned imperfections do not allow for providing the exact values of heating demands, which may cause the use of wrong solutions in the designing of heat supply systems.

In this entry, new and improved methods are presented for more precise solution of heating problems.

INTRODUCTION

This entry was prepared based on summarizing the results of a research work, accomplished by the author, within a program of heating efficiency. For providing energy and fuel savings in heating systems, first the exact values of heating loads of buildings are needed. The use of the developed method provides the best possible accuracy in heating-loads value, as it takes into account the impact of more factors. Particularly, this entry highlights that depending on the thermal properties of buildings' envelopes, the duration-of-heating season for each building is different, even in the same climatic conditions. The proposed method is designed for heating experts and it can be used by students.

Method for Heating Load of Building

The heating load is the quantity of heat needed to provide the given inside temperature of a building under the design outside temperature, $t_{out.d}$ (°C), of given climatic conditions. As a rule, it is more convenient and more exact to calculate the heating load referring to 1 m³ of a building. This value is called the specific heating demand, q_{hd}, W/m³, which can be determined by Eq. 1:

$$q_{hd} = q_{hl} + q_v + q_{inf} - q_d, \quad (1)$$

where

q_{hl} total specific heat lost (W/m³)

Keywords: Heating loads; Seasonal loads; Heating designs; Specific heating demand; Design temperature; Heating degree days.

q_v specific quantity of heat required for heating outside fresh air, supplied into the building for ventilation (W/m³)

q_{inf} specific quantity of heat lost for heating outside fresh air, penetrating into the building through gaps of windows and doors (W/m³)

q_d specific quantity of heat gain in the building from lighting, electrical devices, and inhabitants (W/m³).

For the calculation of the specific values indicated above, we suggest the following equations:

$$q_{hl} = (t_{in} - t_{out.dsg})\left\{2k_w\left(\frac{(1-\mu)}{b} + \frac{1}{a}\right) + \frac{2k_{wd}\mu}{b} + \frac{k_r}{h}\right\}, \quad (2)$$

$$q_v = 0.181(t_{in} - t_{out.dsg}), \quad (3)$$

$$q_{inf} = \frac{4.012\mu(a+b)(t_{in} - t_{out.dsg})}{ab}, \quad (4)$$

By substituting the formulas in Eqs. 2–4 for the sum (Eq. 1), the following equation is obtained for determining the value of specific heating demand for any kind of building:

$$q_{hd} = (t_{in} - t_{out.dsg})\left\{\left[\frac{2k_{wd}\mu}{b} + \frac{2\left(\frac{1-\mu}{b} + \frac{1}{a}\right)}{\frac{1}{\alpha_{in}} + \frac{\delta_w}{\lambda_w} + \frac{\delta_{ins}}{\lambda_{ins}} + \frac{1}{\alpha_{out}}}\right] + \frac{1}{h\left(\frac{1}{\alpha_{in}} + \frac{\delta_r}{\lambda_r} + \frac{\delta_{ins}}{\lambda_{ins}} + \frac{1}{\alpha_{out}}\right)} + 0.181 + \frac{4.012\mu}{b}\right\} - q_d. \quad (5)$$

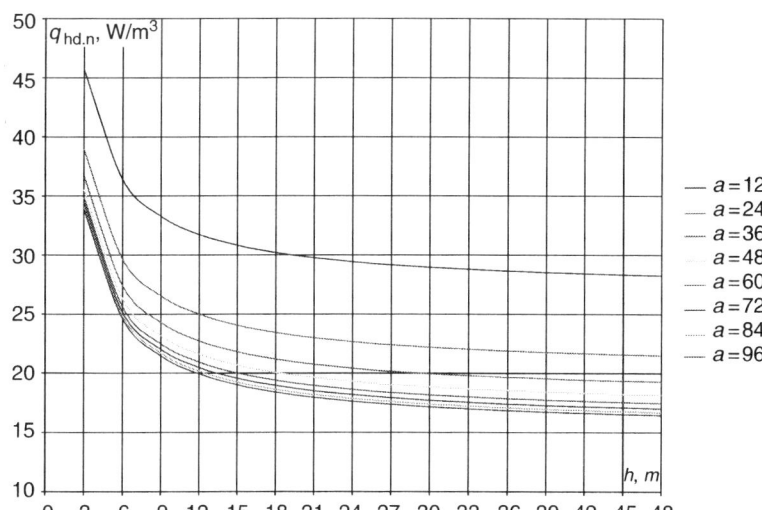

Fig. 1 Values of specific heating demand q_{hd} (W/m³) depending on sizes a and h, m, of buildings.

In Eq. 5, the meanings of values are the following:

- $t_{out.dsg}$ outside design temperature (°C)
- k_{wd} heat transfer coefficient of windows (W/m² °C)
- μ glazing rate of the building
- δ_w and δ_r thickness of the construction material of the walls and ceiling of the building (m)
- λ_w and λ_r heat conductivity of the construction material of the walls and ceiling of the building (W/m °C)
- δ_{ins} thickness of the insulation material covering the walls and ceiling of the building (m)
- λ_{ins} heat conductivity of the insulation material covering the walls and ceiling of the building (W/m °C)
- $\alpha_{in} = 8$ W/m² °C heat convection coefficient on the inside surfaces of the walls and ceiling of the building
- $\alpha_{out} = 20$ W/m² °C heat convection coefficient on the outside surfaces of the walls and ceiling of the building.

These values are taken from the building plan and section:

a, b, h length, width, and height of the building (m)

The glazing rate, μ, of the building is determined by the following fraction:

$$\mu = \frac{\sum F_{wd}}{ah}, \text{ or } \mu = \frac{\sum F_{wd}}{bh},$$

where F_{wd} is the total surface of windows on the vertical surfaces of the building (m²).

For finding the absolute value of heating demand, Q_{hd}, W, the specific value q_{hd} (W/m³) is multiplied by the volume, V_b (m³), of the building:

$$Q_{hd} = q_{hd} V_b \qquad (6)$$

The value of internal heat gains, q_d, in the building from lighting, electrical devices, and occupants has considerable impact on the heating and cooling demands of the building. The total value of q_d depends on the electric power of lighting and domestic devices and on the numbers of occupants simultaneously at home. As a result of the analyzed data resulting from a survey of the inhabitants of 1000 apartments in the city of Yerevan under the direction of the author by the methods of mathematical statistics, the following formulas for defining internal heat gains' specific values for winter ($q_{d.w}$) and summer ($q_{d.sum}$) periods have been carried out:

$$q_{d.w} = 0.0016 S^3 - 0.0622 S^2 + 0.576 S + 2.0566 \qquad (7)$$

$$q_{d.sum} = 0.0044 S^3 - 0.1047 S^2 + 0.6551 S + 1.9572, \qquad (8)$$

where S is the number of stories of the building.

The variation of values of specific heating demands, q_{hd}, depending on the length, a, and width, b (m), of the building is represented in Fig. 1. The diagram shows that the specific heating demands, q_{hd}, for bigger buildings are less, which indicates that the energy efficiency of heating for big buildings is higher.

Method of Seasonal Heating Demand of Buildings

At the present time, the methods for determining the seasonal heating demands of buildings are based on the use of heating degree days or heating degree hours and, therefore, do not take into account the effect of day solar radiation on the surfaces of buildings. Consequently, the values of seasonal heating demands turn out significantly higher. The useful impact of solar radiation on heating demand can be evaluated by the help of radiation temperature, t_R (°C), which is calculated by the following formula:

$$t_R = t_{out} + \frac{Ip}{\alpha_{out}}, \tag{9}$$

where

t_{out} outside air temperature (°C)
I intensity of solar radiation on surfaces of building envelope constructions (W/m²)
p rate of solar radiation absorption by construction surfaces.

The radiation temperature, t_R (°C), on a given surface is formed in a day period; therefore, it should be evaluated by day outside temperatures, $t_{out.d}$, of the heating period. Based on this approach, the method for determination of seasonal heating demands of buildings is divided into two parts: day, $q_{hd.d.seas}$ (Wh/m³), and night, $q_{hd.n.seas}$ (Wh/m³), seasonal heating demands. The total seasonal heating demand will be the sum of daytime and nighttime heating demands:

$$q_{hd.seas} = q_{hd.d.seas} + q_{hd.n.seas} (\text{Wh/m}^3). \tag{10}$$

As the solar radiation intensively lights the south-oriented walls and roof of a building, the heat lost through these constructions for day should be determined by the difference of inside, t_{in}, and radiation, t_R, temperatures. For the constructions of other orientations, heat lost from the building should be determined by the difference of inside, t_{in}, and outside current day temperatures, $t_{out.d.i}$.

Based on this concept, the specific value of total seasonal heat loss through all constructions of the building for the day period is determined by the following equation:

$$q_{hl.d.seas} = \sum_{i=Z_{t_{out.st.d}}}^{Z_{t_{out.dsg}}} Z_{t_{out.d.i}} \left\{ (t_{in} - t_{out.d.i}) \left\{ k_w \left(\frac{2(1-\mu)}{b} + \frac{2}{a} \right) \right. \right.$$
$$\left. + \frac{k_r}{h} + \frac{2k_{wd}\mu}{b} \right\}$$
$$\left. - \left[\frac{I_s p_s k_w (1-\mu)}{\alpha_{out} b} - \frac{I_r p_r k_r}{\alpha_{out} h} \right] \right\}, \tag{11}$$

where

$t_{out.d.i}$ current day temperatures between heating period beginning temperature at day $t_{out.st.d}$ and design temperature $t_{out.dsg}$ of heating period (°C)
$Z_{t_{out.d.i}}$ duration (h) of each outside day temperature, $t_{out.d.i}$, occurring between day heating season starting temperature, $t_{out.st.d}$, and heating design temperature, $t_{out.dsg}$,
I_s and I_r average intensity of solar radiation on south wall and roof surfaces (W/m²).

Besides heat lost, there is direct heat gain, $q_{wd.d.seas}$ (Wh/m³), through windows by solar radiation. The following formula can be used for determining specific heat gain, $q_{wd.d.seas}$ by day through south windows:

$$q_{wd.d.seas} = \sum_{i=Z_{t_{out.st.d}}}^{Z_{t_{out.dsg}}} Z_{t_{out.d.i}} \frac{\mu I_s n_1 n_2 n_3}{b}, \tag{12}$$

where n_1, n_2, and n_3 are rates of reduction of radiation penetration through windows due to reflection, the window's frames, and dust on glazed surfaces.

By adding the seasonal specific values of $q_{wd.d.seas}$, $q_{v.d.seas}$, $q_{inf.d.seas}$ (Wh/m³) for heating of ventilation and infiltration air, and substituting the seasonal specific quantity of heat gain, q_d (Wh/m³), the following equation for determining day seasonal specific heating demand, $q_{hd.d.seas}$ (Wh/m³) for a building is obtained:

$$q_{hd.d.seas} = \sum_{i=Z_{t_{out.st.d}}}^{Z_{t_{out.dsg}}} Z_{t_{out.d.i}} \left\{ (t_{in} - t_{out.d.i}) \left\{ k_w \left(\frac{2(1-\mu)}{b} + \frac{2}{a} \right) + \frac{k_r}{h} + \frac{2k_{wd}\mu}{b} + 0.181 + \frac{4.012\mu}{b} \right\} \right.$$
$$\left. - \left[\frac{I_s p_s k_w (1-\mu)}{\alpha_{out} b} + \frac{I_r p_r k_r}{\alpha_{out} h} + \frac{\mu I_s n_1 n_2 n_3}{b} + q_d \right] \right\}. \tag{13}$$

At night period of the heating season, the heat losses through all external constructions of a building take place under the difference of outside night current temperatures, $t_{out.n.i}$ and inside temperature t_{in}. Therefore, the seasonal specific value of heat lost of a building, $q_{hl.n.seas}$ (Wh/m³) can be determined by the following equation:

$$q_{hl.n.seas} = \sum_{i=Z_{t_{out.st.n}}}^{Z_{t_{out.dsg}}} Z_{t_{out.n.i}} (t_{in} - t_{out.n.i})$$
$$\times \left\{ k_w \left(\frac{2(1-\mu)}{b} + \frac{2}{a} \right) + \frac{k_r}{h} + \frac{2k_{wd}\mu}{b} \right\}. \tag{14}$$

By adding the seasonal specific values of $q_{v.n.seas}$ and $q_{inf.n.seas}$ (for heating of ventilation and infiltration air under night temperatures [Wh/m³]) and substituting the

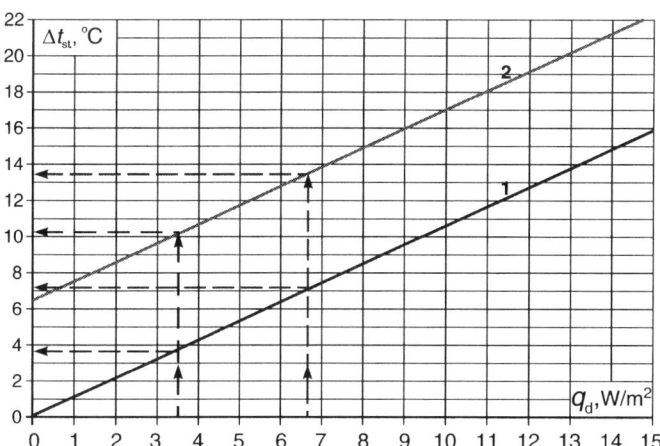

Fig. 2 Difference $\Delta t_{st} = t_{in} - t_{out.st}$ (°C) between inside t_{in} and heating season starting $t_{out.st}$ temperatures in case of different values of internal heat gains, q_d, in a building having sizes $a = 48$ m, $b = 12$ m, and $h = 15$ m (1 for nighttime, 2 for daytime period of heating season).

seasonal specific quantity of heat gain, q_d (Wh/m³), the following equation for determining of night seasonal specific heating demand, $q_{hd.n.seas}$ (Wh/m³) for a building is obtained:

$$q_{hd.n.seas} = \sum_{i=Z_{t_{out.st.n}}}^{Z_{t_{out.dsg}}} Z_{t_{out.n.i}}(t_{in} - t_{out.n.i}) \\ \times \left\{ \left\{ k_w \left(\frac{2(1-\mu)}{b} + \frac{2}{a} \right) + \frac{k_r}{h} + \frac{2k_{wd}\mu}{b} \right. \right. \\ \left. \left. + 0.181 + \frac{4.012\mu}{b} \right\} - q_d \right\},$$

(15)

where

$t_{out.n.i}$ current night temperatures occurring between night heating season's starting temperature, $t_{out.st.n}$, and heating design temperature, $t_{out.dsg}$, (°C)

$Zt_{out.n.i}$ duration (h) of each outside night temperature, $t_{out.n.i}$, occurring between night heating season's starting temperature, $t_{out.st.n}$, and heating design temperature, $t_{out.dsg}$.

To calculate seasonal daytime specific heating demands $q_{hd.d.seas}$, in Eq. 13 is substituted the values of durations $Zt_{out.d.i}$ of daytime current temperatures $t_{out.d.i}$ which occur between temperatures $t_{out.st.d}$ and heating design temperature $t_{out.dsg}$. Seasonal nighttime specific heating demands $q_{hd.n.seas}$, are calculated by substituting in Eq. 15 the values of $Zt_{out.n.i}$ and nighttime current temperatures $t_{out.n.i}$ which occur between temperatures $t_{out.st.n}$ and heating design temperature $t_{out.dsg}$. For this reason, it is necessary to have the durations $Zt_{out.i}$ of each temperature $t_{out.i}$ for each climatic zone, which can be established by special climatologic investigations. For example, for the conditions of Yerevan City, Armenia, the following empirical equations have been obtained:

daytime: $Z_{t_{out.d.i}} = 0.00003 t_{out.d.i}^5 + 0.0006 t_{out.d.i}^4$
$- 0.0133 t_{out.d.i}^3 - 0.25 t_{out.d.i}^2 + 3.92 t_{out.d.i} + 61$

nighttime: $Z_{t_{out.n.i}} = 0.000006 t_{out.n.i}^5 + 0.0014 t_{out.n.i}^4$
$- 0.031 t_{out.n.i}^3 - 0.59 t_{out.n.i}^2 + 9.15 t_{out.n.i} + 142$

For finding out the duration $Zt_{out.d.i}$ (h) of any current daytime temperature $t_{out.d.i}$, within the heating season, the value of that temperature should be substituted in the presented above, first empirical equation. The same kind of procedure should be done for finding out the duration $Zt_{out.n.i}$ (h) of any current nighttime temperature $t_{out.n.i}$, using the second empirical equation.

Heating period's starting daytime $t_{out.st.d}$ and nighttime $t_{out.st.n}$, temperatures are those under which the building is heated enough by internal heat gain, q_d. This means that if the outside temperature is lower than $t_{out.st.d}$ or $t_{out.st.n}$, the building needs to be heated, as the value of heat lost, q_{hl}, becomes higher than q_d.

The values of starting temperatures $t_{out.st.d}$ and $t_{out.st.n}$ can be found analytically, assuming that in Eqs. 5 and 13, the value of current outside temperature, $t_{out.i}$, is equal to $t_{out.st}$. As a result of such a procedure, the following formulas are obtained:

$$t_{out.st.n} = t_{in} - \frac{q_d}{2\left(\frac{k_{wd}\mu}{b} + k_w \left(\frac{1-\mu}{b} + \frac{1}{a} \right) \right) + \frac{k_r}{h} + 0.181 + \frac{4.012\mu}{b}}$$

(16)

$$t_{out.st.d} = t_{in} - \frac{\frac{I_s p_s k_w (1-\mu)}{\alpha_{out} b} + \frac{I_r p_r k_r}{\alpha_{out} h} + \frac{\mu I_s n_1 n_2 n_3}{b} + q_d}{2k_w \left(\frac{(1-\mu)}{b} + \frac{1}{a} \right) + \frac{k_r}{h} + \frac{2k_{wd}\mu}{b} + 0.181 + \frac{4.012\mu}{b}}$$

(17)

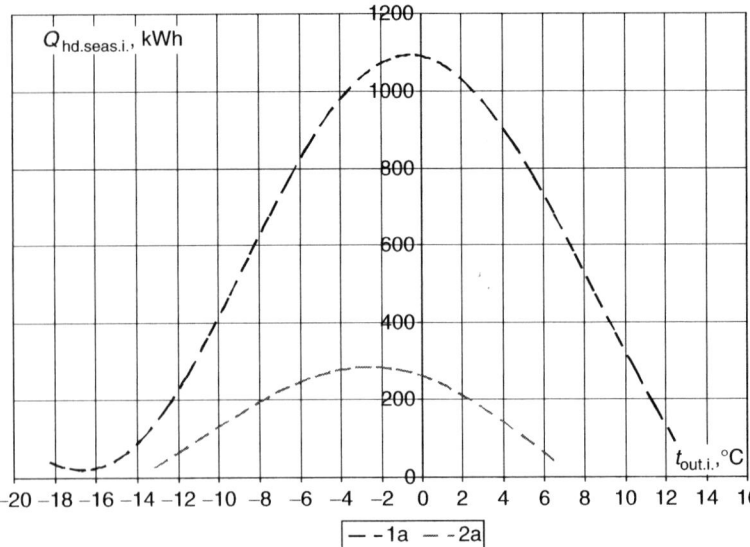

Fig. 3 Daytime and nighttime seasonal specific heating demands for an individual house in the climatic conditions of Yerevan, Armenia; (1a) nighttime seasonal specific heating demands, $q_{hd.n..i}$ (kWh/m^3); (2a) daytime seasonal specific heating demands, $q_{hd.d..i}$ (kWh/m^3).

The Eq.16 permits to find the temperature $t_{out.st.d}$, under which starts the nighttime heating season for any kind of building. By the Eq. 17 the temperature $t_{out.st.d}$, under which starts the daytime heating season, is determined. The impact of internal heat gains q_d, W, on the heating seasons' starting $t_{out.st.d}$ and $t_{out.st.d}$, temperatures is significant, which can be seen from the diagram in Fig. 2. The diagram shows that the growth of q_d, increases the difference $\Delta t_{st} = t_{in} - t_{out.st}$, between inside t_{in} and heating season's starting temperature $t_{out.st}$. This means that in the buildings with higher internal heat gains q_d, the heating starts at lower outside temperatures, and lasts shorter.

Using the collection of methods provided above, simulation software has been developed that calculates the exact values of seasonal heating demands for residential buildings.

The results of the calculations of day and night seasonal heating demands for individual house in Yerevan climatic conditions are represented by Fig. 3.

The diagram proves that for the examined individual house, the night seasonal specific heating demand, $q_{hd.n..i}$, becomes zero ($q_{hd.n..i} = 0$) under outside temperature $t_{out.n.i} < +15°C$. The day seasonal specific heating demand, $q_{hd.d..i}$, becomes zero ($q_{hd.d..i} = 0$) under outside temperature $t_{out.d.i} < +8°C$. This means that the heating of the house at night should start when the outside temperature $t_{out.n.i} < +15°C$, and at day it should start when the outside temperature $t_{out.d.i} < +8°C$.

The sum of all day and night current heating demands under all possible outside temperatures represents the total seasonal specific heating demand for the house. Multiplying total seasonal specific heating demands by the volume of the house will obtain the absolute value of seasonal heating demand.

The experimental calculations of seasonal heating demands for the same building by existing and suggested methods prove that the correct value of heating demand, defined by the help of new method, in 25%–30% is less, than in case of calculation by existing method. This difference is explained by consideration of solar energy impact and by more correct approach to the determination of heating degree hours for each building.

CONCLUSIONS

1. The suggested methods in this entry can be applied for accurate calculation of seasonal heating demands of any kind of building that possess any combination of thermal and physical properties of constructions in any climatic conditions. For this purpose, it is necessary to conduct climatologic investigations and develop empirical equations for determining the total duration of current daytime and nighttime temperatures in given climatic conditions.

2. Each building has its own heating season's beginning temperature, regardless of climatic conditions. The same building in various climatic conditions has the same heating season starting temperature, but the heating season period of a building depends on the total duration of temperatures having values between the heating season starting temperature, $t_{out.st.n}$, and the heating design temperature, $t_{out.dsg}$, in a given area.

3. The accurate values of seasonal specific heating demands allow determining and planning accurate values of fuel consumption for heating purposes.

BIBLIOGRAPHY

1. Turner, W.C., Ed. *Energy Management Handbook* 6th Edition; Fairmont Press: Atlanta, GA, 2006.
2. American Society of Heating, Refrigerating and Air-Conditioning Engineers, Inc. *ASHRAE Handbook of Fundamentals*; ASHRAE: Atlanta, GA, 2005.
3. American Society of Heating, Refrigerating and Air-Conditioning Engineers, Inc. *HVAC Systems and Equipment*; ASHRAE: Atlanta, GA, 2004.
4. Goldberg, N. Air conditioning heating and ventilating. *Marks' Standard Handbook for Mechanical Engineers* 10th Ed.; McGraw Hill: New York, 1996; [Chapter 12.4].
5. Sawer, H.J.; Howell, R.H. *Principles of Heating, Ventilating and Air Conditioning*; ASHRAE: Atlanta, GA, 1994.

Run-Around Heat Recovery Systems

Kamiel S. Gabriel
University of Ontario Institute of Technology, Oshawa, Ontario, Canada

Abstract

Run-around heat-recovery systems are often used to recover heat from the exhaust air in building ventilation systems, particularly in cold climates. In a typical heat-recovery system, an ethylene glycol and water solution is used as a "coupling fluid" to prevent the system from freezing. The design of a run-around heat-recovery system involves consideration of the heat-transfer rates between the fluids. A challenging problem is how to significantly increase those rates between the fluids, while using less energy for heating and cooling loads in buildings. Due to the heat-transfer characteristics of this coupling fluid and the operating and capital cost factors, the typical overall effectiveness of such systems is only about 50%. Studies show that two-phase, gas–liquid coupling fluids have much higher convective heat-transfer rates than single-phase flows at the same mass flow rates.

INTRODUCTION

In industrialized countries, a large portion of the energy consumed is discharged into the ambient. The mining and manufacturing industries account for a large percentage of waste heat. In particular, the iron and steel industry produces a large amount of waste energy with a broad range of temperatures.

In order to reduce the consumption of oils and fossil fuels, emphasis has been placed on the utilization of alternative energy sources, as well as energy conservation. The latter could be achieved through the enhancement of heat-transport systems (thus, reducing the production of waste heat), or by utilizing waste heat that has been already produced to pre-heat incoming air into buildings. Waste energy recovery can be achieved in many ways. Some examples are:

1. Using a pre-heater to transfer heat to another working fluid
2. Conversion into other forms of energy (e.g., electrical generators using steam)
3. Heat storage for control of supply-demand cycles between the source of waste heat and the consumer
4. Using run-around heat-recovery systems employing coupling fluids that absorb heat from the source (waste energy) and transfer it back into usable space

This entry will focus on the last method: run-around heat-recovery systems.

Keywords: Heat-recovery systems; Energy conservation; Heat-exchangers; Gas–liquid flow.

RUN-AROUND HEAT-RECOVERY SYSTEMS

The technique of waste heat-recovery involves heat transfer from the hot fluid (from which the energy is to be extracted) to the cold fluid (which is the energy absorbent). This system is comprised of two air-to-liquid heat exchangers, or coils, that are thermally connected by a pumped liquid. In the Heating, Ventilation, and Air Conditioning (HVAC) industry, an ethylene glycol and water solution is used as the "coupling fluid" to prevent the system from suffering freeze damage in cold winter conditions. This system is favorable for applications that involve remote air streams or require complicated configurations of air ducting, such as those found in the heating and ventilating systems of large buildings.

Due to the heat-transfer characteristics of this coupling fluid and the operating and capital cost factors, the typical overall effectiveness of such systems is only about 50%.[1] The performance of the run-around system depends on many parameters, such as coil geometry, air flowrates, air temperatures, fan power requirements, glycol flowrates, pump power requirements, heat losses in the system, and operating schedules. Computer simulations were conducted by several researchers to investigate the system's performance.[2–4] The results demonstrate that a major change in the coupling fluid is needed to facilitate significant improvement in the overall performance of the system.

Studies related to two-phase, gas–liquid flows in tubes have shown that the heat-transfer coefficients associated with mixtures of liquids and vapors (or gases) have much higher values than those with single-phase flow at the same mass flowrates.[5–7] Rezkallah and Sims[7] developed a general correlation for heat transfer in two-phase, gas–liquid flow in a vertical tube. However, there had been little or no heat transfer data available in a tube configuration similar to that of a typical heat exchanger coil until the work published by Zeng, Rezkallah, and Besant.[8] In addition to obtaining a large set of heat-transfer and pressure-drop data about typical heat-exchanger coils on the supply and exhaust sides of a run-around heat-recovery loop, the researchers[8] also examined the effect of temperature-dependent properties on the performance of the heat-recovery system.[9]

The results have shown that the temperature-dependent properties may further alter the performance of run-around systems if the coupling fluid experiences large temperature cycles in the run-around loop. Since run-around heat-recovery systems using liquid aqueous-glycol coupling fluids are usually designed and operated with low overall heat-transfer units, high Reynolds numbers must be maintained and the maximum overall system effectiveness is limited to 50%.[1] Therefore, pumping rates in a run-around loop need to be carefully selected, taking into account the glycol concentration and the supply air inlet temperature, if the very significant effects of temperature-dependent properties are to be avoided.[9] If, on the other hand, coupling fluids with freeze protection and much better heat-transfer characteristics can be found, the number of transfer units can be raised, resulting in a much higher maximum overall effectiveness. One fluid that has this potential is a liquid–gas mixture described by Zeng, Rezkallah, and Besant.[8]

Fig. 1 shows a schematic of a typical heat-recovery system. It consists of two major subsystems; an air supply system and a circulating fluid (usually known as the "coupling fluid"). In cold climates, the coupling fluid is typically an aqueous glycol solution to prevent the system from damage by freezing. In both the supply and exhaust ducts, air is supplied by fans. The air flows through a normal duct system. For the best performance of the system, the air flowrates (supply and exhaust) need to be balanced. This can be monitored using air pressure transducers.

HEAT-TRANSFER COEFFICIENTS

The overall thermal conductance, UA, for each coil is calculated from:

$$\frac{1}{UA} = \frac{1}{(\eta hA)_h} + R_{sh} + R_{sc} + R_w + R_c + \frac{1}{(\eta hA)_c} \quad (1)$$

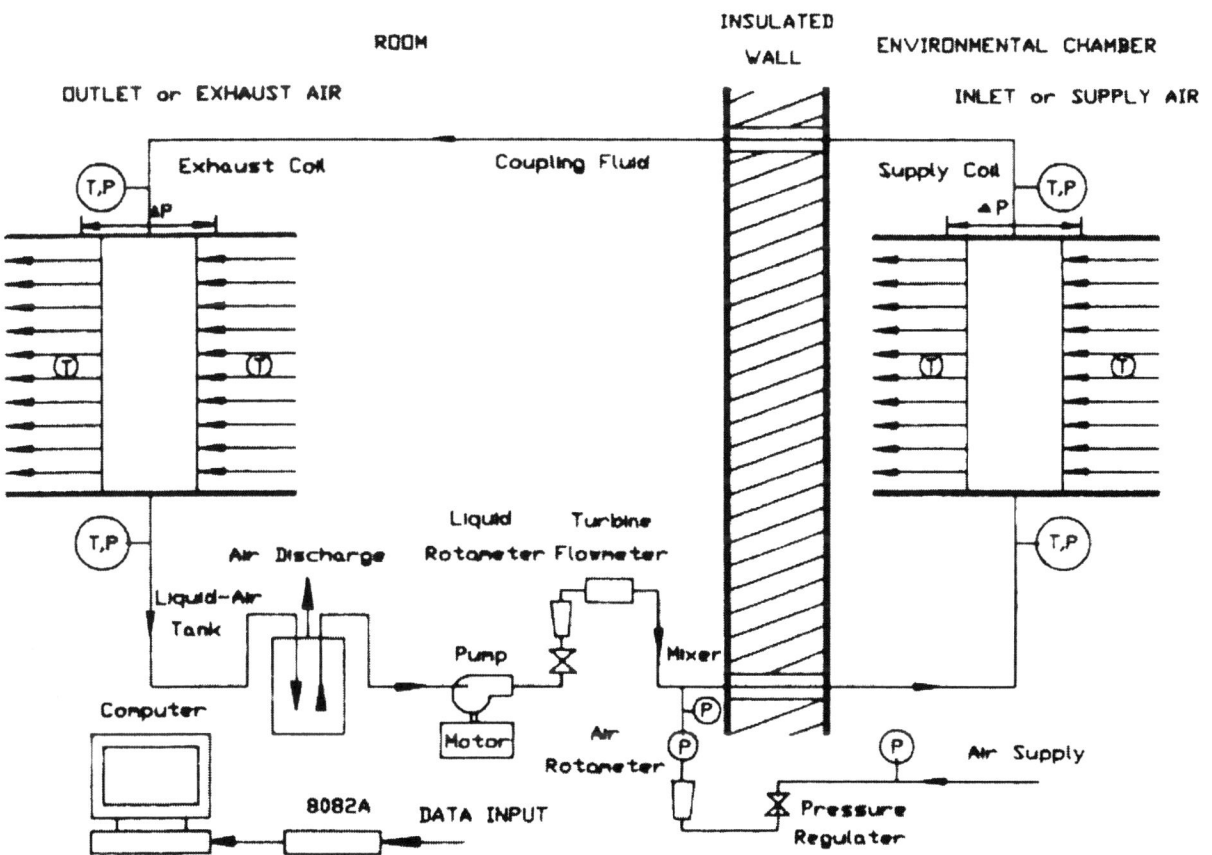

Fig. 1 A schematic of a typical run-around heat-recovery system.

where U is the overall heat-transfer coefficient (W/m² K); A is the tube cross-section area; c and h refer to the cold and hot fluids, respectively; R_{sh} and R_{sc} are the fouling resistances on the hot and cold sides, respectively; and R_w and R_c are the total tube wall thermal resistance and contact resistance of fins and tube, respectively. All resistances are in K/W.

A parameter known as the number of transfer units, N, is commonly used in the equations for calculating the effectiveness of a coil from:

$$N = \frac{UA}{C_{min}} \quad (2)$$

The heat capacity ratio, Cr, is estimated from:

$$Cr = \frac{C_{min}}{C_{max}} \quad (3)$$

where C_{min} and C_{max} are the minimum and maximum heat capacities of the cold and hot fluids (W/K).

Fig. 2 is a plot of the thermal effectiveness, ε_s, against the number of transfer units for heat capacity ratios ranging from 0.58 to 0.97. The experimental data of Zeng, Rezkallah, and Besant[8] are also shown on the plot.

The heat-transfer rate, q, is calculated from:

$$q = \varepsilon q_{max} = UA(LMTD)F = C_h(dT_h)$$
$$= C_c(dT_c) \quad (4)$$

where F is used as a correlation factor with the Logarithmic Mean Temperature Difference (LMTD) method,[10] and is equal to 1.0 in most cases. From Eq. 4, UA can be calculated and the heat transfer coefficients for single-phase liquid flow (h_L), and two-phase flow (h_{TP}) can then be obtained.

Since the minimum heat capacity in the system is that of the supply air, no dramatic changes could be made to the

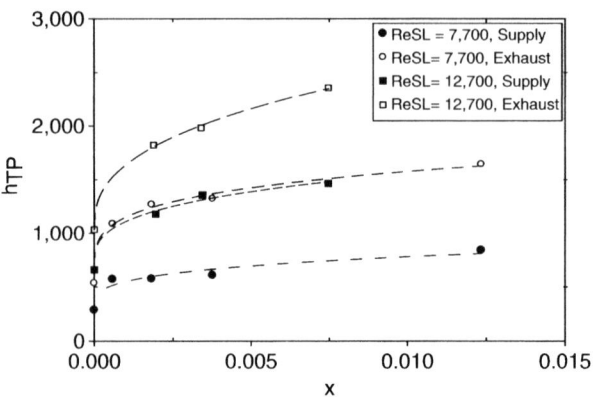

Fig. 3 Two-phase heat transfer coefficients as a function of gas quality using water–air coupling fluid, air supply volumetric flow rate; $Q_{as} = 21$ m³/min. air exhaust volumetric flow rate; $Q_{ae} = 22$ m³/min.

value of N. However, the number of transfer units in each coil and, hence, the overall effectiveness of the run-around system can be significantly increased if the overall heat transfer coefficient U is increased. Fig. 3 is a plot of the two-phase heat-transfer coefficients against the gas quality, x (defined as the ratio of the air mass flow rate to the total mass flow rate of the water–air mixture), for the supply and exhaust coils at liquid flow Reynolds numbers of 7700 and 12,700. It can be easily observed that as the air quality increases, the heat-transfer coefficients increase, yielding a better overall thermal effectiveness of the run-around system. Increasing the liquid flow rate (or Reynolds number) also produces higher heat-transfer coefficients. The heat-transfer coefficient for the supply coil increased by approximately 100% (at $x = 0.004$) when the Reynolds number was increased by 65%. It can be also concluded from the results in Fig. 3 that the exhaust coil heat-transfer coefficient is larger than that of the supply coil.

INFLUENCE OF INLET SUPPLY AIR TEMPERATURE

In general, higher heat-transfer coefficients are associated with lower inlet supply air temperatures. As shown in Fig. 4, the relationship tends to be linear.

INFLUENCE OF AIR FLOW RATE THROUGH THE DUCT

Fig. 5 shows two sets of data for the heat-transfer rate for 35% glycol/water–air coupling fluid at two different duct air flow rates. The results are those of Zeng, Rezkallah, and Besant.[8] It can be observed that higher heat-transfer rates are always associated with lower duct air flow rates;

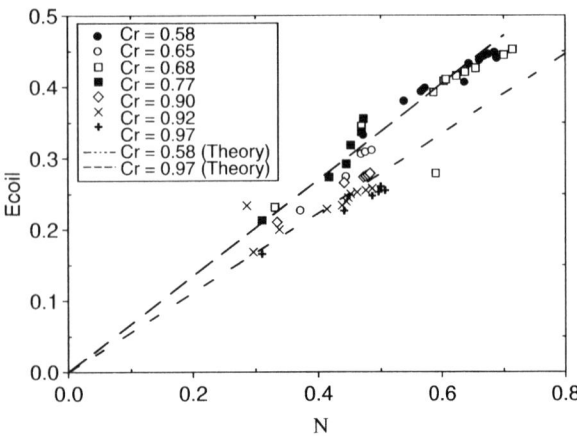

Fig. 2 Coil effectiveness as a function of number of transfer units with Cr as a parameter.

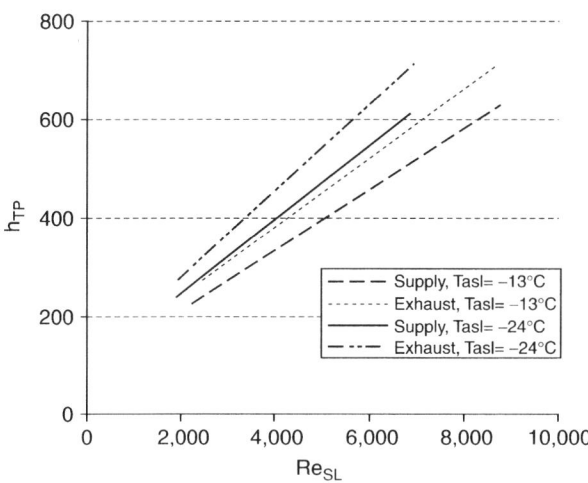

Fig. 4 Two-phase heat transfer coefficients as a function of the liquid Reynolds number using 35% glycol/water–air coupling fluid with $x = 0.0007$, Air supply volumetric flow rate; $Q_{as} = 21$ m³/min, Air exhaust volumetric flow rate; $Q_{ae} = 22$ m³/min.

e.g., at $Re_{SL} = 2500$, h_{TP} decreased by 16% when the duct air flow rates were increased. This is partly attributed to the significant change in viscosity of the glycol/water solution as a result of the change in operating temperature.

INFLUENCE OF CHANGES IN THE OPERATING TEMPERATURE OF ETHYLYNE GLYCOL

Experimental and numerical studies were performed by Zeng, Besant, and Rezkallah[9] to examine the performance of run-around systems when the coupling fluid properties vary with temperature. For a 50% aqueous glycol solution, it was shown that the temperature

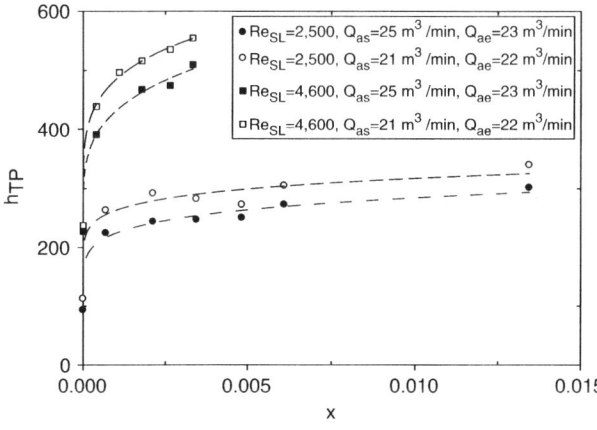

Fig. 5 Two-phase heat transfer coefficients as a function of gas quality using 35% glycol/water–air coupling fluid, Air supply volumetric flow rate; $Q_{as} = 21$ m³/min, Air exhaust volumetric flow rate; $Q_{ae} = 22$ m³/min.

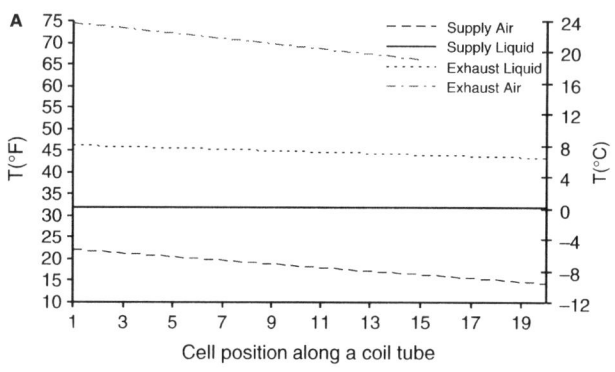

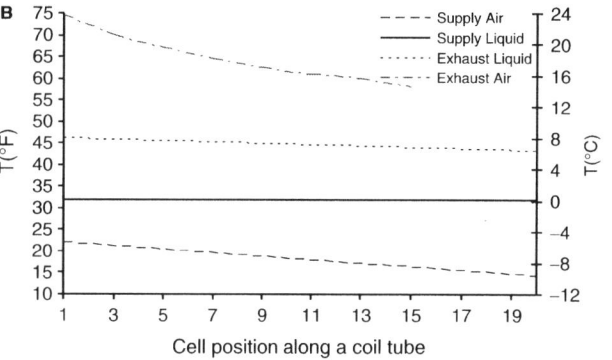

Fig. 6 Temperature distribution in a run-around system (A) Properties independent of coupling fluid temperature (B) Properties dependent on the local coupling fluid temperature.

variations in the glycol solution produced small changes in the supply and exhaust air temperatures. Both the supply and exhaust air temperatures have shown a non-linear variation with position along the coil tube. In the case where properties were independent of the coupling fluid temperature, the relationship was linear. These results are shown in Fig. 6.

OPERATING FEATURES OF THE TWO-PHASE FLOW RUN-AROUND SYSTEM

(Refer to Fig. 1 for the component identification)

The coupling fluid is pumped around the run-around system at a rate controlled by the flow control by-pass valve using a pump(s) sized to provide a coupling liquid flow rate such that the thermal capacity ratio for each coil is equal to one when no flow is by-passed.

The coupling liquid flow is divided equally to each tube in each coil using flow divider valves (not shown in the figure).

Pressurized gas is injected into the coupling liquid flow in each tube of each coil in such a manner that the gas flow is distributed equally in each tube. This rate of gas injection will result in a mass ratio of at least 0.0001 between the gas and the coupling liquid.

The rate of coupling liquid flow and gas flow in each coil is adjusted to maximize the total heat exchange between coils or, in the event that the required load is met, to reduce the total heat exchange between the coils.

The mixed gas and coupling liquid is separated after they flow through each coil.

CONCLUSIONS

Run-around heat-recovery systems are very effective means of reducing the heating costs of buildings during winter months, particularly for northern climates, such as Canada. A recovery system can also reduce the size and capital cost of supporting HVAC equipment, such as boilers and radiators. A practical system for recovering this lost energy is a run-around, or liquid-coupled, heat recovery system.

Heat transfer rates in run-around heat-recovery systems of the type commonly used in HVAC systems are greatly enhanced when two-phase, gas–liquid flow through the coils is utilized. The overall effectiveness of the system increases when the two-phase mixture replaces the commonly used single-phase aqueous glycol solution. Moreover, by using the liquid–gas coupling fluid, the capacity of the coupling fluid could be matched to that of the exhaust and supply air. This leads to a substantial enhancement of the heat transfer process in such systems, resulting in higher efficiency and revenue saving.

REFERENCES

1. Forsyth, B.I.; Besant, R.W. The design of a run-around heat-recovery system. ASHRAE Trans. **1988**, *94* (II), 511–531.
2. Custer, J.B. The coil energy recovery loop. ASHRAE Trans. **1976**, *88*, 998–1007.
3. Wang, J.C.Y. *Practical Thermal Design of Run-Around Systems*, Proceedings of the Solar Energy Society of Canada Energex 8th Conference, ISES: Regina, Canada, 1982; 641–645.
4. Forsyth, B.I.; Besant, R.B. The performance of a run-around heat-recovery system using aqueous glycol as a coupling liquid. ASHRAE Trans. **1988**, *94* (II), 532–545.
5. Collier, J.B. *Convective Boiling and Condensation*, 1st Ed.; McGraw Hill Company: U.K., 1972; 382–406.
6. Michiyoshi, I. *Two-Phase Two-Component Heat Transfer*, Proceedings of the 6th International Heat Transfer Conference, Washington, DC, 1978, 1, 219–233.
7. Rezkallah, K.S.; Sims, G.E. An examination of correlations of mean heat-transfer coefficients in two-phase two-component flow in vertical tubes. AICHE Symp. Ser. **1987**, *83*, 109–114.
8. Zeng, Y.Y.; Rezkallah, K.S.; Besant, R.W. Heat transfer measurements in two-phase two-component flow inside heat exchanger coils of a run-around heat recovery system. *ASME Publication HTD*; Two-phase flow in energy exchange systems; 1992, 1–10.
9. Zeng, Y.Y.; Besant, R.W.; Rezkallah, K.S. The effect of temperature-dependent properties on the performance of run-around heat recovery systems using aqueous-glycol coupling fluid. ASHREA Trans. **1993**, 551–562.
10. Incropera, F.P.; DeWitt, D.P. *Fundamentals of Heat and Mass Transfer*, 5th Ed.; Wiley: U.S.A., 2002.

Savings and Optimization: Case Studies

Robert W. Peters
Department of Civil, Construction & Environmental Engineering, University of Alabama at Birmingham, Birmingham, Alabama, U.S.A.

Jeffery P. Perl
Chicago Chem Consultants Corp., Chicago, Illinois, U.S.A.

Abstract

Since mass and energy balances are coupled, process energy can be optimized only by conducting detailed energy assessments, which investigate both process and energy analyses. A number of plant-wide assessments have been conducted on a variety of industries, including aluminum, chemicals, forest products, glass, metal casting, mining, petroleum, and steel, among others. While a number of industrial organizations have had plant-wide assessments performed at their facilities in each of the above industries [reported on the DOE website], one industrial organization was selected from each type of industry reported above, and the results and recommendations of the energy assessments are described in this chapter as case histories. Examples of recommendations from the industrial energy assessments include: improve waste heat recovery, consolidation of cooling/chiller operations, improved process recycling operations, lower temperature operation, use of variable speed drives on process equipment, implementing more efficient plant lighting systems, reducing scrap/wastage rates, and upgrading equipment with more energy efficient systems. Potential energy savings in terms of payback periods, capital savings, and reductions of Btu's and kWh/yr are reported, for the various assessment recommendations presented.

The U.S. Department of Energy (DOE) has an industrial technology program (ITP) support available. Elements of this program include free technical assistance as well as project co-funding opportunities. The program identifies four broad areas in the Save Energy Now arena and can provide free energy audits to facilities around the US. These audits assist companies in charting facilities energy use, identify areas of concern by matching energy bills with similar facilities and recommending changes along with estimated pay back periods to allow project prioritization

Although this entry is meant to complement our preceding article "Energy Savings and Optimization in the Chemical Process Industry" in this encyclopedia, the examples also apply to a wide variety of applications wherever energy is used in any manufacturing process.

U.S. DOE CASE STUDIES

DOE has published a number of best practices of industrial technology energy optimization case studies at their website: www1.eere.energy.gov/industry/bestpractices/printable_versions/case_studies.html. The write-ups are based on the case studies provided in the DOE Energy Efficiency and Renewable Energy (EERE) website: www1.eere.energy.gov/news/progress_alerts/progress_alert.asp?aid=183. The following case studies involving plant-wide assessments are available at that website, divided according to industrial category:

Aluminum

- Alcoa: Lafayette Operations Energy Efficiency Assessment (PDF 235 KB);
- Alcoa: Plant-Wide Energy Assessment Finds Potential Savings at Aluminum Extrusion Facility (PDF 464 KB);
- Commonwealth Aluminum: Manufacturer Conducts Plant-Wide Energy Assessments at Two Aluminum Sheet Production Operations (PDF 810 KB);
- Pechiney Rolled Products: Plant-Wide Energy Assessment Identifies Opportunities to Optimize Aluminum Casting and Rolling Operations (PDF 417 KB).

Chemicals

- 3M: Hutchinson Plant Focuses on Heat Recovery and Cogeneration during Plant-Wide Energy-Efficiency Assessment (PDF 710 KB);

Keywords: Case studies; Energy efficiency; Payback period; Energy efficiency assessment; Plantwide assessment; Energy optimization; Energy savings; Process system analysis; Waste heat recovery; Variable-speed drive.

- Akro Nobel Morris Plant Implements a Site-Wide Energy Efficiency Plan (PDF 341 KB);
  - Plant–Wide Assessment (PWA) Summary: $1.2 Million in Savings Identified on Akro Novel Assessment (PDF 174 KB);
- Bayer Polymers: Plant Identifies Numerous Projects Following Plant-Wide Energy Efficiency Assessments (PDF 863 KB);
- Dow Chemical Company: By-Product Synergy Process Provides Opportunities to Improve Resource Utilization, Conserve Energy, and Save Money (PDF 218 KB);
- Formosa Plastics Corporation: Plant-Wide Assessment of Texas Plant Identifies Opportunities for Improving Process Efficiency and Reducing Energy Costs (PDF 1 MB);
- Neville Chemical Company: Management Pursues Five Projects Following Plant-Wide Energy-Efficiency Assessment (PDF 1.0 MB);
- Rohm and Haas: Chemical Plant Uses Pinch Analysis to Quantify Energy and Cost Savings Opportunities at Deer Park, Texas (PDF 607 KB);
- Rohm and Haas: Company Uses Knoxville Plant Assessment Results to Develop Best Practices Guidelines and Benchmark for Its Other Sites [Revised July 2003] (PDF 765 KB);
- Solutia: Massachusetts Chemical Manufacturer Uses SECURE Methodology to Identify Potential Reductions in Utility and Process Energy Consumption (PDF 979 KB); and
- W.R. Grace: Plant Uses Six Sigma Methodology and Traditional Heat Balance Analysis to Identify Energy Conservation Opportunities at Curtis Bay Works (PDF 906 KB).

Forest Products

- Appleton Papers Plant-Wide Energy Assessment Saves Energy and Reduces Waste (PDF 220 KB);
  - PWA Summary: $3.5 Million in Savings Identified in Appleton Assessment (PDF 176 KB);
- Augusta Newsprint: Paper Mill Pursues Five Projects Following Plant-Wide Energy Efficiency Assessment [Revised July 2003] (PDF 699 KB);
  - PWA Summary: $1.6 Million in Savings Identified in Augusta Newsprint Assessment (PDF 176 KB);
- Blue Heron Paper Company: Oregon Mill Uses Model-Based Energy Assessment to Identify Energy and Cost Savings Opportunities (PDF 394 KB);
- Boise Cascade Mill Energy Assessment (PDF 297 KB);
  - PWA Summary: $707,000 in Savings Identified in Boise Cascade Assessment (PDF 176 KB);
- Caraustar Industries Energy Assessment (PDF 188 KB);
  - PWA Summary: $1.2 Million in Savings Identified in Caraustar Assessment (PDF 176 KB);
- Georgia-Pacific Palatka Plant Uses Thermal Pinch Analysis and Evaluates Water Reduction in Plant-Wide Energy Assessment (PDF 352 KB);
  - PWA Summary: $2.9 Million in Savings Identified in Georgia-Pacific Assessment (PDF 176 KB);
- Georgia-Pacific: Crossett Mill Identifies Heat Recovery Projects and Operational Improvements That May Save $6 Million Annually (PDF 797 KB);
- Inland Paperboard and Packaging, Rome Linerboard Mill Energy Assessment (PDF 280 KB);
  - PWA Summary: $9.5 Million in Savings Identified through Inland Assessment (PDF 176 KB);
- Weyerhaeuser Company: Longview Mill Conducts Energy and Water Assessment that Finds Potential for $3.1 Million in Annual Savings (PDF 400 KB).

Glass

- Corning, Inc.: Proposed Changes at Glass Plant Indicate $26 Million in Potential Savings (PDF 177 KB); and
- Anchor Glass Container Corporation: Plant-Wide Energy Assessment Saves Electricity and Expenditures (PDF 200 KB);
- PWA Summary: $1.6 Million in Savings Identified in Anchor Assessment (PDF 181 KB).

Metal Casting

- AMCAST Industrial Corporation: Energy Assessment (PDF 295 KB);
  - PWA Summary: $3.6 Million in Savings Identified in AMCAST Assessment (PDF 176 KB);
- Ford Cleveland: Inside-Out Analysis Identifies Energy and Cost Savings Opportunities at Metal Casting Plant (PDF 266 KB).

Mining

- Alcoa World Alumina: Plant-Wide Assessment at Arkansas Operations Reveals More than $900,000 in Potential Annual Savings (PDF 464 KB);
- Coeur Rochester, Inc.: Plant-Wide Assessment of Nevada Silver Mine Finds Opportunities to Improve

Process Control and Reduce Energy Consumption (PDF 300 KB); and
- Kennecott Utah Copper Corporation: Facility Utilizes Energy Assessments to Identify $930,000 in Potential Annual Savings (PDF 307 KB).

Petroleum

- Chevron: Refinery Identifies $4.4 Million in Annual Savings by Using Process Simulation Models to Perform Energy-Efficiency Assessment (PDF 293 KB);
- Martinez Refinery Completes Plant-Wide Energy Assessment (PDF 241 KB); [Equilon]
  — PWA Summary: $52 Million in Savings Identified in Equilon Assessment (PDF 172 KB);
- Paramount Petroleum: Plant-Wide Energy-Efficiency Assessment Identifies Three Projects (PDF 816 KB);
  — PWA Summary: $4.1 Million in Savings Identified in Paramount Petroleum Assessment (PDF 173 KB); and
- Valero: Houston Refinery Uses Plant-Wide Assessment to Develop an Energy Optimization and Management System (EOMS) (PDF 308 KB).

Steel

- Jernberg Industries, Incorporated: Forging Facility Uses Plant-Wide Energy Assessment to Aid Conversion to Lean Manufacturing (PDF 987 KB);
- Full PWA Report: An Assessment of Energy, Waste, and Productivity Improvements for North Star Steel Iowa (PDF 2.32 MB);
- North Star Steel Company: Iowa Mini-Mill Conducts Plant-Wide Energy Assessment Using a Total Assessment Audit (PDF 608 KB); and
- Weirton Steel: Mill Identifies $1.4 Million in Annual Savings Following Plant-Wide Energy-Efficiency Assessment (PDF 555 KB).

Supporting Industries

- Metaldyne: Plant-Wide Assessment of Royal Oaks Finds Opportunities to Improve Manufacturing Efficiency, Reduce Energy Use, and Achieve Significant Cost Savings (PDF 676 KB).

Other

- Aluminum/Alcoa: IAC Energy Assessment of Spanish Fork Plant (PDF 174 KB); and
- Metlab: Plant-Wide Assessment (PDF 640 KB);
- Utica Corporation: Plant-Wide Energy Assessment Report Final Summary (PDF 225 KB).

Examples of a plant energy optimization assessment from these various major industries are summarized below.

Aluminum

Commonwealth Aluminum (now Aleris Rolled Products): Commonwealth Aluminum conducted plant-wide energy assessments at its aluminum sheet rolling mills in Lewisport, Kentucky, and Uhrichsville, Ohio. The Lewisport facility utilizes direct-chill ingot castings, and the Uhrichsville facility employs continuous casting technology. The direct-chill process makes use of reversing hot mills to homogenize, scalp, and break down large cast ingots. Additional processing includes hot reduction using multistand mills, coiling, annealing, cold reduction, and finishing. The continuous-casting operation has molten metal fed from an adjacent scrap casting operation and from melted in-house scrap to the continuous-casting unit, producing an aluminum strip. The as-cast strip is fed to a multistand hot mill and then goes to coiling, annealing, cold reduction, and finishing operations; this process eliminates the scalping, homogenization, and reversing hot mill operations that are part of the direct-chill process.

The assessments focused on analyzing the processing procedures involved in converting scrap feed material to finished coiled sheet products. Specifically, the assessments involved looking at the role of process technologies; the impact of operating procedures; and the effects of alloy composition, temper, sheet width, sheet thickness, and coil weight and size to determining energy requirements. The assessment goals were: (i) identify levels of energy consumption inherent in the direct-chill and continuous casting processes; (ii) identify the best practices at each plant and recommend potential methods to transfer those practices; and (iii) define opportunities for improving future energy consumption by investing in new equipment, changing operating practices, and optimizing production strategies. The assessment team reviewed: (1) process heating, with an emphasis on melting, homogenization, annealing, and painting processes; (2) motors and pumps (including those used for hot and cold rolling operations), saws and shears, material-handling equipment, and environmental control systems; (3) compressed air systems (including locating and repairing air leaks); (4) steam systems, with an emphasis on steam usage for process and plant heating and cooling; and (5) application of new processes and equipment, including energy contracting and scheduling, waste heat recovery, lighting, and heating, ventilating, and air-conditioning equipment. Ten potential projects were identified by this assessment

Table 1 Projects identified from Commonwealth Aluminum's Plantwide Assessment

	Annual projected savings				
Project	Fuel savings (MMBtu/yr)	Electricity savings (kWh/yr)	Cost savings ($/yr)	Capital cost ($)	Payback period (years)
1. Convert the Lewisport mill to the continuous-casting process	546,000	30,900,000	8,000,000	25,000,000	3.1
2. Upgrade melter/holder furnaces at Newport mill	153,000	N/A	918,000	5,000,000	5.4
3. Modify the production strategy to accommodate market demands	83,000	N/A	500,000	None	Immediate
4. Implement the best practices for melting operations	46,000	N/A	273,000	None	Immediate
5. Improve melt stirring processes at Newport mill	33,000	N/A	200,000	400,000	2.0
6. Upgrade soaking pits at Lewisport mill	128,000	N/A	770,000	500,000	0.6
7. Improve annealing operations	22,000	N/A	130,000	None	Immediate
8. Optimize compressed air system	N/A	6,800,000	238,000	248,000	1.0
9. Use infrared imaging technology for process diagnostics	50,000	N/A	300,000	None	Immediate
10. Improve waste heat recovery	Unknown	Unknown	Unknown	Unknown	Unknown
Total	1,061,000	37,700,000	11,329,000	31,148,000	2.7 (mean)

team to improve the efficiency of processes at the two rolling mills.

Estimates indicate that implementing all ten projects could result in annual cost savings exceeding $11 million, corresponding to energy savings of ~1 million MMBtu and 38 million kWh annually. The annual projected savings for the ten projects are summarized in Table 1, below. After the table, brief discussions of the projects are presented.

1. *Convert the Lewisport mill to the continuous-casting process*: The assessment team recommended converting the Lewisport mill from direct-chilling casting to continuous casting integrated with the existing multistand hot mill. A 56% reduction in energy usage was estimated, allowing the mill to significantly reduce energy consumption, improving the sheet recovery from cast material and reduce working capital by eliminating the need for the scalping operation and ingot homogenization using soaking pits and tunnel furnaces.
2. *Upgrade melter/holder furnace at Newport mill*: The Newport melter/holder furnaces have dual purposes–receiving and holding molten metal delivered from the adjacent IMCO plant, and melting heavy gage and primary metal. Replacing three of four existing melters/holders with two furnaces in which melting and holding functions are optimized was recommended by the assessment team. The third melter/holder would be retained as a spare.
3. *Modify the production strategy to accommodate market demands*: The assessment team recommended that the mills minimize or eliminate less-than-capacity operations and optimize operations for which multiple processing devices exist, such as melting, melter/holder and holding furnaces, homogenization furnaces, annealing furnaces, and finishing processes.
4. *Implement best practices for melting operations*: The mills should define and implement best practices for melting operations, including minimizing time that furnace doors are open, improving burner firing, improving skimming, controlling exhaust gas temperature, improving maintenance of furnace sensors and controls, and improving staff communication.
5. *Improve melt stirring processes at Newport mill*: The assessment team recommended that the Newport mill upgrade its melt stirrer units to consume less energy and generate less dross.
6. *Upgrade soaking pits at Lewisport mill*: Recommendations to improve the efficiency of soaking furnaces included: (i) Replacing dry ceramic seals with more thermally efficient water seals; (ii) Upgrading the programmable logic controllers (PLCs) for the furnace; and

(iii) Using variable-speed recirculation fans and reverse air flow to mimic tunnel furnaces.

7. *Improve annealing operations*: Measures to improve the annealing processes included:
   - The Lewisport mill should evaluate methods employed at Newport for controlling annealing operations by using thermocouples on each coil interfaced with a PLC. The system permits focusing on the slowest heating coil while preventing overheating in other coils. This will increase energy savings and resulting in improved energy consumption.
   - Door seals should be improved to minimize heat losses at both mills.
   - The Lewisport mill should implement the Newport mill's practice of pulling coils from the furnaces at higher temperature to reducing annealing time and energy consumption.

8. *Optimize compressed air system*: Both mills should employ measures to upgrade and optimize their compressed air systems to improve system efficiency and maintenance practices, eliminate leaks, and substitute other energy sources for compressed air wherever possible.

9. *Use infrared imaging technology for process diagnostics*: The use of infrared imaging techniques is a powerful detecting heat losses and maintenance problems associated with thermal processes and electrical operations. These devices should be installed at both plants.

10. *Improve waste heat recovery*: Opportunities exist for improving waste heat recovery at both mills, primarily from melting, homogenization, annealing operations and paint line exhausts. Other specific opportunities should also be determined. Uses for recovered heat could include furnace air preheating, regenerative burners, steam generation, and hot water generation for applications within the mills or at nearby sites. The assessment team was unable to provide meaningful estimates of costs or savings associated with waste heat recovery because specific data were not available, although the team believed that these opportunities would be significant.

Potential savings: total capital costs to implement all the projects were estimated to be ~$31 million. Implementing the projects with payback periods of 2 years or less was estimated to yield more than $2.4 million in annual cost savings.

Chemicals

3M Chemical Company: 3M Chemical Company performed a plant-wide energy-efficiency assessment at its Hutchinson, Minnesota, plant to identify opportunities for its plant's operations and utility requirements. This plant produces Scotch brand cellophane tape, and other consumer, office, industrial, and electrical supplies and tapes. The plant covers 1.3 million ft^2 consisting of two buildings (north and south buildings) and has more than 1600 employees. The plant spends about $7 million annually on energy-related expenses. Four separate implementation packages representing various combinations of energy-efficiency projects were identified from the assessment. The packages were developed separately because some of the individual efficiency improvement measures involving the thermal oxidizer units were mutually exclusive. One package (having the shortest payback period) was selected for implementation based on relative aggregate payback periods of the individual packages; this package included projects for chiller consolidation, and a heat recovery boiler for two of the plant's thermal oxidizers to produce low-pressure steam, which would offset the fuel requirements of plant steam production. Although the assessment team did not recommend immediate implementation of the packages involving combustion turbine-based cogeneration to destroy volatile organic compounds (VOCs), research on the technical feasibility of these measures was encouraged; this could help manage plant emissions while simultaneously improving energy efficiency. Energy efficiency improvements identified through the assessment would provide additional benefits by improving the plant's cooling efficiency and improving the reliability of chilled water production. The plant's productivity was also expected to improve because process loads will not need to be reduced if a chiller fails. The heat recovered for steam production will improve the thermal efficiency of the thermal oxidizer, reducing boiler fuel consumption and environmental emissions. Using a steam turbine to reduce steam pressure will offset the plant's retail electricity requirements.

The annual projected savings for the six projects are summarized in Table 2. After the table, brief discussions of the projects are presented.

Individual energy-efficiency measures identified during the plant-wide assessment and scheduled to be implemented as a package are described briefly below.

Cooling-chiller consolidation: The total capacity of the chilled water system is 7640 tons each for the north and south plants. By consolidating the chiller capacity of both plants (by interconnecting the individual chilled water distribution systems serving the plants), electric energy savings could be realized. The newer and more energy efficient chillers at the north plant could be used

Table 2 Projects identified from 3-M Chemical Company's Plant-wide Assessment of its Hutchinson, MN, Facility

	Annual projected savings				
Project	Electricity (kWh/yr)	Natural gas (MMBtu/yr)	No. 6 fuel oil (MMBtu/yr)	1st-Year avoided energy expense ($)	Projected capital cost ($)
Chiller consolidation	1,552,750	N/A	N/A	87,420	292,546
Air Compressor cooling, North Plant	609,000	N/A	N/A	22,168	65,240
Air Compressor cooling, South Plant	393,750	N/A	N/A	34,287	170,775
Thermal oxidizer heat recovery boiler	N/A	38,093	172,557	772,191	913,275
Steam turbine	3,166,000	N/A	N/A	163,999	604,035
Relative humidity	N/A	695	3,145	14,200	0
Total	5,721,500	38,788	175,702	1,094,265	2,045,970

for base loads; these chillers could serve larger loads for longer periods of time, lowering operation costs. The chiller consolidation was estimated to provide an energy savings of more than 1.5 million kWh/yr.

Cooling-air compressor cooling north plant and south plants: The air compressors at both the north and south plants use chilled water for cooling; the nominal chilled water demand is 145 and 75 tons, respectively. The cooling towers of the chilled water system could be used as the primary cooling method for the air compressors–the potential energy savings equals the difference between chiller operation and operation circulating pumps. The electrical requirements were estimated to be reduced by 609,000 kWh/yr and 393,750 kWh/yr in the north and south plants, respectively, by implementing this measure.

Head recovery-thermal oxidizer heat recovery boiler: The 2L thermal oxidizer has a regenerative/recuperative cycle, thereby reducing the exhaust temperature and effectively eliminating the unit as a potential source for heat recovery. Units 1L and 3L, however, are suitable candidates for heat recovery applications. Two applications were considered–an oil-to-air heat exchanger to preheat supply air and makeup air, and a heat recovery boiler for low-pressure steam production. Because the payback period of the heat recovery boiler is considerably shorter than for the heat exchanger (1.2 years vs. 8.1 years), the heat recovery boiler was the recommended option. Using a reheat boiler as the heat recovery mechanism also mitigates the heat recovery limitations of an oil-to-air heat recovery system. With an oil-to-air system, the energy reduction annually is limited to the load of the specific air-handling units. Heat recovered from the thermal oxidizers to produce low-pressure steam has greater potential energy savings because the steam can serve loads throughout the plant. Annual steam production from the heat recovered from the two thermal oxidizers was estimated to be at least 164,000 million pounds annually; estimated savings are more than 210,000 MMBtu/yr due to the reduced need for natural gas and fuel oil.

Cogeneration-steam turbine: Steam is produced at a nominal pressure of 220 psig, but steam pressure is reduced to 125 psig and 15 psig for process loads and humidification, respectively. Using the pressure drop to drive a steam turbine and an electric generator (rather than reduction through a pressure reducing valve) could provide an offset in the retail electrical service requirement of the plant. Replacing the existing pressure-reducing valve with a steam generator provides an opportunity for cogeneration. More than 3 million kWh/yr in electricity were estimated to be saved by implementing this methodology.

Steam system-relative humidity Project: Steam provides part of the plant's relative humidity. From corporate guidelines, a relative humidity of 50% is necessary for the types of materials and processes used at this plant. Reanalysis has indicated that an acceptable range lies between 35 and 40%. Steam production can be reduced to correspond to this allowable range; no capital investment is necessary for this project.

Potential savings: The plant was estimated to save 5.7 million kWh/yr in electricity and 214,499 million Btu/yr in natural gas and fuel oil by implementing the energy savings measures.

Forest Products

Blue Heron Paper Company: Blue Heron Paper Company and Pacific Simulation performed a model-based energy assessment to analyze effluent flow and heat load reduction, fresh water usage minimization, and process energy usage reduction at Blue Heron's paper mill in Oregon City, Oregon. Blue Heron was originally known as the Hawley Pulp and Paper Mill, which began operations in 1908. Times Mirror purchased the facility in 1948 (known as Publisher Paper Company). Smurfit Stone Container Corporation bought the mill in 1986, operating it for about 14 years. The mill was sold in 2000 to its employees and KPS to form the Blue Heron Paper Company. The mill produces ~650 tons of paper daily, making newsprint and specialty paper on three paper machines. Approximately 60% of the fiber furnished to the paper mill is cycled fiber from old newsprint and magazines. Residual wood chips from sawmill waste are an additional primary source of fiber supplied to the thermomechanical pulping (TMP) lines, which convert them to pulp. Six natural, gas-fired utility boilers produce steam for the mill. Two lines of TMP pulping operate at Blue Heron. Each line is a primary and secondary refiner to separate wood chips into pulp fibers suitable for making paper. Rotating plates compress the stock, creating friction thus producing heat and steam. The steam is recovered in one of the lines and is converted into clean steam using a reboiler and thermocompressor.

The deinking pulping process involves pulping, debris screening, and flotation deinking. In pulping, the stock is mixed with steam-heated water to form pulp at a given temperature. Debris is removed from the pulp via screening and cleaning operations, after which the pulp is washed. The pulp slurry is refined and mixed with surfactants to float the ink particles. The slurry is cleaned and screened again to facilitate as much ink removal as possible. Water is filtered from the pulp and returned to the front of the deinking process for reuse.

Five processes were investigated during the energy-efficiency assessment: the deinking facility, three paper machines, the entire white water system, the TMP mill, and the steam system. Opportunities for effluent flow and heat load reduction, fresh water usage minimization, and process energy usage reduction were investigated by building full-mill mass and energy balances. A full-mill balance was required due to the complex system of water and steam reuse at the facility. Data were collected for the various processes to generate a steady-state model representing the base case. After establishing, the base case was a "summer" case and an "average" case were considered. The summer case, based on the highest recorded incoming water temperatures for the previous summer months, was used to determine the effluent heat loading conditions. The average case was used to calculate the overall annual economic benefits for the projects being evaluated.

After establishing the base cases, several possible projects were analyzed and quantified in simulations. Potential projects analyzed included the use of additional heat exchangers and/or cooling towers, modification of process conditions and process flows, and process equipment modifications. Fifteen different projects were evaluated during the plant-wide assessment; of those, 14 were considered potentially feasible (although some projects are considered to be alternatives to other projects). The projects listed in Table 3 exclude those considered to be alternatives to more economically attractive cases.

Individual energy-efficiency measures identified during the plant-wide assessment scheduled to be implemented as a package are described briefly below.

1. *Recycle vacuum pump sea water on paper machines No. 1 and No. 4; heat shower water*: Liquid ringed-seal vacuum pumps were currently installed on all three paper machines. The heat generated by creating the required vacuum is absorbed by the vacuum pump seal water, thereby providing steam that can be recycled and used as a heat source. Because the seal water temperature may be increased to the point where it flashes to steam under the high vacuum, it is possible to recycle up to 90% of the water. The seal water can then used to heat incoming filtered water for use in the paper machines, either by utilizing heat exchanger or by using a cooling tower, to dissipate excess heat for additional water reuse options. This requires recycle pumps, a heat exchanger or cooling tower for each paper machine, or a combination of components, depending on which machine is being considered for modification. Space availability is also a factor.

2. *Heat shower water for paper machines No.1 and No.4*: This activity involved recycling 75% of the vacuum pump seal water from the No.4 paper machine, and routing the flow through the No.1 paper machine to use as vacuum pump seal water. This water can then routed to the de-ink process water clarifier showers to reduce filtered water usage and to decrease the net amount of steam required in the paper mill. The recirculated vacuum pump seal water for each paper machine can be used to heat water in the paper mill. The recirculated vacuum pump seal water for each paper machine can be used to heat water required for the paper machine. The costs for this project activity result from the acquisition of heat exchangers and piping for the water. The estimated capital cost is $1.0 million, but it could be higher because of equipment logistics. In addition to the benefits in reduced relative heat load, and energy cost reduction, this processing alternative reduces the overall effluent flow by approximately 1.6 million gallons per day.

3. *Recover heat from Uhle box effluents on No. 4 paper machine*: Uhle boxes help remove water from the forming felt as it traverses from the forming section toward the dryer section. The Uhle

Table 3 Blue Heron Assessment Results

Project description	Fuel savings (MMBtu/yr)	Electricity savings (kWh/yr)	Cost savings ($/yr)	Capital cost ($)	Payback period (yrs)
1. Recycle vacuum pump and water for paper machines No. 1 and No. 4; heat shower water	N/A	40	Included in total of project directly below	103,000	Included in total of project directly below
2. Heat shower water for paper machines No. 1 and No. 4	124,000	N/A	315,000	1,000,000	3.5
3. Recover heat from Uhle box effluents on No. 4 paper machine	32,000	N/A	100,000	110,000	1.1
4. Recover heat from deinking effluent to No. 1 paper machine	37,000	N/A	125,000	375,000	3.0
5. Reduce operating temperature in deinking plant pulpers	87,000	N/A	230,000	10,000	Immediate
6. Recover heat from vacuum pumps, Uhle boxes, and (TMP) wastewater; cool effluent with incoming filtered water	74,000	N/A	1,150,000	2,000,000	1.7
7. Heat shower water with reboiler steam and vacuum pump seal water; heat de-inking pulpers with TMP waste water	254,000	900	1,050,000	2,500,000	2.4
Total	608,000	940	2,970,000	6,098,000	2.1 (aggregate average)

box downstream and in the proximity of the steam box removes a combination of shower water and water from the sheet. The water's temperature is between 115 and 120°F. The type of heat recovery system is economical and technically feasible for the No. 4 paper machine. The capital cost is estimated at ~$110,000 with significant heat recovery, creating a project return on investment of 1.1 years. Heat recovery is more practical when applied to the Uhle box stream coming off the felt. The stream has the steam box on it and thus has the hottest Uhle box flow.

4. *Recover heat from de-ink effluent to No. 1 paper machine*: The de-ink plant's combined effluent stream at a temperature of ~120°F at a flow of 600 gal/min (gpm). These streams can be collected, passed through a new heat exchanger in the de-ink plant, and cooled with filtered water. The warm filtered water would displace steam that heats shower water on the paper machines. The heat exchanger itself could recover 9 MMBtu/hr at average conditions. The benefit in steam economy is only 3.5 MMBtu/hr due to some heat being lost through the effluent system, and the shower water flow being only 300 gpm. This stream is severely contaminated, so designing for minimal fouling risk increases the estimated capital expense to ~$375,000.

5. *Reduce operating temperature in de-inking plant pulpers*: The de-inking plant uses live steam to heat water in the pulper supply tank to a temperature of 125°F. The energy in the steam input varies from 9 MMBtu/hr in summer to 16 MMBtu/hr in winter; almost all of this heat is added to the mill effluent. The de-ink staff are considering operating at reduced temperatures; each 5°F decrease in temperature could save ~4 MMBtu/hr of steam use. Little capital expense is required for this modification.

6. *Recover heat from vacuum pumps, Uhle boxes, and TMP wastewater; cool effluent with incoming filtered water*: Options that could reduce energy consumption include combining recycling and cooling of vacuum seal water, heating paper machine shower water with TMP hot wastewater, recovering heat from Uhle boxes on the No. 3 and No. 4 paper machines, and cooling the effluent with incoming fresh water.

7. *Heat shower water with reboiler steam and vacuum pump seal water; heat de-ink pulpers with TMP wastewater*: The following operational changes would be required:
   - Preheating paper machine shower water with vacuum pump seal water on a partially closed cycle;
   - Heating paper machine shower water to operating temperature by direct injection of clean

steam from the reboiler at 11 psig, thereby reducing power consumption in the steam compressors; and
- Heating de-ink pulpers by replacing some of the white water from the central wastewater from the TMP mill economizer.

This alternative requires less equipment than any other alternative considered that has comparable energy economy, due to much of the heat recovery is accomplished without heat exchangers. The effluent flow is reduced by 1.2 million gallons/day. The net steam consumption to heat stock and shower water drops to ~65 MMBtu/hr. A reduction in electricity consumption of ~900 kWh can be achieved.

Glass

Anchor Warner Robins: Plant-wide energy assessments were conducted at the Anchor Warner Robins, Georgia and Jacksonville, Florida plants identifying opportunities that can result in significant annual energy savings. Anchor Glass Container Corporation is the 3rd largest manufacturer of glass containers in the U.S. The Warner Robins facility has two furnaces and eight bottle-forming machines that produce more that 4 million bottles per day for approximately 360 days/yr. The facility contained 880,000 ft^2, comprised of 360,000 ft^2 for bottle production, 500,000 ft^2 for the finished-goods warehouse, 12,400 ft^2 for plant utilities, and 19,000 ft^2 for offices and miscellaneous space. Typical electric and gas loads are about 12.5 megawatts/day and 4 million ft^3/day. The Jacksonville plant was reduced in size during the past few years, consuming about half as much energy as the Warner Robins plant. Both facilities produce similar products. The primary materials used in manufacturing glass are sand, soda ash, limestone, cullet, and various corrugated packaging materials.

Anchor's furnaces are primarily gas-fired furnaces with electric booster heat. They melt more than 800 tons of glass/day. The furnaces are equipped with heat recovery regenerators that recover a portion of the waste energy from the 2800°F furnaces using a cycling air flow system. Fluctuations in the temperature differential between the heat recovery masses and the air streams during each 20-min cycle limit the effectiveness of the heat recovery process. Additional equipment that uses notable quantities of electricity include air compressors and vacuum pumps [7900 (hp) total, typically 6050 operating hp], cooling and furnace air blowers (3200 hp total), and lighting. The cooling water system, conveying and packing machinery, raw materials handling equipment, and a limited amount of space conditioning equipment also consume energy. Waste minimization and continued environmental compliance are goals for Anchor Glass. Environmental regulations pertaining to glass manufacturing include the disposal of checker slag, furnace residue removed during furnace rebuilds, furnace bricks containing chromium, and production waste. Additional regulations pertain to the discharge water used to clean machines, cooling water, dust produced by the batch mixing process, and air emissions from furnaces.

A systems approach was used to perform the plant-wide energy efficiency assessments. Opportunities were identified, evaluated, and prioritized for energy savings. Maintenance and operating procedures were also reviewed for their impact on energy efficiency. Three areas were considered in performing the energy assessments: inputs to plant processes, plant process efficiency, and process outputs including waste and heat products. Emphasis was placed on processes and systems identified to have the greatest energy savings potential:

- Cogeneration—installation of gas turbines with waste heat recovery systems;
- Waste heat recovery—recovery and reuse of waste heat from furnaces;
- Motor analysis—motor efficiency improvements;
- Lighting systems—cost-effective lighting improvements; and
- Variable speed drive (VSD) analysis—the installation of VSDs on selected process equipment, particularly blowers.

The results from the site-wide energy efficiency assessment are summarized in Table 4 (for both the Warner Robins and Jacksonville plants). The projections of energy and capital savings are preliminary, based upon the data collected and analyzed during the energy assessment; the data have not been validated through definitive engineering analysis and field testing.

Individual energy-efficiency measures identified during the plant-wide assessment scheduled to be implemented as a package are described briefly below.

Table 4 Potential energy savings summary

Action	Estimated energy savings/yr	Simple payback on capital required (yrs)
Increasing heat recovery	220,388 MMBtu	1.0
Compressed air	2,056,250 kWh	1.2
VSD cooling water pumps	524,600 kWh	1.8
VSD furnace air blowers	808,400 kWh	1.7
VSD machine cool blowers	560,720 kWh	1.8

1. *Thermal Cycle Efficiency Improvement Options*: Heat is introduced to the glass furnace via direct natural gas over the glass melting tank, and electric booster heat arcing between electrodes immersed in the glass melt. Direct firing has lower cost fuels, but the over-melt firing process is much less efficient (50–60%) in delivering heat to the melt. The electric arc is nearly 100% efficient in delivering heat to the glass melt, but the cost of electricity is high relative to other fuels. To recover some of the heat lost through inefficiency of the direct firing process, the glass furnaces employ checker brick regenerators to capture and return some of the waste heat. The regenerators are characterized by their fluctuating effectiveness over the charge/discharge cycle; the regenerators are more effective at the beginning of the cycle because of greater air/regenerator temperature differences, but performance significantly decreases toward the end of the cycle. Further, significant heat is lost up the stack, providing an opportunity to improve process efficiency. Possible methods to improve the glass melting process efficiency include:
   - Generate steam from the stack gas waste to drive a turbine;
   - Recover stack gas waste heat into the incoming air through air-to-air or intermediary heat exchangers; and
   - Optimize charge/discharge cycle times to improve regenerator effectiveness.

   The assessment team considered recovery of the waste heat from the furnace stacks, to produce steam driving a turbine/generator and/or preheat incoming furnace combustion air. Their analysis indicated that the best way to recover the waste heat energy was by incorporating an air-to-air heat exchanger that transfers part of the heat between the air streams exiting and entering the regenerators.

2. *Compressed Air System Efficiency and Improvements*: Significant amounts of compressed air are used at two different pressures in container production processes at their facilities; high-pressure air (~100 psig) is used for operating typical air-powered controls and tools, and low-pressure air (~50 psig) is used in the bottle-blowing process machinery. A flow rate of ~16,000 and 11,000 ft^3/min of high-pressure air is used at the Warner Robins and Jacksonville plants, respectively. The energy assessment indicated that the system operates between 97 and 103 psi in the air compressor room at the Warner Robins plant; a refrigerated dryer rated at 11,000 cfm provides air-drying. The air supply in the plant supply header ranges from 82 to 85 psig. Air leaks were estimated for ~20% of the total air consumption. Efforts should be focused on reducing or eliminating the leaks to minimize air use. Opportunities identified to improve the efficiency of the high-pressure system included:
   - Remove system bottlenecks and add a higher-pressure storage system to reconfigure the compressor room to a lower supply pressure;
   - Provide advanced system controls to optimize compressor operators and energy efficiency; and
   - Reduce leaks in the system to lower flow requirements.

Adding the higher-pressure air receiver storage capacity will allow sufficient air storage so that the compressor can shut down for 10–20 minutes when unloaded, rather than cycling between loaded and unloaded modes.

Motor management and efficiency: The application and operation of electric motors were reviewed, resulting in the following recommendations:

- Economic guidelines should be in place so that initial cost does not drive repair vs. replacement decision-making;
- Larger motors that are sometimes underloaded for their application and are less efficient should be replaced with premium efficiency motors when a payback of 2.5 years or less is possible;
- An overall motor inventory and replacement plan should be developed; and
- Energy-efficient motors should be used for all new or replacement motors when expected annual operating time exceeds 4,000 hr.

Pump and blower VSD application: These facilities use pumps and blowers for various process functions, but three general applications were considered in the assessment: cooling water pumping, process cooling blowers, and furnace air blowers. These pumps currently run at constant speed with either no flow control or with valves for pumps and variable inlet vanes for blowers. Potential options to improve energy efficiency include:

- Use VSDs on the cooling tower water pumps to control pump speed and modulate tower water flow and pressure;
- Use VSDs on the glass furnace air and stack draft blowers to control blower speed as a means of modulating air flow; and
- Use VSDs on the glass forming and hot gas handling machinery air cooling blowers to control blower speed as a means of modulating air flow.

Plant lighting: Improvements in efficiency of plant lighting systems include the following options:

- Provide for expanded use of natural lighting
- Where convenient, convert the plant's outdoor lighting from utility lighting to wall packs or plant-supplied lighting
- Install motion sensors in equipment rooms, warehouse facilities, and other plant areas that are not frequently occupied
- Have lighting performance contractors conduct no-cost reviews of plant lighting to identify possible improvement opportunities.

Plant energy purchase optimization: The assessment team evaluated the potential for fuel substitution (including load management on the electric boost for real-time electricity prices and operational considerations). Break-even guidelines were recommended for consideration of energy purchase optimization and fuel substitution to permit better fuel management for these facilities.

Potential savings: Estimated total potential energy savings were 220,000 MMBtu/yr and approximately 4 million kWh/yr for electricity if all the project recommended were implemented. The associated capital required to achieve the fossil fuel savings was estimated to be $800,000, while the capital required to achieve the project electricity savings was estimated to be $250,000. Average simple payback periods calculated for the primary recommendations ranged from 1 to 2 years.

Metal Casting

AMCAST Industrial Corporation: A plant-wide energy assessment was performed to identify energy savings opportunities at AMCAST Industrial Corp.'s Wapakoneta, Ohio, manufacturing facility. The assessment highlighted process performance and its impact on overall energy and cost savings. The plant was founded in 1866, and is a major supplier of aluminum permanent-mold cast suspension components for the automotive industry; it also serves the construction industry and other industrial sectors. The plant employs nearly 390 people, and processes 20 to 25 million pounds of aluminum yearly.

Twelve separate projects were recommended to improve the efficiency of the plant's production processes and decrease the plant's energy consumption. The aggregate potential of these projects could save the plant $3.7 million, which would lead to a 3-month. payback period. Implementation of these twelve projects has the potential to reduce the plant's CO_2 emissions by more than 11 million pounds per year.

The plant performs heat treating and liquification of aluminum. The raw material is aluminum ingots. Natural gas-fired reverb furnaces melt the aluminum, which is transferred to hold furnaces adjacent to each low-pressure permanent mold machine via heated ladles. After casting, flash and scrap parts are sent back to the jet-melt furnace. Cast products are then solution and age-heat treated in special ovens, trimmed, inspected, and shipped to customers or sent to other locations for additional processing. Primary waste streams include aluminum dross, recyclable aluminum flash, de-burring material, metal shavings, and cooling wastes.

The facility's energy use, from transformers and switches to production lines, was monitored during the energy assessment. Data drawn from the operations were downloaded to data loggers, where it was evaluated utilizing software and documented. The information collected included:

- Thirty-day snapshot of peak kW demand;
- Peak daily kW demand (based on 30 minute readings);
- Metered values by phase (current, voltage, and power factor);
- Circuit-loading summary;
- Energy and power summaries; and
- Harmonic and disturbance analysis.

Opportunities to improve the reliability of the plant's motor systems, and cases where equipment should be added or replaced, were identified during the plant energy assessment. Energy data were used to calculate the expected savings and payback. The initial focus was on identifying and minimizing end-use loads. The distribution system was examined for inefficiencies and savings potential. After the end use and distribution systems were analyzed, the problem areas were examined for savings opportunities. In most cases, end-use and distribution savings directly influenced the recommendations for modifying the energy source.

The following recommendations were established from the plant-wide energy assessment:

- *Electrical systems*: Three-shift operations allowed the opportunity to move noncontinuous operations with large electrical demand to off-peak periods.
- *Lighting*: Levels were recorded and compared to recommended lighting levels. Options recommended included:
  — Disconnecting obstructed lights or lights in overly lit areas;
  — Making better utilization of task lighting and skylights; and
  — Replacing T12 lamps and magnetic ballasts with T8 lamps and electronic ballasts in the offices.
- *Motor drive systems*: Motor repair and replacement policies were examined, and options recommended included:
  — Use premium efficiency motors;
  — Install variable speed drive (VSDs) on the casting machines; and

— Replace belts on the motor drives with notched v-belts.
- *Compressed air system*: The compressed air system was examined for savings potential from using outside air, reducing header pressure, reducing pressure loss in the distribution system, optimal staging of compressors, and effective use of cooling air and water.
- *Processing heat*: Several options were identified:
  — Optimizing combustion efficiency of the ovens and furnaces;
  — Using waste heat and pre-heat combustion or space-heating air; and
  — Using solar walls to preheat process and space-heating air.

Once the data collection and analysis was complete, the assessment team developed twelve separate recommendations for projects. The team validated and supported the implementation of a number of these initiatives. The project recommendations resulting from the assessment are summarized below:

1. Switching to aluminum-titanate riser tubes to replace the Dense Fused Silica (DFS) material. The initiative results in reduced maintenance; less scrap, less downtime; and better product quality. It is estimated that savings in reduced scrap could exceed $1 million per year.
2. Using electric infrared heaters instead if existing gas torches to preheat permanent molds would further reduce scrap and save ~$850,000 annually.
3. Improving tooling design, repair and tool maintenance could lead to annual savings of ~$730,000.
4. Reducing scrap rate by addressing quality and scrap controls and increasing the number of quality control personnel to reduce rejects, could provide a projected net savings of $470,000 per year.
5. Using exhaust heat from reverbatory melting furnaces in heat treating furnaces, was estimated to provide an natural gas savings of $157,000.
6. Using exhaust heat from heat treating furnaces in aging ovens was estimated to provide an annual savings of $93,000.
7. Implementing other energy savings opportunities for reverbatory furnaces (such as capturing flue gas losses, radiation losses, and eliminating air leakage from furnace openings, etc.) was estimated to provide an energy savings of $64,000 annually.
8. Relocating jet melt furnaces and improve jet melt process flow by moving the furnaces closer to the jet melt- recycling unit would result in savings in energy, labor, and maintenance projected to total $89,000 per year.
9. Installing VSDs on casting machines was estimated to provide savings of $37,000 per year.
10. Using blowers to pressurize the cast hold furnaces was estimated to provide an annual energy savings of $11,000.
11. Using plant cooling tower water to cool the compressed air system instead of city water was estimated to provide $10,000 in savings of reduced municipal water fees.
12. Using exhaust heat from heat treating furnaces for scrap preheating (recapturing heat for reuse) was estimated to provide a savings of $3,300 annually.

Potential savings: Implementation of the above twelve projects was estimated to provide an annual energy savings of $3.5 million. Additionally, reductions in the utility costs could provide another $200,000 in savings. Since the total project costs are estimated to be ~$1.0 million the simple payback would be slightly more than 3.0 months. The estimated cost savings are summarized in Table 5 below. The payback periods for the twelve projects ranged from 0 to 29 months. Maximizing the plant's motor system energy efficiency would also reduce the amount of pollutants that the plant generated. Implementation of the twelve projects was estimated to reduce CO_2 emissions by 11,000,000 pounds.

Mining

Couer Rochester, Inc.: Coeur Rochester, Inc. mining operation in Rochester, Nevada, conducted a plant-wide energy efficiency assessment, in which five energy-savings opportunities were identified. Using energy cost tracking, process systems analysis, and characterization of the primary energy-consuming equipment to identify systems with the greatest energy savings potential, the assessment team identified five energy savings projects, focusing on improving the pumping systems, capturing waste heat, and upgrading lighting fixtures. It was estimated that 11 million kWh in annual electricity savings could be realized if the five projects were implemented, yielding an estimated cost savings of ~$813,000 per year. The required capital investment was estimated to be $260,000, resulting in an average simple payback of 4 months for all five projects.

Table 5 Wapakoneta estimated cost savings

Category	Estimated savings
Cost savings	$3.6 million/yr
Plant savings realized	$6.0 million/yr
Electrical energy savings	672,000 kWh/yr
Natural gas savings	9,000 MMBtu/yr

Table 6 Cost Savings Estimated for Projects Identified during Coeur Rochester Plant-wide Energy Assessment

Project Description	Annual Electricity Savings (kWh/yr)	Implementation Cost ($)	Annual Cost Savings ($/yr)	Simple Payback Period (yrs)
Replace 1250-hp barren pump motor with a 600-hp motor	4,520,000	41,000	339,000	0.1
Replace 1250-hp phase 4 pregnant pump motor with a 750-hp motor	3,205,000	69,000	240,000	0.3
Install a regenerative system on downhill conveyor	1,920,000	115,000	144,000	0.8
Install adjustable-speed drive on 150-hp well pump No. 2	544,000	16,000	41,000	0.4
Install efficient lighting fixtures	648,000	19,000	49,000	0.4
Total	10,838,000	260,000	813,000	0.3 (average)

Coeur Rochester, Inc., is a subsidiary of Coeur d'Alene Mines Corporation. This facility opened in 1986, and is the company's largest producer and the largest primary silver mine in North America, producing 60,000 oz of gold and 6 million oz of silver annually. The ore body is blasted, and the ore is fed into a series of crushers to yield ore that is less than 3/8 inch in size. Conveyors transport this ore to the heap; a heap leaching process employing cyanide solution is used to capture the metals from the ore. The cyanide solution is pumped to the top of the heap and is distributed across the ore; the cyanide solution percolates down through the ore and captures the precious metals. The mine uses the Merrill-Crowe process to recover the metal from the cyanide solution. That process involves precipitation with zinc dust and is particularly applicable for gold ores with high silver content. Before gold is precipitated with zinc dust, all solids must be removed from the pregnant solution. The Merrill-Crowe plant has three principal stages: (i) deaeration to prevent gold from dissolving again; (ii) zinc dust precipitation; and (iii) precipitate filtration, in which zinc dust and gold are mixed with sulfuric acid to dissolve the zinc; the resulting mix is filtered, and the remaining solids are smelted into a bullion bar.

The assessment team tracked monthly energy costs, performed a process system analysis, and identified major energy consumers. The assessment involved the following activities:

- Performing a comprehensive review of product flow through the mine, from raw materials to final product, and establishing efficiency benchmarks;
- Developing a database of energy-consuming equipment;
- Identifying potential projects to reduce energy use and demand, and developing preliminary cost estimates for the projects;
- Prioritizing potential projects based on probability of implementation;
- Performing a detailed cost/benefit analysis for selected projects; and
- Developing implementation strategies for selected projects.

The above approach uncovered systems with the greatest potential for energy savings. The assessment team selected a number of projects for further study that involved major energy consumers, offered the greatest savings potential, and met a one-year simple payback criterion. Projects recommended for cost-savings potential included.

- Replace 1250-hp barren pump motor with a 600-hp motor;
- Replace 1250-hp phase 4 pregnant pump motor with a 750-hp motor;
- Install a regenerative system on downhill conveyor;
- Install adjustable-speed drive on 150-hp well pump No. 2; and
- Install efficient lighting fixtures.

Energy savings potentials for implementing these five projects are summarized in Table 6.

Individual energy-efficiency measures identified during the plant-wide assessment scheduled to be implemented as a package are described briefly below.

1. *Replace 1250-hp barren pump motor with a 600-hp motor*: The four-stage barren pump circulates 5500 gallons/minute (gpm) of cyanide solution to the top of the phase 2 and phase 4 heaps. Phase 2 requires 1300 gpm and phase 4 receives 4200 gpm. A 600-hp premium efficiency motor could replace the existing 1250-hp barren pump motor.
2. *Replace 1250-hp phase 4 pregnant pump motor with a 750-hp motor*: The phase 4 pregnant pump transfers metal-laden cyanide solution from the base of phase 4 to the processing plant. The head required is 500 ft, determined by calculating the pressure loss and elevation gain, using a 25 psi pressure drop through the clarifiers in the

processing plant. One bowl could therefore be eliminated from the existing pump configuration. A 750-hp premium efficiency motor could replace the existing 1250-hp pregnant pump motor. Additionally, an adjustable-speed drive could be installed to accommodate the change in flow requirements from phase 4 to the plant.
3. *Install a regenerative system on downhill conveyor*: A downhill conveyor transfers crushed ore to the base of phase 4 at an average rate of 1200 tons/hr. The energy created by the falling ore is currently dissipated as heat in the conveyor system's brakes; this energy can be recovered and used to generate electricity.
4. *Install adjustable-speed drive on 150-hp well pump No. 2*: Well pump No. 2 provides water to the processing plant for use in the plant and for use in suppressing road dust. Flow requirements fluctuate. A valve on the discharge line currently controls the flow rate. Installing an adjustable-speed drive on the 150-hp motor would allow better flow control and reduce the pump operating costs.
5. *Install efficient lighting fixtures*: Metal halide lighting fixtures are currently used throughout the shop and in warehouse spaces. Installing T-5 high-bay fixtures offer significant potential to provide cost savings and improve overall lighting performance.

Potential savings: Implementation of the five projects was estimated to yield a cost savings of ~$813,000 per year. The required capital investment was estimated to be $260,000, resulting in an average simple payback of 4 mo for all five projects.

Petroleum

Valero Energy Corporation: Valero Energy Corporation performed a plant-wide energy assessment at its Houston, Texas refinery, involving an energy systems review to identify the primary natural gas and refinery fuel gas users, electricity, and steam-producing equipment, and cooling water systems, plus develop an energy optimization and management system. Valero Energy Corporation is headquartered in San Antonio, Texas, and has ~20,000 employees. Estimated annual revenues exceed $50 billion. The company owns and operates 14 refineries throughout the U.S., Canada, and the Caribbean. The refineries have a combined throughput capacity of more than 2 million barrels per day (approximately 10% of the total U.S. refining capacity). Valero bought the facility in Houston in 1997. This crude oil processing facility produces a wide range of petroleum products, such as gasoline, diesel fuel, kerosene, asphalt, jet fuel, sulfur, No. 2 and No. 6 fuel oil, and liquefied petroleum gas. It produces ~60,000 bbl/day of gasoline and 35,000 bbl/day of distillates. Processing capabilities primarily involve medium sour crudes and low-sulfur residual oils. The Houston facility is composed of several process units equipped with gas- or refinery gas-fired equipment. The units process raw crude oil into various finished products by means of distillation and chemical reactions. The units include the following:

- *Crude complex* consisting of atmospheric and vacuum distillation towers with two lightends removal towers;
- *Fluidized catalytic cracking complex* that thermally cracks heavy gas oil material into lighter products (such as gasoline and diesel).
- Hydrotreating/reforming complex that processes kerosene, diesel, and naphtha in four desulfurization units; the desulfurized naphtha is reformed in a fixed catalyst bed reformer to produce a high-octane gasoline blend component;
- Sulfur removal complex recovering sulfur gas streams in an amine contacting/regeneration unit and a sulfur plant before using the gas streams as refinery fuel gas; and
- Alkylation complex that produces a high-octane gasoline blending component using a sulfuric acid alkylation unit; an isooctane unit converts butane streams to isooctane for gasoline blending.

Refinery fuel gas is used in process heating and steam production. Four fixed boilers and waste heat recovery equipment (two cogeneration units) produce steam. Each process unit requires cooling water that is supplied by six refinery cooling water towers to remove low-level heat. The refinery's three flare systems vent and burn light gases produced by the process equipment. The facility has access to three water sources: the Coastal Water Authority (CWA) supply, City of Houston municipal water, and well water.

The energy assessment addressed ways to reduce water use and environmental emissions. Fourteen projects were identified for potential implementation at the facility. If all the projects identified were implemented, the potential energy savings to the Houston refinery was estimated to be 1.3 million MMBtu (fuel) and more than 5 million kWh (electricity). Total annual cost savings were estimated to be ~$5 million. The plant-wide assessment consisted of two primary activities:

- *Energy systems review*: The assessment team reviewed the primary energy systems (natural gas, refinery fuel gas, electricity production, steam, and cooling water). Major natural gas and refinery fuel gas users, electricity- and steam-producing equipment, and cooling water systems were identified. Energy-use data was collected to identify major energy-consuming equipment and processes.
- *Energy optimization and management system development*: Assessment of data and energy system information were gathered to develop a computer model of the primary refinery processes and the energy

production and distribution systems. The model can be used to determine the most efficient loading of individual pieces of equipment

Energy savings potentials for implementing these fourteen projects are summarized in Table 7 below.

Individual energy-efficiency measures identified during the plant-wide assessment scheduled to be implemented as a package are described briefly below.

— *Upgrade steam system insulation*: Using a computer model, the assessment team identified several uninsulated or underinsulated steam system lines and equipment; the upgrading the steam system insulation should be implemented to reduce energy losses.

— *Install new air compressor*: Two rented diesel air compressors with a new steam-driven compressor would result in saving ~108 gallons/day of diesel fuel and $160,000/yr in rental fees.

— *Install CO control system*: Two Bambeck CO control systems on the crude unit's atmospheric and vacuum heaters were recommended by the assessment team. By reducing the excess oxygen from 3.5% to 1.5%, energy savings will result. Nitrous oxide (NO_2) and CO emissions would be lower due to the lower flame temperatures. The Bambeck CO control system uses a CO analyzer on the heater flue gas, a stack damper control system, heater draft instrumentation, and computer software to minimize excess oxygen.

Table 7 Estimated Project Cost Savings for Valero's Houston Refinery

Project Description	Cost Savings ($/yr)	Capital Cost ($)	Pay-back Period (yr)	Fuel Savings (MMBtu/yr)	Electricity Savings (kWh/yr)
1. Upgrade steam system insulation	168,100	100,000	0.6	51,906	N/A
2. Install new air compressor	200,000	400,000	2.0	5,125	N/A
3. Install CO control system	247,700	401,000	1.6	77,406	N/A
4. Install cooling tower automatic blowdown control system	130,000	240,000	1.8	N/A	N/A
5. Incorporate an energy optimization and management system	2,200,000	270,000	0.1	687,500	N/A
6. Inspect, repair, and maintain steam traps	656,000	100,000	0.2	205,000	N/A
7. Install boiler automatic blowdown control system	100,000	180,000	1.8	N/A	N/A
8. Replace boiler house boilers	161,000	3,000,000	18.6	50,313	N/A
9. Install flare gas recovery compressor	420,000	1,000,000	2.4	131,250	N/A
10. Install evaporative coolers on cogeneration units	174,700	249,000	1.4	−17,035	4,987,000
11. Clean the CWA water supply line	13,000	6,000	0.5	N/A	N/A
12. Assign an energy coordinator	200,000	130,000	0.7	N/A	N/A
13. Turn off outdoor lighting during daylight hours	6,000	N/A	0	N/A	175,000
14. Install liquid ring vacuum compressor	306,000	450,000	1.5	95,625	N/A
Total	4,980,500	6,526,000	1.3 (mean)	1,287,090	5,162,000

— *Install cooling tower automatic blowdown control system*: Instrumentation and controls on four cooling towers to automatically control each tower's conductivity (which provides a measure of the solids in the circulating water system) were recommended. High conductivity reduces makeup water costs, chemical use for water treatment, and downstream processing costs, but it increases cooling water system fouling and corrosion while reducing heat transfer. Low conductivity has the opposite effects. Maintaining the conductivity within a suitable range is important for overall performance and cost of the cooling water system. Makeup water and chemical costs were estimated to be reduced by $130,000.

— *Incorporate an energy optimization and management system*: An energy optimization and management system (EOMS) should be developed to use in assessing, implementing, and tracking process unit energy changes and new project implementation effects on the refinery. Implementing EOMS was estimated to reduce annual energy use by 2%–6%. The simulation and optimization software optimizes the purchase, supply, and use of fuel, steam, and power at the refinery, based on process unit energy demands and system constraints caused by equipment or environmental regulations. The software analyzes conditions such as supply contract variability, alternative fuel options, optimum loading of steam boiler equipment, motor vs. turbine driver decisions, and importing vs. exporting of steam, fuel, and power. The software performs the following functions for the refinery:
  ○ Facilitates optimal planning of utility equipment;
  ○ Assists in optimal operation of the utility plant and associated equipment;
  ○ Provides real-time information on site-wide energy performance, utility costs, and revenue; and
  ○ Provides real-time information for use in prioritizing maintenance tasks;

— *Inspect, repair, and maintain steam traps*: Leaking or failed steam traps cause significant energy losses and create process problems. Leaking or failed steam traps were recommended to be repaired and implementation of a maintenance program.

— *Install boiler automatic blowdown control system*: Installing instrumentation and controls on three steam boilers was recommended for automatically controlling boiler blowdown conductivity. High conductivity reduces boiler feedwater expenses and chemical use, but it increases boiler system fouling and corrosion, while reducing heat transfer. Low conductivity has the opposite effects. Controlling conductivity within a suitable range is important to overall boiler system operation and costs. Boiler feedwater and chemical feed were estimated to be reduced by ~$100,000 annually.

— *Replace boiler house boilers*: Replacing the four 1940's-era 50,000 lbs/hr induced draft boilers with a new 200,000 lb/hr, boiler was recommended, to improve operating efficiency and reduce fuel gas usage. Additionally, NO_x and other emissions would be reduced, providing additional economic and environmental benefits.

— *Install flare gas recovery compressor*: Installing a compressor and associated equipment was recommended in order to recover flare gases from three existing flare systems. Recovered gas can be amine-treated to remove hydrogen sulfide and routed to the refinery fuel gas balance drum, thereby reducing natural gas makeup to the gas balance drum (except when there is excess recovered fuel gas because one or both off-refinery fuel gas purchasers are shut down).

— *Install evaporative coolers on cogeneration units*: Installing intake air coolers was recommended to increase the cogeneration of electric power or to sell to the local power grid. The cogeneration units will probably be replaced due to their aging components, inefficiency, and NO_x emissions.

— *Clean the CWA water supply line*: Cleaning a raw water supply line was recommended to reduce pressure drop and increase flow. This program has already been implemented; the system now supplies the added water needed to meet refinery demands without having to bring in more expensive municipal water.

— *Assign an energy coordinator*: Prior to conducting the plant-wide assessment, no one was responsible for monitoring refinery energy use and developing energy projects. A full-time energy coordinator and support personnel were recommended. This team would be responsible for such activities as ensuring that steam trap surveys are conducted, developing energy use simulation models, reviewing flare system losses, developing energy projects, and monitoring energy use.

— *Turn off outdoor lighting during daylight hours*: Lighting was being left on unnecessarily during the day. Operations personnel were reminded to turn off these lights when they were not needed. The team also found that most areas in the refinery are equipped with photocells. Several photocell switches had become dirty or had accidentally been painted over, so they were in need of repair or replacement.

— *Install liquid ring vacuum compressor*: Installation of a liquid ring vacuum compressor on the vacuum tower overhead system was recommended to replace the third-stage vacuum steam jets and water cooler. This project was estimated to reduce steam use at 285 psi by ~6,000 lbs/hr. Additional energy savings should result from improved suction pressure on a downstream compressor.

Potential savings: The projects recommended were estimated to save Valero Energy Corporation about

$5 million annually, while reducing environmental emissions.

Steel

Jernberg: To streamline its manufacturing processes at its Chicago, Illinois, facility, Jernberg chose to convert from its traditional batch methods to lean manufacturing. Jernberg was founded in 1937; it was the country's first independent press forging company. The company's forging facility located in south Chicago, produces 170 million lbs of forged parts annually. Jernberg produces a wide variety of gears, yokes, hubs, and other parts for the automobile and motorcycle industries. A plant-wide energy efficiency assessment was conducted to address energy-intensive processes requiring change. Efficiencies of the primary support systems (e.g., compressed air, cooling water, etc.) were also investigated. A process simulation model was used to evaluate the impacts of converting existing batch production to a lean manufacturing operation; it employed a systems approach to identify trends and energy use distribution for plant equipment. Seven projects were identified that were estimated to save Jernberg more than 64,000 MMBtu/yr in fuel and more than 6 million kWh/yr in electricity.

The primary raw material used in forging operations is steel bar. The bar stock is often preheated with a natural gas-fired burner to drive off moisture and to facilitate shearing. The heated bar is fed into one of seven shear presses or saws, where it is cut into billets of 6–15 in. lengths. The billets are deposited into metal boxes as they leave the shear press. The billets are transported by forklift to one of ten forging lines, where the billets are manually fed (charged) into a conveyor that feeds the billets into a pass-through induction furnace. The heated billet is ejected from the furnace discharge and drops down a feed chute that serves the forging press. At the bottom of the chute, a worker lifts the billet from the chute and places it on the bottom die. The forging ram is activated and compresses the billet. The pressed billet is lifted and placed on the second stage die where it is rammed into its final shape. Once the billet has been forged, the part is ejected from the press and conveyed to a trim press where flash is removed from the hot part.

About half of the parts that are produced are also heat-treated, which is performed in one of five batch-style heat treat furnaces. Several heat treat methods are used, including water-quench, oil-quench, normalizing, and annealing. Many parts undergo finishing operations such as shot blasting, drilling, grinding, magnaflux testing, dimensional testing, and ultrasonic testing.

The lean manufacturing methodology optimizes equipment utilization, maintenance programs, information flow through the plant, and workplace organization. Lean manufacturing requires plant personnel to assess ways to eliminate waste, including work-in-progress inventory, unnecessary movement of parts throughout the plant, overproduction, machine downtime, etc. Primary benefits of lean manufacturing include more efficient overall utilization of the plant, reduced inventory (and associated carrying costs), improved product quality, and improved equipment reliability. Jernberg traditionally produced forged parts using batch processes, which can length a production cycle time by 3 to 5 times or longer. The energy assessment team performed a process simulation to evaluate the impacts of converting existing batch production to a lean manufacturing operation. Heat-treating was identified as the primary bottleneck in the process. Using controlled cooling instead of batch heat treating largely eliminates that problem as well as provide significant energy savings. A systems approach was used to evaluate the plant's energy consumption. Historical data over a two-year period was evaluated to identify trends and energy use distribution for plant equipment; this allowed identification of systems to be targeted for further evaluation. Once systems and equipment were identified and targeted, recommendations were developed that were both technically and economically feasible. These included recommendation that supported conversion to lean manufacturing. The efficiencies of the primary support system were also evaluated, such as for compressed air and the cooling water loop.

The assessment team identified seven projects during the energy efficiency assessment. If all the projects were implemented, Jernberg estimated that it could save more than 64,000 MMBtu/yr in fuel and more than 6 million kWh/yr in electricity. Total annual cost savings were estimated to be ~$791,000. Total implementation costs were estimated to be ~$2,000,000 for all the projects. The specific projects recommended are listed below:

- Repair recuperator;
- Recover waste heat from cooling tower loop;
- Eliminate billet reheats;
- Install air compressor controls;
- Convert heat treat processes to controlled cooling;
- Replace air compressors; and
- Reduce forge press downtime.

Individual energy-efficiency measures identified during the plant-wide assessment scheduled to be implemented are described briefly below.

Repair recuperator: The existing recuperator on heat treat furnace No. 5 is not being used; it was taken out of service during a previous furnace rebuild operation. As a result, air at 1400°F is exiting the stack during furnace operation. By repairing the recuperator and returning it back to service, the stack exit temperature could be reduced to 400°F, and the recovered heat could be used to preheat the combustion air in the furnace, saving an estimated 1812 MMBtu/yr. The implementation cost was estimated to be ~$25,000, assuming that the recuperator would need to be repaired before returning it back to service. This project was estimated to provide an annual cost savings of $9,400; the resulting simple payback period would be ~2.7 yr.

Recover waste heat from cooling tower loop: Fourteen cooling towers provide process cooling for the induction heating furnaces, air compressors, and hydraulic systems on each forging line. Jernberg uses ~12,000 MMBtu of natural gas for space heating annually. Most areas of the plant require space heating (except for the forging area, which relies on waste heat from the process to warm the area during the winter). Of the many cooling loops used in the plant, only the loop serving the north air compressors is used consistently enough to provide a steady source of heat for the plant. This loop rejects 3650 MMBtu of heat from two Ingersoll-Rand compressors. Using this waste heat to warm the final processing department and displace the existing gas-fired unit heaters that are only 80% efficient would save 4,652 MMBtu a year (estimated). Annual cost savings are estimated to be $23,700 per year. The estimated cost to implement this project is $25,000, resulting in a simple payback period of 1.1 years.

Eliminate billet reheats: The induction heaters operate on an open-loop control basis to take room-temperature billets and heat them to 2300°F prior to delivering them to the forging process. When a downstream process goes down, the heated billet is diverted to totes where it is allowed to cool to room temperature before being reheated. The furnace has no automatic controls, so an operator must shut down the charging system that feeds the billet stock into the furnace to reduce waste heating of billets when the line goes down. A closed-loop has been proposed for each line to allow the induction furnaces to react to downstream variations and speed up or slow-down as needed to match production requirements more closely. This approach would tie in with the lean manufacturing process, as it would allow all processes in the forging cell to balance. Billet heating is by far the most energy-intensive step in the manufacturing process. Using a closed loop approach would eliminate the reheating step, providing an estimated energy savings of 4.1 million kWh/yr of electricity, with an annual cost savings of ~$247,700. The cost to retrofit all ten lines was estimated to be $500,000 (which includes the installation of controls and queue sensors as well as programming logic to operate the furnaces) providing a simple payback period of ~2 years.

Install air compressor controls: At the plant, Ingersoll-Rand compressors are base loaded while Fuller Rotary vane compressors operate partially loaded in throttle modulation. Throttle modulation is an inefficient mode of control. Plant personnel turn compressors on and off as needed. This generally causes more compressors to be on than are really needed, particularly during shift changes and when process lines are shut down early. An integrated control system would minimize the human factor from compressor control, and ensure that the vane compressors are base loaded when online, using the reciprocating compressors as swing machines. The electronic controls allow the compressors to maintain pressure at ±1 psig, allowing the compressor setpoint to be reduced safely within 2 psig of the plant's minimum pressure requirement. The combined energy savings associated with pressure reduction and capacity control was estimated to be 1.8 million kWh annually with a cost savings of ~76,900/yr. Using an implementation cost of $270,000, the simple payback period was estimated to be 3.5 years.

Convert heat treat processes to controlled cooling: Forged parts are allowed to cool a minimum of 24–48 hours; they are then reheated to 1500°F in the heat treat furnaces. Controlled cooling takes the hot forged parts and carefully cools then in a closed chamber. The cooling profile makes certain that the parts' microstructure is properly developed to ensure proper hardness, which is accomplished by using the residual heat from the forge process. A type of controlled cooling is currently used in the plant by utilizing vanadium alloyed steels. Implementing controlled cooling on other parts was estimated to save $7700 MMBtu annually by utilizing the process heat that is currently wastes for heat treating of parts. This system would readily fit in with the lean manufacturing approach due to the forged parts being fed via conveyor directly through the cooling chamber. Estimated costs to implement this project was $1.1 million, which includes $600,00 to model the cooling profiles for the parts, plus $500,000 to construct two chambers that provide the cooling process. An annual natural gas savings was estimated to be $352,000 yielding a simple payback period of 3.2 years. This project was deemed to be highly important, since it would eliminate a production bottleneck in the heat treat department and would increase the plant's capability such that raw materials could be converted into finished product within one work day.

Replace air compressors: Four existing Fuller air compressors are at the end of their service life. The machines are two-stage rotary vane compressors. Due to their age and wear, maintenance cost on these compressors is high (~$10,000/yr for materials only) and the machines' capacity has degraded because of increased blowby between the vanes and the chamber wall. These compressors are water-cooled; there are additional water costs associated with cooling tower evaporation and drift. Replacing these aged compressors with new, two-stage rotary screw compressors was recommended to increase compressor specific power from 4.3 cfm/bhp to 5.3 cfm/bhp at the application conditions. This improvement in efficiency was estimated to save 143,100 kWh/year with a cost savings of $74,000/yr. The cost of four new compressors was estimated to be $280,000, resulting in a simple payback period of 3.8 years.

Reduce forge press downtime: When dies are changed in the forging process, new dies are heated using a natural gas torch that is sandwiched between the die cavities for about 20 minutes. When die changes are required during production hours, this time represents lost productivity. If an infrared die heating station is installed, the same die heating process could be accomplished in 4 to 5 minutes, enabling the die to be changed and return the press back in to production more quickly. Some energy savings would be realized by eliminating the torch. The primary savings will occur as a result of less downtime, thereby reducing billet heater losses. This savings was estimated to be ~$7,200 annualy. The cost of an infrared die heat station was estimated to be $5,000, yielding a simple payback period of 0.7 yr.

Potential savings: With implementation of the seven projects, Jernberg estimated that it could save more than 64,000 MMBtu/yr in fuel and more than 6 million kWh/yr in electricity. Total annual cost savings were estimated to be ~$791,000. Total implementation costs were estimated to be ~$2,000,000 for all the projects.

Geothermal Plant

Alaska's first geothermal plant was dedicated at Chena Hot Springs, Alaska [see website: www1.eere.energy.gov/news/progress_alerts/progress_alert.asp?aid=183]. This plant is the first to use a new technology making electricity generation possible at lower temperatures. The plant makes geothermal power plants feasible in many more locations than today's high temperature technologies. The technology is a pure cycle organic Rankine device to convert waste heat and liquid streams in to power to increase system efficiency of distributed generation devices. At Chena Hot Springs, the system will offset 160,000 gal of diesel fuel annually, representing a cost savings of ~$384,000.

BIBLIOGRAPHY

1. U.S. Department of Energy, Energy Efficiency and Renewable Energy, U.S. DOE website: http://www1.eere.energy.gov/industry/bestpractices/printable_versions/case_studies.html (accessed January 2007).
2. U.S. Department of Energy, Energy Efficiency and Renewable Energy, U.S. DOE website http://www1.eere.energy.gov/news/progress_alerts/progress_alert.asp?aid=183 (accessed January 2007).

Savings and Optimization: Chemical Process Industry

Jeffery P. Perl
Chicago Chem Consultants Corp., Chicago, Illinois, U.S.A.

Robert W. Peters
Department of Civil, Construction & Environmental Engineering, University of Alabama at Birmingham, Birmingham, Alabama, U.S.A.

Abstract

For the chemical process industry (CPI) to remain competitive in the global environment, both environment, safety and occupational health (ESOH) and process design must be integrated and simultaneously optimized. A practice-based approach to chemical process optimization is outlined emphasizing these concerns. Raw-material usage and energy consumption must be balanced. Mass and energy balances are coupled in order to use each in the most efficient manner possible. Process energy is optimized by consideration of process heat integration, low- and high-temperature sources, mechanical systems (e.g., microturbine power generation, pumping, compressed air, hydraulic systems, pneumatic transport, and chillers), electrical systems, water, utilities, waste-to-energy opportunities, and facility energy integration. These compounds can be applied to CPI unit operations, such as distillation, liquid extraction, drying, freeze crystallization and drying, crystallization, filtration, chemical reactors, etc. Several case studies integrating mass and energy optimization are reported related to process energy optimization. A checklist is presented for optimizing process energy systems.

ENERGY UTILIZATION AND SAVINGS IN THE PROCESS INDUSTRY

Introduction

The business of the chemical process industry (CPI) is turning raw materials into finished products, usually on a large scale. The CPI includes a wide variety of manufacturing sectors including, food, pharmaceutical, plastic, oil refining, and building materials to name only a few. All have as a common requirement a need to balance raw material usage with energy consumption as well as Occupational Health and Safety Administration (OSHA) and Environmental Protection Agency (EPA) regulatory compliance. Product quality assurance requirements must also be met. It will be helpful to review some of the regulations that affect process energy considerations.

Keywords: Chemical process industry; Environment, Safety and Occupational Health (ESOH); Material and energy balance; Energy optimization; Process heat integration; Heating, ventilation, and air conditioning (HVAC); Variable-frequency drive (VED); Unit operations; Combined heat and power; Cost; savings; Electricity; Natural gas; reaction engineering; Building energy management; RCRA; Occupational Health and Safety Administration (OSHA); Environmental Protection Agency (EPA); Process troubleshooting; Low-quality heat; High quality heat; Turbine; Power.

U.S. EPA Regulations

Resource Conservation and Recovery Act of 1976 (RCRA), United States Environmental Protection Agency (EPA) RCRA was crafted exactly during the time of the first (1973) energy crisis in the United States. This happened when oil-exporting countries embargoed and severely curtailed oil exports and raised prices. As a result, the U.S. Department of Energy (DOE) began an extensive program to promote alternative energy sources including novel methods to recover more of the 60%–80% of crude oil left behind after conventional drilling and pumping operations. In light of these contemporary activities, RCRA was written with energy prominently mentioned in the opening paragraphs.

Selected Excerpts from the original RCRA statute:

> Section 1002 (d) ENERGY—The Congress finds that with respect to energy:

1. Solid waste represents a potential source of solid fuel, oil or gas that can be converted into energy;
2. The need exists to develop alternative energy sources for public and private consumption in order to reduce our dependence on such sources as petroleum products, natural gas, nuclear and hydroelectric generation; and
3. Technology exists to produce usable energy from solid waste.

Objectives and National Policy

Section 1003 (a) OBJECTIVES—to promote protection of health and environment and to conserve valuable material and ENERGY resources...

We can see elements of this philosophy on U.S. EPA boiler industrial fuel (BIF) regulations allowing companies to burn used oil onsite to recover energy value, while maintaining clean air compliance.

US Department of Energy Program

The federal government has an excellent program to assist both commercial and residential end users of energy. Information is available for US DOE as well as US EPA Energy Star program on the web. Much information has been developed including efficiency of equipment and appliances as well as actual energy project cost and effectiveness study results. We have presented this information at the end of this article and our following article.

COMBINED ENERGY AND ENVIRONMENT, SAFETY AND OCCUPATIONAL HEALTH (ESOH) CONSIDERATIONS

As we have seen from the very inception of US EPA, both energy and environment have been linked. For the design engineer, it is imperative that both ESOH and process design be simultaneously optimized. Sustainability captures a global concept and includes such elements as raw material and prime energy resource availability into the planned project future. Health and Safety also are of primary concern up front in any process design. We lump all such considerations together as ESOH.

When making process changes to reduce energy consumption, take care to avoid negative impact on the product. This might entail re-testing for intermediate and end point specification. International Standards Organization (ISO) considerations may also be involved regarding total quality assurance. OSHA management of change (MOC) regulations will also come into play to ensure worker safety is maintained or improved, and always in compliance. The EPA risk management program (RMP) requirements also will come into play. The details of these OSHA and EPA programs are beyond the scope of this article, but their importance is not. The law of unintended consequences is be best managed by understanding the interplay of elements involved in production.

MATERIAL AND ENERGY BALANCE... WHERE IT ALL BEGINS

The very first course of study for chemical engineering students is Material and Energy Balances (MEB). When designing a chemical process, a form of "double entry accounting", evolved regarding materials processed and energy consumed. These two elements are always coupled and provide the designer with a convenient method to both design the process as well as double check and optimize the use of both materials and energy.

A simple example is that of boiling water. Here approximately 1000 Btu is required to evaporate a of water. A process designed to distill water, therefore, must account for both providing this energy (source) and dissipating this heat (sink). The heat going into a process minus the heat leaving must be equal to that consumed by reaction requirements or non-steady state energy buildup. So a source such as a boiler must be selected to supply needed heat and a sink such as a condenser must be selected to remove the heat. This calculation can also be used to provide extremely close digital control of the process through coupled feed forward and feedback control loops, with the calculated energy needs (feed-forward) being fine-tuned by the feedback measurement of actual conditions, as tempered by adaptive controls. Detailed knowledge of the process MEB information also identifies energy inefficiencies and can become the basis for improvement in operation.

The material balance alone can provide useful information about the health of the process. Material entering the process minus that leaving must be equal. Differences can arise from non-steady state conditions such as accumulation in a tank. Any unaccounted difference might be considered lost material and may serve as the basis for environmental permit violation as well as bottom line process economic loss.

An energy balance alone can also be useful. For example, if you cannot account for 20,000 Btu of supplied heat, it may indicate a loss of 20 lb of water or poor flow or temperature measurement. A typical steam energy balance problem might involve inadequate or missing condensate return, poor check valves, traps, etc. Similar problems come up in chemical processes such as plugged or leaking heat exchanger tubes, partially shorted pump motor windings, broken mixing blades, poorly distributed or destroyed mass transfer packing, deactivated catalysts, and buildup of harmful byproducts (such as scale and chemicals) that reduce reactor efficiency or selectivity.

So it becomes apparent that while independent material and energy balances can be useful, they really shine when ever coupled as a double check of process or utility health.

If our simple water-boiling example became a solvent distillation instead the unaccounted energy could be tied to solvent loss with attendant environmental implications or worker health impacts, so the MEB is extremely important, not just in design, but as a barometer of process health and efficiency as well. The attempt to address environmental losses through waste minimization must therefore be accompanied by equal consideration to energy optimization. In some instances, energy will be the primary consideration (leading indicator), and in others, material will be.

We will focus more on energy as a leading indicator. An unexpected benefit of process energy review and optimization provides the designer an opportunity to reevaluate the entire system.

PROCESS ENERGY OPTIMIZATION

The overall process of turning raw materials into finished products is energy intensive. Processes consume mechanical and electrical energy as well as water, natural gas, oil and other combustible commodities. Waste byproducts are also produced that must be either treated or disposed of, often at the loss of their energy contents.

Process Heat Integration

Chemical manufacturing is comprised of physical unit operations to separate and concentrate chemicals without changing them, while reactors change raw chemicals into desired products such as, plastics, fuels, pharmaceuticals and other products. Unit operations such as distillation and drying consume large amounts of energy and cooling water. Reactors commonly produce heat (some are endothermic, or energy consuming) and use cooling water to maintain temperature. Mechanical, electrical, power, and heat engine utilities exist to meet process needs. In addition, facilities need heat in winter and cooling in summer, operations that both consume energy. Through careful management and design, the various energy sinks are matched to sources to minimize outside requirements of electricity, gas and oil.

Low- and High-Temperature Sources: Low versus High-Quality Energy

Thermodynamic engines require large temperature differences to produce power, and they still have relatively low energy conversion efficiency. The key to high overall thermal efficiency is to couple a steam-driven electric turbine to a process that requires low-quality heat. With the advent of the microturbine and utility electricity buyback regulations, individual facilities can now consider power generation. When you couple in transmission line loss of up to 30%, it should make on-site generation projects increasingly more attractive. Those that have small process heat requirements might consider locating near or attracting onsite manufacturers that have high heat demand. The same arguments hold for plant chillers and facility HVAC air conditioning system combinations.

Boiler heat exhaust can be used for instance to create high-pressure electrical turbine steam. However, many streams requiring cooling, such as unit operations and reactors and do not meet these differential temperature requirements. In these instances, we look for streams requiring preheating to match against streams requiring cooling.

Facility HVAC equipment, such as chillers and boilers can often exhaust/supply low quality energy to chemical processes at a very high efficiency. When designing a chemical process, engineers typically place collection vessels before and after to take up surges. In fact, the concept is similar to an electrical surge protector. Surge capacity must be engineered if a facility generates electricity for use on- and off-site, or when integrating facility heat and cooling loads with process stream energy demands.

ELECTRICAL, MECHANICAL, AND OTHER UTILITY DESCRIPTIVE CHECKLIST

It is difficult sometimes to separate these elements from the process itself, but it is helpful, as most plants are divided this way. Let us examine some common areas of energy interest. These topics are covered elsewhere in this encyclopedia, but they deserve mention here to identify the method and manner that they interact with and serve the process.

Mechanical

Microturbine Power Generation

Recent advances in electric generation technology have brought this down to the personal home size. In chemical processes, one can generate power from waste heat to steam. The key to success lies in either using generated power for your plant, selling it to the power company or both. Remember to factor in low-quality waste heat that may exit the turbine. This can be captured as preheat for the process or to preheat boilers or hot water heaters or even as seasonal facility heating.

Pumping

Virtually no chemical process is without some sort of pump. Whether liquid, solid, or gaseous, a variable-speed pump or blower versus a fixed-speed pump and valve arrangement can reduce or eliminate expansion/throttling losses. Here the ubiquitous variable-frequency drive (VFD) can take the place of energy consuming throttle type valves and fixed-speed motors. The applications of course must be matched, as would be the case with any design. A concomitant benefit is the reduction of one control element, i.e. a mechanical valve positioner for each loop. While we are on this subject, up front design of pumping systems is critical to long-term energy reduction due to the gross expense of replacing inefficient hydraulic systems.

Compressed Air

The typical specification for compressed air has a broad tolerance. This leads to inefficiencies arising from leaking systems and improperly high or low pressure settings. Poorly maintained line filters can introduce moisture or compressor lubricants and other undesirable elements into a process that might also adversely affect energy consumption. Air compressors also create waste heat that should be integrated whenever possible, but is often overlooked.

Hydraulic Systems

Piping is one of the most commonly overlooked system elements regarding energy. Typically, design engineers are forced to prioritize physical layout restrictions over all else. Still, overall pumping energy requirements can be modeled and used to at least identify potential energy savings.

Pneumatic Transport

Many chemical processing facilities transport solids by conveying them in air or nitrogen. Frictional losses are great in these systems and are supplied by compressors that themselves must be integrated into overall energy conservation plans.

Chillers

One of the largest consumers of plant energy is the chiller. Whether serving process heat dissipation loads or air conditioning work space, much heat can be recovered. Advances in technology have led to more efficient units that although costly, can lead to short payback periods, particularly at end-of-life replacement.

Electrical

- Motors—Choose energy efficient units and ensure they are matched to loads and set to turn off when not needed. Often running against closed valves can damage pumps and waste energy.
- Process Control—Modern Programmable Logic Controls (PLC) can allow a greater degree of flexibility than ever before. In fact, PLC reevaluation can lead to additional savings.
- Variable-Speed Drives—Do not use without proper design and selection, but can achieve significant savings.
- Variable-Speed Fans (VSF)—Air moving is another hidden energy gobbler. Often fans are either on or off, and use of VSF can save energy just as with VFD on motor driven pumps.
- Micro/Macro Electric Generating Turbines—Can be used to couple high to low quality heat source and sink.
- Power Topping Cycles—Particularly useful when multifuel capability or large combustible waste streams are available.

Water

- Once Through Cooling—Many facilities use water from rivers, wells, or municipal supplies. Often, the water is discarded after a single use—hence the term once through. Every attempt must be made to identify process or HVAC interchange opportunities. Lawn watering with this "Gray Water" is also a cost savings item if storage is possible. Avoid sprinkler operations during the rain by use of available rain detection equipment coupled to PLC.
- Process Feed Water—Many processes require water as feed and once through water can be substituted or augmented by clean water for this purpose. Interconnections are a bit more complex but should be considered in every conservation program.
- Cooling Towers—These can be used to reduce once through cooling water requirements. They must be maintained and can also be incorporated into Gray Water plans to save additional resources and money.

Electricity, Natural Gas, and Fuel Oil Supply

Periodic Cost Review

Review all energy procurement contracts to assure best possible pricing and structure. The US industry is undergoing a period of constantly changing de-regulation and restructuring, so these reviews should now as regular as those used for purchasing any other manufacturing supply.

Periodic Demand Charge Review

All electrical equipment briefly more energy during startup than while running at steady state. Utilities commonly levy an extra. "Demand Charge" for this. These charges can be avoided by load shifting to off hours, use of electrical load leveling technology, or just negotiations with the utility supplier. These charges can change even when base line rate does not and so require additional periodic review.

Waste to Energy

- Waste Energy and Material Evaluation.
- Review of all waste streams for energy recovery opportunity.
- Review of all waste streams for material recovery or waste exchange.

Facility Energy Integration

- HAVC Interconnection Review.
- Peak Demand Hours of Operation Review.
- Lighting.

PROCESS ENERGY CONSERVATION EXAMPLES

There is no end of examples, and we commend the reader to look at other sources, such as, the US Department of Energy. DOE maintains a website of process energy savings examples described later. We have provided several of these in the following article.

Chemical process engineering is comprised of physical and chemical manufacturing steps. The physical steps have become known as "Unit Operations" and are comprised of activities that make no change to the individual chemical constituents. Chemical manufacturing steps come under the category of "Reaction Engineering" that involve chemical changes to individual constituents. These are described in general below, and are followed by some common industrial examples.

Chemical Process Unit Operations

Distillation

This is the process of physically separating two or more components by the application of heat. The difference in component boiling point serves as the "separation factor" and as a result, this process uses both heat for boiling and cooling for condensing. As a result, this is one of the most energy intensive of all unit operations.

Liquid Extraction

Here pumping is often the largest source of energy and pressure drop through packed beds are the culprit. The main purpose of extractions is to transfer material from one stream to another, so mass transfer is the controlling design parameter. Continuous pumping of any type is a hidden energy drain so select low pressure-drop packing up front. Indeed this consideration is so new that certainly advances in packing may come about that will improve both energy and mass transfer.

Drying

This is an extremely common unit operation. In this application, controlled heat and humidity reduce or remove moisture from products. Wallboard and food products are two important examples. Low quality heat from other sources can assist in drying, while high moisture bearing waste hot air streams might preheat a process. Always remember surge requirements and design accordingly.

Freeze Drying and Crystallization

Ice undergoes a direct phase change from solid to gas. This process, termed sublimation, of course requires heat. Freeze-drying is employed where higher temperature evaporation is undesirable due to product breakdown or bacterial or enzymatic activity. Moisture is the primary culprit in food spoilage so this method greatly extends product shelf life. Refrigeration equipment is employed that should lend itself to heat recovery and integration, but must be coordinated closely with the process. In addition, vacuum pump equipment should also be included in energy optimization plans.

Crystallization

Water can be purified by alternating freeze-thaw cycles, taking advantage of the large difference in solubility of salt in water with temperature and phase change. As with drying, the focus will again be on refrigeration equipment and associated heat rejection.

Filtration

This process separates solids from liquids as well as liquids from other liquids. In solid-liquid filtration, pumping through a filter media causes retention of solids on the media. This is accomplished by either application of positive pump pressure or negative vacuum, both consuming energy. Liquids comprised of varying molecular size and structure components can be separated by ultrafiltration, a process that requires pump pressure and hence energy.

Chemical Process Reactor Operations

- Heat of Reaction, Exothermic–Heat is evolved and may need to be dissipated,
- Endothermic–Heat is absorbed and may need to be supplied.

Neutralization–A process of adding acid or base to produce a neutral product. These are generally exothermic, require heat dissipation and are ripe for heat recovery.

Polymerization–A very common process for convert individual monomer molecules into longer chain polymers with more desirable properties. Plastics are a good examples of polymers. Catalysts are often used and these processes are prime candidates for energy conservation.

Combustion–A very important class of chemical reactions, this is used to reduce non-recoverable organic waste materials to ash and heat through chemical reaction known as oxidation. Combustion is also employed to create steam and provide process and facility heat.

Whether exothermic or endothermic, energy integration is essential for modern reactor design. Most reactions have optimal temperature ranges so reliance on waste heat requires topping from an alternate source. Again, a modern programmable logic controller (PLC) can easily integrate two heat/cooling sources for on reactor.

Combined Heat and Power: Putting it All Together

Adsorption Chillers and Process Heat Recovery

These can be used to combine high and low quality energy to maximize efficiency. On their own, modern high efficiency chillers already save over older systems. Whenever the waste heat of compression can be recovered to provide process heat in, for instance, a distillation tower or dryer, the savings can often double that of the increased compressor efficiency alone.

Integrated Steam Generation from Waste Process Heat

Here again by combining low and high quality heat, maximum efficiency is obtained. In fact, not combining these two can lead to a decision not to invest as cost recovery time may be well beyond corporate policy.

Integrated Electric Power Generation from Waste Process Heat

Electric generation costs are comprised of nearly 33% line transmission losses. In addition, electric utilities do not always have a use for the low quality waste heat produced. Chemical process manufacturers should always explore the option at least of adding electrical topping cycles that are matched to process heat loads. Always remember to factor in demand charges as well when making investment decisions.

Whenever considering energy upgrades in Combined Heat and Power (CHP), remember that you can minimize break-even cost time for end of life equipment replacements, so make a special effort to integrate process and utility management groups fully in the capital equipment decision-making process.

Process Energy Examples

Distillation Energy Reduction

In this case, we examine 1000 Pounds/Hour of water distillation. This requires an approximately 1 million BTU/Hour heat source, typically a boiler, and a 1 million BTU per hour heat sink, typically a condenser. We will assume cooling water will be used in the condenser and that steam will be made in the boiler.

A common method of reducing distillation energy load is through mechanical vapor recompression (MVR). Simply stated, MVR places a compressor in the overhead vapors to raise the dew point significantly above the reboiler temperature. In this manner, heat is supplied entirely by the addition of work to the compressor through latent heat from overhead vapor condensation in the reboiler. In some instances, this can completely remove the need for an overhead condenser or cooler thereby saving either once through cooling water or reducing the load to the external cooling tower. Pre-heaters typically protect the turbine from harmful condensed liquid.

Another method, multiple effect evaporation (MEV), involves reducing the operating pressure in successive evaporator stages. With MEV, the same principle of MVR, i.e., condensing one stages' overhead against the boiling section of another stage is applied. In either case, the heat load is reduced by as much as 75%, with similar benefits to cooling water load. Of course, this introduces additional equipment cost and operational complexity and becomes part of the energy optimization equation. With a modern PLC, the control methodology to manage complex systems is now readily and inexpensively available—a fact that should make many retrofits of aging equipment very successful.

The above are common seawater desalination methods. They are also highly applicable to simple organic distillation purification systems. Typically, this is limited to systems where top and bottom tower temperatures are similar as with desalination and some solvent separations. As discussed earlier, MEB are highly coupled, and changes in energy flux within a distillation column will affect composition at both exits. The difficulty of process quality control will aid in screening distillation candidates.

Waste to Energy

Municipal Solid Waste (MSW) consists essentially of garbage, for lack of a better term. MSW in turn consists of environmentally recoverable materials (ERM) such as metal, paper, and plastic. MSW stripped of most ERM can still have a relatively high BTU content. Incinerating this material is not preferred environmentally as it requires extensive stack gas treatment and is not trusted by the public. Another process, pyrolysis can be used to produce a combustible gas, but a hybrid is generally preferred and is discussed below.

Classic pyrolysis involves heating organic material in the absence of oxygen. This process produces a combustible gas by-product, has very low particulate emissions, but produces a byproduct carbon char that still contains recoverable Btu. Newer methods incorporate a small, controlled quantity of air to minimize char production. The resulting medium-Btu gas produced can provide process heat through hot water or steam generation, or in some instances, a direct fuel supplement. The resulting pyrolysis ash is often disposed as MSW itself at greatly reduced volume, or in cement kilns where metal content is not a problem. The relatively low operating temperature allows the recovery of metals, including aluminum, as well as glass for recycling. Heat recovery from produced gas is also possible.

An additional benefit of this type of system may lie in its air permitting classification—i.e., not as an incinerator, as the extremely low flow of gases greatly reduces air particulate emissions common to incinerators. The low temperature also allows easier recovery of metal and glass waste. Pre-sorting is minimized, and materials come out clean and easier to recycle.

Fish Processing Filter and Evaporator

The processing of seafood produces many useful byproducts e.g., bone, protein, and oil. Much water is used throughout and often must be removed to recover these products. This is typically done with large evaporators and centrifuges that serve to concentrate the solids to a desired level. In addition, high-Btu-content oil is produced that can be burned to generate evaporator heat or electricity depending upon facility needs and layout. Large processors can produce thousands of gal/day of fish oil saleable for both nutritional value or blended for clean fuel power generation.

Waste Heat Recovery

Facility energy managers are in general familiar with methods used to intertwine HVAC components to recover waste energy from the various components on the utilities side of the house. The very same principles apply to interacting with chemical processes. In particular, be on the lookout for low quality heat, e.g., turbine extraction steam or reboiler condensate that is difficult to reuse in HVAC. Chemical processes often have stream preheating needs that can recover a high percentage of BTU content not thermodynamically available to work engines. HVAC, and air conditioning and plant operational energy integration is challenging, requires coordination and can be rewarding.

Checklist to Review for Process Energy Optimization

Avoiding startup and shutdown excursion is important in any process. During startup and shutdown, processes commonly run off spec and must be wastefully recycled. You may need to consider supplemental fuel to keep an electrical generator running when process output is low.

Without a good relationship with utility providers, energy optimization can be difficult. This concept is generic to virtually any energy optimization scheme designed to capture non-steady waste energy. The key to success here is proper contracts with the utility and grid integration, onsite and offsite.

Review neighboring energy and raw material needs. You might be able to sell energy and chemical by-products just like any other commodity. It takes a little effort, but can be worthwhile in reducing permit and reporting requirements as well as attendant good neighbor status. A material is not a waste if it is still useful. Sometimes reclassification might be required, but EPA and its many State equivalents tend to be very supportive of all legitimate efforts to reduce waste production while conserving raw materials, and, as we have seen, energy.

Process Energy Checklist

- Establish Process/Utility Engineering Working Group—Necessary to ensure overall facility energy optimization.
- Conduct an MSW Audit—Make sure you are not throwing away Btu and recyclables such as metal, plastic, and glass. Get everyone on-board.
- Utility Bill Review—Identify and develop baseline costs for all process energy use including gas, electric, coal, etc.
- Determine Electric Company policy on purchase of onsite generated electricity.
  — Look for financial incentives for equipment purchase or Operations and Maintenance (O&M)
  — Utilize off-peak demand charge and rate structure to save on electrical operations
  — Consider micro-turbine power generation
- Identify Gas Company incentives
  — Look for financial incentives for equipment purchase or O&M
- Identify high quality thermal streams that might be directly tapped or easily upgraded to generate electricity.
- Utilize electric vehicles for plant operations and charge only at night when demand charges are lowest.
- Distillation and drying are high thermal process energy consumers that can often be modified to capture low quality heat, generate electricity, reduce cooling water demand or provide space heating. Multiple effect evaporation is an established method that should be explored.
- Waste Exchange Review—It is not waste until you declare it waste. If your process by-product is useable as a feedstock in some other company's process, then you have a product, not a waste. This will be saleable! If you have already declared by products a waste, they can always be "de-listed" with EPA. The process may

take some doing, but can reap financial and PR benefits as well.
- Load Shifting—Incorporate energy demand charge as function of time into all process design optimization and change programs. The City of Chicago constructed a refrigeration process that makes ice at night and provides district cooling during the day in downtown Chicago. Off-peak electric rates are usually quite a bargain. Factor this into any processing scheme.
- Leading Energy Indicators—Select metric feedback elements to monitor energy performance.

Office HVAC and Lighting

Although this topic is covered in greater detail elsewhere in this encyclopedia, it is worth discussing here as older incandescent bulbs produce more heat than light. Newer compact fluorescent bulbs are at least four times as efficient and last five times longer, saving on energy and maintenance with payback periods of a year or so. Replacement of old magnetic ballast T12 fluorescent bulbs with new style T8 or T5 with electronic ballasts can have payback periods of 2-3 years in upgrade mode or less than a year in the case of end of life replacement. Cooling and heating in large buildings is often accomplished with boilers and chillers with mixing of air taking place in air handlers. Make sure you are not balancing steam against chilled water to maintain office or plant air temperature set points. Several basic ideas to use in saving energy in large or small facilities include:

- Use area motion detectors to actuate light and in some cases zone HVAC if available.
- Use setback thermostats programmed for heating and cooling for several seasons.
- Consider off site control via telephone or internet of HVAC and lighting for large buildings.
- Store process gray water for lawn sprinkling.
- Examine process heaters and chillers for interconnection, at least for topping service.

Building Energy Management and Improvement

Often overlooked in industrial environment, this is area is ripe for energy integration with low quality waste process heat or excess refrigeration for cooling. The off peak ice machine concept can also work here to help reduce demand charge. Consider replacing older copiers and laser printers that require high temperature for operation for replacement with energy star labeled products. The EPA has computer programs to quantify equipment replacement benefits. Computer energy conservation modes must be set by corporate IT departments to assure the proper sleep or hibernation required to save energy when machines are left unused for even a relatively short time. Screen savers of the past do not save energy. The speed of modern computer processors and hard drives makes hibernation or sleep modes very effective. Newer chips can also be set to run slower, consume less energy, and produce less heat. Photocopiers are huge energy wasters and must go to sleep within a reasonable time. Be careful not to shut down important elements such as fax and network printers now common in multifunction office machines.

Government Energy-Related Programs

DOE Programs

The DOE has industrial technology program (ITP) support available. Elements of this program include free technical assistance and project co-funding opportunities. The program identifies four broad areas in the Save Energy Now arena and can provide free energy audits to facilities around the United States. These audits assist companies in charting facilities energy use, identify areas of concern by matching energy bills with similar facilities and recommending changes along with estimated pay back periods to allow project prioritization.

EPA Energy Star Program

EPA and DOE have interlocking missions here, but EPA manages this program for both agencies. The program includes product labeling with energy use and efficiency, electronic automatic energy reduction for computers and other related class of electronics as well as industrial integration. Websites of both offer consumers and energy engineers alike access to a wealth of useful and growing information.

In our article on "Energy Savings and Optimization Case Studies," which follows this one, we have selected several representative cooperative industrial examples from the US Department of Energy Web Site.

Six Sigma Methods: Measurement and Verification

Carla Fair-Wright
Maintenance Technology Services, Cameron's Compression Systems Division, Houston, Texas, U.S.A.

Michael R. Dorrington
Cameron's Compression Systems Division, Houston, Texas, U.S.A.

Abstract

"If you can't measure, you can't control—if you can't control, you can't manage—if you can't manage, you can't contain costs." As a process-oriented activity, energy management begins with a thorough assessment of use and culminates in the implementation of an effective energy management plan. Applying the five process improvement steps of a Six Sigma DMAIC (Define, Measure, Analyze, Improve, and Control) activity provides a way to define, measure, analyze, improve, control, and verify energy use and savings. The straightforward approach of this process provides a cost-effective means of optimizing facility energy and cost efficiencies.

These efficiencies are transformed into energy savings based on improvements to both facility process and utility-related activities, with process-related activities encompassing all product manufacturing events, and utility-related activities including automated building systems and all other activities not directly related to manufacturing events.

The DMAIC process for each begins with a well-defined problem statement that is supported by examination of existing data to formulate the project premise. Combining "tribal knowledge" with sound data analysis highlights the pathway to process improvement and long-term savings.

NOMENCLATURE

DMAIC Acronym for Define, Measure, Analyze, Improve, and Control, a five-phase, Six Sigma project-focused approach to continuous process improvement

Tribal knowledge Tacit knowledge held by individuals or groups based upon their individual and collective experiences

Stage gates A management review of the process used to screen and pass projects as they progress through their various stages of development

ROI Return on investment = net income/investment, where net income = expected earnings

MSCF Million standard cubic feet

Energy audit A systematic survey conducted to measure and record energy consumption, used to identify opportunities to reduce energy use and cost

Keywords: DMAIC; Tribal knowledge; Stage gates; ROI analysis; MSCF; Energy audit; Process capability ratios; Special-cause variation; Common-cause variation; Run chart; Process variation; DPMO; Project charter; As-is process flow; Capability analysis; Cause-and-effect matrix; Failure modes and effect analysis; Risk priority number; Measurement system analysis; Graphical analysis; Hypothesis testing; Regression analysis; Cost-benefit analysis; Predictor variable; Response variable; Correlation coefficient; Coefficient of determination; *p*-value; KPIV; KPOV; Control plan.

Process capability ratios Capability indices used to measure short-term and long-term process standard deviation

Special-cause variation The rate or magnitude of change in an output resulting from an unexpected event

Common-cause variation The rate or magnitude of change in an output resulting from expected or everyday events

Run chart A graphical display of data representing process outputs over a given period of time, used to determine stability of a process

Process variation The quantifiable difference between individual measurements, represented in a designed sequence of operations or events

DPMO Acronym for defects per million opportunities

Project charter A formal document that defines the mission of the project, project scope, financial benefit, measurement metrics, time frame and resource requirements

As-is process map The current flow of events in a process containing the inputs and outputs of each process step

Capability analysis A measure of how a process is performing, visually represented in terms of a bell-shaped curve, in terms of how much variation exists in the process and how centered it is within the limits of the process

Cause-and-effect matrix A chart used to rank the relationship between process inputs and outputs of a designed sequence of operations or events

Failure modes and effects analysis FMEA; a document that provides a systematic technique to identify and analyze potential failure modes by quantifying them according to the severity of risk they present

Risk priority number RPN; a numerical indicator used to identify the severity of risk

Measurement system analysis MSA; an evaluation used to determine the precision of a given response

Graphical analysis An investigation of how a process or event is performing by representing the data in diagram form

Hypothesis test A statistical method to determine if there are differences between two or more process outputs

Regression analysis A statistical technique used to investigate and model the relationship between dependant and independent variables

Cost-benefit analysis A technique used to analyze competing alternatives based upon comparison of total cost to total return

Predictor variable A value that can be used to predict the value of another variable

Response variable A value that represents the outcome of a specified process or experiment

Correlation coefficient r; statistic that describes the strength of a relationship between two variables

Coefficient of determination r^2; the square of the correlation coefficient, and a measure of the strength of the association between two variables

***p*-value** Probability value; a number that reflects the probability that a statistical result happened by chance, causing a hypothesis to be either accepted or rejected

KPIV Acronym for Key Process Input Variable

KPOV Acronym for Key Process Output Variable

Control plan A document that ensures processes are operated and monitored consistently so that the desired result meets the defined requirements

Cp Process capability; represents the spread of the process. A Cp Value of 2 is representative of a 6σ process (0.002% rejects)

Cpk Process capability of non-centered process; is an indication of how the process is aligned to the process mean. Cp and Cpk are equal in a centered process

INTRODUCTION

We live and work in an ever-increasingly volatile and complex business environment that has begun to recognize energy use and efficiency as strategic elements of our overall business planning process. Under deregulation, energy providers pass on a higher percentage of energy cost increases to the consumer in the form of cost adjustment factors. These added costs, the efficiencies of the energy conversion processes we employ, and environmental effects are what Six Sigma practitioners would define as critical inputs (X-values). Outcomes—or Y-values—are a function of these critical inputs and stated in terms of our monthly utility expense. By applying Six Sigma methodology to the issue, we are able to define energy use as a process and employ a useful set of metrics to measure, improve, and control variation in our monthly utility bills.

ENERGY MANAGEMENT—PURPOSE AND GOALS

The goal of energy management planning is to optimize both the consumption energy and its cost. This article offers an overview on how to apply Six Sigma methodology to continuously improve energy consumption processes using DMAIC principles. Six Sigma methodology also benefits energy managers by adding stage-gate approvals into the energy management plan development process.

SIX SIGMA—BACKGROUND

Established in the 1980s at Motorola and brought into mainstream use by GE in the mid-1990s, Six Sigma employs data-based decisions built around baseline measurements to reduce waste and process variation. This fundamental objective is often achieved using the five-step DMAIC process (Define, Measure, Analyze, Improve, and Control). Since most energy use projects are site specific, this article will serve to guide users in the application of Six Sigma methodology rather than developing a specific case study. Additionally, sample sets of data have been employed to simplify data analysis while offering an acceptable level of confidence in the results.

FINANCIAL CONSIDERATIONS AND INCENTIVES

Capital expenditures for facility energy infrastructure projects are rated based upon financial criteria that seek to establish their expected rate of return. While financial analysis of large energy infrastructure projects is often complex, smaller straightforward projects generally stress the expected return on investment (ROI). This article

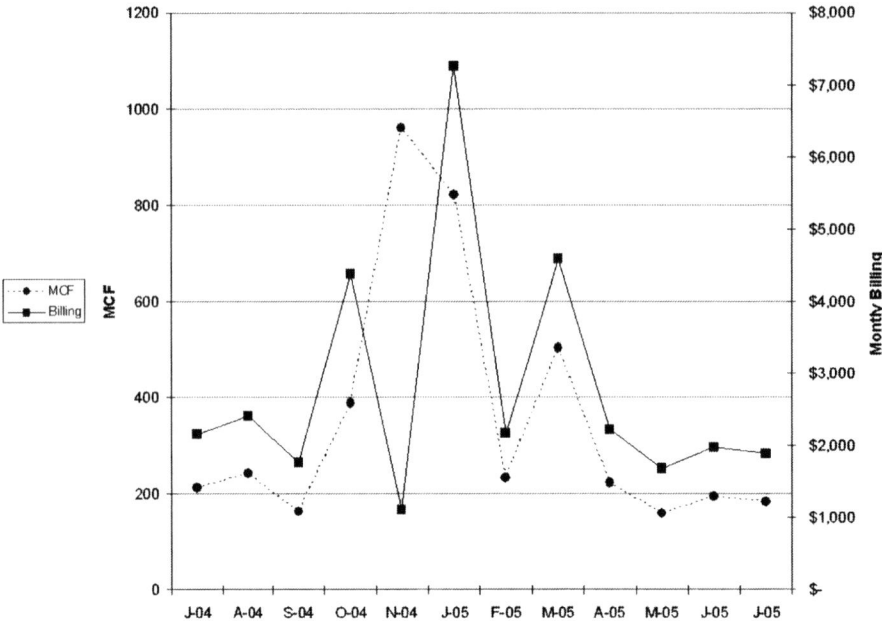

Fig. 1 Energy usage.

employs ROI Analysis to forecast the expected rate of return.

It should be noted that the Energy Policy Act of 2005 provides substantial tax incentives to encourage improving energy efficiency in both new and existing buildings. A major aspect of the act is the ability of facility managers to obtain new energy equipment and apply any derived cost savings to energy utilization.

DATA SOURCE

Research data used in this article were obtained from the monthly service billings of a small aftermarket repair facility in Houston, TX., that employs a single analog utility meter to record energy consumption (Fig. 1). This facility was selected based upon the simplicity of its metering scheme and the ability to readily obtain a sample set of data that fit the reporting purposes of this article. Larger facilities will employ sub-metering or data-recording devices that allow users to categorize energy use by process segment. The ability to link energy use to individual machine tools or processes enhances the ability to analyze data and verify the impact of specific improvements.

In practice, one of the first steps taken in evaluating energy use is an energy audit. Usage data are gathered from a variety of sources including metering devices, sensors, and utility bills, and combined with maintenance information to identify energy cost reduction opportunities. Self-assessment and assessments conducted by university-based industrial assessment centers are two low-cost means of performing an energy audit.

SETTING ENERGY CONSUMPTION TARGETS

Determining the need for improvement requires establishing current usage (baseline) data that can then be compared with desired objectives or best industry practices. Performance against the desired outcome can then be measured in terms of its process capability ratios (Cp and Cpk).

SIX SIGMA—THE DMAIC PERSPECTIVE

Overview

As a rigorous methodology that uses data-based decisions and statistical process control, Six Sigma provides firms with a robust platform to minimize variability, defects, and waste in the processes we employ to convert goods and services into revenues. In practice, Six Sigma represents how well a process is performing and how often a defect is likely to occur—the higher the sigma value, the better the process is performing.

Since energy savings programs are essentially project activities, we can readily apply DMAIC methodology to characterize and optimize desired benefits. The five phases of this process are as follows:

- Define—Stating the desired need in the form of a problem statement.
- Measure—Determining the baseline and target performance values.
- Analyze—Relating process inputs (critical X's) to process outputs (Y's).

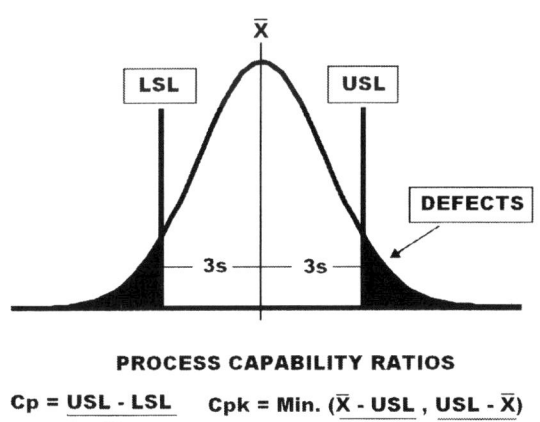

Fig. 2 Process capability.

- Improve—Identifying improvements to optimize outputs.
- Control—Establishing control procedures and accountability.

Variation and Mean Shift

Every process includes both common-cause variation and special-cause variation. Common-cause variation occurs naturally in any process, while special-cause variation is nonrandom and the result of an action or a series of actions. Process improvement seeks to reduce special-cause variation, shifting the mean to center and stabilizing the process (Fig. 2). The capability of the process to consistently perform within specification limits relies on achieving both. A common tool employed to display both stability and variation is the run chart, which allows users to quickly gauge the amount of variation in a process. Points outside the process upper and lower limits, unequal dispersions around the mean, and pattern variations are all indicators of a process variation requiring further investigation.

Defect Prevention

A process, product, or result may contain a single or multiple number of defects before it is considered defective or out of specification. Defect prevention results from error proofing the activity. The common defect measure is expressed in terms of parts per million or defects per million opportunities (Table 1). A process operating at Six Sigma levels produces no more than 3.4 defects per million operations, meaning that the process is operating at a 99.99966% defect free rate. For energy savings projects, a defect may represent any point outside of the run chart control limits.

Data-Based Decisions

Six Sigma converts data into useful information that allows users to base their decisions on facts rather than opinion. Data-based solutions may result in improvements that match expert opinion, but data-driven solutions coupled with statistical process control remove errors while adding process control capability—all resulting in higher probability of a successful outcome.

DMAIC METHODOLOGY

Define

During this initial project phase, the problem statement is created in the form of a project charter. Project charters establish the project's scope and improvement objective; detail the financial impact of the project in a business case; and forecast the project's expense, savings, and required resources.

Measure

Understanding the Process

Before a process can be improved, it must first be measured to establish a baseline for comparison and targeted improvement objectives. Baseline data (typically in kilowatt-hours or million standard cubic feet) may come from utility bills, meter readings, or data recording devices. Following data collection, key process variables are identified by creating a top level as-is process flow to help narrow the focus of the improvement effort to those key process steps having the highest levels of special cause variation.

Process Capability

The demonstration of process performance is accomplished through capability analysis, by comparing

Table 1 Defects per million opportunities yield and Sigma

Sigma (σ) value	Accuracy (%)	Defects per million opportunities (DPMO)
1	30.85	691,500
2	69.15	308,500
3	93.32	66,800
4	99.38	6,210
5	99.977	233
6	99.999	3.4

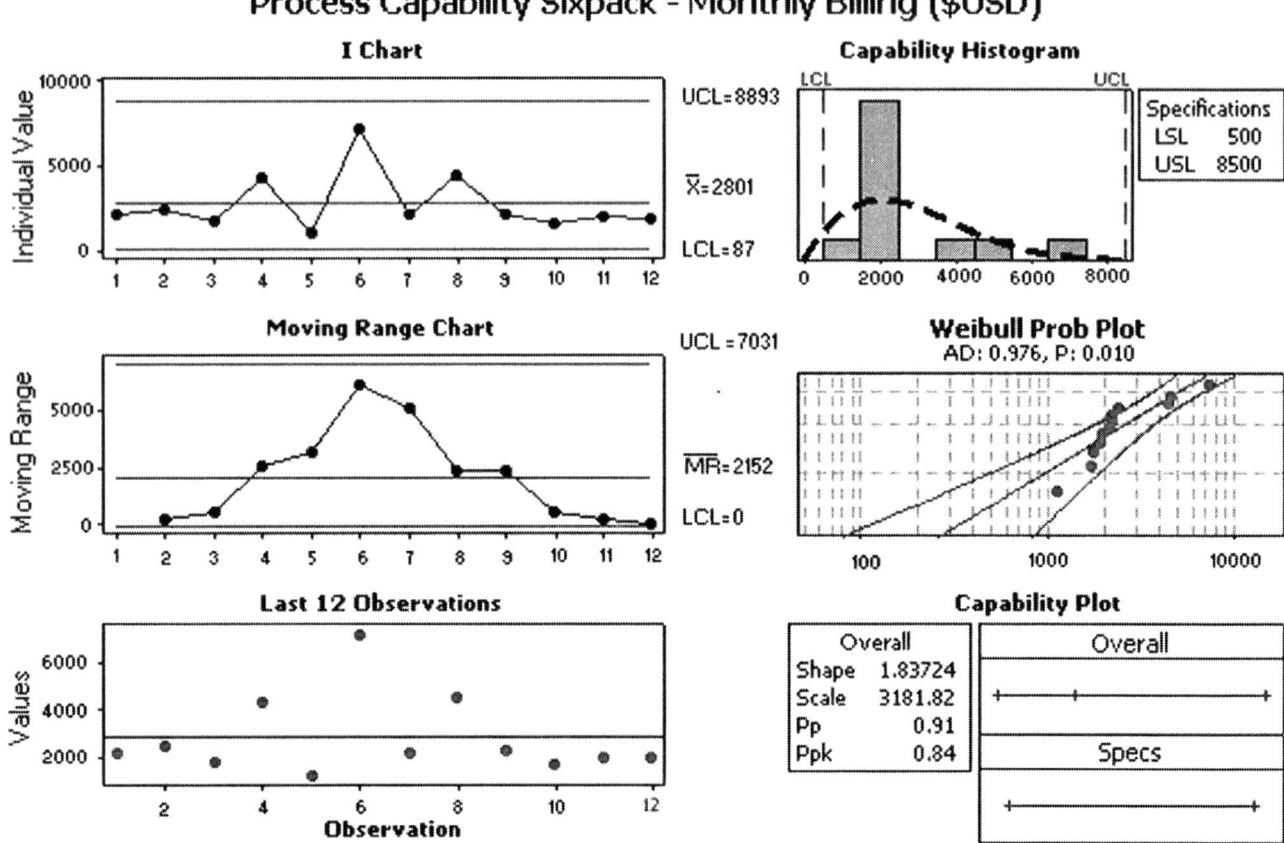

Fig. 3 Energy usage process capability.

the width of the process variation to the width of the acceptable process tolerance limits, or by measuring the defects present in the existing process. The standard measures of capability are defects per million opportunities (DPMO) and process capability ratios (Cp and Cpk).

Analyze

Non-Normal Data

Before performing any data analysis, one should check data for normality, since standard capability indices contain an assumption of normality. With energy usage programs, data is likely to be non-normal due to a number of factors including monthly utility fees and surcharges. The data used in this article is non-normal data that was transformed before use (Fig. 3).

One strategy is to make non-normal data resemble normal data by using a Box-Cox transformation function to convert the data closer to a normal distribution (Fig. 4). The underlying technique of the Box-Cox transformation is based on developing a normality plot using the correlation coefficients for the defined transformation parameter (λ, the lambda value). The relatively high p-value taken from the fitted line plot on Fig. 4 indicates that the transformed data is normal and capable of being analyzed using Six Sigma statistical tools. The capability charts indicate that significant variation occurred between periods 3 and 10, signifying that further investigation of these eight time periods is required. These periods should be analyzed with reference to the process step inputs and outputs developed in the as-is process map to identify any observable key input or output variables.

Sources of Variation

Once as-is process capability is defined, it is necessary to understand cause and effect relationships. A cause-and-effect matrix is the tool used to rank relationships between the key input and output variables. When complete, the cause-and-effect matrix is linked to a failure modes and effect analysis (FMEA) by transferring the highest-ranked input variable or variables onto the FMEA. This analysis is used to identify and force rank by criticality, the potential failure modes, and their effects and causes, using a risk

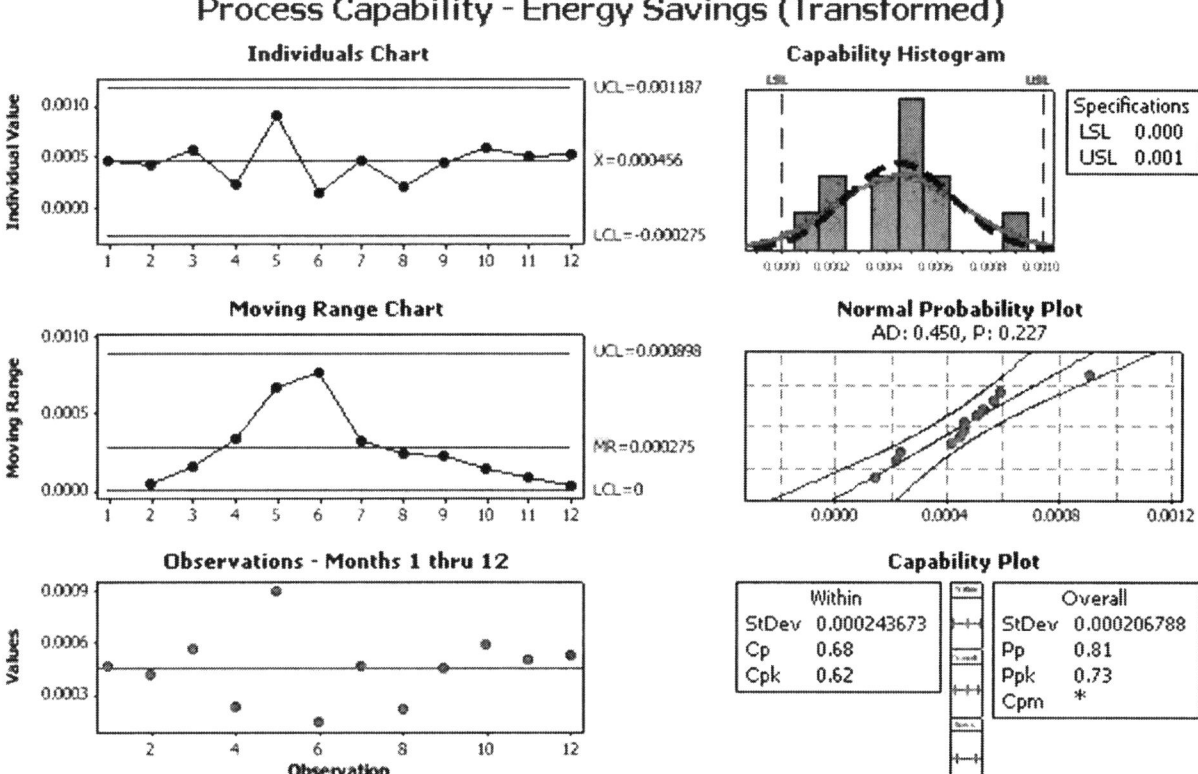

Fig. 4 Energy usage process capability transformed.

priority number (RPN). Only the highest-rated factors qualify for investigation.

Measurement Error

Accuracy of the measurement system can be the predominant reason for process variation. Employing measurement system analysis (MSA) will allow the firm to evaluate the accuracy and precision of the measurement system. Where measurement systems are inadequate, it is recommended that they be upgraded to a level consistent with the project requirements.

Corrective Actions

Completion of the FMEA requires statement of the corrective action needed for process improvement. For each action taken, it is necessary to recalculate the RPN to validate process improvement. A variety of Six Sigma statistical tools are available to confirm the impact of the corrective actions, including graphical analysis, hypothesis testing, and regression analysis. As a supplement to the FMEA, a cost-benefit analysis should be developed to define the optimal level of planned investment.

IMPROVE

Improvement seeks to optimize outputs (Y's), eliminate or minimize defects and variation, and statistically validate new process operating conditions.

Variable Relationships

The governing principle behind variable relationships is expressed in terms of $y = f(x)$, wherein key output (y) is stated as a function (f) of the key inputs (x). A common statistical tool used to examine this relationship is regression analysis, which correlates the degree of association between the predictor variable (x) and the response variable (y) in terms of r, the correlation coefficient. Where $r > 0.8$ or < -0.8, a significant relationship exists; for values between -0.8 and 0.8, no significant relationship exists. The amount of variation in the output that is explained by the regression model is R^2, the coefficient of determination. In terms of energy analysis, a strong correlation indicates that the output (y) or energy use is defined by the input values (x) being modeled. The final term dealing with correlation in the regression is the p-value, which indicates whether there is a correlation between the variables. A p-value can also be used to determine normality, which can be seen by the fit of the variables to a fitted line plot (Fig. 3).

For the facility cited in this article, it was determined that utility fees and surcharges were the critical inputs. Regression analysis indicated a weak correlation to the various input factors of shop load, utilization factors, maintenance, and seasonality, due in part to the limited amount of data and unpredictable increases in utility energy fees and surcharges. Based upon the age of the facility and a cost-benefit analysis of the upgrades needed to operate the facility at optimal energy levels, it was determined that the best alternative was to switch to a lower-cost energy provider as the means to reduce energy costs.

New Process Capability

Following implementation of the recommended actions, it is necessary to establish the new relationships of key process input variables (KPIVs) and key process output variables (KPOVs) by running a second capability analysis using data from the improved process. This information is then compared to the original baseline data to establish the level of improvement obtained. The final steps in defining the new process capability require updating the process map to reflect implemented changes and validating project cost savings.

CONTROL

The final project phase documents, sets up control criteria and assigns accountability for sustaining the gains achieved through process improvement. Control plans are intended to permit operating processes consistently on target with minimum variation by defining and monitoring critical success factors (KPIVs and KPOVs); as well as the employed measurement methods, frequency, and decision rules governing corrective action. It is important to remember that the control plan is a living document and part of a continuous process improvement activity.

CONCLUSION

The primary advantage of Six Sigma as a continuous process improvement tool, is its ability to provide a set of metrics that can then be used to understand and correct process variation. Six Sigma is a project-focused activity that adapts well to time-dependent activities. The five-phase DMAIC model is commonly applied to characterize and optimize industrial processes. Understanding how energy is consumed in the processes we employ requires that we first establish a method of measurement to identify how energy is being consumed and in what quantities. Organizing data into useful information is accomplished in the measurement phase of the DMAIC process, which establishes the baseline against which all future improvements will be compared. What prevents decline of the improved process is the control plan, which is established to identify ongoing surveillance requirements and corrective action plans.

RESOURCES

The American Society for Quality http://www.asq.org/

The DOE Building Technologies Program http://www.eere.energy.gov

International Performance Measurement and Verification Protocol http://www.ipmvp.org

BIBLIOGRAPHY

1. Watson, J.J. Satisfaction through six sigma. *Engineered Systems*, March 2003; 1–2. http://www.findarticles.com/p/articles/mi_m0BPR/is_3_20/ai_98780226 (accessed November 2005).
2. Muller, M.R.; Papadaratsakis, K. *Self-Assessment Workbook for Small Manufacturers*. Version 2.0, Department of Energy (October 2003).
3. The Black Belt Memory Jogger™ *Six Sigma Academy*, 1st Ed.; 2002.
4. Introducing Information Systems for Energy Management, GPG 231, Action Energy, http://www.thecarbontrust.co.uk/energy/pages/home.asp.

Solar Heating and Air Conditioning: Case Study

Eric Oliver

K. Baker

M. Babcock
EMO Energy Solutions, LLC, Vienna, Virginia, U.S.A.

John Archibald
American Solar, Inc., Annandale, Virginia, U.S.A.

Abstract

Mechanical cooling is one of the largest end uses of electricity, and it contributes significantly to peak demand and on-peak energy consumption. The ability to offset electricity consumption during peak usage periods by eliminating mechanical cooling can save costs, reduce strains on transmission grids, and free up much-needed peak period capacity for electric generation utilities. Solar energy use, long encouraged for offsetting space heating and daylighting energy consumption, can now be utilized for offsetting cooling consumption. By using solar heat to regenerate a desiccant dehumidification system, and utilizing direct and indirect evaporative cooling to condition superheated dry supply air, occupied spaces can be cooled using no grid energy input. A demonstration project has been designed to cool an inspection station at the U.S. Pentagon in Arlington, Virginia, and is expected to provide 65°F, moderate humidity air at summer design conditions.

INTRODUCTION

An inspection station at the Pentagon Central Plant currently uses a through-wall heat pump for cooling and heating. This station was chosen for a demonstration project to test the application of space cooling using only solar thermal, desiccant dehumidification, and indirect evaporative cooling. By using collected rainwater for evaporative cooling and photovoltaic panels with battery back-up to run peripheral equipment, this system will have the potential to operate completely off grid, with no purchased electricity or associated emissions. Cool air will be provided directly to the station using a four-stage process:

Stage 1 : An American Solar Inc. tile roofing system will be used to heat ambient air up to 200°F.

Stage 2 : Supply air for the station will be dehumidified and superheated using a small desiccant wheel. The wheel will be regenerated using the solar heated air.

Stage 3 : The superheated air will be cooled using a direct air–air heat exchanger with ambient air.

Stage 4 : The hot air will be further cooled using two stages of indirect evaporative cooling, first with saturated ambient air and second with a portion of the supply air diverted into a second indirect evaporative cooler.

Based on a psychrometric analysis, it can be shown that 65°F 60% relative humidity (RH) air can be delivered from Washington, DC, design ambient conditions of 85°F dry bulb (DB)/75°F wet bulb (WB).

COOLING SYSTEM METHODOLOGY

The cooling process designed utilizes three separate air streams: the supply or primary air stream, the secondary air stream, and the regenerative air stream (RAS). Further details of each of the equipment used will follow later in this abstract. Fig. 1 is a schematic of the design.

The primary air stream is drawn from outside air through air filters and is mixed with the exhaust air from the second evaporative cooler (IEVAP-2). This mixing process reduces the enthalpy and temperature of the incoming air during design conditions. The primary air then passes through the dehumidification side of the desiccant wheel (DESC-1). The air is then passed through an air-to-air heat exchanger (HX-1) with the regenerative air to cool the primary air. In the next stage, the primary air is further cooled by passing through an indirect evaporative cooler (IEVAP-1). Finally, the primary air passes through a Maisotsenko-type indirect evaporative cooler (IEVAP-2) to provide the final conditioning for the supply air. During this final stage, the primary air is divided into the primary air stream and an exhausted secondary air stream.

---

* This entry originally appeared as "Solar Heating and Air Conditioning at a Pentagon Security Inspection Station" in *Energy Engineering*, Vol. 101, No. 4, 2004. Reprinted with permission from AEE/Fairmont Press.

Keywords: Heating, Ventilating and Air Conditioning (HVAC); Solar heating; Cooling; Case studies.

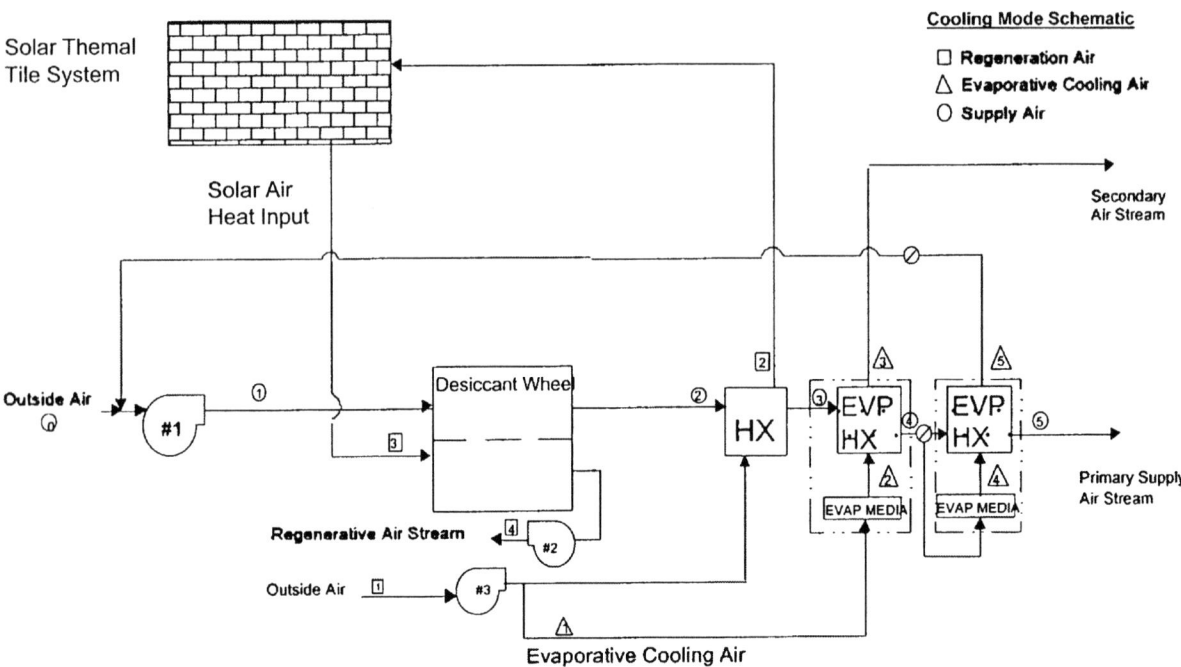

Fig. 1 System schematic.

The secondary air stream feeds IEVAP-1. The secondary air stream is intended to maximize the removal of sensible heat from the primary air and thereby reduce the primary air temperature. To this end, the airstream is designed with twice the airflow as the primary air.

Similar to the primary air stream, the RAS is drawn through air filters from outside. This airstream is first passed through HX-1 to recover heat from the superheated and dried primary air stream. This preheated air is then sent to the solar roof tile system to capture the solar energy and raise the RAS temperature. The solar heated regeneration air passes through the regeneration side of the DESC-1 to enable the desiccant to release and exhaust the absorbed moisture into the air stream.

PSYCHROMETRIC ANALYSIS

Fig. 2 illustrates the psychrometric path of the primary air stream. It begins at State 1—a high temperature, high humidity, and high enthalpy condition. The exhaust from the IEVAP-2 is saturated with water vapor, but has a lower temperature-humidity ratio and enthalpy than the outside air. The mixing of the airstreams reduces the overall enthalpy at State 2. State 3 occurs after the desiccant wheel; as expected, the moisture content is greatly reduced while the temperature is significantly increased. It is important to note that the total enthalpy of the air is actually increased from State 2 to State 3. States 4–6 are the result of sensible-only cooling through the heat exchange with other air streams in HX-1, IEVAP-1, and IEVAP-2. The overall result is a strategy that first shifts the enthalpy of the airstream from latent to sensible energy, and then uses heat exchange with original or tempered outside air to remove the sensible energy and produce cool air without the mechanical cooling.

As Fig. 3 shows, the psychrometric path for the RAS is simple. It starts at State 1, the outside air conditions, and gains sensible heat from the heat exchange at HX-1 to arrive at State 2. The air stream is further heated sensibly by the solar irradiance captured from the solar tile system, and reaches the regeneration temperature at State 3. Once the RAS has passed through the desiccant wheel, State 4 represents the cooler, wetter exhaust air containing the additional moisture from the desiccant.

SOLAR ROOF TILE SYSTEM

American Solar, Inc. has patented a technology to provide solar-heated air using a tile roof system. The solar thermal tile system is designed to provide dual functionality: to serve as the roof of a space and to provide solar heated air. This system translates a portion (∼35%) of the solar insulation available in a given region into preheated air, which can be utilized for heat exchange, thermal storage, space heating, and in this case, desiccant regeneration air. Radiant energy is absorbed by black-painted metal absorbers and heats the air surrounding the absorbers. Transparent tiles over the absorbers contain the heated air, and allow a fan system, at a plenum at one end of the roof, to collect the solar-heated air.

Stagnation tests show that a typical system can achieve internal air temperatures of >200°F (94°C) and more than 130°F (72°C) above ambient temperature during the

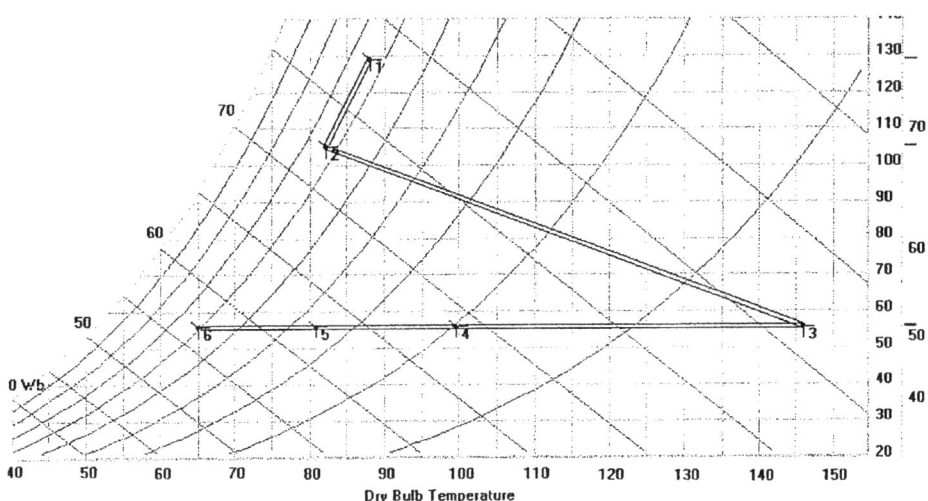

Fig. 2 Supply air stream psychrometric path.

summer months. An air flow test with an early prototype showed outlet air temperatures of 160–180°F (71–82°C) are possible. Higher temperatures are expected with optimal orientation, improved materials (i.e., absorber, glass tile, etc.), and optimal air flow given a system's configuration.

For this particular application, the preheated solar air generated from the solar roof will be used to regenerate a high temperature substrate desiccant, which will be utilized to dry the primary air stream. To achieve the temperatures need to regenerate this particular desiccant, the complete solar roof system had to be designed in three stages:

Stage 1 : This roof section includes about 150 ft² of dark-colored standing seam panels, set at a tilt of 22.9° and facing north. The panels will heat up, and preheated air from underneath the seam panels will flow to the next roof section.

Stage 2 : This roof section includes about 150 ft² of the ASI tile system, set at a tilt of 22.9° and facing south, that will boost up the temperature coming off the standing seam panels and deliver it to the next roof section.

Stage 3 : This roof section includes about 90 ft² and faces south with an angled reflecting surface opposite it (to provide additional insulation). Stage 3 takes the incoming air from Stage 2 and increases the temperature of the air further. The air from Stage 3 is then delivered to the desiccant wheel for regeneration.

DESIGN AND CONSTRUCTION STRATEGY

The overall goal of this showcase project was to create an innovative and fully integrated cooling system which would also have the capability to provide solar-heated air for space heating in the winter. To provide power for peripheral equipment, such as fans and the water pump, several photovoltaic shingles will be installed in the roof system. This particular application will include battery a

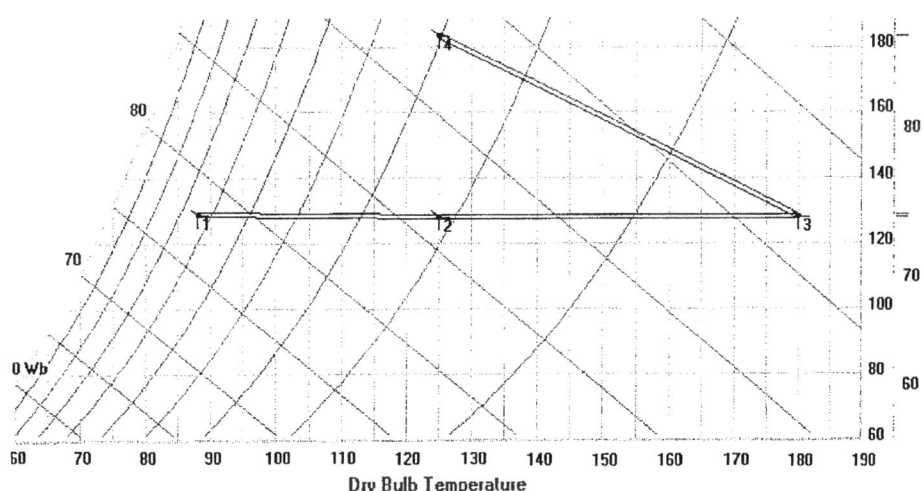

Fig. 3 Regeneration air stream psychrometric path.

backup and backup grid connection for battery charging, but the showcase is meant to demonstrate the potential for off-grid applications. Highlighted design features and modifications for test bed purposes include:

- The ability to alter and test the impact of different face velocities across the solar roof to increase the regeneration air temperature.
- The ability to adjust the relief damper position for the regeneration air to properly pressurize the solar roof.
- The end user's ability to adjust the temperature when the solar cooling system is not providing adequate supply air temperatures during muggy days or when there is little to no (night) solar isolation available.

Some of the challenges and system constraints include:

- The minimal amount of space available for routing ducts and mounting heating, ventilating and air conditioning (HVAC) equipment underneath the solar roof and above the 9' carport ceiling limit.
- The integration of the solar system with the evaporative cooling system.
- The integration of the PV shingles and the ASI tile system into a single entity.
- Minimizing pressure losses through the supply system to minimize fan power consumption and energy losses.

CONCLUSION

There is a great potential to provide spot cooling for remote locations using solar-regenerative desiccant dehumidification, heat exchange, and indirect evaporative cooling. The economics of such an installation will be greatly dependent on climate, energy rates, benefits of avoided grid extension, and other factors where adding mechanical cooling to a location is costly. The potential impact on operating costs, utility generation and transmission demand constraints, and environmental improvement from reduced emissions is significant.

BIBLIOGRAPHY

1. ASHRAE 1993 Fundamentals.
2. ASHRAE 2000 HVAC Systems and Equipment.
3. Watt, J.R.; Brown, W.K. *Evaporative Cooling Handbook*, 3rd Ed.; Fairmont Press: Lilburn, Georgia, 1997.
4. Archibald, J. *A New Desiccant Evaporative Cooling Cycle for Solar Air Conditioning and Hot Water Heating*, Proceedings of the ASES Annual Conference, 2001.
5. U.S. Patent #5,651,226 to Archibald dated July 29, 1997.
6. Coolerado Coolers, www.coolerado.com
7. Novelaire Technologies, www.novelaire.com

Solar Thermal Technologies

M. Asif
School of the Built and Natural Environment, Glasgow Caledonian University, Glasgow, Scotland, U.K.

T. Muneer
School of Engineering, Napier University, Edinburgh, Scotland, U.K.

Abstract

Solar thermal technologies are amongst the most diverse and effective renewable energy technologies. They range from low-temperature (<70°C) and simple in operation technologies such as solar space heating and solar cooking to high-temperature (>200°C) and sophisticated ones like solar air conditioning and solar thermal power generation. Solar thermal technologies, similarly, have a broad economic bandwidth amongst them. The most successful application of solar thermal technologies so far has been in the form of solar water heating. In this entry, a brief overview of the prominent solar thermal technologies has been presented in which the technology fundamental and operational principles of these technologies have been precisely discussed.

INTRODUCTION

The sun is a source of energy on Earth, and in a sense, it is a source of life; as energy is the most important commodity for life besides providing light and heat to the planet. The reaction between the sun's energy and Earth's atmosphere determines weather patterns and rainfall, and Earth's tilt towards the sun creates the seasons. Its role in photosynthesis helps plants to grow and its role in biodegradation helps complete the natural cycle of ecosystems. The sun also sources several other forms of energy on the planet: wind power depends on the sun's impact on atmospheric movement as it creates wind patterns; through photosynthesis, sun contribute to bioenergy (wood and other organic materials); and fossil fuels indirectly owe their creation millions of years ago to solar energy.[1]

Solar energy is one of the most promising renewable technologies. It is abundant, inexhaustible, environmentally friendly, and widely available. Solar energy has the potential not only to play a very important role in providing most of the heating, cooling, and electricity needs of the world, but also to solve global environmental problems. Solar energy can be exploited through solar thermal and solar photovoltaic (SPV) routes for various applications. While SPV technology enables direct conversion of sunlight into electricity through semiconductor devices called solar cells, solar thermal technologies utilize the heat energy from the sun for a wide range of purposes.

Solar thermal technologies are quite diverse in terms of their operational characteristics and applications—they include fairly simple technologies such as solar space heating and solar cooking as well as complex and sophisticated ones like solar air conditioning and solar thermal power generation. Solar thermal technologies have also a broad bandwidth in terms of their economical standing. Solar water heating and solar space heating, for example, are very cost effective and are regarded among the most economical renewable energy technologies while high temperature technologies such as solar thermal power generation and solar air conditioning are on the higher economic bandwidth. Solar thermal technologies on the basis of their working temperature can be classified into the following three types:

- Low-temperature technologies (working temperature <70°C)—solar space heating, solar pond, solar water heating, and solar crop drying.
- Medium-temperature technologies (70°C< working temperature <200°C)—solar distillation, solar cooling, and solar cooking.
- High-temperature technologies (working temperature >200°C)—solar thermal power generation technologies such as parabolic trough, solar tower, and parabolic dish.

In the coming sections, this entry provides an overview of these technologies, briefly highlighting their technology fundamental and operating principles.

SOLAR SPACE HEATING

Solar energy can be used to accomplish heat for comfort in buildings. The application, referred to as solar space

Keywords: Solar thermal technologies; Solar radiation; Concentrated collectors; Parabolic trough; Sorption.

heating, is used to optimize the reduction of auxiliary energy consumption in such a way that minimum overall cost is obtained. In combination with conventional heating equipment, solar heating provides the same levels of comfort, temperature stability, and reliability as conventional systems. A building that includes some arrangement to admit, absorb, store, and distribute solar energy as an integral part is also referred as a solar house. A solar space-heating system can consist of a passive system, an active system, or a hybrid of the two.

Passive Space Heating

In passive space heating, the buildings are designed or modified so that they independently capture, store, and distribute solar heat throughout the building without using any electrical or mechanical equipment. Inherently flexible passive solar design principles typically accrue energy benefits with low maintenance risks over the life of the building. The design does not need to be complex, but it does involve knowledge of solar geometry, window technology, and local climate. Passive solar heating techniques generally fall into three categories: direct solar gain, indirect solar gain, and isolated solar gain, as shown in Fig. 1.

Direct Solar Gain Design

Direct gain passive designs typically have large windows with predominately equatorial aspects. In this design, solar radiation directly penetrates and is stored in the building's inherent thermal mass, in materials such as concrete, stone floor slabs, or masonry partitions that hold and slowly release heat.

Indirect Solar Gain Design

In indirect solar passive design, a glazed heat collector, also referred as Trombe wall, collects and stores solar radiation during the day. A Trombe wall consists of an 8–16″ thick masonry wall coated with a dark, heat-absorbing material and covered by a single or double layer of glass, placed from about 3/4″ to 6″ away from the masonry wall. Heat from the sun is stored in the air space between the glass and the dark material, and conducted slowly to the interior of the building through the masonry through the conduction and convection mechanisms.

Isolated Solar Gain Design

In isolated solar passive systems, an extra highly glazed unheated room—a sun-space or conservatory—is added to the south side of the house. Solar gains always make sun-spaces warmer than the outside air, and this reduces heat losses from the house and warms any ventilation air that passes through the sun-space. When solar gains raise the sun-space above house temperature, the heat collected can be let into the house by opening communicating doors and windows.[2]

Active Space Heating

In active space heating of buildings, additional electrical and mechanical equipment is incorporated to circulate solar heated water or air. The main components of an active system are the heat collectors, storage tanks or pebble bed storage, heat exchangers, heat emitters, fans/pumps, connecting pipes or ducts, and controls. Active solar heating systems can be designed to provide the same levels of control of condition in the heated spaces as conventional systems. With indoor temperature essentially fixed at or little above a minimum, load estimations can be done by conventional methods. Passively heated buildings in many cases are not controlled within the same narrow temperature ranges.[3]

Hybrid Solar Space Heating

Solar space heating system can also be of hybrid nature, combining both the passive and active modes. For example, in a hybrid system, a roof-space collector accomplishes passive collection of solar energy that can be actively distributed in the house using a fan and associated ductwork.

SOLAR WATER HEATING

Water heating is an essential feature of energy requirements in industrial and commercial sectors in general and in domestic sector in particular. A solar water heater consists of two main elements—the collector and the water storage tank, which respectively have the functions of absorbing solar radiation and transferring it to the water, and storing the water for usage. The collectors in solar water heaters can be broadly classified into two categories—flat plate and evacuated tube. A flat plate collector consists of an absorber plate that absorbs solar energy while a glazing above it is used to reduce convective heat loss. An evacuated tube collector consists of tubes with vacuum maintained between the tubes and glazing for better protection against convective heat loss.

Solar water heaters come in three main types: thermosyphon, built-in-storage, and forced circulation. There are two operating principles for solar water heaters: passive system, which relies on natural circulation of water (such as thermosyphon, built-in-storage types); and active system, which uses an external element such as an electric pump to circulate the water (such as the forced circulation type). Another criterion that distinguishes solar water heaters is the

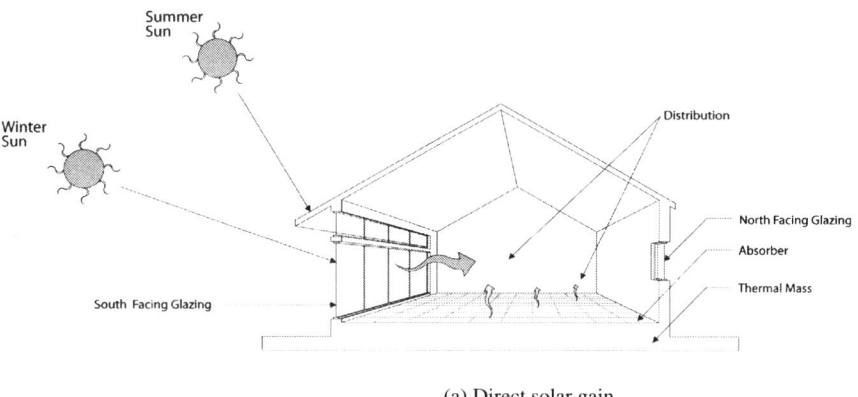

(a) Direct solar gain

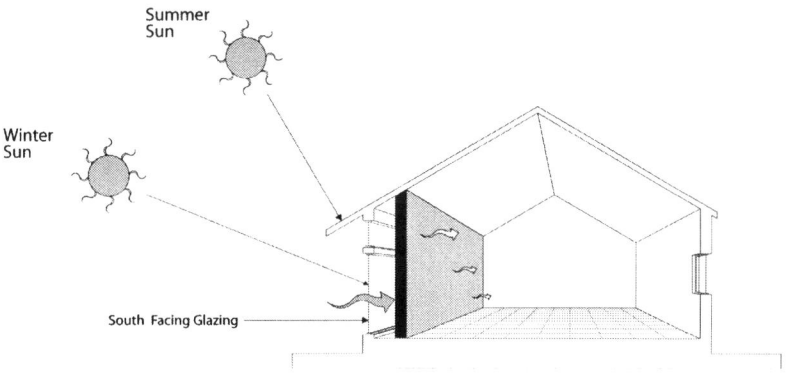

(b) Indirect solar gain (Tormbe wall)

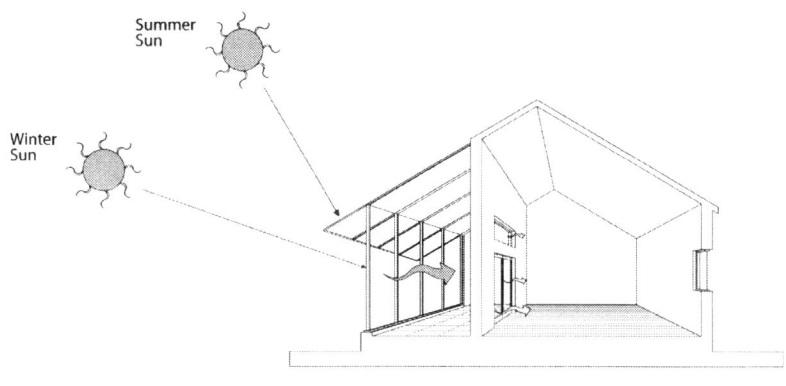

(c) Isolated solar gain (sunspace)

Fig. 1 Passive solar space heating principles.

way they transfer heat to water. Again, there are two types: direct system, in which the collector itself transfers heat to water; and indirect system, in which a heat-transfer fluid, circulating in collector in a closed loop, transfers heat to water through a heat exchanger. Fig. 2 shows an indirect active solar water heater.

The efficiency of a solar water heater depends upon its design and the available solar radiation. In this entry, solar water heating has been classified as a low-temperature thermal technology because most of its application is in residential sector where it operates at a temperature of $\leq 70°C$. Also, in industrial applications, solar water heaters are used as preheaters and to hold supply water at almost the same temperature for further heating by conventional means. Solar water heaters are a cost-effective technology—the payback period of solar water heaters can be as low as 3 years while having a service life of more than 20 years. Owing to its technical and economical viability across the world, solar water heating is one of the most established and efficient application of solar energy. Among solar thermal technologies, solar water heating holds the greatest market share and the highest market growth rate.

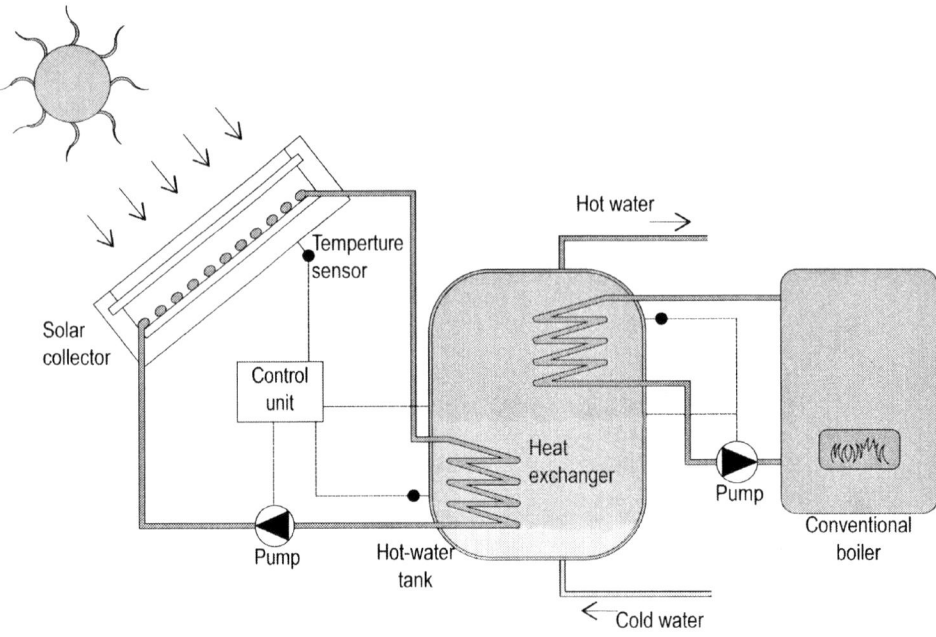

Fig. 2 Indirect active solar water heating system.

SOLAR PONDS

Solar ponds are naturally occurring salt gradient lakes that collect and store solar energy. A solar pond contains salt water with increasing concentrations of salt, hence the density of the solution. When solar radiation is absorbed, the density gradient prevents heat in the lower layers from moving upward by convection and leaving the pond. This results in an increased temperature at the bottom of the pond and a near atmospheric temperature at the top of the pond. The phenomenon of solar ponds was first discovered in 1902 by von Kalecsinsky, who reported that the Medve Lake in Transylvania, containing nearly saturated NaCl solution at a few meters depth with almost fresh water at its surface, had a bottom temperature of 70°C.

A solar pond has three distinctive zones. The top layer is the surface zone that has a low salt content and is at atmospheric temperature. It is also called the upper convective zone (UCZ), as shown in Fig. 3. The bottom layer has a very high salt content and is at a high temperature, 70°C–90°C. This is the zone that collects and stores solar energy in the form of heat and it is called the lower convective zone (LCZ). There is an intermediate insulating zone with a salt gradient. It establishes a density gradient that prevents heat exchange by natural convection, and hence it is called the nonconvective zone (NCZ). In this zone, salt content increases with depth, creating salinity.

Solar ponds can be broadly classified into two main types: nonconvective and convective. In nonconvective solar ponds, the heat loss to environment is reduced by suppressing natural convection normally by using salt stratification. While in convective ponds, heat loss to environment is reduced by covering the pond surface with

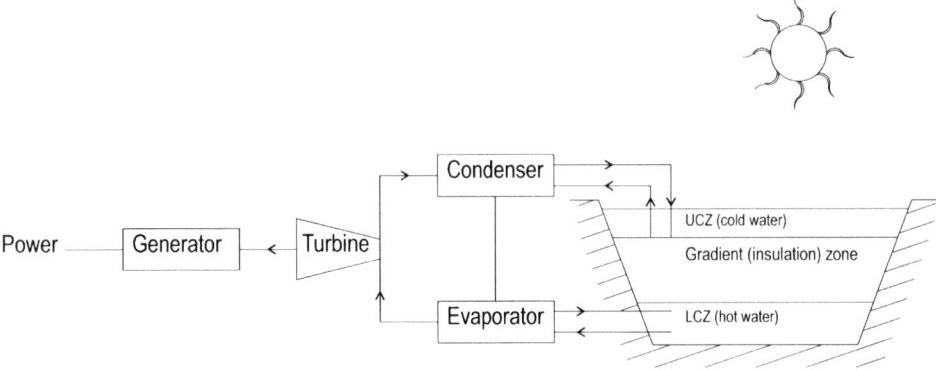

Fig. 3 Schematic of solar pond power generation system.

a transparent material. The heat trapped in the solar ponds can be used for many different purposes, such as industrial process heating, the heating of buildings, desalination, and to drive a turbine for generating electricity.

The first artificial solar pond was developed in Israel in 1958. Since then, many countries such as Australia, the United States, China, India, Iran, Italy, and Mexico have constructed solar ponds, mostly for research and development purposes. During the last decade, significant success in operational practices and applications of solar pond technologies has been achieved.[4]

SOLAR CROP DRYING

Drying is the oldest technique used to preserve food. Until around the end of the 18th century when canning was developed, drying was virtually the only method of food preservation. Solar energy is the main driving force that utilizes warm air to dry food. In drying, the moisture from the food is reduced to a certain level—as low as 5%–25% depending on the type of food-to prevent decay and spoilage in an environment free of contaminations such as dust and insects. Successful drying depends on[5]:

- Enough heat to draw out moisture, without cooking the food
- Dry air to absorb the released moisture
- Adequate air circulation to carry off the moisture

Solar drying can be carried out in open air under the sun by simply spreading the material on a clean surface or in particularly designed solar dryers. Solar dryers, however, exhibit many advantages over open air drying. Firstly, solar dryers are more efficient because they require lesser drying time and area. Secondly, the product is protected from rain, insects, animals, and dust, which may contain faecal material. Thirdly, faster drying reduces the likelihood of mold growth. Fourthly, higher drying temperatures mean that more complete drying is possible, and this may allow much longer storage times (only if rehumidification is prevented in storage). Finally, more complex types of solar driers allow some control over drying rates. Solar dryers can be made in many different designs depending upon various factors, i.e., the type of produce, scale of operation, and local economical and environmental conditions. In terms of their operational mode, solar dryers can be broadly classified into two main types, active and passive dryers, which can both be further subclassified into direct (in which the produce is directly heated from sun) and indirect types (in which the produce is not directly exposed to sun).

Almost all types of food—for example, vegetables, fruits, milk, herbs, spices, meat, and fish—can be dried by solar energy. The advantages of solar food drying are numerous. Dried foods, for example, are tasty, nutritious, lightweight, easy to prepare, and easy to store and use. The energy input is less than what is needed to freeze or can, and the storage space is minimal compared with that needed for canning jars and freezer containers.

SOLAR DISTILLATION

Solar distillation is a process that utilizes solar energy to purify water through evaporation and condensation processes. The process is also referred as water desalination when solar energy is used to purify water from saline water. Solar water distillation is a solar technology with a very long history. Installations were built over 2000 years ago, although they were to produce salt rather than drinking water. Documented use of solar stills (the distillation unit) began in the 16th century. An early large-scale solar still was built in 1872 that spread over an area of 4600 m^2 capable of producing 23,000 liters of drinking water for a mining community in Chile. Mass production occurred for the first time during the World War II when 200,000 inflatable plastic stills were made to be kept in life-crafts for the U.S. Navy.[6] In addition to their use in obtaining drinking water, solar stills are also suitable for the production of distilled water if there is appreciable demand for it in industry, laboratories, and medical facilities or to fill lead acid batteries.

Solar stills come in different designs; however the main features of operation are the same for all of them. In its simple form, water can be placed in an airtight basin that has a sloped transparent cover normally made of glass or plastics, though glass is preferred for its high transparency. The basin is coated with a black lining to maximize absorption of solar radiation. The incident solar radiation is transmitted through the glass cover and is absorbed as heat by the black surface in contact with the water to be distilled. The water is thus heated and gives off water vapor. The vapor condenses on the glass cover, which is at a lower temperature because it is in contact with the ambient air, and runs down into a tray where it is fed to a storage tank, as shown in Fig. 4. The economic viability of solar stills is determined to a critical degree by the design, the construction, the materials employed, and the local market conditions.

SOLAR COOKING

A solar cooker or solar oven harnesses solar energy to cook food. The solar cooker was first developed by a Swiss scientist Horace de Saussure in 1767.[7] Solar cookers are now being used in many countries across the world, especially in remote areas of poor countries. Solar cookers accomplish free cooking with environment friendliness as they only capitalize solar energy. Solar cooking can be very helpful in reducing the deforestation and pollution

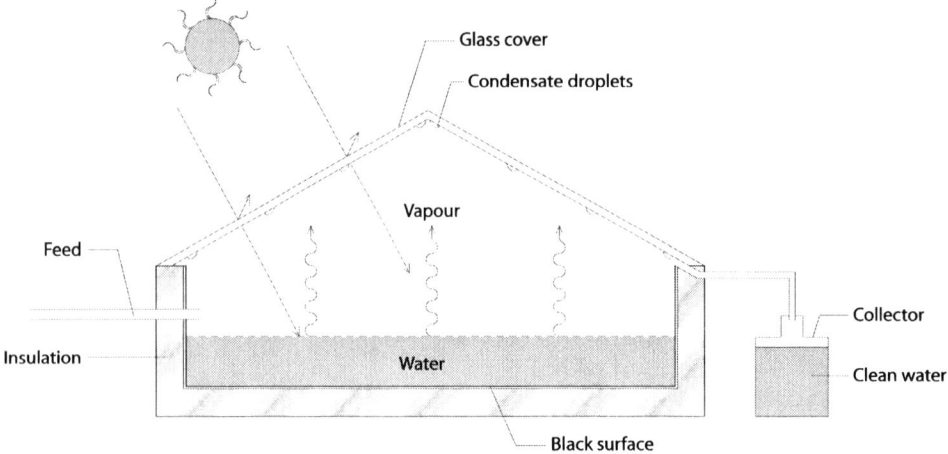

Fig. 4 Schematic diagrams of a solar still.

that originate from consumption of wood, and animal and agricultural residues for cooking in remote areas that lack access to electricity and gas. Solar cookers are capable of performing various types of cooking phenomenas, i.e., frying, baking, and boiling. The maximum achievable temperature depends on the intensity of the available solar radiation and the design and size of the solar cooker. Solar cookers come in a wide range of designs, which can be categorized under the following three major types.

Solar Box Cookers

A solar box cooker consists of an insulated box with a transparent top and a reflective lid. It is designed to capture solar radiation and make use of the greenhouse effect to cause heat to accumulate inside. The top is removable to allow food pots to be placed inside. Temperatures in a typical box cooker can reach above 200°C, but the temperatures achieved obviously depend on the size and design parameters of the cooker and the location of use.

Solar Panel Cookers

The solar panel cooker is the simplest solar cooker, and it consists of multiple simple reflectors arranged to focus solar radiation onto a covered black pot enclosed in a clear heat-resistant plastic bag or other transparent enclosure, such as glass bowel.

Solar Parabolic Cookers

Parabolic solar cookers, also called concentrated cookers, consist of a concave disk that focuses the light onto the bottom of a pot that is arranged at the focal length of the disk, as shown in Fig. 5. These are the most efficient types of solar cookers.

SOLAR COOLING/AIR CONDITIONING

Solar thermal energy can be used for cooling and dehumidification. Collectors play a critical role in extracting the energy from solar radiation to operate the cooling device. The collectors used in solar thermal cooling could be of various types, such as low-temperature

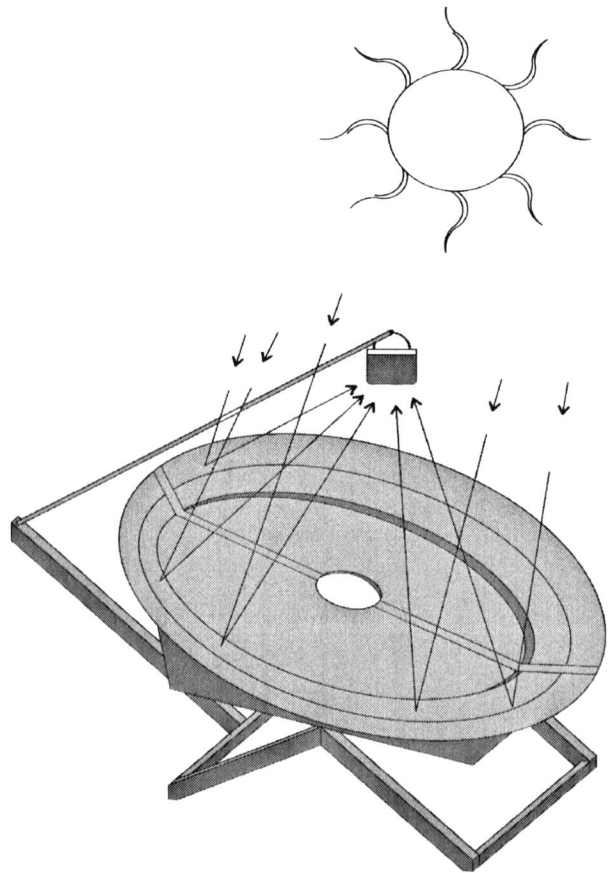

Fig. 5 Parabolic type of solar cooker.

flat plates and high-temperature evacuated tubes and concentrators. The basic principle behind solar thermal cooling is the thermochemical process of sorption—a liquid or gaseous substance is either attached to a solid, porous material (adsorption) or it is taken in by a liquid or solid material (absorption). The heat transfer fluid is heated in the solar collectors to a temperature well above ambient and used to power a cooling device—a type of heat-actuated pump. The heat transfer fluid may be air, water, or another fluid; it can also be stored in a hot state for use during times of no sunshine. Heat extracted by the cooling device from the conditioned space and from the solar energy source is rejected to the environment using ambient air or water from a cooling tower.[8]

The solar thermal cooling process can be broadly classified under open cycle systems and closed cycle systems. Open cycle systems are those in which the refrigerant is in direct contact with atmosphere and is discarded from the system after providing the cooling effect and new refrigerant is supplied in an open-ended loop. In closed systems, on the other hand, the refrigerant is not in direct contact with the atmospheric air. Open and closed cycle systems can further be distinguished according to the type of sorbent used, which can be in a liquid or a solid form. The three main designs of solar thermal cooling technologies that have gained the most attraction include solar adsorption, solar absorption, and solar desiccant. The key features of these designs are provided in Table 1.[9]

SOLAR THERMAL POWER GENERATION

Solar thermal power generation systems start with capturing heat from solar radiation. Direct solar radiation can be concentrated and collected by a range of concentrating solar power technologies to provide medium- to high-temperature heat. This heat then operates a conventional power cycle—for example, through a steam turbine or a Stirling engine to generate electricity. Solar thermal power plants can be designed for solar-only or hybrid operation, where some fossil fuel is used in case of lower radiation intensity to secure reliable peak-load supply. Five distinct solar thermal power generation concepts are available:

- Solar pond
- Solar chimney
- Solar parabolic trough
- Solar central receiver or solar tower
- Solar parabolic dish

Solar pond and solar chimney are nonconcentrated types of technology. In this section, the three concentrated types of solar thermal technologies—solar parabolic trough, solar central receiver, and solar parabolic dish—are discussed, as they have received the greater degree of attention over the years due to their favorable technical and commercial characteristics. These technologies can be used to generate electricity for a variety of applications, ranging from remote power systems as small as a few kilowatts (kW) up to grid-connected applications of 200–350 megawatts (MW) or more.[10]

Solar thermal power generation systems have three essential elements needed to produce electricity: a concentrator (to collect and focus solar radiation), a receiver (to convert concentrated solar radiation into heat), and an engine cycle (to generate electricity). Some systems also involve a transport or storage system. Solar collectors have a crucial role to play in the whole system and can be mainly classified into two types: concentrating and nonconcentrating. They are further categorized on the

Table 1 Overview of processes for thermally powered cooling and air conditioning

Solar thermal cooling system design	Adsorption refrigeration	Absorption refrigeration	Desiccant air conditioning
Solar collector	Vacuum tube collector, flat plate collector	Vacuum tube collector	Flat plate collector, solar air collector
Coolant circulation process	Closed refrigerant circulation systems	Closed refrigerant circulation systems	Open refrigerant circulation systems (in contact with the atmosphere)
Process basic principle	Cold water production	Cold water production	Air dehumidification and evaporative cooling
Sorbent type	Solid	Liquid	Solid
(Refrigerant/sorbent)	Water–silica gel ammonia–salt	Water–water–lithium bromide, ammonia–water	Water–silica gel water–lithium chloride–cellulose
Typical operating temp.	60°C–95°C	80°C–110°C (one step) 130°C–160°C (two step)	45°C–95°C

Parabolic Trough

The parabolic trough systems consist of large curved mirrors or troughs that concentrate sunlight by a factor of 80 or more onto thermally efficient receiver tubes placed in the trough's focal line, as shown in Fig. 6a. A thermal transfer fluid, such as synthetic thermal oil, is circulated in the tubes at focal length. Heated to approximately 400°C by the concentrated sun's rays, this oil is then pumped through a series of heat exchangers to produce superheated steam.[12] The steam is converted to electrical energy in a conventional steam turbine generator, which can either be part of a conventional steam cycle or integrated into a combined steam and gas turbine cycle, as shown in Fig. 7. Parabolic trough power plants are the only type of solar thermal power plant technology with existing commercial operating systems.

It is also possible to produce superheated steam directly using solar collectors. This makes the thermal oil unnecessary and also reduces costs because the relatively expensive thermo oil and the heat exchangers are no longer needed. However, direct solar steam generation is still in the prototype stage.

Central Receiver or Solar Tower

In solar thermal tower power plants, hundreds or even thousands of heliostats (large individually tracking mirrors) are used to concentrate sunlight onto a central receiver mounted at the top of a tower, as indicated in Fig. 6b. A heat-transfer medium in this central receiver absorbs the highly concentrated radiation reflected by the heliostats and converts it into thermal energy to be used for the subsequent generation of superheated steam for turbine operation. To date, the heat transfer media demonstrated include water or steam, molten salts, liquid sodium, and air. If pressurized gas or air is used at very high temperatures of about 1000°C or more as the heat transfer medium, it can even be used to directly replace natural gas in a gas turbine, thus making use of the excellent cycle of modern gas and steam combined cycles.

Parabolic Dish

A parabolic dish system uses a parabolic concave mirror to concentrate sunlight onto a receiver located at the focal point of the mirror, as highlighted in Fig. 6c. The concentrated beam radiation is absorbed into the receiver to heat a fluid or gas (air) to approximately 750°C. This fluid or gas is then used to generate electricity in a small piston, Stirling engine, or a microturbine attached to the receiver. These systems stand alone, and they are normally used to generate electricity in the kilowatts range.[13]

(a) Parabolic trough

(b) Central receiver or solar tower

(c) Parabolic dish (Stirling engine)

Fig. 6 Solar thermal power generation technologies.

basis of their concentrator optical properties and the operating temperature that can be obtained at the receiver. Most of the techniques for generating electricity from heat need high temperatures to achieve reasonable efficiencies. Concentrating systems are hence used to produce higher temperatures. Table 2 shows the operational characteristics of concentrated collectors.[11]

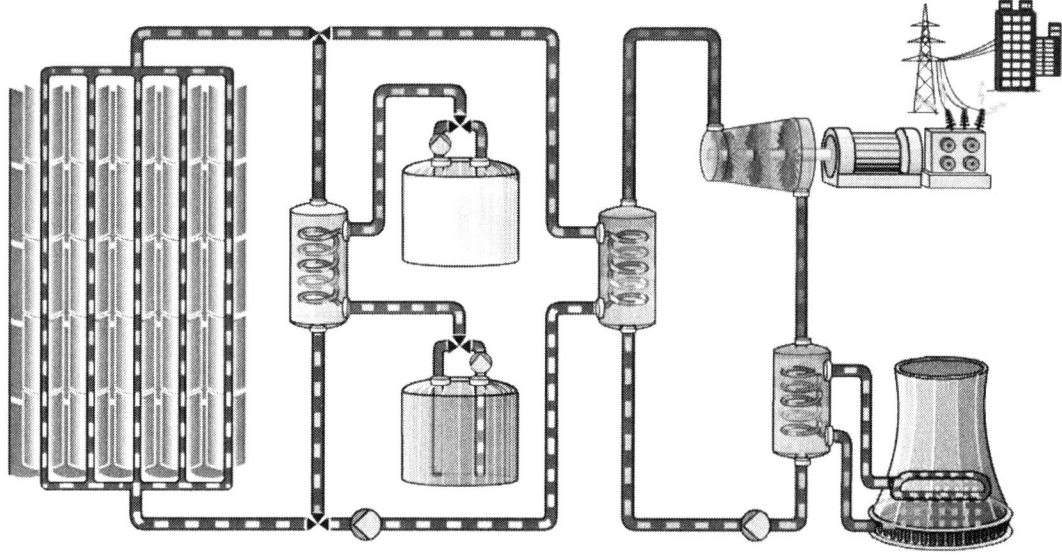

Fig. 7 Schematic of solar parabolic trough power plant.

SOLAR THERMAL TECHNOLOGIES—MARKET GROWTH AND TRENDS

Solar thermal technologies, in general, like SPV and other renewable energy technologies, are rapidly growing. The two key growth areas in solar thermal technologies, however, are solar water heating and solar thermal power generation. Solar water heating is among the fastest growing renewable technologies. As it was reported in the year 2003, solar water heaters received a 21% share of the total investment, $22 billion, in the renewable energy sector worldwide.[14] In 2004, China, which presently holds almost 80% of the installed solar thermal collector area, found its market growth by 30% as it installed well over 10 million m² of new solar thermal collector areas—equivalent to 7 GWth. There are similar growth trends across the world, especially in European countries.

The solar thermal power market has experienced a relative state of stagnancy since the early 1990s. However, new opportunities are now opening up for solar thermal power as a result of the global search for clean energy solutions. Both national and international initiatives are supporting the technology, encouraging commercialization of production. Recently, commercial plans in Spain and the United States have led a resurgence of interest, technology evolution, and potential investment. Some developing countries, including Morocco, India, Egypt, and Mexico have also planned projects with multilateral assistance. Examples of specific large solar thermal projects currently under construction or in advanced permitting and development stage around the world include:

- *Spain*. Over 500 MW solar capacity using steam cycle (4×10–20 MW solar tower and 12×50 MW parabolic trough)
- *Morocco*. 220-MW Integrated Solar Combined Cycle (ISCC) plant with 30 MW solar capacity (trough)
- *United States*. 50-MW solar capacity with parabolic trough in Nevada using steam cycle, preceded by a 1-MW parabolic trough demonstration plant using Organic Rankine Cycle (ORC) turbine in Arizona
- *United States*. 500-MW Solar Dish Park in California, preceded by a 1-MW (40×25 kW) test and demo installation

Table 2 Characteristics of typical concentrated solar collectors

Solar collector technology	Typical operating temperature (°C)	Concentration ratio	Tracking	Maximum conversion efficiency (Carnot) (%)
Solar fresnel reflector technology	260–400	8–80	One-axis	56
Parabolic trough collectors	260–400	8–80	One-axis	56
Heliostat field + central receiver	500–800	600–1000	Two-axis	73
Parabolodial dish concentrators	500–1200	800–8000	Two-axis	80

- *Italy*. 40 MW solar capacity integrated into existing combined cycle plant (trough)
- *Mexico*. 291-MW ISCC plant with 30 MW solar capacity (trough)
- *Algeria*. 140–150 MW ISCC plant with 25 MW solar capacity (trough)

A scenario of what could be achieved by the year 2025 was prepared by Greenpeace International, the European Solar Thermal Industry Association, and International Energy Agency (IEA) SolarPACES projects. From the current level of just 354 MW total installed capacity, the rate of annual installation by 2015 will have reached 970 MW, thus reaching a total installed capacity of 6454 MW. By 2025, 4600 MW will come on stream each year. According to this projection, by 2025, the total installed capacity of solar thermal power around the world will reach over 36,000 MW. It is also projected that by 2040 more than 5% of the world's electricity demand may be satisfied by solar thermal power.[12]

CONCLUSIONS

Solar thermal technologies operate by converting solar radiation into heat, which can be either directly utilized in various applications such as solar space heating, solar water heating, and solar air conditioning, or can be transformed into electricity to serve any purpose similar to conventional electricity. The key element in all solar thermal technologies is the collector, whose function is to gather the heat of solar radiation. Collectors normally come in three different types: flat plate, evacuated tube, and concentrated, and they operate in a wide range of temperatures, i.e., from less than 50°C to more than 1200°C. Solar thermal technologies normally operate in passive or active modes. Different types of solar thermal technologies are gaining huge attention across the world depending upon their technical and economical viability. Solar water heating, for example, in 2003, received a 21% share of the total investment made in renewable energy sector worldwide. Solar thermal power generation is also expected to grow at a healthy rate in coming years, as it is projected that by 2040 more than 5% of the world's electricity demand could be satisfied by solar thermal power.

REFERENCES

1. http://www.mtpc.org/index.asp (accessed on March 2006).
2. Muneer, T. Solar energy. *KEMPS Engineering Year Book*, Miller Freeman: New York, 2000.
3. Duffie, J.; Beckman, W. *Solar Engineering of Thermal Processes*, 2nd Ed.; Wiley: New York, 1991.
4. Akbarzadeh, A.; Andrews, J.; Golding, P. Solar pond technology: A review and future directions. *Advances in Solar Energy*, Earthscan: London, 2005.
5. Whitfield, D.E. *Solar Drying*, International Conference on Solar Cooking, South Africa, November 26–29, 2000.
6. Solar distillation, Technical Brief, Intermediate Technology Development Group.http://www.itdg.org/?id=technical_briefs (accessed on March 2006).
7. http://en.wikipedia.org/wiki/Solar_oven (accessed on March 2006).
8. Grossman, G. Solar cooling, dehumidification, and air-conditioning, solar thermal power generation. In *Encyclopedia of Energy*, Vol. 5; Elsevier: Amsterdam, U.K., 2004.
9. http://www.solarserver.de/solarmagazin/artikeljuni2002-e.html (accessed on March 2006).
10. Concentrating Solar Power: Energy from Mirrors, report Produced by NREL 2001. Available at: http://www.nrel.gov/docs/fy01osti/28751.pdf.
11. Luzzi, A.; Lovegrove, K. Solar thermal power generation. In *Encyclopedia of Energy*, Vol. 5; Elsevier: Amsterdam, U.K., 2004.
12. Concentrated solar thermal power—now!, Report by European Solar Thermal Industry Association and Greenpeace, 2005 http://www.greenpeace.org/raw/content/international/press/reports/Concentrated-Solar-Thermal-Power.pdf.
13. Quaschning, V. Solar thermal power plants. Renewable Energy World **2003**, *June*.
14. Annual investment in renewable energy: 1995–2003. Renewable Energy World **2005**, *February*.

Solar Water Heating: Domestic and Industrial Applications

M. Asif
School of the Built and Natural Environment, Glasgow Caledonian University, Glasgow, Scotland, U.K.

T. Muneer
School of Engineering, Napier University, Edinburgh, Scotland, U.K.

Abstract

Solar water heating is one of the most successful applications of solar thermal technologies. It provides environmentally clean energy and has enormous potential within domestic and industrial sectors. Solar water heaters are categorized into three main types: thermosyphon, built-in-storage, and forced circulation. This entry provides the fundamentals of solar water heating technology and a description of parameters that determine the performance of a solar water heating system. Economics of solar water heating and the market trends around the world have also been discussed. Results from a specific life cycle assessment (LCA) undertaken on built-in-storage solar water heater are also presented.

INTRODUCTION

Energy is imperative for human life. The accomplishments of civilization have largely been achieved through the increasingly efficient and extensive harnessing of various forms of energy to extend human capabilities and ingenuity. One of the biggest challenges to mankind in the 21st century is to develop methods of generating and using energy that could meet the rapidly growing energy demands while protecting the global ecosystem. Renewable energy resources such as solar energy, wind power, biomass, and geothermal energy are abundant, inexhaustible, and environmentally friendly. Solar energy is one of the most promising types of renewable energy that has the potential to meet the energy demand of the entire planet. Solar water heating, one of the oldest and the most successful applications of solar thermal technologies, utilizes solar energy to heat water without producing any harmful emissions into the environment.

Solar water heating, besides its domestic role, has a wide array of applications within the commercial sector (e.g., swimming pools, laundries, hotels, and restaurants) and the industrial sector (e.g., food and beverages, processing, and textile industries). Around the world, water heating accounts for as much as 15%–25% of the total energy consumed in the domestic sector. In the United States and United Kingdom, for example, water heating, respectively, consumes 18 and 23% of the domestic energy.[1,2] In the industrial sector, water heating may account for a significantly higher share of energy usage. In the textile sector, for example, water heating can account for as much as 65% of the total energy used during processes such as dyeing, finishing, drying, and curing.[3] Solar water heating systems can also be used for large industrial loads and for providing energy to district heating networks.

A solar water heating system essentially consists of a collector and a water storage tank. The collector absorbs solar radiation and transfers it to the water stored in the tank. Residential, commercial, and a considerable number of industrial applications often require hot water that is at a temperature of less than 60°C. The required hot water temperature, however, could vary depending upon the type of activity especially within the textile sector where it often needs to be near the boiling level. Modern solar water heaters can accomplish these temperatures. However, taking into account various factors such as unpredictable weather conditions (rain or overcast sky) and a solar heating system's size limitation in case of high demand, solar water heaters require a conventional heating system as a back up. A typical domestic solar water heating system can provide up to two-thirds of the total hot water requirements cutting down on fossil-fuel energy costs and also reducing the associated environmental impacts. There is a significant environmental penalty associated with the combustion of hydrocarbon gas with 14,000 g of CO_2 and 65 g of NO_x emitted per GJ of energy released. This can be reduced by using solar water heater.

This entry provides fundamentals of technology for solar water heating and a description of parameters that determine the performance of a solar water heating system. Economics of solar water heating and the market trends around the world are also discussed. Results of a life cycle assessment (LCA) study of a built-in-storage solar water heater are also presented.

Keywords: Solar energy; Solar water heating; Flat plate collectors; Evacuated tube collectors; Renewable energy; Life cycle assessment.

SOLAR WATER HEATER DESIGNS

Solar water heaters can be categorized into three main types: thermosyphon, built-in-storage, and forced-circulation. Thermosyphon and built-in-storage types are also regarded as passive systems as they rely on the natural circulation of water. The forced-circulation type of heater, on the other hand, is regarded as an active system as it incorporates an external element such as an electric pump to circulate the water. Solar water heaters transfer heat to the water in two ways: a direct system in which the collector itself transfers the heat to water, and an indirect system in which a heat-transfer fluid, circulating in the collector in a closed loop, transfers the heat to water through a heat exchanger.

Thermosyphon Solar Water Heater

The thermosyphon system operates on the principle that cold water has a higher density than warm water, and so being heavier, will sink down. Therefore, the collector is always mounted below the water storage tank, so that cold water from the tank reaches the collector via a descending water pipe (see Fig. 1). The collector absorbs solar radiation and transforms it into thermal energy and induces that energy to the water inside. As the temperature of the water in the collector rises, its density decreases. A circulation is thus established which enables the tank water to be progressively heated. The temperature of water at any point in the circulation determines its corresponding density. If this density variation is plotted against the height contour of the circuit, the magnitude of the driving force for water circulation is obtained. Owing to the low flow rates encountered within such systems, the driving force is balanced against the friction offered by the fluid circuit. The flow rates can thus be estimated. A thermosyphon system's storage tank must be positioned well above the collector, since otherwise the cycle can run backwards during the night, cooling down all the heated water. Furthermore, the cycle does not work properly at very small height differences. To avoid flow reversals, it is recommended that the bottom of the tank should be at least 300 mm above the top of the collector. This requirement makes the system inconvenient, particularly if the collectors are to be roof-mounted. Freezing may occur within the collectors during winter in a direct system. The solution is to employ a heat exchanger within the storage tank, with an anti-freeze solution added to water. This will, however, result in reduced system efficiencies and increased system cost.

Built-in-Storage Solar Water Heater

A "built-in-storage water heater," also referred as an "Integral collector-storage system", combines a flat-plate collector and storage tank in one unit. Built-in-storage water heaters can be further classified into plain and finned types.

Plain Built-in-Storage Water Heater

The construction of a built-in-storage water heater involves a rectangular box-like structure with the top face painted black and enclosed behind a single or double sheet of glass. The back surface and sides are properly insulated and the entire assembly is tilted at a suitable inclination. The built-in-storage water heater possesses several advantages over the other types. First, during the daytime, it operates at a higher efficiency owing to the fact that primarily no heat losses occur during water circulation. Second, in comparison with the thermosyphon and forced-circulation systems, the intimate contact of water with the absorber plate results in better heat transfer. Built-in-storage heaters are also compact in size and

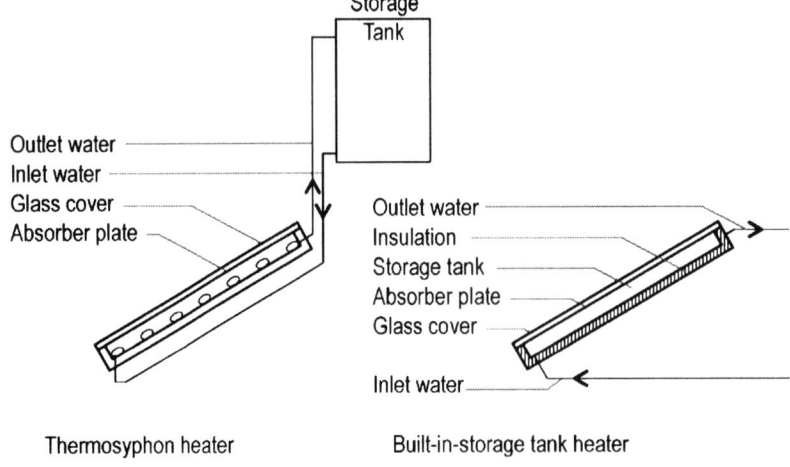

Fig. 1 Schematic diagrams of thermosyphon and built-in-storage solar water heaters.

cheaper due to their simplicity of construction. Based on extensive measurements of a number of designs, Muneer has reported the effects of storage volume/collector area ratio, the number of glazings, and the mode of operation on the performance of built-in-storage heaters.[4] Kreider and Kreith have presented design details of a freeze-tolerant, built-in-storage solar water heater with embedded heat exchanger coil which uses a refrigerant to extract heat.[5] Also in this respect, Davidson and Hammonds have recently undertaken more work.[6,7]

Finned Built-in-Storage Water Heater

The finned type of built in storage solar water heater differs from the plain type only in terms of fins that are incorporated within the thermal collector plate. Fins in this design play a dual role: firstly, they act as a support for the top absorber plate, and thus avoid the bulging due to hydrostatic pressures exerted by the stored water within the heater. Secondly, the fin attempts to enhance the heat transfer process from the absorber plate to the innermost layers of water. Note that within the plain heater the fact that heat is being transferred from a heated plate at the top to a cooler body of water residing underneath is an inefficient convection process. On the other hand, the vertically placed fins have a better opportunity to transfer heat as shown in Fig. 2.[8]

Forced Circulation Solar Water Heater

In a forced-circulation solar water heater, water is actively pumped from the storage tank through the collectors and back into the tank. An electronic controller, a small pump, valves, and other components are needed for proper operation and maintenance. With a forced-circulation water heater, the collector and storage tank can be installed independently, and no height difference between tank and collector is necessary. Two temperature sensors monitor the temperatures in the solar collector and the storage tank. If the collector temperature is higher than the tank temperature by a certain amount, the control starts the pump, which moves the heat transfer fluid in the solar cycle; "switch-on" temperature differences are normally between 5 and 10°C. If the temperature difference decreases below a second threshold, the control switches off the pump again. Fig. 3 shows the schematic of a forced-circulation water heating system. In this design, the problem of placing the solar water tank above the collectors is removed. The water in tank A is the feed water for tank B which has a built-in auxiliary heater. In a more compact version of the forced-circulation system, the two tanks A and B are combined in a single unit. However, this reduces the overall efficiency of the system as the maximum exploitation of solar energy is inhibited.

TYPES OF THERMAL COLLECTORS

The solar thermal collector is at the heart of a solar water heater as it absorbs the solar radiations and then transfers the captured heat to water in the storage tank. Solar thermal collectors may broadly be classified as flat-plate or evacuated-tube collectors.

Flat-Plate Solar Collector

The main components of a flat-plate solar collector include a transparent front cover, a collector housing,

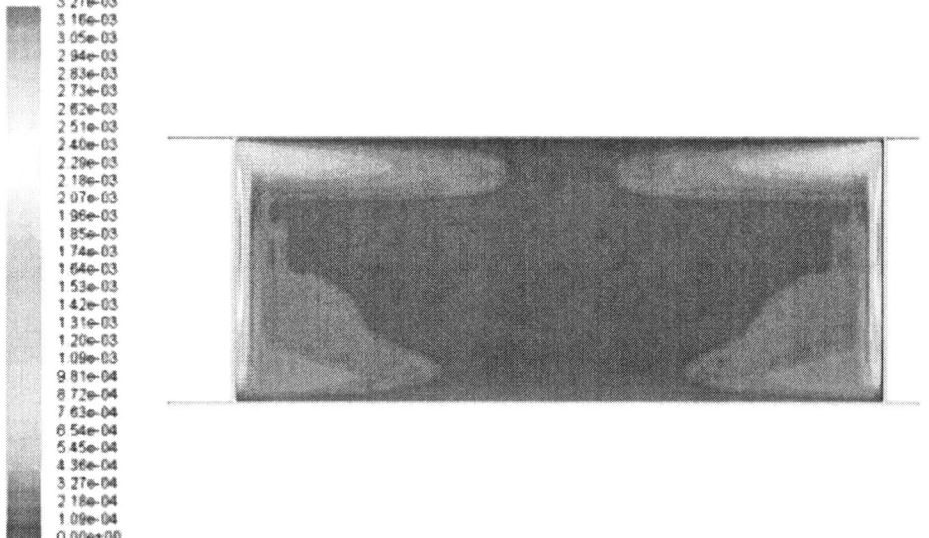

Fig. 2 Computational fluid dynamics (CFD) velocity distribution raster plot for a finned heater. The jets issuing from the left and right fins demonstrate their effectiveness in heating the otherwise quiescent body of water shown.

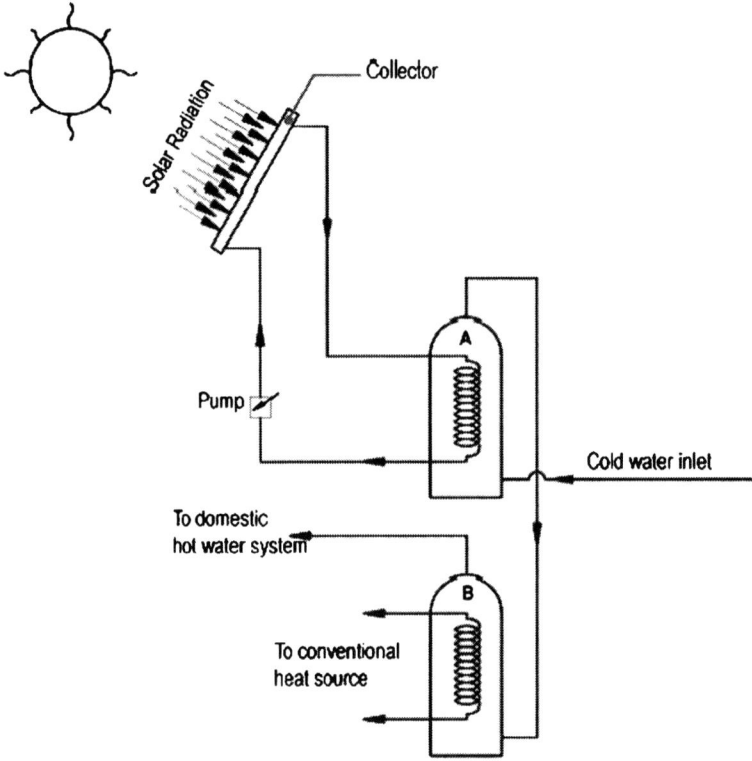

Fig. 3 Forced-circulation solar water heater with indirect cycle.

and an absorber. The absorber is usually made of metal materials such as aluminum, copper, or steel. The collector housing can be made of plastic, metal or wood, and the glass front cover must be sealed so that heat does not escape, and dirt, insects, and humidity do not get into the collector itself. The collector housing is highly insulated at the back and sides, keeping heat losses low. However, there are still minor heat losses from the collector, mainly due to the temperature difference between the absorber and ambient air, and these are subdivided into convection and radiation losses. The former are caused by air movements, while the latter are the product of the exchange of heat by radiation between the absorber and the environment. A sheet of glass covers the collector as it faces the sun, and this helps to prevent most of the convection losses. Furthermore, it reduces heat radiation from the absorber into the environment in a similar way that a greenhouse does. However, the glass also reflects a small part of the sunlight, which does not then reach the absorber at all.

The heat-loss coefficient from any given collector determines its thermal efficiency. This value, in turn, is obtained from the heat loss from the collector top (U_T) and the overall loss coefficient (U_L), i.e., the sum of heat loss from the top, sides, and bottom of the collector housing. For one commercially available single-glazed collector with a selectively coated absorber, U_T and U_L are 3.37 and 3.4 W/m^2-K, respectively.[9]

Eq. 1 is a simplified model for obtaining U_T

$$U_T = \left[\frac{N}{C/T_{pm}[T_{pm} - T_a/(N+f)]^e} + \frac{1}{h_w} \right]^{-1}$$
$$+ \frac{\sigma(T_{pm} + T_a)(T_{pm}^2 + T_a^2)}{(\varepsilon_p + 0.00591 N h_w)^{-1} + 2N + f - 1 + 0.133\varepsilon_p/\varepsilon_g - N}$$

(1)

Evacuated Tube Solar Collector

Evacuated tube solar collectors normally come in two types, coaxial-tube and heat-pipe designs. In the former design, a coaxial heat exchange pipe carrying a counter-flowing fluid is embedded within the absorber. The heat exchange pipe then feeds into a header. To enable maximum exploitation of solar energy, each vacuum tube is pivot mounted, enabling for the optimal orientation. The vacuum in the glass tubes ensures minimal convective heat loss while the selective coating helps suppress radiative losses. The top-loss and overall-loss coefficients for the collector presently under discussion are quoted as 1.37 and 1.375 W/m^2-K.[9]

The design of the collector shown in Fig. 4 is based on the heat pipe principle. Basically a heat pipe is a closed container that employs an evaporating-condensing cycle.

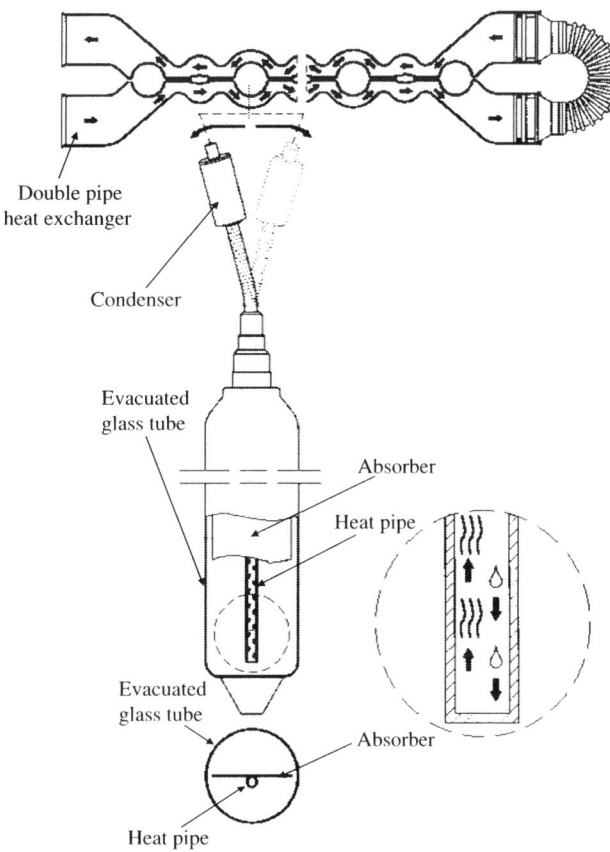

Fig. 4 Evacuated tube solar collector design based on the heat pipe principle.

The transfer of heat from the absorber plate occurs via an efficient and fast heat conductor bearing a low heat capacity. The heat-pipe functions like a thermal diode. It accepts heat from an external source, uses this heat to evaporate the liquid (latent heat) and then releases latent heat by reverse transformation (condensation) to the heat sink. This process is repeated continuously by a return feed mechanism of the condensed fluid back to the heat zone. Typically, evacuated heat pipe solar collectors use a sealed steel pipe, containing alcohol. The pipe is then attached to a black copper fin that fills the tube (absorber plate). Protruding from the top of each tube is a metal tip attached to the sealed pipe (condenser). These tubes feed into a heat exchanger manifold. Under solar flux, the alcohol is heated and hot vapor rises to the top of the pipe. Water or glycol flows through the manifold and picks up the heat from the tubes. The heated liquid circulates through another heat exchanger and gives off its heat to water that is stored in a solar storage tank. The maximum operating temperature of a heat pipe is the critical temperature of this heat transfer medium. Since no evaporation/condensation above the critical temperature is possible, the thermodynamic cycle is interrupted when the temperature of the evaporator exceeds the critical temperature. The top-loss and overall-loss coefficients for the above collector are quoted as 1.34 and 1.346 W/m²-K.[9]

SYSTEM DESIGN

The rule of thumb for a solar water heating system storage design is to use an 80 l/m² collector area for tropical locations, 60 l/m² for the United States, Central Europe, and Canada, 40–45 l/m² for the British Isles and Scandinavian countries.[1] Losses in the generation of domestic hot water maybe summarized as draw-off pipe losses, storage losses, losses in primary circulation pipes and those encountered during the heating up of the hot water generator. For a realistic plant simulation, these losses should be taken into account. In this section, basic mathematical models are presented which enable system simulation and design evaluation of solar water heating systems.

The Hottel–Whillier–Bliss Equation

An important relationship, which essentially represents the steady-state energy balance of any solar heat collector with an absorber area of A_c, is the Hottel–Whillier–Bliss equation (Eq. 2). This equation enables the calculation of useful energy gain (Q_u) as a function of the fluid inlet temperature, T_i and ambient temperature T_a.

$$Q_u = A_c F_R [I_{G,TLT}(\tau\alpha) - U_L(T_i - T_a)] \qquad (2)$$

Eq. 1 enables the computation of U_T. It is customary to add 5%–10% to the value of U_T to obtain the total heat loss coefficient, U_L. Other terms used in Eq. 2 are explained below.

The Transmission-Absorption Product, $(\tau\alpha)$

A solar radiation incident on any inclined collector is composed of direct (beam), sky-diffuse, and ground-reflected components. These transmitted components, passing through the collector glazing, may be estimated if the corresponding transmittances τ_b, τ_d, and τ_g are known. Each of these is a function of their respective irradiance incidence angles. Furthermore, part of the transmitted radiation is reflected back from the absorber plate to the cover system, which is, in turn, absorbed and reflected back to the collector plate. Therefore, as Duffie and Beckman have suggested, the transmittance-absorbance product $(\tau\alpha)$ should be thought of as a symbol representing a property rather than a straightforward product of the two fundamental properties, τ and α. Duffie and Beckman[10] have presented a graphical solution for obtaining $(\tau\alpha)$ from $(\tau\alpha)_n$.

Fin Efficiency (F')

The fin efficiency factor, F' represents the ratio of the actual energy gain to the energy gain that would result if the entire absorbing surface was maintained at the fluid temperature. Duffie and Beckman have presented the physical model for obtaining F' which is shown to be a strong function of the absorber fin heat transfer efficiency, U_L, and other less significant parameters.[10] For most practical designs, F' may be taken to have a value between 0.8 and 0.9.

Heat Removal Factor (F_R)

The collector heat removal factor, F_R, is the product of F' and the flow factor, F''. F_R is equivalent to the effectiveness of the solar collector's heat exchange process. The flow factor is obtained from Eq. 3, which, in turn, enables the estimation of F_R.

$$F'' = (mC_p/A_c U_L F')[1 - \exp(-mC_p/A_c U_L F')] \quad (3)$$

Measured Collector Performance

The basic method of obtaining collector performance is to operate the collector in a steady-state mode with coincident measurements of solar radiation, fluid flow rate, temperature gain, ambient temperature, and wind speed. In Eq. 2 the useful solar heat gain was written in terms of collector parameters. The efficiency of the collector is defined as

$$\eta = Q_u/I_{G,TLT} \quad (4)$$

Eqs. 2 and 4 may be manipulated to express the efficiency as a linear function of the ratio of the temperature differential between the collector plate and the ambient air and the collector irradiation. Thus,

$$\eta = F_R(\tau\alpha) - F_R U_L[(T_i - T_a)/I_{G,TLT}] \quad (5)$$

The European practice is to base the collector test results on $T_{p,av}$ which is the average of the inlet–outlet fluid temperature. Therefore, in Eq. 5 T_i is replaced with $T_{p,av}$.

Eq. 6 may be simplified as

$$\eta = a - bX \quad (6)$$

The X variable represents the term contained within the square brackets of Eq. 5.

Contrary to the dictum of Eq. 5, measured data show a slight non-linear variation of the data points, and as such, second-order models have been suggested for the above efficiency variation. For most practical purposes, though, a straight line fit should suffice.

System Simulation

Eqs. 1–5 enable a fairly representative, hourly simulation of solar water heating plants to be undertaken. A full-blown simulation is beyond the scope of this entry. A detailed simulation exercise would require collector thermal capacitance and tank temperature stratification effects to be taken into account. Hourly (and where applicable, temperature-dependent) estimation of F_R, $(\tau\alpha)$, and U_L may be undertaken with the hot water draw-off profile also incorporated within the simulation program.

ECONOMICS OF SOLAR WATER HEATING

Solar water heaters provide energy with economy. The economic payback period of a solar water heating system, however, is a function of various factors such as system efficiency, local weather conditions (i.e., the level of solar irradiance available), and the cost of fossil fuels. The initial cost of a solar water heater is higher than that of a gas water heater or an electric water heater that varies from region to region. A solar water heater can, however, be much more economical over the lifetime of the system than heating water with electricity, fuel oil, propane, or natural gas, depending upon the price of fuel sources. According to the findings of the Florida Solar Energy Center, the payback period for a well-designed and properly installed solar water heater could be between 4 and 8 years.[11] After the payback period, savings can be accrued over the life of the system, which ranges from

Table 1 Life cycle assessment of a built-in-storage solar water heater

Entity	Quantity (kg)	Embodied energy (MJ)	Carbon released (kg)	Monetary costs (USD)
Stainless steel	33	1155	20.13	108
Glass	11	340	6.70	5
Glass wool	2	40	1.75	2
Rubber	0.1	15	0.28	1
Timber	20	44	0.8	4
Total		1744	39.7	140
Savings/year		3509	69.6	20
Payback period		166 days	156 days	6.1

15 to 40 years, depending on the system and how well it is maintained. Muneer and Asif estimated the payback period for solar water heating incorporated within textile industries in Pakistan to be 6 years. More recent work has concluded that, with more efficient designs, the payback period can be reduced to just over 3 years.[8]

LIFE CYCLE ASSESSMENT (LCA) OF SOLAR WATER HEATERS—A CASE STUDY

Solar water heaters use solar energy as the fuel and hence are environmentally friendly, as they do not generate any toxic emissions during their operation. There are, however, environmental burdens associated with solar water heaters due to the materials used and the fabrication that is involved. In order to improve the thermal performance of a built-in-storage system, Muneer and Asif developed a finned device employing stainless steel, with a collector area of 1 m^2 and capacity of 80 l. They conducted rigorous testing of the heater for a complete year to determine its performance under varying weather conditions in Pakistan. They also conducted LCAs to investigate its energy and environmental performance. Their findings are presented in Table 1.

Table 2 Installed solar thermal collector area for European countries (Muneer 2004)

Country	1	2	3	4	5
Germany	420	615	900	3.71	45.1
Greece	195	181	160	2.98	283.4
Austria	141	168	169	2.34	288.9
France	16	23	38	0.55	9.3
Spain	33	40	50	0.45	11.4
Denmark	22	27	40	0.32	60.6
Italy	16	18	17	0.31	5.5
Switzerland	30	26	27	0.26	36.1
Portugal	5	6	8	0.25	25
Netherlands	30	32	35	0.21	13.5
Sweden	9	18	13	0.21	23.9
United Kingdom	9	10	11	0.21	3.5
Finland	9	10	10	0.03	5.9
Belgium	2	2	3	0.02	2.4
Ireland	1	0.3	0.3	0.003	0.9
Total	914	1171	1488	11.9	

Column labeled 1: thousands of m^2 installed in year 1999. Column labeled 2: thousands of m^2 installed in year 2000. Column labeled 3: thousands of m^2 installed in year 2001. Column labeled 4: millions of m^2 installed, cumulative. Column labeled 5: cumulative installed m^2/thousand population.

The annual average daily incident solar irradiation on a 1 m² collector area of the heater, under the test conditions for the above site was 4.8 kWh. The average daily energy yield of the heater was found to be 2.65 kWh, thus demonstrating an efficiency of 55%.[12] It was found that over a service life of 20 years, the heater would provide 19.4 MWh of energy. The heater consequently has a potential of saving 1393 kg of CO_2 emission.

As part of this work, a life cycle cost assessment was also carried out. Compared to furnace oil systems, typically used within industrial settings, the solar water heater was found to have a payback period of 6 years.[8] Further work in this regard showed that the performance of the system could be significantly improved by employing aluminum collectors. Owing to the fact that aluminum is cheaper and has better thermal properties than stainless steel, the payback period of this type of system was found to be just over 3 years.

USER SURVEY AND MARKET GROWTH

The Energy Technology Support Unit (ETSU) undertook a survey of domestic users of solar water heaters. Users from all regions of the United Kingdom were surveyed on a number of issues ranging from the types of systems in use to economic viability and technical reliability of their systems. The majority of respondents (70%) registered their response as "very satisfied." Any initial dissatisfaction was commonly reported as being due to installation problems. It was also observed that post-1986 improvement in manufacturing technology had resulted in increased user satisfaction. On an average, respondents had paid £2500 for their system. The overall conclusion was that the systems had shown a high reliability, with 85% of the systems sold since 1974 still in use.[1]

Since the early 1970s, the efficiency and reliability of solar water heating systems have increased significantly while the cost has dropped. Improvements to materials, a rating system for consumers, and more attractive designs, have all helped to make systems more successful. Particularly, over the last ten years there has been a significant increase in the use of domestic solar water heaters across the world. In 2003, solar water heaters received 21% share of the total investment, U.S. $22 billion, in the renewable energy sector worldwide.[13] In 2004, China's solar water heating market grew by 30% as it installed well over 10 million m² of new solar thermal collector area—equivalent to 7 GWth. In terms of installed capacity per capita, Israel, Greece, and Austria are the world leaders. The solar water heating market in France grew by over 30% per year during the last 3 years. In Spain, the solar water heating market has grown at a rate of 15% over the last 6 years while an annual growth target of 65% has been set until the year 2010. Similarly, other European countries are also experiencing an upward trend as shown in Table 2.[14] Reported figures indicate a healthy growth in 2004 (compared with 2003) in several countries, even though the total market is still small. Belgium grew by 62%, Estonia by 67%, Hungary by 50%, Ireland by 67%, Malta by 41%, Portugal by 67%, and Slovenia by 64%. In the United States, solar water heating is becoming increasingly popular in domestic and swimming pool applications.[15] From 1996 to 2004, for example, the Hawaiian Electric Company (HECO) has seen more than 25,000 solar water heater (SWH) systems installed within its customer base.[16]

CONCLUSIONS

Solar water heating is an energy efficient, cost effective and environmentally friendly renewable energy technology that has a huge potential in domestic and industrial applications. For most countries around the world, solar water heaters can provide up to two-thirds of the total hot water requirements while cutting down the energy cost and the associated environmental impacts. Due to its favorable characteristics, solar water heating is becoming increasingly popular across the world. In 2003 approximately 4.6 billion USD were invested in solar water heating, accounting for 21% of the total investment made within the renewable energy sector. China, a leading player in the solar water heating market, saw its solar water heating market grow by 30% in the year 2004. On the other hand, several European countries are experiencing an even healthier growth. On the basis of a case study presented in this entry, it may also be noted that the payback periods for built-in-storage heaters using stainless steel and aluminum may be expected to be, respectively, around 6 and 3 years.

REFERENCES

1. Muneer, T. Solar Eenergy In *KEMPS Engineering Year Book*; Miller Freeman: Kent, England, 2000.
2. http://www.toolbase.org/tertiaryT.asp?DocumentID=3216&CategoryID=949 (accessed October 2005).
3. Vijayaraghavan, S. In *Encyclopaedia of Energy*, Vol. 5; Academic Press: London, 2004.
4. Muneer, T. Effect of design parameters on performance of built—in storage solar water heaters. Energy Conversion Manag. **1984**, *3*, 277–281.
5. Kreider, J.F.; Kreith, F. Non-Concentrating Solar Thermal Collectors Solar Energy Handbook. In *Solar Energy Handbook*; McGraw-Hill: New York, 1981.
6. Liu, W.; Davidson, J.H.; Kulacki, F.A. Thermal characterisation of prototypical integral collector storage system with immersed heat exchangers. J. Solar Energy **2005**, *127*, 21–28.

7. Cruza, M.S.; Hammond, G.P.; Reisa, A.S. Thermal performance of a trapezoidal-shaped solar collector/energy store. Appl. Energy **2002**, *73*, 195–212.
8. Munner, T.; Maubleu, S.; Asif, M. Prospects of solar water heating for textile industry in Pakistan. Renew. Sust. Energy Rev. **2006**, *10*, 1–23.
9. *Product Range*, Viessmann Werke GmbH & Co KG, Germany, 1997
10. Duffie, J.; Beckman, W. *Solar Engineering of Thermal Processes*, 2nd Ed.; Wiley: New York, 1991.
11. *Heat your Water With the Sun: A Consumer Guide*; U. S. Department of Energy; Washington, DC, U.S.A.; 2003.
12. Asif, M.; Muneer, T. Life cycle assessment of solar water heating in Pakistan; *Building Services Engineering Research and Technology*; in press.
13. Annual investment in renewable energy: 1995–2003; *Renewable Energy World*; February 2005.
14. Muneer, T. *Solar Radiation and Daylight*, 2nd Ed.; Elsevier Butterworth Heinemann: Oxford, U.K.; 2004.
15. Solar Thermal Systems on the Up; *Renewable Energy World*; July 2005.
16. http://go.ucsusa.org/clean_energy/renewable_energy/page.cfm?pageID=1595 (accessed October 2005).

Solid Waste to Energy by Advanced Thermal Technologies (SWEATT)

Alex E. S. Green
Green Liquids and Gas Technologies, Gainesville, Florida, U.S.A.

Abstract

Solid waste (SW), now mostly wasted biomass, could fuel approximately ten times more of the United States's increasing energy needs than it currently does. At the same time, it would create good nonexportable jobs and local industries. Twenty-four examples of wasted or underutilized solids that contain appreciable organic matter are listed. Estimates of their sustainable tonnage exceed two billion dry tons. Now usually disposal problems, most of these SWs can be pyrolyzed into substitutes for or supplements to expensive natural gas (NG). The large proportion of carbon-dioxide-neutral plant matter in the list would reduce greenhouse problems. Pyrolysis—heating to high temperatures without oxygen—converts such SW into a medium-heating-value gaseous fuel, usually with small energy expenditure. With advanced gas cleaning technologies the pyrogas can be used in high-efficiency gas turbines or fuel-cell systems. This approach has important environmental and efficiency advantages with respect to direct combustion in boilers and even air-blown or oxygen-blown partial combustion gasifiers. Because pyrolysis is still not a predictive science, the Clean Combustion Technology Laboratory (CCTL) has used an analytical semiempirical model (ASEM) to organize experimental measurements of various product $\{C_a H_b O_c\}$ yields vs temperature (T) for dry ash, nitrogen, and sulfur free (DANSF) feedstock having various weight percentages (wt%) of oxygen [O], and hydrogen [H]. With this ASEM, each product is assigned five parameters (W, T_0, D, p, q) in a robust analytical $Y(T)$ expression to represent yields vs temperature of any specific product from any specified feedstock. Patterns in the dependence of these parameters upon [O], [H], a, b, and c suggest that there is some order in pyrolysis yields that might be useful in waste-to-energy conversion (WEC) systems to optimize their throughput. An analytical cost estimation (ACE) model is used to calculate the cost of electricity (COE) vs the cost of fuel (COF) for a SW integrated gasifier combined cycle (IGCC) system for comparison with the COE vs COF for a natural gas combined cycle (NGCC) system. It shows that at high NG prices, SW can be changed from a disposal-cost item to a valuable asset. Comparing COEs when using other SW-capable technologies are also facilitated by the ACE method. Implications of this work for programs that combine conservation with waste-to-energy conversion in efforts to reach Zero Waste are discussed.

SOLID FUELS AND SOLID WASTE

In 1940, when Britain was fighting a ruthless and apparently unstoppable Hitler, Winston Churchill offered only "blood, sweat, toil, and tears" to unite Britain's political factions. At this time in our history we are excessively (60%) reliant on foreign sources for our liquid fuels and are increasingly importing our gaseous fuels (now >15%). Our country is shedding blood in its efforts to stabilize regions of the globe that supply these premium fuels. Yet the United States is well endowed with solid fuels in the form of coal and oil shale, and substantial quantities of renewable but wasted solids. In this paper, in continuation of a long search for alternatives to oil,[1–10] our focus is on converting our solid waste to energy by advanced thermal technologies (SWEATT).

Keywords: Solid waste; Pyrolysis; Gasification; Cost of electricity; Pyrolysis products.

Table 1 is a list of the United States's abundant supply of wasted solids or SW whose organic matter can be made into gaseous and liquid fuels. With recent high natural gas (NG) prices and for technical reasons that will become obvious, this entry will concentrate on advanced thermal technologies (ATT) for the conversions of SW to gaseous fuels. Advanced thermal technologies conversions of coal to liquid and gaseous fuels involve similar technical considerations, but coal to liquid or gas technologies have the attention of many government, business, and engineering personnel. Solid waste to energy by advanced thermal technologies has the attention of only a few in the United States.

In the United States, most of the categories in Table 1 would now be called "biomass" in part because "solid waste" has a bad public image, bringing to mind old incinerators belching black smoke. However, with advances in thermal technologies and gas cleanup systems now being successfully applied in Japan and the European Union (EU),[11] SWEATT deserves a new image. It not only addresses the United States's very urgent need for

Table 1 Wasted solids that could be used as a component of the united state's primary energy supply

Waste type	Million dry tons
Agricultural residues	~0.98
Forest under-story and forestry residues	~0.40
Hurricane debris	~0.04
Construction and deconstruction debris	~0.02
Refuse derived fuels	~0.10
Urban yard waste	~0.02
Food serving and food processing waste	~0.07
Used newspaper and paper towels	~0.02
Used tires	~0.05
Energy crops on under-utilized lands	~0.05
Ethanol production waste	~0.02
Anaerobic digestion waste	~0.01
Bio-oil production waste	~0.01
Waste plastics	~0.03
Infested trees (beetles, canker, spores)	~0.02
Invasive species (cogon-grass, melaluca)	~0.02
Plastics mined when restoring landfills	~0.03
*Bio-solids (dried pelletized sewage sludge)	~0.04
*Poultry and pig farm waste	~0.02
*Water plant-remediators (algae, hydrilla.)	~0.01
*Muck pumped to shore to remediate lakes	~0.01
Manure from cattle feed lots	~0.01
Plants for phyto-remediation of toxic sites	~0.01
Treated wood past its useful life	~0.01
Total	~2 billion dry tons

Table 1 is a list of potential local sources of useful nonconventional fuels cited in the author's conference presentations with recent emphasis on sources available in Florida. Items marked with * help in water remediation and the ~ denotes estimated values.

alternative fuels, but also could mitigate air and water pollution problems. The large carbon dioxide-neutral plant matter components in Table 1 can help in greenhouse mitigation. The great diversity of physical and chemical characteristics in Table 1 implies that the world needs omnivorous feedstock converters (OFCs) to change these solid fuels into much more usable liquid or gaseous fuels.

Fig. 1 is a conceptual illustration of an OFC adapted from several previous CCTL papers[8–10] in which a SW pyrolyzer–gasifier–liquifier is co-utilized with a natural gas combined cycle (NGCC) system, as discussed in "ASEM and Pyrolysis."

PROPERTIES OF SOLID FUELS AND SOLID WASTE

The declining resources of domestic liquid and gaseous fuels are the greatest energy problems facing the United States and many other countries today. Yet, the United States is particularly well endowed with solid fuels in the form of various coals, peat, biomass, and SW. Table 2 shows major ranks of coals as well as of peat, wood and cellulose, and their ultimate and proximate analyses as measured by industry for over a century. The numbers listed in columns 2, 3, and 4 essentially apply to ideal carbon, hydrogen, and oxygen (CHO) materials by correcting measurements to be dry, ash, sulfur, and nitrogen free (DASNF). For such material $[C] = 100 - [O] - [H]$ so $[C]$ becomes a variable dependent upon the values of $[H]$ and $[O]$. Column 5 gives the higher heating value (HHV) in MJ/kg as measured with standard bomb calorimeters after allowing for the minor components.

Fig. 2a is mainly a plot of $[H]$, the wt% of hydrogen (solid diamonds with values read on the left scale) vs $[O]$, the wt% of oxygen, for 185 representative DANSF CHO materials taken from ultimate analysis data available in the technical literature. The bottom scales give conventional coal ranks, some potential names for the biomass region, and some names that might foster more friendly discussions between the coal and biomass communities. This $[H]$ vs $[O]$ coalification plot shows that apart from the anthracite region, all natural DANSF feedstock have $[H]$ values that are close to 6%. The near constancy of $[H]$ together with the relationship $[C] = 100 - [O] - [H]$ for DASNF feedstock imply a linear decline of $[C]$ with increasing $[O]$. The smooth trend of properties with $[O]$ provide strong reasons for regarding peat and biomass simply as lower-rank coals or the various coal ranks simply as aged forms of biomass. The diagram suggests that the natural solid fuels could be ranked simply by $[O]$ to replace the different ranking systems of various countries (a Tower of Babel!). Using 34-O for peat—called "turf" in Ireland—might help temper the "turf wars" in fuel sector competitions and in energy/environmental confrontations on the use of our available solid fuels.

Higher heating values of various fuels measured with calorimeters usually are reported with proximate analyses. Approximate HHVs in MJ/Kg for the seven representative CHOs are given in Column 5 in Table 2. Column 6 gives

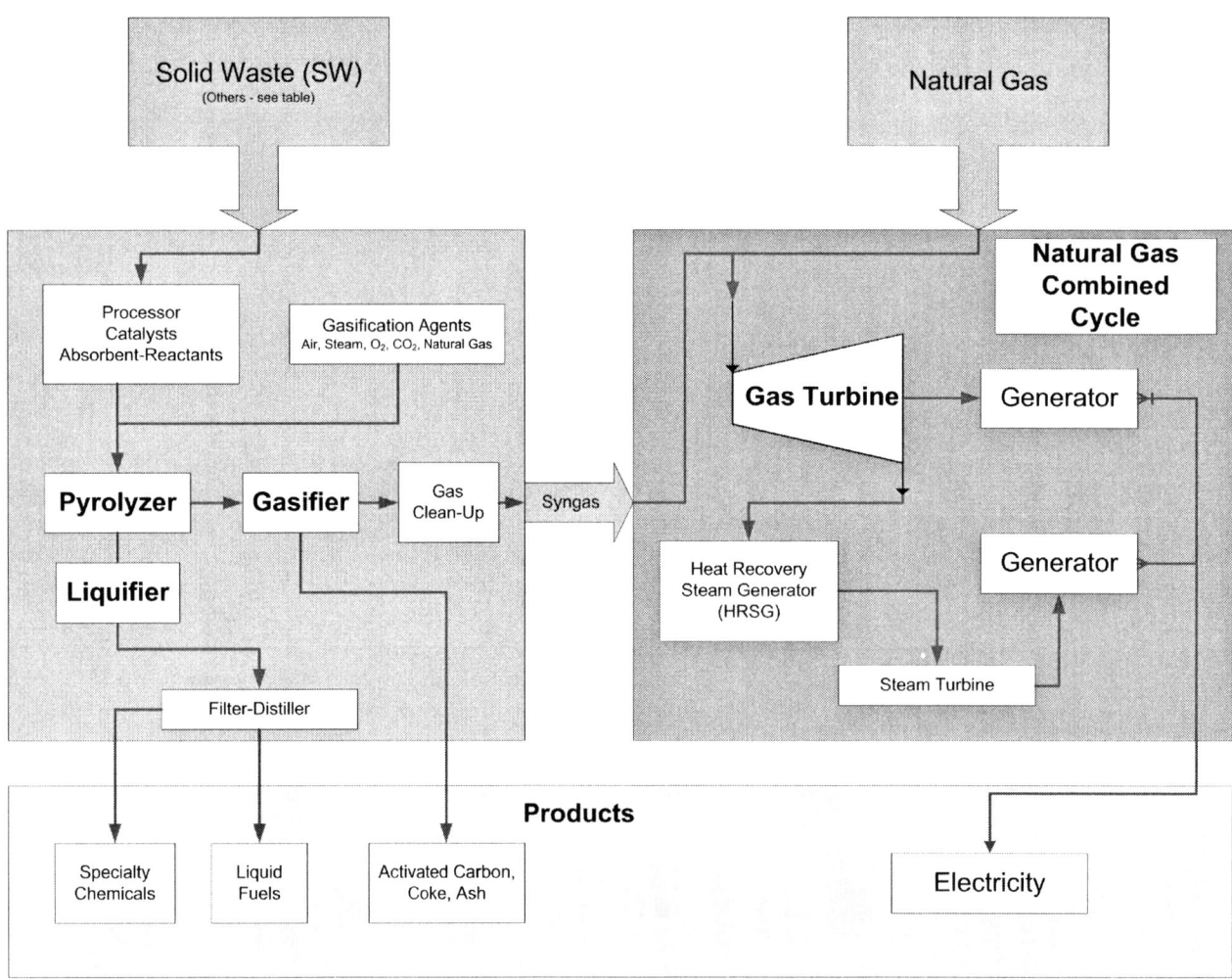

Fig. 1 Diagram of the omnivorous feedstock converter (OFC) illustrating the addition of a solid waste (SW) system to an existing natural gas combined cycle (NGCC) plant to create an effective SWCC system.
Source: Adapted from Ref. [10].

representative total volatiles, V_T, as determined by an American Standard Test Measurement Method (ASTM). A solid sample is heated (pyrolyzed) in an inert atmosphere using a platinum crucible at 950°C for 7 min. The wt% loss between the sample and its char is the total volatile yield. Then the balance from 100% represents the weight of the fixed carbon (FC) plus ash. When this residual is burned, the remainder is the ash wt%. An empirical analytical formula is given in the caption to represent general trends of total volatiles along nature's coalification curve. Note the rapidly increasing trend in V_T from low [O] materials to high [O] materials. The numbers in Column 7 of Table 2 represent $FC = 100 - V_T$, the fixed carbon for pure CHO materials after all volatiles are driven off.

Columns 8 and 9 of Table 2 give the relative physical density and relative energy density of the various natural solid fuels, which are important factors in handling and transportation costs. Columns 10 and 11 relate to the reactivity and H and OH free radical generated in the combustion of these various solid fuels, which have strong influence on rates of reaction. The last column gives the proposed quantitative [O] ranking system for solid fuels that lie along nature's coalification curve.

The HHV of a solid fuel is, perhaps, the most important variable in solid fuel use and solid fuel conversion to liquid and gaseous fuels. A simple form of Dulong's formula, suitable for mental calculations, that captures the main trends is

$$HHV = [C]/3 + 1.2[H] - [O]/10 \text{ in MJ/Kg} \quad (1)$$
$$\text{where } M = 1,000,000$$

Divide this number by 2.3 to get the corresponding HHV in MBtu, where $M = 1000$. Note that the carbon

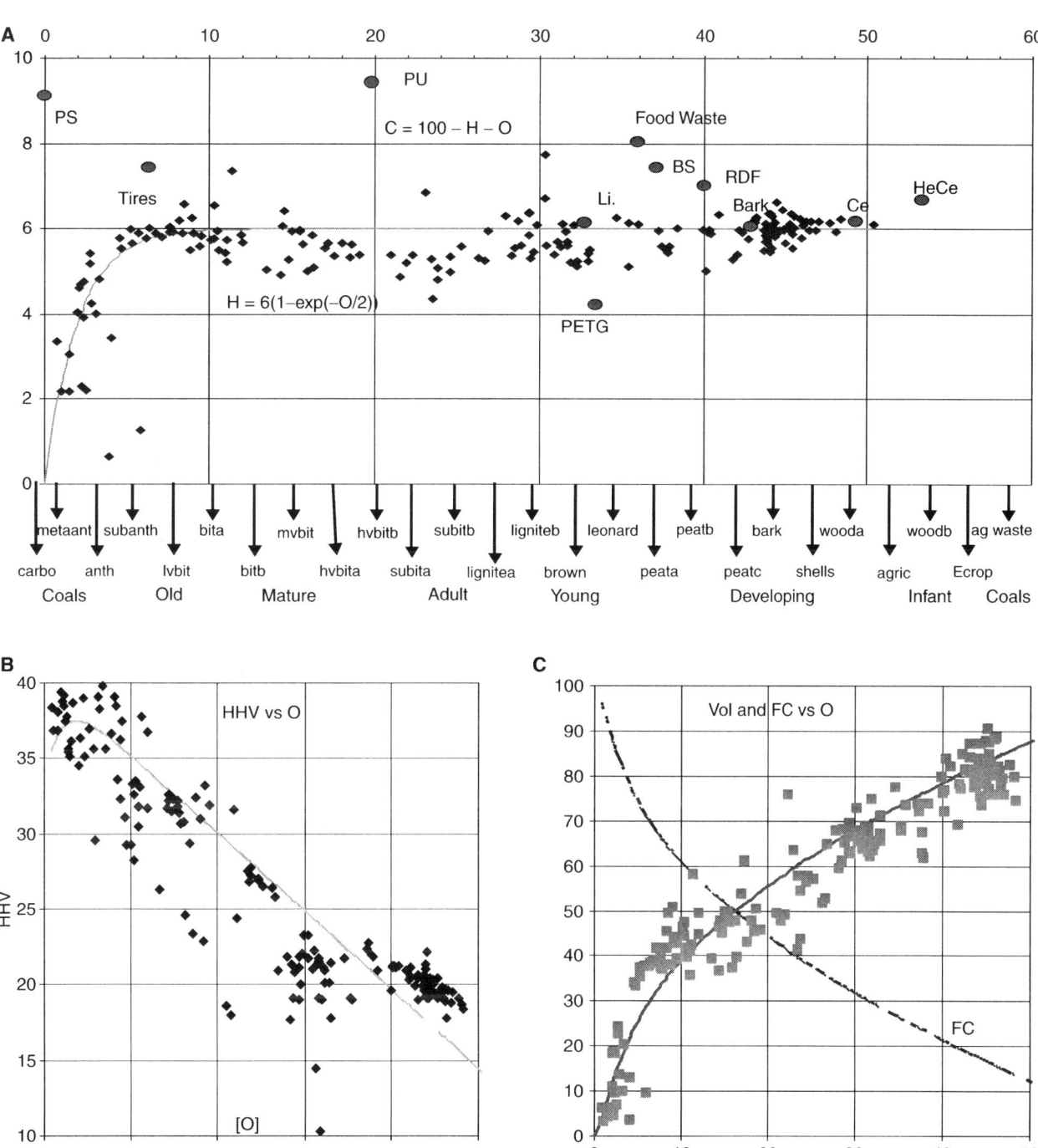

Fig. 2 (A) Weight percentages of hydrogen [H] vs [O] for 185 DANSF carbonaceous materials (black diamonds) vs oxygen wt%. Classification labels are given at the bottom scale and [O] values on top scale. Adapted from Ref. [4]. (B) Higher heating values (HHV) of 185 carbonaceous materials (corrected to DANSF) vs [O]. The smooth curve represents HHV = ([C]/3 + 1.2[H]–[O]/10). (C) Total volatile weight percentages vs [O] for 185 DASNF carbonaceous materials (squares) from proximate analysis. The curve through the data points satisfies $V_T = 62([H]/6)([O]/25)^{1/2}$. The analytic fixed carbon (FC) is shown.

Table 2 Properties of fuels along natures coalification path

Name	Ultimate analysis			Proximate analysis				Other properties			
	C	H	O	HHV	VT	FCCh	Dens	E/vol	RelchR	H,OH Rad	O-Rank
Anthracite	94	3	3	36	7	93	1.6	58	1.5	v. low	3-O
Bituminous	85	5	10	35	33	67	1.4	49	5	low	10-O
Sub bitum	75	5	20	30	51	49	1.2	36	16	med	20-O
Lignite	70	5	25	27	58	42	1	27	50	interm	25-O
Peat	60	6	34	23	69	31	0.8	18	150	high	34-O
Wood	49	7	44	18	81	19	0.6	11	500	v. high	44-O
Cellulose	44	6	50	10	88	12	0.4	9	1600	v v.high	50-O

energy term ([C]/3) usually is much larger than the hydrogen energy contribution (1.2[H]). Oxygen contributes negatively in part because the more [O] implies less [C] and in part because of the subtractive term −[O]/10. The larger points on Fig. 2a give the [H], [O] positions of lignin (6.1,32.6), cellulose (6.2,49.4), and hemi-cellulose (6.7,53.3), the three main components of all plant matter. Also shown in Fig. 2a are the [H] and [O] coordinates of several materials that are present in SW. These depart substantially above and below the coalification curve. Not shown are petroleum and polyethylene, which would lie at [14.2, 0].

Fig. 2b shows the pattern of HHVs vs [O] along the coalification path. The DuLong formula given in the caption is a simple compromise between those used in the coal and biomass sectors. Fig. 2c shows the total volatiles (V_T) for the CHO materials vs [O] mostly for materials close to the coalification path. These values are determined by standard proximate analysis procedures that measure the weight loss of a sample after exposure to 950°C for 7 min in an anoxic medium. It should be obvious that in the high [O] region pyrolysis is substantially equivalent to gasification. Detailed studies point to the fact that small departures of [H] from the coalification curve have large impacts on volatile content.

The three diagrams in Fig. 2 all indicate the importance of the [O] in determining the fuel properties of natural substances. CCTL studies indicate that the [H] dimension is also very important in determining volatile content and that small deviations of [H] from the smooth coalification path have a large impact on the volatile release. Table 3 gives a compact list of heating values of various wastes, fuels, and plastics in units of MBtu/lb.

GLOBAL AND U.S. PRIMARY ENERGY SUPPLIES

Fig. 3a presents an overview of the world total primary energy supply (see the International Energy Agency Web site) at the opening of the millennia. Among the major sources, combustible renewables and waste (CRW, mostly biomass) need only be doubled to be competitive with coal and NG, and tripled to be competitive with petroleum. Already, the category CRW is almost a factor of 2 greater than nuclear. On the other hand, wind and solar must grow by factors of more than 100 to become major global energy supplies. This global total primary energy supply (TPES) picture is not representative of the industrial world, particularly the United States today. Fig. 3b shows the subdivisions of the U.S. TPES in 2005, in quadrillion Btu or quads (see the U.S. Energy Information Agency Web site).[12] Because total consumption is now very close to 100 quads, the numbers might also be considered to be

Table 3 Heating values of MSW components, fuels, and plastics in 1000 Btu/lb

Component	As recd	Dry	Component	Dry
Wastes			Fuel	
Paper and paper products			*Hydrocarbons*	
Paper, mixed	6.80	7.57	Hydrogen	60.99
Newsprint	7.97	8.48	Natural gas	20.00
Brown paper	7.26	7.71	Methane	23.90
Trade magazines	5.25	5.48	Propane	21.52
Corrugated boxes	7.04	7.43	Ethane	22.28
Plastic-coated paper	7.34	7.70	Butane	21.44
Waxed milk cartons	11.33	11.73	Ethylene	21.65
Paper food cartons	7.26	7.73	Acetylene	21.50
Junk mail	6.09	6.38	Naphthalene	17.30
			Benzene	18.21
Domestic wastes			Toluene	18.44
Upholstery	6.96	7.48	Xylene	18.65
Tires	13.80	13.91	Naptha	15.00
Leather	7.96	8.85	Turpentine	17.00
Leather shoe	7.24	7.83		
Shoe, heel, and sole	10.90	11.03	*Oils*	
Rubber	11.20	11.33	No. 1 (Kerosene)	19.94
Mixed plastics	14.10	14.37	No. 2 (Distillate)	19.57
Plastic film	—	13.85	No. 4 (VL residual)	18.90
Linoleum	8.15	8.31	No. 5 (L residual)	18.65
Rags	6.90	7.65	No. 6 (residual)	18.27
Textiles	—	8.04		
Oils, paints	13.40	13.40	*Alcohols*	
Vacuum-cleaner dirt	6.39	6.76	Methanol	10.26
Household dirt	3.67	3.79	Ethanol	13.15
Food and food waste			Plastics	
Vegetable food waste	1.79	8.27		
Citrus rinds and seeds	1.71	8.02	Polyethylene	19.73
Meat scraps (cooked)	7.62	12.44	Polystyrene	16.45
Fried fates	16.47	16.47	Polyurethane	11.22
Mixed garbage 1	2.37	8.48	Polyvinyl chloride	9.78
			PVC (pure resin)	7.20
Coals			Polyvinylidene chloride	4.32
Low-vol. bituminous	—	15.55	Polycarbonate	13.31
Med.-vol. bituminous	—	15.35	Cellulose	7.52
High-vol. bituminous	12.25	14.40	Polypropylene	20.22
Subbituminous	9.90	12.60	Polyester	12.81
Lignite	7.30	11.45		
Anthracite	—	14.00		

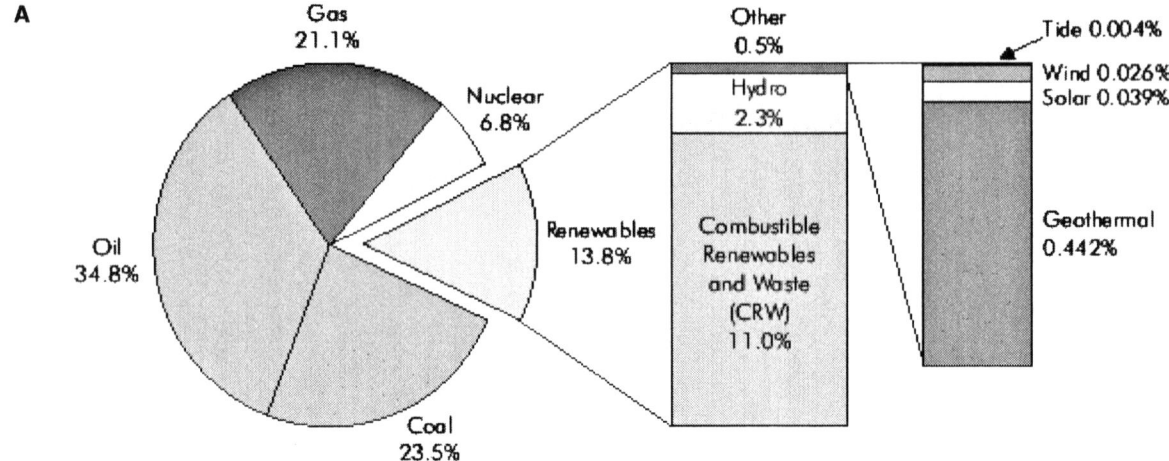

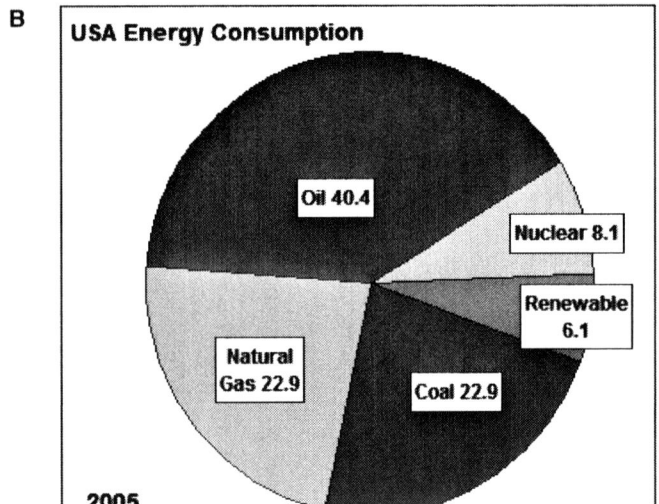 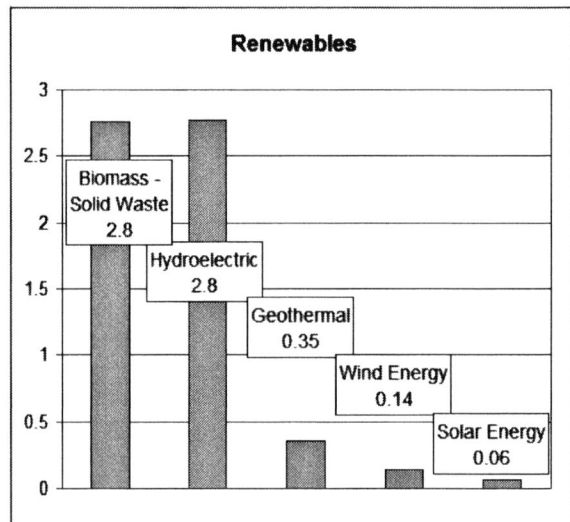

Fig. 3 (A) Total primary energy supply for the globe at the millennium (IEA Web site), (B) (Left) total 2005 annual U.S.A. energy consumption of primary energy sources in quads (Right) renewables.

approximate percentages of U.S. energy consumption. It is seen that more than 40% of our energy consumption is in the form of petroleum, consumed mainly in our transportation sector. Without doubt the biggest energy problem faced by the United States today is the need to find alternatives to oil.[1–3] In the 1970s and early 1980s, the United States focused heavily on alternatives to oil in the utility sector. The alternatives first were pulverized coal plants and, in the late 1980s and 1990s, NGCC systems. At this time, the United States' focus should in part be on the developing alternatives to NG for electricity generation via the use of ATT. It should be noted, however, that ATT can also make significant contributions to the solution of our liquid fuel problem in the transportation sector.[3]

It is important to differentiate secondary energy supplies (SES) from the primary energy supplies (PES) shown in Fig. 3. Secondary energies supplies include steam, syngas, reactive chemicals, hydrogen, charges in batteries, fuel cells, and other energy sources that draw their energy from PES. If an SES is converted to another type of energy—say, mechanical energy—via a steam turbine, the mechanical energy becomes a tertiary energy supply (TES). This TES can be converted to electrical energy via magnetic generators, in which case the electricity is a quaternary supply (QES). In the case of electricity, the many conversions are usually justified because electricity can readily be distributed by wire and has so many uses as a source of energy for highly efficient electric motors, illumination systems, home appliances, computers, etc.

A debate is under way in many communities as to whether increasing electricity needs should be met with solid fuels—particularly coal—via conventional steam and steam turbine generator systems or via conversion to a gaseous fuel to fuel integrated gasifier combined cycle (IGCC) systems. Granting that the steam turbine route has had many advances over the past century, our thesis is that converting the solid fuel to gaseous fuel is the ATT route of the future. The ATT route is driven not only by environmentally acceptable waste disposal needs and increased needs for electricity, but also by the need for liquid and gaseous fuels. A number of petroleum resource experts recently advanced the date when the globe's supply of oil and NG will run out. The prices of oil and NG now reflecting this drawdown are already high enough that conversion of organic matter in SW to liquid and gaseous fuels makes economic sense. We should recognize, however, that for the most part, cartels—not free markets—govern fuel prices. Thus, we should not abandon alternative fuels efforts whenever cartels, for their interests, lower prices.

The SWs listed in Table 1, mostly consisting of biomass, now constitute a minor component ($\sim 2.8\%$) of the United States' annual TPES. This wasted material, however, could in the near term become a major ($>25\%$) component comparable to coal and NG, both now at about 23%. Essentially, the United States now consumes about 100 quadrillion Btu, only about 2.8% of which currently come from SW. The other renewables—hydroelectric (2.8%), geothermal (0.35%), wind (0.14%), and solar (0.06%)—have much further to go than SW before becoming a major primary energy source in the United States. Because SWEATT is based on locally available SW, it would also create good nonexportable local industries and jobs while mitigating serious U.S. energy import and waste disposal problems.

An Oak Ridge National Laboratory study[13] estimates the sustainable supply of the first few biomass categories in Table 1 at about 1.4 billion dry tons. The remaining categories should readily bring the total sustainable U.S. SW available to more than two billion dry tons. Assuming a conservative HHV of 7500 Btu/lb, a simple calculation shows that with SWEATT technologies similar to those that are now in place in Japan, U.S. SW contribution to its primary energy supply could reach the 25% level.

ADVANCED THERMAL TECHNOLOGIES

The largest SW-to-energy systems in operation today are direct combustion municipal solid waste (MSW) incinerators[14] with capacities in the range of 1000–3000 tons per day. In such mass burn systems, the organic constituents of the SW are combusted (in a sense, converted) into the gaseous products CO_2 and H_2O. These have no fuel value but can be carriers of the heat of combustion, as in coal and biomass boiler-furnace systems. Along with the flame radiation, these gases transfer heat to pressurized water to produce pressurized steam that drives a steam turbine-driven electric generator. The steam also can serve as a valuable secondary energy supply (SES) to distribute heat for heating buildings, industrial processes, etc. The production and use of steam, along with the steam engine, launched the Industrial Age, and various steam-driven systems have reached a very high level of refinement, including in waste-to-energy systems.[14]

Solid waste to energy by advanced thermal technologies systems do not involve direct combustion and the use of the heat released to raise steam; rather, the SW is first converted to a gaseous or liquid fuel. Then this fuel serves as an SES that can be combusted in efficient internal combustion engines, combustion turbines, or (in the future) in fuel cells, none of which can directly use solid fuels. Over the past century, automotive and aircraft developments have pushed internal combustion engines (ICE) and gas turbines (GT) to very high levels of efficiency. Furthermore, with the use of modern high-temperature GTs in NG-fired combined cycle (NGCC) systems, the heat of the exhaust gases can be used with a heat recovery steam generator (HRSG) to drive a steam turbine. Alternatively, the HRSG can provide steam for combined heat and power (CHP) system that can effectively make even greater use of the original solid fuel energy.

If one considers the United States' heavy dependence on foreign sources of liquid and gaseous fuels, the most challenging technical problem facing us today should be recognized as the development and implementation of efficient ways of converting our abundant domestic solid fuels to more useful liquid and gaseous fuels. In view of the diversity of feedstock represented in municipal or institutional SW, any successes in SWEATT would advance this more general quest. In effect, the United States and the world need an OFC such as is illustrated in Fig. 1. Here, the right block represents a typical gas-fired combined cycle system, whereas the left block represents a conceptual omnivorous conversion system that can convert any organic material into a gaseous or liquid fuel.

GROSS COMPARISONS OF ATT OUTPUTS

First, we will consider the gross nature of the output gas from biomass and cellulosic-type material, the major organic components of most solid-waste streams. Apart from minor constituents such as sulfur and nitrogen, the cellulosic feed types are complex combinations of carbon, hydrogen, and oxygen in combinations such as $(C_6H_{10}O_5)$ that might serve as the representative cellulosic monomer.

Advanced thermal technologies systems may be divided into (1) air-blown partial combustion (ABPC) gasifiers, (2) oxygen-blown partial combustion (OBPC) gasifiers, and (3) pyrolysis (PYRO) systems. The three approaches for converting waste to a gaseous fuel have many technical forms, depending on the detailed arrangements for applying heat to the incoming feed and the source of heat used to change the solid into a gas or liquid.

Let use "producer gas" as a generic name for gases developed by partial combustion of the feedstock with air, as in many traditional ABPC gasifiers that go back to Clayton's coal gasifier of 1694. We will use "syngas" for gases developed by partial combustion of the feedstock with oxygen, as in OBPC gasifiers, which are mainly a development of the 20th century. We will use "pyrogas" for gases developed by anaerobic heating of the feedstock, such as in indirectly heated (PYRO) gasifiers. Our objective is to replace NG that has HHV $\sim$ 1000 Btu/ft^3 = 1 MBtu/ft^3 (here, $M=1000$).

When an ABPC gasifier is used with cellulosic materials (cardboard, paper, wood chips, bagasse, etc.), the HHV of biomass producer gas is very low (100–200 Btu/ft^3) for two reasons: (1) The main products are CO that has a HHV of 322 Btu/ft^3 and CO_2 and H_2O that have zero heating values and (2) the air nitrogen substantially dilutes the output gas.

The syngas obtained from biomass with an OBPC gasifiers is better—$\sim$ 320 Btu/ft^3 because it is not diluted by the atmospheric nitrogen. It is still somewhat lower than the feedstock molecules, however, because of the partial combustion. The oxygen separator is a major capital-cost component of an OBPC gasifier.

With a PYRO system, the original cellulosic polymer is first broken into its monomers, leading to some CO, CO_2, and H_2O along with paraffins (CH_4, C_2H_6, and C_3H_8...), olefins (C_2H_4, C_3H_6,...) and oxygenated hydrocarbons: carbonyls, alcohols, ethers, aldehydes and phenols, and other oxygenated gaseous products. Cellulosic pyrogas can have heating values in the 400 Btu/ft^3 range.

Hydrocarbon plastics such as polyethylene and polyolefins in general are among the most predominant plastics in many SW streams. Thus, one might use (C_2H_4) as representative of the monomers in the plastic component of MSW or refuse-derived fuels (RDF). Polyethylene pyrolysis products include H_2, olefins, paraffins, acetylenes, aromatics (Ar), and polynuclear aromatics (PNA). On a per-unit weight basis, all but H_2 have gross heating values in the range 18–23 MBtu/lb ($M=1000$), similar to oil, whereas H_2 has a gross heating value of 61 MBtu/lb. On a per-unit volume, basis all polyethylene pyrolysis products have gross heating value ranging from 1 to 5 MBtu/ft^3, whereas H_2 is 0.325 MBtu/ft^3 = 325 Btu/ft^3. Natural gas typically is about 1 MBtu/ft^3. Thus, we would expect the pyrogas from polyethylene to have a gross heating value comparable to or greater than that of NG, and much greater than that of cellusosic pyrogas.

In summary, because cellulosic feedstock is already oxygenated as compared with pure hydrocarbon plastics, its pyrogas, syngas, and producer gas will all have considerably lower heating values than the corresponding gases from hydrocarbon feedstock. From the viewpoint of maximizing the HHV of SW-derived gas, PYRO gasification scores better than OBPC gasification, which scores better than ABPC gasification.

ASEM AND PYROLYSIS

Proximate analyses of coal and biomass measured for more than a century provide extensive data on total volatile content. A predictive method for identifying the molecules in these volatiles is still not available, however, despite the fact that such knowledge could provide a fundamental understanding of humankind's oldest technology: the use of fire. For control and application of a pyrolysis system, it would be useful to have at least an engineering-type knowledge of the expected yields of the main products from various feedstock subjected to anaerobic thermal treatment.

In most attempts to find the systematic of pyrolysis yields of organic materials such as coal and biomass, including the initial CCTL studies,[15–20] it has been customary to characterized the feedstock by its atomic ratios $y=$ H/C and $x=$ O/C. In its recent studies,[21–27] the CCTL has found it more advantageous to work with the weight percentages [C], [H], and [O] of the feedstock after correcting to dry, ash, sulfur, and nitrogen free (DASNF) conditions (i.e., pure CHO materials). These were attempts to find some underlying order of pyrolysis yields of any product $C_aH_bO_c$ vs the [O] and [H] of the DASNF feedstock and the temperature (T) and time (t) of exposure. In organizing CCTL pyrolysis data as well as data in the literature, the CCTL has developed an analytical semi-empirical model (ASEM) that has been useful for several applications of pyrolysis.[19–27] Some progress has been made in including the time dimension, but much more work remains. When the time dimension is not a factor, the yields of each product for slow pyrolysis (or fast pyrolysis at a fixed time) are represented by

$$Y(T) = W[L(T:T_o,D)]^p[F(T:T_o,D)]^q \quad (2)$$

where

$$L(T:T_o,D) = 1/[1 + \exp((T_o - T)/D)] \quad (3)$$

and

$$F(T:T_o,D) = 1 - L(T)$$
$$= 1/[1 + \exp((T - T_o)/D)] \quad (4)$$

Here, $L(T)$ is the well-known logistic function that is often called the learning curve. Thus, its complement $F(T)=1-L(T)$ might be called the forgetting curve. For engineering applications, this curve-fitting approach provide a more robust and convenient means for organizing pyrolysis data than traditional methods that use

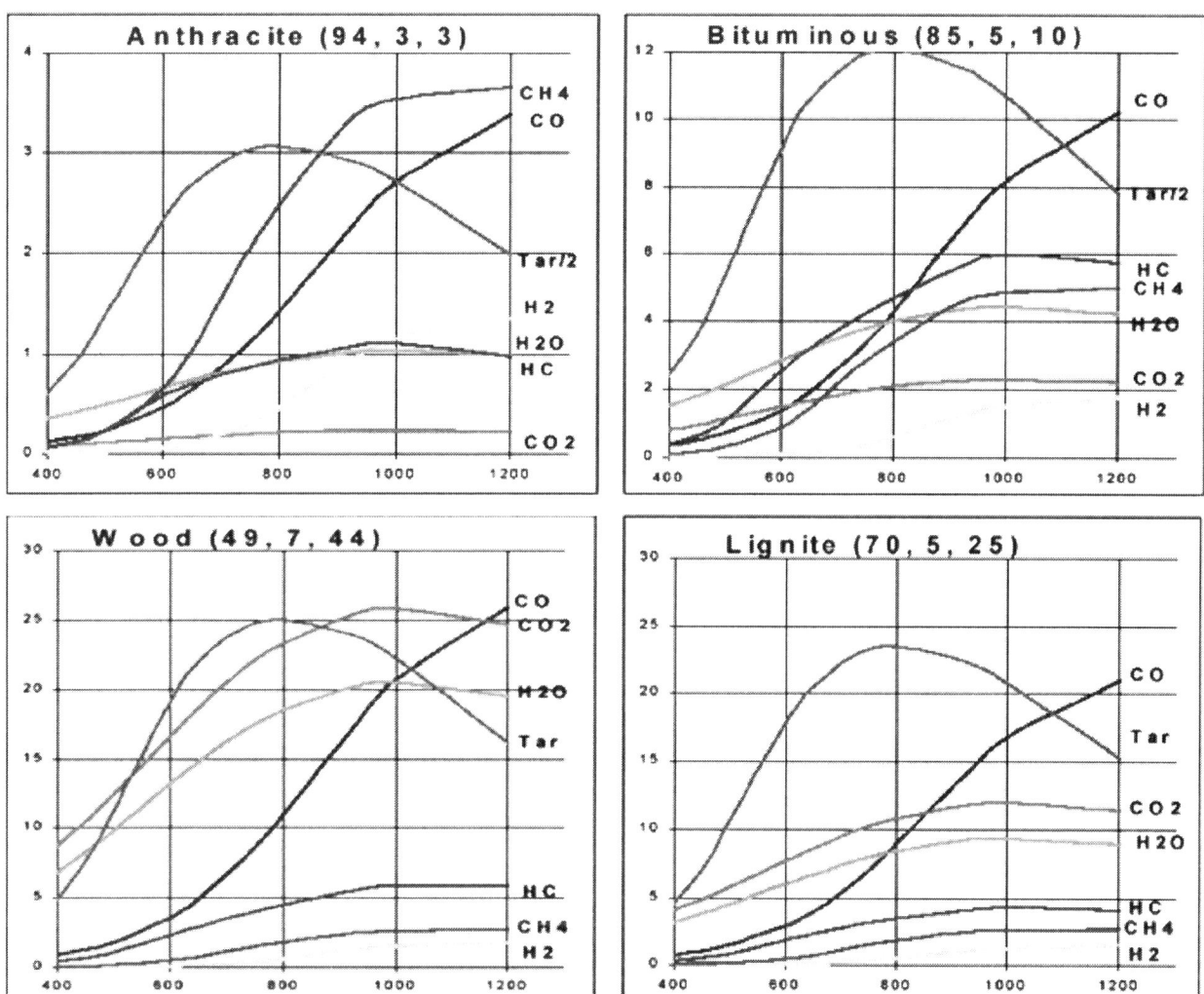

Fig. 4 Wt% yields vs temperature (in °C) from pyrolysis of anthracite, bituminous, lignite, and wood with ([C], [H], [O]) as shown. HC represents C2 and C3 gasses, BTX, phenol and cresol.
Source: Adapted from Ref. [10].

conventional Arrhenius reaction rate formulas.[28] In the ASEM, each product is assigned five parameters (W, T_0, D_0, p, q) to represent its yield-vs-temperature profile. The objective has been to find how these parameters depend on the [H] and [O] of the feedstock and the a, b, c of the $C_aH_bO_c$ product for the data from particular types of pyrolyzers. Studies by Xu and Tomita (XT)[29] that gave data on 15 products from 17 coals at six temperatures have been particularly helpful in revealing trends of the parameters with [O] and [H]. In applying the ASEM to the CCTL data collection, the XT collection, and several other collections, a reasonable working formula was found for the yield of any abc product for any [O], [H] feedstock. It was given by

$$Y(C_aH_bO_c)$$
$$= W_{abc} z^\alpha h^\beta x^\gamma [L(T : T_o, D)]^p [F(T : T_o, D)]^q \quad (5)$$

where $z = [C]/69$, $h = [H]/6$, and $x = [O]/25$, and the parameters α, β, and γ, T_o, D, p, and q were found to have simple relationships to the feedstock and product defining parameters [H], [O], a, b, and c. The final ASEM formulas that fit the data could then be used to extrapolate or interpolate the XT results to any [H], [O] feedstock and temperature. Fig. 4 gives an overview of the interpolated and extrapolated $Y(T)$ outputs for a selection of products for four representative feedstock along nature's coalification path.

Because hundreds or even thousands of organic products of pyrolysis have been identified in the literature, to go much further, some comprehensive organization of these products is needed. Toward this goal, the CCTL has grouped products into the families shown in Table 4, along with the a, b, and c rules that connect these groups. This list can be subdivided into pure hydrocarbons (i.e., C_aH_b) and the oxygenates (C_aH_bO, $C_aH_bO_2$, $C_aH_bO_3$, etc). Isomers (groups with identical a, b, and c) can differ in

Table 4 Organization of functional groups by family

Families	a	b	c
Paraffins	j	$2a+2$	0
Olefins	$j+1$	$2a$	0
Acetylenes	$j+1$	$2a-2$	0
Aromatics	$5+j$	$4+2j$	0
Polynuclear	$6+4j$	$6+2j$	0
Aldehydes	$j+1$	$2a$	1
Carbonyls	j	$2a$	1
Alcohols	j	$2a+2$	1
Ethers	$j+1$	$2a+2$	1
Phenols	$5+j$	$4+2j$	1
Formic acids	j	$2a$	2
Guaiacols	$6+j$	$6+2j$	2
Syringols 1	$7+j$	$8+2j$	3
Syringols 2	$8+j$	$10+2j$	4
Sugars 1	$4+j$	10	5
Sugars 2	$5+j$	$10+2j$	5

a, b, and c are the subscripts in $C_aH_bO_c$, where $j = 1, 2, 3...$.

detailed pyrolysis properties and, hence, parameters. We use $j = 1, 2, 3$, etc. to denote the first, second, third, etc. members of each group or the carbon number (n). In the CCTL's most recent studies[21–27] of specific feedstock pyrolysis, formulas have been proposed and tested for the dependence of the W, T_0, D_0, p, and q parameters on the carbon number of the product within each group. This makes it possible to compact a very large body of data with simple formulas and a table of parameters.

The case of polyethylene is an example of such a study. It is not shown in Fig. 2a, as it is far removed from the coalification curve, having the position $[H] = 14.2$ on the $[O] = 0$ axis. Without oxygen in the feedstock, the pyrolysis products are much fewer, and the ASEM is much simpler to use than with carbohydrates. Thus, only the first five rows of Table 4 are needed to cover the main functional groups involved in organizing the pyrolysis products of polyethylene.

Fig. 5 gives an ASEM-type summary of the product yields vs temperature based on fits to the experimental data of Mastral et al.[30,31] at five temperatures that were constrained to satisfy approximately mass, carbon, and hydrogen balances. When the parameter systematics are identified, the ASEM representation can be used to estimate the pyrolysis product of polyethylene pyrolysis at any intermediate temperature or at reasonable extrapolated temperatures. The experimental data was available only up to 850°C, but the extrapolations to 1000°C were constrained in detail to conform to mass, carbon, hydrogen, and oxygen balance.

Fig. 5 also shows extrapolations to 6000°C that may be of interest if one goes to very high temperatures—by plasma torch heating, for example. Here, we incorporate a conjecture that at the highest temperatures, carbon and hydrogen emerge among the products at the expense of the C1–C2 compounds, as well as aromatics and PNAs components.

Although we have already found that an ASEM can begin to bring some order and overview into pyrolysis yields, clearly, we have a long way to go. When the time dimension is important, the overall search is for a reasonable function of seven variables: [H], [O], a, b, c, T, and t. Einstein's special relativity dealt with only four variables: x, y, z, and t.

SW–IGCC VS NGCC AND ACE

Before World War II almost every town had its own gas works, mainly using coal, as a feedstock. After World War II cheap NG became available and became a major PES for home heating and cooking, as well as for industrial purposes. In the 1980s factory-produced NGCC became available, and NG became a baseload fuel source for many electric utilities, hastening the drawdown of U.S. domestic supplies. In the past four years NG prices have risen to some three to seven times greater than they were when most of these NGCC facilities were built. Thus, pursuing SWEATT is very timely. For most biomass and plastic feedstock, pyrolysis is substantially equivalent to gasification.

The economic feasibility of using a gasifier in front of a gas-fired system can be examined with simple arithmetic and algebra using an analytical cost estimation (ACE) method.[6–10] Analytical cost estimation takes advantage of the almost-linear relationship shown in many detailed cost analyses of the cost of electricity (COE = Y) vs the cost of fuel (COF = X) for many technologies, i.e.,

$$Y(X) = K + XS \qquad (6)$$

Here, Y is given is in cents/kWh, and X is given in dollars/MMBtu. In Eq. 6, S is the slope of the $Y(X)$ line in cents/kWh/$/MMBtu or 10,000 Btu/kWh. S relates to the net plant heat rate (NPHR) via

$$S = \text{NPHR}/10{,}000 \text{ or efficiency via}$$

$$S = 34.12/\text{Eff} \qquad (7)$$

For modern coal plants $S \sim 1$, although supercritical pulverized coal (SCPC) plants are reaching toward 0.9[6–10]. Essentially, the parameter $K = \text{COE}$ if the fuel comes to the utility without cost.

In previous studies,[6–10] we assigned $K_{ng} = 2$ as a reasonable zero fuel cost parameter for, say, a 100 MW NGCC system.[32,33] This low number reflects the low capital costs of the factory-produced gas turbines and

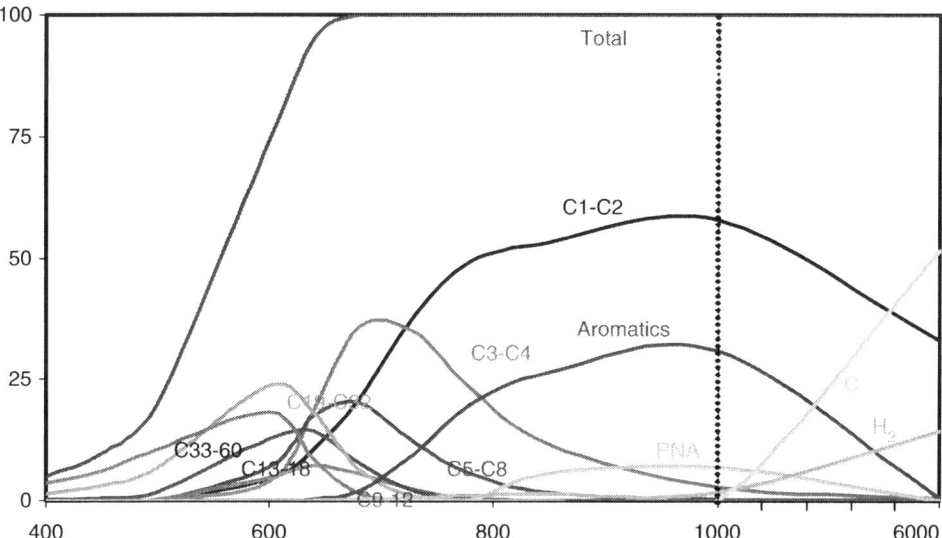

Fig. 5 Yields vs temperature for polyethylene in various hydrocarbon groups.
Source: Adapted from Ref. [23].

steam turbines in NGCC systems reasonable om and en costs. A slope $S_{ng} = 0.7$ is a reasonable assignment reflecting the high efficiency of recent NGCC facilities.

For a SW integrated gasification combined cycle (SW–IGCC) system, K_{sw} generally would be higher than K_{ng} because the capital costs and operating cost must include the gasifier and gas cleanup system. The value of S_{sw} is also higher than S_{ng} because we must first make an SES producer gas, syngas, or pyrogas, which involves some conversion losses. $S_{sw} = 1$ is a reasonable ballpark slope for an up-to-date SW–IGCC system. The X_{sw} for a SW–IGCC system that would compete with a NGCC system at various X_{ng} must satisfy

$$K_{sw} + X_{sw}S_{sw} = K_{ng} + X_{ng}S_{ng} \qquad (8)$$

By algebra, it follows that the SW fuel cost X_{sw} that would enable a SWCC system to deliver electricity at the same cost as a NGCC system paying X_{ng} is given by

$$X_{sw} = (K_{ng} - K_{sw})/S_{sw} + X_{ng}(S_{ng}/S_{sw}) \qquad (9)$$

In what follow, all X numbers are in dollars/MMBtu, and all Y and K numbers are in cents/kWh. Let us use Eq. 9 with $K_{ng} = 2$, $S_{ng} = 0.7$, $S_{sw} = 1$ and $K_{sw} = 4$ as a reasonable ballpark numbers based on several SWCC analyses.[6–10] Then the first term in Eq. 9 is -2. Now when the $X_{ng} = 2$ to generate SWCC electricity at the same cost, the SW provider must deliver the fuel at a negative price—i.e., pay the tipping fee -0.7. If X_{ng} is near 6, however, as it was in 2004 and in the spring and summer of 2006, the SWCC utility could pay up to 2.4 for the SW fuel. If the X_{ng} is 12, the SWCC facility could pay 7.1 to the SW supplier. This X_{sw} price is much higher than that of coal, the delivered price (X_c) of which these days usually is in the 2–3 range. This simple cost comparison is illustrated in Fig. 6, which shows the opportunities for SWCC systems when NG prices are above, say, \$5/MMBtu. The results are slightly less favorable if the K_{sw} were higher—say, $K = 5$.

The conclusion, however, that at high NG prices SWCC electricity becomes competitive with NGCC electricity would be similar. It is conceivable that K_{sw} could be held as low as 2 cents/kWh by retrofitting a NGCC system stranded by high NG prices. In this case, the first term in Eq. 9 vanishes, and the competitive $X_{sw} = (S_{ng}/S_{sw})X_{ng}$. This illustrates the main point that at high NG prices with an ATT system, SW can be a valuable PES. Indeed, this simple algebraic exercise establishes the feasibility of a new paradigm in which SW (mostly biomass but here meaning all solids that are now wasted) becomes a potentially valuable marketable asset.

As described above, the values of K and S are the key factors in determining the COF_{sw} to be used in a SW–IGCC that would be competitive on a COE basis with the COE using a NGCC system at the available COF_{ng}. The ACE method can be extended to the use of SW

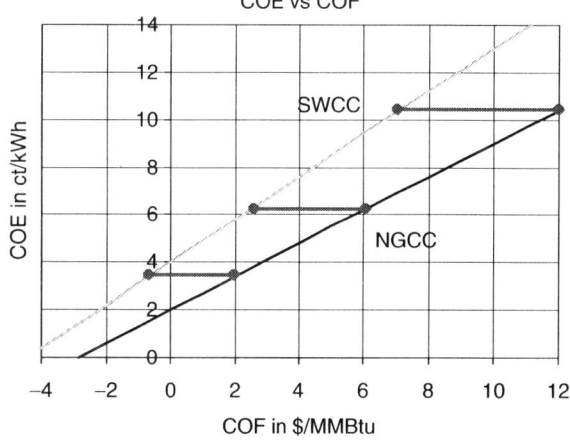

Fig. 6 COE vs COF for SWCC and NGCC at $X_{ng} = 2, 6, 12$.
Source: Adapted from Ref. [10].

or biomass with other technologies if we can identify the K and S for each technology.

The CCTL has applied the ACE method to a large body of COE-vs-COF calculations on biomass use presented in an Antares Group, Inc. report (AGIR)[34] for several technologies. It is reasonable to apply these results to most of the SW listed in Table 1, particularly in small communities that have recycling programs involving residential separation of waste that would minimize the cost of making RDF.

The technologies investigated in the AGIR when 100 tons per day forest thinning was available include a biomass-integrated gasifier combined cycle (B–IGCC) system, a B-IG simple cycle (B-IGSC) system, a BIG internal combustion (B-IGIC) system, a biomass-gasification–coal cofiring BIG–CC of system, a direct co-firing of biomass and coal in a coal-steam boiler BCoSt, a direct use of biomass in a feedwater heat recovery arrangement (FWHR), direct use of biomass in a stoker fire boiler steam turbine (SFST) system. and direct firing in a combined heat and power plant (CHP) with a steam market at $6/MMBtu. For each technology, it was possible to approximately represent the tabulated COE vs COF in Eq. 6 and to evaluate K and S. Then, using reasonable analytical forms for $K(P)$ and $S(P)$, one can make COE comparisons at various power levels and fuel costs. The most interesting result of this ACE digest of the massive AGIR tables is the fact that with slight extrapolations to higher power levels, the results in several important cases were opposite those for lower power levels. Because the assumed 100 tons per day of forest residue was rather low for many areas, these changes in conclusions were important.

Thus far, we have focused on the competition between NG-fueled technologies and SW-fueled technologies. Competition of SW-generated electricity with coal-steam-generated electricity appears to be a bigger problem. If one includes the more expensive scrubber cost in the Ks and externality cost in the coal X's,[35] however, the SW–IGCC route should fare well. Coal burning is a major issue in many communities, yet when one projects technology directions around the globe, it is clear that the Gasification Age is returning.[36]

COMPONENTS OF ACE (CACE)

The CCTL is in the process of examining other economic COE-vs-COF analyses to quantify a more detailed formulation of ACE in which K is broken into components $K = K_c + K_{om} + K_{en}$, where c stands for capital costs, om for operating and maintenance costs, and en for environmental costs. At this time, establishing the magnitudes of these components for various technologies and power levels is at the cutting edge of utility economic analyses, and there are large disagreements—particularly on K_{en}. Future fuel costs (X) in the COE term XS in many important cases, however, probably represents the greatest source of uncertainty. Accordingly, it is foolish to belabor estimating K factors with great precision when X can range over wide limits. In this component form of ACE, Eq. 6 is replaced by

$$Y = K_{cr}(P_r/P)^\alpha + K_{omr}(P_r/P)^\beta + K_{er}(P_r/P)^\gamma + XS_r(P_r/P)^\delta \qquad (10)$$

where K_{cr}, K_{omr}, K_{er}, and S_r are established on the basis of a detailed analysis at a reference power level P_r, and α, β, γ, and δ are scaling parameters intended to reflect the tendency of per-energy-unit cost to go down as the power goes up (economy of scale). These scaling parameters might be established on the basis of a broad set of studies for each technology.

Table 5 lists CACE parameters extracted from a detailed analysis, "Options for Meeting the Electrical Supply Needs of Gainesville," prepared by ICF Consulting.[35] Here, the final cost of electricity is given in 2003 cents/kWh. The third case, NGCCc, has been added to allow for the contingency that with new offshore drilling or the increased development of liquefied natural gas

Table 5 CACE applied to the ICF consulting study

Tech	Pr	K_0	K_{om}	K_{en}	Xr($/MMBtu)	S_o	COEr(ct/kWh)
NGCCa	220	0.598	0.234	−0.170	6.10	0.68	4.81
NGCCb	220	0.598	0.234	−0.170	11.34	0.68	8.37
NGCCc	220	0.598	0.234	−0.170	4.00	0.68	3.38
SCPC	800	1.491	0.299	0.714	1.91	0.93	4.28
CFB-CB	220	2.531	0.261	0.618	1.41	1.05	4.89
CFB-B	75	2.845	0.261	0.039	1.67	1.39	5.47
IGCC	220	2.2	0.196	0.407	1.41	0.86	4.02

(LNG) importing capabilities, NG might return to the $4/MMBtu level of 2003.

The final column shows that at the reference power levels without the NGCCs case, the IGCC scores the lowest COE, as the ICF report (ICFR) concluded. The value of the ACE analysis is that with a bit of algebra, anyone can easily consider other fuel cost projections and other power levels (with assumed values of α, β, γ, and δ). Based on previous CCTL exploratory work and economy-of-scale investigations, the author estimates that for costly field-erected facilities, $\alpha = \beta = 0.3$ are reasonable choices. With factory fabrication of jet and steam turbines, however, these parameters may not follow the usual economy-of-scale pattern and be somewhat smaller in magnitude. Assigning a value for γ is a wide-open question because environmental costs and the methods of incorporating them into the cost of electricity are highly debated issues.[36] Reasonable values for δ are also somewhat difficult to find. For NGCCs, the author tentatively assigns close to zero or a very small value (~ 0.1), perhaps because the development of highly efficient aeroderivative turbines has proceeded on a wide range of power levels.

In concluding this section, it should be clear that the age of making gas has returned and that the time has come to develop a national gas strategy[37] to facilitate the earliest implementation of new gasification systems.

BIOFUELS

Biochemical Conversion by Fermentation

Fermentation, a major form of biochemical conversion, uses bacteria in the presence of oxygen to break down biodegradable organic material into liquid fuels such as ethanol. The ethanol thrust is an extension of the commercial beer, wine, and alcohol industries' processing of sugar- and starch-based feedstock such as corn and sugarcane. An important development,[38] however, has extended these capabilities to cellulose, greatly expanding the mass of biomass that can be transformed by the fermentation route. Ethanol lends itself to conventional automotive liquid fuel storage, although at only 0.6 times the energy density of diesel or gasoline. As Brazil has demonstrated, an automotive fleet can be largely fueled by ethanol. An aircraft fleet will require a higher-energy-density fuel, however.

Biochemical Conversion by Anaerobic Digestion

In anaerobic digestion, bacteria convert biodegradable waste to methane gas, a technology that is an extension of the phenomenon of flatulence of animals.[39] Landfill gas is a product of anaerobic digestion. Although a good-quality gas can be achieved by anaerobic digestion, escaping methane can be a problem, because one molecule of methane is some 20 times more damaging as a greenhouse gas than carbon.

At this time, neither the aerobic nor the anaerobic biochemical conversion is capable of processing lignin (approximately 25 wt% of plant matter) or any plastics except biodegradable plastics. The main disadvantage of bioconversion is that the reaction times are weeks, whereas thermochemical conversion can take place in minutes. Thus, the volume required for large-scale biochemical processing is very much larger than for ATT processing. The fact that the residue from biochemical conversion can be a good feedstock for thermochemical processes[1] suggests that bioconversion and thermal conversion should work together. Estimates of SW from alcohol and methane production are included in Table 1.

Bio-Oils

The esters of vegetable oils are renewable alternative fuels that potentially can serve as direct replacements for diesel fuels in compressed ignition engines (CIE).[39] Oils from soybeans, sunflower seeds, safflower seeds, cotton seeds, peanuts, and rapeseeds, as well as used oil from restaurants, are under considerable investigation as replacements for diesel. Waste from bio-oil programs have been included in Table 1 because only the seeds of the plants are used for bio-oil crops; the rest of the plant becomes SW amenable to serving as an input of a SWEATT program.

RECYCLING AND SWEATT

Although our confrontational society has a tendency to view waste-to-energy as a threat to recycling programs, the opposite may be true. Recycling programs in a community can serve to sort the various components of municipal or institutional SW into categories that lend themselves to maximizing the return on these components. If, for example, newspaper at a given time has no recycling market but must be disposed of at a cost, it could be used as high-energy dry feedstock for ATTs. The same is true for plastics recycling. Thus, the marketplace would be decisive as to whether to recycle via the materials route or the energy route. A recycling community should be able to go the SW–IGCC route with less capital costs than one that does not have waste separation at the source.

The advantages of a biomass alliance with natural gas (BANG) have been described previously.[6–10] Gasification systems that mainly use cellusosic (biomass) inputs produce a low- or medium-heating-value fuel that will result in the derating of a NG designed turbine-generator. By coutilizing the biomass pyrogas with NG, one can ensure that the input energy requirement matches the output needs at least until the maximum rating of the generator is required. At that point, the firing could be entirely on NG. In a solid waste alliance with natural gas (SWANG), an additional option becomes available when the SW comes from a recycling community. Then the utility might prepare and store high-energy plastics for increased use during times of high electricity demand as a means of following peak loads without calling on the full use of NG.

ATT FOR LIQUID FUEL PRODUCTION

Pyrolysis/gasification technologies followed by gas cleanup can greatly reduce emissions of pollutants such as NO_x and SO_x, as well as toxics such as mercury and arsenic. Advanced thermal technologies can treat nearly the entire organic fraction of MSW and, in general, can treat a more heterogeneous feedstock, including high-energy-content plastics.[11] Although this paper focuses on gaseous fuel generation, ATTs for liquid fuel (condensable gas) production are closely related. Considerable research and development work is under way on distillation technologies to refine such liquid fuels for transportation applications, adding a major driver for the ATT route.

It should be noted that Table 1 does not list oil shale or tar sands in the United States that could substantially increase the available SW tonnage that could be used to address our need for transportation fuels. A 2005 Rand study[40] shows that with in-situ thermal treatment, domestic oil shales could substantially lower our oil import problem. Another route would be to convert our coal to liquid fuels, as South Africa has done for many years. A third route would involve the use of methane hydrates to produce methane for use in NG-fueled vehicles.

CONCLUSIONS

The main conclusion of this paper is that the United States has very large sustainable supplies of now-wasted solids that have an annual energy potential comparable to that of our current use of coal and also of NG. With ATT, this SW could, in the near term, multiply its contribution to our national energy supply by a factor of about 10. Robust technology that can handle MSW or RDF also should be able to handle almost any of the categories listed in Table 1. Agricultural and forestry residues are two of the major SW supply components in this list, and many of the other materials are greenhouse-neutral plant matter.

By utilizing thermochemical processes to convert the lignin and plastic content of SW and biochemical residues, we could get much closer to goal of zero SW. Solid waste to energy by advanced thermal technologies also would further reduce the final volume of the waste, and practically all contaminants can be destroyed by high temperatures. Thus, cooperation between biochemical and thermochemical programs would clearly be in the national interest.

Japan, a country with an outstanding sustainability record, is the global leader in SWEATT. With more than 60 pyrolysis and thermal gasification systems now in operation, Japan has established the technical and environmental feasibility of these systems. This should allay the concerns of environmentalists and risk-averse utility decision makers in the United States.

In final summary, our main conclusions are

- The United States is excessively reliant on imported oil (60%) for its liquid fuels.
- The United States is increasingly reliant on imported NG fuel (now >15%).
- The United States is well endowed with solid fuels: wasted solids, coal, and oil shale.
- The organic matter in SW can be converted to useful gaseous or liquid fuels.
- In most cases, ATT provides the fastest and most efficient conversion method.
- Co-use of domestic fuels such as SW and NG can overcome some problems.
- Solid waste to energy by advanced thermal technologies generates lower emissions than combustion waste-to-energy systems.
- Thermal conversion of solid fuels to gaseous and liquids fuels has a long history.
- Utilities are experienced with high-temperature processes in the production of steam.
- Advanced thermal technologies (PABC, POBC, and PYRO) are extensions of high-temperature steam making.
- Conversion to gaseous fuels is essential for SW powering of fuel cells.
- There are many environmental benefits attendant to SWEATT.
- Many areas of engineering research will be needed to optimize SWEATT.
- Cooperation of stakeholders would accelerate the implementation of ATT.
- Conservation and SWEATT together is the fastest realistic path to zero waste.

ACKNOWLEDGMENTS

The author would like to thank his many coauthors in this 30-year quest for alternatives to oil; their contributions are incorporated into this article.

Glossary

ABPC: Air Blown Partial Combustion
ACE: Analytical Cost Estimation
AGIR: Antares Group Inc. Report
Ar: Aromatics
ASEM: Analytical Semiempirical Model
ATT: Advanced Thermal Technologies
BTU: British Thermal Units
CACE: Component Analytic Cost Estimation
CCTL: Clean Combustion Technologies Laboratory
CHP: Combined Heat And Power
CIE: Compressed Ignition Engines
DANSF: Dry Ash, Nitrogen And Sulfur Free
EU: European Union
FC: Fixed Carbon
GT: Gas Turbines
HHV: Higher Heating Values
HRSG: Heat Recovery Steam Generator
ICE: Internal Combustion Engines
IGCC: Integrated Gasifier Combined Cycle
MSW: Municipal Solid Waste
NGCC: Natural Gas-Fired Combined Cycle
NPHR: Net Plant Heat Rate
OBPC: Oxygen Blown Partial Combustion
OFC: Omnivorous Feedstock Converter
PES: Primary Energy Supplies
PNA: Polynuclear Aromatics
PYRO: Pyrolysis Systems
QES: Quaternary Energy Supply
quads: Quadrillion BTUs
RDF: Refuse Derived Fuels
SES: Secondary Energy Supplies
SW: Solid Waste
SWANG: Solid Waste Alliance with Natural Gas
SWEATT: Solid Waste To Energy By Advanced Thermal Technologiess
TES: Tertiary Energy Supply
TPES: Total Primary Energy Supply
VT: Volatiles
WEC: Waste to Energy Conversion

REFERENCES

1. Green, A., Ed. *Alternative to Oil, Burning Coal with Gas*, University Presses of Florida: Gainesville, FL, 1981.
2. Green, A. et al. Coal-water-gas: an all American fuel for oil boilers. Proceedings of the Eleventh International Conference on Slurry Technology, Hilton Head, SC, 1986.
3. Green, A. Ed. Solid fuel conversion for the transportation sector, In FACT. Vol. 12. Proceedings of Special Session at International Joint Power Generation Conference, San Diego, CA; ASME: New York, 1991.
4. Green, A. A Green Alliance of Biomass and Coal (GABC). Appendix F. *National Coal Council Report; May 2002*, Proceedings of the 27th Clearwater Conference, March 2003.
5. Green, A.; Feng, J. A Green Alliance of Biomass and Natural Gas for a Utility Services Total Emission Reduction (GANGBUSTER). Final Report to School of Natural Resources and the Environment, 2003.
6. Green, A.; Hughes, E.; Kandiyoti, R. Proceedings of the first international conference on co-utilization of domestic fuels. Int. J. Power Energy Syst. **2004**, *24* (3).
7. Green, A.; Smith, W.; Hermansen-Baez, A. et al. Multidisciplinary Academic Demonstration of a Biomass Alliance with Natural Gas (MADBANG). Proceedings of the International Conference on Engineering Education, Gainesville, FL, October 2004.
8. Green, A.; Swansong, G.; Najafi, F. Co-utilization of domestic fuels biomass gas/natural gas. In *GT2004-54194*; IGTI: Vienna, Austria, 2004; [June 14–17].
9. Green, A.; Feng, J. Assessment of technologies for biomass conversion to electricity at the wildland urban interface. Proceedings of ASME Turbo Expo, Reno–Tahoe, NV, 2005.
10. Green, A.; Bell, S. Pyrolysis in waste to energy conversion. Proceedings of the NAWTEC 14 Conference, Tampa, FL, May 2006.
11. California Integrated Waste Management Board. Conversion Technologies Report to the Legislature [draft], Available at: www.ciwmb.ca.gov/Organics/Conversion/Events.
12. Energy Information Administration. *January 2006 Monthly Energy Review*, DOE/EIA-0035(2006/01); Available at: www.eia.doe.gov/emeu/mer/contents.html (accessed on January 2006).
13. Perlack, R.; Stokes, B.; Erbach, D. Biomass as Feedstock for a Bioenergy and Bioproducts Industry: The Technical Feasibility of a Billion-Ton Annual Supply, ORNL/TM-2005/66. Oak Ridge National Laboratory: Oak Ridge, TN, 2005.
14. Stultz, S., Kitto, J., Eds. *Steam* 40th Ed.; Babcock and Wilcox: Barberton, OH, 1992.
15. Green, A.; Peres, S.; Mullin, J. et al. Co-gasification of domestic fuels. Proceedings of the IJPGC; ASME: Minneapolis, MN, 1996.
16. Green, A.; Zanardi, M.; Mullin, J. Biomass Bioenergy **1997**, *13*, 15–24.
17. Green, A.; Zanardi, M. Int. J. Quantum Chem. **1998**, *66*, 219–227.
18. Green, A.; Mullin, J. J. Eng. Gas Turbines Power **1999**, *121*, 1–7.
19. Green, A.; Mullin, J.; Schaefer; G. et al. Life Support Applications of TCM-FC Technology. Presented at: 31st ICES Conference, Orlando, FL, July 2001.
20. Green, A.; Venkatachalam, P.; Sankar, M.S. Feedstock Blending of Domestic Fuels in Gasifier/Liquifiers. Presented at: TURBO EXPO 2001, Amsterdam, 2001.
21. Green, A.; Chaube, R. Pyrolysis Systematics for Co-utilization Applications, Presented at: TURBO EXPO 2003, Atlanta, GA, June 2003.
22. Green, A.; Chaube, R.A. Int. J. Power Energy Syst. **2004**, *24* (3), 215–223.
23. Green, A.; Sadrameli, S.M. Analytical representations of experimental polyethylene pyrolysis yields. J. Anal. Appl. Pyrolysis **2004**, *72*, 329–335.
24. Sadrameli, S.; Green, A. J. Anal. Appl. Pyrolysis **2005**, *73*, 305–313.
25. Green, A.; Feng, J. Systematics of corn stover pyrolysis yields and comparisons of analytical and kinetic representations. J. Anal. Appl. Pyrolysis **2006**, *76*, 60–69.

26. Feng, J.; Green, A. Peat Pyrolysis and the Analytical Semi-Empirical Model. J. Energy Sources, **2006**. In press.
27. Feng, J.; YuHong, Q.; Green, A. Analytical model of corn cob pyroprobe-FTIR data. Biomass Bioenergy **2006**, *20*, 486–492.
28. Gaur, S.; Reed, T. *Thermal Data for Natural and Synthetic Fuels*; Marcel Dekker: New York, 1998.
29. Xu, W.C.; Tomita, A. Effects of coal type on the flash pyrolysis of various coals. Fuel **1987**, *66*, 627–631. 632–636.
30. Mastral, F.; Esperanza, J.; Garcia, E. et al. J. Anal. Appl. Pyrolysis **2002**, *63*, 1–15.
31. Mastral, F.; Esperanza, J.; Berruco, E. et al. J. Anal. Appl. Pyrolysis **2003**, *70*, 1–17.
32. Liscinsky, D.; Robson, R.; Foyt, A. et al. Advanced technology biomass-fueled combined cycle. Proceedings of ASME Turbo Expo, ASME: Atlanta, GA, 2003.
33. Phillips, B.; Hassett, S. Technical and Economic Evaluation of a 79 MWe (Emery) Biomass, Presented at: IGCC Gasification Technologies Conference, San Francisco, CA, 2003.
34. Antares Group, Inc. *Assessment of Power Production at Rural Utilities Using Forest Thinnings and Commercially Available Biomass Power Technologies*; Antares Group, Inc.: Landover, MD, 2003.
35. ICF Consulting Options for Meeting the Electrical Supply Needs of Gainesville report submitted March 6, 2006.
36. Roth, I.F.; Ambs, L.L. Incorporation externalities into a full cost approach to electric power generation life-cycle costing. Energy **2004**, *29* (12–15), 2125–2144.
37. Rosenberg, W.; Walker, M.; Alpern, D. *National Gas Strategy*; Harvard University Press: Cambridge, MA, 2005.
38. Ingram, L.; Alterthum, F.; Conway, T. Ethanol production by *Escherichia coli* strains co-expressing zymomonas pdc and adh genes Us Patent No. 5,000,000, 1991.
39. Wilkie, A.C.; Smith, P.H.; Bordeaux, F.M. An economical bioreactor for evaluating biogas potential of particulate biomass. Bioresour. Technol. **2004**, *92* (1), 103–109.
40. Bartis, J.; LaTourrette, T. et al. *Oil Shale Development in the United States: Prospects and Policy Issues*; National Energy Technology Laboratory: Pittsburgh, PA, 2005.

Space Heating

James P. Riordan
Energy Department, DMJM+Harris/AECOM, New York, U.S.A.

Abstract
Nothing seems more welcoming than the coziness of a blazing fireplace on a cold winter night. The technological developments leading up to this inviting scene are the result of over 2000 years of developments. The technological advances within the past 100 years and in our future will shape new industries, create better energy awareness, and mark changes in our lifestyles.

INTRODUCTION

Since early human history, mankind has used fire for protection, lighting, cooking, and generating warmth in cold climates. Fire, utilizing many miscellaneous fuels, is still the main heat source used for space heating worldwide. Over time, the fuels and methodology have varied considerably. Many sources of fuel are available worldwide. Some of the sources that are utilized are wood, coal, oil products, natural gas, geothermal, biomass, peat, solar, wind, and hydro-produced electric energy. The recent use of efficient building design using high R-value insulation and updated technologies has allowed us to redefine industry standards for space heating.

This entry will look at many historical methodologies used for space heating and the improvements made to the processes along the way. Next, it will investigate the current technologies in use, and how they relate to existing energy conservation and indoor air-quality standards. Last, the need for more energy-efficient equipment and better public energy awareness will be addressed.

HISTORICAL INFLUENCES

Thousands of years ago, as mankind explored and populated the continents, heat from fire became an invaluable commodity. It allowed communities to cook, protect themselves, and provide light, and it was also used for the processing of metals, leading to the evolution of tools. As technology advanced, these tools were utilized to build improved structures for habitation and public use. Early habitats used fire pits to radiate heat by a combination of convection, conduction, and radiation in an attempt to keep the inhabitants relatively comfortable. The problems incurred with this type of system were the labor involved in supplying the wood fuel and how to get rid of the smoke. Many early huts were built with holes in the roofs to allow the smoke out, but this also allowed any inclement weather to come in. The dangers involved with this method are not only the risk of fire due to the building materials, but also the danger of lingering smoke that could cause suffocation.

The Romans developed a method for heating buildings that is evident from excavations of ancient villas and bathhouses. This Roman system of generating radiant heat is called a hypocaust. Modern variations of this radiant heat-type system are still in use today. One of these villas, at Newport on the Isle of Wight in Great Britain, is a good example of such a system. The structure's finished floor is supported by numerous tile columns. Under the finished floor is a chamber or crawl space. The crawl space is usually no more than 2 ft high. Apparently, this was found to be the most efficient design. At one end of the building is a furnace or a fire pit where wood is burned, generating hot air. The pit is usually large enough to allow for a reasonably sized fire while also allowing airflow around the fire and into the crawl-space chamber. The walls of the structure include flues, which are scattered around the structure and made of box-shaped tiles stacked one upon another. These flues ascend from the crawl space through the walls, allowing the hot exhaust gases to escape above the rooftop level. This rush of hot air through the crawl space and up the stacks generates what we know today as a stack draft. This negative pressure pulls more heated air from the furnace through the crawl space, heating the floor. The walls are also heated by these multiple hot stacks. Naturally, the floors and walls closest to the furnace

Keywords: ASHRAE: American Society of Heating, Refrigerating and Air-Conditioning Engineers; ASME: American Society of Mechanical Engineers; BTU: British thermal unit—energy required to raise 1 lb of water 1°F; Conduction: The transmission of energy through a nonmoving medium; Convection: Heat transfer through liquid or gas by circulation; Energy Policy Act 1992: Federal legislation adapted enacting many energy standards; Geothermal: Relating to the internal heat of the earth; Heating Degree Day: Traditional procedure for calculating fuel consumption; Radiation: Energy radiated or transmitted by waves or rays; R-Factor: Insulating factor=h-ft^2-F/BTU; Space Heating: The study and science of heating habitable structures; Stack Draft: Negative stack pressure caused by evacuation of hot gas; U-Factor: Inverse of R-factor (1/R); U=BTU/h-ft^2-F.

Fig. 1 Roman bath (Courtesy Bath and North East Somerset Council, England).
Source: From http://www.romanbaths.co.uk/

are the warmest; those farther away do not receive as much heat. This was an early form of radiant heating.[1]

In today's culture, bathing is a private activity. In Roman culture, communal bathing and public bathhouses were widespread leisure activities and fundamental staples of Romans' lives. Many bathhouses were privately owned (called balneae), but there were also many public bathhouses (called thermae) (see Fig. 1). Public bathhouses were accessible by all levels of society—by both men and women—for a trivial fee. The remains of the Stabian Baths in Pompeii are a good example of a public bathhouse. Many bathhouses were lavish in their construction and decoration. Usually, they allowed woman to use them in the morning hours, while men were allowed entry from about 2:00 P.M. until sunset. These bathhouses also provided facilities for sports and recreation, and acted as a community center for cultural and intellectual interaction. Many included gardens, libraries, entertainment areas, and lecture halls.

A typical visit usually started in the warm room (tepidarium), which had heated walls and floors. From there, the client normally would proceed to the hot bath (caldarium) for a leisurely bath. This hot bath normally would be located closest to the furnace. After this, the client would proceed to the cold room (frigidarium) and soak in a cold pool of water. Many bathhouses offered steam rooms and dry heat similar to a sauna (laconicum). Needless to say, these were widely used public buildings handling large occupancies.[2]

The thermal demand for a private-residence bathhouse would be moderate. The thermal demand for a public bathhouse must have been tremendous. The furnace fires would require regular attention from many servants, and obviously, this was extremely labor intensive. The furnace(s) would require frequent addition of quantities of wood and removal of the hot ashes. The wood burned would be mainly small branches up to about 3 in. in diameter and up to 3 ft long. Logs were not effective, as they burned too slowly and blocked the airflow into the crawl-space chamber. The height of the flame was also restricted to about half the height of the chamber to allow for proper airflow.[3]

HEARTHS AND FIREPLACES

During the Middle Ages, only slight improvements were made in space heating. The open hearth preceded the fireplace, and saw wide use in Saxon times and later. The hearth was usually bordered by stone or tile, and smoke rose through a rooftop louver. A "couvre-feu," or fire cover, made of tile or china was placed on the hearth at night to reduce the fire risk. The disadvantages of an open hearth were that the warmest locations in the building were closest to the fire. Smoke also became a fire and health issue. The fireplace, which provided heat both directly and by radiation, saw wide use later during this period. The stones in the back of the hearth and the walls opposite the

Fig. 2 Uncovered medieval fireplace (Courtesy of Chichester District Council, England).
Source: From http://www.chichestertoday.co.uk/viewarticle2....

fireplace were usually made extra thick to radiate stored heat after the fire had burned low.

When multiple-story structures began utilizing fireplaces, the danger of a fire became critical, as most roofs were made of wood. Hearths and fireplaces began to be located close to a perimeter wall where the exhaust smoke could be funneled into the wall, allowing it to vent properly.[4] Near the end of the 12th century, fireplaces began to be constructed with projecting hoods to control the smoke, allowing for a shallow recess. The flues were routed through the walls to discharge above the rooftop level (see Fig. 2).

During this period, space heating was still dependent on the capacity of the household and its ability to keep the fires fed with fuel. Many ornate beds of this age included feather mattresses, fur coverlets, and curtains to reduce drafts. Bed curtains were commonly used into the early 20th century to reduce drafts and maintain personal warmth.

Three major improvements to fireplaces occurred in the 1600s and 1700s. In 1678, Prince Rupert, Duke of Bavaria, invented a system of hinged baffles in an attempt to extract more heat from the fire; it had limited success and was still prone to smoking. Later, Benjamin Franklin produced a design (the Franklin stove; see Fig. 3), which radiated heat but required the exhaust gases to travel up and over a hollow riser extracting convection-heated air. This was somewhat more successful and was later improved upon by David Rittenhouse, one of Franklin's contemporaries. In the late 1790s, Count Rumford came up with a design incorporating a continuously variable flue damper and a sloping fireback. This sloping fireback provides a larger surface area, allowing greater heat radiation. This design is the basis of many fireplaces used today.[5]

A very efficient fireplace design used in many northern climates is the Finnish Masonry Stove (see Fig. 4). The Finnish stove is a more complicated design using a long

Fig. 3 Franklin stove (Courtesy of The Franklin Institute).
Source: From http://en.wikipedia.org/wiki/Franklin_stove

Fig. 4 Finnish stove (Courtesy of Peter Moore Masonry, Inc.). Source: From http://www.vtbrickoven.com/masonry/masonry.html

horizontal flue between the stove section and the flue. Convection-heated air is directed around the masonry sections and into nearby rooms. It is quoted that these stoves capture up to 85% of the BTUs burned in the combustion process for radiated heat. Conventional wood stoves capture about 45%–60%.[6]

Odd fact: up until 1900, the Danish war office was heated by a single furnace in the basement of the building. The furnace was used to heat cannonballs. These cannonballs were carried to every office and placed red hot into a metal bin in the fireplace alcoves to radiate heat.[7]

PRESSURE VESSEL DEVELOPMENT

The capability to heat water and generate steam has been known since ancient times. Its viable use has been seen only since the Industrial Revolution. In 1678, Thomas Savery patented the Savery Pump, which used steam to draw water out of mines.[8] It worked only to a depth of 80 ft, and boiler explosions were common due to lack of advancement in pressure-vessel design. Thomas Newcomen vastly improved upon this design with his Newcomen Atmospheric Engine—which, although inefficient, was used for about 60 years.[9] James Watt and his partner, Matthew Boulton, later improved on the process with the use of a separate condenser. The work done by the Boulton-Watt engines was not by steam pushing a piston but by condensing steam creating a vacuum. They later improved the engine by utilizing the steam to push a piston.[10] The boilers used for this were atmospheric pressure boilers. The technology developments required to drive these engines also led to the development of improved boiler designs.

Using boilers as a heat source is commonplace today. In the 18th and early 19th centuries, boilers were temperamental, and catastrophic explosions frequently occurred. The potential applications for steam power seemed unlimited, but the method of safely harnessing and controlling this energy was not sufficiently developed. Early steamships and locomotives were plagued with pressure limitations to maintain safe operation. In 1865, the Mississippi River steamship Sultana exploded, taking over 1200 lives.[11] At this historical point, there were no boiler codes in use anywhere in the United States or elsewhere. In March 1905, in Brockton, Massachusetts, the Brockton shoe-factory boiler exploded. This catastrophic explosion resulted in 58 deaths and 117 injuries. Massachusetts immediately set about to form a Board of Boiler Rules whose charge was to write a boiler law for the state. This state law was published in 1908.

By 1911, a total of 10 states and 19 metropolitan areas had enacted their own separate boiler laws. The requirements of these individual laws varied greatly. Essentially, a boiler built in one state could not be used in another state due to the differences in the individual codes. In 1911, the American Society of Mechanical Engineers (ASME) appointed a committee to come up with a standard boiler code that could be accepted by all states. The intent of this committee in establishing this code was to protect the public and also to reduce the clutter of miscellaneous state regulations. This was done by bringing boiler and pressure-vessel regulations under one national code. The first ASME Boiler and Pressure Codes were published in 1914. Although accepted as a national code within the United States, it is commonly used by many countries as an international standard.[12]

Since 1914, boiler development has consisted primarily of fire-tube and water-tube boilers. Most early boilers, including those that powered steam locomotives, were fire-tube boilers. A fire-tube boiler is essentially a tank chamber with internal tubes that are immersed in water (see Fig. 5). Hot combustion gases flow through these tubes, boiling the surrounding water. The produced steam accumulates above the water level and is used for the

Space Heating

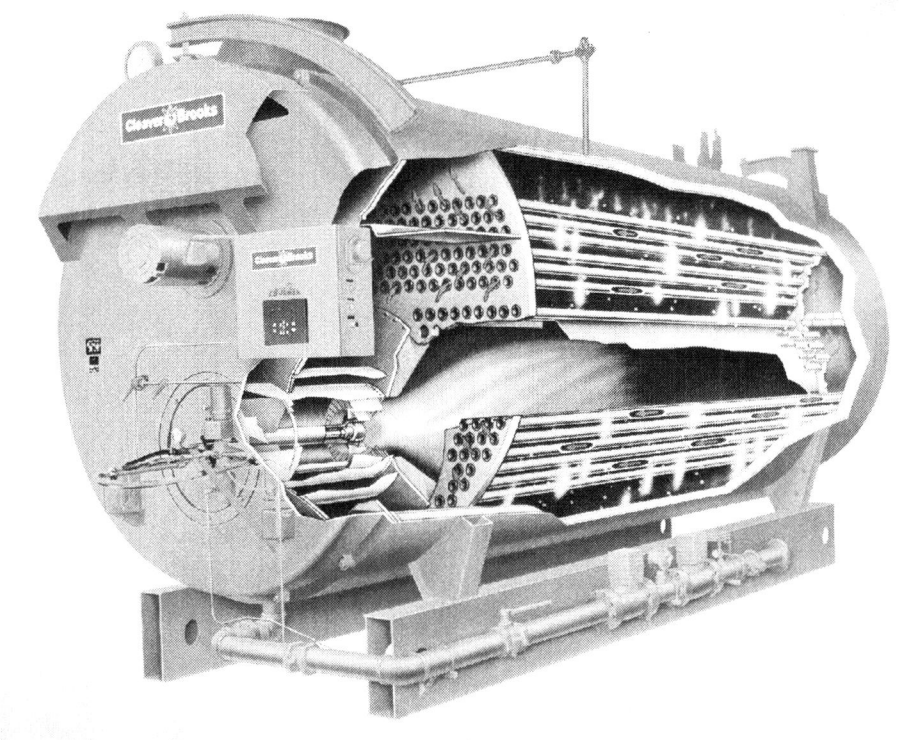

Fig. 5 Fire-tube boiler (Courtesy of Cleaver-Brooks).
Source: From http://www.elizabethtowngas.com/BusinessCustomers.CommercialBoilers.aspx

heating process. Steam traps in the system allow the steam to condense by losing heat to the surrounding air via radiators. Then this condensate is gravity fed or pumped back to the boiler. A water-tube boiler is essentially a tank with internal tubes (see Fig. 6). The tubes contain water, and the hot exhaust gases are passed around the tubes, producing steam. The steam is collected in a steam drum at the top of the boiler for process heat. The condensate is returned the same way as in the fire-tube system.[13]

HEATING SYSTEMS DESCRIPTION

Steam and hot water (hydronic) heating systems utilize different piping arrangements. Steam systems allow the steam to retain high heating values per unit of mass. This is why many large complexes—such as universities, hospitals, and manufacturing facilities—have a central heating facility that utilizes a steam-type system. This system configuration is more complicated, but it alleviates the need for individual building boilers. Another system commonly seen in larger applications is a high-temperature hot water system. This type of system heats water well above its normal boiling point but maintains a liquid state by sustaining high system pressure. Standard hydronic or hot water heating systems are much less complicated and tend to be more centralized. Due to the relatively low temperatures and pressures involved, hydronic systems are much less costly to maintain and are more commonly used in small-building and residential applications.

FUELS

The fuel used for boiler applications has varied widely. Early boilers and industrial drive engines consumed primarily coal and wood. During the 18th century through the early 20th century, these fuels were plentiful and a relatively inexpensive fuel source. In the late 19th century and early 20th century, with the oncoming of improved transportation, petroleum fuels and natural gas took over as the fuels of choice. They were inexpensive, readily available, and easy to store and transport. Wood and coal are still used to this day in many areas of the country. Many metropolitan areas still have remnants of old underground gas systems underneath their streets. Although no longer in use, some of these conduits were made of wood and were used to supply different forms of gas for street lighting and other applications. The recent limitations and petroleum price increases imposed on this country have caused a sudden emergence of the use of high-efficiency wood and coal stoves in residential heating applications. Coal is still used in many large industrial applications due to its availability, low cost, and high fuel-heating values. Of course, current emissions standards

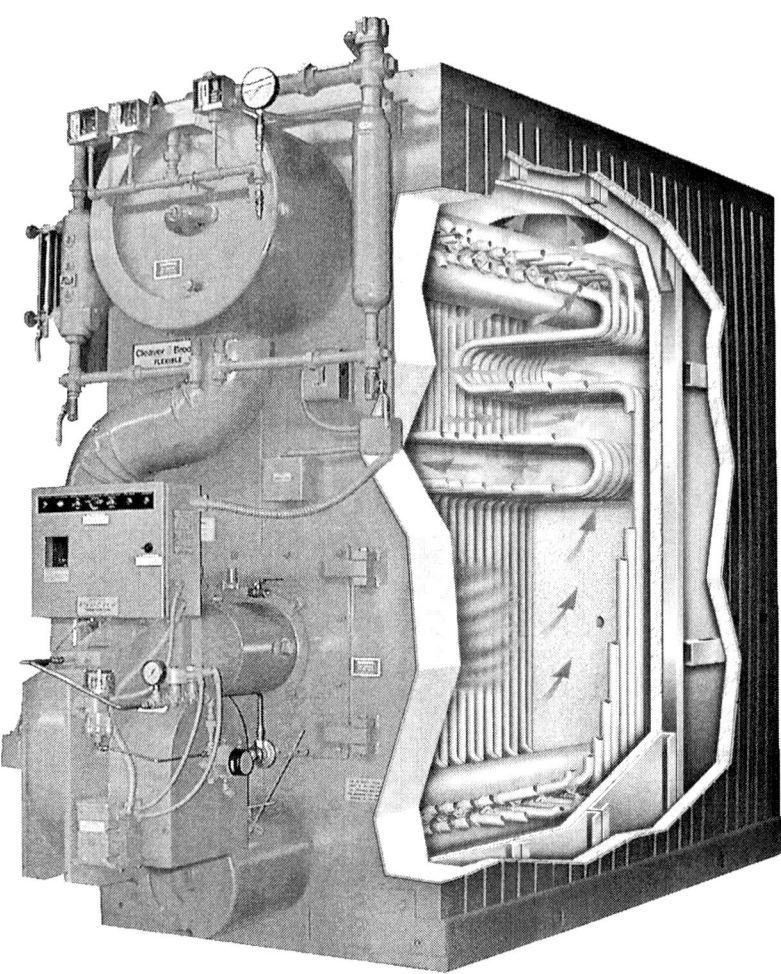

Fig. 6 Water-tube boiler (Courtesy of Cleaver-Brooks).
Source: From http://www.elizabethtowngas.com/BusinessCustomers.CommercialBoilers.aspx

require that special emissions monitoring and treatment remain in effect.

HABITAT DEVELOPMENT

A discussion of space heating would not be complete without reviewing the development of habitat structures and the improvements in the materials of their construction.

Early habitats were constructed primarily of earth, wood, or stone. The roofs were thatched straw, with large openings to allow hearth-fire smoke to escape. The doorways and window openings were rough openings that allowed occupants and light to enter. These doors and window openings may have been covered with animal hides or wood. The structure's main function was to offer some level of protection from predators and the elements. Energy efficiency was not primary on the minds of the occupants. The actual building materials acted as the rough insulation. Drafts causing heat losses were required to allow fire smoke to escape. In moderate climates, the need for protection from the ambient weather would be low to moderate. In the colder latitudes, this became more of a priority. The level of heating required was supplied by maintaining a hearth fire along with heavier clothing, as the climate dictated. The majority of the structures are considered primitive by today's standards.

Later, during the Roman and Middle Ages periods, doors, windows, and shutters came into wider use; these devices reduced air infiltration but did not alleviate it. Glass windows were expensive commodities that were handmade for each application. They saw use more for decorative purposes in wealthier households and churches than for heating efficiency. Many windows leaked and required shutters just to keep the elements out.

In many northern latitudes and during the American colonization period, many log huts or cabins were built. This is only natural, with wood being both plentiful and available. The use of these heavy wood timbers reduced much of the need for thermal insulation. Because wood is fibrous, it acts as a natural insulating thermal barrier.

The openings between the rough-hewn timbers required filling with soil or clay to eliminate drafts, of course. Again, window development did not proceed much past earlier times, and although expensive, it was more readily available. Again, shutters were commonly used for protection from both the elements and predators.

INSULATION DEVELOPMENT

It is only within the past 100 years that the benefits of building insulation have been realized from the standpoint of building comfort and fuel-cost savings. The ideology of efficient insulation practice is to suppress the flow of heat energy from warm areas to cold areas as much as is technically and economically feasible within the guidelines of current air-quality standards. Under normal conditions and in accordance with heat-transfer practices, heat energy will migrate from areas of higher energy (warmer) to areas of lower energy (colder). The rate at which this happens is defined by the thermal conductivity or U-factor of the material. The U-factor is the inverse of the R-factor, or insulating value of the material, in question. Insulating materials that saw use in early structures essentially consisted of the building's structural materials. Later, as frame construction developed and the use of insulation began to develop, whatever would stop the flow of heat was used. This would come to include mineral wool, fiberglass, Styrofoam, and Mylar/aluminum-foil boards.

Insulators can be organic, inorganic or fibrous, cellular, reflective, rigid, soft, or granular. New residences may include many types for different applications. Each type is suited for a specific task, and each type comes in many forms: batts, blankets, loose fill, spray foam, rigid panels, and radiant barriers. Vapor shielding utilized in many homes consisted of single or multiple layers of tar paper; this has steadily been replaced by fiberglass-type blankets. These fabrics have become relatively inexpensive and easy to install, and they essentially eliminate air infiltration where they are used.

Many federal agencies and trade associations have enacted regulation to ensure public safety relating to the use of insulation products. Some of these agencies are the Consumer Product Safety Commission (CPSC), the American Society of Testing Materials (ASTM), and the Federal Trade Commission (FTC). Their regulations and findings are published and easily available for public review.

SPACE HEATING CALCULATION METHODOLOGY

The calculations used to determine acceptable heating comfort standards are in accordance with American Society of Heating, Refrigerating and Air-Conditioning Engineers (ASHRAE) and local codes, and include cooling requirements and special building applications. These will not be discussed here. This section will focus primarily on heating calculations.

Upon determination of the location of the structure, the ASHRAE heating and wind design conditions table needs to be reviewed. The heating dry-bulb temperature columns in this table are shown in 99.6% and 99% columns.[14] What this indicates, for example, is that 99.6% of the time, the temperature will remain above the listed low temperature. Depending on local codes, this would be your low outside ambient air design temperature. At some points, the temperature could fall below that point, but statistically, this time would be minimal. Considering the desire to maintain an indoor air temperature of 70°F, the difference between the two is the maximum design temperature differential (Δt). The structure's design and material selection should use energy-efficient practices. The materials specified should have high insulating values (R-values) and, thus, low heat-loss coefficients (U-values). Utilizing the design Δt and the U-factors, the heat loss is calculated essentially for each square foot of the building's exposed surface area for a maximum design-heating hour. This would include windows, doors, skylights, foundation losses, and all outside exposed surfaces. Also taken into account are the heat load and gain from internal equipment, solar radiation/time-of-day incidence, and surface barrier conditions. Detailed descriptions of these location-sensitive calculations can be found in most Heating, Ventilation and Air Conditioning (HVAC) manuals. Using the design heating load determines the sizing of the boiler (or other device) required. The equipment efficiency losses must also be taken into account to meet the net heating requirements.

A quick simplified example of this type of problem is as follows:

- A one-story residential structure located in Central Islip, New York, is designed with wall R-values of 12. The sloped roof is not insulated, but the internal ceiling is insulated to an R-12 value. The building is built on a slab foundation. Assume that the slab, doors, and windows have a heat-loss total = 7000 BTUs/h for this example. The dimensions of the building are 30 ft by 50 ft with 8-ft-high ceilings. Using a 99.6% heating design DB temperature of 11°F and a boiler efficiency of 80%, how many BTUs of fuel per hour (Q) are required to feed this boiler at the design temperature?
- 1st step: $(70°F - 11°F) = 59°F$
- 2nd step: $R = 12$ (h×ft²×°F)/BTUs, so $U = 1/R$; $1/12 = 0.08333$ BTUs/(h×ft²×°F)
- 3rd step: surface area (A) = $(30' \times 2 \times 8') + (50' \times 2' \times 8') + (30' \times 50') = 2780$ ft²

- 4th step: $Q = UA\Delta t$; 0.08333 BTU's/(h × ft^2 × °F) × 2,780 ft^2 × 59°F = 13,668 BTUs/h boiler output + 7000 BTUs/h (slab/door/windows) = 20,668 BTUs/h.
- 5th step: accounting for equipment efficiency; Solution = (20,668 BTUs/h boiler output)/(80% boiler efficiency) = 25,835 BTUs/h boiler input.

This result uses a standard boiler efficiency of 80%. Using a high-efficiency condensing-type boiler would raise that efficiency to about 90%, reducing fuel consumption. Of course, the current cost of a residential-sized condensing boiler is about double that of a cast iron-type boiler. This is a very straightforward problem; in larger buildings with more complicated duct systems, many other considerations would need to be accounted for.

Estimating annual fuel consumption can be calculated using Heating Degree Day methodology.[16] This method assumes that solar and internal gains for a structure will offset heat loss when the daily mean temperature is 65°F. The formula for this is

$F = (24(DD)qCd)/(n(ti - to)H)$,

where F = Quantity of fuel required, units depend on H
DD = Degree Days for the period desired (from local weather stations)
q = Design heat loss (BTUs—Step 5 of the previous problem)
Cd = Interim correction factor[15]
n = System efficiency
H = Heat value of the fuel

Example: using the above solved problem and the 5-year local average of 6000 HDDs, how many gallons per year of No. 2 oil will be consumed at a heating value of 130,000 BTUs/gallon?

- Step 1: Cd = 0.6, q = 20,668 BTUs/h, n = 80%, H = 130,000 BTUs/gallon, (ti − to) = 59°F
- Step 2: F = (24 × 6000 HDDs × 20,668 BTUs/h × 0.6)/(0.8 × 59°F × 130,000 BTUs/gallon)
- F = approximately 291 gallons of No. 2 oil per year

This result does not account for domestic hot water production or occupancy anomalies. These simplified problems are for demonstration purposes only.

CURRENT TRENDS

Recent historical events have changed the energy market dramatically. Fuel, once cheap and plentiful, has become limited in supply, causing costs to rise dramatically. The effect on lifestyles and economies has been drastic. The effect on technology has been an upswing in energy-efficient technologies and practices. Automobiles with high-efficiency engines and hybrid technologies have replaced powerful but inefficient V8 engines. Lighting has seen more use of fluorescent fixtures, which produce more lumens per watt than the older incandescent bulbs. More efficient condensing-type boilers have seen more use in the marketplace. Condensing boilers capture more heat out of the exhaust, resulting in lower exhaust temperatures and higher overall efficiencies. Many building codes have been modified to reflect the need for "green buildings." These "sustainable" or "high-performance" structures have a reduced impact on Earth's resources. They are designed to be comfortable yet resource efficient, and they will conserve energy, water, and raw materials. They minimize waste both in design and operation, and the life cycle of the building is also taken into account.[17]

Space heating improvements have been evident in the field of solar heating. These systems can be passive, active, or a combination of the two. Passive solar heating systems should take advantage of the design features of a habitat. If used, homes are built of materials such as tile and concrete. These materials absorb heat during the day and release this stored energy in the evening when it is needed. An active system consists of a collection device that absorbs the solar radiation, and in conjunction with fans and pumps, it transfers and distributes this heat. Some active systems may utilize an energy-storage system to provide heat when the sun has gone down. The two basic systems use either liquid or air to store this heat. The liquid systems heat either water or an antifreeze solution in a hydronic collector. Air-based systems use an air-to-water heat exchanger, dissipating the heat into the living space. During the summer months, the system is used to heat water for domestic use. Many of these systems can provide 30%–70% of the heating requirements of most homes.[17]

Another geographically limited means of space heating is geothermal energy. Iceland has a high concentration of volcanic activity and geothermal energy, which is put into wide use for heating and the production of electricity. The energy is so inexpensive that the sidewalks in Reykjavic and Akureyi are heated in the wintertime. About 17% of the country's electricity (219 MW) and 87% of the nation's heating and hot water requirements are supplied by this means.[18]

Improvements in window technology have improved dramatically over the past 100 years. The old single-pane glass windows were literally thermal holes in buildings. Their R-values ranged around 1.0, meaning that they allowed about 19 times more heat to escape per square foot than the average wall. With new insulated windows, there is a gaseous or vacuum barrier between the two sheets of glass, and the frames have been improved to essentially eliminate air infiltration. R-values in the range of 4.0 and higher have been met with current construction practices. This is a vast improvement over the older single- and double-pane storm-type windows. The insulating value of a double- or triple-pane window is a product of the air space between the panes and the type of spacers used.

Metal has been avoided for use as spacers, as it conducts heat and cold very well. Rubber, foam, and fiberglass have replaced metal in most applications.

The frames and materials have progressed where many architectural styles and many materials are available. Some of the materials are aluminum, wood, aluminum clad, vinyl clad, wood–plastic composites, vinyl, and fiberglass. Since the Energy Policy Act of 1992, the National Fenestration Rating Council (NFRC) has worked on a national window-standard testing procedure. It has also set the standard for rating windows according to solar heat gain, air infiltration, and condensation potential.[19]

Originally, most HVAC controllers were pneumatic. Mechanical engineers used their experience with steam and air to control hot or cold air. In time, with standardization, relay ladder logic began to replace pneumatic controls, which in turn were replaced by electronic switches. Modern computerized control systems can now be accessed by Web browsers with the appropriate safety passwords.[20] These controls can remotely control temperature, humidity, and pressure. Many of these systems also include lighting, fire, and other security controls. The use of programmable time-of-day controls allows any building to maintain comfortable set points in accordance with ASHRAE standards for daytime-occupied and evening and weekend operation. In most applications, the building temperatures will be allowed to drop during nonoccupied hours, generating fuel savings. This set point must take into account ASHRAE standards and individual building requirements. Shortly before the building opens, heat will be supplied to bring the comfort level up to normal. At the end of the day, the temperature will be allowed to drop back slowly to the evening set point. Residential use of this technology works much the same way. Occupancy sensors have also been used for lighting and to maintain lower set points in larger unoccupied areas.

CONCLUSION

With the current world energy situation, energy conservation and high efficiency are now household words. Technological advances are being investigated to increase efficiencies and cost effectiveness in many fields of study. This is being attempted while maintaining the lifestyles that we are accustomed to. The public need to be more aware of how it uses energy to reduce dependency on foreign fuels. It is usually during times of adversity that some of the more intuitive insights are realized and great achievements are accomplished.

In the middle of difficulty lies opportunity
-Albert Einstein

REFERENCES

1. http://www.romans-in-britain.org.uk/inv_central_heating.htm (accessed May 2006).
2. http://www.vroma.org/~bmcmanus/baths.html (accessed May 2006).
3. http://www.romans-in-britain.org.uk/inv_central_heating.htm (accessed May 2006).
4. http://www.castlewales.com/life.html (accessed April 2006).
5. http://en.wikipedia.org/wiki/Kitchen (accessed April 2006).
6. http://www.vtbrickoven.com/masonry/masonry.html (accessed June 2006).
7. http://www.secretlifeofmachines.com/secret_life_of_central_heating.shtml (access May 2006).
8. http://www.egr.msu.edu/~lira/supp/steam (accessed May 2006).
9. http://www.egr.msu.edu/~lira/supp/steam (accessed May 2006).
10. http://www.egr.msu.edu/~lira/supp/steam (accessed May 2006).
11. http://www.hsb.comabout.asp?id-50 (accessed April 2006).
12. http://www.memagazine.org/backissues/february2000/features/setting/origin.html (accessed May 2006).
13. http://www.elizabethtowngas.com/BusinessCustomers.CommercialBoilers.aspx (accessed April 2006).
14. ASHRAE Climatic design information. In *1989 ASHRAE Handbook—Fundamentals—I-P Edition*; American Society of Heating, Refrigeration and Air-Conditioning Engineers, Inc.: Atlanta, GA, 1989; 26-6–26-21 [Ch 26-6].
15. ASHRAE. In *1989 ASHRAE Handbook—Fundamentals—I-P Edition*; American Society of Heating, Refrigeration and Air-Conditioning Engineers, Inc.: Atlanta, GA, 1989.
16. http://www.moea.state.mn.us/greenbuilding/index.cfm (accessed May 2006).
17. http://www1.eere.energy.gov/solar/sh_basics_space.html (accessed May 2006).
18. http://www.en.wikipedia.org/wiki/Geothermal_power_in_Iceland (accessed June 2006).
19. http://www.healthgoods.com/EducationHealthy_Home_Information/Doors_and_Windows (accessed May 2006).
20. http://en.wikipedia.org/wiki/HVAC_control_system (accessed May 2006).

Steam and Hot Water System Optimization: Case Study

Nevena H. Iordanova
Senior Utility Systems Engineer, Armstrong Service, Inc., Orlando, Florida, U.S.A.
Rick Cumbo
Operations and Maintenance Manager, Armstrong Service, Inc., Orlando, Florida, U.S.A.

Abstract

In 2001, Armstrong Service performed an audit in a large food processing plant.
The audit focused on the steam and hot water systems improvements, as well as their monitoring, control and sustainability of improvement results.
The objective of this engineered audit was to:

- Improve the reliability of the existing systems, including generation, distribution, use, and condensate return,
- Improve system performance to achieve better efficiency,
- Reduce the steam usage, venting, condensate drain and heat loss from the system,
- Reduce emissions and effluents to the environment,
- Improve monitoring and measurement of the system parameters and assure long lasting results.

The presentation will highlight:

- The engineering analysis of the existing system. The boiler house operates with three natural gas fired boilers consuming up to 310,000 MMBTU/year. Saturated steam is generated at 115 psig and distributed to the plant at an average flow rate between 20,000 and 50,000 lbs/h to multiple process users.
- Identifying and root cause analysis of the deficiencies/problems that cause energy losses,
- Providing the necessary remedial actions to improve the steam and condensate system to reduce energy use and maintenance cost. The overall assessment and analysis resulted in 19 conceptual energy conservation measures (ECMs), with potential to reduce up to 30% of the steam cost. After being presented to the plant, 9 out of 19 proposals were approved for further development. In 2002, the projects were implemented and the plant steam to product efficiency was improved by more than 10%.
- Monitoring/managing the system and sustaining the results.

Additional to the instrumentation and meters installed to collect data on critical points for each individual project, a web enabled monitoring and reporting energy optimization system (EOS) was installed. It made the process of data acquisition, monitoring, recording, trending, reporting, and control possible not only for the plant operation and maintenance personnel, but also accessible from a standard web browser for a control and management prospective.

INTRODUCTION

Improving steam system efficiency will contribute significantly to the profitability for every plant. Nearly half of industrial energy is used to generate steam. There are proven energy savings techniques that capture significant energy saving benefits upon implementation. Some of them are industry wide; others are specific for each plant.

Steam systems consist of several components such as:

- Steam generation
- Steam distribution
- Steam usage
- Condensate collection and return

A comprehensive steam system optimization addresses these interrelated areas, identifies the problems, and recommends all necessary corrective measures to eliminate them.

OVERVIEW OF THE STEAM SYSTEM

Steam Generation Cost

The annual steam generation cost at the Plant is approximately $1,766,000, consuming 310,000 MMBTU/

☆ © Armstrong Service. Reprinted with Permission.

Keywords: Steam systems; Steam system optimization; Food processing heat; Manufacturing steam use; Steam traps; Condensate collection; Waste heat recovery; Boiler blowdown; Steam cookers.

Table 1 Steam generation cost

Utilities	Units/yr	Unit costs Units	$/unit	Incremental steam cost $/yr	$/klb steam
Fuel	309,860	MMBTU	5.24	1,623,666	5.86
Water	21,600	kgal	1.96	42,349	0.15
Sewer	4,324	kgal	4.04	17,469	0.06
Chemicals				31,092	0.11
Electricity	1,289	MWh	40	51,564	0.19
Average steam cost				1,766,139	6.37

year. This cost represents the total cost that is required to generate an average of 25,000 lbs/h of steam at 115 psig. The various costs that make-up the steam generation costs are shown in Table 1.

Steam Generation System

There are three boilers in service in the facility:

- Murray built in 1960 rated at 60,000 #/h
- Union built in 1966 rated at 125,000 #/h
- Murray built in 1972 rated at 100,000 #/h.

All of the boilers use natural gas as their primary fuel. Typically, one or two of the boilers are running with one other boiler in standby (Fig. 1).

Steam Utilization

The main steam users are the product cookers, space heaters and the steam jet ejectors on the product deaerators (DAs). An extremely large amount of heat is being wasted by the venting of steam from the vacuum jet ejectors (see ECM#3). Steam is also used to a lesser extent for sterilization, in the jacketed reheat vessels and for water heating.

Steam Trapping

Results from the steam trap survey performed during the audit are listed in ECM #5. In addition to the defective traps, there are several other notable problems with steam traps at the facility. A large amount of steam is being wasted in order to drain condensate from some of the product equipment (see ECM #6).

Condensate Return System

Condensate is piped back to a main condensate receiver (CR) in the production area and then pumped back to the

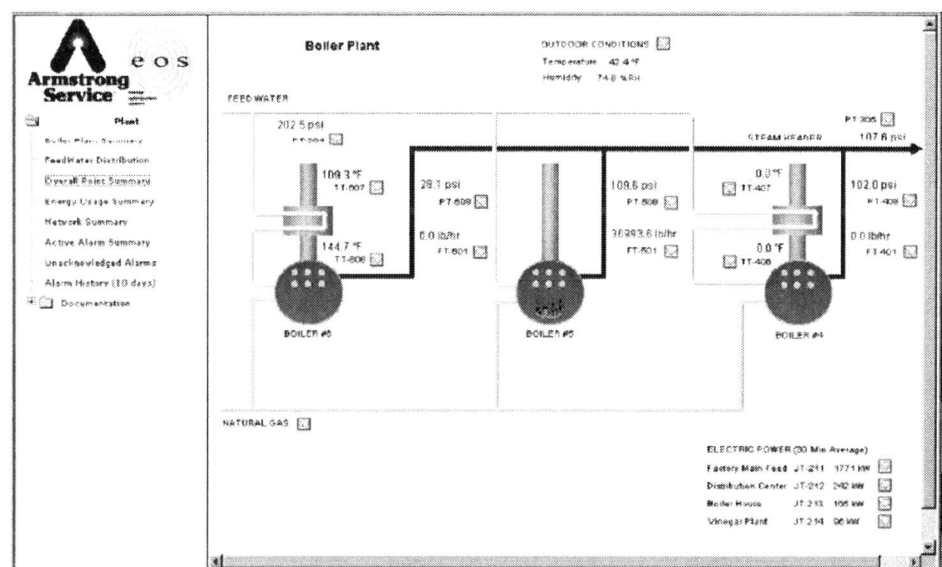

Fig. 1 Steam generation system.

boiler house. Flash steam off this CR is piped back to the boiler house and used to heat makeup water to the DA.

Waste Heat Recovery

Some of the waste heat is recovered by the use of the flash steam in the DA and the use of economizers on several boilers.

SAVINGS OPPORTUNITIES

A thorough review of the system confirmed that there are energy saving potentials in all the areas of the steam system: steam generation, distribution system, utilization, and in condensate return system. The following paragraphs highlight nine energy savings opportunities identified at the site and describe the measures proposed and implemented in order to realize these savings.

ECM 1: INSTALLATION OF A STACK ECONOMIZER

Current System Description

The boiler with the highest design capacity, 125,000 lbs/h, is currently being used as an alternate base load boiler, averaging about 40% of the operating time per year. This boiler is usually used during periods of expected higher steam demand. The boiler does not have an economizer and the flue gas temperature is around 520°F.

ECM Description

Armstrong Service, Inc. (ASI) proposed to install one fin tube stack economizer, which would preheat the boiler feedwater with the heat recovered from the flue gas (Fig. 2). The installation of the economizer will improve the boiler efficiency by reducing the exhaust flue gas temperature down to 285°F. After the installation of the economizer, the boiler will be the primary boiler for the plant and will be used 90% of the time.

The boiler is normally fired on natural gas. However, it has the capability of being fired on Fuel Oil #6. Therefore, an automatically operated soot blower will be installed to help maintain the economizer.

Proposed Energy and Utility Savings

In FY 2002, the steam generation at the plant was 277,151 Mlb. The energy savings expected from the economizer installation were based on 94% of the present steam load, as the load was expected to be reduced down to approximately 260,000 Mlb after the repair of the major non-design vents and leaks.

More details about the data used and the saving calculations are included in Table 2.

By reducing the fuel consumption, total carbon emissions were reduced by 100 tons/year.

The simple payback for this project was 3.5 years.

ECM 2: REPLACEMENT OF BLOWDOWN HEAT RECOVERY EQUIPMENT

Current System Description

All boilers are operated with a continuous blowdown. Each unit has a manual valve, which is used for adjustment. A blowdown heat recovery unit was previously installed in the Boiler House in the DA penthouse. It consisted of a flash tank and a heat exchanger. In the flash tank the blowdown flashes to a lower pressure equal to the DA pressure and supplements the low pressure (LP) steam required at the DA. The heat exchanger recovers the heat from the hot liquid by preheating boiler make-up water (MUW).

The blowdown heat recovery unit has been disconnected and abandoned in place, due to persistently occurring leaks at the heat exchanger and the excess of LP steam to the DA from the plant CR vent line.

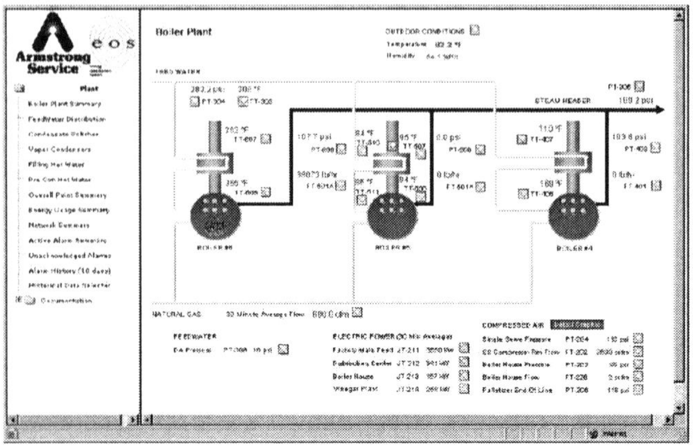

Fig. 2 Proposed economizer.

Table 2 Savings from economizer installation

Economizer savings calculation		Existing condition		Proposed modification	
Hours of operation: 24 h/day 6 days/week 50 weeks/year 7200 h/year		B#5	40%	B#5	90%
		B#6	60%	B#6	10%
		B#5	B#6	B#5	B#6
		Existing	Existing	New	Existing
		No Econ	W/Econ	W/Econ	W/Econ
Input	Unit				
Fuel selection	Type of fuel	NG	NG	NG	NG
Fuel cost	$/MMBTU	5.24	5.24	5.24	5.24
Maximum design firing rate	MMBTU/h	125	80	125	80
Steam load	lb/h	36,001	36,001	36,001	36,001
Steam pressure	psig	115	115	115	115
Feedwater temperature	°F	220	220	220	220
Ambient temperature	°F	80	80	80	80
Flue gas temperature	°F	520	350	285	350
Intake air temperature	°F	100	100		
Flue gas analysis based on measurement (% vol—dry basis)					
	O2	3.5	3	3.5	3
	CO	0	0	0	0
	Combustibles	0	0	0	0
Annual operating hours	h	2,880	4,320	6,480	720
Heating value	Btu/ft^3	1,001.707	1,002	1,001.71	1,002

		Before	After		
Summary of Results					
Fuel consumption	MMBTU/yr	314,706	308,020		
Fuel cost	$/yr	$1,649,062	$1,614,027		
Carbon emissions	ton/yr	4,716	4,615		
Savings					
Fuel consumption	MMBTU/yr	6,686			
Fuel cost	$/yr	$35,035			
Carbon emissions	ton/yr	100			

ECM Description

ASI proposed to replace the blowdown heat recovery system. The waste heat in the blowdown water was used to preheat boiler makeup water (Fig. 3). The flash tank was not reused and the blowdown water was recovered at a high pressure in a shell and tube system of counter flow heat exchangers. The blowdown discharge temperature would be 170°F, due to the MUW temperature coming in at 140°F.

Proposed Energy and Utility Savings

Based on the conductivity readings from the feed water and the boiler water, the average blowdown rate was calculated on 9.4%.

More details about the data used and the saving calculations are included in Table 3.

By reducing the fuel consumption, total carbon emissions will be reduced by an estimated 70 tons/year.

The simple payback for this project was 1.5 years.

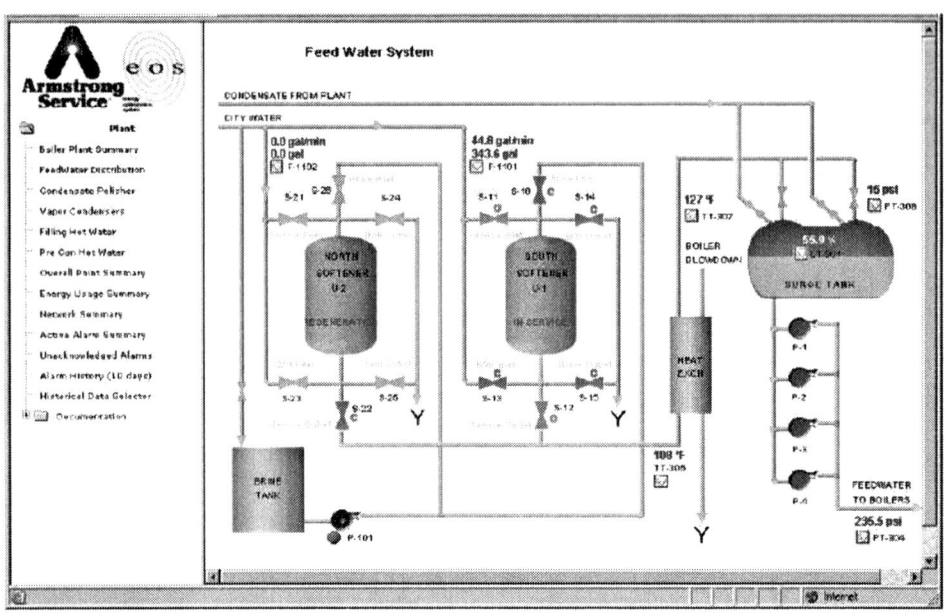

Fig. 3 Blowdown heat recovery.

ECM 3: REPLACEMENT OF PRODUCT DEAERATOR STEAM JET EJECTORS WITH MECHANICAL VACUUM PUMPS

Current System Description

Deaeration is one of the steps of the food product processing. The product is sprayed in a DA at 205°F at a rate of 60 gpm. Due to the vacuum, part of the water is evaporated and part of the oxygen is removed. The product then leaves the DA at 190°F through a process pump.

The plant has a total of eight (8) existing product DAs. The vacuum is drawn through barometric condensers by steam jet ejectors (exhausters) designed to maintain up to 15′ Hg vacuum. Each steam ejector is rated for a dry air suction capacity of up to 70 lb/h. The arrangement of the existing system is shown in Fig. 4.

ECM Description

ASI proposed to replace some of the steam jet ejectors with electrically driven liquid ring (LR) vacuum pumps (VP) and the direct water-cooled condensers with indirect

Table 3 Blowdown heat recovery saving

Annual steam generation	259,204	Mlb/yr
Annual operating hours	7200	hr/yr
Average steam generation	36,001	lb/h average
Blowdown	9%	
Blowdown quantity	3240	lb/h
Pressure	115	psig
Blowdown heat	318.7621	Btu/lb
Blowdown temperature to drain	180	°F
Heat recovered	0.55	MMBTU/h
Steam generation efficiency	85%	
Fuel savings	4,687	MMBTU/yr
Sewage increase	−401	Mgal/yr
Heat savings steam equivalent	4232	klb
Annual dollar savings	$22,937	$/yr

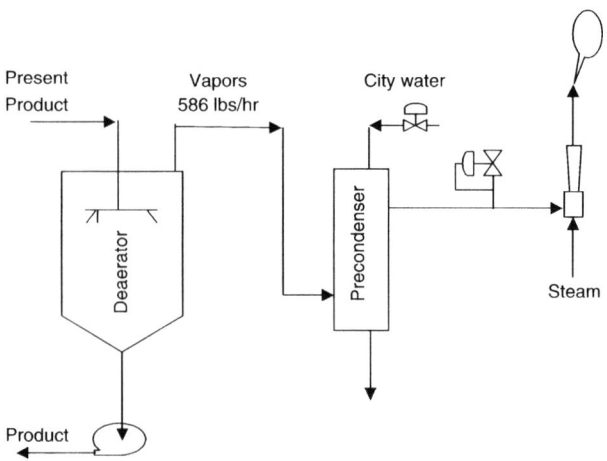

Fig. 4 Existing equipment configuration.

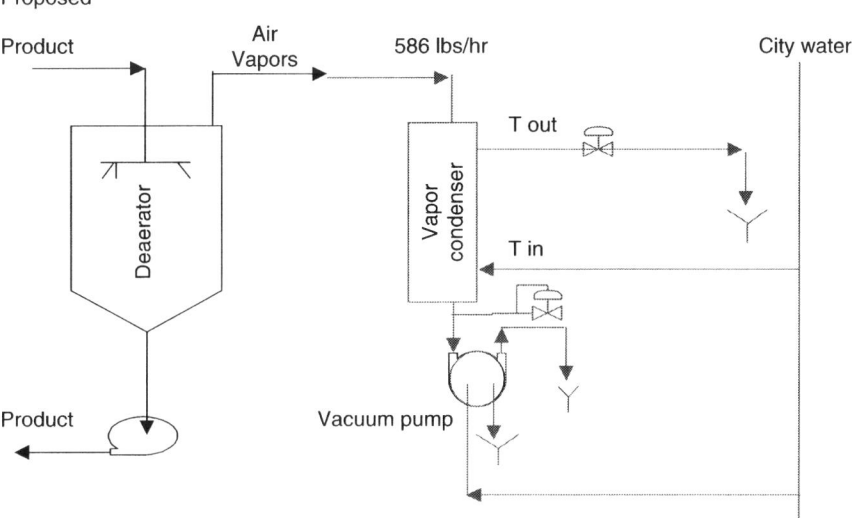

Fig. 5 Proposed equipment configuration.

condensers at three of the DAs. The VPs will eliminate the need for steam use in the DAs and the indirect condensers will eliminate the water discharge to industrial waste sewer (IWS). A simplified schematic of the proposed piping diagram is shown in Fig. 5.

Energy and Utility Savings and Environmental Impact

Under the current conditions each steam ejector was using 357 lb/h of steam and each condenser discharged approximately 15 gpm of cooling water to the IWS.

Details about the data used and the savings calculations for one of the DA are included in Table 4. Table 5 shows the summary of the savings from all the reconfigured DAs.

By reducing the steam usage, total carbon emissions were reduced by 100 tons/year.

The installation of the new design for the product DAs needed to be completed around the work schedule of the current product DAs. The simple payback for this project was 4 years.

ECM 4: REPAIR AND/OR REPLACE RELIEF VALVES AND CONTROL VALVES

Current System Description

There are pressure relief valves and atmospheric control/relief valves located downstream of the control valves on the steam inlet line for multiple cooking kettles. These relief valves are connected to common vents, which are then piped through the wall to the outdoors. Steam is being discharged from these vents approximately 60%–70% of the time (Fig. 6).

An ultrasonic check of the control valves proves that six of the atmospheric valves, all of the pressure relief valves and seven of the steam supply valves are leaking steam.

ECM Description

ASI proposed to replace and/or repair the leaking valves.

Energy and Utility Savings and Environmental Impacts

The unnecessary steam venting was driving up the cost of steam. The boiler fuel usage increased, as more fuel had to be used to supply the additional steam load. The steam lost to the atmosphere increases the amount of MUW required, as it is not recovered as condensate. The additional MUW requires preheating and water treatment/chemicals. The energy savings for this ECM result in 5,223,000 lbs of steam per year and the details can be seen in Table 6.

By reducing the steam venting, the total carbon emissions were reduced by 189 tons/year.

The simple payback for this project was 2.8 years.

ECM 5: STEAM TRAP REPAIR AND REPLACEMENT

Current System Description

ASI conducted a steam trap survey at the plant and 180 traps were tested. The survey and computer analysis revealed that approximately 25% of the in-service traps were defective. The defective traps were wasting steam

Table 4 Utility consumption comparison

Present operation		
Deaerators		Present
Product flow	gpm	XX
Product temperature In	°F	205°F
Product temperature Out	°F	190°F
Vacuum maintained in DA	In HG	10–15
Precondensers		
Water wasted in condensers	gpm	14.8
City water temperature	°F	65
Hot well temperature	°F	130
Steam jets (Exhausters)	psig	100
Steam used	lb/h	357
Utilities required to operate DA present		
City water + MUW (city)	gpm	15.6
Sewer	gpm	14.8
Steam	lb/h	357
Fuel	Btu/h	474453
Proposed operation		
Vacuum liquid ring pumps		
El pump power required	HP	5
Pump water seal flow	gpm	1
Precondenser		
City water In	°F	65
City water Out	°F	105–120
Calculated condenser heat load	Btu/h	500,000
Savings from one DA		
Fuel MMBTU/yr		1358
City water (all = cooling, seal, MUW)	Mgal/yr	−1796
Industrial sewer	Mgal/yr	2195
Chemicals	$/yr	$143
Electricity	kWh/yr	−10675

Table 5 Utility savings—ECM 3

Type	Units	Units/yr	$/yr
Fuel	MMBTU	5,333	27,947
City water	1000 gal	−7,053	−13,828
Waste water	1000 gal	8,620	34,825
Chemicals	Lot	1	562
Electricity	KWh	−41,930	−1,677
Total	$		47,828

Fig. 6 Leaking relief valves vent.

in excess of 2000 lb/h. In addition to the traps wasting steam, 3.7% of these traps were failed in a cold plugged and/or flooded conditions. These traps are not losing steam but rather causing a loss of heat, water hammer, corrosion, and possibly damage to heat transfer equipment. The following is a break down of defective traps as a percentage of the total in-service traps:

Blow-Thru	21.5%
Cold plugged	3.7%
Total	25.2%

ECM Description

The existing steam traps were very old and many of them need to be replaced. Installing the proper replacement trap will improve the systems efficiency. ASI proposed to replace/repair/modify any failed traps or incorrect piping configurations.

Table 6 Utility savings—ECM 4

Type	Units	Units/yr	$/yr
Fuel	MMBTU	7112	37,266
City water	1000 gal	627	1,229
Chemicals	Lot	1	668
Total	$		39,163

Table 7 Utility savings—ECM 5

Type	Units	Units/yr	$/yr
Fuel	MMBTU	4,028	21,107
City Water	1,000 gal	362	709
Chemicals	Lot	1	385
Total	$		22,201

Energy and Utility Savings and Environmental Impacts

By replacing the defective steam traps, 3,012,000 lbs of steam were saved. The details about the savings can be seen in Table 7.

By reducing the steam venting, total carbon emissions were reduced by 60 tons/year.

The simple payback for this project was 1.8 years.

ECM 6: CORRECT STEAM BLOWTHROUGH RATE ON COOKER KETTLES

Current System Description

The cooking kettles are steam jacketed type vessels or have steam heating coils. The steam heating coils in nine of the cooking kettles use float traps (liquid drainers). Steam is supplied to two helical copper coils in the bottom of the vessel. In order to heat up the coils as quickly as possible, a bypass valve ahead of the each trap is open to the condensate return line. During the cooking phase a large amount of steam flow passes directly through this valve and into the return line.

ECM Description

ASI proposed to replace the existing liquid drainers with steam traps that had been designed with a fixed restricted orifice near the top of the body, built into it to bleed off the steam and non-condensable. This would allow the steam bleed rates to be adjusted to achieve maximum performance with minimal steam blow flow.

Table 8 Utility savings—ECM 6

Type	Units	Units/yr	$/yr
Fuel	MMBTU	6,343	33,239
City water	1,000 gal	569	1,115
Chemicals	Lot	1	607
Total	$		34,961

Energy and Utility Savings and Environmental Impacts

By installing the appropriate steam traps for this application, 4,743,000 lbs of steam were saved. The savings were calculated based on 28 psig steam pressure and 60,000 h of operation. The details about the savings can be seen in Table 8.

By reducing the steam venting, the total carbon emissions were reduced by an estimated 95 tons/year.

The simple payback for this project was 4.1 years.

ECM 7: PRODUCT COOLING HEAT RECOVERY AND WATER RE-USE

Current System Description

The plant has multiple product lines equipped with coolers (plate heat exchangers) to precool the product from 190 to 85°F before the fillers. City and/or well water is used as the cooling media. After the coolers, the cooling water (ranging from 145 to 170°F) is discharged to a clear water sewer (CWS) sump. Every shift, samples of the cooling water are taken and tested for the purity and pH of the water. If the samples are contaminated with product, the water is diverted to a common header and then discharged to the IWS.

At the same time, cold city water is used as an ingredient in all feedstock preparation and cooking processes. For many of the water users, heating and/or cooking takes place after the mixing of the ingredients. Steam at 115 psig is used as the heating media.

In the boiler house, cold city water is used as boiler MUW.

ECM Description

ASI proposed to use the hot water discharge from the existing product coolers (155°F average used for savings) for make up-water to the product users. The remainder of the hot water discharge would be used as boiler MUW.

The newly proposed system was designed in details (Fig. 7). After the plant approval, the ECM was implemented and the performance of the equipment and the savings are monitored through the energy optimization system (EOS) system since then (Fig. 8).

Hot water discharge, from the coolers, which use only city water, was collected into two separate tanks. From there, the hot water is pumped to the different users as needed. The pumps are equipped with variable frequency drives (VFD) to efficiently handle the large variations of the flow rates. Each pump has equivalent capacity back-up

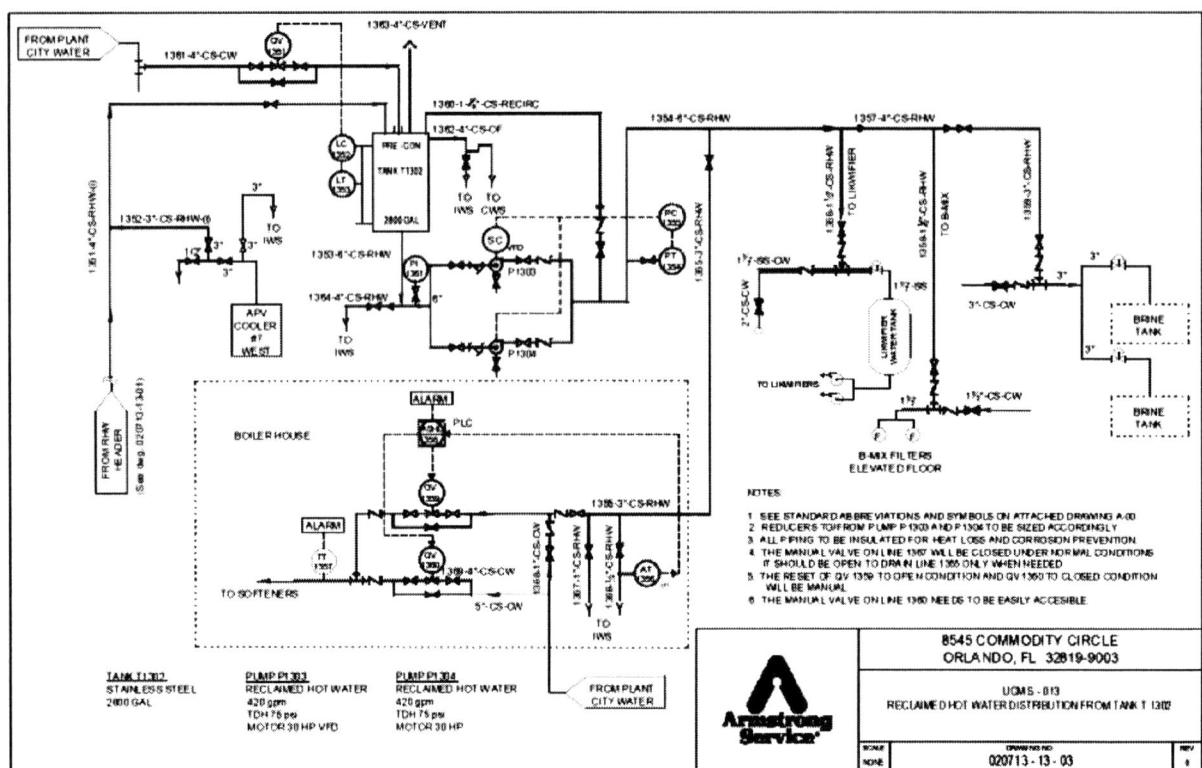

Fig. 7 ECM 7 sample P&ID—TANK 1.

to assure process reliability. Both tanks will have a city water back-up system to assure water availability at any time.

In case of interruption of the Boiler MUW supply, the transition to city water supply, as MUW, is automatic.

Energy and Utility Savings and Environmental Impacts

The implementation of this ECM saved 36,212,000 lbs of steam, shortened the process heat-up time and increased the yield rate without any change of the process equipment.

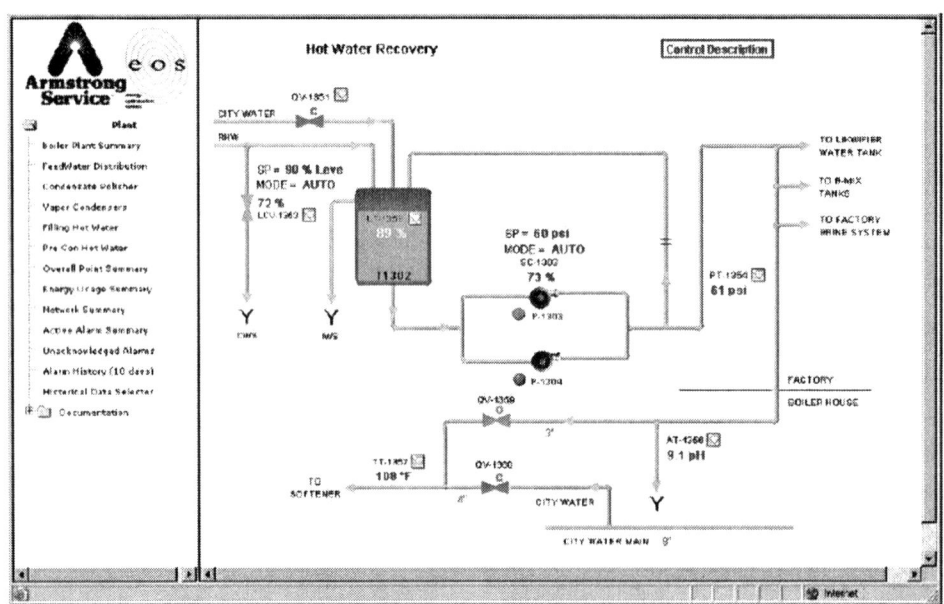

Fig. 8 EOS monitoring screen—ECM 7.

Table 9 Utility savings—ECM 7

Type	Units	Units/yr	$/yr
Fuel	MMBTU	37,857	198,371
City water	1,000 gal	40,525	79,455
Waste water	1,000 gal	−600	−2,424
Electricity	KWh	−232,128	−9,285
Total	$		266,117

Table 10 Utility savings—ECM 8

Type	Units	Units/yr	$/yr
Fuel	MMBTU	1,964	10,293
City water	1,000 gal	173	340
Chemicals	Lot	1	185
Total	$		10,817

The savings were calculated based on city water temperature 65°F, average for the year, heated up to 155°F. The details about the savings can be seen in Table 9.

By reducing the steam usage, the total carbon emissions were reduced by 567 tons/year.

The simple payback for this project was 2.4 years.

ECM 8: ISOLATE ABANDONED LINES

Current System Description

The leak through a strainer ahead of an abandoned valve has been wasting steam for 14 years. Steam is currently being unnecessarily vented through a 2′ pipe, next to an entrance door.

ECM Description

ASI proposes to isolate the leaky strainer and valve.

Energy and Utility Savings and Environmental Impacts

The implementation of this ECM saved 1,443,000 lbs of steam. The details about the savings can be seen in Table 10.

By reducing the steam venting, total carbon emissions will be reduced by an estimated 29 tons/year.

The simple payback for this project was 0.1 years.

ECM 9: CONDENSATE RETURN IMPROVEMENT

Current System Description

The Condensate Return System is a closed system. The condensate from the plant is collected in a common CR. The CR has a pressure equalizing vapor line connecting it to the DA. Normal pressure maintained at the CR ranges from 10 to 20 psig, which is equal to the DA. The liquid condensate at the same pressure is pumped through a pump to the DA.

On average, 50% of the generated steam is returned as condensate (30–60 gpm). Due to the nature of the cooking process, occasionally the condensate gets contaminated and needs to be dumped. The contaminations are caused by hardness from cooling water

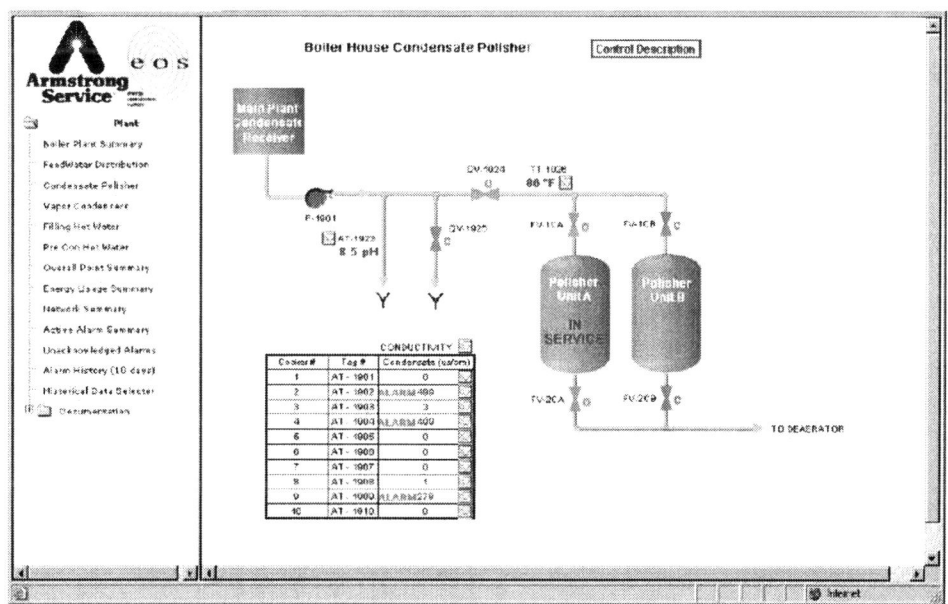

Fig. 9 EOS monitoring screen—ECM 9.

Table 11 Utility savings—ECM 9

Type	Units	Units/yr	$/yr
Fuel	MMBTU	10,699	56,064
City water	1,000 gal	5,326	10,435
Waste water	1,000 gal	5,006	20,212
Chemicals	Lot	1	5,681
Total	$		92,392

crossovers and/or by pH and conductivity from ruptured product coils. The condensate is drained at an estimated 1800 h/year and involves several hours of labor for each occurrence.

There was no arrangement for a fast and effective determination of the location of the contamination source. In addition, there was no way to separate the contaminated condensate from clean condensate. Due to this, all of the collected condensate was dumped. The condensate continued to be dumped until the source of contamination was found and repaired.

Table 12 Measurement and verification gross summary table

ECM	ECM title	Goals and objectives of M&V plan	Baseline measurements	Energy savings calculation	Verification measurement tool	Potential EOS monitoring points
1	Install boiler stack economizer	Calculate savings due to boiler stack exhaust/boiler feedwater pre-heat	Stack temperature log, hours of operation	Preheat feed water	EOS monitoring	Economizer flow, temperature in, temperature out
2	Blowdown heat recovery	Calculate savings due to reusing steam a 2nd time and preheating boiler feed water	Flow of condensate	Blowdown water heat recovery; Btu content of blowdown	EOS monitoring	Heat exchanger flow, temperature in, temperature out
3	Install vacuum pumps	Calculate savings due to steam reduction; increased electrical consumption	Hours of operation; steam flow; water flow; electrical energy	City water flow rate; electrical energy usage	EOS monitoring	Electrical metering; water flow; vapor condensor temperature in, temperature out
4	Relief and control valve replacement	Calculate savings due to steam reduction	Hours of operation; steam flow	Steam plume; condensate return	N/A	N/A
5	Steam trap replacement	Calculate savings due to steam trap replacement and system optimization	Test steam traps for leakage	N/A	Annual trap survey	N/A
6	Steam bleed on product cookers	Eliminate live steam from condensate return system	Boiler house deaerator pressure	Steam plume from DA	EOS monitoring	DA tank pressure

(Continued)

Steam and Hot Water System Optimization: Case Study

Table 12 Measurement and verification gross summary table *(Continued)*

ECM	ECM title	Goals and objectives of M&V plan	Baseline measurements	Energy savings calculation	Verification measurement tool	Potential EOS monitoring points
7	Hot water reuse—product cooling	Confirm reduction in city water use and heating due to pre-heated water re-use	Hours of operation; steam flow; city water flow	Reduced city water consumption, Btu content of hot water reclaimed	EOS monitoring	Heat exchanger flow, tank temperature, city water flow measurements
8	Isolate the abandoned lines	Calculate savings due to steam reduction	Steam flow	Steam plume	Visual	N/A
9	Condensate Return	Calculate savings due to steam reduction; volume of quality condensate return	Hours of cooker operation; condensate return flow; city water	Reduced city water consumption; Btu content of condensate	EOS monitoring	Polisher (conductivity/pH/flow/temperature)

ECM Description

ASI proposed to establish a monitoring system, which helps to identify the source of contamination and prevent continuous and larger than required dumping. ASI also proposed to install a Polisher, which assures good condensate quality before it is sent to the DA (Fig. 9).

A common pH probe automatically monitors and controls a pair of kettle cookers. An alarm is sent to the control room and an operator defines which one of the Cookers is leaking and manually closes the valves. By implementing this project, the draining of condensate was minimized to maximum two cookers for a short period of time.

A new condensate header was built, parallel to the Product cookers, to handle the contaminated condensate. The contaminated condensate was directed to a Flash Tank. The liquid was then discharged to the closest IWS location. A Condensate Polisher was installed in the boiler house to remove hardness from the condensate, which is

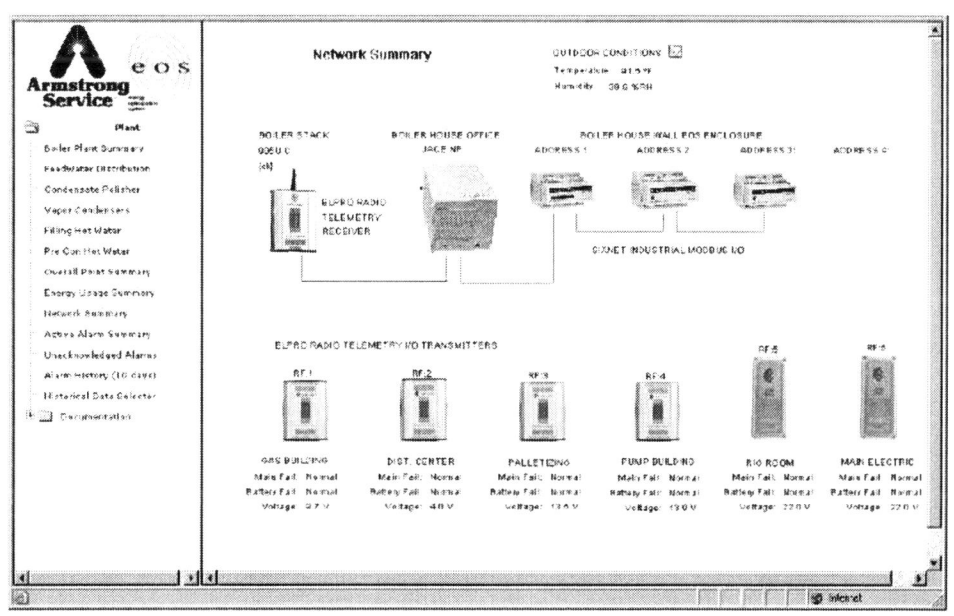

Fig. 10 EOS network summary.

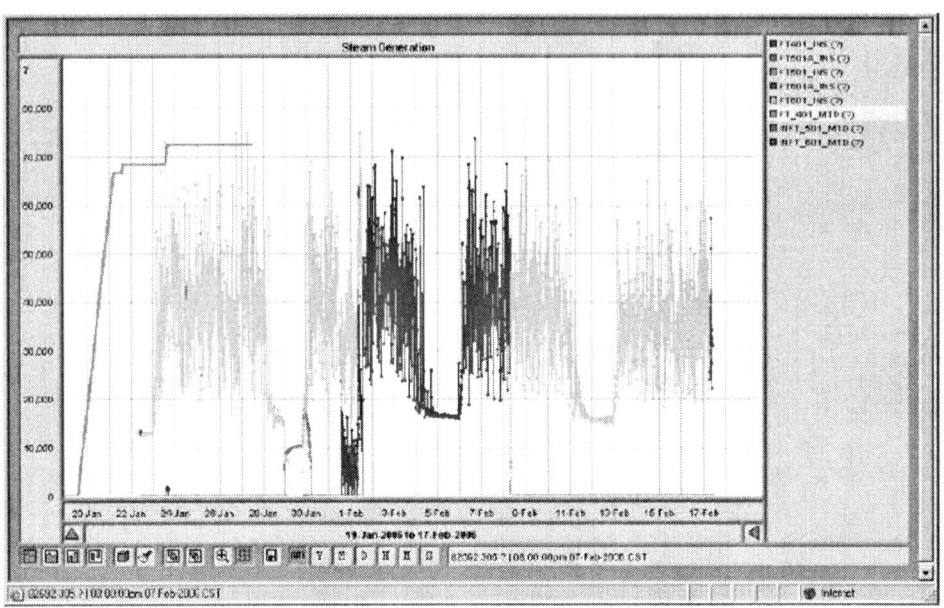

Fig. 11 EOS historic data selection.

coming from different cooling water crossover sources throughout the plant.

If the contaminations are extreme the condensate would be manually dumped to the IWS.

Energy and Utility Savings and Environmental Impacts

The implementation of this ECM saved 9,635,000 lbs of steam. The details about the savings can be seen in Table 11.

By returning the condensate, the total carbon emissions were reduced by an estimated 160 tons/year.

The simple payback for this project was 4 years.

MEASUREMENT AND VERIFICATION PROTOCOL

Before any of the above ECMs were implemented it was mutually agreed with the plant how the savings will be measured and/or calculated. Some of the savings were

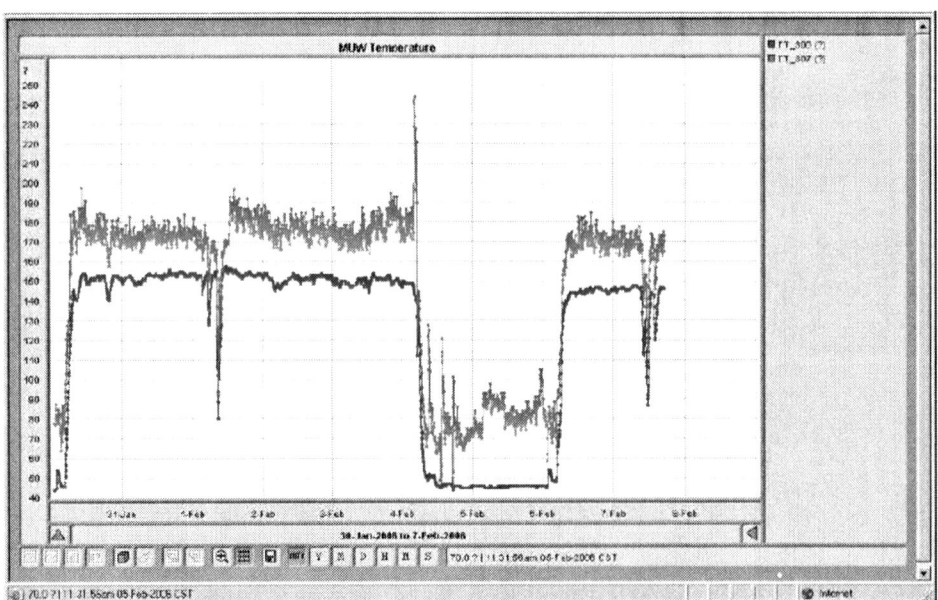

Fig. 12 EOS project performance evaluation.

Table 13 Summarized list of energy conservation measures

ECM	ECM title	Savings, $/yr	Payback/ yr	Carbon emissions, tons/yr
1	Installation of a stack economizer	35,000	3.5	100
2	Blowdown heat recovery	22,900	1.5	70
3	Install vacuum pumps	46,300	4.0	100
4	Repair relief and control valve	65,600	2.8	189
5	Steam trap repair/replacement	22,200	1.8	60
6	Steam bleed on cooker kettles	35,000	4.1	95
7	Product cooling hot water reuse	266,100	2.4	567
8	Isolate the abandoned lines	10,800	0.1	29
9	Condensate return	87,400	4.0	160
	Total savings/emissions	566,000		1,370

calculated based on the readings from the EOS and others were based on stipulated savings. Table 12 provides information on each proposal and the approach taken to establish a proper M&V protocol.

ENERGY OPTIMIZATION SYSTEM

An EOS was installed in the boiler house for monitoring/managing all the utility systems. The EOS was installed in addition to the instrumentation and meters installed to collect data on critical points for each individual project. Summary of the network is shown in Fig. 10.

Based on the existing or newly installed meters, with the EOS, historic data selection is easily manageable and presentable in a chart and/or a table format, depending on the user's preference. Fig. 11 shows the steam generation flow for the period of January 20th to February 17th, 2006.

By the installation of this web enabled monitoring and reporting system, it was possible to monitor the performance of the equipment, at any time and virtually from anywhere. Fig. 12 shows the performance of the heat recovery system (ECM 7) and the blowdown heat recovery (ECM 2). Multiple standard or custom tailored reports/charts could be generated to serve any purpose.

The performance of the equipment and systems is reported to the plant on monthly basis and used to generate invoices for each utility. An overall report is prepared on annual basis and provided to the plant for performance evaluation.

CONCLUSION

By implementing the above nine energy conservation measures, the plant captured $566,000 (at 2002 utility costs) in annual energy savings, and reduced the total carbon emission to the environment with 1370 tons annually.

The identified opportunities to save fuel, steam and condensate are summarized in Table 13. When implemented they lead to 27% overall steam system efficiency improvement.

The installation of EOS made the process of data acquisition, monitoring, recording, trending, reporting and control possible not only for the plant operation and maintenance personnel, but also accessible from a standard web browser for a control and management prospective.

BIBLIOGRAPHY

1. Design and operating data from 2002 to 2005 client's facilities.
2. Turner, W.C. *Energy Management Handbook*, 1992.
3. Armstrong International Inc., *Steam Conservation Guidelines for Condensate Drainage*, Handbook N-101, 2005.

Steam Turbines

Michael Burke Wood
Ministry of Energy, Al Dasmah, Kuwait

Abstract
This article reviews the history of steam turbine development for power generation and marine propulsion. It describes factors affecting thermodynamic and fluid dynamic efficiency and mechanical design principles. Alternative design technologies are described and some comparisons are given. Steam turbines for various applications and probable future developments are described.

INTRODUCTION

The steam turbine has been the predominant prime mover for electricity generation for a century. Its efficiency depends on steam conditions, steam cycle, and turbine fluid dynamic efficiency. Materials properties and mechanical design determine possible steam conditions and turbine rating. Fluid dynamic efficiency is determined by blade and steam path design. These factors and their interaction are described. The changing role of the steam turbine to combined cycle applications, with uncertainty regarding the scope for fossil-fired applications using advanced steam conditions and for new nuclear power plants, is discussed.

HISTORY

Hero of Alexandria's "aoelipile" demonstrated steam-driven rotary motion in 200 B.C. It was driven by the thrust of two tangential outlet pipes and was therefore a reaction turbine. In Italy in 1629, Branca proposed fixed steam jets impinging on rotating blades—the impulse turbine. Both machines were without practical application. Until 1900, motive power was dominated by steam engines. In 1784, James Watt dismissed the turbine as a competitor. He reasoned that steam at boiler pressure would escape to the atmosphere at 2000 ft/s; and because blade speeds of half that were required to extract a useful proportion of its energy, it was impossible to construct a machine to withstand the resultant centrifugal loading. In 1837 in the United States, William Avery used Hero's concept to build machines up to 5 ft in diameter, with peripheral speeds of 900 ft/s, to drive woodworking machinery. They were abandoned because of control difficulties and unreliability. Gustav de Laval in Sweden used a similar device to power a centrifuge, then in the 1890s turned to single-stage impulse turbines. These used a convergent-divergent nozzle to fully expand the steam and had rotor speeds of up to 30,000 rpm. The ascendancy of the steam turbine began when Charles Parsons solved the problem of high velocities by applying pressure compounding to divide pressure drop between a large number of reaction stages along the axis of the turbine. In a year from 1884, he patented and demonstrated the first practical turbine generator. In 1886 in France, Professor Rateau pressure compounded the impulse turbine. In the United States, electrification proceeded rapidly, and in the mid-1990s, George Westinghouse acquired U.S. rights to Parsons' patent and installed the first commercial turbines in 1899. Charles Gordon Curtis addressed the steam velocity problem by velocity compounding; i.e., directing steam from a single row of nozzles onto several rows of moving impulse blades with intervening fixed guide blades. Vertical-axis 5-MW Curtis Turbine Generators were installed in Chicago in 1903. In the United Kingdom, the most rapid development was marine. In 1897, Parsons raced through a fleet review by Queen Victoria in his launch, Turbinia. It was capable of 34 knots, and outpaced the fastest Royal Navy patrol boats. The implications were clear, and by 1905, turbine power was exclusively selected for British warships. It was also used for liners from 1906 until the end of their era.

After 1910, development was rapid, aided by the acceptance of alternating current, which permitted efficient blade design in turbines direct coupled to generators operating at 3600 rpm (60 Hz) in the United States and 3000 rpm (50 Hz) elsewhere. By the 1920s, efficiency was improved by condensers, superheating, and feed heating. In marine applications, multiple reduction gearing allowed diameter and speed to be optimized, resulting in major increases in efficiency. Impulse turbines had the advantages of high power density, smaller rotors, and greater tolerance of rapid temperature changes. They were affected by blade vibration but came to dominate this field after World War II. Reheat was adopted generally at this time, and turbine generator size and efficiency continued to increase. Supercritical steam cycles with steam pressure and temperature conditions of 340 bars/650°C and double

Keywords: Steam turbine; Rankine cycle; Blades; Losses; Mechanical design.

reheat were introduced in the 1960s. Materials difficulties caused reduction to 240 bars/560°C, typical of supercritical units now in service in the United States. Steam turbines operating at up to 250 bars/600°C are used in Europe and Japan. Nuclear reactors require low-speed, wet-steam turbines of up to 1500 MW operating at 75 bars/295°C. In the last decade, combined cycles have become a major steam turbine application. In the marine field, the liners were eliminated by airlines soon after World War II. Rising fuel costs and large, very efficient diesel engines eliminated steam turbines from merchant ships by 1980, despite attempts to compete using very advanced 210 bars/600°C, 14,000-rpm designs. Naval propulsion is now predominantly gas turbine and diesel except for nuclear powered capital ships and submarines.

STEAM CYCLES

The Carnot cycle, shown in Fig. 1, is the most efficient fluid power cycle. It illustrates principles and the reasons for more complex cycles. Processes are as follows: 1–2, reversible isothermal heating in the boiler; 2–3, isentropic expansion in the turbine, 3–4, reversible isothermal condensation; and 4–1, isentropic compression. The area under 1–2 represents heat from the boiler, and that under 3–4 represents heat rejected to the condenser. The difference between the two areas is the work produced. Carnot Efficiency is defined as $(T_{max} - T_{min})/T_{max}$. In a steam cycle, the condenser cooling medium fixes T_{min}; and increasing T_{max}, the temperature at which heat is supplied, maximizes efficiency. This cycle is impractical because isothermal heat supply is only possible in the two-phase region, limiting maximum cycle temperature to below the critical temperature of 374°C. Expansion of saturated steam gives poor steam quality at turbine exhaust, which would cause erosion of the turbine blades. A two-phase compressor is impractical. These difficulties are overcome by the Rankine cycle, shown in Fig. 2, in which heat is supplied at constant pressure from 2 to 3 in successive economizing, evaporating, and superheating. Superheating increases the average heat supply temperature and steam quality at turbine exhaust. Complete condensation from 4 to 1 necessitates only a liquid phase boiler feed pump. Fig. 2 also indicates the effect of irreversibility due to the fluid friction and heat loss in a real cycle. Fuel costs are many times capital costs; and therefore, a complex cycle is justified to maximize efficiency. The values of maximum cycle temperature permitted by creep properties of boiler and turbine materials result in considerable wetness after expansion. For a given maximum temperature, raising the pressure—hence, also raising boiling temperature and average heat supply temperature—increases efficiency. It has the disadvantage of increasing leakage losses and displacing the expansion line to the left, further increasing exhaust wetness. To reduce wetness, steam is returned to the boiler after partial expansion and reheated, as shown in Fig. 3. This further increases average heat supply temperature and efficiency by 4%–5%. Two stages of reheat are sometimes used in supercritical plants. In the economizer, heat is added at low temperature. Efficiency is therefore further improved in the Regenerative Rankine Cycle (Figs. 4 and 5) by feed heaters, heated by steam

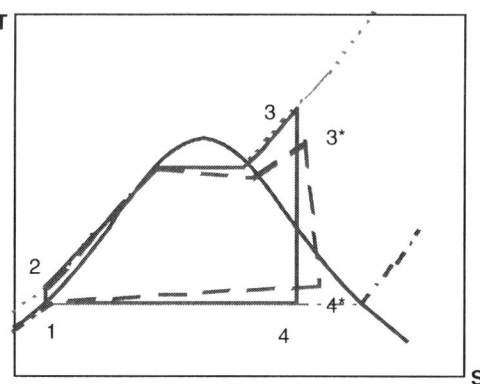

Fig. 2 Rankine cycle.

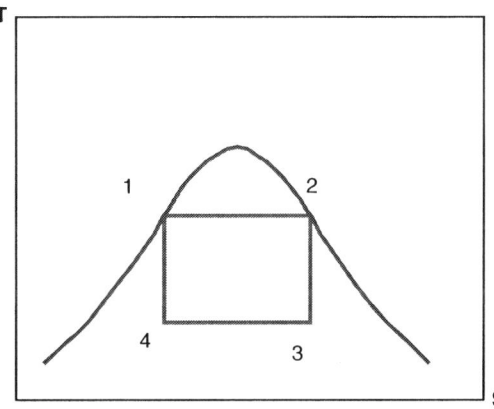

Fig. 1 Carnot vapour cycle.

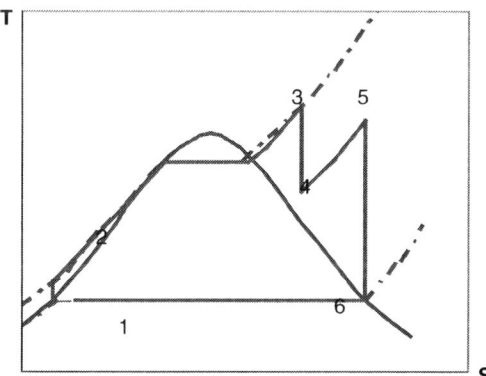

Fig. 3 Reheat rankine cycle.

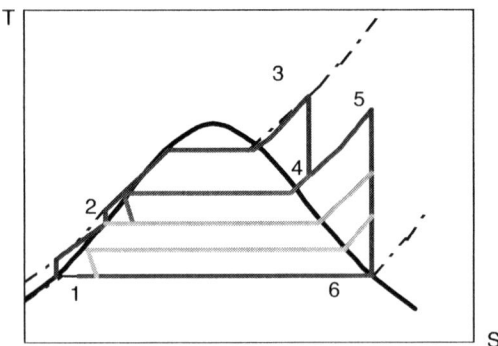

Fig. 4 Regenerative reheat rankine cycle.

extracted from the turbine. Fig. 5 is simplified. In real plants, up to 10 feed heaters are used. The overall energy efficiency of the steam cycle is given by the heat rate, defined as the heat supplied by the boiler (in Btu) divided by the generator output (in MWhr).

The use of a two-phase working fluid has major advantages. The efficiency of energy conversion in heat engines is a function of the pressure ratio. In petrol engines, this is about 12:1; in a diesel engine, 18:1; and in a typical gas turbine, 20:1. In the steam cycle, because of condensing, exhaust pressure is very low, and this ratio is at least 4000:1. The power required for the pump or compressor in the cycle is related to the change in volume. In the Rankine cycle, pumping takes place in the liquid phase; therefore, power requirements are 2 orders less than in a cycle using air as a working fluid. Energy release per unit mass of working fluid is very high, increasing the power density of the heat engine.

STEAM PATH

Steam is expanded in fixed blades or nozzles attached to the casing, and its enthalpy is converted to kinetic energy then to mechanical energy by moving blades attached to the rotor. In the United States, blades are referred to as buckets. The efficiency of the conversion is the ratio of the actual turbine work in process 3* to 4* (Fig. 2) to the isentropic turbine work in process 3–4. This ratio depends on turbine design and losses but is greater than 90% in modern machines. A row of fixed blades and the following row of moving blades is called a stage. The combination of rotors and casings between bearings is a cylinder. This may include two flow paths. Specific volume, and therefore required steam path flow area, increases by a factor of 2000 from inlet to exhaust for subcritical turbines, and by a factor of 3000 for supercritical turbines. This is accommodated by increasing the blade height and rotor diameter from stage to stage through the steam path from the inlet through high- (HP), intermediate- (IP), and low-pressure (LP) cylinders; and, if necessary, by double-flow IP and several double-flow LP cylinders.

Blades and Seals

In a pure impulse stage, pressure drop occurs across the fixed blades and enthalpy is converted to kinetic energy.

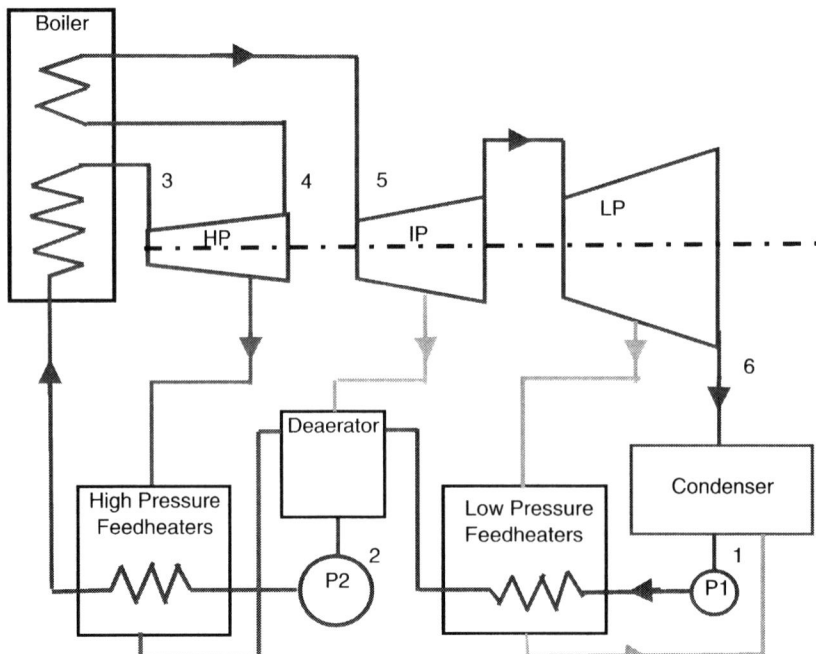

Fig. 5 The regenerative reheat rankine cycle.

Moving blades change the direction of the steam so that energy is transferred to the rotor entirely by change of momentum. Only Hero's concept is pure reaction. All blades have an impulse component. In reaction stages, the flow path between rotating blades is shaped to cause further pressure drop, expansion, and kinetic energy release, which is transferred to the rotor by reaction. The percentage of enthalpy drop across the moving blades is the degree of reaction. Also, in any blade, centrifugal effects in the space between fixed and moving blades increase pressure drop across the latter, and therefore the degree of reaction with radius. All blades have some degree of reaction, varying from a minimum of 5% in "impulse" to 50% in "reaction" designs. These differences have a marked effect on HP and IP design. In impulse design, stationary blades are arranged in rigid diaphragms with a labyrinth seal to the rotor. The diameter of the seal, and therefore leakage area, is minimized. Between seals, the rotor diameter is increased to form a wheel to which the rotating blades are attached. This construction is called "diaphragm and wheel." The pressure drop across moving blades is small, but sufficient to require a tip seal. In reaction design, it is greater and requires higher tip seal duty per moving blade row. It also causes an axial thrust on the rotor; therefore, annular area between the fixed and rotating blade seals must be a minimum, precluding diaphragm and wheel construction. Rotors of cylindrical or slightly tapering conical form, referred to as drum rotors, are used. Thrust is balanced either by a large annular rotor surface subject to steam pressure, a dummy piston; or by combining opposed flows in the same cylinder. For the same enthalpy drop, a reaction turbine requires about twice as many stages as an impulse turbine, but the two concepts have remained competitive through their development histories to present, both having very high efficiencies. Reaction stage efficiencies are higher, but differences in overall turbine efficiency are evidently small, and could only be identified from proprietary information. Diaphragm and wheel construction has the advantages of compactness and lower transient thermal stresses. In LP blades, differences between the two concepts diminish. The last stage can produce up to 10% of total turbine output and is most important in turbine efficiency. Steam volume is large and blades are up to 1.2 M long in 3000 rpm machines. There is a large change in blade speed from root to tip, whilst the axial steam flow is more nearly constant. To align with the resultant steam flow, the angle of the blade is twisted from near axial at the root to near tangential at the tip. Increases in the degree of reaction from root to tip due to centrifugal effects are large, and a marked change in blade profile is required. Blades with these features are known as vortex blades. Full analysis of LP blades is complex and has only become possible in the last two decades with the advent of computational fluid dynamics, correlated with experiments. Their optimization is one of the major recent advances in turbine design.

Losses

Turbine efficiency is determined by fluid dynamic, leakage, wetness, and leaving losses. Fluid dynamic losses due to surface friction and secondary flows are small in modern HP but more significant in impulse stages and LP blades, due to higher steam velocities. Curved nozzles, which lean in the tangential direction, have been developed to reduce secondary flows. Leakage losses due to steam bypassing the tips of moving and stationary blades are more significant in HP, particularly in reaction stages. In LP stages, when steam is wet, water droplet velocities are less than steam velocities and cause drag on, and erosion of, moving blades. Stage efficiency reduces by approximately 1% for 1% additional wetness. Within the steam path, kinetic energy leaving one stage is used in the next, but that leaving the last stage is lost in the condenser. "Leaving loss" is proportional to the square of exhaust velocity. The greater the diameter of the last stage blades and the number of exhaust flows (and therefore the greater the exhaust area), the lower the leaving loss. This is one of the most important issues in determining turbine size, cost, and efficiency. The design of the duct between the turbine exhaust and the condenser—the exhaust hood—is also important. Modern designs can produce a backpressure below condenser pressure, so increasing turbine output.

MECHANICAL DESIGN

Turbines must be designed for temperatures of up to 600°C; pressures of up to 300 bars; transient thermal stresses (mainly during startup and shutdown); and fluid dynamically, and mechanically induced vibration, corrosion, and erosion. In HP cylinders, centrifugal loading essentially depends on peripheral velocity, which is defined by blade efficiency. Optimum designs for pressure containment and minimum thermal stresses are obtained by minimizing rotor diameter and casing thickness, and therefore by achieving high rotational speed. For power generation, gearing is impractical; therefore, optimum rotational speed is either 3000 or 3600 rpm, depending on grid frequency. Impulse designs have advantages with regard to thermal stresses. They have a smaller rotor body, which reduces maximum values, which also do not occur at stress concentrations due to blade attachment details. For LP rotors with large blades, centrifugal loading depends on speed of rotation; therefore, 1500 or 1800 rpm allows the largest blades and exhaust area. For these reasons, until about 1970, large turbines were optimized by cross compounding with a high-speed HP/IP shaft and a low-speed LP shaft, requiring two generators and very large turbine halls. Developments in LP blade design now enable much more compact machines of over 800 MW to be built with 2 LP cylinders,

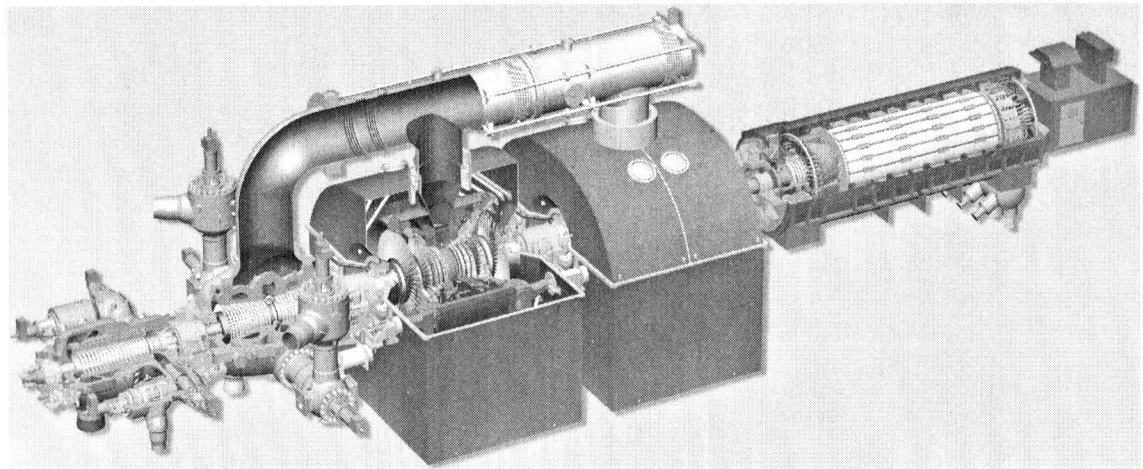

Fig. 6 Arrangement of 4 cylinder up to 1200 MW, 300 bar, 600°C, reheat steam turbine. (Courtesty of Siemens AG.)

and up to 1100 MW with 3 LP cylinders, both using a single high-speed shaft.

Creep properties of high-temperature sections, tensile strength of LP blades and rotors, and toughness to prevent failure of rotating components are the most important material issues. Turbines are large capital investments and must operate reliably for more than 30 years—200,000 h with minimum possible outages. In the past, design was evolutionary; but modern practice is increasingly influenced by engineering analysis of thermal, mechanical, and chemical operating and fault conditions; and by materials development. Extremely high reliability has been achieved. Overhaul intervals approaching 10 years are possible, justified by condition monitoring, and machines' lives extended to more than 40 years. Figs. 6 and 7 show a large modern reaction turbine with a single-flow HP cylinder, a double-flow IP, and two double-flow LP cylinders. Condensers are beneath the LP cylinders. This configuration is used for ratings to over 800 MW and supercritical steam conditions approaching 600°C and 300 bars with a single reheat. Rigid couplings connect all rotors, including the generator. One LP casing is anchored to the foundations to minimize expansion movement relative to the condensers. Other casings are supported on sliding feet to allow free longitudinal expansion. A single thrust bearing is positioned to minimize axial differential expansion between the rotors and casings. Bearings are set at different levels, matching the deflected forms of the rotors to minimize rotating bending stresses. High pressure and IP cylinders have inner and outer casings. The inner casing, which carries the stationary blades, contains the difference between the stage pressure and the cylinder outlet pressure. The outer casing contains the outlet pressure. This allows the casings to be thinner, minimizing

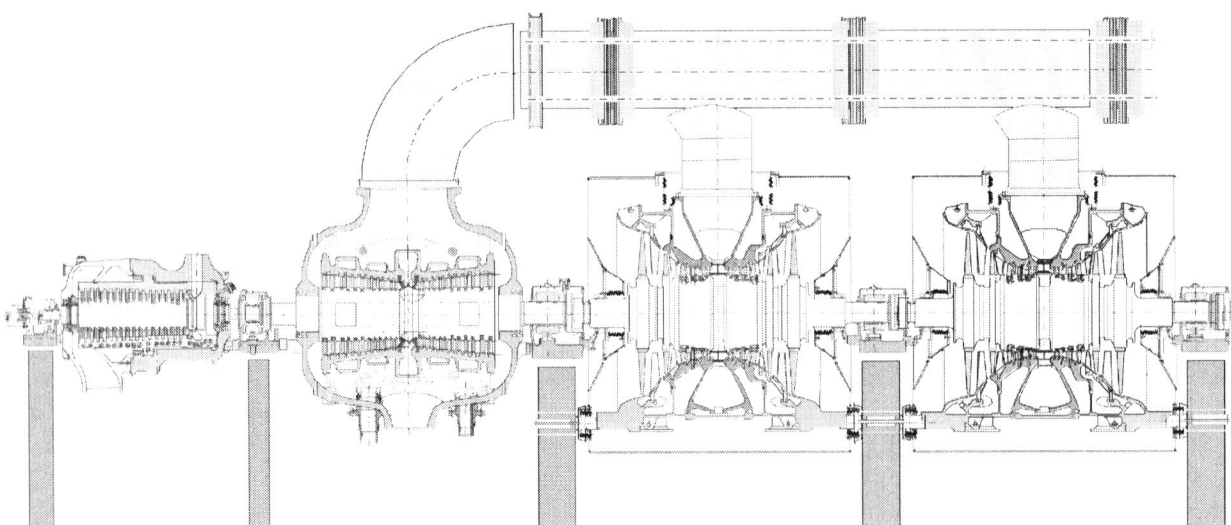

Fig. 7 Section of 4 cylinder up to 1200 MW, 300 bar 600°C, reheat steam turbine. (Courtesty of Siemens AG.)

thermal stresses. In other designs, pressure loading on the casing is further reduced by use of an impulse first stage and a separate inlet nozzle box containing the large pressure drop across the first-stage nozzle. This also reduces the number of stages at some cost in efficiency. Most turbines have horizontal bolted joints in both inner and outer casings to permit assembly and maintenance. Massive flanges with a very high ratio of bolt to flange area are required. Because the casing heats up more rapidly than the flanges during start up, rates must be limited to minimize thermal stress, distortion, and thermal fatigue. In impulse designs, diaphragms reduce leakage caused by casing distortion. In some designs, inner casing bolts are replaced by shrunk-on rings requiring complex assembly and disassembly. High pressure and IP Casings are steel castings; generally, 1%–2% chromium alloys for sub-critical turbines and 9%–12% Cr for supercritical turbines. Low-pressure casings have an inner nodular cast-iron casting carrying the blades, supported in a fabricated steel outer casing including the exhaust hoods. Fig. 8 shows assembly of an LP cylinder. High pressure and IP rotors are one-piece forgings in similar materials to those of the casings, alloyed to maximize toughness. Low-pressure rotors for fossil-fired applications are also normally one-piece forgings. Those for wet steam nuclear turbines comprise a shaft with shrunk-on disks to carry the blades. Compressive stresses in this design increase its resistance to stress corrosion cracking. Because of the high centrifugal loading due to very long blades, low-nickel, chrome, vanadium alloys with very high strength and toughness are used.

Blades are generally 12% chromium or higher steel alloy forgings. They are attached to the rotor by a mechanical connection consisting of a lobed male detail on the blade and an identical female detail on the rotor. The design most commonly used for LP blades is shown in Fig. 9. Similar concepts oriented circumferentially are used for attachment to drum rotors. Multiple pinning through a forked blade root and the disk, over which it is fitted, is also used. Mechanical design of last-stage blades is one of the major factors controlling turbine output and configuration. Length is determined by centrifugal loading. Titanium is now used to increase the material-strength-to-weight ratio. Blade vibration and avoidance of high-cycle fatigue are also major considerations. In the past, blades were rigidly connected in groups to control vibration. Two design approaches are now in use. The blade on the left of Fig. 9 is free standing, designed to ensure that no modes of vibration are excited in operation. The alternative is continuous circumferential connection, suppressing some modes and providing damping. Fatigue of inlet HP blades is also important when steam is admitted over only part of the circumference, and steam forces on the blade vary as it enters and leaves the admission zone.

Sealing is by "labyrinth packings." These consist of several thin metal disks, spaced axially and mounted on the casing with minimum clearances from the shaft. Resistance to leakage is provided by the small clearance gaps in series, and by creating a tortuous flow path. Materials are chosen to prevent damage to the shaft,

Fig. 8 Low pressure cylinder assembly. (Courtesty of Siemens AG.)

Fig. 9 Low pressure rotor. (Courtesty of Siemens AG.)

should a rub occur. Glands, seals between casings and rotors, are required to prevent outward steam leakage from HP and IP cylinders and inward air leakage to LP cylinders, which would reduce condenser vacuum. These consist of several labyrinth packings in series. Steam at IP is introduced between two packings. This controls the pressure gradient along the seal, minimizing leakage. Leakage increases the heat rate, and is minimized by intermediate leak-off paths routed to the feed system. Steam which passes through the whole assembly is captured by a gland steam condenser, maintained at subatmospheric pressure.

Hydrodynamic white metal journal bearings are used. For LP cylinders, bearings with length greater than diameter are required to obtain suitable bearing pressures. When the turbine is shut down, HP and IP rotors remain at temperatures at which permanent creep bending under self-weight would occur. To prevent this, it is essential to rotate the turbine slowly until it cools sufficiently. This is referred to as "barring." Under these conditions, the hydrodynamic oil film is not developed, and alternative pressure lubrication is normally provided.

TURBINE CONTROL

Valves in the steam chests, which can be seen in Figs. 6 and 11, control speed and load. If these are operated in parallel, the turbine is "throttle governed," and steam flow and pressures through the turbine are proportional to load. The disadvantage is that there is a loss of efficiency at part load due to throttling in the valve. Operating the valves sequentially can reduce this. At part-load, when only one valve is used, throttling is less. This is "nozzle governing." Throttling losses can also reduced by operating the boiler at variable pressure; i.e., sliding pressure control, used in combined cycle steam turbines. "Bypass governing," whereby steam is admitted partway down the steam path, can be used if the boiler must operate at constant pressure. Overspeed in the event of load loss and consequent damage is prevented by rapid closure of the valves. In this situation, energy stored in steam in the reheater and feed systems must be isolated from the turbine by further rapidly acting valves.

TURBINE TYPES

The machine shown in Figs. 6 and 7 is typical of those used in a modern fossil-fired plant. The predominant fuel is coal, but lifetime fuel costs are still several times capital cost; therefore, a supercritical plant is generally built. The most common size is about 600 MW, but up to 1200 MW is possible with two LP cylinders. For smaller sizes, the configuration shown in Fig. 10 is attractive and widely used. It is suitable for loadings up to 500 MW, with subcritical steam conditions of 177 bars and up to 600°C. The cylinder on the left includes opposed-flow HP and IP turbines. Steam enters near the center, expands to the left in the HP, is reheated to approximately the same temperature, re-enters near the center, expands to the right in the IP, then exhausts to the LP via the outer casing. This configuration confines high temperatures to the center, reducing thermal stresses and allowing more rapid starting. It is more compact and economic than separate HP and IP cylinders.

Pressurised Water (PWR), Boiling Water (BWR), and Canadian deuterium uranium (CANDU) nuclear plants generate saturated steam at approximately 75 bars and 295°C. It is dried, admitted to the HP turbine, and expanded until it is about 10% wet. It is then exhausted to water separators and superheated in a steam-to-steam heat exchanger before admission to the LP turbine. Low-pressure superheating increases average heat supply temperature and efficiency and limits wetness to about 10% at exhaust to the condenser. Such steam turbines drive four-pole generators and rotate at half the speed of conventional machines. For a given output, flow areas are doubled and velocities are halved. This results in very large, multicylinder turbines but has the advantage of reducing blade erosion by water droplets in wet steam and reducing leaving loss. Machines of up to 1500 MW with three opposed-flow LPs have been built.

Because of lower heat rates, capital costs, and construction times than conventional steam plant, combined cycle gas turbine (CCGT) is now the largest application of steam turbines. Gas turbines exhaust into individual heat recovery steam generators (HRSG's). One or more supply steam to the steam turbine. If only one is used, the gas and steam turbines are arranged on a common shaft, connected by a clutch to allow the gas turbine to be started first. To obtain maximum heat recovery, up to three steam pressures are used in the HRSG, and steam is admitted to the turbine at three positions along the steam path, increasing steam flow from inlet to exhaust. No feed heaters are used, and steam extraction is only for the deaerator. The importance of last-stage blade efficiency is increased. Fig. 11 shows a typical two-cylinder machine for combined cycle application. It is capable of up to 250 MW, with steam conditions of up to 600°C and 170 bars. It is designed for installation on a flat foundation with an end-mounted condenser.

Steam turbine generators are frequently combined with process plants. This is called cogeneration. If the process requirements and generation requirements are fixed, steam is produced at a higher pressure and expanded in a back pressure turbine driving a generator, then exhausted at the pressure required for the process plant. If the steam required for the process is less than that for generation, a pass-out or extraction turbine is used, process steam is extracted from the turbine after partial expansion, and the remainder fully expanded before exhaust to the condenser.

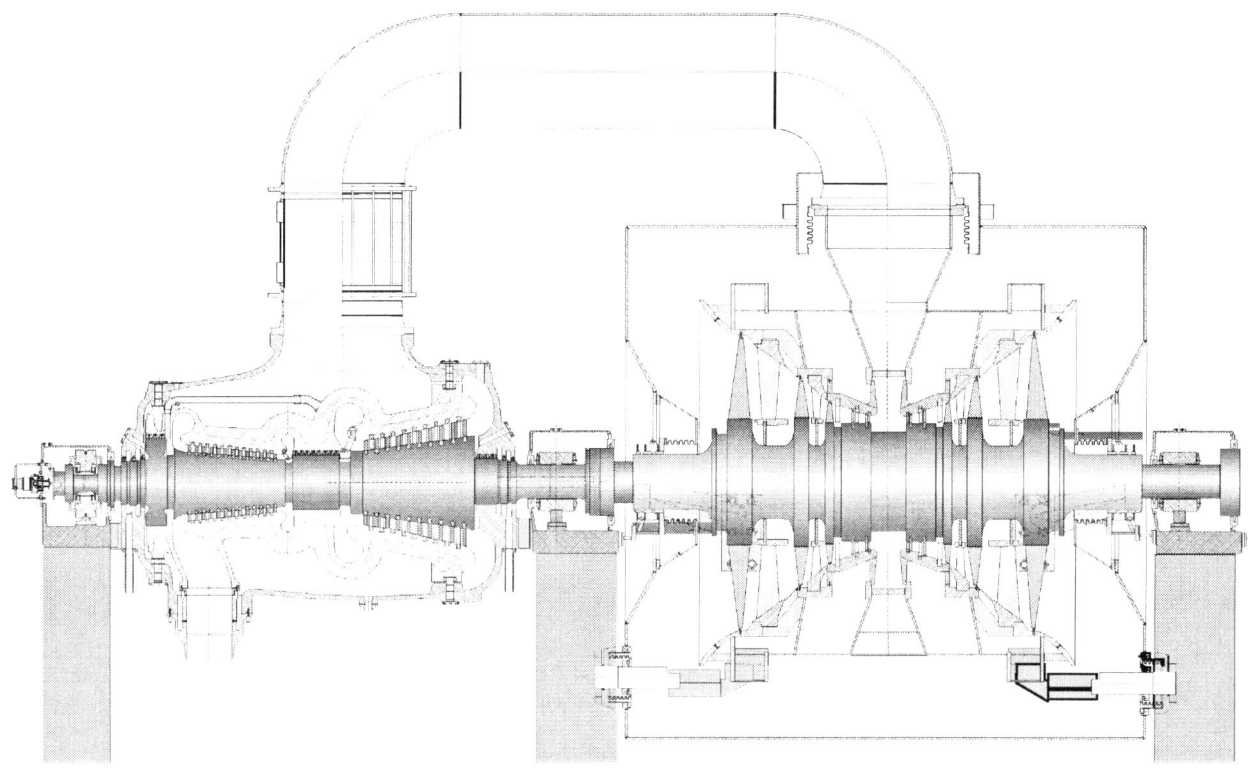

Fig. 10 Arrangement of 2 cylinder up to 500 MW, 177 bar, 600°C, reheat steam turbine. (Courtesty of Siemens AG.)

DEVELOPMENT

Steam turbine technology is thoroughly evolved, and the limits set by material properties and fluid dynamic efficiency are well defined. Applications depend on other technologies. Combined cycles require relatively simple machines with emphasis on maximum exhaust efficiency. Developments will probably be the more complex steam

Fig. 11 Arrangement of 2 cylinder up to 250 MW, 177 bar, 600°C, reheat steam turbine, suitable for combined cycle applications. (Courtesty of Siemens AG.)

path requirements of advanced gas turbines using steam-cooled gas turbine blades and integrated gasifiers. Coal remains a major energy source, and the supercritical steam plant with exhaust conditioning is an economically and environmentally competitive technology for burning it. Related developments are supercritical steam turbines for steam conditions above 700°C, longer last-stage blades, and units with larger outputs and fewer cylinders. Steam turbines are essential for power production from water-cooled nuclear reactors, and steam conditions may increase and unit ratings may reduce in the future.

CONCLUSION

The steam turbine is an elegant concept, capable of converting the energy of steam into mechanical work solely by rotary motion. Condensation of steam enables exceptionally high pressure ratios and energy extraction. Its characteristics as a prime mover are ideal for electricity generation. A steam power plant is capable of over 40% overall energy efficiency, firing a wide range of fuels. The turbine's mechanical elegance results in low maintenance requirements and long service life. For these reasons, it has remained predominant in this application for a century. In this period, efficiency and power density have increased by factors of more than 4, and remarkably, two alternative steam turbine technologies have developed competitively throughout. It had a major role and underwent major development in marine propulsion. This ended because of changes in shipping economics and large, highly efficient diesel engines. Sole predominance in power generation is now ended by large gas turbines, but only combination with the exceptional energy recovery of the steam tubine enables efficiency over 55% to be achieved in combined cycles. These can be adapted to all fuels by the use of gasifiers. Applications of large turbines for nuclear reactor power cycles and coal-fired steam plants with potential for power outputs of up to 1700 MW will continue; but their relative importance depends on a resurgence of nuclear power in response to global warming, the availability of gas to fire CCGT, and advanced clean coal technologies, including gasifiers.

ACKNOWLEDGMENT

Siemens, A.G., for comments and provision of the illustrations included in this article.

BIBLIOGRAPHY

1. Cengel, Y.A.; Bowles, M.A. Vapour and combined power cycles, 4th Ed. In *Thermodynamics, an Engineering Approach*; McGraw-Hill: New York, 2002; 541–547, [Chapter 9].
2. Kearton, W.J. Flow of steam through impulse and impulse—reaction turbine blades. In *Steam Turbine Theory and Practice*; Pitman: London, 1966; 172–227, [Chapters 7–8].
3. El Wakil, M.M. Turbines, 1st Ed. In *Powerplant Technology*; McGraw-Hill: New York, 1984; 173–210.
4. Kehlhofer, R.H.; Warner, J.; Nielsen, H.; Bachmann, R. *Components, Combined-Cycle Gas Steam Turbine Power Plants*, 2nd Ed.; Penwell: Oklahoma, 1999; 155–188.
5. Nicholas, D.G.; Burn, R.E. Marine steam turbines at the centenary. Proceedings of First Parsons International Turbine Conference, University of Dublin, 1984 The Parsons Press: Dublin, Ireland, 1984, 17–36.
6. Mitchell, J.M.; Grant, J. Design and fluid dynamic development of steam turbine blading. Proceedings of First Parsons International Turbine Conference, University of Dublin, 1984, The Parsons Press: Dublin, Ireland, 1984, 81–91.

Sustainability Policies: Sunbelt Cities

Stephen A. Roosa
Energy Systems Group, Inc., Louisville, Kentucky, U.S.A.

Abstract
It has been suggested that "sustainability can provide a qualitative measure of the integrality and wholeness of any given system.[1] This article focuses on evidence indicating that a positive relationship may exist between the adoption of sustainability as a local goal and the rates of local energy policy adoption in Sunbelt cities. The evidence suggests that policies responding to energy and environmental issues may provide long-term solutions to achieving urban sustainability.

The fundamental question becomes "What evidence is there that cities with sustainability as a stated goal have higher rates of local energy policy adoption?" In this article, it is asserted that cities with sustainability as a local goal are more likely to adopt certain energy related policies. It is also determined that there is variability in the implementation of energy related policies in Sunbelt cities. The author concludes that Sunbelt cities with sustainability as an urban goal have higher rates of energy policy adoption when the three selected policies are studied as indicators.

INTRODUCTION

Cities have common energy concerns that impact their urban environments. Cities seeking to incorporate policies that lead to sustainability generally consider energy policy to be a critical component of their urban agendas. These concerns are manifested in urban policies and programs designed to achieve results that include reducing energy usage with the potential of improving the environment. The purpose of this research is multifold:

- Determine which Sunbelt cities have sustainability goals.
- Provide a descriptive comparison of select energy-related policies in Sunbelt cities.
- Identify the specific types of policies that are being adopted and pursued.

In this article, the concept of sustainability is defined and cities with sustainability agendas are identified. The 25 largest cities in the Sunbelt (noted in the attached tables) have been selected for consideration. This research provides an assessment of the selected cities based on the energy-related policies they have adopted. To provide evidence of policy adoption, three locally adoptable energy-related policies are considered. The three policies considered are (1) city operated energy efficiency programs, (2) local governmental energy program support, and (3) Energy Star™ program participation.

Keywords: Energy Policy; Sustainability; Local Energy Policies; Sunbelt.

URBAN SUSTAINABILITY

Sustainability is a broadly defined concept that has a variety of meanings. Urban sustainability refers to an idealized model of urban development that attempts to address concerns about urban growth, patterns of urban development, and issues that arise as urban development occurs. According to Beatley,[2] the four principles of urban sustainability include (1) the principle of urban management, (2) the principle of policy integration, (3) the principle of ecosystems thinking, and (4) the principle of cooperation and partnership. How cities choose to manage energy policy involves each of these principles. Energy policy requires management, needs to be integrated with other urban policies, impacts many ecosystems, and requires cooperation and partnership to be successfully pursued.

For the purposes of this article, sustainable development is defined as the ability of physical urban development and urban environmental impacts to sustain long term inhabitation by human and other indigenous species while providing (1) an opportunity for environmentally safe, ecologically appropriate physical development; (2) efficient use of natural resources; (3) a framework which allows improvement of the human condition and equal opportunity for current and future generations; and (4) manageable urban growth. Non-sustainable urban development is the antithesis of sustainable urban development; it implies growth that is environmentally unsafe, consumes resources inefficiently, degrades the human condition, is characterized by persistently unmanageable development, and fails to value social equity. The energy policies cities choose to adopt are critical to the success or failure of urban sustainability programs. Inappropriate use of energy can cause substantial damage to economies, the environment and create the need for corrective policies. Energy is a resource that begs to be used efficiently. Gross consumption of available energy resources can cause a tragedy of the commons, preventing future use of the resource. Urban growth can be impacted by dependence on fossil fuels.

COMPARING SUNBELT CITIES AND THEIR POLICIES

Why are Sunbelt cities and their sustainability policies particularly interesting? Sunbelt cities are significant centers for urban population growth and development in the United States, generally outpacing their non-Sunbelt counterparts. Many of the selected cities in this broadly defined region (e.g., Las Vegas and Phoenix) are among the fastest growing cities in the United States. A few are leaders in implementing innovative policies. Other cities used in this study, including Atlanta, Austin, Phoenix, Nashville, and Oklahoma City, are their state capitals, making them centers of statewide decision making. With new investment and construction, Sunbelt cities have the opportunity to select from a range of newly available technologies when growing their cities. Their policies will ultimately impact the design of the cities, their future energy usage, and the sustainability of their urban areas.

The cornucopia of policies available to urban regimes to achieve reductions in energy use might include transportation system policies, energy management programs, organizational memberships, and policies designed to improve the environment. However, this examination focuses on three selected indicators: (1) city-operated energy efficiency programs, (2) local government energy program support, and (3) Energy Star program participation. The selected policies require local initiative to implement and sustain. The energy-related policies selected have broad applicability and are available for adoption in some form by all urban areas considered in this study. The first category of policy indicators focuses on whether or not sustainability is an adopted local goal.

SUSTAINABLE DEVELOPMENT AS A POLICY GOAL

This indicator poses a fundamental question: Is sustainable development or urban sustainability a stated goal of the city? Ancient southwestern monuments such as Mesa Verde and Chaco Canyon provide testimony that cities in the southwest may have been abandoned as a result of environmental mismanagement and changes in local environmental conditions. Many southwestern mining towns grew to become boomtowns, only to go bust and ultimately become ghost towns after their resources were no longer considered exploitable. Perhaps even today, we are growing new throw-away cities. If sustainability is not an identified goal of the urban agenda, then it would seem unsurprising if it is not achieved.

The use of the language of sustainability in local policymaking is a relatively recent phenomenon with variable interpretations. To consider this indicator, it must be accepted that there are multiple definitions of sustainability and that the various definitions are subject to a wide range of interpretations. Ignoring the conundrum of interpretations, this policy indicator gauges only whether sustainable development or sustainability has become a stated urban goal. However, the contextual interpretation of the concept of sustainability when searching databases is discriminately limited for the purposes of this article. For example, if a city's only stated "sustainable" policy is to "maintain a sustainable tax base," then the term is judged to be misapplied and the application discredited.

Of the 25 cities in this study, a total of seven cities (28%) have established sustainability as a primary urban goal. These cities are Jacksonville, El Paso, Long Beach, Albuquerque, Atlanta, Mesa, and Tulsa. How cities implement sustainability policies varies. In 1985, Jacksonville initiated its "Quality of Life in Jacksonville" program.[3] The pledge of Atlanta's City Council President to "create an efficient, vibrant and sustainable city" includes a 2002 energy conservation initiative.[4] In Tulsa, urban sustainability is a primary administrative goal, strongly supported by the city administration. Long Beach is atypical in that its 2010 Citywide Strategic Plan identifies "becoming a sustainable city" as a primary strategic goal. There is a statement in the *Vision of the Mesa 2025 General Plan* to support the city "as a sustainable community in the 21st century."

Seven additional cities (28%) have identified programs to support sustainable building policies, have demonstration projects underway, or have established land use requirements to promote sustainable development. San Antonio has a program to support community revitalization with a goal structure that includes "sustaining a strong urban system."[5] The City of Tucson lists a developer-driven project for the new community of Civano to develop a model sustainable community. The goal of the Civano project is "to create a new mixed use community that attains the highest feasible standards of sustainability, resource conservation, and development of Arizona's most abundant energy resource—solar—so that it becomes an international model for sustainable growth."[6] In Los Angeles and Austin, sustainable building programs or guidelines have been established for new construction.

While 14 cities (56%) have identified sustainability as a goal or have related sustainable development policies, the remaining 11 (44%) Sunbelt cities have not established sustainability as an urban goal. It is possible to speculate that for this set of Sunbelt cities, being perceived as having a local goal of "being sustainable" may be unimportant, not considered a priority, counter to the goals of the local regime, or under consideration but not yet implemented.

Table 1 Sustainability as a local policy

Sunbelt city	Sustainability as a local policy goal
Los Angeles, CA	Yes, sustainable building program
Houston, TX	No
Phoenix, AZ	Yes, found in land use plan
San Diego, CA	Yes, goal of Environmental Service Department
Dallas, TX	No
San Antonio, TX	Yes, specific program goal
Jacksonville, FL	Yes
Austin, TX	Yes, established sustainable building guidelines
Memphis, TN	No
Nashville/Davidson, TN	No, excluded in planning mission statement
El Paso, TX	Yes, included as goal in city mission statement
Charlotte, NC	No, focus is on "smart growth"
Fort Worth, TX	No, focus is on "smart growth"
Oklahoma City, OK	No
Tucson, AZ	Yes, Adopted Sustainable Energy Code
New Orleans, LA	No, stated as goal of utility
Las Vegas, NV	No
Long Beach, CA	Yes, sustainability is the primary urban goal
Albuquerque, NM	Yes, Sustainable Community Development
Fresno, CA	No, excluded as goal in planning mission
Virginia Beach, VA	No, excluded from vision statement
Atlanta, GA	Yes, administrative goal of Mayor's office
Mesa, AZ	Yes, included in General Plan for 2025
Tulsa, OK	Yes, administrative goal of Mayor's office
Miami, FL	Yes

Among these cities are Houston, Dallas, Fresno, and Las Vegas. Table 1 provides a summary of cities and indicates which have sustainability agendas.

Other cities in this category, including Charlotte and Fort Worth, have programs with a focused development policy effort based on a "smart growth" agenda. However, policies associated with achieving smart growth agendas are not necessarily sustainable development initiatives. Having identified cities with sustainability as a local goal, three types of energy-related policies will be considered.

CITY OPERATED ENERGY EFFICIENCY PROGRAMS

Cities are owners of many and varied municipal facilities. The types of buildings owned by cities include courthouses, office buildings, educational facilities, fire and police stations, sewage treatment facilities, emergency action and preparedness centers, libraries, public health facilities, training facilities, and subsidized housing. These facilities collectively consume significant amounts of energy. Urban regimes and their administrators may view energy use as a matter of serious concern worthy of attention, as an uncontrollable overhead expense, as an unavoidable but manageable cost, as a concern only if publicly scrutinized, or as inconsequential.

The idea of civic engagement and the ethic of institutional stewardship have been linked to improving sustainability.[8] It seems logical that city governments adopting sustainable policies would be concerned with the costs and impacts of energy in their buildings and facilities. This policy indicator gauges whether the city government has an internal energy efficiency program. The actions taken by local administrations would likely be manifested in policies that support energy efficiency improvements such as the installation of energy-saving technologies, building envelope and architectural improvements, equipment replacement, and adoption of building standards among others. In order to implement facility improvements, partnerships such as performance contracts might be considered useful.[7]

This energy policy indicator provides a gauge of the importance of sustainability to local policy planners. Does the city government feel energy conservation and energy efficiency in its own buildings is important enough to warrant concerted effort? This research indicates that the administrations of most Sunbelt cities feel that energy management programs are important. In this sample of 25 Sunbelt cities, 19 (76%) have initiatives to improve energy efficiency in public buildings, while only six (24%) lacked such programs. What is striking is the wide range of approaches that cities have chosen to employ. Sample efforts implemented by Sunbelt cities include the following:

- Having a departmental division in city government for Energy Conservation and Management (e.g., San Diego).
- Hiring a City Energy Manager and implementing recommended improvements to manage and reduce energy use.

- Establishing a written energy policy for government owned buildings.
- Requiring energy assessment surveys of city-owned buildings to determine economically appropriate actions and alternatives to reduce energy use.
- Mandating the use of "Green Building" construction techniques or incorporating standards such as those required by Leadership in Energy and Environmental Design (LEED) for new construction.
- Participating in packaged programs such as the U.S. Department of Energy's (USDOEs) Rebuild America Program.
- Using energy-saving performance contracts (ESPC) as a vehicle for facility improvements with third party financing, subsidized by energy savings and cost avoidance.

The most popular policy effort among the sampled cities (with seven cities participating) was found to be the adoption of the principles and requirements of the USDOE-sponsored Rebuild America program. Rebuild America is a "network of community-driven voluntary partnerships that foster energy efficiency and renewable energy in commercial, government and public housing programs" that "works to overcome market barriers that inhibit the use of the best technologies."[9] This program is geared toward policies that reduce facility energy use while lowering the costs of energy. Among those participating in Rebuild America are the four largest Sunbelt cities (i.e., Los Angeles, Houston, Phoenix, and San Diego).

Phoenix has budgeted over a million dollars annually through 2005 to directly fund capital-intensive energy conservation improvements. Dallas, Austin, and Long Beach have adopted Green Building or LEED construction standards for city-owned buildings. San Diego's Environmental Services Operations Station administration building has carports in its parking lot with rooftop photovoltaic panels that generate 91,500 kWh per year.[10]

With over two-thirds of the city below sea level, New Orleans has concerns about rising sea levels, which threaten to displace its urban residents. As a result, in October 2001, New Orleans adopted a unique policy to reduce the threat of global warming. Greenhouse gas emissions have been profiled and municipal emission reduction targets have been mandated through 2015. Their research revealed that municipal buildings were responsible for approximately 35% of the CO_2 emissions released from municipal operations. Mitigation measures, justified by energy savings, were implemented in a number of buildings including City Hall, the court complexes, the public library, police headquarters, the airport, and others. These measures include mechanical system upgrades, installation of energy efficient lighting systems, tree planting, installation of LED new traffic signals, establishment of building energy codes for city buildings, and measures to reduce the urban heat island effect. Despite this epochal and precedent setting policy initiative, it is obvious that the actions of one city will not resolve the problems associated with global warming.

Atlanta's Energy Conservation Program exemplifies those programs that offer tangible and measurable financial returns. Atlanta's program included efforts to schedule policy workshops, perform utility rate assessments for over 600 municipal accounts, perform energy audits, and develop an internal employee energy conservation program. Within one year after initiating the program, the city had projected savings of nearly $500,000 from these initiatives.[4] In addition, the city has established a policy goal to reduce energy consumption by an additional 10% by 2010 and has appointed an Energy Conservation Coordinator.[4]

LOCAL GOVERNMENTAL PROGRAM SUPPORT

This policy indicator gauges policy support by the city government for local energy conservation, energy efficiency, and alternative energy programs. This measure asks if the local governments are active in promoting energy related programs within their respective communities.

Local energy conservation efforts can be supported by citizen actions, organizational support, corporations, utilities, local governments, other governmental bodies, or by other means. Local participation and involvement are central to the idea of sustainable cities.[1] Local governments have the primary economic means and leadership infrastructure to direct the orientation of community energy policies should they desire to assume such a role.

Among the local government entities, 14 of the 25 cities in the sample offer some sort of policy or program to support energy conservation or alternative energy, or to provide incentives for complementary technologies for new construction in their local communities. An approach that is used by seven cities is to provide financial incentives or support for community projects involving new building construction that incorporates green building technologies. The City of San Antonio has established the Metropolitan Partnership for Energy, a partnership of the city government and the community at large. The partnership has established an energy council, educational programs, facility and infrastructure improvements, equipment conservation measures, fleet conservation standards, and procurement requirements. Tucson and Las Vegas are among those Sunbelt cities that have adopted building code requirements for energy-efficient construction.

Other cities are less committed and subsidize less extensive programs. While the City of El Paso has a program, perusal of the city budget indicates that only $7500 is allocated annually. Fresno's energy policy provides for weatherization assistance for the homes of senior citizens. There are 11 cities among those sampled that lack any active energy conservation program, alternative energy policy, or support for similar local initiatives.

Combining the results from researching local governmental energy policies or programs, it was determined that (1) a total of 14 cities (56%) meet the requirements of this policy measure and have established policies or programs supported by local government, and (2) eleven cities (44%) lack policies supported by the local government.

ENERGY STAR™ PARTNER

This indicator asks whether or not the city participates as a member in the Energy Star™ program. Energy Star™ partners include manufacturers, retailers, utilities, builders, and governments. While partnership is voluntary, there are commitments to which members must agree. To become a partner, organizations must (1) sign a memorandum of partnership committing the organization to continuous improvement of energy efficiency; (2) measure, track and benchmark energy performance; (3) develop and implement a plan to improve energy performance; and (4) educate staff and the public about the partnership and achievements of the program.[11] For urban governments, the opportunity as an Energy Star™ partner is to use the label to support equipment purchasing decisions, to improve energy planning strategies, and to make better decisions concerning energy-related facility improvements.

Energy Star™ is a voluntary labeling program started in 1992 that is cosponsored jointly by the U.S. Environmental Protection Agency (USEPA) and the USDOE. The focus of the labeling program concerns buildings and the energy-consuming equipment within them. Office products, mechanical equipment, lighting systems, electronics, appliances, and other products are labeled, indicating that they can be promoted as being energy efficient. The Energy Star™ label has been extended to include new construction including homes, commercial structures and industrial buildings. According to its website, "through its partnerships with more than 7000 private and public sector organizations, Energy Star™ delivers the technical information and tools that organizations and consumers need to choose energy-efficient solutions and best management practices."[11] Energy Star™ also offers a building energy performance rating system which has been used for over 10,000 buildings throughout the United States. By leveraging private and governmental partnerships, Energy Star™ has proven to be among the most cost-effective programs sponsored by the U.S. government.

For policymakers in other cities, meeting the partnership requirements might be viewed as being too costly to support and implement. A commitment to specify energy-efficient equipment might be associated with higher initial costs. A fulltime energy engineer might be required to baseline energy usage targets and establish goals. Partnership requirements might also be viewed as potentially intrusive for city administrations who consider it politically ill-advised or otherwise undesirable to advertise their ever-increasing energy costs. City administrations agreeing to measure and track energy performance, implement a plan, and improve energy performance may be subject to public scrutiny should they fail to meet objectives. As a result, some local administrators may not adopt the Energy Star™ program due to the perceived potential for political risk.

Due perhaps to these and other considerations, only 10 of the 25 sampled Sunbelt cities (40%) have become Energy Star™ partners. These cities are Los Angeles, Houston, San Diego, Dallas, Fort Worth, Tucson, Las Vegas, Albuquerque, Atlanta, and Miami. Consider the case of Mesa, which is among those cities that has established sustainability as a primary urban goal, yet has not chosen to be an Energy Star™ partner. On the other hand, Las Vegas, Houston, Fort Worth, and Dallas are examples of cities that have not adopted sustainability goals, but happen to be Energy Star™ partners.

Table 2 provides a summary of the selected policy indicators and identifies which Sunbelt cities have adopted each of the policies considered in this assessment.

FINDINGS

To determine if cities with sustainability as a local goal tend to adopt more energy-related policies, the Sunbelt cities are considered in two groups. It was stated that 14 (56%) of the Sunbelt cities have identified sustainability as a goal or have related sustainable development policies. These cities have adopted an average of 2.07 of the three energy-related policies that were considered. The remaining 11 (44%) Sunbelt cities have not established sustainability as an urban goal. These cities have adopted an average of only 1.18 of the three selected policies.

There are five cities that have adopted sustainability as a local goal (Los Angeles, San Diego, Tucson, Albuquerque, and Atlanta) and have also adopted all three of the considered energy-related policies. In addition, there are two cities that have not adopted sustainability as a local goal (Houston and Las Vegas) that have adopted all three of the considered policies. Far more cities from the sample that have adopted sustainability as a local goal have adopted all three energy related policies. Alternatively, three cities that have not adopted sustainability as a local goal have not chosen to adopt any of the three considered policies (Charolette, Virginia Beach, and Oklahoma City).

This evidence suggests that cities that have identified sustainability as a goal or have implemented related sustainable development policies are more likely to adopt energy-related policies than those that do not.

CONCLUSION

In this research, it was found that the majority of Sunbelt cities have adopted sustainability as an urban goal. It was

Table 2 Energy policy indicators

Sunbelt city	Policies for city buildings	Sponsored by local government	U.S. DoE energy star partner	Total policies
Los Angeles, CA	Yes	Green Building Initiative, Green LA	Yes	3
Houston, TX	Yes	Rebuild America, LEED Program	Yes	3
Phoenix, AZ	Yes	Yes, capital improvement projects	No	2
San Diego, CA	Yes	Green Building, Rebuild America	Yes	3
Dallas, TX	Yes	None	Yes	2
San Antonio, TX	Yes	Metro Partnership for Energy	No	2
Jacksonville, FL	Yes	None	No	1
Austin, TX	Yes	Green Building Program	No	2
Memphis, TN	Yes	None	No	1
Nashville/Davidson, TN	Yes	None	No	1
El Paso, TX	No	Yes	No	1
Charlotte, NC	No	None	No	0
Fort Worth, TX	Yes	None	Yes	2
Oklahoma City, OK	No	None	No	0
Tucson, AZ	Yes	Model Energy Code (1994)	Yes	3
New Orleans, LA	Yes	None	No	1
Las Vegas, NV	Yes	Adopted Model Energy Code	Yes	3
Long Beach, CA	Yes	Yes	No	2
Albuquerque, NM	Yes	Yes, included in 1994 strategic plan	Yes	3
Fresno, CA	No	Senior Citizen Weatherization	No	1
Virginia Beach, VA	No	None	No	0
Atlanta, GA	Yes	Yes	Yes	3
Mesa, AZ	Yes	None	No	1
Tulsa, OK	Yes	None	No	1
Miami, FL	No	Green Building Program	Yes	2

also determined that Sunbelt cities vary in their approaches to implementing energy-related policies. In addition, three specific locally adoptable energy-related policies were considered in detail: (1) city-operated energy efficiency programs, (2) local governmental energy program support, and (3) Energy Star™ program participation. These policies are qualitative indications of the types of programs being pursued by the 25 sampled Sunbelt cities.

The specific energy-related policies adopted by each city were discussed in detail, and the sorts of policies in effect were identified. Evidence was provided that in many Sunbelt cities, policies are in effect to manage and reduce urban energy use. While these policies can be categorized as indicators of energy policy, local policy efforts, and organizational memberships, it is clear that there are variations in the themes of how these policies are locally defined and placed into practice.

Also notable are the examples of cities that use energy policy and energy conservation goals in their agendas as a means of achieving sustainability. Atlanta's program includes an internal energy conservation initiative. Tucson is developing a sustainable community based on use of solar energy. Mesa has created a planning agenda based on sustainability and is among those that have established an energy conservation program for city owned buildings.

It was discovered that the sampled Sunbelt cities vary broadly in their selection and application of policies. Certain cities, including Atlanta, Los Angeles, and San Diego, aggressively pursue multifaceted policies and focus their resources and agendas accordingly. On the other hand, most Sunbelt cities are more selective and limited in their policy choices. Cities such as Charlotte, Virginia Beach, and Oklahoma City are among those that tend not to adopt energy-related local policies.

Finally, it was determined that cities that have identified sustainability as a goal or have related sustainable development policies have substantially higher energy policy adoption rates than those that do not, when three select energy-related policies are used for comparative purposes. This suggests a positive relationship between the adoption of sustainability as a local governmental goal and the implementation of local energy related policies in Sunbelt cities. This research suggests that cities with sustainability as a local goal are more likely to adopt certain energy-related policies. Individuals and organizations seeking ways to get energy-related policies adopted by local governments in the Sunbelt might benefit by promoting sustainability as an urban goal.

REFERENCES

1. Bell, S.; Morse, S. *Sustainability Indicators, Measuring the Immeasurable*; Earthscan: London, 1999; pp. 103–124.
2. Beatley, T. *Green Urbanism: Learning from European Cities*; Island Press, Sage Publications: Washington, DC, 2000; 17.
3. Portney, K.E. *Taking Sustainable Cities Seriously*; The MIT Press: Cambridge, MA, 2003; 213.
4. City of Atlanta Online. *Energy conservation program history*, www.ci.atlanta.ga.us.mayor.energyconservation, 2003; 1.
5. City of San Antonio. *City of San Antonio Public Works*, www.sanantonio.gov, 2002.
6. City of Tucson. *Civano impact system memorandum of understanding in implementation and monitoring process*, June 26, 1998; 1.
7. Presidents Council on Sustainable Development. *Sustainable America—A New Consensus*; U.S. Government Printing Office: Washington, DC, 1996.
8. Roosa, S.A. A discussion of the economic aspects of implementing energy conservation opportunities. In *Solutions for Energy Security and Facilities Management Challenges*; Association of Energy Engineers: Atlanta, 2002; pp. 323–325.
9. United States Department of Energy. *Weatherization and intergovernmental program—rebuild America*, http://www.rebuild.org/index.asp, 2003.
10. City of San Diego. *City of San Diego Mayor Dick Murphy commissions the city's first "energy independent" municipal building utilizing solar power*, Environmental Services News Release, October 18, 2002; 1.
11. United States Environmental Protection Agency. *Join Energy Star—improve your energy efficiency*, www.energystar.gov, 2004.

Sustainable Building Simulation

Birol I. Kilkis
Green Energy Systems, International LLC, Vienna, Virginia, U.S.A.

Abstract
As the need for sustainable development increases, building simulation is becoming more crucial, and it is heading towards new challenges, dimensions, and concepts beyond the building envelope. Buildings are not isolated entities, which are not just responsible for about 35% of the total annual energy demand, but dynamically interact with their environment whose affected perimeter may be far wider than thought. In order to expand the building simulation perimeter to the entire impacted environment, all relevant variables must be factored in on a common base with a uniform metric. Among many attempts in this direction, exergy analysis establishes a uniform metric on all grounds and promises to make building simulation tools to more environmentally conscious.

INTRODUCTION

The objective of this entry is to highlight the importance of building simulation for analyzing and designing truly green and high performance buildings for a sustainable future. It also describes the need for new simulation tools that can cover a wider energy utilization and environment window, including the second law of thermodynamics, in a wide scope beyond the building envelope. More than ever, we need to know how efficient, stable, safe, functional, comfortable, and environmentally sustainable a building is under different indoor and outdoor conditions, functional and occupational requirements, and various equipment and system dynamics. Existing building simulation (BS) tools generally apply to a single building in a narrow window of the environment. These simulation and modeling tools allow us to design with methods to analyze, optimize, and control a building by calculating and compiling the crucial information regarding the overall performance of the building as accurately, precisely, and completely as possible, so that energy savings and a comfortable indoor environment are achieved in the building envelope with minimum possible cost. Only a few of the latest tools, however, include an environmental footprint analysis of a building, and these are rather limited.

NEED FOR BUILDING SIMULATION

People spend about two-thirds of their time indoors, with diverse and often conflicting needs relating to indoor air quality. According to several standards, including American Society of Heating, Refrigerating, and Air-Conditioning Engineers (ASHRAE) Standard 90.1 and ASHRAE Standard 90.2,[1,2] the primary targets are energy efficiency and indoor air quality, which conflict. According to the U.S. Department of Energy (DOE) Energy Efficiency and Renewable Energy Office (EERE), there are 81 million buildings in the U.S., of which 75% were built before 1979 and need a substantial retrofit or replacement of their Heating, Ventilating, and Air-conditioning (HVAC) systems. Fifteen million more buildings are expected by 2010. As a result, residential and commercial buildings in the U.S. are responsible for about 39% of the annual U.S. primary energy consumption, more than 70% of the total electric power consumed,[3] and close to 40% of CO_2 emissions (Fig. 1).[4] Buildings, which use natural gas for HVAC and domestic hot water, produce 20% of the total CO_2 gas emissions, estimated to be responsible for 60% of the greenhouse effect.

Consequently, buildings are the focal point of the sustainability quadrilemma of energy, economy, environment, and people. In order to resolve this quadrilemma, new building and retrofit designs must establish an optimum balance among all elements of environment, economy, people's needs, and rational use of energy resources. If it were possible to utilize the abundant, unused, low-exergy renewable and waste energy resources, all existing buildings could be heated and cooled. However, conventional HVAC systems cannot couple with low-exergy energy resources directly—because of their temperature incompatibility with low-exergy energy resources—unless HVAC equipment is oversized, or the resource temperature is conditioned. Both measures are cost-, energy-, and space-intensive, which greatly diminish the advantages of renewable and waste energy resources. This complexity requires developing new HVAC systems, with an outreach to the environment beyond the building envelope and a special emphasis on building-environment-energy source

Keywords: Building simulation; Exergy efficiency; Low-exergy buildings; Sustainable buildings; Sustainable development.

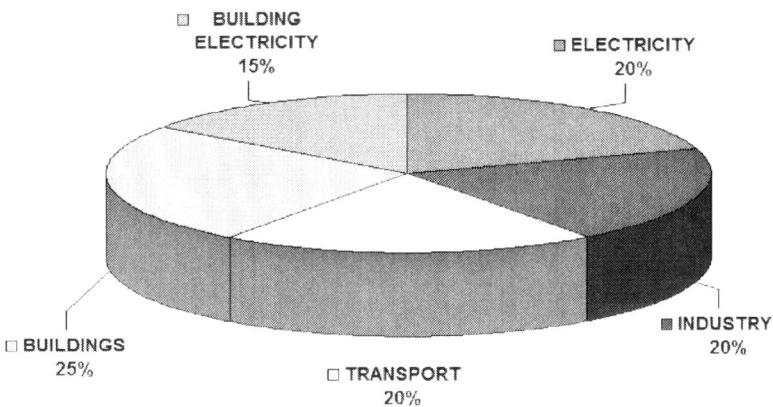

Fig. 1 Carbon emissions from fossil fuels in different sectors.
Source: From Refs 3 and 4.

relations. Today's green buildings may not be truly green unless the simulation window expands beyond the building envelope, with a clear understanding of exergy. Fig. 2 shows sample layers and scale of ideal building simulation windows for sustainability. Today there are almost 300 building simulation tools in 21 countries. However, in spite of this large availability and easy access, they do not yet address the overall picture; they are limited to the building envelope or its close vicinity.

ENERGY AND EXERGY ASPECTS OF BUILDINGS

"The absurdity of cutting butter with a chainsaw is immediately obvious to anyone."[5] On the same token, a conventional HVAC system uses high-grade energy in low-grade space heating or cooling and degrades the original energy resource, most often fossil fuel. According to Dincer and Rosen,[6] "Many scientists and engineers suggest that the impact of energy-resource utilization on the environment is best addressed by considering exergy. The exergy of a quantity of energy or a substance is a measure of the usefulness or quality of the energy or substance or a measure of its potential to cause change. In other words, exergy of any flow or resource of energy is the potential of useful work that is available, and a conventional HVAC system wastes most of it."[7] In fact, exergy is not only wasted but destroyed, because exergy flow is irreversible. Therefore, it is no surprise that the exergy efficiency of existing HVAC systems are less than 10%.[8] An earlier study showed that this efficiency is 5% on average for Swedish homes.[9] The shortcomings of the definition of energy efficiency are particularly apparent for tasks in which fossil fuels are used just for low-temperature heating. Since fossil fuels burn at very high flame temperatures—up to 2000 K[10]—the useful work potential (exergy) of fossil fuels is high. When fossil fuels are used for hot water heating, space heating, or even industrial steam production, most of the exergy is destroyed in these processes.[11] Indoor space heating furnaces have an estimated exergy efficiency of 6%, and heat pumps, when combined with conventional HVAC systems are not much better at 9%.[12] On the other hand, thermal efficiency of HVAC systems has almost reached a saturation point, well above 90% on average, except for thermal energy transport and distribution losses. Therefore, according to Annex 37,[13] the priority must now be shifted towards exergy efficiency in improving buildings, starting from their root causes. Fortunately, most of the root causes can be detected by developing next-generation building simulation codes that address the exergy efficiency at different scales—most importantly the HVAC system, which must be related to the environment and to the primary energy at its source, including the

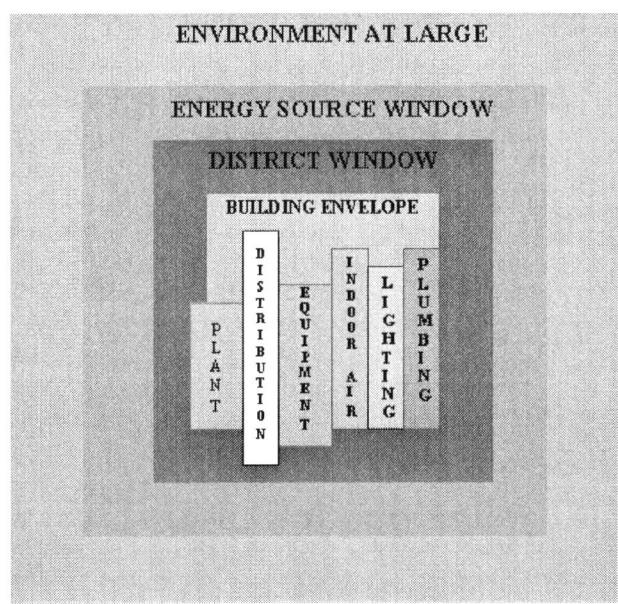

Fig. 2 Sample layers and scale of ideal building simulation windows.

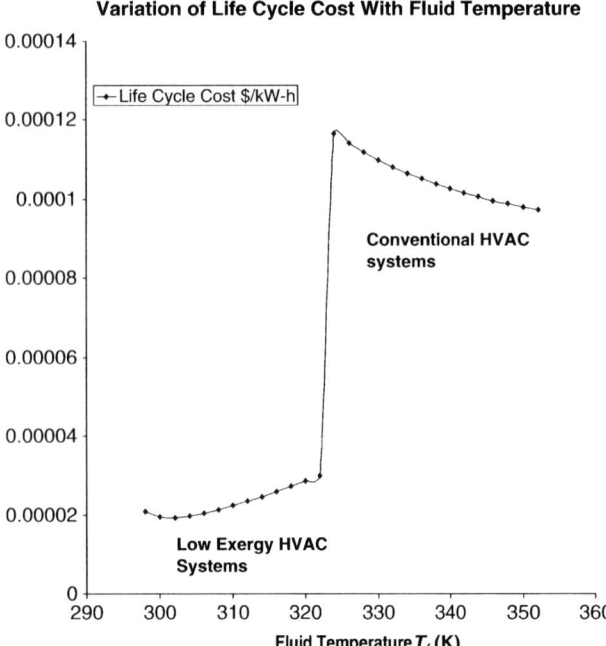

Fig. 3 Heating, ventilating, and air-conditioning life cycle cost including the cost of exergy destruction.
Source: From Ref. 16.

exergetic importance of thermal energy storage systems.[14,15] The rational exergy efficiency (ψ) is the ratio of the minimum exergy required by a given HVAC load to the actually available exergy of the energy source used in satisfying that load.[16]

$$\psi = \frac{\varepsilon_{min}}{\varepsilon_{act}} \qquad (1)$$

The minimum amount of exergy required to satisfy a unit heating load for an indoor space at a dry-bulb air temperature T_a, in reference to the temperature of the environment T_g[17]:

$$\varepsilon_{min} = \left(1 - \frac{T_g}{T_a}\right) \quad \{T_g \neq T_a\} \qquad (2)$$

If the reference temperature of the environment is the ground temperature of 278 K (5°C) in winter, and the indoor air design temperature is 291 K (18°C), the ideal ε_{min} for an HVAC system directly utilizing ground heat is 0.0447. For a conventional HVAC system, the available exergy of the used energy source relates to the resource temperature and T_g. For a natural gas-fired furnace with a flame temperature of 1273 K, the available unit exergy is:

$$\varepsilon_{act} = \left(1 - \frac{278}{1273}\right) = 0.7816 \qquad (3)$$

The exergy efficiency using Eq. 1 is then 0.057 (6%). This means most of the exergy is destroyed. If the same indoor space could be directly heated by an energy source at 302 K (29°C), Ψ could be much higher: 56%. These statements and equations are also valid for electric-based chiller or heat pump space cooling, if electric power is generated in a thermal plant. Fig. 3 illustrates the contrast between low-exergy HVAC systems and conventional HVAC systems as a function of operating fluid temperature.[16] The contrast is a result of including the cost of exergy destruction in the life cycle cost (LCC) analysis. Generally, low-exergy HVAC systems are a hybrid of radiant and convective heat transfer equipment, which can operate at moderate fluid temperatures.[18]

ILLUSTRATIVE EXAMPLES

Fig. 4A–D illustrate the importance of a wider building simulation window for sustainability.

1. *A Conventional HVAC System with a Central Power Plant.* In Fig. 4A, according to a conventional building energy simulation tool covering only the building envelope, the building seems to be energy efficient and may comply with most of the energy codes. Once a wider simulation window is used, covering the plant, the transmission lines, and the environment, it becomes apparent that the exergy efficiency is less than 10%.
2. *A Ground Source Heat Pump (GSHP) Building HVAC System with a Central Power Plant.* According to the same small building simulation window, the ground source heat pump in Fig. 4B makes sense only from the building owner/operator's side, because of a high COP. From a wider window, the overall efficiency remains almost the same and the exergy efficiency is still below 10%. The small window does not reveal how low the exergy efficiency is and does not show the ways and means to increase the exergy efficiency.
3. *A Decentralized Micro Combined Heat and Power (MCHP) Plant with a Conventional HVAC System.* In Fig. 4C, a wider simulation window shows that the overall energy efficiency is improving substantially, because power transmission losses are eliminated. Because the decentralized plant utilizes the primary energy resource for electric power generation on location, and captures most of the waste heat to provide thermal energy to the building, exergy efficiency has also improved. The exergy efficiency may further increase if a low-exergy HVAC system is installed, because waste heat can be more effectively utilized in the building.
4. *A Green Power (MCHP) System with Low-exergy HVAC System.* In Fig. 4D, both the energy and exergy efficiencies are high. Yet, this solution cannot be visible in a narrow simulation window.

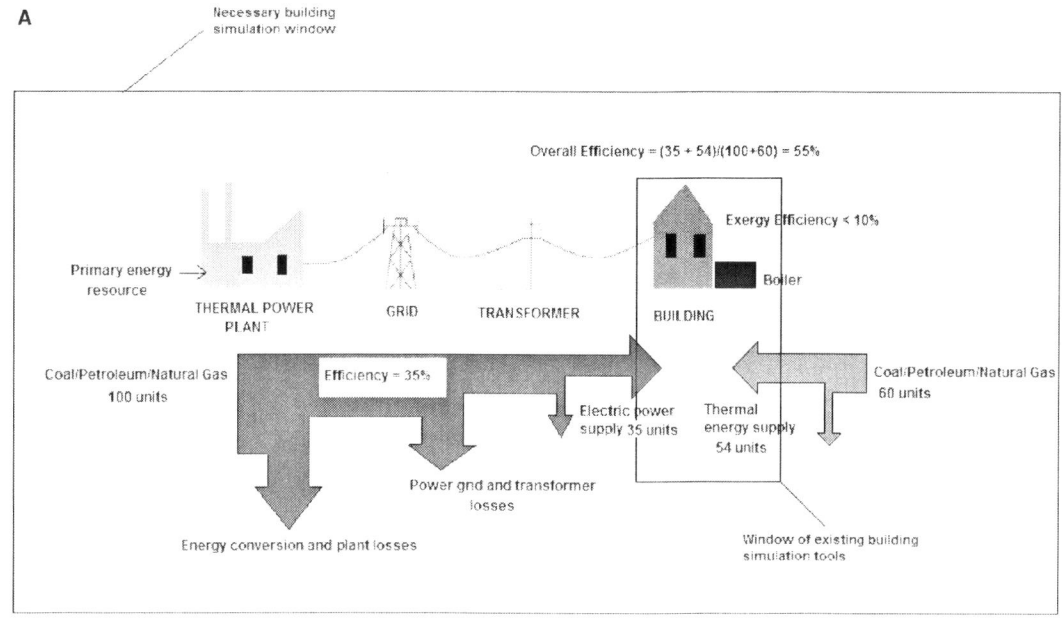
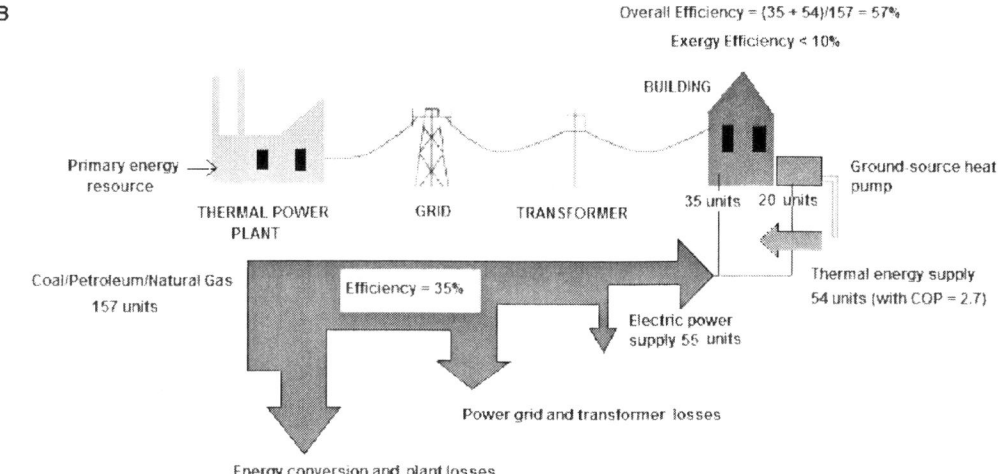
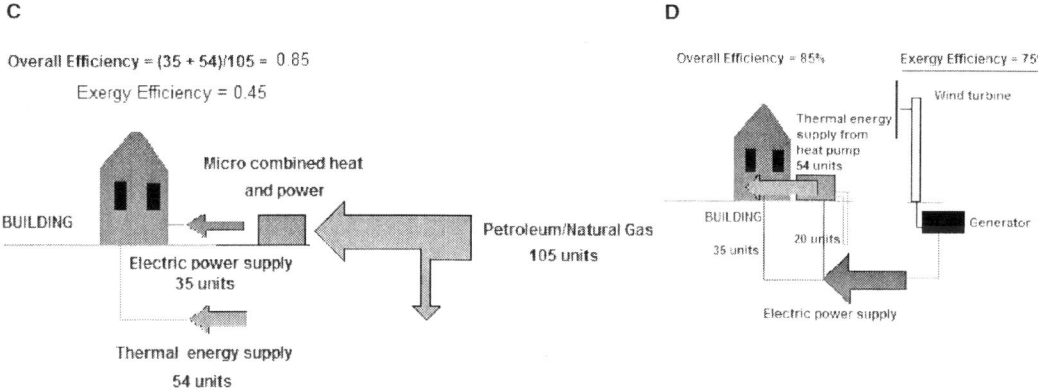

Fig. 4 (A) A centralized energy and power system with buildings using conventional heating, ventilating, and air-conditioning plant and equipment. (B) A centralized energy and power supply system with ground source heat pumps. (C) A decentralized micro combined heat and power system. (D) A decentralized, green energy system.

We can now conclude that whether it is thermal efficiency or exergy efficiency, or both, the simulation window must be far wider than the building envelope for sustainability. The building simulation window may only be reduced to building scale if decentralized energy systems with high efficiency green energy components are considered, as in Fig. 4D.[19]

MAJOR COMPONENTS OF BUILDING SIMULATION

Current theory of building energy simulation is based upon load and energy calculations developed by ASHRAE.[20] The selection of a simulation program for a given task depends on the project requirements, time, and cost of the analysis, experience of the user, and availability of suitable simulation tools and data.[20] Keeping in mind that future BS tools must include exergy analysis on a macro scale, we nevertheless currently need to select a BS tool best suited for a specific project. During the selection process, one needs to consider the following two requirements: (1) algorithms and data must be from reliable, well-established, published sources and (2) validation of the tool must be possible with available validation packages.[21] It is desirable to use a tool with an open structure or open end for future collaborative, modular versions. Expertise and training required must be compatible with the resources and background of the users. In addition, one must ask the following questions before the final decision:

- Who uses this tool? How many active users are there? Which institutions and countries use it?
- Is there a discussion group for this tool?
- What is the target audience? Does it suit your background, expertise, and position?
- What are the main features? Does it satisfy your needs?
- Is this tool compatible with other tools, programs, and databases? If yes, which programs are compatible? Can you bundle them?
- What are the required input and output data? How extensive are they, and what is their format and scope? Are there databases that can be readily used, or are databases already provided in the tool from recognized resources, and are they code approved?
- How many default inputs does the program tend to provide? Generally, more default inputs make it user-friendlier but substantially less accurate, depending on how the defaults are prepared and the assumptions involved.
- What is the computer platform, and which programming language is used?
- How much support can you get, and for how long can you get it? Can you get live support? Is the support free? How much and how detailed are educational materials, if provided?
- What are the costs, licensing options, terms, and conditions?
- Is a trial version available? What is the trial period and is it supported?
- Does the program check necessary code compliances? If yes, which codes?
- Will upgrades be available? What are the terms and conditions?
- What are the speed, capacity, and technical coverage of the program? Does it provide optimization and knowledge base tools?

The following basic components of building simulation must be covered:

- All sensible and latent loads
- Heat losses and gains from the envelope
- Internal gains and losses
- Electrical loads
- Water supply loads
- Waste management loads
- Ventilation loads for acceptable indoor air quality (IAQ)[22,23]
- Human comfort requirements and human behavior in controlled indoor environments
- Radiant asymmetry
- Mean radiant temperature
- Air velocity
- Wet-bulb and dry-bulb air temperature
- Relative humidity
- Mean radiant temperature
- Operative temperature
- AUST (Area Averaged Uncontrolled Surface Temperature)
- Convective and radiant heat transfer split
- Asymmetric thermal radiation
- Draft
- Vertical air temperature difference
- Warm or cold floors
- Hot ceiling
- Heat stress
- Comfort analysis[24,25]
- Duct losses[26]
- Domestic hot water demand modeling and supply[27]
- CBR (chemical, biological, and radiological) attack risk assessment and simulation[28]
- Fragility analysis (earthquakes, etc.)[29]
- Environmental ingress (like moisture penetration from foundations)
- Solar gains
- Shading
- Lighting simulation

INTEGRATED BUILDING DESIGN SYSTEM

Integration of simulation into the building design process can ensure that important data and information for each major design decision are provided in a timely fashion. By establishing design links and exchange between architecture and engineering, an integrated building design system (IBDS) can be developed.[30] Some researchers have taken the initiative to develop more efficient and flexible use of simulation tools. The COMBINE (Computer Models for the Building Industry in Europe) project[31] and the AEDOT (Advanced Energy Design and Operation Technologies) project in the U.S.[32] are typical examples. While design, simulation, operation, and control functions are becoming integrated, all equipment and systems must be bundled around a common protocol. Future integrated building simulation models must know exactly how each piece of equipment or sub-system behaves in the building through BACnet protocol,[33] and the equipment must be architectured accordingly.

AN OVERVIEW OF CURRENT BS TOOLS

According to G. Augenbroe,[34] "A broad range of simulation software applications has become available for a variety of building performance assessments over the last three decades …" The maturation of building simulation into a recognized and indispensable discipline for all professions—involved in the design, engineering, operation, and management of buildings—has now become the imminent challenge. Two key aspects dominate this evolution process: (1) attaining an increased level of quality assurance and (2) offering efficient integration of simulation expertise and tools in the overall building process. Major shortcomings of current simulation tools include:

- Input is usually lengthy and often cumbersome to prepare and compile.
- Parametric studies are not always possible.
- Outputs are generally too voluminous and need further analysis and interpretation.
- The learning curve may be too long and overly frustrating for the novice user.
- The user interface is generally overlooked and not user-friendly.
- The software is not flexible enough to test all possibilities.
- The software does not allow for the design and testing of new components. Off- the-shelf types of equipment must be used. This limitation reduces creative design opportunities.
- Most programs require long run times and a large memory. Personal Computer (PC) versions are available, but the compromises involved must be carefully weighed against the ease and simplicity of using them.
- Lifecycle analyses that include and recognize the exergy efficiency and the cost of exergy waste do not exist.
- Hybrid HVAC systems cannot easily be modeled.
- Optimization tools on a wide simulation window are not yet available.

Building simulation tools are available in the whole building level, equipment and component level, system level, retrofit level, and green building levels. Here only a very small number of available simulation tools are sampled. U.S. DOE EERE[35] provides a comprehensive and up-to-date listing and detailed information. Their Building Energy Software Tools directory lists almost 300 simulation programs categorized under: (1) whole-building analysis (load calculation, renewable energy, retrofit analysis, sustainability, and green buildings), (2) codes and standards, (3) materials, components, equipment, and systems, and (4) other applications.[36] A short list of typical BS tools is provided.

1. *APACHE*. Thermal design (heating, cooling, and latent load calculations). Equipment sizing, codes and standards checks, dynamic building thermal performance analysis, systems and controls performance, and energy use.[37]
2. *APACHE–HVAC*. Flexible and versatile system HVAC and controls modeling. Integrated simulation of building and HVAC systems.[38]
3. *BLAST*. The zone models of BLAST (Building Loads Analysis and System Thermodynamics) are based on the heat balance method. It performs hourly simulations of buildings, air handling systems, and central plant equipment. The output may be coupled to the LCCID (Life Cycle Cost in Design).[39]
4. *BSim2002*. Comprises different programs like a graphic model editor, thermal and moisture building simulation tool, dynamic solar and shadow simulation, daylight calculation tool, and compliance checker. Computer-aided design import and building integrated Photovoltaic system.[40]
5. *BEA*. Building Energy Analyzer is a system-screening tool to evaluate a variety of commercially available HVAC and power generation options. Uses the DOE-2.1E computational engine, includes a life cycle cost analysis module, and handles complex utility rates structures.[41]
6. *BUS^{++}*. New generation platform for building energy, ventilation, noise level, and indoor air quality simulations. A network assumption is adopted, and both steady state and dynamic simulations are possible.[42]

7. *DOE-2*. This is an hourly, whole-building energy analysis program that calculates energy performance and life cycle cost of operation. Can be used to analyze energy efficiency of given designs or efficiency of new technologies. Other uses include utility demand-side management and rebate programs, development and implementation of energy efficiency standards, and compliance certification. Training and expertise required.[43]
8. *EE4 CODE*. Used to determine the compliance of a building to Canada's Model National Energy Code for Buildings (MNECB). EE4 CODE may also be used to perform noncompliance energy analyses and thus to predict the annual energy consumption of a building and to assess the impact of design changes, based on DOE-2.1E.[44]
9. *EED*. A program for borehole heat exchanger design in a ground source heat pump system (GSHP) and borehole thermal storage. In very large and complex tasks, EED allows for the retrieval of the approximate required size and layout before initiating analyses that are more detailed.[45]
10. *EnergyPlus*. This is a whole-building energy simulation program that builds on BLAST and DOE-2. Includes advanced simulation capabilities, including time steps of less than an hour, modular systems simulation modules that are integrated with a heat balance-based zone simulation, and input and output data structures tailored to facilitate third-party interface development. EnergyPlus Version 1.2.2 was released in April 2005. EnergyPlus and weather data for more than 900 locations worldwide can be downloaded at no cost from the EERE home page.[46]
11. *EZDOE*. This tool is an easier-to-use personal computer version of DOE-2. EZDOE calculates the hourly energy use of a building and its lifecycle cost of operation, given information on the building's location, construction, operation, and heating and air conditioning system.[47]
12. *Right Suite-Residential*. All-in-one HVAC software performs residential loads calculations, duct sizing, energy analysis, equipment selection, cost comparison calculations, and geothermal loop design.[48]
13. *TRACE 700*. Follows the algorithms recommended by ASHRAE. Used for assessing the energy and economic impacts of building-related selections, such as architectural features, comfort-system design, HVAC equipment selections, operating schedules, and financial options.[49]
14. *TRNSYS*. TRNSYS (TRaNsient System Simulation Program) includes a graphical interface, a simulation engine, and a library of components that range from various building models to standard HVAC equipment, to renewable energy and emerging technologies. TRNSYS includes a method for creating new components that do not exist in the standard package.[50]
15. *VisualDOE*. A Windows interface to the DOE-2.1E energy simulation program. Users construct a model of the building's geometry using standard block shapes, using a built-in drawing tool, or importing DXF files. Building systems are defined through a point-and-click interface. A library of constructions, fenestrations, systems, and operating schedules is included, and the user can add custom elements. Up to 99 alternatives can be defined.[51]
16. *ESP-r*. Developed by the Energy Simulation Research Unit, University of Strathclyde, this is a general simulation tool that can be used to address a broad range of thermal performance problems, most often used for buildings.[52]
17. *HVAC Solution*. Allows users to graphically design and specify HVAC equipment, picking objects like boilers, pumps, fan coils, and air handlers. Using drag-and-drop methods, an HVAC system can be built. Once the system is built, the tool automatically renders equipment schedules and export schematics.[53]
18. *DUCTSIZE*. Calculates optimal duct sizes using the static regain, equal friction, or constant velocity method. Data entry can be accomplished manually or taken graphically from either Drawing Board or AutoCAD. A library of fan data for noise calculations is built into the program.[54]
19. *Hydronics Design Studio*. Assists in analyzing the thermal and hydraulic performance of modern hydronic heating systems in residential and light commercial buildings. The professional version performs tasks like heating load analysis, series baseboard circuit analysis, piping heat loss estimating, expansion tank sizing, radiant circuit analysis, injection mixing simulation, buffer tank simulation, and fuel cost comparisons.[55]
20. *FLOVENT*. Calculates airflow, heat transfer, and contamination distribution for built environments. FLOVENT uses techniques of Computational Fluid Dynamics (CFD) packaged in a form that addresses the needs of mechanical engineers involved in the design and optimization of ventilation systems.[56]
21. *ArchiPhysics-Solar*. A passive solar energy system for free running buildings. Shows the adaptive comfort level, as well as indoor and outdoor air temperatures for given building geometry, location, and glazing.[57]

22. *MOIST*. This program predicts the combined transfer of heat and moisture in multi-layer building construction. Inputs hourly weather data and predicts the moisture content and temperature of the construction layers as a function of time of year. It can be used to develop guidelines and practices for controlling moisture in walls, flat roofs, and cathedral ceilings.[58]
23. *Daylight*. This program calculates the daylight factor distribution in a room. A user-friendly program by Archiphysics.[59]
24. *BREEZE*. This is a tool for estimating ventilation rates, developed by Building Research Establishment (BRE) in England.[60]

SUSTAINABLE DESIGN AND GREEN BUILDINGS

There are currently some desktop tools available for architects and engineers for sustainable design of buildings. However, these tools perform life cycle environmental impact assessment based solely upon building materials and related items.[61–63] The energy balance, exergy balance, and thermo-physical interactions of the building with the environment are not included.

1. *ATHENA*. This is an environmental impact estimator program developed by ATHENA Sustainable Materials Institute. The program has a large database of building materials and performs an environmental impact assessment.[64]
2. *BEES*. The BEES (Building for Environmental and Economic Sustainability) software enables the use of cost-effective, environmentally preferable building products. The tool is based on consensus standards and includes actual environmental and economic performance data for nearly 200 building products.[65]
3. *RETScreen*. Developed by the Natural Resources of Canada, the RETScreen International Clean Energy Project Analysis Software is a decision support tool that can be used worldwide to evaluate the energy production, life-cycle costs, and greenhouse gas emission reductions for various types of energy efficient and renewable energy technologies. Also includes product, cost, and weather databases, as well as a detailed online user manual.[66]
4. *GBS*. Seamlessly links architectural 3-D CAD building designs with energy analysis. Green Building Studio (GBS) enables architects to quickly calculate the operational and energy implications of early design decisions. GBS uses the DOE-2 simulation engine to calculate energy performance and generates geometrical input files.[67]

More information can be found at the Building Energy Simulation Tools (BEST) Web site,[68] which lists the available tools and their main features and provides Internet sites and references. The most up-to-date information about the energy design tools can be obtained from the International Building Performance Simulation Association.[69]

CONCLUSION

With the ever-increasing need for high performance buildings with truly green features, it is virtually inevitable that future building simulation tools will become an integral part of other macro-scale simulation tools for the environment, the energy sector, and the community. In achieving this goal, building simulation tools must be more compatible, open-structured, and open-ended for greater modularity, compatibility, and data exchange. Finally and most importantly, we must include exergy-based optimization in the broadest simulation window possible within these future tools.

REFERENCES

1. ASHRAE *Standard 90.1-2004—Energy Standard for Buildings Except Low-Rise Buildings, SI Edition (ANSI approved; IESNA Co-sponsored)*; ASHRAE: Atlanta, GA, 2004.
2. ASHRAE *Standard 90.2-2004—Energy-Efficient Design of Low-Rise Residential Buildings (ANSI approved)*; ASHRAE: Atlanta, GA, 2004.
3. Torcellini, P.A.; Judkoff, R.; Crawley, D.B. High-performance buildings. ASHRAE J. **2004**, *46* (9), S4–S12.
4. DOE, U.S. Department of Energy. *Emissions of greenhouse gases in the United States-2 carbon dioxide emissions*, Report No. EIA/DOE-0573, 1998.
5. Simpson, M.; Kay, J. Availability, exergy, the second law and all that, http://www.fes.uwaterloo.ca/u/jjkay/pubs/exergy/, copyrighted in 1989.
6. Dincer, I.; Rosen, M.A. A worldwide perspective on energy, environment and sustainable development. Int. J. Energy Res. **1998**, *22*, 1305–1321.
7. Rosen, M.A.; Dincer, I. Energy and exergy analyses of sectoral energy utilization: an application for Turkey. In *Proceedings of first International Trabzon Energy and Environment Symposium*, Karadeniz Technical University; Tarbzon Turkey, 1996; Vol. 3, 1059–1065.
8. Kilkis, I.B. An exergy aware optimization and control algorithm for sustainable buildings. IJGE **2004**, *1* (1), 65–77.
9. Wall, G. Exergy—a useful concept. *Chalmers Biblioteks Tryckeri*; Göteborg: Sweden, 1986.
10. Dincer, I.; Cengel, Y. Energy, entropy and exergy concepts and their roles in thermal engineering. Entropy **2001**, *3*, 116–149.
11. Ralph, T. *Half Life: Nuclear Power and Future Society*; Ontario Coalition for Nuclear Responsibility: Ottawa, 1981; 178.

12. *American Institute of Physics Efficient Use of Energy*; American Institute of Physics: New York, 1975; 49–50.
13. IEA and ECBCS *Guidebook to IEA ECBCS Annex 37—Low Exergy Systems for Heating and Cooling of Buildings*; VTT: Finland, 2003.
14. Dincer, I.; Rosen, M.A. *Thermal Energy Storage, Systems and Applications*; Wiley: West Sussex, 2002.
15. Dincer, I.; Rosen, M.A. Energetic, environmental and economic aspects of thermal energy storage systems for cooling capacity. Appl. Thermal Eng. **2001**, *21*, 1105–1117.
16. Kilkis, I.B. A road map for emerging low-exergy HVAC systems. Int. J. Exergy **2005**, *2* (4), 348–365.
17. Dincer, I.; Rosen, M.A. Thermodynamic aspects of renewables and sustainable development. Renew. Sust. Energy Rev. **2005**, *9*, 169–189.
18. Kilkis, I.B.; Suntur, A.S.R.; Sapci, M. Hybrid HVAC systems'. ASHRAE J. **1995**, *37* (12), 23–28.
19. Kilkis, S. *Development of a Rational Exergy Management Model to Reduce CO_2 Emissions with Global Exergy Matches*, Honors Thesis, Georgetown University, 2007.
20. ASHRAE *ASHRAE Handbook—HVAC Applications*; ASHRAE: Atlanta, GA, 2003.
21. ASHRAE *Standard 140-2004—Standard Method of Test for the Evaluation of Building Energy Analysis Computer Programs (ANSI Approved)*; ASHRAE: Atlanta, GA, 2004.
22. ASHRAE *Standard 62.1-2004—Ventilation for Acceptable Indoor Air Quality (ANSI Approved)*; ASHRAE: Atlanta, GA, 2004.
23. ASHRAE *Standard 62.2-2004—Ventilation and Acceptable Indoor Air Quality in Low-Rise Residential Buildings (ANSI Approved)*; ASHRAE: Atlanta, GA, 2004.
24. ASHRAE *Standard 55-2004—Thermal Environmental Conditions for Human Occupancy (ANSI Approved)*; ASHRAE: Atlanta, GA, 2004.
25. ASHRAE *ASHRAE Handbook—HVAC Systems and Equipment*; ASHRAE: Atlanta, GA, 2004.
26. ASHRAE *Standard 152-2004—Method of Test for Determining the Design and Seasonal Efficiencies of Residential Thermal Distribution Systems (ANSI Approved)*; ASHRAE: Atlanta, GA, 2004.
27. Tiller, D.K.; Phil, D.; Henze, G.P. *Online domestic hot water end-use database*, ASHRAE research project 1172 report; ASHRAE: Atlanta, GA, 2004.
28. Arvelo, J.; Brandt, A.; Roger, R.P. *An Enhanced Multizone Model and its Application to Optimum Placement of CBW Sensors*; ASHRAE: Atlanta, GA, 2002.
29. Wen, Y.K.; Ellingwood, B.R. The Role of Fragility Assessment in Consequence-Based Engineering, Earthquake Spectra, EERI, August, 2005.
30. Lewis, M. Integrated design for sustainable buildings. Building for the future: a supplement to. ASHRAE J. **2004**, *46* (9), 22–30.
31. COMBINE. http://erg.ucd.ie/combine.html.
32. AEDOT. http://apc.pnl.gov:2080/0projects_and_capabilities/aedot/html/aedot.html.
33. ASHRAE *Standard 135-2004—BACnet®—A Data Communication Protocol for Building Automation and Control Networks (ANSI Approved)*; ASHRAE: Atlanta, GA, 2004.
34. Augenbroe, G. Trends in building simulation. Building Environ. **2002**, *37*, 891–902.
35. U.S. DOE EERE. www.eere.energy.gov.
36. http://www.eere.energy.gov/buildings/tools_directory/subjects.cfm/pagename=subjects/pagename_menu=whole_building_analysis/pagename_submenu=energysimulation.
37. APACHE. Thermal Design (Heating, Cooling & Latent Load Calculations) http://www.ies4d.com.
38. APACHE–HVAC. http://www.ies4d.com.
39. BLAST. http://www.bso.uiuc.edu.
40. BSim2002. http://www.bsim.dk.
41. BEA. http://www.interenergysoftware.com.
42. BUS^{++}. Pekka.Tuomaala@vtt.fi.
43. DOE-2. http://simulationresearch.lbl.gov.
44. EE4 CODE. http://ee4.com.
45. EED. http://www.buildingphysics.com.
46. EnergyPlus. http://www.eere.energy.gov/buildings/energyplus/ or EnergyPlus Web site.http://www.energyplus.gov.
47. EASYDOE. http://www.elitesoft.com.
48. Right Suite-Residential, http://www.wrightsoft.com.
49. TRACE 700. http://www.trane.com/commercial/software.
50. TRSYS. http://sel.me.wisc.edu/trnsys/downloads/download.htm.
51. VisualDOE. http://www.archenergy.com/products/visualdoe/.
52. ESP-r. http://www.strath.ac.uk/Departments/ESRU/Ad_Env_Sim.html.
53. HVAC Solution. http://www.hvacsolution.com.
54. DUCTSIZE. http://www.elitesoft.com.
55. Hydronics Design Studio. http://www.hydronicpros.com.
56. FLOVENT. http://www.flovent.com.
57. ArchiPhysics-Solar. http://www.archiphysics.com/programs/freerunner/freerunnner.htm.
58. MOIST. http://www.bfrl.nist.gov/863/moist.html.
59. Daylight. http://www.archiphysics.com/programs/daylight/daylight.htm.
60. BREEZE. http://www.bre.co.uk/.
61. Gowri, K. Desktop tools for sustainable design. ASHRAE J. **2005**, *47* (1), 42–46.
62. Grumann, L.D., Ed., *ASHRAE GreenGuide*, ASHRAE: Atlanta, GA, 2003; 190.
63. GBC. LEED Reference guide, Version 2.0, U.S. Green Building Council, 2001.
64. ATHENA. www.athenasmi.ca.
65. BEES. http://www.bfrl.nist.gov/oae/software/bees.html.
66. RETScreen. http://www.retscreen.net/.
67. Green Building Studio GBS. http://www.greenbuildingstudio.com/.
68. Building Energy Simulation Tools (BEST). http://www.arch.hku.hk/research/BEER/best.htm#Simulation%20Methods.
69. International Building Performance Simulation Association. http://next1.mae.okstate.edu:80/ibpsa/.

BIBLIOGRAPHY

Additional On-Line Resources

1. *IBPSA*. The International Building Performance Simulation Association (IBPSA), (http://www.ibpsa.org) is a nonprofit international society of building performance simulation researchers, developers, and practitioners, dedicated to

improving the built environment. Since 1985, IBPSA has been organizing international conferences on building simulation every two years. PDF versions of all papers from the 1985–2003 Conferences are available online. http://www.ibpsa.org/m_events.asp.

2. *BLDG–SIM*. This is a mailing list for users of building energy simulation programs. This allows engineers, architects, and others in the building design trade to compare alternative designs and select the design that is cost justified. The Web page for this mailing list is located at http://www.gard.com/ml/bldg-sim.htm.

3. *Center for Buildings and Systems-TNO-TU/e*. The institute's multidisciplinary approach is expected to result in innovative insights in the field of lighting, indoor climate control, noise, health, and comfort, but also with regard to sustainable energy. Their studies include building simulation, acoustics, lighting, renewable energy systems, health, comfort, and productivity. http://www.kcbs.nl.

4. *IEA ECBCS*. The mission of Energy Conservation in Buildings and Community Systems is to facilitate and accelerate the introduction of energy conservation and environmentally sustainable technologies into healthy buildings and community systems, through innovation and research in decision-making, building products and systems, and commercialization. http://www.ecbcs.org.

5. *LowEX (Low Exergy Systems for Heating and Cooling of Buildings)*. The LowExNet has been founded because of the work of the International Energy Agency (IEA) Annex 37: "Low Exergy Systems for Heating and Cooling in Buildings." During an expert meeting in Kassel, Germany in September 2003, the decision was made to form a network on the issues of new ways for energy systems in buildings and to continue the work of the IEA Annex 37, which is to seek various low exergy system solutions. http://www.lowex.net/english/inside/links.html.

6. *ZUM (Zentrum für Umweltbewusstes Bauen e.V. Center for Sustainable Building)*. This institution provides several downloadable programs for building simulation with emphasis on sustainability. www.zub-kassel.de.

Ashrae Guides

1. Advanced Energy Design Guide for Small Office Buildings.
2. Humidity Control Design Guide for Commercial and Institutional Buildings.
3. Practical Guide to Noise and Vibration Control for HVAC Systems (SI).
4. Application Guide: Indoor Air Quality Standards of Performance.

Selected Books

1. Clark, J.A. *Energy Simulation in Building Design*, 2nd Ed.; Heinemann: Butterworth, 2001.
2. Waltz, J.P. *Computerized Building Energy Simulation Handbook*; Fairmont Press: Lilburn, Georgia, 2000.
3. Taplin, H. *Boiler Plant and Distribution System Optimization Manual*, 2nd Ed.; Fairmont Press: Lilburn, Georgia, 1997.
4. Curl, R.S. *Building Owner's & Manager's Guide: Optimizing Facility Performance*; Fairmont Press: Lilburn, Georgia, 1998.
5. Petchers, N. *Combined Heating, Cooling & Power Handbook: Technologies & Applications*; Fairmont Press: Lilburn, Georgia, 2002.

Sustainable Development

Mark A. Peterson
Sustainable Success LLC, Clementon, New Jersey, U.S.A.

Abstract
Sustainable development includes all business and community planning and operating decisions with due consideration for: (1) people—employees, customers, shareholders, community residents, or anyone that is involved or affected; (2) planet—material and energy resource management that does not hurt the environment; and (3) profits—or economics or prosperity. Sustainable development takes a different, more caring look at how people interact with themselves and how their activities affect the planet and the general well being of life for sustained economic growth.

INTRODUCTION

This article defines sustainable development and its three basic aspects. Because sustainable development is a relatively new concept, a short history and description of the drivers that lead to sustainable development are described. Then, a sustainable energy future is presented. To be sustainable as a society requires cooperation and collaboration rather than command and control management. A very key aspect of sustainable development, social synergy, is covered. Also, because sustainable development is relatively new but essential to build a better and viable future, children from the earliest age through college need to learn to understand and apply the concepts of sustainable development. A paragraph is included that describes current efforts in the United States to incorporate education in sustainable development (ESD) in K-12 and college curricula.

Sustainable development encompasses stewardship of many areas of human and planetary life. In business, one of the key motivators is to implement sustainable development measures to be profitably successful indefinitely. To do this requires that businesses show their due diligence to both society and the environment while maximizing profits. Sustainability reporting assists businesses in assessing their efforts. Sustainability reports are both management and public relations tools. An innovative, very effective, time-tested form of sustainable development, "renting a service" rather than "selling a product," is covered. Then, sustainable development for community vitality is described. As the sizes, types, and socioeconomics of communities vary considerably, so do all of the related aspects of evolving them to be sustainable. Reference is made to a web site that thoroughly describes all aspects. Some of the subjects covered on the web site are briefly summarized in this article. The final section covers the vast, opening market of sustainable development in developing countries.

Keywords: Sustainable; Sustainability; Development; Triple bottom line; World; Business; Communities; Success; Social; Responsibility.

WHAT IS SUSTAINABLE DEVELOPMENT?

Sustainable development is development that meets the needs of the present without compromising the ability of future generations to meet their needs. Sustainable development has three aspects:

1. Social (people).
2. Environmental (planet).
3. Economic (profits)/prosperity.

All development affects all three aspects. All three aspects are interdependent. Thus, being mindful of these interdependencies in management and leadership decisions will result in the best overall solution—a win–win–win solution that maximizes success and minimizes any negative social, environmental, and economic costs. This is called managing the triple bottom line of people, planet, and profits. This is also called whole systems thinking[1,2] because all relevant factors are considered as a whole. The role of engineers is to help their clients be successful. This requires integrated whole systems thinking that covers all related liabilities that a company or community (their client) may have and provides the most efficient and profitable solution to the challenge. Often, the best whole system solution is also the most efficient and most sustainably profitable.

The environmental (planet) aspect is significantly affected by energy consumption and management, including: the entire national power infrastructure and distribution, transportation, plus the construction and renovation of all residential, commercial, and industrial facilities.

HISTORY, ENVIRONMENTAL DEGRADATION, AND NATIONAL SECURITY

During the last century, while fossil fuels were abundant and cheap, those fuels fulfilled a majority of our energy conversion needs. The mounting problem is that combustion emissions have fouled the environment in a number of ways, resulting in increases in respiratory illnesses, mercury pollution, and a rise in global temperatures. The quantity of easily retrieved fossil fuels is significantly depleted. Coal is still relatively abundant, but it does not burn cleanly. Technologies need to be developed to both mine the coal safely and to burn it cleanly. Regarding petroleum, many countries that are not friendly to the United States control most of the remaining easily extractable sources. Many national security advisors have indicated the urgency of severing our dependence on foreign oil as a part of an overall strategy for the security of the United States and as a means to prevent oil-related conflicts.[3,4] It is also apparent that there is a need to protect and allow the environment to regenerate. The effect of using fossil fuels extensively and inefficiently is that we are simultaneously poisoning the environment and ourselves. Nuclear power emissions are clean, but the nuclear power industry has significant obstacles such as storage of radioactive wastes for many thousands of years. In addition, there are security concerns to safeguard radioactive material from being stolen for production of atomic weapons.

Governors and mayors are taking action to implement clean energy technologies. In response to clear signs of increased cost from continuing to use fossil fuels and scientific evidence showing that by burning fossil fuels we are initiating a possibly devastating global warming trend[5] that could flood coastal cities, disrupt the food chain, and change climate patterns significantly, many states have taken the initiative and enacted renewable energy portfolios to fund the transition to renewable energy resources. Many remaining states are in the process of developing their own renewable energy portfolios. These renewable energy portfolios provide significant state- and utility-sponsored financial incentives for the commercial, industrial, and residential use of renewable energy systems and fuels. On the city level, many mayors from major cities around the world have made commitments to cut greenhouse gas emissions to slow the rate of global warming.[6] Many of these cities are coastal and could be severely impaired or destroyed from rising sea levels from global warming. So, civic action to switch to cleaner energy options is beginning in earnest.

THE ENERGY FUTURE

What this means to energy engineers is that petroleum-derived fuels are on their way out over the next half century. Hydrogen (where the hydrogen is derived from renewable energy sources), ethanol, biodiesel, and other forms of renewable fuels are on their way in. Direct and indirect conversion of solar energy, including wind, biomass, wave/tidal power, and small-scale hydroelectric power will increasingly be part of the energy infrastructure that energy engineers will design and build. The bottom line with energy is that it needs to be relatively nonpolluting and indefinitely available. It is a very dynamic time for energy engineers as the entire, worldwide energy picture transitions to clean renewable technologies. This will eventually add a lot of stability to the world economy, the world political environment and to everyone's lives. The stability will come from the fact that renewable energy technologies can be used to tap the natural energy resources that are available everywhere. Stability will also come, as the environment regenerates, the climate stabilizes and resources remain available for our sustenance.

HOW WE SOCIALLY AND PROFESSIONALLY INTERACT WITH EACH OTHER DETERMINES OUR DEGREE OF SUCCESS

Another aspect of sustainability has to do with how well we interact and collaborate. In the past, management of most activities was by a top-down hierarchy. Now humanity is evolving and it is driven by high levels of sophistication in technology and communications, which has resulted in individual knowledge and skill level increases. Plus, many families are now structured such that individual adults and children are taking more independent responsibilities for the many aspects of their lives. This has all created a desire by many people to be more intimately involved in solutions rather than just letting someone else do the thinking. In this new economy, teamwork and open communication are important to bring all stakeholders together, whether in a business or community, to manage by consensus and cooperation. Everyone affected should have the opportunity to be involved in the solution, even if just by being informed as the planning and decision-making are in process. The outcome then is one that promotes efficiency, for the simple reason that all persons affected are involved, which promotes enthusiasm and "buy-in." Typically with this process, more work and time are invested up-front such that all aspects are considered and thus everything proceeds more efficiently down-stream.[7]

CREATIVE, COOPERATIVE, DESIGN AND PLANNING TEAMWORK[2]

The ASHRAE GreenGuide recommends "integrated design teams" that have all of the design, economic, planning, and other related disciplines involved up-front to create better designs. If this process is not used, design typically proceeds in a series of "handoffs" that tend to compound problems, as each succeeding team designs "around" any incompatibilities that the previous designers have already finished. This adds unnecessary complexities and inefficiencies, which increase construction and life-cycle costs. Through coordination and the collaboration of designers, architects, engineers, and key players, an integral design can be created that functions as an efficient system and not as a collection of parts that are force-fit together. This approach has enabled design teams to design very energy efficient, comfortable, aesthetically and environmentally friendly buildings at or less than the conventional price per square foot of traditionally designed buildings. So, working as a team from the beginning is the most efficient way of designing because potential conflicts are resolved up-front rather than later at a higher cost. Effectually, when people are creating and they know that their contributions are respected, superior planning and designs are achieved. The upfront work of coordinating planning and brainstorming sessions with many people of diverse backgrounds can be a challenge. However, the results and life-cycle costs are almost assuredly optimal.

EDUCATION

The United Nations has declared the decade from 2005 to 2014 as The Decade of Education for Sustainable Development.[8]

There are many national teams around the world that have taken the lead to work across the educational spectrum—including public and private education, primary and higher education, independent, charter, and home-schooling—to incorporate ESD in their curricula. The U.S. Partnership for The Decade of Education for Sustainable Development[9] was formed to facilitate implementation of ESD in the United States.

Effective education can demonstrate the inter-relationship and interdependence of people, planet, and profits in all life activities. We have the technology to transition to a clean energy future and to manage materials in a cyclic manner. We know that pollution is causing global environmental change. We also know that teamwork and cooperation get better results than working in a hierarchical or isolated manner. Education in sustainable development will show that the current "consume and throw away" economy no longer works for the benefit of humanity and life. Rather, there is a need for cyclic, whole-systems thinking that integrates all relevant factors into the best, longest lasting results.

SUSTAINABILITY REPORTING

Realizing the importance of being "sustainable" and understanding that they are good long-term investments, many companies are developing sustainability reports. These reports are strong management tools that show how well a company is progressing with their sustainability, their corporate social responsibility (CSR), and their goals to continually improve. They provide an openness that stakeholders (employees, stockholders, regulatory agencies, customers, and community leaders) expect so they know if a company is a good place to work, a good investment, or a good neighbor. These reports are available to the public (free download).[10] For engineers, preparation of sustainability reports involves the collection and analysis of energy and environmental data. Demonstration of energy savings plans and associated pollutant emission reductions can have significant public relations and market share value. This can promote higher sales and flow of investor capital as companies prove their social responsibility and long-term viability.

"High 5!—Communicating your Business Success through Sustainability Reporting—A Guide for Small and Not-So-Small Businesses," from the Global Reporting Initiative, describes the benefits of sustainability reporting:

"Sustainability reporting has many advantages that benefit different areas of your business. Some benefits are purely financial while others deal with customer or employee satisfaction. Sustainability reporting helps organizations identify and address their current and potential risks, saving time and money in the short and long term. As the public becomes more aware of your efforts, customer loyalty and credibility of your business will greatly increase. When you take a deeper look into your daily business operations through sustainability reporting you will be able to discover new opportunities.

Businesses continuously seek to generate income and acquire a competitive edge by identifying new market opportunities and determining current and potential risks. When your organization embraces a sustainability perspective, i.e., simultaneously addressing social, environmental and economic issues, you can benefit from cost-savings and improvements in product quality and employee performance.

When considering how to improve financial performance, many organizations only look at financial aspects, such as the cost of purchasing goods, personnel costs, or tax payments. However, working on environmental and social issues can also positively affect your financial bottom line.

Sustainability reporting is the way to identify these potential benefits and realize the economic gains. It helps you achieve your business goals by setting up a continuous improvement process based on target setting and progress measurement. All in all, sustainability reporting helps you to acquire that competitive advantage."

RENTING VS BUYING (A SUSTAINABILITY INNOVATION)[11,12]

Here is an example of systems thinking to ensure that a product has minimal environmental impact plus high social and economic value:

In today's consumer/throw-away economy, typically a product is manufactured and sold. There is producer incentive to minimize the use of labor and material resources put into a product, thus saving on cost. The product is sold as cheaply as possible to maximize sales. The product eventually wears out and is disposed of. A product that breaks and is disposed of soon after its warranty period is best for sales, so that a consumer will go out and buy a replacement. This is obviously a wasteful scenario, which is prevalent in commerce today. However, that is changing.

A more efficient scenario that many companies have successfully deployed for years is to manufacture and rent a product. Then, there is an incentive to maximize the utility and life of the product, thus ensuring an income to the owner/renter for as long as possible into the future. (This is what sustainable development is all about—maintaining economic flow for as long as possible into the future.) When this product is manufactured, due care in manufacturing processes and sufficient material are used to maximize durability, reliability, and longevity. Customer satisfaction is high from having a reliable product, which also builds name recognition and increases desirability. The product is designed to be easily maintainable, again, to maximize longevity, and also to facilitate dismantlement of the product when it has come to the end of its useful life. Thus, the parts can be easily remanufactured and reused or segregated for efficient recycling of the raw materials. So, maximum utility and income is achieved from the product, and most of the resources that went into manufacturing and maintaining the product throughout its life are recycled with minimal impact to the environment.

So, engineers with sustainable development in mind might be thinking of the "rented" product scenario, which is inherently efficient, rather than the "sold" product scenario, which is inherently wasteful. For example, Interface, Inc., the largest carpet manufacturer in the world and a company that has committed itself to be as sustainable as possible, is using this renting concept in its products. They rent carpet tiles. As the carpet eventually wears, tiles are replaced and recycled to make new tiles.

COMMUNITY DEVELOPMENT TO DECREASE ENERGY COSTS[13]

Over the last three-quarters of a century, there was an assumption that gasoline and diesel fuel would be cheap and plentiful, indefinitely. Thus, urban sprawl developed because of the low cost of owning and operating one or more cars. This is not the case anymore! Now, with the realization that cars are expensive to both purchase and operate, there are efforts by many cities to re-establish neighborhoods that have all of the amenities needed for occupants all within a short distance that can be covered on foot, with a bicycle, or with convenient public transportation. These cities are excluding automobiles from certain areas and some are charging admission fees for cars to enter semirestricted areas. This has helped to revitalize many city commercial districts because of the park-like feeling of being in these areas without the noise and exhaust from cars. With the economic benefits that have been realized by these arrangements, more and more urban and suburban cities are pursuing these types of commercial district renovations in their cities.

Obviously, society has made a major investment that established urban/suburban sprawl. That investment now needs to be made sustainable. Consequently, there will be major investments in clean fuels such as biodiesel, ethanol, and renewably derived hydrogen to power the huge fleet of vehicles in the United States. A remaining economic burden to maintain suburban living is the maintenance of roads. However, that may be relieved as convenient, modern, and cleaner public transit busses, trains, and other guided vehicles are developed into transportation networks, thus decreasing the number of cars on the road. Energy engineers will be integral to this transition in transportation.

So, the paradigm for energy engineers will evolve as community structures around urban and suburban cities change and improve for energy efficiency. There will probably be a tendency to create nodes, where residents have a majority of the amenities nearby. Then, clean, comfortable, energy-efficient public transportation and express lanes for cars will connect each node.

OVERVIEW OF THE MANY ASPECTS OF SUSTAINABLE COMMUNITIES INFRASTRUCTURE AND NATURE

The web site http://www.conservationeconomy.net/INDEX.CFM (courtesy of the Ecotrust) summarizes the many and various aspects of sustainable communities and preserving our natural assets. Many social factors that play a role in the culture and systems changes associated with sustainable development are thoroughly explained on the web site. Engineers should know these social interactions, which are cohesive and essential to sustainable communities. Readers are encouraged to visit the web site to get a flavor of all of the aspects of sustainability or to narrow in on aspects that are of particular interest to them. The following is a narrative summary of some of the chapters on the web site:

A Conservation Economy describes how social capital, natural capital, and economic capital can be synergetic and sustainable.

Social Capital covers fundamental needs, which include: the strong need for local sources of food; accessible, healthy shelter; healthy environment and access to healthcare; plus access to knowledge about the interconnectedness of us to our environment and to each other. The section on community discusses collaborative processes that honor: social equity, which promotes prosperity for all; security from fear and violence; recognition of the wealth and strength in our cultural diversity and establishment of a will to preserve it; plus, local celebrations to honor a sense of place with the community and environment. It proceeds to describe the importance of enjoying beauty and play, to relieve stress from our busyness; learning to welcome transitions that improve communities as a whole; and establishment of civic society where all residents can manage their communities collaboratively.

Natural Capital includes the atmosphere, biosphere, and earth. To sustain it requires ecological land use, which includes connected wild lands, in which indigenous animals, plant and people can coexist together. Protected core reserves can be set aside for native plants and animals to thrive, without interference. Wildlife corridors can connect reserves such that animals may migrate freely and parks can be established to act as buffer zones between developed and undeveloped areas. Productive rural areas can be re-established through: sustainable agriculture, which does not cause runoff of pollutants into streams and is more in harmony with natural processes; sustainable forestry, which thins rather than clear-cuts stands of trees; sustainable fisheries, which establish quotas, such that species are not depleted; and eco-tourism, to provide natural getaways and education for people that want to learn more about sustainability and to be close to nature. Compact towns and cities, with human scale neighborhoods, green buildings, convenient transit access, ecological infrastructure, and urban growth boundaries will create healthy, vibrant communities.

Economic capital in healthier communities and commerce will tend to expand through synergies with social and environmental capital, thus building prosperity.

ENERGY ENGINEERING FOR DEVELOPING COUNTRIES[14]

Approximately four billion of the six and one half billion people on the planet live in extreme poverty, where a poor sanitation infrastructure results in disease, there is minimal economic productivity, and there is significant economic burden on governments and aid agencies. However, recently, it has been found that given some of the modern necessities such as water wells, electronic communication, and dependable energy, people that are impoverished can quickly and enthusiastically become productive to their communities and not be an economic burden.

The use of small photovoltaic power systems in villages has literally energized villages into minieconomic zones of relative wealth and flow of capital, thus creating self-sufficiency rather than dependence. Just providing electricity for lighting, water pumps, and small power tools can tremendously boost the productive capabilities of a village.

The establishment of cell phone repeaters and wireless infrastructure is much less expensive than running miles of telephone cable. A limited number of cell phones in villages have also promoted prosperity because farmers and merchants can effectively communicate and market their products.

As developing countries continue to develop, water, energy, and communication will be vital to their success. So, energy engineers will be a key part of this process. Many multinational and small companies and financial institutions are tapping into bringing these four billion people into the world of commercial success and out of the world of poverty. So, engineers will play an important part in this next major step toward a sustainable world.

CONCLUSION

Though we are just now seeing the start of a transition to sustainable development, it is clearly economically, ecologically, and socially advantageous to choose this means to success, prosperity, well-being, stability, and peace. We have an abundance of all of the material and energy resources we need. We have the ability, knowledge, and conscience to do the best for humanity, all of life, and the planet. We have the inspiration and insight to transition to a sustainable world. Now all it takes is the willingness to accept change and make the transition to sustainability. The outcome will be a world that is more prosperous and stable than it is now. Sustainable development is a goal to embrace and make part of our daily decisions.

REFERENCES

1. Waage, S. *Ants, Galileo, and Gandhi—Designing the Future of Business Through Nature, Genius, and Compassion*; Greenleaf Publishing: Sheffield S3 8GG, U.K., 2003; [www.greenleaf-publishing.com].
2. Stecky, N. *Introduction to the ASHRAE Greenguide for Leed, Globalcon 2005 Proceedings CD*, https://www.aeecenter.org/store/detail.cfm?id=895&category_id=6.
3. Woolsey, R.J.; Hunter, L.; Amory, L. *Energy Security: It Takes More Than Drilling*, http://www.csmonitor.com/2002/0329/p11s02-coop.html.
4. Woolsey, R.J.; McFarlane, B. Conference Report: *Renewable Energy in America: The Call for Phase II, Session 3: Using Renewable Energy to Meet Our National Security Needs*; December 2004, http://www.acore.org/pdfs/04policy_report.pdf.

5. http://www.insnet.org/ins_headlines.rxml?cust=2&id=1234.
6. http://www.insnet.org/ins_headlines.rxml?cust=2&id=1248.
7. Anderson, C.; Katharine, R. *The Co-Creator's Handbook*, www.globalfamily.net.
8. http://portal.unesco.org/education/en/ev.php-URL_ID=27234&URL_DO=DO_TOPIC&URL_SECTION=201.html.
9. http://www.uspartnership.org.
10. http://www.CorporateRegister.com.
11. Williard, B. *The Sustainability Advantage—Seven Business Case Benefits of a Triple Bottom Line*; New Society Publishers: British Columbia, V0R 1X0, Canada, 2002; [www.newsociety.com].
12. Doppelt, B. *Leading Change Toward Sustainability—A Change-Management Guide for Business, Government and Civil Society*; Greenleaf Publishing: Sheffield S3 8GG, U.K., 2003; [www.greenleaf-publishing.com].
13. http://magma.nationalgeographic.com/ngm/data/2001/07/01/html/ft_20010701.3.html.
14. http://www.csrwire.com/print.cgi?sfArticleId=3709.

Thermal Energy Storage

Brian Silvetti
CALMAC Manufacturing Corporation, Fair Lawn, New Jersey, U.S.A.

Abstract

Often unnoticed, elements of thermal energy storage (TES) technology can be found in many common products. As an engineered system, however, thermal storage has matured into an extensively developed technology primarily for commercial comfort and process cooling. Commercial utility rates and the unbalanced cooling demand of this sector make it an ideal candidate for TES. Although water is the predominant storage material, methods of its use vary across a diverse selection of latent and sensible heat equipment types, each with unique performance and operational characteristics. Additionally, systems can be selected and designed to address any fraction of the cooling load in partial or full storage strategies. In manipulating the contribution of chiller and storage to the cooling load, control logic must be consistent with equipment capabilities and capacities, utility energy and demand rate structures, and integrated load requirements.

INTRODUCTION

Thermal energy storage is one of the world's oldest energy management technologies. For millennia, the mass of shelters has served to moderate extreme temperature excursions by gradually absorbing and releasing heat. And virtually all thermal storage reviews recount the centuries old practice of harvesting ice from rivers and lakes for use throughout the following warmer seasons. Indeed, ice was a major North American export in the 19th century.

One of the unique historical demonstrations of stored cooling occurred on a hot July day in 1620, when the Dutch scientist, Cornelis Van Drebbel, chilled the Great Hall of Westminster Abbey for the benefit of King James I. Van Drebbel reportedly combined snow saved from the previous winter, water, potassium nitrate, and salt in a prescient precursor to the ice cream maker.

There are dozens of more contemporary examples of thermal storage. These include the familiar domestic hot water heater or the household refrigerator icemaker, as well as many lesser known uses, such as in athletic equipment, medical braces and wraps, thermal barriers for fire protection, temperature stabilization of electronic components, and space exploration components.

However, central plant commercial cooling applications have evolved into the most extensively developed sector of thermal storage technology. Professional organizations have published an impressive library of design literature.[1] Equipment and system standards have been developed,[2,3] and manufacturers provide performance data, installation and maintenance instructions, and of course, a wide variety of available products. Other than a broader discussion of storage materials and brief introductions to a selection of additional TES technologies, commercial cooling application and design will dominate the following discussion.

THERMAL STORAGE MATERIALS

Thermal energy is stored by adding or removing heat from a substance. The change experienced by the substance helps define the type of storage system. Although materials such as clathrates and hydrides can store thermal energy through a change in composition, our attention focuses on the more conventional sensible and latent heat properties of materials.

Sensible Heat Storage

When a substance experiences a rise or fall in temperature as heat is added or removed, it is referred to as sensible heat storage. The amount of heat required to raise or lower the temperature of liquid water, its specific heat, is 1 Btu/lb/°F (1 cal/g/°C). That is, 1 Btu will raise or lower the temperature of 1 lb of water 1°F, an unusually high value. High density, another advantageous property of water, improves volumetric efficiency. The density of water is 62.4 lbs/ft^3 (1 g/cm^3). Virtually any material can be used for sensible heat storage, but water, concrete, masonry, rock, earth, and ceramic brick are some of the more common. In fact, the structure itself is sometimes used as a thermal storage medium through precooling or solar heat gain.[4,5]

Keywords: Full storage; Partial storage; Ice storage; Latent heat; Cooling storage; Sensible heat; Thermocline; Chiller; Demand charge; Secondary coolant; Off-peak; TES.

Latent Heat Storage

When liquid water and ice are in thermal equilibrium at 32°F (0°C), the addition or removal of limited quantities of heat does not change the temperature. The addition of heat will change the phase of some of the water from solid to liquid, and the removal of heat will change the phase of some of the water from liquid to solid. This is an example of latent heat of fusion (solid/liquid), the form of latent heat most commonly applied in thermal storage.

Materials employed for thermal storage in this manner are often referred to as phase change materials (PCMs). Water has an extremely high latent heat capacity. It takes the addition or removal of 144 Btu to change the phase of 1 lb of water (80 cal/g) at 32°F (0°C). Water is by far the most widely applied PCM for cooling TES. Its safety, physical properties, high heat capacity, and chemical stability are all important characteristics, but appropriate fusion temperature and negligible cost secure its popularity.

Other latent storage materials have been commercialized or researched over the years for both heating and cooling applications.[6] Sodium sulfate decahydrate (Glauber's salt), one of many inorganic salt hydrates, was perhaps the most intensely studied and typifies both the promise and pitfalls often encountered. Its low cost, high density (specific gravity 1.46), high latent heat capacity (108 Btu/lb, 60 cal/g), and fusion temperature of 89°F (31.7°C) make it apparently ideal for active (mechanically assisted) and some passive heat storage applications. The performance of salt hydrates has proven difficult to perfect because of the tendency to form undesirable hydrates (e.g., sodium sulfate dihydrate), segregate into separate constituents or supercool (undercool). Supercooling is the tendency of most liquids to remain a liquid, even as the temperature drops below the fusion point, until some nucleating event precipitates the formation of solid crystals. Once nucleated, the entire mass of material typically rises rapidly back to its customary fusion temperature.

Mixtures of materials at their eutectic, or lowest freezing point concentration, are also effective storage materials, often employed below 32°F (0°C). For example, 23.3% sodium chloride in water exhibits a uniform freezing temperature at −6°F (−21°C). Some salt hydrates and eutectics are available for thermal storage applications that use a variety of proprietary or patented techniques for improving performance. Rolling drum containers, thixotropic mixtures, gelling agents, mechanical stirring, clever container geometry, microencapsulation, and other chemical additives have all been used to prevent segregation and/or induce nucleation.

Other materials include paraffins, fatty acids, and their eutectic mixtures. These are chemically stable, and the ability to tailor the fusion temperature of paraffins by controlling the carbon chain length is an attractive feature. They typically have lower heat capacities, densities, and thermal conductivities than the water-based materials, and they sometimes experience comparatively high volume changes with phase transition. They are employed in some specialized applications referred to earlier. There have also been several attempts over the years to infuse paraffins into building materials (gypsum wallboard, for instance) to enhance the storage capacity of the building structure.

Although rarely applied, heat transfer to or from many solids at specific temperatures can produce a change in their crystalline structures, another form of latent heat. Some materials will even undergo a series of structural changes (e.g., hexatriacontane) at different temperature plateaus.

TES FOR COOLING APPLICATIONS

Benefits

Many factors encourage the use of thermal storage. Most applications, not only the classic examples like churches, dairies, and theaters that exhibit brief but also intense cooling loads, benefit from substantial reductions in refrigeration equipment and electrical infrastructure. But the principal basis for many TES benefits is energy cost reduction, particularly with the current emphasis on Leadership in Energy and Environmental Design (LEED®) certification.

In addition to the recognized benefits of avoiding the construction of additional generation and transmission facilities, there is also substantial evidence that power generated off-peak is produced more efficiently and with fewer emissions than the peaking power it replaces.[7]

Engineers often install excess capacity to meet unexpected cooling loads or to compensate for mechanical failure, an inefficient and expensive practice. Rather than invest in chiller capacity that will rarely be used, and may produce even higher electrical demand, the same objectives can be achieved with TES while realizing continuing economic benefits. A selection example is included under "System Design."

Commercial Utility Rates and TES

The commercial customer utility bill is usually comprised of two main components. The first, the energy charge, is based on the amount of electrical energy consumed, or kWh. This energy charge is often reduced during off-peak hours, providing the first incentive for TES. The second component is referred to as a "demand" charge and is based on the rate (kW) at which a customer consumes energy, often for the highest 15-min demand period during on-peak hours. A demand ratchet may extend the impact of high demand to subsequent months. The demand component may be incorporated in an energy charge that is reduced for customers with a flatter electrical demand profile, rather than explicitly identified.

To illustrate, a customer that operates a 100-W light bulb for 10 h will consume the same amount of electric energy as a customer who burns ten 100-W light bulbs for 1 h. However, the customer who consumes that energy over the 10-h period will pay one-tenth of the demand-related utility cost. It is quite common for a commercial customer's demand charges to contribute more than 50% of the utility bill.

Refrigeration equipment can constitute 40% or more of a commercial building's peak electric demand, with little or no nighttime cooling load, representing a significant opportunity for energy cost savings with TES. In a very real sense, TES applications that reduce on-peak electrical usage are storing electricity as much as heating or cooling capacity.

Cooling Storage Equipment

Cooling storage devices encompass a wide variety of equipment designs with unique characteristics. Although far from comprehensive, cool storage systems can usually be classified within one of the following groups, as adopted by most manufacturers and ASHRAE.[8] Because storage systems must meet not only the peak instantaneous load (tons), but also provide the integrated cooling capacity over the design period, an essential measure of capacity is the ton-hour (0.0127 GJ).

Ice Storage-HX Coil Builds and Melts Ice

Fig. 1 illustrates a typical design. The complete assembly is usually factory manufactured and provided in thermally insulated, modular units that can be combined to provide the needed capacity (Fig. 2). Cylindrical plastic tanks and rectangular galvanized or stainless steel tanks are available. Plastic tanks provide water containment, while metal tanks include an additional flexible or metal liner to achieve water containment.

An antifreeze solution, usually 25% ethylene glycol/75% water[9] and referred to as a secondary coolant, is circulated through the tubular heat exchanger (HX) that is distributed throughout most of the tank's internal volume. The HX is completely immersed in water that usually occupies about 80% of the tank volume. The remaining space is reserved for the HX and to allow for the expansion of the water as it freezes, decreasing in density by approximately 9%. The expansion of the water into the space above the HX is often used as a measure of ice inventory. Heat exchangers are constructed of plastic, usually polyethylene or polypropylene, or galvanized steel. Copper heat exchangers circulating refrigerants are a more recent development in TES systems with smaller capacities. The spacing between the tubes of the HX is determined by the diameter of the tubes themselves, desired rates of charging (freezing) and discharging (melting), and heat transfer properties of the HX and coolant.

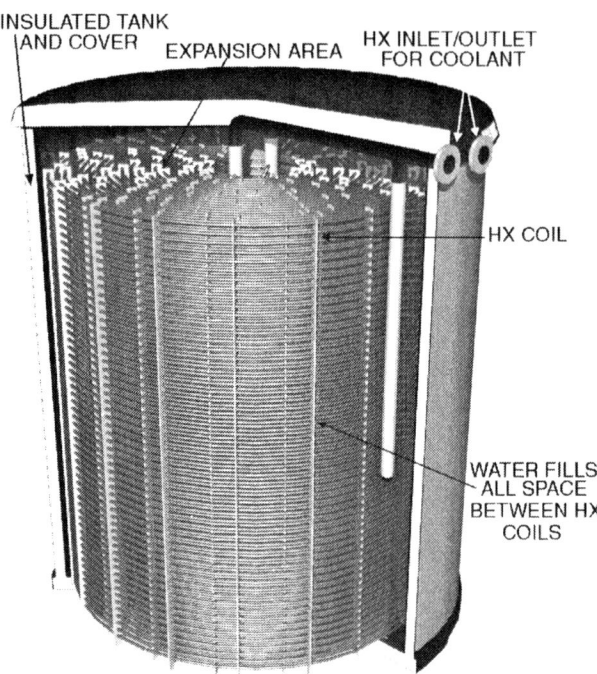

Fig. 1 Modular ice storage tank—charge and discharge through HX. (Courtesy CALMAC Manufacturing and Nihon Spindle.)

Note that the water in the tank is not hydronically connected to the building-cooling loop and does not circulate. All of the heat transfer is accomplished by circulation of the coolant through the HX. The HX usually operates as part of the pressurized, closed building loop while the tank itself is at atmospheric pressure.

Standard chillers (excluding absorbers) provide the cooling source for freezing the water. During the charging cycle, the chiller cools the secondary coolant to a range of perhaps 20°F (−6.7°C) to 27°F (−2.8°C), the lower temperatures reached at the end of the charge cycle. Coolant is circulated through the HX until all the water between the HX tubes is frozen (Fig. 3).

During the discharge, the same coolant is circulated between the cooling load and the storage tank. As the ice surrounding the HX is melted, the 32°F (0°C) liquid water/ice surface gradually retreats from the HX tube surface (Fig. 3). Therefore, the performance, as reflected in the temperature of the coolant leaving the HX, varies as the ice is melted. Some manufacturers publish charts to describe the variable performance due to this characteristic. This design sometimes employs alternate PCMs.

Ice Storage-HX Coil Builds Ice-Surrounding Water Melts Ice

This design shares some features of the internal melt system. Tanks are typically large, site-built, rectangular concrete structures that operate at atmospheric pressure (Fig. 4). The HXs are usually galvanized steel coils,

Fig. 2 14,000 tn-h, 5000 ft² modular ice storage system, cooling 1,000,000 ft² district system. (Courtesy CALMAC Manufacturing.)

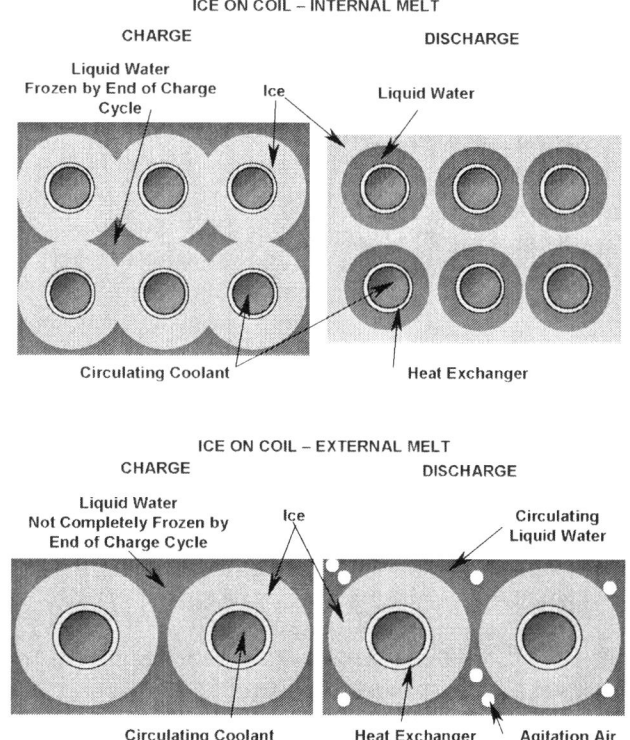

Fig. 3 Comparison of charging and discharging processes from interior and exterior of ice on coil.

Fig. 4 50,000 tn-h district cooling storage system under construction. (Courtesy CALMAC Manufacturing.)

similar to the internal melt coil, but often with wider tube spacing.

Circulating cold coolant through the HX coils, as in the internal melt design, forms the ice. The major difference is that some liquid water is always left surrounding the ice formed on the HX tubes (Fig. 3). This water is the heat transfer fluid for the discharge cycle, and ice thickness must be carefully controlled to ensure that a flow passage is always available for the circulating water while still achieving the required ice inventory. Because the liquid water is in continuous direct contact with the ice, it is available at consistently low temperatures and high rates of discharge.

Various methods are used to interface the atmospheric pressure tank with a pressurized building-cooling loop when necessary. For instance, an intermediate HX or pressure sustaining valves (see Fig. 5 for stratified chilled water) can be placed between the two systems.

The velocity of the water within the tank is extremely low. Compressed air is typically distributed through separate piping located under the bottom layer of coils to augment convection and promote even melting and freezing of the ice throughout the often considerable dimensions of the tank.

The refrigeration equipment often contributes to the cooling load during the discharge cycle (see "System Design"), but separation of the chiller coolant (antifreeze solution) from the water loop must be maintained. Rather than simply continue to circulate the coolant through the ice-encased storage coils, improved efficiency can be realized by adding an additional HX between the chiller coolant loop and the building water loop. Sometimes the actual refrigerant is redirected to a separate evaporator where it cools the circulating water.

Although the chiller/coolant method of ice-build has become common, direct expansion or liquid-overfeed of refrigerants, circulated directly through the ice coils, is applied in certain process applications. Water is the only PCM currently used in external melt systems.

Ice Storage-PCM in Sealed Containers or Encapsulated

These systems also employ standard chillers circulating a secondary coolant. However, the PCM, usually water, is contained within sealed plastic containers. Insulated tanks are filled with the containers, and coolant circulates through the tank and around the containers.

Tank configurations include horizontal or vertical cylinders, both pressurized and at atmospheric pressure. The useable volume of PCM depends on the shape of the container, the geometry of the tank, and whether the containers are individually positioned or allowed to pack randomly.

Many container shapes have appeared over the years, including cylindrical tubes and flat trays, but current products are mainly spherical. Additional features, intended to enhance heat transfer or accommodate expansion of the water, include flexible surface geometry, internal voids, or collapsible internal chambers.

Encapsulated systems are available with alternate PCMs, usually salt hydrates and eutectics, but other materials are possible. Supercooling must be prevented by including a nucleating agent or device.

Ice Storage-Harvester

Refrigerant is directed through a collection of vertical plates or tubes while water constantly flows around the external surface. Ice accumulates on the surface, usually to a thickness less than $\frac{1}{2}$ in. Repositioning an arrangement of valves introduces compressed, hot refrigerant gas into

Thermal Energy Storage

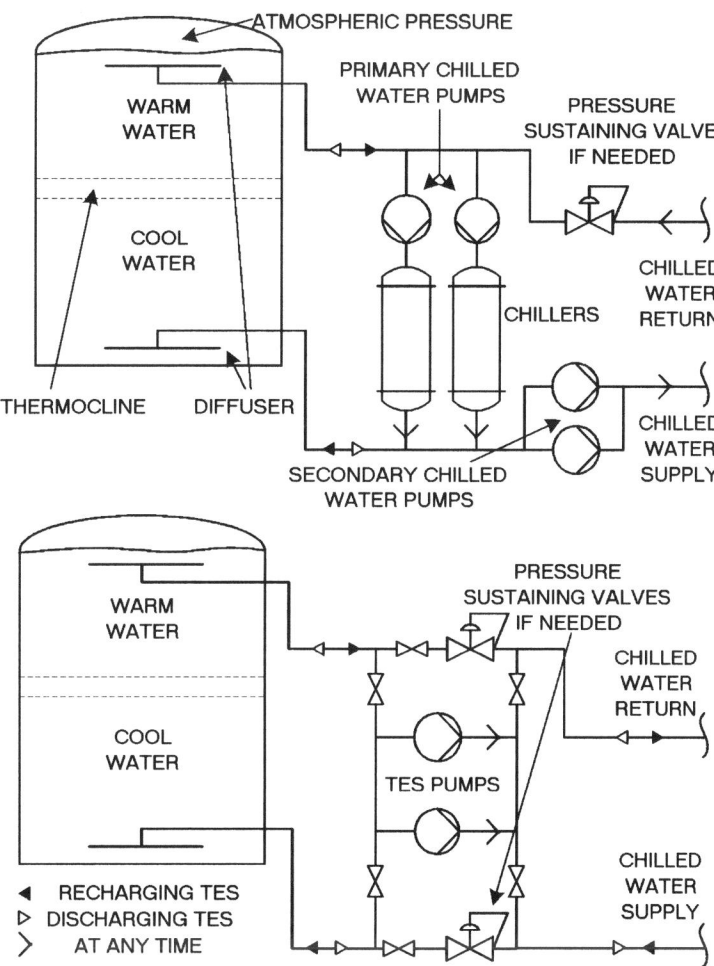

Fig. 5 Chilled water storage piping schematics Top: TES near chiller water plant. Bottom: TES near or remote from chilled water plant. (Designs courtesy The Coolsolutions Company.)

the plate. The ice separates from the surface and drops into an atmospheric pressure steel or concrete tank below. This procedure is repeated sequentially over a series of many plates or tubes, so that the process is essentially continuous.

An alternate approach circulates coolant or refrigerant around an HX that has water combined with a low concentration of glycol flowing on the opposite surface. A combination of mechanical disturbance and the fluid's reduced tendency to adhere to a surface when frozen eliminate the need for hot gas defrost.

Harvesters are less common in comfort conditioning applications today but are still used in process and industrial applications. As in the external melt design, the direct contact between the circulating water and the ice provides low temperatures and high discharge rates.

Stratified Chilled Water Storage

Chilled water is a sensible heat storage method. The available capacity is proportional to the volume of water and the temperature change it experiences. Cost considerations dictate that the same tank, usually a steel or concrete vertical cylinder, be used to store both the warm water returning from the cooling load and the cold water circulated to the cooling load. It is critical that the warm and cool water are prevented from mixing within the tank.

The preferred approach that has evolved is the simple stratified water system. Because warm water is less dense than cooler water, it will float on cooler water below. However, the density of water at 58°F (14.4°C) is only about 0.05 lb/ft^3 (0.8 kg/m^3) less than water at 40°F (4.4°C). Maintaining separation is now commonly achieved through careful design of the water diffusers located in the top and bottom of the tank. The goal is to prevent disruption of the stable but delicate density relationship as water enters and leaves.

Two nondimensional flow parameters have usually been accepted as crucial to proper performance. The Froude number is the ratio of inertial to gravitational forces (related to the Richardson number), and the

Reynold's number is the ratio of inertial to viscous forces. Research, with sometimes inconsistent results, is still evaluating the influences of these and other parameters.[10] Designers, using a combination of analytical and empirical approaches, have nonetheless arrived at a successful design process.

Fig. 5 represents chilled water storage systems part way through a charge or discharge cycle. During charge, flow exits the top of the tank, is cooled by the chiller, and returns to the tank bottom. During discharge, the flow reverses, and water exits the bottom of the tank, is distributed to the load, and returns to the top. Note the area between the warm and cold layers. Known as a thermocline, this short cylinder of water, usually 1–3 ft thick, exhibits a vertical temperature gradient between the cold and warm water temperatures.

Because the maximum density point occurs at 39.2°F (4°C), water is usually cooled to no less than 40°F (4.4°C), and chiller efficiencies are very good during the charging cycle. Additives are sometimes employed to lower the temperature of maximum density and increase the storage capacity. Increasing the system return temperature also expands the storage capacity. However, the water must return to the tank from the cooling system at the design temperature because any reduction represents lost capacity.

Large chilled water storage systems are very economical. Examples include a district cooling application with a 17.6 million gallon (66.6 million liter) tank, 160,000 tn-h total capacity, and a peak discharge rate of 20,000 tn.

Cool Storage System Design

All equipment types require some unique design elements, but an ice storage system will introduce most of the important issues.

Load Profile

The first step is to determine the cooling load profile for the entire design day. In a nonstorage system, the peak load is the primary influence on chiller selection. For storage systems, both the peak load and the total combined load for the design period will determine chiller and storage capacities. Furthermore, chillers may operate at different capacities at different times of the day, and storage performance may vary with system geometry, control sequence, storage inventory, and system temperatures. For this reason, ARI has published a format for specifying operating temperatures, modes, loads, etc. on an hour-by-hour basis.[11]

An assumed load profile for a typical office building is depicted in Fig. 6 (top). The peak load is 1000 tn and the total integrated cooling load over a 12-h period is 10,000 tn-h.

Chiller and Storage Capacity

Because all of the cooling originates with the chiller, we first calculate chiller capacity by equating the full load chiller operating hours to the total cooling load. The calculations are quite simple, requiring only careful assignment of the hourly operating modes, consistent with the utility tariff, building occupancy, and the planned strategy.[12] There are a number of options available to the designer. Possible modes of operation for any particular time include charging (ice-making), charging with cooling, cooling with a combination of storage and chiller, cooling with chiller only, cooling with storage only, and system off.

$$\text{Total ton hours} = \text{Chiller day capacity} + \text{Chiller ice capacity} \quad (1)$$

Full storage is the easiest to calculate and the simplest to implement. It also provides the greatest operating cost savings, but the highest initial investment. The entire on-peak cooling load is provided from storage, and the chiller only operates to produce ice at night, off-peak.

In this simplest of cases, the day capacity is zero, because the chiller is off. The ice-making capacity, at this point unknown, will be less than the chiller's nominal daytime capacity. However, the relative daytime and ice-making capacities are easily determined for a particular chiller category. A centrifugal chiller, for instance, might produce 70% of its nominal capacity when in the ice-making mode, and 100% during the day. We usually want to describe the required chiller capacity in terms of its conventional rating and will therefore include a factor of 0.7 for hours operated at ice-making conditions. This is a capacity modifier, not an efficiency adjustment. Whether the efficiency is the same, better, or somewhat worse during charging will depend on a number of factors, such as nighttime condenser temperatures, arrangement and type of components, and choice of air or water cooled chiller.[13] With 12 h available for charging storage, Eq. 1 becomes:

$$10,000 \text{ tn h} = 0 + (NC \times 12 \text{ h} \times 0.7)$$

$$NC = \frac{10,000 \text{ tn h}}{12 \text{ h} \times 0.7} = 1190 \text{ tn}$$

where NC is the nominal chiller size. Of course, the required storage capacity is the entire 10,000 tn-h.

Another more common approach is to operate the chiller throughout the day, producing ice at night and directly contributing to the load during the day. At its limit, this results in the smallest possible chiller capacity, and equipment costs are often competitive with nonstorage systems.

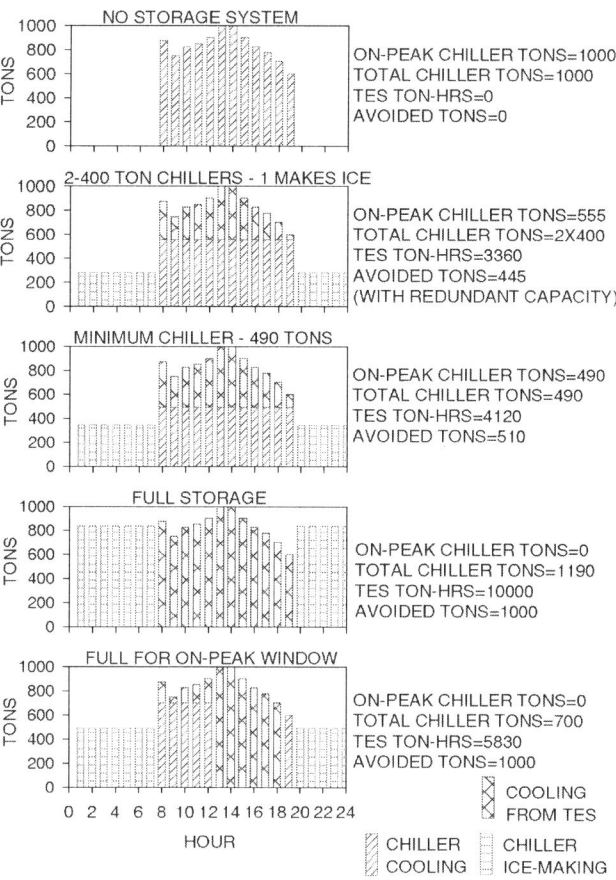

Fig. 6 Alternative TES selections for the same cooling load profile.

$$10{,}000 \text{ tn h} = (NC \times 12 \text{ h} \times 1) + (NC \times 12 \text{ h} \times 0.7)$$

$$NC = \frac{10{,}000 \text{ tn h}}{(12 \text{ h} + 8.4 \text{ h})} = 490 \text{ tn}$$

The amount of storage becomes:

$$\text{Storage ton hours} = NC \times \text{Ice making hours} \\ \times \text{Capacity factor} \quad (2)$$

$$\text{Storage ton hours} = 490 \text{ tn} \times 12 \text{ h} \times 0.7 = 4118 \text{ tn h}$$

Note that this partial storage chiller is only 49% of the peak load and 41% of the full storage selection. Storage capacity is also reduced to 41% of the full storage requirement.

There is any number of intermediate selections. For instance, a common conventional design might include three 400-tn chillers for our 1000-tn building. An alternative is to select two 400-tn chillers and storage equal to the ice-making capacity of one of the machines, or about 3360 tn-h for our example. Similar redundancy is provided in the event of a chiller failure, with the added benefit of permanent operating cost savings.

Other multiple chiller options are possible. Certain utilities experience high, but relatively short, afternoon demand peaks, and a common alternative in these areas treats a 4–6 h period of the day as full storage. Other applications, churches for instance, may operate on multiple-day cycles, or industrial applications may experience a series of brief but intense cooling loads over the period of several hours. Fig. 6 depicts some possible selections and the on-peak chiller reductions that are possible with the more common daily cycles.

System Layout and Control

The final step is to construct a simple schematic representation of the proposed system and verify the control logic. A common partial storage design places the storage and chiller in series (Fig. 7). Our example positions the chiller upstream of the storage relative to the cooling load, but the reverse is sometimes preferred. Remember that in the internal melt system, the only fluid circulating is the freeze-protected coolant; the storage water/ice never leaves the storage tanks.

Note the modulating three-way valve located at the storage system. This valve automatically controls coolant flow through the storage tanks to maintain the desired

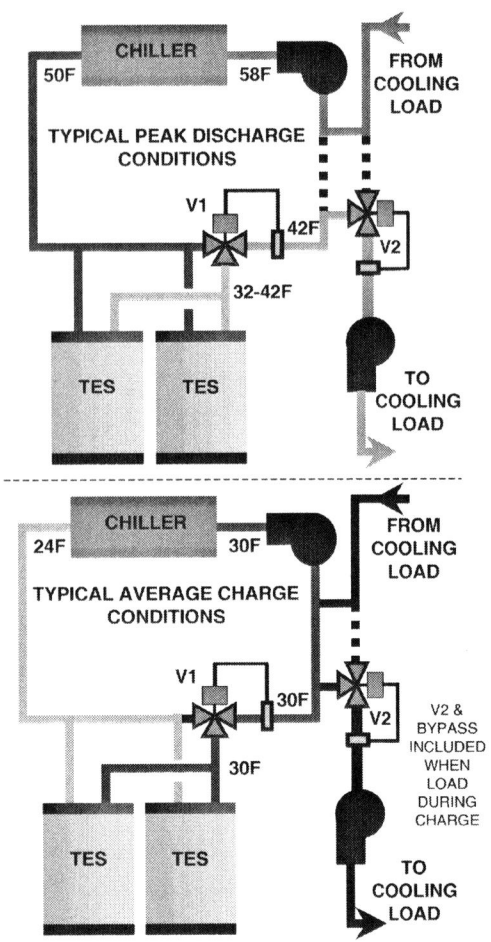

Fig. 7 Ice storage system schematic—charge and discharge.

coolant temperature throughout the discharge period as ice is melted.

During charging, the chiller operates at the lower ice-making temperatures. The three-way valve directs all flow through storage. With no direct cooling needed during the ice-making period, the load can simply be bypassed. If there is a cooling load during ice-making, a second three-way valve and bypass is sometimes included (dashed lines) so that warm coolant returning from the load can be blended with the cold ice-making coolant before being sent to the cooling coils. The ability to efficiently meet small nighttime loads is often seen as a major storage advantage. A separate chiller often serves substantial nighttime load.

Discharge logic is dependent on the utility rate, equipment limitations, cooling loads, and the level of on-site supervision. The chiller can simply be set to provide the design supply temperature (e.g., 42°F [5.6°C]). Storage will only be discharged if the smaller chiller cannot meet the load. The amount of ice melted each day is minimized, referred to as chiller priority. Alternately, on cooler than design days, the chiller leaving temperature can be raised to increase the load on storage (storage priority) and take advantage of additional demand or energy cost savings.

Of course, the danger is that an error in the estimated total cooling load may exhaust storage, leaving only an undersized chiller to carry the load later in the day. Even unsophisticated systems, however, successfully employ this strategy by incorporating adequate margins in the level of chiller loading.[14]

Verify Initial Assumptions

Once the control and operating sequences have been established, return to the original selection calculations to insure that chiller and/or storage contribution, device operating temperatures, and modes of operation remain consistent with original assumptions. There are many alternative designs. Some include parallel flow arrangements, additional heat exchangers, or chillers. By applying the proposed logic on an hour-by-hour basis, the designer is assured of reliable performance.

BRIEF REVIEW OF OTHER TES OPTIONS

There are, of course, many other interesting and promising areas of TES technology. In some cases, precooling of the building structure and contents provides significant storage.[5] Research in this area is identifying and quantifying relevant design parameters, such as the location and type of auxiliary mass, effects of space dimensions, control methods, and potential benefits.

Underground thermal storage systems are used for both cooling and heating. Ground water from aquifers (ATES) is transferred between a warm well and cold well, extracting or adding heat either directly or through a heat pump. Alternatively, a closed loop array of vertical heat exchangers is placed into boreholes (DTES). Some issues that influence the choice or design of these systems include the balance between seasonal heating and cooling loads of the building, local geology, topography, hydrology, extraction and reinjection flow rates, and impact on neighboring water use (wells).[15]

Some winter-peaking utilities promote off-peak heat storage. A common design distributes an array of electric resistance heaters throughout a matrix of high-temperature ceramic (or other refractory) bricks within an insulated cabinet. Temperatures in excess of 1400°F (760°C) can be reached, providing large storage capacities in a small volume utilizing only sensible heat effects. A fan circulating air through passages within the storage device may assist discharge.[16]

Heat storage, combined with passive solar techniques, can be very effective if properly applied with regard to local climate, space geometry, etc. The Trombe wall, popularized in the mid-1960s, provides insight into several solar heating principles,[4] and, of course, the simple hot water tank is a common storage vessel for active solar systems.

CONCLUSION

Progress continues in many areas of TES. Unique applications, like combustion turbine inlet air-cooling that recovers capacity normally lost during hot weather, are continually being developed.[17] Efforts are currently underway to describe more precisely TES performance in energy modeling software, motivated largely by increasing interest in the U.S. Green Building Council (USGBC) LEED®[18] certification program. Thermal storage is a uniquely diverse area of energy technology with both a long history of reliable use and the emerging potential of new and interesting developments.

REFERENCES

1. Dorgan, C.E.; Elleson, J.S. *Design Guide for Cool Thermal Storage*; ASHRAE: Atlanta, GA, 1993.
2. ANSI/ASHRAE/Standard 150-2000. *Method of Testing the Performance of Cool Storage Systems*, ASHRAE: Atlanta, GA, 2000.
3. ARI Standard 900-98. *Thermal Storage Equipment Used for Cooling*, Air-Conditioning and Refrigeration Institute: Arlington, VA, 1998.
4. Torcellini, P.; Pless, S. *Trombe Walls in Low-Energy Buildings: Practical Experiences (Preprint)*; National Renewal Energy Laboratory: Golden, CO, 2004.
5. Braun, J.E. Load Control Using Building Thermal Mass. J. Solar Energy Eng. **2003**, *125*, 292–301.
6. Zalba, B.; Marin, J.; Cabeza, L.; Mehling, H. Review on thermal energy storage with phase change: materials, heat transfer analysis. Appl. Thermal Eng. **2003**, *23*, 251–283.
7. Tabors Caramanis & Associates. *Source Energy and Environmental Impacts of Thermal Energy Storage*, California Energy Commission: Davis, CA, 1995.
8. Elleson, J.; Kooy, R.; Bahnfleth, W. *2003 ASHRAE Handbook, HVAC Applications, Chapter 34 Thermal Storage*, ASHRAE: Atlanta, GA, 2003.
9. Gatley, D.P. *Cool Storage Ethylene Glycol Design Guide*, EPRI TR-100945; Electric Power Research Institute: Palo Alto, CA, 1992.
10. Bahnfleth, W.P.; Musser, A. Thermal performance of a full-scale stratified chilled-water thermal storage tank. ASHRAE Trans. **1998**, 377–388.
11. ARI Guideline T-2002. Specifying the Thermal Performance of Cool Storage Equipment, Air-Conditioning and Refrigeration Institute: Arlington, VA, 2002.
12. Silvetti, B. Application fundamentals of ice-based thermal storage. ASHRAE J. **2002**, *February*, 30–35.
13. MacCracken, M.M. Thermal energy storage myths. ASHRAE J. **2003**, *September*, 36–43.
14. Elleson, J.S. *Successful Cool Storage Projects: From Planning To Operation*; ASHRAE: Atlanta, GA, 1997.
15. http://www.worldenergy.org/wec-geis/publications/default/tech_papers/17th_congress/4_1_16.asp (accessed June 2005).
16. http://www.electricenergyonline.com/IndustryNews.asp?m=&id=38585 (accessed July 2005).
17. Andrepont, J.S. Thermal energy storage technologies for turbine inlet cooling. Energy-Tech. **2004**, *August*, 18–19.
18. *LEED Green Building Rating System For New Construction & Major Renovations (LEED-NC) Version 2.1*, U.S. Green Building Council: Washington, DC, 2002.

Thermodynamics

Ronald L. Klaus
VAST Power Systems, Elkhart, Indiana, U.S.A.

Abstract
The foundational ideas of thermodynamics, especially the implications of its two main laws, are presented, and their implications for energy engineering are discussed. Special attention is given to the implications of the second law in the design of processes that are thermodynamically efficient. The application of thermodynamics to practical engineering problems requires accurate estimation of thermodynamic properties of fluids. A survey of the methods used for such estimations is also presented.

NOMENCLATURE

All thermodynamic properties have the units of energy or energy per unit time except where otherwise specified.

C_p	Heat capacity at constant pressure (energy/temperature)
e^{Ch}	Chemical exergy
e^{Ph}	Physical exergy
G	Gibbs free energy, or Gibbs function
H	Enthalpy
M	General thermodynamic property
M_i	Partial molal property of specie "i"
n	total number of moles
n_i	number of moles of specie "i"
P	Pressure (force/area)
P^V	Vapor pressure (force/area)
Q	Heat or heat flux added to a process (energy or energy per unit time)
Q_{ideal}	Heat or heat flux added to a completely reversible process (energy or energy per unit time)
R	Universal gas constant (energy/temperature)
S	Entropy (energy/temperature)
T	Temperature
T_0	Reference temperature or temperature of the surroundings
U	Internal energy
V	Volume (length^3)
W	Work or power produced by a process. Usually does not include work done by or on fluids entering or leaving the process (energy or energy per unit time)
W_{ideal}	Work or power produced by a perfectly reversible process (energy or energy per unit time)
W_{lost}	Lost work or power (energy or energy per unit time)
W_{total}	Total work or power produced by the process (energy or energy per unit time)
x_i	mole fraction of specie "i"
x	vector containing all mole fractions
yH_2O	Mole fraction of water vapor in a vapor phase
Z	Compressibility factor
Δ	Difference in the value between two states or differences between the properties of outlet and input streams
ΔH_f	Enthalpy of formation
ΔS_f	Entropy of formation (energy/temperature)

Subscripts

f	Formation
o	Base or reference state
C	Cold sink
H	Hot source
rev	Reversible

Superscripts

*	Ideal gas or ideal gas state
id	Ideal solution

INTRODUCTION

Thermodynamics is concerned with the interaction between matter and energy on a macroscopic level. It is therefore the basic science that underlies the engineering of the use of energy in all its forms. It is governed by two basic laws with far-reaching implications. The first concerns the conservation of energy, and the second puts limits on the amount of energy that can be converted into work over and above what the first law might be thought to imply.

The actual performance of practical process equipment can be predicted by application of these two laws. That, in turn, depends on the ability to estimate the thermodynamic properties of the streams that enter and leave this equipment. The exact equations needed to calculate these properties can be derived from the formalism of thermodynamics. Certain simplifications, such as the ideal

Keywords: Combustion; Energy; Enthalpy; Entropy; Exergy; Gibbs- free energy; Ideal gas; Irreversibility; Lost work; Power generation; Phase equilibrium; Thermodynamic properties.

THE FIRST LAW OF THERMODYNAMICS

Formulation

The first law of thermodynamics is a statement of the principle of conservation of energy for a closed system—namely, one in which there is no transfer of mass across the system boundaries.

$$\Delta U = Q - W_{total} \quad (1)$$

This equation ignores certain effects such as differences in fluid velocities, which become important only at very high velocities, and differences in fluid elevations, which become important only when there are greatly varying elevations. U is a thermodynamic property called internal energy. It accounts for energy due to the motion of molecules in a fluid and the exchanges of energy between them. The Δs refer to differences in the property value between the final and initial states of the system. U, Q, and W all have units of energy. However, this equation can also have another interpretation for open flow systems at steady state. Such systems have input and outlet streams, and possibly an exchange of heat or work with the surroundings (Fig. 1). However, the stream flows and property values throughout the system are all considered constant over time. In this case, the first law is a rate equation. ΔU is the difference between the sum of U times the mass flow rate for all the output streams minus that same sum for all the inlet streams. Q and W are rates of energy transfer. The arrows in Fig. 1 indicate the direction of the flow of work, W, and heat, Q, when these quantities are positive.

When applying Eq. 1 to a flow process, W_{total} includes the work done in moving fluids into and out of the process. To eliminate these effects, another thermodynamic property, namely enthalpy, H, is defined,

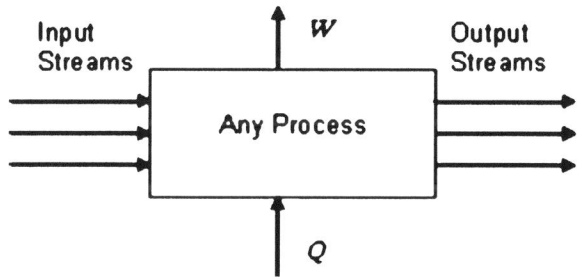

Fig. 1 Schematic representation of any thermodynamic process.

$$H = U + PV \quad (2)$$

where P is pressure and V is volume.

For a flow process at steady state, this changes the first law to

$$\Delta H = Q - W \quad (3)$$

In Eq. 3, W is the work produced by the process apart from that involved in moving fluids into and out of the process. It is often called the shaft work.

In these equations, it is important to make a distinction between intrinsic and extrinsic properties. The former are properties per unit mass or per mole, whereas the latter are total values of the property for the mass flow of the given process stream. These and other thermodynamic relationships can be applied to either, provided that they are applied consistently.

U, H, P, and V are examples of thermodynamic state functions—that is, their intrinsic values depend only on the state of the fluid (usually its temperature, pressure, and composition) and not on the process used to bring the fluid to that state. On the other hand, quantities like Q and W are not state functions. When a fluid makes a transition from one state to another, the values of Q and W associated with that transition are dependent on the "path" over which that transition was made.

Enthalpies of Formation and Reaction

Enthalpy cannot be calculated in an absolute sense. It requires the definition of a reference state, at which it is taken to be zero. Enthalpies in all other states are referenced to this state. The most rigorous definition of a reference state is to take it as comprising the constituent atomic species in their normal molecular configurations at some temperature and pressure—typically 25°C and 1 atm. Normal molecular configuration means the normal configuration the atomic molecules take in nature, such as O_2, H_2, N_2, etc. Thus, these molecules would be taken to have zero enthalpy at 25°C and 1 atm. The phases of these elemental species also have to be stated. They are generally taken to be the normal phase of the element at the stated conditions. The reference states for gases are usually taken to be in their "ideal gas state" (ig).[1]

With these ideas as a starting point, the enthalpies of compounds can be built up by considering real or imaginary reactions through which they are formed from their constituent elements. For example, the enthalpy of formation of methane would be the enthalpy change associated with the following reaction:

$$C(s) + 2H_2(ig) = CH_4(ig) \quad (25°C, 1\ atm)$$

where s indicates solid and ig indicates the ideal gas state. Such chemical reactions are usually accompanied by the absorption or release of a certain amount of heat. If the reaction is carried out at constant temperature and pressure,

Table 1 Enthalpies of formation of some common compounds

Chemical specie	Formula	State	MW	ΔH_f (T = 298.15°C)	ΔH_f (non dim.)
Methane	CH_4	IG	16.04	−74.87	−32.968
Ethane	C_2H_6	IG	30.08	−83.8	−36.901
Propane	C_3H_8	IG	44.10	−104.7	−46.104
n-Octane	C_8H_{18}	IG	114.23	−208.4	−91.767
n-Octane	C_8H_{18}	L	114.23	−250.3	−110.217
Carbon dioxide	CO_2	IG	44.01	−393.51	−173.278
Carbon monoxide	CO	IG	28.01	−110.53	−48.671
Water	H_2O	IG	18.02	−241.826	−106.486
Water	H_2O	L	18.02	−285.830	−125.863

Units of ΔH_f = kJ/gmole; Non-dimensionalization, $\Delta H_f/RT_0$ where T_0 = 273.15 °K.

and if no work is done, according to the first law, this heat is the enthalpy difference of the reaction and defines the standard enthalpy (or heat) of formation of the resulting compound.

Standard enthalpies of formation are widely tabulated. An excellent source is the collection by the National Institute of Standards and Technology (NIST).[2] Enthalpies of formation for some common species are given in Table 1. They are given both in the units that appear in the NIST collection and in nondimensionalized form. The latter permits their use in any system of units simply by multiplying by RT_0 in the desired units. T_0 is taken to be 0°C (= 273.15 °K = 459.67 °R), and R is the Universal Gas Constant. Note that °R represents degrees Rankine and °K represents degrees Kelvin.

Many different units are used in energy calculations. The values of some useful conversion factors and the values of the universal gas constant appear in Table 2.

These enthalpies of formation can be combined to produce enthalpy changes for any chemical reaction whose species have known enthalpies of formation. For example, in the combustion reaction

$$CH_4 + 2O_2 = CO_2 + 2H_2O$$

the enthalpy change may be calculated as follows:

$$\Delta H = \Delta H_f(CO_2) + 2\Delta H_f(H_2O) - \Delta H_f(CH_4)$$
$$- 2\Delta H_f(O_2)$$
$$= (-393.51) + 2(-285.830) - (-74.87) - 0$$
$$= -890.3 \text{ kJ/gmole}$$

If ΔH is positive, the reaction is called endothermic, whereas if it is negative, it is called exothermic. Because this particular reaction is a combustion reaction, the absolute value of its enthalpy change is also called the enthalpy change (or heat) of combustion.

Application to Combustion

The goal of a power cycle is to convert the chemical energy of a fuel into usable mechanical energy that can produce useful shaft work or drive a generator to produce electricity (Fig. 2). What one hopes to do is to convert as much of the fuel's chemical energy into useful work. The hope for this ideal is reflected in what is called the heating value of the fuel. It is taken to be the negative of the enthalpy change of the combustion reaction through which a fuel is completely oxidized at the so-called ISO (International Organization for Standardization) conditions—namely, 15°C and 1 atm pressure.[3] Thus, it is the negative of the enthalpy change of the following kind of reaction:

Fuel + aO_2(ig) → bCO_2(ig) + cH_2O(ig or l)
+ Any other products(15°C, 1 atm)

where a, b, and c are coefficients that depend on the particular fuel.

It is unfortunate that ISO conditions differ from the reference state at which enthalpies of formation are usually tabulated. Nevertheless, the latter can be adjusted to ISO conditions, and values for some common fuels appear in Table 3.

Heating Values, Heat Rates, and Power Cycle Efficiency

One area of confusion about heating values comes because of the choice of the phase of the water produced as a combustion product. If the product water is considered to be a liquid, its enthalpy will be lower than that if it is considered to be a vapor. The former choice leads to

Table 2 Some useful conversion factors and constants

Type	Value	Conversion
Length	1 m	3.28084 ft
		39.3701 in.
Mass	1 kg	2.20462 lbm
Force	1 N	1 kg m/s^2
		0.224809 lbf
Pressure	1 bar	10^5 N/m^2
		10^5 Pa
		0.986923 atm
		14.5038 psia
Pressure	1 atm	14.6960 psia
Volume	1 m^3	35.3147 ft^3
Density	1 g/cm^3	62.4278 lbm/ft^3
Energy	1 J	1 N m
		1 m^3 Pa
		10^{-5} m^3 bar
		10^{-3} kW s
		9.86923 cm^3 atm
		0.239006 cal
		5.12197×10^{-3} ft^3 psia
		0.737562 ft lbf
		9.47831×10^{-4} Btu
Power	1 kW	10^3 J/s
		239.006 cal/s
		737.562 ft lbf/s
		0.94783 Btu/s
		1.34102 hp

Values of the universal gas constant, R

8.314 J/gmole K = 8.314 m^3 Pa/(gmol K) = 83.14 cm^3 bar/(gmole K)

82.06 cm^3 atm/(gmole K)

1.987 cal/(gmole K) = 1.986 Btu/(lbmole R)

0.7302 ft^3 atm/(lbmole R) = 10.73 ft^3 psia/(lbmole R)

1545 ft lbf/(lbmole R)

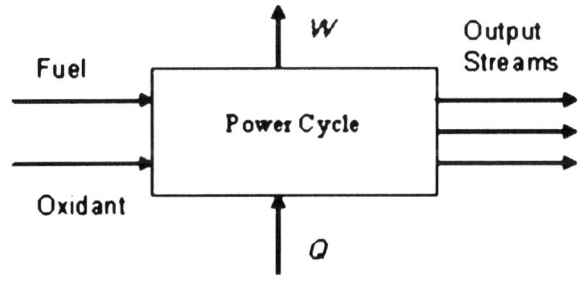

Fig. 2 Schematic representation of a power cycle.

a higher heating value (HHV), whereas the latter leads to a lower heating value (LHV). If the chemical composition of the fuel is known, the heating value is the mole fraction average of the heating values of the individual constituents divided by the average molecular weight of the fuel. Where the composition of the fuel is not known, accurate heating values are determined experimentally.

The heat rate is one often-tabulated measure that is used to describe the effectiveness of a machine—such as an internal combustion engine or a gas turbine—used to convert the chemical energy of a fuel into usable mechanical or electrical energy. Several definitions go into it.

Heat Energy Input(Btu/s)

= First Law Available Chemical Energy

= Heating Value of a Fuel (Btu/lb)

× Fuel flow rate (lb/s)

Heat Rate(Btu/kWh)

= Heat Energy Input (Btu/s)

× 3600 (s/h) / Power Produced (kW)

The heat rate is based on a particular choice of either HHV or LHV of the fuel. The higher the heat rate, the more fuel is being used per kW of power produced and therefore the less efficient the machine. The heat rate is directly related to the efficiency.

Efficiency

$$= \frac{\text{Power Output (kW)} \times 9.47831 \times 10^{-4} \text{ Btu/J}}{\text{Heat Energy Input (Btu/s)}}$$

Comparison with the previous equations yields

Efficiency

= Conversion Factor (Btu/kWh)/Heat Rate (Btu/kWh)

The conversion factor's role is strictly to convert the units. In the units shown it is equal to

Conversion Factor

$= 9.47831 \times 10^{-4}$ Btu/J × 1000 J/kW-s

× 3600 s/h

= 3412.19 Btu/kWh

Table 3 Heating values of some common fuels

Fuel	HHV		LHV		LHV/HHV
	Btu/lbm	Btu/gal	Btu/lbm	Btu/gal	
No. 2 oil	19,580	142,031	18,421	133,623	0.9408
No. 4 oil	18,890	146,476	17,804	138,055	0.9425
No. 6 oil	18,270	150,808	17,312	142,901	0.9476
Diesel fuel	19,733		18,487		0.9368
Hydrogen	61,007		51,635		0.8464
Methane	23,876		21,518		0.9012
Typical natural gas	22,615		20,450		0.9019
Propane	21,653		19,922		0.9201
Butane	21,266		19,623		0.9227
Gasoline	19,657	121,808	18,434	114,235	0.9379
Reformulated gasoline	19,545	120,103	18,304	112,377	0.9365
Methanol	11,274	73,882	10,115	66,289	0.8972
Anthracite coal	14,661		14,317		0.9765
Bituminous coal	14,100		13,600		0.9645

Source: From Based on calculations by the author and data from The Association of Energy Engineers (see Refs. 4–6).

When the power-generating equipment is used to produce electricity, the efficiency is often called the electrical efficiency.

The efficiency calculated by the above equations is based on the hope of converting the entire heating value of the fuel into useful work. It is unfortunate that this has come to be the standard because there is a further limitation imposed by the second law of thermodynamics. It tells us that nature will not allow conversion of this much chemical energy into useful work, even if the power cycle were constructed as perfectly as possible under the most ideal conditions.

THE SECOND LAW OF THERMODYNAMICS

Reversibility

Before stating the second law specifically, it is necessary to introduce the concept of reversibility. Most processes produce changes in the substances on which they operate. Reversibility has to do with how easy it would be to undo the changes that are produced in a substance through some process. Specifically, reversibility means that a change that a process makes on a substance could theoretically be undone (reversed) with only an infinitesimal change in the conditions in the process.

For example, suppose that heat is exchanged by allowing heat to flow from a hotter substance to a cooler one. For this change to be reversible, temperatures at every point where heat transfer takes place must differ only infinitesimally. Thus, if we raised the temperature profile of the cooler substance by just an infinitesimal amount, the heat could be transferred back to the hotter substance.

As another example, suppose that a gas is compressed in a cylinder by moving a piston in such a way that it reduces its volume. For the change to be reversible, the pressure the piston exerts must be only differentially greater than the pressure of the gas. In this case, if the piston pressure were reduced only infinitesimally, the gas could be expanded back to its original state, and all the work of compression would be recovered. On the other hand, if a pressure were applied that differed substantially from that of the gas in the piston, the change brought about would be irreversible, because a differential change in the applied pressure could not reverse the effects of the compression.

Formulation

The second law of thermodynamics has a certain mystique to it because unlike most physical laws, it is not a conservation law (e.g., conservation of mass, energy, momentum, etc.). Instead, it puts certain limits on what nature allows in spite of the great conservation laws. It is also not as intuitive as the other great laws. Therefore, much has been written to try to explain it. However, it can be applied rigorously based on the following three-part formulation. This formulation refers to a given substance of constant mass that experiences changes from an initial to a final state through the possible exchange of heat and work with its surroundings.

Part 1: There is a thermodynamic state function, called entropy, S, which is defined by the following equation:

Thermodynamics

$$dS = \frac{dQ_{rev}}{T} \quad (4)$$

where dS is the differential change in entropy and dQ_{rev} refers to a differential amount of heat transferred to or from a substance in a reversible manner. That is, to calculate finite entropy differences, this equation needs to be integrated along a reversible path.

Stating that entropy is a state function is not a trivial claim. What it means is that no matter what reversible path is chosen, the net change in entropy will be the same, provided that the initial and final states are the same.

Part 2: No matter how a change of state is brought about in a given substance,

$$\Delta S_{substance} + \Delta S_{surroundings} \geq 0 \quad (5)$$

The differences implied by the Δs are differences between the end state and the initial state of both the substance and surroundings. Notice that there is such a thing as a change in the entropy of the surroundings as well as the substance. It too can be calculated by Eq. 4.

Part 3: The inequality in the previous equation becomes an equality if, and only if, the process used to bring about the change is completely internally reversible and if the process also exchanges heat with the surroundings in a completely reversible manner.

Thus, this third part becomes a criterion of reversibility for a process or piece of equipment. Furthermore, the greater the inequality, the greater the irreversibility.

In general, the surroundings are considered to be at a constant ambient temperature, T_0. Thus,

$$\Delta S_{surroundings} = \frac{Q_{surroundings}}{T_0} = -\frac{Q_{substance}}{T_0} \quad (6)$$

where the two Qs are heat transfers to the surroundings and substance, respectively, which are the same in magnitude but opposite in sign. This equation can be combined with Eq. 5 to give

$$Q \leq T_0 \Delta S \quad (7)$$

Along with the first law for flow processes, this gives

$$W \leq T_0 \Delta S - \Delta H \quad (8)$$

In these equations, the subscripts have been dropped and all quantities refer to the substance, not the surroundings. Furthermore, although these equations were derived for changes in a substance of constant mass (closed system), as with the first law, they can also be applied to a steady flow system. In this case, the Δs refer to differences in the fluxes of the thermodynamic properties between the output and inlet streams.

The implications of these two equations are very significant. They mean that in a flow process, if the inlet and outlet conditions of the various streams are fixed, there are an upper bound to the amount of heat that can be transferred to the process and an upper bound on the amount of work it can produce. The latter claim has an impact on power generation systems because it imposes a limit on the amount of power that can be produced from a given amount of fuel, regardless of its energy content. Furthermore, because the entropy change of such processes is usually negative, another implication of Eq. 7 is that a certain amount of heat must be rejected to the surroundings regardless of how well the process is designed internally.

The Carnot Engine

Some of the implications of the second law can be illustrated through the so-called Carnot engine. This is an imaginary reversible engine in which heat, Q_H, is transferred from a high-temperature source at T_H into a reversible engine, which produces work, W, and discards heat, Q_C, into a cold sink—possibly the environment—at T_C (Fig. 3). From the first law,

$$W = Q_H - Q_C$$

From the second law,

$$\frac{Q_C}{T_C} - \frac{Q_H}{T_H} \geq 0$$

If the Carnot efficiency is defined as the amount of work that can be produced from the high-temperature heat (i.e., W/Q_H),

$$\frac{W}{Q_H} = \frac{Q_H - Q_C}{Q_H} = \frac{T_H - T_C}{T_H} \quad (9)$$

The last term on the right is of particular interest. It means that the Carnot efficiency depends on only the

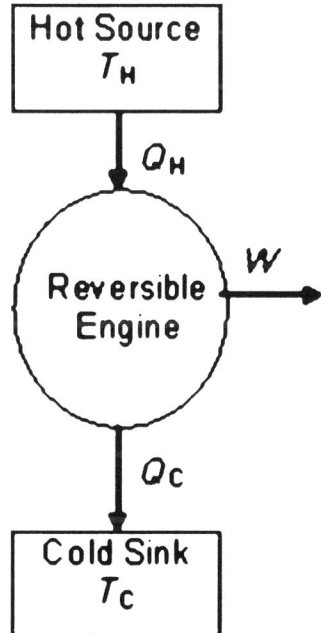

Fig. 3 The Carnot engine.

temperatures of the two heat reservoirs. All the heat in a high-temperature source can never be converted to work, even if the apparatus is perfectly reversible. The higher the temperature of the hot source relative to the cold sink, the higher the Carnot efficiency.

Ideal Work, Lost Work

The second law gives a means to measure the irreversibility of a process or piece of equipment. That, in turn, is a measure of its departure from an ideal design—one which makes changes to substances in a way that most preserves their work-producing potential. This analysis begins by imagining a perfectly reversible process and defining W_{ideal} and Q_{ideal} to be the limiting values defined by Eqs. 7 and 8 (see Fig. 4).

$$W_{ideal} = T_0 \Delta S - \Delta H \tag{10}$$

$$Q_{ideal} = T_0 \Delta S \tag{11}$$

These ideal values can be determined from the enthalpies and entropies of the various streams entering and leaving the process alone, without reference to any of the internal details of the process.

The level of irreversibility in a real process or in some part of it can be quantified through what is called lost work.[7] Because the actual work that can be obtained from any real process is always less than that what would have resulted from a completely reversible process, this leads to the following definition:

$$W_{lost} = W_{ideal} - W \tag{12}$$

By combining this with Eqs. 10 and 3, this becomes

$$W_{lost} = T_0 \Delta S - Q \tag{13}$$

A real process—or any subprocess—can be represented as shown in Fig. 5. The W_{lost} stream is not an actual work output. However, showing it in this manner is a reminder that in any real process a certain amount of the potential to do work has been lost due to the irreversibilities of the process. Furthermore, the lost work of all the subprocesses within the overall process sum up to the lost work of the process as a whole,

$$W_{ideal} = W + \sum W_{lost} \tag{14}$$

where the summation is carried out over each of the subprocesses. When analyzing a process, Eq. 13 can be applied to each piece of equipment. Then the W_{lost} for each of these can be summed according to Eq. 14. Such an analysis brings out where the greatest irreversibilities occur and therefore shows where the greatest opportunities are for improvement.

Exergy

Lost work is related to the modern concept of exergy. Lost work gives a quantitative measure of the irreversibility of a process or subprocesses within it. Furthermore, the term "lost work" seems quite descriptive. However, it is also possible to assign values—perhaps even monetary values—to streams as well as equipment. This can be done through the concept of exergy.

The exergy of a stream is the maximum amount of work that could theoretically be extracted from that stream if its temperature and pressure were reduced to that of the environment and if the concentrations of its chemical species were brought to that of the environment. It should be clear that from a power-production point of view, the economic value of a stream is related to its exergy because the production of work is usually more valuable than the production of heat alone.

Exergy analysis is based on the idea that it costs something in terms of equipment, power, adding costly streams, etc. to raise a stream's exergy—and hence its value. If what it costs to raise the value of a given stream is less than the increased value of the stream, it is a desirable thing to do.

Exergy is generally divided into four parts: physical, chemical, kinetic (having to do with the velocity of the fluid), and potential (having to do with its height above some datum level). The first two of these parts are of primary interest here. Physical exergy is the negative of the ideal amount of work that could be extracted from a stream if its temperature and pressure were reduced to ambient conditions. That is,

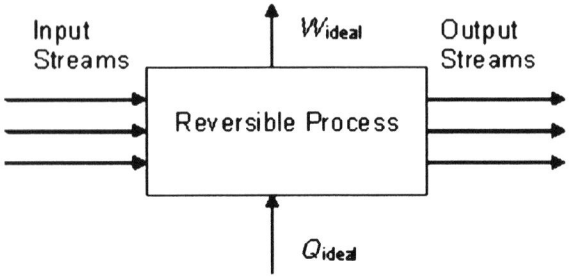

Fig. 4 Schematic representation of a reversible thermodynamic process.

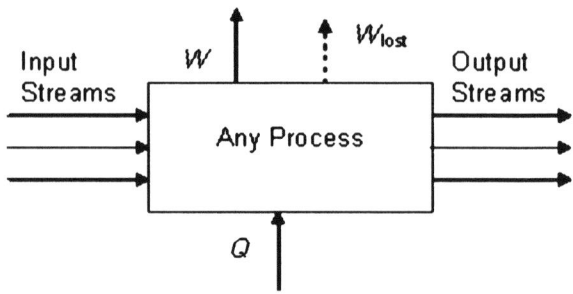

Fig. 5 Schematic representation of any process showing lost work.

$$e^{Ph} = \Delta H - T_0 \Delta S \quad (15)$$

In Eq. 15 and the following equations, the Δs refer to differences between the present state of the fluid and its state at ambient conditions.

Chemical exergy, e^{Ch}, also has to be added because the exergy, say, of a fuel at ambient conditions ought to be higher than that of its eventual combustion products at those conditions because it has the potential of producing work in being transformed into its products. This can be accounted for in an equation similar to the one above, but this time involving properties of formation at ambient conditions.

$$e^{Ch} = \Delta H_f - T_0 \Delta S_f \quad (16)$$

Exergy is not quite a thermodynamic state function because it depends on environmental temperature. Also, unlike enthalpy or internal energy, it is not conserved in a process. Nevertheless, it is a useful measure of the power-producing value of a process stream. In any real processes, every process step results in a net loss of exergy.

The relationship between exergy and lost work is very straightforward—namely,

$$\text{Lost Work} = \sum \text{Exergies of incoming streams} - \sum \text{Exergies of outgoing streams} \quad (17)$$

where the summations are carried out over all incoming and outgoing streams, respectively.

Reversibility Revisited

The preceding analysis shows how the second law can be used to provide a measure of the irreversibility of a process or part of a process. Every irreversibility causes a loss in the potential to convert energy into work. In this era of diminishing supplies of cheap fuels and increasing demands for electric power, this is an important consideration. Furthermore, irrisibilities in one part of a process reduce the amount of work that can be produced—or else increase the amount of work that is needed—in the rest of the process, even though the source of the irreversibility may be far removed from the place in the process where work is produced or consumed.

The following is a list of some process steps that introduce irreversibility into processes, including power cycles:

- Heat transfer across a nonzero temperature difference
- Mixing of fluids of differing pressures, temperatures, or compositions
- Pressure drops that do not recover work
- Chemical reactions that do not take place at equilibrium conditions
- Flashing of liquids when pressure is reduced
- Two-phase contact and mass transfer between fluids whose various species have concentrations that differ from their equilibrium values
- Temperature and pressure shocks when fluids enter equipment at different temperatures or pressures from those that are present within the equipment

The most thermodynamically efficient designs—those that most preserve the work-producing potential of the fluids in the process—are those that strive for the greatest reversibilities in the various pieces of process equipment. However, there is another side to this question. The more reversible a piece of equipment is, the greater its capital cost is apt to be. Thus, in the practical design of equipment there is often a trade-off between reversibility and capital cost. Processes need to be evaluated not only for their thermodynamic efficiency, but also for their thermoeconomic effectiveness.

CALCULATION OF THERMODYNAMIC PROPERTIES

The Rigorous Equations

Process equipment cannot be designed or evaluated apart from the ability to estimate the thermodynamic properties of the entering and exiting streams. The following are rigorous equations for the calculation of some of the important thermodynamic properties of pure or constant-composition substances[8]:

$$dH = C_p dT + \left[V - T\left(\frac{\partial V}{\partial T}\right)_P \right] dP \quad (18)$$

$$dS = \left(\frac{C_p}{T}\right) dT - \left(\frac{\partial V}{\partial T}\right)_P dP \quad (19)$$

$$G = H - TS \quad (20)$$

C_p is the heat capacity at constant pressure, defined by

$$C_p = \left(\frac{\partial H}{\partial T}\right)_P \quad (21)$$

G is called the Gibbs free energy or Gibbs function and has importance in equilibrium calculations. Eq. 20 can be considered its definition. By differentiation and comparison with the other two, it can be shown that

$$dG = VdP - SdT \quad (22)$$

These are differential equations that emphasize that these properties need reference values from which differences can be calculated. To obtain values for the property differences for real substances, the pressure–volume–temperature (PVT) behavior of the fluid needs to be inserted into the equations, which would then be integrated to produce the final results.

The Ideal Gas

One way of inserting PVT behavior is through equations of state. One of the simplest models for gases is that of the ideal gas, whose equation of state on a molar basis is:

$$PV = RT \tag{23}$$

This equation reflects the fact that the molecules in the gas do not interact in any way, although they may have very complex energy interactions within a given molecule.

When this equation is inserted into the rigorous equations, one obtains

$$dH^* = C_p^* dT \tag{24}$$

$$dS^* = \left(\frac{C_p^*}{T}\right) dT + \left(\frac{R}{P}\right) dP \tag{25}$$

$$dG^* = RT\, d\ln P - S^* dT \tag{26}$$

where the asterisk designates the properties as those of an ideal gas.

The integrated forms of Eqs. 24 and 25 are

$$\Delta H^*(P2, T2; P1, T1) = \int_{T1}^{T2} C_p^* dT \tag{27}$$

$$\Delta S^*(P2, T2; P1, T1)$$

$$= \int_{T1}^{T2} \left(\frac{C_p^*}{T}\right) dT + R \ln\left(\frac{P2}{P1}\right) \tag{28}$$

The heat capacity has been left behind the integral sign because even for most ideal gases, it varies with temperature. However, neither ideal gas heat capacities nor enthalpies vary with pressure, but ideal gas entropies do. Ideal gas heat capacity data is often presented in the form of an analytical equation. One common form used by NIST[9] is

$$C_p^* = A + BT + CT^2 + DT^3 + \frac{E}{T^2} \tag{29}$$

NIST tabulates the constants (A, B, etc.) for many compounds.

Real Fluids

Real fluids may be represented by the following general equation of state:

$$V = Z(P,T)RT/P \tag{30}$$

where Z is the compressibility factor, which can be a function of both P and T. It is common to report PVT data in the form of the compressibility factor. For an ideal gas, $Z = 1$. Thus, the compressibility factor is a measure of the deviation from ideal gas behavior.

When Eq. 30 is substituted in Eqs. 18 and 19, one obtains

$$dH = C_p dT - \left[\frac{RT^2}{P}\left(\frac{\partial Z}{\partial T}\right)_P\right] dP \tag{31}$$

$$dS = \left(\frac{C_p}{T}\right) dT - R\left(Z + T\frac{\partial Z}{\partial T}\right)\frac{dP}{P} \tag{32}$$

One consequence of these equations is that thermodynamic properties of real fluids can be calculated from their ideal gas–heat capacity (as a function of temperature) and their PVT behavior. This follows from the fact that to get from one state to another, these equations can first be integrated to zero pressure (which requires only PVT data), then integrated at zero pressure to the final temperature (which requires only ideal gas heat capacity data), and then integrated back to the end-point pressure.

Several approaches have been used to apply these equations for practical calculations. A first approach is to take known ideal gas-heat capacity and PVT data, and use them directly in these equations. Such data are not often available, and even when they are, this is a tedious process. Nevertheless, it has been done for a number of well-studied chemical species, most notably water. This is the basis of the steam tables, which have been put into analytical form for easy computer calculations.[10]

A second approach is through generalized correlations. An early attempt to describe the PVT behavior of many fluids was begun by Pitzer[11] through the use of his so-called acentric factor. His correlations have been extended so that they can be used to predict the more important thermodynamic properties of real no-polar fluids in both vapor and liquid phases.[12]

A third approach is to insert an equation of state into Eqs. 31 and 32 so that they may be integrated analytically. One excellent equation of state for low-density gases is the virial equation

$$Z = 1 + \frac{B(T)}{V} + \frac{C(T)}{V^2} + \frac{D(T)}{V^3} + \cdots. \tag{33}$$

where B, C, D, etc. are functions of temperature only and are called the second, third, fourth, etc. virial coefficients. There is a great deal of data for the second virial coefficients.[13] Where data are not available, correlations have been developed to estimate them.[14,15] Third virial coefficients are usually not available, but prediction methods have been proposed.[16] Very little information is available for higher-order coefficients. Other equations of state have also enjoyed success in predicting properties of real fluids, both in the vapor and liquid phases. Among the more successful are the Soave-Redlich-Kwong[17] and Peng-Robinson[18] equations.

Solutions

The properties of solutions—whether they are gas or liquid mixtures—are not merely the mole fraction averages of the properties of the pure components. For any thermodynamic property, M, the exact equation for the properties of mixtures is

$$M(P,T,\bar{x}) = \sum_i x_i \overline{M_i}(P,T,\bar{x}) \tag{34}$$

where $\overline{M_i}$ is called the "partial molal property of i" and $\bar{x}$ is a vector containing all of the compositions of the various constituents of the solution. The summation is carried out over all constituents. $\overline{M_i}$ is the property of constituent i at pressure, P, and temperature, T, as it exists in solution of composition $\bar{x}$. In general, this is different from its value as a pure substance.

The most general equation through which partial molal properties may be evaluated is

$$\overline{M_i} = \left(\frac{\partial(nM)}{\partial n_i}\right)_{P,T,n_j} \tag{35}$$

where n is the total number of moles and n_i is the number of moles of specie "i".

The partial derivative implies that the extensive property, nM, is differentiated with respect to the particular specie of interest, i, with pressure, temperature, and all of the other constituent amounts held constant. If properties of the various possible mixtures are known or can be approximated analytically, this equation provides a means to calculate their partial molal properties.

An important approximation for solutions is that of so-called "ideal solutions," whose properties are calculated as follows,

$$V^{id} = \sum_i x_i V_i \tag{36}$$

$$H^{id} = \sum_i x_i H_i \tag{37}$$

$$S^{id} = \sum_i x_i S_i - R \sum_i x_i \ln x_i \tag{38}$$

$$G^{id} = \sum_i x_i G_i + RT \sum_i x_i \ln x_i \tag{39}$$

where the superscript, "id," designates these as the properties for ideal solutions. All ideal gases are ideal solutions. The ideal-solution approximation is useful in some situations, but most liquids depart from it significantly. Notice also that properties involving entropy have a mixing effect (second terms on the right side of Eqs. 38 and 39) even if they are ideal solutions. This reflects the fact that the entropy of a solution is always higher than the combined entropies of its constituents.

The true properties of solutions can be calculated from their PVT behavior. For humid air, for example, the mixing effect is often neglected. However, one recent study done by M. Conde Company[19] includes virial coefficient information, including information for mixtures, sufficient to produce very accurate psychrometric properties of air–water mixtures.

Phase Equilibrium

Phase equilibrium can also be predicted from thermodynamic properties. The way this is done goes beyond the scope of this entry. However, several ideas can be stated here. One aspect of phase equilibria—namely pure-component vapor pressures—has been widely measured and can be used as a starting point for multicomponent phase equilibrium estimates. Vapor pressure data is often correlated by the Antoine equation,

$$\log_{10}(P^V) = A - \left(\frac{B}{T+C}\right) \tag{40}$$

where A, B, and C are constants. Such constants for many compounds also appear in the NIST collection.[20]

One special case of interest in energy calculations is the two-phase equilibrium between (1) pure liquid water and (2) a vapor phase one of whose components is water vapor. If the vapor phase is considered to be an ideal gas (and thus an ideal mixture), and if certain other assumptions are made, the composition of water in the vapor phase can be estimated by

$$y_{H_2O} = \frac{P^V(T)}{P} \tag{41}$$

where the vapor pressure can be calculated from the Antoine equation. This is a useful approximation that can be refined if PVT data for the vapor mixture are available.

CONCLUSIONS

This entry attempts to give an introduction to some of the basic principles of thermodynamics and how they relate to various energy engineering applications. The two main laws of thermodynamics govern the performance of any energy-conversion process. Their application requires the ability to estimate the thermodynamic properties of the various process streams. Such estimates are also based on the formalism that thermodynamics provides. Thus, thermodynamics provides the basic framework within which the design and evaluation of all energy-conversion equipment and processes must be performed.

ACKNOWLEDGMENT

Support for this work by VAST Power Systems, Elkhart, Indiana, is gratefully acknowledged.

REFERENCES

1. The ideal gas state is difficult to describe briefly. It is a fluid with the same heat capacity as the real fluid's heat capacity at very low pressure. However, unlike the real fluid, it obeys the ideal gas law. For low-pressure calculations, the difference between the ideal gas state and the real fluid at low pressure is not important.
2. http://webbook.nist.gov/chemistry/form-ser.html (accessed July 2005).
3. ISO conditions for air also include a relative humidity of 60%.
4. Notes from short course: Capehart, B.L.; Salas, C.E. *The Theory and Practice of Distributed Generation and On-Site CHP Systems*; The Association of Energy Engineers: Atlanta, GA, June 2003.
5. Bosul, Ulf. *Well-to-Wheel Studies, Heating Values and the Energy Conversation Principle*, http://www.efce.com/reports/E10.pdf (accessed July 2005).
6. http://www.chpcentermw.org/pdfs/toolkit/7c_rules_thumb.pdf (accessed July 2005).
7. Van Ness, H.C.; Abbott, M.M. Thermodynamics. In *Perry's Chemical Engineers' Handbook*, 7th Ed.; Perry, R.H., Green, D.W., Eds.; McGraw Hill: New York, 1997; 4-1–4-36; However, note that because of a different sign convention their formulas for ideal work have the opposite signs from those in this article.
8. The equations can be found in many thermodynamics text books. A concise useful summary is found in: Van Ness, H.C.; Abbott, M.M. In *Perry's Chemical Engineers' Handbook*, 7th Ed.; Perry, R.H., Green, D.W., Eds.; McGraw Hill: New York, 1997; 4-1–4-36.
9. http://webbook.nist.gov/chemistry/form-ser.html (accessed July 2005).
10. http://www.cheresources.com/iapwsif971.pdf (accessed July 2005). A free downloadable Excel add-in which embodies the equations is also available on this Web site.
11. Pitzer, K.S. *Thermodynamics*, 3rd Ed.; Mc-Graw Hill: New York, 1995; App. 3.
12. Lee, B.I.; Kessler, M.G. A generalized thermodynamic correlation based on three-parameter corresponding states. AIChE J. **1975**, *21* (3), 510–527.
13. Daubert, T.E.; Danner, R.P. *Physical and Thermodynamic Properties of Pure Chemicals Data Compilation*; Taylor & Francis: Bristol, PA, 1994.
14. Tsonopoulos, C. An empirical correlation of second virial coefficients. AIChE J. **1974**, *20* (2), 263–272.
15. Liley, P.E.; Thomson, G.H.; Daubert, T.E.; Buck, E. Physical and chemical data. In *Perry's Chemical Engineers' Handbook*, 7th Ed.; Perry, R.H., Green, D.W., Eds.; McGraw Hill: New York, 1997; 2-355–2-358.
16. De Santis, R.; Grande, B. An equation for predicting third virial coefficients of nonpolar gases. AIChE J. **1979**, *25* (6), 931–938.
17. Soave, G. Equilibrium constants from a modified Redlich-Kwong equation of state. Chem. Eng. Sci. **1972**, *27*, 1197–1203.
18. Peng, D.-Y.; Robinson, B.B. A new two-constant equation of state. Ind. Eng. Chem. Fund. **1976**, *15*, 59–64.
19. http://www.mrc-eng.com/Downloads/Moist%20Air%20Props%20English.pdf (accessed July 2005).
20. http://webbook.nist.gov/chemistry/form-ser.html (accessed July 2005).

Tradable Certificates for Energy Savings

Silvia Rezessy
Energy Efficiency Advisory, REEEP International Secretariat, Vienna International Centre, Austria, Environmental Sciences and Policy Department, Central European University, Nador, Hungary

Paolo Bertoldi
European Commission, Directorate General JRC, Ispra (VA), Italy

Abstract

Recently tradable certificates for energy savings have attracted the attention of policy makers as a tool to stimulate energy efficiency investments and deliver energy savings. While such schemes have been introduced in different forms in Italy, France and Great Britain and considered in other European countries, there is an ongoing debate over their effectiveness and applicability. The paper describes the concept and main elements of schemes that involve tradable certificates for energy savings (TCES) and how these have been put into practice in Italy, France and Great Britain. The entry discusses some key design and operational features of TCES schemes such as scope, additionality rules, measurement and verification. The implications of different certificate trading rules and how these can affect the actual structure of the certificate market are also explored.

INTRODUCTION

Market-based instruments (MBIs [MBIs are public policies that make use of market mechanisms with transferable property rights to distribute the burden from a policy. We recognize the difference between policy instruments that are well positioned to harness market forces to achieve a certain policy goal (such as renewable energy quotas or renewable portfolio standards [RPS]) and the market instruments (namely, carbon allowances, and green and white certificates), the latter being a much narrower concept representing just a tradable commodity. This differentiation is not so important in the context of the present entry, and in the text, we refer to complex policy tools/portfolios that include trading of financial commodities (such as certificates or allowances) as that aim at MBIs.) that aim at bringing sustainability to the energy sector have been implemented to promote electricity from renewable energy sources and to cut harmful emissions. Quota systems (also known as RPS) coupled with tradable green certificate (TGC) schemes have been developed and tested in several European countries to foster market-driven penetration of renewable energy sources. Another well-known and widely analyzed type of MBI is the tradable emission allowance.

To stimulate energy-efficiency investments and to achieve national energy savings targets, the attention of policy-makers in Europe has recently been attracted by the possibility of introducing energy savings obligations on certain types of market players coupled with tradable certificates for energy savings (Tradable Certificates for Energy Savings [TCES], or white certificates). Such schemes have been introduced in different forms in Italy, Great Britain, and France, and are being considered in other European countries.

This entry describes the concept and main elements of a TCES scheme and compares and analyses how these have been put into practice in Italy, Great Britain, and France. The entry builds on a sequence of publications by the authors, giving specific details on the still fairly short implementation track record of TCES schemes, as well as qualitative comparison of the TCES scheme with other policy tools promoting energy efficiency and the possibilities and dangers associated with attempts to integrate existing MBIs into the energy sector.[2–5]

SCHEMES WITH TRADABLE CERTIFICATES FOR ENERGY SAVINGS

Energy efficiency is a well-established option to decouple economic growth from the increase in energy consumption and thus reduce greenhouse gas (GHG) emissions by cutting the amount of energy required for a particular amount of end-use energy service. Apart from being a sound part of the environmental and climate change agenda, increased energy efficiency can contribute to meeting widely accepted goals of energy policy such as improved security of supply, economic efficiency, and increased business competitiveness coupled with job creation and improved consumer welfare. Alongside environmental and economic sustainability, the other main driver of energy policy in the European Union (EU) is to restructure electricity and gas markets. Many energy-efficiency advocates and policy-makers have

Keywords: Energy saving obligations; Tradable certificates for energy savings; Energy suppliers.

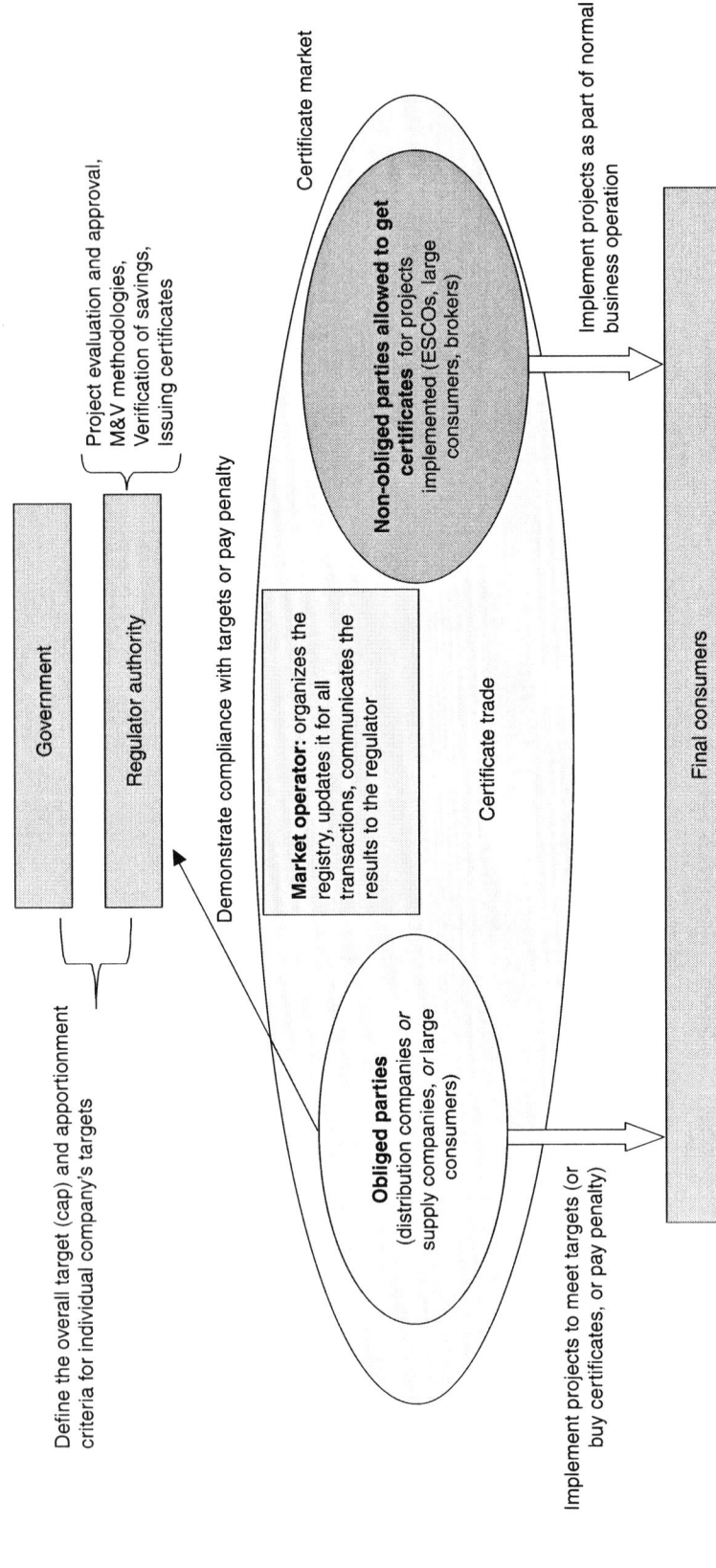

Fig. 1 A policy portfolio with mandatory savings targets and white certificates, and a summary of roles of actors and relationships among them.

called for legislation introducing energy efficiency and energy services as natural complements to the electricity and gas market liberalization. Otherwise, market failures in the energy sector would lead to lower levels of investment in energy efficiency than is socially optimal, with the final outcome being additional cost to the economy due to an imbalance between the supply side and demand side in the energy sector.

A possible market-based policy portfolio could comprise energy-savings quotas for some category of energy operators (distributors, suppliers, consumers, etc.) coupled with a trading system for energy-efficiency measures resulting in energy savings. The savings would be verified by the regulator and certified by means of the so-called white certificates (tradable certificates for energy savings). In Italy certificates are called Energy Efficiency Titles, while in France they are referred to as Certificates of Energy Savings.

A TCES portfolio involves the following basic elements[2,3,5,7,12,13]:

- The creation and framing of the demand. Tradable certificates represent a meaningful option only if there is interest in buying/selling them.
- The tradable instrument (certificate), representing the savings, conferring property rights to the holder, and providing the rules for trading.
- Institutional infrastructure and processes to support the scheme and creation of the market. An aspect that is often overlooked is that markets do not function in a vacuum. These activities include, for instance, measurement and verification (M&V), evaluation methods and rules for issuing certificates, a data management and certificate tracking system, and a registry.
- Cost-recovery mechanisms, in some cases.

Fig. 1 illustrates the roles and summarizes the responsibilities of different categories of actors involved in the design and operation of a TCES policy portfolio.

GENERAL CHARACTERISTICS OF ENERGY-SAVING OBLIGATIONS AND TRADING SYSTEMS IN EUROPE

Variations of the policy mix described above have been introduced in Italy, Great Britain, and (since July 2006) France. In the Flemish region of Belgium, savings obligations are imposed on electricity distributors without any trading option (certificate or, as in Great Britain, obligation trading). The first operational scheme in the world with a white-certificate trading element has been introduced in New South Wales (NSW), Australia. It is, however, a GHG trading system that has an end-use energy efficiency element and is left out of the present discussion.

Box 1. Tradable certificate for energy savings (white certificate)

A White certificate is an instrument issued by an authority or an authorized body providing a guarantee that a certain amount of energy savings has been achieved. Each certificate is a unique and traceable commodity that carries a property right over a certain amount of additional savings and guarantees that the benefit of these savings has not been accounted for elsewhere.

In Italy, command-and-control measures (energy savings targets in primary energy consumption for electricity and gas grid distribution companies with more than 100,000 customers as of the end of 2001) are combined with market instruments (tradable certificates for energy savings issued to distributors and energy service companies) as well as with elements of tariff regulation (a cost-recovery mechanism via electricity and gas tariffs and multiple-driver tariff (MDT) schemes (MDTs essentially constitute tariff regulation schemes linking the evolution over time of allowed revenue with cost drivers such as the number of customers, grid lengths, and energy sales.[17]) to reduce the disincentives for regulated electricity and natural gas companies to promote end-use energy efficiency among their customers) or dedicated funds in some circumstances. Over the 5 years of the current phase of the scheme, 3 million tons of oil-equivalent (Mtoe) cumulative primary energy savings are projected to be realized, of which 1.6 Mtoe is by electricity distributors and 1.3 Mtoe is by natural gas distributors. At least half of the target set for each single year is to be achieved via a reduction of electricity and gas end-use consumption (referred to as the 50% constraint, to which each distributor is subject). The remaining share can be achieved via primary energy savings in all the other end-use sectors. Energy-saving projects contribute to the achievement of targets for up to 5 years (with only some exceptions). Only savings that are additional to spontaneous market trends and legislative requirements are considered. After a long process of designing and elaborating elements, the Italian scheme finally became operational in January 2005.[12,14–16]

In Great Britain, the Energy Efficiency Commitment (EEC) runs in 3-year cycles from 2002 to 2011. It replaced the Energy Efficiency Standards of Performance (EESOP), running from 1994 until 2002, which established the principle of pooled spending on energy efficiency for domestic consumers. The EEC-1 program required that all gas and electricity suppliers with 15,000 or more domestic customers deliver a certain quantity of fuel standardized energy benefits by encouraging or assisting customers to take energy-efficiency measures in their homes. The overall savings target was 62 fuel standardized TWh (Energy savings are discounted over the lifetime of the measure and then standardized according to the carbon content of the fuel saved.) (lifetime discounted), and the total delivered savings reached 86.8 TWh.[10] In EEC-2 (2005–2008), the threshold for obligation was increased to 50,000 domestic customers. The target has been increased to 130 TWh; however, due to the carrying over of savings from EEC-1, in 2005 more than a quarter of this target has already been achieved. Suppliers must achieve at least half of their energy savings in households on income-related benefits and tax credits. Projects can be related to electricity, gas, coal, oil, and Liquefied Petroleum Gas (LPG). Suppliers are not limited to assisting their own customers only and can achieve improvements in relation to any domestic consumers in the Great Britain. Carbon-benefits estimations take into account the rebound effect—the likely proportion of the investment to be taken up by improved comfort—by adjusting the benefits to comfort factors. In addition, dead-weight factors are considered to account for the effect of investments that would be made anyway. At present, certificate trading is not a feature of the scheme in Great Britain.

The French system, introduced in 2006, envisages that all electricity, gas, LPG, (Domestic fuel excludes transport usages.) oil, cooling, and heating fuel for stationary applications suppliers that supply over 0.4 TWh/year will have to meet a target of energy savings. It excludes plants under the EU Emission Trading System and fuel substitution between fossil fuels, as well as energy savings resulting only from measures implemented just to comply with current legislation. Apart from these, no additional restrictions on compliance are foreseen. To meet the obligation, savings can be made in any sector and with any energy source or carrier. The following details related to the obligation parameters are known at the time of finalizing this entry: the total target for the first 3 years (2006–2008) will be 54 TWh (in final energy), cumulated over the life of the energy-efficiency actions with a 4% discount rate. The target apportionment is a two-step procedure: first, the total obligation is divided among different energies; then the obligation for a particular energy is apportioned among respective energy suppliers included in the system. The expected cost of action is below 20 Euro/MWh.[1]

Energy-efficiency obligations without certificate trading are also in place in the Flemish region of Belgium. Regional utility obligations were introduced in 2003 and are imposed on the electricity distributors. Currently, 16 electricity distributors are covered by the obligation. The annual target is 0.58 TWh; eligible actions refer to residential and nonenergy-intensive industry and service, and can involve saving fuel from any sources. Separate targets are set for low-voltage clients (<1 kV, mainly residential) and high-voltage clients (>1 kV). For the low-voltage clients, the target is 10.5% of electricity supplied over the 6 years from 2003 to 2008, and for high-voltage users (>1 kV), the target is 1% per annum for each over the same period. The reason for the higher than 1% per annum target for the low-voltage users is because of the Flemish Parliament's decision to provide free vouchers for the head of every family in 2004 and 2005, which can be exchanged via the electricity distributor for either an energy-saving Compact Fluorescent Lamps (CFL), a low-flow shower head, or an energy meter. In 2006 and 2007, it is planned that the other family members will receive a voucher for an energy-efficient light bulb.[9] There is a discussion going on in the Netherlands about the introduction of a white-certificate scheme.

ENERGY-SAVING OBLIGATIONS AND TRADING SYSTEMS: EUROPEAN EXPERIENCES TO DATE

Below, details on and first experiences with the following parameters of the existing European schemes with energy-saving obligations and energy-saving trading elements are reviewed: (a) eligible projects allowed; (b) institutional infrastructure and processes to support the scheme; and (c) certificate delineation, trading rules, and tools to stabilize the market. A comprehensive discussion of these and other design and operational features is available in Bertoldi and Rezessy.[3]

Eligible Projects

In Italy, projects in all end-use sectors are eligible. At least half of the target set for each single year should be achieved by reduction of the supplied energy vector—i.e., electricity and gas uses, a.k.a. the 50% constraint.[12] The remaining share can be achieved via primary energy savings in all of the other end-use sectors. There is an illustrative list of eligible projects. Energy savings projects contribute to the achievement of targets for up to 5 years (with only some exceptions). Early experience in Italy shows that a significant share of savings certified in autumn 2005 is coming from cogeneration, district heating, and public lighting projects. There are numerous submissions for certification of projects following the deemed savings verification method, which has very minor data requirements (see more details later in this entry). A surplus of certificates and banking of certificates are expected in Italy because the regulator has to evaluate and certify the savings from eligible projects starting from 2001, when the decrees were passed.

In Great Britain, only activities concerning domestic users are eligible. At least 50% of the energy savings must be targeted to customers who receive income-related benefits or tax credits (a.k.a. the priority group). Projects can be related to electricity, gas, coal, oil, and LPG. Suppliers can achieve improvements in relation to any domestic consumers in the United Kingdom. A nonexclusive list of measures is included within the illustrative mix for EEC 2005–2008. Measures that are related to the reduction of energy vectors other than the one supplied by the obliged party are allowed. Experience from EEC-1 in Great Britain shows that a significant share (56%) of the 86.8 TWh of savings delivered in the period 2002–2005 came from building insulation (wall and loft). CFLs accounted for a quarter of the savings achieved, followed by appliances (11%) and heating measures (9%).[10] CFLs accounted for the largest number of projects undertaken (almost 40 million measures related to CFL installation in EEC-1), followed by wet and cold appliances.[8] All but two suppliers—which went into administration and administrative receivership—achieved their targets; six suppliers exceeded their targets in EEC-1 and carried out their additional savings to EEC-2. Suppliers can receive a 50% uplift on the savings of energy-efficiency measures that are promoted through energy service activities. This uplift is limited to 10% of the overall activity. Of the six major suppliers with an EEC target, three submitted schemes that would take them over the 10% threshold if take up had been as forecasted. In reality, the energy services uplift was only 3.6% of all insulation activity. There is uplift on innovative technologies as well.

Apart from plants under the EU Emission Trading System (ETS) directive, fuel substitution between fossil fuels, and projects resulting just from measures implemented only to conform to current legislation, no other restrictions on compliance are foreseen in the French scheme. Any economic actor can implement projects and get savings certified as long as savings are above 3 GWh over the lifetime of a project. Actions must be additional relative to their usual activity, and there is a possibility to pool savings from similar actions to reach the threshold. All energies (including fuel) and all of the sectors (including transports and excluding installations covered by ETS) are eligible. Certification of projects implemented by bodies, which do not have savings obligation, is allowed, but only after considering the impact of a project on business turnover. If impact on business turnover is identified, certification of savings is allowed only for innovative products and services. "Innovative" product in this discourse means that efficiency is at least 20% higher compared with standard equipment and market share is below 5%.

Institutional Infrastructure and Processes to Support the Scheme

A sound institutional structure is needed for a white-certificate system to function, including administrative bodies to manage the system, as well as processes such as verification, certification and market operation, transaction registry, and detection and penalization of noncompliance.

Under the EEC in Great Britain, the regulator Office of Gas and Electricity Markets (OFGEM) manages project evaluation and approval, verifies savings, and manages the data. In Italy, the regulator Authorita per l'energia elettrica e il gas (AEEG) implements the scheme; the marketplace is organized and managed by the electricity market operator Gestore del Mercato Elettrico (GME) according to rules and criteria approved by AEEG. GME issues and registers certificates upon specific request by AEEG, organizes market sessions, and registers bilateral over-the-counter contracts according to rules set by AEEG.[12] In France, certificates are issued by the Ministry of Industry, and the French Agency for Environment and Energy Management and Association Technique pour l'Environnement et l'Energie (ATEE) are in charge of the definition of standardized actions.

To determine the energy savings resulting from an energy-efficiency activity, the eventual energy consumption has to be compared with a baseline (reference situation) without additional saving efforts. The choice of the reference scenario—in terms of reference consumption and conditions—raises some challenges, such as determining the relevant system boundary, minimizing the risk of producing leakage, establishing the practicality and cost effectiveness of a baseline methodology, and treating no-regret measures in the baseline determination (We are indebted to Ole Langniss for these comments). Additionality refers to certification of genuine and durable increases in the level of energy efficiency beyond what would have occurred in the absence of the energy-efficiency intervention—for instance, due only to technical and market development trends and policies in place. While in practice, projects tend to have a mix of public and private benefits, costs of disaggregating these benefits and precisely accounting for the exact share of no-regret measures in a larger action may be prohibitively high.

One way of overcoming this problem would be to place an objectively defined discount factor on investments, which accounts for these private benefits. One possibility is to use minimum efficiency requirements or current sale-weighted average efficiency levels. Furthermore, the electricity price and the effects of the EU ETS and other policies in place (such as taxation or standards) may also be accounted for in the baseline to ensure genuine additional savings. In Great Britain, a discount factor of 3.5% over the lifetime of the measure is applied, while in France, the discount factor is 4%. However, in both the British and French schemes, the savings are cumulated over the lifetime of equipment, and the discount factor refers to actualizing the annual savings for different measures with different life spans.

In Great Britain, saving estimations take into account the likely proportion of the investment to be taken up by improved comfort (comfort factors adjustment of carbon benefits; see earlier discussion) as well as dead-weight factors to account for the effect of investments that would be made anyway.

In Great Britain, the Department for Environment, Food, and Rural Affairs (DEFRA) requires suppliers to demonstrate additionality. Concerns have been raised that energy suppliers can claim toward their EEC target the total energy savings that flow from a partnership project regardless of the actual financial contribution made by the supplier.

In Italy, savings have to go over and above spontaneous market trends or legislative requirements[14,16]; the business-as-usual trend will be adjusted with time. The nature of the additionality check differs for project types (e.g., installation of efficient equipment may be evaluated on the basis of difference with the national average installed or with what is offered in shops). For projects that are based on the deemed savings and engineering verification approach (see details below), a case-by-case additionality check is performed by the regulator.

The Italian scheme uses three valuation (M&V) approaches: (a) a deemed savings approach with default factors for free riding, delivery mechanism, and persistence; (b) an engineering approach; and (c) a third approach based on monitoring plans whereby energy savings are inferred through the measurement of energy use. In the latter case, all monitoring plans must be submitted for preapproval to the regulatory authority AEEG, and they must conform to predetermined criteria (e.g., sample size, criteria to choose the measurement technology).[14,16] In practice, most of the projects submitted to date have been of the deemed saving variety. There is ex-post verification and certification of actual energy savings achieved on a yearly basis (e.g., in the case of Combined Heat and Power [CHP], the plant operator has to prove that the plant has run a certain number of hours, etc).[11] In principle, the metering approach is a more accurate guarantee of energy saved than the standard factors approach (the latter cannot verify details such as location and operating hours of installed CFLs), but in practice it can be difficult to identify the actual savings (e.g., in households, there is only one meter for all electricity usage, which increases each year due to growth in appliances and can fluctuate with changing household numbers, lifestyle, weather, etc). It may be reasonable for large installations or projects, but it may result in high monitoring costs for smaller projects.[14,16]

In Great Britain, the savings of a project are calculated and set when a project is submitted based on a standardized estimate taking into consideration the technology used, weighted for fuel type and discounted over the lifetime of the measure. There is limited ex-post verification of the energy savings carried out by the government, although this work would not affect the way energy savings are accredited in the current scheme; the monitoring work affects the energy savings accredited in future schemes.

Energy savings can be determined by metering or estimating energy consumption ahead of time and comparing this with consumption after the implementation of one or more energy-efficiency improvement measures, adjusting for external factors such as occupancy levels and level of production. Certificates, therefore, can be issued either ex-post (representing the energy saved over a certain period) or ex-ante (representing the estimation of the energy to be saved over a certain period). With regard to ex-post certification, there are different options: the saved energy resulting from an energy-efficiency measure could be measured at the end of a predetermined period (e.g., after 1 year) or over the lifetime of the project (which has to be assessed accurately). The latter option will make the system more comparable to a green certificate. The certificate has a unique time of issue attached to it; it indicates the period over which energy has been saved, the location where energy has been saved, and who is saving the energy (initial owner of the certificate). Ex-post certification, however, will probably increase validation efforts and verification costs. Alternatively, for projects that can be monitored through a standard savings approach, certificates can be granted in advance (ex-ante) of the actual energy savings delivery. This will mitigate liquidity constraints of project implementers and allow them to finance new projects. If underperformance is detected at the end of the lifetime of the measure, the underperforming project owner should be asked to cover the shortage with certificates purchased on the spot market.

Depending on the design of the scheme, the role of the regulator may or may not include the issues of certificates and the verification of savings. For instance, third parties may be licensed to evaluate and approve projects, verify savings, and issue certificates. Then the role of the regulator would be to accredit third parties and audit their performance. With the cost of compliance being one of the major issues raised about the implementation of white-certificate schemes, this potentially reduces the overall cost. It is not as crucial which body issues the certificates, provided that these are based on verified data, which can come from the energy regulator (as is the case in Italy) or from a certified verifier.

Certificate Delineation, Trading Rules, and Tools to Stabilize the Market

The certificate is an instrument that provides a guarantee that savings have been achieved. Each certificate should be unique, traceable, and at any time have a single owner. A certificate needs to be a well-defined commodity that carries a property right over a certain amount of additional savings and guarantees that the benefit of these savings has not been accounted for elsewhere. Property rights must be clear and legally secured, as it is unlikely that trades will occur if either party is unsure of ownership.[6]

Minimum project size may be applied for certification of savings to reduce transaction costs and encourage pooling of projects.[12] The size of a certificate also has important implication on the number of parties that can offer certificates for sale (unless other restrictions apply). In Italy, certificates are expressed in primary energy saved, and the unit is 1 ton of oil equivalent (toe). In France, certification is allowed only above a threshold of 3 GWh of savings over the lifetime of a project.[1]

The validity and any associated intertemporal flexibility embodied by banking and borrowing rules, the rules for ownership transfer, the length of the compliance period, and the expectations of market actors about policy stability and continuity will all influence the market for white certificates. A long certificate lifetime and banking increase the elasticity and flexibility of demand in the long term. To mitigate the uncertainties about the achievement of the quantified policy target within the prespecified time frame, banking for obliged parties may be allowed only after they achieve their own targets. In Italy, certificates are valid for up to 5 years, with a few exceptions.[12] In Great Britain, suppliers can carry over to EEC-2 all of their excess savings from measures implemented under EEC (this refers to measures rather than savings). In France, it has been proposed that the certificates' validity be at least 10 years. Borrowing is discouraged, because it makes the attainment of a target uncertain and is against the ex-post logic of the white-certificate scheme as applied in Italy, for instance.

Rules defining trading parties are also important for market liquidity. Provided that administrative and monitoring costs are not disproportionate, as many parties should be allowed in the scheme as possible, because this enhances the prospects of diversity in marginal abatement costs and lowers the risks of excessive market power.[13] Parties that may be allowed to receive and sell certificates include obliged actors, exempt actors, Energy Service Companies (ESCOs), consumers, market intermediaries, Non Governmental Organizations (NGOs), and even manufacturers of appliances. A key benefit of allowing many parties into the scheme is that new entrants may have the incentive to innovate and deliver energy-efficiency solutions that have a lower marginal cost.

In Italy, certificates are issued by the electricity market operator upon request of the regulator AEEG to all distributors and their controlled companies and to ESCOs. Certificates are tradable via bilateral contracts or on a spot market organized and administered according to rules set out jointly by AEEG and the electricity market operator. There are three types of certificates—for electricity savings, for gas savings, and for primary energy savings—and they are fully fungible.

In France, any economic actor can make savings actions and get certificates as long as the savings are at least 3 GWh over the lifetime of a measure. Certificates are delivered after the programs are carried out but before the realization of energy savings.[1]

In Great Britain, there are no certificates in the strict sense of the word. The scheme covers obliged parties, and no other party can receive verified savings that can be used to demonstrate compliance with the savings target. Suppliers may trade among themselves either energy savings from approved measures or obligations with written agreement from the regulator. There has been little interest in trading to date because energy savings can be traded only after the supplier's own energy-saving target has been achieved. Suppliers are also allowed to trade excess energy savings into the national emission trading scheme as carbon savings. However, the linking of carbon savings to the national emission trading scheme was never formalized. Suppliers have been allowed to carry savings over from EEC-1 to EEC-2, and this is what all suppliers that exceeded their target have chosen to do.

CONCLUSION

This entry discussed the general concept and the key elements of a system with energy-saving targets and tradable certificates for energy savings. It provided an up-to-date review of white-certificate schemes as implemented in three European countries, discussing some key design and operational features such as projects, implementer, and technology eligibility, and it pointed out key issues such as additionality, baseline setting, and M&V. This entry also explored the implications of different certificate trading rules and how these can affect the actual structure of the certificate market. Because the implementation track record of white-certificate schemes is very limited, it remains to be seen whether this policy portfolio will perform as expected, at what cost this will be achieved, and whether it can co-exist with and complement other MBIs in the energy sector to pave the road to a sustainable energy future.

ACKNOWLEDGMENTS

The authors would like to acknowledge the contribution of numerous reviewers who have provided comments, suggestions, updated country information, and clarifications at various stages of the writing of the report "Tradable certificates for energy savings (white certificates): theory and practiceon which this entry is based."[3]

REFERENCES

1. Baudry, P.; Monjon, S. *The French Energy Law and White Certificates System*, RECS open seminar, Copenhagen, DK, 25 September, 2005.
2. Bertoldi, P.; Rezessy, S. Tradable certificates for energy efficiency: the dawn of a new trend in energy policy?

American Council for Energy Efficient Economy (ACEEE) Summer Study on Energy Efficiency in Buildings, ACEEE: Asilomar, CA, 2004.

3. Bertoldi, P.; Rezessy, S. Tradable certificates for energy efficiency (white certificates): theory and practice In *EUR 22196 EN*; Institute for Environment and Sustainability, DG Joint Research Centre, European Commission: Ispra, VA, 2006.

4. Bertoldi, P.; Rezessy, S.; Bürer, M.J. Will emission trading promote end-use energy efficiency and renewable energy projects? *Summer Study on Efficiency in Industry of the American Council for Energy Efficient Economy*, ACEEE: New York, 2005.

5. Bertoldi, P.; Rezessy, S.; Langniss, O.; Voogt, M. White, green and brown certificates: how to make the most of them? *Summer Study of the European Council for Energy Efficient Economy*, ECEEE: Mandelieu, 2005.

6. Jaccard, M.; Mao, Y. Making markets work better. In *Energy for Sustainable Development. A Policy Agenda*; Goldemberg, J., Ed.; UNDP: New York, 2002; 41–76.

7. Langniss, O.; Praetorius, B. How much market do market-based instruments create? An analysis for the case of "white" certificates. *European Council for Energy Efficient Economy (ECEEE) Summer Study*, ECEEE: Saint Raphael, France, Stockholm, 2003.

8. Lees, E. *How Increased Activity on Energy Efficiency Lowers the Cost of Energy Efficiency*; RECS seminar: Copenhagen, DK, 2005.

9. Lees, E. Summary of workshop. *Bottom-up Measurement and Verification of Energy Efficiency Improvements: National and Regional Examples, DG TREN, European Parliamnet*, ECEEE: Brussels, 2005.

10. Mansero, S. *Energy Efficiency Commitment. An Example From the U.K.*; RECS International: Copenhagen, DK, 2005.

11. Oikonomou, V.; Patel, M.; Mundaca, L.; Johansson, T.; Farinelli, U. A quallitatiive analysis of white, green certificates and EU CO2 allowances. *Phase II of the White and Green Project*, Utrecht University, Copernicus Institute: Utrecht, 2004.

12. Pavan, M. What's up in Italy? Market liberalization, tariff regulation and incentives to promote energy efficiency in end-use sectors. *American Council for Energy Efficient Economy (ACEEE) Summer Study*, ACEEE: Asilomar, CA, 2002.

13. Pavan, M. Expectation from Italy: suggestions to focus the project proposal. *Workshop on Organising a Task on White Certificates Under the IEA-DSM Implementing Agreement*, CESI: Milan, 2003.

14. Pavan, M. The Italian White Certificates System: Measurement and Verification Protocols. *EU and ECEEE expert seminar on Measurement and Verification in the European Commission's Proposal for a Directive on Energy Efficiency and Energy Services*, Brussels, 2004;Available at http://www.eceee.org/european_directives/EEES/monitoring_evaluation/ESD21SeptPavan.ppt, (accessed on December 2006).

15. Pavan, M. *Italian Energy Efficiency Obligation and White Certificates: Measurement and Evaluation*, Copenhagen, DK, 23 September, 2005.

16. Pavan, M. Italian Energy Efficiency Obligation and White Certificates: Measurement and Evaluation. *European Parliament Workshop on Case studies of Current European Schemes for the Measurement and Verification of Energy Efficiency Improvements*, Brussels, 2005;Available at http://www.danskenergi.net/attachments/8564/1146129527/marcellapavan.pdf, (accessed on December 2006).

17. Politecnico di Milano, *Final report of SAVE project, DSM pilot actions, DSM bidding and removal of DSM disincentives in restructured electricity markets*, 2000.

Transportation Systems: Hydrogen-Fueled*

Robert E. Uhrig
University of Tennessee, Knoxville, Tennessee, U.S.A.

Abstract

The term "hydrogen economy" is the title of a recent book (Rifkin, J. *The Hydrogen Economy*, Tarcher/Putham Publisher, ISBN 1-58542-193-6, 2002.) but the concept of using hydrogen as fuel for transportation systems has been advocated by environmentalists and others for at least three decades. There is no universally accepted definition of "hydrogen economy," but it is generally viewed as the replacement of the vast majority of petroleum fuels used by transportation vehicles of all kinds (automobiles, trucks, trains, and aircraft) with hydrogen that is burned in internal-combustion engines, external-combustion (jet) engines, or preferably, used in fuel cells to more efficiently generate power for transportation. This chapter reviews the hydrogen economy from a basic energy engineering viewpoint to identify potential impediments and opportunities and to quantify the tasks involved in terms of energy and materials required in bringing the hydrogen economy into existence.

PRESENT TRANSPORTATION ENERGY SITUATION

Today, the United States uses over 20 million barrels (a "barrel" as used in the petroleum industry, is a volume of 42 gallons, which is equal to ~ 159 l) of oil per day (Mbbl/day), of which about 13 Mbbl/day are used for transportation of all kinds—cars, trucks, aircraft, trains, and military vehicles. If each barrel of oil contains 5.8 million British thermal units (MBtu) [6.12×10^9 Joules (J)], then the transportation energy to be replaced is about 75.4×10^{12} Btu/day [79.5×10^{15} J/day]. Because the energy content of hydrogen is about 51,600 Btu/lb [120×10^6 J/kg], the amount of hydrogen required to replace transportation fuel would be about 1.46×10^9 lb/day [664×10^6 kg/day]. If we assume that half of the transportation energy goes to fuel cells that are twice as efficient as heat engines while the remaining half is burned in combustion engines with a 20% increase in efficiency from current engines, then the required hydrogen is reduced to about 0.97×10^9 lb/day [441×10^6 kg/day], or about 177 million tons per year [~ 161 million tonnes (a 'tonne" is a metric ton, defined as 1000 kg or about 2200 pounds) per year]. This compares with the current production of about 50 million tons [~ 45.5 million tonnes] of hydrogen per year worldwide for all purposes, including fertilizers, upgrading of hydrocarbon fuels, and chemical-industry feedstock.

The Challenges of Hydrogen

Like electricity, hydrogen is an energy carrier, not an energy resource, and thus it must be extracted from other sources such as water or hydrocarbon fuels. Hydrogen is difficult to store, particularly on transportation vehicles, and difficult to distribute from one location to another. Hydrogen also burns with an almost invisible flame and is subject to special handling regulations by state and national codes for safety reasons. Clearly, the use of hydrogen as a transportation fuel has many engineering challenges that may be difficult to address. Despite this, there is a particular focus on hydrogen as an energy resource because it can be converted into electricity for transportation using fuel cells with an efficiency that is about twice as high as the conversion in conventional thermodynamic heat engines. Other benefits include the very significant reduction of pollutants and greenhouse gases being emitted into the atmosphere by transportation vehicles using hydrogen as a fuel.

HYDROGEN PRODUCTION

The two primary sources of hydrogen today are the decomposition of water by electrolysis and themochemical water-splitting processes and extraction from hydrocarbon fuels by chemical processing. Extracting hydrogen from water is more energy intensive than extracting it from hydrocarbon fuels. Coal gasification and partial oxidation of heavy oils are also demonstrated technologies for producing syngas (CO and H_2) from which hydrogen can be separated, but the large amount of CO_2 that is also

---

*The material in this article was originally published as *Engineering Challenges of the Hydrogen Economy* by Robert E. Uhrig in the Spring 2004 issue of *The Bent of Tau Beta Pi*, quarterly publication of the Engineering Society, Volume XCV, No. 2, Pages 10–19, 2004. That original article has been condensed and modified to reflect changes that have occurred since its publication.

Keywords: Hydrogen economy; Electrolysis; Steam methane reforming; Sulfur iodine process; Hydrogen storage; Hydrogen transportation and distribution; Hydrogen safety.

produced is a greenhouse gas and hence an undesirable by-product.

STEAM-METHANE REFORMING

Today, the vast majority of hydrogen is produced by steam methane reforming (SMR) of natural gas (which is about 95%–98% methane—CH_4) followed by a water–gas shift reaction. The two-step SMR process can be represented by

$$CH_4 + H_2O + \text{Heat} \xrightarrow{\text{Catalyst}} CO + 3H_2$$

Steam reforming reaction

$$CO + H_2O + \text{Heat} \xrightarrow{\text{Catalyst}} CO_2 + H_2$$

Water–gas shift reaction

The hydrogen comes from the methane and the steam. The SMR reaction that takes place at 300 psi [2.07 MegaPascals] and 1562°F [850°C] is endothermic with the required heat energy normally produced by combustion of some of the methane and the exothermic heat of the water–gas shift reaction. A well-designed SMR plant will yield hydrogen having about 75%–80% of the energy of the methane supplied. Unfortunately, SMR produces CO_2 in both the methane combustion and in the water–gas shift reaction. Recent work in Japan has demonstrated the feasibility of substituting high-temperature heat from a gas-cooled nuclear reactor to replace the heat supplied by the combustion of methane. This reduces the amount of methane required by about 20%–25% and eliminates the CO_2 produced by its combustion but not the CO_2 produced by the water–gas shift reaction.

The major disadvantages of steam methane reforming in the United States is the inadequacy of domestic supplies of natural gas, which produces wide fluctuations of price and the fact that it produces significant carbon dioxide that goes into the atmosphere. Reduction in the amount of CO_2 produced and the virtual elimination of pollutants are major benefits of using hydrogen as a transportation fuel in a "hydrogen economy." However, using a method such as steam methane reforming that emits CO_2 to produce hydrogen compromises the environmental benefits of using hydrogen as an automotive fuel. Perhaps more importantly, the demand for natural gas is growing rapidly primarily because of its increasing use as fuel for high-efficiency combined cycle gas turbine plants to generate electricity, as evidenced by the fact that since 2000 the average amount of natural gas used for electrical generation has exceeded the total amount used for home heating even though the vast majority of new homes use natural gas for heating. To produce hydrogen for the mature U.S. hydrogen economy of about 177 million tons per year [$\sim 161 \times 10^6$ tonnes/year], using SMR would require about 22.8×10^{12} standardized cubic feet [~ 707 billion normalized cubic meters] of natural gas per year in the United States.[10] This exceeds the average of about 21.0×10^{12} standardized cubic feet [651 billion normalized cubic meters] of natural gas per year currently used in the United States for all purposes (feedstock for chemical processes, upgrading hydrocarbon fuels, fertilizer, home heating, and generation of electricity) today. Gas utility executives are already planning importation of liquefied natural gas (LNG) to help meet the need in the United States. However, there are serious public concerns about the safety of LNG terminals. Doubling the U.S. demand for natural gas to meet the needs of the hydrogen economy is generally viewed as not being a viable alternative. This leads to the conclusion that SMR is not likely to be used to produce large quantities of hydrogen for the hydrogen economy.

ELECTROLYSIS OF WATER

Electrolysis of water is a mature technology used to produce high-purity hydrogen with an overall efficiency of about 25%. Commercial units of 10 MWe capacity are available today that can be coupled together to produce a facility of any desired size. On average, 1 MW of electricity using today's technology will produce about 1040 pounds (lb) [473 kilograms (kg)] of hydrogen per day and 8320 lb [3747 kg] of oxygen per day from about 10,000 lb [~ 4550 kg] of water per day (including $\sim 6\%$ overall processing loss of water).[9] Hence, to produce 177 million tons [~ 161 million tonnes] of hydrogen per year for a mature hydrogen economy would require about 932 Gigawatts of electricity (GWe), i.e., 927 new nonpolluting 1000 MWe power plants, more than the total generating capacity of the United States today, and about 1700 million tons [~ 1545 million tonnes] of water per day.

Use of "Spinning Reserve" to Produce Hydrogen

One of the useful features of conventional electrolysis (producing hydrogen using electricity only) is the ability to instantaneously disconnect the electrolyzer from the utility electric service and use this available electricity for other purposes. This allows the amount of capacity reserved by the utility to instantaneously pick up the electrical load dropped because of plant or transmission system failure (commonly called "spinning reserve") to be used for electrolysis. In effect, this increases the useful generating capacity of the utility by the amount of spinning reserve used for electrolysis. The additional cost to the utility for this additional capacity is only the cost of fuel and maintenance because the capital charges are already covered. For instance, a utility with a total generating

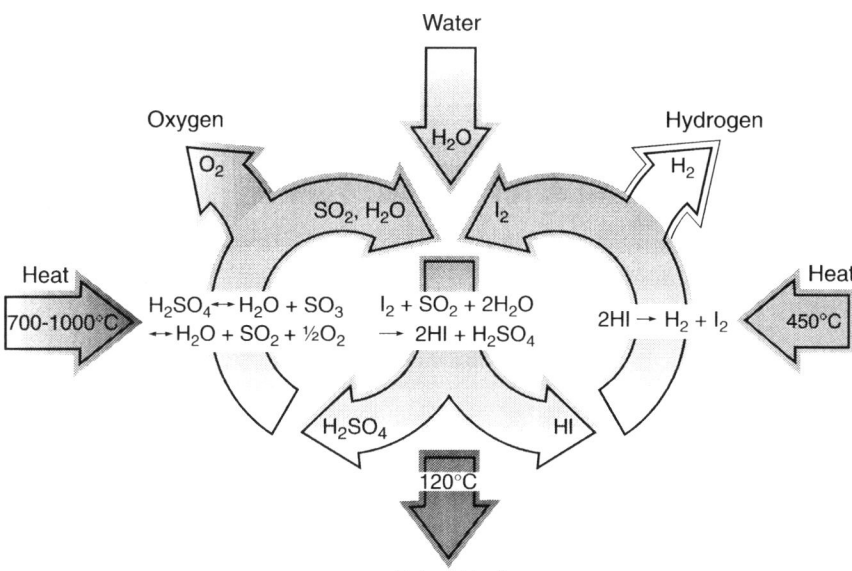

Fig. 1 Schematic diagram of the sulfur iodine process of producing hydrogen courtesy of oak ridge national laboratory. Source: From OECD Nuclear Energy Agency (see Ref. 3).

capacity of 12,000 MWe with its largest plant having a 1200 MWe output would have to carry 1200 MWe of "spinning reserve" to pick up the customer load in the event that the 1200 MWe plant trips offline. Hence, the utility's useful capacity is only 10,800 MWe. If this utility were able to use this spinning reserve to generate hydrogen by electrolysis while still being able to use it for an emergency, the only additional cost would be operating costs (fuel and maintenance). Although cost accountants might want to assign the capital costs equally to all activities, the rationale for not assigning capital costs to the electrolysis is that this capacity cannot be used for any other activity and hence does not increase the cost of electricity produced for other uses.

A study of the price of electricity on the Alberta, Canada Open Market by Atomic Energy of Canada Limited showed that there was an underlying low cost with many cost spikes. These spikes were 10–20 times as high but occurring for only about 5% of the time. Hence, for about 95% of the time, the average cost of electricity is below the overall average cost.[5] The ability to instantaneously switch off the hydrogen electrolyzers during peak loads could significantly reduce the cost of producing hydrogen by electrolysis. Similar pricing patterns exist in the United States and may exist in other countries.

High-Temperature Electrolysis

Electrolysis is more energy intensive than other methods of producing hydrogen. The energy of the hydrogen produced by electrolysis is about 75% of the energy of the electricity used, which in turn, typically has about one-third of the energy of the fossil or nuclear fuel used to generate the electricity. Hence, the overall efficiency of electrolysis is about 25%. There are two ways to improve this overall efficiency: (1) use more efficient high-temperature power plants to generate the electricity and (2) use high-temperature electrolysis.

The use of very high-temperature heat (800°C–900°C) with electricity for electrolysis is being investigated for electrolysis of steam. Substitution of high-temperature heat energy for some of the electrical energy in electrolysis has been demonstrated to decrease the electricity needed and appears to be potentially competitive with other hydrogen production processes under some circumstances. About 30% of the energy is heat that is less expensive than electrical energy. The overall efficiency of high-temperature electrolysis is 45%–50%, higher than conventional electrolysis.[3] Using high-temperature cycle electrical generation with high-temperature electrolysis could further improve the efficiency.

THERMOCHEMICAL CRACKING OF WATER

"Water-Splitting"—The Sulfur–Iodine Process

Thermochemical cracking of water is generally viewed as the primary way that nuclear energy will be used to produce hydrogen on a large scale. Thermochemical cycles are Carnot cycle-limited, which indicates that high temperatures can improve the conversion efficiency. A comprehensive study of potential thermochemical cycles for hydrogen production reactions by the International Atomic Energy Agency (IAEA) included metal oxide systems, metal-halide family processes, sulfur family processes, and numerous hybrid systems incorporating portions of two or more of the basic systems.[4] Only a few of these cycles have survived the comprehensive

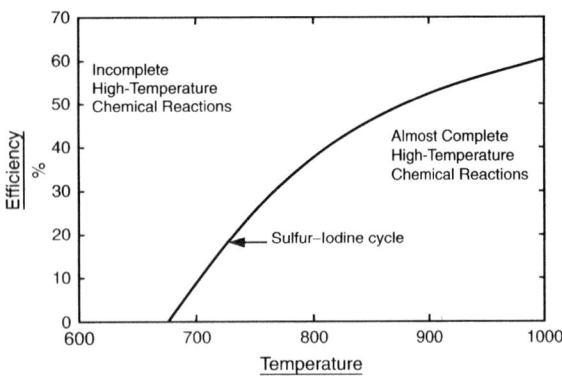

Fig. 2 Efficiency of Sulfur–Iodine cycle as a function of temperature (°C). Courtesy of General Atomics.
Source: From Ref. 8.

reviews and studies of the past two decades. When all of the steps involved in any given thermochemical process are summed up, the result is a closed system represented by

$$H_2O + \text{high-temperature heat} \rightarrow H_2 + \frac{1}{2} O_2 + \text{low-temperature heat}$$

In theory, this reaction could be implemented directly, but it would require a very high temperature (at least in the 2500°C–3000°C range), which is well beyond the capability of any known nuclear reactor.

General Atomics (GA) has studied the potential use of helium-cooled nuclear reactors to produce hydrogen for over three decades. After reviewing 922 references and 115 cycles, some 25 cycles were investigated in detail using thermodynamic calculations and preliminary block flow diagrams. Only two were selected for detailed design studies: the sulfur–iodine (S–I) cycle and the Adiabatic UT-3 cycle. General Atomics chose the S–I process for further detailed study and analysis. The fact that the S–I cycle involved only fluids (liquids and gases), the higher efficiency of the S–I cycle, and the compatibility between the energy requirements of the S–I cycle and the characteristics of its helium-cooled reactors were some of the reasons for this choice.[8]

Sulfur–Iodine (S–I) Cycle

Probably the most studied cycle is the sulfur–iodine cycle, one of three sulfur cycles being considered today. All three sulfur cycles have a common thermal decomposition of sulfuric acid that requires a very high temperature (~900°C, 1652°F) and a catalyst. The basic sulfur–iodine process that has been studied extensively by GA since the 1970s[7] and more recently at Oak Ridge National Laboratory and in Japan is shown schematically in Fig. 1.[3] The high-temperature reaction shown on the left side of Fig. 1 is represented by

$$H_2SO_4 \overset{(600°C)}{\leftrightarrow} SO_3 + H_2O \overset{(850°C \text{ and a catalyst})}{\leftrightarrow} SO_2 + H_2O + \frac{1}{2} O_2$$

The double-headed arrows indicate that the reactions can go either way depending upon temperature, pressure, and material balance. This high-temperature reaction should go almost to completion for efficient hydrogen production. This is shown graphically in Fig. 2, where the efficiency approaches 60% when the temperature is 1000°C (1832°F).[8]

The H-I process has two lower temperature reactions— (1) the hydrogen-generating reaction shown on the right side of Fig. 1 (where only about 16% of the HI is decomposed with the remainder being recycled), represented by[2]

$$2HI \overset{(350°C–450°C)}{\leftrightarrow} I_2 + H_2$$

and (2) the Bunsen reaction shown in the middle of Fig. 1 that involves recycle of the chemicals and the addition of water, represented by

$$I_2 + SO_2 + 2H_2O \overset{(120°C)}{\leftrightarrow} 2HI + H_2SO_4$$

This reaction requires excess iodine so that the hydrogen iodine and sulfuric acid separate into two separate liquids to be recycled in the process. The management of the heat recovery throughout these processes is critically important to the high efficiency of all of the sulfur cycles.

The advantages of the S–I cycle are (1) all of the chemistry reactions have been demonstrated, at least on a laboratory scale, (2) only fluids (liquids and gases) are involved in continuous processes, and (3) all chemicals are recycled. While there have been extensive studies of the various cycles in the United States, the only experimental testing of the whole SI cycle has been in Japan, where successful production of hydrogen has been demonstrated on a bench scale using a nonnuclear high-temperature heat source.

Two other sulfur cycles that are being studied are: (1) the "Hybrid Sulfur" Cycle (known as the GA 22, Westinghouse, and Ispra Mark 11 cycles) and (2) the "Ispra Mark 13 Cycle." Like the S–I cycle, both of these cycles involve the thermal decomposition of sulfuric acid that requires a high temperature (~900°C) and a catalyst but does not involve iodine. In the Hybrid Sulfur cycle, the two low-temperature processes of the S–I cycle are replaced by electrolysis that takes place at 90°C (194°F), and can be represented by

$$SO_2 + 2H_2O \overset{\text{Electrolysis}}{\leftrightarrow} H_2SO_4 + H_2$$

Electrolysis is also used in the Ispra Mark 13 cycle in conjunction with a second chemical process involving bromide to enhance the electrolysis.

Reducing the Temperature Required for Cracking Water

Detailed studies have concluded that peak temperatures in the S–I cycle need to be at least 850°C (1562°F) and preferably 900°C (1652°F) to drive the SO_3 decomposition to near completion.[3] Hence, the reactor outlet temperature would need to be about 950°C (1742°F) to 1000°C (1832°F). The High-Temperature Test Reactor (HTTR) helium-cooled reactor in Japan has a 950°C (1742°F) outlet temperature, the same as the German AVR pebble bed test reactor in the 1970s.

Studies of the process flow sheets show that there are strong economic and other incentives to lower the temperature and increase the pressure at which SO_3 dissociates—the exact opposite of the conditions required by thermodynamic considerations. After the high-temperature dissociation reaction, all the chemicals must be cooled to near room temperature, the SO_2 must be separated out and sent to the next chemical reaction, and the unreacted H_2SO_4 (formed by recombination of SO_3 and H_2O at lower temperatures) must be reheated back to high temperatures. Unless the chemical reactions go almost to completion, the energy losses in separations and in heat exchangers to heat and cool all the unreacted regents (H_2SO_4) could result in an inefficient and uneconomical process.[2]

Recent work at Oak Ridge National Laboratory (ORNL) has been directed at accelerating the SO_3 disassociation into SO_2 and O_2 at lower temperatures and higher pressures by the use of an inorganic separation membrane.[3] It appears to be possible to lower the required temperature to about 700°C (1292°F) by separating and removing the SO_2, H_2O, and O_2 products from the SO_3 disassociation. If the reaction products are removed, the remaining SO_3 (with a catalyst and heat) will disassociate more rapidly. If the reaction products can continue to be selectively removed, the chemical reaction can be driven to completion at a lower temperature, estimated to be ~700°C (~1292°F). The membrane operates with high pressure on one side and lower pressure on the other and this pressure difference dives the separation process. Preliminary results of tests at ORNL have been encouraging.[3]

Lowering the high-temperature requirements for the SI processes has many advantages. Foremost of these advantages is that currently designed high-temperature reactors, i.e., some existing high-temperature gas-cooled reactors as well as liquid metal (lithium or lead) cooled fast reactors, could be used for the nuclear production of hydrogen. Furthermore, lower temperatures significantly enhance the safety of both the reactor and the hydrogen production facilities by reducing peak temperatures in structural materials to levels where there is a better understanding of their mechanical properties, particularly strength and creep behaviors.

Calcium Bromide Cycles (UT-3 and Others)

A family of lower temperature calcium–bromide cycles for producing hydrogen with a 750°C (1382°F) peak temperature is also being studied. The high-temperature steps are well understood, but they involve solid reagents, calcium oxide (CaO), and calcium bromide ($CaBr_2$). The dimensions of the solids change size when the chemical reactions occur and there is the potential for dust. Several options are being studied as methods of producing hydrogen and recycling the bromide, including (1) the Iron Cycle process (UT-3) that was developed at the University of Tokyo in Japan, (2) electrolysis that has a low efficiency, and (3) a cold plasma process that provides energy as a replacement for electricity in electrolysis.[3]

Status of the Development of Thermochemical Cycles in the United States.

The United States Department of Energy's Nuclear Hydrogen R&D plan has as its goal to define the hydrogen production R&D necessary to demonstrate nuclear hydrogen on a production scale by 2017. Currently, they are supporting studies of several of the thermochemical cycles discussed in this chapter as well as experimental work on some of them. The S–I cycle is currently being developed experimentally by three organizations: Sandia National Laboratory (SNL), GA, and the French Commissariat a l'Energie Atomique (CEA). Sandia National Laboratory is building the facilities to study and test the high-temperature thermal decomposition of sulfuric acid using a nonnuclear source of high-temperature heat, GA is investigating the hydrogen generating reaction, and CEA is investigating the Bunsen reaction that involves recycle of the chemicals and the addition of water.[8] When this work is completed, a prototype system integrating these three processes in a single facility will be undertaken. Current DOE plans are to demonstrate both the S–I thermochemical water-splitting and high-temperature electrolysis processes on a production scale at Idaho National Laboratory (INL) by 2017. The INL facility is expected to provide 60 MWt of high-temperature heat for hydrogen production and to demonstrate passive safety while maintaining the purity of the hydrogen to meet fuel cell standards.

HYDROGEN STORAGE

Implementation of the hydrogen economy will require facilities for storing, transporting, and distributing hydrogen to refueling facilities throughout the country.

Historically, hydrogen has been stored as a high-pressure gas (~3000 psi) or a cryogenic liquid (~20°K). Although there are experimentations with liquid hydrogen by some automakers, most current discussions about storage of hydrogen on automotive vehicles involve gaseous hydrogen at 5000 psi or possibly 10,000 psi. The liquefaction of hydrogen consumes about 30% of the energy stored, and there is also a continual loss of energy due to thermal conduction through the insulated walls whether the liquid is stored in a service station or a vehicle on or off the road. Even so, there may be applications for liquid hydrogen in heavy vehicles and long-range aircrafts, where weight is critical.

Storage of Hydrogen as a High-Pressure Gas

An official goal for on-board hydrogen storage is to achieve a 300-mi range with a tank no larger than current automobile fuel tanks. The importance of the goal is reflected in the general belief that failure to meet this goal was a major impediment to battery-powered electric cars—a technology in which the direct storage of electricity in batteries is much simpler than converting electricity to hydrogen, distributing it nationally, dispensing it to vehicles, and then converting hydrogen to electricity in a fuel cell to drive an electric motor to propel a vehicle.

On-board storage tanks for gaseous hydrogen on vehicles, made with filament-wound carbon fibers and lined with aluminized polyester bladders, have been approved for use up to 5000 psi (34.5 MP) by United States and up to 10,000 psi (69 MP) by German authorities. Generally, these tanks are cylindrical, and more than one are sometimes used because of the required volume and the low-volumetric-energy density of hydrogen. Even though the higher efficiency of fuel cells partially compensates for this low-energy density, the large size of the tank(s) required for a 300-mi range is a concern and researchers still seek alternatives. Technologies under serious consideration are metallic hydrides and alanates where hydrogen is adsorbed onto interstitial surfaces. Reports on these two options from a recent conference are summarized below. Adsorption of hydrogen by carbon nanotubes was considered as an option a few years ago, but it has not lived up to its initial promise for hydrogen storage.

Metal Hydrides for Storage of Hydrogen

Hydrogen is a highly reactive element and it will form hydrides and solid solutions with hundreds of metals and alloys as well as form chemical compounds or complexes with many other elements. Metal hydrides are formed by hydrogen bonding to a metal with metallic, ionic, or covalent bonds. Hydrogen is usually bound in the interstitial sites, and it can be removed by applying heat to the metal hydride. Many intermetallic compounds and solid solutions can readily absorb and desorb hydrogen gas at room temperature and atmospheric pressure, but these hydrides can reversibly store only 1–3 weight-percent hydrogen, which is not enough for application to vehicle storage. Covalent and ionic hydrides (e.g., MgH_2 and LiH, respectively) are capable of storing 7–12 weight-percent hydrogen, but they must be heated to above ~325°C (617°F) to release the hydrogen at atmospheric pressure. This temperature is much higher than the ~75°C (167°F) waste heat that is available from a proton-exchange-membrane fuel cell. Recent work with catalyzed complex hydrides containing mixed ionic-covalent bonding can reversibly store more than 4% hydrogen with operating temperatures below 125°C (257°F). Generally, the volumetric energy density of hydrogen stored in metal hydrides is comparable to that of liquid hydrogen. The primary problems with metal hydrides for vehicle applications are their heavy weight and high cost.[1]

Alanates for Hydrogen Storage

Alanates, such as sodium alanate ($NaAlH_4$), are aluminum alloys that contain hydrogen. Sodium alanate undergoes two-step decomposition into sodium hydride (NaH), hydrogen, and aluminum, where the gas temperature and pressure determine the equilibrium quantities of reactants and products that release a total of about 5.6% hydrogen. Pure alanates react slowly at ambient conditions, but recent work indicates significant improvement in sorption kinetics by adding transition metal additives. Many other alanates show promise as a storage medium, but they are at an early stage of their development.[1]

HYDROGEN DISTRIBUTION

Because of its low volumetric energy density, hydrogen is at a disadvantage when compared to other gaseous fuels with higher volumetric energy densities (e.g., methane or propane) and liquid fuels (gasoline or ethanol). The power to perform the pumping of hydrogen is reported to be about a factor of 4.6 greater than for methane over a range of high pressures. The total cost of distribution for an equal amount of energy in the form of hydrogen is estimated by the International Energy Agency to be 15 times that of liquid hydrocarbons. However, the economies of scale associated with large central hydrogen facilities may partially compensate for the extra distribution costs.

There are perhaps two basic but different configurations of a distribution system that may evolve over the next three decades—with many combinations of the two and other variants. Only the two basic configurations will be discussed here.

Hundreds of Large Hydrogen Plants with a National Distribution Pipeline Grid

In this scenario, there are many clusters of large hydrogen plants distributed around the country with an interconnected pipeline grid to distribute the hydrogen to control centers, which in turn carry hydrogen to millions of service stations through a distribution pipe grid. Most likely, the plants would be thermochemical water-splitting plants with nuclear power plants providing the heat. This arrangement is directly analogous to the current electric transmission and distribution grids in the United States.

If we accept the required hydrogen production capacity as about 177 million tons (161 million tonnes) per year for the hydrogen economy (this is based on present-day U.S. transportation sector oil demand), which is equivalent to about 145 billion normalized cubic feet (Mncf/day) [4.50 billion standard cubic meters per day (Mscm/day)], and that today's world-class hydrogen plants each produce about 184 Mncf/day [5.7 Mscm/day], then 790 world-class hydrogen production plants would be required for the hydrogen economy. These plants might be grouped in 197 clusters of four plants per cluster or an average of about four clusters per state. Clearly, more clusters would be needed in the more populous states and fewer plants would be needed in the other states. It can be readily shown that the 184 Mncf/day (5.7 Mscm/day) output of hydrogen is equivalent to $\sim$775 MWt. Hence, if each world-class plant is operating at 50% efficiency (typical of thermochemical water-splitting plants), we would require $\sim$1550 MWt per plant or 6200 MWt for each four-unit cluster and a total of $\sim$1225 GWt of heat energy for 790 plants. A crude calculation indicates that if the 197 four-unit clusters of plants were laid out uniformly on a square grid throughout the United States, the clusters would be about 80 mi apart and would require about 90,000 mi of interconnecting pipe. These interconnecting pipes would be analogous to the high-voltage transmission lines for electricity. If a hydrogen distribution grid were installed with connections at 10-mi intervals on the interconnecting pipes, an additional 725,000 mi of smaller distribution pipe would be required.

Millions of Small Hydrogen Plants Located at Hydrogen Service Stations

The other basic arrangement for distribution of hydrogen to service stations is a distributed array of small hydrogen generators, probably electrolysis units, of sizes selected to provide adequate hydrogen for a local area's need and using electricity from the most economical source. In the hydrogen economy, the total amount of new electrical capacity required for electrolysis would be about 930 GWe—more than doubling the current electrical generating capacity of the United States. The electrical grid would also have to be doubled in size to carry the needed electricity.

The alternative configuration for providing the required power is a distributed array of windmills or photovoltaic generators. Some 930,000 new 1 MWe windmills would be required if they ran 24 h a day. With an average availability factor of about 30%, the number of windmills would have to be increased significantly. The other option, some 93,000,000 new 10 kWe photovoltaic electric generating units, is also burdened with a low-availability factor—only 20% and significant additional capacity would be required. Clearly, the ultimate source of power to provide hydrogen for the hydrogen economy would be some combination of all the options discussed above. There may be others, including fusion plants and fermentation-or solar-driven photo-biological methods that are currently being studied.

Hydrogen at Service Stations

There are several issues of concern associated with dispensing hydrogen into automotive vehicles. The first is that the connection from the station to the vehicle would be complex (with sensors to assure secure attachment without leakage). The valves and other hardware used to control the flow of very high-pressure gases are inevitably complex and expensive. To deliver hydrogen to vehicle tanks at 5000 psi will require a continuous delivery pressure of perhaps 7000 psi to keep the refilling time reasonable. As the station uses its supply of hydrogen, the pressure will drop. To keep the pressure high enough to refuel vehicles in a reasonable time would require a compressor to deliver hydrogen into the vehicle.

EPILOGUE

The National Environmental Policy Act, passed by Congress in 1969, mandated that before a major project can be undertaken, an environmental impact statement (EIS)—a comprehensive assessment of the benefits of the project and its impact upon all aspects of the environment—must be prepared and presented to the authorizing authority. Nothing remotely resembling an environmental impact statement required for the construction of a nuclear power plant or other large projects (an EIS for off-shore oil drilling had $\sim$32,000 pages) has been undertaken for the hydrogen economy. It would seem that, given the extraordinary magnitude of the requirements for power, water, fuels, and infrastructure of all sorts, that at least a scoping study of the environmental and economic consequences of the hydrogen economy should be undertaken now.

REFERENCES

1. Bowman, R. Jr.; Sanrock, G. *Hydrides for Energy Storage and Conversion Application*, Proceedings of GLOBAL 2000 Atoms for Prosperity, American Nuclear Society Meeting, New Orleans, LA, November 16–20, 2002.
2. Forsberg, C. *Production of Hydrogen Using Nuclear Energy*, Presentation to National Research Council/National Academy of Science: Washington, D.C., January 22, 2003.
3. Forsberg, C.; Bischoff, B.; Mansur, L.; Lee Trowbridge, L. *Nuclear Thermo-Chemical Production of Hydrogen with a Lower-Temperature Iodine-Westinghouse-Ispra Sulfur Process*, Second Information Exchange Meeting on Nuclear Production of Hydrogen, OECD Nuclear Energy Agency: Argonne, IL, October 2–3, 2003.
4. IAEA. International Atomic Energy Agency document. Hydrogen as an Energy Carrier and Its Production by Nuclear Power, IAEA-TECDOC-1085, May, 1999.
5. Miller, A.; Duffy, R. *Hydrogen and High Temperature Reactors: a Source Map*, Proceedings of GLOBAL 2000 Meeting, at the Winter American Nuclear Society Meeting, New Orleans, LA, November 16–20, 2003.
6. Rifkin, J. *The Hydrogen Economy*, Tarcher/Putnam Publisher, ISBN 1-58542-193-6, 2002.
7. Schultz, K. *Efficient Production of Hydrogen from Nuclear Energy*, Presentation to the California Hydrogen Business Council, June 27, 2002.
8. Schultz, K. *Fueling the Hydrogen Economy: Hydrogen Production using Nuclear Energy*, Presentation at World Nuclear Association Meeting, October, 2003.
9. Stuart, S. *Nuclear-Electrolysis Synergies: For Neat Hydrogen Production and Cleaner Fossil Reserves Harvesting*. ANS President's Special Session: Hydrogen Systems, ANS Meeting: Reno, NV, November 11–15, 2001.
10. Uhrig, R. *Engineering Challenges of the Hydrogen Economy*. The Bent of Tau Beta Pi, Vol. XCM, No. 2, Spring, 2004.

Transportation: Location Efficiency

Todd A. Litman
Victoria Transport Policy Institute, Victoria, British Columbia, Canada

Abstract
This chapter describes the concept of location efficiency (also called smart growth), which refers to land-use development patterns that maximize the ease with which people can obtain desired goods, services, and activities, and that minimize the need for physical travel. Location efficiency includes factors such as land-use density, land-use mix, roadway connectivity, and transportation system diversity. Improving location efficiency helps reduce transportation costs, including road and parking facility costs, consumer costs, traffic risk, energy consumption, pollutant emissions, and other environmental impacts. It also benefits people who are transportation disadvantaged and increases housing affordability.

INTRODUCTION

Land use and transportation are two sides of the same coin. Transportation affects land use, and land use affects transportation. Decisions that affect one also affect the other. As a result, it is important to coordinate transportation and land-use planning decisions so they are complementary. Land-use planning can help create more efficient transportation systems that reduce per-capita vehicle travel and energy consumption, for example. This is referred to as *location-efficient* development or smart growth.

To understand how land use affects travel patterns it is useful to consider the concept of accessibility. Accessibility (or just access) is the ability to reach desired goods, services, activities, and destinations—together called opportunities. A stepladder provides access to the top shelf in your kitchen. A store provides access to goods. Libraries, telephones, and the Internet provide access to information. Paths and roads provide access from one destination to another by walking, cycling, automobile, and bus.

Access is the ultimate goal of most transportation, excepting the small portion of travel in which movement is an end in itself (e.g., cruising, historic train rides, horseback riding, and jogging). Even recreational travel usually has a destination, such as a resort or a campsite. (Mobility as an end in itself is discussed later in this chapter).

Four general factors affect accessibility:

1. Mobility—that is, physical movement. Mobility can be provided by walking, cycling, public transit, ride sharing, taxi, automobiles, trucks, and other modes.
2. Mobility substitutes, such as telecommunications and delivery services. These services can provide access to some types of goods and activities, particularly those involving information.
3. Transportation system connectivity, which refers to the directness of links and the density of connections in a path or road network.
4. Land-use patterns—that is, the geographic distribution of activities and destinations. When real estate experts say "location, location, location," they mean "accessibility, accessibility, accessibility."

LAND-USE IMPACTS ON TRANSPORTATION

Land-use factors (also called spatial development, community design, urban design, or the built environment) can affect transport activity in several ways.[1] When worksites are dispersed and located in areas without good walking and cycling facilities, for example, most employees will drive, but if the same businesses locate in commercial centers with good walking and cycling facilities, and with good transit services, a significant portion of employees will use alternative modes.

Planners increasingly realize the importance of integrating transportation and land-use decisions to increase accessibility and improve travel options, thereby reducing the amount of motor vehicle travel required to meet people's needs and serve economic activities. This can help achieve a variety of planning objectives, including reduced congestion, energy conservation, pollution and emission reductions, infrastructure cost savings, increasing household affordability, and improving economic opportunities for disadvantaged populations.

Specific land-use factors that affect transportation are described in the following sections.[2]

Keywords: Land use; Smart growth; Location; Energy.

Density

Density refers to the number of people or jobs in a given area. Increased density tends to reduce per-capita automobile ownership and use, and to increase use of alternative modes. Fig. 1 shows how per-capita vehicle mileage tends to decline with density in U.S. urban areas. Many other studies find similar results.

Increased density tends to reduce per capita vehicle travel.

Density at both origins and destinations affect travel behavior. One study found that increasing urban residential population density to 40 people per acre increased transit use from about 2 to 7%, while increasing densities in commercial centers to 100 employees per acre resulted in an additional 4% increase in transit use, to an 11% total mode share.[3] Both work trips and shopping trips are affected by population and employment densities.

Land-Use Mix

Mixed land use (such as locating appropriate businesses and public services in or adjacent to residential areas) can reduce per-capita vehicle travel. It tends to reduce the distances that residents must travel for some services and allows more use of walking and cycling for such trips. It also can reduce some employees' commute distances (some residents may obtain jobs at nearby businesses), and employees who work in a mixed-use commercial area are significantly more likely to commute by alternative modes.

Roadway Design

Roadway design can affect travel behavior in several ways. A connected road network provides better accessibility than a conventional hierarchical road network with a large portion of dead-end streets. Increased connectivity can reduce vehicle travel by reducing travel distances between destinations and by improving walking and cycling conditions, particularly where paths provide shortcuts, so walking and cycling are relatively faster than driving. This also supports transit use.

Transit Service Quality

Per-capita automobile ownership and motor vehicle mileage tend to decline as transit service quality in an area improves. Transit service quality includes the convenience, frequency, comfort, and security of transit vehicles and stations or stops, as well as the quality of walking conditions in nearby areas.

Site Design and Building Orientation

Some research indicates that people walk more and drive less in areas with traditional pedestrian-oriented commercial districts where building entrances connect directly to the sidewalk than in areas with automobile-oriented commercial strips where buildings are set back and separated by large parking lots (Moudon, 1996).[5] This type of building orientation improves pedestrian access and creates a more attractive pedestrian environment.

Cumulative Impacts

The transportation effects of density and clustering, land-use mix, transit access, street design, and building design tend to be cumulative. As an area becomes more urbanized (more dense and mixed activities, higher land prices, and less parking), transportation diversity tends to increase, with fewer trips by automobile and a greater portion of trips by walking, cycling, and public transit.

Holtzclaw[6] finds that average vehicle ownership, vehicle travel, and vehicle expenditure per household decline with increasing residential densities and proximity to public transit, holding constant other demographic factors such as household size and income. The formulas below summarize his findings. An online calculator, This View of Density Calculator (www.sflcv.org/density), uses this model to predict the effects of different land-use patterns on travel behavior (Fig. 2).

This figure illustrates how density and transit accessibility affect household vehicle mileage. The Transit Accessibility Index (TAI) indicates daily transit service nearby.

Ewing, Pendall, and Chen[7] developed a sprawl index based on 22 specific variables related to land-use density, mix, street connectivity, and commercial clustering. The results indicate a high correlation between these factors and travel behavior: a higher sprawl index is associated with higher per-capita vehicle ownership and use, and lower use of alternative modes. Other studies also find

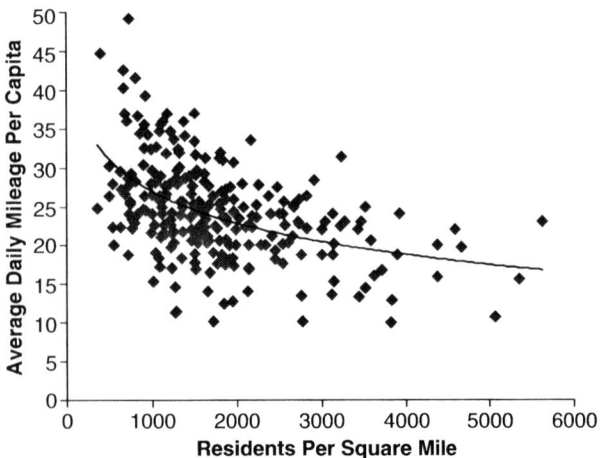

Fig. 1 Density vs vehicle travel for U.S. urban areas. Source: From FHWA, 2005 (see Ref. 4).

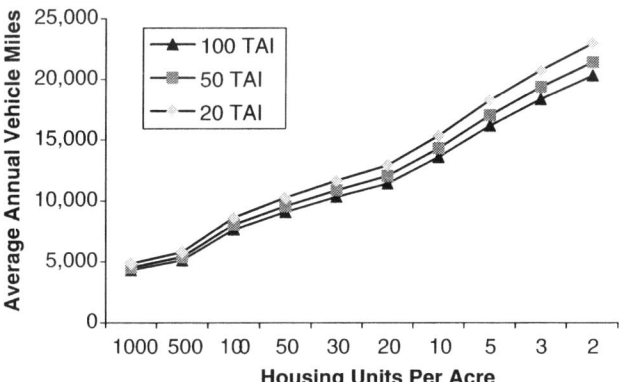

Fig. 2 Annual VMT per household.
Source: From Based on National Resources Defense Council (see Ref. 6).

LOCATION EFFICIENCY

Location-efficient development (also called smart growth) refers to more compact and mixed development located in compact centers designed for walking, cycling, and transit. Table 1 compares the two types of land-use patterns.

Per-capita motor vehicle travel tends to be significantly higher in automobile-dependent areas compared with the same demographic and businesses activity in Smart Growth locations.[10] Table 2 summarizes typical reductions in Vehicle Miles of Travel (VMT) resulting from various Smart Growth developments. This also indicates that location efficiency significantly reduces per-capita vehicle travel. Because of these impacts, planners are increasingly including land-use management strategies to achieve transportation planning objectives, including improved accessibility, reduced infrastructure costs, reduced accidents, and energy conservation.

that per-capita vehicle travel is significantly lower in higher-density, multimodal urban neighborhoods than in automobile-oriented suburban neighborhoods.

Lawton[8] found that average daily motor vehicle miles per adult decreased from 19.8 in the least urbanized residential neighborhoods to 6.3 in the most urban neighborhoods, due to fewer and shorter automobile trips. Even modest land-use changes can provide significant vehicle travel reductions if they are reinforced by other mobility strategies, such as commute trip reduction programs (which encourage commuters to use alternative modes) and parking pricing or cash-out (travelers can choose to receive cash instead of parking subsidies).

As a result, residents of more urbanized areas drive significantly fewer miles and rely more on alternative modes than residents of suburban and rural communities, as indicated in Fig. 3.

Urban residents drive less and use transit, cycling, and walking more than elsewhere.

HOW LOCATION-EFFICIENT DEVELOPMENT IS IMPLEMENTED

Location-efficient development is implemented by developers, usually with support and encouragement from local governments. It often is implemented as part of Smart Growth and New Urbanist planning.[11] In practice, location-efficient development consists of redeveloping older urban residential neighborhoods and commercial centers, creating new transit-oriented suburban neighborhoods, and improving walking conditions, cycling conditions, and transit services. It also can involve the application of mobility management (also called transportation demand management), which includes various policies and programs that encourage people to reduce their automobile travel and use alternative travel options.[13,10] These programs include incentives for more compact, infill development, and parking management to reduce the amount of parking required at each destination.

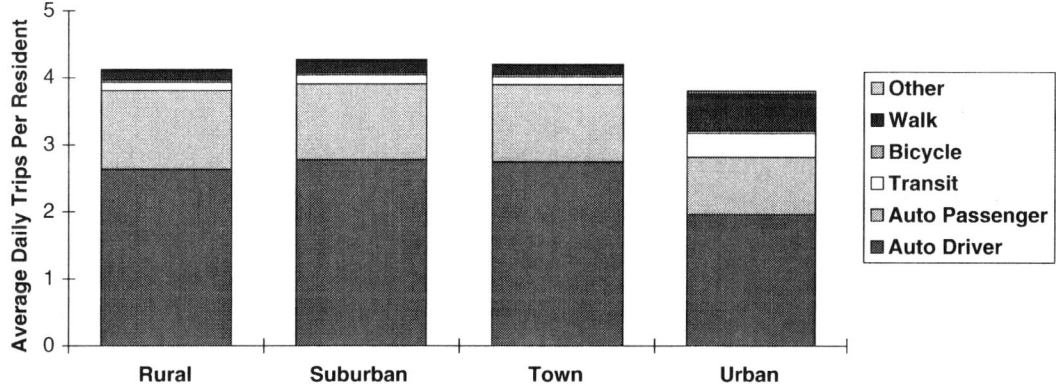

Fig. 3 Average daily trips per resident by geographic area.
Source: From NPTS (see Ref. 9).

Table 1 Comparing automobile dependency and smart growth

	Automobile dependency	Smart growth
Density	Lower-density, dispersed activities	Higher-density, clustered activities
Growth pattern	Urban periphery (greenfield) development	Infill (brownfield) development
Land-use mix	Homogeneous (single-use, segregated) land uses	Mixed land use
Scale	Large scale. Larger buildings and blocks, wide roads. Less detail, because people experience the landscape at a distance, as motorists	Human scale. Smaller buildings, blocks, and roads. Careful detail, because people experience the landscape up close, as pedestrians
Public services (shops, schools, parks)	Regional, consolidated, larger. Requires automobile access	Local, distributed, smaller. Accommodates walking access
Transport	Automobile-oriented transportation and land-use patterns, poorly suited for walking, cycling, and transit	Multimodal transportation and land-use patterns that support walking, cycling, and public transit
Connectivity	Hierarchical road network with numerous loops and dead-end streets, and unconnected sidewalks and paths, with many barriers to nonmotorized travel	Highly connected roads, sidewalks, and paths, allowing relatively direct travel by motorized and nonmotorized modes
Street design	Streets designed to maximize motor vehicle traffic volume and speed	Streets designed to accommodate a variety of activities. Traffic calming
Planning process	Unplanned, with little coordination between jurisdictions and stakeholders	Planned and coordinated between jurisdictions and stakeholders
Public space	Emphasis on the private realm (yards, shopping malls, gated communities, private clubs)	Emphasis on the public realm (streetscapes, pedestrian environment, public parks, public facilities)

Source: From Victoria Transport Policy Institute (see Ref. 12).

One strategy for encouraging households to choose more accessible locations is to offer location-efficient mortgages (LEMs), which means that lenders recognize the potential savings of a more accessible housing location when assessing a household's borrowing ability. It considers transportation and housing costs together, so vehicle cost savings are treated as additional income that can be spent on a mortgage. This gives home buyers an added incentive to choose location-efficient residences and tends to encourage more infill development as opposed to more automobile-dependent development at the urban periphery.[15–17]

Location-efficient mortgages are implemented by residential mortgage lenders, often with the support and encouragement of government agencies such as Fannie Mae and the Canadian Mortgage and Housing Corporation. Lenders use a model to determine which locations have lower transportation costs and, therefore, can qualify for higher mortgage payments. The following factors can be considered in such models:

- Residential density
- Land-use mix—that is, the number of public services within convenient walking distance (schools, shops, parks, medical services, pharmacy, etc.).
- Proximity to high-quality transit (such as a rail transit station or a bus line with frequent service).
- Quality of walking and cycling conditions.

Table 2 Infill VMT reductions

Location	Description	VMT reduction (%)
Atlanta	138-acre brownfield, mixed-use project	15–52
Baltimore	400 housing units and 800 jobs on waterfront infill project	55
Dallas	400 housing units and 1,500 jobs located 0.1 mile from Dallas Area Rapid Transit (DART) station	38
Montgomery county	Infill site near major transit center	42
San diego	Infill development project	52
West palm beach	Auto-dependent infill project	39

Source: From Center for Clean Air Policy (see Ref. 14).

- Car-share services within convenient walking distance (vehicle rental services designed to substitute for automobile ownership).
- Parking management (reduced parking requirements and renting parking spaces separately from building space, so residents who do not own an automobile are not forced to pay for parking they do not need).

BENEFITS AND COSTS

Location-efficient development can provide several benefits:

- Consumers benefit from more housing, shopping, and transportation choices, and from financial savings. Nondrivers in particular benefit from having housing options designed for maximum accessibility, as well as financial savings from reduced driving and parking costs. Per-household transportation expenditures tend to be lower for residents in such areas. Residents of cities with high levels of transit ridership tend to spend significantly less per capita on transportation than residents of more automobile-dependent cities,[2] as illustrated below. Similarly, McCann[18] found that households in more automobile-dependent communities on average spend more than 20% of their household budgets on transportation (more than $8,500 annually), whereas those in communities with more diverse transportation systems spend less than 17% (less than $5,500 annually), representing thousands of dollars in annual savings.
- By reducing per-capita vehicle ownership and use, location-efficient development tends to reduce per-capita traffic congestion and delays, road and parking facility costs, traffic crashes, pollution, energy consumption, and sprawl.
- By improving walking and cycling conditions, an increased portion of travel involves physical activity, which provides health benefits.
- Developers can benefit from having more design flexibility, including more opportunities for infill development and reduced parking costs. Also, LEMs increase the amount a household can spend on housing. It creates new markets and financing options. The New Urbanist movement promotes this type of development from an industry perspective.
- Regional economies tend to benefit when consumers shift their transportation expenditures from vehicles and fuel to transit services or general consumer goods.

There may be costs associated with higher population density in urban neighborhoods.[19] Higher density may increase congestion intensity (i.e., when people drive, they face greater congestion delay, although because they drive shorter distances and have alternative modes, they face less congestion delay per capita). Some households that choose location-efficient housing that has limited parking eventually may purchase additional motor vehicles if their needs change or they become wealthier, thus increasing local traffic and parking problems. This may require parking management. Some location-efficient housing includes resident covenants that restrict vehicle ownership. Urban infill may also cause displacement of lower-income households (gentrification).

EQUITY IMPACTS

Location-efficient housing and location-efficient mortgages tend to increase equity by allowing households that own fewer than average automobiles to avoid paying for parking they don't use, and by increasing housing options for lower-income households and nondrivers. Residential parking requirements reflect suburban, middle-class car ownership rates that are excessive for many households, particularly those with lower incomes. This is both unfair and regressive. Location-efficient development is optional, so consumers will choose it only if they consider themselves to be better off overall.

Location-efficient development and location-efficient mortgages tend to benefit lower-income households by providing financial savings and improving affordable transport and housing options.

Location-efficient development is most appropriate in urban neighborhoods that have good access (services and activities are easily available by walking and transit). It can be implemented by regional or local governments, or by not-for-profit organizations or individual businesses.

BEST PRACTICES

Here are some specific recommendations for implementing location-efficient development:

- It should include a variety of land-use and transportation features that improve access and mobility options, such as pedestrian and cycling improvements, transit improvements, and mixed land use.
- It should include a range of housing types and prices, so that people in various life-cycle stages and income classes can choose such housing.
- Parking requirements should be reduced or eliminated for location-efficient housing. Rather than being included with housing, parking should be rented separately, so that households pay only for the amount of parking they actually use.
- Parking should be managed to prevent spillover problems.

REFERENCES

1. Frank, L.; Kavage, S.; Litman, T. Promoting Public Health Through Smart Growth: Building Healthier Communities Through Transportation and Land Use Policies. Smart Growth BC (www.smartgrowth.bc.ca), 2006.
2. Litman, T. *Evaluating Public Transit Benefits and Costs*; VTPI (www.vtpi.org), 2005.
3. Frank, L.; Pivo, G. Impacts of mixed use and density on utilization of three modes of travel: SOV, transit and walking. Transportation Res. Rec. **1995**, *1466*, 44–55.
4. FHWA. Urbanized areas: selected characteristics. Highway Statistics 2004. U.S. Federal Highway Administration, (www.fhwa.dot.gov/policy/ohim/hs04/xls/htm72.xls), Table 72, 2005.
5. Moudon, A.V.; et al. *Effects of Site Design on Pedestrian Travel in Mixed Use, Medium-Density Environments*, Washington State Transportation Center, Document WA-RD 432.1, 1996. Available at: www.wsdot.wa.gov/ppsc/research/onepages/WA-RD4321.htm
6. Holtzclaw, J. Using Residential Patterns and Transit to Decrease Auto Dependence and Costs. National Resources Defense Council (www.nrdc.org), funded by the California Home Energy Efficiency Rating Systems, 1994.
7. Ewing, R.; Pendall, R.; Chen, D. Measuring Sprawl and its Impacts. Smart Growth America (www.smartgrowthamerica.org), 2002.
8. Lawton, K.T. The Urban Structure and Personal Travel: An Analysis of Portland, Oregon Data and Some National and International Data. E-Vision 2000 Conference (www.rand.org/scitech/stpi/Evision/Supplement/lawton.pdf), June 2001.
9. NPTS. National Personal Transportation Survey. U.S. Department of Transportation (www-cta.ornl.gov/cgi/npts), 1995.
10. Litman, T. *Land Use Impacts on Transportation*; VTPI (www.vtpi.org), 2006.
11. Arigoni, D. Affordable Housing and Smart Growth: Making the Connections. Subgroup on Affordable Housing, Smart Growth Network (www.smartgrowth.org), and National Neighborhood Coalition (www.neighborhoodcoalition.org), 2001.
12. Victoria Transport Policy Institute Online TDM Encyclopedia, 2005. Available at: www.vtpi.org.
13. CCAP. *Transportation Emissions Guidebook: Land Use, Transit & Travel Demand Management*; Center for Clean Air Policy, 2005. Available at: www.ccap.org/trans.htm
14. SGN. *Getting To Smart Growth: 100 Policies for Implementation*, and *Getting to Smart Growth II: 100 More Policies for Implementation*; Smart Growth Network (www.smartgrowth.org) and International City/Country Management Association (www.icma.org); 2002 and 2004.
15. Goldstein, D.B. Making Housing More Affordable by Correcting Misplaced Incentives in the Lending System. National Resources Defense Council (www.nrdc.org), 1996.
16. Hoeveler, K. Accessibility vs mobility: The location efficient mortgage. *Public Investment*. American Planning Association (www.planning.org) and Center for Neighborhood Technology (www.cnt.org/lem), 1997.
17. Russo, R. Planning for Residential Parking: A Guide For Housing Developers and Planners. Non-Profit Housing Association of Northern California (www.nonprofithousing.org) and the Berkeley Program on Housing and Urban Policy (http://urbanpolicy.berkeley.edu), 2001.
18. McCann, B. Driven to Spend. The Impact of Sprawl on Household Transportation Expenses, Surface Transportation Policy Project (www.transact.org), 2000.
19. Litman, T. *Evaluating Transportation Land Use Impacts*; VTPI (www.vtpi.org), 2004.

BIBLIOGRAPHY

1. Carman, E.; Bluestone, B.; White, E. Building on Our Heritage: A Housing Strategy for Smart Growth and Economic Development. Center for Urban and Regional Policy, Northwestern University (www.curp.neu.edu); October 2003.
2. Cervero, R. Built Environments and Mode Choice: Toward a Normative Framework. Transportation Res. D **2002**.
3. City of Orlando, Florida. Applicability of Vehicle Miles of Travel to Transporstation Planning, 1998. Available at: www.cityoforlando.net/planning/Transportation/ddc.htm
4. Ewing, R. *Transportation and Land Use Innovations; When You Cant't Build Your Way Out of Congestion*; Planners Press: Chicago, IL, 1997; [www.planning.com].
5. Friedman, B.; Gordon, S.; Peers, J. Effect of neotraditional neighborhood design on travel characteristics. Transportation Res. Rec. **1995**, *1466*, 63–70.
6. Institute for Location Efficiency (www.locationefficiency.com) is an organization that works to encourage implementation of location-efficient development.
7. Jia, W.; Wachs, M. Parking Requirements and Housing Affordability: A Case Study of San Francisco. Research Paper 380, University of California Transportation Center (www.uctc.net), 1998.
8. LGC. Creating Great Neighborhoods: Density in Your Community. Local Government Commission (www.lgc.org), U.S. Environmental Protection Agency and the National Association of Realtors, 2004. Available at: www.lgc.org/freepub/PDF/Land_Use/reports/density_manual.pdf.
9. Litman, T. *Parking Requirement Impacts on Housing Affordability*; VTPI (www.vtpi.org), 2003.
10. Nelson, A.C. et al. The Link Between Growth Management and Housing Affordability: The Academic Evidence. Brookings Institution Center on Urban and Metropolitan Policy (www.brook.edu/dybdocroot/es/urban/publications/growthmang.pdf), 2002.
11. Nelson/Nygaard Consulting. Housing Shortage/Parking Surplus. Transportation and Land Use Coalition (www.transcoalition.org/southbay/housing_study/index.html), 2002.
12. Non-Profit Housing Association (www.nonprofithousing.org) provides a variety of materials to support development of affordable housing, including location-efficient development.
13. Reconnecting America. Hidden in Plain Sight: Capturing the Demand for Housing Near Transit. Center for

Transit-Oriented Development; Reconnecting America; Federal Transit Administration (www.fta.dot.gov), 2004.
14. Regulatory Barriers Clearinghouse (www.huduser.org/rbc), created by the U.S. Department of Housing and Urban Development, was created to support state and local governments and other organizations seeking information about laws, regulations, and policies affecting the development, maintenance, improvement, availability, and cost of affordable housing.
15. Scheurer, J. Car-Free Housing in European Cities: A New Approach to Sustainable Residential Development. World Transport Policy Practice **1998**, *4* (3) [wwwistp.murdoch.edu.au/publications/projects/carfree/carfree.html].
16. The Affordable Housing Design Advisor Web site (www.designadvisor.org), sponsored by the U.S. Department of Housing and Urban Development, provides information on developing more affordable housing, redeveloping urban communities, and implementing Smart Growth.
17. The San Francisco Planning and Urban Research Association. Reducing Housing Costs by Rethinking Parking Requirements, 1998. Available at: www.spur.org.
18. West Coast Environmental Law Foundation. Smart Bylaws Guide, 2004. Available at: www.wcel.org/issues/urban/sbg.

Tribal Land and Energy Efficiency

Jackie L. Francke
Geotechnika, Inc., Longmont, Colorado, U.S.A.

Sandra B. McCardell
Current-C Energy Systems, Inc., Mills, Wyoming, U.S.A.

Abstract

As energy service companies move beyond their traditional markets, technical competence must be combined with other skills, such as an understanding of different types of institutions and the ability to work with individuals of varying backgrounds and even cultures. In the United States, Tribal communities are one such market. This paper discusses important factors beyond engineering and economics that should be considered when working with Tribal communities, and presents an overview of experiences and recommendations for successful projects. Although the focus is on Tribal communities, the conclusions drawn are also applicable to rural communities and to projects in countries other than the United States—particularly those in developing countries where cultural traditions remain strong and energy service companies are as yet uncommon.

INTRODUCTION

The traditional markets for energy efficiency services are institutional customers, educational institutions of all kinds, and government entities—all institutions with large facilities having high utility bills, significant air handling requirements, and a reasonably uniform building configuration. As one moves outside these traditional markets to those that have historically been underserved, however, life can become both more complicated and more interesting for the energy service provider.

High on the list of underserved markets are facilities in developing countries and (inside the United States) rural and Tribal facilities, which provide challenges that are technical, operational, and above all cultural. To succeed in these types of projects, a company must include cultural considerations in its project implementation. The authors, with experience in implementing projects successfully in these varied environments, present projects on Tribal lands as a model for some of the approaches needed to accomplish energy efficiency programs in these areas.

As Tribal governments and communities gain and establish control of their natural resources, economies, and communities, they are faced with the challenges of developing and maintaining not only skilled labor pools, but also Tribal infrastructure and even sustainable economies. Often, all rural communities—and Tribal communities in particular—must perform and maintain necessary community services such as law enforcement, health services, and government functions. With the limited number of people in these areas, this can mean that opportunities for economic growth fall by the wayside.

Communications technologies and widespread travel are among the trends causing the world to shrink, so that communities that have until now been considered out of the mainstream are turning to urban resources and businesses to gain more knowledge and obtain services that will introduce them to new technologies, improve economic sustainability, and empower local government officials. For service companies working with these communities, it is important to remember that technology and engineering expertise may take a back seat to knowledge, understanding, and awareness of the community's culture. Working with Tribal governments and communities presents technical challenges, of course—but more important are the challenges of cultural differences and geographic location.

ENERGY EFFICIENCY ON TRIBAL LANDS

Background

There are more than 500 federally recognized Tribes in the United States. There are reservations in almost every state, and they range in size from less than one acre to many thousands of square miles, from cities to forests to beaches to deserts. The Tribes themselves are extremely varied, with different histories, cultures, languages, religions, and traditions. The United States recognizes the right of these Tribes to self-government and supports their Tribal sovereignty and self-determination. This means that each Federally recognized Tribe possesses the right to form its own government, to develop and enforce its own laws, to

Keywords: Energy efficiency; Tribal; Rural; Culture; Community.

tax, and exclude persons from Tribal territories, among other rights and responsibilities.

As a side note, it is often believed that, with the advent of Indian Gaming Laws that allow casinos on Tribal lands, all Tribes have adopted this approach to economic development and, therefore, are highly developed and profitable enterprises in themselves. But like all generalizations, this one is incorrect. Many Tribes have resisted gaming for cultural or other reasons, and even where gaming is present, the methods of distributing economic benefits vary widely.

In 2005 and 2006, a series of energy efficiency audits was performed on residential and commercial buildings located on Tribal lands. Under a U.S. Department of Energy grant directed by the Council for Energy Resource Tribes (CERT) based in Denver, Colorado, the purposes of the energy audits were to identify energy conservation opportunities for the Tribes and to establish alliances between teams of private experts and Tribes to develop energy efficiency.

The energy audits were performed on Tribal lands located in California, New Mexico, Oklahoma, and South Dakota by a team consisting of Indian and non-Indian personnel. The team traveled to the Tribal communities and worked closely with Tribal leaders and community members to identify areas that would improve the energy efficiency of the buildings. The results and recommendations were summarized in each case in a report, with copies submitted to Tribal leaders for future implementation.

The recommendations and scenarios discussed in this paper are based on the observations and experiences drawn from the performance of these energy audits on Tribal lands. The challenges the Tribal communities face with regard to geographic location, access to information, lack of access to subject-matter experts, and cultural awareness are similar to those faced by any small rural community in the United States—although the cultural gap can be wider (Fig. 1).

Tribal Governments

Like many small rural communities, Tribal communities are often too small to be considered cities. Instead, as small towns and villages, they usually entrust government and community decisions to an elected board, council, or chapter, which may be made up of commissioners, community elders, or volunteers. Frequently, as in small towns, the positions are part time (though with full-time responsibility), and usually, a president, chairperson, or mayor functions as the chief executive officer. Naturally, in smaller communities such as these, the holding of political office is not a career path, but something one does on the side, as an obligation to the community or as a family responsibility, after achieving a good reputation in another field (Fig. 2).

Fig. 1 Cultural and geographic challenges can be enormous in the open rural areas of the West. Where distances are so vast, the sense of community becomes doubly important.
Source: From Current-C Energy Systems, Inc.

With such a small governing council or board managing a small community where few resources are available, and with both limited funding and time, the primary goal of the board usually is meeting the most critical needs of the community. These include water supply, law enforcement, fire protection, health services, and the school system. Therefore, when opportunities arise for changes that may bring economic growth (energy efficiency improvements, capacity building in the energy arena, and renewable energy projects, for example), the community may have difficulty assessing the opportunities. The first issue here is the lack of a base of shared understanding such as that developed in larger cities by, for example, public relations and advertising campaigns; followed by a lack of skilled community experts, the

Fig. 2 In Tribal communities, people, places, and other components of the world are often given more respect than is true in the urban culture. "Shiprock" on the Navajo Reservation.
Source: From Geotechnika, Inc.

inadequacy of funding, and generally insufficient or inaccurate information. These community challenges require the Tribes to rely on outside resources to fill these voids and provide community opportunities. Evidently, therefore, there is a great possibility for service providers to work with Tribal communities—but it is important first to learn to think outside the traditional "energy service company" box. When working with Tribal communities, businesses must first gain a sense of understanding of the culture, the community and its priorities, and the way that particular government works. Only in this way will any project be successful and lead to future work in the community.

Observations

Cultural Awareness

As noted in the Background section earlier in this chapter, each Tribal community is different, with its own cultural values and traditions. In the context of developing and implementing a project, these differences in turn lead to differences in decision-making, leadership styles, and other business-related characteristics. It is important to understand that what may be observed in one village may not be valued in another. One must observe, learn, respect, and acknowledge the customs and values of each Tribal community.

A few general comments can be made, however, to highlight the types of situations and sensitivities of which any company operating on Tribal lands should be aware. These are not meant to be typical, as there is no such thing, but they are indicative.

It is not uncommon, for example, when working with Tribal communities that meetings be opened with a blessing/prayer led by a community leader or elder before any introductions and certainly before the business part of the meeting can start. When introductions begin, the focus is not on one's own individual accomplishments, but on one's family, heritage, and connection to the community. This stems from the importance of humility in many Indian cultures. Each person is considered to be equal to the others and no better. Therefore, in working with small Tribal communities, it is important to be humble and to focus importance on the Tribal elders, council members, and community members. Those who present a sense of their own importance or take themselves too seriously may violate Tribal cultural values and might be considered untrustworthy by the community.

Another common issue for those who have not worked with Tribal communities is that they must also understand the culture's ways of showing respect for another person. If, during a presentation to the community, the presenter does not see the direct eye contact to which he or she is accustomed, which in urban business culture usually means that the audience is involved, he or she should not feel as though the audience is disinterested in the topic at hand. In many Native American cultures, indirect eye contact may be considered a sign of respect.

Probably the most misunderstood value of American Indian culture (and the cultures of many developing countries) is the meaning of time. For one working with Tribal communities, patience and understanding of the meaning of time are of vital importance. In many American Indian cultures, time is not measured by the clock, but by when the situation feels appropriate. In many instances, Tribal community meetings may run late into the night or may not begin until all appropriate people are present. As in any small community where relationships are more important than deadlines, the necessity of dealing with daily issues may delay Tribal leaders; their commitment to the community is paramount, however, and resolutions of any local issues must be addressed. Everywhere in the world, the rhythms of small-community life are different—not necessarily more relaxed, as the workloads of these local leaders often put those of corporate chiefs to shame—and have a different focus. It is the community that is important. Therefore, service providers should be prepared to wait, if need be, and to change their schedules based on unforeseen circumstances. On the other hand, they should not take this as an excuse to be late themselves; that would be a strong sign of disrespect.

While working on Tribal lands, it is not uncommon to encounter elders and/or Tribal members who speak their native language as their primary language. While performing the energy audits, we had the opportunity to perform an energy audit on the home of an elderly Tribal couple not fluent in English. As in all the audits, a preaudit interview with the occupants was performed to gain a better understanding of energy usage in the home. During the interview, an interpreter was utilized. This situation also provided the team the opportunity to learn a few words in the local language. Most important was "Thank you"—a phrase that was returned with a smile at the end of the interview. Such opportunities should be appreciated, as they provide insight into the culture, show interest and respect, and provide to energy efficiency work a spice and flavor that are not often found in an office building or a hospital.

These examples are neither typical nor exhaustive, of course; with so many different cultures and so many different communities, it would be impossible for them to be so. If service providers are interested, genuine, respectful, and quietly professional, and if they watch and listen for cues from those around them, cultural awareness will develop.

Sense of Community

Business culture is often modeled on the culture of a big city—fast-paced, goal-oriented, with one-dimensional relationships and a separation between professional and personal life. Rural communities, including Tribal ones, are the centers of small-town life—even now, in the 21st century. Community life is important; relationships are of long standing; decisions are often made for reasons that are difficult for those from outside the community to understand; and discussions often appear to go back generations.

To work in such places, this sense of community must be understood and appreciated, and any projects must enhance that community to be successful. In addition, the traditional small-town value of "My word is my bond" is alive and strong in small towns and Tribal lands. When working in a community that has a shared memory going back for generations, it is important for service providers to say what they will do—and do what they say. Trust, which develops slowly in communities that are inherently wary of outsiders, is critical for success.

The clan system prevalent in the American Indian culture makes the community, in essence, a circle of family. This circle provides community members a sense of family support and a strong set of relationships. When accepted into this family, an outside service provider may be expected to assist the community in areas outside the project objectives, such as locating sources of information for the Tribal communities. In a community, boundaries are fluid, and when one joins the community, those fluid boundaries surround everyone.

Long-Term Orientation

In Tribal communities, an issue related to both the sense of community and the sense of time mentioned earlier is the long-term orientation taken to decision-making and project implementation. For most service providers, the process of proposing a project, having it accepted, and then implementing it is most familiar—and preferred. In Tribal communities, that process is often broken down into multiple steps on a time scale that is defined by other constraints, such as the needs of the community, Tribal leaders' lack of time, or insufficient funding. When a Tribal leader agrees that it is a good idea to implement a project, it is tempting to assume that he or she means now, or at least within the next few months. Unfortunately, the real meaning may be (as mentioned earlier) "when the time is right"—by which time the service provider may have given up and moved on to other projects.

It is important, therefore, for each party to understand the other's time scales and the steps that must be taken to move the project along one step at a time (with the associated approximate costs of those steps).

Geographic Location and Coordination

With an audit team of six, transporting equipment and personnel to each Tribal community required significant planning and preparation for each energy audit to be effective, successful, and within budget guidelines. On average, travel entailed one full day of travel by car to reach the sites, with three days on-site performing the energy audits.

To ensure that each trip was successful, documentation on energy usage pertaining to each building was requested in advance; meetings were scheduled; and a Tribal community point of contact was established, most often a Tribal leader. Because time runs differently in most American Indian cultures, this preparation required persistence and patience, as well as understanding. The possibility presented by using energy efficiency programs to bring new technology and opportunities to the community may be overridden by other daily priorities. We discovered that where possible, it was good to include the possibility of staying one extra day, just in case the schedule was changed during the course of the audit. (We always used that extra time).

Unlike most cities, where business representatives have the capability to travel to and from clients and conduct business in one day, rural areas and Tribal locations usually must be the single focus for that day or for several days. Therefore, additional planning, patience, and persistence play an important role (Fig. 3).

Technical Issues

In a standard commercial building, a school, or a hospital, the contact person for the energy audit is likely to be an engineer, an energy manager, or a technician familiar with the equipment in that particular building. In Tribal

Fig. 3 In rural and Tribal areas, a "community" often consists of a very few buildings in large land areas.
Source: From Current-C Energy Systems, Inc.

communities, there certainly are some such individuals, but it is the norm that those responsible for energy issues, construction, and maintenance also have significant other responsibilities. Given that situation, it is simply not possible for these individuals—no matter how dedicated and professional they may be—to be familiar with the opportunities provided by energy efficiency audits and enhancements. Neither is it often appropriate to recommend the latest, fanciest equipment unless the staff will be effectively trained on it and unless it can be serviced throughout its life.

As engineers, we are trained to recommend the best alternatives. On Tribal lands and in rural communities, however, the best technical solution may not be the best all-around solution. Cultural concerns, experience levels, power availability and reliability, time available for maintenance, training needs, service availability, and many other such factors must be taken into consideration for successful project implementation.

During our audits, we found that the most useful information often came from informal discussions, rather than from our carefully constructed audit forms and questionnaires. We also found that Tribal expertise resided in surprising areas (because many Tribal officials and personnel perform multiple roles), and in our implementation plans, we worked to use that expertise for training, project management, and other activities.

Access to Information and Administrative Notes

Unlike most urban communities, small Tribal communities may not have access to digital Internet services, but some may utilize dialup technology. Because of this, additional effort and time should be allocated in transferring information to the Tribal communities. Paper copies may need to be transmitted by mail or fax, depending on the content.

With this limited Internet access, most small communities are at a disadvantage when conducting research on such mainstays of business life as project opportunities and funding information. As an outside resource, a service business may add value by becoming the portal to that Internet-based information for Tribal communities.

During the energy audits, each visit was initiated with an on-site meeting. The meeting was conducted not only to build interest in and commitment to the audit, but also to convey and disseminate information regarding the energy efficiency audits and any new technologies available to the community. Although information was initially transmitted during the pretrip planning, in some instances, the point of contact did not receive the information or did not have the opportunity to review it due to other community concerns of higher priority. In addition, the meeting provided the community leaders the opportunity to discuss concerns and to question auditors on areas that they might be able to support during their visits (Fig. 4).

Therefore, a business expecting to work with Tribal communities should not rely on Internet access and should be understanding of Tribal leaders who may not consider the project to be a priority. Because of the limited time and higher priorities that come with being a Tribal leader, transferred information should be summarized and outlined, followed by contact with the Tribal officials to ensure that any questions, comments, or concerns he or she may have are addressed.

Funding Differences

With a typical energy service project in a city or a town, it is easy to determine who owns the facility, who is responsible for its construction and maintenance, and who would derive benefit from any energy efficiency measures installed. With buildings on Tribal lands, however, the picture is not so clear.

Ownership is a concept that is perceived differently among Tribes, with shared ownership by the entire Tribe, a family, or another group being the norm rather than the exception. In addition, because the laws on Tribal lands are often different from those off the reservations, banks and other financing institutions have historically been uncomfortable with the collateral or the foreclosure procedures required. The result has been that many facilities on reservations and other Tribal lands are inadequate and in

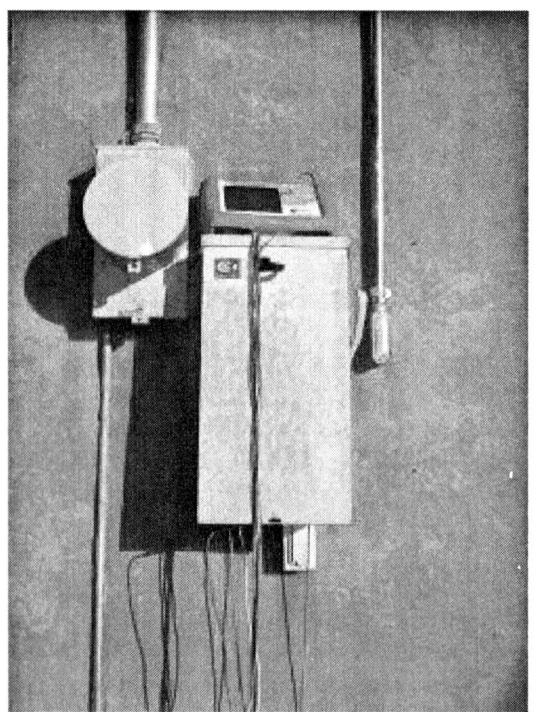

Fig. 4 Tradition meets technology: a power-quality meter determines that the wiring in a traditional adobe structure is inadequate and that the transformer is likely undersized.
Source: From Current-C Energy Systems, Inc.

need of substantive repair or replacement. From an energy efficiency point of view, that presents an opportunity—if funds can be secured.

Often, the first place to look for funds that can be used to implement energy efficiency projects on Tribal lands is the federal government, which under its trust responsibility and as part of its current mandate is encouraging such projects. In addition, several states are focusing on increasing energy efficiency on the Tribal lands within their borders.

There are other instances in which funds are available for either energy efficiency or renewable energy programs from internal Tribal funds produced by casinos or other economic development activities, but government grants and loans form the largest proportion of funds available for such projects. It is important, therefore, for any service provider interested in working with Tribal entities to be familiar with the process of obtaining these grants to move projects forward.

Teamwork and Humility

If one were to develop an equation for the proportion of a successful Tribal project that is related to technical and engineering excellence, and the proportion that is attributable to the development of a project appropriate to the local culture, the importance of technical merit would likely be well under one-third of the project.

Service providers must understand, therefore, that in projects developed on Tribal lands or for rural areas (particularly those in developing countries), as possibly nowhere else, project design and development cannot succeed unless they are true team efforts, in which the service provider/consultant listens to and follows the lead of local Tribal leaders. Technical excellence on its own will go nowhere; only a project that meets the needs (technical, cultural, economic, environmental, and political) of the Tribe and its leaders, and that can be implemented within local constraints, can be accepted and successful.

For this team effort to be possible, service providers must leave the "expert" mantles they are accustomed to wearing in the urban business environment and work hard to be team members, catalysts, and learners—not experts. That requires a fair amount of both flexibility and humility.

One concrete example is that introductions should not focus on one's career accomplishments, but on one's family, heritage, and community. Focusing on one's accomplishments may be interpreted as showing that the presenter believes himself or herself to be better than the audience. In addition, initiating one's introduction with a welcome in the native Tribal language may be interpreted as a sign of respect but should be carried out only after relationships have been established. A sense of humility is important in becoming a part of the community.

Project Success

It should be evident that although the technology and economic opportunity in a recommended project may be unsurpassed, the success of any project on Tribal lands is very much dependent on a number of factors unrelated to engineering and economics. One must be persistent, humble, understanding, and giving (Fig. 5).

Tribal meetings should be initiated with protocols and processes to which community members are accustomed. This may include saying blessings/prayers, contacting a Tribal elder for sponsorship or introductions, working with Tribal leaders, and presenting introductions in a manner consistent with Tribal values. In addition, one must be prepared to present to the community meetings at a time deemed appropriate by the meeting coordinator, rather than at a time measured on the clock. This requires patience, as one may not be called on for a period of time.

Over time, when trust and relationships have been established, one becomes a part of the community. In this capacity, service providers then should be willing to support the community in areas that may not pertain to the project. In any small community, the lines that urban business culture draws around a project do not exist, and if one is part of a community, one helps as one can. Also, in most small communities that lack access to information, one may become a portal of information for the community by informing Tribal leaders of upcoming opportunities, new technologies, and additional information related to community needs or goals.

As stated previously, small Tribal communities are disadvantaged (in terms of access to services) by their locations. In working with these communities, one must be well prepared logistically and highly organized before

Fig. 5 Projects and technologies should be appropriate and meet needs, which may call for innovative and creative approaches. There are great renewable energy resources on Tribal lands, but their development sometimes requires different thinking about organizational structure, financing, management, and training.
Source: From Geotechnika, Inc.

starting a project; it is often not possible to just call up a supply house and ask it to make a delivery. For a project to proceed well, persistent (but polite) phone calls, faxes, and emails may be required to ensure that key players are available for support and that information has been transmitted to the appropriate contacts prior to on-site visits.

Because some small Tribal communities are tight–knit circles of families, the success of any project requires community involvement. The key players in the community may dictate the processes and goals of the project, but community Tribal members need to be motivated and informed of the benefits to themselves and to the community. Community members must feel that they are part of the decision-making process, which can be achieved by disseminating information at community meetings and events. If the community members are motivated, Tribal leaders will become interested and make the project a priority.

CONCLUSION

As our audit teams discovered, the success of any project in a small Tribal community is very much dependent on factors other than engineering and economics.

While providing services to small Tribal communities, the authors learned much about the challenges that Tribal leaders and community members face on a daily basis. Location and access to information prove to be challenges when seeking out services required by the community. Services provided by outside organizations are critical to the viability of small Tribal communities, because they provide assistance in introducing new technologies and in fostering economic sustainability. This is particularly true in remote areas, because subject-matter experts may not be available in the area.

As a result, businesses seeking to work with small Tribal communities should be prepared to become part of the community, learning and understanding their values and customs. They should become culturally aware of the community in which they want to work, and they should learn what is important to the members and Tribal leaders (Fig. 6).

Service providers should understand the concerns and challenges of the community and support Tribal leaders by becoming a portal of information leading to interesting opportunities, new technologies, and other pertinent information. They should also empathize with community leaders and members by lending support where needed.

Overall, and most important, one must be persistent, patient, and humble for the project to succeed. As stated earlier, the technology and opportunity may be unsurpassed from the standpoint of engineering and economics, but if humility, understanding, empathy, and cultural awareness do not exist, neither will the project.

Fig. 6 Tradition and a sense of community are values that are becoming more respected in mainstream culture as well. In some Tribal areas where the cultural tradition is remaining in one place rather than practicing a nomadic lifestyle, traditional buildings are very well adapted to the climate and the area, and in fact are being studied by some architects as models for the architecture of the future.
Source: From Geotechnika, Inc.

About the Authors

Jackie Francke is a Tribal member of the Navajo Nation Tribe, a mining engineer, and president and principal engineer of Geotechnika, Inc. Francke has more than 15 years of experience working in the field of instrumentation, monitoring, and data collection on projects related to energy efficiency, mining, civil, and environmental engineering. Having grown up on the reservation, Francke understands the challenges Tribes face on a daily basis, and hopes to educate and assist businesses that are looking to work with Tribal communities.

Sandra McCardell is president and founder of Current-C Energy Systems, a small woman-owned firm specializing in efficient and sustainable energy applications in institutional, industrial, and community settings. Her experience over 25 years has ranged from consumer-products companies to international development to real estate to engineering. With a strong background in business and significant experience working in other countries and cultures, she particularly enjoys her projects with Tribal entities.

Underfloor Air Distribution (UFAD)

Tom Webster
Fred Bauman
Center for the Built Environment, University of California, Berkeley, California, U.S.A.

Abstract
The purpose of this paper is to provide an overview of the principles, features, benefits, and limitations of the building conditioning technology called underfloor air distribution (UFAD) and the closely related task/ambient conditioning (TAC).

INTRODUCTION AND BACKGROUND

Recent trends in today's office environment make it increasingly more difficult for conventional centralized heating, ventilation, and air-conditioning (HVAC) systems to satisfy the environmental preferences of individual office workers using the standardized approach of providing a single uniform thermal and ventilation environment. Since its original introduction in West Germany during the 1950s, the open plan office, containing modular workstation furniture and partitions, is now the norm. Thermostatically controlled zones in open plan offices typically encompass relatively large numbers of workstations in which a diverse work population having a wide range of preferred temperatures must be accommodated. Modern office buildings are also being impacted by a large influx of heat-generating equipment (e.g., computers and printers) whose loads may vary considerably from workstation to workstation. Offices are often reconfigured during the building's lifetime to respond to changing tenant needs, affecting the distribution of within-space loads and the ventilation pathways among and over office partitions. Compounding this problem, there has been growing awareness of the importance of the comfort, health, and productivity of individual office workers, giving rise to an increased demand among employers and employees for a high-quality work environment.

Underfloor Air Distribution

In the 1970s, underfloor air distribution (UFAD) was introduced into office buildings in West Germany as a solution to cable management and heat load removal issues caused by the proliferation of electronic equipment throughout the office.[13] In these buildings, the comfort of the office workers had to be considered, giving rise to the development of occupant-controlled localized supply diffusers to provide task conditioning. Some of the first UFAD systems in Europe used a combination of desktop outlets for personal comfort control and floor diffusers for ambient space control.[12]

Prior to the 1990s, office installations using underfloor systems were found primarily in South Africa, Germany, and other parts of Europe. The technology was not commonly used in North America before about 1995, in part due to the downturn in office building construction beginning in the mid-1980s. Japan did not experience this same downturn, and as a result, significant growth in UFAD technology was observed during this period. Between 1987 and 1995, more than 250,000 m^2 (2.7 million ft^2) of office space in more than 90 buildings was installed with UFAD systems in Japan.[14]

However, in the late 1990s, growth for raised floor installations in the United States was dramatic, and designers and manufacturers predicted that 35% of new offices would use raised floors by 2004. Half of these installations were expected to incorporate UFAD technology. This rate of increase slowed in 2003 due to the economic downturn and reduced office construction, but has revived since then. Installation data shows that since 2005, 12% of new commercial offices used raised floors, and 45% of those projects used UFAD systems.[11] We estimate that as of 2006, at least 400 UFAD projects have been built in North America.

Task/Ambient Conditioning (TAC)

During recent years, an increasing amount of attention has been paid to air distribution systems that individually condition the immediate environments of office workers within their workstations to address the issues outlined above. As with task/ambient lighting systems, the controls for the task components of TAC systems are partially or

---

[*] A version of this report was originally published by the U.S. Department of Energy, Office of Federal Energy Management, under the title *Technology Focus: Alternative Air Conditioning Technologies: Underfloor Air Distribution (UFAD)*, DOE/EE-0295, 2004.

Keywords: Constant air volume; Stratification; Swirl diffuser; Task ambient condition; Underfloor air distribution; Underfloor plenum; Variable air volume.

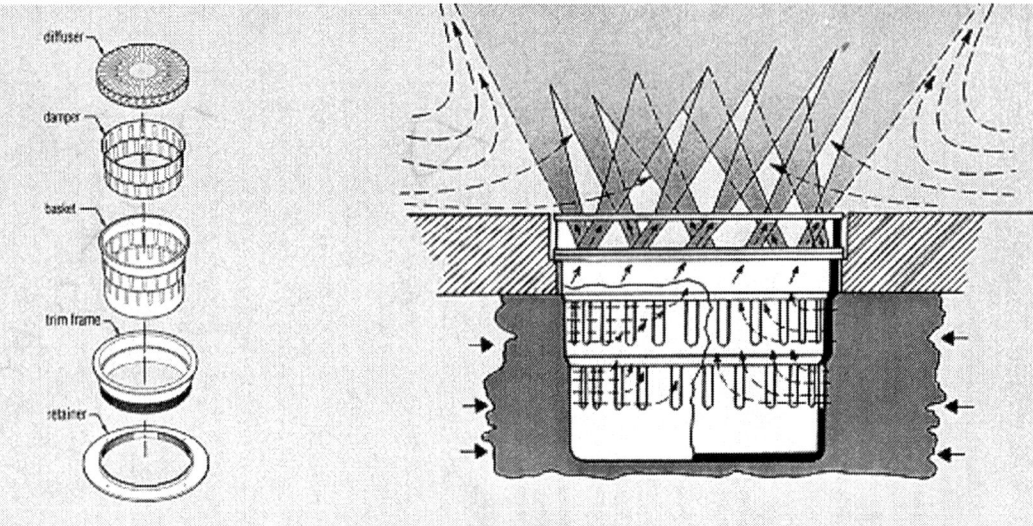

Fig. 1 Illustration of typical swirl diffuser used in underfloor air distribution (UFAD) systems.

entirely decentralized and under the control of the occupants. Typically, the occupant has control over the speed and direction, and in some cases the temperature, of the incoming air supply. Variously called "task/ambient conditioning," "localized thermal distribution," and "personalized air conditioning," these systems have been mostly installed in open-plan office buildings to which they provide supply air and (in some cases) radiant heating directly into workstations.

TECHNOLOGY DESCRIPTION

UFAD Systems

Underfloor air distribution systems use the open space (underfloor plenum, also used for distribution of other services such as power and computer cabling) between a structural slab and the underside of a raised floor system to deliver conditioned air to supply diffusers (Fig. 1) located at or near floor level within the occupied zone (up to 6-ft [1.8-m] height) of the space. The supply diffusers can provide some degree of individual control over the local thermal environment, depending on diffuser design and location. Additional supply diffusers provide ambient environmental control in nonwork areas. Active diffusers (for purposes of this paper) are defined as those with local means of volume adjustment (such as an integral variable speed fan or damper) that is amenable to automatic zone control (in addition to means for occupant control). Passive diffusers, although they may have means for occupant adjustment, are combined with terminal or system elements to achieve zone control. Systems designed with all fan-assisted active diffusers typically use zero-pressure plenums. Passive diffusers require pressurized plenums. The majority of UFAD systems currently being deployed have pressurized plenums with either active or passive diffusers.

TAC Systems

Task/ambient conditioning systems can be distinguished from standard UFAD systems by their higher degree of personal comfort control provided by the localized supply outlets. Task/ambient conditioning supply outlets use direct velocity cooling to achieve this level of control, and are therefore most commonly configured as fan-driven (active), jet-type diffusers that are part of the furniture or partitions or from floor outlets. A majority of TAC systems deployed have used UFAD to distribute conditioned air to fan-driven diffusers located in workstation furniture and partitions. Fig. 2 shows an illustration of such a TAC system.

For further information on a complete range of TAC systems, refer to Bauman et al. and Loftness et al.[6,10] Because few TAC systems are currently (as of 2006) being deployed, the remainder of this report will be focused only on UFAD systems.

PRINCIPLES OF OPERATION OF UFAD SYSTEMS

Figs. 3 and 4 illustrate the fundamental differences between traditional overhead and UFAD systems, respectively. As shown in Fig. 3, overhead systems (in office buildings, these are predominantly variable air volume (VAV) all-air distribution systems) employ an extensive array of ductwork and terminal devices to provide supply air through the ceiling-mounted diffusers. Often referred to

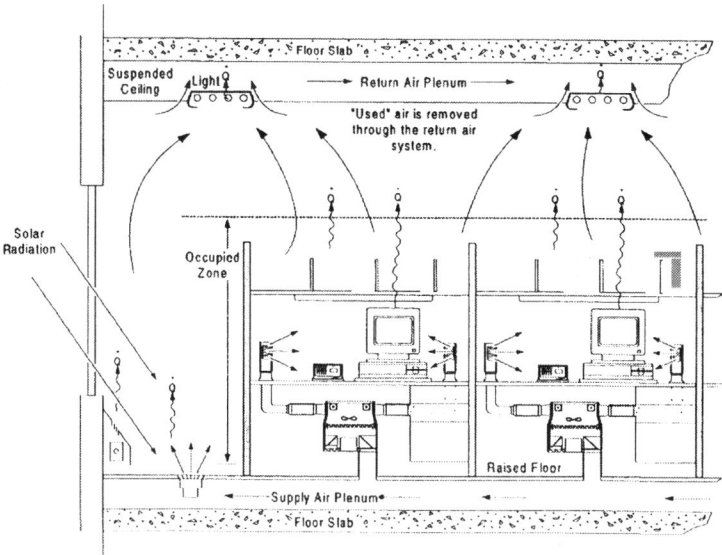

Fig. 2 Task/ambient conditioning (TAC) system.

as mixing ventilation systems, these systems are designed to promote complete mixing of supply air with room air, thereby maintaining the entire volume of air in the space at the desired temperature setpoint. Space air is typically returned to the AHU through an open ceiling plenum that also contains various other systems for lighting, electrical, communications, and fire protection.

Underfloor air distribution systems turn this concept upside down and have the following characteristics:

- Supply air, including at least the minimum required volume of outside air, is filtered and conditioned to the required temperature and humidity by a conventional AHU and passed through a minimum amount of ductwork to an underfloor plenum. The underfloor plenum is formed by installation of a raised floor system, typically consisting of 0.6×0.6 m^2 (2×2 ft^2) of concrete-filled steel floor panels positioned 0.3–0.46 m^2 (12–18 in.) above the concrete structural slab of the building. The raised floor system also allows all cable services, such as power and communication, to be located in the plenum and provides easy access for modifications and maintenance.
- Individual office workers can control their local thermal environment to some extent (typically by adjusting the volume and/or trajectory of the supply air entering the space), giving them the opportunity to fine-tune the thermal conditions in their workstation to

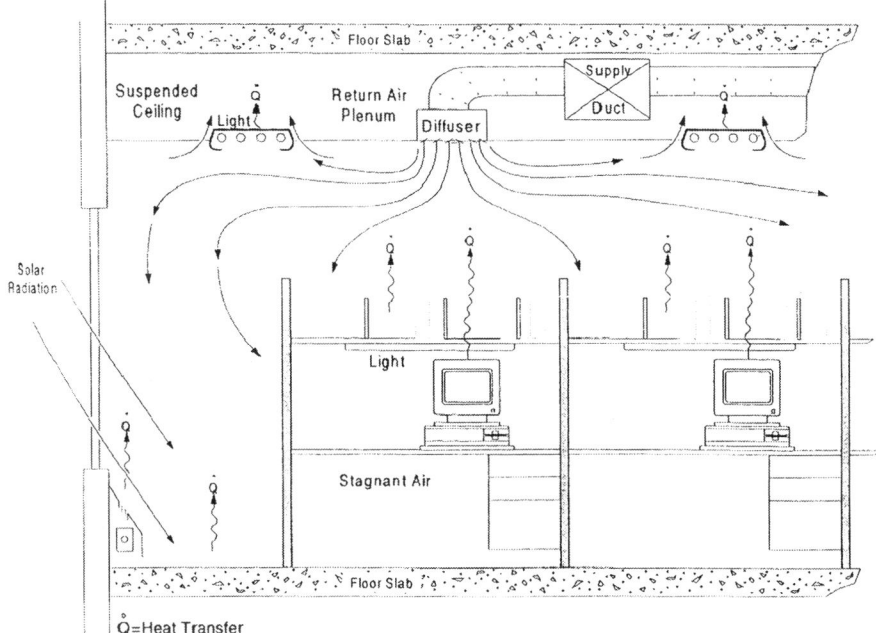

Fig. 3 Overhead system.

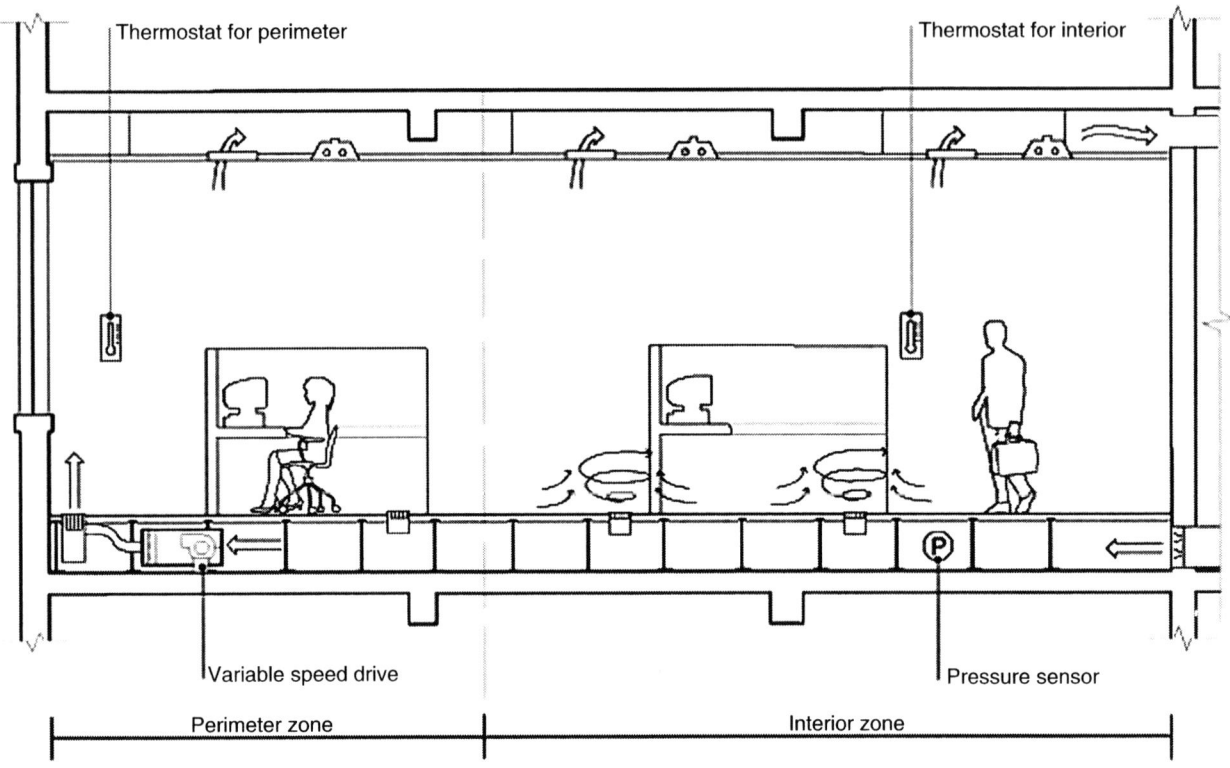

Fig. 4 Pressurized plenum underfloor air distribution (UFAD) system.

their personal comfort preferences. Different supply outlet configurations may be used depending on the conditioning requirements for a particular zone of the building, as discussed below.

- Underfloor air distribution systems benefit from a floor-to-ceiling airflow pattern that takes advantage of the natural buoyancy produced by heat sources in the office to efficiently remove heat loads and contaminants from the space. Air is returned from the room at ceiling level through recessed lighting fixtures and return grilles through a conventional ceiling return plenum or return grilles located high in the space when a return plenum is not used.
- Because the air is supplied directly into the occupied zone (up to 6-ft [1.8-m] height), supply outlet temperatures are generally maintained above 17°C–18°C (63°F–64°F) to avoid uncomfortably cool conditions for the nearby occupants and to minimize cool temperature near the floor.

There are a wide variety of approaches being used to provide a combination of individual and automatic zone control for UFAD systems.[2] The authors have identified over a dozen variations of these systems. Typically, these systems use VAV or constant-air-volume (CAV) methods for general zone control (i.e., overall zone control other than local occupant control).

- Six types of diffusers are currently being offered:
  — Fan-assisted, active
  — Variable area, active
  — Swirl, passive
  — Swirl, active
  — Linear bar grille, passive
  — Linear bar grille, active
- The heating and cooling loads of perimeter zones are handled by fan-powered constant volume or variable volume terminal reheat units located in the underfloor plenum (similar to those shown in Fig. 4). Passive or active diffusers are located in occupied areas of the zone, normally within about 4.5 m (15 ft) from the exterior walls. Passive diffusers are generally supplied by a (series) fan-powered mixing box or fan coil unit, or a VAV box either connected to the diffuser by ducting or by supplying air to a partitioned area of the plenum where the diffusers are located.
- Interior zones are generally large zones each controlled by one thermostat, but with diffusers located near the occupants within or close to their workstations.

Fig. 4 shows a schematic diagram and Fig. 5 shows a typical swirl diffuser layout for a pressurized plenum UFAD system.

This system is being commonly applied to office buildings due to its simplicity and cost savings. Although

Fig. 5 Typical interior swirl diffuser layout for underfloor air distribution (UFAD) systems.

this floor-based air distribution system provides somewhat limited individual comfort control for occupants, it still retains the same flexibility and potential energy-saving benefits associated with the others.

POTENTIAL OF UFAD APPLICATIONS

Benefits

Improved Thermal Comfort for Individual Occupants

Occupant thermal comfort is perhaps the area of greatest potential improvement in that UFAD systems potentially can accommodate individual differences. In today's work environment, there can be significant variations in individual comfort preferences due to differences in clothing and activity level (metabolic rate), as well as differences in the local heat gains and losses. By allowing personal control of the local thermal environment, UFAD systems could potentially satisfy virtually all occupants, including those out of thermal equilibrium with their surrounding ambient environment, as compared with the 80% satisfaction quota targeted in practice by existing thermal comfort standards.[1]

Improved Air Movement and Ventilation Effectiveness; Cleaner Environment

Some amount of improvement over conventional uniformly mixed systems is expected by delivering the fresh supply air near the occupant and at the floor.

Improved Occupant Satisfaction and Increased Worker Productivity

Underfloor air distribution systems have the potential to increase the satisfaction and productivity of occupants, because occupants have the ability to individually control their workspace environments. The financial implications of such improvements can be extremely large, as salary costs typically make up at least 90% of all costs (including construction, operation, and maintenance) over the lifetime of a building.

Energy Savings

Energy savings over conventional overhead systems are predominately associated with two factors: (1) cooling energy savings from economizer operation at higher supply air temperatures and (2) fan energy savings due to reduced static pressure requirements and potentially lower airflow requirements due to optimized stratification. Additional details of how these savings are achieved can be found in Bauman and Bauman and Lehrer.[2,3]

LIMITATIONS

Among the items that limit the widespread acceptance and application of UFAD technology are the following.

New and Unfamiliar Technology

For the majority of U.S. building owners, developers, architects, engineers, and equipment manufacturers, UFAD systems still represent a relatively new and unfamiliar technology. The decision to select a UFAD system will initially require changes in common practice, including new procedures and skills in the design, construction, and operation of such systems, as well as changes in responsibilities of the various installation trades. This situation creates some amount of perceived risk to designers and building owners. Significant progress has been made in these practices over the past several years.

Perceived Higher Costs

Many designers are concerned about the higher first costs of the raised flooring required for UFAD systems. However, as described above, there are many factors associated with raised floor systems that contribute to reducing life-cycle costs in comparison to traditional air distribution systems. (See the "Cost Effectiveness" section below). In a recent study of UFAD costs, we found that there are several ways that UFAD systems can be tailored to reduce the cost differential. Furthermore, the difference depends significantly on the quality of the traditional VAV system being used for comparison.[16]

Limited Applicability to Retrofits and Certain Building Types/Areas

The installation of UFAD systems and the advantages that they offer are most easily achieved in new construction. Some of the key system features are not always suitable for retrofit applications (e.g., access floors cannot be installed in existing buildings with limited floor-to-floor heights). Although widely applicable, there are some building types and areas within buildings where access floors and UFAD are not appropriate. These areas are generally those in which spillage has the potential to occur, including bathrooms, laboratories, cafeterias, and shop areas.

Lack of Information, Design Guidelines, and Evaluation Methods

Although in recent years there has been an increased number of publications on UFAD technology, only very recently has there been effort aimed at providing a complete understanding of the underlying fundamental fluid mechanics and thermal issues to allow for the creation of design and simulation tools. New guidelines and research results are presented in several sources.[4,5,9,15,17] System commissioning procedures, operating sequences, and control techniques are likewise under development.

Potential for Higher Building Energy Use

As with any space conditioning system, a poorly designed and operated UFAD system has the potential to use more energy than that of a well-designed conventional system. Other factors that can influence the energy comparison with conventional systems are number of perimeter fan-powered units, supply plenum thermal decay in UFAD systems, and airflow requirements.

Thermal Discomfort

Underfloor air distribution systems are perceived by some to produce cold floor, and because of the close proximity of supply outlets to the occupants, the increased possibility of excessive draft exists. However, we are finding that thermal comfort problems are more likely to be associated with inadequate control and operating strategies and inexperience of operators with UFAD technology.

Plenum Leakage

There have been reports of problems with leakage from the supply plenums either into the space (through floor penetrations and floor panel leakage) or through bypassing the occupied space to the return or outside the building altogether. This latter type of leakage, if excessive, is a serious problem, because it causes increased fan energy use. This is probably the single most serious problem found upon start-up of UFAD projects to date and can be traced to poor design and construction practices related to the leakage sources.

Virtually all of the issues listed above are actively being researched (refer to References) or addressed by design and construction professionals and equipment vendors in response to market demand. Some issues suggested by critics of UFAD technology have turned out not to be problems, and many of the problems that do occur can be traced to inadequate design, construction, and operating practices that have resulted from lack of knowledge and experience. Problems like these are typical for early installations of any nascent technology.

COST EFFECTIVENESS

Cost considerations will be different depending on whether the installation represents new or retrofit construction. Total first costs (shell and core plus tenant improvement) for UFAD systems using raised flooring will likely be somewhat higher than those for a conventional system. Preliminary results from research studies[16] have shown total building first costs of pressurized UFAD systems in new construction (including raised floor and structural differences) to be about $3.50 per gross square foot (gsf) greater than typical practice conventional VAV systems. However, if a raised floor system has already been justified for other reasons, such as improved cable management, or a high quality conventional system is the standard of comparison, the cost premium is effectively eliminated altogether, and under some circumstances, there can be savings in the range ~$2–7/gsf. In new construction, UFAD can lead to reducing floor-to-floor heights, thus reducing structural costs. Furniture-based and active diffuser-based systems will generally cost more than other solutions.

Operating costs can be reduced in accordance with the energy-saving strategies discussed above. With the improved thermal comfort and individual control provided by UFAD systems, occupant complaints requiring

Table 1 Example churn rates

Group	Churn (%)	Office plan (% office/open/bullpen)
Services	37	35/55/10
Manufacturing	40	34/58/8
Institutional/fovernment	23	67/20/13

response by facility staff can be minimized. Underfloor air distribution systems using raised flooring provide maximum flexibility and significantly lower costs associated with reconfiguring building services (when changes are being made in the office layout) due to churn, and thus reduce life-cycle costs substantially.

First cost for retrofits, generally the bulk of construction activity, is most likely greater than those for new construction. Also, as indicated previously, some facilities are not amenable to retrofit by UFAD systems.

In order to determine total life-cycle cost differences between UFAD systems and conventional designs, the following factors must be considered in addition to first cost:

- *Churn.* This is the cost associated with relocating personnel and it is defined as the ratio of total workplace moves in a year to the total number of building occupants. These figures vary widely by industry type and building activities. Results of a study by IFMA[8] are shown in Table 1.

  As shown, government facilities have significantly lower churn than other industries, which is also reflected in the much lower percentage of open plan space. This indicates that the benefits of reduced churn costs in federal facilities may be limited. It should be noted that the cost of churn can vary considerably depending on the extent of the reconfiguration; i.e., simply moving to a new cubicle is much different that reconfiguring the layout of cubicles or offices. The high proportion of private offices in government facilities could drive these costs significantly if the reconfiguration involves more than office to office moves.

- *Operations and maintenance (O&M).* This item includes the costs of maintenance and repair as well as energy. While energy savings estimates are limited due to the lack of appropriate capabilities in energy simulation programs and data from monitored projects, indications are that savings in annual HVAC system energy can be in the range of 0%–20% depending on system design and weather conditions. Maintenance costs are expected to be less than conventional systems due to the ease of access to the distribution system. However, commissioning/startup costs may be greater because the location and operation of the diffusers may require fine-tuning to optimize the occupant interaction benefits.

- *Productivity and health.* The savings associated with productivity and health benefits are difficult to measure and require considerably more research. However, recent studies[7] indicate that work performance improvements of 0.5%–5% may be possible if the indoor environmental quality is improved.

SUMMARY AND CONCLUSIONS

Underfloor air distribution systems have significant potential advantages compared with traditional VAV systems. Rarely has there been a space conditioning technology that promises the combined benefits of improvements in thermal comfort, energy efficiency, and productivity and health. Underfloor air distribution technology, like all nascent technologies, is being advanced both in theory and practice by researchers, designers, manufacturers, and early adopter owners who are working to bring the design, operation, and costs to the point where they can be more easily and reliably applied. Underfloor air distribution technology may someday displace overhead VAV as the system of choice for space conditioning. Although the use of UFAD systems in particular is becoming more common in the private commercial sector, the overall potential for UFAD in federal facilities may be limited by low churn that reduces life-cycle cost benefits. In addition, the overall federal building stock is not as amenable to UFAD installations as the private sector, due to the higher cost of retrofits (i.e., a greater ratio of fixed private offices).

However, in those situations where it is appropriate, there are many compelling reasons to consider UFAD for the space conditioning solution.

ACKNOWLEDGMENTS

Much of the material for this report was derived from work done at the Center for the Built Environment (CBE) at the University of California, Berkeley. The authors would like to gratefully acknowledge the support CBE and its industry partners, CBE sponsor the National Science Foundation (NSF), the Regents of the University of California, and the DOE/FEMP New Technology Development Program for providing funding for this report. CBE's current Industry Partners include: Armstrong World Industries; Arup; California Department of General Services; California Energy Commission; Charles M. Salter Associates Inc.; Flack & Kurtz Inc.; HOK; Pacific Gas & Electric Company; Price Industries; RTKL Associates Inc.; Skidmore; Owings & Merrill, LLP;

Stantec; Steelcase, Inc.; Syska Hennessy Group; Tate Access Floors, Inc.; Taylor Team: CTG Energetics, Guttmann & Blaevoet, Southland Industries, Swinerton Builders, Taylor Engineering; Trane; U.S. Department of Energy; U.S. General Services Administration; Webcor Builders; York International Corporation.

REFERENCES

1. ASHRAE. ANSI/ASHRAE Standard 55-2004. In *Thermal Environmental Conditions for Human Occupancy*; American Society of Heating, Refrigerating, and Air-Conditioning Engineers, Inc.: Atlanta, 2004.
2. Bauman, F. *Underfloor Air Distribution (UFAD) Design Guide*; ASHRAE, American Society of Heating, Refrigerating, and Air-Conditioning Engineers: Atlanta, 2003.
3. Bauman, F.; Lehrer. D., [Underfloor Technology Website]. Center for the Built Environment (CBE), Available from: http://www.cbe.berkeley.edu/underfloorair/, 2006.
4. Bauman, F.; Jin, H.; Webster, T. Heat transfer pathways in underfloor air distribution (UFAD) systems. ASHRAE Trans. **2006**, *112* (Part 2).
5. Bauman, F.; Webster, T.; Jin, H. Design guidelines for underfloor air supply plenums. HPAC Eng. **2006**, *July*.
6. Bauman, F.S.; Arens, E.A. *Task/Ambient Conditioning Systems: Engineering and Application Guidelines*; Center for Environmental Design Research, University of California: Berkeley, 1996.
7. Fisk, W.J. Health and Productivity Gains from Better Indoor Environments and Their Relationship with Building Energy Efficiency, LBNL-45484, Lawrence Berkeley National Laboratory, July 31, 2000.
8. IFMA. *Benchmarking*; International Facilities Management Association: Atlanta, GA, 1997.
9. Jin, H.; Bauman, F.; Webster, T. Testing and modeling of underfloor air supply plenums. ASHRAE Trans. **2006**, *112* (Part 2).
10. Loftness, V.; Brahme, R.; Mondazzi, M.; Vineyard, E.; MacDonald, M. Energy savings potential of flexible and adaptive HVAC distribution systems for office buildings—Final report. Air Conditioning and Refrigeration Technology Institute 21-CR Research Project 605-30030, June. Full report available at www.arti-21cr.org/research/completed/index.html, 2002.
11. Reynolds, B. Personal communications, Product manager, Tate Access Floors, Inc.: Jessup, MD, 2006.
12. Sodec, F. Air Distribution Systems Report No. 3554A, Krantz GmbH & Co.: Aachen, West Germany, September 19, 1984.
13. Sodec, F.; Craig, R. The underfloor air supply system—the European experience. ASHRAE Trans. **1990**, *96* (Part 2).
14. Tanabe, S. *Task/Ambient Conditioning Systems in Japan*, presented at Workshop on task/ambient conditioning systems in commercial buildings, San Francisco, May 4–5, 1995.
15. Webster, T.; Bauman, F. Design guidelines for stratification in underfloor air distribution (UFAD) systems. HPAC Eng. **2006**, *June*.
16. Webster, T.; Benedek, C.; Bauman, F. *Underfloor Air Distributsion (UFAD) Cost Study: Analysis of First Cost Tradeoffs in UFAD Systems*, Center for the Built Environment, University of California, Berkeley, September 2006.
17. Webster, T.; Bauman, F.; Reese, J.; Shi, M. *Thermal Stratification Performance of Underfloor Air Distribution (UFAD) Systems*, Proceedings, Indoor Air 2002, Monterey, CA, June 2002.

BIBLIOGRAPHY

1. Bauman, F.; Webster, T. Outlook for underfloor air distribution. ASHRAE J. **2001**, *43* (6), 18–27.

Utilities and Energy Suppliers: Bill Analysis

Neil Petchers
NORESCO Energy Consulting, Shelton, Connecticut, U.S.A.

Abstract

Once the energy rate structures for a customer have been examined and been understood, the next step in understanding how utility costs are determined is to perform a utility bill analysis. Knowing how a customer is charged for the energy it uses each month is an important piece of the overall process of energy management at a facility. This article discusses how electric bills and gas bills for large commercial, industrial, and institutional customers are calculated. Average costs, peak load costs, and time-of-use (TOU) costs are presented and evaluated. Electric costs are found as demand costs per kilowatt per month, and energy costs per kilowatt hour. The utility cost data are used initially to analyze potential energy savings opportunities, and will ultimately influence which of these opportunities are recommended.

INTRODUCTION

The ability to calculate utility bill impacts is fundamental to the evaluation of any energy system. An effective method for determining the energy operating cost impact of a proposed option is to compare the calculated annual utility bills with and without the option in place. The difference represents the energy operating cost impact associated with the option. To do so, one must carefully identify not only the aggregate change in fuel or electricity usage associated with the proposed option, but when the changes occur with respect to the various time periods and other billing factors integral to the rate schedule.

The average unit cost for utilities defined as the total annual cost divided by the annual consumption, is a tempting simplification of utility rate structures. It allows easy calculation of energy cost impact from changes in energy usage patterns. In some cases, this simplistic approach yields accurate answers. More typically, however, application of average unit costs gives results that are inaccurate and can, at times, be very misleading.

For any proposed technology application that would change the number of fuel or electricity units consumed, the cost impact can rarely be accurately determined by merely multiplying the change in units by the average cost per unit for the total facility usage. This is because incremental costs are usually quite different from average costs. With most currently available utility rates, the addition or subtraction of usage during peak periods may have several times the cost impact as the addition or subtraction of the same amount of usage in off-peak periods. Therefore, one must consider the weighted average cost of the increase or decrease in consumption units associated with a proposed application.

CALCULATING UTILITY BILLS

Utility bills can typically be broken down into the following basic components:

- A customer or minimum service charge
- An energy or commodity charge
- A demand or maximum level of service charge
- Power factor penalties
- Adjustments such as taxes levied by state, county, and city authorities
- Surcharges or credits associated with specific orders established by various regulatory authorities
- Fuel cost adjustments that reconcile the actual cost of fuel used or delivered by the utility, with the estimated cost used in the most recent rate proceeding to set the energy or commodity charge.

These basic components are expressed and calculated in many ways using various units of measure. Combined, they comprise the total utility bill with each contributing in different ways to the weighted average cost. To calculate a utility bill, one must carefully read the rate tariff inclusive of all rate riders and adjustment clauses. One must also know the current values for items that vary such as fuel adjustment charges. Based on this information, one should be able to calculate the utility bill exactly. If such calculations

---

[*] This entry originally appeared as "Utility Bill Analysis" (Chapter 22) in *Combined Heating, Cooling & Power Handbook*, by Neil Petchers. Published by Marcel Dekker, Inc., New York, NY, 2003; reprinted with permission from AEE/Fairmont Press.

Keywords: Utility bill analysis; Utility cost analysis; Utility bills; Electric bills; Gas bills; Electric costs; Gas costs; Energy bills; Energy costs; Average cost of power.

do not equal the utility bill exactly, either a mistake has been made in the computation or a piece of information is missing.

Utility rate spreadsheets are useful in performing such computations. Once a spreadsheet is built, it can be used to calculate costs for any usage pattern under a given rate or set of rates. It can also be used to quickly calculate cost savings from energy efficiency improvements.

TYPICAL GAS BILL CALCULATION

The following is a sample utility bill calculation for a given natural gas usage profile for natural gas service. This is a typical declining block rate structure, where different levels of usage are billed at different unit costs. In this case, the billing unit is 100 cubic feet (Ccf) of natural gas. Hundred cubic feet is a commonly used volume of gas for billing purposes. Other commonly used billing units are 1000 cubic feet (Mcf), therm (100,000 Btu), and million Btu (MMBtu). The rate schedule is shown in Fig. 1. Refer to Chapter 5 for details on the energy or heat content of natural gas billing units. Assuming that the gas usage for the month was 47,500 Ccf, the bill would be calculated as follows:

The commodity rate is first adjusted to account for the purchased gas adjustment (PGA) and the demand side management (DSM) surcharge. To each rate block, $0.0097 is subtracted to account for the PGA and $0.0020 is added to the commodity rate to account for the DSM surcharge. The resulting net commodity rate is:

Block	Ccf	Per Ccf
First	10,000	$0.4297
Next	20,000	$0.4145
All over	30,000	$0.4023

Therefore,

For the first 10,000 Ccf, the charge is:	$4,297.00
For the next 20,000 Ccf, the charge is:	$8,290.00
For the final 17,500 Ccf, the charge is:	$7,040.25
The total commodity charge is:	$19,627.25
Add the service charge:	$80.00
The total pre-tax utility bill is:	$19,707.25
Add state and city taxes of 5.8%:	$1,143.02
Total bill for the month is:	$20,850.27

Note that if the customer used no gas during the billing month, the bill would have been only the service charge of $80.00 plus the state and city taxes for a total monthly bill of $84.64.

TYPICAL ELECTRIC BILL CALCULATION

Fig. 2 is a sample electric rate tariff for a seasonally differentiated electric rate. In this example, assume that the customer's electricity usage in August was 200,000 kWh and the peak demand was 1655 kW. The energy rate is adjusted to account for the fuel adjustment charge (FAC) and the nuclear decommissioning surcharge. To the base rate of $0.05890/kWh, $0.00123 is subtracted to account for the fuel adjustment and $0.00074/kWh is added to account for the surcharge. The resulting net energy charge is $0.05841. Note that the FAC varies each month and can be either positive or negative.

The monthly charge for energy is therefore:

200,000 kWh x $0.05841 per kWh	= $11,682.00
The monthly charge for demand is:	
1,655 kW x $10.02 per kW	= $16,583.10
Add the service charge:	$71.29
The resulting total pre-tax utility bill is:	$28,336.39
Add state and city taxes of 5.8%:	$1,643.51
The total bill for the month is:	$29,979.90

If the customer had used no electricity during the billing month, and the highest demand in the previous 11 months

---

Monthly service charge: $80.00

Minimum monthly charge = The service charge

Commodity rate:

Block	Ccf	Per Ccf
First	10,000	$0.4374
Next	20,000	$0.4222
All Over	30,000	$0.4100

Purchased gas adjustment (PGA):	($0.0097)
Demand-side management (DSM) charge:	$0.0020

State tax: 4.3%
City tax: 1.5%

Fig. 1 Sample natural gas rate tariff.

	Summer Period*	Other Periods
Monthly service charge	$71.29	$71.29
Demand charge	$10.02/kW	$8.53/kW
Energy charge	$0.05890/kW	$0.05242/kWh

Minimum monthly charge: the customer charge plus $4.57 per kW of the highest billing demand established during the 12 months ending with the current month.

Fuel adjustment charge (FAC):	($0.00123 per kWh)
Nuclear decommissioning surcharge:	$0.00074 per kWh
State tax:	4.3%
City tax:	1.5%

*Summer Period is defined as the billing months of June, July, August, and September. All other billing months are defined as "Other Periods."

Fig. 2 Sample electric rate tariff.

is assumed to be 1890 kW, the pre-tax bill would be only the service charge of $71.29 plus a minimum (demand ratchet) charge of:

1890 kW × $4.57 per kW = $8637.30

for a total of $8708.59. Adding on the state and city tax of 5.8% results in a final bill of $9213.69 for the month. This extreme example illustrates the importance of accounting for all elements of the rate structure. Had the demand charge been ignored, the bill calculation would have been grossly underestimated.

DETERMINING THE WEIGHTED AVERAGE COST OF POWER

In the following pages, three electric rate examples are discussed. To keep the analysis manageable, certain billing factors, such as customer charges, taxes, and power factor, have been excluded. These examples, which represent typical industrial, institutional, and large commercial electric rate structures and clearly demonstrate the relationship between varying consumption load profiles and electricity costs, are based on the following three rate structure types:

- Rate 1. A seasonal time-of-use rate
- Rate 2. A conventional seasonal (CONV) rate
- Rate 3. A four-tier seasonal real-time pricing (RTP) rate.

The three rates presented here are representative of current rates in many parts of the country (between $0.05 and $0.06/kWh for baseloaded usage, inclusive of demand charges). However, they should not be used to evaluate specific technology applications. One must always use the rates charged by the local utility. It must be noted that electric rates vary dramatically across the country. In fact, neighboring utilities in the same state often have significant differences between the types of rates offered and rate levels. These differences will likely increase as the utility industry continues to undergo restructuring. While all the three rates have fairly similar costs for baseloaded usage, they have very different structures.

Rate 2 is referred to as a conventional electric rate because it has historically been the most common type of rate. It is, however, being increasingly replaced by time-of-use (TOU)-differentiated rates designed to send market price signals that shape consumer usage patterns and better reflect the cost to serve. Because the usage charge per kilowatt hour does not vary with time of use and because peak demand charges are more moderate, Rate 2 price signals do not strongly drive usage away from peak periods or attract usage in off-peak periods to the extent TOU rates do.

In the two standard (i.e., Rate 1 and 2) rates, a demand charge combines generation, transmission, and distribution system capacity charges, although each is charged separately in many rate structures. Many TOU rate structures will use varying demand charges in each rate period. In addition to peak demand charges, this TOU rate charges for excess demand in off-peak periods. The peak usage charges also include some allocation for capacity costs. But, in many rate structures, costs between peak and off-peak usage are not nearly so differentiated. In those rate structures, a larger portion of the various capacity costs are embedded in demand charges. Rates 1 and 2 have demand ratchets, which can only be set in summer months, as part of their seasonal differentiation. Many rate structures do not have ratchets and some have ratchets that can be set in any month. The RTP rate combines all capacity and commodity costs into usage charges, differentiated by four rate periods.

WEIGHTED AVERAGE COST FOR REPRESENTATIVE OPERATING LOAD PROFILES

The simple average price of electricity or gas for a given facility can be calculated by dividing the annual cost by the annual usage in billing units (e.g., kilowatt hour or 100 cubic feet). This yields an average cost per kilowatt hour or 100 cubic feet. However, average cost calculations provide limited and often misleading information about the actual incremental cost of a particular end-use or load profile. The weighted average cost for specific usage profiles may vary dramatically. In fact, it could be several times greater with one rate structure compared with another.

To demonstrate this important concept, a table reflecting the price of purchasing electricity under various usage profiles is presented for each of the three example rates. Each of these tables lists ten different power usage profiles that might be associated with usage of a certain device or, perhaps, the usage of an entire facility.

The individual profiles in Tables 1–3 show the annual usage, in kilowatt hour, for each profile, the weighted average incremental cost of a kilowatt hour, and the annual cost of consuming power under specific load profiles for a theoretical kilowatt device. Explanation of how the various profiles in Tables 1–3 are calculated and how they relate to various types of usage follows the three electric rate examples.

While different rate designs result in widely varied costs under different usage profiles, the weighted average cost for the baseloaded kilowatt usually is fairly similar for a given utility's cost structure. The rate structures and costs used in these three rate examples could all realistically be offered by one utility. The baseloaded cost per kilowatt of capacity requirement is based on

Table 1 Billing effect of 1 kW, with usage under different usage profiles operating on electric Rate 1 time-of-use (TOU)

Profile number	Period of use	Annual kilowatt hour (kWh)	Average cost ($/kWh)	Annual cost ($)	Annual load factor (LF) (%)
1	1 ratchet setting kilowatt hour per summer month	4	24.870	99	0.1
2	Baseload (BL) summer peak (no ratchet set)	700	0.137	96	8.0
3	50% LF summer peak (w/ratchet)	350	0.351	123	4.0
4	BL summer peak and shoulder (w/ratchet)	1400	0.134	187	16.0
5	6 month cooling profile (w/ratchet)	1870	0.103	193	21.3
6	12 month cooling profile	3379	0.077	260	38.6
7	50% LF 12 months peak	1043	0.169	176	11.9
8	BL 12 months all rate periods	8760	0.056	494	100.0
9	Mixed use (MU) 12 months all rate periods	4755	0.068	321	54.3
10	BL 12 months off-peak and shoulder (no peak demand)	6674	0.038	254	76.2

Notes: Load factor (LF): ratio of actual use vs. maximum potential use in all or certain rate periods; Baseload (BL): 100% LF, or the maximum use in rate period(s); Mixed use (MU): usage based on 80% peak; 60% shoulder and 40% off-peak usage; Cooling profile: demand based on 100% in 2 summer months, 85% in 2 summer months, and 60% in non-summer months. Summer usage based on 80% peak, 60% shoulder, and 30% off-peak. Non-summer usage based on 48% peak, 36% shoulder, and 18% off-peak.

continuous usage every hour of the year, with the total usage being 8760 kWh/kW of demand. This type of usage is shown for each rate example as Profile 8. This particular profile is illustrated graphically in Fig. 3. As shown, 1 full kilowatt hour is consumed in each of the 24 h in each day in each of the 12 months of the year, producing a volume of 100% of usage for one kilowatt of demand. Hence, the terms baseloaded kilowatt and 100% load factor (LF) are applied. A decrease in usage volume per kilowatt of demand corresponds to a decrease in LF. The weighted average cost per kilowatt hour for the baseloaded kilowatt is $0.0600 in the CONV rate, $0.056 in the TOU rate, and $0.056 in the RTP rate. Since the weighted average cost for baseloaded usage is close, comparison of these three rates clearly demonstrates the cost impact of rate design on various types of usage patterns.

Profile 4 in each of the rates is based on a total annual usage of only 1400 kWh for the 1 kW device. All of this usage is in the peak and shoulder rate periods during the four ratchet-setting summer months. As a result, in each of the three rates, the weighted average cost per kilowatt hour is significantly higher than the weighted average cost of the baseload (BL) usage associated with Profile 8, which also includes off-peak usage. The weighted average cost is so much higher, because it is more expensive to provide power during peak periods than off-peak periods. This is

Table 2 Billing effect of 1 kW, with usage under different usage profiles operating on electric Rate 2 conventional (CONV)

Profile number	Period of use	Annual kilowatt hour (kWh)	Average cost ($/kWh)	Annual cost ($)	Annual load factor (LF) (%)
1	1 ratchet setting kilowatt hour per summer month	4	22.850	91	0.1
2	Baseload (BL) summer peak (no ratchet set)	700	0.108	76	8.0
3	50% LF summer peak (w/ratchet)	350	0.312	109	4.0
4	BL summer peak and shoulder (w/ratchet)	1400	0.116	163	16.0
5	6 month cooling profile all rate periods (w/ratchet)	1870	0.097	181	21.3
6	12 month cooling profile all rate periods	3379	0.075	252	38.6
7	50% LF 12 months peak	1043	0.147	154	11.9
8	BL 12 months all rate periods	8760	0.060	522	100.0
9	Mixed use (MU) 12 months all rate periods	4755	0.066	315	54.3
10	BL 12 months off-peak	6674	0.048	318	76.2

Table 3 Billing effect of 1 kW, with usage under different usage profiles operating on electric Rate 3 real-time pricing (RTP)

Profile number	Period of use	Annual kilowatt hour (kWh)	Average cost ($/kWh)	Annual cost ($)	Annual load factor(LF) (%)
1	1 kWh per summer month (peak)	4	0.750	3	0.1
2	Baseload (BL) summer (peak)	700	0.201	141	8.0
3	50% LF summer (peak)	350	0.234	82	4.0
4	BL summer (peak)	1400	0.152	213	16.0
5	6 month cooling profile	1870	0.102	190	21.3
6	12 month cooling profile	3379	0.074	249	38.6
7	50% LF 12 months (peak)	1043	0.142	148	11.9
8	BL 12 months	8760	0.056	493	100.0
9	Mixed use (MU) 12 months	4755	0.066	316	54.3
10	BL 12 months (base)	6674	0.039	259	76.2

Notes: Peak: indicates that kWh are first charged to the highest rate block and then successively to lower rate blocks; Base: indicates that kWh are first charged to the lowest rate block and then successively to higher rate blocks; Cooling: profile kWh are allocated between rate blocks to correspond to the allocations used in the time-of-use (TOU) rate examples; Mixed-use: profile kWh are allocated between rate blocks to correspond to the allocations used in the TOU rate example.

reflected in the rate structures, though to varying degrees. The BL usage profile of Profile 8 blends this high-cost peak usage with low-cost off-peak usage.

The annual usage for Profile 4 is illustrated graphically in Fig. 4. Note that usage is only shown during the four summer months and during hour 6–21 of each day, which correspond to the peak and shoulder periods (6 A.M.–9 P.M.) from Monday to Friday. Hence, this figure only represents the usage during the normal five-day workweek.

In comparison to Profile 8, which shows an annual consumption volume of 8760 kWh, Profile 4 only shows a volume of 1400 kWh for the same 1 kW of peak demand. As will be shown below in the computations provided in the detailed discussion of each rate profile, the LF for Profile 4 is only 16% since only 1400 of a possible 8760 kWh are consumed over the course of the year.

In contrast, Profile 10 is based on a total annual usage of 6674 kWh, with all usage in the off-peak and shoulder rate periods. As a result, in each of the rates, the weighted average cost per kilowatt hour is even lower than the weighted average cost of the BL usage profile (Profile 8). In this case, since it is far less costly to provide electricity in the off-peak period, the result is a very low weighted average cost. The utility now has this extra capacity

Fig. 3 Rate structure Profile 8, the baseloaded kilowatt.

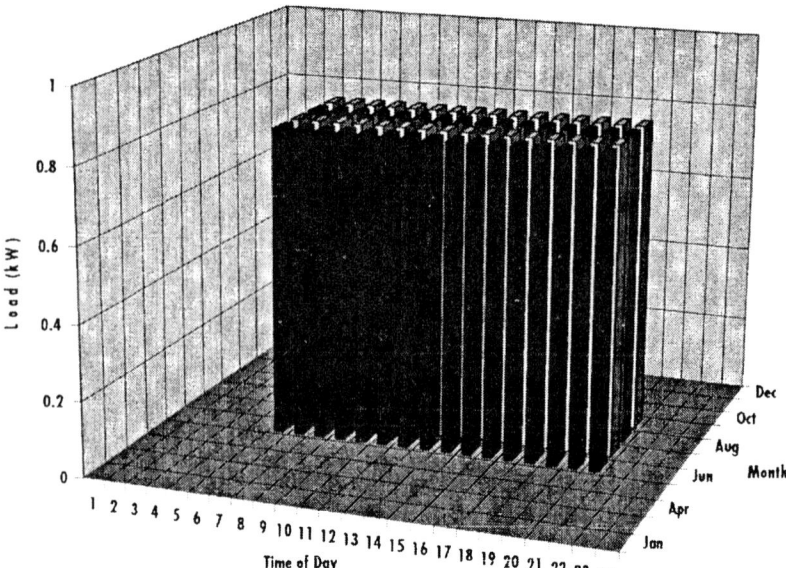

Fig. 4 Rate structure Profile 4, peak and shoulder summer usage Monday–Friday.

available, during the peak periods, that it can sell at the much higher rate to balance the sale of this low-cost usage.

The ten profiles listed in each of the Tables 1–3 were created to match the hours in each of the rate periods specific to the TOU rate structure. To allow for a reasonable basis of comparison, the same rate periods have been imposed on the rate structure used in the RTP rate. While the RTP rate has been calibrated for this purpose, it would also be appropriate to view these rate structures with respect to their own natural rate blocks.

For example, the RTP rate has a total annual base period, or lowest cost rate block of 3020 h/year, while in the TOU rate there are 4588 annual hours in the off-peak rate block. To make the RTP rate compatible with this load profile, it was assumed that the 4588 annual hours would be composed of 3020 h from the base period, with the remaining 1568 h assigned to the intermediate, or next lowest, cost rate period. The rest of the profiles for the RTP rate were calibrated to the TOU rate structure in a similar manner.

The table (i.e., Tables 1–3) accompanying each rate provides a calculation of the annual electric bill that would result from each of the ten usage profiles under each of the rate structures. Following the three rate descriptions are explanations for each of the profiles.

ELECTRIC RATE 1 (TOU)

Electric Rate 1 is a seasonal TOU rate. The basic tariff is summarized in Fig. 5. Under this rate, shoulder and off-peak demand charges are assessed only for demand levels that exceed that of the peak periods. For example, if summer peak usage was 1000 kW and shoulder usage was 1500 kW, the demand charge would be:

$$(1000 \text{ kW} \times \$12.00) + (500 \text{ kW} \times \$6.00)$$

$$= \$15,000 \text{ per mo.}$$

Ratchet Adjustment

Under this rate, if one month had a peak demand of 2000 kW and the next 11 months had a peak demand of 1000 kW, each of those next 11 months' demand charges would be based on 80% of the 2000 kW figure, or 1600 kW The impact over the course of a year on the customer's bills would be an additional 600 kW in monthly billable kilowatt charge over the actual demand. Over a period of one year, the customer would pay for

Electric Rate 1 (TOU) Seasonal, TOU Electric Rate		
	Summer (4 months)	Non-summer (8 months)
Demand Charges	($/kW)	
Peak	$12.00	$8.00
Shoulder Excess	$6.00	$4.00
Off-Peak Excess	$3.00	$3.00
Energy Charges	($/kWh)	
Peak	$0.068	$0.058
Shoulder	$0.058	$0.048
Off-peak	$0.032	$0.032

Fig. 5 Electric Rate 1 tariff summary.

6600 kW in additional demand charges. This is calculated as:

$$[(2000 \text{ kW} \times 0.80) - 1000 \text{ kW}] \times 11 \text{ months}$$
$$= 6600 \text{ kW}$$

Specific Hours of Operation

Peak:
  10:00 A.M.–6:00 P.M., Monday–Friday
4 summer months at 40 h/week (700 h)
  8 non-summer months at 40 h/week (1386 h)
Shoulder:
  6:00 A.M.–10:00 A.M. and 6:00 P.M.–10:00 P.M., Monday–Friday
4 summer months at 40 h/week (700 h)
8 non-summer months at 40 h/week (1386 h)
Off-peak:
  10:00 P.M.–6:00 A.M., Monday–Friday, all day Saturday and Sunday
  4 summer months at 88 h/week (1540 h)
  8 non-summer months at 88 h/week (3048 h)

ELECTRIC RATE 2 (CONV)

Electric Rate 2 is a conventional seasonally differentiated Commercial, Industrial and Institutional (CI&I) rate. The basic tariff is summarized in Fig. 6. Specific hours of operation are:

  Summer–4 months (17.50 weeks totaling 2940 h)
  Non-summer–8 months (34.64 weeks totaling 5820 h).

ELECTRIC RATE 3 (RTP)

Electric Rate 3 is a simplified real-time-pricing rate. There are several ways in which this developing rate is offered by electric utilities to customers. In one common approach, the pricing is established per rate block based on an

Electric Rate 2 (TOU)
Seasonal, TOU Electric Rate

Summer (4 months)	Non-summer (8 months)
Demand Charges ($/kW)	($/kW)
$10.00	$8.00
Energy Charges ($/kWh)	($/kWh)
$0.051	$0.046

Fig. 6 Electric Rate 2 tariff summary.

Rate Block	Unit Cost Total ($/kWh)	Summer Hours	Non-summer Hours	Total Hours
1. Base	$0.025	700	2,320	3,020
2. Intermediate	$0.038	1,200	3,200	4,400
3. Peak	$0.170	1,000	300	1,300
4. Power pool peak	$0.740	40	0	40
Total		2,940	5,820	8,760

Fig. 7 Real-time pricing (RTP) rate pricing blocks.

analysis of the utility's costs associated with the dispatch of various generation stations in the stack. Each day, the utility informs the customer of which hours will be applied to each rate block.

In this case (Fig. 7), four blocks have been assigned: base, intermediate, peak, and power pool peak. The power pool peak refers to costs incurred as a result of the utility requiring peak power from the pool. Since each hour of the year is assigned to a rate block based on actual (real time) dynamic conditions, there is no established schedule. A good approximation can be made based on experience, however. For the purpose of demonstrating the workings of this rate, hours have been assigned for winter (36.64 weeks) and summer (17.50 weeks) to each of the four rate blocks.

Fig. 8 shows the maximum cost in each rate block for a baseloaded kilowatt Fig. 9 shows the annual hours of operating and cost per kilowatt hour for various combinations of rate blocks.

ELECTRIC RATE COMPARISONS AND CONCLUSIONS

The period during which power is used affects operating costs as much as, or more than, the amount of power used. This concept becomes critical in energy use planning as price differentiation by time of use increases.

While accountants may look at electric operating costs in terms of the average cost per kilowatt hour, energy use planners must look at the incremental costs of individual

Rate Block	Summer	Non-Summer	Annual
1. Base	$17.50	$58.00	$75.50
2. Intermediate	$45.60	$121.60	$167.20
3. Peak	$170.00	$51.00	$221.00
4. Power pool peak	$29.60	$0.00	$29.60
Total	$262.70	$230.60	$493.30

Fig. 8 Maximum cost per rate block.

Annual and Average Costs			
Rate Block	Hours	Ann. Cost ($)	Weighted Avg. Cost ($ kWh)
3+4 Summer	1,040	$199.60	*$0.1n'*
2+3+4 Summer	2,240	$245.20	$0.109
1+2+3+4 Summer	2,940	$262.70	$0.089
1+2 Annual	7,420	$242.70	$0.033
1+2+3 Annual	8,720	$463.70	$0.053
3+4 Annual	1,340	$250.60	$0.187
2+3+4 Annual	5,740	$417.80	$0.073
1+2+3+4 Annual	8,760	$493.30	$0.056

Fig. 9 Operating hours and cost per rate block combinations.

end uses and various consumption profiles to understand price impact. Energy planners audit facilities to develop incremental cost-usage profiles associated with individual equipment, systems, and activities. These audits are done in much the same manner as the ten profiles presented in the preceding pages.

Planners look at seasonal end uses, such as cooling, and understand that the relevant weighted average cost per kilowatt hour may be several times greater than the facility's overall average cost per kilowatt hour, especially if a ratchet adjustment is in effect. They look at baseloaded operations and consumption blocks and understand that the costs may be lower than the facility's average cost. They look at identical devices, in a multiple-unit system, that run the same amount of hours per year, and they understand that if they operate with different load profiles, their operating cost may be dramatically different.

Evolving RTP electric rate structures may extend this discrete differentiation to every hour of the year, or perhaps even every minute. An important benefit of RTP rates is that one peak hour of extraordinary usage might not have the dramatic cost impact that a rate with a high demand charge and ratchet adjustment would have. This type of rate flexibility is well suited for electricity purchase strategies that involve a mix of on-site power generation or purchase of non-utility-generated power along with the purchase of utility provided power. In the event of retail purchase of non-utility-generated electricity, traditional, TOU or RTP type rate structures may be applied to transmission and distribution services, while some type of RTP structure would be applied to the usage for the purpose of commodity transaction.

With this understanding, energy planners develop strategies to minimize operating costs and optimize productivity. Efficiency improvements and alternative energy source options are considered with respect to these incremental costs. Self-generation and electricity displacement strategies should be evaluated in the same

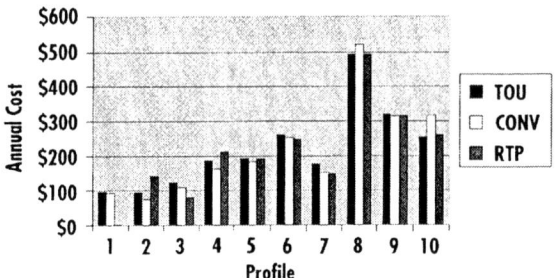

Fig. 10 Annual operating cost for each load profile.

manner. Electric cost savings opportunities should be considered on the basis of incremental cost per kilowatt hour, as well as the total usage. Elimination of 1 kW of low-LF usage should produce larger cost savings per kilowatt hour than the elimination of 1 kW of high-LF usage, even though it may not produce greater aggregate energy savings.

Following are explanations for each of the profiles and a discussion of the type of equipment usage or facility characteristics that would result in each profile. Also included are examples of how LF, annual cost, and weighted average cost were calculated.

DETERMINING THE WEIGHTED AVERAGE COST FOR VARIOUS LOAD PROFILES

For individual equipment or an entire facility operating with anyone of the load profiles presented in Tables 1–3, the total annual usage and cost are based on the total input power (kilowatt) of the equipment (or the connected load of the facility) times the usage and cost of 1 kW as presented in each table entry. In all cases, it is assumed that the facility has only one billing meter that measures consumption and demand of all connected loads.

The annual and weighted average costs per kilowatt hour for the ten sample load profiles under each of the example utility rates are summarized in Figs. 10 and 11. Note that because of the extremely low LF for Profile 1,

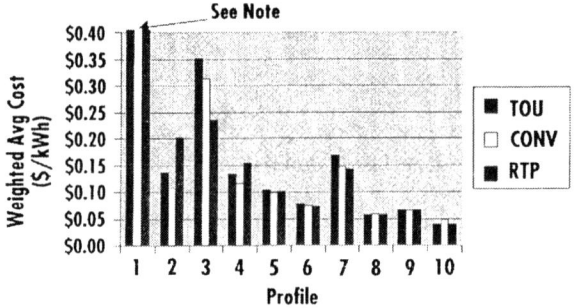

Fig. 11 Weighted average cost ($/kWh) for each load profile.

TOU and CONV average costs shown in Fig. 11 are off the scale (>$20/kWh).

EXPLANATION OF TEN SAMPLE LOAD PROFILES

Profile 1 is based on a device rated at 1 kW, operated only one hour during each summer month June, July, August, and September) when the entire facility is already operating at its highest peak demand level. This added load increases the peak demand level for the month by 1 kW Therefore, there is a peak demand charge for that 1 kW in each of the 4 summer months. It also adds 1 kW to any applicable peak demand ratchet level. Under Rate 1, for instance, the effect of this 1 kW is a monthly demand charge based on 0.8 kW in each of those 8 non-summer months.

Thus, the 4 kWh produce a total annual billable demand charge based on 10.4 kW. In addition, there is a usage charge for the 4 kWh totaling $0.27. Based on Electric Rate 1, the annual cost for using the 4 kWh is:

$(1 \text{ kW} \times \$12/\text{kW} \times 4) + (0.8 \text{ kW} \times \$8/\text{kW} \times 8)$

$+ (4 \text{ kWh} \times \$0.068/\text{kWh})$

$= \$99.47$

The weighted average cost per kilowatt hour is:

$\$99.47/4 \text{ kWh} = \$24.87/\text{kWh}$

If that same 1 kW device is operated for 4 h in the off-peak rate period, then no demand charges apply and the consumption charges are lower. The total annual cost of the 4 kWh is only $0.13, and the weighted average cost of 1 kWh is $0.032/kWh. While the first profile is an extreme example, it demonstrates how significant the impact of demand charges and ratchet adjustments can be in a given rate structure.

A large chiller used for space cooling, for example, may operate at peak capacity only a few hours during the entire year. In many cases, peak cooling demand coincides with the facility's maximum electric use period and establishes not only a demand peak for the month, but also an increased ratchet demand level for the year. If, at the hour of maximum monthly electric usage (the peak demand hour), the chiller consumes 1 extra kilowatt hour to satisfy load, that 1 kWh will increase the billing peak demand by 1 kW.

Taking the most extreme case of Profile 1 in Rate 1, the use of only 4 kWh at the peak demand period of the peak demand month would cost $99. This could actually be the case with a resting lab or a university that holds commencement in the summer and experiences an extraordinary peak load on only one day per year. If the facility is able to somehow shed these 4 kWh at the same time that the chiller or testing equipment consumes these ratchet-setting kilowatt hour, it would save $99.

At first glance, the idea of 4 kWh costing $99 might seem absurd. However, an understanding of how electric rate structures operate shows that the incremental cost per kilowatt hour can actually vary by several thousand percent. In fact, with a typical demand window of 15 min, this equipment need only set a peak for 15 min, consuming only 0.25 kWh to cause the facility to endure the full 10.4 kW of annual demand charge. Under the RTP rate, no such dramatic costs would be incurred. Since there is no demand charge impact, the costs only consist of the peak cost per kilowatt hour times the hours of use. In this example, the most expensive kilowatt hour of the year would cost $0.74 and the 4 kWh would, therefore, cost $2.96. However, it is important to note that in a real-time market, the cost for these 4 kWh could be quite a bit higher, conceivably as high as $99, though perhaps not likely.

Profile 2 is based on a device rated at 1 kW and operated as a BL in all 4 summer month peak hours. Under Rate 1 (the TOU example consisting of 700 h/year based on a 40-h/week rate period extending for 17.5 weeks), the annual cost for using those 700 kWh is:

$(1 \text{ kW} \times \$12/\text{kW} \times 4) + (700 \text{ kWh} \times \$0.068/\text{kWh})$

$= \$96$

The weighted average cost per kilowatt hour is:

$\$96/700 \text{ kWh} = \$0.137/\text{kWh}$

The annual LF is:

$700 \text{ h}/8760 \text{ h} = 8\%$

In this scenario, demand charges total $48, slightly more than half of the annual cost. It is assumed in this case that peak demand is sufficiently high in the winter months so that there is no ratchet in effect. Based on Rate 1, if, for example, an electric motor with an input power rating of 100 kW operates at full load in each of the 700 h in this profile (represented in Table 1, Profile 2), the annual cost would be $9600. This can also be calculated by multiplying the input power (100 kW) by the total full-load hours of operation (700 h) by the weighted average cost per kilowatt hour of $0.137.

Under the RTP rate, assuming that the 700 h corresponding to the TOU rate summer peak would be composed of the full 40 h of the power pool peak block and 660 h of the peak block, the total cost would be $141 and the weighted average cost would be $0.20/kWh. Based on this example, operating the same 100 kW electric motor during the same 700 h would result in an annual cost of $14,350.

In Profile 3 the 1 kW device operates with a 50% LF during the 40 h/week peak rate period over the 4 summer months rather than at a 100% LF. This results in a usage of

350 kWh over the 700 total hours in this rate period and a total annual LF of 4%, rather than 8% with Profile 2. Usage over the 17.5 week summer period and annual LF, respectively, are calculated as follows:

(1 kW × 17.5 weeks) × (40 h/week) × (0.5 LF)

= 350 kWh

$$\frac{700 \text{ h} \times 0.5 \text{ LF}}{8760 \text{ h}} = 4\%$$

Using the TOU rate, the total annual cost is reduced compared with Profile 2 due to reduced usage. Therefore, demand charges as a percent of the total cost increase. In Profile 3 (with ratchet adjustment), demand charges represent 84% of the total cost. The impact of reduced usage with constant demand charges is an increased weighted average cost per kilowatt hour. The weighted average cost per kilowatt hour for Profile 3 increases to $0.351.

For this profile with only summer peak usage, the traditional non-time-differentiated rate is less costly than the TOU rate, and the RTP rate is the lowest, with a weighted average cost of $0.234. The reason is that with such a low LF load of 4%, the impact of demand charges, inclusive of ratchets, drives up costs dramatically for the other rate structures.

Profiles 2 and 3 are realistic examples of the cost of cooling equipment operation during peak periods. In many cases, these operating profiles are only a portion of a cooling unit's total energy usage. In other cases, they may represent the total operation. In facilities with multiple cooling units, one unit is often predominantly used as a peaking unit. Peak cooling loads often correspond to the TOU peak electric rate period (10:00 A.M.–6:00 P.M., Monday–Friday) due to ambient temperature profiles, increased productivity, and increased internal gains from people and equipment. In single shift C&I operations, the peak period coincides with most of the operating hours of the facility. In those cases, profiles such as Profiles 2 and 3 may also be representative.

Electric usage with profiles of the type listed in Profiles 1–3 are often targeted for elimination or reduction by various load shedding or alternative energy source technologies. Peak-shaving generators and fuel- or steam-powered cooling are two commonly applied technologies for eliminating these blocks of electric usage.

Profile 4 is similar to Profiles 2 and 3, in that it reflects the higher cost of power resulting from seasonal differentiation and significant demand charges. With 100% usage in Summer Peak and Shoulder periods, this profile has greater usage, with a LF of 16%, as compared with 4% in Profile 3. Under the TOU rate, this produces a weighted average cost per kilowatt hour of $0.134, as compared to $0.351 for the lower LF usage of Profile 3.

In this profile, demand charges represent a lower percentage of the total cost than in Profile 3 because each unit of demand is spread over a greater usage base. However, demand charges still represent a significant portion of the total cost. Profile 4 also includes the effect of demand ratchets. Extending usage to the shoulder periods in this profile partially integrates lower-cost power into the profiles and results in increased LF and decreased weighted average cost.

Under the RTP rate, assuming the 1400 h in Profile 4 would include 40 power pool peak hours, 1000 peak hours, and the balance of 360 h intermediate, the annual cost and weighted average cost, respectively, are:

(40 kW × $0.74) + (1000 kWh × $0.17) + (360 kWh × $0.038) = $213.28

$$\frac{\$213.28}{1400 \text{ kWh}} = \$0.152/\text{kWh}$$

Profiles 5 and 6 refer to mixed use cooling season profiles. While Profiles 2–4 are all based on a 1 kW device running either all of the time or with a 50% LF in a given rate period, Profiles 5 and 6 are based on a 1 kW cooling device running with a load that varies between each month and rate period. These profiles were designed to represent typical space cooling loads served by an electric vapor compression system.

During the 4 summer months, it is assumed that the equipment operates with a LF of 80% peak, 60% shoulder, and 30% off-peak, based on the TOU rate structures. In July and August, it is assumed that the full 1 kW peak demand is set. In June and September, it is assumed that the peak demand impact of the 1 kW equipment is 0.85 kW In the non-summer months, it is assumed that the equipment operates with a LF of 48% peak, 36% shoulder, and 9% off-peak and the demand impact is 0.60 kW in each month. These profiles were then calibrated and applied to the CONV and RTP rate. The difference between the two profiles is that Profile 5 is based on 6 months of operation and profile 6 is based on year-round operation. Notice that profile 6 has a far lower weighted average cost per kilowatt hour than Profile 5 as more lower cost non-summer usage is blended in and there is no ratchet impact.

Profile 7 has peak usage every week of the year and demand charges in every month. This type of profile might be targeted for elimination with peak shaving power generation or replacement of baseloaded electric-driven equipment with fuel- or steam-driven equipment operated in the peak periods.

Profile 7 is representative of a load profile that might result from electric-driven equipment operated with a 50% LF during the peak period only. Under the TOU rate, this corresponds to 20 h of operation per week. There is a significant increase in weighted average cost for Profile 7 for all rate types. This is due to the fact that demand charges remain the same, but are spread over only half the usage as the LF is only 11.9%. Under the RTP rate, the weighted average cost is relatively high because as LF is decreased, a greater percentage of the usage is assumed to fall in the highest cost rate block.

Profile 8 (BL 12 months, all rate periods) represents the baseloaded kilowatt, or 1 kW baseloaded every hour of the year. The annual usage of 8760 kWh has an annual LF of 100%.

With this profile, the fixed investment in electric generation and distribution capacity is spread over the maximum possible annual usage. As shown in Profiles 1–7, different rate structures result in widely varied costs under different usage profiles. However, the weighted average costs for the baseloaded kilowatt usually are fairly similar for different rates under a given utility's cost structure. The rate structures and costs used in the TOU, CONV and RTP rates could realistically be offered by one utility.

While with the profiles with lower LF, peak usage produced much higher costs for the RTP rate, compared with the CONV rate, the higher LF off-peak usage produced much lower costs with the RTP rate. These opposing trends are roughly canceled out with continuous BL usage, though traditional rate structures, such as Rate 2, commonly produce slightly higher baseloaded kilowatt hour costs. Thus, a three-shift facility with a very high LF would choose the TOU or RTP rate structures.

The annually baseloaded kilowatt is the load profile most often targeted for prime mover-driven power generation and mechanical service applications that employ heat recovery (cogeneration cycles). The weighted average cost per kilowatt hour of the baseloaded kilowatt is often the benchmark for determining the feasibility of such applications. While the cost per kilowatt hour is lower than most of the other load profile entries, the total annual cost is the greatest.

Profile 9 represents a profile that might result from annual operation of individual electric motors or other process equipment. It might also result from the combined operation of multiple equipment in a facility. Facilities rarely have absolutely flat loads. Therefore, the BL, or 100% LF load profile, is not necessarily representative of the weighted average cost of power. Some type of mixed use profile, such as Profile 9, is generally more representative of the weighted average cost. This is an aggregate of varying individual components, such as lights and motors, each with a different load profile.

Sometimes, equipment does operate with a LF of 100%, either in a specific rate period or in all rate periods. More often, equipment operates under varying load or under full load for intermittent periods. Profile 9 is an example of such operation. The LF of this profile is about half that of Profile 8. The total annual cost is lower due to significantly lower use, but is greater than half the cost of Profile 8, because the weighted average cost per kilowatt hour is greater. This is due to the greater relative impact of demand charges (the dollar value of which remains the same as in Profile 8).

Profile 10 represents extensive off-peak non-demand setting usage. It has no demand charge at all. Notice that with no demand charges, the weighted average costs are lower for the TOU rate than the CONV rate. This is a result of the price signals of the TOU differentiated rates, which greatly emphasize usage in off-peak and shoulder periods—which the traditional rate does not do. Also note that only a lower cost excess demand charge can be assessed to these usage profiles in the TOU rate example, while a full demand peak can be set in the CONV rate. This would increase costs still further. Under the RTP rate, these average weighted costs are produced by blending the lowest cost rate block with the successively higher cost rate blocks.

As Profile 10 shows, TOU rates offer relatively low-cost power for about three quarters of the hours of the year. When an excess demand charge is assessed to shoulder usage, or if standard shoulder-period demand charges are in effect, the costs will be slightly greater, although still significantly lower than during the costly peak period.

The RTP rate offers relatively low-cost power for about 85% of the annual hours. The weighted average annual cost for the 3020 h of base block usage and 4400 h of intermediate block usage is only $0.033/kWh. This is balanced by a much higher unit cost in the peak and power pool peak blocks, which comprise about 15% of the total annual hours, but slightly more than half the total annual cost. The weighted average cost is only a usage cost, but it also reflects imbedded fixed costs, a large portion of which are included in the demand charges associated with the other two rates. Table 4 provides a comparison of Profiles 7, 8, and 10 for the CONV (2) and RTP (3) rates.

Notice that the sum of total costs for Profiles 7 and 10 approximately equal Profile 8, the annually baseloaded kilowatt The total annual costs of Profiles 7 and 10 are somewhat dose, compared with the stark contrast in usage. The weighted average cost per kilowatt hour for Profile 7 is about three times that of Profile 10.

MATCHING ENERGY TECHNOLOGY ALTERNATIVES WITH ELECTRIC RATES AND USAGE PROFILES

Following is a brief discussion of the fuel- and steam-powered technology applications that should be considered for use in eliminating electricity purchases associated with the 10 representative load profiles presented in Tables 1–3.

Table 4 Comparison of Profiles 7, 8, and 10 for conventional (CONV) and real-time pricing (RTP) rates

Rate number	Profile number	Usage (kWh)	Annual load factor (LF) (%)	Average cost per kWh ($)	Total cost ($)
2	7	1043	12	0.147	154
2	10	6674	76	0.048	318
2	8	8760	100	0.060	522
3	7	1043	12	0.142	148
3	10	6674	76	0.039	259
3	8	8760	100	0.056	493

Electric Peak Shaving Generation Applications

The primary focus here is on peak demand and usage charge savings. Thus, the emphasis is on eliminating costly peak demand charges resulting from poor LF or, in the case of RTP rates, eliminating the usage in the highest cost rate blocks. As shown above in the RTP rate, the total annual cost is about the same for a load with a 15% LF occurring in the highest cost rate blocks as a load with an 85% LF occurring in the lowest cost rate blocks. Since peak shaving applications will have relatively low annual hours of operation, low capital cost is emphasized more heavily than optimum thermal efficiency and simple energy costs, and heat recovery is not commonly used. Representative load profiles that might be targeted for peak shaving generation include Profiles 1–3, and 7. Other potential technology applications include load shedding control, battery storage and thermal energy storage (TES).

Electric Cogeneration Applications

The primary focus of power generation applications that employ heat recovery is on overall system thermal fuel efficiency and durability, since equipment run times may range from several thousand hours to continuous operation. The object is to minimize the use of purchased electricity while simultaneously eliminating other internal fuel usage via heat recovery. The 8760 h load profile associated with Profile 8 is ideal for cogeneration in most cases. Other high LF load profiles, such as Profile 9, may also be targeted for elimination with on-site application of cogeneration technologies. In some cases, notably with highly stratified rate structures, it may not be economical to generate power on site in periods when the lowest cost power is available from the utility. In such cases, load profiles with somewhat less than 100% LF may be appropriately targeted for elimination. Other potential technology applications include combined cycles and steam injection cycles, along with a host of other standard BL energy conservation measures.

Single- Unit, Year-Round Mechanical Drive Applications

The primary focus here is on satisfying end use requirements with a single unit that has the lowest life cycle cost. Single-unit systems may operate several thousand hours or more annually. In some cases, prime mover-driven systems using heat recovery may be less costly to operate than electric-driven systems in all use periods. In other cases, prime mover-driven systems may be more costly in some periods (i.e., off-peak), but less costly to operate on the average (mixed use) due to significant savings in peak and shoulder periods. System efficiency and durability are particularly emphasized for applications with significant run-time requirement. Heat recovery will be increasingly cost-effective with increased hours of operation. In single or two-shift operations, the unit may operate only in costly peak periods or in peak and shoulder rate periods. Representative load profiles that might be targeted for elimination with single-unit prime-mover mechanical drive systems include Profiles 8 and 9. Other potential technology applications include building and process automation systems and variable volume distribution systems (i.e., air, water, steam, etc.).

Multiple-Unit Mechanical Drive Mixed (Hybrid) System Applications

The primary focus here is overall system optimization with use of electric units in off-peak periods and non-electric units during peak and shoulder periods, or a variation with some baseloading of equipment. Equipment may operate anywhere from several hundred hours to several thousand hours annually. Thermal fuel efficiency (and heat recovery potential) of non-baseloaded individual units may be sacrificed for lower capital costs.

A cogeneration cycle unit may be baseloaded and electric units used for the remaining off-peak load and non-electric units used for the remaining peak and shoulder loads. Alternatively, an electric unit may be baseloaded and non-electric units used for peaking duty only. In one- or two-shift, five-day operations, a single unit might experience a similar operating profile as would a peaking unit in a three-shift, seven-day operation. Profile 7 might be targeted for elimination with use of a prime mover-driven mechanical drive system. Other potential technology applications include building and process automation, peak shaving, load shedding control and variable volume distribution systems (i.e., air, water, steam, etc.).

Single-Unit Seasonal Cooling Applications

The primary focus here is satisfying cooling requirements with a single unit with the lowest life cycle cost. Equipment may operate anywhere from several hundred hours to a few thousand hours annually. Heat recovery is a viable option, but has less impact than in year-round, single-unit applications, because operating hours are typically lower. Heat recovery becomes more important with higher hours of operation and fuel costs. In single- or two-shift operations, the unit may operate predominantly in costly peak periods or peak and shoulder rate periods. Representative load profiles that might be targeted for elimination with use of a non-electric-driven cooling systems include Profiles 2–6. Other potential technology applications include peak shaving, load shedding control, battery storage and TES.

Multiple-Unit Cooling (Hybrid) Applications

The primary focus here is to minimize system operating costs with use of electric units in off-peak periods and non-electric units in peak and shoulder periods, or a variation with some baseloading of equipment. Electric peak demand charges are a critical consideration. Equipment may operate anywhere from two hundred to several thousand hours annually. Thermal efficiency (and heat recovery potential) on non-baseloaded individual units may be sacrificed for lower capital costs. This is similar to strategies for year-round, multiple-unit mechanical drive systems. However, optimization strategies will differ somewhat due to the greater electric unit costs with seasonal pricing and ratchet potential. In one- or two-shift, five-day operations, a single unit might experience a similar operating profile as would a peaking unit in a three-shift, seven-day operation. Representative load profiles that might be targeted for elimination with use of non-electric-driven cooling equipment as part of a mixed system application include Profiles 2–4. Other potential technology applications include peak shaving, load shedding control, battery storage and TES.

CONCLUSION

As shown in this article, the weighted average cost of a kilowatt hour is highly variable. Different technology applications must be matched with different cost scenarios and electric load profiles. The aggregate energy cost for any particular application is the real determinant of operating cost savings potential. While savings of $0.30 or $0.60/kWh are attractive targets, savings must accrue over enough hours to provide sufficient payback on the investment in alternative energy equipment. Baseloaded cogeneration applications, for example, emphasize overall thermal fuel efficiency, effective heat recovery, and durability. These systems will save less per hour of operation than other technology applications in many cases, but will accrue savings over a greater number of operating hours. Conversely, peak shaving applications may run fewer hours, but target the most expensive electricity usage. In these applications, thermal fuel efficiency is generally less of a concern than low capital costs, because the key is elimination of low LF high-cost power purchases.

BIBLIOGRAPHY

1. Capehart, B.L.; Wayne, C.T.; William, J.K. Understanding energy bills. *Guide to Energy Management* 5th Ed.; The Fairmont Press, Inc.: Lilburn, GA, 2006; [Chapter 3].
2. Mahoney, J.F.; Barney, L.C. Strategies for electric utility bill reduction when alternative rate structures apply. Eng. Economist **1994**, *Spring*.
3. Petchers, N. Utility rate structures, In *Combined Heating, Cooling and Power Handbook*; Marcel Dekker, Inc.: New York, 2003; [Chapter 21]. Petchers, N. Utility bill analysis; In *Combined Heating, Cooling and Power Handbook*; Marcel Dekker, Inc.: New York, 2003; [Chapter 21].
4. Turner, W.C. Electric and gas utility rates for commercial and industrial customers. *Energy Management Handbook* 5th Ed.; The Fairmont Press, Inc.: Lilburn, GA, 2005; [Chapter 18].

Utilities and Energy Suppliers: Business Partnership Management

S. Kay Tuttle
Process Heating, Boilers and Foodservices, Duke Energy, Charlotte, North Carolina, U.S.A.

Abstract
This paper explores the role of the utility and the industrial or commercial facility engineer in developing a business partnership especially as it relates to the quality of electric power. The industrial or commercial facility engineer needs a thorough knowledge of the electrical power quality needs of equipment and processes critical to the business's financial success. Information relating to the operation of the electric utility system is essential to understanding what the utility can reasonably deliver and determining the best solution for bridging any gaps between process requirements and available power quality. The utility needs a thorough understanding of the equipment and processes of businesses they serve as well as a thorough knowledge of the power quality requirements of that equipment. Also, the utility should have available the best options for each process to resolve any gaps between the quality of power supplied and that required by the process. The utility has an obligation to its business partners to explain the operations of the utility system and to make available solutions to power quality problems or needs. Some utilities now offer to install power quality monitors, diagnose power quality problems, perform wiring and grounding assessments, and procure and install mitigation equipment within commercial and industrial facilities. When an opportunity arises for the facility engineer and the utility power quality engineer to work together to prevent or resolve a power quality issue, each should be prepared with the knowledge, information and resolve to do so.

INTRODUCTION

It is imperative for utilities and their customers to form a partnership in order for both to optimize their businesses. Electricity and productivity in industry are inherently linked. For a utility to remain profitable, it must have productive customers, and for a customer to be productive, he or she must have reliable, high quality power. Partnership implies trust. The utility must trust customers and disclose the reliability and quality of power on the electrical system, and the customers must trust the utility and inform the utility engineers regarding their processes. Working together both can meet their business objectives.

SHARING INFORMATION WITH CUSTOMERS ABOUT THE UTILITY SYSTEM

Before the utility can share information regarding typical system performance, the utility must first monitor and analyze the system. In an effort to characterize the utility distribution system, Duke Power and other electric utilities funded a study to determine what quality of service a distribution served customer might expect. This study called Distribution Power Quality (DPQ) was conducted by Electric Power Research Institute (EPRI) and focused on voltage sags, transients, and harmonics. The results for voltage sags are presented in Fig. 1. This information is extremely valuable to customers when specifying new equipment or selecting the optimum power quality mitigation devices. For example, the data indicates that the great majority of voltage sags are 6–10 cycles in duration and from 60% to 90% remaining voltage. A customer can use this information when ordering new equipment by requesting equipment that can ride through these voltage sags. If the equipment manufacturer does not have the capability to test the ability of the new equipment to ride through voltage sags, the utility can do so. Or, if mitigation equipment is needed, this information is valuable when choosing which mitigation equipment to apply.

UNDERSTANDING CUSTOMER PROCESSES

A utility gains valuable information by analyzing its system, but that is only part of the information needed to make businesses more productive. A detailed understanding of customer processes and what causes these

---

* This entry originally appeared as "Building Business Partnerships with Your Electric Utility: A Power Quality Perspective" in *Energy Engineering*, Vol. 101, No. 3. Reprinted with permission from AEE/Fairmont Press.

Keywords: Power quality; Power disturbances; Sags; Utility partnerships; Equipment sensitivity.

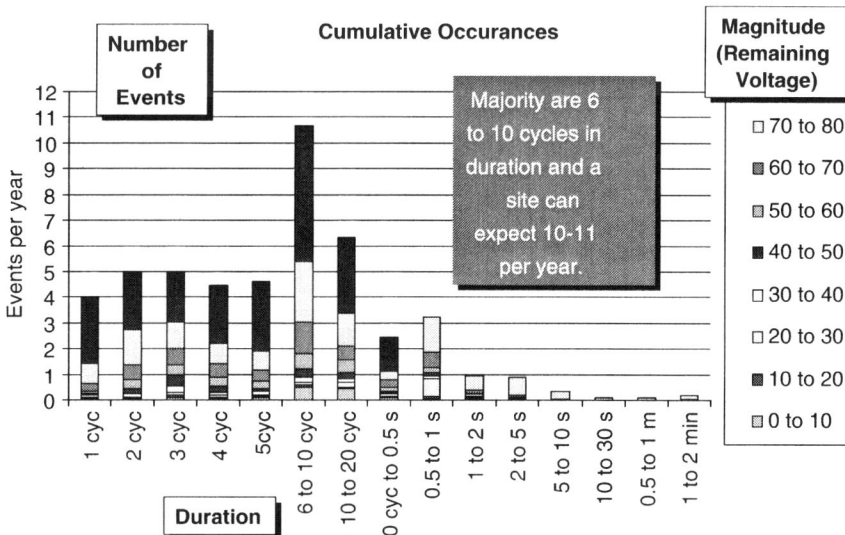

Fig. 1 Voltage sag data from the distribution power quality (DPQ) study.

processes to unexpectedly shutdown is also needed. Voltage sags are a common cause of process interruptions for many industrial customers. Voltage sags cause equipment and therefore, entire processes to shut down, impacting productivity as well as product quality. In an effort to resolve many of the problems caused by voltage sags, utilities provided funding to EPRI to perform an extensive testing of processes prone to shutdowns and to determine effective solutions. The information is in a web-based tool called the Industrial Design Guide (IDG).

EPRI and the participating utilities prioritized industries for study each year. To date, plastics and polymers processing, metals fabrication, semiconductor fabrication, pharmaceuticals, printing and publishing, and healthcare and hospitals have been addressed. The plastics and polymers processing industry was first completed. Processes within this industry were identified early and included pipe extrusion, plastic jacket extrusion, thermoforming, blown film—bubble, and injection molding. Each of these processes was tested in order to find the weak links within the process. One objective was to identify the most economical and effective power quality solutions and to do so required, identifying what components within the process caused the process to shut down. A process diagram of blown film is shown in Fig. 2. The equipment shown in dark gray is possible weak links and can be viewed in even more detail within the IDG.

A tool that enables component level testing was developed several years ago. This tool is connected to and powers the process being tested. The process is then subjected to controlled voltage sags of different magnitudes and durations, originating at various points on the voltage sinewave while the response of process components is being monitored. The result is detailed information about the ride through capabilities of various components during voltage sags. A sample of the results from such a test is shown in Fig. 3. This figure shows that for a sag to 87% nominal voltage for one cycle or longer, the photo eye, the most sensitive component in the process, will drop out causing the process to shut down. The next most sensitive component in this example is the double-pole, double-throw, DPDT, relay. A voltage sag to 75% of nominal voltage lasting two cycles or longer will cause this relay to dropout.

Once the process is characterized to the component level, solutions for each of the components can be identified as well as the cost effectiveness of each solution. With this information, the customer can make an informed business decision about the acceptable number of shutdowns per year and the cost he or she is willing to pay to achieve this level of ride-through.

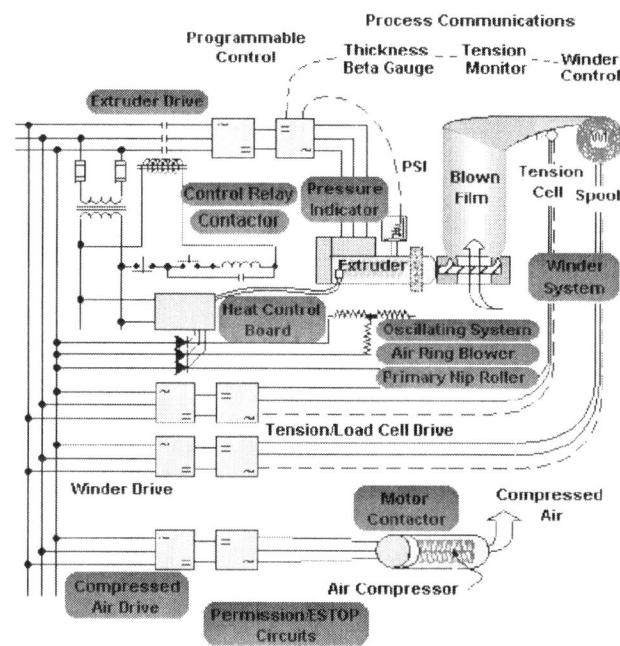

Fig. 2 Diagram of blown film process and possible weak links.

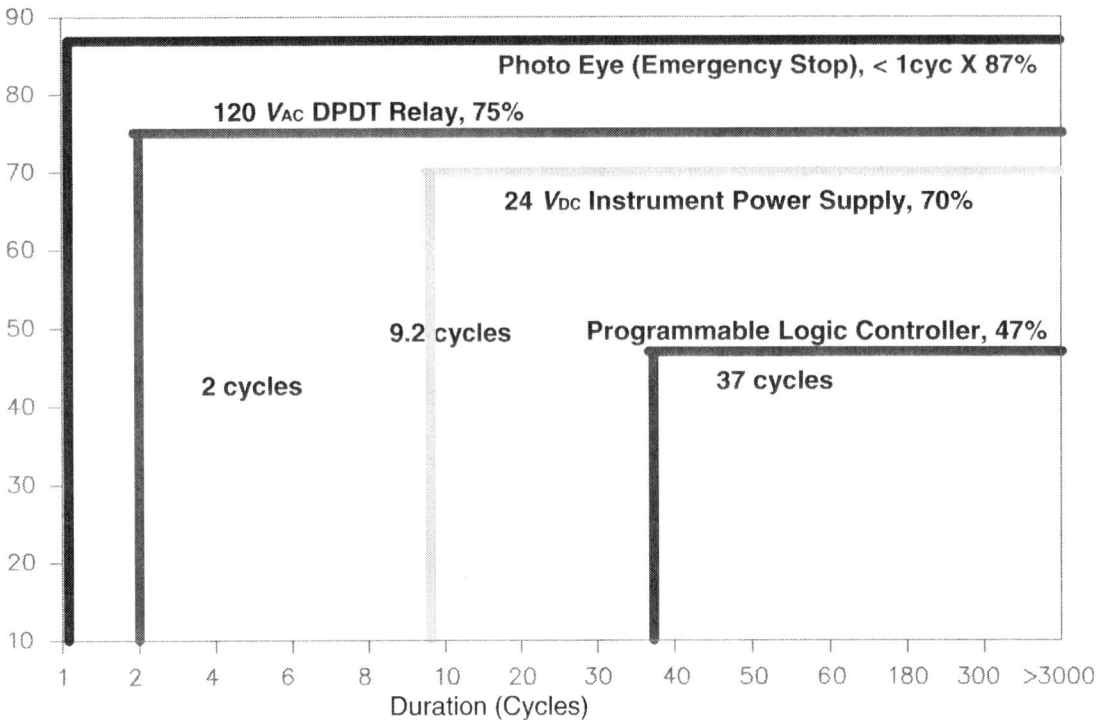

Fig. 3 Process component sensitivity.

Fig. 4 shows contour lines generated from the DPQ Study. For example, a distribution served customer with equipment sensitive to voltage sags in the highlighted area can reasonably expect between 10 and 15 process interruptions a year.

Fig. 5 overlays the process component sensitivity found from sag testing with the distribution system characterization curves from the DPQ Study. By examining the contour lines that the equipment sensitivity plot is within, a customer can predict how often a particular component

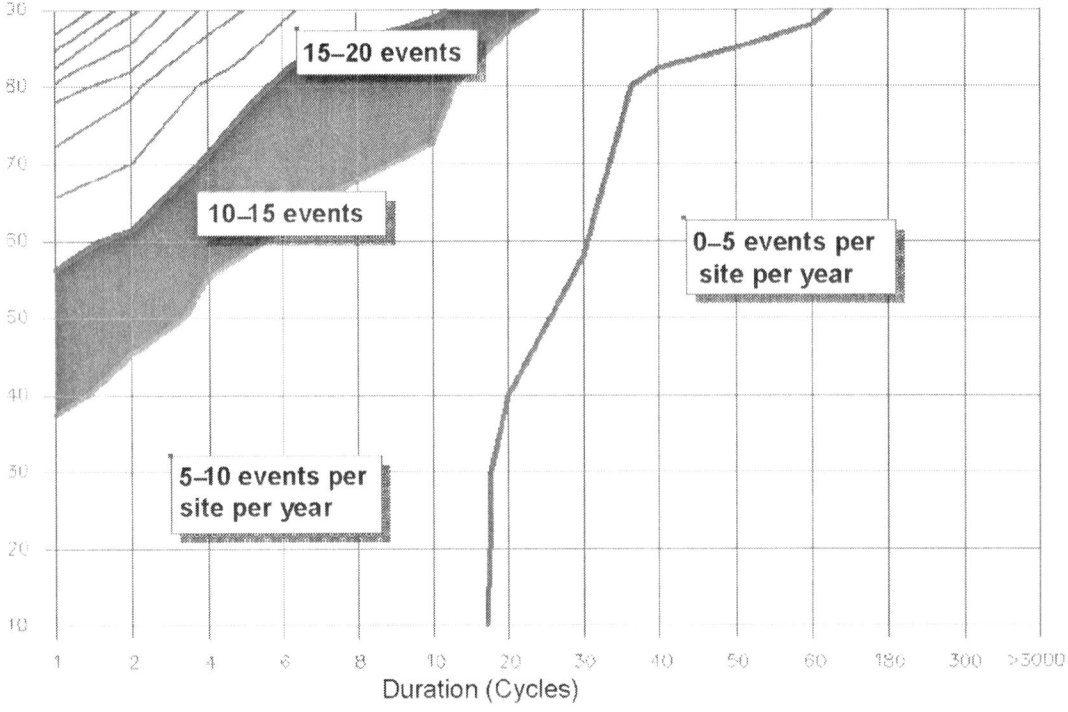

Fig. 4 DPQ study distribution system characterization curves.

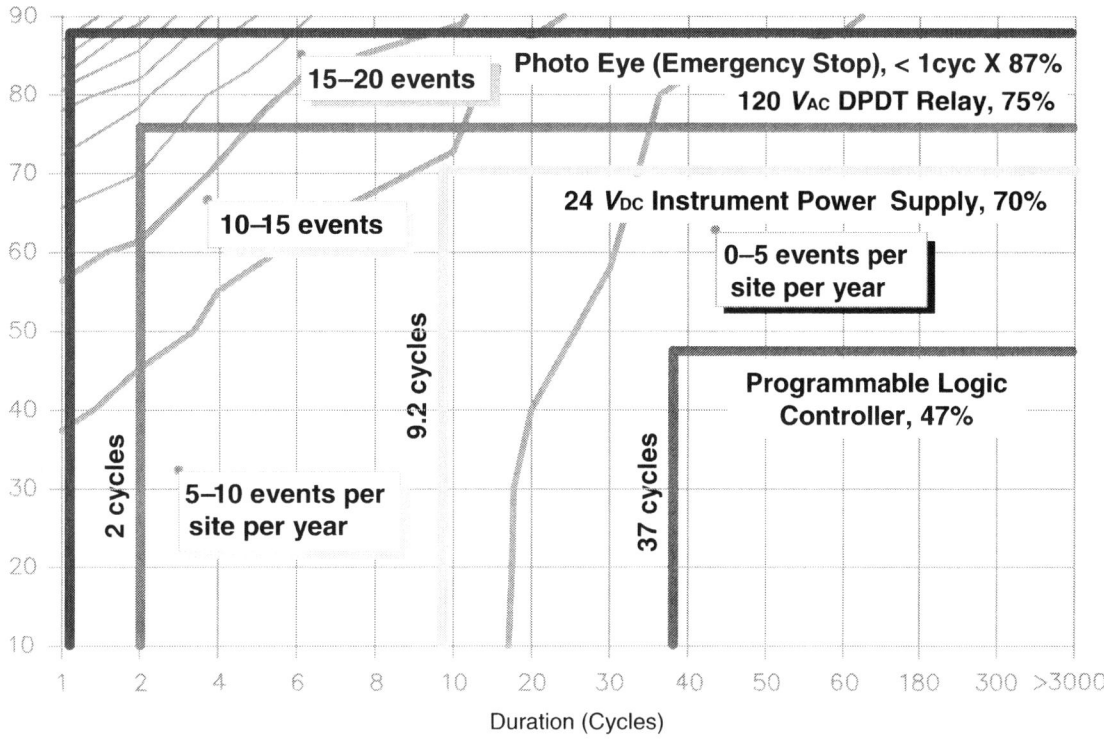

Fig. 5 Compatibility.

will result in process shutdowns each year. If the power supply were the weakest link in the process, a distribution served customer could expect between five and ten process interruptions a year. Unfortunately, the DPDT relay and the photo eye would cause far more disruptions.

Power Quality solutions can be applied at the facility level to protect the entire facility, at the process level to protect a process, at the equipment to protect the piece of equipment, or at the component level for example to hold in a relay or contactor. In general, power quality solutions applied at the facility level are the most expensive while solutions applied at the component level are the least expensive. There are situations however when each is appropriate. For example, if the cost of a process interruption is minimal, then perhaps improving the power supplies would be all that could be economically justified. However, if the cost of a process interruption is substantial, then a process DVR, dynamic voltage restorer, or perhaps even a flywheel would be the right business decision. Armed with this information and the cost of each solution, the customer can select and justify the appropriate level of voltage sag ride-through (Fig. 6).

TRACKING CUSTOMER REQUESTS

Each year customers contact their utility for help, resolving reliability or power quality problems. If the customer and the utility already have a good working relationship, the utilities are usually the first contacted. Sometimes the customer first contacts the equipment manufacturer and then at the recommendation of the manufacturer, contacts the utility. Unless there is an obvious problem with the equipment, the utility usually receives a call. This gives the utility the unique advantage of knowing what problems a particular business or industry is consistently experiencing. With this information, the utility can track and trend the data to determine what problems should be studied to find effective and economical solutions.

One tool used by utilities is the PQ Database. This database was developed by EPRI for utilities to use to store and sort power quality inquiry data. When a customer contacts a utility to request assistance with a power related problem, the description of the problem is entered into the database. As the information about the problem is gathered, it too goes into the database. Monitoring data is often collected and saved, and the actual resolution to the inquiry is also stored. This data can then be sorted in a variety of ways. Samples from a few such sorts are shown. Tables 1–3 illustrate such data collected by Duke Power's PQ engineers from 1997 to 2000.

Table 1 shows equipment affected by power quality events. Table 2 shows the responses of the equipment to the power quality events, and Table 3 shows the causes of the power quality events.

Once the data is collected and sorted, it quickly becomes obvious which equipment is experiencing the

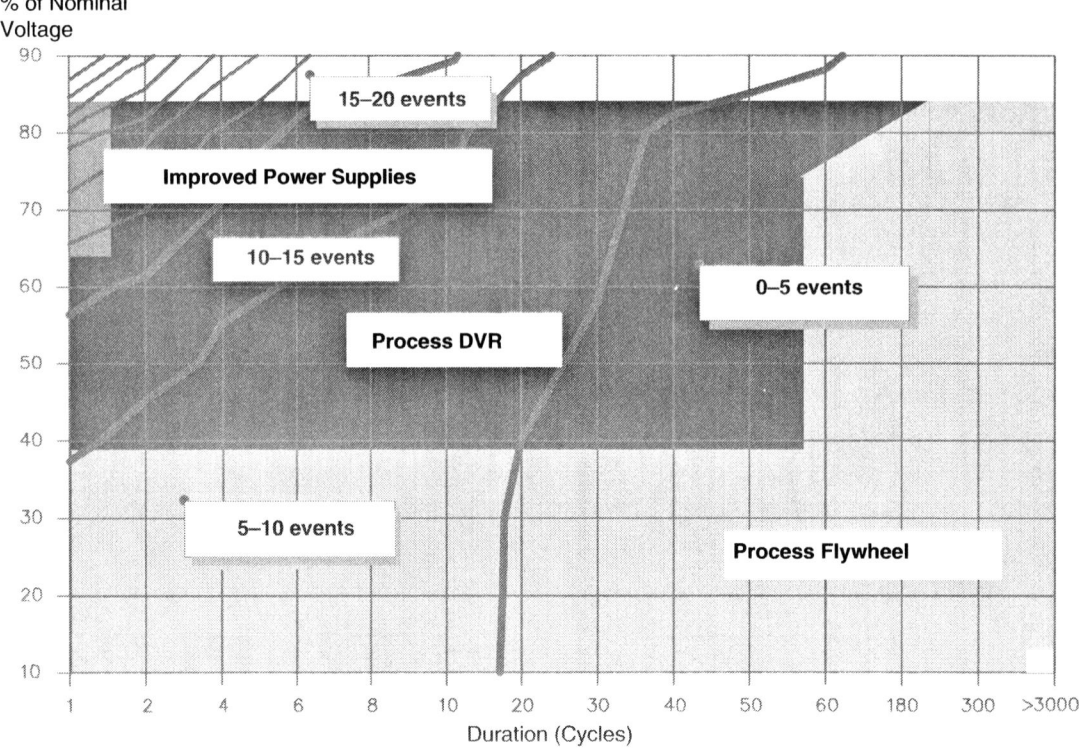

Fig. 6 Selecting a solution.

most problems and what problems are most often encountered. Many utilities contribute to a collaborative effort, to research ways to better solve common power quality problems. Often solutions exist, and it is just a matter of identifying the most appropriate solution and where the solution would best be applied. Sometimes existing solutions are not appropriate and a new solution is developed, applied, and tested. This effort requires collaboration between utilities, a research and development team, customers, and equipment manufacturers.

DEMONSTRATION PROJECTS

When ideal solutions to existing power quality problems do not exist, new solutions that solve the identified problems can be designed, developed, and tested. One example is the low speed flywheel. This flywheel was specified by EPRI, developed by Active Power, and installed and tested at Shaw Industries, a customer served by Duke Power. This demonstration project required collaboration and cooperation of EPRI, Active Power,

Table 1 Affected equipment from duke power PQ database

Affected equipment	Count	Percent
Plant equipment—in general	317	14.2
AC drive	204	9.2
Process controls	152	6.8
Motor	140	6.3
Other-affected equipment	97	4.4
PC	84	3.8
Breaker	73	3.3
LANS–Networks, Mainframes	62	2.8
UPS	61	2.7
Chiller, HVAC	61	2.7

Table 2 Equipment response

Affected equipment	Count	Percent
Equipment/component failure	501	27.5
Drop-out	334	18.3
Erratic operation	232	12.7
General-customer request	207	11.3
Nuisance tripping	154	8.4
Energy usage monitoring	65	3.6
Cycling—on and off	52	2.9
Lights dimming/flickering	42	2.3
PQ characterization	31	1.7
Fuse expulsion	29	1.6

Table 3 Problem causes

Affected equipment	Count	Percent
Duke system; fault, operation	209	13.4
Customer equipment degradation	155	9.9
No PQ problem	153	9.8
Other—customer system problems	148	9.5
Lightning strike	115	7.3
Grounding/bonding	94	6.0
Capacitor switching	91	5.8
Equipment specification	71	4.5
Problem cause no found	68	4.3
No electrical problem	61	3.9

Shaw Industries, and Duke Power. The flywheel was applied to a process line that extrudes filaments to produce carpet backing. This facility was served by a distribution circuit and the process shut down frequently due to voltage sags and momentary interruptions.

The key to successful implementation of the flywheel as a power quality solution included:

- Characterizing the electrical system,
- Determining the sensitivity of the process components,
- Matching the solution to the process line power requirements as well as to the utility distribution system relaying schemes,
- Including specific and accurate language in specifying the solution,
- Meticulously diagramming the process,
- Properly installing the solution, and
- Monitoring the results.

Shaw Industries realized both productivity and efficiency gains from the operation of the flywheel.

UTILITY/CUSTOMER FORUMS

Some utilities have customer groups that meet regularly to discuss power quality and reliability issues, as well as utility standards and changes, and customer or industry challenges. This forum offers a unique opportunity for the utility and the customer to find ways to help each other. Many power quality opportunities are discovered this way. Without this interaction, the utility could only guess what their customers needed and customers could only guess how and why utilities implemented policies and changes. One such group, the Power Quality Issues Forum was established at Duke Power in 1992. This group meets quarterly and participates in activities that include: visiting the Clemson University Electrical Engineering Laboratory to see utility funded power quality research and development efforts; attending PQA, a yearly power quality conference hosted by EPRI; traveling to EPRI-PEAC to see equipment and product testing and research; sponsoring power quality training to stay current on industry initiatives; and visiting demonstration sites such as the installation of the flywheel at Shaw Industries.

WORKSHOPS

As information is gained from demonstration projects, IDG work, SagGen testing, equipment testing done by EPRI-PEAC, and research and development conducted by universities and funded by utilities, this information must be shared with utility customers. One effective way of doing so is by holding workshops, conferences, and seminars. Some specific examples include the Industrial Technology Information Exchange conference sponsored by Duke Power yearly, for the past two years. This conference highlights new electric technology applications, power quality and reliability efforts, and innovative ways to reduce energy costs. Based on customer feedback from the first year, the conference was held twice in 2001 in both North and South Carolina and participation increased dramatically. Another successful way to share relevant power quality information is conducting workshops and seminars. Such workshops offered by utilities focus on wiring and grounding for enhanced equipment performance, effectively applying surge suppression, improving process ride-through, selecting effective power quality solutions, improving drive performance, selecting appropriate monitoring equipment for the application, designing effective lightning protection systems, and solving problems with ungrounded electrical systems.

TRACKING COSTS OF PQ PROBLEMS

In order for a customer to make an informed business decision about power quality solutions, it is imperative to know the cost to the business when a power quality problem occurs. This cost should include scrap, unproductive labor, lost productivity, missed orders, risk, and recovery. A process for tracking these costs should be implemented if not already in place.

SPECIFYING NEW EQUIPMENT FOR RIDE-THROUGH

In order to minimize costly power quality solutions, ride-through specifications can be included when purchasing new equipment. Ideally, one would characterize the

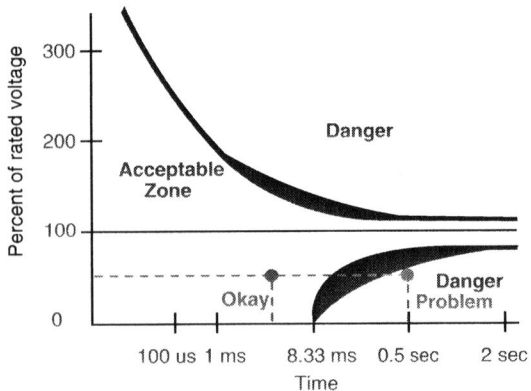

Fig. 7 The Information Technology Industry Council (ITIC) curve.

electrical environment to know exactly what specifications to include, but this is not always feasible. A generic ride-through curve can be included which would allow a process to ride-through most voltage sags. Fig. 7 illustrates the Information Technology Industry Council (ITIC) Curve. This curve was developed to represent the sensitivity of most computer equipment. The region within the two curves represents conditions where computer equipment should operate without disruption or damage. A similar curve has also been developed for the semiconductor industry.

KEEPING ABREAST OF NEW PQ SOLUTIONS

Utility customers can keep abreast of new power quality solutions in a variety of ways. Power quality magazines have articles describing new solutions. Utilities offer seminars and conferences outlining the features of power quality solutions. Literature from equipment testing is available through EPRI utility members. With the tendency for business and industry to reduce personnel, it is not always practical to stay informed about power quality issues. In which case, one can contact utility power quality experts when power quality problems occur or when specifying new equipment.

EDUCATING UTILITY ON PROCESSES

In order for utilities to help when power quality problems do arise, it is imperative that the utility power quality engineers be familiar with the processes in question. This can only be accomplished if the utility customer experiencing the problem feels comfortable sharing process information with the utility. Much information regarding processes is confidential and proprietary, and the utility must respect that confidentiality. It is much easier to expeditiously solve a power quality concern if the utility power quality engineer already has a good understanding of the process equipment. This requires resources and commitment from the utility management. Power quality expertise takes years to acquire technology, equipment, and power quality solutions are constantly changing, requiring time and effort to remain informed.

CONCLUSION

By actively participating in a partnership, both the electric utility and utility customers can significantly improve their business results. This requires collecting and sharing information regarding utility system operation, studying and sharing process requirements—including sensitivity to electrical disturbances, and evaluating solutions to power quality problems.

Information can be effectively shared through customer forums, conferences, and workshops. A utility, customer partnership requires commitment from all involved, but the benefits are well worth the effort.

BIBLIOGRAPHY

1. *Industrial Design Guide*: Version 3.0, EPRI: Palo Alto, CA, 2001; 1005912.
2. *Power Quality Applications Guide for Architects and Engineers*, EPRICSG: Palo Alto, CA, 1999; TR-113874.
3. *Duke Power Energy Solutions for the Information Age Workshop*, Charlotte: NC, May 23, 2002.

Utilities and Energy Suppliers: Planning and Portfolio Management

Devra Bachrach Wang
Natural Resources Defense Council, San Francisco, California, U.S.A.

Abstract

Throughout the United States, most customers continue to receive service from their hometown utility companies, regardless of the status of retail competition in their state's electric industry. Recent turmoil within the industry has focused attention once again on a crucial responsibility of those utilities: electric-resource portfolio management. Effective portfolio management requires a fully integrated approach to identify customers' electric service needs and to select demand- and supply-side alternatives to meet those needs through a portfolio that minimizes the total cost and environmental impacts and has an acceptable level of risk. To enable effective portfolio management, regulators must align utilities' financial incentives with customers' interests. Traditional regulation creates a substantial financial disincentive for utilities to meet customers' energy service needs through cost-effective energy efficiency or other demand-side resources, but regulators and governing boards can and should eliminate this disincentive.

Every utility's current resource mix and its cost of providing energy is largely inherited from past portfolio managers' investment decisions. Long-term, integrated resource planning provides an opportunity for portfolio managers to work with their regulators or governing boards, customers, and other stakeholders in an inclusive and transparent manner to create a plan to meet future energy service needs. The resulting long-term plan should enable stakeholders to understand what the resource mix will be in ten or twenty years, the expected costs of the resource plan, and its environmental impact. Long-term plans provide a common basis for both the utility and the regulator upon which to found subsequent resource investment decisions.

The long-term planning process begins with a forecast of customers' demand and an accounting of the resources that are available to meet that demand. The portfolio manager should conduct a comprehensive analysis of the costs, risks, and environmental impacts associated with all resource options, including both demand-side and supply-side resources, that could be included in the portfolio to meet customers' needs. A number of potential resource portfolios should be tested against the portfolio manager's primary objectives. The final long-term plan should document this analysis and describe the elements of the preferred resource portfolio.

PROVIDING ENERGY SERVICES: THE ROLE OF UTILITIES IN SOCIETY

The overriding purpose of any utility or energy service provider is to meet its customers' energy service needs in an affordable, reliable, and environmentally sensitive manner (Throughout this entry, I often refer to the portfolio manager as a utility, since utilities are the most common entity to perform the portfolio management function; however, as I discuss in a later section, other entities also perform this function). While this may sound simple in concept, achieving these goals in practice is difficult and requires expertise in managing a diverse portfolio of resources to achieve sometimes conflicting objectives.

Managing a portfolio of demand- and supply-side resources is complex in part because the services that energy utilities provide customers vary widely. Utility customers do not purchase electricity from a utility for the sake of the electricity itself. Instead, customers pay the utility in order to receive the services that the electricity can provide. Customers want light to read by, cool and comfortable homes in the summer, and clean clothing, all for a reasonable price. As such, customers generally care more about the total bill they will pay to receive those services than about the rate for each kilowatt-hour of electricity consumed. In order to best meet customers' needs, utilities must focus on the broader scope of providing the energy services that customers want, rather than just the energy itself as a commodity.

Moreover, many of the services that utility customers demand, along with access to shelter, food, and water, are considered essential to achieve even a basic quality of life in modern times. As the single largest source of air pollution in the United States, the electricity industry must also be a key contributor in the effort to provide all people with another basic human right: access to clean and healthy air to breathe. Thus, how utilities provide their

Keywords: Portfolio management; Integrated resource planning; Risk management; Long-term plan; Resource procurement.

customers with energy services is fundamentally intertwined with a broad range of social and political issues.

THE BENEFITS OF PLANNING AND PORTFOLIO MANAGEMENT

Perhaps nothing in recent history demonstrated the need for long-term planning and portfolio management in the electric industry as clearly as the California energy crisis of 2001. (In 2002, the California Legislature enacted Assembly Bill 57, returning the utilities to the role of portfolio managers.[1] The California Public Utilities Commission has adopted several subsequent decisions providing guidelines for the utilities' portfolio management activities.[2,3]) That experience showed forcefully that simply relying on a spot market to meet customers' needs will not achieve the industry's overriding goals of providing customers with energy services in an affordable, reliable, and environmentally sensitive manner. To achieve these goals, utilities and other service providers must assemble a diverse portfolio of resources that will be robust in the face of the many uncertainties prevalent in the electric industry. Such a portfolio is likely to include demand- and supply-side resources with a mix of long-, medium-, and short-term commitments, owned and contracted resources, and supply-side resources with a variety of fuel sources, along with financial instruments to help manage any remaining risks.[4] Careful diversification along a number of different dimensions will help to avoid the classic pitfall of putting all the portfolio manager's "eggs in one basket" and can provide protection against risks, including those related to fuel prices, future loads, fuel supply availability, and future environmental regulations.

Who is Responsible for Portfolio Management?

Throughout the United States, the most common electric-resource portfolio manager is the hometown utility. This includes publicly owned utilities that are governed by local boards, as well as investor-owned utilities that are regulated by state Public Utilities Commissions. These boards and regulators are responsible for guiding the utilities' portfolio management and long-term investment activities.

In states that have restructured their electricity industry and established a competitive retail market, energy service providers (ESPs) also perform the portfolio management function on behalf of their customers, while the management of the distribution system remains with the utility. Most ESPs are required to sign up each individual customer that they serve; however, a few states, including California, Massachusetts, Ohio, and Rhode Island, have enacted legislation that enables local governments to become portfolio managers, commonly known as community choice aggregators (CCAs), by providing energy services to utility customers in their jurisdictions on an "opt-out" and aggregate basis.

The Long-Term Planning Process

Long-term planning is an essential responsibility of portfolio managers. A long-term view is necessary because the electricity industry is characterized by the need for capital-intensive investments with sometimes long lead times. These investments are often lumpy in nature, and portfolio managers' investment decisions must be made based on uncertain information. In addition, many new resources will continue generating or saving electricity for thirty to forty years or more, so the costs and benefits of investing in a particular resource must be analyzed over an extended time period. Moreover, the various resources from which a portfolio manager can choose to meet its customers' needs expose it and its customers to various types and levels of risks, which must be analyzed in the context of the overall portfolio of resources.

Portfolio managers should conduct integrated analyses of various resource portfolios and investment options on a regular basis, for example, every two to three years. These long-term plans can build off of the experience utilities and regulators gained with Integrated Resource Planning over the past quarter-century. Regular long-term planning processes enable a portfolio manager to compare resource alternatives in a manner that captures interactive portfolio effects. Without long-term, integrated planning, a utility that analyzes procurement options one by one is likely to "miss the forest for the trees." Each individual investment decision may seem like the best decision, but the *additive* effect of the decisions and the impact on the overall portfolio would not be considered without true long-term plans. The preferred resource plan generally has, among other factors, the lowest lifecycle cost (i.e., the lowest anticipated long-term revenue requirement) and is most robust in the face of various risks.

Long-term plans provide a portfolio manager and its regulators or governing board, customers, and other stakeholders with a common roadmap toward the future. Importantly, the process of developing a long-term plan provides a crucial opportunity for portfolio managers to engage all of these stakeholders in a collaborative, public process to understand the risks and benefits of alternative portfolios and to seek a general consensus on the best path forward. (Montana's guidelines for electricity procurement provide a good example of the policies governing, and the processes used for, portfolio management.[5]) Such a public process not only benefits the various stakeholders who have an opportunity to help determine the future of the portfolio, it may also help the portfolio manager to avoid potentially lengthy and adversarial regulatory hearings.

While long-term plans must delve into the technical details of load forecasts and the characteristics of resource alternatives discussed further below, a plan should also be able to answer key policy-level questions. For example, a long-term plan should enable policymakers and the public to understand what the portfolio manager's resource mix will be in ten or twenty years under its preferred plan and how it will differ from the present resource mix. The plan should also clearly present the expected costs of the resource plan (and the impact on average customer bills over time), the accompanying risks, and its impact on pollution emissions.

Long-term plans (or Integrated Resource Plans) serve as common guidebooks for both the utility and the regulator, so that subsequent resource decisions are founded upon common understandings and assumptions. These plans may assist utilities in making a strong case for cost recovery, and they help regulators ensure that the energy system is on an appropriate path toward the future.

UTILITIES' FINANCIAL INCENTIVES MUST BE ALIGNED WITH CUSTOMER INTERESTS TO ENABLE EFFECTIVE PORTFOLIO MANAGEMENT

Effective portfolio managers should analyze both demand- and supply-side resources to meet their customers' needs. In many instances, it is more cost-effective for a utility to help customers use energy more efficiently than to buy power (or build a power plant) to serve inefficient uses of energy. However, traditional regulation creates a substantial financial disincentive for utilities to meet customers' energy service needs through cost-effective energy efficiency or other demand-side resources.

This disincentive arises because once a utility's rates are set, its revenues are tied to the volume of electricity that is sold. If actual annual electricity sales diverge from the forecast that is used to set the authorized revenue requirement, the utility will either under- or over-recover the fixed-cost element of its revenue requirement. While this problem is most often discussed in the context of investor-owned utilities, publicly-owned portfolio managers, including municipal utilities and CCAs, can face similar "conflicts-of-interest" with demand-side resources if their governing boards use a similar process to set rates. Fortunately, this disincentive can be eliminated through the use of modest, regular "true-ups" in rates to ensure that any fixed costs recovered in kilowatt-hour charges are not dependent on sales volumes. This solution has been implemented successfully by a number of regulators to remove utilities' disincentives for investments in energy efficiency and other demand-side resources. (For a complete discussion of the conflicts of interest that portfolio managers can face relative to demand-side resources, and the solutions to it, see the reference section.[6])

In addition to eliminating these disincentives, development of the most reliable, affordable, environmentally responsible energy service portfolio requires, among other things, a balanced, performance-based incentive system that provides the portfolio manager with risks and rewards to the extent it achieves (or does not achieve) these objectives. More than a decade ago, the national association of regulatory utility commissioners (NARUC) urged its members to "ensure that the successful implementation of a utility's least-cost (investment and procurement) plan is its most profitable course of action."[7] The resolution framed the term "least-cost" over an extended time period. Congress endorsed NARUC's objective in the National Energy Policy Act of 1992, for both electric and gas utilities, although the final decision remains with state regulators.[8] NARUC's stated objective remains an important prerequisite to enable effective electric-resource portfolio management, although in most states, current regulatory incentives do not achieve this objective.

KEY ELEMENTS OF A LONG-TERM PORTFOLIO MANAGEMENT PLAN

The process of developing a long-term plan begins with a forecast of customers' demand and an inventory of the resources already at the portfolio manager's disposal to meet that demand. Next, the portfolio manager should conduct a comprehensive analysis of the costs, risks, and environmental impacts associated with all available resource options, including both demand-side and supply-side resources, that could be added to the portfolio to meet customers' needs. Finally, the portfolio manager should test a number of potential resource portfolios against its primary objectives, conduct a risk analysis under various scenarios, and assemble a plan for an optimal portfolio.[9]

Forecasting Demand and Existing Resources

Forecasting customers' energy service needs (both energy and peak-load demand) is the first step in long-term planning. The changes in demand over time are a function of many factors, including population, economic conditions, weather, and how energy is used by customers (known as "end-uses"). Portfolio managers can use both econometric and end-use forecasting techniques to project future customer demand.[10] Econometric models use a "top-down" approach to project future demand by extrapolating trends based on past relationships between demand and key metrics such as population, weather, and economic growth. End-use forecasting models produce "bottom-up" forecasts based on the penetration and energy consumption of various identifiable end-uses such as refrigeration, lighting, air conditioning, and other electrical equipment.

Together, these types of models can capture the key variables affecting changes in customer demand over time. However, since forecasts are inherently uncertain, portfolio managers should also conduct sensitivity analyses to understand the full range of possible future demand scenarios. These sensitivity analyses might, for example, analyze a low load-growth scenario under which population and economic growth estimates are at the low end of the range, and conversely, a high load-growth scenario with higher possible population and economic growth forecasts. In states with retail competition, these scenarios might also include forecasts of various levels of customers switching providers. This forecasting and scenario analysis process often results in a "jaws" forecast of future demand, defining both the best estimate of future demand as well as the potential range of future demand.

Once a portfolio manager has forecast the range of possible demand, the next step is to assess the existing resources that it either owns or has under contract, and to compare these available resources to the demand forecast. The difference between the two will determine the portfolio manager's remaining need for additional demand- or supply-side resources. During the planning process, the portfolio manager should also assess whether the portfolio could be improved by replacing or re-powering any of the existing resources.

Analysis of Resource Alternatives

To fill that remaining need, the portfolio manager should analyze the costs, risks, and environmental impact associated with the full range of potential resource options. These options include energy efficiency, distributed generation, renewable resources, thermal resources (such as natural gas-fired plants and integrated gasification combined-cycle coal plants), transmission, and more.

Energy efficiency is the most cost-effective, reliable, and environmentally friendly resource available to portfolio managers. Assessing the potential for energy efficiency resources requires an analysis of the various end-uses (i.e., how customers use energy), how much more efficient those end-uses could be, and what level of efficiency is achievable through voluntary programs that provide incentives and information to customers to improve their efficiency or through mandatory standards that set the minimum level of required efficiency. (California's recent analysis of the potential for cost-effective energy efficiency provides a good example of this type of study in potential.[11])

Determining what portion of that energy efficiency potential is cost-effective requires an analysis of the total cost to society of procuring the energy savings; the total resource cost (TRC) test is the cost-benefit test most commonly used by regulators and portfolio managers to determine what level of energy efficiency is cost-effective.[12] The TRC test accounts for the total cost to society of acquiring efficiency resources, including the incremental cost to customers (if any) of implementing more efficient technologies or practices and the cost to the portfolio manager of offering the programs. Since energy efficiency resources are comprised of many smaller resources, portfolio managers often shorten the analysis of determining what portfolio of programs is cost-effective by comparing the costs of the efficiency resources to a static estimate of avoided costs (including avoided costs of generation, transmission, distribution, and environmental pollution), rather than putting the energy efficiency programs into the iterative portfolio analysis alongside the supply-side resources.

Assessing supply-side options requires an analysis of the costs, attributes, and risks associated with each resource. Every resource's fixed and variable costs should be assessed either over the lifetime of the resource or over some fixed period, often thirty years. In order to allow all resources to compete on a level playing field in this assessment, portfolio managers must use accurate operating and cost assumptions for each resource. For fossil-fueled resources, forecasting fuel prices (with a sensitivity analysis) is a critical element of this cost assessment. In analyzing renewable resources, assumptions regarding capacity factors and integration costs must be realistic in order to accurately reflect the cost of these resources.

Different resources have different operating characteristics, which may make them more or less valuable to a portfolio manager. For example, some resources are designed to operate at a relatively constant output to serve baseload demand, whereas other resources are dispatchable and can ramp up and down quickly to follow load or to meet peak demand. The portfolio manager must consider these attributes in order to design a portfolio that enables it to maintain reliable service.

Each resource exposes the portfolio manager to certain risks while mitigating other risks.[13] For example, natural gas-fired resources expose the portfolio manager to fuel price risks, whereas renewable resources should help mitigate exposure to this risk in a portfolio. Conversely, some types of renewable resources are intermittent and expose the portfolio manager to reliability risks, whereas natural gas-fired resources are more dispatchable and can help mitigate this risk. Conventional coal-fired generation is particularly vulnerable to financial risks associated with the potential regulation of greenhouse gases. These are just a few examples; all of the significant risk attributes of each resource must be well understood so that the various portfolios of resources can be tested against the variety of risks and designed to minimize both risks and costs.

Finally, resources have widely varying environmental impacts. By analyzing the environmental profile of each type of resource, the portfolio manager can assess the projected environmental impact of various portfolio options to help select a portfolio that meets the objective of providing energy services in an environmentally responsible manner. This information is also necessary to assess the financial risk exposure due to pollution emissions, as we discuss further below.

Determining the Optimal Portfolio of Resources

The final steps in assembling a long-term plan are to test a number of potential resource portfolios to determine their total long-term costs, conduct a risk analysis of those portfolios under various scenarios, and select an optimal portfolio that best meets the portfolio manager's objectives. While optimization models have been used to help determine the preferred plan, most utilities construct dozens of potential portfolios "by hand." These portfolios should span a wide spectrum of possible options. The operation of these portfolios is then modeled over the timeframe addressed by the plan (often 30 years) and the portfolios' total long-term revenue requirements, environmental impacts, and other metrics are compared. As the analysis illuminates the components of the portfolios that are more or less desirable, new portfolios may also be constructed and tested.

This application of "base case" assumptions in the initial analysis should produce useful information, but given the numerous risks in the electric industry, it is essential to conduct a risk analysis to test how robust each portfolio is in the face of various uncertainties. There are generally at least three different types of risks: (1) risks that can be quantified and for which historical experience can inform assessments of the future risk (e.g., load forecasts, natural gas price risk, etc.); (2) risks that can be quantified but for which no historical experience can inform the assessment (e.g., future regulation of carbon dioxide emissions); and (3) risks that cannot be easily quantified, but can be assessed qualitatively (e.g., a change in FERC's market design, public acceptance of new resource siting, etc.) PacifiCorp's 2003 Integrated Resource Plan provides a good discussion of these different types of risks and develops a framework for analyzing the risks.[14]

Utilities have traditionally emphasized the first type of risk listed above in their analyses. However, the other two types of risks are no less significant or real; even if they can't be quantified based solely on historical experience, they can often be quantified and incorporated into the integrated resource analysis. The financial risk associated with future regulation of carbon dioxide emissions is a prime example of the type of risk listed in the second category above that utilities have historically failed to assess or mitigate.[15] As the electric industry becomes more sophisticated with risk management, a number of utilities are beginning to conduct comprehensive risk assessments. Leading examples include PacifiCorp, Idaho Power, Puget Sound Energy, and Pacific Gas & Electric.[16–19] The Northwest Power and Conservation Council is a leading public sector practitioner.[20]

Based on these cost, risk, and environmental performance analyses, the portfolio manager should select a preferred portfolio that best meets its objectives. The final long-term plan should outline the elements of that preferred portfolio, in addition to documenting the full analysis conducted to assemble the preferred portfolio.

TURNING A LONG-TERM PLAN INTO ACTION

The principal value of a long-term plan is determined by its translation into the actual procurement of resources by the portfolio manager. Long-term plans lay out the elements of a preferred portfolio over ten to thirty years or longer, but circumstances within the industry may change over time and affect the underlying assumptions that led to the selection of the preferred plan. As such, long-term plans should be living documents that are updated regularly, as needed. In the intervening years, the procurement of resources should be consistent with the long-term plan. However, if conditions have changed enough to warrant deviations from the plan, the long-term plan should provide a common basis for stakeholders and regulators to understand what has changed and how those changes affect procurement.

Different portfolio managers have utilized their long-term plans in resource procurement in varying ways. Some include a near-term action plan within the long-term plan itself, laying out the steps to be taken in the near future to work toward the long-term vision described in the plan. In some cases, utilities will build or contract for each specific type of resource described in the plan. In other cases, utilities will solicit competitive bids and determine whether to procure more or less of a certain type of resource based upon the offers that are received. Either way, it is important for the portfolio manager to remain flexible to respond to actual circumstances while still utilizing the guidance that is provided from the full portfolio analysis produced in the long-term plan.

CONCLUSION

Portfolio management is essential to enable utilities and other energy service providers to meet customers' energy service needs in an affordable, reliable, and environmentally sensitive manner. Long-term integrated resource planning is a crucial tool for portfolio managers to balance costs, risks, and other objectives in an industry that is characterized by widely varying resource options and long-lived investments. Recent turmoil within the electric industry has put a spotlight on the benefits of portfolio management and, in particular, the tools it provides for risk management. Regardless of what type of entity is responsible for meeting customers' energy service needs, or a region's electric industry structure, portfolio management and long-term planning enables all stakeholders to work together to create a common roadmap toward a better energy future.

REFERENCES

1. California Assembly Bill 57 (Wright, Chapter 835, Statutes of 2002).
2. California Public Utilities Commission (CPUC) Decision 03-12-062, December 18, 2003.
3. CPUC Decision 04-12-048, December 16, 2004.
4. Harrington, C.; Moskovitz, D.; Shirley, W.; Weston, F.; Sedano, R.; Cowart, R. *Portfolio Management: Protecting Customers in an Electric Market That Isn't Working Very Well*; Regulatory Assistance Project for the Energy Foundation and the Hewlett Foundation: Montpelier, VT, July 2002.
5. Administrative Rules of Montana, Default Electric Supplier Procurement Guidelines, Title 38, Chapter 5, Sub-Chapter 82, section 38.5.8201 et seq., December 31, 2003, http://161.7.8.61/38/38-6151.htm.
6. Bachrach, D.; Carter, S.; Jaffe, S. Do portfolio managers have an *inherent* conflict of interest with energy efficiency?. The Electricity Journal **2004**, *17* (8), 52–62.
7. Moskovitz, D. *Profits and Progress Through Least-Cost Planning*; The National Association of Regulatory Utility Commissioners: Washington, DC, 1989.
8. 16 USC section 2621(d)(8).
9. Biewald, B.; Woolf, T.; Roschelle, A.; Steinhurst, W. *Portfolio Management: How to Procure Electricity Resources to Provide Reliable, Low-Cost, and Efficient Electricity Services to All Retail Customers*; Synapse Energy Economics prepared for the Regulatory Assistance Project and the Energy Foundation: Cambridge, MA, October 10, 2003, www.raponline.org/Pages/Feature.asp?select=15.
10. Tellus Institute and Institute of International Education. *Best Practices Guide: Integrated Resource Planning for Electricity*. Prepared for United States Agency for International Development: Washington, DC. Available at: http://telluspublications.stores.yahoo.net/bespracguidi.html.
11. Rufo, M.; Coito, F. *California's Secret Energy Surplus: The Potential for Energy Efficiency*. Xenergy Inc. for the Energy Foundation and the Hewlett Foundation, 2002, www.energyfoundation.org/energyseries.cfm.
12. California Public Utilities Commission. *California Standard Practice Manual: Economic Analysis of Demand-Side Programs and Projects*. October 2001, www.cpuc.ca.gov/static/industry/electric/energy+efficiency/rulemaking/eeevaluation.htm.
13. Bachrach, D.; Wiser, R.; Bolinger, M.; Golove, W. *Comparing the Risk Profiles of Renewable and Natural Gas Electricity Contracts: A Summary of the California Department of Water Resources Contracts*, LBNL-50965; Lawrence Berkeley National Laboratory: Berkeley, CA, April 2003.
14. PacifiCorp. 2003 *Integrated Resource Plan*, pp. 4–7, www.pacificorp.com.
15. Bokenkamp, K.; LaFlash, H.; Singh, V.; Wang, D. Hedging carbon risk: protecting customers and shareholders from the financial risk associated with carbon dioxide emissions. The Electricity Journal, **2005**, *18* (6), 11–24.
16. Idaho Power Company. 2004 *Integrated Resource Plan*, July 2004, www.idahopower.com/energycenter/2004irp.htm.
17. PacifiCorp. 2004 *Integrated Resource Plan*, 2004, p.155, www.pacificorp.com/Navigation/Navigation23807.html.
18. Puget Sound Energy. *Least Cost Plan*, April 2005, www.pse.com/about/supply/resourceplanning.html.
19. Pacific Gas and Electric Company. *Prepared Testimony*, R.04-04-003, Order Instituting Rulemaking to Promote Policy and Program Coordination and Integration in Electric Utility Resource Planning, July 9, 2004.
20. Northwest Power and Conservation Council. *The Fifth Power Plan: A guide for the Northwest's Energy Future*, Pre-publication Draft, Document 2004-18, January 2005, www.nwcouncil.org/energy/powerplan/draftplan/Default.htm.

Utilities and Energy Suppliers: Rate Structures

Neil Petchers
NORESCO Energy Consulting, Shelton, Connecticut, U.S.A.

Abstract

This article discusses how electric, gas, and other regulated utilities charge commercial, industrial, and institutional customers for their energy services. In today's utility industry, there are numerous factors which make up a consumer's monthly electric or gas bill. This article contains sections that will individually describe each important component of most common electric and gas rate structures. A full understanding of each component will then allow a customer to comprehensively read and understand a monthly utility bill. Types of loads, cost allocation methods, billing factors, rate design strategies, common electric rates, and common gas rates will all be discussed.

INTRODUCTION

Natural gas and electric utility rates are determined through a regulatory process under the jurisdiction of a public utility commission (PUC). The utility periodically applies for a rate case hearing to request rate adjustments in response to changes in economic factors, such as supply, demand, distribution, and market forces that affect the utility's cost to serve. Rates are designed so that the utility can recover sufficient funds to cover costs and, in the case of investor-owned utilities, generate a reasonable return on investment for its stockholders. The PUC first determines the level of allowable investment by the utility in its facilities (referred to as the rate base) and its level of operating expenses. It then determines the rate of return the utility may earn on its investment and adds to it the operating expenses. This sets the total revenue requirement, which is equal to the total cost to serve. Rates are then designed to allocate revenue requirements among the various customer classes and subcategories within the classes.

Typically, customer classes include residential, commercial, industrial, institutional, and municipal. They may be subdivided into many rate classes or lumped together into broader rate classes. Commercial and industrial (C&I) as well as institutional (CI&I) customers, for example, are commonly lumped together.

General rate classes include categories, such as residential, small CI&I, and large CI&I. Residential rate classes are typically limited to single-family dwellings and multifamily dwellings metered separately from one another. Master-metered multifamily dwellings can be treated either as a separate rate class or as part of a commercial rate class.

Two common distinctions between customer classes are size of load and usage profile, although distinctions are made between CI&I classes based on other criteria as well. Time-of-use (TOU) and load factor are, in many cases, even more significant factors than size of load. Additional rate classes sometimes exist for cogenerators, nonutility generators, and other end-use- or equipment-specific categories. Another major distinction is between firm and nonfirm service. Many utilities are also able to offer special contracts that are customer-specific.

In assigning rate classes, utilities attempt to determine commonalties among customers within a class. This includes the assignment of load behavior. Contribution to a probability of a peak, for example, is an extremely important factor. Common characteristics within the class contribute to when the load occurs.

TYPES OF GAS AND ELECTRIC UTILITY LOADS

Consumer electric and gas loads may be characterized as peak, off-peak, baseload, and seasonal. Each is rated differently by the utilities, produces different costs to the end-user, and has different implications for the system.

PEAK AND OFF-PEAK LOADS

The broadest and most critical distinction is between peak and off-peak loads. Peak loads are those that occur during the time periods in which the utility must supply the maximum amount of energy and, therefore, use the highest amount of system capacity. Increases in peak load are closely tied to the need for investment in facility expansion. For electric utilities, peak loads often necessitate the use of the less fuel-efficient peaking plants. During peak periods, the rate of transmission and distribution system line losses also increases, further adding to supply requirements. For these reasons, peak capacity is the most expensive to purchase and carries with it the burden of increased capital cost and decreased fuel and delivery system efficiency. For gas utilities, peak loads that must be served on a firm basis determine the capacity requirements of the local distribution system piping network, the amount of gas supply and transmission capacity that must be reserved on interstate pipelines, and/or the amount of gas storage capacity required.

Off-peak loads are those that occur during the time of day, week, month, or year when the utility uses a relatively small amount of its total system capacity. Consequently, off-peak loads do not contribute to the need for facility expansion. For electric utilities, off-peak loads are usually served by the most efficient electric generation plants and distributed with relatively low line losses. For gas utilities, off-peak loads are generally served by a low-cost source of gas supply. For all of these reasons, off-peak capacity is the least expensive to purchase and carries with it the benefits of operating cost-efficiency and minimized capital cost impact.

In addition, many utilities further segment loads into intermediate categories, sometimes referred to as shoulder- or intermediate-peak periods. These loads and their effects lie between the peak and off-peak loads.

BASELOADS

Baseloads, from an end-user perspective, are those that occur all of the time (i.e., a 100% load factor). The quantity that can be referred to as baseload is defined as the baseload rate times every hour in a given period (i.e., day, month, and year). Baseloads are served, in theory, by the portion of the utility's capacity that is used constantly. Baseloads are often considered optimal loads because the fixed costs, associated with capacity-related investments, are spread over the maximum potential units of sales, and, in the case of electric utilities, the life-cycle costs of highly efficient power plants are at the lowest possible rate.

In many rate structures, baseloads carry a higher weighted average cost than off-peak loads. This is because off-peak load is a component of baseload that is separated from the peak component. Capital investment-related costs are frequently stripped away and added to the peak constituent. Variable operating costs, which are typically lower than the average due to increased efficiency, are broken out and assigned incrementally to off-peak load. Baseload costs are thus a composite of a fixed load experienced continuously in both periods.

Different rate structures separate or stratify costs to varying degrees between peak and off-peak. Traditional rate structures tend to have a relatively low differentiation of costs in different rate periods. On the other hand, more progressive rate structures have a wider range of prices. They attempt to assign costs to a greater number of finite blocks of usage based on their varying impact on capital and operating costs, with the result that the cost of a kilowatt-hour (kWh) of electricity or thousand cubic feet (Mcf) of gas at any given time may be far different than the average cost.

SEASONAL LOADS

Utility systems are generally either summer or winter peakers, referring to the season during which their peak-capacity requirements are highest. Most of the nation's electric utilities are summer peakers, because summer cooling loads outweigh the winter-heat-load component. In most regions of the country, heating loads are mostly served by direct on-site fuel use, such as gas or oil.

Even though fuel costs (driven by supply and demand) are even greater in the winter, the capacity-related capital cost component and the inefficiency of electric generation peaking plants produce a greater effect on seasonal electric costs than do their fuel cost component. Summer peakers, therefore, will often construct seasonally differentiated rates that are higher in the summer than in the winter.

However, not all electric utility systems are summer peakers. Some are winter peakers, some are balanced, and some (due to a high growth rate) set a new peak every season. Winter peakers are found in certain northern regions where cooling loads are modest and/or electric heat is predominant. Winter peakers or balanced systems may also be found in moderate climatic regions that are more conducive to the use of electric heat. Balanced systems are more likely for utilities with a heavy industrial base in which temperature-related end uses are minor, compared with the baseload process end uses.

Almost all gas utilities are winter peakers, due to the preponderance of space heating loads. Northern climate local distribution companies (LDCs) tend to have the most dramatic winter peaking load profiles. The LDCs that serve a significant industrial sector and/or provide large quantities of gas to electric generation plants tend to have more balanced loads, as do LDCs in warmer climates.

LOAD FACTOR AND USAGE PROFILE

The load factor is one of the most significant elements of a customer usage profile and utility's operating profile. For the utility system, it is the measurement of actual energy output over a period vs potential output over that period, based on the full capacity of the system. If, for example, an electric utility has 10,000 MW of total capacity and generates 120,000 MWh over a 24-h period, its load factor is 50% [120,000 MWh/(10,000 MW × 24 h)]. If this utility had a 100% load factor, it could generate the same 120,000 MWh over 24 h with only 5000 MW of capacity. Hence, an incremental load with a 100% load factor will produce a significantly lower revenue recovery requirement than one with a 50% load factor, because the capital cost associated with the additional 5000 MW of capacity is eliminated.

If, for example, a gas LDC has a maximum daily distribution capacity of 100,000 Mcfd and a 100% annual load factor, it would distribute 36.5 million Mcf (100,000 Mcfd × 365 days) per year. If, in actuality, it distributed only 14.6 million Mcf per year, the annual distribution system load factor would be 40% (14.6 million Mcf/36.5 million Mcf).

A load factor of 100% is the theoretical ideal operating state for utilities. In this ideal state, the utility's investment in capacity is spread over the maximum amount of potential output, and the revenue requirement per unit (kilowatt-hour or thousand cubic feet) sold is minimized. In reality, however, a 100% load factor is not an attainable goal. Some margin of excess supply, transmission, and distribution capacity is necessary for maintenance and emergency backup. It is also not possible to balance consumer loads perfectly.

There are, however, some conditions that produce utility load profiles approaching a 100% load factor. For example, electric utilities that operate as part of regional power pools and/or have sufficiently low operating costs to allow for export of all excess capacity can approach a 100% load factor. Utilities that are capacity constrained may also need to operate all of their plants at 100% load factor, and purchase power to meet the rest of their load requirements. Gas utilities that predominantly serve heavy industrial loads, for instance, may have a high load factor.

In order to improve system load factor, utilities often seek to market electricity or gas in low-load periods at a relatively low cost. They may also provide incentives, through various conservation and load management programs, to promote elimination or shifting of peak loads.

The utility considers load factor both on a discrete minute-by-minute basis and an hourly, daily, weekly, monthly, and yearly basis. Other common measurement periods are seasons, normal workdays, workweeks, and weekends. Sophisticated dispatch modeling is used to determine when to bring additional capacity (i.e., electric generation plants or gas storage) on- and off-line in response to load fluctuations.

Electric utility systems (or regional power pools) have a dispatch stack order. Generating stations are arranged in the order in which they will be brought on- and off-line in response to changing load requirements. Typically, plants in the stack are classified as baseload, intermediate, and peak-load plants. Baseload plants are generally the most cost efficient to operate, or have the lowest operating cost, including fuel costs per unit output. Intermediate (or swing-load) plants run much of the time, often under varying loads. Peak-load or peaking plants are typically the least costly plants to construct, but also the least cost-efficient to operate. The LDCs also have a type of dispatch order. Supply sources, including storage reserves and liquid natural gas facilities, are arranged in the order in which they will be used.

Load factor is one of the most important determinants of the cost to serve a given facility. Consider two electric utility customers that have the same monthly load of 72,000 kWh. One of the customers uses 100 kWh every hour of the month (720 h). Under a typical demand/commodity-type rate, this customer would have a peak demand of 100 kW and a monthly load factor of 100%. The other customer uses 75 kWh every hour of the month, except for 1 h every day, when a certain process requires 600 kWh. This customer would have a peak demand of 600 kW and a monthly load factor of 17% [72,000 kWh/(600 kW × 720 h)]. While, in this example, the monthly usage is exactly the same for both customers, the cost to serve the customer with the 17% load factor would likely be several times that of the customer with the 100% load factor. To serve the customer with the 17% load factor, the utility would have to reserve 600 kW in generation, transmission, and distribution capacity as opposed to 100 kW for the other customer. If both customers paid the same amount for electricity on a per kilowatt-hour basis, much of the charges paid by the customer with the 100% load factor would be to support the investment required to serve the other customer.

Now consider a third customer that uses 72,000 kWh/month. This customer operates a night and weekend shift factory that uses 200 kWh every hour for half of the hours every month (360 h). Therefore, this customer has a maximum demand requirement of 200 kW and a monthly load factor of 50% [72,000 kWh/(200 kW × 720 h)]. In this case, since the entire load occurs in the utility's off-peak period when load requirements are low, the utility does not require any additional capacity to serve this customer, except for the local wires and transformers connected directly to the facility. Additionally, this customer's loads can be served by the utility's most efficient generation plant. When the power is required by the customer (i.e., the specific usage profile), in this example, it is even less costly to serve the off-peak customer with the 50% load factor than it is to serve the customer with the 100% load factor.

A parallel example can be drawn with three gas utility customers that all consume the same amount of gas on an annual basis. The first customer has a 17% load factor, based on a winter heating load requirement. The second customer has a 100% load factor, based on a continuous industrial process requirement. The third customer has a 50% load factor, based on a continuous non-winter-month gas-fired cooling load requirement. Similar to the three electric utility customers, in this hypothetical example, the customer with the 17% load factor will likely be the most costly to serve and the customer with the 50% load factor will likely be the least costly to serve, with the 100% load factor customer falling somewhere in the middle.

The conclusion that can be drawn from these examples of customers with the same exact daily, monthly, or annual usage is that the load factor and, even more importantly, the specific load profile are key determinants of the actual cost of service. The main reason is the impact of the load profile on fixed cost requirements and how fixed costs are recovered with respect to overall usage requirements.

In addition to load profile, the magnitude of overall load is an important cost factor. There is an economy of scale in serving customers that requires large amounts of energy. Costs of running lines or pipes to a facility, installing service and meters, billing every month, and providing other goods and services are less when averaged over a large, rather than a small volume.

ALLOCATION OF COSTS

Utility allocation of fixed costs and the associated setting of rate levels is a process that is partly scientific and partly subjective. One major consideration in cost allocation and rate design is the desire for rate stability. All usage charges should contribute to fixed costs in some way, and demand charges should serve to mitigate against significant distortions in cost allocation resulting from varying load profiles.

If a utility placed all capital cost recovery in a single period of use, such as peak summer, the rate might be too unstable. The market might overreact with a rush toward solutions, such as peak shaving and, in short order, cost recovery would be insufficient. If all usage in that single usage period disappeared, the capital costs would not disappear. The utility system would still have to be financially supported. Of course, loads do not disappear all at once. Hence, as they change, the utility rate structures must change with them.

Fixed cost-related charges are neither arbitrary nor capricious, but they are also not a perfect scientific cost allocator. The underlying theory behind demand charges and other fixed cost rate components is that they better approximate real costs. If, for example, all customer electric bills under demand rates were broken down into incremental usage costs, the cost per kilowatt-hour might range from $0.02 to $0.90, or even higher, depending on the severity of the impact of demand charges. This wide range of price differentiation is said to allow rates to more closely represent the utility's actual cost to serve. As this process continues to evolve, along with advances in measurement and market communication technology, such determinations may be made dynamically, based on actual events, as opposed to predicted events.

Clearly, the movement in today's energy market is toward more precise cost allocation strategies. Factors that have contributed to this movement include the following:

- The need to protect utility revenues by having rates that are competitive with energy alternatives. More competitive, diversified rates have been constructed by electric and gas utilities to protect revenues and maximize sales during periods when electricity or gas is less costly.
- The desire of utilities to influence customer usage patterns in order to create higher system load factors. Price differentiation directs equipment investment choices and operating schedules toward customer load profiles that improve system load factor and lower the average cost to serve. If all similar energy units cost the same, load growth would tend to move away from a high load factor as the prevailing forces of weather and the normal work week directed loads to peak periods. Cost differentiation provides incentives to counter these forces through careful end-use planning.
- The directive from regulators to redesign cost allocation processes and avoid the need for additional capacity. Careful review of utility management decisions by regulators has directed utilities away from the business of building new plants much in advance of load growth.

In addition to least-cost planning, regulators have demanded more precise cost allocation in an effort to make rates more fair. Regulators have also provided utilities with incentives for activities other than construction. A preferred rate of return for investments in conservation and load growth reduction activity is an example. Many demand-side management programs involve shifting usage from the utility peak period to an off-peak period. In order to make such programs effective, utilities must have rate structures that charge according to when energy is used instead of just how much energy is used.

The imperative of the regulated least-cost planning activities has increasingly been replaced by competitive market-based imperatives. As a result, utility rate design is driven more and more by competition and less by static regulatory planning.

On the other end of the spectrum from a single fixed-price usage charge is a completely market-driven discrete usage charge that changes from hour to hour, or even minute to minute. In this scenario, gas and electricity usage charges continually vary with the price being established on almost an instantaneous basis. This price must be market competitive and reflect all embedded fixed and variable costs. This type of pricing would be applied to both the commodity side and the transmission and distribution sides, either separately or as a bundled price.

While the market is still not fully mature enough or unencumbered by imperfection to function with such pricing, the technology of metering and transmitting gas and electricity is approaching the level of sophistication necessary for such a market. Moreover, the forces of competition are moving the market in this direction.

However, today's market does not yet function in this manner. Instead, some utility rates are designed to approximate this type of instantaneous or real-time pricing (RTP). To do so, rate designs use a mix of various rate components. They can be classified into several general categories on the basis of energy-use characteristics. Each category may constitute one rate or several categories may be aggregated to determine a rate.

BILLING FACTORS

Utilities use a number of billing factors in their rate structures. Some of these factors are clearly stated in all utility bills, while some are considered in every transaction but not necessarily identified separately in all bills. Others are considered and applied, depending on the particular characteristics of the utility or the customer class under a given tariff. The most common billing factors are as follows:

- *Basic, or customer, service charge.* A fixed charge is assessed to each customer based on costs related to connection, metering, billing, service maintenance, etc. Typically, basic service charges are greater for rates designed to serve large users because of the greater cost associated with larger pipes, regulators, and metering equipment. Facilities with multiple services under the same rate may have summarized billing with only one basic service charge. Facilities with multiple services under different rates will often pay a basic service charge for service under each rate.
- *Minimum charge.* This is the lowest bill a customer is required to pay for service on a given rate schedule during each billing period, regardless of actual usage. In most cases, this is equal to the customer charge. However, for larger customers, it can include other charges, such as demand charges or charges established by individual contracts between the customer and the utility.
- *Commodity charge.* This is a charge based on energy usage or the number of energy units actually consumed by the facility during each billing period. Commodity charges generally include the utility's incremental operating costs, plus some contribution to fixed costs.
    — Common natural gas billing units are: cubic feet (cf), hundred cubic feet (Ccf), thousand cubic feet (Mcf), and therm (100,000 Btu or 105,480 kJ). While a therm is a specific quantity of energy, a cubic foot of gas has varying British thermal unit or kilojoule levels. Typically, 1 cf of pipeline quality natural gas contains about 1000 Btu, but this can vary by several percentage, depending on the specific source of natural gas.
    — Common electricity billing units are: kilowatt-hours and kilovolt-amperes.
- *Demand, or maximum level of service, charge.* This charge is based on a customer's peak rate of consumption and takes into account the utility's required investment needed to serve that load. Measurement of demand, or rate-of-use, is somewhat like a car speedometer. However, a demand meter does not measure rate-of-use instantaneously, but averages it over a discrete utility-selected interval. This interval, or "demand window," for electric utilities is usually 15 or 30 min. The interval for gas utilities is usually the highest daily consumption or, in some cases, it may be measured as the peak usage for 1 month. Demand can be measured in multiple-use periods, and charges can be differentiated for each use period.

  Demand charges are assessed in several different ways. They may vary by season, TOU, and level of use:

    — Seasonal variation produces greater charges for peak demand during months in which the utility experiences its highest peak demand. With gas utilities, demand charges are often applied only in winter months.
    — The TOU variation can be applied in several different ways. In many cases, demand is only measured by electric utilities during a peak period. In other cases, demand is measured during several different periods, such as peak, shoulder, and off-peak. The charge per unit of demand varies with each period, peak charges being the highest.
    — Level-of-use variation produces different charges for increasing blocks, or steps, of demand. For example, the first 500 kW or 500 Ccf of demand will be billed at one rate, the next 1000 kW or 1000 Ccf of demand will be billed at another rate, and so on.
    — Minimum demand charges are sometimes set at a given level for each rate tariff. In some cases, they may be based on a percentage of the customer's connected load or main transformer or gas meter size.
- *Ratchet adjustments.* A demand ratchet adjustment sets minimum monthly billable demand at a certain percentage (usually 60%–100%) of the highest month's peak demand in a given preceding period (typically 11 or 12 months). When the utility has a severe summer or winter peaking system, the ratchet may be based on the highest month's peak demand only in the peaking season. The minimum monthly demand charge assures the utility of cost recovery for investment in peak

capacity, even if a customer does not require that peak capacity in a given month. In some cases, this preceding period may be as long as several years or for the life of a contract. Ratchets serve as a mechanism to approximate the true cost of service and to distribute that cost impact on the customer over several months.

Consider the example of an electric utility with an 80% demand ratchet adjustment. If 1 month had a peak demand of 2000 kW and the next 11 months had a peak demand of 1000 kW, each of the proceeding 11 months' demand charges would be based on 80% of the 2000 kW figure or 1600 kW. The result would be an additional annual billable 600 kW/month, totaling 6600 kW additional billable kilowatt over the next 11 months. Fig. 1 provides an example of the impact of an 80% ratchet on an 11-month rolling basis.

The reasoning behind ratchet adjustments is that monthly demand charges alone do not sufficiently compensate the utility for the impact of a customer who sets a large peak once per year rather than every month. The utility must have sufficient capacity to meet the annual highest peak hour of demand at all times and must incur the cost of building or securing contracts to purchase the needed capacity. A customer does not need that peak demand every month must still pay for a large portion of the demand in order to reserve it for the peak month.

- *Energy adjustment charge (EAC)*. This mechanism is designed to pass increasing or decreasing fuel costs per billable unit directly to the customer. Energy adjustment charges may vary on a monthly, quarterly, or annual basis. It is normally based on a 12-month rolling average of fuel costs and used to maintain rate stability between rate-case proceedings. Common terms used to express EAC are: fuel cost adjustment, fuel adjustment charge (FAC), purchased power adjustment, purchased gas adjustment (PGA), and levelized gas adjustment. Fig. 2 is an illustration of a FAC to monthly electricity charges.

- *Other adjustment charges*. A utility bill may also include additional adjustment charges. Many electric companies have a nuclear capacity adjustment or nuclear plant decommissioning charge. Some companies also have conservation adjustment mechanisms or sales adjustments. Gas utilities have shrinkage or retainage charges. These are applied as a percentage increase to transported gas volumes at delivery points for transportation customers. This charge reflects lost or unaccounted for gas volumes that arise from the transportation of gas over the LDCs pipeline network.

- *Taxes and fees*. These are added charges that the utility bears as operating expenses and passes on to individual customers. These may include gross receipt taxes or sales taxes. They may also include surcharges for special regulatory commission approved programs that are added to the bill. Such charges may be shown on the bill as itemized charges, or they may be embedded in the derivation of other utility charges and are thus not easily spotted on the bill.

RATE DESIGN STRATEGIES

The purpose of rate design is to recover revenue requirements associated with the costs incurred by utilities to provide services to the customer classes, while also recognizing the different energy use profiles of customers between and within the customer classes. Rate design also sends price signals to consumers and can influence energy-use decision makers to favor more cost-efficient energy use profiles. Price signals mean that a particular utility's rate design makes it clear when it is more or less costly to purchase a unit of energy.

If all customers had the same usage pattern, fixed and variable costs could be readily integrated into one simple cost recovery mechanism—a usage charge. To determine an average price per kilowatt-hour or thousand cubic feet, the total revenue requirement would be divided by all of

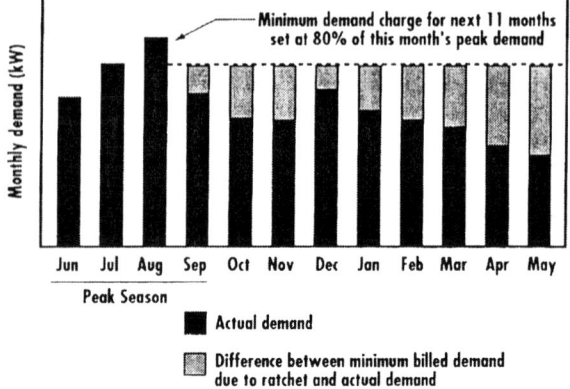

Fig. 1 Impact of 80% demand ratchet.

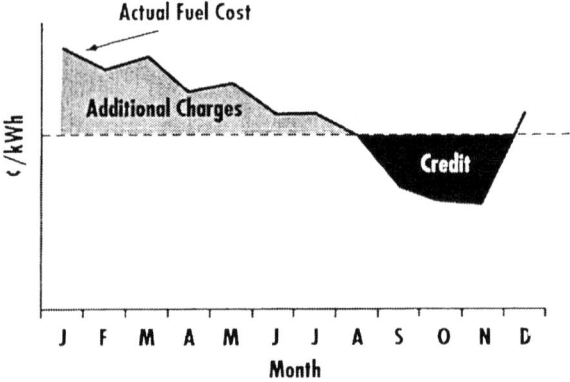

Fig. 2 Representative illustration of fuel adjustment charge to monthly electricity charges.

the kilowatt-hour or thousand cubic feet to be sold. In reality, however, customers do not have the same usage patterns and, as demonstrated in the above load factor examples, distinctions must be made as to how varying customer load patterns affect system cost.

BLOCK RATES

Block rates are rates in which the charges for a unit of service vary with consumption. The billing period's consumption levels within a rate are often broken down into blocks, or steps, with different charges for each block. There are several different common types of block rates:

- *Declining block rates.* Historically, the most common type of rates, declining block rates have lower usage charges as levels of consumption increase. The reasoning behind these rates is that increasing customer usage reduces the cost to serve the customer on a per-unit basis. These rates have been phased out in many utilities because they are believed to encourage greater consumption and discourage conservation. Fig. 3 illustrates a natural gas declining block service rate.
- *Increasing, or inverted, block rates.* These rates charge more per unit as levels of consumption increase. They are relatively uncommon with natural gas service, but they do show up with electric services. They are used by utilities that are supply-constrained, or used for conservation purposes to discourage increased consumption. Fig. 4 illustrates a natural gas inverted block service rate.
- *Sliding block rates.* These rates use a peak demand value times a multiplier to determine the first-step size. They encourage greater load factors or, in other words, more constant levels of energy usage by the customer. This type of rate is common today in the electric industry for C&I customers. Fig. 5 illustrates a natural gas sliding block service rate.

In some cases, the utilities charge the same unit price regardless of the level of consumption, meaning there is

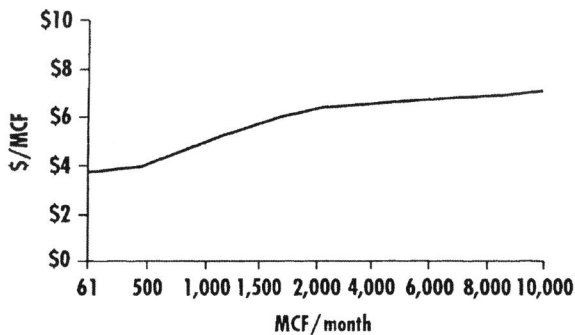

Fig. 4 Representative natural gas inverted block service rate.

only one block. This is somewhat common for residential customers.

Many utilities use a combination of block-pricing structures. They are often used to reflect a seasonal energy-cost differential. A utility may use an inverted block rate in the season of highest consumption and a flat or declining rate in the off-peak season.

DEMAND/COMMODITY RATES

These rates consist of two basic components: a demand charge and a commodity charge. The demand charge is typically based on the peak hourly or daily volume in the billing period. The commodity charge is based on the units of energy consumed. In many cases, a ratchet penalty is used. As previously discussed, the ratchet penalty is typically based on a certain percentage of the peak demand incurred within the previous year on an 11- or 12-month rolling basis. This type of rate somewhat more accurately reflects the cost to serve than flat commodity rates, because it takes into account peak-day requirements. Such rates encourage customers to maintain a high load factor.

In many cases, a sliding block rate structure is used to blend demand and usage charges into stepped rates based on utilization (or load) factors. Usage is billed at varying

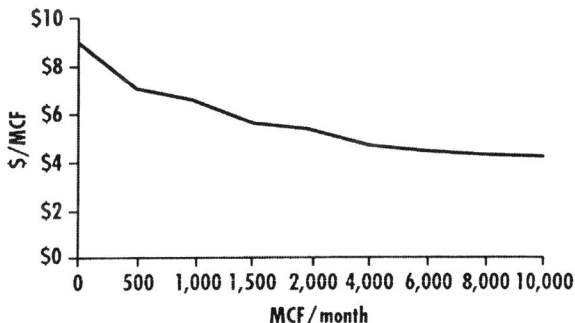

Fig. 3 Representative natural gas declining block service rate.

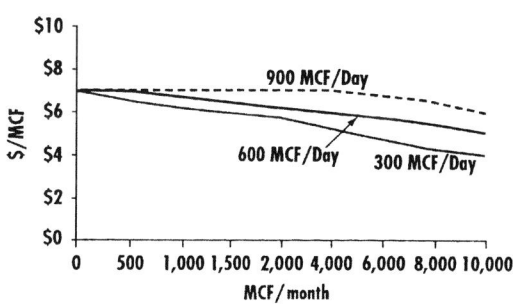

Fig. 5 Representative natural gas sliding block service rate.

rates based on hours of use of demand. For example, a 30-day billing cycle has 720 h. If the peak demand on an electric utility bill is 1000 kW, a 100% load factor would represent 72,000 kWh, or 720 h of use of demand. A 25% load factor would represent 18,000 kWh, or 180 h of use of demand. This type of rate schedule breaks the hours of the month into different blocks. With a declining block rate structure, in each subsequent block or hours of use of demand, the rate is lower. For example, the first 180 h of use of equivalent full-load demand (18,000 kWh) would be billed at a certain rate per kilowatt-hour. The next 180 h of use of demand (from 18,000 to 36,000 kWh) would be billed at a lower rate per kilowatt-hour, and so on. This type of rate rewards high-load factor customers. If the customer's peak demand increases, more kilowatt-hour are billed at the higher rate and vice versa.

TIME-OF-USE RATES

The TOU rate structures are often designed to differentiate among months of the year (seasonal rates), days of the week, or hours of the day. These rates assign greater costs to peak usage periods, to discourage consumption, and lower costs to off-peak periods. Commodity cost differentiation may be augmented by peak demand charge differentiation. These TOU rates are more reflective of the cost to serve than standard block rates and provide price signals that direct consumers toward off-season or off-time usage. The level of differentiation will often be greater for utilities with poor annual load factors, such as those with loads that are predominantly heating or cooling. Typically, higher summer rates will be used by electric utilities and higher winter rates will be used by gas utilities.

- *Seasonally differentiated rates.* These are the type of TOU rate designed to assign greater cost to consumption, on a per-unit basis, for corresponding rate blocks in peak usage months. Commodity cost differentiation may be augmented by peak season demand charges and ratchets. These rates provide price signals that influence the consumers toward off-season usage.
- *Off-season rates.* These services are provided to customers in specific nonpeak months. They are designed for customers who do not require gas service in peak months, but do not have the alternative energy sources typically required for standard interruptible sales service. Similar to interruptible rates (IR), which are discussed later, these rates provide lower cost gas, because they do not contribute to the high cost of maintaining peak facilities. These rates may also be attractive to customers that have alternative energy sources, but cannot use them in certain months because they must meet seasonal emissions standards.

END-USE RATES

End-use rates are designed to provide incentives to customers to install and operate specific equipment that uses the utility's energy. This usually includes space heating, water heating, and cooling equipment. These rates generally have some basic equipment requirements and often need separate metering. Specialized heating rates are generally used by electric utilities. These are sometimes combined into an all-electric rate that is offered to customers using electric heating, water heating, air conditioning, cooking, etc. Specialized cooling equipment rates are more commonly offered by gas utilities.

End uses selected for specific rates can often be placed in given rate periods or seasons. In some cases, end-use rates are combined with TOU rates. The lowest rates typically are offered for end uses employed during the off-peak season or off-peak time of day or week. More moderate rates may be offered for baseload process end uses, such as water heating, cogeneration, or year-round industrial processes.

Many PUCs prefer that price signal tools be applied uniformly to usage characteristics rather than to specific end uses. However, some utilities feel that their typical rates already accomplish this and that further distinctions are needed to attract (or discourage) certain loads.

NONFIRM SERVICE RATES

Nonfirm rates include IR, standby rates, and load management rates. With multiple energy source options increasingly available, nonfirm service is becoming increasingly popular:

- *Interruptible rates.* These rates, which are commonly offered to CI&I customers, are designed for customers whose entire load, or large blocks of load, can be dropped at virtually any time. The primary benefit to utilities is that it reduces the need to guarantee service during peak demand periods. The utility can pass on savings, which come predominantly from reduced fixed-capacity costs, to customers. These rates are also used to attain a competitive advantage and/or attract loads based on lower costs. Interruptible service may involve commodity or distribution components.

    There are many conditions under which a facility can withstand such interruptions. Examples include:
    — Customers with dual-fuel burners capable of operation on natural gas and alternative fuels, such as propane or oil, or distribution systems supplied by propane–air mixtures can easily withstand interruption of natural gas service by switching over to their alternative burner fuel.
    — Customers with on-site electric power generation capacity can go off-line and generate their own

power to serve the entire facility or selected circuits within the facility.

— Customers with equipment that can operate on either electricity or fuel or steam, can simply use the non-electrical equipment during periods of interruption. This would be the case with a dual-drive mechanical service device that had both an electric motor drive and a prime mover drive, or with mixed energy source (hybrid) multiple unit systems that feature both electrical and nonelectrical powered units.

Electric utility IR are often referred to as utility-controlled peak shaving. Customers are required to reduce their demand on the utility system completely, or to some predetermined level when necessary. Contracts may be designed with a set rate-break on demand or usage, a flat annual or monthly fee paid by the utility, or a specific payment rate for each period of interruption. Significant penalties for failure to interrupt are also common.

The LDC IR are typically based on the customer's ability to use an alternative fuel, such as oil or propane with the natural gas commodity charge set or negotiated based on the price of the alternative fuel. Natural gas distribution services (in cases where a customer purchases gas from a seller other than the LDC) may be purchased on a nonfirm basis. Pricing by the LDC may be based on competitive alternatives similar to interruptible commodity pricing or may be offered at a firm cost-of-service-based price.

- *Load control, or load management, rates.* These rates are a type of IR that gives the utility direct control over specific loads (sometimes via radio wave signals) during system peak periods. This strategy is most commonly used by electric utilities for residential water heating and air-conditioning customers, but can be used for commercial, industrial, and agricultural customers. Load control is intended to minimize customer inconvenience by selecting loads that can most easily be eliminated or cycled.
- *Standby rates.* Utilities offer these rates to customers that have their own source of energy but require service from the utility on an intermittent basis. These rates are more common to electric utilities and may be categorized as maintenance rates for power supplied by the utility when the customer prearranges downtime for generation equipment maintenance; supplementary rates used by self-generators that regularly require additional power from the utility; and backup rates used by self-generators in the event of an unexpected system outage.

NEGOTIATED AND SPECIALTY RATES

Negotiated rates may be cost-based, designed to allow the utility to compete with alternatives, used to support economic development or business recovery, or implemented to permit unique arrangements, such as interruptible service.

Cost-based negotiated rates are designed for customers whose usage and characteristics vary considerably from the average of the rate class or have realistic competitive alternatives to the utility offering the rates. Most commonly, this is only done with large CI&I customers. In many cases, the utility has the ability to negotiate rates down to a level equal to or, more commonly, slightly above its short-term marginal cost. The regulatory justification is that the rest of the utility's customers will benefit from such contracts as long as the negotiated rate charged to the particular customer covers the incremental variable cost of service and provides some contribution to fixed costs. These rates provide the utility with a maximum degree of flexibility to market their product to customers with special needs or competitive alternatives. Sometimes, particularly when the contract period is longer than 5 years, these rates may be designed to recover long-term marginal costs. With these rates, a larger minimum demand charge is required.

- *Economic development rates.* Utilities commonly offer these rates to provide economic incentives for businesses to locate or expand into their home service territory and/or into economically depressed areas. They are often based on a schedule in which rates are initially discounted and then phased into a standard rate over a period of several years.
- *Business retention rates.* These rates are designed to retain customers with either competitive options from other energy sources, self-generation capabilities, or an interest in moving to another state or service territory. Business recovery rates are designed to retain customers in financial difficulty.
- *Conservation and load management rates.* Many utilities offer these rates to customers that meet certain equipment or building envelope thermal efficiency standards and/or operating standards. These rates may be designed with a simple percentage rate break on usage based on achieving a certain level of conservation and efficiency. They may also involve a reduction in charges based on some type of load-control incentive, or they may be based on a combination of both. Rate design also may include incentive mechanisms for shifting load from peak to off-peak periods.
- *Special contract rates.* In cases when it is in the best interest of all parties (i.e., the utility, the customer, and the rest of the utility's customers) and when the unique conditions of the situation cannot be met under standard rules, utilities may develop special contracts with individual customers. Examples are a very large cogeneration application or a customer who makes year-round third-party gas purchases and is willing to

make volumes available to the utility for resale during peak periods. These contracts often require individual PUC approval, which can be a lengthy process.

Examples of other types of specialty rates are compressed natural gas rates for natural gas-fueled vehicles and rates that support the introduction of new technologies.

COMPETITIVE ENERGY RATES

Competitive energy rates give utilities the maximum flexibility to sell power or natural gas in competitive situations. For example, customers considering a gas or steam technology application, such as cogeneration, are often presented with some type of competitive energy rate by their electric utility in an effort to keep the full load on the utility system. Competitive pricing may be offered down to some small level above the incremental avoided cost.

This pricing structure is also used for peak usage by utilities with some excess system capacity. There is some concern that this practice skews market choices, keeps load at the expense of conservation opportunities, discounts opportunity costs, and, in the long run, leads to additional capacity needs at the expense of the rest of the rate payers.

WHOLESALE (OFF-SYSTEM) SALES

Another area for incremental-cost commodity sales is on the wholesale market. This market, while often less stable than on-system sales, can be very profitable for utilities if the plant supplying power or the reserved gas pipeline capacity is in the rate base. A utility with excess capacity in a given period can sell gas or electricity to other utilities or, in some cases, to customers outside of their service territory. The prevailing logic is that excess capacity should be marketed whenever possible as long as variable costs are recovered and some contribution, however small, is made toward fixed cost recovery. With the potential for capacity release, however, LDCs can instead sell excess capacity rights to others rather than use the capacity for the purpose of off-system sales.

REAL TIME PRICING

The RTP is an emerging utility rate strategy that goes a step further than demand charges and varying usage charges in allowing for differentiation of costs that better reflect the utility's actual incremental cost. Real-time pricing rates typically do not have a demand component. Instead, kilowatt-hour, or thousand cubic feet, consumption is priced by the hour. For example, an hourly RTP structure may charge $0.90/kWh at noon on a Wednesday in August and only $0.02/kWh at midnight on a Sunday in March. Currently, RTP is being used by numerous electric utilities.

The theory behind RTP is that if customers are told in advance of the utility's anticipated system and price conditions, customer demand will respond most directly to price changes. That is, a decrease in consumption as the price rises and an increase in consumption as the price falls. Typically, customers are given a schedule of hourly prices one day in advance. In some cases, the cost per kilowatt-hour is fixed for several categories (i.e., off-peak, utility-peak, or regional power pool-peak), but the hours during which they are applied are varied and communicated by the utility to customers on an hourly, daily, or weekly basis.

The procedures used to design rates that differentiate usage and demand charges by time and season of use come close to approximating what the utility determines is proper hourly cost allocation. The RTP accomplishes this with greater certainty and fewer complications. Taking the example of a TOU rate with a peak demand period of the typical 9-to-5 workweek, a peak demand may be set at 9 A.M. by a facility. This peak may have no impact on the utility peak, but is charged as if it were set at the utility system peak hour. With RTP, if in fact this peak had no impact on the utility peak, it would be priced at a far lower level than a peak that did have an actual impact on the utility peak.

The concept behind RTP is that pricing reflects real, almost instantaneous, market conditions instead of predicted market conditions. While TOU rate blocks are bins which approximate what actual costs are in different periods, RTP more closely reflects the actual value of electricity (or gas) at any given point in time. NonRTP rates are, therefore, based on probability of occurrence, rather than occurrence.

Another type of pricing that more closely represents real events is ambient temperature-based pricing. For example, when the outside temperature falls below 30 or 20°F (-1 or $-7°C$), natural gas pricing could automatically shift to a higher rate. Currently, there are many interruptible gas rates that base interruptions on temperature. A similar strategy could be employed for summer electricity pricing, based on rising temperature. While this is not an instantaneous pricing mechanism, it is one based on real events as opposed to predicted events. When the event occurs, the pricing schedule is in effect.

Electric utility dispatch modeling has become an increasingly precise process. Utilities can identify where each incremental kilowatt-hour comes from and its value. Large facilities can then perform the same modeling of in-house usage. Furthermore, the cost of telemetry is decreasing, while capabilities are increasing. As alternative electricity purchase options become available, RTP may become a mainstream sales pricing tool. Consumers may elect to purchase certain blocks from the utility, generate certain blocks on site, and purchase certain blocks from other sources through retail wheeling, all based on real-time price signals.

RATE RIDERS

In addition to various rate design options, rate riders are special charges or programs integrated into rate schedules that modify the structure based on specific customer qualifications. Riders are used to account for unique conditions or to give the utility added flexibility to apply rates without dozens of additional tariffs. Riders may include: negotiated competitive-energy riders, interruptible riders, standby riders, buy-back riders, conservation and other load-control riders, end-use riders, and other types of discounts, such as an electric utility discount for customers receiving service at a voltage above the standard voltage.

COMMON NATURAL GAS RATE SCHEDULES

Actual natural gas rate schedules include many of the same rate-design strategies previously mentioned. These rates are offered to customers who meet specific criteria. Commonly used natural gas rate schedules are:

- *Firm sales rates.* These are the highest priority of service offered by LDCs. Gas is made available throughout the year, on an uninterrupted basis, to serve customers' needs. Typically, there are several categories of firm sales rates, such as residential, small commercial, and large CI&I. Rate design may include block rates, seasonally differentiated rates, demand charges, etc.
- *Firm transportation rates.* These provide uninterrupted transportation service, through the LDCs distribution system, of natural gas purchased directly by the customer. The LDC takes on the obligation to deliver to the customers' facilities gas that has been delivered to the LDCs gate station by an interstate pipeline.
- *Dual-fuel firm rates.* These provide service to customers who have the option of using an alternative fuel source, but who, at any time, may request firm delivery of gas from the LDC. Since the LDC must stand ready to serve, and may have significant investment in, distribution facilities, supply contracts, or storage capacity, it may require purchase of some level of guaranteed volume or include a demand-charge component. Because this type of service has the potential to negatively affect the LDCs load factor if the customer only uses gas services during peak periods, the rate may be expensive.

- *Interruptible sales rates.* A utility can sell to its interruptible sales customers spot market gas or excess gas that it purchased as a reserve for its firm sales service customers. Selling gas on a commodity basis only allows gas to be priced competitively with oil, propane, and other energy sources. Therefore, customers benefit from lower costs.

  Prices are often negotiated competitively and indexed each month to the alternative fuel or energy source (e.g., No. 6 oil, No. 4 oil, No. 2 oil, propane, or electricity). A rate offering, commonly known as the standard offer, may be made to the entire group with the same energy-source alternative. Individual customers with better purchasing capabilities may negotiate price and be permitted to lock in a rate for a longer or shorter period of time. Prices may also vary with notice period. The shorter the notice period the lower the gas cost.

  The LDC may have the ability to negotiate downward to a certain floor. In many cases, the floor is set a few cents above the LDCs actual supply cost. This competitive approach is accepted by PUCs, because it holds down overall rates by maintaining gas sales that would otherwise be lost to alternative fuels. Rate-case proceedings may include an agreement that a large portion of the marginal revenues from these sales flow back and reduce the rates of firm-rate customers, often via the PGA. In a sense, this flow-back of revenues provides compensation for the use of facilities (fixed costs) that is amortized through cost recovery via firm rate charges.
- *Interruptible transportation rates.* These services are similar to interruptible sales rates, except they relate solely to distribution rather than to both sales and distribution. The LDCs ability and/or need to interrupt are tied to local distribution constraints, not to supply constraints.
- *Cogeneration, air conditioning, and other end-use rates.* These services are offered for specific equipment applications and, because they have a predictable effect on the LDCs load factor and cost, they are typically grouped under individual rate schedules. These rates are designed to be attractive to customers, because they are more closely tailored to actual energy use profiles of the applications, often improving LDC load factor.

COMMON ELECTRIC RATE SCHEDULES

Electric rates use a mix of various rate components, and rate schedules include one or more of the rate design strategies discussed earlier. These components typically consist of an energy charge, a demand charge, a ratchet clause, a FAC, surcharges for factors such as conservation or nuclear plant decommissioning, power factor (PF) charges, and taxes. Often, rates are offered to customers who meet specific criteria, such as type of facility or type of equipment used. Some of the most common non-residential electric rate schedules include:

- *General service (GS) rates.* General service rates are typically used by most small commercial customers. Rate design may include block rates, seasonally differentiated rates, and demand charges. Generally, these rates place a greater emphasis on usage than on demand and are less differentiated than large customer power rates. In some cases, these rates are available without demand metering, using higher usage charges instead. Typically, the availability of GS rates is limited to customers whose demand does not exceed a particular specified level.
- *General service TOU (GST) rates.* General service TOU rates are generally used by small C&I customers with multiple shift operations. They commonly consist of peak and off-peak periods and often register peak demand only during the peak period. They are not time-differentiated as much as are large customer power TOU rates, but they do allow customers to benefit from lower costs for extended use in off-peak periods. They are also often used by customers with electric heat or some type of thermal storage. Rate design may include block rates and seasonally differentiated rates.
- *General service heating (GSH) rates.* General service heating rates are common end-use-specific rates. Typically, they are GS rates that are available only to electric heating customers whose heating-related usage makes up a certain minimum portion of total usage. They are often used by summer-peaking utilities to build winter load. They have a greater degree of seasonal differentiation than standard GS rates, with depressed winter rates compensating for extensive usage. They also may have a TOU component to allow for the use of domestic hot water or heating thermal storage.
- *Street lighting (SL) rates.* Street lighting rates are end-use-specific, typically offered to states, cities, or other municipalities, and sometimes to large campus-type facilities.
- *Large power (LP) rates.* Large power rates are the traditional rates offered to larger CI&I customers and virtually always include monthly or fixed-contract demand charges and may use rate blocks and seasonal differentiation. Typically, the design is somewhat similar to GS rates, except that there is usually increased emphasis on demand charges.
- *Large power TOU (LPT) rates.* These rates are becoming more predominant for large CI&I customers. Typically, they consist of two, three, or four rate periods, such as peak, shoulder, or off-peak. They may register demand only during peak or all rate periods or have fixed-contract demand charges. Off-peak usage may be handled with varied charges or as peak usage with charges in the off-peak periods only for demand in excess of peak demand. Usage charges are varied by rate period. Rate design may include block rates and seasonally differentiated rates.

  These rates are more stratified than traditional LP rates. They are advantageous for facilities with high-load factors and extended hours of operation, which can offset costly peak usage with inexpensive off-peak usage. They are also attractive for facilities that use thermal storage or some type of peak-shaving technology.
- *Real-time pricing rates.* Typically, RTP rates do not have a demand component. Instead, kilowatt-hour may be priced by the individual hour or, in some cases, charges may be fixed, but the hours in which different charges are applied will vary. In either case, the utility communicates these varying costs or hours of application on an hourly, daily, or weekly basis to customers. Often, these rates are used by facilities with alternative energy sources in place or with the ability to shed loads on a regular basis. The RTP rates may become increasingly common as electric rates become even more sensitive to market competition. As opposed to demand-based TOU rates, RTP rates are thought to more closely reflect the actual discrete price of power at a given hour or even minute.
- *Transmission (T) rates.* Currently, T rates are typically used for special cases, such as wheeling, in which power is either bought from or sold to a party other than the local utility. Rate design is based on the use of the utility's transmission facilities only. Under the National Energy Policy Act of 1992 (EPAct 92), electric utilities are required to more clearly define transmission rates, as well as identify available capacity and known restraints. Over the long term, it is anticipated that the advent of retail wheeling and the further unbundling of electric rates will result in transmission/distribution rates being available to all customer classes.
- *Interruptible rates.* These rates and rate riders are designed for customers that have blocks of load (or all of their loads) that can be dropped at any (or almost any) time. Commonly, this includes the use of standby generation as a load-shedding technology. Interruptible rates may be designed with a set rate break on demand or usage, or may consist of a flat annual or monthly fee paid by the utility to the customer with a specific payment rate for each period of interruption. Rate design includes different steps, or levels, of availability, with 100% availability receiving the most beneficial treatment. Rates also vary with notice period. The shorter the notice needed for an interruption, the higher the rate discount.

Further refinement of IR involves differentiation of services between nonfirm electricity sales and nonfirm transmission/distribution services. Facilities with on-site energy alternatives can benefit from the ability to withstand sales and transmission interruptions. Facilities with alternative electricity purchase options, via retail wheeling, can benefit from the ability to withstand sales interruptions, but may still require firm transmission/distribution services.

- *QF rates and rate riders*. Many utilities have special QF rates or rate riders for self-generators. In some cases, these are elective rates (or riders), while in other cases, they are required. These riders often include rate designs that emphasize high peak demand charges, such as TOU rates. These special QF rates also usually include mandatory provisions that require the self-generator to purchase a form of backup services to provide a payment stream to the host utility for providing a reliability service by virtue of the physical connection. The provision of backup service is generally desired by the customer to prevent interruptions. The charge for backup is supported by the argument that a rate recovery mechanism is needed to prevent self-generating facilities. from taking advantage of utility capacity supported by other rate-paying classes, in the event of an outage of self-generation equipment, or during periods of planned system maintenance or for purchasing supplementary power.

  Rate restrictions are often put in place to limit rate options available to self-generators. A self-generator forced to be on a highly demand-sensitive rate with a ratchet penalty clause could end up with nearly a full year's worth of demand charges for a single outage. This is often considered excessive recovery. On the other hand, a self-generator allowed to be on a low-demand, highly usage-sensitive rate may pay a minimal amount, which is often considered insufficient cost recovery.

- *Standby rates*. These rates are offered to self-generators requiring power from the utility when their own or alternative energy supply is inadequate or unavailable. Standby rates are often riders which affect several rates offered by a utility. Many utilities have special rate recovery treatment for providing standby (QF backup) power to self-generators when their own or alternative supply is inadequate or unavailable. Standby charges are a type of demand (or insurance) charge paid to the utility to reserve replacement capacity and energy if a system failure or normal maintenance interval takes the on-site generator out of service. These standby rates may be offered as separate rates or rate riders.

There are three general types of standby rates offered by electric utilities: backup, maintenance, and supplementary rates. Backup and maintenance rates are offered to provide power when a self-generator's system is fully or partially out of service. Maintenance rates are offered for use during prearranged downtime and backup rates are offered for unanticipated downtime. Supplementary rates offer power for regular use and are offered to partial-requirements customers that may require purchased power in addition to their own self-generated power.

Standby rates for backup and maintenance are usually based on a monthly charge per kilowatt of capacity reserved. There is also a commodity charge for actual energy usage during the down-time period. These standby demand charges are less costly per kilowatt than the actual demand charges on a given full service rate, but are paid for on a take-or-pay basis, regardless of whether additional power is ever required.

Maintenance service rates provide convenience in that they allow self-generators to perform routine service and overhaul during peak demand setting periods. An alternative is to perform maintenance in off-peak periods. However, this is not always possible, particularly for lengthy overhauls. Backup service rates are somewhat like an insurance policy. By purchasing this capacity insurance for a given amount of kilowatt on a monthly basis, self-generators avoid ratcheting and/or full demand charges that might otherwise result from outages.

Consider an example in which the full service rate demand charge is $18/kW/month and the standby charge is $8/kW. The facility pays this charge regardless of whether backup power is used. In this example, the annual fee of $96/kW would be a wise investment only if the facility experienced peak setting outages more than 5 month/year, since 5 months' demand charges would only cost $90/kW.

The cost of standby service varies widely. Some utilities require self-generators to purchase standby insurance. Other utilities offer it as an option, while some have no provision for standby power at all. The alternative is the use of standard rates. Key questions in cost allocation are: "What are true costs?" and "What is a fair and reasonable price for such capacity insurance?"

The logic behind this particular cost allocation is that the utility must stand ready to serve these loads when needed. Cost allocation is based on a determination of the impact on generation and distribution capacity requirements. The cost-of-service analyses take into account all self-generators and the real probabilities of peak demand impact resulting from random system outages. If, for example, each of 100 self-generators were to set a peak once a year at different times, what would the real impact be on the capacity requirements of the utility?

For cases in which standby service is not mandatory, but offered as an option, customers must make the determination whether to take this type of insurance or take their chances on standard rates. Customers may also elect to secure standby power for a portion, rather than all, of their self-generation load.

In cases where standby charges are mandatory and very high, the cost may be sufficient to make projects uneconomical. In some cases, a change to mandatory requirements, resulting from rate case proceedings years after a system has been installed, may provide sufficient incentive to abandon a project due to the evaporation of savings critical to successful economic operation.

One hypothetical example of such prohibitive effects is a system with three generation units with required standby charges for the full connected load at 66% of the standard demand charge. In this case, the cost of standby service is equivalent to that of the system operating on a standard rate and experiencing the highly unlikely occurrence of a peak-setting outage in every single month for two out of the three units. Add to this, the potential of being forced onto an uneconomical rate, and a self-generator could end up with no savings at all. While this example is extreme, it helps to explain why self-generation has been underdeveloped in certain utility service territories.

Real-time pricing rate structures may offer an effective means of allocating costs for standby power. Real-time pricing is an attempt to reflect short-term costs so that consumers may make short-term purchase decisions. These same varying short-term prices could be made available to QFs. California, Florida, and Virginia are a few of the states that currently have QF purchase power pricing tied to variants of RTP. For example, one rate structure on file with a state commission provides payment for a QFs energy sales at the corresponding marginal cost (i.e., system lambda, $/MWh) of power the host utility experiences. In California, a forecast of marginal costs are the primary input in determining an RTP pricing structure for as-available energy.

MEASURING ELECTRIC DEMAND

The measurement of demand is fundamental to most electric rate structures. It is a tool that allows a utility to differentiate capital cost requirements for serving customers of varying usage patterns. By measuring and billing for demand, the utility can assign costs more fairly to customers.

As opposed to gas utilities, the demand interval for electric utilities is extremely short, usually 15 or 30 min. It is not an instantaneous measurement, but an average of discrete measurements over time. If, for example, a facility experienced a rate-of-use pattern of 200 kW for 5 min, 300 kW for 5 min, and 700 kW for 5 min, a 15-min demand interval meter would register an average demand of 400 kW. The demand recording meter would log 400 kW and reset only when a higher level of demand was reached. Sometimes, utilities set peak demand by averaging peaks of a few demand intervals over the billing period.

Many customers use demand monitoring and load-shedding techniques to minimize the impact of peak demand billing. These facilities often attempt to synchronize their operations with the utility demand interval and use intermittent load shedding to reduce their average rate-of-use during intervals when a large surge of power is required for a period less than the full demand interval.

From the utility perspective, this load shedding technique may partially defeat the purpose behind demand metering, which is to charge for peak capacity requirements. A sliding demand interval is sometimes used to more accurately measure the impact of peak demand. With a sliding demand interval, for example, a 15-min demand interval may be broken into smaller intervals of 5 min. These smaller intervals are averaged and then added together, as in the previous example, to set the peak demand for the entire 15-min interval.

In some cases, utilities simply use smaller demand intervals, starting as low as 5 min. More common, however, is the use of the typical 15- or 30-min interval with a clause in the rate schedule that states that in cases of rapidly fluctuating loads or other special conditions in which the established demand measurement time interval does not equitably compensate the utility, demand may be based on the peak for a shorter period.

Traditional rates often use only one peak demand measurement for billing. Some rates call for demand measurement only in certain peak periods. The rationale is that individual facility peaks in utility off-peak periods have no real impact on capacity requirements. The TOU rates, however, measure peak demand in several periods. Demand charges may be set at a different rate for each period. For example, peak demand might be billed at $20/kW during peak periods and $5/kW during off-peak periods. In some cases, off-peak period peak demand may only be billed for the portion that exceeds peak period peak demand. This is referred to as excess demand billing.

INTEGRATING POWER FACTOR INTO DEMAND BILLING

Utility generation is measured in volt-amperes, or apparent power, while most customer meters are measured in watts, or real power. In alternating current (a.c.) circuits, watts (power) are equal to volts (potential) times amps (current) only when the wave-forms of voltage and amperage are in phase. This is an ideal condition that does not exist in electric distribution systems. Many types of equipment, such as induction motors, require more apparent power than the amount of real power consumed because their inductive impedance causes current and voltage to be out of phase.

The difference between apparent power volt-amperes and real power watts is called volt-amperes reactive. This is the component of volt-amperes that circulates back and forth between the utility and the equipment, but is not consumed by the load. It is, however, partially consumed by distribution losses.

The PF is the ratio of W:VA. A facility with a PF of 0.75, requiring the same wattage as another facility with a PF of 1.0, for example, will be more costly to serve, because the utility will require one-third more system capacity (1.00 W/0.75 PF = 1.33 VA) to serve the facility with the PF of 0.75.

To more accurately allocate capacity costs through demand charges, some utilities measure and bill demand charges based on kilovolt-amperes rather than kilowatt. This shifts the cost for maintaining nonproductive capacity, or a PF of less than 1.0, to the customer and acts as an added incentive to improve the facility's PE.

Many utilities simply institute a penalty for a lagging PE For every increment below a required minimum PF, a charge is levied against the facility. The minimum allowable PF is typically in the range of 0.80–0.90. Many utilities establish this penalty on the rate schedule, but often do not invoke it.

Another way utilities establish a PF penalty is to specify a maximum free kilovolt-amperes reactive as a percentage of the maximum kilowatt of demand. Utilities then bill for all metered kilovolt-amperes reactive above this level. There are several other ways to build PF into billing, such as increasing the peak demand measured by a certain percentage for every percentage, the PF is below a specified level.

Many utilities currently do not penalize for lagging PF, and many that do only impose modest penalties. In those cases, from the customer's perspective, the advantages of a higher PF and the benefits of investment in capacitors and other higher PF equipment are savings from reduced internal line losses, down-sized equipment, and avoided billing penalties. To encourage such customer actions, utilities attempt to set PF penalties at levels that will have sufficient economic impact. Refer to Chapter 24 for a detailed technical description of PE.

METERING POINT AND TRANSFORMER OWNERSHIP

Another factor that is often an element in electric utility rate structures is metering point and transformer ownership. Utility distribution voltage is almost always greater than the voltage required inside a facility. The main transformer brings the voltage down to a suitable level for the service entrance at the facility. Typically, the utility owns the transformer and meters usage on the customer (low-voltage or secondary) side of the meter.

These factors will figure into rate design. If the customer owns the transformer, the utility saves on capital and maintenance costs and can pass those savings on to the customer. If power is metered on the high side of the transformer, the utility saves on transformer-related power losses and can pass those savings on to the customer to compensate for losses now occurring on the facility side of the meter.

SEPARATION OF COMMODITY AND DISTRIBUTION FUNCTION

Overtime, the forces of deregulation are moving the jurisdiction of regulated utility rate structures toward the transmission and distribution functions and away from the commodity sale function, which is falling under the control of free market forces. However, in today's market, the rate structures presented above still widely apply as the utilities continue to sell a bundled commodity to their customers or sell unbundled services, notably transmission and distribution under regulated cost-of-service-based pricing.

With the advent of electricity wheeling and gas brokering, with open access to transmission and distribution service, opportunities for utilities to make off-system sales have become subject to fierce competition. With open access for the transmission of power for wholesale sale in interstate commerce, the utility must compete for wholesale market share with cogenerators and other independent power producers, many of which can generate power at a very low cost. Gas brokers present similar challenges to gas utilities.

From the consumer perspective, the trend of competitive pricing can be very attractive. Currently, competitive pricing is most beneficial to large consumers, notably those with favorable load profiles or competitive options. Such customers often face a win–win situation in which they install operating cost-reducing alternatives or reap savings from a competitive energy rate break or, in some cases, both.

Over time, it is anticipated that an increasing number of consumers will be the beneficiaries of competitive pricing. However, there is also concern that the embedded cost in utility capacity that is no longer cost-competitive will drive up prices to customers stranded without strong competitive options.

The overwhelming long-term trend in the electric and the gas industries is one of deregulation, fierce market competition, unbundling of services, and increased consumer choice in energy services. Historically, natural gas companies mostly sold and delivered purchased gas to customers at a price determined by a regulated rate structure that included a bundled host of services. Electric utilities sold electricity that they had generated or purchased under similar arrangements.

While far from fully evolved, the current trend is toward a condition under which the local utility's prime function is the transmission and/or distribution of natural gas or electricity, but not necessarily the sale of the energy commodity. The purchase and sale of natural gas and electricity is becoming more of an independent market function in which the utilities are but one broker among many. It is very conceivable that over the next decade, this trend will be all encompassing, inclusive of the residential market, with newly developed rate structures that are compatible with the evolving conditions surrounding consumer transactions. Though the trend continues at an accelerated rate, utilities still provide both sales and distribution functions to most customers.

CONCLUSION

The concepts behind the design of natural gas and electric utility rate structures are extremely similar in that they are developed through the same regulatory process. Both have been greatly affected by the movement toward deregulation and are moving more toward market-based competitive rate structures. While the pace of change varies from state to state and utility to utility, the commodity component is being separated from the transmission and distribution component, necessitating significant changes in cost recovery and, therefore, rate design.

Finally, the trends toward some variation on the concept of RTP are apparent, as is the trend toward increased customer choice in the selection of utility rates and energy services in general. Still, the basic concepts of fixed and variable cost recovery shall continue to apply. An understanding of these concepts allows gas and electricity consumers to understand utility rates and select rates and services that most effectively meet their energy-use needs.

BIBLIOGRAPHY

1. Berg, S.V. *Innovative Electric Rates*, Lexington Books, D C Heath and Company: Lexington, MA, 1983.
2. Berg, S.V.; Tschirhart, J. *Natural Monopoly Regulation: Principles and Practices*; Cambridge University Press: New York, NY, 1988.
3. Bonbright, J.C.; Danelson, A.L.; Kamerschen, D.B. *Principles of Public Utility Rates*, Public Utility Reports, Inc.: Arlington, VA, 1988.
4. Capehart, B.L.; Turner, W.C.; Kennedy, W.J. *Guide to Energy Management*, 5th Ed.; The Fairmont Press, Inc.: Lilburn, GA, 2006; [Chapter Three, Understanding Energy Bills].
5. Enholm, G.B.; Malico, J.R. *Electric Utilities Moving Into the 21st Century*, Public Utility Reports: Arlington, VA, 1994.
6. Hirsh, R.F. *Power Loss—The Origins of Deregulation & Restructuring in the American Electric Utility Stystem*; MIT Press: Cambridge, 2002.
7. Petchers, N. *Combined Heating, Cooling & Power Handbook*; Marcel Dekker, Inc.: New York, NY, 2003; [Chapter 21, Utility Rate Structures, and Chapter 22, Utility Bill Analysis].
8. Turner, W.C. *Energy Management Handbook*, 5th Ed.; The Fairmont Press, Inc.: Lilburn, GA, 2005; [Chapter 18, Electric and Gas Utility Rates for Commercial and Industrial Customers].
9. The Changing Structure of the Electric Power Industry 2000: An Update; October 2000, US Energy Information Agency, Washington, DC.
10. Introduction to Rate Design for Electric Utilities, Professional Development Course, Electric Utility Consultants, Inc. Denver, Co., 2006.

Walls and Windows

Therese Stovall
Oak Ridge National Laboratory, Oak Ridge, Tennessee, U.S.A.

Abstract
Energy travels in and out of a building through the walls and windows by means of conduction, convection, and radiation. The walls and windows, complex systems in themselves, are part of the overall building system. A wall system is composed of multiple layers that work in concert to provide shelter from the exterior weather. Wall systems vary in the degree to which they provide thermal resistance, moisture resistance, durability, and thermal storage. High-tech windows are now available, which can resist radiation heat transfer while still providing light and visibility. The combination of walls and windows within the building system can be adapted to meet a wide range of environmental conditions, recognizing that the best building envelope system for one climate may not be the first choice for another location.

INTRODUCTION

The building envelope protects you from the weather, separating the indoor air that you have paid to heat or cool from the outdoor air. The walls, roofs, and windows form the major surfaces of this envelope. Energy travels through these surfaces along many paths and in many forms, such as unintended air leakage through a wall or sunlight streaming through a window. We can improve the energy efficiency of the building envelope if we carefully consider all of these energy pathways.

Here, we are focusing on energy efficiency, but keep in mind that each part of the building envelope performs multiple jobs under challenging conditions. The roof keeps out rain, hail, and snow; bakes in the hot summer sun; freezes in the cold winter night; and must be sturdy enough to survive a workman's boots. The walls must repel the rain, hold up the roof, stop the wind, and provide a rigid support for windows and doors. The windows have to let in the light, allow ventilation when they are open, and keep out drafts when they are shut. Any change we make to the building envelope to conserve energy must account for these multiple functions and the complex interactions between the envelope components.

Resistance to heat transfer is often expressed as an *R*-value. For a complete description of R-values and their use, please refer to the DOE Insulation Fact Sheet (www.ornl.gov/roofs+walls/insulation).

BUILDING TYPES

Buildings fall into two main classes: high-rise and low-rise. The high-rise buildings are typically custom engineered with structural steel frames. Low-rise buildings are typically divided into commercial, low-rise multifamily, and single-family buildings. The high-rise, commercial, and low-rise multifamily buildings are more likely to have low-slope (often mislabeled as "flat") roofs, whereas the single-family houses are more likely to have pitched or steep-sloped roofs.

Construction methods can be roughly grouped according to the portion of the assembly performed on site and the portion performed at a factory. With today's engineered wood products and premade trusses, few buildings are strictly built on site; but we still refer to a stick-built building as the one where the greatest part of the assembly takes part on the construction site. At the other end of the spectrum are manufactured buildings that can be moved from one site to another with relatively little effort. In the middle are factory-built modular buildings and panelized construction. A factory-built modular building typically includes one or more modules that are placed on a permanent foundation. In some modular buildings, the windows are installed after the modules have been installed. A panelized building is closer to the site-built model, but will have major wall, floor, or foundation sections prebuilt and delivered to the site.

Every building must conform to local building codes, which can limit the material choices or construction methods. Many local building codes now include energy conservation clauses or incorporate the Model Energy Code or the International Energy Conservation Code.[1]

WALLS

The walls make up most of the exterior surface area of many buildings and therefore the energy transported through these surfaces is very important. Wall issues vary according to the building type. A framed building is constructed with a wood or metal skeleton that provides structural support to both the building and all the other wall components.

Keywords: Building envelope; Wall; Window; Thermal energy; Heat transport; Solar heat gain coefficient; Thermal mass; Adobe; Structural insulated panel; Exterior insulation finish system; Masonry walls.

A framed wall is characterized by numerous parallel heat paths and multiple layers of different materials. A non-framed building uses bulk material, such as masonry or adobe, to provide the structural support. This type of building is characterized by a more homogenous heat path and relatively few layers of materials.

For either type of building, the connections between the wall and the roof and between the wall and the foundation are important construction details from an energy conservation standpoint. These connections can provide unintended air passageways and may be overlooked in the overall insulation scheme. An Air Drywall Approach is an effective way to limit air leakage from walls, roofs, and windows. Here, a rubber gasket is fitted along the perimeter of the window frame and along the exterior wall's base board and ceiling plate to compensate for openings that occur as the wood changes shape with time. The gasket, once placed, seals against the drywall gypsum board and makes an airtight barrier.

Whether the wall is built on a frame or constructed from masonry, there is a wide selection of exterior siding choices. These include brick; wood, fiber cement, or vinyl siding; or an exterior insulation finish system (EIFS). For all of these facades, repelling rain and wind is often more complex than it looks. For example, a brick wall looks like a solid surface, but the mortar joints provide capillary paths for moisture, especially when subjected to wind-driven rain. Similar pathways exist for other cladding materials. For this reason, air gaps are often provided behind the outermost wall layer. Depending on the vent/drain arrangement, this may provide a true pressure-equalized rain screen or a simple break in the capillary path. Old-fashioned wood lap siding is a good example of a simple rain screen, as shown in Fig. 1. The small air gap behind each wood layer is well vented and drained to the outside, so that the air pressure within the air gap is equal to the air pressure outside the wall. This pressure equalization reduces the moisture moving into the air space, and therefore reduces the amount of moisture available to penetrate the rest of the wall.[2,3]

Proper moisture management is important for energy conservation for two reasons. First, unmanaged moisture must be removed by additional ventilation, which entails the energy load needed to heat and cool the additional air mass. Second, moist building materials will always have a higher thermal conductivity than dry materials. When wet, some insulation materials become matted and lose the greater part of their insulating value.

Every wall system stores energy. This storage quality is called "thermal mass." The thermal mass can reduce the amount of heat lost or gained through the wall whenever the outdoor temperature varies above and below the indoor temperature on a daily basis. This occurs during the spring, summer, and fall for most of the United States, and during the winter as well for the southern regions. The relative benefit of thermal mass is therefore determined by both the wall properties and the climate. A wood-framed wall has very little thermal mass compared with a masonry wall. One study compared the energy consumed by a house with traditional wood-framed walls with a house constructed with masonry walls for six cities. Depending on the location and wall thickness, the more massive wall reduced the household energy use by an amount equal to increasing the traditional wall's thermal resistance by 10–50%, with the savings greatest in Denver and Phoenix and least in Miami.[4]

Although we typically think of the exterior walls when we think about energy losses, the interior walls are also important. Air enters the wall cavity through a number of penetrations, some visible and some not. Holes made in the drywall to accommodate electrical outlets and plumbing connections also allow uncontrolled airflow. Other gaps are often present at the floor–wall and floor–ceiling connections. Therefore, it is important to provide a continuous top plate above every wall cavity. This top plate separates the interior wall cavity from the attic space, thus preventing a free flow of conditioned air from your house into the attic.

In addition to the energy used to heat and cool a building, energy is also embodied in the building materials and expended in the construction process. Among low-rise residential buildings, studies have shown that many of the building elements, including gypsum drywall, roofing materials, and carpeting, are common to all the wall types. But the wall type still makes a significant difference in the overall energy embodied in the building, with a

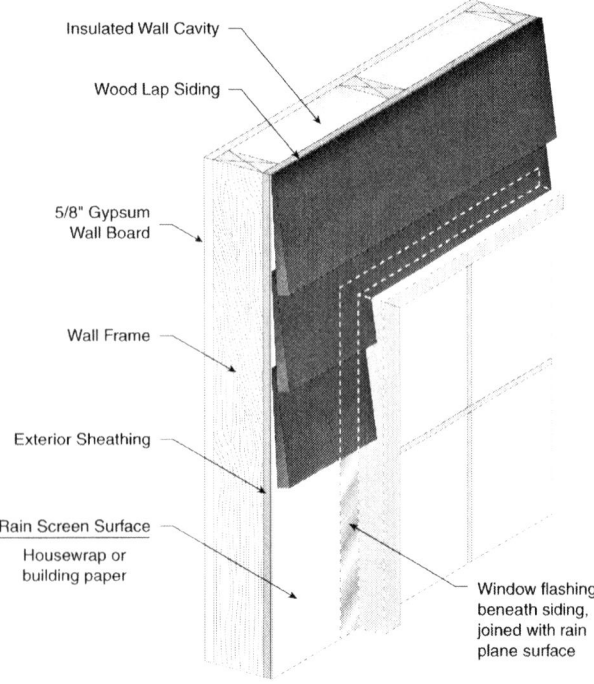

Fig. 1 Simple rain screen within a typical wood lap siding wall.

wood-framed house containing about 15% less embodied energy than either a metal-framed or concrete house.[5,6]

Wood-Framed Walls

Wall construction methods and materials vary somewhat according to the local climate and natural resources, but most walls in the United States are made from wood framing with insulation between the studs, drywall on the inner surface, and exterior sheathing layer(s) (Fig. 2). The wall studs used are either nominal 2×4, with a 3.5-in. cavity depth, or nominal 2×6, with a 5.5-in. cavity depth. Cavity insulation options for new construction include batts, a blown-in mixture of a foam binder and loose-fill insulation, blown-in loose-fill insulation secured by nets, or blown-in foam insulation.[8] In commercial buildings, high-density batts are sometimes used to provide both thermal insulation and acoustical buffering. In retrofit situations, professional installers can blow loose-fill insulation into the wall cavities by drilling a series of holes through the interior or exterior facade between each pair of adjacent studs.

Energy flows through the wood studs more easily than through the surrounding insulation. Most walls contain much more framing than you would think, as shown in Fig. 3. In addition to the studs, there is additional wood framing around each window, around each door, at each building corner, where the wall sits on the foundation, where the roof sits on the wall, where an interior wall meets an exterior wall, and between floors in a multistory building. When you combine all of these thermal "short circuits," a wall (in a one-story ranch style house) filled with R11 insulation provides an overall performance of only R9–R10. If you replace that insulation with foam in the wall cavity, thereby increasing the cavity insulation from R11 up to about R18, you get an overall performance of R13 or an increase that is about half of the increased insulation value. One way to improve the thermal performance of a wall is therefore to place insulation between the studs and the exterior surface of the wall. For

Fig. 3 Framing lies behind a significant portion of the wall area in many houses.
Source: From ASHRAE Special Publications (see Ref. 9).

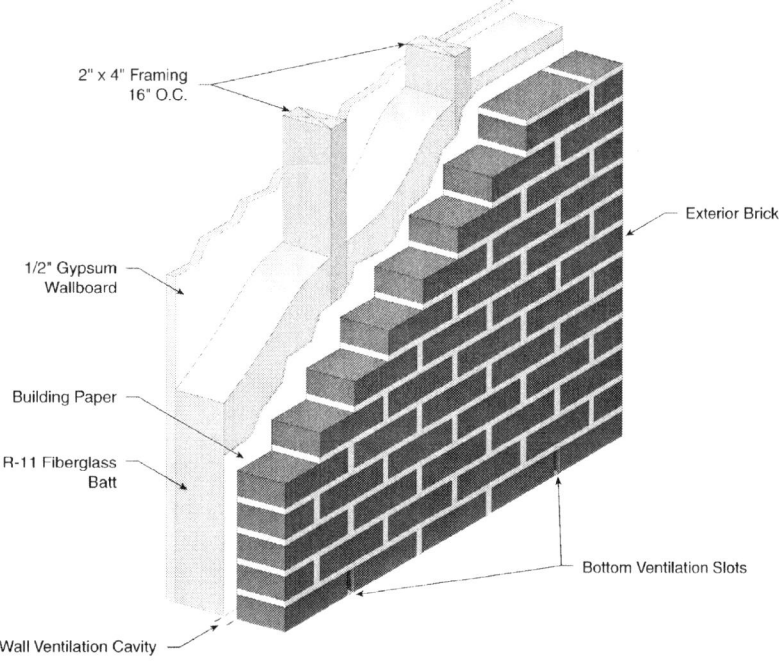

Fig. 2 Typical wall structure for a wood-framed wall with brick cladding.
Source: From ASHRAE Special Publications (see Ref. 7).

example, adding only 1 in. (R5) of foam sheathing to the R10 wall brings its overall *R*-value up to R14 (Fig. 4).

Advanced framing techniques are available which reduce the amount of lumber required. These methods reduce the energy losses through the framing and allow more room for insulation. Many of these techniques also provide for improved air sealing.[10]

The wall sheathing provides a flat uniform surface to support the exterior air barrier, vapor retarder, and siding. If a wood product is used for the sheathing, it will also provide the structural stiffness needed at the building corners. If a foam insulation product is used instead, some form of additional bracing will be necessary at the corners. Sometimes a layer of foam is placed on top of a layer of wood product sheathing. This greatly improves the wall's thermal resistance because that layer covers the thermal short circuits provided by the wood frame. However, this configuration can require extra care during the finishing process, with longer nails needed to fasten siding materials. Also, specialty brick ties will be needed if a brick decor is selected for finishing the wall.

Steel-Framed Walls

Steel-framed walls share many of the characteristics of a wood-framed wall, but the steel components themselves have a very high thermal conductivity. Therefore, most steel-framed walls are built with a layer of foam sheathing to break that thermal pathway. There is also research underway to produce complex steel shapes to provide the structural support needed while providing a longer heat transfer pathway, and therefore greater thermal resistance.[11] Some metal-framed products also include an integral foam insulation element for the same reason. Connections between the walls and a steel-framed roof can be problematic from an energy point of view, especially if the steel framing extends out in the eaves. Such arrangements act like the fins on a heat exchanger and can cause excessive energy consumption if appropriate insulation arrangements are not included in the design. Steel-framed walls are most attractive where the heating loads are modest and where insect damage is more challenging.

Masonry Walls

Masonry walls can refer to walls with a masonry veneer, such as brick or stone, or to a wall where the masonry also provides the structural support, such as a poured concrete or concrete block wall. Such buildings are relatively resistant to the corrosive environment common near the ocean. Masonry also provides thermal mass that can both save energy and improve the interior comfort level in a hot climate with daily temperature swings. Aside from this thermal mass effect, masonry veneer walls share the same energy characteristics as other wood- or steel-framed walls.

Concrete block and poured concrete

Full masonry walls are often used for foundations or basements. Whole houses built from masonry are popular in the warmer climates where termites and other insects are more populous. A full masonry wall can be built from concrete blocks or by pouring concrete into forms. For a load-bearing wall, steel reinforcement rods, or rebar, are used with either method to add strength. For a reinforced concrete block wall, reinforcing rods are placed vertically in the block cavities which are then filled with mortar. Steel wires or mesh are laid in the horizontal mortar joints to resist shear stress. In a poured concrete wall, the steel reinforcing rods are positioned both vertically and horizontally within the forms before the concrete is poured. Because these steel rods tend to be perpendicular to the heat transfer direction, they have little effect on the overall wall thermal resistance.

The thermal characteristics of a masonry wall vary, depending on the whether blocks or poured concrete is used and on the density of the concrete. But in general, the thermal resistance of the masonry portion of such walls will be very small, ranging from R1 to R2 for an 8-in. thick wall, even if the cores of a hollow concrete block are filled with perlite or vermiculite.[12] Foam board insulation, with a thermal resistance in the range of 5*R*/in. is often used to increase the total thermal resistance of a masonry wall and can be placed on the inside and/or the outside surface. The foam board must be covered by some material with an appropriate fire rating, such as gypsum board. One study has shown that the thermal mass is more effective when the concrete is in good thermal contact with the interior building, i.e., when the insulation is placed on the outside of the wall.[13]

Autoclaved aerated concrete can be used in place of ordinary concrete for low-rise buildings. This type of concrete is much lighter than standard concrete, available

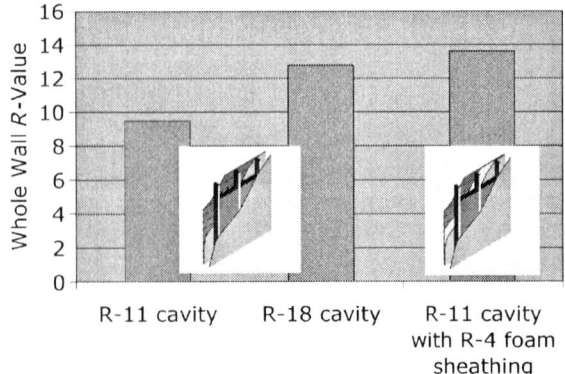

Fig. 4 Whole wall *R*-values for wood-framed (nominal 2×4) walls on 16-in. centers with interior gypsum and exterior wood siding in a one-story ranch house.

in a variety of sizes and shapes, and may be reinforced. The autoclaved aerated concrete has a much higher R-value (1.25R/in.) than the standard concrete (0.05R/in.).[3] The overall R-value of a wall built with this material will depend on the shape and thickness of the concrete and on the thermal resistance of other wall components, such as air cavities. Several walls tested with aerated concrete (no facings applied), both in the traditional hollow concrete block and solid block forms, had R-values between 6 and 9.[14]

Precast concrete

Walls can be made from reinforced concrete slabs that have been precast into the desired shape. Unless insulation is applied, these walls will have approximately the same thermal resistance as a site-poured concrete wall, i.e., about R1–R2 for an 8-in. thick wall. This construction method is more often used for larger buildings, such as apartment buildings or hospitals, but is also used for residential buildings. In the smaller buildings, the precast panels were first used for basement walls, but precast panels can be used for all exterior walls. The panels are precast and cured in the factory, thus avoiding weather limitations associated with pouring and curing concrete at the building site. The factory environment also allows the production of concrete that is stronger and more water resistant than site-poured concrete. Some of the panels are cast against foam insulation to improve the wall's R-value. In addition to the steel reinforcing, these walls can be produced with cavities for electrical wiring and rough openings. Some of the panels have been designed to provide the appearance of bricks and limestone, so that a new building will blend into an existing urban neighborhood. A crane is usually used to place the precast panels on top of a bed of crushed stone. The panels are then connected in place using weld joints or bolts and sealants. Installation for a typical residential unit can be completed in a single day.[15,16]

Insulated concrete forms

An insulated concrete form (ICF) wall is composed of a set of joined polystyrene or polyurethane forms that are filled with reinforced concrete, as shown in Fig. 5. The joining methods vary from one manufacturer to another. Some have interlocking foam panels; others are linked with ties made from polypropylene or steel. The foam panels become a permanent part of the wall, providing a continuous layer of insulation on both the inner and outer wall surfaces. The concrete portion of the wall is reinforced with rebar positioned inside the forms before the concrete is poured to provide needed strength. The ICF walls can be covered on the outside with light-weight stucco, brick, or wood or vinyl siding. Gypsum board fastens onto the interior side. The attachment methods for

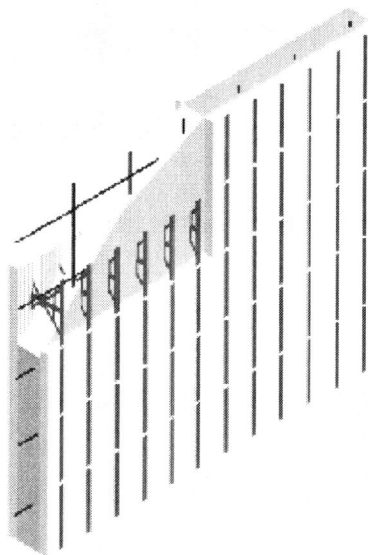

Fig. 5 Insulated concrete form (ICF) wall.

these facing materials vary from one manufacturer to another. In the laboratory, the simple steady-state R-value of unfinished ICF walls varied from 12 to 18.[14] The time-varying effective energy performance of these walls is also determined by the temperature profile within the wall, which is improved by the walls' well-insulated thermal mass. Therefore, the ICF wall design provides both greater thermal mass and a higher thermal resistance than a typical 2×4 frame wall. Although the airtightness of any wall system tends to vary depending upon the expertise of the construction crew, an ICF wall is generally more airtight than a wood-framed wall.

A variant of the ICF wall uses forms made from a polystyrene–cement composite. The composite walls may have a greater thermal mass, but provide a significantly lower steady-state thermal resistance (about R8 for one 10-in. thick configuration).[14]

Adobe

In parts of the southwest, adobe walls are popular because they absorb and store the daytime heat until it can be released to the cooler night air. Traditional adobe bricks are sun-dried, not fired, mixtures of clay, sand, gravel, water, and straw or grass. (For fired or stabilized bricks that are made to look like adobe, see the previous section on concrete masonry). The traditional production method produces a brick that swells and shrinks depending on its fluctuating water content. Adobe walls are relatively fragile and must be sealed with a protective covering.

The low compressive strength of the adobe bricks leads to the use of very thick (10–30 in.) walls that are seldom more than two stories high. These thick solid walls provide significant thermal mass. The thermal mass effect is especially important because the steady-state thermal resistance of these walls is not very great. A 14-in. thick wall would have a thermal resistance of from R2 to R10, depending on the density and water content of the wall. Insulation can of course be added to the interior or exterior face of the wall if covered with an appropriate coating.[17]

Exterior Insulation Finish Systems

The EIFS can be placed on a wood- or steel-framed wall or a masonry wall. In this system, a layer of polystyrene board insulation, one or more inches thick, is applied to the wall which is then covered with multiple coatings that produce the finished appearance of stucco. This wall system provides a continuous cover of insulation that breaks all the thermal short circuits associated with the framing materials and has the thermal advantage previously shown in Fig. 4 for foam sheathing. The EIFS system has been and continues to be one of the most popular exterior claddings for commercial and institutional buildings. However, residential construction jobs often do not have the same high level of quality control and job oversight. This difference led to moisture-related problems in residential EIFS walls where moisture seeped into inadequately sealed window openings and became trapped within the walls.

Subsequent building failures led to the development of two classes of EIFS: barrier and drainable. Both classes are used in the commercial building class, but only the drainable system is allowed for residential construction in many locations. In the barrier class, the outer finish layer is designed to be the one and only weather-resistive barrier on the wall. In the drainable class, some form of spacer is placed between the polystyrene board insulation and a second weather-resistive barrier is located atop the wall's structural sheathing layer. This drain plane allows any moisture that seeps into the wall to drain safely out of the wall.[3]

Structural Insulated Panels (SIPS)

The SIPS walls are made by sandwiching foam insulation, typically 4–6 in. thick, between two sheets of a wood product, thus providing structural support, insulation, and exterior sheathing in a single panel. Each manufacturer specifies the proper method and materials to use when joining adjacent panels. Because the panels themselves have such a high thermal resistance, these joints are critical in maintaining a high thermal resistance for the whole wall. Walls have been measured with R-values of about R14 for a wall with a 3.5-in. thick foam core and about R22 for a wall with a 5.5-in. thick foam core.[14] The wall sections are relatively light and the exterior walls of a building can often be completed in a day. Two variations of the system include the use of metal sheets for the exterior skin and the use of alternative insulation materials.

Straw Bale Walls

Exterior walls can be made from stacked straw bales. The straw is a natural insulation material, but must be protected from the weather by an exterior surface, often a stucco-type finish applied over a wire mesh. Gypsum board can be used on the interior surface using a number of methods.[3] Experiments have shown that it is very important not to leave any air space between the straw and the surfacing material. Such air gaps work in concert with the hollow straw tubes to set up convection loops that may cut the overall thermal resistance from around R50 down to around R16.[18]

WINDOWS

Transparent glass was first used for windows during Roman times. Technology has gradually advanced, improving the smoothness, strength, and clarity, and increasing the maximum size of manufactured glass. More recently, double-pane windows became popular as energy costs rose and the use of air conditioning became more common. By the mid-1990s, nearly 90% of all residential windows sold had two or more layers of glass.[19] The windows in older homes are being upgraded to double-pane windows as well; from 2000 to 2001, about nine million homeowners spent $15 billion on window and door replacements.[20] From 1990 to 2003, the replacement window market made up from 38% to 52% of the total window market.[21] However, in 2004, two-thirds of residential buildings and about half of the non-residential building stock still had single-pane windows. The potential energy savings in this population is great, especially with the new selective coating techniques for glass.

Energy Transport

The primary function of windows is to let light and air into a building; so it is not surprising that the windows can be very challenging from an energy conservation point of view. Energy travels through a window by all three energy transfer phenomena, as shown in Fig. 6.

The radiation portion of the energy transport includes short-wave radiation, or ultraviolet, that tends to fade the colors in fabric and paint; visible radiation, or light, that we desire; and long-wave radiation or heat. Long-wave radiation is a normal part of the solar spectrum. It helps to heat our buildings during the winter but increases our air conditioning load during the summer. Some values for the solar heat gain through a few prototypical windows are shown in Fig. 7. Long-wave radiation also occurs between any two surfaces, traveling from the warmer surface to the cooler surface. Because the outside environment is warmer than the inside surfaces in the summer and the reverse in

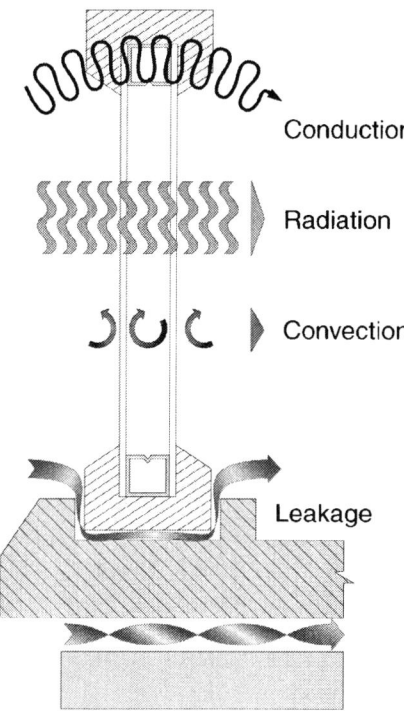

Fig. 6 Energy traveling through windows.

the winter, this long-wave radiation travels through the window and increases both our winter heating and summer cooling energy use. Newer windows have a "low-e" coating to reduce long-wave radiation and thus reduce the energy losses.

The conduction portion of the energy transport includes the heat that travels through the window frame and heat conducted through the glass pane(s) and through any gas between the panes. The energy that travels through the window frame and sashes is a complex function of the frame material(s), the shape (including any hollow cavities), and the exposed surface area. Each material used in windows has positive and negative qualities. In general, wood has a lower thermal conductivity than metal or plastic. However, wood is more likely to change in shape during its lifetime, so that sealing air leakage out of a wood window over a long time period may be more problematic. Also, wood is more susceptible to moisture damage and must be kept painted or varnished. Metal frames will conduct more heat than wood, but the exposed area can be reduced because the metal is a stronger material. Some of the best performance is found in windows that combine multiple materials. For example, a metal- or vinyl-clad window frame will require less maintenance than a wood window, but will conduct less heat than a solid metal or solid plastic frame. Some special gases, such as argon, have a lower thermal conductivity than air and are sometimes used to fill the gap between panes in multiple-pane windows.

The convection portion of the energy transport includes exterior air movement, or wind, across the glass surface; gas movement between the panes in a multilayer window; and air movement across the interior face of the window. The gas movement within a multilayer window is determined by the thickness of the gap, the height of the window, the gas temperature, and the temperature difference between the two panes of glass. Closed draperies or shades can reduce the air flow across the inside pane of glass.

The energy carried by air leakage falls into two major categories. As Fig. 6 shows, some air leaks around the window frame itself. This leakage path should be sealed when the window is installed, although caulking around the frame of an existing window can also reduce the air leakage. Air also leaks through any moving joint in a window. These joints must be sealed by weather stripping or special gaskets that are built into the windows themselves.

Storm windows have been used for a long time and were very popular in northern climates before the introduction of multiple-pane windows. Storm windows require annual installation and storage and there are more glass surfaces to clean. Storm windows can be mounted on either the inside or outside of the window

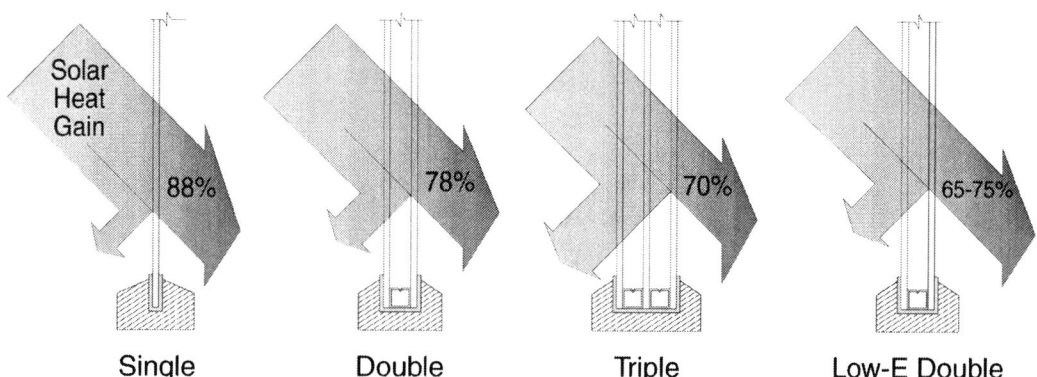

Fig. 7 Solar heat gain through various windows.

frame.[22] Tests have shown that the energy savings can be substantial when storm windows are added to an existing single-pane window. However, replacing an existing single-pane window with a modern double-pane window will save more energy than the addition of a storm window.[23,24]

Window Rating Systems

Considering the complexities of energy transport through windows, it can be difficult for consumers to compare one window with another. Fortunately, there are two important tools available to help with window selection. The National Fenestration Rating Council (NFRC) has developed a rating system that includes a standard label.[25] The U.S. Departments of Environmental Protection and Energy have cooperated in the production of an Energy Star label for windows.[26]

The NFRC label (Fig. 8) shows four values: the U-factor, the solar heat gain coefficient (SHGC), the visible transmittance, and air leakage. Manufacturers may also choose to show a value for condensation resistance. The two most important factors used to rate the energy efficiency of a window are the U-factor and the solar heat gain coefficient. The U-factor is the inverse of the R-value (the label that is quoted for insulation), so that a lower U-factor indicates a slower rate of heat transfer for any given temperature difference.[27] The NFRC U-factor ratings for windows sold nowadays range from 0.2 to 1.2.[28] The SHGC ranges from 0 to 1 and measures how much heat from the sunlight incident upon a window will enter the building. Windows with lower SGHC ratings do a good job of blocking this heat.

The Energy Star label is available to windows that have been certified and labeled by NFRC and that meet special standards. The standards vary among the four NFRC climatic regions because of the trade-off between desirable winter heating and undesirable summer heating. The four regions used by the Energy Star program for windows are shown in Fig. 9 and Table 1 shows the required U-factors and SHGCs for each region.[29]

Future Improvements

Researchers are working on a portfolio of window designs with automatic energy-saving features. Some forms of "smart" windows with switchable glazing have become commercially available, albeit at a relatively high cost. These windows vary the amount of light and heat transmitted based upon an electric current, which is usually programmed to respond to either the temperature or the amount of sunlight hitting the window. The electrochromic window uses a multilayer electrically conductive film where ions are moved from one layer to another by a short electrical signal. In one layer, the ions allow only 5% of the sunlight through the window; when the ions move to the other layer, 80% of the sunlight is transmitted. This system has the advantage that once the change from one state to another has been made, no electrical energy is required to maintain that state. Another switchable glazing, the suspended particle display (SPD) places a solution containing suspended particles between two glass panes. When an electrical charge is applied, the particles align and light is transmitted through the window. Without an electrical charge, the particles move about randomly, blocking up to 90% of the light. Other switchable windows are designed to provide privacy by changing from transparent to translucent, but are not effective at saving energy.[30,31] Another proposed window design includes sensors that automatically raise or lower a

Fig. 8 National Fenestration Rating Council (NFRC) label.

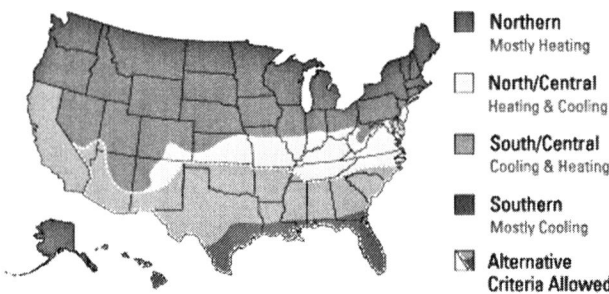

Fig. 9 Energy Star's four regions used for window rating program.

Table 1 Energy Star window criteria

Climate zone	U-factor	Solar heat gain coefficient (SHGC)	
Northern	≤0.35	Any	
North/Central	≤0.40	≤0.55	
South/Central	≤0.40	≤0.40	Prescriptive
	≤0.41	≤0.36	Equivalent performance (excluding CA)
	≤0.42	≤0.31	
	≤0.43	≤0.24	
Southern	≤0.65	≤0.40	Prescriptive
	≤0.66	≤0.39	Equivalent performance
	≤0.67		
	≤0.68	≤0.38	
	≤0.69	≤0.037	
	≤0.70		
	≤0.71	≤0.36	
	≤0.72	≤0.35	
	≤0.73		
	≤0.74	≤0.34	
	≤0.75	≤0.33	

blind enclosed between two panes of glass based upon the outdoor temperature and solar radiation. This design admits solar radiation when it will help heat the house, but lowers the blind to block solar radiation when it will increase the air conditioning load.[32]

CONCLUSION

Walls and windows are often selected to achieve a desired appearance. Considering today's emphasis on energy conservation and overall sustainability, it is important to consider their thermal characteristics as well. Selecting more energy-efficient walls and windows will permit the building designer to specify a smaller heating and cooling system, so that often the total building cost is little more than that of a standard building. When you consider the reduced cost of heating and cooling the building during its lifetime, any added investment during construction is returned many times over.

OTHER USEFUL GUIDES

Graphic Guide to Frame Construction: Walls, by *Rob Thallon*, Published by Taunton Press, 2000, ISBN 1-56158-353-7, # 070470 (http://www.taunton.com/finehomebuilding/pdf/Framing%20Walls.pdf) http://www.taunton.com/finehomebuilding/pdf/Grading%20and%20Drainage.pdf

REFERENCES

1. http://www.energycodes.gov/ (accessed February 12, 2007).
2. Ontario Association of Architects, *The Rain Screen Wall*, found at http://www.cmhc-schl.gc.ca/en/inpr/bude/himu/coedar/loader.ffm?url=/commonspot/security/getfile.cfm&PageID=70139, other related publications from the Canadian Mortgage Housing C... available at http://www.cmhc-schl.gc.ca/en/ (accessed February 12, 2007).
3. Toolbase Services, a consortium including the U.S. Department of Housing and Urban Development and the National Association of Home Builders (NAHB) provides a wealth of information about current and future building materials and housing systems, NAHB Research Center, Maryland, 2005, found at www.toolbase.org (accessed February 13, 2007).
4. Kosny, J.; Kossecka, E.; Desjarlais, A.O.; Christian, J.E. In *Dynamic Thermal Performance of Concrete and Masonry Walls*, Thermal Envelopes VII, Conference Proceedings, Clearwater Beach, FL, December 1998; ASHRAE Special Publications, 1998.
5. Lippke, B.; Wilson, J.; Perez-Garcia, J.; Bowyer, J.; Meil, J. CORRIM: life-cycle environmental performance of renewable building materials. Forest Prod. J. **2004**, *54* (6), 8–19 http://www.corrim.org/reports/pdfs/FPJ_Sept2004.pdf, (accessed February 13, 2007).
6. Brown, M.A.; Southworth, F.; Stovall, T.K. *Towards a Climate-Friendly Builtt Environment*; Pew Center on Global Climate Change: Arlington, VA, 2005; available at

http://www.pewclimate.org/document.cfm?documentID=469 (accessed February 13, 2007).

7. Stovall, T. K., and Karagiozis, A., *Airflow in the Ventilation Space Behind a Rain Screen Wall*, Thermal Envelopes IX, Conference Proceedings, Clearwater Beach, FL, December 2004; ASHRAE Special Publications, 2004.

8. U.S. Department of Energy, Technology Fact Sheet: Wall Insulation, October, 2000, http://www.eere.energy.gov/buildings/info/documents/pdfs/26451.pdf (accessed February 13, 2007).

9. Kosny, J.; Mohiuddin, S. A. *Interactive Internet-Based Building Envelope Materials Database for Whole-building Energy Simulation Programs*, Thermal Envelopes IX, Conference Proceedings, Clearwater Beach, FL, December 2004; ASHRAE Special Publications, 2004.

10. Toolbase Services, a consortium including the U.S. Department of Housing and Urban Development and the National Association of Home Builders (NAHB) provides a wealth of information about current and future building materials and housing systems, NAHB Research Center, Maryland, 2005, found at www.toolbase.org

11. Kosny, J. *Steel Stud Walls; Breaking a Thermal Bridge*. Home Energy July 2001, http://www.homeenergy.org/article_preview.php?id=249&article_title=Steel_Stud_Walls:_-Breaking_the_Thermal_Bridge (accessed February 13, 2007).

12. ASHRAE Handbook of Fundamentals, 2005 Chapter 25.

13. Kosny, J.; Petrie, T.; Gawin, D.; Childs, P.; Desjarlais, A.; Christian, J. Thermal Mass—Energy Savings Potential in Residential Buildings, August 2001, http://www.ornl.gov/sci/roofs+walls/research/detailed_papers/thermal/index.html (accessed February 13, 2007).

14. Advanced Wall Systems Hotbox Test *R*-value Database, Oak Ridge National Laboratory, http://www.ornl.gov/sci/roofs+walls/AWT/Ref/TechHome.htm

15. NAHB Research Center, ToolBase Services, "PATH Technology Inventory, Precast Concrete Foundation and Wall Panels," Upper Marlboro, MD, 2005, http://www.toolbase.org/Technology-Inventory/Foundations/precast-concrete-panels (accessed February 13, 2007).

16. Precast/Prestressed Concrete Institute, "Single Family Housing," http://www.pci.org/markets/projects/single.cfm (accessed February 13, 2007).

17. Thermal conductivity estimated considering the reported thermal conductivity of unfired clay bricks found at http://www.constructionresources.com/products/ and the measured thermal conductivity of a wall built from rammed earth bricks reported at http://www.ornl.gov/sci/roofs+walls/AWT/Ref/TechHome.htm (accessed February 13, 2007).

18. Christian, J.E.; Desjarlais, A.O.; Stovall, T.K. In *Straw Bale Wall Hot Box Test Results and Analysis*, Conference Proceedings, Clearwater Beach, FL, December, 1999; Thermal Performance of the Exterior Envelopes of Buildings VII, 1999; 275–285.

19. Carmody, J.; Selkowitz, S.; Heschong, L. *Residential Windows, A Guide to New Technologies and Energy Performance*; W.W. Norton & Company: New York, 1996.

20. Joint Center for Housing Studies of Harvard University, *Remodeling, Measuring the Benefits of Home Remodeling*; Harvard University: Cambridge, MA, 2003.

21. U.S. Department of Energy, 2004 Buildings Energy Databook, http://buildingsdatabook.eren.doe.gov/default.asp

22. U.S. Department of Energy, A Consumer's Guide to Energy Efficiency and Renewable Energy, Storm Windows, http://www.eere.energy.gov/consumer/your_home/windows_doors_skylights/index.cfm/mytopic=13490

23. Klems, J.H. In *Measured Winter Performance of Storm Windows*, ASHRAE 2003 Meeting, Kansas City, MO, June 28–July 2, 2003; LBNL-51453.

24. Turrell, C., Storm windows Save Energy, Home Energy Magazine Online, July/August 2000, http://homeenergy.org/archive/hem.dis.anl.gov/eehem/00/000711.html

25. The NFRC has developed a number of informative fact sheets regarding window issues. You can find them at www.nfrc.org/factsheets.aspx (accessed February 13, 2007).

26. www.energystar.gov/index.cfm?c=windows_doors.pr_windows

27. The *U*-factors used by the NFRC and the Energy Star program have units of Btu/h-ft^2-°F.

28. www.nfrc.org/label.aspx (accessed February 13, 2007).

29. http://www.energystar.gov/index.cfm?c=windows_doors.win_elig

30. Fanjoy, R. Smart Windows Ready to Graduate, HGTV Pro. com, found at http://www.hgtvpro.com/hpro/dj_construction/article/0,2619,HPRO_20156_3679648,00.html (accessed February 13, 2007).

31. Toolbase Services, a consortium including the U.S. Department of Housing and Urban Development and the National Association of Home Builders (NAHB) provides a wealth of information about current and future building materials and housing systems, NAHB Research Center, Maryland, 2005, found at www.toolbase.org

32. Kohler, C.; Goudey, H.; Arasteh, D. A First-Generation Prototype Dynamic Residential Window, LBNL-56075, Lawrence Berkeley National Laboratory, Berkeley, CL, October 2004, available at http://www-library.lbl.gov/docs/LBNL/560/75/PDF/LBNL-56075.pdf (accessed February 13, 2007).

Waste Fuels

Robert J. Tidona
RMT, Incorporated, Plymouth Meeting, Pennsylvania, U.S.A.

Abstract

Opportunity fuels are those wastes or process byproducts that have significant energy content and could be used to provide energy to generate electricity but have not traditionally been used for that purpose. Burgeoning interest in the use of opportunity fuels to offset purchased traditional fossil fuels has focused on the combustion, material handling, and environmental permitting challenges associated with their exploitation. A number of industries have been taking advantage of opportunity fuels since long before that moniker was coined. Other industries are relative newcomers to the field. Some industry analysts have indicated the potential for as much as 100 GW of electric generation from opportunity fuels associated with distributed generation facilities. Although such estimates are not unfounded, a number of challenges are associated with a potential application of the use of opportunity fuels. These challenges include defining the fuel sufficiently for equipment vendors and regulatory agencies. Characteristics to be established are fuel particle size, shape, and propensity to bridge, agglomerate, stick to equipment surfaces, leach, or contain hazardous components. Other important considerations are the ease and cost of transporting the fuel (if required); its ability to be co-fired with other traditional or waste fuels; the ability to obtain performance guarantees from boiler/furnace/steam generator vendors with the desired opportunity material; and pollutant emissions from combustion, gasification, and handling. The answers to these issues, along with the typical design issues revolving around complex multifuel steam plants with on-site electric generation, must be obtained in the course of any feasibility analysis of opportunity fuels.

INTRODUCTION

The price of natural gas has been steadily rising throughout the 1990s and the first half-decade of the new millennium. From a benchmark NYNEX price averaging about $2–$4/MMBtu in the early 1990s to a 2006 level of approximately $10–$14/MMBtu, the spot market price has doubled over the past decade. Perhaps even more significant, the *perception* that we are in the midst of a significant upward trend in natural gas prices has grown increasingly strong. Increased demand from electric utilities, which have become more dependent upon this relatively clean fossil fuel over the past decade to fuel their gas turbine peaking plants and natural gas-fired steam boilers, is creating "shortages" at peak usage times, thereby putting upward pressure on the price of gas. The upward spike in the price of natural gas is illustrated in Fig. 1.

The relative ease of environmental permitting of natural gas or light distillate oil combustion devices, whether for electric generation or not, has made these fuels the popular choice for industrial, commercial, and utility boilers and furnaces nationwide. In many cases, to obtain a permit in a timely fashion, alternative fuels that might be more economically attractive, but are as yet unproven for specific applications, often lose out to these more traditional clean fossil fuels. Natural gas and distillate oil have become the "fuels of least resistance" from an environmental permitting standpoint, making them more valuable and driving up prices.

The upward trend in fossil-fuel prices, exacerbated by the shortages created by the Gulf hurricanes of 2005, has driven many companies to begin seriously considering alternative and opportunity fuels to supplement or replace their current fossil fuels. This article examines the current status of these fuels and their potential.

Because this topic crosses many disciplinary lines, a number of terms are defined below for purposes of the ensuing discussion.

EXAMPLES OF ALTERNATE AND OPPORTUNITY FUELS

Alternate fuel is a general term that can apply to any fuel not normally used by a process, boiler, or furnace but that can be used to provide energy. These fuels can be utilized either with or without changes to burner, fuel, and fuel residue handing, or pollution control systems to supplement or replace the use of fossil fuels. Numerous

Keywords: Alternate fuel; Opportunity fuel; Waste fuel; Industrial energy recovery; Municipal waste; Solid waste; Cogeneration; Distributed generation.

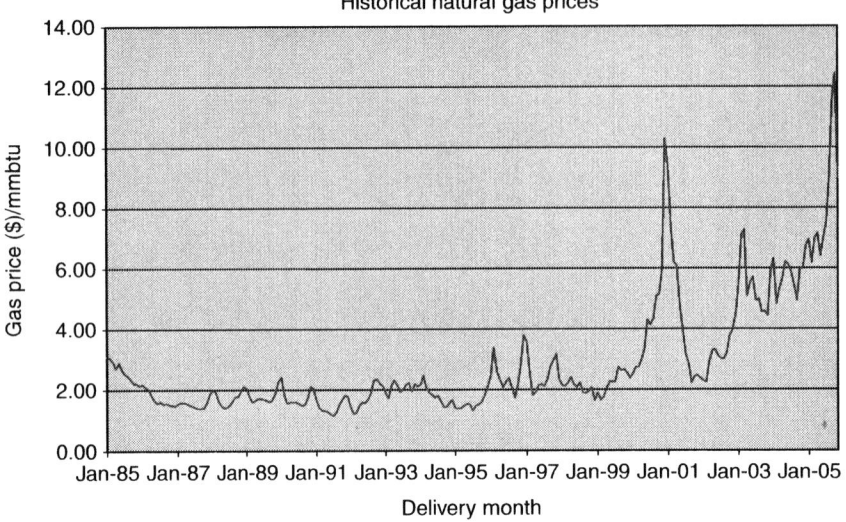

Fig. 1 Historical natural gas prices.
Source: From RMT, Incorporated internal report, December 2005.

examples of alternative and opportunity fuels are listed below:

- Agricultural wastes: corn cobs, cotton seed hulls, rice hulls, peanut shells, sugar cane waste, coconut husks.
- Anaerobic digester gas: gas consisting of approximately two-thirds methane and one-third carbon dioxide, along with a few percent nitrogen and small quantities of other gases, including oxygen, hydrogen, and hydrogen sulfide. This gas is produced by the anaerobic digestion process, which is often associated with sewage treatment, animal waste, vegetable oil, or alcohol mills. Anaerobic digestion is the process of using microorganisms that live only in the absence of oxygen to decompose organic materials into more valuable, or more easily disposed, components. The digestion process can also be used for synthetic natural gas from organic wastes or from algae.
- Bagasse: the fibrous material remaining after the extraction of the juice from sugar cane. It is used as a fuel and as a mix in making lightweight refractories.
- Black liquor: from pulp and paper plants, the organic residue from the wood digestion process that is burned in recovery boilers to eliminate the wood organics and to recover sodium sulfate for reuse in the wood digestion process. It has a heating value of approximately 4000 Btu/lb.
- Blast furnace gas: from steel mills, a gas, low in heat content (typically 100 Btu/ft^3 or less), recovered from a blast furnace as a byproduct and used as a fuel.
- Carpet waste: surplus cuttings and rejected product from manufacturing, as well as postconsumer carpet waste.
- Coalbed methane: methane that diffuses out of seams in coal mines and is channeled to the surface by mine shafts.
- Coke oven gas: obtained as a byproduct of coke production processes (mostly hydrogen and methane) with a heating value typically just under 600 Btu/ft^3. Coke is the solid, combustible residue left after the destructive distillation of coal or crude petroleum.
- Food processing wastes: include animal carcasses, animal fats, citrus rinds, peanut shells, and corn husks.
- Industrial VOCs: volatile organic chemicals, usually emitted in the manufacture of products.
- Industrial organic sludges: heavy liquids with integral suspended and possible dissolved solid phases.
- Landfill gas: approximately 50% methane and 45% CO_2, generated by the anaerobic decay of organic materials.
- Municipal solid waste (MSW): includes trash; yard wastes; garbage; household waste (including paper, wood, and carpet waste); textile waste; plastic and other synthetic packaging materials; and minor amounts of metals, soil, and other noncombustibles.
- Petroleum coke: coal (carbon black) that has been heated and partially oxidized.
- Refuse-derived fuel (RDF): fuel made from municipal solid waste, typically in pellet form.
- Tape manufacturing waste: depending on the product line, can be paper- or cloth-backed waste, either coated with adhesive or noncoated, and can include cardboard and plastic packaging materials, unusable end rolls, trimmings, waste from web breaks, and off-specification product.

- Textile wastes: cuttings and rejected product from manufacturing.
- Tires: whole, chipped, shredded postconsumer waste.
- Tire-derived fuel: fuel made from postconsumer tires, usually in pellet form.
- Used oil: defined in Title 40 of the Code of Federal Regulations (CFR) Part 279, Standards for the Management of Used Oil, as "any oil that has been refined from crude oil, or any synthetic oil, that has been used and as a result of such use is contaminated by physical or chemical impurities." Used lubricants, hydraulic fluids, and heat transfer fluids are examples. Used oil typically is contaminated or mixed with dirt, fine particles, water, or chemicals, all of which affect the performance of the oil and eventually render it unusable. By itself, it is not a hazardous waste, even though it may exhibit hazardous waste characteristics.
- Wastewater treatment sludge: heavy liquid with integral suspended and possible dissolved solid phases, usually dried by mechanical or heat exchanger method.
- Wood products industry:
  — Sanding/finishing dust: waste material from sanding or other finishing process, such as grinding and sawing
  — Hogged (wood) fuel: wood logs that have been reduced in size in a shredding machine called a hogger
  — Bark: stripped from the surface of logs, usually with a high moisture content (often at or above 50%).

A partial summary of the key characteristics of opportunity fuels is contained in Table 1.

MATERIAL HANDLING AND WASTE DISPOSAL

Blending of waste materials with other wastes or with conventional fuels can be the key to ultimate success in utilizing opportunity fuels. Blending can take place either en route to day storage or just prior to introduction into the energy conversion device. The target attributes of the blend include consistency of thermal energy content, particle sizing, material handling characteristics, bulk density, stickiness, and ash content.

Typical sizing of fuel specified by stoker boiler vendors is 2 in. maximum particle size in any dimension, with uniform sizing generally desired. An occasional large particle can be tolerated, provided that the material is relatively unlikely to agglomerate or accumulate in pipes or ducts. A very light material can be expensive to store due to its low bulk density and correspondingly large required storage vessels. A very heavy material can cause problems with elevating or pneumatic conveyors.

Standard disposal of solid waste materials is by landfill or municipal incinerator. The costs of these disposal options may range from $10 to $40 per ton of waste for landfilling and up to $80 per ton for incineration. Often, these costs can be eliminated in large part by co-firing the wastes in a boiler or furnace. Only the remaining noncombustible ash components and any residue remaining from the neutralization of acid gases in the boiler exhaust gases (see the next section for more details) will still need to be landfilled.

ENVIRONMENTAL CONSIDERATIONS

Combustion Processes

Federal, state, and in some cases county and/or city regulations that apply in most cases include Title 40 of the CFR Part 60 New Source Performance Standards (NSPS); 40 CFR Part 63 National Emission Standards for Hazardous Air Pollutants (NESHAP), specifically the Industrial, Commercial, and Institutional Boiler and Process Heater Maximum Achievable Control Technology (ICIB/PH MACT) rule; and process-based MACT standards.

- Gaseous emissions: NO_x, SO_2, CO, HCl, VOC, Hg.

Note that SO_2, HCl, and Hg emissions are primarily associated with initial concentrations of sulfur, chlorine, and mercury in the waste material prior to thermal processing, whereas NO_x, CO, and VOC occur due to the presence (or lack thereof) of oxygen and the mixing, or lack thereof, of fuel and air in the combustion process.

- Particulate emissions:
  — Total particulate matter (PM)
  — Total selected metals (TSM): includes 6 metals that are known health threats.
- Opacity.

NonCombustion Processes

Gasification

Gasification is gaining interest as an alternative to combustion-based energy generation, particularly among industrial concerns. The advantage of gasification is that a relatively clean, gaseous fuel may be generated from a solid waste stream, eliminating the need for expensive particulate control equipment at the downstream end of the process. In addition, for the production of electricity, gasification may permit direct conversion to power, averting the need for a steam generator and turbine. Some cleanup of the gas is typically required prior to being introduced into a burner. Permitting considerations are dependent upon the particular application and location but

Table 1 Opportunity fuel performance matrix

Opportunity fuel	Availability	Heating value	Fuel cost	Equipment cost	Emissions/ environment	Combined heat and power (CHP) potential	Rating	Limitations
Anaerobic digester gas	●	◐	●	◐	●	●	5.0	Need anaerobic digester
Biomass gas	●	◐	◐	○	●	●	4.0	Gasifiers extremely expensive
Black liquor	○	◐	●	◐	◐	◐	3.0	Most BL already used up by mills
Blast furnace gas	○	○	●	◐	◐	○	2.0	Limited availability, low Btu
Coalbed methane	◐	●	●	●	◐	◐	5.0	Coal mines—lack CHP demand
Coke oven gas	○	◐	●	◐	◐	◐	3.0	Availability—most already used
Crop residues	◐	◐	○	◐	●	◐	3.0	Difficulty in gathering/transport
Food processing waste	◐	◐	●	◐	●	◐	4.0	Limited market, broad category
Ethanol	●	◐	◐	◐	●	◐	4.0	Currently only used for vehicles
Industrial volatile organic chemicals (VOC's)	○	○	◐	◐	◐	◐	2.0	Must be used w/ NG turbine
Landfill gas	●	◐	●	◐	●	◐	4.5	Landfills—little demand for CHP
MSW/Refuse-derived fuels (RDF)	●	○	●	○	◐	◐	3.0	Low heating value, contaminants
Orimulsion	○	●	◐	◐	◐	◐	2.5	Orimulsion not available in U.S.
Petroleum coke	●	●	●	◐	○	○	3.5	Many contaminants; large apps
Sludge waste	●	○	●	○	◐	○	2.5	Low heating value, contaminants
Textile waste	◐	◐	◐	◐	◐	○	3.0	Must be cofired; larger apps
Tire-derived fuel	●	●	●	◐	◐	◐	4.0	Best suited for large apps
Wellhead gas	◐	●	●	◐	●	◐	4.5	Oil/gas wells—no CHP demand
Wood (forest residues)	●	◐	◐	◐	●	◐	4.0	Fuel can be expensive
Wood waste	●	◐	●	◐	◐	◐	4.5	Waste may have contaminants

Key: ●, excellent/not an issue; ◐, average/could become an issue; ○, poor/major issue.
Source: From Resource Dynamics Corporation (see Refs. 1 and 2).

can be somewhat reduced from combustion-based equipment. A more detailed discussion of gasification technologies may be found in "Gasifiers."

Gasification, however, is not a universally applicable technology and can be expensive, relative to combustion. Further, there is limited full-scale operating experience with this technology.

Anaerobic Digestion

Anaerobic digestion is the process of using microorganisms that live only in the absence of oxygen to decompose organic materials into more valuable or more easily disposed-of components. The digestion process can also be used for synthetic natural gas from organic wastes or from algae. As with gasification processes, because anaerobic digestion does not involve combustion, the time and effort spent meeting environmental permitting requirements tends to be reduced.

TECHNOLOGIES

Retrofit Applications

Existing Solid Fuel Boilers

Many of the best opportunities for energy recovery are to be found in industry. The pulp-and-paper and petroleum refining industries have long utilized numerous waste streams for energy. There are more than 7000 MW of biomass-fired energy systems in place in the United States. Portland cement plants have been co-firing tires in their kilns since the 1970s.

Co-firing is a promising technology for burning opportunity fuels within conventional coal-fired boilers. Fig. 2 shows an example of a commercially available burner for this application.

Waste Heat Recovery Steam Generators

Often, wastes are burned in incinerators, and the hot flue gas is exhausted to the atmosphere. The use of waste heat boilers and steam generators is gaining popularity as traditional fuel costs rise and state-of-the-art exchanger design and control systems provide added margins of safety.

Burner/Engine Redesign

Occasionally, an excellent opportunity for the reclamation of the energy potential from wastes may be obtained by a redesign of an existing burner or engine. These opportunities include the use of bagasse in Combined Heat and Power (CHP) plants and other relatively high-Btu waste gas streams, such as landfill gas and coke oven gas.

Sludge Drying

Some chemical or biological sludges contain large quantities of moisture after initial dewatering steps. It is sometimes cost effective to reduce the moisture levels in these sludges prior to burning them in a boiler or furnace. Many dewatering processes are available. They include physical or mechanical means such as centrifuges or filter presses, as well as thermal drying systems. Typically, the mechanical-type dewatering systems are the initial step in sludge drying. The high heat transfer rates possible in specialized thermal drying equipment (such as hollow screw or paddle type dryers) can provide enhanced drying capability, resulting in relatively high solids content, thus further reducing the need for fuel to evaporate the moisture in a combustion process. During the process of drying some materials, proper consideration must be given to the potential for the development of a "sticky" phase in which the material becomes non-Newtonian and can agglomerate or plug the handling system.

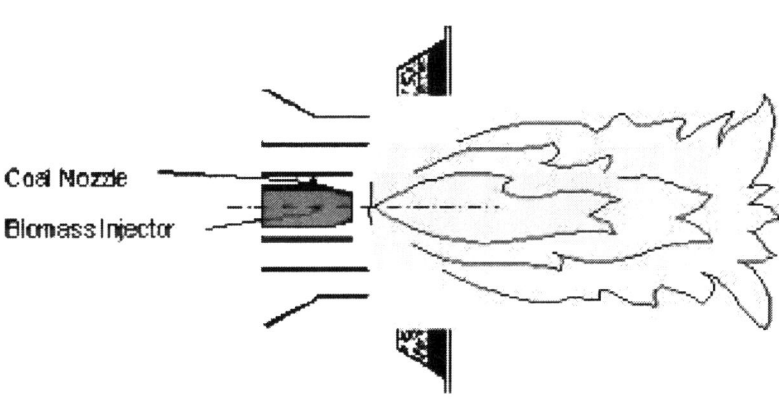

Fig. 2 Biomass co-firing through air.
Source: From Foster Wheeler Corporation (see Ref. 3).

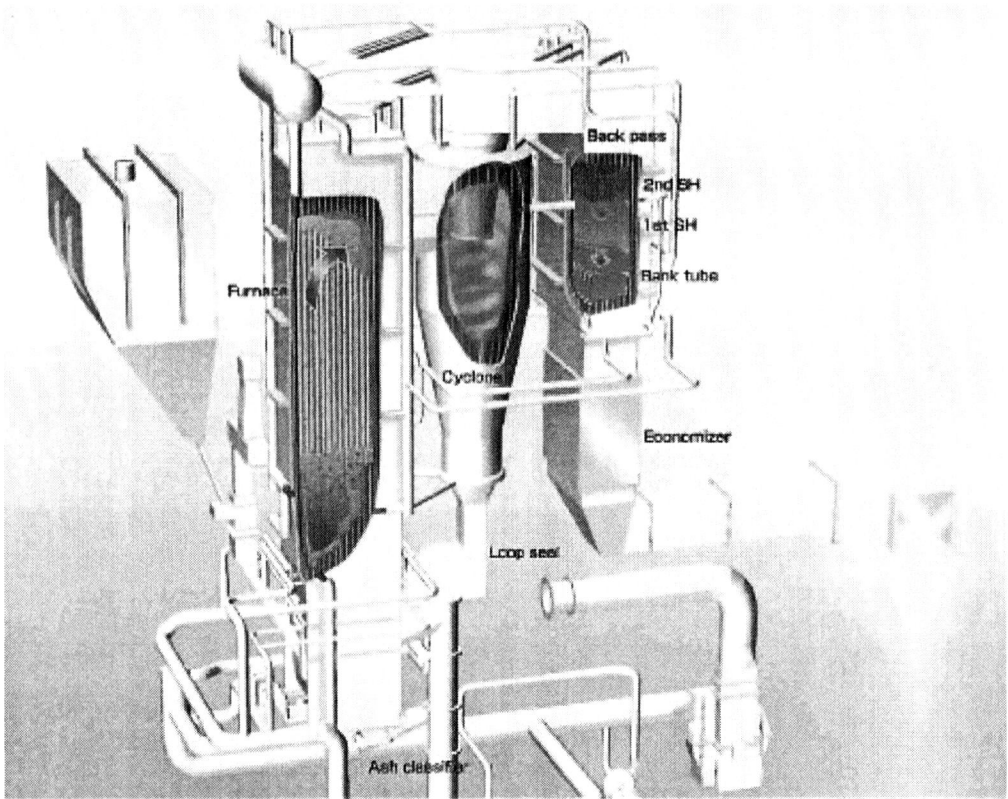

Fig. 3 Coal/biomass—fired circulating fluid bed boiler.
Source: From Takuma Company, Ltd. (see Ref. 4).

New Equipment

Circulating Fluid Bed (CFB) and Bubbling Fluid Bed (BFB) Boilers

Fluidized bed boilers have been in use for many years. They are well known for their capability to burn diverse materials efficiently and with low emissions. Compared with the older stoker or underfed boiler designs of the past, they can achieve greater burnout of the carbonaceous feed materials at relatively low emission rates of NO_x and PM.

As their name implies, Circulating Fluid Bed (CFB) boilers achieve lower and more uniform combustion-zone temperatures and higher heat transfer rates by recirculating the bed material. Consequently, CFB boilers generally have lower NO_x emissions and can burn many types of fuels. Their chief advantage over the lower-velocity, fixed fluid bed boilers is that they can achieve higher energy densities and, consequently, smaller footprints. It is expected that these types of boilers will have an efficiency and flexibility advantage for larger steam and electric generation applications.

Some commercial suppliers of CFB boilers are

- Foster Wheeler
- Babcock and Wilcox
- Kvaerner Power
- Alstom
- Takuma

Fig. 3 shows an example of a coal/biomass circulating fluidized bed boiler.

Bubbling fluid bed boilers (BFBs) have been in use longer than CFBs and are more suited for lower-heating-value materials, such as waste sludges and biomass fuels containing large amounts of moisture. Because they do not recirculate the bed material as in CFBs, they tend to be somewhat lower in cost for the same duty requirement, though their physical size is larger. Turndown is expected to be greater for BFBs than CFBs.

Examples of suppliers of BFB boilers are

- Babcock and Wilcox
- Kvaerner Power
- Austrian Energy
- Energy Products of Idaho

Gasifiers

Due to rising natural gas and distillate oil prices, there is intense interest in the United States regarding alternate fuels. Where space or air pollution control constraints make the application of large combustion devices such as CFBs and all the ancillary equipment accompanying them problematic, gasifiers offer energy recovery that can be permitted in a relatively short period and that, depending on the syngas cleanup requirements, requires a relatively small installed footprint.

The following Web site provides an excellent description of the major gasification processes: www.eere.energy.gov/biomass/large_scale_gasification.html (Ref. 5).

There are several widely used process designs for biomass gasification. In staged steam reformation with a fluidized-bed reactor, the biomass is first pyrolyzed in the absence of oxygen; then the pyrolysis vapors subsequently reform to synthesis gas, with steam providing added hydrogen, as well as the proper amount of oxygen and process heat that comes from burning the char. With a screw auger reactor, moisture (and oxygen) are introduced at the pyrolysis stage, and process heat comes from burning some of the gas produced in the latter stage. In entrained flow reformation, both external steam and air are introduced in a single-stage gasification reactor. Partial oxidation gasification uses pure oxygen, with no steam, to provide the proper amount of oxygen. (Using air instead of oxygen, as in small modular uses, yields producer gas including nitrogen oxides rather than synthesis gas.)

Fig. 4 shows a simplified process diagram of a fluidized bed gasifier.

Commercial suppliers of gasifiers include

- Energy Products of Idaho (EPI)
- Pyromex
- Primenergy
- Chiptec
- Interstate Waste Technologies
- Omnifuels Technology, Inc.
- Precision Energy Services
- PRM Energy Systems, Inc.
- Thermogenics, Inc.

Anaerobic Digesters

Anaerobic digestion is gaining in popularity in the United States due to its reduced processing cycle times and energy requirements. This process eliminates the need for expensive pumps required in aerobic (oxygenated) processes.

Anaerobic sludge digestion is the destruction of biological solids using bacteria that function in the absence of oxygen. This process produces methane gas, which can be used as an energy source and can make anaerobic digestion more economically attractive than aerobic digestion. In essence, the larger the treatment plant, the greater the economic incentive to use anaerobic digestion. Anaerobic digestion is considerably more difficult to operate than aerobic digestion, however. As such, the decision to use anaerobic digestion must take into consideration the operational capability of the installation.

There are two conventional processes: mesophilic, which takes place at ambient temperatures typically between 20 and 40°C, and thermophilic, which takes place at elevated temperatures, typically up to 70°C. The residence time in a digester varies with the type and amount of feed material and the temperature. In the case of mesophilic digestion, residence time may be between 15 and 30 days; the thermophillic process is usually faster, requiring only about 2 weeks to complete, but thermophilic digestion is more expensive, requires more energy, and is less stable than the mesophillic process.

Typically, two types of organisms are found in digester sludge:

- Saprophytic bacteria, which are acid formers that break down complex solids to volatile acids
- Methane fermenters, which produce methane, carbon dioxide, and inert materials from the volatile acids that are produced by the saprophytic bacteria

The efficiency of the digesters is typically measured by the amount of volatile solids reduction between the raw sludge and the digested sludge. Many variables affect the volatile solids reduction in the digestion process. The major variables are sludge type, digestion time, digestion temperature, and mixing.

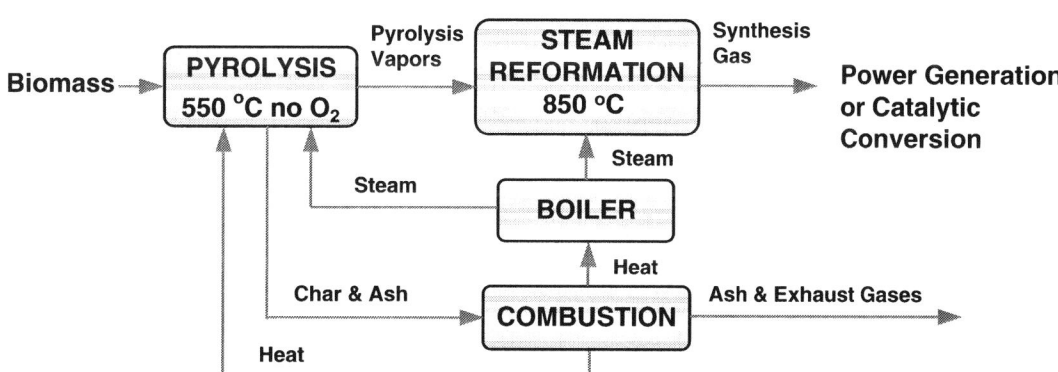

Fig. 4 Biomass gasification via staged steam reformation with a fluidized bed gasifier.

Applications for digesters fall into two major categories: wastewater treatment and farm-animal wastes.

Funding for anaerobic digester projects is available in some states. Details of one such program can be found on the New York State Energy Research and Development Administration (NYSERDA) Web site: www.nyserda.org/programs/pdfs/digestergrantlist.pdf (Ref. 6).

Commercial suppliers of anaerobic digesters include

- Arrow Ecology
- CCI US Corporation
- EcoCorp, Inc.
- Onsite Power Systems

POTENTIAL AND ECONOMICS

Reduced-Cost Alternate and Opportunity Fuels

Commercial Waste Fuel Suppliers

Biomass fuels are available in all parts of the country, particularly where the forest-products industry has a presence. Commercial suppliers are finding many new customers that are seeking alternatives to higher-priced traditional fossil fuels for boiler plants and power generation facilities. Mixed wood fuels from pulping, hogging, and debarking operations, as well as sanderdust, sawdust, and furniture manufacturing waste, can be sourced through a number of commercial suppliers.

The relative costs of many opportunity fuels may be estimated through information readily available on the World Wide Web at sites like the North Dakota State University's Fuel Cost Comparison site: www.ext.nodak.edu/extpubs/ageng/structu/ae1015.htm (Ref. 7).

Recycling Centers

Recycling of many manufacturing and postconsumer wastes for use as opportunity fuels has begun in earnest. As an example, nearly 5 billion lb/yr of carpet waste is generated by the carpet manufacturing industry. Much of this waste material is produced in Georgia. Commercial "reprocessors" of this waste material are pelletizing, baling, compressing, and gasifying it for its ultimate energy utilization.

Disposal Cost Reduction

Landfill Cost

Landfill of wastes can cost in the range of $10–$40 per ton. Typically, the greater the hauling distance, the greater the cost of disposal. This operating expense, coupled with the loss of usable thermal energy ("opportunity cost"), has often been ignored in an era of relatively inexpensive fossil fuels. This is no longer the case in many areas. Landfills themselves are aggressively seeking cost mitigation through energy utilization of landfill gas methane. As landfill costs climb due to the diminishing available land, increased environmental restrictions on groundwater cleanliness, and increased transportation costs as landfills are sited more remotely from population centers, conversion of waste to energy (WTE) is gaining more attention.

Trash Collection and Hauling

Just as landfill costs are spurring interest in WTE, increased trash hauling and collection costs are adding fuel to the fire. Industries are considering their options when it comes to paying rental costs for numerous waste containers on their manufacturing sites and the emptying of those containers. Such costs can run thousands of dollars per day at many manufacturing facilities.

Combined Heat and Power (CHP) for Increased Efficiency

Thermal Load Following (Bottoming) Cycle

An example of a thermal load following WTE is a process steam boiler with a backpressure turbine generator. This type of facility typically uses steam throughput rates from about 40,000 to several hundred thousand pounds of steam per hour. Steam pressures typically range from 150 psig to several thousand pounds per square inch gauge. Electric output can range from 100 kW up to about 20 MW.

Limitations include the necessity to use all steam or vent the excess that cannot be used by the process or the turbine. Facilities with greatly varying steam loads can experience problems with upset conditions caused when loads change rapidly and without warning. The turbines can trip offline, and process steam pressures and temperatures may fluctuate.

These problems can be alleviated by the use of a condensing turbine at the cost of additional capital. Larger systems with electric generation capacities of several MW or more will usually have a condensing turbine. The flexibility offered by the condensing turbine allows a boiler to run base loaded, which permits the use of large fluidized bed boilers designed for a multitude of fuels. Extraction turbines allow the tapping of steam slipstreams at various pressures for process use.

Electric Load Following (Topping) Cycle

An example of an electric load following cycle is a central steam boiler plant, which generates electricity to supply its users' base electrical loads via a condensing turbine.

The steam that is used in the turbine can be reused for process or heating needs at the exhaust pressure of the turbine or at intermediate pressures extracted from the turbine at certain points in the expansion from turbine inlet to turbine exhaust pressure.

Advantages include the ability to accommodate variable steam loads simply by condensing more or less steam. This also provides flexibility during startup and transient steam load conditions, and can reduce fluctuations in steam flow, temperature, and pressure to process users at such times.

The disadvantages are added initial cost and complication of the steam flow system and turbine due to the addition of the condenser; increased turbine manufacturing cost; and increased piping, wiring, and installation costs.

Sizes of condensing turbines range from about 5 to 100 MW of electrical output.

The potential for generation from CHP plants burning opportunity fuels has been estimated at more than 100 GW.

Table 2 shows a chart comparing estimated capital and operating and maintenance costs to generate electricity utilizing several opportunity fuels.

Wood fuels in particular have been studied for their potential use in CHP, including both thermal and electric load following applications. Table 3 illustrates the costs associated with their construction, operation, and maintenance.

Integrated Coal or Waste Gasification/Combined Cycle (IGCC) Power Plants

An alternative to conventional boilers in which fuels are burned to release energy that is then used to make steam is gasification wherein a solid fuel is converted to a gaseous fuel stream by means of thermal energy addition—generally in an oxygen-deficient atmosphere.

The integrated coal gasification/combined cycle is one version of this concept. There are many others, and the concept is growing in popularity among operators of large industrial and institutional boiler plants because of the relative ease of permitting this type of facility compared with installing a traditional boiler.

The advantage of this type of system is generally regarded as being its ability to convert coal or a waste byproduct (such as a solid waste or sludge) to usable energy, typically in the form of steam and/or electricity.

Project Funding Opportunities

Biomass Programs

- U.S. Department of Energy (DOE) grant program:
  The DOE provides grants for projects that have the objective of furthering the use of biomass fuels.

- U.S. Department of Agriculture loan guarantee program:
  The USDA has a loan guarantee program for projects that expand the use of agricultural byproducts as energy sources.

- State-funded programs:
  The following is a useful link to a U.S. DOE Energy Efficiency and Renewable Energy (EERE) Web site that gives a great deal of information about programs run by individual states: www.eere.energy.gov/state_energy_program (Ref. 8).

Renewable Energy Resources

The general category of renewable energy resources includes biomass and other opportunity fuels, in addition to traditional solar and wind energy sources.

Fig. 5 shows two possible projections for the future composition of the U.S. renewable energy portfolio.

Fig. 6 is a projection of the cumulative savings of natural gas resulting from the increasing use of renewable energy resources.

Cogeneration

The DOE supports a series of regional CHP Application Centers to provide application assistance, technology information, and education to architects and engineering companies, energy services companies, and building owners in eight states (Illinois, Indiana, Iowa, Michigan, Minnesota, Missouri, Ohio, and Wisconsin).

DOE's Office of Industrial Technology (OIT) coordinates two programs—the Combustion Program and Steam Challenge Program—to support the use of CHP technologies. The Combustion Program's goal is to foster research and development opportunities for combustion systems used to generate steam and heat for manufacturing processes, and the heat-processing materials from metals to chemical feedstocks. The Steam Challenge Program provides technical assistance to industrial customers to improve the efficiency of their steam plants and increase awareness of benefits from cogeneration systems. OIT predicts opportunities for future CHP development for any industrial business where boilers are installed.

The Office of Distributed Energy Resources coordinates the CHP Initiative to raise awareness of the energy, economic, and environmental benefits of CHP and to highlight barriers that limit its increased implementation.

Greenhouse Gas Emission Reduction

Projects in developing countries that seek to reduce the emissions of greenhouse gases (GHGs), including CO_2, methane, nitrous oxide, hydrofluorocarbons (HFCs), and SF_6 can obtain financial support through a Kyoto Protocol

Table 2 Equipment and maintenance average costs [not including combined heat and power (CHP) equipment]

Fuel	Type of cost	Steam turbine[a]($)	Gas turbine($)	Recip engine($)	Microturbine($)	Fuel cell($)	Stirling engine($)
Anaerobic digester gas[b]	Equipment ($/kW)	2150–3500	1800–3600	1900–3200	2650–4400	4800–7500	2400–3500
	Maintenance ($/kWh)	0.007–0.022	0.008–0.021	0.015–0.043	0.025–0.035	0.21–0.043	0.009–0.013
Biomass gas[c]	Equipment ($/kW)	1700–2800	1300–2550	1500–2550	2150–3500	4100–6500	2100–3000
	Maintenance ($/kWh)	0.007–0.022	0.005–0.016	0.01–0.029	0.017–0.027	0.017–0.038	0.009–0.015
Coalbed methane	Equipment ($/kW)	1000–1600	600–1400	800–1400	1400–2300	3500–5500	1500–2000
	Maintenance ($/kWh)	0.005–0.015	0.004–0.01	0.008–0.022	0.015–0.02	0.015–0.03	0.008–0.01
Landfill gas	Equipment ($/kW)	1250–2000	900–2100	1000–1700	1750–2900	3900–6000	1500–2000
	Maintenance ($/kWh)	0.016–0.019	0.007–0.018	0.014–0.04	0.024–0.032	0.02–0.04	0.008–0.01
Tire-derived fuel	Equipment ($/kW)	1000–1600	NA	NA	NA	NA	NA
	Maintenance ($/kWh)	0.006–0.019					
Wellhead gas	Equipment ($/kW)	NA	NA	NA	1550–2500	NA	NA
	Maintenance ($/kWh)				0.024–0.032		
Wood (forest residues)	Equipment ($/kW)	1250–2000	NA	NA	NA	NA	NA
	Maintenance ($/kWh)	0.008–0.023					
Wood waste	Equipment ($/kW)	1300–2100	NA	NA	NA	NA	NA
	Maintenance ($/kWh)	0.008–0.024					

[a] Including boiler costs.
[b] Including digester costs.
[c] Including gasifier costs.

Table 3 Comparisons of electric, thermal, and combined heat and power facilities

	Size (MW)	Fuel use (green ton/yr)	Capital cost (million $)	O&M[a] (million $)	Efficiency (%)
Electrical					
Utility plant	10–75	100,000–800,000	20–150	2–15	18–24
Industrial plant	2–25	10,000–150,000	4–50	0.5–5	20–25
School campus	NA	NA	NA	NA	NA
Commercial/institutional	NA	NA	NA	NA	NA
Thermal					
Utility plant	14.6–29.3	20,000–40,000	10–20	2–4	50–70
Industrial plant	1.5–22.0	5,000–60,000	1.5–10	1–3	50–70
School campus	1.5–17.6	2,000–20,000	1.5–8	0.15–3	55–75
Commercial/institutional	0.3–5.9	200–20,000	0.25–4	0.02–2	55–75
Combined heat and power (CHP)					
Utility plant	25 (73)[b]	275,000	50	5–10	60–80
Industrial plant	0.2–7 (2.9–4.4)	10,000–100,000	5–25	0.5–3	60–80
School campus	0.5–1 (2.9–4.4)	5,000–10,000	5–7.5	0.5–2	65–75
Commercial/institutional	0.5–1 (2.9–7.3)	5,000	5	0.5–2	65–75

[a] Operating and maintenance.
[b] Sizes for the CHP facilities are a combination of electrical and thermal; the first figure is electrical and the figure in parentheses is thermal. 1 MW = 3.413 million Btu/h.
Source: From www.fpl.fs.fed.us/documents/techline/wood_biwnass_fas_energy.pdf (see Ref. 11).

program known as the Clean Development Mechanism. Joint Implementation is another mechanism built into the Kyoto Protocol that funds GHG reduction projects in countries with economies in transition, including the former Soviet-bloc nations.

The potential for the reduction in GHG emission in the United States has been modeled based on adherence to proposed national renewable energy standards. Fig. 7 illustrates that a 15% reduction in U.S. power plant carbon dioxide emissions could be achieved due to the displacement of traditional fossil fuels by renewable energy resources through 2020.

CONCLUSION

Opportunity fuels are those wastes or process byproducts that have significant energy content and could be used to

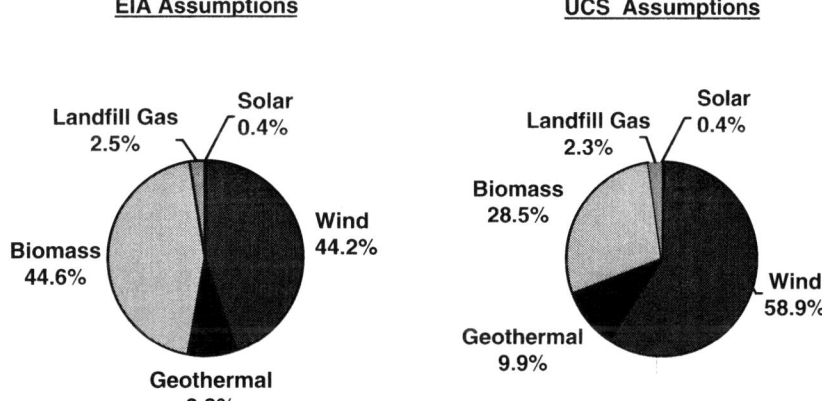

Fig. 5 Two predictions for the U.S. renewable generation mix in the year 2020.
Source: From Ref. 9.

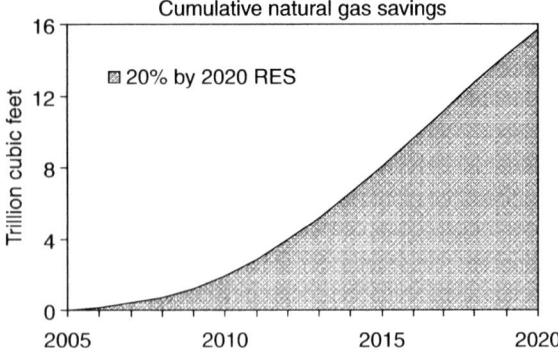

Fig. 6 Renewable energy conserves natural gas supply.
Source: From NREL Energy Analysis Seminar (see Ref. 10).

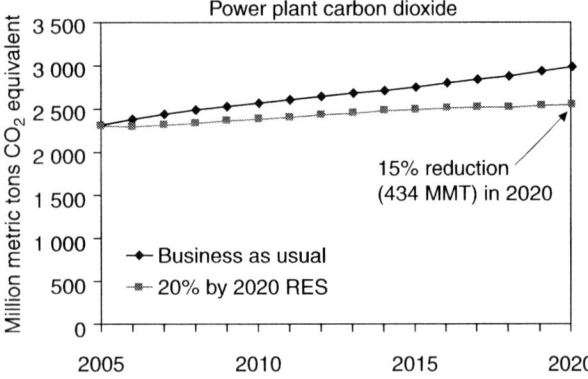

Fig. 7 Renewable energy reduces emissions and environmental compliance costs.
Source: From NREL Energy Analysis Seminar (see Ref. 10).

provide energy to generate electricity but have not traditionally been used for that purpose.

Relative to the combustion of traditional fossil fuels, burning or otherwise thermally processing opportunity and waste fuels offers the promise of reduced energy supply cost, with attendant environmental benefits including reductions in the amount of material going to landfills and a possible net reduction in carbon dioxide emissions.

Current interest in the utilization of these low-cost fuels for both thermal and electrical energy production is strong and is expected to increase due to the continuing upward trend in natural gas and oil prices. Further enhancing the appeal of prospective projects involving the use of these fuels and the technologies that exploit them is the potential for reduced environmental permitting time. Shorter project implementation times reduce costs and increase the rate of return on investment.

Glossary

Alternate fuel: Alternate fuel is a general term that can apply to any fuel not normally used by a process, boiler, or furnace, but that could be used either with or without appropriate changes to burner, fuel and fuel residue handling, or pollution control systems.

Co-firing: Co-firing is the combustion of a waste fuel with a traditional fossil fuel or fuels in a burner, boiler, or furnace—generally, as a low-cost substitute for another conventional fuel.

Industrial energy recovery: Industrial energy recovery is the reuse of energy exhausted in process waste streams or steam condensate, typically to heat reactants, feedwater, or combustion air.

Municipal waste: Municipal waste consists of a variety of materials (some combustible and some not), typically including trash; yard wastes; garbage; household waste (including paper, wood, and carpet waste); textile waste; plastic and other synthetic packaging materials; and minor amounts of metals, soil, and other noncombustibles.

Nonhazardous waste fuel: A nonhazardous waste fuel is a combustible organic material that is a byproduct of a manufacturing process; a waste packaging or shipping material; a component of municipal solid waste such as paper waste, yard waste, and household and office waste; and any other such materials other than those classified by a government regulatory authority as hazardous that are capable of being burned or co-fired in a boiler or furnace.

Opportunity fuel: An opportunity fuel is one that could be used economically as a source of energy to generate electric power but is not normally used for this purpose.

REFERENCES

1. Lemar, P. Jr. *Combined Heat and Power Market Potential for Opportunity Fuels*; Resource Dynamics Corporation: Vienna, VA, 2005.
2. Lemar, P. Jr. *Market Potential for Opportunity Fuels in DER/CHP Applications*, CHP Subcontractors Coordination Review Meeting, Washington, DC, April 14, 2005.
3. http://www.fwc.com/pow_services/parts/burners/Co_Firing_of_Biomass.cfm Foster Wheeler Corporation (accessed January 2006).
4. http://nett21.gec.jp/jsim%5F3/html/doc%5F458.html Takuma Company, Ltd (accessed November 2005).
5. http://www.eere.energy.gov/biomass/large_scale_gasification.html (accessed January 2006).
6. http://www.nyserda.org/programs/pdfs/digestergrantlist.pdf (accessed December 2005).
7. http://www.ext.nodak.edu/extpubs/ageng/structu/ae1015.htm (accessed December 2005).
8. http://www.eere.energy.gov/state_energy_program/ (accessed January 2006).
9. *Renewable Energy Can Help Ease Natural Gas Crunch*, UCS, March 2004.
10. Clemmer, S. *Renewable Energy Modeling Issues in the National Energy Modeling System*, NREL Energy Analysis Seminar, Washington, DC, December 9, 2004.
11. http://www.fpl.fs.fed.us/documnts/techline/wood_biomass_for_energy.pdf (accessed January 2006).

BIBLIOGRAPHY

Relevant materials are too numerous to mention, however a few highlighted references are provided below.

1. Summary of the Proceedings of the International Conference on the Co-Utilization of Domestic Fuels, University of Florida Conference Center, Gainesville, FL, February 5–6, 2003.
2. Gielen, D.; Unander, F. *Alternative Fuels: An Energy Technology Perspective*, International Energy Agency Workshop on Technology Issues for the Oil and Gas Sector, Paris, France, January 13–14, 2005.
3. Tillman, D.A.; Johnson, D.; Miller, B.G. *Analyzing Opportunity Fuels for Firing in Coal-Fired Boilers*, 28th International Technical Conference on Coal Utilization and Fuel Systems, Clearwater, FL, March 10–13, 2003.

Waste Heat Recovery

Martin A. Mozzo
M and A Associates Inc., Robbinsville, New Jersey, U.S.A.

Abstract
Many industrial, commercial, and institutional uses of energy result in excessive rates of waste heat rejection. Heat rejection is typically inherent in process uses; however, it may be utilized to meet other needs. Recovering and reusing rejected heat is known as waste heat recovery. Waste heat is usually recovered in the forms of steam, hot water, or hot air. The recovery medium is dependent on the quality of the waste stream, the potential use for the waste heat at the host, and the cleanliness of the waste stream. In this entry, the energy engineer is introduced to issues that should be considered in the economical and technical evaluation of waste heat recovery potential. These issues include: (1) the quality of the waste heat stream; (2) the calculation of the availability and applications of waste heat; and (3) the types of heat recovery equipment available.

INTRODUCTION

In many industrial and commercial energy applications, only a portion of the energy input is used in the process. The remainder of the useful energy is rejected to the environment. This rejected energy may potentially be recaptured as useful energy through waste heat recovery. Not all rejected energy can be recovered due to quality, usefulness in a host's load profile, and/or economical reasons that may make its recovery infeasible. This entry will serve as a guide to waste heat recovery in order to provide a framework for the energy engineer to develop waste heat recovery projects. It will discuss the concept of quality vs quantity, engineering concerns in waste heat recovery, sample calculations for waste heat recovery, and the types of waste heat recovery equipment that can be used.

THE CONCEPT OF QUALITY VS QUANTITY

While at first glance, a waste heat recovery project may appear feasible, this may not always be the case. The concept of quality vs quantity plays an important role. For the purposes of this entry, there are three classifications of waste heat. These are: (1) high-grade waste heat, generally 1000°F and above; (2) medium-grade waste heat, generally in the range of 400°F–1000°F; and (3) low-grade waste heat, generally below 400°F. Typically, the higher the grade of waste heat, the better the application for a successful and economical waste heat recovery project. This is referred to as the "quality" of the waste heat. It is better to have marginal amounts of high quality waste heat than large quantities of lower-grade waste heat.

Keywords: Waste heat recovery; Equipment; Heat conservation; Heat transfer.

There are numerous reasons why higher quality waste heat recovery provides a better application. One reason is that the waste heat stream has a much higher temperature than the recovery medium, resulting in higher temperature differences, and, thus, waste heat recovery equipment sizes can be smaller. This is a result of the higher heat transfer efficiency between the waste heat stream at a higher temperature to a lower temperature heat recovery medium. Second, the recovery medium can take on many forms, namely air, steam, or hot water, whichever is best suited for the host facility, whereas, with lower quality heat recovery, the form may be very limited, especially with low-grade waste heat. For example, if you have a waste heat stream that is at 400°F, steam is not a viable heat recovery medium, as the temperature difference will be too low. Finally, cost and savings become important issues in regard to the quality of the waste heat recovery system. Generally, the higher the quality of waste heat, the lower the capital costs for waste heat recovery equipment. This is because better heat transfer efficiencies will be realized with heat recovery equipment. Additionally, the higher the quality of the waste heat stream, the higher the savings usually are. Thus, with lower costs and higher savings, simple payback periods will be shortened, a result which is generally well-received by management.

A word of caution should be offered regarding the quality of waste heat. There are vendors who will reduce the grade of the available waste heat so that their equipment can be used in a heat recovery application. One such vendor manufactured only low temperature heat transfer equipment, which he was trying to sell as waste heat recovery equipment. His proposals included the dilution of high-grade waste heat streams by mixing the high temperature waste heat stream with ambient air in order to bring the temperature down to levels that his equipment could handle. His argument was that the total Btu content of the waste heat stream was not changed, just the temperature. Unfortunately, in reducing the temperature and quality of the waste heat stream, the size of the heat recovery equipment had to be increased to recover the waste energy, and the type of waste heat recovery medium was limited. This results in higher costs, lower savings, and longer-term payback periods.

There is one instance where the dilution of a waste heat stream may be justified. Steel is usually used in the manufacture of waste heat equipment, and there is an upper temperature limit for the waste stream where steel can be used. A substitution of other materials, for example, ceramics, can be made; however, the capital costs may become too prohibitive. In this instance, the dilution of the waste heat stream may be justified to bring the waste stream temperature down for the safe operation of the waste heat recovery equipment. It should, however, never be diluted simply to enable the use of a particular piece of heat recovery equipment. The primary reason is that by diluting the waste stream, energy savings may decrease and equipment costs will definitely increase; thus, payback periods will increase, perhaps significantly.

ENGINEERING CONSIDERATIONS

There are several engineering factors that must be evaluated when considering and designing a waste heat recovery system. In terms of steps, these are: (1) quantifying the waste heat stream; (2) determining the value of the waste heat stream; (3) evaluating the best form of heat recovery for the host facility; (4) determining the host site heat load profile; (5) determining the grade of waste heat; (6) determining the cleanliness and quality of the waste stream; and (7) selecting the proper waste heat recovery equipment by considering size, location, and maintainability.

The first step that should be executed is to quantify the waste heat stream by determining how many Btu/h are in the waste stream. The equation to calculate this is as follows:

$$Q = M \times \text{specific heat} \times \text{delta temp} \qquad (1)$$

where Q = total heat flow rate of waste stream in Btu/h; M = mass flow rate in Lb/h; specific heat = for air, 0.24 (Btu/Lb/°F); delta temp = $(T_{upper} - T_{lower})$ in °F.

The specific heat (Cp) changes as the temperature of the air rises. For example, at 1000°F, Cp is 0.26. Since this value is higher than that of Eq. 1, using the lower value, rather than correcting for temperature, provides a conservative calculation of the total heat rate in the waste stream.

The mass flow rate (M) is calculated as follows:

$$M = \rho \times V \qquad (2)$$

where M = mass flow rate in Lb/h; ρ = density of the waste stream in Lb/ft^3; V = volumetric flow rate in ft^3/h (in standard cubic feet).

Note that the value of Q is not the total amount of waste heat that will be recovered, but, rather, the total amount of waste heat that is ideally available for recovery. Not all of this waste heat will be recovered, or even can be recovered. The total amount that will be recovered will be determined by numerous other factors, such as the cleanliness of the waste stream and the form of recovery (i.e., high pressure, superheated steam, saturated steam, or hot water). This step is necessary in determining if there are sufficient volumes available for waste heat recovery.

One common mistake committed when quantifying the amount of waste heat available is assuming values for flow and/or temperatures without making measurements. Too often, actual conditions vary from assumed conditions, a variation that can often cause disastrous results in a waste heat recovery project. The cost of making actual field measurements is a small price to pay to obtain reliable data.

After the quantity of the waste stream is correctly determined, step two is to find the dollar value of the waste heat stream to determine how much capital cost a potential project can bear, or if it even makes economical sense. In all waste heat recovery projects, the heat recovered will displace a medium, such as steam, which would have to be generated using another piece of equipment, such as a boiler. This equipment likewise has a related efficiency, and the heat output is always less than the heat input. To determine the dollar value of the waste heat stream, use Eqs. 3 and 4 below:

$$\text{Value} = Q \times \text{unit cost} \qquad (3)$$

where value = the dollar value of the waste heat stream, per hour; Q = total heat flow rate of waste stream in Btu/h as calculated from Eq. 1; unit cost = unit cost of the waste stream in dollars/Btu.

$$\text{Unit cost} = \frac{\text{fuel cost}}{\text{efficiency}} \qquad (4)$$

where unit cost = dollars/Btu; fuel cost = cost for fuel displaced in dollars/Btu; efficiency = efficiency of unused equipment. For example, a steam boiler @ 75%.

The third through fifth steps tend to overlap and be interrelated, and are discussed together. In order for waste heat recovery to be acceptable to a host facility, its use must be consistent with current energy usage at the facility. To best apply waste heat recovery, the engineer needs to examine current energy usage patterns, heat loads for the site, and the quality of the waste heat that is available. The load factor of the waste heat recovery stream is the first thing to determine. For example, the waste heat recovery stream may be exhaust air flows from a process furnace that is periodic in nature, rather than a continuous operation. In this case, the best potential would be to return the waste heat to the process in some manner that allows the usage to follow the waste heat stream generation. Examples include utilization in a drying operation or as pre-heated combustion air for the process furnace. In another situation, the process furnace may run continuously and be of a high enough grade to generate steam; however, if the host facility only uses steam for heating in the winter, the loads do not match and, thus, steam is not a good recovery medium. On the other hand, there may be a need for large quantities of hot water in the facility, and the loads may match up. A thorough evaluation of both the source and operation of the waste heat and potential uses to recover the waste heat must be performed.

The sixth step, determining the cleanliness and quality of the waste stream, is important. If the waste heat stream is the product of the combustion of a natural gas-fired or fuel oil-fired operation, then the gases should be relatively safe for waste heat recovery. The important issue here would be the condensation of acidic liquids in the products of combustion gas stream, especially if fuel oil is used. This condensation would happen if the temperature of the waste stream is allowed to drop to low temperatures, typically below 300°F. The waste stream could also contain a burn-off from the product itself, which could condense on the waste heat recovery equipment and cause blockages. Here, it is best to maintain the exit temperature significantly above the condensation temperature to avoid the formation of acids or other potential hazards from the waste stream.

If the waste heat stream contains the products of combustion and the products of the process, great care must be exercised when utilizng the waste heat. The selection and design of waste heat recovery equipment could lower the waste heat exhaust temperature below acceptable levels, resulting in condensation or blockage problems with the heat recovery equipment. When unsure of the condensation temperatures and the effect on the heat recovery equipment, simple tests must be conducted to obtain this information.

Some actual examples wherein the waste heat stream was dirty and created problems follow. The first was an air-to-air heat recuperator used in a process operation. The products of combustion gases were dirty, creating a buildup on the heat exchanger. A soot blower was used to remove the buildup on a periodic basis; however, this was not totally successful. During actual heat recovery operations, the soot blowers could not remove the buildup adequately. A solution to this problem occurred accidentally. The operation was a 24 h/five days a week, and by mistake one weekend, the soot blower was left on, while the waste heat recovery equipment was shut off. Because the buildup and the heat exchanger were composed of different materials, the heat exchanger and the buildup thermal contraction rates were different. As a result, the soot blower was able to break up the buildup on the heat exchanger surfaces. This method was then used to clean the heat exchanger each week. The efficiency of the heat exchanger did decrease somewhat during the week, though not significantly. During the weekend, the buildup was blasted off the heat exchanger surface, thus increasing the heat exchanger's efficiency at the beginning of another week.

The second example did not turn out as well. In this scenario, the quality of the waste heat stream was low, and the products of combustion, combined with product burn-off, created a dirty waste stream. As a result, a serious problem quickly developed. With the waste heat recovery system in operation, gas products in the waste heat stream condensed on the heat exchanger surfaces, and within a very short period of time, the heat exchanger was completely blocked. The only corrective action was to remove the waste heat recovery equipment from service and steam clean it, a very costly operation that eventually resulted in the failure of the project.

The seventh step in the design of a heat recovery system is the selection of the waste heat recovery equipment. Descriptions of several different types of equipment are given in another entry in this encyclopedia. Selection of the proper piece of equipment will be based on the quality of available waste heat; the recovery fluid that can be generated, i.e., steam, hot water, or air; the location of the equipment; and the maintenance capabilities of the host facility. While the first two criteria should be obvious, the last two are sometimes neglected.

The location where the equipment is to be placed is important. It obviously should be close to the waste heat stream, yet should not interfere with the other equipment involved in the generation of waste heat, for example, a process furnace. The equipment also must not be squeezed into an area where maintainability is an issue. Sufficient access for maintenance and overhaul must be included since these pieces of equipment will require periodic maintenance. Accessibility to perform maintenance is not just desirable, but mandatory.

SAMPLE CALCULATIONS

This section describes some sample calculations that are used in the design of waste heat recovery systems. The first step is to calculate the amount of waste heat that is available for recovery. This is done using Eq. 1.

Example—Waste heat is available in the form of hot products of combustion from a natural gas-fired furnace. Assume that there are no contaminants in the waste heat stream. Measurements indicate that there are 14,000 available actual cubic feet per minute (ACFM) flowing from a process at 1000°F. The actual volumetric flow must first be converted to standard cubic feet per minute (SCFM) at 60°F, using the following equation:

$$\text{SCFM} = \text{ACFM} \times (T_{\text{absolute}} + 60)/(T_{\text{absolute}} + T_{\text{actual}}) \qquad (5)$$

where SCFM = standard cubic feet per minute; ACFM = actual cubic feet per minute; $T_{absolute} = 460°F$; T_{actual} = actual gas temperature.

Using the data available here,

SCFM = $14,000 \times (460 + 60)/(460 + 1000)$

= 4,986 ft^3/min

Since the product of combustion is essentially air, a good approximation of the density of the gases is similar to that of air at standard conditions, or $\rho = 0.074$ Lb/ft^3.

Using Eq. 2, we can calculate the mass flow rate for the waste stream as follows:

$M = \rho \times V$

= 0.074 Lb/ft^3 × 4,986 ft^3/min × 60 min/h

= 22,139 Lb/h

The next step is to calculate how much waste heat is available for transfer to the heat recovery system. The upper temperature has been given as 1000°F. The lower temperature will be determined by the medium used for heat recovery, that is, steam, hot water, or air, as well as the cleanliness and condensation temperature of the waste stream.

To continue the sample calculations, we will use three different applications of heat recovery mediums: 125 PSIG steam, the hot water required at 350°F, and pre-heated combustion air at whatever final temperature is available. Each of these applications is analyzed separately.

The first heat recovery application medium is saturated steam at 125 PSIG (140 PSIA). From steam tables, we find the enthalpy for saturated liquid (h_f) to be 324.82 Btu/Lb. and for saturated vapor (h_g) 1193.0 Btu/Lb. Thus, we need 868.18 Btu for every pound of steam (1193.0−324.82). The temperature of steam at this pressure is 353°F. For this temperature of steam, we can take the waste stream down to approximately 400°F without the risk of condensation (since condensation usually does not occur above about 325°F). Using Eq. 1, we can calculate the available waste heat as follows:

$Q = 22,139$ Lb/h × 0.24 Btu/Lb/°F × (1000 − 400)

= 3,188,041 Btu/h = 3.188041 MMBtu/h.

This waste heat stream's value can be calculated using Eqs. 3 and 4. Assuming we have natural gas priced at $8.00/MMBtu in a steam boiler with an efficiency of 75%, the value becomes:

Unit cost = $8.00/MMBtu/0.75 = $10.67/MMBtu

Value = $10.67/MMBtu × 3.188,042 MMBtu/h

= $34.00/h

To obtain steam mass flow, we must use Eq. 6 below:

$$M_s = \frac{Q}{(h_g - h_f)} = \frac{3,188,041 \text{ Btu/h}}{(1193.0 - 324.82) \text{Btu/Lb}}$$

= 3,672 Lb/h. (6)

The second heat recovery application medium will be hot water leaving the heat recovery equipment at 250°F, using the same waste stream. Water will be supplied to the heat recovery equipment at 200°F. With the temperatures involved, we can take the waste stream down to approximately 300°F, provided this low temperature does not cause any condensation issues. For the purposes of this example, we have determined that condensation occurs below 325°F. We will use 350°F as our lower limit. Using Eq. 1, we can calculate the available heat as:

$Q = 22,139$ Lb/h × 0.24 Btu/Lb/°F × (1000 − 350)

= 3,453,711 Btu/h.

In this case, we need to determine the mass flow rate, expressed as gallons per minute (GPM), of hot water we can heat from 200 to 250°F. Eq. 7 below provides the methodology:

$$Q' = 500 \times V \times (T_{upper} - T_{lower}) \tag{7}$$

where Q' = total heat recovered in Btu/h; V = volumetric flow rate of water in GPM. 500 is the constant used to convert GPM to Lb of water per hour.

To calculate the volume of water required for this system, we calculate:

$$V = \frac{Q'}{500 \times (T_{upper} - T_{lower})}$$

$$= \frac{3,453,711 \text{ Btu/h}}{500 \times (250 - 200)} = 138 \text{ GPM}$$

Thus, our waste heat recovery system will generate approximately 138 GPM of hot water raised from 200 to 250°F.

The third heat recovery application medium is an air-to-air recuperation system. The waste stream is the same as in the previous example, providing 3,453,711 Btu/h (assuming that 350°F will be the lower limit that we can go to without condensation issues). The recovery medium will be air applied as pre-heated combustion air with an input temperature of 75°F. In this system, we need to provide a set volume of combustion air to the waste heat recovery equipment, so we are interested in finding the final temperature of the pre-heated combustion air.

Using Eq. 1:

$$Q = M \times 0.24 \times (T_{upper} - T_{lower})$$

$$3,453,711 \text{ Btu/h} = 50,000 \text{ Lb/h} \times 0.24 \text{ Btu/Lb} -°F \times (T_{upper} - 75)$$

Solving for T_{upper} we get 362.8°F.

HEAT RECOVERY EQUIPMENT

There is a variety of equipment manufactured and/or sold as heat recovery equipment. The following is a list of some more common types of equipment.

Waste heat steam recovery—This piece of equipment can be either a water-tube or fire-tube boiler that uses the hot waste heat gas stream to heat boiler feed water to generate either low-pressure or high-pressure steam. Typically, this piece of equipment is called a heat recovery steam generator (HRSG). The unit can be a heat recovery-specific piece of equipment; however, some boiler manufacturers will sell their boilers without a burner package and call them HRSGs. This type of boiler can lose some of its effectiveness as a steam generator; however, it still is an acceptable piece of equipment. The reason for the decrease in recovery effectiveness is solely based on the possibility that the waste heat stream temperature will be somewhat lower than the burner flame temperature.

Recuperator—This piece of equipment is generally an air-to-air heat exchanger that transfers heat from the waste heat gas stream to air on the recovery side of the heat exchanger. This air can be used as pre-heated combustion air, or host make-up air in the facility.

Shell and tube heat exchanger—This piece of equipment consists of a bundle of tubes within a steel shell. It is usually used for water-to-water heat recovery. One of its uses may be to recover heat from a boiler condensate blow down on the shell side to heat up boiler feed water on the tube side.

Fin-tube heat exchanger—This type of heat exchanger uses air (usually the waste heat stream) blowing across finned coils that contain water. A typical application for this type of heat exchanger is using boiler products of combustion gases to preheat boiler feed water. Plugging of the finned coils could be a problem if the fins are closely spaced, the exhaust gases are dirty, or the condensation temperature (of the gases) is approached.

Plate and frame heat exchanger—This piece of equipment consists of two frames sandwiching thin plates. It is usually used with fluids of relatively low temperature, which flow through alternate plates. Since the plates are thin, heat transfer is usually good; however, both fluid streams need to be relatively clean or the exchanger will plug up.

Heat wheels—These are typically used in low temperature applications, such as the exhaust of environmental (space) air coupled with the introduction of outside air. The two ducts, outside and exhaust air, must be adjacent to one another so that the wheel turns through both ducts. The wheel will turn at low revolutions per minute (RPM) collecting the waste heat or cooling, depending on what season it is, and exchange the heat or cooling to the outside air being introduced to the facility.

Heat pipes—This equipment consists of a pipe heat exchanger with the interior containing a coolant. The coolant is alternately vaporizing and condensing between an exhaust and outside air stream, exchanging cooling and heating between the two air streams. The exhaust air and outside air ducts must reside together in the same manner as a Heat Transfer Wheel.

Run-around coils—These are similar to heat pipes in operation. They cool or heat exhaust air and outside air streams. Their typical construction is a finned water coil with air blown across the coil. The advantage of the run-around coil is that the exhaust air and outside air ducts can be physically separated. The disadvantage is that a pump, with pumping power, is required. Additionally, if the coils are subject to freezing conditions, then a water/glycol solution must be used. This solution will decrease the effectiveness of the heat transfer.

SUMMARY

The purpose of this entry has been to provide the energy engineer with some general guidelines to applying waste heat recovery. There are numerous applications in the industrial, institutional, and commercial sectors where waste heat recovery can be used cost effectively. The best applications are situations wherein the waste heat stream is of high quality, produced for many hours of the year, preferably 24/7, with a heat load that matches the waste heat availability. Care must be taken by the engineer to ensure that the waste heat stream will not block or destroy the waste heat recovery equipment.

There are also numerous pieces of equipment that can be used for waste heat recovery. These include steam boilers, hot water boilers, recuperators, coil heat exchangers, shell and tube heat exchangers, plate and frame heat exchangers, heat transfer wheels, heat pipes, and run-around coil systems.

BIBLIOGRAPHY

1. Thumann Albert, P.E.; C.E.M; Metha, D.P. *Handbook of Energy Engineering*; The Fairmont Press, Inc.: Lilburn, GA, 1997.
2. Turner, W.C. *Energy Management Handbook*, 3rd Ed.; The Fairmont Press, Inc.: Lilburn, GA, 1997.

Waste Heat Recovery Applications: Absorption Heat Pumps

Keith E. Herold
Department of Mechanical Engineering, University of Maryland, College Park, Maryland, U.S.A.

Abstract

Absorption refrigeration/heat pump technology is a heat-driven system for transferring heat from a low temperature to a high temperature. The major current application is in the building cooling market where such machines provide a natural-gas-fired cooling option which is particularly popular in markets where electric costs are high in comparison with natural gas costs. Another set of applications exists where the absorption machines are fired by waste heat, typically acting as a bottoming cycle for an integrated energy system. Examples include turbine inlet air cooling, engine cooling jacket heat recovery, and energy recovery in chemical plants. At current energy prices, waste heat has little value and is often discarded even though it has thermodynamic potential. Projections of the depletion of fossil fuel reserves and projections of the environmental implications of wasteful burning of fossil fuels seem to point toward a future where the value of waste heat would increase relative to other commodities. In such a future, it is expected that waste-heat-driven absorption technology would become more widespread as a natural bottoming cycle for many high temperature processes. This entry summarizes the potential for the use of absorption cycles for waste heat recovery.

INTRODUCTION

Absorption chillers and heat pumps are machines that transfer heat from a low temperature to a high temperature.[1–4] As we know from the second law of thermodynamics, such a process requires an energy input. In the case of absorption technology, the energy input is in the form of a high temperature heat transfer into the system that drives the process. Absorption machines typically also require an electricity input to power the pumps and the control system, but the primary energy input is heat transfer. As such, absorption machines have an important role in waste heat utilization. In the current economic climate, characterized by relatively low energy costs, waste heat utilization does not get much attention. However, a trend toward more conservative use of energy appears to be gaining momentum based on economic and environmental concerns. This trend is expected to lead to resurgence in the demand for waste-heat-fired absorption technology.

THERMAL DESIGN FUNDAMENTALS

The ability of a heat transfer to do work (including providing refrigeration) depends on its temperature according to the well-known Carnot law

$$W_{max} = Q\left(1 - \frac{T}{T_o}\right)$$

where W_{max} is the maximum possible work output, Q is the heat transfer, T_o is the ambient temperature and T is the temperature at which the heat transfer occurs. Note that the temperatures must be expressed in absolute temperature units (e.g., K). For heat transfer processes that occur over a range of temperatures, the temperature T can be replaced by an average temperature to give a reasonable estimate. The fact that the same amount of heat transfer can do more work if it occurs at a higher temperature comes directly from the second law of thermodynamics and is one of the keystones of thermal design. One of the corollaries is that it is important to provide temperature matching between a heat source and a process that requires heat, so as to avoid wasting the potential to do work.

Combustion temperatures can be quite high, easily above 1000°C. However, the strength of the metals used in the construction of energy systems usually limits high temperatures to values of less than about 500°C. This value depends on the materials and on the mechanical stresses that the system experiences during operation. Temperature matching at the high temperature end would argue for an approach closer to combustion temperatures, but the materials issues are significant, intertwining both economics and materials science.

One of the requirements of the second law of thermodynamics is that all thermal power cycles must reject a portion of the high temperature heat input as heat

---

Keywords: Absorption chiller; Heat pump; Waste heat; Cogeneration; Combined heat and power (CHP); Turbine inlet air cooling.

at the low temperature of the cycle. Another way to say this is that the heat transfer input cannot be converted completely into work. The energy input as heat is only partially converted to work and the remainder of the energy must be rejected as low temperature heat. This low temperature heat is often referred to as waste heat (in general, any heat that is not used in a process can be called waste heat). Depending on the thermal efficiency of the power cycle, the rejected energy can be significant. For example, if the power cycle is 33% efficient, then 67% of the input energy is rejected as heat. This waste heat is a resource, but it is important to realize that its value is in its ability to do work, which is influenced both by its temperature and by the energy content. If the waste heat is at ambient temperature, then it has zero value even though it may be a significant energy flow.

BOTTOMING AND TOPPING CYCLES

In the design of a thermal power cycle, the thermal efficiency can, in general, be maximized by introducing the input heat at the highest possible temperature and rejecting heat at the lowest possible temperature. However, design details often restrict the ability to do this (e.g., materials issues at the high temperature end) and the waste heat is often rejected at a temperature sufficiently far above ambient so as to be useful. Examples include: the waste heat from an automobile engine where the exhaust gases and the jacket cooling fluid are both at temperatures well above ambient; the waste heat from a combustion turbine.

In the practice of thermal design, the idea of using a heat transfer stream twice (or more times) as it cascades down from high temperature to ambient, is often discussed in terms of topping and bottoming cycles. When the waste heat from a power cycle is used to drive an absorption chiller, the power cycle is called the topping cycle and the absorption cycle is the bottoming cycle. Both cycles are driven by the same heat transfer as it cascades down in temperature. Although in theory, an absorption cycle could be inserted as a topping cycle, materials considerations limit the range of temperatures for the most common working fluids limiting the capability of an absorption machine for use as a bottoming cycle. The absorption cycle must also reject heat, but if the temperature of the rejected heat is close to ambient, than it has little value even if a large energy flow rate is present.

The bottom line for practical waste heat utilization is that the temperature of the waste heat must be sufficiently different from the ambient temperature to make the stream look attractive. Beyond that, the particular technology considered for utilizing the waste heat must be well matched to the waste heat and to the load. For example, a standard (so-called single effect) absorption cycle requires a heat input of around 120°C and a cooling tower to reject the heat. The machine provides chilled water as the output. If all three heat sources/sinks exist in a given application, then it is a good match for the absorption machine.

ABSORPTION HEAT PUMP CONFIGURATIONS

There are a number of useful ways to categorize absorption chillers and heat pumps. These include:

- Working fluid systems
- Heat source and sink configurations

In a sense, the heat source and sink configuration is more fundamental than the working fluid system, although in a practical sense these categories overlap and influence each other.

Heat Source and Heat Sink Configurations

Many heat source and sink configurations are possible with absorption technology if simple cycles are combined into more complex combinations. There are many publications about such cycles.[5–7] However, as an introduction to the subject, only the simplest cycles are considered here. For this purpose, it is assumed that the absorption machine interacts with three external heat sources or sinks at three distinct temperature levels.

Historically, absorption cycles are categorized as either Type I or Type II as shown in Figs. 1 and 2. Type I refers to the configuration found in building

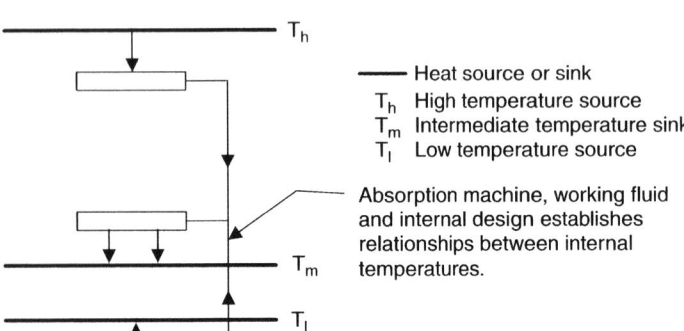

Fig. 1 Type I absorption heat pump. Acts as a chiller (refrigerator) if T_m is the ambient temperature sink. Acts as a heat pump if T_l is the ambient temperature.

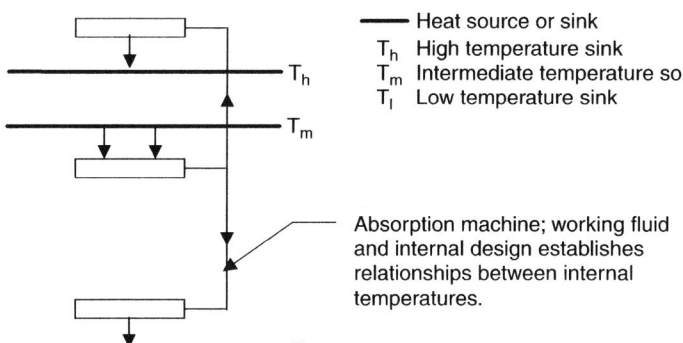

Fig. 2 Type II absorption heat pump. Also called a heat transformer or a temperature boosting heat pump.

air-conditioning applications and is the most well-known configuration. Although a Type I machine can act as either a chiller or a heat pump, the chiller application is, by far, more common due to the lack of competing technologies for the heat-driven chiller market. For a building's chilling application, the absorption machine will reject heat to ambient at the intermediate temperature (typically via a cooling tower). However, the same machine can be used for heating purposes by connecting the lowest temperature to the ambient. For both of these Type I configurations, the heat flow directions are the same as shown in Fig. 1. It is noted that the heat transfer out of the system at the intermediate temperature is equal to the sum of the heat transfers into the system at the high and low temperatures. Also, the maximum rate of refrigeration/heat pumping is limited by the second law of thermodynamics and depends on the temperatures of the three heat sources and sinks.

Type II heat pumps, as shown in Fig. 2, have been built and run as research machines but are not commercially available (because of the low current value of waste heat). In comparison with the Type I configuration, all of the heat transfers occur in the opposite directions in a Type II machine. The heat is input to the machine at the intermediate temperature, and a portion of the input energy is pumped out of the system at the highest temperature. The remaining energy is rejected to ambient at the lowest temperature. The purpose of a Type II heat pump is to upgrade the temperature of a heat transfer. An example application might be in a chemical plant where a component must be cooled at 100°C and elsewhere in the plant a component requires heat at 120°C. In such a case, the plant owner could burn some fuel to increase the temperature of the available heat by 20°C but a Type II heat pump provides an option that supplies the needed heat without requiring additional fuel. As was noted for the Type I heat pump, it is noted here that the sum of the energy out at the high and low temperatures must equal the input energy. Furthermore, the second law of thermodynamics puts constraints on the portion of the input heat that can be upgraded, which depends on the temperatures of the three sources and sinks.

Absorption Working Fluids

Absorption working fluids consist of a refrigerant and an absorbent. In general, the absorbent has a much higher boiling point than the refrigerant so that the refrigerant can be separated readily from the absorbent by heating. Over the years of absorption system development (which goes back to the 1800s), many working fluid pairs have been tried. A large number of thermophysical properties interact in a working absorption chiller/heat pump, and the performance of a particular working fluid pair can be derailed by many properties. Thus, the process of selecting working fluids is bound by many constraints. Although many fluid pairs can be made to work, there are only two pairs that have emerged in commercial systems. These two are: water/ammonia and lithium bromide/water.[8-11] In each of these pairs, the absorbent is listed first, followed by the refrigerant. Note that the water plays the role of absorbent in one case and refrigerant in the other.

Both water and ammonia make good refrigerants because they each have a high latent heat, resulting in a low refrigerant flow rate for a given heat transfer rate. Although water has twice the latent heat of ammonia, water suffers from two major drawbacks which limit its use: water's low vapor pressure, resulting in sub-atmospheric (vacuum) internal pressures and in-leakage of air; and the freezing point of water limits its use to temperatures above 0°C. On the other hand, the vapor pressure of ammonia is, in some sense, too high such that ammonia is restricted to applications with temperatures below 150°C to avoid high pressure on vessel walls. Many other refrigerants have been considered for use in absorption cycles, including the refrigerants used in vapor compression refrigeration, but water and ammonia are preferred because they are paired with effective absorbents.

A key property of an effective absorbent is that it must have a high affinity for the refrigerant (in other words, there needs to be a molecular attraction that makes mixing occur readily). Other beneficial properties include low vapor pressure and maintenance of a liquid phase over the operating temperature range of interest. The water/ammonia pair remains liquid over the full concentration range and over a wide range of useful temperatures but the vapor pressure of water is sufficiently high so as to require special design features (i.e., a rectifier) to remove water vapor from the refrigerant after it is boiled out of solution. This extra component must be cooled resulting in a reduction in thermal performance for a water/ammonia machine. Lithium bromide is a good absorbent for water and has a very low vapor pressure, but it has limited solubility leading to crystallization (solid hydrate formation) that tends to plug up flow passages under conditions of high concentration. Crystallization can be avoided by maintaining low heat rejection temperatures; this implies that lithium bromide/water machines must operate with a cooling tower.

Lithium bromide/water systems utilize a surfactant as a heat and mass transfer enhancement factor to enhance internal transfer processes.[12–14]

Corrosion is an issue for both of the common working fluid pairs. In most cases, corrosion inhibitors are found to effectively limit corrosion to acceptable levels.

HOW DOES ABSORPTION REFRIGERATION WORK?

The easiest way to understand absorption refrigeration technology is to first understand vapor compression technology as shown schematically in Fig. 3. A vapor compression cycle consists of just four components: a compressor, a condenser, a throttle, and an evaporator. Mechanical energy is input to the compressor to raise the pressure of the refrigerant vapor, allowing it to condense into liquid in the condenser (accompanied by heat rejection). The liquid then passes through a throttle which lowers the pressure. A throttle is just a flow restriction, which can take the form of an orifice, a valve, or a capillary tube. Due to the drop in pressure across the throttle, a portion of the liquid evaporates and this cools the remaining liquid. The liquid then enters the evaporator where it absorbs heat from the outside of the device while it evaporates at low temperature. The cooling effect comes from the low pressure maintained in the evaporator by the compressor.

A similar schematic of an absorption refrigeration machine is shown in Fig. 4. Similarities to the vapor compression refrigeration machine can be seen on the left side including the condenser, throttle, and evaporator. The difference in the two technologies comes in how the compression process is accomplished. In the absorption machine, compression is accomplished via a thermal compressor consisting of the absorber, a solution heat exchanger, and the desorber (sometimes called the generator). The compression process goes as follows. First the vapor from the evaporator is absorbed into the liquid absorbent accompanied by heat rejection. Then the refrigerant-laden absorbent is pumped up to the higher pressure level in the desorber by a liquid pump. By the application of heat to the desorber, the refrigerant is then boiled out of the absorbent and sent to the condenser. The liquid absorbent leaving the desorber is hot and is brought into regenerative heat transfer contact with the solution leaving the absorber to minimize the heat requirements in both the absorber and the desorber.

One of the features of the thermal compressor is that the increase in pressure is accomplished while the refrigerant is in its liquid phase. In contrast to the compression of a vapor, the compression of a liquid requires very little energy because the small volume changes that the liquid experiences do not store much energy. This implies that the electrical energy input required by an absorption machine is minimal. Another aspect of the thermal

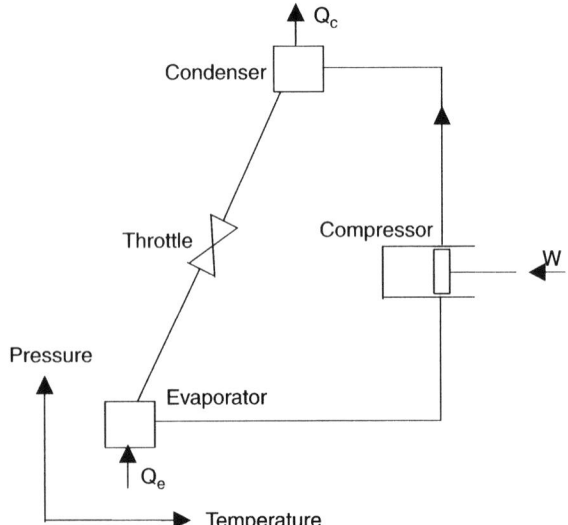

Fig. 3 Schematic of a vapor compression refrigeration/heat pump cycle.

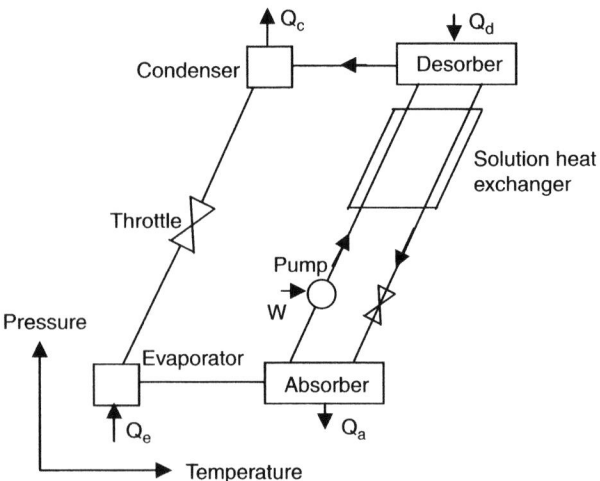

Fig. 4 Schematic of an absorption refrigeration/heat pump cycle.

compressor is the fact that the energy that must be rejected by the absorber is similar in magnitude to the energy that must be supplied to the desorber. Furthermore, this energy is somewhat greater than (on the order of 20% greater) the latent heat of the refrigerant due to the energy interactions between the absorbent and the refrigerant (associated with the affinity of the absorbent for the refrigerant; i.e., the heat of mixing). The presence of the heat of mixing further limits the thermal performance of an absorption machine.

ABSORPTION MACHINE PERFORMANCE

As with other refrigeration technologies, the thermal performance of an absorption machine is defined in terms of a coefficient of performance as

$$\eta_I = \frac{Q_e}{Q_d} \qquad \eta_{II} = \frac{Q_a}{Q_d + Q_e}$$

$$\eta_{vc} = \frac{Q_e}{W}$$

where Q_e is the evaporator heat (i.e., the refrigeration effect), Q_a is the absorber heat, Q_c is the condenser heat, Q_d is the desorber heat (i.e., the high temperature heat transfer), and W is the compressor work input. Typical coefficient of performance values are given in Table 1. The value listed in parentheses for the double effect machine is a gas-fired value that is lower than the machine value due to gas burner losses.

The range of values shown for the single effect, Type I machine represent differences in performance between the two standard working fluid pairs with water/ammonia having lower performance due to the energy penalty of the rectifier that is needed to strip water out of the refrigerant after the desorber. The values shown for the double effect machine are only lithium bromide/water values since the temperatures for double effect make water/ammonia less practical. The Type II value is a value obtained by the author on a research machine (unpublished).

The value given for the vapor compression technology represents a typical value for an application comparable to the absorption chiller applications. It is evident that the vapor compression technology has a high coefficient of performance but it must be understood that this is a somewhat biased comparison because the Carnot factor for the work input has a value of 1. A fairer comparison results if a power cycle efficiency factor is included in the calculations so that both technologies are compared based on a fuel input. If this is done with a power cycle efficiency of 33%, then the coefficient of performance of the vapor compression machine is reduced to 1.67. Although this comparison is more favorable to the absorption technologies, the conclusion is that vapor compression technologies, combined with high-efficiency electricity generation, are hard to beat for refrigeration. However, for the waste heat applications that are the focus of this article, the comparison to vapor compression technology is not relevant. If the input heat for the absorption machine is waste heat, then it implies that any output from the absorption machine is over and above what would otherwise have been produced.

BUILDING COOLING APPLICATIONS

The use of absorption chillers for building cooling applications is widespread, with a market share of only a few percent in the U.S., but up to 50% in Japan, China, and South Korea. The primary technology that is utilized in these applications is the direct-fired, double-effect lithium bromide/water chiller. The primary driver in the building cooling market is the ability to provide building cooling with a gas or oil input. All of these markets are characterized by relatively high electricity costs in comparison to fuel costs. In the U.S., absorption chillers have significant market share in markets such as New York City where the electric infrastructure is expensive to upgrade but where natural gas is available. In Japan, government policy requires gas-fired cooling in an attempt to control electricity growth. In China and Korea, urban growth outpaced power plant construction leading to limited availability of electric power in some cities.

Building cooling using direct-fired absorption chillers makes sense economically in these locations but makes little sense thermodynamically. In particular, much of the potential of the fuel is lost when it is burned and then utilized at such low temperatures. The combustion process occurs at temperatures approaching 1000°C, while the heat input to the absorption machine is needed at around 250°C. Thus, there is a thermodynamic opportunity to extract more useful work from the fuel than is obtained when it is used for direct firing of an absorption chiller. In particular, it would be possible to run a topping cycle (e.g., a power cycle) that would make use of the high temperature combustion heat and which would be mated with the absorption chiller so that both machines would use the same energy input as it cascaded down in temperature.

Table 1 Typical coefficient of performance values for various absorption technologies, including vapor compression technology for comparison

Cycle configuration	η
Type I	
Single effect	0.5–0.72
Double effect	1.3 (1.1)
Type II	
Single effect	0.3
Vapor compression	
Typical	5

This is not a new idea, but it is not widely used today due to various economic realities (i.e., small-scale power generation is difficult to integrate into a grid-connected electricity network, and small power generating equipment is less efficient and costly to maintain).

Since absorption technology has a unique role in energy integration, it is expected that the application mix for absorption chillers will shift over time from the current mix dominated by direct-fired machines to a future application mix dominated by waste heat fired machines.

TURBINE INLET AIR COOLING

Combustion turbines, fired by natural gas, are widely used for electric power generation due to a range of features including:

- Modular design
- Low initial cost compared to coal-fired power
- Relative ease of power modulation

These machines are modified gas turbines and operate at high rotational speeds, transferring the energy released from combustion into rotating power to drive a generator. These systems ingest ambient air at their compressor inlet which is then fed to the combustion chamber. After combustion, the hot gases are expanded through a power generating section. The power output is strongly dependent on the mass of air ingested which is a function of the turbine's rotational speed and the density of the air. The performance of the turbine is maximized when it operates at its maximum speed. However, because the air density varies with ambient conditions (temperature and pressure), the power output also varies. This is problematic in many applications because the electric demand often peaks when the ambient temperature is high (when electric usage is dominated by electric-drive cooling systems). Thus, these combustion turbines lose capacity when it is most needed.

One option for leveling the power output is to provide inlet air cooling.[15] This can be done in a number of ways, including evaporative cooling (useful in dry climates where water is available), vapor compression cooling (which can be electric- or steam-driven), or absorption cooling.[16,17] Absorption cooling is a good match in applications where the waste heat from the turbine is not otherwise used. When waste heat is available free, the only costs are the capital cost of the absorption chiller and the additional pressure drop due to the cooling coil in the inlet air stream. Turbine inlet air cooling by absorption chillers can be implemented with chillers based on either water/ammonia or lithium bromide/water working fluids with the latter being more prevalent due to the availability of standard chiller packages designed for this application.

One interesting aspect of the inlet air cooling application is that the thermal efficiency of the gas turbine system is not as strongly influenced by the inlet air temperature as is the system capacity (i.e., power output). One of the key material design issues in such a system is the maximum temperature that the turbine blades experience. On the operational side, this maximum temperature is governed largely by the ratio of the fuel and air flow rates. Thus, when the ambient temperature is low, the reduced air flow rate requires a reduced fuel flow rate. The efficiency of the system is largely a function of the maximum temperature (assuming other design factors such as turbine losses are fixed). Thus, material strength considerations force the operator to reduce fuel flow rate at low ambient. When the machine is always operated at the same high temperature, then the thermal efficiency is essentially a fixed value. This is still true when you add the inlet air cooling. The benefit of the inlet air cooling comes primarily from capacity increases and not from efficiency increases.

COMBINED HEAT AND POWER

Cogeneration is a term that implies the simultaneous generation of heat and power from a fuel source. Many examples exist in practice,[18,19] including the power system here at the University of Maryland where the University recently updated its antiquated and inefficient steam generation system with a combined system that generates sufficient electric power for the entire campus plus the steam for building heat with the same amount of fuel as was previously used. The system is based on combustion turbines with heat recovery steam generators. At present, the University does not use absorption chillers although their use would be a natural way to leverage the existing steam distribution system in the summer months (the University did use absorption chillers 20 years ago in an era when the technology required considerable user expertise to avoid maintenance problems but has since replaced all of those chillers with electric-drive machines). Many such installations exist around the world where various levels of energy integration are used at campus or municipal scales.

Recently, the term CHP (for combined heat and power) has come into use to describe energy integration at the individual building scale. In the CHP scenario, several energy technologies would be cascaded (possibly including electricity generation, heating, cooling, dehumidification, and/or hot water). Absorption chillers have a natural bottoming cycle role in such integration schemes. It remains to be seen if the CHP trend proves economical since integration implies more complexity and, hence, more maintenance and operator attention. However, the concept of energy integration existed long before the term CHP was introduced. Additionally, the potential benefits in reduced fuel usage and reduced pollutant production are expected to become more important as current resource and environmental conservation trends continue. The concept of cogeneration, implying the integration of energy systems, is well known to energy engineers. Many cities and large campuses around the world have various levels of energy integration. On a municipal scale, it is easier to solve the integration issues and to negotiate reasonable terms with the power utility. A similar idea has caught on recently to provide energy integration at the building level (under the name CHP, for combined heat and power). Absorption chillers often play a key role in energy integration schemes because they represent one of the few heat-driven refrigeration technologies. Such integration schemes have a great potential for improving the overall energy utilization by utilizing the full temperature range of a combustion process. Better utilization implies longer-lasting fuel supplies and a lower rate of environmental degradation. Thus, energy integration is expected to become more important as the economics of energy use responds to the twin drivers of global warming and fossil fuel depletion.

SOLAR AIR CONDITIONING

Absorption chillers are a natural match to solar thermal energy and solar absorption cooling has a long history of research and demonstration projects. Solar energy suffers from the same economic problem as waste heat; that is, current fuel costs make both options un-economic. Solar thermal energy collection is a demonstrated technology but it requires large collector arrays due to the low intensity level of the solar energy. Furthermore, the temperature level required by standard absorption machines is at a level where concentrating collectors are needed, and these collectors are expensive to build and maintain. All of these problems have been dealt with and the technology has been demonstrated at numerous sites, and on various scales. At the end of each project, the result was that the economic constraints hold the technology back and not the technical details.

Solar-driven absorption air conditioning is a proven technology, but until the economic environment changes significantly, it is not expected to gain any market share.

CONCLUSIONS

Absorption technology has a natural supporting role in energy integration as a bottoming cycle. As one of the most widely demonstrated, heat-fired energy technologies, waste heat-driven absorption cooling is expected to grow in importance as the value of waste heat grows in relation to other resources. Particularly, when the environmental costs associated with energy are properly charged, waste heat takes on a higher value than current practice would imply. It remains to be seen how the environmental penalties will be assessed to energy users, but increases in energy costs seem likely. When the cost of fuel use rises, then the economic incentive to utilize waste heat rises as well. Although this scenario implies many challenges for us as energy users, it also implies a brightening future for waste heat-driven absorption technology.

REFERENCES

1. Bogart, M. *Ammonia Absorption Refrigeration in Industrial Processes*; Gulf Publishing Co: Houston, TX, 1981.
2. ASHRAE Chapter 1, Thermodynamics and Refrigeration Cycles. *ASHRAE Fundamentals*; ASHRAE:Atlanta, 2005.
3. Herold, K.E.; Radermacher, R.; Klein, S.A. *Absorption Chillers and Heat Pumps*; CRC: Boca Raton, FL, 1996.
4. Herold, K.E.; de los Reyes, E.; Harriman, L.; Punwani, D.V.; Ryan, W.A. *Natural Gas-Fired Cooling Technologies and Economics*; Gas Technology Institute: Chicago, IL, 2005.
5. Alefeld, G.; Radermacher, R. *Heat Conversion Systems*; CRC: Boca Raton, FL, 1994.
6. Ziegler, F.; Kahn, R.; Summerer, F.; Alefeld, G. Multieffect absorption chillers. Int. J. Refrig.-Rev. Int. Froid **1993**, *16* (5), 301–311.
7. Ziegler, F.; Alefeld, G. Coefficient of performance of multistage absorption cycles. Int. J. Refrig.-Rev. Int. Froid **1987**, *10* (5), 285–295.
8. Bosnjakovic, F., translated by Blackshear, P.L. Jr, *Technical Thermodynamics*, Holt, Rinehart, and Winston: New York, 1965.
9. Ibrahim, O.M.; Klein, S.A. Thermodynamic properties of ammonia–water mixtures. ASHRAE Trans. **1993**, *99* (Part 1), 1495–1502.
10. Yuan, Z.; Herold, K.E. Specific heat measurements on aqueous lithium bromide. Hvac&R Res. **2005**, *11* (3), 361–375.
11. Yuan, Z.; Herold, K.E. Thermodynamic properties of aqueous lithium bromide using a multiproperty free energy correlation. Hvac&R Res. **2005**, *11* (3), 377–393.
12. Zhou, X.; Yuan, Z.; Herold, K.E. Phase distribution of the surfactant, 2-ethyl-hexanol in aqueous lithium bromide. Hvac&R Res. **2002**, *8* (4), 371–381.
13. Yuan, Z.; Herold, K.E. Surface tension of pure water and aqueous lithium bromide with 2-ethyl-hexanol. Appl. Thermal Eng. **2001**, *21* (8), 881–897.
14. Kulankara, S.; Herold, K.E. Theory of heat/mass transfer additives in absorption chillers. Hvac&R Res. **2000**, *6* (4), 369–380.
15. TICA, *Turbine Inlet Air Cooling*, 2006, http://www.turbineinletcooling.org/index.html
16. Erickson, D.C.; Anand, G.; Kyung, I. Heat-activated dual-function absorption cycle. ASHRAE Trans. **2004**, *110* (Part 1), 515–524.
17. Erickson, D.C. Power fogger cycle. ASHRAE Trans. **1995**, *111* (Part 2), 551–554.
18. Erickson, D.C. Waste-heat-powered icemaker for isolated fishing villages. ASHRAE Trans. **1995**, *101* (Part 1), 1185–1188.
19. Brant, B.; Brueske, S.; Erickson, D.C.; Papar, R.A. New waste-heat refrigeration cuts flaring, reduces pollution. Oil Gas J. **1998**, *96* (20), 61–65.

Water and Wastewater Plants: Energy Use

Alfred J. Lutz
AJL Resources, LLC, Philadelphia, Pennsylvania, U.S.A.

Abstract
Energy efficiency in existing and new wastewater treatment facilities (WWTF) can be improved by benchmarking energy use for different types of plants. Specific applications involve dissolved oxygen (DO) control, the use of adjustable-speed drives (ASDs), methane gas, heat recovery, and renewable energy. Typical savings estimates are shown to be in the range between 10 and 30%. Challenges in implementation and operation are highlighted, along with possible solutions.

INTRODUCTION

Wastewater treatment is a mature industry that has adopted tried-and-true solutions to specific challenges. In the past, the majority of these systems have performed at reasonably satisfactory levels. There is significant potential for improvement, however, in economics and in energy efficiency—both in existing and new wastewater treatment facilities (WWTF). Explanation and discussion of the potential for savings is directed toward benchmarking WWTF energy use and to specific applications in dissolved oxygen (DO) control, including the use of adjustable-speed drives (ASDs). The use of methane gas, heat recovery, and renewable energy are also discussed. Challenges in implementation and operation are highlighted, along with possible solutions.

SUMMARY

A number of opportunities in WWTF extend beyond common commercial and industrial energy efficiency and energy cost control measures. These site- and industry-specific measures often account for the bulk of the energy savings in comprehensive energy audits and retrofit projects.

Most of these measures entail control of processes on specific variables, not unlike the control of heating ventilation and air conditioning (HVAC) systems on temperature. Unfamiliarity with the application of specialized equipment and challenges in application, however, can limit their recommendation.

Nevertheless, many wastewater systems can be operated more efficiently, and the opportunities to do so must be pursued for the benefit of municipalities and private concerns that operate these systems.

Opportunities may exist in several areas. Analysis at many facilities has shown that DO control and the conversion of other speed control technologies to modern ASDs can save significant amounts of the total facility energy; typical savings estimates range between 10 and 25%. Significant additional savings can be gained by utilizing methane generated in the process, where applicable.

By gathering the necessary information and identifying appropriate partners, energy service providers and municipalities themselves can concentrate efforts on implementing the largest and most cost-effective energy efficiency measures.

GOALS OF WASTEWATER TREATMENT

The desired result in wastewater facility operations is the treatment of water to control water pollution. Certain standards and parameters are contained in operating permits for a facility; these typically are regulated and enforced by various governmental agencies. There can be substantial penalties for noncompliance.

The primary goal of WWTF is to comply with permit requirements concerning effluent water quality. These concerns—and not energy use—are paramount. Changes to systems and processes can involve savings and reduced manpower requirements but also involve some risk—especially if there are no existing areas of noncompliance. There can be substantial resistance to any changes in the way a plant operates, at many different levels.

Nevertheless, the savings opportunities are too great to wait for a major plant upgrade or total redesign to implement cost-effective energy measures, especially at older plants.

WASTEWATER TREATMENT FACILITY OPERATIONS

There are several types of treatment plants in operation, with different types of equipment. The decision to use

Keywords: Wastewater treatment; Energy use; Benchmarking; Efficiency; Dissolved oxygen; Adjustable-speed drives.

various types of equipment and plant processes is based upon several factors, including the characteristics of waste treated, variability of flows, levels of treatment needed, facility location, access to land, and proximity to the community served. A discussion of all types of plants and operations is beyond the scope of this article. The basic principles of wastewater treatment should be understood, however, to appreciate the variables and the effects of changes that may be considered.

Setting boundaries on the process, for this discussion consider first the inputs—raw wastewater, oxygen, and chemicals—and the outputs—treated effluent water and dewatered solids (sludge).

Several processes and functions cause the transformation from the input to the output stage. Although flow by gravity is desirable, most WWTF rely on a substantial number of pumps to move the wastewater to, through, and from the plant. Influent wetwells may be required. A typical first process is primary screening and settling operations. This can be followed by an aeration process, where oxygen is mechanically introduced into the wastewater or diffused into the wastewater. The oxygen is needed by microorganisms in the biological processes of transforming the incoming wastewater.

Chemicals are added to the water for various purposes. Important processes use chemicals to "coagulate" and "flocculate" solids, binding them together so that they fall to the bottom of the tank and form a sludge blanket. Then this sludge is pumped to holding, storage, or processing operations.

Some of this sludge is sent back to the aeration process. The microorganisms and bacteria responsible for the processing of wastewater possess biological activity that is based on the age of the organism, so the microorganisms must be refreshed continually. Return activated sludge is "recycled" to keep the process fresh and alive. Varying recycle rates are recommended for different types of plants, and each plant fine-tunes this measure, based on incoming flow and transit time through the plant. Some form of variable flow technology usually is employed to allow optimal recycle rates.

This return of activated sludge is a critical part of the process. The process can be "upset" and can become unstable; the process can be unable to produce good-quality effluent if too much or too little sludge is returned. A large rain event can create excess flow through the plant (due to infiltration and inflow into sewer lines or to the presence of combined wastewater/stormwater systems). In this case, the active sludge may be washed out of the plant.

Waste sludge that is not returned to the process is handled in various ways. It may be thickened and dewatered using belt presses, vacuum filtration, and centrifuges. In some cases, sludge drying beds are employed. In larger plants, sludge may also be incinerated; in this case, electricity or heat for other plant processes may be generated. The sludge also may be sent to closed digesters for aerobic and anaerobic digestion. Methane will be generated in this process. This methane can be used to produce heating for the digester operations, space heating, hot water, and electricity. In many cases, however, this methane is flared on site due to technical, maintenance, or other operational issues.

The primary output of the plant is treated water, which can be chlorinated or disinfected as needed prior to discharge. Secondary outputs (ash from sludge incineration and processed dewatered sludge solids) typically are landfilled or land-applied.

Some important terms that define the qualities and the rate of flow of influent raw water and treated effluent water are:

- DO
- Biochemical oxygen demand (BOD)
- Total suspended solids (TSS)
- Millions of gallons per day (MGD or mgd)

It is normal to measure chlorine residuals in the effluent if chlorine is used in the final disinfection process. Other measurements may entail ammonia and phosphorous.

TYPES OF TREATMENT SYSTEMS AND OPPORTUNITIES

Many facilities use a form of the activated sludge process, which, as described above, returns live microorganisms to the aeration treatment tank.

The aeration treatment tank can be aerated in several ways:

- Mechanical aeration—Large motors drive impellers to agitate water in the basin, forcing air contact with wastewater.
- Course bubble diffusion—Blowers pressurize air in headers along the bottom of the aeration basin; headers have large diffuser openings; bubbles typically are 20 mm in diameter; and agitation can be quite violent.
- Fine bubble diffusion—Blowers pressurize air in headers along the bottom of the aeration basin; headers possess ceramic heads or membrane diffusers; bubbles typically are 2 mm in diameter; contact areas are increased; and agitation is minimal.
- Pure oxygen systems—Liquid oxygen is converted to pressurized gas and fed into covered aeration tanks; motors are used for mixing and distributing oxygen throughout the wastewater.

Extended aeration refers to a system in which the aeration times are longer due to system design. This method may be preferable when there are low organic loadings (low BOD) and when space is available.

Trickling filters may also be used; basically beds of a solid, rock, or plastic medium, they provide acceptable contact time for oxygen uptake and treatment, at a low energy cost.

Advanced systems encompass nitrification, denitrification, phosphorous removal, and filtration methods, among other processes. Many municipalities currently have permit discharge requirements that dictate some form of advanced treatment.

This article concentrates on facilities utilizing secondary treatment only. Many of the results contained herein are relevant and can be applied to facilities with advanced design.

FACILITY ENERGY USE

Different types of plants will require different amounts of energy for a measure of treatment. One metric for energy use—kWh/mg—has been used to describe the energy intensity of WWTF. This metric—the kWh needed to treat each million gallons (mg) of wastewater—may be useful when correlated to a size range (e.g., 10–50 mgd) and type of treatment (e.g., secondary treatment using mechanical aeration). It may also be useful to correlate the kWh/mg with similar input and output parameters in the water treated.

From field studies that concentrated on smaller plants (primarily with average flows between 1 and 20 mgd) in the northeastern United States, the overall average for plant energy use was determined to be 1356 kWh/mg (as depicted in Table 1).

Table 1 Energy use in specific wastewater plants

Plant type no.	Average millions of gallons per day (MGD)	Peak/design MGD	kWh/mg
1	12.0	17.0	891
1	5.0	7.1	1062
1	1.0	1.4	2174
1	5.8	10.0	1596
2	17.3	20.0	1860
2	5.8	7.0	1542
3	1.5	2.75	1193
3	8.9	10.75	1317
3	62.0	80.0	943
4	12.5	16.0	980
Average	13.2	17.2	1356

Plant type legend: 1, surface aeration; 2, course bubble diffusion; 3, finebubble diffusion; 4, high purity oxygen.

It is interesting to note the range of usage values by the types of plant. Energy use in activated sludge plants with surface aerators appears to be somewhat higher than energy use in plants with fine bubble aeration. Course bubble diffusion systems, however, appear to yield the highest energy consumption per unit of wastewater treated.

The last two columns in Table 2 show other metrics of energy use; KW/mg represents plant peak kW per average mg, whereas kW/Design mg represents maximum kW for design flows. These metrics are instructive in considering energy demand as well as consumption, and in making observations on how a plant operates relative to its loading (i.e., how variable energy use is in comparison to variable flows and how energy intensity tracks plant design flows). From these values, it appears that surface aeration may result in the highest kW energy demand for all types of plants studied.

Note, however, that the results in Table 2 are taken from a small sample size. The figures obtained should be taken as a guide and used as a frame of reference only. Further work needs to be done to compile a robust data set. This article presents a methodology and approach that can be used to create this body of data.

The use of kWh/mg and kW/mg for wastewater treatment is a valuable metric—an energy use index that can be used to set targets for energy use and identify the plants that have the greatest opportunities for savings. The results also could be used with a portfolio of plants, similar to the U.S. Environmental Protection Agency's ENERGY STAR program. One of the goals of that program is to benchmark facilities across a portfolio of commercial or institutional properties operated or managed by a single entity, to identify the best candidates for energy efficiency upgrades.

Many studies contain backup information regarding the value of kWh/mg. An Electric Power Research Institute (EPRI) study[1] cites a figure of 1212 kWh/mg for plants using secondary treatment. The same study cites figures of:

- 955 kWh/mg for plants with trickling filters
- 1322 kWh/mg for plants using activated sludge (no breakdown for type of aeration process)
- 1541 kWh/mg for plants with some advanced treatment (but without nitrification)
- 1911 kWh/mg for plants with advanced treatment and nitrification

It is noted that some facilities operate at significantly higher levels of energy intensity; some advanced or extended aeration plants with nitrification have recorded intensities that exceed 3000 kWh/mg. Results indicated in Tables 1 and 2 pertain primarily to facilities that utilize only secondary treatment processes.

Size of facilities can be important when using kWh/mg as a measure of energy intensity or retrofit potential.

Table 2 Energy use ranges by type of aeration process

Plant type	Value type	kWh/mg	kW/mg	kW/design mg
All Types	Average	1356	94.0	67.9
	Maximum	2174	182.0	130.0
	Minimum	891	51.4	40.2
Surface Aeration	Average	1431	119.8	80.4
	Maximum	2174	182.0	130.0
	Minimum	891	71.8	50.7
Coarse Bubble	Average	1701	97.6	83.0
	Maximum	1860	116.5	100.8
	Minimum	1542	78.6	65.1
Fine Bubble	Average	1151	71.6	50.4
	Maximum	1317	83.3	64.1
	Minimum	943	53.9	41.8

Studies conducted by the United Nations[2] and others indicate that results vary for different sizes of plants. In ranges of 1–10 mgd, for example, results from Germany indicated energy use for activated sludge plants of approximately 1600 kWh/mg. For plants smaller than 1 mgd, consumption was approximately 1900 kWh/mg, whereas for plants larger than 10 mgd, consumption was approximately 1200 kWh/mg. This large variation illustrates the importance of considering plant size when using kWh/mg as a comparator.

Interestingly, unit consumption was reported as sometimes increasing with plant size, perhaps due to higher standards of treatment. Other countries in this study, including Canada and the United Kingdom, showed very low energy consumption figures in wastewater treatment. Again, this may be for different levels of treatment required or the use of other treatment technologies.

Electric Power Research Institute also reports a sizable increase in energy intensity in facilities that are below the 5 mgd range, as compared with larger facilities.[1]

Amounts of energy use by process for a typical activated sludge secondary treatment plant are estimated[1] as follows:

- Aeration: 47%
- Sludge processing: 32%
- Pumping: 16%
- Balance of plant: 5%

These estimates of energy use by process are only approximate figures. One particular case study of a 34.1 mgd plant has shown energy use in aeration systems as high as 73% of total plant energy.[4] Field investigation and measurement are required to determine reasonable and valid estimates of energy use at individual facilities.

These percentages can be used, however, to indicate the processes in which it is possible to find large savings opportunities—namely, aeration, sludge processing, and pumping.

ENERGY EFFICIENCY MEASURES

Large opportunities exist for efficiency in WWTF (see Fig. 1). Many of the larger systems will interact with smaller systems, and opportunities can be implemented concurrently. A detailed analysis should entail all processes and systems within the facility.

Dissolved Oxygen Retrofits and Control

Dissolved oxygen is required due to the BOD of the incoming untreated wastewater. Dissolved oxygen is introduced into the wastewater by several means. The most common techniques are through the use of mechanical surface aeration or aeration blowers.

Table 3 shows estimated savings at several WWTF studied. The savings are attributed to different measures affecting aeration systems, given as a percentage of total facility kWh consumption. The percentage of kWh savings

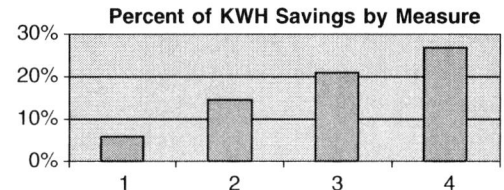

Fig. 1 Example of savings due to energy efficiency.

Table 3 Energy savings from aeration system modifications

Plant type no.	Pre-retrofit KWH/yr	Aeration system savings (%)	Savings from all measures (%)	Notes
1	3,901,200	7	31	Dual speed motors
1	1,938,300	4	34	—
1	793,600	6	20	Weir control
1	3,377,900	0	25	—
2	11,747,400	18	26	Fine bubble and downsizing
2	3,265,000	14	20	Fine bubble and downsizing
3	652,900	23	23	Dissolved oxygen (DO) Control and adjustable-speed drives (ASDs)
3	4,278,960	0	6	Process control savings
3	21,340,000	15	27	Downsizing and guide vane control
4	4,471,400	21	27	ASDs on aeration motors
Average	5,576,666	11	24	—

from all energy efficiency measures identified, as a percentage of total kWh consumption, also is given to illustrate the contribution of aeration system modifications in a comprehensive energy efficiency program.

In mechanical surface aeration systems (plant type no. 1), a number of motors will be used to drive large impellers. These motors agitate the surface of the water, continually entraining oxygen. Motors may be left on continuously or energized/deenergized manually or automatically, dependent on operator experience or lab results. There are several possible retrofits to these systems.

A seemingly simple retrofit would be the conversion from single-speed to two-speed motors. There is also a possibility of using ASDs. Some considerations include oil circulation in gear boxes at reduced speed; tank geometry and the possibility/consequences of short-circuiting of wastewater flows, impacting treatment; location of motor control centers and electronic variable-speed drive placement to prevent damage due to severe ambient conditions and to minimize impact on facility power quality; and the ability of existing motors and their insulation systems to handle reduced speeds.

It is also possible to reduce energy use of surface mechanical aerators by changing the submergence of the driven impellers through weir control or other means. In this case, motor loading will decrease. Again, short-circuiting may need to be evaluated. In general, automatic control should yield the greatest savings, although a manual weir control system could be implemented. This strategy has been reported in a case study to yield savings of 12%–15% in total plant energy.[4]

With automatic control systems, sensors that measure control parameters are needed. Dissolved oxygen is the most common control parameter. A number of DO sensors are available. Newer technologies include self-cleaning sensors. These can be important; sensors lose accuracy as they are subject to fouling in unclean environments. Sensor location is important; there typically are many sensors per tank. Routine calibration also is essential to process control and achieving energy savings.

The inputs from DO sensors and the output signals to weirs, motors, or ASDs can be fed into the plant's computerized control system.

The Environmental Protection Agency (EPA) study cites that the use of better DO control can save significant amounts in older facilities, as shown in three case studies: 12.5% in a 90 mgd plant, 14.5% in a 53 mgd plant, and 57% in a 36 mgd plant.[3] These results were measured in pounds of BOD removed per kWh used, which ranged from a low of 0.88 lb/kWh to a high of 5.95 lb/kWh. Other studies concur that savings from automatic control of 8% to more than 20% are possible.

A common aeration technique involves the use of blowers and pressurized air distributed through tanks via headers. The blowers may be the centrifugal or the positive-displacement type. Normally, valves or guide vanes would be used to regulate airflow to the manifolds and headers. The valves or vanes generally would be controlled on inputs from DO sensors. In some cases, there is the ability to optimize these controls.

In other cases, instead of throttling flow via valves, which introduce friction and lead to energy losses, ASDs can be used. Positive-displacement blowers and centrifugal blowers lend themselves to the use of ASDs.

The application to centrifugal blowers requires that turndown ratios be regulated to prevent blower surging. Airflow must be maintained above the minimum standard cubic feet per minute (SCFM) required by the diffusers in the system.

In addition to ASDs to control blower energy use, stepped motors can be considered for new facilities, and possible motor downsizing should be considered in existing facilities. It was forecast that facilities in a California program would be able to obtain 39 and 63% energy savings by using smaller blowers in aeration systems.

Course bubble systems can be inefficient due to large bubble size, low oxygen transfer efficiency, the requirements for blowers, and violent agitation of water in the aeration basins. Many plants have been retrofitted to fine bubble diffusion systems. A range of 9%–40% savings on aeration system energy use by converting from course to fine bubble diffusion has been documented.[1]

Various systems have been used for fine bubble diffusers. Ceramic disks were some of the first available systems; membrane systems are also available. There are efficiency advantages to membrane systems; oxygen transfer efficiencies of 13 vs 10% have been cited.[5] Membrane systems may also have the ability to be cleaned by "pulsing," or rapidly changing airflow. Cleaning and maintenance are important considerations in the selection of diffuser heads for use in specific facilities.

Adjustable-Speed Drives

There are many pumping systems within a wastewater treatment facility. These include influent wetwell pumps, return-activated sludge pumps, waste-activated sludge pumps, effluent pumps, and plant water systems.

The need to regulate variable flows and speeds at many stages in the process derives from the need for stable plant operations to produce high-quality treated effluent. Because pumping also consumes a large quantity of energy, better control of all pumping processes should be considered. In many cases, a reduced cycling schedule or better pump impeller selection for changed conditions may be relatively simple, low-cost improvements. The use of variable-speed pumping and ASDs to replace older flow control technology, although having possible higher capital costs, has proved to be advantageous both in reducing energy use and in improving overall process control.

It is possible to replace certain existing speed control technologies with electronic or mechanical ASDs. With addition of control system components, process control can be optimized.

Eddy current drives, for example, have been used to reduce speeds and flows mechanically through return sludge pumps in many plants. Flow generally is set at a certain recycle percentage. During periods of high or low flow, that percentage may require manual adjustment. By linking a process controller with an influent flow meter, an ASD will be able to select a variable optimized setting automatically—a process known as flow pacing.

Similarly, larger pumps may have been controlled with liquid rheostat-type controls. These controls can regulate flow but do not match the efficiencies of electronic ASD.

Table 4 shows estimated savings due to application of modern electronic variable-speed technologies, in kWh and as a percentage of total facility kWh consumption. As in Table 3, the percentage of kWh savings from all measures as a percentage of total kWh consumption also is given to illustrate the contribution of speed control modifications to a comprehensive energy efficiency program.

Care must be taken to locate these drives with the correct enclosures, provide a suitable operating environment, and check compatibility with driven motors and pumps. Analysis of the pump curves is essential.

Energy savings calculations must take into consideration the efficiencies of existing systems at various flow ranges. In many cases, the existing motor systems draw less than the motor full load horsepower. This can lengthen the payback on energy-saving retrofits. Detailed study

Table 4 Energy savings from speed control modifications

Plant type no.	Pre-retrofit consumption kWh/yr	Speed control savings	kWh Savings — % of total (%)	Total savings of all measures (%)
1	3,901,200	607,300	16	31
1	1,938,300	450,669	23	34
1	793,600	64,800	8	20
1	3,377,900	409,629	12	25
2	11,747,400	107,900	1	26
2	3,265,000	119,961	4	20
3	652,900	—	0	23
3	4,278,960	—	0	6
3	21,340,000	1,304,058	6	27
4	4,471,400	—	0	27
Overall	5,576,666	437,760	8	24

supplemented by field measurements should be used to verify the efficacy of these retrofits.

Sludge Processing and Methane Gas Recovery

Sludge processing consumes a significant amount of energy. Dewatering and digestion of the sludge can be an energy-intensive process. One must consider not only the energy used in processing sludge, but also the energy value of the byproducts of this process. Sludge can be incinerated, and methane gas produced in digestion can be used. Both byproducts can generate electricity and heat, which can be used in plant operations. Conversion from vacuum filters to more efficient technologies should be evaluated. Solar sludge drying, for example, also may prove to be cost effective.

Methane is formed in the anaerobic digestion of sludge. In some cases this valuable fuel is wasted, being flared to the atmosphere. Some facilities flare the total quantity of gas generated, whereas others use methane as a source of process, space, and water heating through on-site boilers.

In many cases it is economical to capture this methane; treat the gas by drying and removing impurities; and use this gas in an internal-combustion engine, in gas turbines, or even as feedstock for fuel cells.

In one instance, in a project funded and financed by an energy service company, a New Jersey municipal authority saved more than $450,000 (estimated as approximately 20% of its aeration costs) by using a gas-fired engine to drive a new blower directly.[6]

In another case, a smaller plant with flows of less than 2 mgd expects to save more than 200,000 kWh/yr, or approximately 30% of total plant electrical energy use. Not only will the system provide some backup power, but also, the savings for this small municipality will be more than $10,000 per year.

Sludge also has been incinerated to reduce volume and landfill costs. On-site incineration generally applies to larger systems. Sometimes it is cost effective to install heat-recovery boilers, which can be used to generate electricity and provide heat for both processes and winter space heating. One such project at a large municipal WWTF with a 40 mgd design flow has been estimated to save $1.2 million annually.

Renewable Energy Measures

There have been a few applications for other renewable or alternative energy sources. Solar sludge drying has potential applications where space and odor problems can be addressed.

Another approach for larger lagoons involves the use of introducing slow-rate laminar flows through motors that can be powered by solar electric (photovoltaic) cells. These systems increase the oxygen transfer to the system at a very low energy cost.

A few municipalities with high elevation drops on the outflow from WWTF (or in related water systems) have the potential to use simple hydroelectric turbines. This measure applies to few systems, as the elevation potential (in feet of head) usually is not sufficient to make this approach economical compared with current power purchase prices and electric tariffs.

Other Energy Cost Control Measures

Other process areas that should be addressed include the optimization of control systems, wetwell level adjustments to increase suction head, and chlorine system modifications. As an alternative to using variable-speed drives or dual-speed motors, some facilities may be better served by downsizing motors to accommodate low flow periods or actual operating conditions. One area where this has been implemented is in the optimization of plant water systems.

Due to the number of motors, it may make economic sense to retrofit using premium high-efficiency motors. In motor replacement, the highest-efficiency motor for a specific application should be selected, as the incremental costs of this motor can yield a short payback, sometimes paying for itself in a few months.

To reduce energy costs, hardware and software changes may be of benefit in improving power factor, limiting peak kW demand, and transferring load to off-peak rates through time-of-use scheduling.

Another area that deserves attention is the use of chemicals to reduce loading and energy requirements in secondary processes. This was implemented at a facility designed for 216 mgd that served a large North American city.[4]

Proper equipment selection and facility design enable significant energy savings. A savings of slightly more than 20% in oxygen uptake—a difference of 1.59 vs 1.26 kg of oxygen per kWh consumed—was reported at different plants within the same facility.[4] This applies to the design of new plants and in substantial expansions/renovations.

In support buildings, HVAC and lighting improvements should be considered. Due to the high hours of use in some plants, facilitywide lighting retrofits usually are cost effective.

CONCLUSIONS

There are enormous opportunities to save energy in existing WWTF and to optimize design in newer facilities. Many plants can save on the order of 10%–30% of energy costs through cost-effective retrofits. Baselines of energy usage have been developed for different types of plants. This benchmark for estimating possible savings in a particular facility may be useful. Process operations with high energy use, particularly the DO process, should be given special attention. Modern speed control technologies, such as electronic or mechanical ASDs, have relevance in reducing operational costs and in improving process control. Both energy and environmental benefits accrue when renewable energies are utilized for some of the energy requirements. In particular, methane recovery and use are particularly appropriate for those facilities with digesters and in the design of new facilities.

REFERENCES

1. Electric Power Research Institute. *Water and Wastewater Industries: Characteristics and DSM Opportunities. EPRI TR-102015*; Burton Environmental Engineering: Los Altos, CA, 1993.
2. United Nations. *Strategies, Technologies, and Economics of Wastewater Management in ECE Countries*; 1984.
3. U.S. Environmental Protection Agency. *Upgrading Existing Wastewater Treatment Plants-Case Histories. EPA-625/4-77-005a*; 1973.
4. Diagger, G.T.; Buttz, J.A. *Upgrading Wastewater Treatment Plants*; Technomic Publishing, Inc.: Lancaster, PA, 1998.
5. Iranpour, R.; Stenstrom, R.K. Relationship between oxygen transfer rate and flow for fine pore aeration under process conditions. Water Environ. Res. **1991**, *73* (3).
6. Polaski, J.; Schmalz, G. Energy-efficient retrofit saves electricity and money for wastewater plant. Water Eng. Manag. **2000**, *147* (1), 28–29.

Water and Wastewater Utilities

Rudolf Marloth
Department of Mechanical Engineering, Loyola Marymount University, Los Angeles, California, U.S.A.

Abstract

In the first half of the 20th century, life expectancy in the United States increased dramatically, primarily because of water treatment, which greatly reduced the incidence of waterborne bacterial infections such as cholera, typhoid fever, and dysentery. The Safe Drinking Water Act of 1974 and its amendments of 1986 and 1996 are the primary pieces of federal legislation protecting drinking water supplied by public water systems. A standard process chain for water treatment consists of grit removal, flash mixing with chemicals, flocculation, sedimentation, filtering, and disinfection.

The Clean Water Act of 1972 and subsequent legislation with the objective of reducing the discharge of pollutants to natural waters imposed standards on the secondary treatment of wastewater. A standard process chain for wastewater treatment consists of preliminary (physical), primary (physical and chemical), and secondary (biological) steps. In secondary treatment, microorganisms consume organic pollutants. Much industrial waste is incompatible with public treatment systems, so it is usually subject to pretreatment.

INTRODUCTION

Municipalities are normally responsible for providing water and disposing of wastewater, either by privately or publicly owned utilities. A water utility is responsible for the source, transmission, treatment, and distribution of potable water. Water usage is metered at the user. A wastewater treatment utility is responsible for collection, treatment, and disposal. Industrial wastewater is often metered. Residential wastewater charges are based on water usage.

Municipalities are also responsible for the collection and disposal of runoff, which consists of stormwater, misplaced or excessive irrigation, domestic car washing, etc. Runoff collects in gutters or holding ponds; then goes to storm sewers; and is dumped into a nearby river or, in coastal cities, directly into the ocean. This water carries with it a variety of pollutants: fertilizer, animal waste, oils, tire dust, etc. For municipalities that do not have combined storm and sanitary sewers, this runoff is dumped without treatment. Worse, in severe storms, combined sewer systems are likely to dump runoff combined with untreated sewage. Furthermore, runoff is not metered, so the municipality bears the cost of installing and maintaining the infrastructure. Possible approaches to reducing pollution of natural waters are (a) enactment of ordinances to make property owners responsible for reducing runoff and its pollutants, (b) diverting light flow or the first flow in a storm to a wastewater treatment plant (WWTP), and (c) installing treatment facilities to make runoff suitable for reuse. (The city of Santa Monica, California recently demonstrated its SMURRF (Santa Monica Urban Runoff Recycling Facility), the first of its kind in the nation.)

WATER

The primary objective of water treatment is to make it safe for human consumption at a reasonable cost. It is possible to produce safe water that has objectionable taste, odor, or color, so a secondary goal is to make the water appealing to the consumer. Turbidity and color are qualities apparent to the naked eye. The former is caused by particles in suspension and is measured in nephalometric turbidity units (NTU) by passing light through a sample. These particles can be removed by settling or filtering. Colloidal particles will not settle in a reasonable amount of time and must be removed by other physical processes. Dissolved substances are removed or transformed by chemical treatment. Color may be caused either by materials in solution (true color) or in suspension (apparent color).[1]

History

Early water treatment focused on what was apparent to the senses: appearance, taste, and odor. These qualities were improved by removing turbidity through filtration or precipitation. It was found much later that particles in the water harbored pathogens, which were largely removed while clarifying the water.

Keywords: Activated sludge; Baffles; Broad street pump; Clean Water Act; Coagulation; Coliforms; *Cryptosporidium*; Disinfection; Filter cycle time; Filtering; Flash mixing; Flocculation; *Giardia*; Head; Hydraulic; Industrial waste; Log removals; Microorganism; Natural waters; Pollutants; Preliminary; Primary; Reactor; Safe Drinking Water Act; Secondary; Sedimentation; Settling velocity; Sewage; Slow sand filter; Sludge; Tertiary; Turbidity; Wastewater; Water.

A slow sand filter (SSF) designed by James Simpson was commissioned in 1829, but it was some time before its full importance was realized. This simple device is essentially a tank filled with sand with water introduced at the top and removed at the bottom. Some time after the filter is started up (a few days to a few weeks), the upper layer of sand becomes coated with a gelatinous biological layer called the schmutzedecke, made up of algae, bacteria, protozoa, and small invertebrates. This sticky layer is biologically active and converts organic matter in the water to water, carbon dioxide, and harmless salts (i.e., it is mineralized). Later, research showed the importance of biological removal and also showed that the SSF was very effective in that respect. In particular, the SSF is effective in removing *Giardia* and *cryptosporidium* oocysts, which are nearly unaffected by chlorination.[1,2]

In the famous Broad Street pump episode of 1854, an outbreak of cholera in the Soho district of London killed more than 600 people. Dr. John Snow, who had theorized that the disease was spread by contaminated water, traced it to water from the Broad Street pump. The likely cause was from a leaking and cholera-infested cesspool located only three feet from the Broad Street well. In fact, cesspits lay under many of the houses in the district.[3] Acceptance of Snow's theory was slow, as was conversion to a sewer system that conveyed wastewater to central plants.

Recently, the Centers for Disease Control and Prevention and the National Academy of Engineering named water treatment as one of the most significant public health advancements of the 20th century.[4] This is rightly so, for in the first half of the 20th century, life expectancy in the United States increased dramatically, primarily because of water treatment, which has greatly reduced the incidence of waterborne bacterial infections such as cholera, typhoid fever, and dysentery. Even so, waterborne disease does occur in this country, the most well-known instance of which is the cryptosporidiosis outbreak of 1973 in Milwaukee. Most episodes are due to contamination of raw or treated water, inadequate treatment, and cross-contamination between sewers and water mains.

Standards and Monitoring

The Safe Drinking Water Act of 1974 and its amendments of 1986 and 1996 are the primary pieces of federal legislation protecting drinking water supplied by public water systems. Primary regulations under the act are for the protection of public health; secondary regulations are for regulations pertaining to taste, odor, and appearance.

The Surface Water Treatment Rule mandates that surface water or groundwater under the influence of surface water must be treated to remove or inactivate 99.9% of *Giardia lamblia* cysts and 99.99% of enteric viruses. These requirements are commonly stated as "log removals," where n-log removal is removal of a $1-10^{-n}$ fraction of the pollutant. Treatment processes using filtration are judged to comply by providing an adequate concentration/contact time (Ct) product, where C is the concentration of the disinfectant in mg/L and t is the time in minutes.[5]

The Enhanced Surface Water Treatment Rule requires a 2-log removal of *Cryptosporidium*. Systems using filtration are granted credit if they meet certain turbidity criteria. The Filter Backwash Recycling Rule requires treatment plants to recycle filter backwash water through the entire process cycle. The Disinfectants/Disinfection Byproduct Rule establishes maximum contaminant levels (MCLs) on total trihalomethanes, haloacetic acids, bromate ion, and chlorite ion.[5]

The most common way of testing the quality of drinking water is the coliform test, as specified by the U.S. Environmental Protection Agency. Coliforms are bacteria that are gram-negative, aerobic and facultative anaerobic, nonspore-forming rods, which ferment lactose with gas formation in 48 h at 35°C. When a sample tests positive for coliforms, it must be tested for fecal coliforms. Fecal or thermotolerant coliforms include all coliforms that can ferment lactose at 44.5°C. It is common to identify *Escherichia coli* uniquely with fecal coliforms.[5]

Sources and Transmission

Large-scale sources are primarily surface water and groundwater. Desalinated ocean water is not yet a major source in the United States, although Spain is a major producer and user of desalinated water. In the past 30 years, the energy required for desalination has fallen from 12 to 3 or 4 kWh/m^3 using reverse osmosis.[6] In a few areas, recycled water (treated wastewater) is used for irrigation. Water is transmitted by way of natural water courses, lined open channels, or pressure pipe. Los Angeles, for example, receives water from the Sacramento Delta via a concrete-lined open channel, from snowmelt in the Sierra via a natural watercourse that is diverted into an iron pipe aqueduct. New York reservoirs supply water to treatment plants through a series of underground tunnels. In some cases, water treatment plants are located right at the site where water is drawn from a river or lake.

Treatment

Treatment is tailored somewhat to the characteristics of the influent and the effluent limits, but in general, large particles are removed first by screening and then by settling. Next are particles that will float or settle with some assistance, such as air floatation or mixing with coagulants. Colloids and dissolved materials are removed last. The water should be nearly clear before disinfection so that contaminants may not hide in turbidity.

Standard water treatment (Fig. 1) consists of coagulation, flocculation (aggregation into a wooly mass), sedimentation, filtration, and disinfection. The size of the

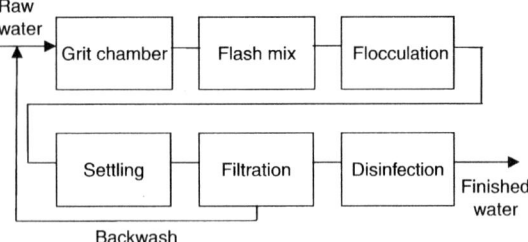

Fig. 1 Water treatment.

units and major equipment are determined by the hydraulic loading. The examples below are based on a plant with an average flow of 24 mgd (million gal per day). Plants are commonly designed for a maximum daily flow that is 1.5 times the average daily flow. Plants whose main supply is groundwater will have a somewhat different process chain because the water will contain less settleable and suspended solids but probably more dissolved metals. Plants that draw from a river should have at their head a coarse screen for tree branches, etc., and a grit chamber for sand and silt, because this kind of grit will cause a serious maintenance problem with pumps.

The coagulation operation consists of the addition and flash mixing of chemicals meant to remove the charge on colloids and suspended solids so that they will aggregate in the flocculation step. Depending on the type of chemical used, flash mixing should occur in about one to five seconds. The addition of more than the optimum amount of chemicals can compensate for less-than-optimum mixing. A potential problem in this step is the clogging of feed lines. Various methods of flash mixing are available, the best of which are diffusion mixing by water jets and inline static mixing. Proper operation at this step requires choosing the right chemicals and quantities in response to the raw water quality and flow rate. Chemicals used in this step include aluminum sulfate (alum), polyaluminum chloride, various iron compounds, polymers, and bentonite.[5]

Flocculation is slow mixing that increases the rate of collisions between particles whose electrostatic repulsive forces have been neutralized in the coagulation step. Now the particles will stick together into sizes that will settle. Flocculation mixing is performed by mechanical mixers, baffles, or other methods. Paddle reels with a horizontal axle perform well. In a reasonable design for two parallel, 12-mgd floc tanks, each of three chambers in a tank might be about 13.5 ft^2 in cross section so that a 13-ft horizontal paddle reel would nearly fill a chamber. Perforated baffle walls separate the chambers to promote good mixing. Proper operation of this step requires continual monitoring of and adjusting for the floc size, and removing scum from the surface of the water, sludge from the bottom of the tank, and algae from the vertical surfaces (walls and baffles). Transfer of the flocculated water to the sedimentation tank must be done at low velocity to avoid breaking up the floc.

Four progressive stages of sedimentation are distinguished. Type I sediment consists of separate, destabilized particles. Type II is made of larger groups of flocculated particles. In Type III, the particles have formed a blanket that initiates hindered settling. Type IV settling is compression of the sludge blanket at the bottom of the tank.

A typical filter bed is made up of a layer of sand and a layer of coal, charcoal, or granular activated carbon. The water from the sedimentation tank is introduced at the top of the filter and moves by gravity down through the media to the underdrain. It is most desirable to have the filter backwashed once per day. Thus, the design depends upon the quality of the raw water, the required throughput, the local climate, and the skill level of the operators. The backwash water is required to be recirculated to the head of the plant.

Filter efficiency is determined by the unit filter run volume (UFRV), which is the ratio of the amount of water processed during a filter cycle to the amount that could be processed if no backwashing were necessary. The effective filtration rate, R_e, is

$$R_e = (\text{UFRV} - \text{UBWV})/T$$

where UBWV (unit backwash volume), gal/ft^2; T, filter cycle time, min; and

$$\text{UFRV} = V_f/A \qquad \text{UBWV} = V_b/A$$

where V_f, volume filtered per filter cycle, gal; A, area of filter, ft^2; V_f, volume of backwash water, gal.

The design filtration rate, R_d, is the maximum filtration rate, which can be achieved only if no backwash were necessary. Then the production efficiency is

$$R_e/R_d = (\text{UFRV} - \text{UBWV})/\text{UFRV}$$

Example 1. Find the production efficiency and the filter cycle time for a filter with UFRV equal to 7500 gal/ft^2, UBWV equal to 200 gal/ft^2, and design filtration rate equal to 5 gpm/ft^2.

Solution. The effective filtration rate, filter efficiency, and filter cycle time are

$$R_e = (5 \text{ gpm/ft}^2)[(7500 \text{ gal/ft}^2) - (200 \text{ gal/ft}^2)]/$$
$$(7500 \text{ gal/ft}^2) = 4.87 \text{ gpm/ft}^2$$

$$R_e/R_d = [(7500 \text{ gal/ft}^2) - (200 \text{ gal/ft}^2)]/(7500 \text{ gal/ft}^2)$$
$$= 0.973$$

and

$$T = [(7500 \text{ gal/ft}^2) - (200 \text{ gal/ft}^2)]/(4.87 \text{ gpm/ft}^2)$$
$$= 1499 \text{ min}$$

Note that the filter cycle time is very close to one day (1440 min). Also, increasing the filtration rate will not necessarily increase the amount of water filtered per day.

Increasing the rate increases the amount of deposition on the filter media, reducing the filter run time.

Most of the water treatment plants in this country disinfect with chlorine. Common alternatives are ozone, chloramine, and ultraviolet light. The disinfectant is sometimes added at the head of the plant to give an adequate concentration-contact time product. When chlorine is used, disinfection byproducts can be formed—most notably, trihalomethanes. To suppress this, ammonia might be added at the end of sedimentation to form chloramines. Then the water is lightly rechlorinated in the clearwell to suppress regrowth of pollutants in the distribution system.

In small municipalities across the heartland, treated water is pumped into elevated tanks emblazoned with the name of the town. These towers serve to keep the water clean, meet surges in demand, and supply even pressure at the tap. Large cities tend to keep the water in open reservoirs, an unfortunate but perhaps necessary practice that leaves the water subject to the reintroduction of various undesirable substances.

Hydraulics

Flow through the plant is described by Bernoulli's equation, where all the components are expressed in terms of head in feet,

$$z_1 + P_1/\gamma + V_1^2/2g = z_2 + P_2/\gamma + V_2^2/2g + H_L$$

and z_i, distance of water level above datum; P_i/γ, pressure head at surface; $V_i^2/2g$, velocity head; H_L, head loss; $i=1$, upstream; $i=2$, downstream.

The head-loss term is made up of entrance and exit losses, pipe friction losses, and minor losses. Hydraulic calculations are best made starting at the clearwell and working upstream.[1]

It is preferable to have the water flow through the plant by gravity. The head losses through the processes are approximately (a) rapid mixing, 1 ft; (b) flocculation, 2 ft; (c) sedimentation, 2 ft; (d) and filtration, 10 ft, for a total of 15 ft. If the water comes into the plant at the level of the clearwell, this head and the friction losses in the lift pipe must be provided by a pump. At the pump discharge, the water horsepower is HP_w; the pump input is the motor brake horsepower, HP_b; and the power required to drive the motor is P, where

$$HP_w = QH/C_1 \qquad HP_b = QH/C_1 e_p$$
$$P = C_2 HP_b/e_m$$

and Q, pump flow rate, gal/min; H, pump discharge head, ft; C_1, constant, (550 ft-lb/s-hp) (60 s/min)/(8.34 lb/gal) $\approx$ 3960 ft-gal/min-hp; e_p, pump efficiency; C_2, constant, 0.746 kW/hp; e_m, motor and drive efficiency.

Example 2. Size the motor, and find the input power required to provide 15 ft of head for a 24-mgd water treatment plant. Ignore plumbing losses. Take the pump efficiency to be 70% and the motor/drive efficiency to be 90%.

Solution. The flow rate is $(24 \times 10^6 \text{ gal/da})/[(24 \text{ h/da})(60 \text{ min/h})] = 16,667$ gal/min. Thus

$$HP_b = QH/C_1 e_p = (16,667 \text{ gal/min})(15 \text{ ft})/[(3960 \text{ ft-gal/min-hp})(0.70)] = 90 \text{ hp}$$

and

$$P = C_2 HP_b/e_m = (0.746 \text{ kW/hp})(90 \text{ hp})/(0.90) = 75 \text{ kW}$$

Because a motor is just as efficient at 75% load as at full load, a 125-hp motor should be installed.

The important hydraulic parameters of a sedimentation tank (Fig. 2) are the hydraulic retention time (HRT), the horizontal velocity of the water through the tank (V_h), and the surface overflow rate (SOR). For a tank with dimensions L, W, D for length, width, and depth, the volume V, is LWD; the surface area, A_s, is LW; and the vertical cross-sectional area is $A_c = WD$. Therefore, the horizontal velocity and the SOR are

$$V_h = Q/A_c = Q/DW \qquad \text{and} \qquad \text{SOR} = Q/A_s$$

The HRT is

$$\text{HRT} = L/V_h = L/(Q/DW) = V/Q$$

A particle with settling velocity satisfying $\text{HRT} = D/V_s$ will reach the bottom of the tank before being swept up and out (as will particles with V_s greater than this). Then

$$V/Q = D/V_s \qquad \text{or}$$
$$V_s = DQ/V = Q/LW = \text{SOR}$$

That is, for a particle to settle, the settling velocity must be equal to the SOR (or greater).

The power (P) required for mixing or flocculation with an impeller in a tank is dependent upon the average

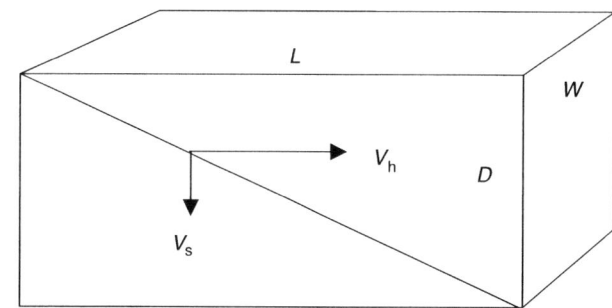

Fig. 2 Sedimentation parameters.

velocity gradient (G) in the fluid, dynamic viscosity (μ) of the fluid, and volume ($\forall$) of the tank: $P = G^2 \mu \forall$. For wastewater treatment, the average velocity gradient for rapid mixing is about 500–1500/s and for flocculation is about 50–100/s.[5]

Example 3. Find the power (P) required to achieve an average velocity gradient (G) of 100/s in a 1 million-gal flocculation tank whose contents are at a temperature (T) of 40°F.

Solution. The dynamic viscosity of water at 40°F is $\mu = 2.359 \times 10^{-5}$ lb-s/ft^2. Therefore,

$$P = (100/s)^2 (2.359 \times 10^{-5} \text{lb–s/ft}^2)(10^6 \text{ gal})$$
$$\times (0.746 \text{ kW/hp})/[(550 \text{ ft–lb/s–hp})(7.48 \text{ gal/ft}^3)]$$
$$= 42.8 \text{ kW}$$

Example 4. Consider a plant with an input flow rate of $Q = 24$ mgd $= 24 \times 10^6$ gal/da. The two sedimentation tanks are 300 ft long, 40 ft wide, and 13 ft deep. Find the HRT, the velocity of the water through the tank, and the SOR.

Solution. The flow rate through each tank is $Q = (24/2 \times 10^6$ gal/da)/[(24 h/da)(60 min/h)] $= 8333$ gal/min $= (8333$ gal/min)/[(7.48 gal/ft^3) (1440 min/da)] $= 1114$ ft^3/min. Therefore, the velocity in each tank is $V = (1114$ ft^3/min)/[(40 ft) (13 ft)] $= 2.14$ ft/min, and the surface overflow rate is SOR $= (8333$ gal/min)/[(40 ft) (300 ft)] $= 0.694$ gal/min-ft^2.

WASTEWATER

Some authors make the following distinction: wastewater is water that has been used for domestic, industrial, or commercial purposes, whereas sewage is more inclusive in that it can include water that has not been used, such as rain runoff. In the past 50 years, sanitation agencies have made a great effort to confine runoff to storm sewers and out of treatment plants, so the term WWTP is appropriate. These facilities are also called publicly owned treatment works (POTWs).

History

The history of wastewater treatment is a sordid one of determined ignorance and apathy. Until 1965, for example, Salt Lake City was dumping raw sewage into a 9-mi open canal that emptied into the Great Salt Lake.[7] In other parts of the world, this kind of practice continues to the present day.

Dr. John Snow in London convincingly linked cholera with the consumption of contaminated waters. This most famous episode in the history of both epidemiology and water treatment occurred in the late summer of 1854. Repeated outbreaks of cholera had occurred between 1831 and 1854 in the industrial cities of England, with little being done to prevent or contain it. The particularly sudden and violent episode in and around Broad Street in September gave Dr. Snow the opportunity to verify his belief that the cause was in contaminated water. When he persuaded the authorities to remove the handle from the Broad Street pump, the spread of the disease was halted. Later, Snow established that wastes from a single infected individual had been dumped into a leaking cesspit near the Broad Street well. After some time, people accepted the fact that fecal contamination of drinking water was a major cause of disease.[3]

With better water supplies and sewer systems, there was a sharp decrease in the incidence of waterborne diseases, even before the agents were identified. After half a century of research, the concept of waterborne disease was established. The cause was known to be microorganisms in the digestive tract and the associated health hazards had been proven. Then work proceeded on two fronts: analytical methods for the detection of fecal pollution and the development of treatment methods and facilities. Research led to publication of the first edition of *Standard Methods* in 1901,[8] and the SSF was an early and very effective method for treatment. (In fact, the newer and faster conventional rapid sand filter does not remove dissolved constituents as effectively.)

Standards and Monitoring

The Clean Water Act of 1972 and subsequent legislation placed increased emphasis on the importance of reducing the discharge of pollutants to natural waters. The minimum national standards now for secondary treatment are the "30/30" rule: a 30-day average of no more than 30 mg/L of BOD$_5$ (5-day biochemical oxygen demand) and TSS (total suspended solids), as well as pH to be between 6.0 and 9.0 at all times. Unfortunately, meeting these standards does not guarantee the absence of disease-causing agents—notably, *Giardia lamblia* and *Cryptosporidium parvum*. Increased sophistication of monitoring techniques is leading to better treatment techniques and stricter standards.[9]

Sources and Collection

Sanitary sewers receive some groundwater infiltration and stormwater. Otherwise, 90% or more of the intended influent is of residential or commercial origin. Industrial users are either direct dischargers dumping into a waterway or indirect dischargers dumping into a

POTW. (Some industrial liquid waste may also be hauled off.)

Influent is collected in closed pipe, mostly by gravity flow. When pump lift is necessary, the flow is into short runs of pressure pipe, eventually returning to gravity flow lines. Because plants are often located next to natural waters, they are typically at the low points of terrain, which keeps pumping to a minimum.

Treatment

Several levels of increasing care are defined. Preliminary treatment is the removal of large items, sand and grit, floatables, and grease. Wastewater typically contains floatable materials—particularly fats, oils, and grease (FOG)—whereas (fresh) water does not. Removal is accomplished purely by physical processes such as screening and gravity, and is intended to protect the plant equipment. Primary treatment is the removal of suspended solids and organic matter, often by the addition of chemicals. Secondary treatment is a biological process that removes organics and suspended solids (and sometimes nitrogen and phosphorus), followed by disinfection. Tertiary treatment removes remaining suspended solids by fine filtering, and may include disinfection and nutrient removal. The standard today for wastewater is full secondary treatment, meaning that all the influent to a plant is given secondary treatment. When a plant cannot handle the flow, partial secondary treatment means that all the influent is given preliminary and primary treatment, but only part of it is given secondary treatment.

A typical process chain for secondary treatment is the activated sludge process shown in Fig. 3.[9] Large items are screened out. Dense noncontaminants are removed purely by gravity in the grit chamber. Primary treatment is a chemical/physical process, whereby small particles agglomerate (flocculate) and gravitate out. Secondary treatment is biological, in which microbes consume the dissolved and suspended organic matter. Disinfection kills most of the remaining contaminants.

Preliminary and primary treatment for wastewater is much like that for water. Preliminary screening of wastewater is necessary because large objects sometimes find their way into the sewer, an unfortunate example being construction debris illegally dumped into a manhole (alternatively, "maintenance hole"). (Note that a manhole cover is round for at least three reasons: (a) It will not fall through the hole, no matter how oriented; (b) it is easy to move by rolling; and (c) it need not be rotated to fit.) Primary treatment for both is nominally flash mix, flocculation, and sedimentation. Wastewater undergoes secondary treatment, in which microbes remove biological pollutants.

Both organic and inorganic particulates may be removed by settling, flotation, or filtering, depending on particle size and density. Carbon filtration is preferable to

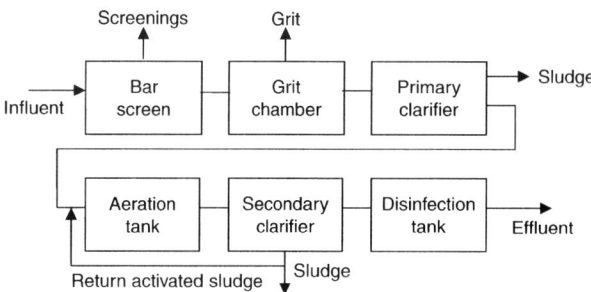

Fig. 3 Activated sludge wastewater treatment.

vaporizing, because the latter merely moves the substance from the water to the atmosphere.

Most reactions in waste treatment are of first order—that is, the rate of reaction is proportional to the concentration of the pollutant.

$$dC/dt = -k\,C$$

where C is concentration of pollutant (mg/L); t is time (min); k is reaction constant (mg/L-min).

Reactors are of three types: complete mix (or batch), plug flow, and dispersed flow. In a complete mix reactor, the reactor is filled; the reaction is allowed to take place; and then the reactor is emptied. The concentration of pollutant is equal throughout the tank. Complete mix reactors are approximately cubical and can be set up in sequence to provide an increasing proportion of removal. In a plug flow reactor, the flow moves through the reactor with the reaction taking place so that the concentration of pollutant is less at the outlet than at the inlet. These tanks are long in proportion to their length and width. Turbulence should be minimized to retain the form of the plug. Most reactors are of the dispersed flow type, intermediate between the other 2.[9]

Liquid and Solids Disposal

By the principle of conservation of mass, treatment does not make the contaminants disappear; it merely separates them from the water that bears them. When most of the contaminants have been removed and the water is sufficiently clean, it may be discharged to a natural waterway, such as a river, lake, or ocean. Most of the processes in Fig. 3 produce residuals. These impurities—such as sediment, sludge, waste washwater, and brine—are left behind to be treated and disposed of in other ways. Large items from screening and grit from the grit chamber are sent to landfill. Sludges from primary and secondary treatment are sent to a digester, which itself produces waste. Waste washwater from the filters is recycled, but because it adds to the throughput volume, it should be kept to a minimum.

Industrial Waste

For both direct dischargers and indirect dischargers, the content of industrial wastewater is regulated. Industrial users are allowed to send the first fraction of runoff from rainstorms to the sanitary sewer, for the purpose of keeping pollutants out of the storm sewers and, ultimately, out of natural waters after which it must be diverted to storm sewers.

Because POTWs are set up to treat organic waste, much industrial waste is incompatible with public treatment systems, so is usually subject to pretreatment. The objectives of such treatment are to prevent interference with the process in the POTW, to prevent pass-through of pollutants to the receiving waters, and to make possible reuse of the effluent and sludge from the POTW.[10]

Some treatment strategies are flow equalization to prevent shock loading to the POTW; solids removal by straining or settling; removal of FOG by dissolved air floatation or centrifuging; neutralization of high-pH or low-pH solutions; and hydroxide precipitation of heavy metals. A notable exception to the last is the removal of Cr^{+6}, which will not respond to hydroxide precipitation. Instead, it is converted by chemical reduction at low pH to Cr^{+3}, which can be removed by hydroxide precipitation. Dissolved inorganics may be removed by hydroxide precipitation, ion exchange, or membrane filtering.[10] It should be noted that diluting industrial wastewater to reduce the concentration of pollutants is not acceptable, and that "dilution of pollution is not a solution, and can lead to prosecution." The most important principle is segregation: keeping the pollutants separated so that they can be treated individually.

Issues

Wastewater treatment plants are designed to treat organic wastes. Other substances (nutrients in fertilizer, pharmaceuticals, etc.) can pass through and cause problems in receiving waters (e.g., algal growth and abnormal growth in fauna). Yet others, such as heavy metals and toxic chemicals, can cause interference (also called upset)—disruption of the process in the biological reactor.

Failed equipment can cause raw or partially treated wastewater to flow into storm drains and then into natural waters. Runoff from heavy storms can flow into sanitary sewers and overwhelm treatment plants. Cities with combined sewers are especially subject to this problem. Inadequately sized treatment plants will discharge partially treated wastewater in times of heavy flow.

Everything removed from wastewater must be disposed of. Sometimes, objections are raised to the release of volatiles into the atmosphere. Although the creators believe that digested sludge is a fertilizer rich in nutrients, others are not convinced.[11]

More radically, some have questioned the wisdom of the whole process of fouling great quantities of cleaned water and then cleaning it again.[12]

CONCLUSION

A basic requirement of human existence is an adequate supply of clean water. Today, very few have access to clean, untreated water. Wealthy societies obtain clean water by treating it, while poor ones often rely on polluted sources and suffer from the resulting waterborne diseases.

The idea that many municipalities draw from surface waters that are used for the disposal of wastewater is sobering, if not chilling. In the United States, the Safe Drinking Water Act mandates water treatment standards, and the Clean Water Act mandates wastewater treatment standards. Observance of these standards has made the practice of having a common source and sink acceptable.

The conventional water treatment process described in this article has been very effective in removing bacterial pathogens. Dechlorination to control disinfection byproducts created by chlorination was instituted as a result of the Safe Drinking Water Act. The most recent major issue is the resistance of *Cryptosporidium* oocysts to chlorination. Membrane filtration is an effective way to remove these and other very small suspended pathogens. New water treatment plants are likely to be based on this technique because of its effectiveness and ease of operation.[5]

The most important developments required in wastewater treatment are (a) building or expanding facilities to provide the capacity to subject all flow to complete secondary treatment, (b) repairing and maintaining the collection system to ensure that all wastewater reaches the treatment plant, and (c) finding practicable ways to dispose of the residual solids.

REFERENCES

1. Reichenberger, J. Unpublished notes. Loyala Marymount University: Los Angeles, 2005.
2. Jesperson, K. *Search for Clean Water Continues*. Available at: www.nesc.wvu.edu/ndwc (accessed 27 November 2006).
3. Judith Summers. *Broad Street Pump Outbreak*. Available at: www.ph.ucla.edu/epi/snow/broadstreetpump.html (accessed 27 November 2006).
4. United States Environmental Protection Agency, Office of Water. *The History of Drinking Water Treatment*. Available at: http://epa.gov/safewater/consumer/pdf/hist.pdf (accessed 28 November 2006).
5. Kawamura, S. *Integrated Operation and Design of Water Treatment Facilities*, 2d Ed.; Wiley: New York, 2000.

6. Graber, C. *Desalination in Spain*. Available at: www.technologyreview.com/spain/water (accessed 27 November 2006).
7. Salt Lake City Department of Public Utilities. *Conduits of Civilization*. Available at: www.ci.slc.ut.us/utilities/NewsEvents/news1998/news9281998-1.htm (accessed 27 November 2006).
8. Greenberg, Arnold, et al., eds. *Standard Methods for the Examination of Water and Wastewater*, 21st Ed.; American Public Health Association (APHA), American Water Works Association (AWWA), and Water Environment Federation (WEF): Washington, D.C., 2005.
9. Tchobanoglous, G.; Burton, F.L.; Stensel, H.D. *Wastewater Engineering, Treatment and Reuse*, 4th Ed.; Metcalf & Eddy Inc., McGraw-Hill: New York, 2003.
10. Water Environment Federation Pretreatment of Industrial Wastes In *Manual of Practice FD-3*; Water Environment Federation: Alexandria, VA, 1994.
11. Voters seek to block sludge, *Los Angeles Times* 2006, January 2. L.A. Fights Kern County Sludge Ban. *Los Angles Times* 2006, August 16.
12. Stauber, J.C.; Sheldon, R.S. *A Brief History of Slime*. Available at: www.prwatch.org/prwissues/1995Q3/slime.html (accessed 27 November 2006).

Water Source Heat Pump for Modular Classrooms

Andrew R. Forrest
Government Communications Systems Division, Harris Corporation, Melbourne, Florida, U.S.A.

James W. Leach
Industrial Assessment Center, Mechanical and Aerospace Engineering, North Carolina State University, Raleigh, North Carolina, U.S.A.

Abstract

This work investigates and recommends design improvements for a water source heat pump system for mobile homes and modular classrooms. It builds on a previous study that tested a 3-ton geothermal heat pump in a modular classroom at Wilson Mills Elementary School in Johnston County, North Carolina. Water stored in flexible plastic bladders resting on the ground underneath the classroom served as the heat source. The bladders were filled with 2000 gal of saltwater. Using Transient Systems Simulation Program (TRNSYS), a model of the original system was constructed and validated by comparing model predictions with measured performance. Transient systems simulation program models of several new designs were constructed to evaluate potential design improvements. The system models were evaluated based on predicted performance for a typical meteorological year, and on other criteria, such as initial cost, maintenance, and portability. This resulted in a new optimized system design in which the water storage volume is reduced to 120 gal, and the predicted electrical energy requirements are about two-thirds of those of an air source heat pump. The predominant design improvement is to replace the bladders with heat exchangers constructed of PVC pipe. Design, costs, and assembly procedures for the PVC heat exchanger are presented in this study.

INTRODUCTION

The purpose of this project is to optimize the operating parameters and design of an earth-coupled heat pump system with aboveground water storage.[1] A computer model was created in a Transient Systems Simulation Program (TRNSYS) to predict the best geothermal heat pump design. Available experimental results from the previous project were compared with predicted results to validate the model. Afterwards, the computer model was used to predict how changes in system parameters would affect the energy use of the water source heat pump.

Background

Two previous studies were performed to develop geothermal heat pump technology specifically for mobile classrooms, homes, and offices. Researchers at Progress Energy Carolinas (formally Carolina Power & Light) conceived the original design and funded the first study, which was conducted by a graduate student at North Carolina State University.[2] A 1-ton water source heat pump was installed on a mobile office building, and the water in the system circulated through a 7000-gal plastic bladder that rested on top of the ground. The bladder was insulated from the atmosphere to provide some freeze protection, and heat strips were installed in the bladder to heat the water if temperatures approached the freezing point.

The second study, which was funded by the North Carolina State Energy Office, was twofold, containing a theoretical analysis and experimental validation of a low cost version of the original design. The second heat pump was installed in a modular classroom in Johnston County, North Carolina at Wilson Mills Elementary School (WMES). The heat pump capacity increased from 1 ton in the original mobile office to 3 ton for the modular classroom, and the volume of the storage bladders decreased from 7000 to 2000 gal. The bladders underneath the classroom were not insulated, and saltwater was used to prevent freezing. Fig. 1 is a schematic of the water source heat pump system at WMES. A photograph of the classroom with the heat pump installed is shown in Fig. 2. The theoretical study[3] predicted that the geothermal heat pump would use approximately one-half the energy that a similar air source heat pump would use for the same classroom.

The experimental part of the study[4,5] compared the performance of the geothermal heat pump system with that

---

*This entry originally appeared as "Water Source Heat Pump for Modular Classrooms" in *Energy Engineering*, Vol. 102, No. 2, 2005. Reprinted with permission from AEE/Fairmont Press.

Keywords: Water source heat pump; Modular classrooms; Mobile homes; TRNSYS; PVC.

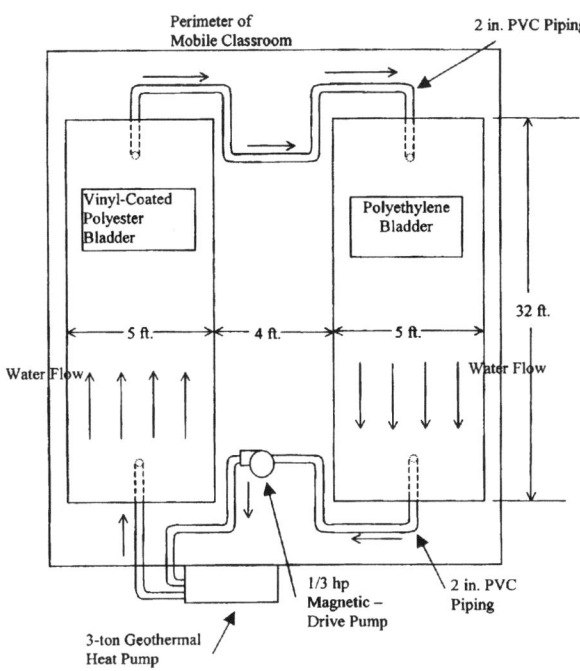

Fig. 1 Layout of original bladder system.
Source: From North Carolina State University and ASHRAE (see Refs. 4 and 5).

The purpose of the current work is to improve the design of the water source heat pump so that it is more energy efficient and reliable. Also, a reduction in the initial cost of the heat pump system and additional system mobility is desired. The work will also demonstrate the performance advantages of the improved design by comparing two water source heat pumps with three air source heat pumps in modular classrooms.

In the first phase of this work, candidate design improvements were evaluated, new heat exchangers were designed and fabricated, and a test site was selected. A TRNSYS model of the classroom and heat pump was developed. The model was calibrated by comparing predicted performance with experimental data from WMES. Wake County Public School Systems agreed to allow their mobile classrooms to be used to demonstrate the water source heat pumps for a period of 1 year. Several potential test sites were considered, and Davis Drive Elementary School (DDES) was selected because the site contained multiple school trailers with similar building loads. Several data loggers were left at selected trailers to confirm this assertion. Measurements from the crawl space from two of these trailers were used to design the PVC heat exchangers.

The second phase of this study is in progress. The PVC heat exchangers and water source heat pump were installed in June 2003 as shown in Fig. 3. Data are being collected to determine the system's actual economic and physical performance. The data will be compared with predicted performance to further validate the TRNSYS model. The design and volume capacity of the bladder are substantially different from those of the PVC heat exchanger, which is evident in Figs. 3 and 4.

Previous Project Problems

The experiment at WMES provided valuable experimental results to validate previous theories. However, several problems with the water source heat pump design became evident as the experiment was being conducted. Initially, of a nearly identical air source heat pump at an adjacent classroom. The experimental results showed that the heat pump did use about one-half as much electricity as the air source heat pump while in heating mode. However, in cooling mode, the geothermal heat pump used about 80% as much energy as the air source heat pump. Subsequent study showed that the ventilation rate in the classroom with the air source heat pump was well below the value recommended by ASHRAE. This changed the building load in favor of the air source heat pump. In addition, the air source heat pump could not maintain the interior of the classroom at comfortable temperature during the hot summer months.

Fig. 2 Modular classroom style where experiments are conducted.
Source: From North Carolina State University and ASHRAE (see Refs. 4 and 5).

Fig. 3 PVC heat exchangers installed at DDES.

Fig. 4 Installed bladder underneath modular classroom at WMES.
Source: From North Carolina State University and ASHRAE (see Refs. 4 and 5).

one of the bladders split at the seam during filling, and required replacement. In addition, the saltwater used for freeze protection was corrosive to the pump and heat pump. The volume of the brine solution (2000 gal) presented disposal and portability problems.

The original bladder that split at the seam was made from a high-density polyethylene material, and cost $250. The bladder that was used as a replacement was constructed of a vinyl coated woven polyester fabric, and cost $750. The $250 polyethylene bladder that did not fail is shown filled with 1000 gal of saltwater in Fig. 4.

With a 2000-gal storage volume, saltwater provided a low-cost option when compared with using antifreeze for freeze protection. However, the saltwater corroded the direct drive pump and required the use of a more expensive, magnetic drive pump. Furthermore, the saltwater destroyed the soft solder copper pipe connections. The leakage of saltwater around the connections caused the base of the heat pump enclosure to rust.

Two thousand gallons of saltwater cannot be easily transported or easily disposed of when the modular classroom is relocated to another site. To ensure freeze protection, salt was added to the water in the bladders to a 15% solution.[4] This volume of saltwater would present a disposal problem at the end of the equipment life.

The North Carolina Energy Office recognizes the need for efficient and economical water source heat pumps suitable for mobile buildings. Therefore, the present work was funded to further develop and refine the system design. One potential design improvement would incorporate low cost solar collectors to heat the storage water.

MODEL EXPLANATION AND VALIDATION

TRNSYS Description

The TRNSYS was chosen as the software to model the geothermal heat pump system for three reasons. Transient systems simulation program is a Fortran language-based program, which allowed for a ground temperature distribution finite difference model and an earth-coupled heat exchanger to be modeled in conjunction with the existing TRNSYS components. Transient systems simulation program is well known for its ability to model solar radiation and solar collectors in thermal systems. The software was used to predict the performance of solar collectors in the geothermal heat pump system. Since the system being modeled in TRNSYS consists of multiple components, it allowed a large, complex system to be broken into smaller parts. Thus, it was convenient to analyze the effects of modifying various parameters in the system.

WMES Experimental Results and Model Predictions

It was necessary to construct a TRNSYS model of the previous experiment to validate the calculations for the new design. The model predicts the heat pump electrical energy requirements based on typical meteorological year (TMY) data. Unknown parameters in the model, such as the heat transfer coefficient between the bladder surface and the ambient air, were adjusted until it predicted most accurately the performance of the previous experiment. The theoretical data and experimental results presented in Fig. 5 are organized to correspond directly to the days that actual heat pump energy use was recorded. The model correlates very closely with recorded data during the spring and fall months. When divergences do occur, the majority of the differences in experimental and theoretical energy use can be attributed to differences in actual degree days and TMY degree days as shown in Fig. 6. The two exceptions occur during the summer months. In the summer 2000, the degree days difference accounts for some of the divergence, while the remainder could be attributed to larger infiltration and internal loads due to increased student activity levels. In summer 2001, the classroom was not used for summer school, as it was the previous summer. The disparity in energy use in the winter months can be explained by differences in degree days. The predicted annual energy use was 9337 kWh, and the actual annual energy use during the previous study totaled 10,783 kWh. This underestimation should be taken into account when viewing the subsequent model predictions for the new designs.

NEW DESIGNS

The initial results from the TRNSYS models showed that the heat pump performance is not very sensitive to the volume of water stored, but is sensitive to the surface area available for heat transfer from the water to the ground and the ambient air. This led us to consider the possibility of replacing the bladders with PVC heat exchangers as shown

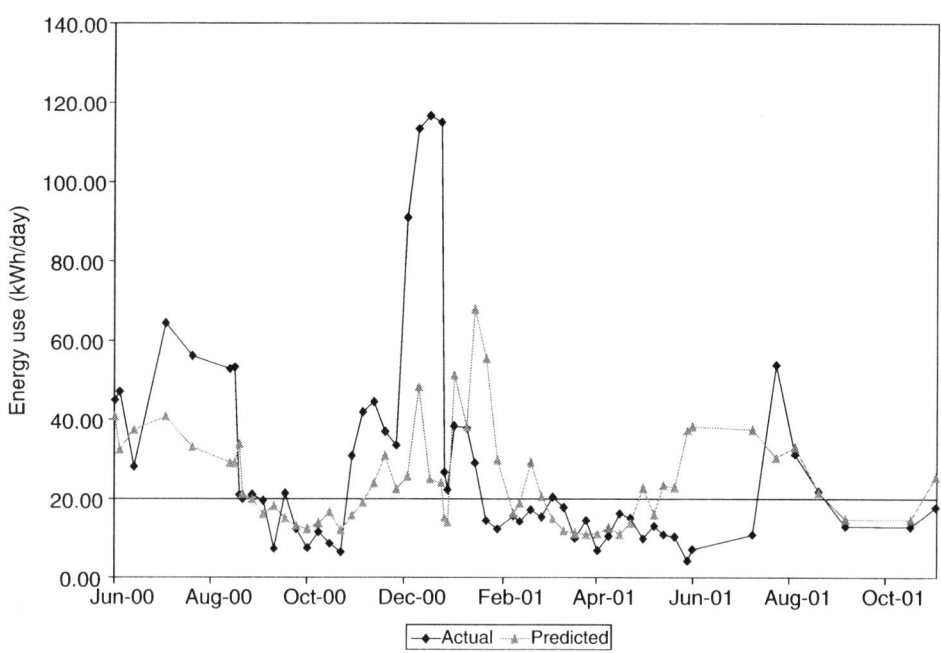

Fig. 5 Energy use comparison for original geothermal heat pump study.

in Fig. 3. Heat exchangers such as this would be more durable than the bladders and would be easier to move. Also the surface area and volume can be optimized independently by changing the diameter of the PVC pipe. The surface area-to-volume ratio for the small diameter pipe in Fig. 3 is much greater than the corresponding ratio for the bladders.

Surface Area and Volume Effects

Parametric analyses were performed to establish the dependence of the annual electrical energy requirements on the volume of the water storage and the surface area available for heat transfer. The heat transfer coefficient for convection to the ambient air was assumed to be constant and independent of pipe diameter. The surface area available for heat transfer to the ground was assumed to be 30% of the total surface area. A simulation with hourly time steps was performed using TMY data.

The trends predicted in Fig. 7 are the fundamental reason for using a heat exchanger design in place of the bladders. The predicted annual energy use of the heat pump in the bladder system (with two 1000 gal bladders and

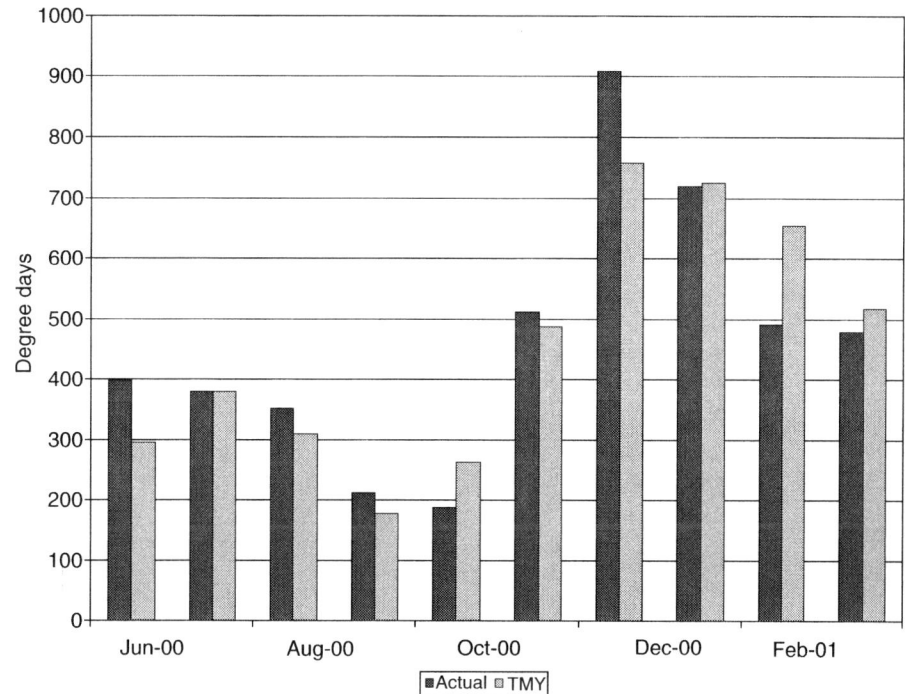

Fig. 6 Degree day comparison for portion of original geothermal heat pump study.

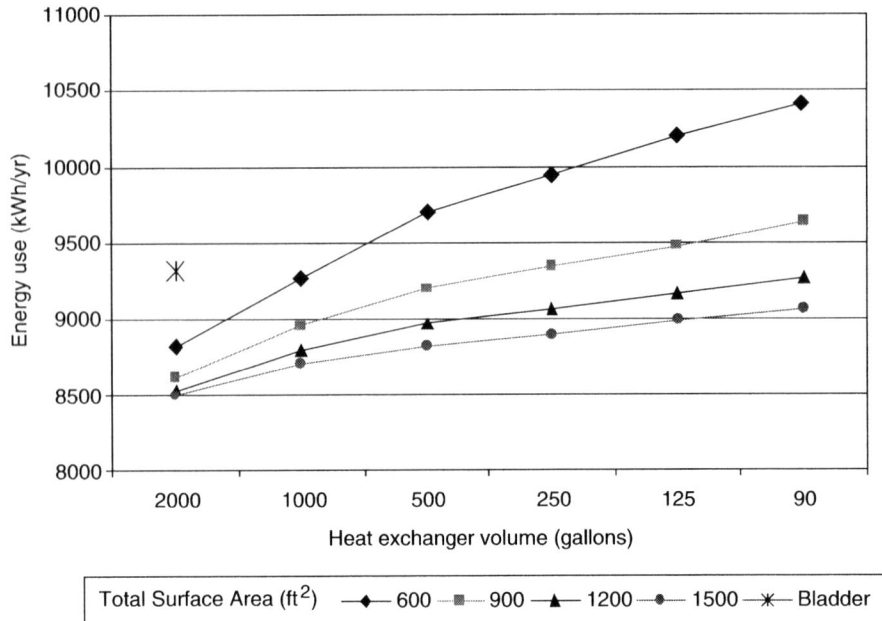

Fig. 7 Water source heat pump and pump annual energy use for various surface areas and volumes.

640 ft² of total surface area) is included for comparison. The difference in energy use is due to different percentages of surface area allocated for heat transfer to air and water between the bladder and heat exchanger models. Furthermore, the theoretical heat transfer coefficient for the heat exchangers to air is twice as large as the heat transfer coefficient for the bladders to air. The model predicts that a heat pump coupled with a heat exchanger having a volume of 120 gal and a surface area of 1200 ft² would require less electricity than one coupled with 2000 gal of bladder storage. The 120 gal storage volume is much more desirable for mobile buildings.

As shown in Figs. 8 and 9, the fluid temperature in heat exchangers with a volume of 120 gal and a surface area of 1200 ft² is fairly close to the ambient air temperature. The fluid in the bladders is warmer in the summer and cooler in the winter than the fluid in the heat exchangers. Therefore, the coefficient of performance (COP) of the heat pump is expected to be greater when the PVC heat exchangers are used. The average amplitude of daily temperature fluctuation in the bladders and PVC heat exchangers was predicted to be about 10 and 20°F, respectively.

Solar Collector Investigation

One of the objectives of this work is to evaluate the economics of using low-cost unglazed solar collectors to improve the heat pump coefficient of performance. The storage water would flow through the collectors before it arrives at the heat pump. The storage water, which is

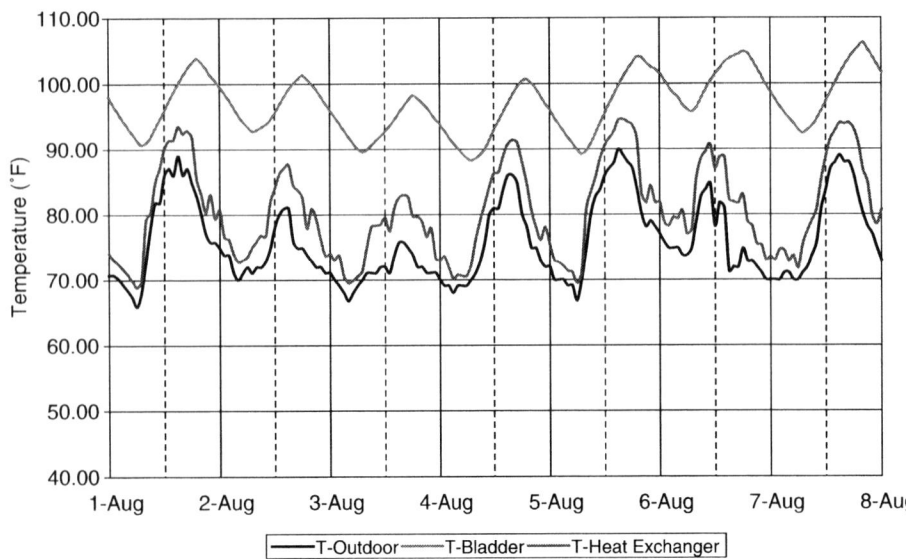

Fig. 8 Typical predicted summer temperatures.

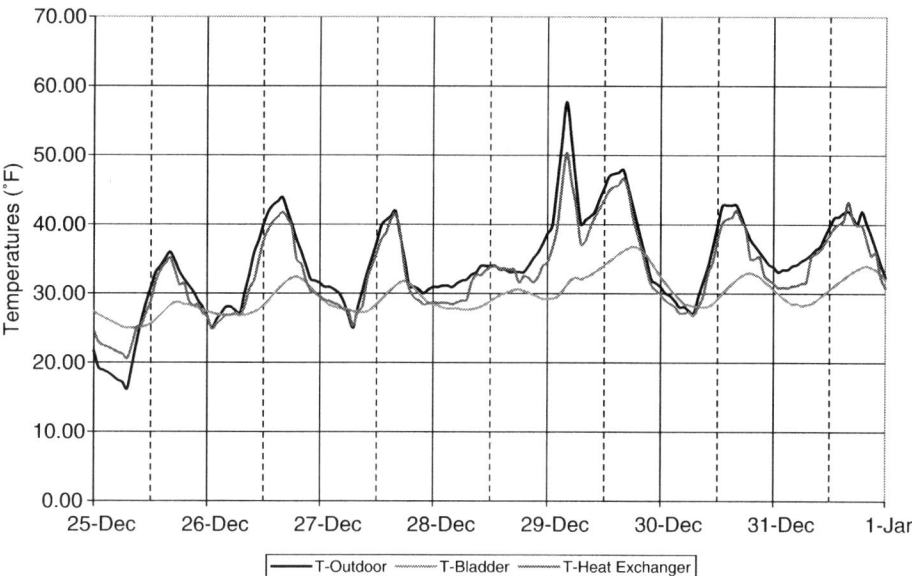

Fig. 9 Typical predicted winter temperatures.

always colder than the ambient temperature in the winter, would be heated by the sun and by convection from the air. During the summer, the collectors could be shaded so that the storage water is cooled by convection.

The interest in modeling solar collectors as an energy savings measure motivated the decision to use TRNSYS. However, the TRNSYS code also proved to be useful for modeling systems without solar collectors. The predicted electrical energy savings yielded by the solar collectors was much less than expected. A system with three 4×8 ft unglazed collectors was modeled. The collectors were plumbed in parallel to prevent excessive pressure drop. The model predicts that solar collectors provide more savings for a system with bladders than for a system with PVC heat exchangers. Still, the system with solar collectors and PVC heat exchangers uses less energy. The larger storage capacity of the bladders enabled them to store the energy gained by the solar collectors. Fig. 10 shows the energy savings from using solar collectors with bladders and heat exchangers. Despite the increased energy savings from using the solar collectors with the bladders as opposed to heat exchangers, the predicted payback period for the three solar collectors being modeled would exceed 10 years. Further evidence of the solar collectors ineffectiveness can be seen by the fraction of time the fluid flows through the collectors. The simplest control strategy causes fluid to flow though the collectors when the sun is shining and the heat pump is heating, or when the sun is not shining and the heat pump is cooling. In this case, the solar collectors are used only about 15% of the time. More complex strategies increase the time that the collectors can be used, but do not have much effect on energy usage.

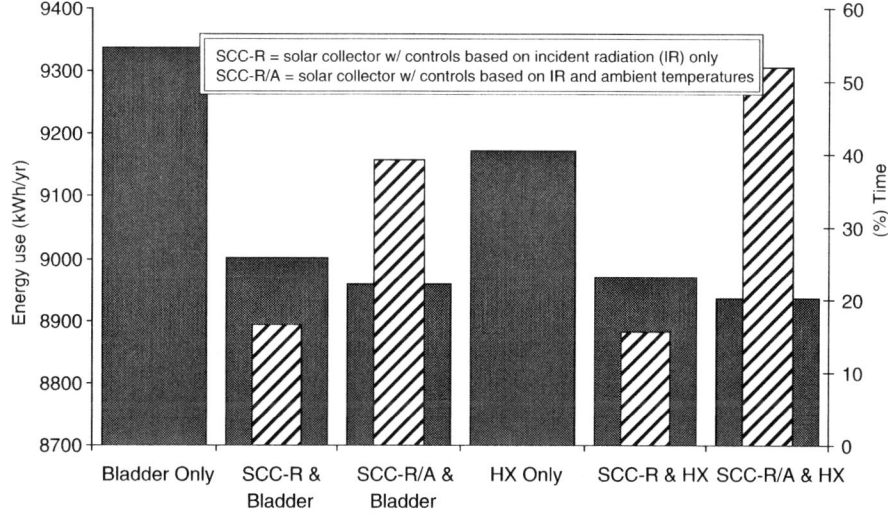

Fig. 10 Geothermal heat pump energy use with solar collectors.

Another issue to be examined is the optimum orientation of the solar collectors. For appearance, it is desirable to have the collectors lie flat on the roof. Measurements were taken at DDES to determine the slope of the classroom roof, β, and the azimuth, γ, of the most southern-facing roof. β was measured to be about 14° and γ was measured to be about 30° east of south. The optimum case for solar collector orientation was determined to be at a slope of 47° and facing south. According to the Florida Solar Energy Center, a solar collector performs best facing south, but a 30° variation in either direction will still allow the solar collector to capture 90% of the maximum solar energy available.[6] The recommended optimum slope that a solar collector should be positioned for winter use is the location's latitude plus 15°. In Raleigh, North Carolina, this would equate to about 51°. The TRNSYS model predicted that energy savings associated with changing the orientation of the collectors are very small. Thus, the solar collectors could lie flat on the roof at DDES.

Fluid Flow Effects

Parametric analyses were performed to determine whether an optimum flow rate exists, and to determine whether it coincides with the heat pump manufacturer's recommended flow rate of 7 gpm. The simulation, which was carried out over the range of flow rates, accounts for improvements in heat pump COP and energy costs to drive the circulation pump. The optimum flow rate for heat pumps with solar collectors and heat pumps without solar collectors was determined to be 13 gpm, which differs from the manufacturer's recommended flow rate, but is not excessive. As shown in Fig. 11, the differences in energy consumption between a system with and without solar collectors dwindle as the flow rates are increased.

Three different types of fluids were modeled for the same heat pump: pure water, 25% ethylene glycol, and 25% brine solution. The system containing the brine solution properties performed the worst. The system containing the ethylene glycol solution yielded about a half percent energy savings when compared with the brine solution. The system containing pure water performed the best (about one-third percent savings over the ethylene glycol solution). The results of the simulations were in line with expectations due to the variations in specific heat of the three solutions.

Thermostat Setback

Another objective of this work is to estimate the energy savings made possible by automated thermostat setback. All mobile classrooms in Wake County, North Carolina are equipped with radio-controlled night setback controls. During the winter months, the thermostat is supposed to be set back to 55°F typically from about 4:00 P.M. to 6:00 A.M. In the summer months, the heat pump is supposed to be turned off during a similar time frame. Some of the night setback thermostat controls were not functioning during our visits to the classrooms. In addition, these controls can be manually overridden with the thermostat inside the classroom. The night setback feature was modeled in TRNSYS to predict the energy savings generated by the use of these controls for a geothermal heat pump system with PVC heat exchangers and for an air source heat pump. The results generated by the TRNSYS model, which are summarized in Fig. 12, show that the night setback controls reduce electrical energy consumption by about 30%.

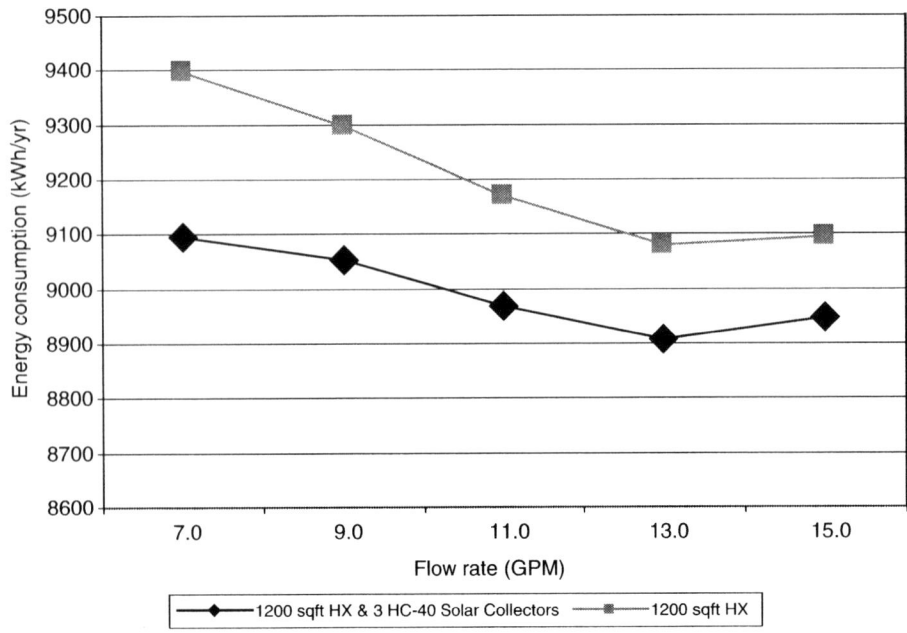

Fig. 11 Heat pump annual energy use for manufacturer's flow rate range.

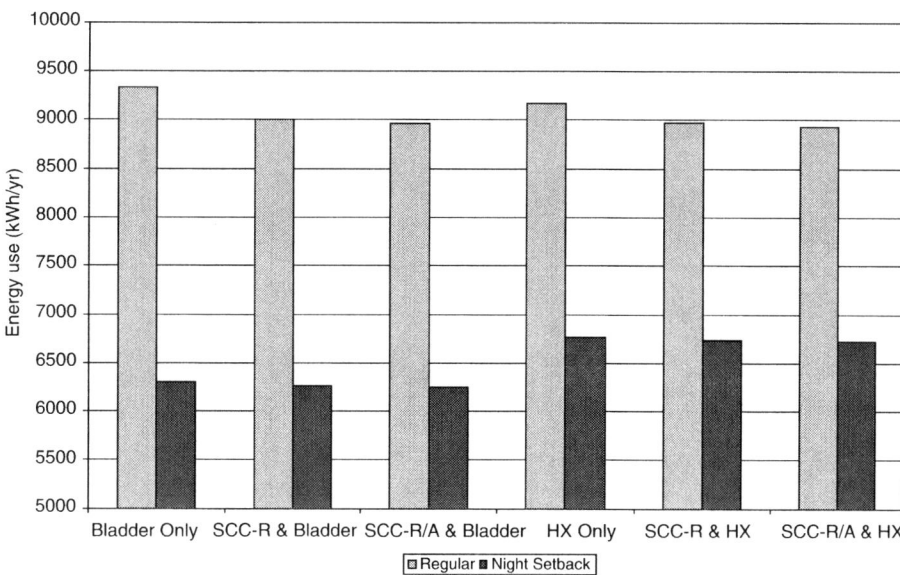

Fig. 12 Energy use with night setback controls.

Optimum System

An optimum system was established based on the analysis presented in the previous sections and on the existing layout of the concrete piers in the classroom crawlspace. The optimum system was determined to have a surface area of 1200 ft^2 and a volume of 120 gal. The flow rate of the ethylene glycol solution in the system was prescribed to be 13 gpm. Fig. 13 compares the daily energy use of the original bladder system and the optimum PVC heat exchanger system. The total predicted electrical energy consumption of the optimized heat pump system was 9069 kWh/yr, a predicted energy savings of 3% compared with the original system. The optimum system used much less energy in the summer months, but used slightly more or about the same throughout the rest of the year.

PVC HEAT EXCHANGER DESIGN

The PVC heat exchangers will be more durable and less likely to leak than the bladders were. The significant reduction in volume of the PVC heat exchangers will allow an ethylene glycol solution to be used instead of a brine solution, eliminating the problems caused by saltwater. A less-expensive, direct drive pump can be used to circulate the ethylene glycol solution. In addition, the ethylene glycol solution can be reused when the mobile classrooms are moved, further reducing the overall cost of the system.

The heat exchanger that evolved as a result of this work is made from PVC pipe, and is comprised of a header and footer, and 0.5-in. PVC pipe (Fig. 14). Space limitations in the crawl space of mobile buildings influenced the design

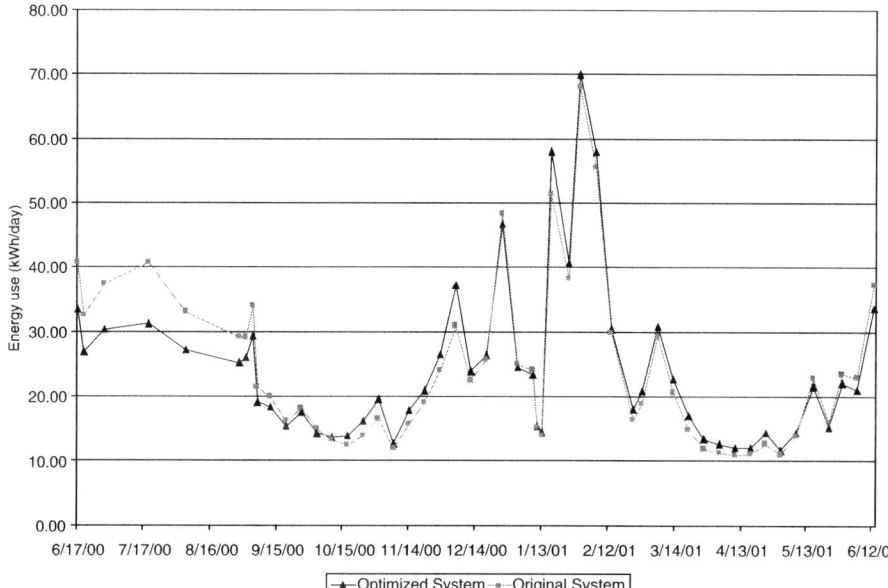

Fig. 13 Annual energy comparison of original and optimized systems.

of the heat exchanger. The heat exchanger was designed to be approximately 2-ft wide, and to run the length of the space it is intended to occupy. The heat exchangers were designed to this width to ease manufacturing and handling of the headers, and to maintain the desired fluid velocity of about 1 ft/s. An additional benefit from using the multiple 2-ft wide heat exchangers would be realized if a replacement were required. The heat exchangers are connected in series with a hose and a clamp fitting.

The available area for heat exchanger placement varies between mobile classrooms. Fig. 15 shows a layout for a typical classroom. This style layout was used at DDES. The piers under some mobile buildings are arranged in lines that extend across the width of the building instead of the length. In this case, the heat exchanger should also extend across the width to fully utilize the available space.

Pressure Drop

The heat exchanger was designed to have minimal pressure drop. The diameter of the headers ensured that the pressure drop in each header could be neglected during the overall pressure drop calculations. Based on 15 or 16 tubes per heat exchanger and an optimum system flow rate of 13 gpm, the velocity through each of the tubes in the heat exchangers is about 0.75 ft/s. The total pressure drop across ten heat exchangers 30-ft long is predicted to be less than 2 ft of water. By comparison, the pressure drop across the water to refrigerant heat exchanger in the heat pump is about 20 ft. A one-third horsepower centrifugal pump will provide the required flow of 13 gpm.

Material Specification

The heat exchanger header shown in Fig. 14 is made of 2.5-in. SCH-40 PVC pipe and 3-in. SCH-80 reinforcement where holes for the risers are to be drilled. The reinforcement is made of quarter sections of the 3-in. SCH-80 PVC pipe. The reinforcement was added to

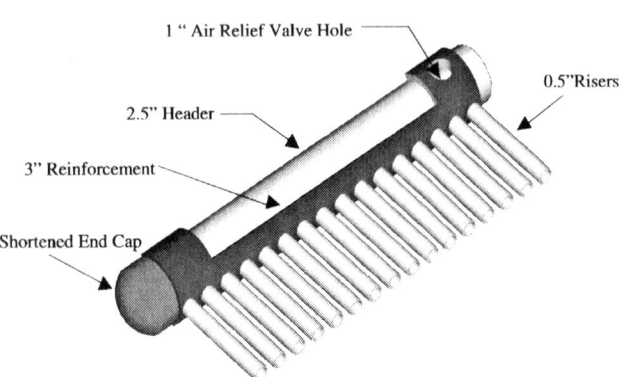

Fig. 14 PVC heat exchanger header design.

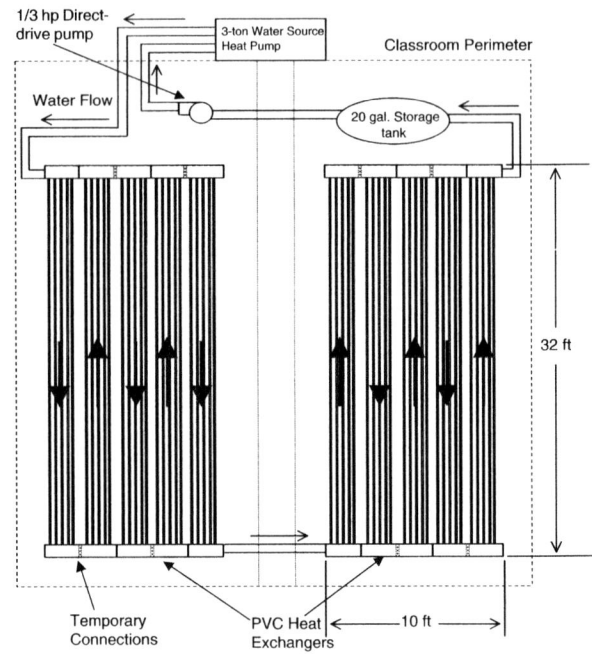

Fig. 15 Layout of PVC heat exchanger system.

increase bonding surface area between the headers and the risers. By adding the reinforcement, the wall thickness was increased by approximately 60%. The inner diameter of the 3-in. pipe is approximately equal to the outer diameter of the 2.5-in. pipe, which ensures a natural fit. This extra step significantly decreases the likelihood that the heat exchanger will leak at the riser holes.

The tubes of the heat exchanger were made from 0.5-in. SDR-13.5 PVC pipe. This thin wall pipe was selected rather than 0.5-in. SCH-40 for its lower cost, lower thermal resistance to heat transfer, and greater volume capacity. The result of using SDR-13.5 PVC instead of SCH-40 PVC yielded a cost savings of 30% and a volume capacity increase of 33%.

An additional 1-in. hole was drilled in the header assemblies so that an adapter could be installed to allow air to be purged from the system. Emersion thermocouples were installed at theses locations so that temperatures could be monitored throughout the network of heat exchangers.

The risers from each header are joined by standard SCH-40 0.5-in. couplings and standard 10 or 20 ft lengths of 0.5-in. pipe. The 20 ft lengths contain a belled end, which reduce the number of couplings required. The use of standard PVC lengths and couplings simplifies the assembly of the heat exchangers during installation, and helps to keep costs low.

The headers are connected to each other via 3-in. lengths of nitrile (Buna-N) hose and stainless steel worm gear hose clamps. Nitrile (Buna-N) hose was chosen because it will not deteriorate under the temperature range expected in the system (20°F–120°F). The hose has an "A" rating for transporting ethylene glycol (antifreeze) solutions.[7] Because the water pressure in the heat exchangers is relatively low, hose connections can be used in place of more expensive PVC unions.

Materials and Labor Cost

North Carolina State University's overall cost to manufacture and install the 1200 ft^2 surface area PVC heat exchangers in one classroom was approximately $1000. This cost represents retail material costs, and a $10/h labor rate. The cost was approximately $500 less than the uninstalled cost of two vinyl-coated polyester bladders. The less expensive and less reliable bladders made from polyethylene cost about $500 less than the PVC heat exchangers.

CONCLUSIONS AND RECOMMENDATIONS

A TRNSYS computer model of a mobile classroom with a water source heat pump was constructed to predict the effects of proposed design modifications. Comparisons of predicted performance with experimental data from a previous project were used to validate the computer model. The computer model was then used to evaluate proposed design modifications. Parametric analyses showed that solar collectors are not cost effective in this system. However, the TRNSYS model made it possible to design PVC heat exchangers to replace large bladders employed in previous designs. The heat exchangers are expected to improve energy efficiency, reduce initial costs, increase mobility (in the event the system is moved), and improve reliability.

The predicted and actual water source heat pump electrical energy use correlated as well as could be expected using actual data and TMY data. Thus, the predictions and conclusions presented in this study should provide significant insight into the actual behavior of the enhanced geothermal heat pump system. The redesigned geothermal heat pump system is currently being tested at DDES and will be monitored for a test period of 1 year.

ACKNOWLEDGMENTS

The authors wish to thank the North Carolina State Energy Office for its technical guidance and financial support. We especially acknowledge the help of our project monitor, Mr Bob Leker, who worked many hours with us to ensure that our analyses are sound, and that our objectives are in line with those of the State Energy Office.

REFERENCES

1. Forrest, A.R. *Optimization of a Geothermal Heat Pump System with Aboveground Water Storage*; North Carolina State University: Raleigh, NC, 2003; [master's thesis].
2. Rich, J.L. *Thermal Performance of a Bladder Tank Water Storage System Coupled in a Closed Loop Series Configuration with a Water Source Heat Pump*; North Carolina State University: Raleigh, NC, 2003; [master's thesis].
3. Smith, K.M. *Geothermal Heat Pumps and Modular Classroom Units*; North Carolina State University: Raleigh, NC, 1999; [master's thesis].
4. Soderberg, E.W. *Test of an Earth-Coupled Heat Pump With Aboveground Water Storage*; North Carolina State University: Raleigh, NC, 2001; [master's thesis].
5. Soderberg, E.W.; Leach, J.W. Test of an earth-coupled heat pump with aboveground water Storage. *ASHRAE Transactions Symposia*; CI-01-7-2; ASHRAE; 600–607.
6. Florida Solar Energy Center. Available at: http://www.fsec.ucf.edu/en/consumer/solar_hot_water/pools/q_and_a/#Orientation (accessed on May 2003).
7. Arimes, T. *HVAC and Chemical Resistance Handbook for the Engineer and Architect*; BCT: Lexington, KY, 1994.

Water-Augmented Gas Turbine Power Cycles

Ronald L. Klaus
VAST Power Systems, Elkhart, Indiana, U.S.A.

Abstract

Conventional gas turbine systems convert only a small fraction (25%–35%) of the available chemical energy in their fuel into useful mechanical or electrical energy. One of the most common ways to improve these power cycles is through the imaginative use of water. Such cycles fall into four families: (1) combined cycles (CCs), which convert some otherwise wasted energy into steam to drive a steam turbine, (2) cycles that inject water into the compressor train and thus reduce the work of compression, (3) cycles that humidify air before combustion in a very thermodynamically efficient manner, and (4) cycles that inject steam or water into the combustor or immediately thereafter. All of these approaches are discussed and certain performance parameters are presented.

NOMENCLATURE

R	Universal gas constant [(energy/mole * temperature)]
T_C	absolute temperature of cold reservoir (^{0}K or ^{0}R)
T_H	absolute temperature of hot reservoir (^{0}K or ^{0}R)
T_o	absolute base or environmental temperature (^{0}K or ^{0}R)
TIT	turbine inlet temperature (any temperature units)
W	work per mole (energy/mole)
γ	ratio of constant pressure to constant volume heat capacities
υ	isentropic efficiency

INTRODUCTION

Conventional gas turbines convert only a small fraction (25%–35%) of the chemical energy in a typical fuel to usable mechanical or electric power. Because of the second law of thermodynamics, not all of that energy can be converted, even with perfect machinery. However, there has been an ongoing effort to develop gas turbine systems that come closer to the maximum second-law efficiencies. One approach has been to use water in imaginative ways to boost thermodynamic efficiency. One of the most comprehensive recent surveys of the literature that is worthy of special mention was done by Maria Jonsson.[1]

Water-augmented cycles fall into four categories: (1) cycles that recover otherwise wasted energy from the exhaust stream and use it to produce steam to power a steam turbine, (2) cycles that inject water into the compressor train to reduce the work of compression and therefore make more of the expander work available for useful purposes, (3) cycles that evaporate water into the air stream, which adds mass to be expanded and also permits better recovery of wasted exhaust gas energy, and (4) cycles that recover heat from the exhaust gas and use it to produce steam or hot water that is injected back into the combustor. Combinations of these ideas have also been proposed. Besides making electric power, each of these kinds of cycles has its own particular capacity to recover heat in the form of hot water and steam, which can be used for heating purposes and also for refrigeration.

Obtaining high efficiency of the energy conversion is only one side of the design problem. In general, the more efficient the energy conversion process, the higher is its capital cost. Thus, the best design has to balance these two issues and produce a power cycle that has the greatest thermoeconomic potential. Optimum design is also affected by plant capacity. Because capital costs are proportional to plant capacity raised approximately to the 0.8 power, larger plants favor higher capital cost expenditures to obtain greater efficiency, whereas the opposite is true for smaller systems.

The economic analysis of these cycles goes beyond the scope of this survey. However, one measure of the capital cost is the specific power, defined as the net power produced divided by the total mass flow through the turbine. The higher this mass flow—and thus the smaller the specific power—the larger the major equipment that handles it and thus the larger its capital cost. Thus, specific power is a rough but useful measure of the capital cost of the power cycle.

BACKGROUND

Carnot Efficiency

One way of quantifying the second-law limitation on the conversion of energy to work is through the so-called

Keywords: Gas turbines; Wet cycles; Recuperation; Brayton cycle; Combined cycle; Compressor cooling; RWI cycle; HAT cycle; STIG cycle; Combustion; Combined heat and power.

Water-Augmented Gas Turbine Power Cycles

Table 1 Carnot efficiencies

T (°F)	Carnot efficiency (%)
1600	74.8
1700	76.0
1800	77.0
1900	78.0
2000	78.9
2100	79.7
2200	80.5
2300	81.2
2400	81.9
2500	82.5

Carnot efficiency.[2] The Carnot efficiency can be calculated to be:

$$\text{Carnot efficiency} = \frac{T_H - T_C}{T_H}$$

Table 1 shows efficiencies for the range of combustion temperatures typical in present gas turbines. These efficiencies are quite high and show that there is considerable room for improvement in efficiency over the typical 25%–35% efficiencies of conventional gas turbines. Notice also that the higher the combustion temperature, the higher the potential efficiency.

Computer Systems for Cycle Calculations

Many computer programs have been developed to analyze various power cycles.[3] Two of the most extensive commercial systems that are specially oriented toward power systems are Simtech's IPSEPro[4] and Thermoflow.[5] The simulations presented here were performed with the latter system. These and systems like them are placing increasing emphasis on economic as well as thermodynamic analysis of these cycles.

Choice of Machines for Simulation

Rather than studying specific machines, the bulk of this entry will deal with a range of "rubber machines" whose pressure ratios and turbine inlet temperatures (TITs) are taken over ranges typical for actual gas turbine systems (see Table 2, which shows ranges of TITs as a function of pressure ratio for a sample of typical gas turbines). For the purposes of this article, results will be presented for three representative machines whose pressure ratios, TITs, and power levels are also shown in that table.

Blade coolant flow is usually considered proprietary information and is difficult to estimate. This entry will use the following empirical equation for blade cooling as a percentage of the main flow entering the expander,

$$\text{Blade cooling \%} = 0.0013(\text{TIT})^{1.286}$$

where TIT is in °C. This equation is based on cooling flow data for approximately 20 gas turbine systems collected by Traverso[6] and correlated against TIT by ourselves. Values for each of the TITs used here appear in Table 3. This is only a crude way to specify blade cooling flow. More accurate blade coolant flow models can be found in Refs. 7 and 8. In addition to the major parameters discussed above, Table 4 lists some of the minor parameters that were used in the simulations.

Table 2 Pressure ratios and TIT's of machines investigated

Pressure ratio	Low TIT (°F)	High TIT (°F)	Number of machines
5	1600	1700	2
10	1700	2100	5
15	1900	2500	7
20	2100	2500	5
25	2200	2300	2
30	2100	2300	3
35	2100	2400	4
		Total number	28

Base machines		
Pressure ratio	TIT (°F)	Net power (kW)
10	1900	4600
20	2200	25500
30	2399	191000

Table 3 Blade coolant flow

Temp (°F)	Temp (°C)	Blade cooling flow (% of main input)
1,600	871	7.6198
1,700	927	8.2481
1,800	982	8.8871
1,900	1,038	9.5364
2,000	1,093	10.1955
2,100	1,149	10.8641
2,200	1,204	11.5420
2,300	1,260	12.2286
2,400	1,316	12.9239
2,500	1,371	13.6275

Table 4 Design parameters for turbine systems

Parameter	Value
Air intake	
Pressure	14.7 psia
Temperature	59°F
Relative humidity	60%
Compressor	
Isentropic efficiency	90%
Mechanical efficiency	99.8%
Inlet pressure drop	1%
Outlet pressure drop	0%
Combustor	
Pressure drop	4%
Heat loss	0.3%
Ratio of inlet fluid pressure to inlet combustor pressure	1.4
Expander	
Polytropic efficiency	85%
Mechanical efficiency	99.8%
Outlet pressure drop	0.5%

Compressor Surge

In a commercial turbine system, the compressors and expander are each designed for a particular combination of gas flow. All the water-augmented cycles affect this balance because they all add additional water to one or more of the gas streams.

This can lead to unstable operation of the compressor, called surge, which can result in catastrophic failure of the turbine system. Thus, for all the simulations of advanced cycles, an appropriate compressor has been matched to the existing expander.

THE CONVENTIONAL BRAYTON CYCLE

The Basic Cycle

Conventional cycles are Brayton cycles (Fig. 1). They consist of four major pieces of equipment: a compressor that compresses air to a high pressure; a combustor in which fuel is added and combustion takes place, producing a high-temperature gas; an expander, in which the gas is expanded to produce mechanical power; and a generator in which the mechanical power is converted to electric power. Some of the power produced in the expander powers the compressor. What remains above that is called the net power.

The advantages of this cycle are:

1. It is simple. There is no additional equipment required beyond the four basic units.
2. It is proven technology. Most of the gas turbine systems in the world are of this design.

The disadvantages are:

1. The adiabatic flame temperature of a stoichiometric mixture of fuel and air is far too high for the expander blades to withstand. Thus, the fuel is mixed with a rather large amount of excess air in the combustion chamber to reduce the TIT to a value that is consistent with the requirements of expander blade integrity (1600°F–2500°F.) Because today's compressors and expanders are quite efficient (~90%), this does not decrease the efficiency of the system greatly. Much of the power required to compress the excess air is recovered in the expander. But because this additional air has to be compressed and expanded, this does add considerably to the capital cost of the turbine system. Furthermore, even with excess air, at the high TITs of modern turbines, a certain amount of the gas produced by the compressor has to be used to cool the hot turbine blades. This results in something of a loss in turbine performance.
2. The exhaust gas exits at a high temperature; thus, a large amount of energy is wasted.
3. Pollutant gases, mainly CO and NO_x, are produced at unacceptably high levels. Additional technology needs to be used to control their levels.
4. The temperature of the gas leaving the combustor is difficult to control accurately. This causes operational problems in the control of the overall system.
5. At off-peak loads, the temperature of the gas entering the turbine drops. This cyclic temperature variation tends to increase corrosion and maintenance.
6. At off-peak electrical loads, the efficiency of the cycle also falls off quite rapidly. Because power generation systems often operate at less than peak load, this has an adverse effect on the average cost of the electricity produced.

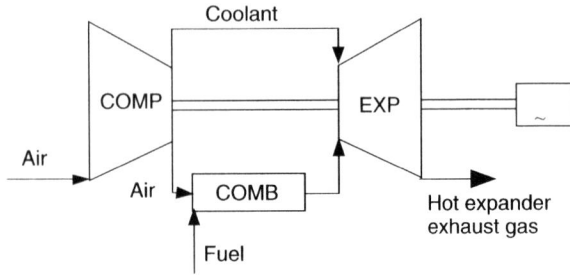

Fig. 1 The Brayton cycle.

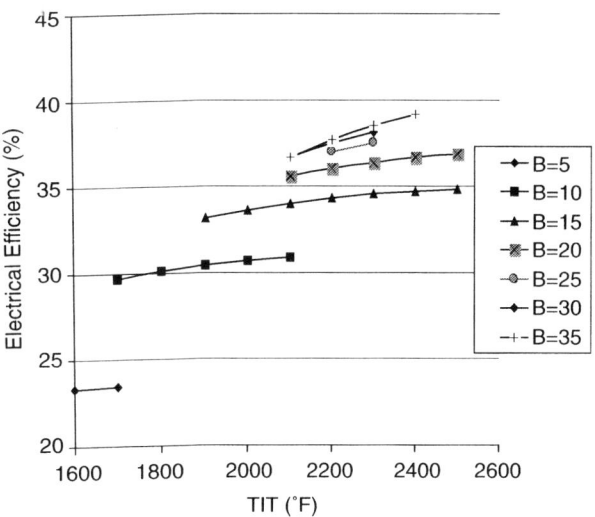

Fig. 2 Efficiencies of Brayton cycles.

The results of simulations for Brayton cycles for all the machines in Table 2 are shown in Figs. 2 and 3. Efficiency is not a strong function of TIT, but it rises as the pressure ratio increases. Specific power is a strong function of TIT, but it actually falls with the pressure ratio. The higher the TIT, the less excess air is needed for cooling, and this drives the specific power up. For larger machines, the best designs are at higher TITs and pressure ratios. For smaller machines, the capital costs are such that lower TITs and pressure ratios are more desirable.

Use of Recuperators

Whenever the expander exhaust temperature is higher than the compressor exhaust temperature, there is an

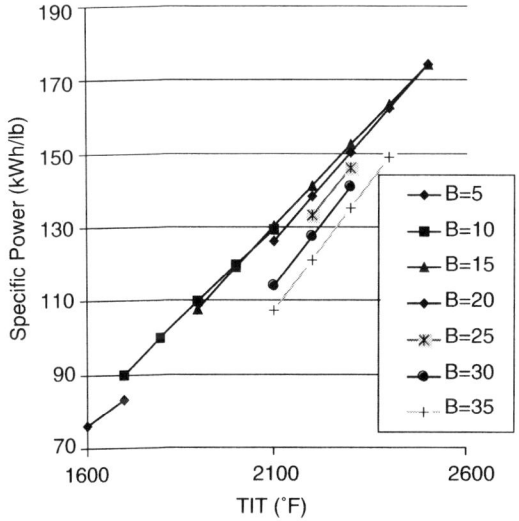

Fig. 3 Specific power of Brayton cycles.

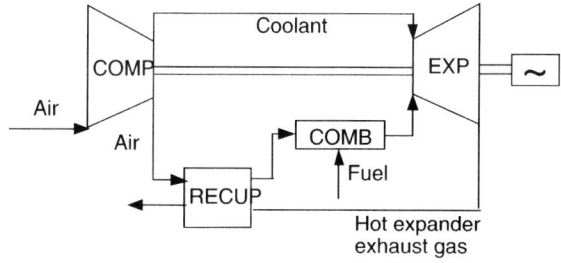

Fig. 4 Recuperated Brayton cycle.

opportunity to recover some of the exhaust heat by heat exchange with the compressed air before it enters the combustor (see Fig. 4). This has the effect of recycling some of the energy in the exhaust stream to enable it to be converted into mechanical energy. The cycles with pressure ratios of 20 and below were eligible for recuperation. In the simulations shown, it was assumed that the recuperators had 90% effectiveness. The effect of their inclusion on efficiency cycles at lower pressure ratios is dramatic (compare Figs. 5 and 2). There is a modest increase in air flow because at a higher temperature it has less cooling effect per unit mass. Thus, specific power drops somewhat (compare Figs. 6 and 3).

Recuperators are gas-to-gas heat exchangers; therefore, they require very large surface areas, and are expensive and difficult to maintain. Where they can have great impact on efficiency—for small machines at lower pressure ratios—they comprise a large fraction of these machines' capital cost. Nevertheless, wherever recuperators can be used in power cycles, and to the extent that they can be made cost effective and trouble free, they represent a good way to improve cycle efficiency. This conclusion

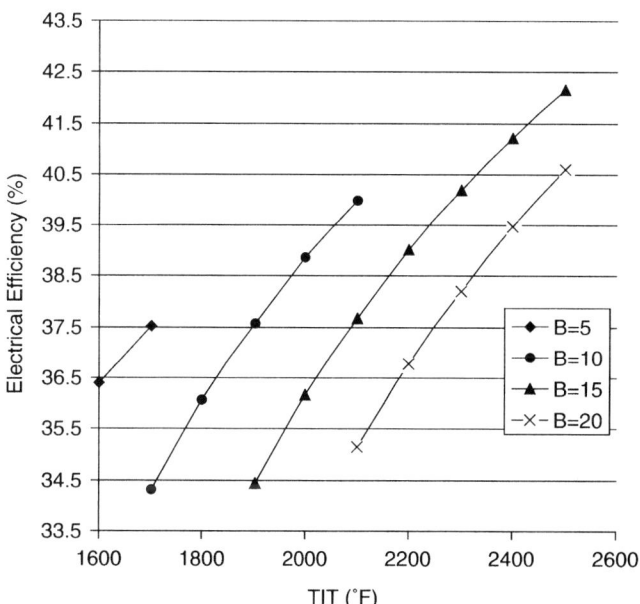

Fig. 5 Efficiencies of recuperated Brayton cycles.

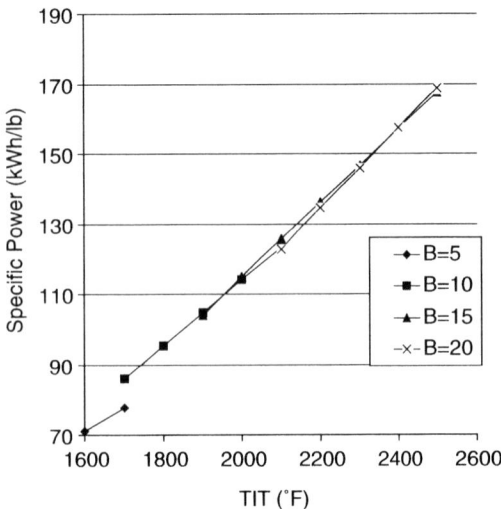

Fig. 6 Specific power of recuperated Brayton cycles.

applies not only to Brayton cycles, but also to most water-augmented cycles.

COMBINED CYCLES

The Basic Idea

The idea behind combined cycles (CCs) is to recover as much of the heat exhausted from a Brayton cycle (called the topping cycle) and use it to produce steam that is used to drive a steam turbine (the bottoming cycle). Fig. 7 shows a simple single-pressure CC. Heat is exchanged between the gas turbine exhaust and the water and steam of the steam turbine through a heat recovery steam generator (HRSG). This consists of three major units—an economizer to heat the colder water to near its boiling tempertuare, an evaporator to evaporate the water, and a superheater to heat the resulting steam to a higher temperature. In actual HRSGs, there is also a deaeration step that takes place before the water enters the evaporator, but in these conceptual simulations, this was omitted. This does not affect the thermodynamic results significantly.

On the water/steam side, cool water is heated, evaporated, and superheated and then sent to the steam turbine, where the steam is expanded to subatmospheric pressure to produce mechanical energy. The steam is then condensed, pumped up to the high pressure, and reheated.

Heat Recovery Steam Generators (HRSGs)

Heat recovery steam generators are widely used to recover heat from hot gas streams. They have an inherent inefficiency in that evaporation takes place at a constant temperature, although it is driven by the hot exhaust gas, whose entry temperature is far above the boiling temperature of the water.

Heat transfer is limited by the so-called evaporator pinch temperature, which is the difference between the gas stream leaving the evaporator and the water's boiling temperature. Water is usually fed to the evaporator at a temperature slightly below boiling to prevent premature flashing. This difference is called the approach temperature of the liquid.

Evaluation of Combined Cycles

Combined cycles (CCS) are discussed extensively in a book by Kehlhofer et al.[9] This study used the midrange of his recommended approach temperature (8°F) and evaporator pinch temperature (12°F). It also uses a superheater pinch temperature of 18°F. These temperature differences are consistently used, wherever possible, in all HRSG designs for all of the cycles.

The advantages of CCs are:

1. There is a significant increase in efficiency ($\sim 15\%$).
2. Conventional gas turbine and steam turbine equipment can be used. The two cycles are somewhat independent, linked only by the HRSG, whose technology is also well known.

The disadvantages are:

1. There is a high capital cost because of the addition of a steam turbine. For this reason, such cycles have generally been implemented for larger power plants.
2. The HRSG is only moderately efficient in recovering gas turbine exhaust heat. The exhaust gas still exits at a relatively high temperature.
3. Steam turbine efficiency drops off at smaller sizes, which compromises the overall efficiency of this cycle at these sizes.

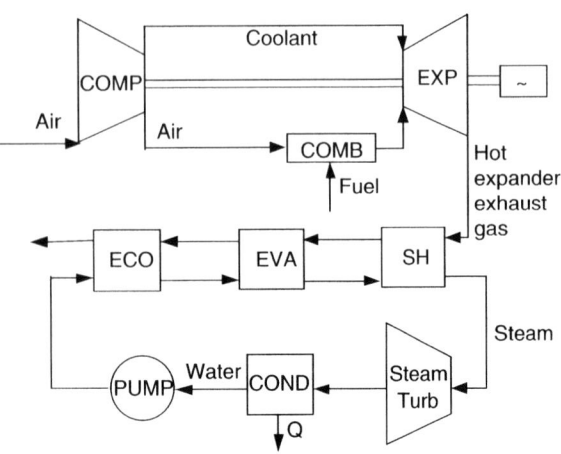

Fig. 7 Single-pressure combined cycle.

4. Disadvantages 1 and 3–6 of the Brayton cycle are unchanged.

The three base machines were simulated using a simple single-pressure steam bottoming cycle operating between two pressures suggested by Kehlhofer,[10] namely, 0.045 bar (=0.6586 psia) and 100.8 bar (=1475.3 psia). Summary results are shown in Table 5. In this table, the three base machines are identified by hyphenated numbers referring to the three most important characteristics of them three base machines, namely, base power, pressure ratio, and TIT. Because of its high efficiency, this is a widely implemented cycle. A number of variations have been developed that operate at up to three different pressures for even greater efficiencies. The high capital cost of this cycle has favored implementation in large power plants. However, the need for two separate turbine systems has motivated the search for cycles that do not require a second turbine but improve on the efficiency of the Brayton cycle. All the subsequent cycles considered here are examples of these.

WATER INJECTION INTO THE COMPRESSOR TRAIN

Quasi-Isothermal Compression

Isothermal compression up to the same pressure is known to require much less work than adiabatic compression to the same pressure. For example, for a compression ratio of 20 and $\gamma = 1.4$, which is its value for air, and v (isentropic efficiency)=0.9 the adiabatic $W/RT_0 = 5.56$, whereas the isothermal value is 3.33. (W/RT_0 is the non-dimensionalized adiabatic work.) Unfortunately, true isothermal compression is not possible to implement; therefore, two approaches, called quasi-isothermal compression, are often used. The first approach cools the gas between compression stages with conventional heat exchangers. The second approach injects water between stages. This is essentially a method of cooling because the absorption of the latent heat of evaporation of the water cools the gas.

Water-Injected Compression

Water injection into compressors is now being widely used both to improve cycle efficiency and to reduce emissions. Some early patents were obtained by Dow Chemical.[11] This idea has been incorporated into the general electric (GE) Sprint system.[12] Because water injection reduces the temperature of the compressor exhaust air, it can be combined with recuperation to produce efficient cycles. This is the idea behind the recuperated water injection (RWI) cycle developed by Rolls-Royce[13] (Fig. 8) and evaluated by Bassily.[14]

The Dutch utility company N. V. Kema took this idea one step further and obtained patents for what it calls the TOPHAT (Top Humidified Air Turbine) cycle with Swirlflash technology, which injects enough water to supersaturate the compressor air stream, thus ensuring a supply of water that evaporates continuously as the air passes through the compressor.[15]

The advantages of the RWI cycle are:

1. Significant gains in efficiency are achieved.
2. Relatively minor changes to the turbine system are necessary.

Its disadvantages are:

1. The amount of water that can be injected is limited by the amount required to saturate the air stream. Thus, the cooling effect is limited.
2. The use of a recuperator to recover some of the exhaust gas waste heat means that the temperature of the air that enters the combustor is higher than for a Brayton cycle, which requires somewhat more excess air for cooling.
3. The amount of waste heat recovered is limited by the temperature of the expanded air.
4. The water injected is lost as water vapor in the exhaust gas. This requires a continual use of treated water to be added into the cycle. There is usually not enough water injected to make it economical to recover it.
5. Although emissions are reduced through this technique, they are probably not reduced sufficiently to meet forthcoming standards.
6. Disadvantages 4, 5, and 6 of the Brayton cycle still apply.

Nevertheless, on balance and considering its relative simplicity, this is a remarkably efficient cycle, and it is surprising that it is not more widely implemented. In the simulations, it was found that water injection before the first compressor did not improve efficiency greatly; neither did the temperature of the injected water have much effect. The latent heat of evaporation dominated. In the simulated cycles, water was heated to 8°F below its boiling point before injection. It was assumed that the injected water led to 95% saturation of the air stream. Summary results are shown in Table 5.

HUMIDIFICATION OF AIR

The third major category of water-augmented cycles has as its main idea the very efficient humidification of the compressed air entering the combustor. The water used is pumped up to the system pressure and therefore does not need to be compressed as a gas. It replaces some of the

Table 5 Summary of performance for various cycles

	Brayton	Brayton w/ Recup.	CC	RWI	STIG-1:1	STIG-2.8:1	FSTIG	VAST	VASTIG
Efficiency									
46-10-19	30.46%	37.57%	44.96%	42.80%	31.44%	33.48%	40.48%	28.01%	35.26%
255-20-22	36.45%	37.21%	49.10%	46.94%	37.74%	40.32%	46.97%	36.46%	43.59%
1910-30-23	38.96%	N/A	49.55%	48.31%	40.37%	43.19%	48.85%	38.07%	46.49%
Power (kW)									
46-10-19	4,600	4,466	6,788	5,329	4,751	5,022	5,896	6,841	7,333
255-20-22	25,500	25,117	34,344	33,422	26,596	28,540	33,455	39,137	41,451
1910-30-23	191,000	N/A	242,953	276,734	200,565	217,785	252,696	305,975	326,852
Specific power (kJ/lb)									
46-10-19	109.8	104.8	146.5	132.0	114.5	123.9	157.7	205.3	228.8
255-20-22	139.9	136.1	172.3	196.6	147.6	163.0	206.4	268.8	295.7
1910-30-23	143.8	N/A	168.8	229.1	152.8	171.0	211.4	289.8	320.8
Water/fuel ratio									
46-10-19			6.63	5.82	1.00	2.80	8.97	8.94	12.74
255-20-22			5.90	6.11	1.00	2.80	7.40	7.52	10.85
1910-30-23			5.15	6.48	1.00	2.80	6.39	7.35	10.15

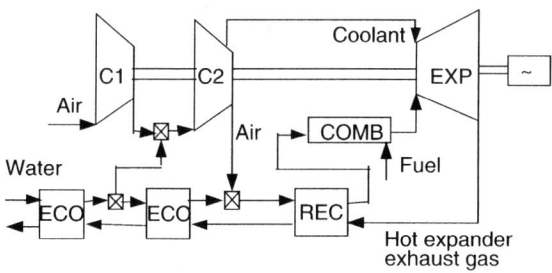

Fig. 8 RWI cycle (pumps not shown for clarity).

excess air and provides extra mass to be expanded. Furthermore, the humidification of the air—done relatively reversibly in a column—involves evaporation of water, which has a cooling effect on this water; therefore, the water can be used for interstage and aftercooling of the gas stream with surface heat exchangers.

This cycle has the most complex configuration of all the cycles considered (Fig. 9). The complexity comes from the pains that are taken to achieve thermodynamic efficiency. The distinctive feature of cycles in this family is the presence of a humidification column in which hot water flows downward, countercurrently, against rising cool air. Thus, the air is warmed and humidified. The configuration shown is a slight modification of the original humidified air turbine (HAT) cycle proposed by Ashok Rao,[16] which was an improvement over several previous schemes.

The entering air stream is both inter-and aftercooled with surface heat exchangers. The intercooling helps reduce the work of compression. It uses both recycled water, fresh feed water, and a heat sink to reduce these temperatures as much as possible. The aftercooling brings the temperature of the air down to condition it for the humidification column. Hot water is passed countercurrently to the air stream and humidifies it. The hot water also warms the air, increasing its capacity to contain water vapor. The cooled water is used for cooling of the air stream, and it is also used in an economizer to capture and recycle expander gas heat. After humidification, the air stream passes through a recuperator and then to the combustor and expander.

A Swedish consortium consisting of Alstom's small turbine group, Lund Institute of Technology, the Royal Institute of Technology, and other educational and industrial organizations did research on this and related cycles.[17] They have proposed several modifications, which they called the evaporative gas turbine (EvGT) cycles.[18] Several additional studies were done by consortiums of companies and the Electric Power Research Institute (EPRI).[19,20] More recently, Hitachi has been building a non-intercooled demonstration plant (2–3 MW) in Japan.[21]

The efficiency of the cycle depends a great deal on the match between TIT and pressure, and also on the choice of temperatures and flow rates of the many streams in the cycle. Because of that, this entry doesn't report particular numbers, but Rao claims that in some configurations 65% efficiency is achievable, and our own calculations bear this out.

The advantages of this cycle are:

1. It has a very high efficiency and is also quite cost effective.
2. The water vapor in the air stream reduces NO_x emissions.
3. It is claimed that at off-peak operation, the TIT can be maintained somewhat.
4. It is claimed that at off-peak electrical load, the efficiency of the cycle stays essentially constant down to 60% of full load and decreases only to 74% of full load efficiency at 20% of full load, whereas in a CC, it can decrease to 59%.[22]
5. Efficiency and power output aren't greatly sensitive to changes in ambient temperature.
6. The amount of makeup water required is similar to what is required for a CC that uses a wet cooling tower.
7. The quality of the makeup water need not be extremely high, because it is not injected into the turbomachinery but evaporated in a saturator column.

The disadvantages are:

1. It is a complex cycle that seems to have deterred implementation.
2. Although the humidification column is similar to devices used in the chemical processing industry,

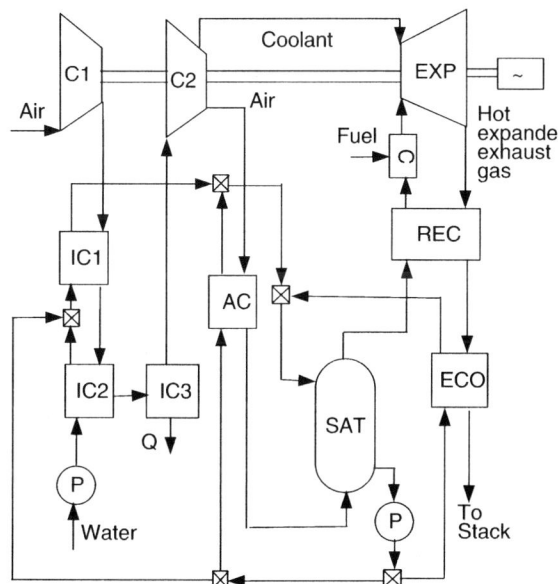

Fig. 9 HAT cycle.

it is not as familiar in the power industry. The column also increases the capital cost of the cycle.
3. Although more water can be injected than in RWI, the amount is still limited to that which saturates the air stream. Thus, excess air is still needed to control the TIT.
4. The cycle incorporates a recuperator, which is a troublesome piece of equipment.
5. The temperature of the gas leaving the combustor is difficult to control exactly.
6. Disadvantages 4 and 5 of the RWI cycle still apply.

Nevertheless, this is an excellent family of cycles—especially for power production less than about 50 MW. It appears that the cycles' complexity has been the main obstacle to their implementation.

INJECTION OF STEAM AND WATER INTO THE COMBUSTOR

STIG Cycle

The last family of cycles consists of those that inject steam or water into the combustion chamber directly or immediately thereafter. This steam/water is produced by recovering heat from the expander exhaust with an HRSG, similar to what is done in the CC. Therefore, it suffers from the same HRSG efficiency limitation as the CC. Steam injection began with GE (steam injected gas turbines—STIG)[23] and Dah Yu Cheng,[24] who is the principal of Cheng Power Systems. The Cheng cycle is essentially very similar to the STIG cycle.

International Power Technologies (IPT)[25] now holds his major patents, which are listed on the Web.[26] International Power Technologies reports 128 systems installed worldwide, mostly in Japan.

"General electric in the mid 1980s had done a detailed study for Pacific Gas and Electric (PG&E) on the STIG and in the intercooled steam injected gas turbines (ISTIG) utilizing a modified LM5000 gas turbine. The steam to fuel ratio for the ISTIG was as high as 3.76 lb/lb fuel (natural gas) while cooling the stack gas to as low as 336 F (i.e., steam generation/injection was maximized). The resulting efficiency of the ISTIG was 54% on an LHV (lower heating value) basis."[22]

The STIG system makes steam in a manner very similar to the CC but injects it directly into the combustion chamber. Fig. 10 incorporates this cycle, although the injection of liquid water shown there is not done in STIG cycles. Cheng implemented his machines using existing turbine systems, especially the Allison (Rolls-Royce) 501-KX, which had a wide surge margin. He does not inject enough steam to lead to surge. Originally, this involved a

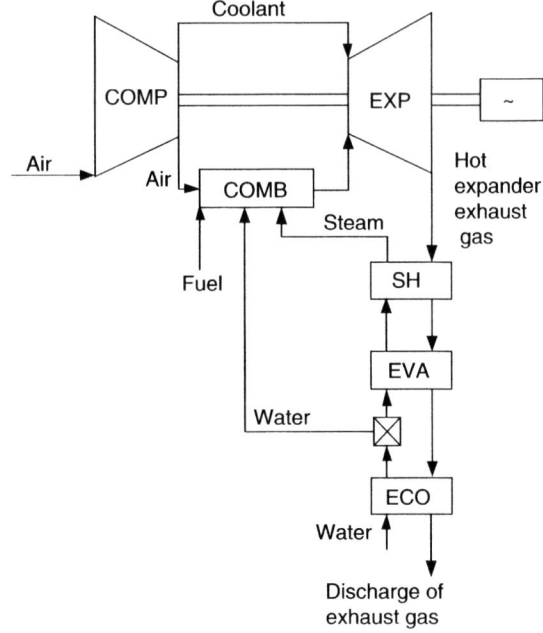

Fig. 10 VASTIG cycle.

steam/fuel ratio of 1:1, but 2.8:1 and higher is now state of the art. International Power Technologies claims that with the use of duct burners to increase the amount of steam that can be produced, it has achieved stable burning and no surge with steam/fuel ratios as high as 10:1.[27] Cheng has also developed a burner that he claims will reduce NO_x and CO emissions to below 2 parts per million (ppm). The Russian company Zorya–Mashproekt has developed what it calls the Aquarius cycle, which adds water recovery to a conventional STIG cycle.[28]

As with the CC, best results occur when steam is produced at the lowest possible pressure consistent with good injection into the combustor. Results from simulations of these systems are reported in Table 5.

The advantages of STIG cycles are:

1. The additional water mass is brought up to pressure as a liquid but is additional mass available for expansion.
2. Some of the excess air needed for cooling is replaced by water vapor.
3. Efficiency is improved.
4. Existing machines with a modified combustor can be used.
5. A burner has been redesigned, which not only permits combustion of the more humid gas, but also reduces emissions considerably.
6. Turbine inlet temperature can be maintained some degree at off-peak operation.
7. Off-peak efficiency is better than for the Brayton cycle.

The disadvantages are:

1. Excess air is still required.
2. There is still heat lost in the exhaust gas.
3. The steam injected is usually lost as steam in the flue gas. This typically requires a continual use of treated water added into the cycle.
4. Control of the steam injection is a technical challenge.

FSTIG Cycle

In the STIG cycles, the superheater pinch temperature (18°F) can be reached, but the large pinch temperature in the evaporator suggests that theoretically, quite a bit more steam or water could be injected. The latter idea was explored by various investigators.[29,30] Urbach called the point of maximum steam injection the Cheng point. We have called this cycle full steam injected gas turbines (FSTIG), and performance results are also shown in Fig. 5.

The advantages of this cycle are:

1. Even more steam is introduced, thus accentuating the first three advantages of the STIG cycle.
2. There is now enough water in the exhaust gas to allow recovery and a water-neutral cycle, if that is desirable.
3. Turbine inlet temperature can be maintained further into off-peak operation.
4. Off-peak efficiency is better than for the STIG cycle.

However, there are some disadvantages (or at least problems) to be overcome:

1. The amount of steam injected will cause some existing turbomachinery to go into surge. Thus, a different compressor may have to be matched with the rest of the cycle or else other means have to be taken to prevent surge—but see the comments above concerning the STIG cycle.
2. The high concentration of water in the combustor exceeds the conventional combustion limit. However, VAST Power Systems (Value Added Steam Technologies) has developed a proprietary method of combustion that permits stable combustion at these high water levels and also reduces emissisons to the 5 ppm level.
3. Some excess air is still required.
4. The control problem is unchanged.

This cycle has not yet been commercially implemented, although it is a promising cycle because it brings steam injection to its limit and therefore improves efficiency.

The VAST Cycle

Water injection into the combustor goes back to 1950, where it was used for thrust augmentation for jet engines.[31] An early water-injection patent was obtained by Lyle Ginter,[32] and VAST Power Systems developed a cycle based on water-only injection into the combustor.[33] In Fig. 10, this corresponds to eliminating the steam injection and the superheater. The best results are obtained when the water is taken to the highest temperature before injection, which requires that it be brought to as high a pressure as is practical. Efficiency is low for such cycles because the water injection is a highly irreversible process. However, the significant reduction in capital costs makes such cycles competitive in cost effectiveness with other cycles of higher efficiency.

Its advantages are:

1. It is possible to eliminate virtually all excess air. This reduces the capital cost of the compressor and expander considerably.
2. Advantages 2–4 of the FSTIG cycle are maintained.
3. Water injection is easier to control than steam injection. Thus, the combustor temperature control point can be set closer to the desired TIT without as much concern for temperature overshoot.
4. With the addition of VAST combustion technology, stable burning can be achieved that reduces emissions to a very low level.
5. The injection of water tends to dampen acoustic vibrations.

The disadvantages are:

1. There is low efficiency.
2. Disadvantages 1 and 2 of the FSTIG cycle are retained.

The VASTIG Cycle

An obvious extension of the two previous cycles is to combine them as shown in Fig. 10. The name, VASTIG, indicates that this cycle is a combination of the VAST and STIG cycles. This idea was envisioned by Guillet[29] and Urbach.[30] As much steam as possible is injected, as in FSTIG. However, the exhaust gas still has heat that can be used to heat additional water in the economizer to produce enough water to replace virtually all the excess air. In many ways, this cycle combines the best of the two previous cycles. Results of the simulation of this cycle are shown in Table 5.

This cycle's advantages are the same as for the VAST cycle, with the additional advantage of higher efficiency

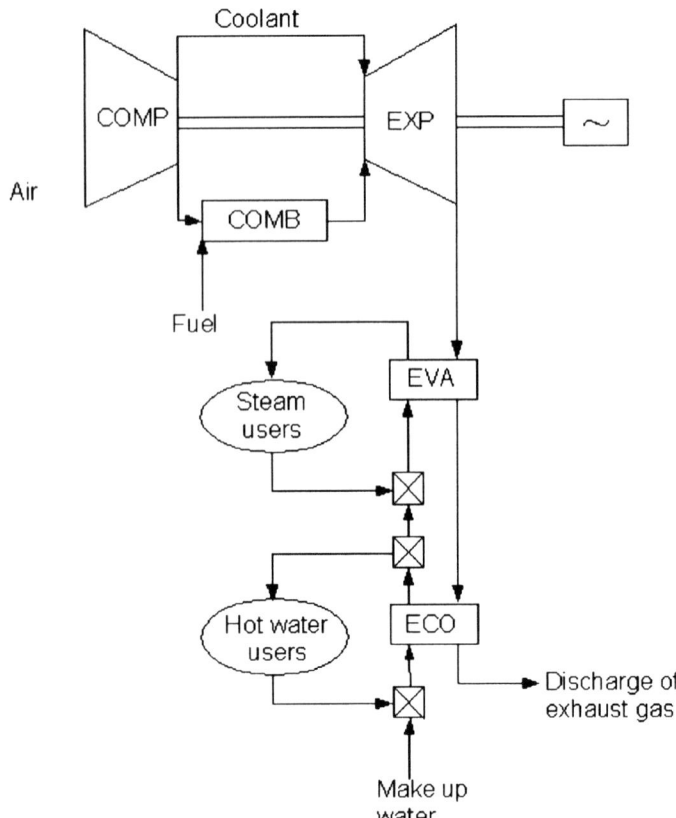

Fig. 11 CHP with Brayton cycle.

and better cost effectiveness. The disadvantages are the same as the first two for the FSTIG cycle.

It was thought that combining the VASTIG cycle with compressor water injection might lead to a very efficient cycle, but the efficiency of such a cycle was only marginally better than that of VASTIG alone.

COMBINED HEAT AND POWER

An idea that is coming into its own is that of producing both electric (or mechanical) power and heat (or refrigeration) with the same apparatus. This is called combined heat and power (CHP) or tri-generation when some of the energy is used for heating and some for chilling or refrigeration. Even the best gas-turbine power cycles recover only part of the chemical energy in the fuel. Some of the rest, which would otherwise be wasted, can be recovered as heat.

One straightforward and state-of-the-art method of doing this is illustrated in Fig. 11. The well-known Brayton cycle is equipped with an HRSG to recover heat from the exhaust gas in the form of steam or hot water. Table 6 shows how much heat can be recovered both in the form of steam and hot water. Such a cycle has a very high overall efficiency when one counts not only the electrical energy, but also the heat energy produced. A great deal of this energy is in the form of less desirable hot water rather than steam. However, there are refrigeration systems than

Table 6 Summary of performance of CHP cycles (pressure = 209.8 psia)

	Conventional Brayton	STIG 1:1	STIG 2.8:1
Electrical efficiency	30.36%	31.43%	33.46%
CHP efficiency	96.37%	95.88%	96.67%
Net power (kW)	4597	4749	5019
Steam heat (Btu/s)	6026	5142	3629
Steam temp. (F)	386	386	386
Steam heat/ total energy (%)	41.98%	35.91%	25.54%
Hot water heat (BTU/s)	3449	4086	5354
Hot water temp. (F)	262	190	171

(Continued)

Table 6 (Continued)

Water heat/total energy (%)	24.03%	28.54%	37.67%
Total heat (Btu/s)	9475	9228	8983
Total heat/total energy (%)	66.01%	64.45%	63.21%

can be driven by hot water; therefore, this is a possible application.

Other cycles can be used for CHP with varying effectiveness. Table 6 also gives results for two STIG cycles. There are increases in electrical efficiency and power production. (Fig. 12)

CONCLUSIONS

Many cycles have been proposed that use water in imaginative ways to improve the efficiency and cost-effectiveness of power-generating cycles. Furthermore, some of these can be used with advantage in CHP systems as well. Cycles like these are only in the beginning stages of implementation, possibly facing the same kind of resistance that many new technologies face. However, world energy considerations and the need to reduce CO_2 emissions are likely to bring increasing pressure on utilities to generate power more efficiently. In the long run, more efficient systems will likely be based on one or more of the water-augmented systems that are just beginning to be implemented or are still on the drawing board.

ACKNOWLEDGMENTS

Helpful comments were made by Dr. David Hagen, VAST Power Systems, and Dr. Ashok Rao, University of California, Irvine. Allan McGuire assisted with some of the simulations. Support of this work by VAST Power Systems is gratefully acknowledged.

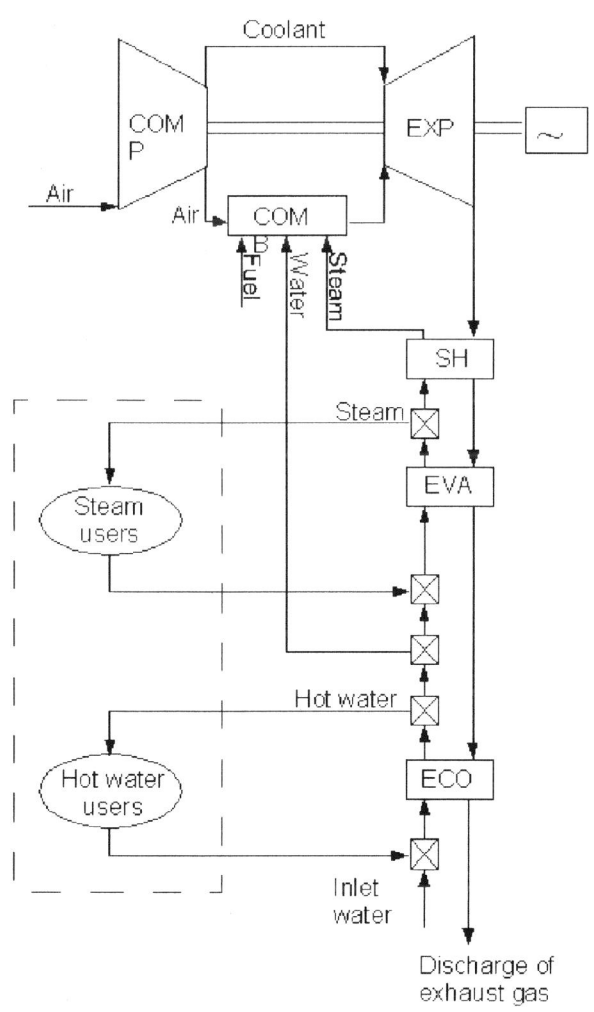

Fig. 12 CHP with VASTIG cycle.

REFERENCES

1. Jonsson, M. *Advanced Power Cycles with Mixtures as the Working Fluid*, Doctoral thesis; Royal Institute of Technology, Dept. of Chemical Engineering and Technology, Division of Energy Processes: Stockholm, Sweden, 2003.
2. Klaus, R.L. Thermodynamics. In *Encyclopedia of Energy Engineering and Technology*; Capehart, B.L., Ed.; Taylor and Francis: NY, 2005.
3. http://www.technology.novem.nl/en/processtools/ (accessed July 2005).
4. http://www.simtechnology.com/IPSEpro/english/IPSEpro.php (accessed July 2005).
5. http://www.thermoflow.com/index.htm (accessed July 2005).
6. Traverso, A.; Massardo, A.F.; et al. *WIDGET-TEMP: A Novel Web-Based Approach for Thermoeconomic Analysis and Optimization of Conventional and Innovative Cycles*, ASME Paper GT2004-54115, Proceedings of ASME Turbo Expo 2004, Vienna, Austria, June 17, 2004.
7. Jordal, K. *Modeling and Performance of Gas Turbine Cycles with Various Means of Blade Cooling*, Doctoral thesis; Division of Thermoal Power Engineering, Dept. of Heat and Power Engineering, Lund Univ.: Sweden, 2001.
8. Torbidoni, L.; Horlock, J.H. *A New Method to Calculate the Coolant Requirements of a High Temerature Gas Turbine Blade*, Proceedings GT2004-53729, ASME Turbo Expo 2004, Vienna, Austria, June 14–17, 2004.
9. Kehlhofer, R.H.; Warner, J.; Nielsin, H.; Bachmann, R. *Combined-Cycle Gas Steam Turbine Power Plants*, 2nd Ed.; PennWell: Tulsa, OK, 1999.

10. Kehlhofer, R.H.; Warner, J.; Nielsin, H.; Bachmann, R. *Combined-Cycle Gas Steam Turbine Power Plants,* 2nd Ed.; Penn Well: Tulsa, OK, 1999; 51–53.
11. Zachary, R.E.; Hudson, R.D. Method and Apparatus for Achieving Power Augmentation in Gas Turbines Via Wet Compression. U.S. Patents 5,867,977, Feb. 9, 1999 and 5,930,990, Aug. 3, 1999, assigned to The Dow Chemical Company.
12. http://www.gepower.com/prod_serv/products/aero_turbines/en/lm6000_sprint.htm (accessed August 2005).
13. Reynolds, G.A. Power Generation Equipment. U.S. Patent 6,293,086, September 25, 2001, assigned to Rolls-Royce, London.
14. Bassily, A.M. Performance improvements of the recuperated gas turbine cycle using absorption inlet cooling and evaporative aftercooling. Proc. Inst. Mech. Eng., Part A: J. Power Energy **2002**, *216* (4), 295–306.
15. http://www.alphapowersystems.nl/Duesseldorf.pdf (accessed August 2005).
16. Rao, A. Process for Producing Power. U.S. Patent 4,829,763, May 16, 1989, assigned to Fluor Corp., now Fluor Daniels.
17. Lindquist, T. *Evaluation, Experience and Potential of Gas Turbine Based Cycles with Humidification,* Doctoral Thesis; Thermal Power Engineering, Lund Inst. Tech.: Sweden, 2002, http://www.lub.lu.se/cgi-bin/show_diss.pl/tec_560.html (accessed August 2005).
18. Bartlett, M. *Developing Humidified Gas Turbine Cycles,* Doctoral Thesis; Dept. of Chemical Engineering and Technology, Division of Energy Processes, Royal Institute of Technology: Stockholm, 2002.
19. Rao, A.D.; Francuz, V.J.; Shen, L.C.; West, E.W. *A Comparison of Humid Air Turbine (HAT) Cycle and Combined Cycle Power Plants,* EPRI Report EI-7300, March 1991.
20. Rao, A.D.; Francuz, V.J.; Mulato, F.J.; et al. *A Feasibility and Assessment Study for FT4000 Humid Air Turbine (HAT),* EPRI Report TR-102156, Sept. 1993.
21. Hatamiya, S.; Araki, H.; Higuchi, S. *An Evaluation of Advanced Humid Air Turbine System with Water Recovery,* Paper GT2004-54031, Turbo 2004, Vienna.
22. Rao, A. Private communication.
23. http://www.gepower.com/prod_serv/products/aero_turbines/en/ (accessed August 2005); lm2500_stig.htm (accessed August 2005).
24. http://www.chengpower.com/ (accessed August 2005). See also Cheng, D.T.; Nelson, A.L.C. *The Chronological Development of the Cheng Cycle Steam Injected Gas Turbine During the past 25 Years,* Paper GT-2002-30119 at ASME Turbo Expo 2002, Amsterdam. A copy of this paper may be found at :http://www.aem.no/download/THE%20CHRONOLOGICAL%20DEVELOPMENT%20OF%20THE%20CHENG%20CYCLE%20STEAM%20.pdf (accessed August 2005).
25. http://www.intpower.com/ (accessed August 2005).
26. http://www.intpower.com/technology/patents.htm. (accessed August 2005).
27. Turley, R. *International Power Technology,* Private communication.
28. See http://www.mashproekt.nikolaev.ua/ and http://www.zorya.com.ua/Eng/Products/IndustrialEquipment/AquariusEng.htm (accessed August 2005).
29. Guillet, R. Natural Gas Stream Turbine System Operating with a Semi-Open Cycle. U.S. Patent 5,271,215, December 21, 1993.
30. Urbach, H.R.; Quandt, E.R. Direct Open Loop Rankine Engine System and Method of Operating Same. U.S. Patent 4.509,324, April 9, 1985.
31. Lundin, B.L. *Theoretical Analysis of Various Thrust-Augmentation Cycles for Turbojet Engines,* NASA Report TR-981; 1950.
32. Ginter, L. Engine System and Thermogenerator Therefore. U.S. Patent 3,651,641, March 28, 1972.
33. Ginter, L. Vapor-Air Steam Engine. U.S. Patents 5,617,719, April 8, 1997 and 5,743,080, April 28, 1998.

Water-Using Equipment: Commercial and Industrial

Kate McMordie Stoughton
Battelle Pacific Northwest National Laboratory, U.S. Department of Energy, Richland, Washington, U.S.A.

Amy Solana
Pacific Northwest National Laboratory, Portland, Oregon, U.S.A.

Abstract

Water is an important aspect of many facets in energy engineering. The companion article of this entry, "Domestic Water-Using Equipment," detailed domestic related water-using equipment such as toilets and showerheads; this entry focuses on various types of water-using equipment in commercial and industrial facilities, including commercial dishwashers and clothes washers, single-pass cooling equipment, boilers and steam generators, cooling towers, and landscape irrigation. Opportunities for water and energy conservation are explained, including both technology retrofits and operation and maintenance changes. Water management planning and leak detection are also included because they are essential to a successful water management program.

INTRODUCTION

In the study of energy engineering, the importance of water management is often overlooked. But water is closely linked to energy. For commercial and industrial facilities, there are many pieces of equipment that utilize both water and energy to perform processes. For example, energy is required to circulate water through systems like cooling towers and to heat water for steam systems and boilers. In the development of a comprehensive energy management plan, it is important to be fluent in water management as well. Efficiency improvements in these types of systems typically work synergistically.

Common commercial and industrial water-using equipment are discussed below, with options for efficiency improvements and guidelines for estimating water consumption when not metered. In addition, leak detection and water management planning are discussed, and references are provided for further information.

DISHWASHERS

Dishwashing uses almost two-thirds of the water consumption in typical food preparation facilities such as restaurants.[1] Washing steps often include a scrapping trough system; a prewash; and a machine that cleans, rinses, and sanitizes. The scrapping trough system pushes garbage off the plates and into the disposer, using 3.0–5.0 gpm. The prewash removes anything left on the plates, using 3.0–6.0 gpm for both automatic and hand sprayer types. The final cleaning step uses 2.8–8.0 gpm.[2]

Retrofit options in the commercial setting are varied because of the multiple steps involved. The scrapping trough system could be used without water or replaced with a conveyor system, because food waste does not need to be disposed of in the sewer system. Prewash systems can be retrofitted with low-flow, high-pressure spray heads and flow reduction valves. A more efficient prewash is a manual high-pressure spray nozzle, which shuts off the flow when the user lets go and typically reduces water use to 1.8–2.5 gpm.[2] The most efficient models reduce flow to 1.6 gpm or less.[1]

A sensor could be installed on the dishwashing machine to ensure water flow only when dishes are passing through, instead of constantly or whenever the belt is in motion, empty or full. Low-temperature dishwashers are available that rely more on chemical agents instead of large volumes of hot water.[3]

Water recycling is another efficiency improvement. Rinse water could be reused for prewash or the garbage disposer. Dishwater could be reused in the garbage disposer. Water from final rinse steps could be reused in initial rinse or wash steps.

Auditing

To estimate the amount of water used in commercial dishwashing, information regarding flow rates of each component and frequency of use are needed. Flow rates of prewashers and sprayer nozzles may be measured as with faucets (see companion article, "Domestic Water-Using Equipment"), and information detailing the wash times and number of wash cycles of all equipment can be gathered. Use patterns need to be collected via interviews,

Keywords: Water; Dishwasher; Clothes washer; Single-pass cooling; Steam boiler; Cooling tower; Irrigation; Audit; Conservation.

Fig. 1 Large commercial clothes washer.

including number of wash cycles per day and operating days per month.

CLOTHES WASHERS

Large commercial clothes washers (as shown in Fig. 1), on average, use between 2.5 and 3.5 gallons of water per dry pound of laundry.[2] A continuous batch (tunnel) washer reuses rinse water, which saves about 60% of the water typically used per load. Ozonated laundering systems use ozone to kill bacteria and disinfect fabric. They require no rinsing, hot water, or detergent, and are ideal for the sterilization of hospital gowns (for example), but are not suited for heavily soiled laundry (unless detergent is added).[4]

Large clothes-washing systems may be adapted to recycle rinse water for the next cycle's wash water, resulting in 25% or more water savings. Waste wash and rinse water can also be treated and then reused, allowing up to 50% water savings. Chemicals may be changed in the washing process to reduce the number of wash and rinse steps. Using only the necessary number of steps will help reduce water and energy use.[2] Find more information on laundry water recycling systems in the resource section of this article.

In any laundry facility or machine, washing only full loads will reduce overall energy and water use. Using cold water whenever possible saves energy by saving heated water.

Auditing

To determine laundry water use, machine frequency of use must be known. Industrial and some commercial laundry facilities usually operate on a regular schedule.

Once the number of loads per year is determined, the amount of water used per load can be estimated or found. Some washers have this value written on the machine. Others, usually in large commercial or industrial facilities, give the capacity in pounds of clothes. For machines with little or no information, manufacturer and model information can be used to obtain water use data. Incomplete data can be combined with typical consumption values based on the type of washer, as previously noted.

SINGLE-PASS COOLING EQUIPMENT

Single-pass cooling (also referred to as once-through or noncontact cooling) is a water-intensive cooling process that passes water once through equipment to remove heat before discharging it. Many different types of equipment may utilize single-pass cooling, including x-ray machines, hydraulic equipment, condensers, air compressors, welding machines, ice machines, air conditioners, and degreasers. These systems often run 24 h a day, 7 days a week, wasting thousands of gallons of water. A mere 2-gpm cooling rate would consume over 20,000 gallons per week alone.

To greatly increase efficiency, retrofit the single-pass cooling system to recirculate the water in a closed-loop system. If a closed-loop retrofit is not feasible, equipment should be operated at peak efficiency. This includes ensuring that the cooling process and water flow are turned off when not needed, accomplished either manually or through an automatic timer. Also, set the flow rate to the minimum allowed by the manufacturer's specifications if possible, and make sure the entering and leaving temperatures are in the specified range to ensure adequate cooling is taking place.

Routinely cleaning the heat exchange coils will maximize heat exchange, saving energy, and water. Consider using the effluent for cooling tower or boiler make-up water, landscape irrigation, or another nonpotable use. Another option is to replace the equipment with an air-cooled system to eliminate all water consumption.

Auditing

Information to gather that will help in the estimation of water and energy consumption includes the brand and model number, equipment time of use, entering and leaving temperatures, and flow rate. With this information, the manufacturer typically can provide the general range of cooling-water flow rates and energy requirements. The bucket-and-stopwatch method can also be used to estimate the flow rate of the cooling water (as explained in the faucet section of the companion article, "Domestic Water-Using Equipment"). Water consumption can be estimated by multiplying the operating hours by the flow rate. Energy consumption can be estimated by multiplying the gallons of water consumed by the energy required to cool 1 gallon to the working temperature, and dividing by the cooling equipment efficiency.

BOILERS AND STEAM SYSTEMS

Boilers and steam generators are addressed elsewhere in this book, but are discussed here because of their significant water use and conservation opportunities. These systems are typically used for process steam requirements, large heating systems, and cooking. Because the water is brought to such a high temperature in these systems, energy loss can be substantial in poorly operated equipment.

The largest water and energy losses are seen in systems without condensate return. In these systems, the condensed steam is sent down the drain after use instead of being recirculated. A continuous flow of fresh makeup water is required, and the system does not take advantage of any residual heat from the condensed steam, resulting in significant water and energy waste—up to 70% of operating costs.[2] Existing boilers and generators can be modified to return steam condensate if they do not already do so. Additional energy can be saved by insulating the condensate return pipes.

Systems with condensate return can still achieve efficiency improvements through proper operation and maintenance (O&M) of steam traps, lines, and blowdown. Steam traps can become stuck open, allowing steam to escape continuously instead of only when necessary as pressure builds. A maintenance program that includes regular steam trap checks and quick repairs helps to reduce these losses. Line leaks can also be moderated in this way.

Blowdown is necessary to control scale buildup inside the system. Phosphate-based chemicals and non-chemical methods are often used to reduce scale growth, but they cannot prevent it. Therefore, some water must be expelled and replaced periodically with make-up water. An automatic blowdown control manages this process to optimize water use, water quality, and boiler conditions.

Tracking the amount of make-up water a boiler consumes can help identify leaks in the system. This is true for systems with or without condensate return. Water meter data can show abnormal increases or decreases in consumption, indicating potential leaks. An increase may signify a leak in steam or condensate lines, while a decrease may signify a leak in a shell-and-tube heat exchanger. The latter type of leak is more problematic, because it means untreated water is entering the system, which can cause a boiler to fail.

Properly sized systems also can reduce water requirements, as can shutting down the system when not in use or during unoccupied hours. System shutdown can be manual or controlled automatically. Simply using the minimum amount of steam or heat required can greatly reduce the amount of water and energy used.

Auditing

Determining the amount of water used by a steam system can be difficult because it requires detailed system data; however, if a water meter is present on the make-up water line, the metered data can provide the consumption instead. Water and energy consumption can be calculated using some or all of the following information, depending on what is available: system capacity and time of use, steam pressures and temperatures, and fuel type and consumption. The operator and operating logs are the best source for most of this data, as well as manufacturer information found on the system and in manuals. Systems with condensate return use different calculations than systems without. Water lost through steam traps, leaks, and blowdown also needs to be identified and included.

COOLING TOWERS

Cooling towers (as shown in Fig. 2) are also discussed elsewhere in this book, but are briefly addressed here because of their considerable impact on water consumption. Because they provide cooling by evaporating water, conservation opportunities are minimal; however, some key O&M techniques can improve efficiency.

Cooling towers evaporate about 3 gpm through evaporation, drift, and blowdown (sometimes called bleed-off) for every 100 tons of cooling, with some variation by climate.[5] Reducing blowdown is the primary way to save water. Additionally, other uses may be found for blowdown water, and reclaimed water from single-pass cooling, reverse-osmosis systems, or other facility processes can be used as make-up water.

A high concentration ratio minimizes blowdown. The concentration ratio, or cycles of concentration, indicates how many times water circulates before being discharged. It is the ratio of the concentration of dissolved solids in the

Fig. 2 Cooling tower.

bleed-off water to the concentration of dissolved solids in the make-up water. The concentration ratio can also be calculated with the volume of the make-up and bleed-off water, if metered.

To increase the concentration ratio and save water, maintain water quality through chemical or other treatment methods:

- Shade cooling towers from sunlight by installing covers, which can significantly reduce biological growth, such as algae.
- Ultraviolet light added through an intense UV lighting module kills microorganisms that can build up and lead to fouling or even Legionnaire's disease.
- Sulfuric (or other) acid treatment lowers the pH of the water, which controls scale buildup created from mineral deposits.
- Sidestream filtration filters a portion of the flow to remove sediment and other impurities.
- Ozonation is a powerful oxidizer that controls scale, corrosion, and biological growth.
- Automatic chemical feed enables larger systems to automatically add chemicals to the system based on the conductivity, which optimally controls corrosion, scale, and biological growth.

Auditing

Metering is the best option to determine cooling tower water consumption. Meters that display total amount of water being used as well as current rate of flow are most useful. Measuring the water conductivity with a conductivity sensor is another way to understand the system. Conductivity measures how well electricity travels through the water, which is a function of the total ionic dissolved solids in the water. The ratio of make-up water conductivity to bleed-off water conductivity should be about the same as the ratio of make-up to bleed-off flow.[5]

LANDSCAPE IRRIGATION

Landscape irrigation can be a significant water consumer at many types of facilities. The amount used depends on the climate, water rates, type of landscape plants and plant health, type of irrigation equipment, and O&M practices. If nonpotable water cannot be used for irrigation, efficiency is essential.

The place to start improving efficiency is the landscape itself. Native turf and plants (an example of native southwestern landscape is shown in Fig. 3) have water and soil requirements that match the local average rainfall and soil type. Using these types of plants can greatly reduce or eliminate irrigation needs.

Healthy plants allow more efficient irrigation because the roots tend to be stronger and the plants tend to be drought and disease resistant. Frequent, shallow watering can result in compacted soil, shallow root systems, and increased disease incidence. Deep, occasional watering promotes deep, healthy roots. For turf, mulching mowers and high

Fig. 3 Native landscape in a dry climate.

blade settings leave a layer of protection to shade root systems, encourage deeper roots, and reduce evaporation.

Soil condition also plays an important role in plant health. Depending on the soil composition and plant type, organic matter can be added to the soil to help retain moisture and to ensure proper nutrients for the plant. Aeration loosens compacted soil, allowing more air and water to reach the roots; ideally, it should be performed at least twice per growing season (as shown in Fig. 4).

Optimized irrigation provides the landscape with the amount of water it actually needs. The evapotranspiration (ET) rate, (ET, the loss of water by evaporation and transpiration, indicates how much water is needed by the plant for the current conditions based on plant type, soil type, and season. Local county extension offices or local university agricultural departments often have information on how to estimate ET rates.) typically provided in inches of water, is compared to the area's rainfall and soil type to

Fig. 4 Turf aeration.

find the amount of irrigation needed. The soil type determines how much moisture can be held at the plant's root base (sandy soil will retain much less water than clay soil) and what types of nutrients might be deficient. Divide the ET by the irrigation equipment's precipitation rate (PR), or flow rate, to calculate the watering time needed. Precipitation rate is typically provided in inches per hour, and can be obtained from the manufacturer.

Appropriate irrigation scheduling can then provide the landscape with the required amount of water. Irrigation scheduled in the early morning helps to reduce evaporation and minimize fungus growth. This time period is typically less windy as well, helping to avoid overspray and evaporation. Overspray to sidewalks and pavement can also be avoided by placing sprinkler heads only where they will evenly distribute water over the desired area.

Irrigation controllers function automatically and allow flexibility in scheduling. Automatic timers are helpful for watering during inconvenient times of day, and for preventing overwatering because of negligence of sprinkler shutoff. A manual override option is important for factors where irrigation is unnecessary, such as rainfall.

Other technologies, including rain, wind, freeze, and humidity sensors, respond to weather conditions and can be connected to the existing control system. An ET controller is a more robust option, which can be programmed to match the plant's exact water needs with the PR of the irrigation system and the local weather conditions.

Efficient irrigation equipment applies water directly to the plants' root base and avoids overspray and evaporation. Low-volume drip systems apply water at a low pressure to nonturf landscape such as flowerbeds and shrubs. Subsurface drip irrigation applies water directly to roots of turf or other plants through a perforated piping system buried underground.

Proper maintenance of irrigation equipment can also save water. System checks for broken heads and line leaks at the beginning and end of each watering season are crucial. Routine spot checks of the irrigation system are also important, especially if landscape is watered early in the morning when broken heads and leaks can go unnoticed. The system's pressure should meet the range of pressures specified by the manufacturer. If the pressure is too high, a pressure-reducing valve might be required, which also needs to be checked periodically for leaks.

Auditing

Metering is the best approach to measuring irrigation water consumption. This way, the consumption can be tracked and monitored to ensure the irrigation needs are as expected. If metered data are not available, system operators should be able to estimate how long the irrigation system runs each day, week, and year. If the system flow rate is unknown, it may be found in manufacturer data. Alternatively, consumption can be estimated using the following typical values:

- Water-thirsty landscape: 25 gallons per square foot per year.[2]
- Water-wise landscape: 5–10 gallons per square foot per year.[2]

LEAK DETECTION

Distribution system leaks can be a large source of water loss, especially in aging infrastructure. Leaks typically occur at joints and connections such as meter boxes, hydrants, and valves; resulting from loose connections, corrosion, improper piping installation, settlement, or overloading. Leak rates have a wide range; a large main break can lose water at 1000 gpm, while a slow service line leak can be as small as 0.5 gpm.[6] These small leaks will most likely grow over time, especially if they are caused by corrosion.

A comprehensive leak-detection program is the best approach to limiting system losses. A good program includes the following steps:

- Estimate the total loss rate through a distribution system audit.
- Identify the areas of highest leak rate.
- Inspect and record the location and condition of service lines, valves, meters, and other connections.
- Prepare a detection-and-repair plan that schedules leak inspections and repairs and analyzes economics.
- Implement the plan and revise when needed.

A simplified technique for finding leaks is called the zone-flow procedure, which examines water use in specific areas of the system.[6] Zones of the distribution system are individually isolated by closing the valves in that area, and flow is recorded for about 24 hour. If flow exists during the nighttime hours but there are no major nighttime water-using processes, leaks are most likely present in that zone.

Leaks are most often pinpointed by using electronic listening devices, which detect the sound of the pressurized water being forced through the pipe wall.[6] Typically, a trained operator is contracted to perform leak-detection services, because it requires specific instrumentation and knowledge of how to determine the exact location of the leak.

Thermal imaging can help find leaks in large distribution lines (16 in. or greater) using multispectral and thermal aerial photography. Areas of cooler temperatures reveal possible water leaks because the saturated area will be cooler than the surrounding soil. This technique works best during warmer seasons and in arid areas for two reasons: (1) the leak is easy to pinpoint because vegetation grows at the point of the leak; and (2) the water table is typically deeper, so the surrounding soil temperatures are warmer than the area of leakage.

WATER MANAGEMENT PLANNING

Water management planning is at the heart of all water efficiency improvements, and sets water conservation as a priority alongside energy efficiency. Ideally, a comprehensive water management plan uncovers the entire cycle of water use at a facility from supply through discharge (as shown in Fig. 5 below).

A tiered approach can help develop a comprehensive water management plan, as described in the following steps:

1. Assess current conditions:
   - Understand and apply water-related regulations and laws.
   - Work with stakeholders to understand issues revolving around different aspects of water management.
   - Gather comprehensive utility information, including rates and at least one year's bills. Include information on rate structure and future water rates.
   - Perform audits to quantify water consumed by equipment, other end-uses, and leaks or losses. Use auditing techniques (described in both water-related articles) to help estimate water consumption at the equipment level.
   - Document current O&M practices and schedules.
   - Categorize and quantify water consumption by end use to help prioritize efforts.
2. Improve operations and increase efficiency:
   - Identify potential operational efficiency improvements and technology retrofits for each major water-using equipment and process.
   - Perform economic analyses to identify feasible projects.
   - Prioritize efficiency improvements based on largest water user and best economic feasibility.
   - Set goals and a schedule for project completion.
   - Ascertain funding to implement projects.
   - Share success.
3. Plan for contingencies:
   - Assess current contingency plans for emergencies such as short term outages and drought.
   - Develop a proactive plan that prepares and develops strategies for contingencies. This includes assigning specific tasks to teams that are responsible for monitoring and mitigating water shortages and outages.

Understanding current consumption patterns is a crucial element to successful water management. Submetering buildings and end uses is the best and most accurate method to allocate water use. When submetering is not feasible, manual measurements or engineering estimates can be used as an alternative.

When deciding which conservation opportunities to pursue, a good resource is the U.S. Department of Energy Federal Energy Management Program's (FEMP) "Water Efficiency Best Management Practices" (BMPs), which are recommendations for efficiency improvements in ten

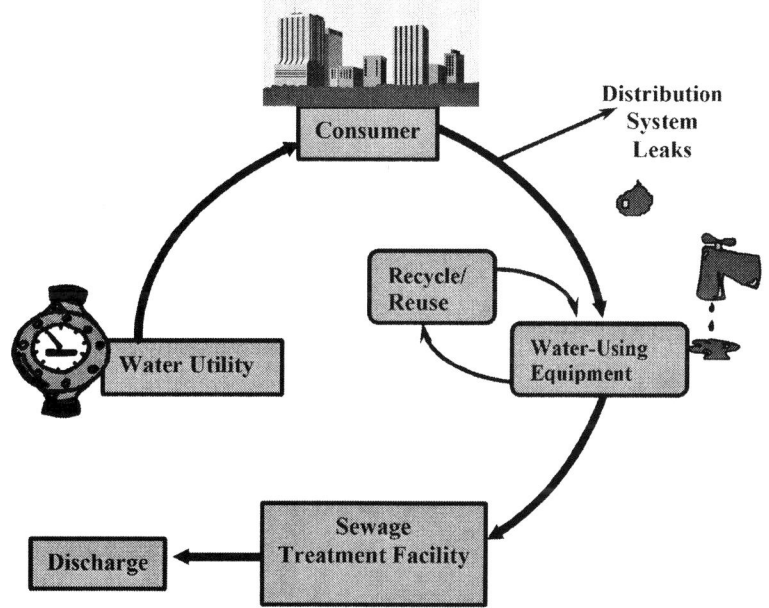

Fig. 5 Facility water cycle.

distinct areas. In the "Resources" section below, a link is provided to the website.

CONCLUSION

Water-using technologies are an important aspect of energy engineering, because their use has such a wide, but often unrealized, impact on energy. Because unrestrained water consumption is becoming more of a concern, more technologies are being developed to help conserve both water and energy. Simple retrofits and proper O&M will make a significant impact on water consumption, especially in large water-consuming systems found in commercial and industrial facilities. These changes begin with organized water management planning and a comprehensive audit; conservation follows.

RESOURCES

General

American Water Works Association
www.awwa.org
Information on AWWA's activities, programs, conferences, publications, and useful links.

WATER EFFICIENCY AND CONSERVATION

Environmental Protection Agency Water Efficiency Program
www.epa.gov/owm/water-efficiency/index.htm
Efficiency tips on agriculture, residential water use, drought management, water recycling, and more.
California Urban Water Conservation Council
www.cuwcc.org
Statewide water conservation awareness program with urban water conservation best management practices.

Water and Energy

State of California Energy Commission Energy and Water Connection
www.energy.ca.gov/process/water/water_index.html
Information specifically tying together energy and water.
Watergy
watergy.org
A division of the Alliance to Save Energy, drawing the links between water supply and distribution and energy.

Federal Guidance

Federal Energy Management Program—Water Efficiency
www.eere.energy.gov/femp/technologies/water_efficiency.cfm
Guidance to federal agencies on water efficiency improvements.
Federal Water Efficiency Best Management Practices
http://www.eere.energy.gov/femp/technologies/water_fedrequire.cfm.
Listing of ten key areas for improving efficiency in water-using equipment and processes.
Energy Center of Expertise—Water Management Guide
www.gsa.gov/gsa/cm_attachments/GSA_DOCUMENT/waterguide_new_R2E-c-t-r_0Z5RDZ-i34K-pR.pdf
A comprehensive handbook for federal facility managers on water conservation.

Clothes Washers

http://www.h2oreuse.com/WhatHow.html
Information on a laundry water recycling system.

Cooling Towers

Federal Technology Alert: Ozone Treatment for Cooling Towers
http://www.eere.energy.gov/femp/pdfs/FTA_OTCT.pdf
Technical bulletin on ozonation for cooling towers.
Labs for the 21st Century Best Practices for Water
http://www.labs21century.gov/pdf/bp_water_508.pdf
Resources on water management in federal laboratories.

Irrigation and Landscape

Irrigation Association
www.irrigation.org
Irrigation best management practices and information on auditor training classes.
Center for Irrigation Technology
cati.csufresno.edu/cit/index.html
Independent research and testing of irrigation technology.
GreenCo
www.greenco.org
Information on creating drought-resistant landscapes through best management practices.
Sustainable Building Xeriscape Sourcebook
www.greenbuilder.com/sourcebook/xeriscape.html
Sustainable building sourcebook on xeriscape principals and construction specifications.

ACKNOWLEDGMENTS

The authors would like to thank Bill Chvala for his technical contributions, Mark Halverson for his technical peer review, and Sue Arey for her editorial review.

REFERENCES

1. http://www.cuwcc.org/sprayvalves.lasso (accessed on January 2006).
2. A Water Conservation Guide for Commercial, Institutional and Industrial Users, New Mexico Office of the State Engineer, Santa Fe, New Mexico, 1999; pp. 38–68.
3. Water Management Guide, U.S. General Services Administration, Kansas City, Missouri, 2000; pp. 66–67.
4. McMordie Stoughton, K.L.; Solana, A.E.; Elliott, D.B.; Sullivan, G.P.; Parker, G.B. In *Update of Market Assessment for Capturing Water Conservation Opportunities in the Federal Sector, PNNL-15320*; Pacific Northwest National Laboratory: Richland, Washington, 2005; p. E.4.
5. McMordie Stoughton, K.L.; Chvala, W.D.; Solana, A.E.; Halverson, M.A. In *Actions You Can Take Now to Save Water and Reduce Costs, PNNL-SA-45992*; Pacific Northwest National Laboratory: Richland, Washington, 2005; pp. 7–9.
6. Water Audits and Leak Detection, Vol. M36, 2nd Ed., American Water Works Association, Denver, Colorado, 1999; pp. 38–40.

Water-Using Equipment: Domestic

Amy Solana
Pacific Northwest National Laboratory, Portland, Oregon, U.S.A.

Kate McMordie Stoughton
Battelle Pacific Northwest National Laboratory, U.S. Department of Energy, Richland, Washington, U.S.A.

Abstract

Water management is an important aspect of energy engineering. This entry addresses water-using equipment used primarily for household purposes, including faucets, showers, toilets, urinals, dishwashers, and clothes washers, and focuses on how this equipment can be optimized to save both water and energy. Technology retrofits and operation and maintenance changes are the main water and energy conservation methods discussed. Auditing to determine current consumption rates is also described for each technology.

INTRODUCTION

In the study of energy engineering, the importance of water is often overlooked, although energy and water are closely linked. Water not only generates energy (as hydropower), but also consumes energy. The energy required to heat water used in sinks and showers, for instance, directly correlates to the amount of water being used. All potable water (water safe for human consumption) is treated and pumped before and after use, requiring energy for the pumping, monitoring, and production and transport of chemicals.[1]

Water is important regardless of its ties to energy. Droughts are having a greater impact on wildlife, animal and fish habitats, and humans; they can easily result in water shortages at the current level of consumption.[2] In addition, rising water rates are a major concern for consumers. Water conservation has become a necessity and should be interrelated with energy efficiency, helping to conserve natural resources.

Common domestic water-using equipment is discussed below, namely faucets, showerheads, toilets, urinals, dishwashers, and clothes washers. These technologies do not consume large amounts of water alone, but their combined daily use significantly impacts available resources. Water consumption can be greatly reduced by installing and using the most efficient appliances and water-using systems. Proper operation and maintenance (O&M) is even more essential to minimizing water and energy use. Options for efficiency improvements and guidelines to determine water consumption for non-metered domestic end uses are addressed in this section.

Keywords: Water; Faucet; Shower; Toilet; Urinal; Dishwasher; Clothes Washer; Audit; Conservation.

FAUCETS

A variety of faucets on the market today fall into two main categories: manually operated or automatic closure. As a result of the Energy Policy Act of 1992 (EPAct 1992), faucets are mandated to remain below a flow of 2.2 gallons per minute (gpm) at 60 pounds per square inch (psi); however, flow rates as low as 0.5 gpm are available. Older faucets flow at about 3–5 gpm at 60 psi. Automatic sensor faucets typically run at 0.5 gpm[3] and use infrared or ultrasonic sensors to turn on the faucet in the presence of a person. These sensors are powered by hard-wired, solar-powered, or hydro-powered electricity. The solar- and hydro-powered versions use batteries that are recharged with ambient light and running water, respectively. Self-closing faucets shut off the water when the knob is released, and typically have a 2.2-gpm flow rate. Metered valve faucets release a preset amount of water; the EPAct 1992 standard is 0.25 gallons per use.[4] Foot pedal faucets are retrofits to existing faucets that allow hands-free operation.

When replacing faucets with new fixtures, use the lowest flow possible for faucets that are not used for filling containers. If sanitation is a concern, consider automatic sensors or foot pedals.

Low- and no-cost options that do not replace the entire faucet also reduce water consumption. Aerators (see Fig. 1) are inexpensive and simple to install, and can reduce the flow of an existing faucet to 2.2 gpm, or as low as 1.0 gpm.[3] Flow restrictors are washer-like devices that fit into a faucet head. The flow valve on the water line can also be adjusted to reduce flow. Regular maintenance, which includes checking for leaks and proper line pressure (30–60 psi) and making appropriate adjustments, will save water. High pressures can waste water and quickly erode fixtures.

Fig. 1 Faucet aerator.

Auditing

Some faucets and aerators have flow rates imprinted on the spout. These numbers can be used if available, but are often not accurate. A simple measuring method is the bucket and stopwatch method, which measures the time that it takes to fill a container of known volume, allowing the gallons per minute to be calculated. A bag with volume increments marked on it can be filled for 5 seconds and the volume read to determine the gallons per minute; these supplies are available through water conservation suppliers.

The flow rates are then multiplied by the amount of time each faucet is used. Building occupancy (hours occupied and number of people, split by gender if possible) determines this. The following average values for faucet run times during an 8-hour work day can be used:

Men: 1.2 min/day
Women: 2.0 min/day[4]

Daily and annual water use per person for common faucet flow rates is summarized in Table 1. The table also shows the savings potential that can be achieved by retrofitting an older faucet with a more efficient one. This table is based on a 40-hour work week.

When determining energy savings associated with faucet water savings, it can be assumed that about 50% of the total flow is hot water.[6] When heating water from 60 (typical delivery temperature) to 120°F, electric hot water heaters use about 0.147 kilowatt-hours (kWh) per gallon, and natural gas heaters use about 0.005 therms per gallon. Energy savings can be estimated using the following equation with these assumptions:

$$\text{Energy saved} = \frac{\text{Water saved (gal)} \times \text{\% hot water} \times \text{energy used (kWh or therms/gal)}}{\text{heater efficiency}}$$

Daily and annual energy use per person is summarized in Table 2, using the same faucet types as in Table 1. This table also shows the energy savings potential from the water savings calculated in Table 1.

SHOWERS

Energy Policy Act 1992 standards require a 2.5 gpm flow rate, but older showerheads are still widely utilized today, at an average of 5–7 gpm. Low-flow showerheads are available with flow rates as low as 2.5 gpm. These do not always perform well, resulting in a somewhat negative reputation. However, water-conserving showerheads offering strong performance are widely available.

The same water conservation options for faucets exist for showers. Flow restrictors are extremely cost-effective, but can result in poor water pressure.[5] Showerheads with temporary cutoff valves are another option (see Fig. 2); water can be shut off while soaping and turned on to rinse at the same temperature. Additionally, using individual control valves in common shower rooms saves water compared with a single controller, which turns all showers on or off at the same time.

Table 1 Estimated water consumption and savings by efficient faucets in commercial facilities

Fixture/ year of installation	Water use (gpm)	Daily consumption male/ female (gal/day)	Annual consumption male/ female (gal/year)	Annual savings—Energy Policy Act (EPAct) standard retrofit male/ female (gal/year)	Annual savings—low-flow retrofit male/female (gal/year)	Annual savings—high efficiency retrofit male/female (gal/year)
High efficiency	0.5	0.6 / 1.0	156 / 260	—	—	—
Low-flow	1.0	1.2 / 2.0	312 / 520	—	—	156 / 260
EPAct standard	2.2	2.6 / 4.4	676 / 1,144	—	364 / 624	520 / 884
Pre-1994	3.0–5.0	3.6–6.0 / 6.0–10.0	936–1,560 / 1560–2,600	260–884 / 416–1,456	624–1,248 / 1,040–2,080	780–1,404 / 1,300–2,340

Table 2 Estimated energy consumption and savings by efficient faucets in commercial facilities

Fixture/ year of installation	Energy use (kWh/min)	Daily consumption male/ female (kWh/day)	Annual consumption male/ female (kWh/year)	Annual savings—Energy Policy Act (EPAct) standard retrofit male/female (kWh/ year)	Annual savings—low-flow retrofit male/female (kWh/ year)	Annual savings—high efficiency retrofit male/female (kWh/year)
High efficiency	0.04	0.04 / 0.07	12 / 19	—	—	—
Low-flow	0.07	0.09 / 0.15	23 / 38	—	—	12 / 19
EPAct standard	0.16	0.19 / 0.32	50 / 84	—	27 / 46	38 / 65
Pre-1994	0.22–0.37	0.26–0.44 / 0.44–0.74	69–115 / 115–191	19–65 / 31–107	46–92 / 76–153	57–103 / 96–172

Fig. 2 Temporary cutoff valve showerhead.

Auditing

Similar to faucets, shower water consumption can be found using the bucket and stopwatch method, or by checking the manufacturer's rate imprinted on the showerhead. Building occupancy may or may not be representative of the number of showers taken daily. This should be confirmed during an interview of facility staff or occupants. Some occupants may not shower every day; some may shower elsewhere, like a health club or workplace.

The length of showers can vary depending on the facility and user. Five-minute showers can be used as a conservative estimate, but interviews may help determine a more accurate length. Daily and annual water consumption per person for common showerhead flow rates is summarized in Table 3. The table also shows the savings potential that can be achieved by retrofitting an older showerhead with a more efficient one. This table assumes a person showers once every day for 5 minutes.

About 60% of a typical shower flow rate is hot water.[6] Otherwise, energy savings is calculated in the same manner as for faucets. Table 4 summarizes daily and annual shower energy use and the energy savings potential based on the water savings in Table 3.

TOILETS

Toilets are the largest indoor water consumer at most facilities.[7] The most common types of toilets on the

Table 3 Estimated water consumption and savings by efficient showerheads

Fixture/year of installation	Water use (gpm)	Daily consumption (gal/day)	Annual consumption (gal/year)	Annual savings—Energy Policy Act (EPAct) standard retrofit (gal/year)	Annual savings—low-flow retrofit (gal/year)	Annual savings—high efficiency retrofit (gal/year)
High efficiency	1.5	7.5	2,738	—	—	—
Low-flow	2.0	10.0	3,650	—	—	912
EPAct standard	2.5	12.5	4,563	—	913	1,825
Pre-1994	5.0–7.0	25.0–35.0	9,125–12,775	4,562–8,212	5,475–9,125	6,387–10,037

Table 4 Estimated energy consumption and savings by efficient showerheads

Fixture/year of installation	Energy use (kWh/min)	Daily consumption (kWh/day)	Annual consumption (kWh/year)	Annual savings—Energy Policy Act (EPAct) standard retrofit (kWh/year)	Annual savings—low-flow retrofit (kWh/year)	Annual savings—high efficiency retrofit (kWh/year)
High efficiency	0.13	0.66	242	—	—	—
Low-flow	0.18	0.88	322	—	—	80
EPAct standard	0.22	1.10	403	—	81	161
Pre-1994	0.44–0.62	2.21–3.09	805–1,127	402–724	483–805	563–885

market are tank toilets (see Fig. 3), typical in residential and light commercial applications, and flush valve toilets (see Fig. 4), typical in commercial applications. Tank toilets flush water through a siphon created by emptying water into the bowl through a flapper valve, whereby water is sucked out of the bowl and down the sewer line. Flush valve (or flushometer) toilets use the line pressure to flush high velocity water through the toilet; line pressure of at least 30 psi is required.

Pressure-assisted toilets contain a chamber with trapped air inside the toilet tank. The trapped air compresses as the chamber fills with water. When the toilet flushes, the compressed air pushes the water out at a high velocity, providing a powerful flush. Dual flush toilets provide two flushing options: a full flush and a partial flush. These toilets are available as tank, pressure-assisted, or flush valve types.

Toilets manufactured between 1980 and 1994 typically consume about 3.5 gallons per flush (gpf); toilets manufactured prior to 1980 consume around 5.0 gpf. Energy Policy Act 1992 mandated that toilets do not exceed 1.6 gpf; these toilets were termed ultra low-flow toilets or ULFs. High efficiency toilets (HETs) have since emerged on the market, defined as consuming at least 20% less than standard 1.6 gpf toilets. High efficiency toilets on the market include 1.0 gpf pressure-assisted and flush valve toilets and dual flush toilets, which use 1.6 gpf for the full flush and between 0.8 and 1.1 gpf for the partial flush. The flushing performance of low-flow toilets has been a consumer concern in the past, but has since improved.

An option for increasing the efficiency of an existing flush valve toilet is a flush valve retrofit kit. This kit involves replacement of the existing valve and toilet bowl

Fig. 3 Inside of tank toilet.

Fig. 4 Flush valve toilet.

with a high efficiency valve (rather than the entire plumbing) and new bowl. Toilets will not have a proper flushing pattern if the bowl is not replaced with the valve retrofit.

Retrofit devices for water efficiency in tank toilets include displacement devices, toilet dams, dual flush adapters, and early closure devices. These devices either displace volume in the tank or change the flushing mechanism to minimize the amount of flush water. Water utilities commonly provide these types of devices as a low cost retrofit option, but flush performance and long term maintenance may be problematic.

Leaks are quite common in toilets and are often inaudible, allowing them to go undetected for months, possibly years. A good O&M program can go a long way to save water, usually at little or no cost. Some simple procedures to include are as follows.

- Periodically check for leaks, at least bi-annually.
- Replace worn valves and flapper valves, matching the proper flush rate for the existing toilet. For example, if a 1.6-gpf tank or flush valve toilet is retrofitted with a 3.5-gpf flapper valve, the toilet will flush with 3.5 gallons.
- Periodically sample flush rates.
- Avoid harsh chemicals dispensed through tank toilets to avoid degrading flapper materials prematurely.

Auditing

Toilet water consumption begins with vintage identification to find the consumption per flush described above. The total water use can be found by understanding the number of people using the toilets and the total number of uses by each person. Daily and annual water use for each toilet vintage is summarized in Table 5, based on common estimates of toilet use frequency for one man and one woman. The table also shows the savings potential that can be achieved by retrofitting an older toilet with a ULF or HET. This table is relevant for workplaces with a 40-hour work week; because men have access to urinals, it is assumed they use toilets once per day and women use toilets three times per day.

URINALS

Urinals are typically used in commercial applications only. There are two basic types of urinals: flush valve (in Fig. 5 with infrared sensor) and no-water. A flush valve urinal works the same way as a flush valve toilet (described above). The no-water urinal is distinctly different, as the name implies. Instead of flushing, it contains a cartridge in the drain with a sealing liquid less dense than urine, allowing the urine to pass into the drain line. The liquid then seals the cartridge and prevents sewer vapors from escaping back into the restroom.

Similar to toilets, the amount of water a urinal uses depends on its manufacture date. Urinals built prior to 1980 use approximately 5.0 gpf. Urinals manufactured between 1980 and 1994 consume 1.5–3.0 gpf. From 1994 forward, EPAct 1992 required urinals to flush no more than 1.0 gpf. Flush valve urinals on the market today use 1.0 or 0.5 gpf.

Older urinal valves can be replaced with efficient valves similar to flush valve toilets. To achieve the proper flush pattern, the urinal should also be replaced with a low volume urinal. However, because urinals only flush liquids, poor flush performance is less of a concern. Checking the flush valves bi-annually can reduce the amount of water lost from broken valves that are stuck open, causing a continuous flush. Periodic leak detection at the pipe and handle connections is an important maintenance practice, as well.

No-water urinals can fairly easily replace existing flush valve urinals. However, a thorough economic analysis, including both the initial cost and the long-term O&M costs, is very important. No-water urinals have no valves to maintain, but the sealant in the cartridge must be

Table 5 Estimated consumption and savings by efficient toilets in commercial facilities

Fixture/year of installation	Water use (gpf)	Daily consumption male/female (gal/day)	Annual consumption male/female (gal/year)	Annual water savings—ultra low-flow toilets (ULFs) retrofit male/female (gal/year)	Annual water savings—high efficiency toilets (HETs) retrofit male/female (gal/year)
HETs	1.0–1.28	1.0–1.28 / 3.0–3.84	260–333 / 780–999	— / —	— / —
1994–present (ULFs)	1.6	1.6 / 4.8	416 / 1,248	— / —	83–156 / 250–468
1980–1994	3.5	3.5 / 10.5	910 / 2,730	494 / 1,482	577–650 / 1,732–1,950
Pre-1980	5.0	5.0 / 15.0	1,300 / 3,900	884 / 2,652	967–1,040 / 2,902–3,120

Fig. 5 Urinal with infrared sensor.

replaced, and some brands require periodic cartridge replacement. Additional training for janitorial and maintenance staff is required.

Other maintenance issues with no-water urinals are still not well understood. Some evidence points to maintenance savings because there may be a decrease in sewer line calcification. With older sewer lines and hard water, uric acid in urine can combine with minerals in water and deposit on the interior of the pipe. Other evidence points to the opposite; no-water urinals can possibly cause line stoppage. At the time of this writing, no major independent study has quantified these issues.

A small demonstration of no-water urinals prior to a large replacement may help ease the adjustment period and solve issues more quickly.

Table 6 Estimated consumption and savings by efficient urinals in commercial facilities

Fixture/year of installation	Water use (gpf)	Daily consumption (gal/day)	Annual consumption (gal/year)	Annual water savings—Energy Policy Act (EPAct) standard retrofit (gal/year)	Annual water savings—high efficiency retrofit (gal/year)	Annual water savings—no-water retrofit (gal/year)
No-water	0	—	—	—	—	—
High efficiency	0.5	1.0	260	—	—	260
1994–present	1.0	2.0	520	—	260	520
1980–1994	1.5	3.0	780	260	520	780
1980–1994	3.0	6.0	1,560	1,040	1,300	1,560
Pre-1980	5.0	10.0	3,900	3,380	3,640	3,900

Auditing

After determining the vintage of the flush valve units and, subsequently, their water consumption per flush, the total water consumption can be estimated based on the number of men in the facility and the average number of uses per day. Daily and annual water use per man for each urinal vintage is summarized in Table 6. The table also shows the savings potential that can be achieved by retrofitting the urinal to a more efficient unit. It is assumed that men use urinals twice per day during an 8-hour work day.

DISHWASHERS

Conventional, standard-size residential dishwashing machines (see Fig. 6) use about 9.5 gallons of water per cycle. More efficient EnergyStar™ machines use less than six gallons per cycle on average.[8] Compact size machines appropriate for smaller loads are also available, and they utilize less water and energy per cycle. In any dishwasher, washing full loads is always most efficient.

Standards currently regulate only energy use in dishwashing machines; however, because about 56% of dishwasher energy use is water heating, conserving energy will most likely conserve water. Federal standards require at least 0.46 cycles per kWh (of machine electrical energy and water heating energy) for standard size machines and 0.62 cycles per kWh for compact machines. EnergyStar™ currently has requirements only for standard-size machines—0.58 cycles per kWh. In 2007, EnergyStar™ will require standard machines to obtain 0.65 cycles per kWh and compact machines to obtain 0.88 cycles per kWh.[9]

Pressure or flow regulators can be installed to limit the water entering the machine in order to match the manufacturer's recommendations.[4] Machines should be checked for proper operation and correct water use.

Auditing

Residential style dishwasher water consumption can be estimated by obtaining the gallons per cycle, found either on the machine or from manufacturer's product data. Use patterns need to be collected via interviews.

CLOTHES WASHERS

A conventional top-loading (vertical-axis) residential machine uses about 40 gallons per load (gpl).[10] Efficient top-loading machines that use a high-pressure spray instead of tub soaking to rinse clothes, and sensors to measure the hot water needs per load are also available. Horizontal-axis machines use gravity to assist in spinning on a horizontal axis, and consume only 15–25 gpl[3,9] because only the bottom third of the machine is filled with water. Horizontal-axis machines are mostly front-loading (see Fig. 7), but top-loading machines are available as well.

Federal standards currently require clothes washers to meet a minimum energy factor (MEF) of 1.04 cubic feet (ft^3) per kWh per cycle, which will be raised to 1.26 ft^3/kWh/cycle in 2007. EnergyStar™ requires 1.42 ft^3/kWh/cycle. The MEF takes water use and dryer energy into account, not just washer energy.[11]

Replacing vertical-axis machines with horizontal-axis machines saves about 40% of the typical water used, plus almost 60% of the typical energy used, not including the dryer energy savings, and about 50% of the typical amount of detergent used.[10] Additional energy savings is achieved as a result of higher-efficiency motors and high spin speeds that result in dryer clothes than with vertical-axis machines.

In apartment buildings or other multi-family residential facilities, common coin-operated laundry instead of individual machines in each apartment has reduced water consumption by about 72%, according to the Multi-Housing Laundry Association study.[4]

Fig. 6 Standard-size residential dishwasher.

In any laundry facility or machine, washing only full loads will reduce overall energy and water use. Using cold water whenever possible saves energy by saving heated water.

Auditing

To determine laundry water use, the frequency of machine use and water consumption per load must be known. Frequency of use will probably have to be estimated using employee or resident interviews. Water consumption may be written on the machine; if not, manufacturer and model information can be used to obtain water use data. Incomplete data can be combined with typical consumption values based on the type of washer, as previously noted.

GENERAL CONSERVATION TECHNIQUES

Education and outreach is a fundamental part of water conservation. A leak of one drop per second can waste 36 gallons of water per day[5]—a maintenance team's quick response can greatly limit this waste. User-friendly mechanisms to report fixture problems in public restrooms can help achieve faster repairs.

Energy is conserved with many water conservation strategies, especially when hot water is saved. It is important to note that lowering the water heater temperature to what is needed, generally no higher than 120°F, is a simple no-cost measure that saves energy without involving water conservation.

Domestic water consumption should utilize water management planning as described in the following entry, "Commercial and Industrial Water-Using Equipment."

CONCLUSION

Water-using technologies are an important aspect of energy engineering because their use has such a wide, but often unrealized, impact on energy. Simple retrofits and proper O&M can have a significant impact on domestic water and energy consumption. Auditing to

Fig. 7 Front-loading horizontal-axis clothes washer.

calculate current consumption provides the baseline from which to start reducing daily water use.

RESOURCES

General

American Water Works Association
www.awwa.org
Information on American Water Works Association (AWWA's) activities, programs, conferences, and publications and useful links

Water Efficiency and Conservation

Environmental Protection Agency Water Efficiency Program
www.epa.gov/owm/water-efficiency/index.htm
Efficiency tips on agriculture, residential water use, drought management, water recycling, and more
California Urban Water Conservation Council
www.cuwcc.org
Statewide water conservation awareness program with urban water conservation Best Management Practices

Water and Energy

State of California Energy Commission Energy and Water Connection
www.energy.ca.gov/process/water/water_index.html
Information specifically tying together energy and water
Watergy
www.watergy.org
A division of the Alliance to Save Energy, drawing the links between water supply and distribution and limited energy resources

Education

Saving Water Partnership
www.savingwater.org

Educational resources about water conservation from Seattle-area utilities

Federal Guidance

Federal Energy Management Program (FEMP)—Water Efficiency
www.eere.energy.gov/femp/technologies/water_efficiency.cfm
Guidance for federal agencies on water efficiency improvements

Energy Center of Expertise—Water Management Guide
www.gsa.gov/gsa/cm_attachments/GSA_DOCUMENT/waterguide_new_R2E-c-t-r_0Z5RDZ-i34K-pR.pdf
A comprehensive handbook for federal facility managers on water conservation

Toilets

Toiletology 101
www.toiletology.com
General information on toilets

How Stuff Works—Tank Toilets
www.home.howstuffworks.com/toilet.htm
Basics of how a tank toilet works

ACKNOWLEDGMENTS

The authors would like to thank Mark Halverson for his technical peer review, Sue Arey for her editorial review, and David Trost and David Gualtieri for their photographic assistance.

REFERENCES

1. http://www.energystar.gov/index.cfm?c=products.pr_protect_water_supplies (accessed December 2005).
2. Rehfeld, B. *Spending. Steps to Take Before the Collective Well Runs Dry*; New York Times: New York, 2005.
3. McMordie Stoughton, K.L.; Solana, A.E.; Elliott, D.B.; Sullivan, G.P.; Parker, G.B. *Update of Market Assessment for Capturing Water Conservation Opportunities in the Federal Sector*, PNNL-15320; Pacific Northwest National Laboratory: Richland, WA, 2005; 38.
4. *A Water Conservation Guide for Commercial, Institutional and Industrial Users*; New Mexico Office of the State Engineer: Albuquerque, NM, 1999; 33–41.
5. *Water Management Guide*; U.S. General Services Administration: Kansas City, MO, 2000; 61–62.
6. http://www.eere.energy.gov/femp/pdfs/watergy_manual.pdf (accessed December 2005).
7. Vickers, A. *Handbook of Water Use and Conservation*; Water Plow Press: Amherst, MA, 2001; 26.
8. Sullivan, G.P.; Elliott, D.B.; Hadley, A.; Hillman, T.C.; Ledbetter, M.R.; Payson, D.R. *SWEEP—Save Water and Energy Education Program*, PNNL-13538; Pacific Northwest National Laboratory: Richland, WA, 2001; 34.
9. http://www.energystar.gov/ia/partners/prod_development/revisions/downloads/dishwashers/ES_Dishwasher_Criteria Letter_finalDec20.pdf (accessed December 2005).
10. http://www.toolbase.org/techinv/techDetails.aspx?technologyID=177 (accessed December 2005).
11. http://www.aceee.org/consumerguide/topwash.htm (accessed December 2005).

Wind Power

K. E. Ohrn
Cypress Digital Ltd., Vancouver, British Columbia, Canada

Abstract
This entry discusses a broad range of topics as an overview of the field of wind power. The topics include current penetration, market shares, technology, costs, governments, and regulation.

INTRODUCTION (WEB PREVIEW)

This article is a broad overview of wind power. It covers a range of topics, each of which could be expanded considerably. It is intended as an introductory reference for engineers, students, policy-makers, and the lay public.

Wind power is a small but growing source of electrical energy. Its economics are well known; there are several large and competent manufacturers, and technical problems are steadily being addressed. Wind power can now be considered as a financially and operationally viable alternative when planning additional electrical capacity. However, as shown below, wind power is only a minor component of present energy sources.

World final energy consumption (2002)[1]: 100%
World final electrical consumption (2002)[2]: 16.1%
World final wind power consumption (2004)[3]: 0.15%

Wind power's main deficiency as a power source is variability. Since wind velocity cannot be controlled or predicted with pinpoint accuracy, alternatives must be available to meet demand fluctuations.

Wind power carries few environmental penalties and makes use of a renewable resource. It has the potential to become a major but not dominant part of the future energy equation.

History

People have used wind to move boats, grind grains, and pump water for thousands of years. Wind-powered flour mills were common in Europe in the 12th century. In the 1700s, the Dutch added technical sophistication to their windmills with improved blades and a method to follow the prevailing wind. Isolated farms in the last century used windmills to generate electricity until the availability of the electrical power grid became widespread.

Past interest in wind power has tended to rise and fall with fuel prices for the predominant method of electrical production—thermal plants burning oil, natural gas, and coal.

CURRENT

Electrical Production

Although small, wind power is a fast-growing part of the energy picture. Since 1990, worldwide installed capacity has grown about 27% (Table 1).[3]

Business is good for the leading manufacturers of wind power devices. Sales have increased; the technology is stable and predictable, with low maintenance and high availability.

Geographical Distribution (Countries)

The European Union had around 72% of installed capacity in 2004, and Germany, Spain, and the United States accounted for 66.1%. Denmark, Spain, and Germany had by far the largest 2004 capacity in terms of MW per million populations, and 10.6% of the world's population had 81.9% of its wind power capacity. In 2004, Denmark produced about 20% of its electrical power from wind power in 2004 (Table 2).[3]

Manufacturing capacity in 2002 was largely confined to this group of countries, with five big vendors accounting for 76% of sales. European Union vendors accounted for 85% of manufacturing market share.[3]

ECONOMICS

Cost per kWh

Wind power is a viable method of producing electricity that is capital intensive with low operating costs. The cost of production compares favorably with traditional fossil fuel or nuclear plant costs.

Keywords: Wind power; Wind energy; Electrical energy; Environment; Wind farm; Wind turbine; Renewable.

Table 1 Capacity growth

Year	Capacity (MW)	Growth rate year-over-year (%)
1990	1,743	13.8
1991	1,983	17.0
1992	2,321	20.7
1993	2,801	26.1
1994	3,531	36.5
1995	4,821	26.6
1996	6,104	25.1
1997	7,636	33.0
1998	10,153	33.9
1999	13,594	27.7
2000	17,357	66.3
2001	28,857	7.9
2002	31,128	26.9
2003	39,500	20.3
2004	47,500	13.8
Average Growth Rate		26.7

Source: Reprinted with permission from European Wind Energy Association (see Ref. 3).

The major cost elements of a modern wind power installation are as follows[3,5,9]:

Capital
    Onshore: 1200–1500 USD/kW
    Offshore: 1700–2200 USD/kW
Operating: usually about 1.5%–2.0% of capital cost per year.[12]

Capital costs include wind capacity survey and analysis, land surveying, permits, roads, foundations, towers and turbines, sensors and communications systems, cabling to transformers and substations, maintenance facilities, testing, and commissioning. By far the largest individual capital cost is the turbine (up to 75%).

Operating and maintenance costs include management fees, insurance, property taxes, rent, and both scheduled and unscheduled maintenance.

Financing costs are a major portion of energy production costs, making them very sensitive to interest rates, incentives and subsidization.

Energy production cost estimates vary considerably. Optimists in the industry, such as the British Wind Energy Association, quote a low of 4.8 USD cents per kWh for an onshore plant in an optimal location. Pessimists, like the Royal Academy of Engineering in the UK, quote up to 13.2 USD cents per kWh for an offshore plant, partly by including a controversial 3.1 USD cents per kWh cost for

Table 2 Capacity distribution

Country	Wind power installed capacity (MW)				One-year (%)	Three-year (%)	Population (millions)	Capacity (MW/Million)	Percent of world capacity
	2001	2002	2003	2004					
Denmark	2,456	2,880	3,076	3,083	0.2	7.9	5.4	570.9	6.4
Spain	3,550	5,043	6,420	8,263	28.7	32.5	40.3	205.0	17.2
Germany	8,734	11,968	14,612	16,649	13.9	24.0	82.4	202.1	34.7
Netherlands	523	727	938	1,081	15.2	27.4	16.4	65.9	2.3
USA	4,245	4,674	6,361	6,750	6.1	16.7	295.8	22.8	14.1
Italy	700	806	922	1,261	36.8	21.7	58.1	21.7	2.6
UK	525	570	759	889	17.1	19.2	60.4	14.7	1.9
Japan	357	486	761	991	30.2	40.5	127.4	7.8	2.1
India	1,456	1,702	2,125	3,000	41.2	27.2	1080.4	2.8	6.3
China	406	473	571	769	34.7	23.7	1306.4	0.6	1.6
Total	22,952	29,329	36,545	42,736	16.9	23.0	3073		89.2

Source: From Global Wind Energy Council—"Wind Force 12" (see Ref. 17).

Table 3 Wind power costs

Wind power costs	Cents (US) per kWh			
	Wind onshore	Wind offshore	Coal CFB	Gas CCGT
RAE [15]	6.8	9.9	4.8	4.0
RAE 2[a]	10.1	13.2	9.2	6.0
EWEA [3]	5.5	8.5		
AWEA (5)	6.6			
AWEA 2[b]	5.4			
BWEA (minimum) [4]	4.8	6.9	4.8	4.8
BWEA (maximum)[c]	6.8	9.1	6.8	5.9
Euro to USD conversion (2004)				1.22
GBP to USD conversion (2004)				1.83

[a]Adds 3.1 cents per kWh for wind power backup capacity and 1.9–4.6 cents/kWh for coal and gas carbon capture.
[b]AWEA figures adjusted to delete 1.8 cents US/kWh production tax credit and are for onshore sites with different average wind speeds.
[c]BWEA figures for a range of site types in November 2004 and 1–2 cents/kWh for carbon capture.
Source: RAE, Royal Academy of Engineering (see Ref. 15); BWEA, British Wind Energy Association (see Ref. 4); EWEA, European Wind Energy Association (see Ref. 3); AWEA, American Wind Energy Association (see Ref. 5).

"standby capacity" required to supply demand when wind power is not available (Table 3).

Capacity Factor

The power generated by a wind turbine depends on the speed of the wind, and on how often it is available. At any given site, this is measured by the capacity factor, or the ratio of actual generated energy to the theoretical maximum. Wind power turbine electrical output rises as the cube of the wind speed. When wind speed doubles, energy output increases eightfold. A typical turbine begins to turn when wind speed is at 9 MPH and will cut out at 56 MPH for safety reasons.

Capacity factors vary by site but are typically in the range of 20%–30% with occasional very good offshore sites reaching 40%. The yearly energy output from a wind farm is given by the following formula:

Output/year(kWh)

$$= [Capacity(kW)] \times [8760 \text{ hours/year}]$$
$$\times [Capacity \text{ factor}]$$

The Site

Power production costs, site size, site design, and energy output and variability will depend mainly on details about the wind. These details include wind speed, wind direction, and the geographical distribution of favorable wind profiles. During analysis of potential sites, most planners use high (60 m plus) anemometer towers—often several of them—to gather at least one year's data per site. These data are usually correlated with national meteorological observations, if these are available and suitable. If not available, it would be prudent to gather site data for a longer period of up to three years.

Investors and regulators are increasingly aware of the crucial nature of wind data in estimating the quantity and timing of potential power production at a specific site. This research is crucial to the financial analysis of a potential wind power venture.

Other site analysis factors are accessibility via road for heavy equipment, electrical grid proximity and capacity, land ownership, and environmental impact.

LOCATION

Favored Geography

Many countries have developed wind charts of broad areas based on meteorological data gathered for weather and aviation purposes. These charts show potential areas for investigation, where wind strength is high and constant over long periods of the year. Once potential sites, and their extent, have been identified, on-site data measurements provide the basis for analysis and modeling of potential energy production for a specific site.

After wind modeling, the site's geographical, environmental, financing, and ownership issues can be explored in detail.

Generally, sites are either onshore or offshore. Onshore sites are cheaper to construct, but have lower capacity factors due to wind turbulence from nearby hills, trees, and buildings. Offshore sites can have more potential energy available due to higher wind speeds and lower turbulence, which also reduces turbine component wear. Good offshore sites can be near high-demand load areas such as coastal cities, which also increases transmission options. Aesthetic and noise concerns are often fewer offshore, and sea-bed environmental concerns can be lower than land-use concerns for an onshore site.

Sizing a Location

Wind farm towers are usually spread over a large area in order to minimize wake losses. A spacing of five rotor diameters is often recommended. In a typical wind farm, the land physically occupied by tower foundations, buildings, and roads is often less than 2% of overall land area.[6] The remaining land is quite suitable for agriculture and other uses.

Limits to Maximum Production

How much capacity exists to generate electricity from wind? Is it possible that we will require more energy than wind can provide? After surveying wind patterns in the United States and applying energy density and extraction calculations, Elliott and Schwartz[14] concluded in 1993 that 6% of the available U.S. land mass could provide 150% of then-current U.S. electrical consumption. Furthermore, the needed land would be sparsely affected by the wind farm installation, with the vast majority of it (95%–98%) unoccupied by tower foundations, roads, or ancillary equipment and suitable for farming, ranching, and other uses. This study excluded land that is environmentally or otherwise unsuitable, such as cities, forests, parks, wildlife refuges, and environmental exclusion areas.

In the European Union, potential wind power capacity is also larger than current electrical consumption.

FUTURE

Projected Growth

Thanks to increasing concern over the environmental effects of greenhouse gas emissions, the rising cost of fossil fuels, and the impending decrease in availability of oil and natural gas, wind power has a bright future.

Current 25%–30% growth rates are likely not sustainable, due to equipment production volume constraints and limits to perceived need for further capacity. Given the Eurocentric, highly clustered nature of current installed capacity, there is significant potential for high-rate growth elsewhere. However, even in European countries like Germany and Denmark, steady growth will be driven by predicted cost reductions in the 10%–20% range and by regulatory and governmental initiatives aimed at reducing emissions from electrical energy production and transitioning to renewable resources.

Projected Cost

Wind power technology is well down the cost improvement curve, with costs having fallen to present levels, below ten cents USD per kWh, from over $1.00 U.S. in 1978. Costs for a medium-sized turbine have dropped 50% since the mid-1980s, reflecting increasing maturity in the market. Cost projections range from a further 9%–17% drop as installed capacity doubles in the near-term future.

Projected Production

With increasing governmental policy support and commitment, growth rates of 15%–20% appear achievable in five to ten years. But there is likely an upper limit to the amount of electrical energy that can be produced from the wind.

Reaching Maximum Production

Production limits for wind power are based on its variable nature. Other types of electrical production capacity will be needed to provide base-load electrical capacity in the event that there is little wind available. Wind power will then become one player in a mix of generating technologies.

The Hydrogen Economy

As wind power becomes a larger portion of electrical supply, occasionally its supply will exceed demand. Rather than simply curtailing wind plant production, it is attractive to think of using this excess electrical power capacity to generate hydrogen via electrolysis. This has the effect of storing wind energy that would otherwise not be harvested. This energy, in the form of hydrogen, can be used directly as a non-polluting fuel or as an input source to fuel cells to produce electricity at a later time.

When there is a significant hydrogen economy, with transmission lines, storage, and fuel cell capacity, this use for wind power will become a very attractive scenario.

Other Issues for the Future

- Learning more about wind and forecasting—predicting the best locations, wind farm output, gusts, and directional shear.[10] This will help reduce financing costs when wind power plant output and impact on the grid are better understood and more predictable.

- Improving the control of demand through incentives around end-user load shedding, rescheduling and simple conservation methods. This could be used to offset wind power production shortfalls as an alternative to other forms of generation.
- Advancing aerodynamics specific to wind turbine blades and control systems.
- Designing extremely large wind tunnels to study wake effects minimization, structural load prediction, and energy output maximization at lower wind speeds.
- Enhancing power system capacity planning models to include wind farm components.
- Re-planting, or upgrading older mechanical and electrical components at existing wind farms.
- Wind farm siting further offshore and on floating platforms.
- Combining wind power and hydroelectric capacity by using surplus wind power to pump water behind dams and so store power that might otherwise be wasted.
- Determining how and whether to allocate full costs of environmental impact to fossil and nuclear plants.

STRENGTHS AND WEAKNESSES

Strengths

Environment

Wind power installations do not emit air pollution in the form of carbon dioxide, sulfur dioxide, nitrogen oxides, or other particulate matter such as heavy metal air toxins. Wind power installations do not use water or discharge any hazardous waste or heat into water. Conventional coal, oil, and gas electric power plants produce significant emissions of all kinds. Nuclear power plants produce dangerous and long-lasting radioactive waste. Greater use of wind power means less impact on health and the environment, particularly regarding climate change due to greenhouse gas emissions.

Renewable

Wind power produces energy from a resource that is constantly renewed. The energy in wind is derived from the sun, which heats different parts of the earth at different rates during the day and over the seasons. Unlike fossil and nuclear plants, the source of energy is essentially inexhaustible.

Costs

Wind power's costs are well known and are dropping to the point at which this technology is very competitive with other means of production. Fuel costs are nil, meaning that fuel costs have no uncertainty. Wind power costs should be more stable and predictable over the lifetime of the plant than power costs for fossil fuel plants.

Local and Diverse

Wind power plants provide energy source diversity and reduce the need to find, develop, and secure sources of fossil or nuclear fuel. This reduces foreign dependencies in energy supply, and reduces the chances of a political problem or natural disaster interfering with and diminishing the supply of electricity.

Quick to Build, Easy to Expand

Wind power plants of significant capacity can be constructed and installed within a year, a much shorter time than conventional plants. The planning time horizon is similar to conventional plants, given the need to accurately survey site wind characteristics and deal with normal environmental and related site issues. This means that capacity can be increased in closer step with demand than with conventional plants. With the right site and design, a wind power plant can be incrementally expanded very quickly.

Weaknesses

Natural Variability

A single wind farm produces variable amounts of energy, and its output is not yet as predictable as a traditional plant. As the geographical distribution and number of wind plants increases, and as research into predicting wind continues, these problems should be minimized, allowing cost-effective and orderly scheduling and dispatch of total grid capacity sources—but it is difficult to see traditional power sources disappearing entirely.

Connection to Grid

As the amount of electrical power supplied by wind power plants increases, concern increases over its effects on the electrical grid.

In order to maintain a reliable supply of electricity that matches demand, utility operators maintain emergency reserve capacity in order to deal with plant outages (failures) and unexpected demand across their entire system. This reserve is in the form of purchased power, unused capacity at conventional plants running below their maximum, or quick-start plants such as gas-fired turbines. Often, conventional plants on the grid are allocated a cost to cover this reserve based on their capacity (large plant, large reserve) and reliability (more outages, more reserve).

The industry is working on ways to determine and allocate this reserve cost for wind power plants. Yet to be agreed upon is the statistical basis for calculating such wind power plant reserve costs. Improvements in day-ahead wind forecasting will greatly reduce the uncertainty around wind plant output, and so decrease the cost burden to provide this reserve.

Several current estimates prepared for U.S. utilities show this reserve cost burden (or ancillary services cost) to increase with the amount of capacity provided by wind power, and to be in the range of 0.1–0.5 cents USD per kWh for penetrations between 3.5 and 29%. In no case was it thought necessary to allocate a reserve equal to 100% of the wind power capacity.[11] German experience is similar,[8] with no additional reserve capacity required for the 14% wind energy share of the national electrical consumption forecast for 2015.

When wind power supplies less than 20% of electrical consumption, these problems are not severe. At larger penetrations, reserves become a major issue. Interestingly, wind power plants may be subject to shutdown or voluntary power reductions in the event of coincident high wind, low demand situations. This is occasionally the case today in Denmark and Spain.

In some cases, wind power sites are situated far from the location of high electrical power demand, placing strain or potential overload on existing transmission facilities. In these cases, there are often cost, ownership and responsibility issues yet to be resolved.

Power quality problems around power factor, harmonic distortions, and frequency and voltage fluctuations are being successfully addressed in modern large production wind farms.[8]

This is one of the most difficult sets of issues facing the future of wind power as it matures from small-scale and local to large-scale penetration.

Local Resource Shortage

In a few places, high-quality wind power sites are not available or are already in production, leaving these places to import electrical power or use traditional sources.

Noise

Noise levels have decreased and are now confined to blade noise in modern units. Generator and related mechanical noise has been effectively eliminated. Noise, however, will always be a significant factor. Blade noise is described as a "whoosh, whoosh" sound, and is in the 45–50 decibel range at a distance of 200–300 m. This noise level is consistent with many national noise level regulations. However, this noise buffer zone adds to the overall land requirement for a wind power plant and so increases costs.

Visual Impact

Onshore wind farms are highly visible due to the height of towers and the size of the blades and generator. The impact of this varies with each person. Each wind plant operator needs to determine the levels of support and opposition from those who live and work within sight of the plant. Offshore plants attract fewer detractors than onshore plants—one of the reasons for their increasing popularity.

Offshore wind plants are less likely to cause unwanted noise since they are far from human habitation. This reduces turbine and blade design constraints and can lead to higher capacity factors.

Bird Impact

Bird deaths are a regrettable reality. The bird death rate at a specific wind farm project is quite variable. Several early wind farms (Altamont Pass, California, and La Tarifa, Spain) caused concern over death rates. The California Energy Commission estimates the death rate at Altamont (5400 turbines) to be 0.33–0.87 bird deaths per turbine per year.[16] The overall recent U.S. national average[13] is 2.3 bird deaths per turbine per year. Prudently located sites are off migration routes and not in nesting, over-wintering, or feeding areas. Their tower designs do not offer nesting or even roosting places. In such locations, death rates are lower, and overall impact is much lower than that caused by other types of human activity.

Since climate change is a very serious environmental problem faced by bird populations, wind power and other renewables are an important part of the solution.

TECHNOLOGY

Overview

Wind turbine design has three major components, and there are large economies of scale in design.

- Tower height: Wind turbine energy output is proportional to the cube of wind speed. Since moving air (wind) is subject to drag and turbulence from its contact with the earth and the objects on the earth, wind speed increases with height (vertical shear). The higher the tower, the more advantage there is for power generation. The tradeoff is between tower costs and increase in power generation. Typically, tower heights are rising, and are currently in the 100 m range. Off shore, vertical wind shear is generally less than onshore, so towers can be shorter, with wave height clearance being the factor that determines tower height (Fig. 1).
- Blade diameter: The power capacity (watts) of a wind turbine varies with the square of its blade diameter, because a blade with a larger diameter has a larger area available for harvesting the wind energy passing through it. The coefficient of performance defines the actual power capacity compared to the

Fig. 1 Typical large wind turbine. Note entrance steps and utility pole at base for scale. (Photo courtesy of Suncor Energy, Inc.)

maximum—how much energy can be extracted from the wind compared to the available energy. Modern wind turbines can achieve a coefficient of performance approaching 0.5, very close to the theoretical maximum of 0.59 derived by Betz.[3,18] This maximum is derived from the concept that if 100% of wind energy were extracted, the wind exiting the turbine would be at zero speed, so no new air could enter the turbine. Larger capacity turbines benefit significantly from economies of scale in foundation and support costs as well as swept area (Fig. 2).

- Controls and generating equipment: The turbine's hub (or nacelle) is the most costly component and contains the generator, gear boxes, yaw controls, brakes, cooling mechanism, computer controls, anemometer, and wind directional vane.

Generators

As the blades turn, they drive a generator to produce electricity. Generating capacity ranges from a few hundred

Fig. 2 Site assembly of large wind turbine nacelles and blades. (Photo courtesy of Suncor Energy, Inc.)

kW to over 3 MW. In older designs, there is usually a 40:1 gearbox to match low, fixed rotor speeds (~30 RPM) to required generator speed (1200 RPM for a 60 Hz output, 6-pole generator). The gearbox often incorporates brakes as a part of the overall wind turbine control system. Generators that operate at low RPM are available and are called direct drive generators. These would eliminate the gearbox.

In more modern designs, rotor speed is variable and controlled to optimize power extraction from the available wind. Generator output is converted to d.c. and then back to a.c. at the required grid frequency and voltage. The conversion equipment is sometimes located at a central part of the wind power plant.

This is an active area for ongoing technical innovation.

Blades

In order to maximize power capacity through size, blades are very long, up to 50 m. To minimize noise, they must turn slowly so as to reduce tip speed, the primary blade noise source. Typical rotation speeds are in the 10–30 RPM range. Blades are increasingly made from composites (carbon fibre reinforced epoxy resins).

Rotor blade aerodynamics[19] have much in common with the aerodynamics of a propeller or a helicopter blade, but they are sufficiently different that the aerodynamics of wind turbine blades is an evolving field. The difference is that wind turbine airflows are unsteady due to gusting, turbulence, vertical shear, turbine tower upstream shadow, yaw correction lag, and the effects of rotation on flow development. For example, at present it can be difficult to predict rotor torque (and therefore power output) accurately for normal turbine operating conditions. Further development of theory and modeling tools should allow the industry to improve rotor strength, weight, power predictability, power output, and plant longevity while controlling cost and structural life.

For a given site wind velocity, the rotor blade's tip has very different air flow than its root, requiring the blade to be designed in a careful twist. The outer third of the blade generally produces two-thirds of the rotor's power. The third nearest the hub provides mechanical strength to support the tip, and also provides starting torque in startup situations.

Each blade generally has lightning protection in the form of a metallic piece on the tip and a conductor running to the hub.

Some manufacturers place Whitcomb winglets at the blade tips to reduce induced drag and rotor noise, in common with aircraft wing design.

In order to control blades during high wind speeds, some are designed with a fixed pitch that will progressively stall in high wind speeds. Others incorporate active pitch control mechanisms at the hub. Such control systems use hydraulic actuators or electric stepper motors and must act very quickly to be effective.

Wind Sensors

Wind turbines incorporate an anemometer to measure wind speed and one or more vanes to measure direction. These are primary inputs to the control mechanism and data gathering systems usually incorporated into a wind turbine.

Control Mechanisms (Computer Systems)

Control systems are used to yaw the wind turbine to face into the wind, and in some designs to control blade pitch angle or activate brakes when wind gets too strong. In sophisticated cases, the controllers are redundant closed-loop systems that operate pumps, valves and motors to achieve optimum wind turbine performance. They also monitor and collect data about wind strength and direction; electrical voltage, frequency and current; nacelle and bearing temperatures; hydraulic pressure levels; and rotor speeds, vibration, yaw, fluid levels, and blade pitch angle. Some designs provide warnings and alarms to central site operators via landline or radio. Manufacturers do not release much detail about these systems, since they are a critical contributor to a wind turbine's overall effectiveness, safety, and mechanical longevity.

ROLE OF GOVERNMENTS AND REGULATORS

Governments play a large part in determining the role and scale of wind energy in our future mix of energy production capabilities.

Subsidies, Tax Incentives

As part of programs to encourage wind power production, the following are used in varying ways[7]:

- Outright subsidies, grants and no-interest loans.
- Tax incentives such as accelerated depreciation.
- Fixed prices paid for produced electrical power.
- Renewable energy quantity targets imposed on power utility operators.

Grid Interconnection and Regulatory Issues

Since many power utility operators are owned by governments, and most are regulated heavily, governments have a role to play in encouraging solutions to grid interconnection issues. There must be a political will to address issues, find solutions, and develop practices and different management strategies that will allow greater penetration of wind power into the electrical supply.

Improving Wind Information

Climate and environmental information is most often collected and supplied by national governments in support of weather and aviation services. Wind atlases are an invaluable resource to the wind plant planning process. National efforts to improve long- and short-term wind forecasting, atmospheric modeling tools, and techniques will benefit wind power projects' ability to forecast power output for long-term and short-term planning purposes.[10]

Environmental Regulation

In this controversial area, government can tighten its regulation of air quality, carbon emissions and other environmental areas. This would have the effect of increasing the apparent cost of conventional thermal electrical power, which is responsible for significant emissions. It is often argued that wind power would already be cost competitive if environmental and health costs were to be fully allocated to conventional oil, gas, and particularly coal-powered plants, or if such plants were required to make investments to significantly reduce emissions.

CONCLUSION

Western societies depend on a steady supply of energy, much of it in the form of electricity. Most of that supply comes from thermal plants that burn oil, natural gas, and coal, or from nuclear plants. Where will our electrical energy come from in the future? How will we keep our environment livable and healthy?

One part of this answer lies in wind power. Its costs are within reason; the technology has matured with some gains yet to be realized; it carries little environmental penalty; and the source of its energy is renewable. As long as the sun heats the earth, wind power will be available.

Wind power will not likely be the complete answer; it is an intermittent source because the wind doesn't always blow. But there is a very large amount of it available for us to harvest. As wind power moves quickly from small-scale to large-scale, its future path depends on governments and regulators as much as it does on technical innovators and manufacturers.

REFERENCES

1. Annual Energy Review 2003, DOE/EIA-0384(2003), Energy Information Administration, U.S. Department of Energy, Washington, DC, September 2004.
2. Key World Energy Statistics - 2004 Edition, International Energy Agency, OECD/IEA - 2, rue André-Pascal, 75775 Paris Cedex 16, France or 9, rue de la Fédération, 75739 Paris Cedex 15, France, 2004.
3. Wind Energy, The Facts, Analysis of Wind Energy In the EU-25, European Wind Energy Association, Brussels, Belgium, 2004.
4. BWEA Briefing Sheet, Wind and the U.K.'s 10% Target, The British Wind Energy Association, http://www.bwea.com/energy/10percent.pdf (accessed July 2005).
5. The Economics of Wind Energy, American Wind Energy Association, http://www.awea.org/pubs/factsheets/EconomicsOfWind-Feb2005.pdf (accessed July 2005).
6. Saddler, H., Dr.; Diesendorf, M., Dr.; Denniss, R.; A Clean Energy Future For Australia, A Study By Energy Strategies for the Clean Energy Future Group, Australia, March 2004.
7. Policies to Promote Non-hydro Renewable Energy in the United States and Selected Countries, Energy Information Administration, Office of Coal, Nuclear, Electric and Alternate Fuels, United States Department of Energy, Washington, DC, February 2005.
8. BRIEFING May 10th, 2005. German Energy Agency DENA study demonstrates that large scale integration of wind energy in the electricity system is technically and economically feasible. http://www.ewea.org/documents/0510_EWEA_BWE_VDMA_dena_briefing.pdf (accessed July 2005).
9. Offshore Wind Experiences, International Energy Agency, 9, rue de la Fédération, Paris Cedex 15, 2004.
10. Milborrow, D., Forecasting for Scheduled Delivery, Windpower Monthly, December 2003.
11. Utility Wind Interest Group—Wind Power Impacts On Electric-Power-System Operating Costs, November 2003.
12. Danish Wind Industry Association, http://www.windpower.org/en/tour.htm (accessed July 2005).
13. Wind Turbine Interactions With Birds and Bats: A Summary of Research Results and Remaining Questions, Fact Sheet, Second Edition, National Wind Coordinating Committee, Washington, DC, November 2004. http://www.nationalwind.org/publications/avian/wildlife_factsheet.pdf (accessed July 2005).
14. Elliott, D.L.; Schwartz, M.N., Wind Energy Potential In the United States, Pacific Northwest Laboratory, PNL-SA-23109, Richland, Washington, USA, September 1993.
15. The Cost of Generating Electricity, Royal Academy of Engineering, http://www.raeng.org.uk/news/publications/list/reports/Cost_of_Generating_Electricity.pdf (accessed July 2005).
16. Developing Methods to Reduce Bird Mortality in the Altamont Pass Wind Resource Area, August 2004, http://www.energy.ca.gov/reports/500-04-052/500-04-052_00_EXEC_SUM.PDF (accessed July 2005).
17. Wind Force 12, Global Wind Energy Council, June 2005, http://www.ewea.org/03publications/WindForce12.htm (accessed July 2005).
18. Proof of Betz Theorem, Danish Wind Industry Association, http://www.windpower.org/en/stat/betzpro.htm (accessed July 2005).
19. National Wind Technology Center, http://www.nrel.gov/wind/about_aerodynamics.html (accessed July 2005).

Window Energy

William Ross McCluney
Florida Solar Energy Center, University of Central Florida, Cocoa, Florida, U.S.A.

Abstract
Modern factory-built windows and window components are available which can greatly reduce that portion of building energy consumption attributable to the windows. The adverse energy consequences of using larger window apertures for good daylighting (thereby realizing its attributes) can be reduced or even eliminated. The result can be buildings with greater comfort, more productive occupants, and substantially reduced purchased energy costs. As we pass the peak of world oil production and approach peaks in other fossil-fuel supplies, considerable energy and dollar savings are possible through the use of the new high performance windows.

INTRODUCTION

Windows have been included in building envelopes since the first buildings were constructed—mainly for the illumination and view that they provide. The addition of glazing provided increased protection from wind and rain. In the past, windows were considered "holes in the insulation." As a result, window areas have generally been kept modest in size, especially in buildings designed for hot or cold climates. Modern window designs and materials have made it possible to minimize—and, in some cases, avoid altogether—the energy costs of windows, enabling builders to utilize larger areas for aesthetic, viewing, and health benefits. In some circumstances, good windows can outperform opaque insulated walls, energy-wise.

The energy performance of a building's window system can be estimated through a variety of engineering calculations. Due to the vagaries of weather and climate, these measures are best performed with the aid of computer simulation, wherein the hourly energy and illumination performances of windows can be assessed quickly with modest precision. In window design, care must be taken to ensure strong energy performance sans excessive glare, overheating, or thermal discomfort.

This article examines the fundamentals of window energy performance and presents some issues surrounding window and shade design and product selection. A separate chapter discusses daylight design more generally.

FUNDAMENTAL QUANTITIES

Radiant flux, Φ, is the time rate of radiant energy flow (unit: watt).

Irradiance, E, is the radiant flux per unit area at a specified point in a specified surface that is incident on, passing through or emerging from that point in the surface, or the average of this quantity over the surface's area (units: watt m^{-2}).

Conductive heat flow, Q, is the flow of heat through a specified distance in a material per unit area and per unit temperature difference. It has units of energy flux per unit area and per unit temperature difference (units: W m^{-2} K^{-1}).

An important characteristic of radiant flux is its distribution over the electromagnetic spectrum, known as a spectral distribution or spectrum. The Greek symbol λ is used to symbolize the wavelength of monochromatic radiation, defined as radiation with only one frequency and wavelength. The symbol for frequency is the Greek symbol ν. The relationship between frequency ν and wavelength λ is shown in the equation

$$\lambda \nu = c \tag{1}$$

where c is the speed of propagation in the medium (more familiarly, the "speed of light," even though "light" is properly applied only to the visible portion of the spectrum). The spectral concentration of radiant flux at a given wavelength λ is given the name "spectral radiant flux" and represented by the symbol Φ_λ (units: W nm^{-1}). The wavelength dependence is represented by the functional notation, thus, $\Phi_\lambda(\lambda)$, where the subscript denotes the concentration of the quantity at a specific wavelength.

Spectral irradiance is the spectral "concentration" of the irradiance E, and its wavelength dependence is denoted by the symbol $E_\lambda(\lambda)$ (units: W m^{-2} nm^{-1}).

Names for the relevant regions of the electromagnetic spectrum are provided in Table 1.[1]

ENERGY BASICS

Heat Transfer

Energy is transferred through windows by the three mechanisms: radiation, conduction, and convection, illustrated in Fig. 1.

Keywords: Windows; Daylight; Glazing; Aperture; Illumination; View; Energy.

Window Energy

Table 1 CIE vocabulary for spectral regions

Name	Wavelength range
UV-C	100–280 nm
UV-B	280–315 nm
UV-A	315–400
VIS	Approx. 360–400 to 760–800 nm
IR-A[a]	780–1,400 nm
IR-B	1.4–3.0 µm
IR-C[b]	3 µm–1 mm

[a]Also called "near IR" or NIR.
[b]Also called "far IR" or FIR.
Source: Excerpts from International Lighting Vocabulary (see Ref. 1).

Radiation Heat Transfer

The primary form of radiative heat transfer through windows is solar energy. Irradiance spectra for the sun and the diffuse sky for a clear day are shown in Fig. 2.

All objects above an absolute zero temperature emit radiation. Those with temperatures much cooler than the sun emit radiation in the far IR (FIR) region. Radiative transfer can also occur in windows over this spectral band. The FIR region is separated significantly from that of the solar spectrum.[2]

The spectral separation between solar and FIR radiation makes the greenhouse effect possible, a process whereby solar radiation is transmitted effectively through clear glass, but the longer-wavelength radiation emitted from objects receiving the transmitted solar radiation does not transmit back through the glass easily. This mechanism enables greenhouses to remain warm on cold winter days when they receive ample solar radiation, without the need for additional heating. Window manufacturers have learned to take advantage of this spectral separation to improve the winter day and night performance of glazing systems for a variety of applications. When it is cold outside and warm inside, significant quantities of FIR radiation can pass from the warm inner glazing across the gas gap in a double pane window to a colder outer pane. Several mechanisms, described below, are available for reducing the magnitude of this component of heat loss through windows.

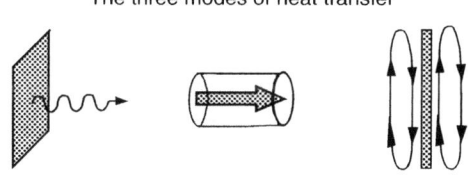

Fig. 1 The three modes of heat transfer.

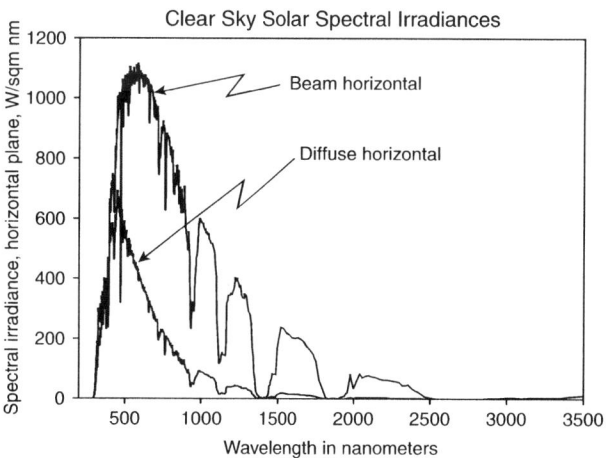

Fig. 2 Spectral irradiance distributions for clear sky direct beam and diffuse sky radiation on a horizontal plane.

Conduction Heat Transfer

Heat is transferred via conduction through material that has at least two surfaces with different temperatures. The heat conducts through the material from the warm surface to the cooler one. The rate of transfer depends upon the magnitude of the temperature difference and the thermal conductivity of the material, k, having units of energy per unit time, per unit area, per unit distance through the material, and per unit temperature difference. For homogeneous materials of specific thickness t, the conductance C of that thickness is a product of the conductivity of the material and the distance t through it. Conductance has units of energy per unit time, per unit area, and per unit temperature difference. Window frames with sections that have lower thermal conductivity are more insulating than those without these thermal breaks. Generally, the thermal resistance R is the reciprocal of the thermal conductance C.

Convection Heat Transfer

A warm surface can transfer some of its heat to the adjacent air by conduction. Once the air in a thin film adjacent to the surface receives this heat, it expands slightly and there is a tendency for the air to rise. This air movement can carry heat to or from the glazing of a single-pane window. If the surface is a vertical one, and if the rising warmed air reaches a boundary (like the window frame around an insulating gas space between the panes of a double pane window), it tends to move laterally. If there is a cooler vertical second pane, the heated air will tend to move downward. As it descends, it will lose some of its heat via conduction into the cooler pane of glass. The higher density of the cooler air will drive it further downward and it will, at the bottom, move laterally from the cooler toward the warmer surface, filling the space left

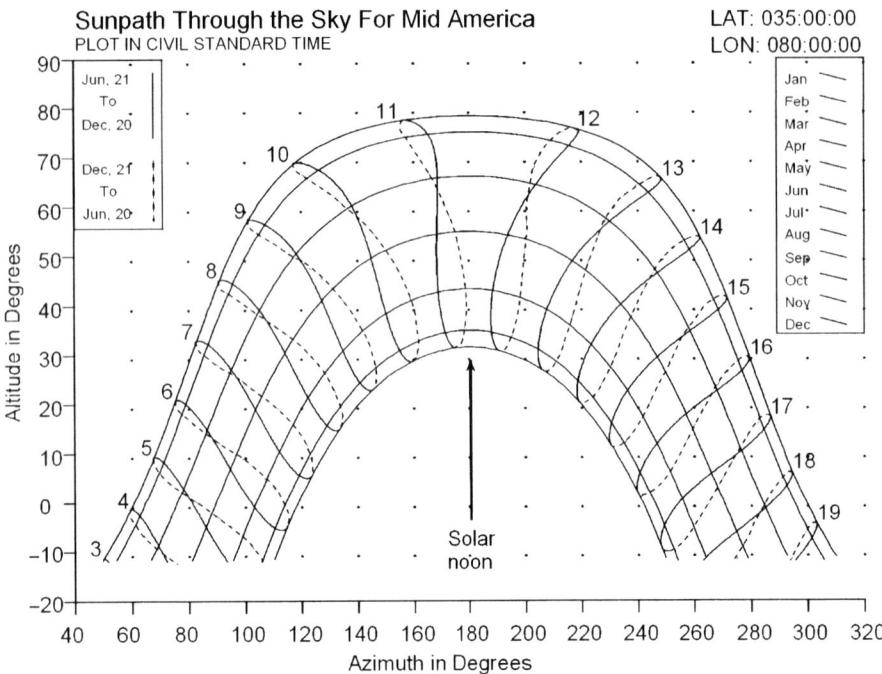

Fig. 3 Sunpath chart for a central U.S. location, in local standard time. Due south is at azimuth 180°. East is at 90°.

by the less dense, rising air next to the warmer pane of glass. This circulation of air from a warmer pane to a cooler one provides the third mechanism for heat transfer through a window—convection.

Air adjacent to a warm pane thus convects to the cooler pane, releases its heat, and repeats the cycle. See Fig. 1. If the air were not moving, it would be a relatively effective insulator since the conductivity of air at atmospheric pressure is quite small. Convective movement of air, however, can "short-circuit" the insulating value of still air and promote heat transfer. Changing the gap width affects the rate of heat transfer by convection. Generally, there is a non-zero optimum gap width for a given gas that maximizes the thermal resistance due to convection.

Double pane windows generally have substantially lower conduction and convection heat transfer. In climates with modest year-round inside/outside temperature differences, single pane windows may be adequate. Multiple pane windows exhibit significantly lower energy costs on buildings in climate regions that experience significant inside/outside temperature differences.

THE PATH OF THE SUN THROUGH THE SKY

The sun moves in a predictable manner each day. During the winter in the northern hemisphere at middle latitudes, it rises south of due east (azimuth 90) and sets south of due west (azimuth 270). In the summer, it rises north of due east and sets north of due west. Plots of solar position vs time on a chart of solar coordinates form what is called a sunpath chart. Fig. 3 shows an example of a sunpath chart for latitude 35 degrees north and longitude 80 degrees west. A variety of sunpath calculators are available on the Web.

ORIENTATION AND SHADING

If the paths of the sun for a given location are known, the building and its windows can be oriented to either capture or prevent the entry of solar heat gain from the direct sun and/or diffuse sky light. Shading devices of many kinds can also be applied to the building to block direct solar beam entry. A variety of devices for exterior application are illustrated in Fig. 4.

Interior shades can also be effective if their solar reflectances on the window-facing side are high, thereby causing them to reflect most of the solar radiation incident back toward—and hopefully through—the window to the outside.

Solar heat loading on cooling equipment can be further reduced when desired by placing buffer spaces like garages, laundry rooms, warehouse space, porches, and utility rooms on the east- or west-facing sides of a building.

When passive solar space heating is desired, moderately large window areas can be added to facades that face directions from which strong direct solar radiation can be expected during certain seasons of the year and times of day. This solar gain can be moderated by judicious building and window orientation and through the use of exterior shading devices and vertical and horizontal fins

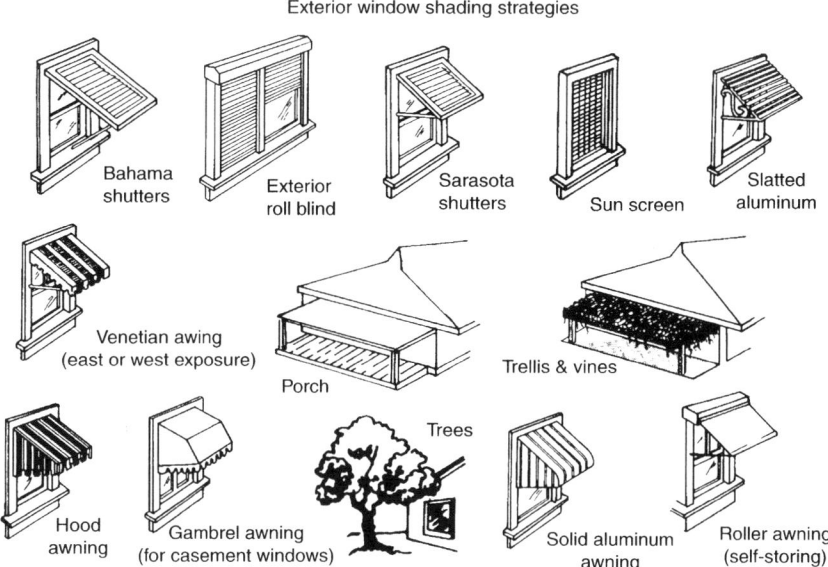

Fig. 4 Window shading options illustrated. Credit: Florida Solar Energy Center, 1679 Clear lake Rd., Cocoa, FL 32922.

and overhangs designed to selectively shade the window when direct sunlight is not desired.

Direct beam entry into a building should be carefully facilitated due to the potential adverse consequences that can result. Bright spots in a generally darker room can be powerful sources of glare and should therefore be avoided or carefully moderated by architectural or interior design features. Localized overheating and the fading of interior furnishings are additional difficulties that can be produced by direct beam entry. Both exterior and interior shades, properly operated through either manual or automatic motorized control, can be effective means of reducing glare and controlling direct beam solar heat gain, while still allowing unobstructed views of the exterior when the sun is not a problem.

WINDOW HEAT TRANSFERS

Solar radiant heat is conducted through a window via several mechanisms, depicted schematically in Fig. 5.

The thermal conductivity of glass is typically about 1.5 $Wm^{-1} K^{-1}$. Transparent acrylic and polycarbonate plastics are sometimes used as glazing materials, and their conductivities (around 0.2 $Wm^{-1} K^{-1}$) are somewhat lower than that of glass, but the inherent thermal conductivity of air in the absence of convection is approximately 0.002 $Wm^{-1} K^{-1}$, a much lower value. Double pane windows with a correctly sized gas gap can therefore significantly reduce the conductive heat transfers through them. Use of a more insulating gas, such as Argon or Krypton, can further reduce conductive heat transfer through windows. Due to convection, however, the thermal resistance of the gas gap is somewhat less than the conductance of still air (or a more insulating gas) with the same gap thickness.

All objects above absolute zero emit radiation. Long-wavelength infrared radiation (often termed "thermal" radiation) is therefore another mechanism of heat transfer in windows (the net heat transfer being from a warm to a cold surface). At the temperatures encountered in glazing systems, this radiation is invisible; it is confined to the long-wavelength (FIR) region of the electromagnetic spectrum.

The combined effects of conduction, convection, and radiation across the gas gap between glazing panes, which results from the temperature difference between inside and outside air, are often lumped together in colloquial descriptions and given the name "conductivity." A double pane glazing system has two panes separated around the edge by a spacer bar, a structural material that keeps the panes the correct distance from one another. The spacer

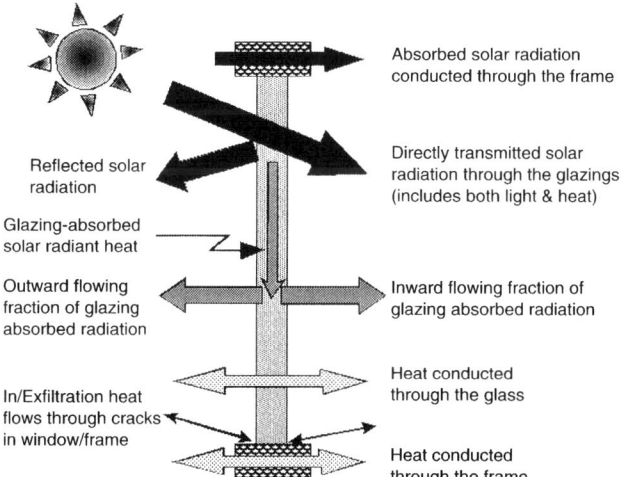

Fig. 5 Heat flows through window glass and window frames.

bar is often made of metal to ensure its strength. The higher thermal conductivity of metal results in higher conduction of heat through the glass/spacer/glass edge than through the center of the glass/gas/glass combination. Insulated glazing units (IGUs) usually have an edge seal around the outside perimeter of the unit. Sealant is often used between the glass and spacer bar, as well.

To prevent condensation inside sealed IGUs, a desiccant material is usually incorporated into the spacer bar, which contains small openings to allow water vapor in the gas gap to be absorbed by the desiccant inside. Some spacer bars are made of materials with higher insulating value to reduce conduction transfer, or they may contain a thermal break, a section of non-metallic, lower conduction material.

The glazing system is set within a frame. The total window unit, including the glazing and frame, is called a sash. The sash fits into another frame that is connected to the wall and holds the sash in place. Due to its use of different materials, the frame around a glazing conducts heat differently from the glazing. The National Fenestration Rating Council, responsible for rating the energy performance of windows and other fenestrations in the U.S., requires different values for the combined conductive/convective/FIR radiative transfer of heat through the center of a glazing system, through its edge, and through the framing members surrounding the glass.[3] The total thermal conductance of the entire sash is an area-weighted average of the frame, edge, and center of the glass values. Frames made of metal often contain thermal breaks to reduce conduction heat transfer through them.

If a window has cracks or other openings that permit the flow of air through these openings between the inside and outside of the building, the air so transferred can carry heat with it, thereby reducing the thermal integrity of the window. Such convective heat flow is termed infiltration or exfiltration, depending upon whether the air is flowing from outside to inside or vice versa. Modern factory-built windows generally exhibit relatively little infiltration and the heat gains or losses via this mechanism are generally much smaller than other heat transfers through the window. Windows in very cold areas, such as close to the poles of the Earth during the winter, can be exceptions to this rule if they are not very carefully constructed to minimize infiltration transfers.

In hot climates and buildings with air cooling equipment, the infiltration of hot and humid outside air poses a special problem. The air conditioner has to both cool the air and remove some of its moisture. At times, it takes more energy to remove the moisture (called a latent cooling load) than to lower the air's temperature (called a sensible cooling load). Windows, such as "jalousie" windows with large infiltration openings, often contribute large latent and sensible cooling loads to the building. Repairing or replacing such windows would be a useful energy conservation strategy.

THE "U-FACTOR"—CONDUCTIVE HEAT TRANSFER

The American Society of Heating, Refrigerating, and Air Conditioning Engineers (ASHRAE) uses the term "U-factor" to designate the overall conductive transfer property of a window: the heat transferred through the window by virtue of the temperature difference between the inside and outside air adjacent to the window. The full value for the window as a whole has units of heat flux per unit area and per unit temperature difference. In the *Systeme International* (SI, the international metric system), the units are $W\,m^{-2}\,°K^{-1}$ and in the inch-pound (IP) system, the common unit is $Btu\,hr^{-1}\,ft^{-2}\,°F^{-1}$. As previously mentioned, National Fenestration Rating Council (NFRC) standards divide the U-factor into frame, edge, and center-of-glass components, using an area-weighted average of these for the total product U-factor. The conductive heat loss Q_c for the whole window is the product of the window's U-factor, its area A, and the temperature difference going from inside t_i to outside t_o.

$$Q_c = UA(t_i - t_o) \qquad (2)$$

LOW-E COATINGS

In a multiple pane glazing system, it is common to number the glazing surfaces from outside in. Thus, the outer surface in contact with the outside air is surface 1 and the next one is surface 2. Surface 2 faces the first gas gap in the glazing system. It is common in multiple pane systems to place a coating on surface 2. Sometimes, surface 3 is used for the coating. Many such coatings are relatively soft microscopic multilayer structures that would be damaged while cleaning the window. These coatings are placed only

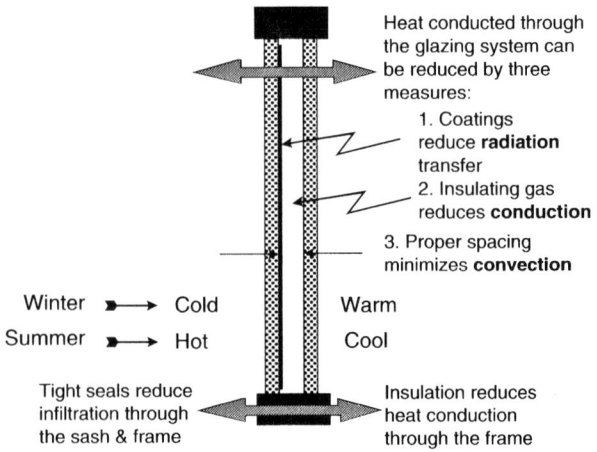

Fig. 6 Double pane windows with insulating gas, special coatings, and insulated frames reduce heat flows. Tight windows eliminate significant infiltration heat transfer.

on one of the inner surfaces in a multiple-pane system, so they can be sealed in the IGU and protected from damage and degradation. Hard coatings with similar properties that can be applied to single pane windows have been developed, but their energy performances are generally not quite as good as for the soft coats. See Fig. 6.

The glass coating has one or two purposes, depending upon its design and construction. The most common purpose is to admit the maximum quantity of solar radiation (over the spectral range from approximately 350–3500 nm) through the glass, while, at the same time, blocking FIR radiation emitted by the warm inner pane from crossing the gas gap and being absorbed by the colder outer pane during cold periods. Such coatings are called low-e coatings because their emissive property over the long-wavelength FIR portion of the spectrum is quite low. Emissivity is the ratio of thermal radiant emission from a surface to the maximum possible emission from the best possible radiator at the same temperature, called a blackbody. Bare glass has an emissivity of approximately 0.94, but low-e coating emissivities from commercial glass companies can be substantially below 0.2. Values as low as 0.09 are common. Such coatings can reduce the emitted long-wavelength radiation from glass by a factor of ten. The effect of this is to reduce the center-of-glass U-factor by reducing the radiative heat transfer component between the panes in the IGU. The U-factor is further lowered by using an insulating gas of the proper thickness between panes.

To distinguish these low-e coatings from a second kind of coating (to be described), they are given the generic name high solar gain low-e coating. An alternative name that has been used is "cold climate low-e coating."

Kirchhoff's Law states that the emissivity of a material is equal to the absorptivity of that material on a wavelength-by-wavelength basis. If the emissivity is low over a spectral range, then the absorptance will also be low over that range. The reflectance is generally high for such surfaces, especially if the coating and/or glass substrate are not very transmissive over the spectral range involved. This is because the sum of the absorptance, transmittance, and reflectance of a surface is 1.0. If two of these properties are low, the other must be high and vice versa. Thus the "low-e" designation applied to window glass refers to the emissivity (low) and reflectance (high) properties of the coated surface over the long-wavelength FIR spectral region of surface-emitted radiation.

The emissivity of a glass surface over the solar portion of the spectrum is not likely to be low because that implies high absorptance and a very dark window, one that would not transmit substantial quantities of solar or visible radiation. A low-e coating placed on surface 3 works by blocking the FIR radiative heat transfer, reducing its emission from the warmer inner pane. When it is placed on surface 2, it takes advantage of the coating's high reflectivity over the FIR, which causes much of the radiated heat from the warm inner pane (surface 3) to be reflected back to that pane from the colder outer pane (surface 2) before the radiant heat can be absorbed by the outer pane and then conducted and convected outdoors as a heat loss from the interior.

It is a quirk of common practice that radiative issues are embedded in a discussion of the conductive properties of a glazing system. The U-factor is but an approximation, a simplification of the more general heat transfer properties of a glazing system for the purposes of simplifying engineering calculations of window energy performance.

The second kind of coating commonly used in IGUs is a modification of the first. In this case, the purpose is to adapt the coating to the needs of buildings with significant cooling periods—when daytime solar radiant heat gain is as important as (or more so than) conductive heat loss on cold winter nights. In this case, the high spectral reflectance characterizing the FIR region for high solar-gain low-e coatings is extended toward lower-wavelength radiation across the near-IR region to the edge of the visible portion of the spectrum, at which point this reflectivity ideally should drop to zero and permit good transmission of visible daylight, the primary purpose of the window.

This coating is called a low solar gain low-e coating because it reflects near IR solar radiation back outdoors before it can be absorbed in the glazing or transmitted directly to the interior. Invisible solar IR radiation only enters a building as heat and is therefore unwanted in hot climate regions. This relatively new glass coating helps

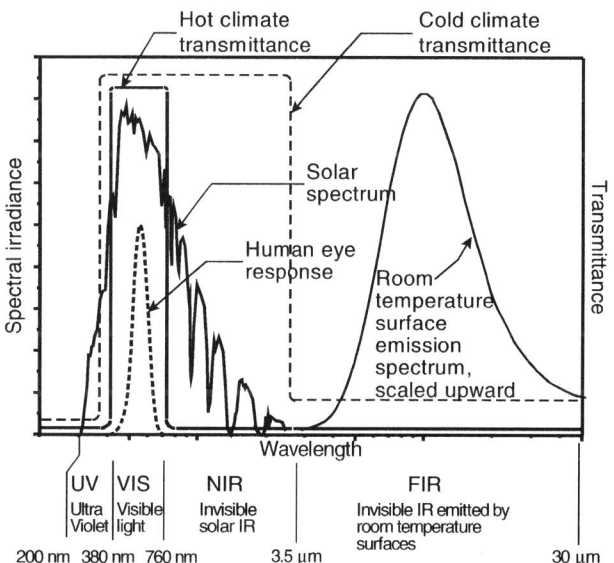

Fig. 7 Spectral irradiances of solar radiation and the infrared emission irradiance from a blackbody at room temperature. The blackbody curve is scaled so that its maximum value is the same as the solar spectrum maximum. Also shown are idealized spectral transmittances of both high and low solar gain low-e coatings.

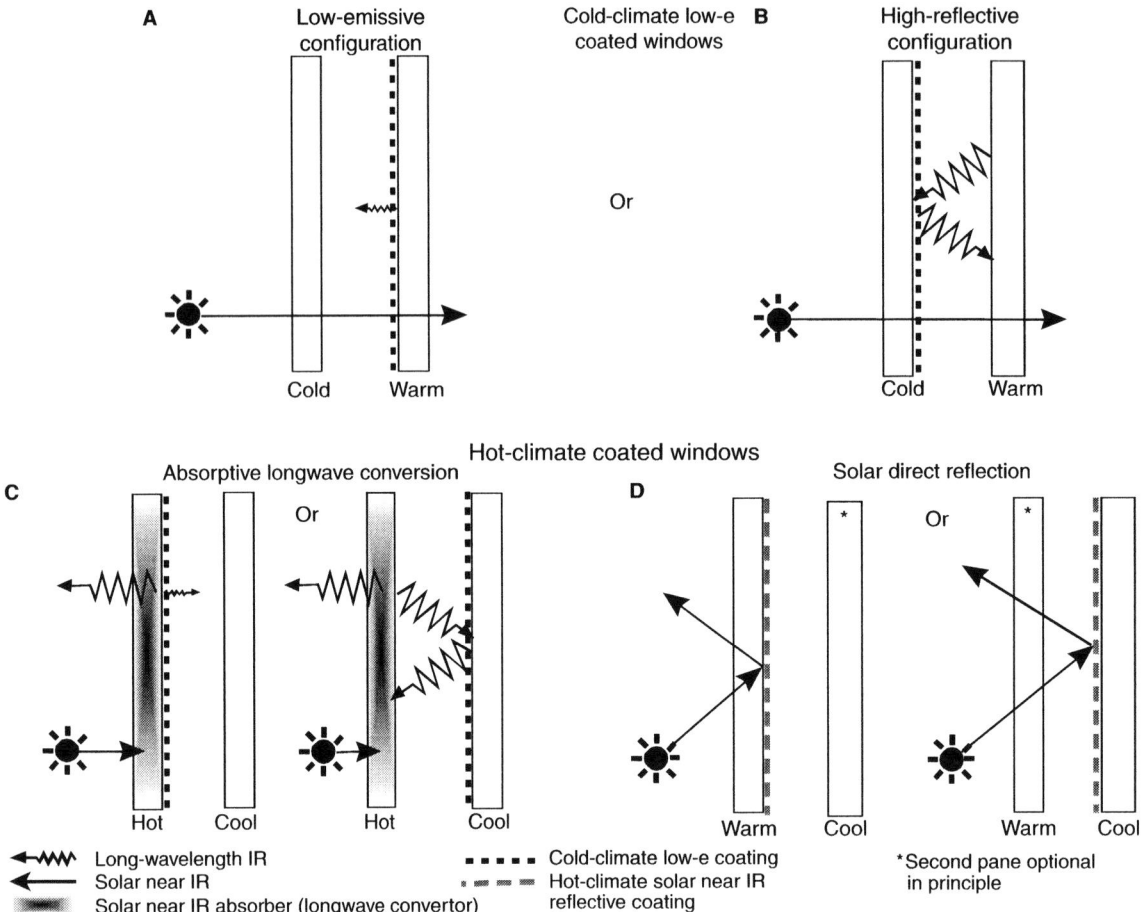

Fig. 8 Coatings and tints for energy efficient glazing systems.

keep a building's interior cool on hot summer days without placing an unnecessary heat burden on the air cooling system, while still admitting ample quantities of natural daylight. The idealized spectral transmittances of the two types of coatings are illustrated in Fig. 7.

Fig. 8 illustrates several ways these two coatings and spectrally selective absorption in the glass can be used to accomplish energy and illumination objectives in both hot and cold climates.

In Fig. 8, A and B describe the two cold-climate coated window configurations. A has the high-solar-gain low-e coating on surface 3 to reduce the wintertime emission of heat from the warmer inner pane to the colder outer one while B has the coating on surface 2 to reflect the long wavelength IR radiation emitted by the warmer inner pane toward the colder outer one. Likewise, parts C and D illustrate the operation of low-solar-gain low-e coatings in reducing summertime solar heat gain through double pane windows. C illustrates the absorptive case in which solar radiation is absorbed within the outer glass and the emission of the absorbed heat toward the colder inner pane is suppressed by a high solar gain type low-e coating on surface 2 by the low emittance of that coating or it is suppressed by reflection from the low-e coating on surface 3, sending the reflected heat back to the outer pane which dissipates it to the outside air. D illustrates the reflective case in which solar radiation is not absorbed by the glass but rather is reflected from a special low-solar-gain low-e coating on either surface 2 or surface 3, sending that radiation back outdoors, without much of it being absorbed within the window glass. The coating in this case, though reflective to the infrared portion of the solar spectrum, is transmissive over the visible portion, thereby allowing good visible transmittance for good view and daylight admission. In some cases, it is less expensive to increase the chemicals responsible for NIR absorption in glass than to make a low solar gain low-e coating. The high solar gain low-e coating generally has fewer layers and—for this and other reasons—is therefore less expensive to manufacture.

CONDENSATION RESISTANCE

Water vapor generally condenses from the vapor phase to liquid droplets when the temperature of the air or a surface in contact with the air drops below what is known as the dew point temperature, which is the temperature below which water vapor mixed with air at a specific humidity level can no longer remain in the vapor phase. The greater the air's humidity, the higher the dew point temperature. Condensation on the surfaces of a window is common with insufficiently insulated windows and most often seen when it is cold outside and warm inside. In buildings well-sealed against cold winter air and filled with relatively warm air, the moisture content indoors tends to rise due to several interior sources of water vapor. If a window is not very well-insulated, the window glass and/or frame can lose heat to the cold outside, dropping their interior surface temperatures. When these temperatures drop below the dew point for the indoor air near the window, condensation forms. Condensation can also occur on air conditioned buildings during humid weather conditions, with the condensation in this case forming on the outside surfaces of the window.

Condensation is annoying, impedes the view, and can be destructive to the window and building materials. The solution is to insulate the window better, keeping surfaces adjacent to the humid air warmer than the dew point temperature of that air.

The Condensation Resistance Factor (CRF) is a measure of the effectiveness of a window or glazing system in reducing the potential for condensation. The higher the CRF, the more energy efficient the window and glazing system, generally. Thus, a low U-factor generally leads to a high CRF, as long as there is not a small "thermal short-circuit" in the window system that allows an interior surface to become too cool.

THE "SOLAR HEAT GAIN COEFFICIENT"—SOLAR RADIANT HEAT GAIN

The solar heat gain property of a fenestration is the sum of two factors. See Fig. 5. The first is characterized by the solar transmittance of the window, T_s, given by

$$T_S = \frac{\int T(\lambda) E_{S\lambda}(\lambda)\, d\lambda}{\int E_{S\lambda}(\lambda)\, d\lambda} \qquad (3)$$

where $E_{S\lambda}(\lambda)$ is the solar spectral irradiance distribution and $T(\lambda)$ is the spectral transmittance. The second is the inward-flowing fraction N_i of the window system's solar absorptance A_s, given by

$$N_i A_S = N_i \frac{\int A(\lambda) E_{S\lambda}(\lambda)\, d\lambda}{\int E_{S\lambda}(\lambda)\, d\lambda} \qquad (4)$$

where $A(\lambda)$ is the spectral absorptance. Added together, the two of these form what is called the Solar Heat Gain Coefficient (SHGC).

$$SHGC = T_s + N_i A_s \qquad (5)$$

This is the ratio of total solar heat gain through the window to the incident solar irradiance on the window. The solar heat gain Q_s through the window is the product of the incident solar irradiance E_s over the plane of the window, the area A of the window, and the SHGC value for the window.

$$Q_s = E_s A\; SHGC\; [\text{watts}] \qquad (6)$$

VISIBLE TRANSMITTANCE, UV TRANSMITTANCE, AND DAMAGE-WEIGHTED TRANSMITTANCE

The human eye can see only wavelengths over the approximate range from 380 to 760 nm. Radiation outside this range is not visible and should not be called light. The eye's spectral response follows a bell-shaped curve, seen in Fig. 1 of the entry on daylighting, below the solar spectrum therein plotted, peaking at 555 nm. The visible transmittance T_v of a glazing system is the ratio of luminous flux transmitted by the window to the luminous flux incident upon it, defined by the following equation:

$$T_V = \frac{\int T(\lambda) V(\lambda)\, d\lambda}{\int V(\lambda)\, d\lambda} \qquad (7)$$

where $V(\lambda)$ is the photopic spectral luminous efficiency function or "V-lambda" weighting function and $T(\lambda)$ is the spectral transmittance of the glazing system. T_V is also known as VT.

The ultraviolet transmittance of a glazing system is also of interest because many people associate the transmission of ultraviolet radiation with fabric fading and other damage to a building's interior. There have been studies of fading and other damaging effects of solar radiation, but no consensus has yet emerged on which portion of the solar spectrum is most responsible or what spectral weighting function is appropriate for assessing the contribution of different solar UV wavelengths to the damage in a single "UV transmittance" figure. For a while, the straight integrated spectral transmittance from 300 to 380 nm was used, given the symbol T_{-UV}. More recently, the spectrum of interest has been extended beyond the UV portion to cover the range from 300 to 700 nm, and a different weighting function was selected. With this system, the $V(\lambda)$ weighting function of Eq. 7 is replaced by another function purported to represent the damaging portions of the solar spectrum. The damage-weighting function is

$$S_{\text{dm,rel}}(\lambda) = e^{3.6-12.0\lambda}\; \lambda\; \text{in}\; \mu\text{m} \qquad (8)$$

The resulting damage-weighted transmittance has the symbol $T_{-\text{dw}}$. The methodology is based on the work of

Jurgen Krochmann in Germany and stems from his studies of the damaging effects of radiation on paintings and other museum artifacts. The Krochmann damage-weighting function was incorporated into ISO/CIE publication 89/3 "On the Deterioration of Exhibited Museum Objects by Optical Radiation" and is referenced by standard NFRC 300 optical properties standard in computing the damage-weighted transmittance T_{-dw}.[3]

ILLUMINATION

Daylight illumination from windows and other fenestrations can be used to displace electric lighting energy use. The conditions for this to be successful are that: (1) the space should be occupied by people (in need of illumination); (2) the building should be designed for ample, high quality illumination from the fenestration systems; (3) the electric lighting should be turned off or dimmed to save energy when daylighting is adequate; and (4) the sun and sky conditions are such that the daylight illumination incident on the building is adequate. Design for daylighting is discussed in the entry on daylighting.

ENERGY PERFORMANCE

Solar radiant heat gain through windows, which is generally desirable during the winter and undesirable in the summer, and conductive heat transfer are the principal sources of unwanted energy costs associated with fenestrations. They can add to the cost of heating the building in winter and cooling it in summer. Computing the energy costs and savings associated with a building's fenestration systems has been eased by a variety of computer simulation tools that are now available. The WINDOW program, coupled with THERM for calculating 2-D frame and edge effects, both developed at Lawrence Berkeley National Laboratory, allows rapid calculation of the instantaneous energy performance characteristics of a window.[4] The results of these calculations can be used in computerized hourly building energy performance computer programs. RESFEN is another LBNL computer program for calculating the heating and cooling energy use of windows in residential buildings over a long time period. All of these LBNL computer programs are available for free download at http://windows.lbl.gov/software/. Additional energy simulation tools are available for determining fenestration contributions to a building's energy performance. They are described at the LBNL web site and some are available there. Additional tools can be found via a Web search.

WINDOW ECONOMICS

Replacing old windows in a building with energy-efficient ones and installing high performance windows during new construction can result in significant yearly energy savings, increased comfort, and improved occupant satisfaction. While these are desirable in residential buildings, better windows can also increase productivity in commercial or office buildings. Occasionally, the productivity enhancements save more dollars than the entire energy bill itself.

Achieving these desirable results can be expensive. (There are exceptions, such as the case where reduced solar heat gain from improved windows allows the air conditioning equipment to be significantly smaller, enough so that the saved equipment costs offset the extra window costs.) In markets driven primarily by least-first-cost goals, it can be difficult to justify the extra costs of better windows. Several accounting tools are available for calculating the longer-term benefits. These include payback time, return on investment, cash flow analysis, life-cycle cost/benefit analysis, and net energy analysis. Using these tools, the designer can capture at least some of the future dollar savings attributable to better windows. But these measures generally fail to include a variety of additional benefits, such as the dollar savings of improved worker productivity, the economic benefits of a healthier society, the values of improved ecosystem health, or the influence of future energy prospects.

CONCLUSION

As we pass the peak of world oil production, entering a subsequent steady decline in oil's availability, and as we approach future peaks in the production of coal and natural gas, the rising price of fossil fuel energy will shorten payback times and improve other indicators of window economic performance. Government actions, such as green taxes and other positive and negative incentives, can speed the adoption of energy-efficient and otherwise high-performance fenestration systems. Large projects offer higher budgets for design and specification, simplifying decisions to incorporate some of these additional benefits in design and product selection. For smaller projects, perhaps the best advice would be to use the most energy-efficient and affordable window system for comfort and as a hedge against future energy price increases.

ACKNOWLEDGMENTS

Portions of the Glossary section have been excerpted, with permission, from "Photometry" by Ross McCluney, in

Encyclopedia of Optical Engineering, Marcel Dekker, New York, 2003.

REFERENCES

1. International Lighting Vocabulary, CIE Publ No. 17.4 IEC Publ. 50(845), Central Bureau of the Commission Internationale de l'Eclairage: Kegelgasse 27, A-1030 Vienna, P.O. Box 169, Austria.
2. McCluney, R. *Introduction to Radiometry and Photometry*; Artech House: Boston, MA, 1994.
3. National Fenestration Rating Council. 8484 Georgia Avenue, Suite 320, Silver Spring, MD 20910, http://www.nfrc.org.
4. Building Technologies Dept., Building 90 Room 3111, Lawrence Berkeley National Laboratory: 1 Cyclotron Road, Berkeley, CA, 94720.

Window Films: Savings from IPMVP Options C and D[*]

Steve DeBusk
Energy Solutions Manager, CPFilms, Inc., Martinsville, Virginia, U.S.A.

Abstract

Building owners and property managers have provided numerous reports of the energy savings obtained from the retrofit installation of energy-control window film to existing buildings. In many cases, substantial energy savings are reported with simple paybacks often within one to four years. However, such claims have not been produced using the rigorous methods prescribed by the International Performance Measurement and Verification Protocol (IPMVP, see www.ipmvp.org).

To provide for more definitive evidence of the savings from the use of energy-control window films, CPFilms undertook a demonstration project to install window film on a mid-sized commercial office building and then measure the resulting energy savings using two separate IPMVP methods. The measurement of savings involved the use of IPMVP Option C (whole-building method using utility bill analysis) and Option D (calibrated simulation). During the twelve months following film installation, Option C measured savings of 8.8% were achieved, which includes savings over both the cooling and heating seasons. This compares favorably to Option D calculated savings of 8.4% that were estimated prior to film installation. Both estimates of energy savings demonstrate a simple payback in less than three years.

PROJECT OVERVIEW

In early 2002, CPFilms completed the installation of 9200 ft^2 of energy-control window film on an eight-story building in Rockford, Illinois (near Chicago). Fig. 1 shows the building chosen for the demonstration project following completion of window film installation. The building's conditioned space is 59,000 ft^2 and the building's windows are single-pane bronze tinted glass. The building is heated and cooled using all-electric room unit ventilators.

The window film chosen for this demonstration project was CPFilms' LLumar® E-1220 Low-E film. As shown in Table 1, addition of this film reduced solar heat gain 67% based on the solar heat gain coefficients (SHGC) of the windows before and after film installation. In addition, the film is constructed with a low-emissivity, or low-e coating that improved the insulating value (U-value) of the single-pane windows by 23%. This improvement in the window's insulating properties provided additional cooling-season energy savings, beyond the reduction in solar heat-gain, and provided heating-season energy savings by reducing heat loss through the windows.

Rockford has over 6000 heating degree days per year (using a 65°F base). This cool-climate location was chosen to demonstrate that, under the proper circumstances, energy-control window films are an attractive option even in cooler climates and are not simply a "warm climate only" technology, especially when utilizing low-e insulating films. The location was also chosen to show that low-e window films, and the improvement in window insulating properties they offer, can provide substantial energy savings during a prolonged heating season, more than offsetting the loss of "free" solar heat. Negotiations with the building owner prohibited other energy-saving measures from taking place in order to isolate the effects from window film installation.

Before film installation, CPFilms utilized DOE-2.1 to evaluate project energy savings using IPMVP Option D calibrated simulation methods. CPFilms then enlisted the efforts of Johnson Controls, a leading national energy service company (ESCO), to independently measure and verify energy savings using IPMVP Option C. The Option C measurement of energy savings was compared to predicted savings from the calibrated simulation as a means of determining the suitability of using DOE-2 in predicting energy savings for future projects. This was deemed important, as normally it is not possible to isolate the energy savings from window film installation. Often multiple energy-saving measures are performed along with window film. As a result, Option D typically is the only viable means for determining energy savings from window film installation.

WINDOW FILM ENERGY SAVINGS USING IPMVP OPTION D

Prior to window film installation, a whole-building energy simulation of the demonstration building was prepared.

---

[*] This entry originally appeared as "Measuring Savings from Energy-Control Window Films Installations Using IPMVP Options C and D" in *Energy Engineering*, Vol. 102, No. 4, 2005. Reprinted with permission from AEE/Fairmont Press.

Keywords: Building envelope; DOE-2; Energy-control films; IPMVP; Low-emissivity; Solar heat gain coefficient; Windows; Window film.

Fig. 1 Demonstration building following film installation.

Table 2 Values used in doe-2 simulation

Building type	Commercial office
Building size	8 floors, 59,000 ft² total conditioned area
Window areas	North—2875 ft²
	South—2875 ft²
	East—1725 ft²
	West—1725 ft²
Window type	Single-pane, 1/4-in., Bronze
Lighting offices	1.0 w/ ft²
Plug load	w/ ft²
Heating temperatures	70°F daytime, 65°F night
Cooling temperatures	75°F daytime, 85°F night
Ventilation rate	20 cfm per person, 250 ft² per person
Cooling system efficiency	1 kW/ton

To ensure that the simulation model would provide an accurate assessment of the energy savings resulting from window film installation, the monthly and annual energy consumption of the simulated building was compared to the building's actual consumption for the two previous years. DOE-2 energy analysis software was chosen as the method for simulating the building's energy performance, based upon the regular use of the program in the research, ESCO, and utility industries as a reliable method for such purposes. Table 2 provides a review of the most pertinent information used in the simulation.

Fig. 2 shows the comparison of monthly and annual electricity consumption for the building using averaged data from 2000 to 2001 compared to the simulation results. As indicated in Fig. 2, the simulation estimate of annual energy use for the demonstration building closely matches actual data for the 2000–2001 period, while matching the monthly fluctuations in energy use reasonably well. As the simulation model uses averaged weather data over multiple years, it is not reasonable to expect simulation and actual monthly energy use to match precisely each month.

Next, the building DOE-2 model was modified to show the effect of adding CPFilms' LLumar® E-1220 low-e window film to the existing windows. As seen in Table 3, the DOE-2 calculations indicated an 8.4% reduction in annual electricity consumption, saving approximately 180,850 kWh annually and producing a simple payback in 2.77 years.

WINDOW FILM ENERGY SAVINGS USING IPMVP OPTION C

Energy savings for the demonstration building were also independently measured by Johnson Controls using IPMVP Option C, employing the analysis of monthly utility bills. This method was deemed appropriate, as the demonstration building is all electric, making it relatively simple to identify the energy savings from window film installation for both the heating and cooling seasons.

As expected, the main factors causing fluctuations in energy consumption for the building, based on a review of prior energy use, were: (1) building occupancy, and (2) weather conditions. As mentioned, other factors affecting building energy consumption (type and amount of lighting, heating, ventilating and air-conditioning (HVAC) equipment efficiency and type, cooling and heating temperatures, etc.) were held relatively stable to enable isolating the savings from window film installation as much as possible. Accordingly, the method for measuring

Table 1 Window thermal and solar properties

Window values	U-value[a] (Btu/ h/ ft²/deg F)	Solar heat gain coefficient
Before film installation	1.09	0.64
After film installation	0.84	0.21
% Improvement with film	23%	67%

[a]National Fenestration Rating Council (NFRC)/American Society of Heating, Refrigerating, and Air-Conditioning Engineers (ASHRAE) Winter night-time conditions, 0°F outdoors, 15 mph wind, 70°F indoors.

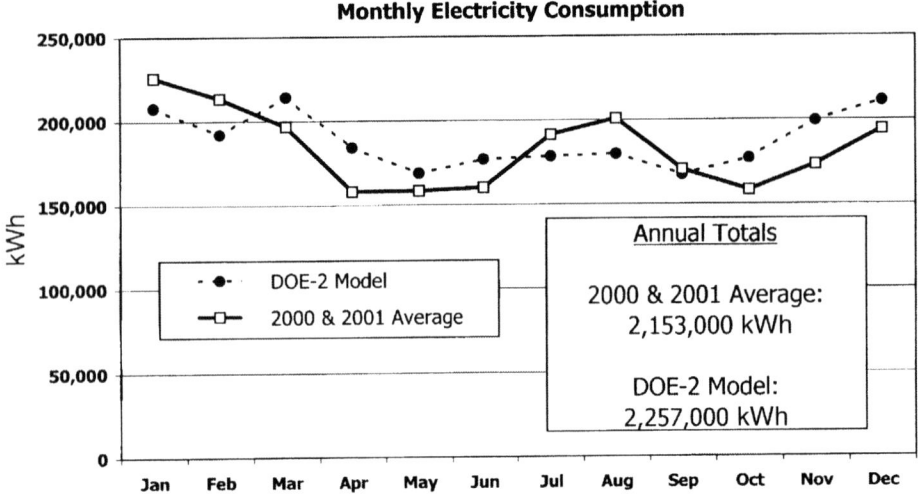

Fig. 2 Monthly and annual electricity consumption (actual vs DOE-2 model).

Table 3 Annual savings using Option D

Existing annual electric consumption and cost based on average of 2000 & 2001	2,153,000 kWh	$138,222
Estimated savings (as a % of existing consumption)	8.4%	
Estimated annual savings	180,850 kWh	$11,610
Simple payback	2.77 years	

energy savings would be to determine the energy consumption over the 12-month period prior to window film installation, after adjusting consumption for occupancy and weather, and compare this figure to the value for the 12 months following window film installation.

Fig. 3 shows the electricity use per thousand ft^2 of occupied space per degree day (kWh/ksqft/DD), before, during and after window film installation. From Fig. 3 it is apparent that during and immediately following window film installation, energy usage began to decrease significantly. As seen in Fig. 4, the average energy consumption for the twelve months prior to film installation was 11.18 kWh/ksqft/DD. Following film installation, the same information was tracked on a monthly basis and the average energy consumption for the twelve months following film installation was 10.19 kWh/ksqft/DD. A comparison of these "before" and "after" values indicated the project's overall savings to be 8.8%.

Table 4 shows the results obtained from both IPMVP methods, in a side-by-side comparison. As indicated in this table, the results of both methods match closely, giving additional confidence to the results obtained.

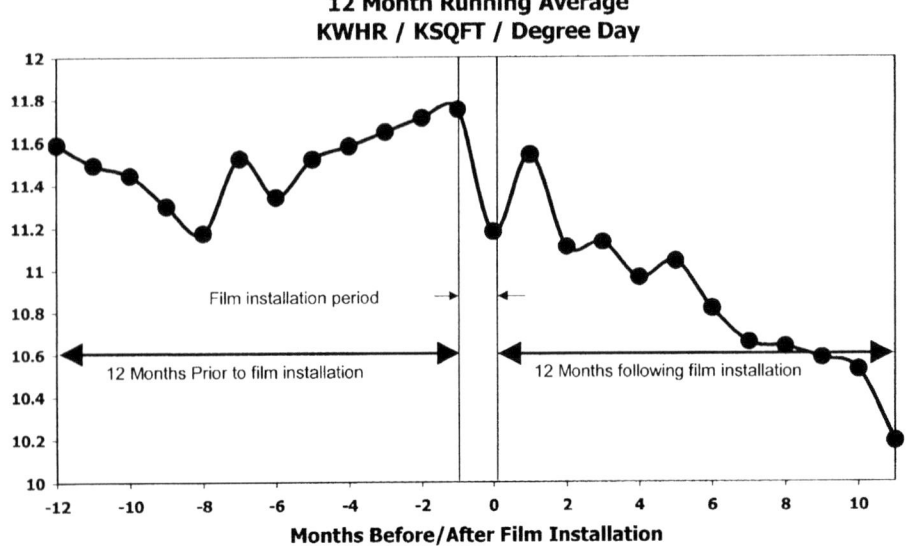

Fig. 3 Building energy use—before and after film installation.

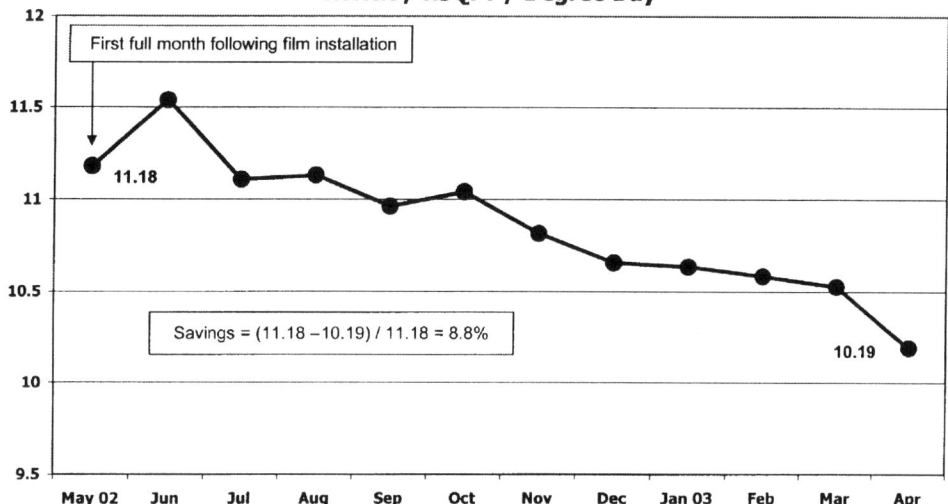

Fig. 4 Determining energy savings after film installation.

Table 4 Comparison of energy savings using Option C and Option D

	Option C	Option D
Annual kWh savings	189,460	180,850
Savings as a % of existing consumption	8.8%	8.4%
Annual dollar savings	$12,163	$11,610
Simple payback (years)	2.65	2.77

CONCLUSION

The demonstration project provided insight into the substantial savings that can be achieved from the installation of energy-control window films. The project also demonstrated that significant energy savings can be obtained even in cool climates, if a significant amount of energy-inefficient window glass is utilized on the building envelope. Energy savings in warmer regions are likely to be greater than those shown in this article, but actual savings will be dependent upon many variables, especially existing glass type and type of window film utilized. These variations can easily be accounted for within a DOE-2 analysis for an accurate assessment.

The project also demonstrated that energy savings from window film installation can be accurately predicted using DOE-2 or other appropriate computer simulation software, if an accurate model of the building is prepared. This is especially important, as directly measuring the energy savings from window film installation using IPMVP methods other than Option D is often difficult, as the many factors affecting energy consumption are not normally held constant as they were in this study. Using the information provided by this project, energy savings from future window film projects can be predicted with confidence using DOE-2.

ACKNOWLEDGMENTS

The author wishes to thank the following for their assistance: Jeff Hongsermeier, owner of Glass Enhancements, Rockford, Illinois, for providing window film installation services and for his assistance in developing this project, Johnson Controls for assistance in project development and for determining Option C savings, and the Fridh Corporation for use of their property.

Window Films: Solar-Control and Insulating

Steve DeBusk
Energy Solutions Manager, CPFilms Inc., Martinsville, Virginia, U.S.A.

Abstract

Solar-control films are used worldwide as an effective means of lowering building energy costs by reducing excessive solar heat gain through windows. Determining energy savings from actual installations is difficult due to the common practice of implementing several energy conservation measures simultaneously and due to annual variations in the many factors that can affect a building's energy usage.

To isolate and quantify the energy-saving benefits of solar films, an energy analysis study on a conventional office building was undertaken using the U.S. Department of Energy (DOE)'s sophisticated DOE-2 energy-simulation software. The study included several types of solar films applied to various glazing systems. Various cities were included in the study to illustrate the energy savings in different climates and to show the effect of differing electricity costs.

Based on the most typical types of installations and on customary installation costs for medium-size commercial projects, the average return on investment (payback) from solar film application was an impressive 2.65 year. These savings were the result of reducing annual electricity kilowatt-hour (kWh) usage by an average of 6.6% and reducing peak summer month kilowatt demand on average by 6.4%.

INTRODUCTION

Solar-control window films are considered in the building industry to be "retrofit" products—that is, products that are applied to existing buildings after construction as opposed to being used in new construction.[1,2]

Solar Film Construction

Solar-control window films typically consist of a thin (0.025-mm, 0.001-in.) polyester film substrate that has a microthin transparent metal coating applied to one side. This metal coating is applied using vacuum-based technologies such as vapor deposition or sputtering. The metal coating may be a single metal, an alloy, a metal-oxide, or a combination of these coatings. A second layer of polyester film is laminated over the metal coating to protect the metal. Onto one side of this laminated composite, an acrylic scratch-resistant (SR) coating is applied to the surface that will face the building interior. This SR coating protects the film during normal window cleaning. On the opposite side of this film laminate, a clear adhesive is applied, which will eventually bond the film to the window glass. This adhesive layer is protected by a removable release liner until just before field application. The film is protected from ultraviolet (UV) degradation by UV absorbers that are added to the polyester film layers, the adhesive layer, or both.

Keywords: Building envelope; DOE-2; Energy simulation; Solar control; Solar films; Window films; Windows; Window systems.

Solar Film Appearance and Properties

The appearance of the film (the color, the level of visible light transmission, and the degree of reflectivity) depends on the metal coatings used. Typical all-metal solar films can be silver-reflective, gray, silver-gray, bronze, or light green in color. Visible light transmissions can vary from very dark (10%) to very light (70%), and visible reflectance can vary from the same reflectance as clear glass (8%) to highly reflective (60%). The ability of a glazing system to reduce solar heat gain is measured by its solar heat-gain coefficient (SHGC). As expected from the variety of films available, the SHGC for solar films can vary significantly, from 0.14 to 0.69, as measured on 6-mm (1/4-in.) clear glass.

Solar Film Benefits

This combination of film properties produces a product that provides several important benefits:

- Reduced cooling energy costs by reducing excessive solar heat gain
- Enhanced reduction in cooling and heating energy costs when low-e type films are used
- Enhanced tenant comfort from improved temperature distribution (less hot and cold spots) and reduced glare
- Uniform building appearance from the exterior—improving tenant retention in leased buildings
- Reduced fading of carpets, drapes, and furnishings due to the UV-blocking ability of films
- Privacy for building occupants when using reflective or dark films

Solar Film Installation Process

The first step in the installation of solar film involves rigorous cleaning of the window glass surface. Next, an application solution is sprayed onto the glass; the release liner is removed from the film; and the adhesive side of the film is carefully placed onto the glass surface. The application solution allows one to move the film on the glass for precise film placement, and it prevents air from becoming trapped between the film and glass. Then the film is carefully trimmed around the perimeter of the window, leaving a 1–3-mm (1/32-in. to 1/8-in.) gap between the film and frame. Water is sprayed onto the film surface, and a rubber squeegee is used to remove the application solution and bond the film to the glass surface.

MOTIVATION FOR SOLAR-CONTROL FILM ENERGY ANALYSIS STUDY

Solar-control films have been used since the early 1960s as an effective means of reducing building energy costs. Unfortunately, it is difficult to determine precisely the cost savings from applying film to a building, due to variations in the many factors that affect a building's energy consumption from year to year (weather changes such as the amount of sunshine and temperature/wind differences, changes in occupancy, upgrades or other changes in building energy-using equipment, changes in maintenance for key equipment, the addition of energy-consuming equipment such as computers, etc). This situation is usually complicated by the fact that building owners usually perform energy-conservation upgrades such as solar-control film application in conjunction with other upgrades, making it impossible to determine the savings from film application alone. Therefore, a means of accurately estimating the energy savings from solar film application was needed.

One of the most accurate and reliable energy simulation software packages available is the DOE-2 energy analysis program.[3] DOE-2, which uses an hourly calculation method, has been validated many times by comparing its results with thermal and energy-use measurements on actual buildings (see http://gundog.lbl.gov).[4] DOE-2 is used worldwide by energy engineers, architects, government organizations, and utilities as a means of estimating the effect of various measures on a building's energy consumption and for developing building energy codes. As a result, it was determined that a reasonable course of action was to use DOE-2 modeling to estimate the energy savings from application of solar-control window film.

SCOPE OF ENERGY ANALYSIS STUDY

The DOE-2 energy study was performed on a conventional (1990s) 10-story office building with a total floorspace of 16,257 m^2 (175,000 ft^2). To gauge the effect of different films, four films were chosen and categorized as "Maximum-Performance," "Maximum-Performance

Table 1 Solar and thermal performance factors for windows and films in study

	No film	Maximum performance film	Maximum performance low-E film	High performance film	High performance low-E film
Solar heat-gain coefficient (SHGC)					
Single clear	0.81	0.23	0.17	0.36	0.28
Dual clear	0.70	0.31	0.27	0.42	0.35
Single gray	0.57	0.27	0.21	0.33	0.26
Dual gray	0.45	0.24	0.21	0.30	0.25
U-values (W/m^2 °C)					
Single clear	6.18	5.81	4.78	5.87	4.69
Dual clear	2.74	2.65	2.35	2.66	2.32
Single gray	6.19	5.81	4.78	5.87	4.69
Dual gray	2.74	2.65	2.35	2.66	2.32
U-values (BTU/h/ft^2/°F)					
Single clear	1.09	1.02	0.84	1.03	0.83
Dual clear	0.48	0.47	0.41	0.47	0.41
Single gray	1.09	1.02	0.84	1.03	0.83
Dual gray	0.48	0.47	0.41	0.47	0.41

Low-E," "High-Performance," and "High-Performance Low-E," based on the film's SHGC on single-pane, 6-mm (1/4-in.) clear glass.

The study was performed on single-pane clear, dual-pane clear, single-pane gray-tinted, and dual-pane gray-tinted window systems. Each film type was analyzed on each of these glazing systems. All windows consisted of 6-mm-thick (1/4-in.-thick) panes, and for the dual-pane units, the panes were separated by a 12-mm (1/2-in.) air space.

The SHGC and Winter Night-Time U-values for all glazing systems contained in the study are shown below. The SHGC shown is at normal solar incidence angle, and the Winter Night-Time U-value is based on American Society of Heating, Refrigerating, and Air-Conditioning Engineers (ASHRAE) standard Winter conditions, using a $-17.8°C$ (0°F) outdoor air temperature, an indoor air temperature of 21°C (70°F), a 24.1 km/h (15 mph) wind speed outdoors, a 0 km/h (0 mph) wind speed indoors, and 0 W/m^2 (0 Btu/h/ft^2) solar intensity. In the analyses, the SHGC was calculated (by DOE-2) for the precise sun angle at each hour of the day; and the window U-value was calculated at each hour based on indoor and outdoor temperatures, outdoor air speed, and solar intensity (Table 1).

The model building was a square building with equal glass area facing north, south, east, and west. The glass area on each of the four building exposures was 557.4 m^2 (6,000 ft^2). Models with film applied used film on the east, south, and west exposures only.

Other model parameters, typical of modern office buildings, used in the study included

- Indoor lighting 10.76 W/m^2 (1.0 W/ft^2)
- Office equipment 11.95 W/m^2 (1.1 W/ft^2)
- Heating setpoint 21.1°C (70°F)
- Heating setback 18.3°C (65°F)
- Cooling setpoint 23.9°C (75°F)
- Cooling setback 26.7°C (80°F)
- Medium-colored blinds used 25% of the time; SHGC of blinds 0.69.
- Windows recessed from building face 15 cm (6 in.) providing partial shading of all windows
- Variable-air-volume (VAV) air-distribution system and air-side economizer
- Heating plant using gas boilers with an efficiency of 80%
- Chillers with full-load efficiency of 0.69 KW per ton (Coefficient of Performance [COP] of 5.1)

The parameters used (such as the use of blinds, recessed windows, the VAV air distribution system, the air-side economizer, and the high-efficiency chiller system) effectively reduce the savings from solar film application. This was desired to provide reasonable and conservative estimates of energy savings from solar film installation, not to create a "best case" scenario.

Electricity costs for each location were determined from the commercial rate schedules published on the Web sites for electric utilities in each city. Rate schedules that applied to buildings with peak kilowatt demands of approximately 1000 KW were used. Both kilowatt-hour and kilowatt demand charges were used. The rate schedules used typically vary the kilowatt-hour and kilowatt charges by time of year and time of day, and are too complex to provide here; however, to provide the reader a general idea of the costs used, Table 2 shows the average costs for each city based on the total annual kilowatt-hour used and the total annual electricity costs for the four building models without film.

RESULTS OF STUDY

Tables 3–5 show the payback, reduction in annual kilowatt-hour usage, and reduction in summer month

Table 2 Average electricity costs by location

City	Average cost ($/kWh)	Electric utility and rate schedule
Boston	0.1220	Boston edison, rate G-3
Chicago	0.0892	Commonwealth edison, rate 6L
Dallas	0.0907	Texas-new mexico power, large general service
Jacksonville	0.0697	Florida power & light, GSLD-1
Los Angeles	0.1336	Southern California edison, TOU-8, large general service
Memphis	0.0604	Memphis light, gas & water, general service GSA
Phoenix	0.0595	Arizona public service co., general service
Toronto (Canada)	0.0553	Toronto hydro-electric, business rates
Washington, D.C.	0.0706	Baltimore gas & electric, schedule GL (option 2)
Overall average	0.0834	

Natural gas was used as the heating fuel and for domestic hot water production at a cost of 70 cents per therm for all locations.

Table 3 Simple payback by location, window, and film type

	Boston	Chicago	Dallas	Jacksonville	Los Angeles	Memphis	Phoenix	Toronto	Washington	Average
Single clear										
Max-perf film (%)	0.89	1.11	0.87	1.18	0.69	1.42	1.10	1.73	1.28	1.14
Max-perf low-E film (%)	0.81	1.00	0.85	1.15	0.69	1.33	1.07	1.43	1.16	1.06
High-perf film (%)	1.12	1.37	1.10	1.53	0.87	1.82	1.38	2.19	1.66	1.45
High-perf low-E film (%)	0.91	1.10	0.95	1.31	0.79	1.48	1.20	1.55	1.30	1.18
Single clear-all film types average (%):										1.21
Dual clear										
Max-perf film (%)	1.46	1.79	1.42	2.03	1.12	2.57	1.77	2.64	2.12	1.88
Max-perf low-E film (%)	1.34	1.65	1.34	1.90	1.08	2.29	1.70	2.29	1.93	1.72
High-perf film (%)	2.01	2.50	2.00	3.02	1.57	3.69	2.47	3.94	3.14	2.70
High-perf low-E film (%)	1.55	1.91	1.60	2.31	1.25	2.68	2.01	2.69	2.30	2.03
Dual clear-all film types average (%):										2.09
Single gray										
Max-perf film (%)	2.09	3.14	2.18	2.81	1.78	3.81	2.39	4.16	2.90	2.81
Max-perf low-E film (%)	1.42	1.94	1.59	2.09	1.36	2.49	1.90	2.34	1.96	1.90
High-perf film (%)	2.62	3.95	2.81	3.70	2.26	4.96	2.99	5.31	3.84	3.61
High-perf low-E film (%)	1.49	1.98	1.73	2.28	1.49	2.61	2.03	2.39	2.08	2.01
Single gray-all film types average (%):										2.58
Dual gray										
Max-perf film (%)	3.92	5.57	3.71	4.63	2.99	6.90	3.86	6.53	4.96	4.79
Max-perf low-E film (%)	2.90	3.56	2.81	3.57	2.32	4.65	3.18	4.26	3.57	3.42
High-perf film (%)	5.60	8.02	5.49	6.82	4.34	10.26	5.20	9.19	7.04	6.88
High-perf low-E film (%)	3.12	3.86	3.19	4.04	2.62	5.11	3.58	4.53	3.99	3.78
Dual gray-all film types average (%):										4.72
Location average (%)	2.08	2.78	2.10	2.77	1.70	3.63	2.36	3.57	2.83	
All location, all window types, all film types (%)	2.65									

Table 4 Reduction in annual kilowatt-hour usage

	Boston	Chicago	Dallas	Jacksonville	Los Angeles	Memphis	Phoenix	Toronto	Washington	Average
Single clear										
Max-perf film (%)	9.5	9.6	12.1	11.4	11.7	11.0	13.6	9.2	10.3	10.9
Max-perf low-E film (%)	10.9	10.9	13.5	12.7	12.9	12.3	15.3	10.5	11.7	12.3
High-perf film (%)	7.5	7.8	9.6	8.9	9.3	8.7	10.7	7.2	8.1	8.6
High-perf low-E film (%)	9.4	9.6	11.8	11.0	11.1	10.8	13.4	9.1	10.2	10.7
Single clear-all film types average (%):										10.6
Dual clear										
Max-perf film (%)	6.2	6.4	7.9	7.1	7.6	6.8	9.1	6.2	6.7	7.1
Max-perf low-E film (%)	7.3	7.4	9.2	8.4	8.8	8.0	10.5	7.3	7.9	8.3
High-perf film (%)	4.5	4.6	5.7	4.9	5.5	4.8	6.6	4.3	4.7	5.1
High-perf low-E film (%)	6.1	6.2	7.6	6.9	7.4	6.7	8.8	6.0	6.5	6.9
Dual clear-all film types average (%):										6.8
Single gray										
Max-perf film (%)	4.2	3.8	5.3	5.2	5.0	4.6	6.7	4.0	4.7	4.8
Max-perf low-E film (%)	5.9	5.4	7.3	7.1	6.8	6.5	9.0	5.7	6.5	6.7
High-perf film (%)	3.4	3.0	4.2	3.9	4.0	3.6	5.4	3.1	3.6	3.8
High-perf low-E film (%)	5.4	5.0	6.6	6.3	6.1	5.9	8.2	5.1	5.8	6.1
Single gray-all film types average (%):										5.3
Dual gray										
Max-perf film (%)	2.6	2.5	3.5	3.4	3.3	3.0	4.6	2.8	3.1	3.2
Max-perf low-E film (%)	3.5	3.5	4.7	4.6	4.4	4.1	6.0	3.8	4.2	4.3
High-perf film (%)	1.8	1.8	2.4	2.4	2.3	2.1	3.5	2.0	2.2	2.3
High-perf low-E film (%)	3.1	3.1	4.0	3.9	3.8	3.6	5.2	3.3	3.7	3.8
Dual gray-all film types average (%):										3.4
Location average (%)	5.7	5.7	7.2	6.8	6.9	6.4	8.5	5.6	6.2	
All location, all window types, all film types (%)	6.6									

Table 5 Reduction in summer peak kilowatt demand

	Boston	Chicago	Dallas	Jacksonville	Los Angeles	Memphis	Phoenix	Toronto	Washington	Average
Single clear										
Max-perf film (%)	11.4	10.6	11.2	10.0	9.7	10.1	11.3	11.4	10.4	10.7
Max-perf low-E film (%)	13.0	12.2	12.9	11.5	11.2	11.8	13.6	13.0	12.3	12.4
High-perf film (%)	8.7	8.2	8.5	7.4	7.4	7.6	8.9	8.7	7.7	8.1
High-perf low-E film (%)	10.9	10.4	11.1	9.9	9.1	10.0	11.7	11.1	10.3	10.5
Single clear-all film types average (%):										10.4
Dual clear										
Max-perf film (%)	6.5	6.8	6.9	5.7	5.7	5.8	7.4	7.6	6.3	6.5
Max-perf low-E film (%)	7.8	8.1	8.4	7.1	6.5	7.0	8.9	9.0	7.6	7.8
High-perf film (%)	4.7	4.9	4.8	3.6	4.0	3.7	5.2	5.2	4.1	4.5
High-perf low-E film (%)	6.6	6.6	6.9	5.5	5.8	5.8	7.4	7.3	6.1	6.4
Dual clear-all film types average (%):										6.3
Single gray										
Max-perf film (%)	5.3	4.4	4.9	4.7	4.6	4.4	5.7	5.3	5.4	5.0
Max-perf low-E film (%)	7.4	6.3	7.3	7.0	6.5	6.7	8.3	7.4	7.5	7.2
High-perf film (%)	4.2	3.4	3.8	3.6	3.4	3.3	4.6	4.0	4.2	3.8
High-perf low-E film (%)	6.8	5.9	6.6	6.3	5.4	6.1	7.7	6.6	6.9	6.5
Single gray-all film types average (%):										5.6
Dual gray										
Max-perf film (%)	2.9	2.8	2.9	3.0	2.3	2.6	3.3	3.8	3.2	3.0
Max-perf low-E film (%)	3.7	4.0	4.3	4.2	3.2	3.8	4.7	5.2	4.4	4.2
High-perf film (%)	2.0	1.7	1.9	1.9	1.6	1.7	2.4	2.7	2.4	2.0
High-perf low-E film (%)	3.5	3.4	3.6	3.7	3.0	3.4	4.2	4.5	3.7	3.7
Dual gray-all film types average (%):										3.2
Location average (%)	6.6	6.2	6.6	5.9	5.6	5.9	7.2	7.1	6.4	
All location, all window types, all film types (%)	6.4									

peak demand for each location and for each film and window combination. Also shown are the averages for each window type and each location. The overall average for all locations, window types, and films is also given.

Following are some general observations concerning the results of the study.

- For all window and film types and all locations, the overall average payback for solar film installation was 2.65 year (see Table 3). The average payback by window type: Single Clear, 1.21 year; Dual Clear, 2.09 year; Single Gray, 2.58 year; and Dual Gray, 4.72 year (in almost 50% of the cases involving Dual Gray windows, the payback was less than 4 year).
- As shown in Table 3, it appears that the payback period is affected more by the cost of electricity than by climate effects. The average payback in Boston (2.1 year), for example, is less than the average payback in Jacksonville, Florida (2.8 year), due mainly to the higher average cost of electricity in Boston compared with Jacksonville (12.2 cents per kWh average vs 6.97 cents). Also, the average payback for Memphis (3.6 year) was more than in Washington, D.C. (2.8 year), even though the climate in Memphis is somewhat warmer, solely due to the lower cost of electricity in Memphis.
- The data also show that solar-control film is not a "warm climate only" product. The average payback for cities not considered to be in the Sun Belt was still less than 3 year on average (Boston, 2.1 year; Chicago, 2.8 year; Washington, DC, 2.8 year). The average payback for Toronto (the coolest climate of all cities considered) was still a very respectable 3.6 year, despite the fact that Toronto has the lowest overall electricity prices of cities in the study.
- Solar-control film has a considerable positive effect on reducing annual kilowatt-hour and summer peak demand, on average reducing annual kilowatt-hour usage by 6.6% and summer month peak demand by 6.4% (see Tables 4 and 5).

CONCLUSIONS

This study clearly indicates that solar-control window film can play a useful and viable role in improving the energy efficiency of many buildings and that window films can be effective in reducing energy costs and energy consumption for buildings in many locations. Excellent energy savings can be provided by this technology—typical 5%–10% reductions in peak demand and annual cooling costs with such savings provided within a reasonable payback period (averaging less than 3 year). Although the focus of this entry was locations in the United States, it has been the author's experience (and it should be apparent) that solar-control window films are applicable to a wide range of locations, climates, and countries.

It is important to note that while providing these important energy-saving benefits, window films are able to provide many other benefits that directly hit the mark of key scoring components for "green building" specification programs, such as the Leadership in Energy and Environmental Design (LEED).[5,6] As such, solar-control window films are able to meet the needs of many different design professionals, from property owner/managers to architects to energy engineers to green-building professionals.

REFERENCES

1. International Window Film Association. Advanced Solar Control Guide, 2003.
2. International Window Film Association. Window Film Information Center. Available at www.iwfa.com (accessed on December 2006).
3. Hirsch J.J. and Associates. Welcome to DOE-2.com. Available at www.doe2.com (accessed on December 2006).
4. Lawrence Berkeley National Laboratories. DOE 2.1E. Available at http://gundog.lbl.gov (accessed on December 2006).
5. U.S. Green Building Council. LEED for Existing Buildings Project Checklist. Available at http://www.usgbc.org/DisplayPage.aspx?CMSPageID=221 (accessed on December 2006).
6. U.S. Green Building Council. LEED: Leadership in Energy and Environmental Design. Available at www.usgbc.org/DisplayPage.aspx?CategoryID=19 (accessed on December 2006).

Window Films: Spectrally Selective versus Conventional Applied[*]

Marty Watts
V-Kool, Inc., Houston, Texas, U.S.A.

Abstract

Compared to conventional applied window film, high light-transmitting spectrally selective film cost-effectively blocks unwanted solar energy from entering windows and reduces air conditioning costs without darkening building interiors, impeding the ability to see through existing glass, or changing the external appearance of a building.

According to the California Energy Commission, as much as 40% of a building's cooling requirements is a function of heat entering through existing glass. Stopping heat at the window is the most effective means of lowering temperatures and reducing heating, ventilation and air conditioning (HVAC) operating costs. In new construction, reducing heat at the window can mean the need for smaller and less-expensive HVAC systems.

The solution to overheating through windows is to specify solar control glass or applied window film, though even the best solar control glass performs no better than the best applied window film. Solar control glass can be selected for optimum energy performance in reference to the geographic orientation of any given building or section of a building. However, even in new construction, the cost of solar control glass often exceeds the cost of standard glass to which a solar control film is later applied.

For existing buildings experiencing problems from heat through windows, the most expensive option is to replace existing glass and frames with a new window system designed to block heat and deal with a building's energy performance needs. Less expensive is keeping existing frames and replacing only the glass. In either case, building managers may be understandably reluctant to replace existing windows or glass whose performance is generally adequate, though not optimum, in the case of blocking unwanted heat.

For all existing glass and in much new construction, applied window film is the least expensive and preferred solution to mitigate the impact of too much solar heat entering windows. Conventional dark and reflective applied window films successfully block a significant amount of solar heat, thereby reducing the use of HVAC systems.

Unfortunately, the same films reduce a significant percentage of visible light through the glass. Most of these films are highly reflective in daylight, giving them a mirror-like appearance when viewed externally. In artificial light and at night, internally they appear mirrored. In the case of retail establishments, visible light is reduced inside the store and shoppers outside cannot see clearly inside.

Most conventional window films transmit less than 35% of visible light, a good 35% less than the 70% necessary to be undetected by the naked eye. The result is that building interiors are correspondingly darkened, often requiring the use of increased illumination. This leads to higher electricity consumption that may increase inside temperatures requiring more air-conditioning. Increased utility costs defeat the major benefit of the film—cost savings.

THE BEST SOLUTION TO OVERHEATING—CLEAR, SPECTRALLY SELECTIVE FILM

Clear, spectrally selective applied window film offers the best ratio of visible light transmission to heat rejection. Spectrally selective refers to the ability of the film to select or let in desirable daylight, while blocking out undesirable heat.

While some manufacturers call their films spectrally selective, the definitive test is, how much visible light the film transmits. Most so-called spectrally selective films transmit no more than 54% of visible light. If a window film looks tinted and not clear, it is not optimally selective in the all-important category of visible light transmission.

The following table shows how different kinds of glass and applied films transmit light and heat.

Building management should consider the following points when evaluating spectrally selective vs. conventional window films.

---

[*] This entry originally appeared as "Comparing the Energy Conservation Capabilities of Spectrally Selective and Conventional Applied Window Film" in *Energy Engineering*, Vol. 102, No. 4, 2005. Reprinted with Permission from AEE/Fairmont Press.

Keywords: Window film; Applied film; Spectrally selective; Visible light transmission; Reduced HVAC operating cost.

Table 1 Window film performance

Type of glass or applied film	Percentage of daylight through glass	Percentage of solar energy through glass	Shading coefficient[a]	Luminous efficacy constant[b]	Percentage of light reflectance interior/exterior
¼″ clear glass	89	77	0.96	0.93	7/7
¼″ clear glass with tinted film	37	64	0.74	0.50	6/6
¼ clear glass with reflective film	37	44	0.51	0.73	18/28
¼″ clear glass with clear spectrally selective film	70	45	0.51	1.37	8/8

[a]The lower the shading coefficient, the lower the solar heat gain.
[b]Luminous efficacy constant, a measurement of a window glass or film's ability to simultaneously block heat and transmit light. (Visible light divided by the shading coefficient). The higher the number, the more efficiently the glass or film blocks heat and transmits light.

How do they Compare in Clarity?

The ideal film would be totally clear, yet, able to significantly block unwanted solar heat and reduce glare. Most dark and reflective films transmit less than 35% of visible light and correspondingly appear unclear. Spectrally selective film, which blocks heat equivalent to the darkest films, transmits 70% of the visible light and in so doing possesses a clear appearance (Table 1).

Data by Southwall Technologies, Inc., Palo Alto, CA, and Lawrence Berkeley National Laboratory, Berkeley, CA.

How do they Compare in Blocking Heat?

Most conventional tinted films transmit over 65% of solar energy, giving them an unacceptable shading coefficient of over 0.70 (the lower the shading coefficient, the lower the solar heat gain). With a shading coefficient as low as 0.51, reflective films block more heat, but many transmit as little as 15% of the visible light. When considering both heat rejection and light transmission, spectrally selective films out perform conventional competitors.

How do they Compare in Mitigating Heat Loss in Cold Weather?

Both conventional and spectrally selective window films are designed to block near infrared or solar heat. However, both conventional and spectrally selective window films will enhance the ability of existing glass to insulate against heat loss by as much as 15%.

How do they Compare in Applicability to Different Types of Glass?

Both conventional and spectrally selective films can be applied to single pane and insulating fixed glass, windows and doors. Always identify existing glass and follow the advice of a qualified film installer.

According to tests conducted by independent laboratories under the auspices of the Association of Industrial Metallizers, Coaters, and Laminators (AIMCAL), applied window film properly installed on insulating glass does not cause seal failure. Accordingly, most window film manufacturers offer an insulating glass warranty in the event of seal failure. For further information on the use of window film on insulating glass consult AIMCAL, Ft. Mill, SC (www.aimcal.com).

How do they Compare in Requiring Special Care?

The best-applied films require no special care. They can be cleaned just like the surface of glass using no abrasives, just soap and water.

How do they Compare in Price?

The price of dark, tinted, and reflective window film ranges from 4 to 6 dollars per installed square foot. Depending on the particulars of the installation and the geographic area, the best spectrally selective applied window film ranges in price from approximately $9 to $12 per square foot installed. Installed prices are volume dependent; on larger projects such superior performing films may be installed for less.

How do they Compare Aesthetically?

Conventional dark and reflective window film changes the appearance of existing glass and therefore the external appearance of a building. Clear, spectrally selective film does not change the appearance of existing glass, allowing its application on the entire building or on as few windows as necessary to deal with a localized overheating problem. For limited applications, spectrally selective film is competitive in price with conventional film.

How do they Compare in Payback?

Less expensive, conventional window films have a shorter payback compared to more expensive, spectrally selective films. However, it's not that simple. It is necessary to add on the cost of extra energy used for lighting due to the inability of conventional film to transmit sufficient visible light. Also, because extra lighting generates additional heat, the use of conventional window film may also increase air conditioning cost.

In reality, the payback for conventional film and spectrally selective film becomes comparable. Given rising electricity and natural gas rates, the rate of payback for spectrally selective film is always improving—averaging less than four years.

Use of Spectrally Selective Applied Window Film at Stanford University

Spectrally Selective Window Film Saves Energy at Stanford University.

While recent summers will be remembered for actual and threatened energy blackouts in California, Stanford University is doing its part to reduce energy use, thanks to the ongoing energy conservation program. Among many measures taken to improve energy efficiency at Stanford is the installation of clear, spectrally selective applied window film in 20 academic and administrative buildings on the 8,000 acre campus south of San Francisco.

At Stanford, the least expensive option to dramatically reduce unwanted solar heat and improve the performance of existing windows is applied window film. Scott Gould, Stanford's energy engineer, contends that since the 80s, window film performance has improved and that films look better and last longer. However, it was clear to decision-makers at Stanford that not all windows films are alike.

Not any film would do.

According to Alan Cummings, associate director of Facilities Operations, the primary reason for window film at Stanford is heat loading and occupant comfort. In collaboration with campus architect David Neuman, Cummings reviewed a variety of applied film products. Their primary objective was to select a window film with high light level transmission and heat load reduction. Equally important was the need for a window film that would not appear reflective. According to Neuman, the traditional architecture of the campus precludes using reflective glass, even on the newest buildings.

Conventional mirrored and tinted window films do prevent some solar heat loading, but cannot transmit high levels of light. Spectrally selective film freely transmits visible daylighting while blocking the near infrared and UV portions of the sun's spectrum. Tinted films may reduce heat gain but darken building interiors; spectrally selective film is virtually clear and so doesn't change the color of existing glass.

Over the past five years, spectrally selective applied film has been applied to selected south-, west-, and east-facing facades on 20 Stanford buildings totaling 120,000 ft^2. They include the Stanford Law School and the Green Earth Sciences Building. The film's energy payback, through lowered air-conditioning bills, is from three to four years, depending on the building, electricity rate, and weather.

The most recent film installation at Stanford took place at Encina Hall, a renovated administration building that was originally constructed as a dorm in 1891 and completely renovated in 1998. Some 6,212 ft^2 of film was applied in June, 2003.

According to a before and after energy audit conducted by V-Kool, Inc., the building has a British thermal unit meter reading (Btu/ft^2/h.) of 225 and 5.71 h a day of peak load. Daily air conditioning requirements to remove heat without the spectrally selected film amounted to 665.57 A/C tons to remove heat at a cost of AC at $66.56 per day. Daily air conditioning requirements with the film installed are 339.44 A/C tons to remove heat at a cost of AC at $33.94 per day.

Energy savings on the building with the film consist of:

1. $32.61 daily in A/C savings;
2. $978.39 in monthly A/C savings;
3. $4,891.95 in annual A/C savings.

Approximate project cost: $43,000. Return on investment: nine years or less, gave increase in the cost of electricity.

The window film installation program at Stanford University is an example of a long-term commitment to energy conservation on a campus with some buildings over 100 years old. To our knowledge, there are more buildings equipped with spectrally selective window film at Stanford University than at any other single institution in the country.

A case in point is the Los Angeles Department of Water and Power's (LADWP) rebate program for window film. It is based on a film's luminous efficacy constant, a measurement of its ability to simultaneously block heat and transmit light. While a very reflective film that blocks more heat than a spectrally selective film earns a 55 cent per square foot rebate from LADWP, a spectrally selective film that blocks less heat but lets in more light receives a higher rebate of 85 cents per square foot. Only spectrally selective films with luminous efficacy constants over 1.0 receive the higher rebate.

How do they Compare in Guarantees?

The best applied films are guaranteed not to peel, discolor, blister, bubble or demetalize for at least 10 years on a commercial installation. Look for a guarantee from the manufacturer in addition to any by the installer.

Where can I Find More Information on Conventional and Spectrally Selective Window Film?

The International Window Film Association, Martinsville, VA, (www.iwfa.com) and the AIMCAL, Ft. Mill, SC, (www.aimcal.com) provide a range of information.

REAL-LIFE INSTALLATIONS OF SPECTRALLY SELECTIVE WINDOW FILM

Both company-owned and franchise properties of the following retailers use spectrally selective window film in selected establishments: Hallmark Cards, Calico Corners, Public Storage, Albertson's, Esprit, McDonalds, Exxon, and Quik Trip convenience stores. Spectrally selective window film is saving energy at the Eldorado Country Club in Palm Springs, the Ontario, CA, convention center, and in such landmark buildings as the former headquarters of Montgomery Ward in Chicago and the headquarters of the American Institute of Architects in Washington, DC.

CONCLUSION

The universal applicability of spectrally selective applied window film makes it particularly cost effective for institutional, commercial, and residential structures in need of solutions to one or more of the following problems: solar overheating; reduction in HVAC size, operation, and cost; reduction in discomfort to building occupants and visitors due to high temperatures; increasing productivity of building occupants through maximizing natural light; utilization of floor space adjacent to windows and fixed glass; some increase in insulation against heat loss in winter; reduction of some glare while maintaining high levels of natural light; reducing internal temperature swings that can damage expensive interior landscaping; increasing a building's energy efficiency without compromising its historic and aesthetic qualities; mitigating the impact from ultraviolet radiation (UV) induced fading of furniture and window treatments; reducing exposure to cancer-causing UV, and increasing the resistance of existing glass to wind-blown debris, earthquake stress, explosions, and forced entry.

Of course, film manufacturers should and will continue research and development of applied films that will block even greater amounts of heat while transmitting high levels of visible light. Other enhancements in applied window film that may one day become available include increased durability, strength, and resistance to environmental and climatic stress, resulting in even greater film life expectancy.

BIBLIOGRAPHY

1. Watts M. Spectrally selective window film reduces heat. *College Planning and Management*, December, 2003. Case study of film use at Eastern Oregon University.
2. Barista D. Sun block, *Building Design and Construction*, December 2003. Review of latest window film technology.
3. Yates P. Seeing the light without feeling the heat. *Construction Canada*, November 2003. The benefits of energy efficient window films.
4. Abate K. Vacuum-sputtered coatings in glazing applications. *The Construction Specifier*, July 2003. Review of how the sputtering technology used in window glass and film manufacturing improves energy efficiency.
5. Yates P. Window films. *Window World*, Fall 2002. A review of the growth of window film use in Canada.
6. Egri J. Caught on film. *California Real Estate Journal*, August 5, 2002. Review of developments in window film.
7. A look at Stanford's conservation plans. *Energy & Power Management*, Coxsackie, NY, USA, October 2001. A review of Stanford's use of spectrally selective window film and other energy saving technologies.
8. Applied window film protects against heat damage to merchandise. *Energy & Power Management*, Coxsackie, NY, USA, November 2000. Reviews the use of window film by three major convenience store chains.
9. *Window Film Magazine*, Safford, VA, USA (http://www.windowfilmmag.com) is the only publication written for and about window film manufacturers, distributors, dealers and installers. It often contains material on the energy benefits and performance of applied window film.
10. *Energy Management Program Window Film Training Guide* published by the Association of Industrial Metallizers Coaters and Laminators, Window Film Committee. Ft. Mill, SC, USA. (http://www.aimcal.org) Excellent review of how window film is made and its energy conservation performance.

Windows: Shading Devices

Svetlana J. Olbina
Rinker School of Building Construction, University of Florida, Gainesville, Florida, U.S.A.

Abstract
A shading device, as an integrated component of a window and facade, protects space from direct sun, overheating, and glare; and provides increased daylight levels, desired privacy, or a view to the outside. This paper presents a classification of shading devices based on their assembly, their material, their position relative to the facade, and the control strategy used for the shading device position adjustment. The paper also introduces the decision-making framework (DMF) that can help in the selection of the most appropriate shading device for a specific building. The DMF is a tool for the analysis of the shading device performance; it is meant to be used by architects, engineers, window manufacturers, and shading device manufacturers.

INTRODUCTION

The proper design of a building with a window as a building component has the goal of providing comfort for the occupants, as well as energy efficiency for the building, by reducing the heating and cooling loads of the building. Direct sun radiation—one of the most significant cooling loads—is admitted through the windows. Therefore, direct sun radiation through the window should be prevented by the appropriate application of shading devices.[1]

The window is part of any conventional facade system, including both single-skin and double-skin facades. Windows as multifunctional systems provide thermal, visual, and acoustic comfort, and affect air quality. The thermal requirements for windows are to protect the building from heat loss in winter and heat gain in summer, and to collect energy in winter. The visual requirements for windows are to provide the occupants with a view, daylighting, privacy, and protection from glare.

The shading device is an integrated component of the window. Proper application of the shading device is especially important in curtain wall systems. Large glass areas can create a greenhouse effect, contribute to overheating, and increase cooling loads. Glass can also cause visual problems with direct and reflected glare.[2] Therefore, the application of shading devices in windows and large glass facades is necessary for controlling sunlight penetration through the glass.

Shading devices regulate heat by maximizing the reception of welcome heat in winter, and excluding excessive heat penetration in summer.[3] The shading device protects the space from direct sun and overheating in summer, reducing the cooling loads of the building by 23%–89%.[2] As a result, the appropriate use of a shading device in a window contributes to energy savings. When designing a window and a shading device for the window, the goal is to achieve a low total energy transmittance while maintaining high light transmission and good transparency.[4]

Advanced shading devices also provide daylight for the interior space. "The maximization of daylight is recognized as one of the key-goals in low-energy design."[5] A shading device as a daylighting system can redirect daylight to spaces where daylight is needed; for example, to spaces at a large distance from the window wall. Use of daylight decreases the use of artificial lighting, decreasing the following:

- Use of electricity
- Internal heat gain from lighting
- Cooling loads

This leads to energy savings for the building. The application of daylighting can decrease energy cost by 30%.

The shading device can also provide the following benefits:

- Protection from glare
- View to the outside
- Privacy
- Collection of sun energy in the double-skin facade
- Thermal insulation during winter nights

"Venetian blinds, draperies and roller shades inside single-pane, clear glass windows, reduce heat losses by

Keywords: Shading device; Shading device classification; Decision-making framework; Shading device selection; Shading device performance; Energy-efficient design; Daylighting.

25%–40% and metallic coated shades may further reduce losses by 45%–58%."[2] However, use of the shading devices in the window can obstruct the view to the outside and limit the amount of daylight that penetrates into the interior space.

The second section of this article presents a classification of shading devices based on their assembly and material, their position relative to the facade, and the control strategy used for the shading device's position adjustment. The third section explains the decision-making framework (DMF) that can help in selection of the most appropriate shading device for a specific building.

CLASSIFICATION OF SHADING DEVICES

Various shading device systems available on the market can be classified based on the shading device's assembly, its material, its position relative to the facade, and the control strategy used for the shading device adjustment.

Shading Device Assemblies and Materials

Various assemblies and materials are used for the shading device systems:

- Architectural solutions—Shading devices are an integral part of the building (e.g., overhangs, fins, brise-soleils, window setback, and light shelves).
- Window treatments—Shading devices are industrially manufactured systems (e.g., awnings, louvers, blinds, roller blinds, solar films, shades, sun screens, drapes, and shutters).[1]

The description of some of the shading assemblies follows:

- Overhangs and fins: fixed architectural shading elements, usually in the form of the balconies or projected spaces. Horizontal overhangs are effective devices for the south orientation, while the vertical fins work better for east–west orientation.[3]
- Brise-soleils: fixed architectural shading elements that consist of horizontal or vertical brise-soleil louvers. They are effective for east–west orientations.
- Awnings: consist of a frame that supports a horizontal or sloped surface on the exterior of windows. Awnings can be made of fabric, plastic, and aluminum. They can be fixed or moveable.[1]
- Louvers and blinds: consist of multiple horizontal or vertical slats. Horizontal devices are the most efficient for south orientation. Vertical devices give the best protection for east–west orientation. Slats can be either flat or curved, fixed or moveable. Louvers are exterior devices made of galvanized steel, anodized or painted aluminum, plastics, or glass (Fig. 1). Blinds are interior or between-glass devices made of painted aluminum, perforated metal, wood, glass, or plastic (Fig. 2).[1,6]

Advanced shading devices not only meet thermal performance requirements but also improve daylight levels in the space. The examples of advanced devices are as follows:

- Light shelves: flat or curved elements that reflect light either outside (exterior light shelf) or inside (interior light shelf). They divide the window into two areas; the upper area provides daylight, while the lower area provides shading.[7]

Fig. 1 Louvers: an example of the automatically controlled exterior shading device, made of glass. Source: From Architectural Press (see Ref. 5).

Fig. 2 Venetian Blinds: typical example of the manually controlled interior shading devices. The slats are made of perforated aluminum.

- Mini light shelves: for example, Okasolar units have a concave and convex shape and are made of a highly reflective light-gauge steel.[8,9] Louvers are fixed at a predetermined angle and spacing to respond to different seasonal conditions. Louvers are installed between the two panes of glass (Fig. 3).
- Prismatic and refraction elements: can be made of acrylic (Fig. 4). They can be installed in the upper part of the window to protect the space from glare and veiling reflections, while the lower part of the window provides the view.[8]

Position of Shading Devices Relative to the Facade

Based on their position relative to the facade, shading devices can be classified into three major groups:

- Exterior devices
- Interior devices
- Between-glass devices

An exterior shading device is installed in front of the facade (Figs. 1 and 4). Examples include overhangs, fins, awnings, louvers, brise-soleils, fabric blinds or screens, and roller blinds.[9] Exterior devices provide better solar

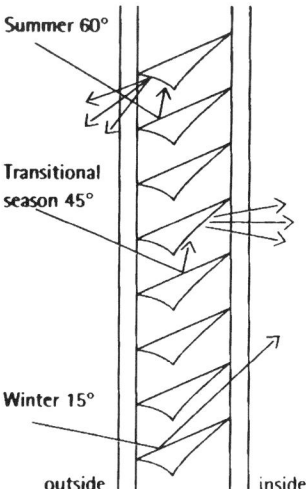

Fig. 3 Mini light shelves: the Okasolar unit is an example of the fixed, between-glass shading device, made of a highly reflective light-gauge steel.
Source: From Birkhauser Publishers (see Ref. 8).

protection in summer than interior devices, because exterior devices block sun radiation before it enters the glass panel and interior space. Shading effectiveness increases 35% by using an exterior shade instead of an interior one. As a result, building cooling loads are reduced. The exterior shading device's maintenance is more complicated and expensive than maintenance of the interior shading device. Exterior devices are more expensive because of the structural and durability requirements.

Interior shading devices are installed in the building's interior space. Examples include Venetian blinds (Fig. 2), traditional roller shades, drapes, and blackout screens.[7] Interior shading captures sun energy that can be used in

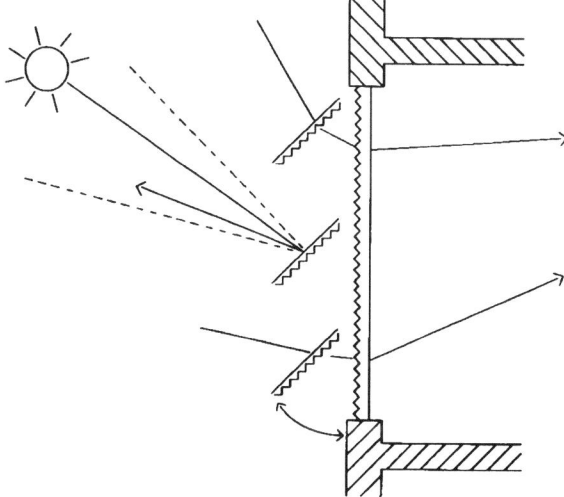

Fig. 4 Prismatic shading elements: an example of a moveable exterior shading device made of acrylic.
Source: From Birkhauser Publishers (see Ref. 8).

winter for space heating. Maintenance of interior devices is easier and less expensive than for exterior devices. Efficient solar protection in summer is difficult to achieve with an interior device, since sunlight enters the interior space and overheats the space between the blinds and the interior glass layer.

A between-glass shading device can be installed in an air cavity in one of two ways:

- Between two panes of glass in a double insulating glass unit (DGU) (Fig. 3).
- Between two facade layers in a double-skin facade.

The between-glass devices are less exposed to dust and dirt, so there is less need for cleaning. If the blinds are moveable and fully automatically controlled, maintenance can be complicated and more expensive, and a more complex window structure can be required. In summer, the between-glass shading device in the DGU usually reflects all sun energy in order to protect the interior from overheating. Any energy absorbed by the shading device contributes to the heating of the glass panes and of the air in the cavity between glass panes. This creates a problem in the DGU. However, in a double-skin facade, this warm air can be exhausted at the top of the facade, so that the facade and interior space can be protected from overheating in the summer.

Control Strategy for the Shading Device Position Adjustment

Shading devices can be either fixed (Fig. 3) or moveable. "Fixed systems are usually designed for solar shading, and operable systems can be used to control thermal gains, protect against glare, and redirect daylight."[6] The use of fixed blind systems requires higher energy consumption than moveable blind systems.[10] Moveable systems follow the dynamic exterior thermal and luminous conditions.[1] Position of the moveable shading device can be adjusted manually (Fig. 2) or automatically (Figs. 1 and 4), depending on the sun position, the sun radiation intensity, and the requirements for interior temperature and light levels. The moveable shading device can have three basic positions: open, partially open/partially closed, and completely closed.

Manually operated systems are generally low energy-efficient, because occupants may or may not operate them "optimally."[6] Occupants very often close the blinds completely to protect the space from overheating and glare, but at the same time the amount of daylight in the space is reduced; therefore, both the use of electric lighting and, thus, the cooling loads are increased. "If the blinds are open when a large amount of solar radiation enters, excessive energy is consumed for air-conditioning... When the blinds are closed on days without solar radiation, the advantage of the view from the window is lost."[11] The occupants will adjust the blinds to protect the space from direct sunlight, but will rarely adjust the blinds again when the direct sunlight is gone, and daylighting can be admitted.[10]

The automated shading device systems optimize energy use and control interior conditions without relying on occupants.[7] Automated systems can achieve savings in both cooling loads and lighting energy.[12] Automated blinds have better thermal and daylighting performance than both fixed blinds and manually controlled blinds. Automated systems "close automatically when the interior becomes too glary or too hot, and re-open later to admit useful light."[10] The use of automated Venetian blinds decreases the energy cost by 30% during the winter and by 50% during the summer. However, automatic systems can produce discomfort in occupants who dislike the feeling of not having personal control over the system.[6] Automated devices are often high-maintenance, and therefore expensive, solutions.[8]

SHADING DEVICE SELECTION

Several criteria should be considered when selecting the most appropriate shading device among the devices available on the market. To make the proper choice of the shading device, the required or desired performance for the shading device and the variables that affect that performance need to be determined. Fig. 5 shows the structure of the DMF for the shading device selection. The user of this DMF can be an architect, engineer, windows manufacturer, or shading device manufacturer. The DMF is an analysis tool that can help its user to select the most appropriate shading device for a building.

The structure of the DMF includes the following:

- Independent, dependent, and shading device variables that influence shading device performance.
- Performance parameters (thermal, visual, acoustic, aesthetic, cost, and control) that are used as criteria for the shading devices' evaluation and selection.
- Relationships and interactions among the variables and performance parameters.

Independent Variables

Independent variables such as climate, location, site, and building type are given to the user of the DMF.

The United States has four major climate zones: hot dry, hot humid, cold dry, and cold humid. For each of these climates, characteristics need to be determined. Climate directly affects the type of heat transfer, Heating, Ventilation, and Air Conditioning (HVAC) conditions, the facade type, shading device variables, the shading device's thermal and visual performance, the operational cost of

Windows: Shading Devices

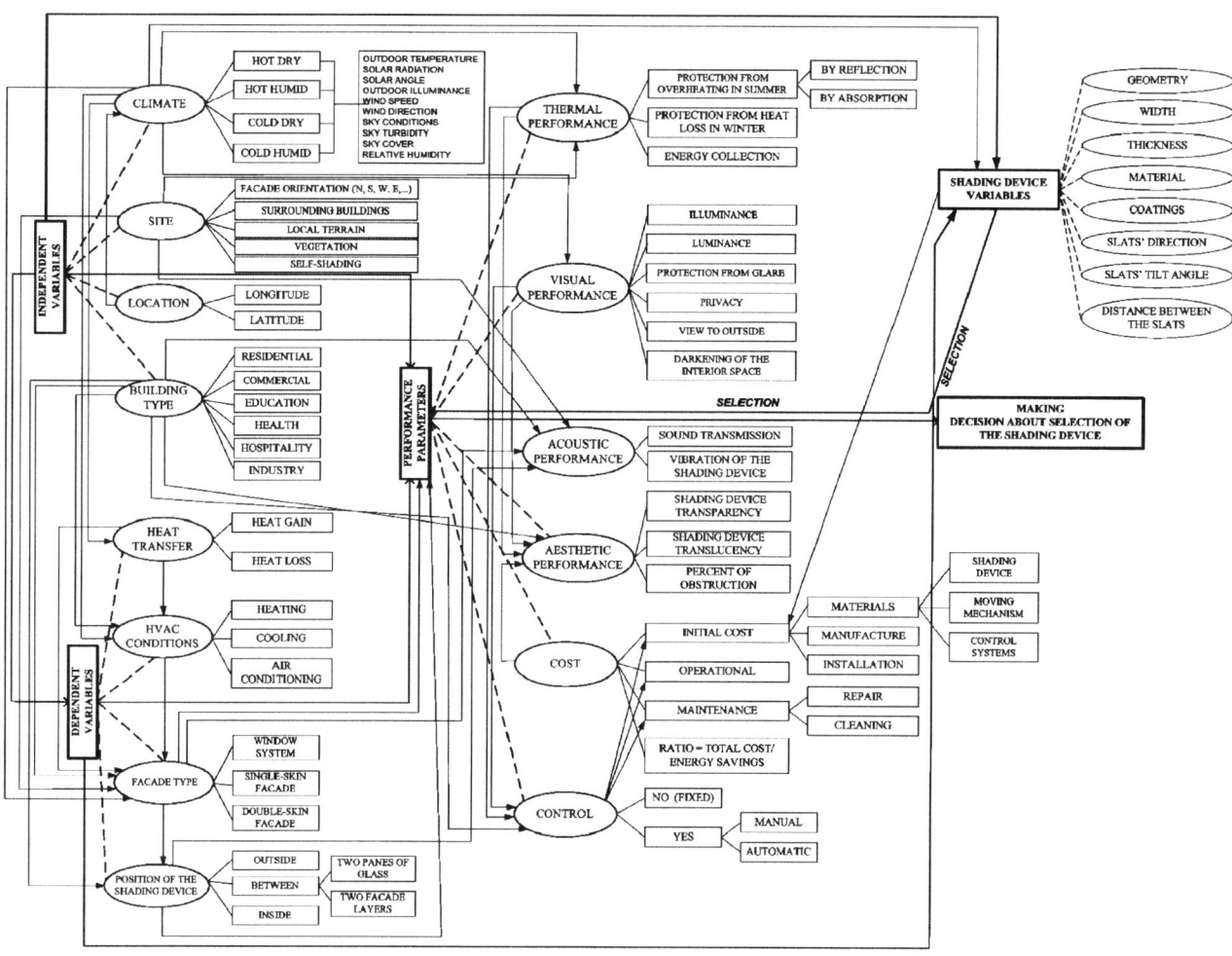

Fig. 5 The decision-making framework (DMF) for the shading device selection: the figure shows the structure of the DMF including variables (independent, dependent, and shading device variables), performance parameters (thermal, visual, acoustic, aesthetic, cost, and control), and the relationships and interactions among the variables and performance parameters.

the shading device, and the control strategy for the devices' adjustment.

The building location is defined by latitude and longitude. Based on the building location, the climate of the region and the microclimate of the particular building's site can be determined. Location indirectly affects heat transfer, HVAC conditions, the facade type, the shading device's thermal and visual performance, and the control strategy.

The building site is given to the designer, who has two choices: to select the position of the building on the site, or to accept the predefined position of the building, as in the case of the dense, urban setting. The site has a strong relationship with the climate and location. The site has direct influence on the facade type, the shading device variables, the shading device's thermal and visual performance, the operational cost, and the control strategy for the blinds' adjustment.

The building type defines a function of the building; for example, residential, commercial, education, or health.

The building type directly influences heat transfer, HVAC conditions, facade type, the position of the shading device, the shading device variables, and the performance of the shading device.

Dependent Variables

The user of the DMF defines dependent variables, such as heat transfer, HVAC conditions, the facade type, and the position of the shading device relative to the facade.

The user of the DMF defines dominating heat transfer conditions in the building, such as heat gain or heat loss. Heat transfer depends on the climate, location, and building type. Heat transfer affects the HVAC conditions, the facade type, the shading device variables, and the values of the performance parameters.

By selecting the HVAC conditions, the user of the DMF decides whether or not there is a need for heating, cooling, or air conditioning—or a combination of any of these systems. The selection of the HVAC conditions is made based on the heat transfer in the building, the climate, and the building type. The HVAC conditions have an impact on the shading device variables, the shading device's thermal performance, and the control strategy for the devices' adjustment.

The selection of the facade type is affected by the climate, site, and building type. The facade type has an effect on the shading devices' variables and on the shading device's thermal, visual, aesthetic, and cost performance. For example, if the double-skin facade is chosen, the shading device can be installed between the two facade layers and function as a solar collector, thus improving the facade's thermal performance.

The position of the shading device is dependent on the climate, site, building type, heat transfer, HVAC conditions, and facade type. The position of the shading device strongly influences the shading device variables. For example, an exterior device should be made of different material than an interior device because exterior devices must be weather resistant. The position of the shading device affects the performance of the shading device, especially the maintenance cost (e.g., it is more expensive to clean an exterior device than an interior one).

Shading Device Variables

This DMF includes the following shading device variables: the shading device's geometry, width, and thickness; applied materials and coatings; and in the case of Venetian blinds, distance between the blinds, the blinds' direction, and the blinds' tilt angle. Shading device variables are independent in the process of the shading device's selection because the shading device variables are already defined by the manufacturer and given to the designer of the building. These predefined variables are used to analyze the performance of the shading device. Shading device variables, together with the independent and dependent variables, directly affect the performance of the shading device.

Performance Parameters for the Shading Devices

Performance parameters considered in this DMF are the thermal, visual, acoustic and aesthetic performance; the cost of the shading devices; and the control strategy for the shading devices' position adjustments. The performance parameters' values depend on the independent, dependent, and shading device variables. There are also interactions among the performance parameters in this DMF.

The thermal performance of the shading device includes protection from overheating in summer, protection from heat loss in winter, and collection of sun energy. Thermal performance strongly depends on the climate, site, building type, heat transfer, facade type, position of the shading device, and shading device variables. In a hot climate, the office building's shading device will be required to provide protection from overheating. The shading device's geometry, materials, and position will be chosen to achieve the required protection from overheating. The level of protection from heat loss during winter nights or in cold climates and the collection of sun energy are also measures of the shading device's thermal performance. The device can absorb solar energy instead of reflecting it, and also collect this energy for application in the building's mechanical systems. There is a strong relationship between the shading device's thermal performance, its visual performance, and its control strategy. The shading device can be designed to provide maximum overheating protection but also to allow the sufficient daylight level in the space. To achieve this goal in the case of Venetian blinds, control systems should provide the blinds' optimum tilt angle.

The visual performance of the shading device includes providing the following desired effects:

- Illuminance
- Luminance
- Protection from glare
- Privacy
- Darkening of the space
- Visual contact to the outside space

Climate and site affect illuminance, luminance, and protection from glare. The building type strongly affects the requirements for providing privacy, darkening of the space, and direct visual contact to the outside space. For example, providing privacy and darkening of the space is often desirable in residential buildings, but not necessarily in office buildings. The visual performance of shading devices depends on the facade type, the position of the device, and the devices' geometry, dimensions, material, direction, and tilt angle. Also, there is the interaction among the thermal, visual, and aesthetic performance, and the control strategy. When selecting the shading device, the user of the DMF needs to understand that good visual performance can be achieved only with a thoughtful control strategy of the shading device's adjustment. The user of the DMF also must consider the impact of such a shading device on the appearance of the facade, i.e., on the aesthetic performance.

Acoustic performance parameters of the shading device include sound transmission and vibration of the blinds. The acoustic performance is significantly affected by the following:

- The building location and site: a higher level of noise occurs in urban areas; therefore, the shading device has to be designed to reduce this noise.
- The building type: different levels of acoustic comfort are required in different types of buildings, and a

specific shading device can help in meeting the acoustics requirements.
- Facade type and position of the shading device: the exterior shading device can vibrate because of wind, resulting in increased noise level. For that reason, structure of the exterior device should be designed properly to avoid the problem of vibration.
- The thermal performance: the shading devices that protect space from heat loss during cold nights can also protect the space from noise, because devices can be made of materials with good thermal and acoustical properties.
- Control strategy: completely closed blinds provide the lowest sound transmission, but at the same time they block the daylight. The control strategy needs to balance the sound transmission, penetration of the daylight, and protection from heat loss or gain.
- The shading device variables: the shading device's material, shape, dimensions, and position need to be selected to achieve the best possible protection from noise.

The shading device has a significant impact on both the exterior and interior appearance of the facade. To achieve an aesthetically pleasing look for the shading device and facade, the device's transparency or translucency and the percent of the window area obstructed by the device need to be considered. The shading device's aesthetic performance is affected by the climate, site, and building type. The appearance of the surrounding buildings also affects the appearance of the analyzed building, and, consequently, the appearance of the shading device. Different aesthetic requirements are imposed for the shading devices installed on different building types. The shading devices can have a different look if they are installed on an office building than they would on a hospital or industrial building. The aesthetic performance of the shading device is influenced by the facade type; the shading device's position; its thermal, visual, and acoustic performance; the cost of the shading device, and the applied control systems. The shading device's transparency or translucency and the percent of the window area obstructed by the shading device depend on the requirements for the protection from overheating, the desired light level in the interior space, and the choice of the shading device's variables.

Total cost of the shading device consists of initial, operational and maintenance cost. Cost analysis should also include calculation of the ratio between total cost and energy savings achieved due to use of the shading devices. The initial cost of the shading device includes the cost of material, cost of manufacture of the system, and cost of installation of the shading device on the facade. The initial cost includes also the cost of the moving mechanisms and control systems if the device is moveable. The operational cost includes the operating cost of the shading device itself and the cost of the heating, cooling, and lighting of the space, which is a result of the application of the shading device on the building. The cost of maintenance includes cleaning and repair costs. The cost of the device is influenced by the climate; location or site; building type; thermal, visual, acoustic, and aesthetic performance parameters; and control systems.

The control strategy for the shading devices' adjustment in this DMF considers two options: shading devices without need for control (fixed) and shading devices controlled manually or automatically. The control strategy for the shading devices' adjustment depends strongly on the following:

- Climate, particularly sun radiation, sun angle, and sky conditions
- The building type, because different building types require different values of the performance parameters for the shading device; and, therefore, different positions and tilt angles of the device
- The facade type
- Heat transfer
- HVAC conditions
- The shading device variables, particularly geometry of the device, dimensions, and materials
- The required thermal, visual, and acoustic performance, and the cost of the shading device

Making the Decision by Using the DMF

The process of making a decision about selection of the shading device by using this DMF includes the following steps:

- Identifying input for the DMF—The user of the DMF prepares the input for
  — The independent, dependent, and shading device variables
  — The required values of the performance parameters that can be taken from active standards, codes, and recommendations
- Testing the shading devices—Separate testing is performed for each type of shading device to analyze thermal, visual, acoustic, aesthetic, and cost performance; and the effect of the applied control strategies. Depending on the nature of the performance parameter, the actual values of the performance parameters can be obtained by experimental testing, computer simulations, and mathematical calculations.
- Obtaining output results of testing—The actual values of thermal, visual, acoustic, and aesthetic performance parameters; the cost of the shading device; and the effect of the control strategy are collected for each type of shading device. The results are organized in an

understandable and useful format and prepared for analysis.
- Making the decision about the selection of the most appropriate shading device for the particular building—Output results of testing are compared to the required values of performance parameters for the shading devices. If the shading device's actual performance meets the requirements of the standards, then the particular device can be considered for further analysis and application on the building. Then the alternative shading devices are compared to each other. The actual values of the performance parameters for each shading device are compared, and the device with the best overall performance is selected for use on the specific building.

CONCLUSION

Shading devices are integrated elements of windows and building facades. They increase the energy efficiency of buildings and improve comfort for the building occupants. Shading devices provide the following benefits:

- Protection from direct sun
- Protection from overheating
- Protection from glare
- Increased daylight levels
- Privacy
- View to the outside space

Different types of shading devices that exist in the market can be classified based on their materials, their assemblies, their position relative to the facade, and the control strategy for the shading device position adjustment. The DMF presented in this paper helps the user, whether an architect, engineer, window manufacturer, or shading device manufacturer, to select the most appropriate shading system among several available systems. The DMF offers the user a tool for an analysis of the shading device's performance.

REFERENCES

1. Schuman, R.; Rubinstein, F.; Papamichael, K.; Beltran, L.; Lee, E.S.; Selkowitz, S. *Technology Review: Shading Systems*; Lawrence Berkley National Lab: Berkley, CA, 1992.
2. Dubois, M. *Solar shading for low energy use and daylight quality in offices. Simulations, measurements and design tools*, Report No TABK-01/1023; Department of Construction and Architecture, Lund University: Lund, Sweden, 2001.
3. Olgay, A.; Olgay, V. *Solar Control and Shading Devices*; Princeton University Press: Princeton, NJ, 1957.
4. Koster, H. *Dynamic Daylighting Architecture: Basics, Systems, Projects*; Birkhauser Publishers: Basel, SW, 2004.
5. Wigginton, M.; Harris, J. *Intelligent Skins*; Architectural Press: Oxford, UK, 2002.
6. Ruck, N.; Aschehoug, Ø.; Aydinli, S.; et al. *Daylight in Buildings: A Source Book on Daylighting Systems and Components*, A Report of IEA SHC Task 21/ECBCS Annex 29; Lawrence Berkeley National Laboratory: Berkley, CA, 2000.
7. Cardomy, J.; Selkowitz, S.; Lee, E.S.; Arasteh, D.; Willmert, T. *Window Systems for High-Performance Building*; W.W. Norton & Company: New York, 2004.
8. Daniels, K. *Low-Tech Light-Tech High-Tech*; Birkhauser Publishers: Basel, SW, 1998.
9. Campagno, A. *Intelligent Glass Facades: Material, Practice, Design*; Birkhauser Publishers: Basel, SW, 2002.
10. Galasiu, A.D.; Atif, M.R.; MacDonald, R.A. Impact of window blinds on daylight-linked dimming and automatic on/off lighting controls. Solar Energy **2004**, *76*, 523–544.
11. Inoue, T.; Kawase, T.; Ibamoto, T.; Takakusa, S.; Matsuo, Y. The development of an optimal control system for window shading devices based on investigations in office buildings. ASHRAE Trans. **1988**, *104*, 1034–1049.
12. Lee, E.S.; DiBartolomeo, D.L.; Selkowitz, S.E. Thermal and daylighting performance of an automated venetian blind and lighting system in a full-scale private office. Energy and Buildings **1998**, *29*, 47–63.

Wireless Applications: Energy Information and Control

Jim Lewis
Obvius LLC, Hillsboro, Oregon, U.S.A.

Abstract

Most owners and managers of commercial and industrial facilities recognize the value of using interval energy information to allocate costs to building users and fine-tune building operations, but find that installation costs are very high in existing buildings. Some of the most promising technological developments in the submetering world are centered on the use of wireless devices to gather data from the meters, communicate it to a central data acquisition location, and forward the interval data to a remote server using wireless Web networks. This entry examines some of the key elements of wireless energy information and control networks and provides an overview of important underlying wireless infrastructures. Particular emphasis is placed on using "mesh" radio networks to communicate raw meter data within a building or campus and using GSM/general packet radio service (GPRS) cellular networks for long-haul communications.

INTRODUCTION

Recent volatility in energy prices has made submetering (i.e., any meter located on the customer side of the primary utility meter) a valuable tool for building operations and management. One of the biggest obstacles to getting this kind of information has been the high cost of running wiring from the meter hardware (gas, electricity, water, etc.) to a data acquisition device in existing buildings. Many radio manufacturers (such as Maxstream and AeroComm) have introduced low-cost radio modules that can be adapted to provide the backbone of a wireless metering system. Using these off-the-shelf radio modules in conjunction with specially designed hardware and software for RS485 communications provides a seamless wireless communications network for both new and existing meters. In addition to this internal communications network, the development of GSM/general packet radio service (GPRS) cellular networks for wireless networking (think personal digital assistants (PDAs) with email capability) provides a convenient means of gathering interval data from locations around the world without the need for connecting to an existing local area network (LAN) or phone line.

A SUBMETERING PRIMER

For most end users, when we use the term "submetering," the immediate image is of the hardware installed on the electrical system to measure kW and kWh. While these devices are certainly an indispensable part of a submetering system, they are only one of the components. A successful submetering system takes the raw data from one or more meters and converts it into useful and timely information to be used for the following:

- Cost allocation to tenants
- Operational analysis and improvement
- Measurement and verification of energy savings
- Benchmarking and accountability

Converting the raw data at the meter level to actionable information requires several steps. Within the metering industry, this process can best be illustrated by looking at five key components:

1. *Meters and sensors*—This is the hardware level where the devices are actually installed in the electrical (or gas or water) system to capture energy consumption for one or more different areas of the building. In the electrical system, these meters might be the traditional round glass meters or they might be meters designed especially for submetering.
2. *Internal communications*—This is the mechanism used to communicate data from the meters and sensors to the data acquisition server (DAS). In traditional applications, this is usually a simple twisted pair of wires connected on an RS485 serial daisy chain.
3. *Data acquisition*—The meters and sensors from level 1 produce industry standard outputs (e.g., pulse or Modbus) that correspond to the real-time outputs being monitored. For example, each pulse from an electrical or flow meter has an assigned value (e.g., 1 pulse = 10 kWh). In order for this raw data to be utilized, it must be captured in a timely manner, time-stamped, and made available for presentation.

Keywords: Submetering; Wireless; RS 485; Modbus; Energy information.

The DAS takes this raw data, time-stamps it, and then sends it to a local or remote server for storage and report generation.
4. *External communications*—Once the data is collected by the DAS, it is sent to a local or remote server for reporting. In traditional metering applications, this communication may use a LAN connection or a modem.
5. *Storage and reports*—Once the data from the meters is gathered at the building level, it still requires further processing to produce user-friendly reports and information. This processing occurs at the user interface level, where information from one or more buildings is collected into a traditional database for reporting and storage.

A detailed discussion of each of these components is beyond the scope of this entry, but most successful submetering systems have the following characteristics:

- "Open" meter protocols that provide the capability for any DAS to connect to any meter or sensor, regardless of what is being measured (e.g., electricity, gas, water, flow, british thermal units (BTUs)). Most meter manufacturers use Modbus remote thermal unit (RTU) as the primary protocol for communications with submeters, but the DAS should also accept pulse or analog values that meet industry standards.
- Nonvolatile storage of data—the DAS should be able to store at least 30 days of data from multiple meters without the risk of data loss in the event of power failure.
- Options for user-selected data intervals—the user should be able to select data intervals to match the utility interval (e.g., 15 min).
- "Open" protocol support at the DAS level that allows the data collected to be sent to any Web server (i.e., nonproprietary format). In most cases, this can be accomplished using standard Internet protocols such as HTTP, FTP, and XML. This allows the end user to select virtually any software program (including custom programs), regardless of the DAS or meter brand.

Fig. 1 (below) shows a graphical representation of a traditional wired submetering system, starting with the meters and sensors at the bottom and moving up to the storage and reports level at the top of the figure.

WHY DO WE NEED WIRELESS SUBMETERING?

As outlined in the previous section, there are readily available wired solutions that provide reliable submetering data using commonly available tools such as serial ports and LANs—so why do we need wireless? The simplest answer is that in many applications (particularly retrofits), the cost of running wire between meters and the DAS is prohibitively expensive, and using wireless components can dramatically reduce both the cost and the installation time. In addition, in many submetering projects, the cost and time involved in securing a network connection or phone line for communicating from the DAS to the Internet becomes a major headache.

Using wireless solutions in submetering not only minimizes the costs of many projects, but it also greatly reduces the disruption of day-to-day operations caused by running wire several hundred feet or more through an existing facility. Wireless also provides a very attractive alternative to trenching between buildings in a campus environment and eliminates the need for coordinating with the IT department for use of an existing LAN connection.

WHICH "WIRELESS" ARE WE TALKING ABOUT?

Ask anyone what the term "wireless" means, and you'll probably find yourself on the receiving end of one of those looks that implies you must be one of the dumbest people on Earth. Wireless, as any fool knows, means no wires, and therein lies the problem with understanding wireless submetering. A short list of wireless terms includes the following:

- Satellite
- Cell phone
- Wi-fi
- Pager
- Proprietary radio
- Bluetooth
- Zigbee

In essence, wireless (or for that matter, wired) communications form the bridge between each of the levels of metering discussed above. We will be examining two different wireless applications in this entry, the first being communications between meters and the DAS and the second the communications of interval data from the DAS to a remote server for storage and reporting.

PART I-METER-LEVEL WIRELESS COMMUNICATIONS

One of the most expensive aspects of adding submetering to an existing facility or facilities is the cost of running wires from multiple locations to the central data acquisition point. This cost can be significant, particularly when wiring must be run between multiple buildings (i.e., campuses or bases). The first level of wireless communications we will consider is the communication of data from meters and sensors wirelessly to the DAS. This

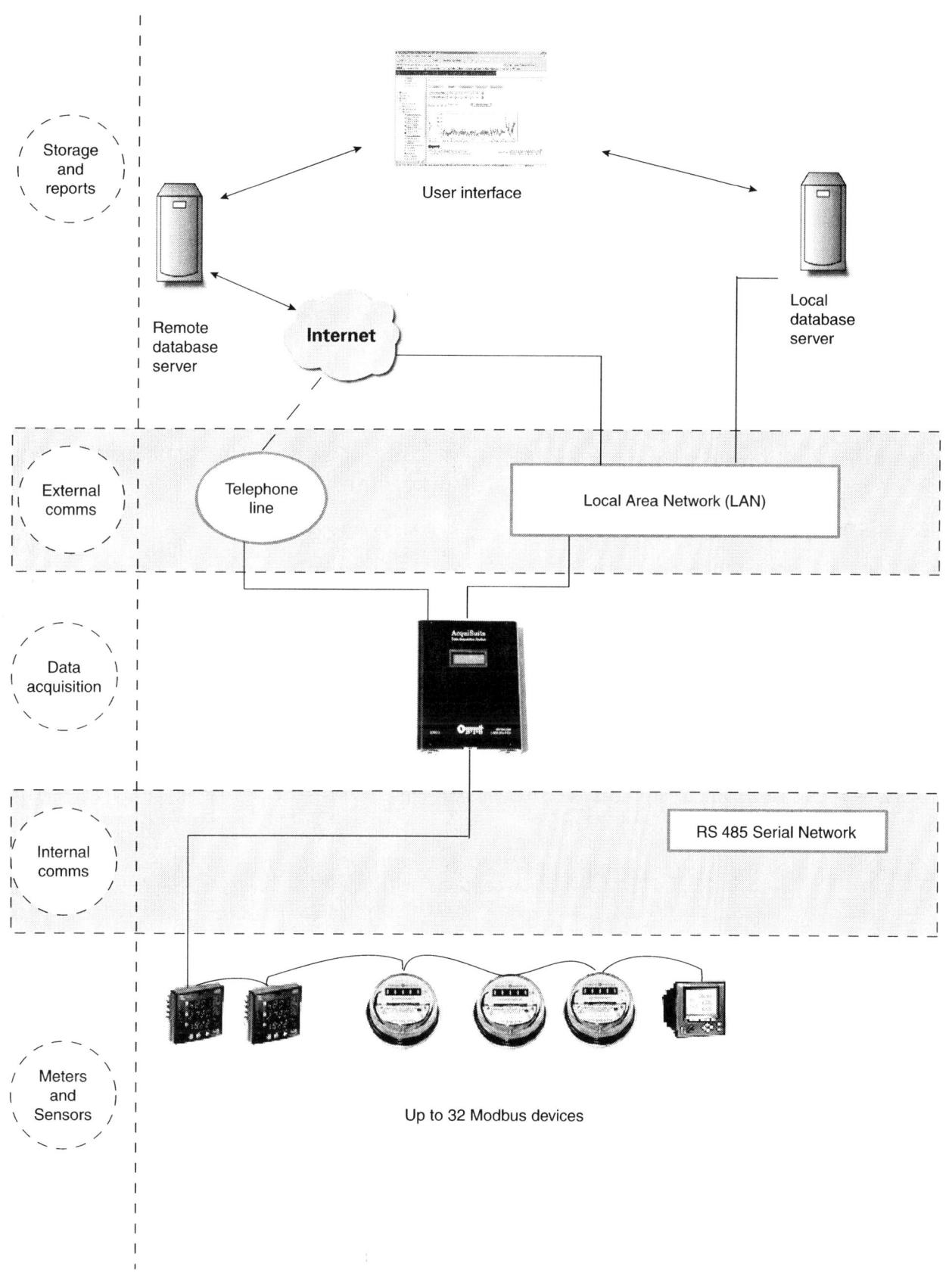

Fig. 1 Overview of metering system components.

level represents the area in which technology is advancing the most rapidly, as electronics developed for a wide variety of wireless communications are being utilized to reduce the cost of submetering.

A BRIEF PRIMER ON WIRELESS TERMS

A detailed technical discussion of wireless technologies is beyond the scope of this entry, so the purpose here is just to provide a brief overview that will serve as a jumping-off point for later discussion. At the most basic level, all wireless communications are based around radio waves, not unlike the over-the-air broadcast of radio and TV signals. (For the younger readers, yes; it is true that in the olden days, TV was limited to a half-dozen channels captured out of the air with a pair of rabbit ears.) There are significant differences between the radios used, but at the heart are some shared attributes that help to determine the suitability for use in wireless metering. Among the most common and important terms associated with radios are the following:

1. *Frequency*—Frequency refers to the time between peaks of the radio signal generated by the radio. In order for two radios to communicate with each other, they must be tuned to the same frequency, whether in a broadcast (one-way) or two-way communication environment.
2. *Power*—Most radios are rated for specific power levels that serve to determine how far a signal can be detected.
3. *Interference*—As anyone who has ever driven under a power line or tried to use a cell phone knows, there are a number of things that can cause interference with a radio signal, and the same is true for wireless metering technologies. Concrete walls, steel panels and other radio sources all provide challenges to the successful communication of metering data.
4. *Throughput*—A combination of the above factors, plus a few others, determines the throughput of the radio system. Basically, this just means how much data can be successfully transmitted in a given period of time, including any repeat requests or other delays.
5. *Repeaters*—Despite the best design of radios, it is very likely that there may be "dead spots" or areas where radio transmission is not successful. This may be due to interference, lack of power, or some other factor; but regardless of the cause, it may be necessary to install repeaters, which are basically designed to relay a lost or weak signal from one point to another.
6. *Mesh networks*—The term "mesh network" refers to networks designed to be self-healing radio networks that automatically configure themselves to optimally send data to one or more other radios without the need for setup and configuration by the user.
7. *Magic*—One of the most interesting characteristics of most radio networks is that they represent a blend of science and art. Most cell phone users have experienced the joy of having a cell phone that always works in a particular location, but suddenly doesn't, and the same magic applies to radios used for metering applications. While there are many tools available to assist in predicting the odds of success in any given location, the reality of wireless communications is that even the best of tools may not prove 100% reliable.

WIRELESS SERIAL COMMUNICATIONS NETWORKS

In order to better understand some of the options available for wireless communications between meters and the DAS, let us take a closer look at the wired version of this same communication. Fig. 2 shows a section of the previous figure that focuses on the internal serial communications networks for submeters.

As Fig. 2 shows, in a wired sensor-level communications network, a twisted pair of wires is run from a serial port (RS485) on the DAS and daisy chained to each of the Modbus meters or devices. The DAS uses plug-and-play technology to detect the type of meter and load the appropriate drivers to interrogate the meter(s) to obtain the desired data. On preselected intervals (typically 15 min or 1 h), the DAS will gather data from the meters that may be limited to consumption (kWh) or may include other parameters such as power factor or total harmonic distortion (THD), if the meter supports those additional functions. The readings are then stored by the DAS until they are "pushed" or "pulled" to a remote server for storage in a standard database (more on that later).

This system is quite reliable and provides an excellent method of communication if the meters and the DAS are all in close proximity. This type of serial network is typically capable of reliable communications up to 4000 ft. On the other hand, if there are multiple locations throughout the facility or campus that make wiring difficult or expensive, wireless nodes near the meters can provide a very cost-effective alternative to wiring.

HOW DOES WIRELESS SERIAL COMMUNICATION WORK?

There are at least a couple of different ways of wirelessly transferring data from meters. One method involves using

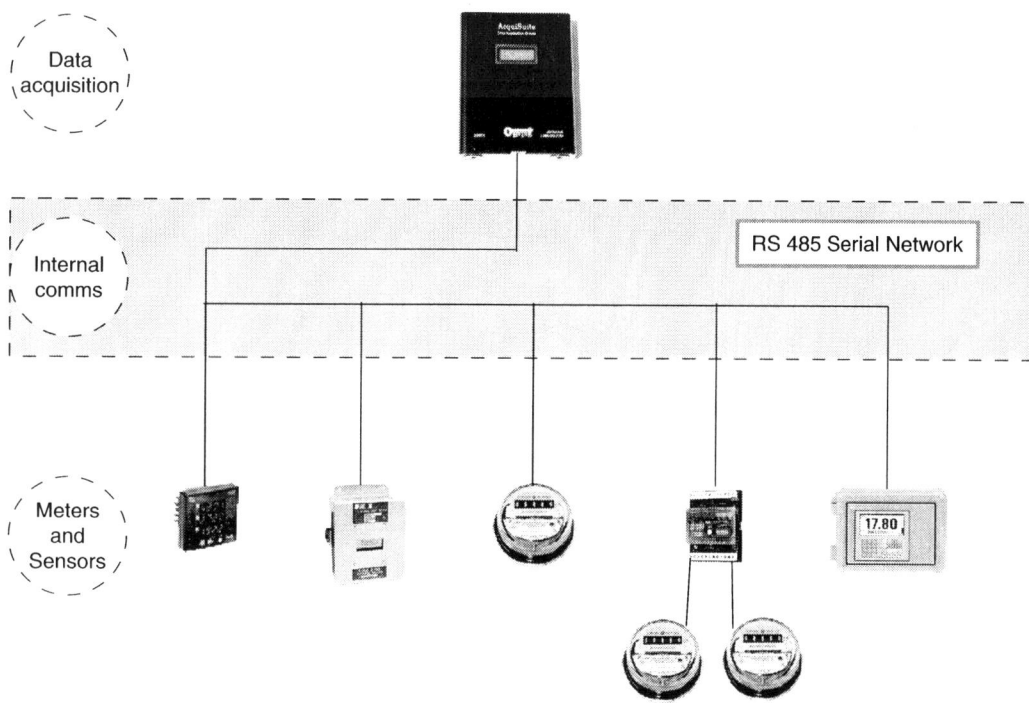

Fig. 2 Internal metering components view.

low-powered battery devices to capture and transmit pulses from meters to the DAS, which represents a very low-cost means of communications, but is not a replacement for the ideal serial network shown above. The wireless nodes function in one-way communication only and cannot provide the reliability or the additional data available from reading and sending Modbus serial data to the network.

The second (and preferred) method is to simply use wireless nodes on either end of the transmission to seamlessly replace the twisted pair of wires in Fig. 2. The newest technologies on the market accomplish this in a mode that is completely transparent to the user (i.e., the data transmitted wirelessly is indistinguishable from the data gathered on a wired network).

Fig. 3 shows a typical system that substitutes wireless Modbus communications for wired networks. In this case, we will assume that the two meters on the left of the graphic are located in one electrical room; the three on the right are located in a second electrical room; and the single meter in the center is the primary meter for the facility, which is located in close proximity to the DAS, and thus can be hard-wired.

In this example, all of the meters can communicate all of their data using the standard Modbus protocol, regardless of whether they are connected wirelessly or via wires. In the case of the remote meters, we have simply wired the RS485 outputs from the meters into a wireless node that puts the Modbus data into a wireless form and sends the data via radio to another node connected directly to the DAS, where the data is once again converted to RS485.

SO HOW DOES THIS MESH NETWORK FUNCTION?

The two keys to making the wireless communications network shown in Fig. 3 actually work are (1) that the individual radio transceivers (or nodes) are specifically designed for transmitting Modbus data and (2) that the radios contained within the transceivers can function as part of a "mesh" network. A brief overview of the functioning of a mesh network will provide a better understanding of the benefits of using wireless mesh networks to replace wired serial communications.

All mesh networks (regardless of the application) share certain common characteristics that make them valuable for replacing or augmenting wired networks. Among the most important are:

- *Self-configuring nodes*—The whole concept of mesh networks is that the radio nodes will be aware of other nodes and will automatically configure the network to optimize throughput of data, without the need of programming or configuration on the site.
- *Multiple routing paths*—Because each node is "aware" of all the other nodes it can communicate with, there are multiple options available for routing the data. As new nodes are added to the network, the network should recalculate the optimal routing paths to take advantage of more efficient routing options.
- *Spread spectrum radios*—This feature allows the radios to work on a variety of predetermined frequencies, minimizing the likelihood of interference

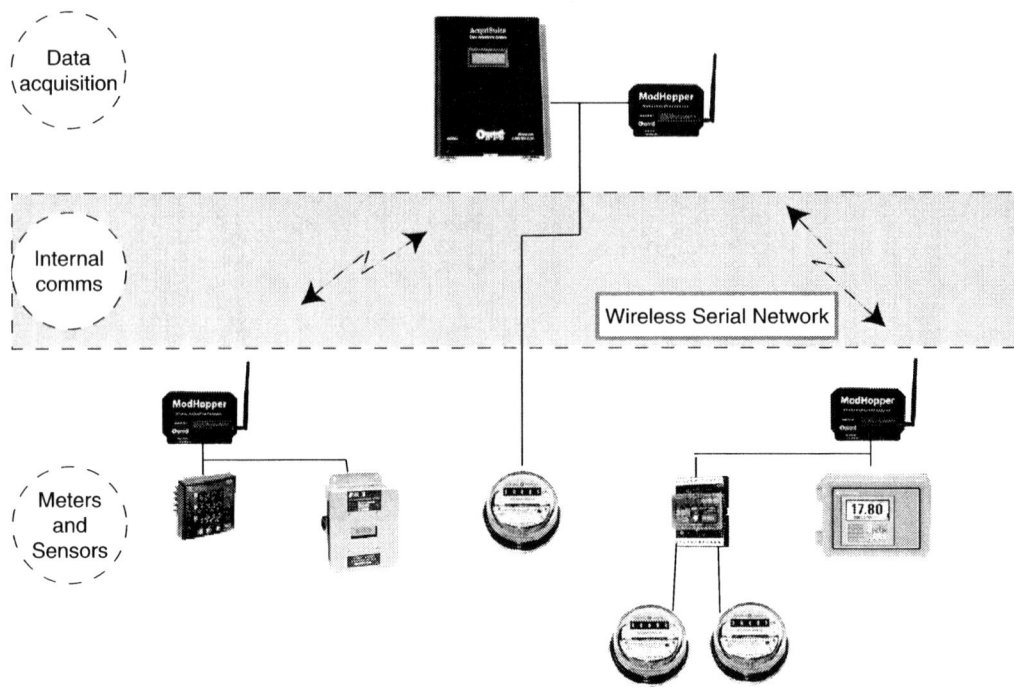

Fig. 3 Metering components with wireless communications.

from other radio sources. Essentially, the radios are designed to "hop" to different frequencies if interference is encountered, until the two radios find a frequency that minimizes interference.
- *ISM band radios*—Transceivers that use the instrumentation, scientific, and medical (ISM) frequencies set aside by the FCC do not need to have a local license for each site. These radios are designed to share the frequency bands with other systems and devices with a minimum of interference.

Fig. 4 shows a diagram of a typical wireless mesh network. Each of the transceivers can accept up to 32 Modbus devices and 2 pulses.

This figure shows a mesh network with eight transceivers and nine meters, all connected to a DAS for data collection. Two of the transceivers are functioning as repeaters, because they have no meters connected and serve only to route traffic to other nodes. The dashed lines between the transceivers show the other nodes that each can communicate with, and most of the nodes can see multiple other nodes, providing multiple routing paths for requests for data from the DAS. The network will automatically figure the optimal routing to reach each of the meters and will adjust these routing paths based on the success rate of transmissions. If other Modbus devices are added to any of the nodes, their presence is detected by the network and optimal routing paths are determined. If other transceivers are added, they will similarly be detected by the network and added to the routing paths.

SUMMARY

The technology available today makes wireless sub-metering a viable alternative to hardwired systems. The installation of these products, whether as the total solution or in conjunction with wired Modbus meters, provides a transparent and cost-effective option when wiring costs are prohibitively expensive.

PART 2—EXTERNAL WIRELESS COMMUNICATIONS

Summary

Just as mesh networking provides an excellent alternative to a wired solution at the sensor communication network level, advances in wireless technology allow the user to send data from the DAS to a remote server without the need for a hardwired connection. The use of GSM/GPRS (cell) modems in the DAS provides a very cost-effective means of communications that piggybacks on the structure built out for cell phone communications.

WIRED EXTERNAL SOLUTIONS

Fig. 5 shows a closer look at the upper part of Fig. 1 and provides some details about how data is typically sent from the DAS to a remote server for logging and reports. The DAS provides several options for communications,

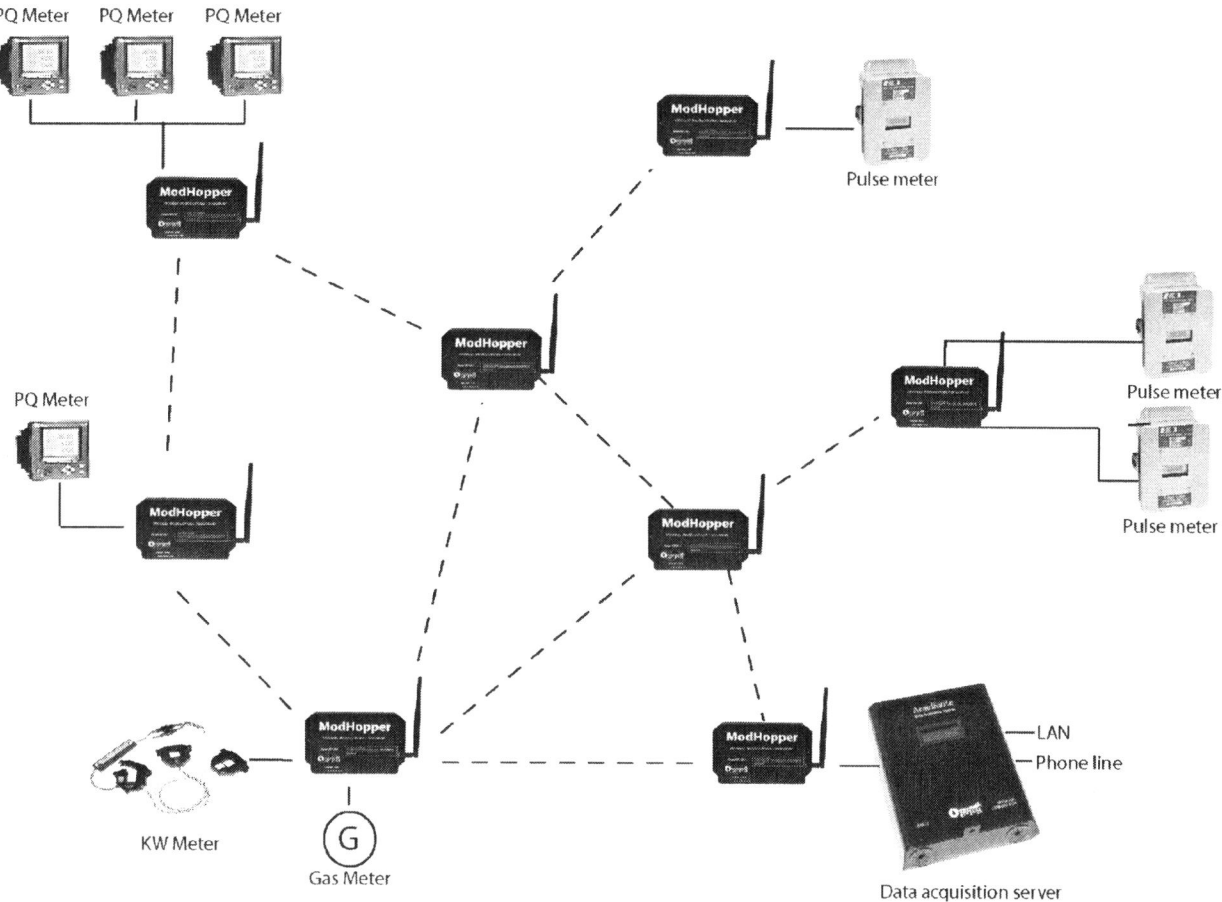

Fig. 4 Typical mesh network metering system.

depending on the needs of the user and available communications options. There are usually two alternatives for connecting to either a remote or a local server:

- *Existing LAN*—If the DAS is located in a facility that has an existing LAN, this provides an excellent mechanism for getting data from the DAS to a server. The DAS is IP-addressable, which means that a network-connected DAS can provide two-way communications and allow the user to see near real-time data in addition to interval information.
- *Phone line*—If there is no LAN connection (or if the IT department cannot or will not make a connection available), the DAS also provides the option to use an on-board modem to dial in or out to the system. If the user only needs to upload data (dial-out), the phone line can be shared with other devices such as fax machines.

Details of the file upload in a wired system are beyond the scope of this entry, but a typical upload session on an existing LAN would have the following elements:

1. The DAS initiates a connection to the remote server via the LAN by accessing a URL (e.g., http://www.obvius.com) where the server is located. The DAS provides a user name and password to the server, which uses this information to access the existing database records.
2. Once the connection is established and verified, the DAS will upload data collected since the last upload, in most cases using HTTP or FTP protocols.
3. Once the data is uploaded and verified, the DAS can optionally get additional information from the server, such as time checks.
4. Finally, the session is terminated.

The process for using a phone line connection is very similar, with the primary difference being that the upload speeds are slower (think dial-up modem vs DSL) and the connection is a PPP session.

WIRELESS OPTIONS FOR EXTERNAL COMMUNICATIONS

There are two basic applications for wireless communication from the DAS—one for inside the building to a LAN and one for totally wireless communications.

We will first look briefly at the simpler of the two, wireless communications within the building.

In this scenario, we are simply substituting a wireless access point for the wired LAN connection that then connects to another wireless access point on the local LAN. Because the DAS is an IP-addressable device like any workstation, there is no special setup required, and the functionality is exactly the same as that shown for the wired LAN connection in Figs. 5 and 6.

The second option for wireless external connectivity uses the existing cell phone networks built around the country to bypass any local connection. The primary reasons for doing this are the following:

- Lack of available LAN or phone lines
- Security concerns from IT personnel about using existing LANs
- High cost of acquiring and maintaining either phone or LAN connections
- Delays in getting phone or LAN connections installed
- Ease of installation

WHAT ADDITIONAL HARDWARE IS REQUIRED?

In the simplest terms, the only real change from the LAN to the phone-connected version is that a cell phone modem is used in place of the typical RJ11 modem. This modem is connected to an antenna mounted on the DAS, but these are the only functional differences between the cell modem version and any other DAS.

HOW DOES IT WORK?

In the olden days (think 1990s), cell phone communication was based on taking analog voice signals from one user and transmitting them to an analog receiver held by another user on the other end. Today, analog transmission is virtually nonexistent—ithas been replaced by digital cell technology (basically a process that converts the analog voice signal to digital for transmission that is decoded on the other end). The evolution from analog to digital is beneficial for data communication because data transmission on the Internet is inherently digital, and thus ideally suited for the digital networks (Fig. 7).

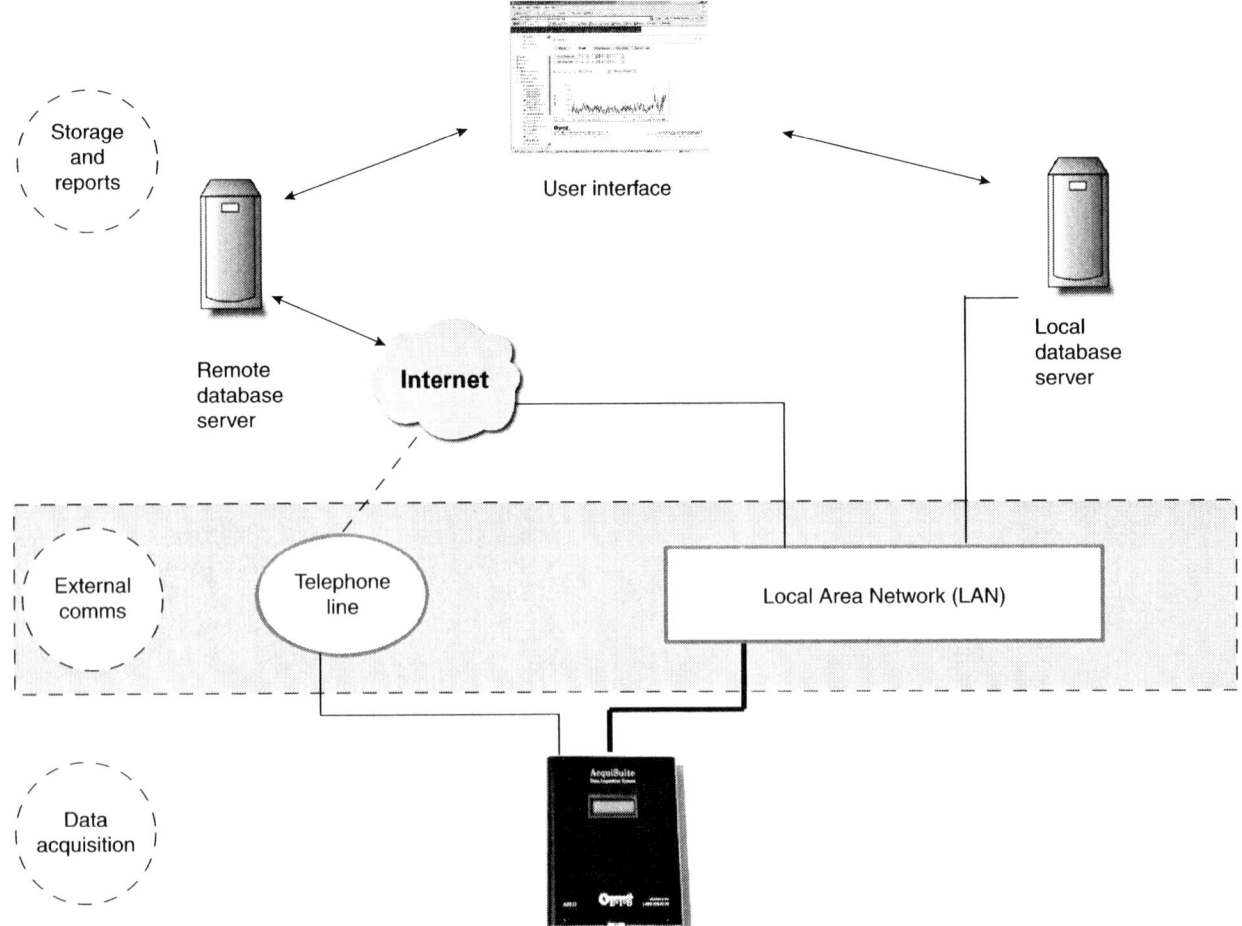

Fig. 5 Wired paths to the Web or a local database.

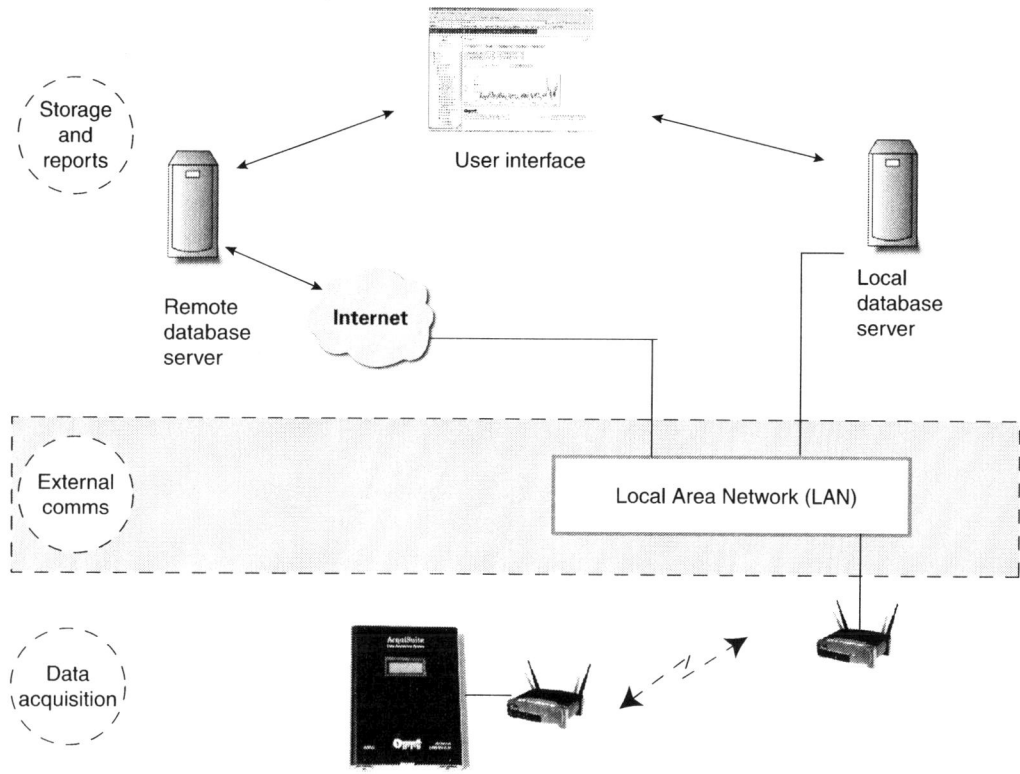

Fig. 6 Wireless Ethernet connection.

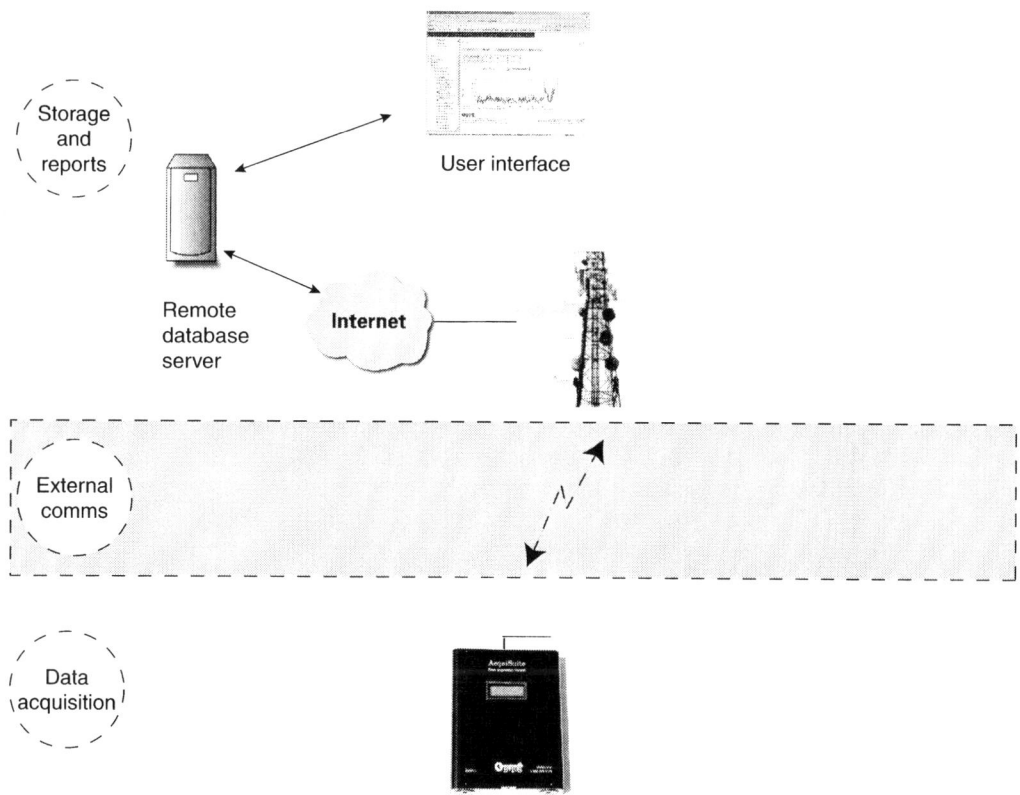

Fig. 7 Cell-based Web connection.

The DAS is functionally the same device used in the hardwired (LAN or phone modem) connection shown earlier in this entry. To use the cell phone system for data transmission, the phone modem module is replaced with a cell modem that connects to the cellular network just like any cell phone. The system relies on the use of the GSM network using GPRS to transmit data packets wirelessly over the cell network. This is the same system and service used by cell providers to allow users to view email and browse the internet using cell phones and PDAs.

Implementing this method of communication requires a SIM card (available from the wireless carrier) and a monthly data plan from the same carrier (e.g., Cingular) that typically runs $20–$30 per month for most DAS installations.

WHY USE THE CELL NETWORK INSTEAD OF A LAN OR PHONE LINE?

In addition to the ongoing monthly costs of cell service, the GSM version of the DAS is also typically 20%–30% more expensive than an identical LAN version, and the upload times are much slower. So, the obvious question is: Why anyone would use the cell system? There are a number of reasons why the GSM version may be preferable in some cases, despite the higher costs:

- *Costs of LAN or phone connections*—In many cases, it may be prohibitively expensive to add a network drop or phone line because of either the end user's policies or physical limitations (long wire runs, firewalls, etc.). In cases like this, the incremental costs of installing GSM may be less than the hardwired costs.
- *IT security concerns*—The LAN-based DAS is designed to function using the existing network without creating any security concerns. Because the GSM DAS has no connection to the existing network or phone system, any concerns are alleviated.
- *Installation delays*—Adding a network drop or phone line frequently involves coordinating with either other departments or contractors. In many installations, the lead times for getting the lines set and activated is the longest part of the installation; whereas the GSM system will be operating without the need for any outside resources the same day it is installed.

The most appealing aspects of the cell-based DAS are that it functions as a self-contained system without relying on existing networks or departments for installation or operation, and it can be up and sending data within minutes of the installation.

WHAT DOES THE FUTURE HOLD?

There are two primary forces driving improvements in the market for wireless submetering systems. First, advances in radio technology (both at the RS485 meter level and at the cellular system level) that are focused on broader markets such as consumer cell phones will be adopted by the energy information industry. Second, as the market for energy information grows, companies (such as Obvius) will emerge that specialize in the development of hardware and software specifically for gathering energy information, which will improve the functionality and ease of installation.

In general, it is reasonable to expect the following changes in the market:

- *Lower costs*—As companies outside the energy information industry (such as cell carriers) expand their offerings and increase the volume of sales, there will be lower costs for radio technologies and for the cellular service.
- *Easier installation*—The increased focus of specialized companies in this market will produce hardware and firmware that is focused on minimizing the time and costs associated with installation.
- *Broader coverage*—As the cellular companies expand and improve the coverage of their networks, it is reasonable to assume that some areas and locations that do not provide cell coverage today will be accessible in the future.

CONCLUSION

As we have seen, the development of application-specific wireless products designed to serve the submetering market has made the implementation of submetering networks much more cost effective. It is now practical to use wireless transceivers to transmit data within existing buildings or from building to building in a campus or base environment. In addition, the explosion of wireless Web options (including GPRS and WiFi) provides managers of submetering networks with wireless options that bypass the issues involved in accessing data via LANs or phone lines. For managers considering the installation of energy information systems, the cost of installing these systems is likely to drop even more in the future as more companies create new solutions and adapt existing technologies to data acquisition.

BIBLIOGRAPHY

1. Brambley, M.R.; Kintner-Meyer, M.; ONeill, P.J. Wireless sensor applications for building operation and management. In *Web Based Energy Information and Control Systems—Case Studies and Applications*; Capehart, B.L., Capehart, L.C., Eds., Fairmont Press: Lilburn, GA, 2005.
2. Standard for Information Technology, Telecommunications and Information Exchange Between Systems, Local and

Metropolitan Area Networks, Specific Requirements, *Part 11: Wireless LAN Medium Access Control (MAC) and Physical Layer (PHY) Specifications: Higher Speed Physical Layer Extension in the 2.4 GHz Band*. EE 802.11-1997. Institute of Electrical and Electronic Engineers: New York.

3. Katipamula, S.; Brambley, M.R.; Bauman, N.N.; Pratt, R.G. *Enhancing Building Operations Through Automated Diagnostics: Field Test Results*; Proceedings of the Third International Conference for Enhanced Building Operations, Texas A&M University, College Station: Texas, 2003.

4. Kitner-Meyer, M.; Brambley, M.R.; Carlon, T.A.; Bauman, N.N. Wireless Sensors: Technology and Cost-Savings for Commercial Buildings. *Information and Electronic Technologies; Promises and Pifalls*. In Teaming for Efficiency: Proceedings 2002 ACEEE Summer Study on Energy Efficiency in Buildings: Aug. 18–23, 2002; American Council for Energy Efficient Economy: Washington, DC, 2003; Vol. 7, pp. 121–134.

5. Kitner-Meyer, M.; Brambley, M.R. Pros & cons of wireless. ASHRAE J. **2002**, *44* (11), 54–61.

6. Kitner-Meyer, M.; Conant, R. *Opportunities of wireless sensors and controls for building operation*. 2004 ACEEE Summer Study on Energy Efficiency in Buildings. American Council for Energy Efficient Economy: Washington, DC, 2004.

7. Roundy, S.; Wright, P.K.; Rabaey, J.M. *Energy Scavenging for Wireless Sensor Networks with Special Focus on Vibrations*; Kluwer Academic Publishers: Boston, MA, 2003.

8. Su, W.; Akun, O.B.; Cayirici, E. Communication protocols for sensor networks. In *Wireless Sensor Networks*; Raghavendra, C.S., Sivalingam, K.M., Znati, T., Eds., Kluwer Academic Publishers: Boston, MA, 2004; 21–50.

9. Ye, W.; Heideman, J. Medium access control in wireless sensor networks. In *Wireless Sensor Networks*; Raghavendra, C.S., Sivalingam, K.M., Znati, T., Eds., Academic Publishers: Boston, MA, 2004; 73–91.

10. Michael, R.B.; Kintner-Meyer, M.; ONeill, P.J. Wireless sensor applications for building operation and management. In *Web Based Energy Information and Control Systems-Case Studies and Applications*; Capehart, B.L., Capehart, L.C., Eds., Fairmont Press: Lilburn, GA, 2005.

Wireless Applications: Mobile Thermostat Climate Control and Energy Conservation

Alexander L. Burd
Galina S. Burd
Advanced Research Technology, Suffield, Connecticut, U.S.A.

Abstract

The annual energy savings (AES) associated with the application of the wireless mobile thermostat for an average two-story residential house in the United States are projected to be approximately in the range of 4.5%–15.7% for heating and 3.1%–8.8% for cooling, respectively, depending on geographic location. The simple payback for the new control system serving space heating and space cooling loads is estimated to be approximately 1.9–3.5 years.

TYPICAL CLIMATE CONTROL ARRANGEMENT IN RESIDENTIAL BUILDINGS

The so-called forced air system with central furnace is the most frequently used system in the United States. The forced air system is utilized for space heating in 56.1 million households out of 101.5 million residential houses (or in 55.3% houses), including apartment complexes. The forced air system is quite extensively used for air conditioning (space cooling) and heating, utilizing the same ductwork to distribute heating and cooling throughout a house. Approximately 47.5 million houses out of 73.7 million single-family houses (or 64.5% houses) equipped with central air conditioning system also utilize a central furnace.[1]

About 30.1 million households (or 40.8%) among 73.7 million single-family houses with space heating and cooling are either two-story (26.9 million or 36.5%) or three-story (3.2 million or 4.3%) houses. The rest of the houses (43.6 million) are single-story houses (about 59.2%). Of these 73.7 million households, only approximately 400,000 households (about 0.55%) utilize two zones, with each zone (most frequently subdivided by floor) served by a dedicated furnace, which has its own stationary thermostat to control heating and cooling.[2]

A vast majority of residential houses with forced air systems have only one thermostat to control both heating and cooling modes of operation.

PREVIOUS AND CURRENT INVESTIGATIONS

The previous conducted investigation established that a significant temperature differential exists between the first and second floors when control of space heating and cooling is implemented from the central stationary thermostat located in the living room on the first floor. The air temperature is also subject to significant variation within each floor.[4] A temperature differential between various locations in the rooms is a function of many factors, such as the distance to the heating/cooling discharge air outlets and the windows, as well as occupancy level, heat gains from appliances, wind direction, orientation toward the sun, etc.

The application of a mobile thermostat allows for control of air temperature on demand at any time in any location of the building. This leads to better indoor climate control and reduces annual thermal and electrical energy consumption.[4,5] The potential AES for an average household in New England were analyzed earlier by Burd and Burd.[6] This paper evaluates the potential AES that could be realized via utilization of the new control system, which features a mobile thermostat in residential buildings at different geographic locations in the United States. Although the application of a mobile thermostat would reduce energy consumption in residential buildings due to multiple factors (including horizontal temperature variation within each floor), in our feasibility evaluation for the developed system, we conservatively considered energy savings that are caused by the vertical temperature variation only. These savings are associated with the temperature differential due to forced and natural convections[3] between floors in multistory buildings (i.e., two- and three-story houses).

INVESTIGATION OF TEMPERATURE DIFFERENTIAL BETWEEN SECOND AND FIRST FLOOR IN A RESIDENTIAL TOWNHOUSE

This investigation was conducted in a typical residential townhouse built in 1986 and located in the state of

Keywords: Residential; Comercial; Wireless mobile thermostat; Space heating; Air conditioning; Control system; Energy conservation.

Connecticut (New England region of the United States). The two-floor townhouse has approximately 92.9 m^2 (1000 ft^2) of total area. The height of each floor is 2.44 m (8 ft). The townhouse has a living room, a kitchen, and a half bathroom located on the first floor and two bedrooms and a full bathroom on the second floor (the smaller bedroom is used as a study room). In addition, the townhouse has a full unfinished basement used as storage. The basement is also heated and cooled via two diffusers supplying the warm and cold air; the forced-air furnace and associated air distribution ductwork are located in the basement. The townhouse is situated in the middle of a four-unit residential building and has easterly- and westerly-oriented outside exposures.

The monitored residential townhouse was built after the new ASHRAE ventilation norms were adapted and, thus, could be rated as a "tight" house. The infiltration rates in ACH (air change per hour) for residential buildings in the United States were adapted from the data published by McQuiston. The ACH indicates the ratio of the hourly infiltration air volume to the volume of the house. These rates vary from 0.51 ACH ("tight" houses) to 1.3 ACH ("loose" houses). Natural gas and electricity are used for space heating and space cooling, respectively. The furnace has an evaporator (cooling coil); the cooling compressor and condenser units of the cooling system are located outside of the house. The air treated in the furnace (heated or cooled) is delivered throughout the house via ductwork, supply diffusers, and registers, and then it is returned back to the furnace via return grills and ductwork. Two people live in the townhouse. The house has one stationary, nonprogrammable thermostat located in the living room, which controls both heating and cooling.

The temperature differential between the second and the first floors (TDSFF) depends on both forced and natural convection. For the purpose of this analysis, in Fig. 1, we considered a separate impact on TDSFF from the forced and natural air convection.

Natural convection is present in one way or the other during the entire heating season. However, its impact is more significant when the forced air space heating system is turned off. Because of that, natural convection impact is more pronounced during relatively warm outdoor air temperatures with fewer hours of space heating operation.

As opposed to natural convection, the impact of forced air convection is more noticeable during lower outdoor air temperatures with frequent space heating system operation. Therefore, the TDSFF, due to the combined effect of the forced air and natural convection, will be somewhat higher during lower outdoor air temperatures, as compared to the high outdoor air temperatures (see Fig. 1).

The combined impact of natural and forced air convection on TDSFF for the entire heating season is represented by line 3 in Fig. 1. The data in line 3 represent the summation of the data in line 1 and line 2.

According to Fig. 1, the average TDSFF during heating season, with an outdoor air temperature of 4.7°C (40.5°F), will be close to about 2.2°C (36°F).

The stationary thermostat set point in the townhouse (for the monitored data used in Fig. 2) was maintained at about 18°C (64.4°F).

WIRELESS MOBILE THERMOSTAT CONTROL SYSTEM

The accuracy of maintaining a desired air temperature in the building greatly depends on the utilized control system.

The closer the control device is to the area where the targeted temperature value is critical to maintain, the more accurate the control becomes. Obviously, the best and most precise control could be achieved by combining various stages of control and, eventually, by utilizing a final (ultimate) control stage when the control device is

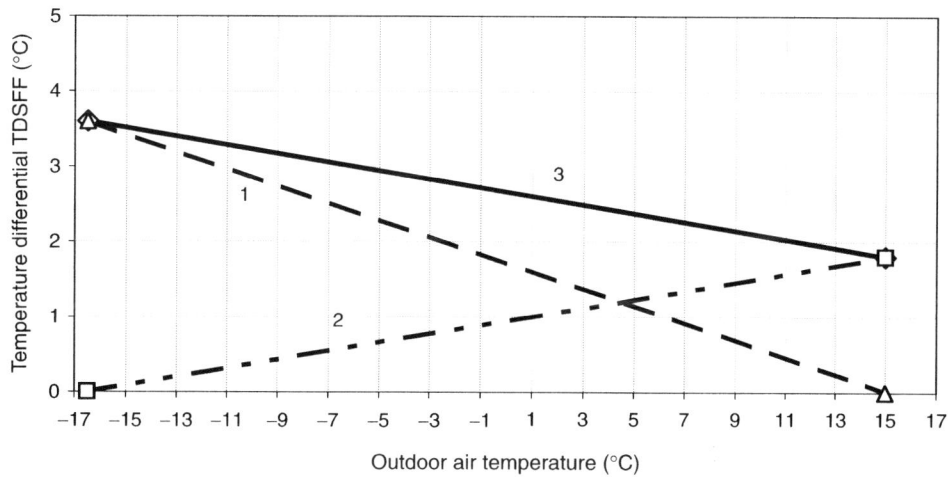

Fig. 1 Variations in outdoor air temperature and temperature differential between second and first floors (TDSFF). 1. Forced air convection impact; 2. Natural air convection impact; 3. Combined impact of forced air and natural air convection.

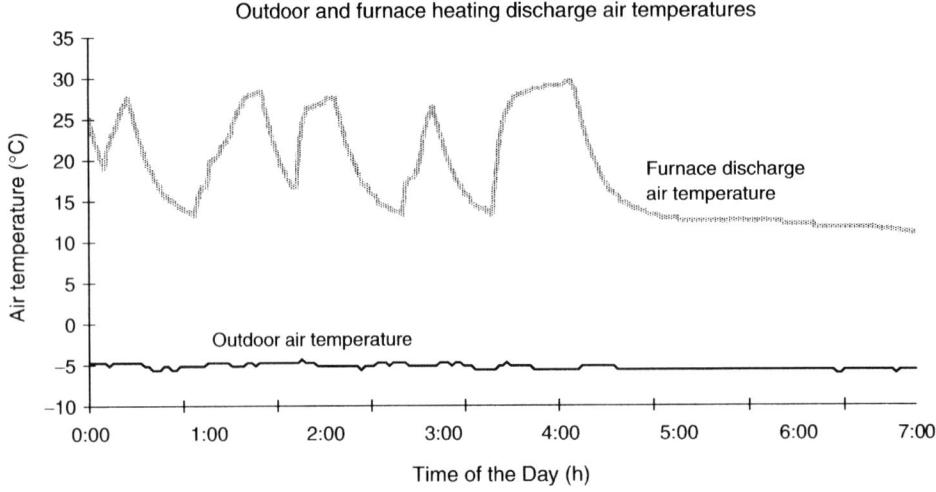

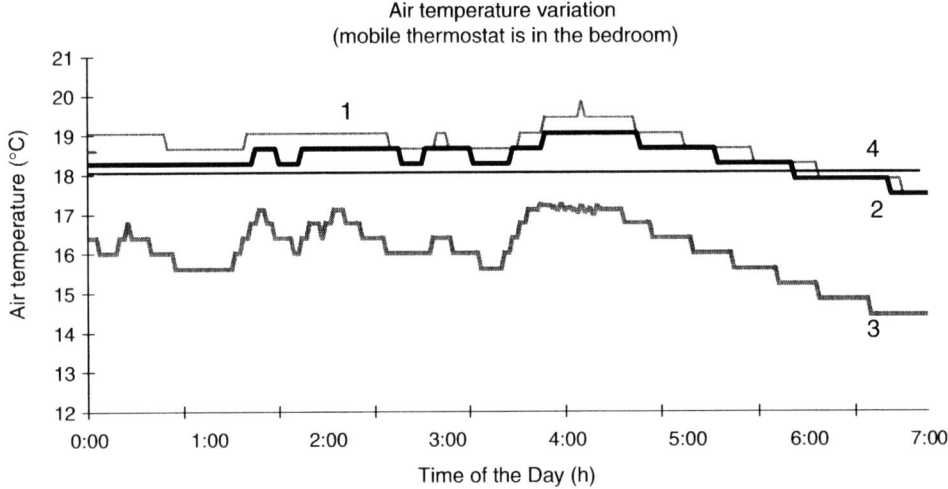

Fig. 2 Mobile thermostat operation during nighttime (heating mode). 1. Study room; 2. Bedroom; 3. Living room (dinner table); 4. Thermostat temperature set point.

located at the place where a desired air temperature value has to be maintained.

An innovative control system that utilizes a wireless mobile thermostat has been developed.[7] The mobile thermostat is a battery-operated device, which facilitates the implementation of the final stage of control to accurately balance the heat loss and heat supply in order to maintain the required air temperature in a particular area of the house. A prototype of the wireless mobile thermostat was manufactured to conduct testing of the thermostat's performance.

The prototype consists of two parts: a receiver and a transmitter. The prototype was installed in the investigated townhouse. The mobile thermostat's transmitter (remote temperature sensing element) was moved to different areas throughout the house (such as the bedroom, living room, study, etc.) to control air temperature as necessary.

The transmitter of the wireless mobile thermostat is capable of sending a remote radio signal to the receiver which, in turn, activates or deactivates the furnace's heating or cooling systems to maintain a set point temperature at the thermostat's transmitter location. This remote signal is proportional to the temperature differential between the wireless mobile thermostat's set point temperature and the air temperature at the thermostat's location. This proportionality is realized via the length of time during which the signal is sent from the thermostat to the receiver. The higher the initial deviation of the air temperature at the thermostat's location from the set point, the longer the time period during which the signal is transmitted from the transmitter to the receiver. This, in turn, would increase the operating time for the heating/cooling unit to heat/cool the area. A reverse procedure would ocurr if the initial temperature at the thermostat's location was closer to the thermostat's set point. Another distinctive feature of the wireless mobile thermostat is its low inertia and ability to react quickly to any temperature changes at the thermostat's location. A detailed presentation of the wireless mobile thermostat system's major elements is given in Burd and Burd.[4]

WIRELESS MOBILE THERMOSTAT OPERATION DURING HEATING MODE

The results of the wireless mobile thermostat's operational tests are shown in Fig. 2.

The upper part of the graph in Fig. 2 shows that at nighttime, the outdoor air temperature was near −6.5°C (20.3°F). The maximum value of the furnace's discharge air temperature measured in the basement was near 30°C (86°F). The furnace cycles on and off.

The mobile thermostat was located on the nightstand in the bedroom, and its temperature set point was 18°C (64.4°F). The lower part of the graph shows that the temperature in the bedroom was maintained close to the thermostat's set point [maximum deviation did not exceed +1°C (1.8°F) and −0.5°C (0.9°F)]. The mobile thermostat overrode the stationary thermostat control and turned the furnace on and off, as necessary (as shown in the upper portion of the graph), to maintain the required temperature at the mobile thermostat's transmitter location.

While the temperature in the bedroom (as well as in the study) closely followed the wireless mobile thermostat temperature set point, the temperature in the living room was maintained at a substantially lower magnitude.

The air temperature difference between the bedroom and the study room was not quite noticeable. On the other hand, the air temperature in the living room was significantly lower—by 1°C (1.8°F) to 4°C (7.2°F)—than the temperature at the mobile thermostat location. This demonstrates the ability of the wireless mobile thermostat to save energy for space heating by maintaining the desired temperature in the occupied rooms upstairs on demand, while keeping a lower temperature on the first floor when it is not occupied.

PROJECTED ANNUAL ENERGY SAVINGS IN RESIDENTIAL BUILDINGS

The projected AES due to the application of the wireless mobile thermostat are shown in Table 1. The energy savings were calculated for a number of the selected U.S. cities, which represent a wide variety of climatic conditions.

The design space heating outdoor air temperature conditions for these cities vary from −29.4°C (−20.9°F) for Bismark (North Dakota) to 7.8°C (46°F) for Miami (Florida). Table 1 indicates that the average (per heating season) outdoor air temperature for the selected cities vary from −1.1°C (30.1°F) for Bismarck (North Dakota) to 14°C (57.2°F) for Miami (Florida). The annual space heating run time for these cities (when the outdoor temperature is lower than the space heating balance temperature) varies from 894 h for Miami (Florida) to 7815 for Seattle (Washington).

The design space's cooling outdoor air temperature conditions differ from 43.3°C (109.9°F) for Phoenix (Arizona) to 29.4°C (84.9°F) for Seattle (Washington). Table 1 also demonstrates that the average outdoor air temperature over the cooling season for the selected cities varies from 26.1°C (79°F) for Portland (Maine) to 31.3°C (88.3°F) for Phoenix (Arizona). The annual space cooling run time for these cities (when the outdoor air temperature is higher than the space cooling balance temperature) varies from 237 h for Seattle (Washington) to 5390 h for Miami (Florida).

The cumulative annual operating time for space heating and cooling ranges from the maximum of 8052 h (91.9% of the length of the entire year) for Seattle (Washington) to the minimum of 6141 h (70.1% of the length of the entire year) for Tampa (Florida).

The application of the mobile thermostat would allow a user to setback the air temperature during the heating period and to setforward the air temperature during a cooling period. We used a simplified engineering method of calculation based on the assumption that the energy consumption for space heating and space cooling can be expressed as a linear function of the temperture differential between the two major parameters, averaged over a heating/cooling season. These parameters are: indoor dry-bulb air temperature set point and oudoor dry-bulb air temperature. This approach assumes that the outdoor dry-bulb air temperature can be used as a defining parameter, impacting energy conservation. This is utilized for the purpose of initial evaluating analysis only. A detailed presentation of the methodology for energy savings calculations is given in Burd and Burd.[6]

Table 1 shows the potential energy savings associated with the utilization of the mobile thermostat. The AES due to lowering or increasing a stationary thermostat's set point temperature by +1°C during the heating or cooling mode of operation were calculated by the formula:

$$\text{AES} = \frac{\Delta T}{\pm (\text{HCS} - T_{\text{AV.OUT.HC}})} \times 100, \%$$

where ΔT, the magnitude of lowering or increasing a set point temperature with the mobile thermostat, °C (°F); HCS, the current heating or cooling set point temperature maintained by the stationary thermostat, °C (°F); and $T_{\text{AV.OUT.HC}}$, the average outdoor air temperature during the heating or cooling season, respectively, °C (°F).

Table 1 denotes that the potential energy savings associated with the 1°C (1.8°F) setback in air temperature via the wireless mobile thermostat during heating season would vary from 4.7% (for Bismarck, North Dakota) to 16.7% (for Miami, Florida). The potential energy savings associated with the 1°C (1.8°F) setforward in air

Table 1 Potential annual energy savings due to mobile thermostat application for space heating and space cooling for selected cities in U.S.A.

No	City	State	North latitude (Degrees)	Design outdoor air temperature for heating (°C)	Average per heating season outdoor air temperature (°C)	Annual space heating system operating time (h)	Potential energy savings for space heating per 1°C thermostat temperature setback (%)	Design outdoor air temperature for cooling (°C)	Average per cooling season outdoor air temperature (°C)	Annual space cooling system operating time (h)	Potential energy savings for space cooling per 1°C thermostat setforward (%)	Annual space heating and space cooling systems operating time (h)	Cumulative percentage of space heating and space cooling systems operating time during a year (%)
1	Bismarck	North Dakota (ND)	47	−29.4	−1.1	6,896	4.7	33.9	27.8	824	25.4	7,720	88.1
2	Chicago	Illinois (IL)	42	−21.1	4.4	6,075	6.4	32.8	27.6	1,193	27.3	7,268	83.0
3	Portland	Maine (ME)	46	−5.6	4.5	7,265	6.5	32.2	26.1	383	45.0	7,648	87.3
4	Buffalo	New York (NY)	43	−16.7	4.5	6,704	6.5	30.0	26.3	616	40.9	7,320	83.6
5	Hartford	Connecticut (CT)	42	−16.7	4.7	6,455	6.5	32.8	27.2	934	30.0	7,389	84.3
6	Baltimore	Maryland (MD)	39	−11.7	6.2	5,717	7.2	33.9	27.5	1,308	27.7	7,025	80.2
7	Charleston	South Carolina (SC)	33	−3.9	6.8	5,703	7.6	34.4	27.2	1,206	30.0	6,909	78.9
8	Seattle	Washington (WA)	47	−5.0	9.1	7,815	9.1	29.4	26.3	237	40.9	8,052	91.9
9	Dallas	Texas (TX)	33	−8.3	9.7	4,079	9.7	37.8	28.6	3,029	21.4	7,108	81.1
10	Phoenix	Arizona (AZ)	34	1.1	11.6	3,305	11.8	43.3	30.2	3,861	15.8	7,166	81.8
11	Tampa	Florida (FL)	28	2.2	13.1	2,262	14.4	33.3	27.4	3,879	28.6	6,141	70.1
12	Miami	Florida (FL)	26	7.8	14.0	894	16.7	32.8	26.9	5,390	32.7	6,284	71.7

Assumptions: 1. Space heating balance temperature is assumed to be 16.7°C (62°F)—below 62°F heating is required; 2. Space cooling balance temperature is assumed to be 23.9°C (75°F)—above 75°F cooling is required; 3. Space heating and space cooling design dry-bulb air temperatures are assumed at their 99.6 percentile.

No	City	State	Annual space heating energy consumption (kWh)	Annual space heating energy savings per 1°C temperature setback (kWh)	Annual space cooling energy consumption (kWh)	Annual space cooling energy savings per 1°C thermostat temperature setforward (kWh)	Cumulative annual space heating and space cooling energy consumption (kWh)	Percentage of space heating energy consumption (%)	Percentage of space cooling energy consumption (%)	Annual space heating and cooling energy savings per 1°C temp. setback for heating and setforward for cooling (kWh)	Cumulative annual relative savings percentage for space heating and space cooling per 1°C temp. setback and setforward (%)
1	Bismarck	North Dakota (ND)	35,350	1,679	459	116	35,809	98.7	1.3	1,795	5.0
2	Chicago	Illinois (IL)	23,007	1,479	617	168	23,624	97.4	2.6	1,647	7.0
3	Portland	Maine (ME)	27,415	1,769	120	54	27,536	99.6	0.4	1,823	6.6
4	Buffalo	New York (NY)	25,298	1,632	212	87	25,511	99.2	0.8	1,719	6.7
5	Hartford	Connecticut (CT)	24,010	1,572	439	132	24,449	98.2	1.8	1,703	7.0
6	Baltimore	Maryland (MD)	19,254	1,392	666	185	19,920	96.7	3.3	1,576	7.9
7	Charleston	South Carolina (SC)	18,358	1,388	567	170	18,926	97.0	3.0	1,559	8.2
8	Seattle	Washington (WA)	20,823	1,903	82	33	20,905	99.6	0.4	1,936	9.3
9	Dallas	Texas (TX)	10,207	993	1,994	427	12,201	83.7	16.3	1,420	11.6
10	Phoenix	Arizona (AZ)	6,795	805	3,450	545	10,244	66.3	33.7	1,349	13.2
11	Tampa	Florida (FL)	3,824	551	1,915	547	5,740	66.6	33.4	1,098	19.1
12	Miami	Florida (FL)	1,306	218	2,323	760	3,629	36.0	64.0	978	26.9

Assumptions: 1. Space heating and space cooling energy consumption changes in direct proportion to the difference between indoor and outdoor air temperatures; 2. Historical energy consumption data for the monitored two-story townhouse in Hartford, CT was assumed to be a base for the calculations 0.000293 conversion factor from Btu to kWh.

temperature via the wireless mobile thermostat during cooling season would vary from 15.8% (for Phoenix, Arizona) to 45% (for Portland, Maine). The projected relative value of energy savings [for 1°C (1.8°F) reset in temperature set point] during the cooling mode of operation is higher than for the heating mode of operation because the temperature differential between the thermostat set point and average seasonal outdoor air temperature is lower for cooling as compared to heating.

The continuation of Table 1 shows annual heating and cooling energy consumption as well as cumulative annual energy consumption for space heating and cooling for the considered cities. The heating and cooling annual energy consumption for the monitored townhouse in Connecticut was calculated based on the actual electrical and gas meters data. The annual heating energy consumption (HEC) for the houses in the selected cities was calculated by the formula:

$$HEC = \frac{HEC_{MH}(T_{INH} - T_{AV.OUT.HSC})}{T_{IN.H} - T_{AV.OUT.MH}} \times \frac{ASHOT_{SC}}{ASHOT_{MH}}, \text{ kWh/yr},$$

where HEC_{MH}, annual energy consumption for heating in the monitored house, kWh (MJ); $T_{IN.H}$, the air temperature inside the house during heating season was assumed to be 20°C (68°F); $T_{AV.OUT.HSC}$, average (per heating season) outdoor air temperature for the selected cities, °C (°F); $T_{AV.OUT.MH}$, average (per heating season) outdoor air temperature for the monitored house, °C (°F); $ASHOT_{SC}$, annual space heating operating time for the selected cities, h; and $ASHOT_{MH}$, annual space heating operating time for the monitored house, h.

The annual cooling energy consumption (CEC) for the houses in the selected cities was calculated by the formula:

$$CEC = \frac{CEC_{MH}(T_{AV.OUT.CSC} - T_{IN.C})}{T_{AV.OUT.MH} - T_{IN.C}} \times \frac{ASCOT_{SC}}{ASCOT_{MH}}, \text{ kWh/yr},$$

where $T_{IN.C}$, the air temperature inside the house during cooling season was assumed to be 23.9°C (75°F); CEC_{MH}, the annual energy consumption for cooling in the monitored house, kWh (MJ); $T_{AV.OUT.CSC}$, the average (per cooling season) outdoor air temperature for the selected cities, °C (°F); $T_{AV.OUT.MH}$, the average (per cooling season) outdoor air temperature for the monitored house, °C (°F); $ASCOT_{SC}$, the annual space cooling operating time for the selected cities, h; and $ASCOT_{MH}$, the annual space cooling operating time for the monitored house, h.

Table 1 (continuation) also indicates the cumulative potential energy savings of a 1°C (1.8°F) temperature setback for space heating and temperature setforward for cooling. For the majority of the selected cities, the HEC far exceeds the cooling energy consumption. This is only not the case for Miami (Florida), where the annual energy consumption for cooling (64%) is higher than for heating (36%).

Table 2 demonstrates that the cumulative annual relative energy savings for heating and cooling with the wireless mobile thermostat vary from 5.0% for Bismark (North Dakota) to 26.9% for Miami (Florida) of the current (baseline) energy consumption for the considered buildings for each °C (1.8°F) of the temperature setback and setforward.

Table 2 illustrates the potential annual energy cost savings per 1°C (1.8°F) in temperature setback and setforward for space heating and space cooling, respectively, for the geographic locations considered in Table 1. The cumulative annual space heating and space cooling energy cost savings range from approximately $102 for Portland (Maine) to $447 for Phoenix (Arizona).

Table 3 shows potential annual energy cost savings due to the wireless mobile thermostat application, considering the results of temperature monitoring and the occupancy schedule in the representative house.[6] Based on the results of temperature monitoring discussed earlier, we assumed that the utilization of the wireless mobile thermostat would allow reduction of the air temperature in the house by 2.2°C (3.96°F) during the heating season, while the upstairs areas are occupied. Based on the results of temperature monitoring,[5,6] we also assumed that the use of the wireless mobile thermostat would allow the user to increase the air temperature of the house by 0.5°C (0.9°F) during the cooling season when it is occupied and occupants are on the second floor. The occupancy schedule assumes that the daily, average, per-week occupied, and nonoccupied time is 14.4 and 9.6 h, respectively. The occupancy schedule also assumes that the occupants spend 30 and 70% of the occupied time in the house downstairs and upstairs, respectively.

Considering the above occupancy schedule, the overall wireless mobile thermostat's daily setback for space heating and setforward for space cooling is assumed to be approximately 0.9°C (1.62°F) and 0.2°C (0.36°F), respectively. These additional savings are once again projected, compared to the existing stationary thermostat located on the first floor of the two-story house. The cumulative annual space heating and space cooling savings vary from $64 for Miami (Florida) to $117 for Phoenix (Arizona). The simple payback for the wireless mobile thermostat would vary from 1.9 years for Phoenix (Arizona) to 3.5 years for Miami (Florida). For the majority of geographic locations considered in the study, the mobile thermostat would have a simple payback of 2.4–2.7 years.

An application of the wireless mobile thermostat in residential buildings with only space heating or space cooling loads might be less advantageous from an

Table 2 Potential annual energy cost savings due to mobile thermostat application per one °C temperature setback for heating and temperature setforward for cooling

No	City	State	Annual space energy heating savings (kWh)	Annual space heating energy cost savings ($)	Annual space cooling energy savings (kWh)	Annual space cooling energy cost savings ($)	Cumulative annual space heating and cooling energy cost savings ($)
1	Bismarck	North Dakota (ND)	1,679	84	459	54	138
2	Chicago	Illinois (IL)	1,479	74	617	73	147
3	Portland	Maine (ME)	1,769	88	120	14	102
4	Buffalo	New York (NY)	1,632	81	212	25	106
5	Hartford	Connecticut (CT)	1,572	78	439	52	130
6	Baltimore	Maryland (MD)	1,392	69	666	79	148
7	Charleston	South Carolina (SC)	1,388	69	567	67	136
8	Seattle	Washington (WA)	1,903	95	82	10	105
9	Dallas	Texas (TX)	993	50	1,994	235	285
10	Phoenix	Arizona (AZ)	805	40	3,450	407	447
11	Tampa	Florida (FL)	551	27	1,915	226	253
12	Miami	Florida (FL)	218	11	2,323	274	285

Notes: 1. The above calculations are conducted for 1°C in temperature setback and setforward for space heating and cooling, respectively; 2. The installed cost of the mobile thermostat for the residential application is assumed to be $220 [10–12]; 3. Cost of natural gas used for space heating was assumed to be $1.46 for 29.3 kWh (100,000 Btu) or $0.0498/kWh; 4. Cost of electricity used for space cooling was assumed to be $0.118/kWh.

Table 3 Potential annual energy cost savings due to mobile thermostat application considering occupancy pattern in the representative house

No	City	State	Potential energy savings for space heating thermostat temperature setback (%)	Annual space heating energy savings (kWh)	Annual space heating energy cost savings ($)	Potential energy savings for space cooling thermostat temperature setforward (%)	Annual space cooling energy savings (kWh)	Annual space cooling energy cost savings ($)	Cumulative annual space heating and cooling energy cost savings ($)	Simple payback period (yrs)
1	Bismarck	North Dakota (ND)	4.5	1,586	79	4.9	89	11	90	2.5
2	Chicago	Illinois (IL)	6.1	1,397	70	5.3	120	14	84	2.6
3	Portland	Maine (ME)	6.1	1,670	83	8.8	23	3	86	2.6
4	Buffalo	New York (NY)	6.1	1,541	77	8.0	41	5	82	2.7
5	Hartford	Connecticut (CT)	6.2	1,484	74	5.8	85	10	84	2.6
6	Baltimore	Maryland (MD)	6.8	1,315	66	5.4	130	15	81	2.7
7	Charleston	South Carolina (SC)	7.1	1,311	65	5.8	110	13	78	2.8
8	Seattle	Washington (WA)	8.6	1,797	90	8.0	16	2	91	2.4
9	Dallas	Texas (TX)	9.2	938	47	4.2	388	46	93	2.4
10	Phoenix	Arizona (AZ)	11.2	760	38	3.1	671	79	117	1.9
11	Tampa	Florida (FL)	13.6	520	26	5.6	372	44	70	3.1
12	Miami	Florida (FL)	15.7	206	10	6.4	452	53	64	3.5

Notes: 1. Assumed average per week temperature setback for heating 0.94°C, 1.7°F; 2. Assumed average per week temperature setforward for cooling 0.19°C, 0.35°F.

economical point of view, as compared to the houses with both loads.

Energy savings for the wireless mobile thermostat will be achieved due to minimization of what we called a "comfort satisfaction safety factor" (CSSF)—when the set point temperature at the stationary thermostat is maintained at a higher level during the heating season and a lower level during the cooling season to compensate for any deviations of the temperature at the occupants' location to ensure their satisfaction with the indoor climate control based on the emperical anticipation. Similarly, the occupant applies the CSSF for the first floor stationary thermostat set point to satisfy the required comfort conditions on the second floor.

The above leads to a logical conclusion that the utilization of the mobile thermostat would always produce energy savings as compared to a baseline energy consumption with a stationary thermostat. Obviously, the CSSF would be different for various users and applications.

The projected savings in residential buildings with a single stationary thermostat serving radiation or convector steam or hot water heating systems with natural air convection may be somewhat lower as compared to the forced air heating systems. This could be due to the reduced temperature differential between the second and first floors for systems with natural air convection. Further investigation would be necessary to verify this assumption.

In addition to individual residential houses, the developed control system could also be used in a variety of heating and cooling applications in commercial buildings, as well. Wireless mobile thermostat could also be instrumental in reducing energy consumption in district energy systems, such as district heating/cooling systems,[8] as well as for various industrial applications—for climate controls in zones served by rooftop units, where manufacturing equipment is frequently moved around the facility,[9] etc.

CONCLUSION

The wireless mobile thermostat—the ultimate stage of climate control—would allow a user to setback and setforward temperatures in the house for space heating and space cooling, respectively. The conducted initial investigation utilizing a simplified model for different geographic locations in the United States showed that the application of a wireless mobile thermostat in the two-story townhouse would allow savings of 4.5%–15.7% of annual space heating energy. Utilization of the wireless mobile thermostat would also allow savings of 3.1%–8.8% of energy for space cooling. The simple payback for the new control system serving space heating and space cooling loads would vary from approximately 1.9–3.5 years.

The mass application of the wireless mobile thermostat, considering the scale of energy consumption in residential buildings, which consume more than 20% of the total energy use—including residential, commercial, agricultural and transportation sectors—could be an important step in energy resource.[10]

The ever-increasing cost of fuel could become a contributing factor for wireless mobile thermostat utilization. In addition, the application of the wireless mobile thermostat will have a positive environmental impact due to emissions reduction at power plants as well as at residential heat generation systems.

REFERENCES

1. Energy Information Administration/Household Energy Consumption and Expenditures, U.S.A., 1997.
2. Energy Information Administration/Annual Energy Review, U.S.A., 1999.
3. ASHRAE Handbook *Fundamentals*; American Society of Heating, Refrigerating, and Air Conditioning Engineers, Inc.: Atlanta, GA, 1997.
4. Burd, A.L.; Burd, G.S. Mobile thermostat for climate control in residential buildings, Proceedings of CIBSE/ASHRAE Joint Conference 2000, paper H-1562, Conference Papers on CD ROM: 1–22, Dublin, Ireland, 2000.
5. Burd, A.L.; Burd, G.S. *Wireless Mobile Thermostat for Climate Control Enhancement in Buildings*. Proceedings of Healthy Buildings 2000, Espoo, Finland, 2000; Vol. 2, 763–768.
6. Burd, A.L.; Burd, G.S. Application of Mobile Wireless Thermostat and Energy Conservation and Comfort in Residential Houses. ASHRAE Transactions, Honolulu, U.S.A., 2002; Vol. 108, Part 2, 202–211.
7. Burd, A.L. Mobile Thermostat to Control Space Temperature in the Building, Patent Number 5,419,489, 1–6, U.S.A., 1995.
8. Burd, A.L. Deferred heat supply for space heating using a capacity-limiting device—a beneficial approach for district heating. ASHRAE Transactions: Research, Boston, U.S.A., 1997; Vol. 103, Part 2, 23–31.
9. Burd, A.L.; Burd, G.S.; De Maio, M. Smith and Wesson: The Story of a Chilled-Water Retrofit. HPAC Engineering, March 2005; Part. 1, 38–47 and July 2005, Part. 2, 24–37.
10. Wisconsin Energy Statistics. Wisconsin Energy Bureau, Department of Admnistration, U.S.A., 1997.

BIBLIOGRAPHY

1. An idea whose time has come...the wireless thermostat! Totaline Star. Replacement Component Division, Carrier Corporation; U.S.A., 1999; 1.
2. ASHRAE Handbook *Systems and Equipment*; American Society of Heating, Refrigerating, and Air Conditioning Engineers, Inc.: Atlanta, GA, 2000.
3. McQuiston, F.C. A study and review of existing data to develop a standard methodology or residential heating and cooling load calculations. ASHRAE Trans., *90* (2A), 102–135.
4. Wireless thermostat. RCI Automation, Specification, 2002.
5. Wireless thermostat. Southern California Air Conditioning Distributors. Specifications, 2002.

Index

2×6 wall construction, 899
3E Plus, 880

A

Abandoned steam line isolation, 1357
Absolute implementation, global temperature adjustment and, 275
Absorption heat pump configurations, 1542–1544
   heat sink configurations, 1542–1543
   heat source configurations, 1542–1543
   working fluids, 1543–1544
Absorption heat pump, 816–817, 1541–1547
   bottoming and topping cycles, 1542
   thermal design fundamentals, 1541–1542
Absorption refrigeration, 1544–1546
   applications of, 1545–1546
   components of, 1544
      compressor, 1544
      condenser, 1544
      evaporator, 1544
      throttle, 1544
   performance of, 1545
Absorption working fluids, 1543–1544
Acceptance based commissioning, 189–190
Acceptance phase, commissioning process and, 196–197
Access, electric power transmission systems, 360
Accidents, nuclear, 1131–1132
Accounting of energy, 1–7
Accreditation, management system standards and, 1030–1031
Accumulated depreciation, rate of return regulation and, 1255
Accumulator, mobile HVAC systems and, 1079
Acid deposition, air quality modeling and, 719
Acid gas removal processes for syngas, 908
Acid rain, 873
Actionable information, interactive access and, 619–622
Active server pages, 537
Active space heating, 1322
Activity based costing, 1042–1048
   advantages and disadvantages of, 1048
   allocation via cost drivers, 1042–1043
   background of, 1042–1043
   calculations for
      cooling and lighting, 1044
      motors and machines, 1045
   example of, 1043–1048
   implementation cost, 1048
   tracing overhead, 1042
Additional energy, 1090–1091
   natural energy, a comparison, 1088–1094
   types, 1090–1091
      chemical fuels, 1091
      non renewable, 1090–1091
      not natural, 1091
      waste products, 1091
Adiabatic mixing, air streams and, 1188

Adjustable speed drive pumps, 1553–1554
   savings from pump modifications, 1553
Adjustable speed drives, energy cost calculations and, 74–75
Adjustment charges, utility billing and, 1502
Adobe masonry walls, 1517–1518
Advanced gas-cooled reactors, 40
Advanced shading devices
   light shelves, 1642–1643
   mini-light shelves, 1643
   prismatics and refractives, 1643
Advanced thermal technologies, 1347
   analytical semiempirical modeling, 1348–1350
   biofuels, 1353
   liquid fuel production, 1354
   output comparisons, 1347–1348
   pyrolysis, 1348
Advanced Transport Reactor, 906–907
Advertising, energy demand-side management programs and, 290
Advisors, independent power producers and, 857
AEC. See alternate energy credit.
AEE. See Association of Energy Engineers.
Aerated concrete, as insulation material, 891
Aerospace uses, heat pipe applications and, 807–808
AFC. See alkaline fuel cells.
Affected sources, emissions trading and, 431
Aging thermal processes, 868
Agricultural products, drying of, 332–336
Agriculture residues, coal and, cofiring of, 490
Agriculture, heat pipe applications and, 810
Air balancing, ventilation system and, 61
Air barriers, 662–663
   airtight drywall approach, 667–668
   bypasses, 664
   leakage details, 664–666
   materials, 662–663
   seal penetrations, 664
Air compressors, accounting for in energy assessment, 71
Air conditioning
   accounting for in energy assessment, 70
   capacity calculations, 1061–1066
   cycles, 1188–1191
      summer hot and
         dry mode, 1190
         humid mode, 1189–1190
      winter, 1190–1191
   electricity usage of, 583
   natural gas and, 1507
   solar, 1326–1327
   solar heating and, case study of, 1317–1320
      cooling system methodology, 1317–1318
      problems with, 1320
      psychrometric analsysis, 1318
      solar roof tile system, 1318–1319
   units, review use of, 521
Air cooled equipment, 532

Air costs, air distribution and, 212
Air density impact, fan system performance and, 1224
Air discharge volume table, 220
Air distribution adjustment
   duct static pressure setpoint reduction, 275–276
   electric demand response and, 275–276
   evaluation of, 276
   fan quantity reduction, 276
   fan speed limit, 276
Air distribution compressed air control systems and, 211–212
   zone management, 211
Air emission reductions
   energy efficiency and, 9–17
   Kyoto Treaty, 9
Air emission standards, 715–717
   vehicles, 717
Air emissions reductions
   carbon sequestration, 14–15
   energy efficiency
      emissions impacts, 11–12
      hydrogen economy, 10–11
      countries focus on, 10
      effectiveness of, 10
   United States
      alternate energy credit, 14
      attempts to regulate state by state, 12–13
      renewable energy credits, 14
Air flow measurement devices, variable air volume and, importance of, 52
Air flow rate, run around heat recovery systems and, 1280–1281
Air flow restrictions, ceiling and roof insulation, 902
Air flow, 116
Air humidification, 1579, 1581–1582
Air leak detection, compressed, 219–225
Air leakage
   details, 664–666
   facility, 662–670
   measuring, web-based compressor management and, 213
Air leaks
   common sealants, 665
   driving forces of, 666
Air Master + 1.0.9, 880
Air movement, underfloor air distribution (UFAD) and, 1467
Air pollution
   air quality modeling, 719–721
   ambient air quality standards, 715–717
   emissions standards, 715–717
   fossil fuels and, 715–721
   transport and dispersion, 718–721
Air production control, compressed air and, 207
Air quality modeling
   acid deposition, 719
   photo-oxidants, 719–721
   regional haze, 719

Air quality
　ANSI/ASHRAE Standard 62.1–2004 and, 51
　coal to liquid fuels and, 169
　compressed air control systems and, 211
　　monitoring of, 211
　indoor, 18–23
　monitoring of, 212
　standards, ambient, 715–717
Air storage
　compressed, 226–235
　distribution, 226–235
Air streams, adiabatic mixing of, 1188
Air traffic control (ATC), Future Air Navigation System, 1, 30
Air usage, reduced, 211
Air zone leakages, 212
Aircraft fuel
　consumption, 24–30
　　available ton-kilometer measurement, 24–25
　　breakthrough gains in efficiency, 29
　　energy costs
　　　advantages of conservation, 25
　　　impact on, 25
　　future of, 29
　　government's role in, 29–30
　　　air traffic control, 29–30
　　　global positioning satellites, 29
　　　incremental gains in efficiency, 29
　　reduction factors, 26–29
　　　drag reduction, 26–27
　　　engine efficiency, 26
　　　engine maintenance, 28–29
　　　flight controls, 28
　　　piloting techniques, 28
　　　pre-flight planning, 27–28
　　　weight reduction, 26–27
　　totals used, 24
　　trends in, 24–25
　jet, 26
　kerosene based, 25–26
　production of, 25–26
　types of, 25–26
　gasoline, 26
Airflow sensors, 58
Air-Krete, as insulation material, 894
Airports
　construction, federally funded, 1020
　Federal funding of, 591
　traffic problems, 588
Air-separation technologies, gasifiers and, 909
Airside heat transfer calculations, mobile HVAC systems and, 1062–1063
Airtight drywall
　advantages and disadvantages, 667
　air barriers and, 667–668
　installation techniques, 668
Airtightness, measuring of, blower door test equipment, 668–669
Alanates, 1446
Alaska coal basins, 158
Alignment tools, energy project management, 563–564
Alkaline fuel cells (AFC), 734–736
Alliance to Save Energy, 513
Allowance allocation, 434–435
　emissions trading and, 432

Allowance assignments, Kyoto protocol and, 141
Allowance, emissions trading and, 432
Allowance/permit allocation
　auction, 432
　baseline choosing, 432
　fixed historical baseline, 432
　historical baseline with updating, 432
Alterative transport fuels vs. gasoline, comparison of, 47–48
Alternate energy credit (AEC), 14
Alternating current induction synchronous electric motor, 349
Alternating current power system, 389
　rotary converter, 389
　Westinghouse, 389
Alternating current synchronous electric motor, 349
Alternative energy sources, regulatory issues, wind power, 1614–1615
Alternative energy
　future of, 43–46
　　economically competitive, 44–45
　　fossil fuel dominance, 45–46
　　renewable electricity generation, 44
　market penetration of, 46
　technologies
　　biomass, 31–34
　　hydrogen, 35–37
　　pricing of, 31–48
　　renewable, 31
　　solar energy, 34
　　types, 31–43
　　wind, 34–35
Alternative fuel, 1523–1525
　cost assumptions, nuclear energy economic issues and, 1105
　technologies
　　fuel cells, 37–38
　　nuclear energy, 38–43
Alternative metrics, 933
Alternative power sources
　tax incentives, wind power, 1614
　wind power, government subsidies of, 1614–1615
Aluminum alloy artificial aging, 868
Aluminum plant assessments, Dept of Energy best practices case studies and, 1283, 1285–1288
Ambient air, 117
　quality standards, 715–717
Ambient energy fraction, 819–820
Ambient temperatures, 521–522
　based utility pricing, 1506
American Council for an Energy Efficient Economy, 513
American energy master planning, 551
　models, management system for energy, 551
American National Standard Institute (ANSI), 1028–1029
American National Standards Institute. See ANSI.
American Society of Heating and Refrigeration and Air Conditioning Engineers Inc. See ASHRAE.
American Society of Heating, Refrigerating and Air Conditioning Engineers (ASHRAE), 438

U factor definition, 1620
Amtrak, failure of, 1020
Anaerobic digesters, 90, 1529–1530
　waste fuels and, 1527
Analytical semiempirical model (ASEM), 1348–1350
Analyze data, Six Sigma and, 1314–1315
Ancillary services, electric demand response and, 284
Andon signals, lean manufacturing and, 473
Annual electricity generation, electric power projects and, 955–957
ANSI (American National Standards Institute), 50–62, 1028–1029
ANSI. See American National Standard Institute.
ANSI/ASHRAE Standard 62.1–2004, 50–62
　analysis of, 50–61
　building renovations, 61
　compliance to
　　indoor air quality procedure, 54, 60
　　ventilation rate procedure, 54, 55–60
　construction process, 61
　definitions as used in, 51
　　indoor air quality, 51
　design documentation procedures, 60
　high humidity, 53–54
　indoor air quality procedures, 60
　minimal system requirements, 51–52
　mold, 53–54
　outdoor air quality, 51
　pressurization flow, 53–54
　procedures, 54
　sensor verification, 61
　system start-up guidelines, 61
　　air balancing, 61
　　operations and maintenance, 61
ANSI/MSE 2000:2005 management system for energy, 1032–1034
　benefits of, 1034–1034
　case study, 1037–1041
　problems addressed, 1035–1037
　process format, 1032–1033
　standard elements, 1033–1034
Anthracite coal, 156
APACHE HVAC software, 1401
Appalachian coal
　basin, 157
　production, 159
Apparent power volt amperes, 1510
Appendix, use of in energy audit report, 79
Appliances, electricity usage of, 583
Application interface, interactive access and, 621–622
Applications, enterprise energy management systems and, 623
ArchiPhysics-Solar software, 1402
Argon, as insulating gas, 1619
Arsenic, as syngas contaminant, 909
Artic Wildlife Reserve oil exploration, 595
　royalties, 596
　nationalization of, 596
Artificial aging of aluminum alloy, 868
ASD technology, electric pump motors and, 353, 355
ASEM. See analytical semiempirical modeling, 1348–1350

Ash content of syngas feedstocks, 907
Ash fusion temperature, syngas feedstocks, 907
ASHRAE, 60–62, 438, 1620
   GreenGuide, 1407–1408
   Standard 62.1–2004, 50–62
   standards, 1363
      Energy Star indoor air quality requirements and, 577
Asian coal to liquid fuel plants, 167
Assets, electricity producers and, 388
Association of Energy Engineers (AEE), 131–135
   certification process, 133–135
   continuing education programs, 135
   salary information, 131–133
ATHENA green building design software, 1403
ATK. *See* available ton-kilometer.
Atomic structure, 1127
Attic radiant barriers, 1227–1232
Attic ventilation
   powered attic ventilator, 900
   vent selection, 899–900
Auction, allowance/permit allocation and, 432
Auditing
   facility energy use, 63–68
   user friendly report, 76–80
Australian coal to liquid fuel plants, 167
Autoclased concrete, as insulation material, 891
Automatic lighting control systems, 978–979
Automation, compressor air control systems and, 210
Automobile technology
   evolution of, 673–675
      communication buses, 674
      computers, 673–674
      engineering analysis and design, 673–675
      microprocessors, 673–674
      quality control, 673
      skilled assembly workers, 675
      smart sensors, 674–675
      standard connectors, 675
      system modules, 674
      wire harnesses, 675
   facility energy controls, 671–679
      building code support, 677–678
      equipment and system modules, 677
      impose standards, 677
      modular buildings, 677
      standardized commissioning of buildings, 678–679
   reasons not found in facility energy controls, 675–677
   transferable to building use, 672–673
      display systems, 672–673
      individual control systems, 672
      operational controls, 672
      options, 673
Autopilot flight controls, 28
Available ton-kilometer (ATK), consumption per, 24–25
Average unit cost, 920
Avoided cost, 919–920
Awnings, as window shading device, 1642
Axial fans, 1220

B

Backup electric rate schedule, 1509
Backup systems, photovoltaic systems and, 1152–1153
Balance supply and demand, 503
Balanced system, 700–701
Ballasts, high intensity discharge, 830–832
Banking, emissions reduction credit and, 432–433
BAS. *See* building automation systems.
Baseline construction, energy benchmarking and, 692–693
Baseloads, 1498
   rate structures of, 1498
Basins, coal, 157–158
Batch dryers, 338
Batch mail applications, 538
Bathhouses, Roman heating designs and, 1358
Batt-attic insulation, 901
   cathedral ceilings, 902
   full, 901
   installation of, 901–902
Batteries
   lead acid, 1155
   nickel cadmium, 1155
Battery technology, 403
Batts, in wall insulation, 896
BCA. *See* benefit cost analysis.
BCR. *See* benefit to cost ratio.
BEA software, 1401
Beadboard, as insulation material, 890–891, 894
BECP. *See* DOE Building Energy Codes Program.
BEES green building design software, 1403
Benefit cost analysis (BCA)
   definition of, 81
   energy efficiency and, 81–85
   net present value calculation, 84–85
   types, 81
      benefit to cost ratio (BCR), 81
      savings to investment ratio, 81
Benefit to cost ratio (BCR), 81
   calculating of, 82–84
      biasing effects, 84
      cost placement, 83
   significance of, 81–82
Benefits
   commissioning existing buildings and, 181
   energy project life cycle costing and, 971–972
Best Practices (BP) program, in industrial energy management, 873
Biasing effects, importance in BCR, 84
Bidding, energy project management planning and, 560–561
Billet heating, 868
Billing factors, utility rate structures and, 1501–1502
Binding energy, 1127
Bio solid gasification, 492–494
Biochemical conversion
   biogas, 90
   liquid fuels, 90–91
Biodiesel, 33–34
Bioethanol, 32–33
   needed resources, 33

Biofuels, 90–91, 1353
Biogas, 90
   anaerobic digestion, 90
Biological sequestration, natural carbon sinks, 129
Biomass, 14–15, 1266–1267
Biomass biochemical conversion, 90–91
Biomass combustion
   co-combustion, 88–89
   problems with, 89
Biomass energy, 31–34, 86–92
   benefits of, 91
      GHG reduction, 91
   combustion, 88–89
      circulating fluidized bed, 88
      stationary fluidized bed, 88
   content, 86–87
      higher heating value, 86–87
      lower heating value, 86–87
   definition of, 86
      chemical composition of, 86–87
   fuel types, 32–34
   modern types, 32
   organic basis, 31–34
   problems with, high bulk volume, 87
   traditional types, 32
   types available, 87
   uses, 31–32
   wood fuels, comparison of, 88
Biomass fuels
   heating values, 482–483
   solid fuel properties and, 482–483
   types, 32–34
      biodiesel, 33–34
      bioethanol, 32–33
      ethanol, 32–33
      methanol, 33
Biomass moisture content, 87
Biomass program funding, 1531
Biomass thermochemical conversion, 89–90
Bipolar direct current transmission, 359
Bituminous coal, 156
Blade materials for turbines, 1382–1383, 1385
BLAST software, 1401
Blended utility rate, 967
Block rates
   declining, 1503
   increasing, 1503
   inverted, 1503
   sliding, 1503
Blow off reduction, compressor air control systems and, 210
Blowdown heat recovery equipment replacement, 1368
   energy and cost savings, 1369, 1370
Blower door test equipment, 668–669
Blowers, 464
Blowing agents, in insulation materials, 891–892
Blown cellulose, ceiling and roof insulation, 901
Blown foam insulation, 897
Blown loose-fill insulation, 897
BMCS. *See* boiler control system management.
Boil water reactors, 39–40
Boiler control systems, 93–102
   management, 96–98
      flame safeguard control, 97–98
      multiple boilers, 98–102

Boiler shutoff, low cost energy efficient improvement case studies and, 519
Boilers, 93–102, 531,1360, 1589
    combustion and, 95–96
    construction of, 93
    control of multiple, 98–102
    efficiency of, 94–95
    energy intensive, 463
    ratings of, 94–95
    types, 93–94
        high pressure, 93
        low pressure, 93
        medium pressure, 93
Boiling limitations, heat pipes and, 805, 806
Boiling water nuclear reactors, 1117–1118
BOOT. *See* build own operate transfer.
Borehole, geothermal heat pump system development and, 758
Boric acid flame retardants, in insulation materials, 891
Bottoming cycles, 1542
Box cooking, solar, 1326
Brayton cycles, basic, 1576–1577
    advantages and disadvantages, 1576–1577
Breathing zone outdoor airflow, 57
BREEM (Building Research Establishment Environmental Assessment Method), 948
BREEZE software, 1403
Brise-soleils, as window shading devices, 1642
Brown coal, 156
BSim2002 software, 1401
BSS. *See* building system simulation.
Btu/sf, 967
Bubble back, as insulation material, 895
Bubble, ERC and, 430–431
Bubbling fluid bed boiler, 1528
Build own operate transfer (BOOT), 571
Building automation systems (BAS), 104–110
    basic features of, 104
        controller-level hardware, 105
        client hardware and software, 106–107
        controller
            communications network, 106
            communications protocol, 106
            software programming, 105
        design issues, 107
            new facility, 107
        direct digital control, 104–119
        energy data collection and, 617–618
        future costs of, 108
        future trends, 108–110
        machine to machine communications, 108–110
        server hardware and software, 106–107
        upgrading of, 107–108
Building codes, incorporating in facility energy controls, 677–678
Building commissioning, 179–187
    monitoring and verification, 179
    new structures, 188–199
Building cooling applications, direct fired absorption chillers, 1545–1546
Building design system, integrated, 1401
Building envelope, 528
    energy efficiency and, 1094

measurements, data collection and, 258–259
Building geometry
    building system simulation process and, 119
    energy simulator software, 111
    energy use
        effect on, 111–115
        geometric ratio, 113
        index, 111
        modeling, 112–113
    factors re energy, 112
Building location, building system simulation process and, 119
Building maintenance, retrocommissioning and, 203
Building materials, building system simulation process and, 119
Building occupancy, daylighting and, 267
Building renovations, ANSI/ASHRAE Standard 62.1–2004 and, 61
Building Research Establishment Environmental Assessment Method (BREEM), 948
Building simulation, sustainable, 1396–1403
Building standards, incorporating in facility energy controls, 677–678
Building system simulation (BSS), 116–123
    applications of, 118–119
    definitions of, 116
    energy flow paths, 116–117
    evolution of, 117–118
        energy flow, 117–118
        frequency domain, 117
        numerical methods, 118
        response factor, 117
        time domain, 117
    physical issues, 116–117
        air flow, 116
        ambient air, 117
        casual heat gain, 116
        control processes, 117
        inter-surface longwave radiation, 116
        moisture transfer processes, 117
        plant interaction with building zones, 117
        shading from sun rays, 116
        shortwave radiation, 116
        solar radiation, 117
        surface convection, 116
    process
        case studies of, 120–123
        input parameters, 119–120
            building geometry, 119
            building location, 119
            building materials, 119
            casual loads, 119
            HVAC control, 120
            infiltration, 120
            internal environment control, 120
            standing devices, 120
            transparent surfaces, 120
            weather data, 119
Building types
    high rise, 1513
    low rise, 1513
    walls and windows and, 1513
Building zones, plant interaction with, 117
Buildings

electric demand response solutions for, 281–283
rating systems for, 948–949
    Building Research Establishment Environmental Assessment Method (BREEM), 948
    Energy Star™, 948
    Green Building Challenge program, 948
standardized commissioning of, 678–679
ventilation problems and, 20–21
Built in storage solar water heater, 1332–1333
    finned, 1333
    plain, 1332–1333
Bulk problem, biomass energy and, 87
Burden of proof, environmental policy and, 627–628
Burner/engine redesign, waste fuel technology and, 1527
Bus circuit breaker arrangements importance of, 359
BUS^{++} software, 1401
Buses, communication, 674
Business retention utility rates, 1505
Bypasses, air barriers and, 664

C

Cables, 357, 357
    capacity determination, 358
    electric power transmission, 358
    transmission limitations, 358
CAC. *See* command and control policies.
Cadmium, as syngas contaminant, 909
CAES. *See* compressed air energy storage.
Calciners, 464
Calcium bromide cycles, 1445
Calculated data, interactive access to, 621
Calculation methodology, space heating and, 1363–1364
Calculations
    residential building heating loads, 1272–1273
    seasonal heating demand, 1272–1273
California Energy Commission, 511
California Statewide Emerging Technology Program, 514
California, electricity deregulation results in, 385
Candela, 998–999
    definition of, 998–999
CANDU. *See* pressurized heavy water nuclear reactors.
Cap and trade emissions trading, 431
Capacity calculations, inverse heat transfer methods and, 1065
Capacity losses, 361
Capacity, electric power transmission systems and, 360
Capital costs, 956–957
    nuclear energy and, 1101–1103
Capital expenditure, reducing energy consumption for drying, 346–347
Capital investment, energy conservation and industrial processes, 464
Capital leasing, 571–572
Capital rationing, 84–85
Capture and storage, carbon, 125–126

Index

Carbon capture and storage (CCS), 125–129
    carbon sources, 125–126
    costs of, 128–129
    geologic sequestration, 127–128
    industrial CO_2 capture, 127
    ocean direct injection, 128
    post-combustion capture, 126
    pre-combustion capture, 126–127
Carbon capture, sequestration and, 125–129, 169
Carbon dioxide, 20
    based DCV, 56
    capture, 127
        storage of, 127
    differentials, steady-state, 52
    emissions
        gasoline fueled vehicles, 788
        hybrid vehicles, 788
        plug-in hybrid vehicles, 788–789
    levels, ventilation rate procedures and, 56–57
    reduction in, 15–17
    regulating of, 11–12
    solid fuel properties and, 484
Carbon emissions, geothermal energy and, 744
Carbon filtration of wastewater, 1561
Carbon sequestration, 125–130
    biological methods, 129
    biomass, 14–15
    carbon capture and storage, 125–129
    definition of, 125–129
    disadvantages of, 15
    future prospects of, 129
    technologies, 14–15
Carbon sources, carbon capture and storage and, 125–126
Carbonyls, metal, as syngas contaminants, 909
Careers, energy engineering and, 131–136
Carnot coefficient of performance, 815
Carnot cycle efficiency, 1171
    thermodynamic temperature, 1171
Carnot efficiency, 1574–1575
Carnot engine, 1427–1428
Carnot Law, 1541
Carnot steam cycle, 1381
Cascading Style Sheets, 537
Cash flows, 932
    IRR and, NPR, 935
Casual heat gain, 116
Casual loads, building system simulation process and, 119
Catalysts, Fischer-Tropsch process and, 164–165
Cathedral ceiling insulation
    batt-attic, 902
    exposed rafting, 902–903
    R-30 batts and, 902
    raised top plate, 902
    raised-heel trusses, 902
    scissor trusses, 902
    soffit air ventilation, 902
Cathedral ceiling insulation, techniques, 902–903
CAV. *See* constant volume of supply air.
CCGT. *See* combined cycle gas turbine.
CCS. *See* carbon capture and storage.
CDD. *See* cooling/heating day adjustment.

CEE premium efficiency motor, 352–353
CEE. Consortium for Energy Efficiency.
Ceiling and roof insulation
    air flow restrictions, 902
    attic
        floor insulation, 900901
        ventilation, 899–900
    batt attic insulation, 901
    blown cellulose, 901
    cathedral ceilings. *See* Cathedral ceiling insulation.
    loose-fill insulation, 900–901
        installation techniques, 900–901
    powered attic ventilator, 900
    recessed lights, 903
    under storage floor, 901
    vent selection, 899–900
Cellulose insulation, 890, 893
    dust inhalation and, 891
    effect on environment, 891
Center for Analysis and Dissemination of Demonstrated Energy Technologies, 515
Center of gravity, aircraft pre-flight planning and, 27
Central chiller plants, cooling system adjustment and, 276
Central receiver, 1328
Centrifugal compressor surge protector, compressed air and, 209
Centrifugal compressors, 464
Centrifugal fans, 1220
Centrifugal pumps, 464, 1213
    theory, 1215
Certification, management system standards and, 1030–1031
CFB. *See* circulating fluidized bed.
CFBC. *See* circulating fluidized bed combustor.
CGI. *See* common gateway interface, 536
Chain reaction, 1128–1129
Chaos, 659
Char reaction, 487
Charting tools, 538
Chauffage, 1136
    contracts, 571
Cheap energy, U.S. transportation and, 588
Chemical fuels, energy form, 1091
Chemical plant assessments, Dept of Energy best practices case studies and, 1283–1284
Chemical process industry
    electrical element, 1305
    electricity supply, 1305
    energy conservation, examples of, 1306
    energy efficiency, 1302–1309
    EPA regulations, 1302–1303
    fuel oil supply, 1305
    material and energy balance concept, 1303–1304
    mechanical element, 1304–1305
        chillers, 1305
        compressed air, 1305
        hydraulic systems, 1305
        microturbine power generation, 1304
        pneumatic transport, 1305
        pumping, 1304
    natural gas supply, 1305

    Occupational Safety and Health Act, 1303
    process energy optimization, 1304
    waste to energy, 1305
    water element, 1305
Chemical reactions, 462
Chicago Edison, infrastructure growth, 391–392
Chilled water storage, 1417–1418
Chiller capacity, 311–312, 1418
Chiller efficiency, higher lift, 309–310
Chiller plants, design comparison, 312
Chiller energy only analyses, 309
Chillers, 531
    chemical process industry and, 1305
    plant retrofitting, 315
    pre-packaged, 314
    series, 310
China's economic growth, impact on U.S. transportation energy, 592
Chloramine, 1559
Chlorine, 1559
Chlorofluorocarbons (CFCs), effect on environment, 891–892
Cholera, 1557, 1560
CHP. *See* combined heat and power.
Churn factor, Underfloor air distribution cost effectiveness and, 1469
CIE vocabulary for spectral regions, 1617

Circuit breakers, 360, 704
    importance of adequacy, 360
Circuit resistance, 360–361
Circulating fluid bed boiler, 1528
Circulating fluidized bed (CFB), 88
Circulating fluidized bed combustor (CFBC), 489
Clausisus statement, 1171
    Carnot cycle efficiency, 1171
Client side programs, web publishing and, 537–538
    Cascading Style sheets, 537
    Dynamic HTML, 537
    extensible markup language, 537
    Java applets, 537
Climate changes, global warming and, 723
Climate control
    common types, 1660
    differences in two story building, 1660–1661
    mobile thermostat, 1661–1663
    wireless mobile thermostat, 1660–1667
Climate policy
    European Union Emissions Trading System, 141
    Kyoto protocol, 139–141
    options for, 137–138
        emission reduction, 137–138
        emissions quota allocation, 138
    regional programs, 141
    U.N. Framework Convention on Climate Change, 139
    voluntary programs, 141
    world motivation for, 137–142
    world response to, 138–139
    worldwide adaptation of, 138
Climate Technology Initiative, 448–449
Closed loop control, electric demand response and, 273

Closed-cell, high-density polyurethane, as insulation material, 891, 894
Clothes washers, 1588
CMMS. *See* computerized maintenance management systems.
Co combustion concepts, 88–89
   direct cofiring, 88–89
   indirect cofiring, 89
   parallel combustion, 89
Coagulation, 1557, 1558
Coal basins, 157–158
   Alaska, 158
   Appalachian, 157
   Great Plains, 158
   Gulf Coast region, 158
   interior, 157–158
   Rocky Mountain states, 158
Coal boiler vs. IGCC, 912
Coal composition, solid fuel properties and, 481
Coal derived syngas contaminants, 907
Coal fired electrical generating station
   exergy analysis and, 648–651
   material flows data, 650
Coal gasifier, 163
Coal mining
   liquid fuels and, 168–169
   underground, 144
Coal power plants, 954
Coal prices in United States, 152–154
Coal production in United States, 143–154
   coal prices, 152–154
   employment, 150
   history of, 143–144, 145
   innovations in, 150–151
   preparation of, 151–152
   productivity, 147–150
      regional basis, 147–149
      regional changes, 149–150
   surface mining, 144, 146
   technology trends in mining, 146–147
   tonnages produced by location, 159, 160
      Appalachia, 159
      Texas, 159
      Wyoming, 159
      western United States, 159
   transporting of, 159, 161
   types of, 144, 146
      underground mining, 144
   uses of, 159–161
      electricity generation, 159–160, 161
      industry, 160
      steel industry, 160–161
Coal supply in United States, 156–161
   basins and types found, 157–158
   coal to liquid fuels and, 168
   mining methods, 158–159
   types, 156–157, 158
      anthracite, 156
      bituminous, 156
      lignite, 156
      subbituminous, 156
Coal to liquid fuels, 163–171
   advantages of, 165
   air quality, 169
      carbon capture and sequestration, 169
   coal gasifier, 163
   coal mining, 168–169

   coal supply in United States, 168
   current and planned world production of, 167
   diesel products, 165–166
      specifications, 165
      utilization of, 166
   electricity requirements, 170
   Fischer-Tropsch process, 163
   manufacturing of, 163
   natural gas issues, 170
   plant employment, 170
   plant land requirements, 169
   plant locations, 168
      Asia, 167
      Australia, 167
      India and Pakistan, 167
      North America, 167
   plant operation of, 168
   products from, 163
   transportation needs, 168–169
      railroad limitations, 168
   waste products, 169–170
   water quality, 169
   water neutral concept, 169
Coal transportation, 159, 161
   coal to liquid fuels, 168–169
Coal
   agriculture residues and, cofiring of, 490
   anthracite, 156
   bituminous, 156
   brown, 156
   heating fuel, 1361
   lignite, 156
   manure and, cofiring and, 490–491
   natural gas vs., 161
   RDF and, cofiring and, 490
   soft, 156
   subbituminous, 156
   wind vs., 957
Coefficient of performance, 819
Cofiring, 490–491
   coal and agriculture residues, 490
   coal and manure, 490–491
   coal and RDF, 490
   fouling during, 491, 493
   NO_x emissions, 491
Cogeneration 1203
Cogeneration applications, load usage and, 1482
Cogeneration district heating and cooling systems, 318–319
   characteristics and technical aspects of, 319
   heat recycling, 318
   use in Europe, 318
Cogeneration natural gas and, 1507
Cogeneration project funding, 1531
Cogeneration Public Utility Regulatory Policies Act (PURPA) of 1978 and, 1202
Cogeneration steam turbines, 1386
Cogeneration systems, combined heat and power and, 177–178
Cogeneration thermal output temperature, combined heat and power and, 176
Cogeneration unit location, combined heat and power and, 175
Coil designs, mobile HVAC systems and, 1061–1066

Cold air retrofit case study, HVAC and, 172–174
Cold gas cleanup technologies, 908–909
Coldfusion, 537
Coliform water test, 1557
Collaboration tools, energy project management and, 563–564
Collection of data, 255–263
Collector performance, solar water heating and, 1336
Color preference index, definition of, 999
Color preference, lighting design and, 999
Color rendering
   definition of, 999
   improvement, definition of, 999
   lighting design and, 999
Color temperature
   definition of, 999
   lighting design and, 999
Combined cycle gas turbine (CCGT), 1386
Combined cycles, 1578–1579
   basics of, 1578
   evaluation, 1578–1579
   heat recovery steam generators, 1578
Combined gas law, 228–229
Combined heat and power, 863–864
   applications
      distributed generation technologies and benefits of, 299
      resale of excess power, 865
      waste heat recovery and, 864–867
   cogeneration systems, 177–178
   cogeneration thermal output temperature and, 176
   cogeneration unit location, 175
   distributed generation and, 303–308
      advantages of, 303–304
      philosophy behind, 304
      users of, 305–307
         energy service companies, 306
      efficiencies
         electric load following cycle, 1530–1531
         integrated coal gasification/combined cycle power plants, 1531
         thermal load following cycle, 1530
         waste fuels and, 1530
   examples of usage, 176–177
   exhaust gas condensation solutions, 177
   Federal support for, 866
   generator voltage, 175
   generators, induction vs. synchronous, 175–176
   industrial
      electric power systems, 175
      plant savings and, 864–867
      processes, 175–178
      thermal considerations, 176
   interconnection standards, 866, 867
   interruptible fuel rates, 177–178
   steam vs. hot water, 176
   waste heat recovery and, 1546–1547
   water augmented gas turbine power cycles and, 1584–1585
Combuster, 489
Combustion analysis, data collection and, 260–261
Combustion capture, 126–127

Combustion of biomass, 88–89
Combustion processes, waste fuels and, 1525
Combustion systems, energy intensive, 463–464
Combustion
    boilers and, 95–96
    circulating fluidized bed combustor, 489
    energy conversion and, 487–489
    first law of thermodynamics and, 1424
    fixed bed combustor, 489
    fluidized bed combustor, 489
    ignition and, 487
    stoker firing, 488–489
    suspension firing, 488
Combustor
    FSTIG cycle, 1583
    injection of steam and water, 1582–1583
    STIG cycle, 1582–1582
    VAST cycle, 1583
    VASTIG cycle, 1583–1584
Comfort control, HVAC systems and, 839
Command and control costs (CAC), 430, 435
    emissions trading vs., 435
Commercial buildings, electric demand response solutions for, 270–278, 281–282
Commercial real time pricing, 1180–1182
Commercial use of energy in U.S., 583
    space vs. energy usage, 583
Commercial utility rates, 1413
Commercial waste fuel suppliers, 1530
Commissioning existing buildings, 710–712
    case study, 180–181, 185–187
    cost and benefits, 181
    definitions, 711
    energy management process, 185
    findings from, 711
    process of, 182–185, 711–712
        implementation and verification, 184
        phase development, 183–184
        team members, 183
            responsibilities of, 183
Commissioning process, 191–193
    acceptance phase, 196–197
    agent selection, 192
    agent skill set, 192
    common mistakes, 199
    phases of, 193–197
        construction, 195–196
        design, 193–195
        installation, 195–196
        predesign, 193
    postacceptance phase, 197
    problems with, 191
    reasons for, 191
    retrocommissioning, 200–206
    success factors, 197–199
    systems to include, 191
    team composition, 192–193
    testing, adjusting and balancing submittal, 195
Commissioning types, 179–180
    continuous, 180
    recommissioning, 180
    retrocommissioning, 180
Commissioning, 179–187
    acceptance-based, 189–190
    definition of, 188–189
    definitions used in, 179–181
    history of, 190
    market acceptance of, 190–191
    new buildings, 188–199
    process-based, 189–190
Commodity charge, utility billing and, 1501
Commodity rates, 1503–1504
Common gateway interface (CGI), 536
Communication architecture, IntelliGrid^{SM} and, 917
Communication buses, automobile technology evolution and, 674
Communications
    controlling of electric demand response and, 283–284
    energy management system problems and, 1034–1035
Community development, design for energy efficiency, 1409
Comparable value, 930
Competitive energy rates, 1506
Complete mix reactor treatment, 1561
Compliance costs, electricity usage and, 401
Compliance period, emissions trading and, 432
Compressed air control systems, 207–213
    air distribution, 211–212
    air quality, 211
    centralization of, 212–213
        human machine interface workstation, 212
    components, 207–209
        air production control, 207
        compressor control, 208
        motor control, 207–208
Compressed air energy storage (CAES), 214–218
    expansion turbine and control, 215–216
    lead-acid battery alternative, 214
    thermal, 214–215, 216
    uninterruptible power supply, 214–215
    vessels, 217
Compressed air leak
    control and prevention, 224–225
    detection
        discharge volume table, 220
        methods, 219–222
        repair and, 219–225
    management of, 223–224
    repairs of, 222–223
        logistical procedures, 223
        methods, 223
        overhead associated with, 223, 225
Compressed air storage
    control and maximization, 233–234
    distribution, 226–235
Compressed air systems, 236–239
    components of, 236–237
        demand side, 238
        supply side, 237
    costs of, 239
    energy balance, 227
    energy flow, 226
        generation efficiency of, 226
    optimization of, 241–245
        controlling demand, 241–242
        minimizing energy losses, 243
        reducing demand, 242
        storing, 242–243
        supply energy reduction, 243–245
    pressure set point, 238
    storage of, 228–233
        calculating
            peak air demand, 231–232
            receiver volume, 232
            usable energy, 231
        combined gas law, 228–229
        permissive start-up time, 232–233
        pneumatic capacitance, 229–230
        usable energy, 230–233
        pressure profile, 230–231
        storage delta, 230–231
    uses of, 236
Compressed air, chemical process industry and, 1305
Compression, 1235
Compressor air control systems
    control types, 208–209
        centrifugal compressor surge protector, 209
        machine protection, 209
        micropressors, 208
        plant safety, 209
        pneumatic, 208
        programmable logic controllers, 208
        protection and safety elements, 208–209
    energy savings, 209–211
        automation, 210
        blow off reduction, 210
        intercooler control, 210
        leak loss reduction, 210
        load scheduling, 210
        networked capacity control, 209–210
        precise pressure regulation, 209
        scheduling, 210–211
        start/stop sequence, 210
        surge controls, 210
        system controllers, 210
Compressor control, 208
Compressor discharge output pressure, 209
Compressor failure, storing compressed air, 242–243
Compressor permission start-up time, 232–233
Compressor surge, 1576
Compressor systems
    common sources of energy waste, 867
    industrial plant savings and, 867
Compressor train, 1579
    water injection, 1579
        quasi isothermal compression, 1579
Compressors, mobile HVAC systems and, 1076
Computerized maintenance management systems (CMMS), 710
Computers, automobile technology evolution and, 673–674
Concentric tube heat exchanger, 801
Conceptual stage, energy project management planning and, 558–559
Concrete block core insulation, 892
    grades of, 892
Concrete block walls, 1516
    reflective insulation, 891
Concrete form insulation systems, 895–896
    cost premium, 895–896
    foams in, 895
    wall shape, 895

Concrete form walls, insulated, 1517
Concrete as insulation material
　　aerated, 891
　　lightweight, autoclaved, 891
Concrete products, lightweight, as insulation materials, 896
Concrete wall insulation, 892
Concrete walls, 1516–1517
　　precast, 1517
Concrete, Air-Krete, 894
Condensate return system, 1367–1368
　　improvement of, 1375–1378
Condensation resistance, windows, 1622–1623
Condenser, mobile HVAC systems and, 1077
Condensors, 1544
　　heat exchangers and, 803–804
　　maintenance of, 804
　　materials used in, 804
　　safe operation of, 804
Conditioned floor area corrections, 2
Conduction heat transfer
　　double pane window, 1618
　　windows, 1617
Conduction, 1161
　　heat transfer and, 823
Conductive heat flow, in window energy performance, 1616
Conductor losses, 360–361
　　circuit resistance, 360–361
Conductor sizing, 701–702
Conductors, 358
Cone density chart, 994
Conformance audits, 1030
Conservation economy, 1409
　　sustainable development and, 1409–1410
Conservation of energy, 1015
　　efficiency vs., 21
Conservation utility rates, 1505
Conservation, aircraft fuel consumption and, 25
Consortium for Energy Efficiency, 513
Consortium for Energy Efficiency. See CEE.
Constant volume of supply air (CAV), 51
Constrained equilibrium modeling, exergy analysis and, 657
Construction guidelines, ANSI/ASHRAE Standard 62.1–2004 and, 61
Construction work in progress, rate of return regulation and, 1255
Construction
　　commissioning process and, 195–196
　　energy project management planning and, 561
Consultants, energy service companies and, 306–307
Consumer good demand, developing countries and energy demand, 500–501
Consumption
　　U.S. energy's, 608
　　worldwide energy resources and, 445–446, 447
Contactors, 704
Contaminant control, 839
Contamination levels
　　direct methanol fuel cells (DMFC) and, 741
　　molten carbonate fuel cells and, 728
　　proton exchange membrane fuel cells (PEMFC) and, 737

solid oxide fuel cells and, 730
Continuous commissioning, 180
Continuous cyclic process, 1170–1171
Continuous dryers, 338
Contract utility rates, 1505–1506
Contracted HVAC services, 709–710
Contracting models, energy service companies in Europe and, 568–572
　　energy savings, 569–571
　　other types, 571–572
Contractors, independent power producers and, 857
Control designs for HVAC facilities, 532–533
Control processes, 117
Control system complexity, energy efficiency and industrial plants, 870
Control
　　granularity of, 273
　　resolution of, 273
　　Six Sigma and, 1316
Controller communications network, building automation systems and, 106
Controller communications protocol, 106
Controller software programming, building automation systems and, 105
Controller, 413–415
　　enthalpy, 415
　　indicating devices, 416
　　input type, 414
　　level hardware, building automation systems and, 105
　　modes, 415
　　modulating, 415–416
　　output devices, 415
　　relative humidity, 414–415
　　system interfaces, 416
　　temperature, 414
　　transducer, 416
　　two-position, 415
　　universal, 415
Controls
　　distributed control systems, 459
　　industrial processes and, 459
Convection heat transfer, in windows, 1617–1618
　　double pane, 1618
Convection, 1161
Conventional Brayton cycles, 1576–1577
　　basic, 1576–1577
　　recuperator usage, 1577–1578
Conversion efficiencies, renewable energy and, 1266
Conversion factors
　　energy and, 603
　　energy units and, 1164
Conversion, uranium and, 1113
Cooker kettles, steam blowthrough rate and, 1373
Cooking, 809
　　solar, 1325–1327
Cool storage system design, 1418–1420
　　chiller and storage capacity, 1418
　　load profile, 1418
　　system layout and control, 1419–1420
Cooling applications, thermal energy storage and, system design, 1418–1420
Cooling costs, activity based costing and, 1044

Cooling equipment, solar heat effect, 1618
Cooling process, energy intensive, 462–463
Cooling system adjustment
　　central chiller plants, 276
　　electric demand response and, 275–276
　　evaluation of, 276
　　increase supply air temperature, 276
Cooling system methodology, 1317–1318
Cooling towers, 246–253, 531–532, 1589–1590
　　performance and rating of, 251–253
　　thermo-fluid dynamic efficiency, 250
　　types, 246–251
　　　　fill type, 249–250
　　　　mechanical draft, 247–248
　　　　natural draft, 247–248
　　　　packing, 249–250
　　　　wet, 246–247
　　　　wet-dry, 250–251
Cooling
　　dehumidification and, 1186–1187
　　sensible, 1186
　　solar, 1326–1327
　　thermal energy storage and, 1413–1418
　　　　commercial utility rates, 1413
　　　　storage equipment, 1414–1418
Cooling/heating day adjustment, 2
Core losses, 360
Corrective action, 1030
Cost accounting, HVAC monitoring and, 522–523
Cost allocation
　　fixed cost related charges, 1500
　　least-cost planning, 1500
　　market driven discrete usage charge, 1500
　　utility rate structures and, 1500
Cost analysis, energy efficiency and, 81–85
Cost based negotiated rates, 1505–1506
　　business retention, 1505
　　conservation and load management, 1505
　　economic development, 1505
　　special contract rates, 1505–1506
Cost benefit analysis, environmental policy and, 628–629
Cost benefits, energy efficiency and, 15–17
Cost centers, energy accounting and,
　　criteria for, 3
　　establishment of, 3
Cost components, electric power projects, 954–955
Cost drivers, 1042–1043
　　first stage, 1042
　　second stage, 1042–1043
Cost effective analysis, environmental policy and, 629–630
Cost metering, web-based compressor management and, 213
Cost minimizing pollution abatements, 433–436
　　allowance allocation, 434–435
　　command and control costs, 435
　　distribution of costs, 434–435
　　emission taxes, 435–436
Cost of aircraft fuel, industry impact, 25
Cost of capital, 1256
Cost of electricity, 959
Cost placement, importance in BCR, 83
Cost savings, energy master planning and, 552–553

Cost savings, radiant barriers and, 1228
Cost
   pumped storage hydroelectricity and, 1210–1211
   steam generation, 1366–1367
Costs
   air, 212
   commissioning existing buildings and, 181
   compressed air leaks and, 223, 225
   energy project life cycle costing and, 971–972
Cotton, as insulation material, 893
Counter flow heat exchangers, 802–803
Course bubble diffusion, 1549
CPFilms LLumar® window film, case study, 1626
   demonstration building, 1627
   DOE-2 simulation values, 1627
CPP. *See* critical peak pricing.
Credits, carbon dioxide CO_2 and, 15–17
Critical peak pricing (CPP), 1175, 1177–1180
   technologies available, 1177–1178
Crop drying, solar, 1325
Cross flow heat exchanger, 801
Cross quality, 421, 422
Cryogenic air-separation technologies, gasifiers and, 909
Crystalline SI photovoltaic (PV) systems, 1148–1149
Culture, energy use and, connection between, 1016
Cumulative sum of differences (CUSUM) analysis, 6
Current flows, 357
CUSUM. *See* cumulative sum of differences.
Cutting tools, 813

D

Damage-weighted transmittance, of windows, 1623–1624
Data analysis, energy balance and, 688
Data collection, 255–263
   building envelope measurements, 258–259
   combustion analysis, 260–261
   electrical energy measurements, 255–257
      current, 255–256
      power factor, 255–256
      voltage, 255–256
   energy information system processes and, 535
   energy management system problems and, 1037
   light measurements, 257–258
      light meters, 258
   low cost data acquisition, 262–263
      software, 262–263
   thermal imaging, 259–260
   ultrasonic leak detection, 261–262
Data compilation, energy balance and, 686–686
   building or location, 685
   equipment type, 685
   meter, 686
   processes used, 686–686
Data management, energy benchmarking and, 692
Data organization, interactive access and, 620

Data presentation, interactive access and, 620
Data sources, energy benchmarking and, 691
Data structuring, energy, 618–619
Data tabulation, energy accounting and, 3–4
   fuel, 4
   needed data, 3–4
Databases, 538
Daylight harvesting, 978
Daylight illumination from windows, as energy saver, 1624
Daylight software, 1403
Daylighting design, 267
Daylighting simulation, 267–268
   scale model testing, 268
   software, 267–268
Daylighting, 264–269
   building occupancy, 267
   illumination basics, 264–267
   word glossary, 269
DCS. *See* distributed control systems.
DCV. *See* demand control ventilation.
DDC. *See* direct digital control.
Deaerator steam jet ejectors, mechanical vacuum pumps, 1370–1371
Decentralized energy
   advantages of, 1259–1260
   renewable, 1258–1264
Declining block rates, 1503
Define, measure, analyze, improve and control, 1310
Define, Six Sigma and, 131
Defining energy project management plan, 559–560
Degree of saturation, 1185
Dehumidification
   cooling and, 1186–1187
   designing a system, 839
   heating and, 1187–1188
   HVAC systems and, 839–840
Delta Score estimator, Energy Star Portfolio Manager software and, 578
Demand billing, 1510–1511
Demand buyback program, 281
Demand charge, utility billing and, 1501
Demand control ventilation (DCV), 56
Demand control, compressed air systems and, 241–242
   reducing demand, 242
Demand forecasting, utility long term planning and, 1493–1494
Demand shifting, electric demand response and, 272
Demand side energy consumption, IEKPI and, 1140
Demand side management, 499
   case studies and
      energy efficiency and, 503–505
      Northern Cyprus, 503–504
      Turkey, 503–504
   energy programs, 286–291
Demand/commodity rates, 1503–1504
Density compensation, gas and steam and, 1057–1059
Density, land use impact and, 1450
Department of Defense, fuel cell program, 511
Department of Energy role in

   revising ASHRAE/IESNA/ANSI Standard 90.1 & 90.2, 439
   setting energy standards, 439
Depreciation, 957, 1256–1257
Dept. of Commerce, 510–511
   National Institute of Standards and Technology, 510–511
   National Technical Information Service, 510
Dept. of Energy best practices case studies, energy savings and, 1283–1301
   aluminum plant assessments, 1283, 1285–1288
   chemical plant assessments, 1283–1284
   forest product plant assessments, 1284, 1288–1291
   glass plant assessments, 1284, 1291–1293
   metal casting plant assessments, 1284, 1293–1294
   mining plant assessments, 1284–1285, 1294–1296
   petroleum plant assessments, 1285, 1296–1299
   steel plant assessments, 1285, 1299–1301
Dept. of Energy, 508–510
   Emerging Technologies, 509
   Federal Energy Management Program, 509
   Inventions and Innovation group, 509
   national laboratories, 510
   Office of Electricity Delivery and Energy Reliability, 509–510
   Office of Energy Efficiency and Renewable Energy, 508–509
   Office of Fossil Energy, 509
   Office of Scientific and Technical Information, 510
DER. *See* distributed energy resources.
Deregulation of electricity, 379–386
   arguments for the consumer, 384
   political reaction, 384–384
   regional transmission operator, 380
   regulator reaction, 384–384
   retail markets, 382–383
   third party involvement, 380
   utility rate structures and, 1511–1512
   utility reaction to, 385–386
   wholesale markets, 381–382
Desiccant dehumidification, case study, 292–293
   accuracy, 294
   computer modeling, 293–294
   field testing, 292–293
   results, 294
Desiccant drying wheels, 795
Design documentation procedures, ANSI/ASHRAE Standard 62.1-2004 and, 60
Design issues, building automation systems and, 107
   new facility, 107
Design phase, commissioning process and, 193–195
DESs. *See* district energy systems.
Developer, independent power producers and, 856
Developing countries
   barriers to energy efficiency, 501
   demand for consumer goods, 500–501

economic growth, 500
energy and, financing of, 501
energy efficiency and, 498–506
implementation of energy efficiency, 501–505
measuring economic development, 501–502
population growth, 500
status of energy in, 501
Dew point temperature, 1184–1185
DHC. *See* district heating and cooling.
Die temperature, 868
Diesel engines
reciprocating, 1238, 1239, 1241, 1242
specifications for, 1238
submodels used, 1239
Diesel products, coal to liquid fuels and, 165–166
Digital devices
electricity quality and, 402
electricity usage and, 402
Dilution, indoor air ventilation and, 19–20
Dimmers, 982
ballast to, 982
integrated, 982
programmed start ballasts, 982–984
Dimming ballasts, lighting control systems, 982
Dimming electronic ceramic ballasts, 830
Dimming electronic metal halide ballasts, 830
Dimming limits, electronic ballast primer and, 837
Dimming system, 982
lighting control systems, 983, 984
Dimming, 831, 996
lighting based demand response and, 277
lighting design and retrofits, 996
switching vs., lighting controls and, 989
Direct cofiring, 88–89
Direct current electric motors, 349
Direct current power system, 389
Direct current systems, 705
Direct current transmission, 359
bipolar type, 359
disadvantages of, 359
inverters, 359
rectifiers, 359
single pole type, 359
thyristors, 359
Direct digital control (DDC), building automation systems, 104–119
Direct fired absorption chillers, advantages and disadvantages, 1545–1546
Direct incentives, energy demand-side management programs and, 291
Direct methanol fuel cells (DMFC), 739–743
applications, 743
contamination levels, 741
principles of, 740
problems with, 741–742
technological status, 742
Direct solar gain design, 1322
Dirty area systems, lighting design and retrofits, 996
Dirty environments, lighting design, 996
Discount rate determination, 932–933

Discount rate issues, nuclear energy economics and, 1104–1105
Discount rate, 930
Dishwashers, 1587–1588, 1603
Disinfection of treated water, 1557, 1559
chloramine, 1559
chlorine, 1559
ozone, 1559
ultraviolet light, 1559
Dispatch stack order, 1499
Dispersed flow, 1561
Dispersion of air pollution, 718–721
Display systems, automobile technology and, transferable to building use, 672–673
Disposal cost reduction
landfill cost, 1530
trash collection and hauling, 1530
waste fuels and, 1530
Disposal, nuclear energy and, 1113–1114
Dissolved oxygen mechanics, 1551–1553
Distillation process, 462
Distillation, 809
solar, 1325
Distributed control systems (DCS), 459
process historian, 459
Distributed energy resources (DER), 404
Distributed generation, 296–308
combined heat and power, 303–308
advantages of, 303–304
philosophy behind, 304
users of, 305–307
energy service companies, 306
definition of, 297
history of, 296–297
markets
existing, 304
future, 304
technologies available, 304–305
technologies, 297–299
benefits of, 299–301
combined heat and power applications, 299
environment, 301
postpone power plant investment, 301
security of supply, 299–300
cost estimates, role of
natural gas, 299
petroleum, 299
fuel cells, 298–299
microturbines, 298
network losses and usage, 300–301
reciprocating engines, 297–298
renewable, 299
simple cycle gas turbines, 298
Distribution of costs, emission trading and, 434–435
Distribution Power Quality (DPQ), 1484
District cooling systems, 309–315
chillers, plant retrofitting, 315
control strategies, 314
cost savings from, 314
design parameters, 313
full-load efficiency improvement, 313–314
part-load efficiency improvement, 314
plant design energy comparison, 312

pre-packaged chillers, 314
series chillers, 310
variable flow, 310–311
chiller capacity, 311–312
water-cooled chiller efficiency, 309
District energy systems (DESs), 316–330
district heating and cooling, 317
nomenclature, 316
District heating and cooling (DHC), 317
benefits of, 317
case studies, 324–330
cogeneration, 318–319
efficiencies of, 320–321
environmental impact of, 320–321
exergy, 321
history of, 319
implementation of, 320
market establishment of, 322
organizations involved, 317
performance evaluation, 322–324
exergy analysis, 322–324
sustainable development, 321–322
renewable energy resources, 322
systems, 320
energy generation, 320
energy transmission, 320
energy use, 320
technical aspects, 319–320
energy generation, 320
energy transmission, 320
energy use, 320
Diversity factor, energy balance and, 688
DMAIC methodology, 1312, 1313–1316
analyze data, 1314–1315
control, 1316
define, 1313
improvement by
new process capability, 1316
variable relationships, 1315–1316
measure the process, 1313–1314
DMAIC. Define, measure, analyze, improve and control, 1310
DMFC. *See* direct methanol fuel cells.
Documentation procedures, ANSI/ASHRAE Standard 62.1–2004 and, 60
Documented processes, 1029
Documenting management systems
guidance documents, 1030
specification standard, 1030
DOE Building Energy Codes Program (BECP), 443
DOE-2 energy study, solar-control film, 1631
DOE-2 software, 1402
Double pane windows, conduction and convection heat transfer, 1618
Downtime
lean manufacturing and, 472
preventive maintenance, 472
set ups, 473
Drag reduction
aircraft fuel consumption and, 26–27
programs to reduce, 27
Dry bulb temperature, 1184
Dry cooling towers, 250
Dry steam, 745
Dryer types, 340
Dryer, mobile HVAC systems and, 1077

Drying
   agricultural and forestry products, 332–336
   definition of, 338–339
   energy audits,
      basic, 344–345
      detailed, 345
   energy consumption of, 339–343
      analyses, 341–343
      reducing, 345–347
         capital expenditure, 346–347
         non capital expenditure, 346
   energy efficiency of, 334–335
   energy intensive, 462
   industry operations, 338–348
      dryer types, 338–339, 340
   new methods and technology, 336
   performance of, 334–335
   reducing electrical consumption, 347
      future of, 347–348
   research on, 335–336
   scheduling, 333–334
   steps in, 332–333
   system design and selection, 335
   thermal, 332
   types, 333
Drywall, airtight barrier, 667–668
DSM. *See* demand side management.
Dual fuel firm rates, natural gas and, 1507
Duct static pressure setpoint reduction, air distribution adjustment and, 275–276
Ductile metal filament lamp, 392
DUCTSIZE software, 1402
Dust inhalation, cellulose insulation and, 891
Duties, 959
Dynamic data sources, 691–692
   automated meter data collection, 691–692
   metering, 691
   temperature and humidity data, 692
   utility bills, 691
Dynamic HTML, 537

E

E Source Companies LLC, 513
EAC. *See* energy account centers.
Economic capital, 1410
Economic development utility rates, 1505
Economic framework, environmental policy and, 628–630
Economic growth, developing countries and energy demand, 500
Economic indicators, energy efficiency and, 501–502
Economics
   energy analysis and, 652–653
   nuclear energy and, 1101–1110
   pumped storage hydroelectricity and, 1210–1211
   solar water heating and, 1336–1337
Edison, Thomas, 379, 389
EE4 CODE software, 1402
EED software, 1402
EEM. *See* energy efficiency measure.
EEMS. *See* enterprise energy management systems.

EERE. *See* Office of Energy Efficiency and Renewable Energy.
Efficiencies
   boilers, 94–95
   energy use and, 1093–1094
   energy use vs. conservation, 21
   IEKPI and, 1141
EIS. *See* energy information systems.
Elastic potential energy, 1163
Electric bulk power operations, 361–363
   control areas, 361–362
   coordination of, 361
   NERC, 361
   open access same-time information system, 362–363
   reliability councils, 361
   transmission capacity, 362
Electric consumption, industrial plants energy efficiency and, 863
Electric demand
   measuring of, 1510
   time frame of, 1510
Electric demand response, 270–278
   air distribution adjustment, 275–276
   ancillary services, 284
   benefits of, 279–280
   closed loop control, 273
   communications, 283–284
   concepts and terminology, 271–273
   controlling of, 283–384
      communications, 283–284
      metering, 283
   cooling system adjustment, 275–276
   cost savings realized, 270–271
   daily peak load management, 272
   definition of, 279
demand shifting, 272
economic issues, 284–285
   energy efficiency, 271
   granularity of control, 273
   HVAC based, 273–274
   lighting based, 276–277
   metering, 283
   rebound, 273
   reduction in service, 272–273
   resolution of control, 273
   resources and programs, 279–285
   shared burden, 273
   solutions for commercial buildings, 281–282
   solutions for industrial facilities, 282
   solutions for residential areas, 283
   zone global temperature adjustment, 274
Electric demand response programs, 280–281
   demand buyback program, 281
   interruptible, 280
   load curtailment, 280–281
Electric energy production, 744–745
Electric energy supply by source, 608
Electric kinetic energy, 1163
Electric load following cycle, 1530–1531
Electric meters, 283
Electric motors, 349–355
   actual efficiency vs. Energy Policy Act standards, 353
   alternating current induction synchronous, 349
   alternating current synchronous, 349

applications of, 350
ASD technology, 353, 355
CEE premium efficiency motor, 352–353
classification of, 349
direct current, 349
efficiency, 352
   CEE premium efficiency motor, 352–353
   EPAct motor cost to premium efficiency motor cost, 353
   equation for, 352
energy consumption, 349, 350
EPAct motor cost to premium efficiency motor cost, 353
evaluation, 353
full load nominal efficiencies, 352
lack of efficiency, 387
nominal efficiencies for, 354
power supply issues, 352
pump usage, 353, 355
   ASD technology, 353, 355
replacement considerations, 351–352
selection criteria, 350–351
service rating of, 351
sizes, 350
starting of, 355
technologies available, 355
type classification, 351
types, 349–352
   alternating current induction synchronous, 349
   alternating current synchronous, 349
   differences in, 349–350
Electric power losses, 360–361
   capacity losses, 361
   conductor losses, 360–361
   core losses, 360
   costs of, 360
   energy losses, 361
   high operating voltage advantage of, 361
   measurements of, 361
   reactive power, 361
      losses, 361
   types, 360–361
      conductor losses, 360–361
      core, 360
Electric power projects, life cycle costing, 953–966
   annual electricity generation, 955–957
   calculation methodology, 955
   capital and noncapital costs, 956–957
   coal power plant, 954
   coal vs. wind, 957
   cost components of, 954–955, 959
   depreciation, 957
   electricity generation, 955–957
   financing charge, 957–958
   interest during construction, 958
   loan repayment, 959
   photovoltaic power system, 954
   project cost, 958
   renewable energy systems incentives, 959–961
   taxes and duties, 959
   variable costs, 958
   wind power plant, 954
   working capital interest, 958–959

Electric power transmission systems, 356–363
　access to, 360
　biological effects, 358
　bulk operations, 361–363
　cable capacity determination, 358
　cable limitations, 358
　cables, 358
　　composition of, 358
　　　high pressure fluid filled, 358
　　　high pressure liquid filled pipe, 358
　capacity, 360
　characteristics of, 357
　　current flows, 357
　　resistance, 357
　components, 357–359
　　conductors, 358
　　three phase transmission lines, 357–358
　coordination of, 361–363
　direct current transmission, 359
　functions of, 356–357
　ground or shield wires, 358
　insulators, 358
　interconnection of, 356
　Kirchhoff's Laws, 360
　limitations of, 360
　　stability disturbances, 360
　North American systems, 357
　overhead cable, 357
　power losses, 360–361
　purposes of, 356
　rating types, 358
　　determination of, 358
　　emergency, 358
　　normal, 358
　ratings of, 358
　　heating factor, 358
　reactance, 357
　reactive power, 356
　real power, 356
　role of, 356
　short circuit duties, 360
　substations, 358–359
　support structures, 358
　transmission components, 357–359
　types, costs, 357
　workings of, 359–361
　workings of, synchronous operation, 359–360
Electric power transmission systems, operation of, 359–361
　access, 360
　capacity, 360
　Kirchhoff's Laws, 360
　power losses, 360–361
　short circuit duties, 360
　synchronous operation, 359–360
Electric power transmission systems, synchronous operation of, 359–360
　advantages of, 350–359
　effect of power loss, 359
　importance of network coordination, 360
　regional effects, 359
Electric rate development, time of use and, 1177
Electric rate schedules, 1507–1510
　general service, 1508
　general service heating, 1508

　interruptible, 1508
　large power, 1508
　rate riders, 1509
　real time pricing, 1508
　standby, 1509
　street lighting, 1508
　TOU large power, 1508
　TOU rates, 1508
　transmission rates, 1508
Electric supply, generation of, 374–378
　fuels, 378
　resources, conventional, 374–376
　renewable, 376–378
Electric utilities, restructuring of, 400–401
Electric utility bill, calculating of, 1472–1473
Electric utility dispatch modeling, 1506–1507
Electrical consumption reduction, drying and, 347
Electrical current, 255–256
Electrical elements, chemical process industry and, 1305
Electrical energy measuring
　data collection of, 255–257
　power quality analyzers, 256
　practice activities for, 256–257
Electrical grid, wind power effects on, 1611–1612
Electrical processes, energy intensive, 463
Electrical resistance space heater, 648
Electrical service, facility guidelines for new structures and, 528
Electrical tariff filings, 380
Electricity
　alternating current power system, 389
　chemical process industry and, 1305
　Chicago Edison, infrastructure growth, 391–392
　commodity, 392
　costs. See electricity pricing programs.
　definition of, 387
　deregulation of, 379–386. See also electricity, deregulation of.
　development in the United States, 387–398
　ductile metal filament lamp, 392
　early history of, 389–392
　efficiencies and, 387
　　small motors, 387
　enterprise, modernizing of
　　achieving goals, 403–406
　Energy Policy Act of 1992 (EPAct '92), 380
　Galvin Electricity Initiative, 404–405
　growth of, 389
　history of
　　growth in 20th century, 392–394
　　market maturity, 394–398
　　nuclear power, 394–395
　　open markets, 395
　　stasis of, 395–398
　hybrid electric vehicles and, 851–852
　impact on society, 387–389
　infrastructure obsolescence, 387–388
　initial usage of, 389–392
　　Edison, 389
　IntelliGrid1 supply system, 402–403
　meters, 392
　nuclear power and, 394–395
　portable storage, 389–390

　power system, alternating current, 389
　producers of, 387
　　assets and investment, 388
　Public Utility Holding Company Act of 1935, 392
　reasons for growth, 389
　regulation of, 379–380
　Rural Electrification Administration, 392
　Societal impact, 387–389
　supply systems, 379–380
　　origins of, 379–380
　　　investor owned utilities, 379–380
　　　Public Utilities Holding Act of 1935, 380
　　　Edison, 379
　tax impact, electric vehicles and, 851–852
　Tesla, Nicola, 389
　transformation of. See electricity transformation.
　turbine generators, 392
　usage. See electricity usage.
　utilities profitability, 392
　utilities, consolidation of, 392
　wind power as source of, 1607
Electricity, deregulation of, 379–386
　arguments for the consumer, 384
　Energy Policy Act of 1992 (EPAct '92), 380
　Federal Energy Regulatory Commission, 380
　future of, 386
　political reaction, 384–385
　Public Utilities Regulatory Policy Act of 1978, 380
　regional transmission operator, 380
　regulator reaction, 384–385
　results in California, 385
　retail markets, 382–383
　utility reaction, 385–386
　wholesale markets, 381–382
　deregulation, results in Pennsylvania, 385
　deregulation, third party involvement, 380
Electricity consumption, worldwide, 499
Electricity costs, average, by location, 1632
Electricity enterprise, future of, 399–406
　modernizing of, 399–403
Electricity generation in U.S., 581–582
　input type, 582
　residential, 583
　coal and, 159–160, 161
　coal vs. natural gas, 161
　renewable, 44
Electricity pricing, 1175
　green, 1262–1264
Electricity pricing programs, 1175–1183
　critical peak pricing, 1175, 1177–1180
　extreme day CPP, 1178
　extreme day pricing, 1175
　green, 1262–1264
　real time pricing, 1175, 1180–1182
　time of use pricing, 1175–1177
　types, 1175
Electricity requirements, coal to liquid fuel plants and, 170
Electricity supply
　global warming and, 724
　world and U.S., 608
Electricity tax impact, plug in hybrid electric vehicles and, 851–852

Index

Electricity transformation
  architecture of, 404
  assistance in achieving, 403–404
  battery technology, 402–403
  distributed energy resources, 404
  fuel cells, 403
  Galvin Electricity Initiative, 404–405
  IntelliGrid1 supply system, 402–403
  key steps needed, 402–403
  organizational initiatives to assist, 403–404
  resistance to, 404
Electricity transmission
  open access, 380
    effects of, 380
Electricity usage
  compliance costs, 401
  digital devices, growth of, 402
    performance concerns, 402
  future costs of, 402
  future demand, 402
  reinvention and transformation of, 399–406
    benefits of, 401
  security costs, 401–402
Electrolysis of water, 1442
  high-temperature, 1443
  spinning reserve concept, 1442–1443
Electrolysis produced hydrogen vehicles, greenhouse gas emissions and, 790–791
Electromagnetic field energy, 1163
Electromagnetic suspension maglev system, 1023
Electronic ballast primer, 830–832
  advantages, 832–833
  ceramic ballasts, 830
  dimming, 831
  dimming limits, 837
  efficiency of, 831–832
  full-light output, 834
  ignition methods, 833–834
  lamp ignition, 831
  lamp performance, 831
  life expectancy, 833
  lumen depreciation, 836–837
  metal halide, 830
  microcontrollers, 834
  problems with, 835–836
  starting behavior, 831
  system efficacy, 834–835
  thermal cut-off, 834
  types, 831
  voltage, 831
Electronic components, heat pipe applications and, 808–809
Electronic control systems
  applications, 416–418
  basics of, 407–419
  characteristics, 407
  components, 408–416
    controllers, 413–415
    sensors, 408–413
  definitions of common terms, 418
  fundamentals, 416
    power circuit supply, 416
Electronics, photovoltaic (PV) systems and, 1154–1155

Elemental analysis, solid fuel properties and, 479
EM programs. *See* Energy management programs.
Em$. *See* emdollar.
Emdollar (em$), 420
Emergency authority, Natural Gas Policy Act of 1978 and, 1084
Emergency lighting, 998
  lighting design and retrofits, 998
Emerging Technologies Program, DOE and, 509
Emergy accounting, 420–428
  arguments with, 427
  definitions of, 420–421
    emdollar, 420
    emjoule, 420
    geobiosphere, 422–423
    H.T. Odum, 420, 421
    net emergy and time, 426
    net energy, 424–426
  transformity
    and quality, 421
    and specific emergy, 421
  unit emergy values, 420–421, 423–424
  yield ratio, 424–426
Emergy inputs, geobiosphere and, 423
Emergy per unit money, 420
Emergy products, 423
Emergy yield ratio (EYR) 424–426
Emissions
  Kyoto protocol targeting of, 140
  waste energy, 659–660
Emission reduction credit (ERC), 430
  offset policy, 430
Emission reduction, 137–138
Emission reduction mandates, 630–631
Emission taxes, emission trading vs., 435–436
Emissions impact
  energy efficiency and, 11–12
  regulating of carbon dioxide, 11–12
Emissions permits, tradable, 630
Emissions quota, allocation of, 138
Emissions reduction credit, 432
  banking of, 432–433
Emissions tax, 630
Emissions trading, 430–436
  allowance/permit allocation, 432
  cap determination, 432
  command and control costs vs., 435
  compliance period, 432
  cost-minimizing pollution abatement, 433–436
  effectiveness, 436
  elements of, 431–433
    affected sources, 431
    enforcement, 433
    inventory, 431–432
    measurement, 431–432
    penalties and enforcement, 433
    spatial trading rules, 432–433
    temporal trading rules, 432–433
    verification of, 431–432
  emission reduction credit, 430
  emission taxes vs., 435–436
  enforcement, 433
  firm incentives, 433

firm incentives, marginal cost of abatement, 433
  measurement, 431–432
  National Ambient Air Quality Standards, 430
  penalties, 433
  thought processes of, 430–431
    command and control policies, 430
    National Ambient Air Quality Standards, 430
  trading rules
    spatial trading rules, 432–433
    temporal trading rules, 432–433
  true-up, 432
  types of, 431
    cap-and-trade, 431
    offset, 431
    project-based, 431
    rate-based, 431
  unit of trade used, 432
Emjoule, 420
EMP. *See* energy master planning.
EMPI. *See* Energy Master Planning Institute.
Empirical determinations, 1144–1145
Employment in energy industry
  coal production in United States and, 150
  coal to liquid fuel plants and, 170
Empower, 421
End to end system management
  air distribution and, 212
End use rates, 1504
  natural gas and, 1507
End use retrofit isolation, 1051
Energy
  additional vs. natural, 1088–1094
  alternative technologies, 31–48
  average unit cost, 920
  avoided cost, 919–920
  carriers and sources, differences in, 1088–1089
  chronology of selected events in history, 610
  conservation vs. efficiency, 21
  consumption by the United States, 608
  conversion factors, 603
  conversion to work, 604
  definition of, 1160, 1162
  developing countries and, 501
    balance supply and demand, 503
    financing of, 501
  development and, 1014–1015
  disruption of, 1258–1259
  diversity of, 601–605
  energy transfer vs. energy property, 1163–1164
  environment, and development, 873–874
  equivalents of, 603, 605
  forms and classification, 1165
  future reserves, 608–610
  global and historical background, 601–615
  global overview, 601–605
  energy diversity, 601–605
  historical background
    fossil fuels and, 606–607
    industrial era, 606
    modern era, 607–608
    post fossil fuel era, 608–610
    preindustrial era, 605–606

importance in wealth generation, 444
new and alternative sources, 880–881
physics of, 1160–1173
   energy conservation, 1164
   energy degradation, 1168–1172
   forms, 1163–164
   heat, 1161
   work, 1160–1161
post fossil fuel era, 608–610
primary sources of, 604
public policy and, 1193–1200
real time costing, 920, 927–928
renewable and decentralized, 1258–1264
sources of, 605
use of, 1091–1094
   efficiency, 1093–1094
   energy conversion technologies, 1091–1092
   selection of, 1092–1093
weighted average unit cost, 920, 921–926
work and heat units, 605, 1162
Energy account centers (EAC), 3
Energy accounting, 1–7
cost centers, 3
   criteria for, 3
data tabulation, 3–4
   fuel, 4
   needed data, 3–4
definition of, 1–2
energy account centers (EAC), 3
energy measurement, 2
Envision software, 3
Fraser software, 3
Global Mvo Asset Manager software, 3
Meter Manager software, 3
method types, 2
   comparison methods, 2
   comparison methods, changing production adjustment, 2
   comparison methods, conditioned floor area corrections, 2
   comparison methods, heating/cooling degree day adjustment, 2
   comparison methods, multiple year monthly average, 2
   comparison methods, present to past, 2
monitoring, targeting and reporting analysis, 4–7
objectives of, 1–2
performance indicators, 4
statistical measuring, 1–2
tools, 2–3
   manual types, 2–3
   software, 3
      Envision, 3
      Fraser, 3
      Global Mvo Asset Manager, 3
      Meter Manager, 3
      Utility Manager, 3
Energy adjustment charge, utility billing and, 1502
Energy analysis, district heating and cooling and, 322–324
Energy and atmosphere
   LEED-C&S, rating system, 943–944
   LEED-CI, rating system, 939
   LEED-NC, rating system, 950

Energy assessments, 63–68
accuracy of, 69–75
   energy and demand balances, 69–70
calculating energy cost, issues with, 71
cost containment projects, 63
levels of, 63–64
detailed, 65–66
   major review areas, 65
pre-assessment procedures, 64
   energy invoice review, 64
   facility layout review, 64
   pollution and waste summaries review, 64
   tariff review, 64
procedures, 64–66
reporting of, 66–68
   consumption data, 67–68
   executive summary, 68
   implementation, 66
   recommendations, 66–67
   physical plant description, 67
savings from, case study, 878
walkthrough, 64
See also energy and demand balances, energy benchmarks.
Energy audits
basic drying type, 344–345
coordination in tribal communities,1459
detailed drying type, 345
drying and, 343–345
funding in tribal communities, 1460
geographic location, 1459
information and tribal communities, 1460
teamwork while in tribal communities, 1461
technical issues in tribal communities, 1459–1460
Energy balance, 681–689, 1142–1143
adjusting, 687
benefits from, 688–689
data analysis, 688
diversity factor, 688
historical usage patterns, 686–687
in energy management programs, 875
input, measuring of, 4
inputs and outputs, 4
load factor adjustments, 687–688
output, energy load inventory, 4
spreadsheet construction, 681–686
   data compilation, 685–686
   equipment energy use data, 681–685
   historical energy use data, 681
utilization factor adjustments, 688
Energy basics, in windows. See Windows, energy basics.
Energy benchmarks, 690–698
analysis using aggregate, 695–698
anomalies, 696
   spikes, 696
   sustained changes, 696–697
seasonal, 696
shift work, 696
determining users, 694–695
   departmental managers, 694
   executives, 694
   facility engineering staff, 694
   tenants, 694
displaying of analysis, 695

external, 690
internal, 690
normalized, 690–692
organizations involved in, 691
preparing of, 691–693
   baseline construction, 692–693
      timeline, 692
   data management, 692
   data sources, 691
      dynamic data, 691
      static data, 691
   scope of, 691
setting of, 690–691
Energy carriers, 1088
Energy Center of Wisconsin, 512
Energy code, model, 438
Energy codes
adoption on state and local level, 440–441
   legislation, 441
   process overview, 440
   regulation, 441
   timing of, 441
energy standards and, differences between, 438
enforcement of, 442
   appointed third party, 442
   local level, 442
   state level, 442
implementation of, 441–442
   timing of, 441–442
local government, municipal code, 441
standards for facilities, 438–443
support of
   DOE Building Energy Codes Program, 443
voluntary programs, 442–443
Energy codes and standards for facilities, 438–443
Energy conservation, 444–457
chemical process industry and, 1306–1309
Climate Technology Initiative, 448–449
energy efficiency, 450. See also energy efficiency.
environmental issues, 446–449, 450
   environmental degradation, 447
   greenhouse gases, 446–447
   solutions, 449
environmental protection vs., 450
examples of, chemical process industry and, 1306–1309
financing of, 450–451
global warming and, 724
implementing of, 453–454
importance of energy efficiency, 455
lack of success, 454–455
lean manufacturing, 467–475
   impact on energy usage, 469–474
   lean traditional vs., 468–469
life-cycle costing, 456, 457
measures to take, 454–456
   parameters, 456
   sectoral, 455
practical aspects, exergy, 450
research and development status, 450–451
sustainable development, 451–453
   definitions of, 451–452
   environmental concerns, 452–453

technical limitations, 450
work-heat-energy principle, 1164–1168
world energy resources, 445–446
See also energy efficiency.
Energy conservation, analysis
energy consumption allocation, 460
industrial processes and, 459–461
Pinch analysis, 460
See also energy efficiency.
Energy conservation, industrial processes, 458–465
capital investment, 464
current applications of, 465
differentiation, 458–459
distributed control systems, 459
effective management of, 464–465
key performance metrics, 465
total quality energy, 465
energy consumption analysis, 459–461, 1141–1142
energy intensive, 461–464
lack of awareness, 465
need and payback, 465
retrocommissioning process and, 202–203, 204
See also energy efficiency.
Energy conservation, opportunities
chemical process industry and, 1306–1309
Climate Technology Initiative, 448–449
Florida's fruit industry, case study, 878
retrocommissioning process and, 202–203, 204
See also energy efficiency.
Energy consumption
dryers and, 339–343
capital expenditure, reducing by, 346–347
reducing of, 345–347
electric motors, 349, 350
industrial plants energy efficiency and, 864
maglev and, 1019–1020
reducing by capital expenditure, drying and, 346–347
reducing by non capital expenditure, drying and, 346
Energy consumption allocation, 460
Energy consumption analyses, 459–461, 1141–1142. *See also* energy efficiency analysis.
Energy consumption data
energy assessments and, 67–68
retrocommissioning process and, 202–203, 204
Energy conversion
bio-solid and coal gasification, 492–494
char reaction, 487
cofiring, 490–491
combustion, 487–489
futuregen, 494–496
general principles of, 489–490
ignition and combustion, 487
pyrolysis, 485–486
volatile oxidation, 486–487
Energy conversion efficiencies, 607
Energy conversion processes
animal waste, 476–496
biomass fuels, 476–496

properties, 478–485
coal, 476–496
Energy conversion technologies, 1091–1092
char reaction, 487
cofiring, 490–491
combustion, 487–489
futuregen, 494–496
general principles of, 489–490
ignition and combustion, 487
pyrolysis, 485–486
volatile oxidation, 486–487
Energy cost savings
adjustable speed drives, 74–75
issues with, 71–75
motor belts and drives, 73–74
motor load factors, 73
motors, high-efficiency, 73
off-peak vs. on-peak use, 71–73
motors, high-efficiency, 73
off-peak vs. on-peak use, 71–73
Operations and Maintenance Best Practices Guide and, 708
See also energy efficiency.
Energy costs
aircraft fuel consumption and, impact on, 25
effect on United States economy, 590
impact on indoor air quality, 22–23
Energy crisis, Public Utility Regulatory Policies Act (PURPA) of 1978 and, 1201
Energy currencies, 1088
Energy data collection
building automation systems, 617–618
creating data history, 618
enterprise energy management systems and, 616–624
meters, 618
sources, 617–618
generated utilities, 617
other sources, 618
purchased utilities, 617
Energy data structuring, 618–619
hierarchy, 619, 620
time intervals, 619
Energy degradation
entropy and exergy, 1168–1172
heat engines, 1171
irreversibility, 1168–1172
reversibility, 1168–1172
Energy demand
electricity consumption, 499
Kyoto Protocol, 499
trends, 499–501
worldwide, 499–501
Energy and demand balances
accuracy of
air compressors, 71
air conditioning, 70
lighting, 70
motors, 70–71
process equipment, 71
energy assessment and, 69–70
results verification, 71
Energy demand-side management programs, 286–291
advertising and promotion, 290
alternative pricing, 291
attributes of, 286

customer education and contact, 290
direct incentives, 291
end-use technology activities, 288
load-shape changes, 287
load-shape objectives, 288
market implementation methods, 289–290
selection of, 286
technology alternatives to use, 288–289
choosing of, 289
trade-ally cooperation, 290
See also energy efficiency.
Energy Design Resources, 514
Energy efficiency, 1093–1094
advanced energy systems, 1093
air emissions reduction, 9–17
effectiveness, 10
emissions impacts, 11–12
hydrogen economy, 10–11
benefit cost analysis, 81–85
building envelopes, 1094. *See also* energy efficient buildings.
chemical process industry and, 1302–1309
community development and, 1409
cost benefits from, 15–17
countries' focus on, 10
definition of, 9
demand side management, 499. *See also* energy demand-side management programs.
developing countries and, 498–506
barriers to, 501
devices, 1093
drying and, 334–335
effect on GHG emissions, 12
electric demand response and, 271
energy conservation and, 455
energy storage, 1093–1094
engineering and, 136
environmental benefits from, 15–17
exergy analysis, 1094
implementation. *See* energy efficiency, implementation.
indoor air quality, 18, 21–22
industrial plants, 862–871. *See also* energy efficiency, industrial plants.
industry classification, links between, 868–869
integrated energy systems, 1093
investments in, 13
leaks and loss prevention, 1093
low cost improvements, 517–523
case studies, 517–523
payback analyses, 517
matching supply and demand, 1093
measuring economic development, 501–502
municipal wastewater treatment plant (WWTP), 15
passive strategies, 1094
practical aspects of, 450, 451
technology information sources, 507–516
associations and organizations, 513–514
Center for Analysis and Dissemination of Demonstrated Energy Technologies, 515
technology information sources. *See* energy efficiencies, technology information sources.

tribal lands, 1456–1462
  and tribal governments, 1457–1458
  worldwide energy demand, 499–501
  See also energy demand-side management programs; energy efficiency, implementation, energy management programs.
Energy efficiency, implementation of
  balance supply and demand, 503
  demand side management, 503–505
  reducing power capacity needs, 503
  technology transfer, 502–503
  See also energy management (EM) programs.
Energy efficiency, industrial plants
  challenges to, 869–870
  combined heat and power, 864–867
  complexity of control systems, 870
  compressor systems, 867
  electric consumption, 863, 864
  energy saving opportunities, 862–869
    combined heat and power, 864–867
    compressor systems, 867
    equipment modernization, 867–868
    variable speed devices, 867
    waste heat recovery, 864–867
  energy savings, 863–864
    combined heat and power, 863–864
    lighting retrofits, 863
  equipment modernization, 867–868
  industry classification, links between, 868–869
  lighting, 870
    retrofits, 863
  monitoring of, 869–870
  pollution reduction, 870
  support of, 869–870
  variable speed drives, 862, 867
  waste heat recovery, 864–867
  See also energy efficient buildings; energy management (EM) programs.
Energy efficiency, technology information sources.
  miscellaneous groups, 515
  Portland Energy Conservation Inc., 515
  state agencies, 511–513
  United States Government Agencies, 508–511
  university programs, 511–513
  utility organizations, 514–515
  See also energy management (EM) program.
Energy efficiency measure (EEM), 967
Energy efficiency programs
  sunbelt cities and, 1391–1392
Energy efficiency ratio, 819
Energy efficient buildings
  electrical service, 528
  envelope, 528
  facility guidelines, 524–534
    for integrated design, 517
    top-down approach, 525
  HVAC, 529–530
  integrated design, 517
  irrigation water use, 527
  lighting, 528–529
  maintenance, 533
  motors and drives, 529

new structures, 527
overall energy use, 527
plumbing, 533
sustain efficiency, 533–534
test and balance, 527
top-down approach, 525
  integrated design, 525–526
  sustainability, 526
  See also energy management (EM) programs.
Energy engineering
  Association of Energy Engineers, 131–135
  careers in, 131–136
  future of, 135–136
    energy efficient economy, 136
    energy security, 135–136
    updating transmission grid, 136
Energy flow, 117–118
Energy forms
  elastic potential, 1163
  electrical kinetic, 1163
  electromagnetic field, 1163
  gravitational potential, 1163
  latent thermal, 1163
  mechanical kinetic, 1163
  sensible thermal, 1163
  thermal, 1163
Energy generation, 320
  district heating and cooling and, 320
Energy glossary, 1173
Energy glow paths, building system simulation and, 116–117
Energy impact on living standards and culture. *See* Living standards and culture.
Energy information system processes, 539–540
Energy information systems (EIS), 535–541
  processes, 535–536
    data collection, 535
    web publishing, 535–536
  utility report cards 539–541
  web publishing, purchase or self design, 538
Energy intensity indicator, 7
  disadvantages of, 7
Energy intensive industrial processes, 461–464
  boilers, 463
  calciners, 464
  centrifugal compressors, 464
  centrifugal pumps, 464
  chemical reactions, 462
  combustion systems, 463, 464
  distillation, 462
  drying, 462
  electrical processes, 463
  fans and blowers, 464
  fired heaters, 464
  flare systems, 463
  fractionation, 462
  furnaces, 464
  incinerator systems, 463
  kilns, 464
  liquid ring vacuum pumps, 464
  mechanical, 463
  melting and fusing, 461
  process cooling, 462–463
  process heating, 461
  steam systems, 463
  vacuum systems, 463–464

Energy invoice review, energy assessments and, 64
Energy leaks, 1093
Energy load inventory, output, 4
Energy loss, 361, 1093
  minimization of, 243
Energy management
  attributes of, 543–544
  commissioning existing buildings and, 185
  determining organization's aptitude for, 546–548
    energy marketing services, 547
    engineering protocol, 547–548
    fiscal protocol, 547
    leadership strength, 547
    leadership intensity, 547
    pride intensity, 547
    replication capacity, 546–547
    visibility of, 546
  fan systems and, 1224–1226
  measurement tools for, 1056–1060
    flow improvement, 1056–1059
    temperature accuracy, 1059–1060
    temperature multiplexing, 1059
  organizational aptitude for, 543–548
    facilitating reasons, 543–544
    management pathfinding for, 545–546
    strategies, 544–545
  prevailing strategies in, 544–545
  See also energy management (EM) programs; energy management systems.
Energy management (EM) programs
  3E Plus, 880
  audit process, 874–875
  available software
    3E Plus, 880
    Industrial Energy Management Software, 880
    Process Heating Assessment and Survey Tool (PHAST 1.1.1), 880
    PSAT: Pumping Assessment Tool, 880
    Steam System Assessment Tool 1.0.0, 880
    Steam System Scoping Tool 1.0d, 880
  current trends, 875
  definition of, 873
  energy balance in, 875
  importance of ECOs, 875
  Industrial Energy Management Software, 880
  Process Heating Assessment and Survey Tool (PHAST 1.1.1), 880
  PSAT: Pumping Assessment Tool 880
  scheme for, 875
  Steam System Assessment Tool 1.0.0, 880
  Steam System Scoping Tool 1.0d, 880
  trends, 875
Energy management systems, 1027–1041
  case study, 1037–1041
  common problems, 1035
    ineffective follow-through, 1036
    lack of communication, 1035
    lack of data, 1037
    lack of focus, 1036
    lack of management commitment, 1035–1036

limited resources, 1037
market reaction, 1037
procedures, 1036
recurring issues, 1036–1037
reliance on single person, 1036
shifting priorities, 1036
drivers of, 1034–1035
establish objectives, 1034
implementation of, 1035
organizational communication, 1034
relationships with suppliers, 1035
team based approach, 1034
See also energy management; energy management (EM) programs.
Energy manager teams, 873.
Japan, 873
Latin America, 874
See also Energy projects management.
Energy master planning (EMP), 549–555
benefits of, 549–550
creating a plan, 552–555
announce success of, 554–555
cost savings proposal, 552–553
create goals, 554
management support, 552
opportunity, 552
organization's energy use, 553
team planning, 553
upgrade implementation, 554
verify savings, 554
current practices, 550–551
failure of, 550–551
leadership, 551
line management accountability, 555
organizational optimization, 550
origins of American approach, 551
Energy Master Planning Institute (EMPI), 552–555
Energy measurement, 2
key functions, 2
other types of measurement, 2
Energy monitoring, 5–6
cumulative sum of differences analysis, 6
performance modeling, 5–6
performance modeling, intensity indicator, 7
steps of, 5
Energy performance indicators, 4, 578
energy utilization index, 4
Energy Policy Act of 1992 (EPAct '92), 380
effects of, 380–381
Order 888, 380
tariff filings, 380
Energy Policy Act standards, electric motors and actual efficiency, 353
Energy production adjustment, 2
Energy project management, 556–565, 874
definitions, 556–558
finance opportunities, 874
reasons to use, 556
skills for
personnel, 564–565
technical skills, 564
skills needed, 564–565
communication skills, 564
team commitment, 564
team composition, 564
tools, 562–564

collaboration and alignment, 563–564
RACI concept, 563–564
scheduling tool, 563
scope of, 562
See also energy project management plan.
Energy project management plan, 558–562
bid preparation, 560–561
conceptual stage, 558–559
construction stage, 561
definition stage, 559–560
delivery stage, 561–562
design stage, 560
See also energy project management.
Energy projects, life cycle analysis (LCA). *See* energy projects, life cycle costing.
Energy projects, life cycle costing
antecedents of, 968
blended utility rate, 967
btu/sf, 967
cost analysis, 967–968
definitions, 967–968, 970–971
energy efficiency measure (EEM), 967
framework of, 969–972
costs and benefits, 971–972
current situation, 969–970
gas station example, 969
HVAC example, 970
definition, 970–971
gas station example, 969
HVAC example, 970
internal rate of return (IRR), 967
LED exit lights example, 970–971
net present value (NPV), 968
payback period, 968
time value of money, 968
total owning costs, 968
Energy property, energy transfer vs., 1163–1164
Energy quality, 421
Energy recovery ventilation (ERV), 52
Energy reduction, compressed air systems and, 243–245
Energy requirement compilations, 1142
Energy requirements, industrial processes and, 458–459
Energy resource distribution, 917
Energy resource utilization data, 1010
Energy resources worldwide, 445–446
fossil fuels, 445
production and consumption, 445–446, 447
statistical analysis of, 445
Energy saving opportunities, 868
Energy saving opportunity gap (ESOG), 503
Energy savings, 569–571
3E Plus, 880
Air Master +1.0.9, 880
available software, 879–880
3E Plus, 880
Air Master +1.0.9, 880
Industrial Energy Management Software, 880
Motor Master (MM+4.0), 879–880
Process Heating Assessment and Survey Tool (PHAST 1.1.1), 880
Steam System Assessment Tool 1.0.0, 880
Steam System Scoping Tool 1.0d, 880

combined heat and power, 863–864
compressed air control systems and, 209–211
as driver in motor optimization projects, 888
guaranteed, 569
Industrial Energy Management Software, 880
industrial plants energy efficiency and, 863–864
lighting retrofits, 863
Motor Master (MM+4.0), 879–880
Process Heating Assessment and Survey Tool (PHAST 1.1.1), 880
recognized validation of, 554
shared, 569
Steam System Assessment Tool 1.0.0, 880
Steam System Scoping Tool 1.0d, 880
tradable certificates, 1433–1439
underfloor air distribution (UFAD) and, 1467
in variable speed drives, 864
Energy savings, case studies
Dept. of Energy best practices case studies, 1283–1301
aluminum plant assessments, 1283, 1285–1288
chemical plant assessments, 1283–1284
forest product plant assessments, 1284, 1288–1291
glass plant assessments, 1284, 1291–1293
metal casting plant assessments, 1284, 1293–1294
mining plant assessments, 1284–1285, 1294–1296
petroleum plant assessments, 1285, 1296–1299
steel plant assessments, 1285, 1299–1301
Florida's fruit industry, case study, 876–877
average equipment energy usage, 877
citrus juice production, 876
company profiles, 876
company profiles, prior to audits, 877
current/new trends, 877, 879
data analysis, 878
descriptors for, 879
energy assessments, 878
energy balance, 877
energy conservation opportunities, 878
energy consumption, 876
energy data, 877
energy distribution, 877
inspection, 876
utility company profile, 876–877
Energy security, 135–136
Energy service companies (ESCO), 566–572
successful projects, 307–308
use of, 306
consultant basis, 306–307
utility based, 307
users of distributed generation, 306–307
Energy service companies (ESCO), in Europe, 566–572
contracting models, 568–572
build own operate transfer, 571
chauffage, 571

energy savings, 569–571
    first out approach, 571
    leasing, 571–572
    other types, 571–572
equity contributions, 568
financing options, 567–568
    internally funded, 567
    third party, 567–568
investment grade audit, 567
measurement and verification of savings, 567, 572
project elements, 566–567
Energy service companies (ESCO), United States, 111
    savings model, 570
Energy service provider companies (ESPC), 566
Energy simulator software, 111
Energy sources, 1088
    net emergy and, 426
    non renewable, unit emergy values and, 425
    See Wind power.
Energy standards
    Department of Energy's role, 439
    development and revision of, 438
        groups involved, 439
    energy codes and, differences between, 438
    management systems and, 1031–1032
Energy Star Building label, 576–577
    applying for, 577
    ASHRAE standards, 577
    indoor air quality requirements, 577
        ASHRAE standards, 577
    professional engineer review and validation, 576–577
    statement of energy performance, 576
Energy Star Partnership, 510
    sunbelt cities and, 1393
Energy Star Portfolio Manager software, 573–578
    additional tools, 577–578
        Target Finder, 577–578
    background of, 573–574
    case studies available, 578
    data input information, 576
    Delta Score Estimator, 578
    eligibility requirements, 575–576
    energy benchmarking tool, 573–574
    Energy Performance Indicators, 578
    Energy Star building label, 576–577
    overview, 574
    required building information, 576
    rulesets, 575–576
        primary spaces, 575
        secondary spaces, 575–576
    Target Finder, 577–578
    tracking water consumption, 578
    using, 574–576
    weather normalization, 576
Energy Star window criteria, 1521
ENERGY STAR, 442
Energy Start™, 948
Energy storage, photovoltaic (PV) systems and, 1155–1156
Energy supplies
    global, 1344, 1346–1347
    primary vs. secondary, 1346
    United States, 1344, 1346–1347

Energy supply and demand, matching of, 1093
Energy supply by source, 607
    electricity, 608
    world and United States, 607–608
Energy system incentives, 959–961
Energy Tax Act of 1978, 1084
    tax disincentives, 1084–1085
Energy transfer, energy property vs., 1163–1164
Energy transmission, 320
    district heating and cooling and, 320
Energy transport systems, 530–531
Energy units, conversion factors, 1164
Energy usage
    impact on life and environment, 873
    square footage vs., 583
    wastewater plants and, 1550–1551
Energy usage patterns, energy balance and, 686–687
Energy usage reduction, 599
Energy use, 320
    building geometry and
        effect on, 111–115
        modeling, 112–113
    culture and society, connection between, 1016
    district heating and cooling and, 320
    environmental impact of, 1013
    modification of, 1015
        conservation, 1015
        efficiency increase, 1015
    past global usage, 1012
    population and, 1010
    projected global usage, 1012
    United States overview, 580–586
        electricity generation, 581–582
        major sectors usage, 580–581
        types used, 581
Energy use data, energy balance, 681–685
Energy use growth patterns, living standards and, 1010–1012
Energy use index (EUI), 111
    energy service company (ESCO), 111
    use of, 111
Energy use modifications
    strategic planning for, 1015–1016
    to improve living standards, 1015
Energy use in the United States
    commercial, 583
    industrial, 584–585
Energy use in U.S. transportation, 585, 588–599
    household vehicles, 585
Energy utilization index (EUI), 4
    energy balance, 4
    normalized performance indicator, 4
    specific energy consumption, 4
Energy waste sources, compressor systems and, 867
Energy wheels, 794–795
    See heat wheels.
EnergyPlus software, 1402
EnergyPort, 918
Energy-resource utilization data, 1010
Enforcement
    of energy codes, 442
    emissions trading and, 433
Engine analysis, reciprocating, 1237–1238

Engine redesign, waste fuel technology and, 1527
Engineering, automobile technology evolution and, 673–675
Engineering review, Energy Star Building label and, 576–577
Engines, heat pipe applications and, 811–812
Enriching, uranium and, 1113
Enterprise energy management systems (EEMS), 616–624
    business applications, 623
    data collection re, 616–624
    definition of, 616–617
    principles of, 617–623
        actionable information access, 619–622
        data collection, 617–618
        data structuring, 618–619
        industry standards, 622–623
        interactive access, 619–622
        measure and verify results, 622
    results, 622
        life-cycle costing, 622
        savings, 622
Enthalpy
    formation, 1423–1424
    reaction, 1423–1424
Enthalpy calculations, two phase heat exchanger condenser and evaporator, 803
Enthalpy controller, 415
Entrained bed gasifiers, 907
    General Electric type, 907
    Shell type, 907
Entrainment, heat pipes and, 805–806
Entropy, energy degradation and, 1168–1172
Environment
    district heating and cooling and, 320–321
    green energy and, 772–774
    minimal impact of geothermal energy, 748
    waste fuels and, 1525, 1527
Environmental agencies, independent power producers and, 856
Environmental benefits, energy efficiency and, 15–17
Environmental concerns
    3-E trilemma, 1147–1148
    sustainable development and, 452–453
Environmental degradation, 447
Environmental impact
    assessment applications, exergy and, 655–660
    energy use and, 1013
    exergy and, 658–660
    geothermal energy and, 746–747
Environmental issues
    consequences of, 448
    energy conservation and, 446–449
        solutions, 449
    nuclear energy and, 1103–1104
    pumped storage hydroelectricity and, 1211–1212
Environmental policy, 625–632
    development of, 625–626
    impact of, 626–627
        air, 627
        burden of proof, 627–628
        economic framework, 628–630

cost-benefit analysis, 628–629
cost-effective analysis, 629–630
land use, 627
water, 627
policy instruments, 630–631
emission reduction mandates, 630–631
emissions tax, 630
liability rules, 630
nonuse benefits, 631
pollution tax, 630
redistributive effects, 631
renewable energy resources, 631
sustainable development, 631
tradable emissions permits, 630
Environmental Protection Agency, 510. *See* EPA.
ENERGY STAR Partnership, 510
Environmental protection, energy conservation vs., 450
Environmental regulation, wind power, 1615
Environmental uses, exergy analysis and, 651–652
Envision software, 3
EPA regulations, chemical process industry and, 1302–1303
EPAact motor cost, premium efficiency motor cost ratio, 353
EPAct '92. *See* Energy Policy Act of 1992.
Equilibrium design, heat pipes and, 805
Equilibrium modeling
constrained, 657
exergy analysis and, 657
Equipment energy use data, energy balance and, 681–685
Equipment modernization
aging thermal processes, 868
aluminum alloy artificial aging, 868
assistance for, 868–869
billet heating, 868
die temperature, 868
energy saving opportunities, 868
industrial plant savings and, 867–868
metal industry, 868
solution heat treatment, 868
Equity contributions by ESCO, 568
Equivalents of energy, 603
ERC. *See* emission reduction credit.
ESCO and ESPC
contrasts between, 567
ESCO. *See* energy service companies.
ESOG. *See* energy saving opportunity gap.
ESP-r software, 1402
Ethanol, 32–33
MTBE, 33
Ethylyne glycol operating temperature, run around heat recovery systems and, 1281
EUI. *See* energy use index.
EUI. *See* energy utilization index.
European energy service companies, 566–572
European Union
adoption of Kyoto Treaty, 13
Emissions Trading System, 141
Evacuated tube thermal collector, 1334–1335
Evaporating process, mobile HVAC systems and, 1073–1074
Evaporator, 1544

heat pipe and, 804–805
mobile HVAC systems and, 1079–1080
heat exchangers and, 803–804
maintenance of, 804
materials used in, 804
safe operation of, 804
Evaporator core identification, 1064–1065
Evolution, automobile technology and, 673–675
Excess energy consumption gap (EXEC), 503
EXEC. *See* excess energy consumption gap.
Executive summary
energy assessments and, 68
user friendly energy audit report suggested format, 78
Exergy, 321, 450, 645–653, 1428–1429, 321
definition of, 645–646, 655
energy degradation and, 1168–1172
environmental impact, 655–660
assessment applications, 655–660
chaos, 659
order destruction, 659
resource degradation, 659
waste energy emissions, 659–660
green energy and, 779
reference environment, 646
related definitions to, 646
sustainable building simulation and, 1397–1398
Exergy analysis, 322–324, 647, 655–656
applications of, 647–653
coal-fired electrical generating station, 648–651
economics, 652–653
electrical resistance space heater, 648
environmental uses, 651–652
thermal storage system, 648
efficiency, 647
energy efficiency and, 1094
geothermal energy and, 748–750
reference-environment modeling, 656–658
constrained-equilibrium, 657
equilibrium, 657
natural-environment-subsystem, 656–657
process-dependent, 657–658
reference-substance, 657
steady flow process, 748
steady state, 748
Exergy balances, 646
Exfiltration, infiltration vs., 844
Exhaust, 1235
Exhaust gas condensation solutions, combined heat and power, 177
Existing building commissioning, 179–187
Expansion, 1235
Expansion turbine and control, 215–216
Expansion valve, mobile HVAC systems and, 1077–1078
Exposed rafting, 902–903
Extensible markup language (XML), 537
Exterior insulation finish system walls, 1518
External communications, wireless applications and, 1655–1656
External energy benchmarks, 690
External wireless applications, 1654–1658
Extreme day CPP, 1178
Extreme day pricing, 1175

Extrusion polystyrene (XPS), as insulation material, 891, 894
Eye anatomy and functions, 993–995
EYR. *See* emergy yield ratio.
EZDOE software, 1402

F

Facilities, building automation system design issues and, 107
Facility air leakage, 662–670
infiltration control, 662
measuring airtightness, 668–669
Facility energy codes and standards, 438–443
Facility energy controls
automobile technology and, 671–679
automobile technology transferable to
display systems, 672–673
individual control systems, 672
operational controls, 672
options, 673
incorporating automobile technology in
building code support, 677–678
equipment and system modules, 677
impose standards, 677
modular buildings, 677
standardized commissioning of buildings, 678–679
lack of automobile technology in, 675–677
Facility energy use auditing, 63–68
energy assessments, 63–68
Facility energy use
benchmarking of, 690–698
energy benchmarks, 690–698
tools to assist in analysis of, 680–689
energy balance, 681–689
Facility guidelines, energy efficient building and, 524–534
Facility intranet monitoring and control, HMI compressor workstation and, 212
Facility layout review, energy assessments and, 64
Facility power distribution systems, 699–706
direct current systems, 705
fault current coordination, 704
power factor, 706
transformer ratings, 705
variable frequency, 705–706
variable speed drives, 705–706
Fan quantity reduction, air distribution adjustment and, 276
Fan speed limit, air distribution adjustment and, 276
Fan speeds, 521
Fan systems, 1220–1226
energy management, 1224–1226
flow control, 1225–1226
low cost options, 1225
maintenance, 1224–1225
laws, 1224
measurements, 1220–1223
flow, 1220–1221
performance, 1222–1223
performance, 1223
air density impact, 1224
curves, 1223–1224
power requirements, 1223

types, 1220
   axial, 1220
   centrifugal, 1220
Fans, 464
Fast breeder nuclear reactors, 1120–1121
Fast neutron reactors, 40
Fast simulation modeling
   focus on, 916
      impact on reliability, 916
      integrated model validation, 916
      multiresolution, 916
   IntelliGrid^{SM} and, 915–916
Faucets, 1596–1597
Fault current circuit breakers, 704
Fault current contactors, 704
Fault current coordination, 704
Fault current energy levels, 703
Fault current energy sources, 703
Fault current interrupting devices, 703–704
Fault current support, 703
Fault current, 703–704
Faults, power system, 702–703
Federal Aid Highway Act of 1956, 1021
Federal Energy Management Program (FEMP), 509, 873
Federal Energy Management Program, Operations and Maintenance Best Practices Guide (O&M BPG), 707
Federal Energy Regulatory Commission (FERC), 380
   transmission open access, 380
Federal Government funding, airports and roads, 591
Federally funded airport construction, 1020
Feedstock for syngas
   ash content of, 907
   ash fusion temperature of, 907
   free swelling index, 908
   moisture content of, 907
   particle size of, 908
   reactivity of, 907
FEMP. See Federal Energy Management Program.
FERC. See Federal Energy Regulatory Commission.
Ferrous electromagnets maglev system, 1022–1023
Fiberglass insulation, exterior rigid, 895
Fiberglass, as insulation material, 890, 893
   effect on environment, 891
Fill type, cooling towers and, 249–250
Filtration, water treatment process and, 1557, 1558–1559
Fin efficiency, 1336
Fin tube heat exchanger, 1540
Financing charges, 957–958
Financing, energy and developing countries, 501
Financing, energy service companies in Europe and, 567–568
Fine bubble diffusion, 1549
Finned built in storage solar water heater, 1333
Finned heat exchangers, 803
Finnish Masonry Stove, 1359–1360
Fired heaters, 464
Fireplaces, 1358–1360
Firm sales rates, natural gas and, 1507

Firm transportation rates, natural gas and, 1507
First Law of Thermodynamics, 1161, 1167, 1423–1426
   combustion, 1424
   enthalpy, 1423–1424
   formulation, 1423
   heat rates, 1424–1426
   heating values, 1424–1426
   power cycle efficiency, 1424–1426
First out contracting model, ESCO and, 571
First stage cost drivers, 1042
Fischer-Tropsch process, 163–165
   catalysts, 164–165
   providers of, 166–167
   South African use of, 163
Fission, 1128
5S's, 474
   self-discipline, 474
   simplification, 474
   sorting, 474
   standardization, 474
   sweeping, 474
Fixed bed combustor, 489
Fixed cost related charges, utilities and, 1500
Fixed history baseline, allowance/permit allocation and, 432
Flame safeguard control, 97–98
Flame temperatures, solid fuel properties and, 484
Flare systems, energy intensive, 463
Flat plate thermal collector, 1333–1334
Flight controls
   aircraft fuel consumption and, 28
   autopilot, 28
   fuel control and management computer, 28
Flight management computer (FMC), 28
Flocculation, 1557, 1558
Floor insulation
   raised floor, 892
   slab-on-grade, 892
   strategies for use, 892
Florida Solar Energy Center, 512
Florida's orange and grape industry, case study of energy savings, 876–877
FLOVENT software, 1402
Flow measurement
   common problems and fixing of, 1056–1057
      orifice plate management, 1057
      straight pipe issues, 1056–1057
   fan systems and, 1220–1221
   flow technologies, comparison of, 1059
   gas and steam
      density compensation of, 1057–1059
      minimizing permanent pressure loss, 1058–1059
      improve accuracy and repeatability, 1056–1059
   temperature, 1059
Flow technologies, comparison of, 1059
Flows data, exergy analysis and, 650
Flue gas volume, solid fuel properties and, 485
Fluidized bed combustor, 489
Fluidized bed gasifiers, 907
   Advanced Transport Reactor, 906–907
Fluidized bed
   circulating, 88
   stationary, 88

Fluids
   heat exchange and, 801
   heat pipes and, 805
   real, 1430–1431
Fluorescent dimming ballast, 984
   lighting control systems, 984
FMC. See flight management computer.
Foam insulation
   blown, 897
   exterior and interior, 895
   strategies for use, 892
Focus and lack of, energy management system problems and, 1036
Foil faced batt insulation, 1231
Foil-faced OSB, as insulation material, 895
Follow through, energy management system problems and, 1036
Footcandle, 999
Footlambert, 999
Forced circulation solar water heater, 1333
Forest product plant assessments, Dept. of Energy best practices case studies and, 1284, 1288–1291
Forestry products, drying of, 332–336
Formation of enthalpy, 1423–1424
Formulation, 1426–1427
Fossil fuels, 445, 606–607
   air pollution, 715–721
      ambient standards, 715–717
   dominance of, 45–46
   global warming, 721–725
   post era, 608–610
Fouling during cofiring, 491, 493
FoxWeb, 537
Fractionation process, 462
Framed wall insulation, interior, 895
Franklin stove, 1359
Fraser software, 3
Free swelling index, of syngas feedstocks, 908
Frequency domain, 117
Frequency, 1652
Fresh air, indoor air quality and, 19
Frosting, heat wheel performance factors and, 797–798
FSTIG cycle, 1583
Fuel cell program, 511
Fuel cell vehicles, greenhouse gas emissions and, 792–793
Fuel cells, 37–38, 298–299, 403, 726–732
   compared to existing technologies, 38
   current use of, 37
   efficiency of, 37
   low temperature, 733–743
   molten carbonate, 727–729
   phosphoric acid fuel cells, 726–727
   solid oxide, 729–732
   technology challenges, 37–38
      durability and reliability of, 37–38
      size, 38
   workings of, 37
Fuel consumption, aircraft and, 24–30
Fuel control and management computer (FCMC), 28
Fuel cost savings, plug in hybrid electric vehicles and, 851
Fuel cycles, nuclear energy and, 1111–1114
Fuel data tabulation, energy accounting and, 4

Fuel ethanol, 1267
Fuel fabrication, uranium and, 1113
Fuel independence, 873
Fuel oil
　chemical process industry and, 1305
　savings of geothermal energy, 744
Fuel operation, nuclear energy and, 1113–1114
Fuel prices, 1364
Fuel properties, 478–485
　gaseous, 485
　liquid, 485
　solid, 479–485
Fuel quantity calculations, pre flight planning and, 28
Fuel supplier, independent power producers and, 856
Fuel Use Act, 1085
Fuels, generation of electricity and, 378
Full load nominal efficiencies, electric motors and, 352
Full-light output, 834
Full-load efficiency, district cooling systems and, 313–314
Furnaces, 464, 531, 809
Fusing, 461
Fusion nuclear reactors, 1123–1124
Fusion, 1130–1131
Future Air Navigation System 1, 30
Future reserves, energy and, 608–610
Futuregen, 494–496

G

Galvin Electricity Initiative, 404–405
　configuration levels, 405
　Perfect Power System, 404–405
Gas compression heat pump, 817
Gas cooled nuclear reactors, 1119–1120
　locations of, 1119–1120
Gas exchange process, reciprocating engine analysis and, 1237–1238
Gas supply curtailment, Natural Gas Policy Act of 1978 and, 1084
Gas transportation, Natural Gas Policy Act of 1978 and, 1084
Gas turbines, 1203
　simple cycle, 298
Gas utilities, seasonal loads, 1498
Gas utility bill, calculating of, 1472
Gas
　density compensation of, 1057–1059
　ideal, 1430
Gaseous fuel properties, 485
Gasification, 89–90, 492–494, 1525, 1527
　feedstock conversion, 906
Gasifiers, 1528–1529
　air-separation technologies in, 909
　　cryogenic, 909
　　high-temperature membrane, 909
　biomass gasification, 912
　contaminants, 908
　economies of scale, 912
　feedstock conversion, 906–907
　　entrained bed gasifiers, 907
　　　General Electric type, 907
　　　Shell type, 907
　　fluidized bed gasifiers, 906–907

Advanced Transport Reactor, 906–907
　moving bed gasifiers, 906
　　Lurgi gasifier, 906
　types of, 906–907
overall plant integration, 910–911
　coproduction, 911
　environmental signature, 911
　IGCC overall block flow, 910
　raw gas in, 908
Gasoline engines, reciprocating, 1239–1240
Gasoline fueled vehicles, CO_2 emissions and, 788
Gasoline,
　aircraft fuel type, 26
　alternative transport fuels vs., comparison of, 47–48
GBS green building design software, 1403
General Electric Co., 390
General Electric gasifiers, 907
General service electric rate schedule, 1508
General service heating electric rate schedule, 1508
Generated utilities, energy data collection source and, 617
Generation energy flow, compressed air systems and, 226
Generation of electricity
　fuels used, 378
　resources
　　conventional, 374–376
　　renewable, 376–378
Generator voltage, combined heat and power and, 175
Generators, induction vs. synchronous, 175–176
Geobiosphere emergy, 422–423
　computing of, 422–423
　inputs to, 423
　products, 423
Geologic sequestration, carbon capture and storage, 127–128
Geometric ratio (GR), 113
　using in energy use reduction, 113
Geometry of buildings, energy use effect of, 111–115
Geothermal electricity, 1266
Geothermal energy, 744–752, 1364
　carbon emissions saved, 744
　case study of, 750–752
　commercial viability of, 745
　direct uses, 744–745
　efficiency of, 745
　electric energy production, 744–745
　environmental impact, 746–747
　　research and development, 746–747
　　standards development, 747
　　technology
　　　assessment, 747
　　　transfer, 747
　exergy analysis, 748–750
　　steady state, steady flow process, 748
　history of, 745, 746
　installed capacity, 744–745
　nomenclature, 744
　performance evaluation of, 748–750
　pollutants in water, 746
　savings in fuel oil, 744

　sustainability of, 747–748
　　minimal environment impact, 748
　　network flexibility, 748
　　renewal ability, 748
　technical aspects of, 745–746
　　fields, 745
Geothermal fields
　dry steam, 745
　hot water, 745
　wet steam, 745
Geothermal heat pump (GHP) systems, 753–759
　definitions, 753–754
　development of, 757
　　load calculation, 757
　　single borehole, 758
　　thermal mass, 757–758
　earth heat exchangers, 756
　HVAC vs., 755–756
　maintenance costs, 755
　organizations, 756–757
　principles of, 754–755
　types, 756
GHG emissions, energy's efficiency effect on, 12
GHG reduction due to biomass energy, 91
GHG. *See* green house gases.
GHP. *See* geothermal heat pump.
Glare, 993
　illumination basics and, 265–267
Glass coatings, 1621
Glass plant assessments, Dept. of Energy best practices case studies and, 1284, 1291–1293
Glass, thermal conductivity of, 1619
Glazing coatings, 1621
Glazing system
　coatings and tints for energy efficiency, 1622
　general description, 1620–1621
　multiple pane, 1620–1622
Global energy supplies, 1344, 1346–1347
Global Mvo Asset Manager software, 3
Global peace, green energy implementation and, 781
Global positioning satellites (GPS), aircraft fuel consumption and, 29
　local-area augmentation system, 29
　wide-area augmentation system, 29
Global temperature adjustment
　absolute vs. relative implementation, 275
　decay of shed savings, 274
　electric demand response and, 274
　evaluation of, 275
　factory vs. field implementation, 275
　impediments to, 275
　implementation of, 274
　mode transitions, 274
Global unrest, green energy implementation and, 781
Global warming
　effects of, 722–723
　　climate changes, 723
　　sea level rise, 722–723
　fossil fuels and, 721–725
　greenhouse gas concentration trends, 723–724

lessening of, 724–725
    conservation measures, 724
    electricity supply efficiency, 724
    non-fossil energy sources, 724–725
    workings of, 721–722
Goal gasification, 492–494
Goal oriented planning, energy master planning and, 554
GPS. *See* global positioning satellites.
GR. *See* geometric ratio.
Granularity of control, electric demand response and, 273
Gravitational potential energy, 1163
Gravity treatment of wastewater, 1561
Great Plains coal basins, 158
Green Building Challenge program, 948
Green building design software, 1403
    ATHENA, 1403
    BEES, 1403
    GBS, 1403
    RETScreen, 1403
Green buildings
    concept of, 948
    rating system, 948–949. *See also* Buildings, rating systems for.
Green electricity pricing, 1262–1264
Green energy, 771–786
    advantages and disadvantages, 774–775
    applications, 779
    environmental consequences, 772–774
    exergetic aspects, 779
    factors need for success, 778
    implementation of, 779–781
        global peace, 781
        global unrest, 781
        sustainable development ratio, 779–781
    need for, 772
    resources, 777–778
    sustainable development, 776–777
        case study, 781–786
    technologies, 777–778
Greenhouse gases (GHG), 9
Greenhouse gas concentration trends, 723–724
Greenhouse gas emissions, 446–447, 787–793
    carbon dioxide
        gasoline fueled vehicles, 788
        hybrid vehicles, 788
        plug-in hybrid vehicles, 788
    electrolysis produced hydrogen vehicles, 790–791
    hydrogen vehicles
        fueled, 789
            fuel cell vehicles, 792–793
            steam methane reforming, 790
        plug-in hybrid electric vehicles, 791–792
        reduction funding, 1531–1532
        steam methane reforming, 789–790
Ground fault detection, 703
Ground source heat pumps, 532
Ground wires, 358
Grounding, 701
Guaranteed energy savings, 569
Guidance documents, 1030
Gulf Coast region coal basins, 158

H
Habitat structures
    evolution of, 1362–1363
    space heating and, 1362–1363
Hardware, client and server, building automation systems and, 106–107
Harvester ice storage equipment, 1416–1417
HCCI. *See* homogeneous charge compression ignition.
HDD. *See* heating/degree day adjustment.
Hearths, 1358–1360
Heat buildup, radiant barriers and, 1231
Heat engines, 1171
    Second Law of Thermodynamics, 1171–1172
Heat exchangers, 801–804, 808, 1540
    analysis of, 801–803
        counter flow, 802–803
        parallel flow configuration, 802
    design considerations, 803
    finned types, 803
    fluids, 801
    two phase, 803–804
    types
        concentric-tube, 801
        cross flow, 801
        shell and tube, 801
    wet cooling towers and, 248
Heat flow, conductive, in window energy performance, 1614
Heat gain
    casual, 116
    decreasing of, 1231
Heat pipe applications, 807–812
    aerospace, 807–808
    agriculture, 810
    cutting tools, 812
    electronic components, 808–809
    engines, 811–812
    exchangers, 808
    medical, 810–811
    ovens and furnaces, 809
    solar thermal applications, 809
    transportation systems, 811
Heat pipes, 804–806, 1540
    applications of, 804
    boiling limitations, 805, 806
    definition of, 804
    entrainment, 805–806
    equilibrium design, 805
    evaporator, 804–805
    sonic limitations, 805, 806
    thermal conductivity, 805
    wick type, 805
    wicking, 805–806
    working fluid, 805
Heat pumps, 814–821
    applications of, 820–821
    Carnot coefficient of performance, 815
    fundamentals of, 814–815
    ground source, 532
    heating and, 815–816
    history of, 814
    performance parameters, 818–820
        ambient energy fraction, 819–820
        coefficient of performance, 819
        energy efficiency ratio, 819
        primary energy ratio, 819
    types, 816–818
        absorption, 816–817
        gas compression, 817
        thermoelectric, 816
        vapor compression, 817–818
Heat rates, 1424–1426
Heat recovery equipment, 1540
    fin tube heat exchanger, 1540
    heat pipes, 1540
    heat wheels, 1540
    plate and frame heat exchanger, 1540
    recuperator, 1540
    run-around coils, 1540
    shell and tube heat exchanger, 1540
    steam recovery, 1540
Heat recovery steam generators, 1578
Heat recovery systems, run around, 1278
Heat recycling, cogeneration district heating and cooling systems and, 318
Heat removal factor, 1336
Heat sink configurations, 1542–1543
Heat source configurations, 1542–1543
Heat storage, 1420
Heat transfer blocking, radiant barriers and, 1227–1228
Heat transfer calculations, 1063
Heat transfer coefficients, 1279–1280
Heat transfer method, 1063–1066
Heat transfer, 822–829
    conduction, 823
        in windows, 1617
    convection, in windows, 1617–1618
    history of, 822–823
    radiation, 828–289
    in windows, 1617
    thermal conductivity, 824–827
    three moles of, 1617
    windows, 1616–1617
Heat units, energy and, 605
Heat values of waste products, 1345
Heat wheels, 794–799, 1540
    construction and materials, 795
    economic factors, 798–799
        life cycle costs, 798
        payback, 798
    environmental factors, life cycle analysis, 798–799
    operation of, 795–796
    performance factors, 796–797
        frosting, 797–798
        pressure drop and leakage, 797
        reliability, 798
Heat, 1161
    First Law of Thermodynamics, 1161
    transfer of, 1161
        conduction, 1161
        convection, 1161
        thermal radiation, 1161
    work units and, 1162
Heater core identification, inverse heat transfer methods and, 1065–1066
Heater, electrical resistance space, 648
Heating capacity calculations, 1061–1066
Heating demand, seasonal, 1273–1276

Heating factor, electric power transmission systems rating and, 358
Heating fuels, 1361–1362
   coal, 1361
   oil, 1361
   petroleum based, 1361
   wood, 1361
Heating loads, residential buildings and, 1272–1277
Heating mode, mobile thermostat climate control and, 1663
Heating oil, 1361
Heating process, mobile HVAC systems and, 1074–1075
Heating values, 1424–1426
   biomass fuels and, 482–483
   solid fuel properties and, 479, 484
Heating
   dehumidification and, 1187–1188
   heat pumps and, 815–816
   Romans
      bathhouses, 1358
      hypocaust, 1357
   sensible, 1186
Heating/cooling degree day adjustment (HDD and CDD), 2
HERS. *See* home energy rating system.
HHV. *See* higher heating value.
HID dimming ballast, 985
   lighting control systems, 985
HID. *See* high intensity discharge.
Hierarchy structure, energy data, 619, 620
High efficiency motors, cost calculations and, 73
High humidity, control of, 843–844
High intensity discharge (HID) electronic lighting, 830–838
High intensity discharge ballasts, 830–832
   electronic ballast primer, 830–832
High operating voltage, advantage, 361
High pressure boilers, 93
High pressure cylinders (HP), 1382
High pressure fluid filled (HPFF) cable, 358
High pressure gas storage, hydrogen and, 1446
High pressure liquid filled pipe (HPLF), 358
High rise buildings, 1513
High solar gain low-e coatings, for windows, 1621, 1622
High voltage alternating current (HVAC), 357
High voltage direct current (HVDC), 359
High voltage direct current lines (HVDC), 357
Higher heating value (HHV), 86–87
Higher lift, chiller efficiency and, 309–310
High-temperature electrolysis, 1443
High-temperature membrane technologies, gasifiers and, 909
Historical baseline with updating, allowance/permit allocation and, 432
Historical energy use data, energy balance and, 681
History of energy, chronology of selected events, 610
HMI workstation. *See* human machine interface.
Home energy rating system (HERS), 442
Homogeneous charge compression ignition (HCCI), 1240, 1242
Hot water (hydronic) heating systems, 1361
Hot water optimization, case study, 1366–1379

Hot water vs. steam, combined heat and power, 176
Hot water, geothermal field and, 745
Hottel Whillier Bliss equation, 1335–1336
Housing growth, U.S. transportation energy use and, 590
HP. High pressure cylinders.
HPFF. *See* high pressure fluid filled cable.
HPLF. *See* high pressure liquid filled pipe.
Human eye anatomy, 994
Human eye functioning, 993–995
Human machine interface (HMI) compressor workstation, 212
   facility intranet monitoring and control, 212
   web-based, 212–213
Humid climates, HVAC systems and 839–846
   mold growth, 839
Humidification, 1186
Humidity
   control of, 843–844
   relative, 19
HVAC based electric demand response, 273–274
HVAC capacity, review of, 519–520
HVAC control, building system simulation process and, 120
HVAC controllers, improvement in, 1365
HVAC solution software, 1402
HVAC summer winter cutover, low cost energy efficient improvement case studies and, 520
HVAC systems
   dehumidification, 839–840
   functions of, 839
      comfort control, 839
      contaminant control, 839
      pressurization, 839
      ventilation, 839
   humid climates and, 839–846
      mold growth, 839
   mobile, 1061–1069
   poor humidity control
      case study, 840–844
      predictability of high humidity, 843–844
      proper pressurization design, 844
      vertical stacking fan-coil air conditioning, 840
HVAC
   cold air retrofit case study, 172–174
   facility guidelines for new structures and, 529–530
      air cooled equipment, 532
      boilers and furnaces, 531
      chillers, 531
      controls, 532–533
      cooling towers, 531–532
      energy transport systems, 530–531
      ground source heat pumps, 532
      hydronic circulating systems, 531
   geothermal heat pump systems vs., 755–756
HVAC. *See* high voltage alternating current, 357
HX coil ice storage equipment, 1414–1416
Hybird vehicles, carbon dioxide emissions and, 788
Hybrid electric vehicles
   current models, 849–850
   efficiency of, 844

   plug in configuration, 847–853
   plug in type, 850–851
      electricity capacity needed, 852
      electricity tax impact, 851–852
      fuel cost savings, 851
      home electrical service needed, 852
      implementation issues, 853
      sales incentives for, 852–853
      utility issues, 852
   responsiveness of, 848
   types, 848
Hybrid solar space heating, 1322
Hybrid vehicles, plug-in, carbon dioxide emissions and, 788–789
Hydraulics, water and, 1559–1560
Hydraulic systems, chemical process industry and, 1305
Hydrid cooling towers. *See* wet-dry cooling towers.
Hydrochloro-fluorocarbons (HFCs), as insulation blowing agent, 892
Hydroelectric generation, 1207–1208
Hydroelectricity, pumped storage, 1207–1212
Hydrogen economy
   advantages of, 11
   motor vehicles, 11
   availability of, 11
   disadvantages of, 11
   wind power and, 1610
Hydrogen fuel service stations, 1447
Hydrogen fuel
   challenges of, 1441
   distribution of, 1446–1147
      plants, 1447
      service stations, 1447
   production of, 1441–1445
      steam-methane reforming, 1442
      thermochemical cracking of water, 1443–1445
      water electrolysis, 1442
   storage of, 1445–1446
      alanates, 1446
      high pressure gas, 1446
      metal hydrides, 1446
Hydrogen fueled fuel cell vehicles, greenhouse gas emissions and, 792–793
Hydrogen fueled transportation systems, 1441–1447
Hydrogen fueled vehicles, 789
Hydrogen vehicles
   greenhouse gas emissions and, 790–791
   high-temperature nuclear reactor, steam methane reforming and, emissions from, 790
   steam methane reforming and, emissions from, 790
Hydrogen, 35–37
Hydrogen, barriers to commercial use, 36
   public acceptance, 36
   safety of, 36
   storage, 36
Hydrogen, energy storage for photovoltaic systems, 1155–1156
Hydrogen, future technology, 36–37
Hydrogen, production of, 36–37
Hydronic circulating systems, 531
Hydronic heating systems, 1361

advantages in, 1361
Hydronics Design Studio software, 1402
Hypertext preprocessor, 537
Hypocaust, 1357

I

IAC (Industrial Assessment Center) program, in industrial energy management, 873
IAQ. *See* indoor air quality.
IAQP. *See* indoor air quality procedure.
Ice storage equipment, 1414–1417
    harvester, 1416–1417
    HX coil, 1414–1416
    PCM, 1414
    stratified chilled water storage, 1417–1418
ICFs. *See* Concrete form insulation systems.
Ideal gas, 1430
IECC. International Energy Conservation Code.
IEKPI. *See* industrial energy key performance indicators.
IGCC vs. pulverized-coal boiler, 912
Ignition methods, electronic ballast primers and, 833–834
Ignition, combustion and, 487
Illumination basics, 264–267
    glare, 265–267
    movement of the sun, 264–265
    photometry, 264
    radiometry, 264
    solar spectrum, 265
Improvement, Six Sigma and, 1315–1316
Incentives
    emissions trading and, 433
    energy demand-side management programs and, 291
Incinerator systems, energy intensive, 463
Increasing block rates, 1503
Incremental pricing, Natural Gas Policy Act of 1978 and, 1083–1084
Independent power producers, 854–861
    history of, 854–855
    plant
        construction, 860
        operation, 860
    project development, 858
    project definition, 858
    stakeholders, 855
        advisors, 857
        contractors, 857
        environmental agencies, 856
        fuel supplier, 856
        government authorities, 856
        lenders, 857
        off taker, 856
        owner/developer, 856
        payment structure, 857–858
        stakeholders, project, 855
        stakeholders, risks, 858, 859
    types, 855
        long-term, 855
        merchant, 855
India's economic growth, impact on U.S. transportation energy, 592
Indian coal to liquid fuel plants, 167
Indicating devices, 416

Indirect cofiring, 89
Indirect solar gain design, 1322
Individual control systems, automobile technology and, transferable to building use, 672
Indoor air quality (IAQ), 18–23
    as defined by ANSI/ASHRAE Standard 62.1–2004, 51
    energy costs, impact on, 22–23
    energy efficiency and
        mutual goals of, 21–22
        relationship between, 21
    Energy Star Building label and, 577
    fresh air, 19
    investigation of, 21–22
        sophisticated testing, 21–22
        walk-through type, 21–22
    measuring of, 51
    procedure (IAQP) compliance to ANSI/ASHRAE Standard 62.1–2004, 54, 60
    sources of problems, 22
    ventilation, 18–19
        advantages, 21
        problems and, 19–20
            dilution, 19–20
            increased outside air, 20
            pollution sources, 20–21
            relative humidity, 19
            tight buildings, 20–21
Indoor environmental quality
    LEED-C&S, rating system, 944
    LEED-CI, rating system, 940
    LEED-NC, rating system, 950
Indoor negative pressures, 844
Induction generator vs. synchronous, combined heat and power, 175–176
Induction systems, 996
    lighting design and retrofits, 996
Industrial Assessment Center (IAC) program, in industrial energy management, 873
Industrial carbon dioxide capture, 127
Industrial drying operations, 338–348
    nomenclature, 338
    types, 338–339, 340
        batch, 338
        continuous, 338
    uses of, 338
Industrial electric power systems, combined heat and power and, 175
Industrial energy key performance indicators (IEKPI), 1140–1146
    cost savings tool, 1146
    demand side energy consumption, 1140
    proper usage of, 1145–1146
    types, 1141–1145
        efficiencies, 1141
        empirical determinations, 1144–1145
        energy
            balances, 1142–1143
            consumption analyses, 1141–1142
            requirement compilations, 1142
        measurements, 1143–1144
        process analyses, 1143
Industrial Energy Management Software, 880
Industrial energy management

3M company, 873
Colombia, 874
Comision Nacional deEnergia (CNE), 874
cost of, 874
Ecuador, 874
Florida, 874
fuel independence, 873
global trends, 872–881. *See also* Industrial energy management.
    acid rain, 873
    economic competitiveness, 873
    governmental energy management policies, 872
    ozone layer effects, 873
    pollution impact, 873
    U.S.A., 872
    U.S. Executive Order 12,123, 872
Japan, 873
Kuznets Curve prediction, 874
Latin America, 874
management support, 874
Peru, 874
secondary benefits of, 873
Walt Disney World, 873
Industrial energy management. *See also* Energy savings, Florida's fruit industry, case study.
Industrial facilities, electric demand response solutions for, 281–283
Industrial key performance indicators, development of, 1145
Industrial motor system optimization projects, in the U.S. *See* Motor optimization projects.
Industrial plants, energy efficiency and, 862–871
    complexity of control systems, 870
    lighting, 870
    monitoring of, 869–870
    pollution reduction, 870
    support of, 869–870
Industrial processes, differentiation, 458–459
    controls, 459
    energy requirements, 458–459
    technology requirements, 459
Industrial processes, energy conservation and, 458–465
    analysis, energy consumption analysis, 459–461
    distributed control systems, 459
    effective management, 464–465
    energy intensive, 461–464
    need and payback, 465
Industrial real time pricing, 1180–1182
Industrial thermal considerations, combined heat and power, 176
    steam vs. hot water, 176
Industrial use of energy in U.S., 584–585
    complexity of, 584
    Manufacturing Energy Consumption Survey, 584
Industrial wastewater, 1562
Industry classification, energy efficiency, links between, 868–869
Industry era, fossil fuels, 606–607
Industry standards, enterprise energy management systems and, 622–623

Industry use of coal, 160
Infiltration control, 662
　air barriers, 662–663
　vapor retarders, 662
Infiltration
　building system simulation process and, 120
　exfiltration vs., 844
Inlet supply air temperature, 1280
Innovation and design process
　LEED-C&S, rating system, 945
　LEED-CI, rating system, 941
　LEED-NC, rating system, 950
Innovations, coal production in United States and, 150–151
Input of energy, balancing of, 4
Input type controllers, 414
Installation, commissioning process and, 195–196
Insulated concrete form walls, 1517
Insulated concrete forms (ICFs). See Concrete form insulation systems.
Insulation blowing agents
　expanding polystyrene, 891–892
　extrusion polystyrene, 892
　hydrochloro-fluorocarbons (HFCs), 892
　open-cell polyurethane, 892
　pentane, 891–892
　polystyrene, 891–892
　polyurethane, 892
Insulation materials
　aerated concrete, 891
　blown foam, 897
　blown loose-fill insulation, 897
　bubble back, 895
　cellulose insulation, 890
　　effect on environment, 891
　closed-cell, high-density spray polyurethane, 891, 894
　comparison chart, 893–895
　concrete from systems, 895–896
　concrete products, lightweight, 896
　containing blowing agents, 891
　containing chlorofluorocarbons (CFCs), effect on environment, 891
　containing pentane, 891–892
　cotton, 893
　environment and, 891–892
　extrusion polystyrene (XPS), 891, 894
　fiberglass, 890, 893
　fiberglass, effect on environment, 891
　foam
　　exterior and interior, 895
　　products, effect on environment, 891
　foil-faced OSB, 895
　interior framed wall, 895
　loose-fill insulation, 897
　metal framing for, 898–899
　mineral wool, 890, 893
　　effect on environment, 891
　molded expansion polystyrene (MEPS), 890–891, 894
　open-cell, high-density spray polyurethane, 891, 894
　paperboard sheathing, 895
　perlite, 893
　polyisocyanurate, 891
　rating system, 890
　reflective insulation, 891
　rock wool, 890, 893
　　effect on environment, 891
　slag wool, 890, 893
　　effect on environment, 891
　strategies for use. See Insulation strategies.
　structured insulation panels, 897–898
　types of, comparison chart, 893–895
　wall, 2 x 4, 896
Insulation materials. See also Insulation blowing materials.
Insulation panels, structured, 897–898
Insulation strategies, 892
　critical guidelines, 892
　floor insulation, 892
　foam insulation, 892
　wall insulation, 892
Insulation, ceilings and roofs. See Ceiling and roof insulation.
Insulation, 1363
　exterior rigid fiberglass, 895
　regulation of, 1363
　R-factor, 1363
　types, 1363
　U-factor, 1363
Insulators, 358
Integrated building design system, 1401
Integrated coal gasification/combined cycle power plants, 1531
Integrated design, energy efficient buildings and, 525–526
Integrated dimmers, lighting control systems, 982
Integrated gasification combined cycle (IGCC), coal- and biomass-based, 906–912
Intelligent on/off devices, lighting control systems, 979
IntelliGrid^{SM}, 914–918
　advantages of, 915
　automation and, 917
　communication architecture, 917
　description of, 914
　energy resource distribution, 917
　EnergyPort, 918
　fast simulation modeling, 915
　power electronics-based controllers, 917
　power flow increases, 917
　value added services, 918
IntelliGrid1 supply system, 402–403
Intensity indicator, energy, 7
Inter surface longwave radiation, 116
Interactive access
　actionable information, 619–622
　application interface, 621–622
　calculated data, 621
　data organization, 620
　data presentation, 620
　enterprise energy management systems and, 619–622
　time and usability, 620
　data, 620–621
　user-specific information, 622
Intercity mobility, maglev and, 1021–1022
Interconnection standards, 866–867
Intercooler control, compressor air control systems and, 210
Interfacing systems, controllers and, 416
Interference using wireless, 1652
Interior coal basins, 157–158
Interior framed wall insulation, 895
Intermediate pressure cylinders (IP), 1382
Intermittent energy sources, 1268
Internal audits, 1030
Internal energy benchmark, 690
Internal environment control, building system simulation process and, 120
Internal rate of return (IRR), 967
International Energy Conservation Code (IECC), 438
International Organization for Standardization (ISO), 1028–1029
International performance measurement and verification protocol (IPMVP), 919
Internet based access, web-based compressor management and, 213
Interruptible demand response programs, 280
Interruptible electric rate schedule, 1508
Interruptible fuel rates, combined heat and power, 177–178
Interruptible rates, 1504–1505
Interruptible sales rates, natural gas and, 1507
Interruptible transportation rates, 1507
Interrupting devices, fault current, 703–704
Intervention, public policy rationale for, 1194
Inventions and Innovation group, DOE and, 509
Inventory reduction, lean manufacturing and, 470
Inverse heat transfer method, 1063–1066
　capacity calculations, 1065
　evaporator core identification, 1064
　heater core identification, 1065–1066
Inverted block rates, 1503
Inverters, 359
Investment analysis techniques, 930–936
　fundamentals of, 930–932
　　alternative metrics, 933
　　cash flows, 932
　　comparable value, 930
　　discount rate, 930
　　determination, 932
　　NPR and IRR
　　　cash flows, 935
　　　comparison, 933
　　simple payback period, 930–931
　　time value of money, 930
　　variable discount rates, 935
Investment grade audit, 567
Investments
　electricity producers and, 388
　energy efficiency and, 13
Investor owned utilities, 379–380
IOF vs. non-IOF facilities, study results
　capital spending vs. re-engineering, 886–887
　energy consumption data, 885
　individual projects, 887
　industries of the future, 885
　U.S. motor systems, 885
Iowa Energy Center, 512
IP. Intermediate pressure cylinders.
IPMVP options C and D, window films, 1626–1629

IPMVP. *See* international performance measurement and verification protocol.
IRR (internal rates of return), in motor optimization projects, 883–884
IRR cash flows, 935
IRR, NPR vs., 933
IRR. *See* internal rate of return.
Irradiance, 1148
Irradiation, 1148
Irreversibility, energy degradation and, 1168–1172
Irrigation water use, 527
ISM band radios, 1654
ISO. *See* International Organization for Standardization.
Isolated solar gain design, 1322

J

Java applets, 537
Java Server Pages, 537
Java Servlets, 537
Jet aircraft fuel, types, 26
Jet engine efficiency, fuel consumption and, 26
Jet engine maintenance, aircraft fuel consumption and, 28–29
Just in Time manufacturing, 468

K

Kelvin-Plank statement, 1170
 continuous cyclic process, 1170–1171
Kerosene based aircraft fuel, 25–26
Key performance indicators (KPI), distinguishing of, 1140–1141
Key performance measurements, 1143–1144
Key performance metrics (KPM), 465
Kilns, 464
Kirchhoff's Laws
 electric power transmission systems and, 360
 importance of, 360
 loop flow, 360
 parallel path flow, 360
 window emissivity calculations, 1621
Korean maglev system, 1025
KPI. *See* key performance indicators.
KPM. *See* key performance metrics.
Krypton, as insulating gas, 1619
Kuznets Curve
 illustration of, 874
 industrial energy management assessment, 874
Kyoto Protocol, 499
 allowance assignments, 141
 emission targets, 140
 flexibility mechanisms of, 141
 implementation of, 140–141
Kyoto treaty
 European Union countries adoption of, 13
 regulating of carbon dioxide CO_2, 11–12
 requirements of, 9–10
 United States Government and, 12–14

L

Lamp ignition, 831
Lamp life, 995–996
Lamp performance, 831
Lamp switching, lighting based demand response and, 277
LAN, 1654–1655
 wireless application vs., 1658
Land development policies, U.S. transportation energy use and, 588
Land use mix, 1450
Land use
 environmental policy and, 627
 impact on transportation, 1449–1451
  cumulative impact, 1450–1451
  density, 1450
  mix of, 1450
  roadway design, 1450
  site design, 1450
  transit service quality, 1450
Landfill cost, 1530
Landscape irrigation, 1590–1592
Large power electric rate schedule, 1508
Latent heat storage, 1413
Latent thermal energy, 1163
Laws, creation of, 1193–1194
Lead acid batteries, 1155
 compressed air energy storage as alternative to, 214
Leadership in Energy and Environmental Design (LEED), 200, 206
Leadership in Energy and Environmental Design for Commercial Interiors. *See* LEED-CI.
Leadership in Energy and Environmental Design for Commercial Interiors. *See* LEED. *See also* LEED-C&S, LEED-CI, LEED-NC.
Leadership in Energy and Environmental Design for Core and Shell. *See* LEED-C&S.
Leadership in Energy and Environmental Design for New Construction. *See* LEED-NC.
Leak detection, 1592–1593
Leak loss reduction, compressor air control systems and, 210
Leakage measuring, air, 213
Leakage, heat wheel performance factors and, 797
Leaks
 air, 212, 664–666
Leaky dampers, 521
Lean manufacturing
 development of, 467
 elements of, 467
 energy conservation and, 467–475
  traditional vs., 468–469
 5S's, 474
 impact on energy use, 469–474
  Andon signals, 473
  downtime, 472
  inventory reduction, 470
  overage reduction, 471
  production flexibility, 473
  single piece flow, 470
  training, 470–471
  visual factory, 473
  workmanship, 470–471
  workplace organization, 474
 Just in Time, 468
 overage reduction, 471
 quality assurance, 470–471
 Toyota, 467, 468
Leasing
 ESCO and, 571–572
 types, 571–572
  capital, 571–572
  operating, 571–572
Least cost planning, 1500
LEED
 discussion of, 945–946
 USBBG and, 945–946
LEED. Leadership in Energy and Environmental Design.
LEED-C&S
 credit categories, 941–945
 energy and atmosphere, 943–944
 indoor environmental quality, 944
 innovation and design process, 945
 materials and resources, 944
 sustainable sites, 942
 water use efficiency, 943
LEED-CI, 937–941
 benefits of, 938
 certification, 938
 energy and atmosphere, 939
 indoor environmental quality, 940
 innovation and design process, 941
 materials and resources, 940
 overview, 937, 938
 point distribution, 938
 rating formats, 938–940
 site selection, 939
 sustainable sites, 939
 technical review, 938–941
 vs. C&S, 937
 water use efficiency, 939
LEED-NC, 947–952
 green buildings, 948
 purpose of, 947–948
 rating system, 948–949
  energy and atmosphere, 950
  indoor environmental quality, 950
  innovation and design process, 950
  materials and resources, 950
  prerequisite categories, 949–950
  sustainable sites, 949
  water use efficiency, 949
Legal issues, net metering and, 1096
Legislation, adoption of energy codes, 441
Lenders, independent power producers and, 857
Level of irreversibility, 1428
LHV. *See* lower heating value.
Liability rules, environmental policy instruments and, 630
Life cycle analysis (LCA). *See* Energy projects, life cycle costing; Life cycle costing.
Life cycle analysis, heat wheels and, 798–799
Life cycle assessment, solar water heating and, 1337–1338
Life cycle cost analysis, 967–968
 calculators and models, 975–976
 uses of, 972–974
  lighting retrofit example, 974–975

Life cycle costing (LCC)
  definition of, 953–954. *See also* Electric power projects, life cycle costing; Energy projects, life cycle costing.
  electric power projects
    analysis of 3–KWP PV system, 965
    annual electricity generation, 955–957
    assumption and input parameter, 964
    calculation methodology, 955
    capital and noncapital costs, 956–957
    coal power plant, 954
    coal vs. wind, 957
    cost components of, 954–955
    cost of electricity, 959
    depreciation, 957
    50–MW wind project, 965
    financing charge, 957–958
    520–MW coal power project, 965–966
    interest during construction, 958
    loan repayment, 959
    photovoltaic power system, 954
    project cost, 958
    renewable energy systems incentives, 959–961
    taxes and duties, 959
    variable cost, 958
    wind power plant, 954
    working capital interest, 958–959
Life cycle costing
  EEMS and, 622
  energy projects
    antecedents of, 968
    blended utility rate, 967
    btu/sf, 967
    definitions, 967–968
    energy efficiency measure (EEM), 967
    framework of, 969–972
      costs and benefits, 971–972
      definition, 970–971
      gas station example, 969
      HVAC example, 970
      situation, 969–970
    internal rate of return (IRR), 967
    LED exit lights example, 970–971
    net present value (NPV), 968
    payback period, 968
    time value of money, 968
    total owning costs, 968
  heat wheels and, 798
Life expectancy, electronic ballast primer and, 833
Light level uniformity, 993
Light measurements, data collection and, 257–258
Light meters, 258
Light shelves, as advanced window shading device, 1642–1643
Light switches, manual, 522
Light water nuclear reactors, 1115–1116
  history of, 1115–1116
  locations of, 1115–1116
Lighting based demand response, 276–277
  dimming, 277
  evaluation of, 277
  fixture/lamp switching, 277
  zone switching, 277
Lighting contractors, 979

Lighting control panel, 981
Lighting controls, 977–992. *See also* Lighting design and retrofits.
  adaptability, 986
  application issues, 986–991
  automatic, 978–979
  commissioning, 990–991
  common strategies, 977
    daylight harvesting, 978
    lumen depreciation, 978
    on/off, 977, 979
    scheduling, 977, 979
    tuning, 977
  complexity, 986
  construction documents, 990
  cost-effectiveness, 985–986
  coverage patterns, 990
  design controls, 986
  dimming ballasts, 982
  dimming system, 983, 984
  ease of use and maintainability, 990
  electrical design, 989
  energy codes, 986
  equipment location, 990
  evaluating options, 985–986
  flexibility, 986
  fluorescent dimming ballast, 984
  HID dimming ballast, 985
  installation, 990
  integrated dimmers, 982
  intelligent on/off devices, 979
  interoperability, 986
  layering control systems, 989
  light levels, 990
  lighting contractors, 979
  local wall switches, 979
  low-voltage controls, 980–981, 982
  maintainability, 986
  microprocessor-based, centralized, 984–985
  occupancy sensors, 981, 982
    coverage patterns, 990
  product selection, 989
  programmability, 986
  programmed-start ballasts, 982–984
  questions to ask vendors, 991
  reliability, 986
  security, 990
  selection table, 987–989
  step-level HID controls, 982
    schematic, 982
  switching vs. dimming, 989
  system dimmers, 982
  typical controls, 978–985
  user education, 991
  utility rebates, 989
  voltage control, 989–990
  wallbox dimmers, 982
Lighting costs, activity based costing and, 1044
Lighting design and retrofits, 993–1000. *See also* Lighting controls.
  case study, Macy's/Brooklyn, 999–1000
  considerations, 993
    color quality, 993
    eye anatomy, 994
    glare, 993
    human eye functioning, 993–995
    retinal rod and cone density chart, 994

    uniformity of light levels, 993
  cost calculations, 997–998
  definitions, 998–999
    candela, 998–999
    color preference index, 999
    color rendering improvement, 999
    color temperature, 999
    footcandle, 999
    footlambert, 999
    lumen, 999
    lux, 999
    mesopic vision, 999
    photopic vision, 999
    scotopic vision, 999
  dimming, 996
  dirty area systems, 996
  emergency lighting, 998
  goal, 993
  induction systems, 996
  lamp life, 995–996
  linear T5 lamped systems, 996
  motion sensors, 995
  scotopic CIE, 995
  scotopic/photopic vision sensitivity chart, 995
Lighting Research Center of Rensselaer Polytechnic Institute, 512
Lighting retrofits, 863
Lighting
  accounting for in energy assessment, 70
  energy efficiency and industrial plants, 870
  facility guidelines for new structures and, 528–529
Light-weight autoclaved concrete, as insulation material, 891
Lignite, 156
Line management accountability, energy master planning success and, 555
Linear induction motor maglev system, 1023
Linear T5 lamped systems, 996
Liquefaction, liquefied natural gas and, 1001–1002
Liquefied natural gas (LNG), 1001
  description of, 1001
  history, 1001
  liquefaction, 1001–1002
  regasification, 1004–1006
  storage of, 1002–1004
  transportation of, 1006–1008
  vaporization, 1004–1006
Liquid fuel production, 1354
Liquid fuel properties, 485
Liquid ring vacuum pumps, 464
Living standards and culture
  definitions of, 1009
  energy and development, 1014–1015
  energy and society, 1014, 1016
  energy use modifications, 1015
  energy-resource utilization data, 1010
  energy-use growth patterns, 1010–1012
  impact of energy use on, 1013–1015
  population and energy use, 1010
LNG. *See* liquefied natural gas.
Load calculation, geothermal heat pump system development and, 757
Load control rates, 1505
Load curtailment demand response programs 280–281

Load factor adjustments, energy balance and, 687–688
Load factor for electricity, dispatch stack order, 1499
Load factor
    measurement tool, 1498–1499
    specific customer profiles and effect on, 1499–1500
    utility rate structure and, 1498–1500
Load inventory, energy, 4
Load management rates, 1505
Load management utility rates, 1505
Load profile, cool storage system design and, 1418
Load scheduling, compressor air control systems and, 210
Load shape changes, energy demand-side management programs and, 287
Load shape objectives, energy demand-side management programs and, 288
Load usage
    cogeneration applications, 1482
    multiple unit cooling applications, 1483
    peak shaving, 1482
    single unit seasonal cooling appliances, 1483
    weighted average cost of power and, 1478–1482
Loads, building system simulation process and, 119
Loan repayment, 959
Local adoption of energy codes, 440–441
Local area augmentation system, global positioning satellites and, 29
Local wall switches, lighting control systems, 979
Location efficiency
    benefits and costs, 1453
    best practices, 1453
    impact on equity, 1453
    implementation of, 1451–1453
    transportation and, 1449–1453
Long term power purchase, 855
Longwave radiation, inter surface, 116
Loop flow, 360
Loose-fill
    attic insulation, 900–901
    insulation, blown, 897
Losses of power, 360–361
Louvers, as window shading device, 1642
Low cost energy efficient improvements, 517–523
    case studies, 517–523
        ambient temperature, 521–522
        boiler shut off in summer, 519
        cost accounting monitoring, 522–523
        dedicated AC units, 521
        fan speeds, 521
        leaky dampers, 521
        manual light switches, 522
        night setback on time clocks, 518
        review
            HVAC capacity, 519–520
            scheduling, 518
            summer-winter cutover, 520
        scheduling and building usage, 518–519
        thermostat location, 520
        cases, 520–521
    payback analyses, 517
Low pressure boilers, 93
Low pressure cylinders (LP), 1382
Low rise buildings, 1513
Low solar gain low-e coating, for windows, 1621–1622
Low temperature fuel cells, 733–734
    alkaline, 734–736
    direct methanol, 739–743
    proton exchange membrane fuel cells, 726
    summary, 742
Low voltage controls, lighting and, 980, 981–982
Low-e glass coatings, for windows, 1621
Lower heating value (LHV), 86–87
Low-voltage controls, lighting control systems, 980–981, 982
LP. Low pressure cylinders.
Lumen depreciation, 978
Lumen, 999
Lurgi gasifier, 906
Lux, 999

M

M & V. *See* measurement and verification.
M & V. *See* monitoring and verification.
M T & R. *See* monitoring, targeting and reporting energy analysis technique.
Machine to machine communications, 108–110
    XML based, 108–110
Machinery, compressor air control protection systems, 209
Macro indicators, technology transfer and, 502–503
Maglev transportation system, 591
Maglev (magnetic levitation) system, 1018–1026
    aid to intercity mobility, 1021–1022
    as transportation system, 1018
    benefits of, 1025
    case for building, 1019–1020
    dedicated right of way, 1022
    impact on society, 1018–1020
    politics and, 1020–1022
        Amtrak, 1020
        Federal Aid Highway Act of 1956, 1021
        federally funded airport construction, 1020
        reduce energy consumption, 1019–1020
    systems in use, 1024–1025
        Korea, 1025
        Nagoya, 1025
        Shanghai Airport, 1024–1025
    technical issues, 1022–2024
    types of, 1018
        electromagnetic suspension system, 1023
        ferrous electromagnets, 1022–1023
        linear induction motor, 1023
        superconductor magnets, 1024
    use in United States, 1025
Magnetic levitation. *See* maglev system.
Main meter M&V procedures, 1051
Maintenance guidelines
    facility guidelines for new structures, 533
    fan system energy management, 1224–1225
    Maintenance Best Practices Guide and, 710
    pumps and energy management, 1218
Maintenance electric rate schedule, 1509
Maintenance programs, Operations and Maintenance Best Practices Guide and, 710
Management, energy management system problems
    ignorance, 1035–1036
    reviews, 1030
    support, energy master planning and, 552
Management system for energy (MSE), 551
    current standards, 1032–1035
        ANSI/MSE 2000:2005, 1032–1034
    defining, by documents, 1030
    definition of, 1027–1028
    energy and, 1027–1041
    energy standards, evolution of, 1031–1032
    key elements, 1029–1030
        corrective action, 1030
        defined management reviews, 1030
        documented processes, 1029
        internal audits, 1030
        ongoing training, 1029–1030
        preventive action, 1030
        responsibilities, 1029
    plan do check act cycle, 1027, 1028
    standardization process, 1028–1031
        American National Standards Institute (ANSI), 1028–1029
        International Organization for Standardization (ISO), 1028–1029
Management systems, standards
    accreditation of, 1030–1031
    American National Standards Institute (ANSI), 1028–1029
    ANSI/MSE 2000:2005, 1032–1034
    certification of, 1030–1031
    current standards, 1032–1035
    defining by documents, 1030
    International Organization for Standardization (ISO), 1028–1029
    registration of, 1030–1031
    types, 1028
Manual light switches, 522
Manufacturing Energy Consumption Survey (MECS), 584
Manufacturing, lean vs. traditional, 468–469
Manure and coal
    cofiring and, 490–491
Marginal cost of abatement, 433
Market driven discrete usage charge, utilities and, 1500
Market implementation methods, energy demand-side management programs and, 289–290
Masonry walls, 1516–1518
    adobe, 1517–1518
    concrete block, 1516
    insulated concrete forms, 1517
    poured concrete, 1516–1517
    precast concrete, 1517
Material and energy balance (MEB) concept, 1303–1304
Material flows data, 650

Materials and resources
  LEED-C&S, rating system, 944
  LEED-CI, rating system, 940
  LEED-NC, rating system, 950
MCFC. *See* molten carbonate fuel cells.
Mean shift, Six Sigma Methods and, 1313
Measure the process, Six Sigma and, 1313–1314
Measured collector performance, solar water heating and, 1336
Measurement and verification (M&V), 567, 572
Measurement and verification (M&V) procedures, 1050–1055
  options available, 1051–1052
    end use retrofit isolation, 1051
    main meter, 1051
    partially measured retrofit isolation, 1051
    whole meter, 1051
  process of, 1052
    establishing baseline, 1053
    periodic status reports, 1054
    plan development, 1052–1053
    post implementation report, 1054
    preconstruction assessment, 1052
  theory of, 1051
Measurement of energy, 2
  tools for, 1056–1060
Measurement protocol, steam system optimization and, 1378–1379
Measurement tools, energy management and, 1056–1060
Measuring emissions, 431–432
MEB. *See* material and energy balance.
MEC. *See* model energy code.
Mechanical aeration, 1549
Mechanical draft cooling towers, 247–248
Mechanical kinetic energy, 1163
Mechanical processes, energy intensive, 463
Mechanical vacuum pumps, 1370–1371
MECS. *See* Manufacturing Energy Consumption Survey, 584
Medicine, heat pipe applications and, 810–811
Medium pressure boilers, 93
Melting, 461
Membrane technologies, high-temperature, gasifiers and, 909
MEPS (molded expansion polystyrene), as insulation material, 890–891, 894
Merchant power producers, 855
Mercury, as syngas contaminant, 909
Mesh network, 1653–1654
  ISM band radios, 1654
  multiple routing paths, 1653
  self configuring nodes, 1653
  spread spectrum radios, 1653–1654
Mesopic vision, 999
  definition of, 999
Metal carbonyls, as syngas contaminants, 909
Metal casting plant assessments, Dept. of Energy best practices case studies, 1284, 1293–1294
Metal framing, for insulation materials, 898–899
Metal hydrides, 1446
Metal industry
  aging thermal processes, 868
  aluminum alloy artificial aging, 868
  billet heating, 868
  die temperature, 868
  energy saving opportunities, 868
  equipment modernization and, 868
  solution heat treatment, 868
Meter, energy balance data compilation and, 686
Meter level, wireless applications and, 1650, 1652
Meter Manager software, 3
Metering, 691
  applications of, 712
  importance of, 712
  net, 1096–1100
  planning, 713
Metering point, 1511
Meters
  electric, 283, 392
  energy data collection and, 618
Methane gas recovery, 1554
Methanol, 33
Methyl-tertiary butyl ester. *See* MTBE.
Micro indicators, technology transfer and, 502–503
Microcontrollers, electronic ballast primer and, 834
Micropressors, 208
Microprocessor-based, centralized lighting control systems, 984–985
Microprocessors, automobile technology evolution and, 673–674
Microturbine power generation, 1304
Microturbines, 298
Milling, uranium and, 1112–1113
Mineral wool
  effect on environment, 891
  as insulation material, 890, 893
Minimum charge, utility billing and, 1501
Mining, uranium and, 1112–1113
Mining methods, 158–159
  surface, 158–159
  underground, 158
Mining plant assessments, Dept. of Energy best practices case studies and, 1284–1285, 1294–1296
Mining technology trends, 146–147
Mobile HVAC systems, 1061–1069
  accumulator, 1079
  airside heat transfer calculations, 1062–1063
  capacity calculations, 1061–1066
  coils design, 1061–1066
  components, 1076–1081
    electrical systems, 1080–1081
  components evaluation, 1066
  compressors, 1076
  condenser, 1077
  configurations, 1075–1076
  evaporator, 1079–1080
  expansion valve, 1077–1078
  history of, 1072–1073
  innovations to, 1066–1069
  inverse heat transfer method, 1063–1066
  nomenclature, 1070–1071
  orifice tube, 1078–1079
  physics of, 1073
    evaporating process, 1073–1074
    heating process, 1074–1075
  pressure switches, 1080
  psychrometric fundamentals, 1061–1066
  receiver/dryer, 1077
  testing procedures, 1061–1066
  thermostats, 1080
  TXV system vs. OT, 1081
  tube-side heat transfer calculations, 1063
Mode transitions, global temperature adjustment and, 274
Model energy code (MEC), 438, 439
  adoption process, 440–441
    groups involved, 440
    timing of, 440
  development and revision of, 440
  International Energy Conservation Code (IECC), 438
  state's right in revising, 439
Model validation, fast simulation modeling and, 916
Modeling
  building geometry and energy use, 112–113
  geometric ratio, 113
Modeling
  energy monitoring performance, 5–6
  IntelliGrid^{SM}, 915
Modes, controller, 415
Modular boilers, control of, 102
Modular buildings, 677
  water source heat pump for, case study, 1564–1573
Modulating controller, 415–416
MOIST software, 1403
Moisture, biomass energy and, 87
Moisture content, of syngas feedstocks, 907
Moisture transfer processes, 117
Mold
  ANSI/ASHRAE Standard 62.1–2004 and, 53–54
  growth, 839, 844, 845, 846
  prevention by pressurization flow, 54
Molded expansion polystyrene (MEPS), as insulation material, 890–891, 894
Molten carbonate fuel cells (MCFC), 727–729
  actual use of, 729
  applications, 729
  contamination levels, 728
  operating principle of, 727–728
  problems with, 728–729
Monitoring and verification (M & V), 179
Monitoring of energy, 5–6
Monitoring, targeting and reporting (M T & R) energy analysis technique, 4–7
  definitions of, 5
  reporting, 7
  targeting, 6–7
Motion sensors, 995
  lighting design and retrofits, 995
Motor belts and drives, cost calculations and, 73–74
Motor control
  compressed air and, 207–208
  primary, 208
  ride through, 208
Motor load factors, energy cost savings and, 73
Motor Master (MM+4.0), 879–880
Motor optimization projects, 883–889
  capital spending vs. re-engineering, 886–887

drivers of, 888
empirical measures, 883–884
energy consumption data, 886
indirect impacts, 887–888
individual project results, 887
industry-wide energy savings potential, 888
internal rates of return (IRR), 883–884
IOF vs. non-IOF facilities, 885
methodology, 883
net present value (NPV), 883–884
simple payback results, 883–884
study results, 41
U.S. motor systems, 884–885
Motor system effectiveness, as driver in motor optimization projects, 888
Motor usage, accounting for in energy assessment, 70–71
Motors, high efficiency, cost calculations and, 73
Motors and drives, facility guidelines for new structures and, 529
Moving bed gasifiers, 906
Lurgi gasifier, 906
MSE. See management system for energy.
MTBE (methyl-tertiary butyl ester), 33
Multiple boilers, control of, 98–102
functional description of, 100–102
modular, 102
Multiple pump theory, 1217
Multiple routing paths, 1653
Multiple unit cooling applications, 1483
Multiple year monthly average comparison method, 2
Multiple zone recirculating system, 57
Multiresolution modeling, 916
Municipal code, energy code adoption and, 441

N

NAAQS. See National Ambient Air Quality Standards.
Nagoya maglev system, 1024–1025
National Ambient Air Quality Standards (NAAQS), 430
National Energy Act of 1978, 1082–1086
components of, 1083–1086
Energy Tax Act of 1978, 1084
Fuel Use Act, 1085
National Energy Conservation Policy Act, 1085–1086
Natural Gas Policy Act of 1978, 1083
Public Utility Regulatory Policies Act, 1084, 1201–1205
National Energy Conservation Policy Act, 1085–1086
National Institute of Standards and Technology (NIST), 510–511
National Technical Information Service (NTIS), 510
Natural capital, 1410
Natural carbon sinks
biological sequestration and, 129
ocean fertilization, 129
terrestrial, 129
Natural draft cooling towers, 247–248
Natural energy, 1089–1090
additional energy, a comparison, 1088–1094

types, 1089–1090
nonsolar related, 1090
solar related, 1090
solar, 1089–1090
Natural environment subsystem modeling, exergy analysis and, 656–657
Natural gas issues, coal to liquid fuel plants and, 170
Natural Gas Policy Act of 1978, 1083–1084
curtailment, 1084
emergency authority, 1084
incremental pricing, 1083–1084
transportation, 1084
wellhead pricing, 1083
Natural gas
chemical process industry and, 1305
coal vs., 161
distributed generation technology and cost estimating, 299
Natural gas rate schedules, 1507
air conditioning, 1507
congeneration, 1507
dual fuel firm rates, 1507
end use, 1507
firm sales rates, 1507
firm transportation rates, 1507
interruptible sales rates, 1507
interruptible transportation rates, 1507
Natural nuclear reactor, 1111
Negotiated rates, 1505–1506
cost based, 1505–1506
NERC (North American electricity reliability council), 361
Net emergy
energy sources, 426
time and, 426
Net energy, 424–426
Net metering, 1096–1100
description of, 1096
implementation of, 1097–1098, 1100
by state, 1097–1100
technologies eligible for, 1098–1100
use of, legal and policy problems, 1096
Net present value (NPV), 968
calculation, 84–85
capital rationing, 84–85
in motor optimization projects, 883–884
Networked capacity control, compressor air control systems and, 209–210
New building commissioning, 188–199
New process capability, 1316
New York State Energy Research and Development Authority, 512
Nickel cadmium batteries, 1155
NIST. See National Institute of Standards and Technology.
Noise level, of wind power, 1612
Non capital costs, 956–957
Non combustion processes
gasification, 1525, 1527
waste fuels and, 1525, 1526
Non firm service rates, 1504–1505
interruptible, 1504–1505
load control, 1505
load management, 1505
standby, 1505
Non renewable energy, 1090–1091

Non renewable energy sources, unit emergy values and, 425
Non-fossil energy sources, global warming and, 724–725
Nonsolar related energy, 1090
Nonuse benefits, environmental policy instruments and, 631
Nonutility generators, electricity deregulation and, 385–386
Normalized energy benchmarks, 690–692
Normalized performance indicator (NPI), 4
Normalized time intervals, 619
North American coal to liquid fuel plants, 167
North American electricity reliability council. See NERC.
Northern Cyprus, 503–504
NO_x emissions, cofiring and, 491
NPI. See normalized performance indicator.
NPR, IRR vs., 933
NPR cash flows, 935
NPV (net present value), in motor optimization projects, 883–884
NTIS. See National Technical Information Service.
Nuclear energy, 38–43
accidents, 1131–1132
competitive cost of, 39
development challenges, 41–43
long term storage, 43
politics, 41–43
safety, 42
spent fuel management, 42–43
waste management, 42–43
disposal of, 1113–1114
economic issues
alternative fuel cost assumptions, 1105
modeling results, 1105–1109
economics of, 1101–1110
capital costs, 1101–1103
discount rate, 1104–1105
environmental issues, 1103–1104
waste disposal, 1103–1104
exposure to, 1132–1133
fuel cycles, 1111–1114
fuel operation, 1113–1114
future challenges, 41
uranium availability, 41
waste product disposal, 41
future technologies, 41
helium cooled very high temperature gas reactor, 41
supercritical water-cooled reactor, 41
Generation IV program, 1121–1123
generation types, 40–41
history of, 38–39, 1125–1127
pebble bed modular reactor, 40–41
power plant types, 1115–1124
boiling water, 1117–1118
fast breeder, 1120–1121
future of, 1121–1124
future of, fusion, 1123–1124
gas-cooled, 1119–1120
light water, 1115–1116
pressurized heavy water, 1118–1119
pressurized water, 1116–1117
principles of, 39
reactor types, 39–41

advanced gas-cooled, 40
boil water, 39–40
fast neutron, 40
pressurized heavy water, 40
pressurized water, 39
technology of, 1125–1133
atomic structure, 1127
binding energy, 1127
chain reaction, 1128–1129
fission, 1128
fusion, 1130–1131
radioactivity, 1127–1128
reactor design, 1129–1130
safety, 1131
waste disposal, 1103–1104
Nuclear energy, challenges of, 41–43
uranium availability, 41
long term storage, 43
politics, 41–43
safety, 42
spent fuel management, 42–43
waste management, 42–43
waste product disposal, 41
Nuclear energy power plant types, Generation IV program, 1121–1123
Nuclear power, 394–395
Nuclear power plant types, 1115–1124
boiling water, 1117–1118
fast breeder, 1120–1121
future of, 1121–1124
fusion, 1123–1124
gas-cooled, 1119–1120
light water, 1115–1116
pressurized heavy water, 1118–1119
pressurized water, 1116–1117
Nuclear power, technology of, 1125–1133
atomic structure, 1127
binding energy, 1127
chain reaction, 1128–1129
fission, 1128
fusion, 1130–1131
radioactivity, 1127–1128
reactor design, 1129–1130
Nuclear reactor types, 39–41
advanced gas-cooled, 40
boil water, 39–40
commercially in use, 40
fast neutron, 40
natural, 1111
pressurized heavy water, 40
pressurized water, 39
uranium history, 1112
Nuclear reactors, commercially in use, 40
Numerical methods, building system simulations and, 118

O

O&M BPG. *See* Operations and Maintenance Best Practices Guide.
OASIS. *See* open access same-time information system.
Occupancy sensors, 981, 982
lighting control systems, 981, 982
coverage patterns, 990
Occupational Safety and Health Act (OSHA), 1303

Ocean carbon sinks, 129
Ocean currents, renewable energy and, 1267–1268
Ocean direct injection of captured CO_2, 128
Ocean sources, renewable energy and, 1267
Ocean thermal differences, renewable energy and, 1267–1268
Odum, H.T., 420, 421
Off peak
energy cost calculations, on-peak vs., 71–73
heat storage, 1420
utility load, 1497–1498
Off season rates, time of use and, 1504
Off taker, independent power producers and, 856
Office of Electricity Delivery and Energy Reliability, 509–510
Office of Energy Efficiency and Renewable Energy (EERE), 508–509, 873
Office of Fossil Energy, 509
Office of Industrial Technologies (OIT), 873
Office of Scientific and Technical Information, 510
Offset emissions trading, 431
Offset policy, 430
bubble, 430–431
Oil companies, nationalization of, 595
Oil consumption in United States, 595
Oil exploration, Artic Wildlife Reserve and, 595
Oil in ground, uncertainty of, 593
Oil production vulnerabilities, impact of, 592–593
Oil reserve uncertainty, 592
Oil shale deposits, 594
disadvantages of, 594
Oil tankers, impact on U.S. energy, 591–592
Oil, U.S. over reliance on, 596–597
OIT (Office of Industrial Technology), 873
On peak energy cost calculations, off-peak vs., 71–73
Open access same-time information system (OASIS), 362–363
Open-cell, high-density spray polyurethane, as insulation material, 891, 894
Open-cell, polyurethane, as insulation blowing agent, 892
Operating expenses, 1256
Operating leasing, 571–572
Operating voltage, electric power losses and, 361
Operational controls, automobile technology and, transferable to building use, 672
Operations and Maintenance Best Practices Guide (O&M BPG), 707–714
commissioning existing buildings, 710–712
definitions, 711
findings from, 711
process of, 711–712
computerized maintenance management systems, 710
definitions, 708
energy savings from, 708
important elements
contracting outside services, 709–710
management support, 709
measuring program quality, 709
program implementation, 709

maintenance programs, 710
major equipment, 713
management practices, 708–710
metering, 712–713
new technologies, 713
operational efficiency, 713
predictive maintenance, 710
purpose of, 707–708
Opportunity fuels, 1523–1525, 1526, 1530
Optimal time intervals, 619
Options, automobile technology and, transferable to building use, 673
Order, 888, 380
Order destruction, exergy and, 659
Organic materials, thin film PV systems and, 1150
Organic matter, biomass energy and, 31–34
Orifice plate management, flow measurement problems and, 1057
Orifice tube, mobile HVAC systems and, 1078–1079
OSHA. *See* Occupational Safety and Health Act.
OT system, TXV vs. mobile HVAC systems and, 1081
Outdoor air quality
ANSI/ASHRAE Standard 62.1–2004 and, 51
outdoor airflow rates, 51
Outdoor airflow rates, 51
constant volume of supply air, 51
variable air volume, 51
Output controller, 415
Output devices, controllers and, 415
Output of energy
balancing of, 4
load inventory, 4
Outside air, value of re indoor air quality, 20
Ovens, 809
Overage reduction, lean manufacturing and, 471
Overhead, activity based costing and tracing of, 1042
Overheating, use of spectrally selective window films, 1637
Owner, independent power producers and, 856
Ozone layer, 873
Ozone, 1559

P

Packing type, cooling towers and, 249–250
PAFC. *See* phosphoric acid fuel cells.
Pakistanian coal to liquid fuel plants, 167
Panel cooking, solar, 1326
Paperboard sheathing, as insulation material, 895
Parabolic cooking, solar, 1326
Parabolic dish, 1328
Parabolic trough, 1328
Parallel combustion, 89
Parallel flow heat exchanger, 802
Parallel path flow, 360
Parallel quality, 421–422
Partially measured retrofit isolation, 1051
Particle size, syngas feedstocks, 908
Part-load efficiency, district cooling systems and, 314
Passive solar heating, windows and, 1618–1619
Passive space heating, 1322

direct solar gain design, 1322
indirect solar gain design, 1322
isolated solar gain design, 1322
Payback analyses, low cost energy efficient improvements and, 517
Payback period, 968
Payback, heat wheels and, 798
Payment structure, independent power producers and, 857–858
PCM ice storage, 1414
Peak air demand, calculating of, 231–232
Peak load management, electric demand response and, 272
Peak shaving, 1482
Peak usage, utility billing and, 1501
Peak utility load, 1497–1498
Pebble bed modular reactor, 40–41
PEMFC. *See* proton exchange membrane fuel cells.
Penalties, emissions trading and, 433
Pennsylvania, electricity deregulation results in, 385
Pentane (polystyrene, expanding)
  insulation blowing agent, 891–892
  smog and, 891
Perfect Power System, 404–405
  goals of, 404–405
Performance contracting, 1134–1139
  benefits of, 1138–1139
  chauffage, 1136
  contemporary usage, 1135
  ESCOs, 1134–1135
  history of, 1134–1135
  types, 1136
    guaranteed savings, 1136–1137
    shared savings, 1136–1137
    workings of, 1135–1138
      deal structuring, 1136
      financial modeling, 1136–1137
      model comparison, 1137
      value chain management, 1137–1138
Performance curves, fan systems and, 1223–1224
Performance indicators, energy and, 4
Performance measurement, fan systems and, 1222–1223
Perl, 537
Perlite, as insulation material, 893
Permanent pressure loss, flow measurement and, 1058–1059
Permit allocation, emissions trading and, 432
Permits
  emissions trading and, 432
  tradable emissions, 630
Personnel skills, energy project management and, 564–565
PES. Primary energy supplies.
Petroleum based products as heating fuels, 1361
Petroleum fuels, end of, 1407
Petroleum plant assessments, Dept. of Energy best practices case studies and, 1285, 1296–1299
Petroleum, distributed generation technology and cost estimating, 299
Phase equilibrium, 1431
Phone line, 1655

wireless application vs., 1658
Phosphoric acid fuel cells (PAFC), 726–727
Photometry, 264
Photo-oxidants, air quality modeling and, 719–721
Photopic vision, 999
  definition of, 999
  sensitivity chart, 995
Photovoltaic electricity, 1267
Photovoltaic power system, 954
Photovoltaic (PV) systems, 1147–1157
  pplications of, 1156–1157
  backup to, 1152–1153
  costs of, 1153
  crystalline SI, 1148–1149
  electronic usage, 1154–1155
  energy storage
    hydrogen, 1155–1156
    lead acid batteries, 1155
    nickel cadmium batteries, 1155
  full solar spectrum usage, 1150–1151
  history of, 1147–1148
  power electronic converters, 1154
  power plant usage, 1151–1153
  satellites, 1147
  solar resources, 1148
  stand alone, 1153
  sunlight concentration, 1151
  thin films, 1149–1150
Physical plant, energy assessments and, 67
Physics of energy, 1160–1173
Piloting techniques, aircraft fuel consumption and, 28
Pinch analysis, 460
Piston devices, reciprocating, 1233–1234
Plan do check act cycle (PDCA), 1027, 1028
Plant construction, independent power producers and, 860
Plant expansion, as driver in motor optimization projects, 888
Plant interaction with building zones, 117
Plant locations, coal to liquid fuels and, 168
Plant operation, independent power producers and, 860
Plant safety, compressor air control systems and, 209
Plant size, coal to liquid fuel plant, 169
Plant valuation, rate base determination and, 1254
Plate and frame heat exchanger, 1540
PLC. *See* programmable logic controllers.
Plenum leakage, 1468
Plug flow, 1561
Plug in configuration, hybrid electric vehicles and, 847–853
Plug in hybrid electric vehicles, greenhouse gas emissions and, 791–792
Plug in hybrid vehicles, CO_2 emissions and, 788–789
Plumbing, facility guidelines for new structures and, 533
Pneumatic capacitance, 229–230
Pneumatic compressor control, 208
Pneumatic transport, chemical process industry and, 1305
Policy instruments, environmental policy and, 630–631

Politics
  nuclear energy development and, 41–43
  U.S. transportation energy use and, 594–595
Pollutants, water, 746
Pollution abatements, cost minimizing, 434–435
Pollution reduction, 870
Pollution sources
  carbon dioxide, 20
  indoor air quality and, 20–21
  volatile organic compounds, 20
Pollution summaries review, energy assessments and, 64
Pollution tax, 630
Polyisocyanurate, as insulation blowing agent, 891, 892
Polystyrene
  expanding (pentane), as insulation blowing agent, 891–892
  extruded, as insulation blowing agent, 892
Polyurethane
  closed-cell, high-density, as insulation material, 891, 894
  insulation blowing agent, 892
  open-cell
    high-density, as insulation material, 891, 894
    insulation blowing agent, 892
Ponds, solar, 1324–1325
Population growth, developing countries and energy demand, 500
Population, energy use and, 1010
Portland Energy Conservation, Inc., 515
Positive displacement pump, 1213–1214
  theory, 1216
Post combustion capture, 126
Post fossil fuel era, 608–610
Poured concrete walls, 1516–1517
Power capacity needs, reducing in developing countries, 503
  energy saving opportunity gap, 503
  excess energy consumption gap, 503
Power circuit supply, 416
Power cycle efficiency, 1424–1426
Power distribution systems, 699–706
  basic principles, 699–701
    balanced system, 700–701
    single phase system, 699–700
    three phase system, 699–700
    unbalanced system, 700–701
  design guidelines
    conductor sizing, 701–702
    grounding, 701
    voltage levels, 702
  fault current, 703–704
  system protection, 702–703
    design guidelines, 702
    ground fault detection, 703
    power system faults, 702–703
Power electronic converters, 1154
  switch mode, 1154
Power electronics based controllers, 917
Power factor, 255–256, 706
  apparent power volt amperes, 1510
  demand billing, 1510–1511
  real power watts, 1510
  volt amperes reactive, 1510
Power flow increases, 917

Power generation, solar thermal technologies and, 1327–1328
Power losses
    regional effects of, 359
    electric power transmission systems and, 360–361
Power plant types, nuclear energy and, 1115–1124
Power plants, photovoltaic (PV) systems and, 1151–1153
Power quality analyzers, 256
Power quality issues, 1484–1490
    cost tracking, 1489
    customer forums, 1489
    Distribution Power Quality (DPQ), 1484
    education re, 1490
    power sags, 1484–1487
        minimizing, 1489–1490
    solutions to, 1490
    workshops, 1489
Power rating, wireless terms and, 1652
Power requirements
    fan systems and, 1223
    pump theory and, 1216–1217
Power sags, 1484–1487
    customers response to, 1487–1488
Power sources
    alternative, government subsidies of, wind power 1614–1615
    wind. *See* Wind power.
Power supply, electric motors and, 352
Power system faults, 702–703
Power usage determination, pump theory and, 1217
Powered attic ventilator, 900
Pre flight planning
    aircraft fuel consumption and, 27–28
    calculation of fuel quantity, 28
    center of gravity, 27
Precast concrete walls, 1517
Precise pressure regulation
    compressed air control systems and, 209
    compressor discharge output pressure, 209
Pre-combustion capture, 126–127
Preconstruction assessment, M&V procedures and, 1052
Predesign phase, commissioning process and, 193
Predictive maintenance, Operations and Maintenance Best Practices Guide and, 710
Preindustrial era, energy and, 605–606
Premium efficiency motor cost to EPA motor cost efficiency, 353
Preparation of coal in United States, 151–152
Present to past energy comparison method, 2
Pressure drop, heat wheel performance factors and, 797
Pressure profile, compressed air storage systems and, 230–231
Pressure sensor, 412–413
Pressure set point, compressed air systems and, 238
Pressure switches, mobile HVAC systems and, 1080
Pressure vessel development
    boilers, 1360

    space heating and, 1360–1361
Pressurization design
    exfiltration vs. infiltration, 844
    HVAC systems and, 844
    improper, case study, 844–846
    smart air syndrome, 844
    understanding of, 844
Pressurization flow
    ANSI/ASHRAE Standard 62.1-2004 and, 53–54
    energy recovery ventilation, 52
    importance in preventing mold, 54
Pressurization, 839
    indoor negative pressures, 844
Pressurized heavy water nuclear reactors, 40
    CANDU, 1118–1119
    locations of, 1118
Pressurized water nuclear reactors, 1116–1117
Pressurized water reactors, 39
Preventive action, 1030
Preventive maintenance, downtime and, 472
Preventive service monitoring, web-based compressor management and, 213
Price cap regulation, 1245–1250
    basic price restriction, 1246–1248
    case studies, 1249–1250
    service baskets, 1248
Price restrictions, price cap regulations and, 1246–1248
Pricing, energy demand-side management programs and, 291
Primary energy forms, 1088
Primary energy ratio, 819
Primary energy supplies (PES) vs. secondary, 1346
Primary motor control, 208
Priorities, energy management system problems and, 1036
Procedures, energy management system problems and, 1036
Process analyses, 1143
Process based commissioning, 189–190
Process cooling, energy intensive, 462–463
Process dependent modeling, exergy analysis and, 657–658
Process energy optimization, 1304
    process heat integration, 1304
    temperature, low and high, 1304
Process equipment, accounting for in energy assessment, 71
Process heat integration, 1304
Process Heating Assessment and Survey Tool (PHAST 1.1.1), 880
Process heating, 461
Process historian, 459
Process reliability, 888
Product cooling heat recovery, 1373–1375
Production flexibility, lean manufacturing and, 473
Production issues, as driver in motor optimization projects, 888
Production, worldwide energy resources and, 445–446, 447
Productivity, coal production in United States and, 147–149
Programmable logic controllers (PLC), 208

Programmed start ballasts
    dimmers and, 982–984
    lighting control systems, 982–984
Project based emissions trading, 431
Project costs, electric power projects and, 958
Project definition, independent power producers and, 858
Project development, independent power producers and, 858
Promotion, energy demand-side management programs and, 290
Protection features, compressor air control systems and, 208–209
Protection, power distribution systems and, 702–703
Proton exchange membrane fuel cells (PEMFC), 726
    application of, 738–739
    contamination levels, 737
    principles of, 736–737
    problems, 737–738
    technological status, 738
Proximate analysis, solid fuel properties and, 479, 480
Prudence concept, rate of return regulation and, 1255
PSAT: Pumping Assessment Tool, 880
Psychrometric analysis, 1318
Psychrometrics, fundamentals of, mobile HVAC systems and, 1061–1066
Psychrometrics, 1184–1191
    air conditioning cycles, 1188–1191
    definition of, 1184
    degree of saturation, 1185
    dew point temperature, 1184–1185
    dry bulb temperature, 1184
    humidity ratio, 1185
    processes, 1186–1188
        air streams, adiabatic mixing of, 1188
        cooling and dehumidification, 1186–1187
        heating and dehumidification, 1187–1188
        humidification, 1186
        sensible
            cooling, 1186
            heating, 1186
    saturation pressure, 1185
    wet bulb temperature, 1184
Public policy
    creation of laws and rules, 1193–1194
    description of, 1193
    energy and, 1193–1200
    factors affecting, 1195–1198
        corporate behavior, 1197
        corporate governance, 1198
        economic conditions, 1196
        industrial conditions, 1196
        input markets, 1196
        institutional conditions, 1195
        international experience, 1195
        international risk experience, 1196
        legitimacy and credibility, 1198
        market structure, 1197
        objectives and priorities, 1195
        policy incentives, 1197
        regulatory governance, 1196–1197

sector performance, 1197–1198
    implementation of, 1198–1199
        regulatory processes, 1198–1199
        rationale for intervention, 1194
Public Utilities Holding Act of 1935, 380
Public Utilities Regulatory Policy Act of 1978 (PURPA), 380
    effectiveness of, 380
Public Utility Holding Company Act of 1935, 392
Public Utility Regulatory Policies Act (PURPA) of 1978, 1201–1205
    cogeneration technology, 1202
    energy crisis, 1201
    focus of, 1202
    implementation of, 1202–1203
    origin of, 1201–1202
    Section, 210,1202
    technological innovations and growth of, 1203
        cogeneration, 1203
        gas turbines, 1203
        solar energy, 1204
        wind, 1204
    unintended consequences of, 1204–1205
Pulverized coal boiler versus IGCC, 912
Pump motors, ASD technology and, 353, 355
    issues with, 355
Pump seals, power usage determination, 1217–1218
Pumped storage hydroelectricity, 1207–1212
    cost and economics, 1210–1211
    costs of, 1207
    efficiency level, 1209
    environmental issues, 1211–1212
    facility for, 1208–1209
    workings of, 1209–1210
Pumping, chemical process industry and, 1304
Pumps, 1213–1220
    electric motors and, 353, 355
    energy management opportunities, 1218–1220
        maintenance, 1218
        retrofitting, 1218–1220
    theory of, 1214–1218
        centrifugal, 1215
        multiple, 1217
        operating characteristics, 1214–1215
        positive displacement, 1216
        power requirements, 1216–1217
        power usage determination, 1217
        pump seals, 1217–1218
    types, 1213–1214
        centrifugal, 1213
        positive displacement, 1213–1214
Purchased utilities, energy data collection source and, 617
Pure oxygen systems, 1549
PURPA. See Public Utilities Regulatory Policy Act of 1978.
PV. See photovoltaic.
Pyrolysis, 89, 485–486, 1348

Q

Quality assurance, lean manufacturing and, 470–471
Quality control, automobile technology evolution and, 673
Quality, emergy definitions of, 421–422
    cross, 421, 422
    parallel, 421–422
Quasi isothermal compression, 1579

R

R Factor insulation, 1363
R values, 2×6 wall construction, 899
RACI concept, 563–564
Radiant barriers, 1227–1232
    airtightness, 1231
    benefits of, 1227
    cost savings, 1228
        claims of, 1228
    decreasing heat gain, 1231
    definition of, 1227
    foil side, 1229–1230
    foil-faced batt insulation, 1231
    heat buildup, 1231
    heat transfer blocking, 1227–1228
    installation, 1230–1231
        airtightness, 1231
        new construction, 1231
        placement, 1231
        safety tips, 1230
    material types, 1228–1229
        costs, 1229
    payback, 1232
    reshingling, 1231
    shingle warranties, 1231
    vs. reflective insulation, 891
Radiant flux
    spectral distribution, in window energy performance, 1616
    in window energy performance, 1616
Radiant heat barriers (RHB), 903–904
    configuration of, 904
    definition of, 903
    mechanism of operation, 903–904
Radiant heat gain, solar, in windows, 1623
Radiation heat transfer, in windows, 1617
Radiation, heat transfer and, 828–829
Radioactivity, 1127–1128
Radiometry, 264
Railroads
    coal to liquid fuel transportation needs and, 168
    ease of use, 591
    U.S. transportation energy use and, 590–591
Raised floor insulation, 892
Raised heel trusses, soffit air ventilation and, 902
Raised top plate, soffit air ventilation and, 902
Rankine steam cycle, 1381
Ratchet adjustments, utility billing and, 1501–1502
Rate base, rate of return regulation and, 1254–1255. See also rate of return regulation.
    accumulated depreciation determination, 1255
    construction work in progress determination, 1255
    plant valuation determination, 1254
    prudence concept, 1255
    working capital determination, 1255
Rate design strategies, 1502–1503
Rate of return
    assessing of, 1252
    formula, 1252–1253
Rate of return regulation, 1252–1257
    advantages and disadvantages, 1253
    cost of capital, 1256
    depreciation, 1256–1257
    formula, 1252–1253
    operating expenses, 1256
    rate base, determining of, 1254–1255
        accumulated depreciation, 1255
        construction work in progress, 1255
        plant valuation, 1254
        prudence concept, 1255
        working capital, 1255
    revenue imputation, 1253–1254
    taxes, 1257
Rate riders, 1507
    electric rate schedule, 1509
Rate structures, utilities and, 1497–1512
Rate-based emissions trading, 431
Ratings
    boilers, 94–95
    electric power transmission systems and, 358
RDF, coal and, cofiring and, 490
Reactance, 357
Reaction of enthalpy, 1423–1424
Reactive power, 356
    electric power transmission systems and, 356
    losses, 361
    production of, 356
Reactive power losses, 361
Reactivity of syngas feedstocks, 907
Reactor design, 1129–1130
Reactor treatment of wastewater, 1561
    complete mix, 1561
    dispersed flow, 1561
    plug flow, 1561
Reactors, nuclear, 39–41. See also nuclear energy; nuclear power plant types; nuclear power technology.
    advanced gas-cooled, 40
    boil water, 39–40
    commercially in use, 40
    fast neutron, 40
    natural, 1111
    pressurized heavy water, 40
    pressurized water, 39
    types of, 39–41
    uranium history, 1112
Real fluids, 1430–1431
Real power watts, 1510
Real power, 356
    electric power transmission systems and, 356
Real time costing, 920, 927–928
Real time pricing (RTP), 1175, 1178, 1180–1182
    commercial basis, 1180–1182
    electric rate schedule, 1508
    example of, 1181–1182
    industrial basis, 1180–1182
    residential, 1180
    See also real time utility pricing.
Real time pricing electric rate schedule, 1508
Real time utility pricing, 1506–1507

ambient temperature based, 1506
electric utility dispatch modeling, 1506–1507
Rebound, electric demand response and, 273
REC. *See* renewable energy credits.
Receiver, mobile HVAC systems and, 1077
Receiver volume, compressed air systems and calculating of, 232
Recessed lights, ceiling and roof insulation and, 903
Reciprocating engines, 1233–1243
Reciprocating engines, analysis, 1237–1238
  compression, 1235
  diesel, 1238, 1239,1241, 1242
  distributed generation technologies and, 297–298
  exhaust, 1235
  gas exchange process, 1237–1238
  gasoline, 1239–1240
  hardware of, 1235–1237
  homogeneous charge compression ignition, 1240, 1242
  intake operating cycle, 1235
  operating cycles, 1235–1237
    compression, 1235
    compression, 1235
    exhaust, 1235
    intake, 1235
  piston devices, 1233–1234
Reciprocating engines, piston devices, 1233–1234
Recommissioning, 180
Rectifiers, 359
Rectisol, in syngas treatment, 908
Recuperator, 1540
Recuperator usage, 1577–1578
Recycling, SWEATT and, 1353–1354
Recycling centers, 1530
Recycling products, sustainable development and, 1409
Redistributive effects, environmental policy instruments and, 631
Reduced air usage, 211
Reduction in aircraft fuel consumption, 26–29
  drag reduction, 26–27
  engine efficiency, 26
  engine maintenance, 28–29
  flight controls, 28
  piloting techniques, 28
  pre-flight planning, 27–28
  weight reduction, 26–27
Reference environment, exergy and, 646
Reference environment modeling, exergy analysis and, 656–658
  constrained-equilbrum, 657
  equilbirum, 657
  natural-environment-subsystem, 656–657
  process-dependent, 657–658
  reference-substance, 657
Reference substance modeling, exergy analysis and, 657
Reflective insulation, 891
  vs. radiant barriers, 891
Regasification, liquefied natural gas and, 1004–1006
Regional coal mining
  changes in, 149–150
  productivity, 147–149

Regional haze, air quality modeling and, 719
Regional transmission operator (RTO), 380
  effect on wholesale customers, 380–381
  results of, 381
Registration, management system standards and, 1030–1031
Regulating rate of return, 1252–1257
  formula, 1252–1253
Regulation
  adoption of energy codes and, 441
  price cap, 1245–1250
  revenue cap, 1245–1250
Regulatory processes, decisions and outcomes subject to, 1199
Reheat rankine steam cycles, 1381
Relative humidity, 19
Relative humidity controller, 414–415
Relative humidity sensor, 411–412
Relative implementation, global temperature adjustment and, 275
Reliability factor, fast simulation modeling and, 916
Reliance on personnel, energy management system problems and, 1036
Relief control valves, 1371
Renewable electricity generation, 44
Renewable energy credits (REC), 14
Renewable energy resources, 453
  district heating and cooling and use of, 322
  environmental policy instruments and, 631
  program funding, 1531
Renewable energy resources program funding, 1531
Renewable energy systems incentives, 959–961
Renewable energy technologies, 31, 1265–1270
  available technologies, 1266–1267
    biomass, 1266–1267
    geothermal electricity, 1266
    ocean sources, 1267
  biomass, 1266–1267
  conversion efficiencies, 1266
  decentralized, 1258–1264
  distributed generation and, 299
  fuel ethanol, 1267
  future of, 1268–1270
    adequacy, 1269–1270
    current energy use, 1268–1269
    efficiency gains, 1269
  geothermal electricity, 1266
  green electricity pricing, 1262–1264
  intermittent sources, 1268
  minimal impact of geothermal energy, 748
  ocean currents, 1267–1268
  ocean sources, 1267
  photovoltaic electricity, 1267
  promotion of, 1260–1264
    green electricity pricing, 1262–1264
    portfolio requirements, 1261–1262
    renewable trust funds, 1260–1261
    system benefits charge, 1260–1261
  resources available, 1266
  solar industrial process heat, 1267
  solar water heating, 1267
  technology development, 1267–1268
    ocean currents, 1267–1268
    ocean thermal differences, 1267–1268
    solar industrial process heat, 1267

  wave energy, 1267
  technologies not cost competitive, 1267
    fuel ethanol, 1267
    photovoltaic electricity, 1267
    solar water heating, 1267
  thermal differences, ocean, 1267–1268
  trust funds, 1260–1261
  types, 1265–1266
  wastewater plant operation and, 1554
  wave energy, 1267
Renewable trust funds, 1260–1261
Rensselaer Polytechnic Institute, Lighting Research Center, 512
Repair of compressed air leaks, 219–225
  tagging of, 222–223
Repeaters, 1652
Reporting functions, energy analysis and, 7
Resale of excess power, combined heat and power applications and, 865
Research and development
  energy conservation and, 450–451
  geothermal energy environmental impact, 746–747
Reserves, energy, 608–610
Reshingling, radiant barriers and, 1231
Residential areas, electric demand response solutions for, 283
Residential building heating loads, 1272–1277
  calculations for, 1272–1273
  seasonal heating demand, calculations, 1273–1276
Residential electricity generation in the United States, 583
  air conditioning, 583
  appliances, 583
  space heating, 583
Residential energy use in U.S., 583
Residential real time pricing, 1180
Resistance, 357
Resistance temperature devices, 409–410
  solid-state, 410–411
Resolution of control, 273
Resource availability, energy management system problems and, 1037
Resource degradation, exergy and, 659
Response factor, 117
Responsibility, management systems and, 1029
Results, measuring and verifying, enterprise energy management systems and, 622
Retail power markets, 382–383
  federal government intervention, lack of, 382
  individual state regulation of, 382
  potential savings of, 382–383
Retinal rod chart, 994
Retrocommissioning process, 180, 201–206
  benefits of, 203
  definitions of, 200–201
  energy conservation benefits, 202–203, 204
  Leadership in Energy and Environmental Design, 200, 206
  maintenance activities, 203
Retrofit isolation, 1051
Retrofitting, pumps and energy management opportunities and, 1218–1220
RETScreen green building design software, 1403

Revenue cap regulation, 1245–1250
Revenue imputation, 1253–1254
Reversibility, 1426
    energy degradation and, 1168–1172
    Second Law of Thermodynamics and, 1429
RHB. *See* Radiant heat barriers.
Ride through motor control, 208
Right of way, maglev and, 1022
Right Suite Residential software, 1402
Rigorous equations, 1429–1430
Risks, independent power producers and, 858, 859
Road infrastructure, strain on, 590
Roads, Federal funding of, 591
Roadway design, land use impact and, 1450
Rock wool, as insulation material, 890, 893
    effect on environment, 891
Rocky Mountain States coal basis, 158
Romans, heating and, 1357–1358
Rotary converter, 389
    direct current power system, 389
Royalties, Artic Wildlife Reserve oil exploration and, 596
RTO. *See* regional transmission operator.
RTP. *See* real time pricing.
Rules, creation of, 1193–1194
Rulesets, Energy Star Portfolio Manager software and, 575
Run around coils, 1540
Run around heat recovery systems, 1278
    air flow rate through duct, 1280–1281
    ethylyne glycol operating temperature, 1281
    heat transfer coefficients, 1279–1280
    inlet supply air temperature, 1280
    techniques of, 1278–1279
    two phase, 1281–1282
R-value, in insulation rating systems, 890

S

Sacramento Municipal Utility District Customer Advanced Technologies Progam, 515
Safe Drinking Water Act of 1974, 1557
Safety features, compressor air control systems and, 208–209
Safety
    hydrogen, 36
    nuclear energy and, 42, 1131
Satellites, 147
Saturation, degree of, 1185
Saturation pressure, 1185
Savings
    dollars vs. energy, EEMS and, 622
    stipulated vs. real, EEMS and, 622
Savings to investment ratio (SIR), 81
Scale model testing, daylighting simulation and, 268
Scheduling
    building usage and low cost energy efficient improvement case studies, 518–519
    compressor air control systems and, 210–211
Scheduling review, low cost energy efficiency improvement, 518
Scheduling tools, energy project management and, 563

Scissor trusses, 902
Scoping tools, 562–563
Scotopic CIE, 995
    in lighting design and retrofits, 995
Scotopic vision, 999
    definition of, 999
    vision sensitivity chart, 995
Scotopic/photopic vision sensitivity chart, 995
Screening treatment of wastewater, 1561
Sea level rise, global warming and, 722–723
Seal penetrations, air barriers and, 664
Sealants for air leaks, 665
Seals, 1382–1382
Seasonal energy consumption
    calculation of, 1273–1276
    energy benchmark analysis, 696
    heating demand, 1273–1276
Seasonal utility loads, 1498
    gas utilities, 1498
Seasonally differentiated rates, time of use and, 1504
SEC. *See* specific energy consumption.
Second Law of Thermodynamics, 1167, 1426–1429
    Carnot engine, 1427–1428
    Clausius statement, 1171
    exergy, 1428–1429
    formulation, 1426–1427
    heat engines and, 1171–1172
    Kelvin-Plank statement, 1170
    level of irreversibility, 1428
    reversibility, 1426, 1429
Second stage cost drivers, 1042–1043
Secondary energy supplies (SES) vs. primary, 1346
Sectoral energy conservation measures, 455
Security, lighting controls and, 990
Security costs, electricity usage and, 401–402
Security of supply, distributed generation technologies and benefits, 299–300
Sedimentation, 1557, 1558
Selective Amine Scrubbing, in syngas treatment, 908
Selenium, as syngas contaminant, 909
Self configuring nodes, 1653
Self-discipline, lean manufacturing and, 474
Sensible cooling, 1186
Sensible heat storage, 1412
Sensible heating, 1186
Sensible thermal energy, 1163
Sensor verification, ANSI/ASHRAE Standard 62.1-2004 and, 61
Sensors
    occupancy, 981, 982
    pressure, 412–413
    relative humidity, 411–412
    resistance temperature devices, 409–410
    solid-state, 410–411
    temperature, 409
    thermocouples, 411
    transmitter/transducer, 411
Sequestration, carbon, 169
Serial communications networks, 1652–1653
Series chillers, 310
Server side programs, web publishing and, 536–537
    Active Server Pages, 537

    ColdFusion, 537
    FoxWeb, 537
    hypertext preprocessor, 537
    Java Server Pages, 537
    Java Servlets, 537
    Perl, 537
Service baskets, price cap regulation and, 1248
Service charge, utililty billing and, 1501
Service ratings, electric motors and, 351
Service reductions, electric demand response and, 272–273
SES. Secondary energy supplies.
Set ups, preventive maintenance and, 473
SFB. *See* stationary fluidized bed.
Shading devices for windows, 1641–1648.
    acoustic performance, 1646
    advanced, 1641
    advanced assemblies, 1642–1643
    architectural, 1642
    assemblies, 1642
        descriptions of various, 1642
    awnings, 1642
    brise-soleils, 1642
    classification, 1642
    control strategy for, 1644
    decision-making framework, 1645, 1647–1648
    heat regulation, 1641
    louvers and blinds, 1643
    materials for, 1642
    overhangs, 1642
    performance parameters, 1646
    position of, 1643
    selection of, 1644–1645
    thermal performance, 1646
    variables, 1646
    visual performance, 1646
    window treatments, 1642
    See also advanced shading devices; windows, shading devices for.
Shading from sun rays, 116
Shanghai Airport Maglev system, 1024–1025
Shared burden, electric demand response and, 273
Shared energy savings, 569, 570, 1136–1137
Shed savings, decay of, 274
Shell and tube heat exchanger, 801, 1540
Shell gasifiers, 907
Shield wires, 358
Shift work, energy benchmark analysis and, 696
Shingle warranties, radiant barriers and, 1231
Short circuit duties, electric power transmission systems, 360
Short circuits
    circuit breakers, 360
    electric power transmission systems and, 360
Shortwave radiation, 116
Shower flow heads, 1597–1599
Silicon, 1150
Simple cycle gas turbines, 298
Simple payback period, 930–931
Simplification, lean manufacturing and, 474
Single borehole, geothermal heat pump system development and, 758
Single pass cooling equipment, 1588–1589
Single phase system, 699–700
Single piece flow, lean manufacturing and, 470

Single pole direct current transmission, 359
Single unit seasonal cooling appliances, 1483
SIR. *See* savings to investment ratio.
Site design, land use impact and, 1450
Site selection
    LEED-C&S, rating system, 939
    LEED-CI, 939
Six Sigma Methods, 1310–1316
    background of, 1311
    data based decisions, 1313
    data source, 1312
    defect prevention, 1313
    DMAIC methodology, 1312, 1313–1316
    energy consumption targets, 1312
    financial considerations, 1311–1312
    mean shift, 1313
    nomenclature, 1310–1311
    overview, 1312–1313
    variation and mean shift, 1313
Skill sets, energy project management and, 564–565
Slab-on-grade floor insulation, 892
Slag wool
    effect on environment, 891
    as insulation material, 890, 893
Sliding block rates, 1503
Slow sand water filter, 1557
Sludge drying, 1527
Sludge processing, 1554
Smart air syndrome, 844
Smart sensors, automobile technology evolution and, 674–675. *See also* sensors.
Social capital, 1410
Society, energy use and, 1016
SOFC. *See* solid oxide fuel cells.
Soffit air ventilation, 902
    raised heel trusses, 902
    raised top plate, 902
Soft coal, 156
Software, energy accounting tools and, 3
    Envision, 3
    Fraser, 3
    Global Mvo Asset Manager, 3
    Meter Manager, 3
    Utility Manager, 3
Software, energy simulator, 111
Sofware, client and server, building automation systems and, 106–107
Solar air conditioning, 1326–1327, 1457
    feasibility of, 1547
Solar cooking, 1325–1327
    box, 1326
    panel, 1326
    parabolic, 1326
Solar cooling, 1326–1327
Solar crop drying, 1325
Solar distillation, 1325
Solar energy, 34, 1204
    future of, 34
    See also solar thermal technologies.
Solar heat, effect on cooling equipment. 1618
Solar heat gain coefficient, 1623
Solar heating, 1364
    air conditioning and, case study of, 1317–1320
    cooling system methodology, 1317–1318, 1618
    features of, 1320
    heat gain coefficient, 1623
    industrial process heat, 1267
    passive, windows and, 1618–1619
    psychrometric analysis, 1318
    solar roof tile system, 1318–1319
    space heating
        active, 1322
        hybrid, 1322
        passive, 1322
    See also solar thermal applications; solar water heating.
Solar heating, passive, windows and, 1618–1619
Solar industrial process heat, 1267
Solar ponds, 1324–1325
Solar radiant heat gain, in windows, 1623
Solar radiation, 117
Solar related energy, 1090
Solar resources
    definitions, 1148
    irradiation, 1148
    photovoltaic (PV) systems and, 1148
Solar roof tile system, 1318–1319
Solar space heating
    active, 1322
    hybrid, 1322
    passive, 1322
Solar spectrum
    illumination basics and, 265
    photovoltaic (PV) systems and, 1150–1151
Solar thermal applications, 809
    cooking, 809
    distillation, 809
Solar thermal technologies, 1321–1330
    cooking, 1325–1327
    cooling, 1326
    crop drying, 1325
    distillation, 1325
    ponds, 1324–1325
    power generation, 1327–1328
        central receiver, 1328
        parabolic dish, 1328
        parabolic trough, 1328
    present use, 1329–1330
    solar space heating, 1321–1322
    water heating, 1322–1323
Solar water heating, 1267, 1322–1323, 1331–1338
    applications of, 1331
    collector performance, 1336
    current use of, 1338
    design simulation, 1336
    design transmission-absorption product, 1336
    designs, 1332–1333
        built in storage, 1332–1333
        forced circulation, 1333
        thermosyphon, 1332
    economics of, 1336–1337
    fin efficiency, 1336
    forced circulation, 1333
    heat removal factor, 1336
    Hottel Whillier Bliss equation, 1335–1336
    life cycle assessment, 1337–1338
    system design, 1335–1336
        fin efficiency, 1336
        heat removal factor, 1336
        Hottel Whillier Bliss equation, 1335–1336
        measured collector performance, 1336
        simulation, 1336
        transmission-absorption product, 1336
    thermal collectors, 1333–1335
    thermosyphon, 1332
    workings of, 1331
Solar-control film
    analysis study
        DOE-2 energy study, 1631
        kilowatt-hour usages, 1634
        motivation for, 1631
        results of, 1632–1635
            kilowatt-hour usages, 1634
            simple payback, 1633
            summer peak hour demand, 1635
        scope of, 1631–1632
        solar performance factors, 1632
        summer peak hour demand, 1635
        thermal performance factors, 1631
    appearance of, 1630
    benefits from, 1630
    construction of, 1630
    energy savings from, 1630
    installation process, 1630
    properties of, 1630
Solid fuel boilers, 1527
Solid fuels, 1340–1341
    biomass fuels, 482–483
    CO_2 emissions, 484
    coal composition, 481
    flame temperatures, 484
    flue gas volume, 485
    heating value, 479, 484
    properties, 479–485, 1341–1344
    proximate analysis, 479, 480
    ultimate/elemental analysis, 479
Solid oxide fuel cells (SOFC), 729–732
    actual use of, 731
    applications, 731–732
    contamination levels, 730
    principles of, 729–730
    problems with, 730–731
Solid waste, 1340–1341
    properties of, 1341–1344
Solid waste to energy, advanced thermal technologies (SWEATT), 1340–1354
Solid-state resistance temperature devices, 410–411
Solution heat treatment, 868
Sonic limitations, heat pipes and, 805, 806
Sorting, lean manufacturing and, 474
South African, Fischer-Tropsch process and, 163
Southern California Edison Energy Centers, 515
Southern California Gas Company Energy Resource Center, 515
Space heater, 648
Space heating, 583, 1357–1365
    calculation methodology, 1363–1364
        ASHRAE standards, 1363
    current trends, 1364–1365
        geothermal energy, 1364
        HVAC controllers, 1365
        price of fuel, 1364
        solar heating, 1364
        window technology, 1364–1365

Finnish Masonry stove, 1359–1360
Franklin stove, 1359
fuels, 1361–1362
habitat structures, 1362–1363
hearths and fireplaces, 1358–1360
history of, 1357–1358
    Romans, 1357–1358
insulation development, 1363
pressure vessel development, 1360–1361
system types, 1361
    hot water, 1361
    steam, 1361
See also solar space heating.
Spatial trading rules, emissions and, 432–433
Specialty rates, 1505–1506
Specific emergy, 421
Specific energy, 420
Specific energy consumption (SEC), 4
    disadvantages of, 4
Specification standard, 1030
Spectral distribution of radiant flux, in window energy performance, 1616
Spectral irradiance, in window energy performance, 1616
Spectrally selective window films
    aesthetic differences from conventional, 1638–1639
    applicability on different glasses, 1638
    case study, Stamford University, 1640
    conventionally applied, 1637–1639
    definition of, 1637–1638
    effectiveness guarantees, 1639
    heat block, 1638
    mitigating heat loss, 1638
    payback, 1639
    price comparison vs. conventionally applied, 1638
    special care needs, 1638
Spent nuclear fuel management, 42–43
Spikes, energy benchmark analysis and, 696
Spinning reserve concept, 1442–1443
Spread spectrum radios, 1653–1654
Square footage, energy usage vs., 583
Stability disturbances, electric power transmission systems and, 360
Stack economizer
    installation of, 1368
    savings from, 1368
Stakeholders, independent power producers and, 855–858
Stand alone photovoltaic systems, 1153
Standard connectors, automobile technology evolution and, 675
Standard of living, energy use modifications and, 1015
Standardization, lean manufacturing and, 474
Standardized commissioning of buildings, 678–679
Standards 90.1 and 90.2, 439
    review and revision process, 439–440
    timing of, 440
Standards development, geothermal energy environmental impact and, 747
Standards, impose on facility energy control suppliers, 677
Standby electric rate schedule, 1505, 1509
    backup, 1509
    maintenance, 1509
    supplementary rates, 1509
Standby rates, 1505
Standing devices, building system simulation process and, 120
Start/stop sequence, compressor air control systems and, 210
State adoption of energy codes, 440–441
State agencies, energy efficiency information sources and, 511–513
    California Energy Commission, 511
    Energy Center of Wisconsin, 512
    Florida Solar Energy Center, 512
    Iowa Energy Center, 512
    New York State Energy Research and Development Authority, 512
Static data, 691
Stationary fluidized bed (SFB), 88
Steady state, stead flow process, 748
Steady-state CO_2 differentials, 52
Steam
    density compensation of, 1057–1059
    dry, 745
    wet, 745
Steam blowthrough rate, cooker kettles and, 1373
Steam cycles, 1381–1382
    Carnot, 1381
    Rankine, 1381
    reheat rankine, 1381
    two phase working fluid, 1382
Steam generation cost, 1366–1367
Steam generators, 1589
Steam heating systems, 1361
    energy intensive, 463
Steam vs. hot water, combined heat and power, 176
Steam line, isolation of abandoned ones, 1357
Steam methane reforming, 789–790
    hydrogen vehicles
        emissions from, 790
        high-temperature nuclear reactor, 790
Steam path, 1382–1383
    blades and seals, 1382–1383
    high pressure cylinders, 1382
    intermediate pressure, 1382
    losses, 1383
    low pressure, 1382
Steam recovery, waste heat, 1540
Steam System Assessment Tool 1.0.0, 880
Steam system optimization
    abandoned line isolation, 1375
    blowdown heat recovery equipment replacement, 1368
    case study, 1366–1379
    condensate return improvement, 1375–1378
    condensate return system, 1367–1368
    measurement and verification protocol, 1378–1379
    product cooling heat recovery, 1373–1375
    relief control valves, 1371
    replacement of deaerator steam jet ejectors, 1370
    stack economizer installation, 1368
    steam blowthrough rate on cooker kettles, 1373
    steam generation cost, 1366–67
    steam trap, 1367, 1371–1373
    steam utilization, 1367
Steam System Scoping Tool 1.0d, 880
Steam trap, 1371–1373
Steam trapping, 1367
Steam turbines, 1380–1388
    blade materials, 1385
    control, 1386
    cycles, 1381–1382
    future development of, 1387–1388
    history of, 1380–1381
    mechanical design, 1383–1386
    mechanical design, stresses on, 1383–1384
    sizes, 1386
    steam path, 1382
    types, 1386
        combined cycle gas turbine, 1386
        congeneration, 1386
Steam utilization, 1367
Steam-methane reforming, 1442
Steel framed walls, 1516
Steel industry use of coal, 160–161
Steel plant assessments, Dept. of Energy best practices case studies and, 1285, 1299–1301
Step level HID controls, 982
    lighting control systems, 982
STIG cycle, 1582–1583
Stoker firing, 488–489
Storage
    compressed air systems and, 228–233
    electricity and, 389–390
    liquefied natural gas and, 1002–1004
    nuclear energy waste and, 43
Storage capacity, cool storage system design and, 1418–1419
Storage delta, compressed air systems and, 230–231
Storage of hydrogen, 36
Storage vessels, compressed air energy storage and, 217
Storing compressed air, compressor failure, 242–243
Straight pipe, flow measurement problems and, 1056–1057
Stratified chilled water storage, 1417–1418
Straw bale walls, 1518
Street lighting electric rate schedule, 1508
Structural insulated panel walls, 1518
Subbituminous coal, 156
Submetering primer, 1649–1650
Substations, 358–359
Substations
    bus/circuit breaker arrangements, 359
    equipment make-up, 359
    functions of, 358–359
Sulfur iodine (S-I) cycle, 1444–1445
Sulfur iodine process, 1443–1444
Sulfur, in syngas, 908
Sun, illumination basics and, 264–265
Sun rays, shading from, 116
Sunbelt cities
    energy efficiency programs, 1391–1392
    energy polices, 1394
    Energy Star partnership, 1393
    government support of sustainability policies, 1392–1393

sustainability and implementation of, 1393–1394
sustainability polices and, 1389–1395
Sunlight, direct beam entry, 1619
Sunlight concentration, photovoltaic (PV) systems and, 1151
Superconductor magnet maglev system, 1024
Supercritical water-cooled reactor, 41
Supplementary electric rate schedule, 1509
Supply air temperature, increasing of, 276
Surface coal mining, 144, 146
Surface convection, 116
Surface mining methods, 158–159
Surface Water Treatment rule, 1557
Surge controls, compressor air control systems and, 210
Suspension firing, 488
Sustainability, energy efficient buildings and, 526
Sustainability policies, sunbelt cities and, 1389–1395
    energy efficiency programs, 1391–1392
    Energy Star partner, 1393
    implementation studies, 1393–1394
    local government support, 1392–1393
    significance of, 1390
    sustainability as a goal, 1390–1391
    urban, 1389–1390
Sustainable building simulation, 1396–1403
    available software, 1401–1404
        APACHE HVAC, 1401
        APACHE, 1401
        ArchiPhysics-Solar, 1402
        BEA, 1401
        BLAST, 1401
        BREEZE, 1403
        BSim2002, 1401
        BUS^{++}, 1401
        Daylight, 1403
        DOE-2, 1402
        DUCTSIZE, 1402
        EE4 CODE, 1402
        EED, 1402
        EnergyPlus, 1402
        ESP-r, 1402
        EZDOE, 1402
        FLOVENT, 1402
        HVAC solution, 1402
        Hydronics Design Studio, 1402
        MOIST, 1403
        Right Suite Residential, 1402
        TRACE 700, 1402
        TRNSYS, 1402
        VisualDOE, 1402
    available software for green buildings, 1403
    components of, 1400
    energy aspects, 1397–1398
    exergy aspects, 1397–1398
    integrated building design system, 1401
    need for, 1396–1397
    reason for, 1398–1400
Sustainable development, 1406–1410
    action being taken, 1407
    ASHRAE GreenGuide, 1407–1408
    community development, 1409
    community planning and design, 1409
    conservation economy implementation, 1409–1410
    cooperation in implementation and use, 1407–1408
    definition of, 1406
    district heating and cooling and, 321–322
    educating the population, 1408
        U.S. Partnership for The Decade of Education for, 1408
        United Nations efforts, 1408
    end of petroleum fuels, 1407
    energy conservation and, 451–453
    environmental policy instruments and, 631
    future energy types, 1407
    green energy and, 776–777
        case study, 781–786
    information sources, 1409
        conservation economy, 1409
        economic capital, 1410
        natural capital, 1410
        social capital, 1410
    reasons for, 1407
    recycling products, 1409
    renewable energy resources, 453
    reporting of, 1408
    use in developing countries, 1410
    workable technologies needed, 453
Sustainable sites, LEED-NC, rating system, 949
Sustained changes, energy benchmark analysis and, 696–697
SWEATT
    coal vs., 1344
    recycling, 1353–1354
    See solid waste to energy by advanced thermal technologies.
Sweeping, lean manufacturing and, 474
Switch mode power converters, 1154
Switching, dimming vs., lighting controls and, 989
Synchronous generator vs. induction, combined heat and power, 175–176
Synchronous operation, 359–360
    electric power transmission systems and, 359–360
Syncrude, 594
Syngas compositions, 907
Syngas contaminants, 908
    arsenic, 909
    cadmium, 909
    coal derived, 908
    mercury, 909
    metal carbonyls, 909
    selenium, 909
Syngas feedstocks, 907
    ash content of, 907
    ash fusion temperature of, 907
    characteristics of, 907
    free swelling index, 908
    moisture content of, 907
    particle size of, 908
    reactivity of, 907
Syngas treatment, 908–909
    acid gas removal, 908
    Rectisol, 908
    Selective Amine Scrubbing, 908
    sulfur, 908
System benefits charge, 1260–1261
System controllers, compressor air control systems and, 210
System dimmers, lighting control systems, 982
System modules, automobile technology evolution and, 674
System start-up guidelines, ANSI/ASHRAE Standard 62.1-2004 and, 61

T

TAB. *See* testing, adjusting and balancing.
TAC. Task/ambient conditioning.
TACAS. *See* thermal compressed air energy storage.
Tar sand extraction, 594
    disadvantages of, 594
    Syncrude, 594
Target Finder, 577–578
Targeting reductions in energy use, 6–7
Tariff filings, electricity and, 380
Tariff review, energy assessments and, 64
Task/ambient conditioning (TAC), 1463–1464
    technology description, 1464
Tax disincentives, Energy Tax Act of 1978, 1084–1085
Tax incentives, Energy Tax Act of 1978 and, 1084
Taxes, 959, 1257
    emissions, 630
    fees, utility billing, 1502
    pollution, 630
Team planning, energy master planning and, 553
Technical information, user friendly energy audit report suggested format, 78
Technical skills, energy project management and, 564
Technological growth, Public Utility Regulatory Policies Act (PURPA) of 1978 and, 1203
Technology assessment, geothermal energy environmental impact and, 747
Technology requirements, industrial processes and, 459
Technology transfer
    feasibility studies, 502–503
        macro indicators, 502–503
        micro indicators, 502–503
    geothermal energy environmental impact and, 747
    implementation of energy efficiency, 502–502
Temperature based utility pricing, 1506
Temperature controller, 414
Temperature measurement
    accuracy of, 1059
    flow measurement and, 1059
Temperature multiplexing, reduce cost, 1059
Temperature sensors, 409
Temperature, process energy optimization and, 1304
Temporal trading rules, 432–433
Terminals, 357
Terrestrial carbon sinks, 129
Tesla, Nicola, 389
Test and balance, facility guidelines for new structures and, 527

Testing, adjusting and balancing (TAB), 196
Texas coal production, 159
Thermal collectors
    evacuated tube, 1334–1335
    flat plate, 1333–1334
    solar water heating and, 1333–1335
Thermal comfort, underfloor air distribution (UFAD) and, 1467
Thermal compressed air energy storage (TACAS), 214–215, 216
    adaptations of, 217–218
Thermal conductivity
    heat pipes and, 805
    heat transfer and, 824–827
Thermal cut-off, electronic ballast primer and, 834
Thermal design
    Carnot Law, 1541
    fundamentals of, 1541–1542
Thermal discomfort, underfloor air distribution and, 1468
Thermal efficiency, turbine inlet air cooling and, 1546
Thermal energy storage (TES), 1412–1421
    cooling applications, 1413–1418
        benefits of, 1413
        chilled water, 1417–1418
        commercial utility rates, 1413
        ice storage, 1414–1417
        storage equipment, 1414–1418
        storage system design, 1418–1420
    off-peak heat storage, 1420
    types, 1412–1413
        latent heat storage, 1413
        sensible heat storage, 1412
    underground, 1420
Thermal energy, 1163
Thermal imaging, data collection and, 259–260
Thermal load following cycle, 1530
Thermal mass of walls, 1513
Thermal mass, geothermal heat pump system development and, 757–758
Thermal radiation, 1161
Thermal storage system, 648
Thermal technologies
    advanced, 1347
    solid waste to energy and, 1340–1354
Thermochemical conversion, 89–90
    gasification, 89–90
    pyrolysis, 89
    technology comparison, 90
Thermochemical cracking of water, 1443–1445
    calcium bromide cycles, 1445
    future of, 1445
    sulfur iodine (S-I) cycle, 1444–1445
    sulfur iodine process, 1443–1444
    temperature reduction, 1445
Thermocouples, 411
Thermodynamic properties, calculation of, 1429–1431
    ideal gas, 1430
    phase equilibrium, 1431
    real fluids, 1430–1431
    rigorous equations, 1429–1430
Thermodynamic temperature, 1171
Thermodynamic wet bulb temperature, 1184
Thermodynamics, 1422–1432
    first law, 1423
    nomenclature, 1422
    second law, 1426–1429
Thermoelectric heat pump, 816
Thermo-fluid dynamic efficiency, cooling towers and, 250
Thermostat cases, review if needed, 520–521
Thermostat climate control, mobile, 1661–1663
Thermostat location, 520
Thermostats, mobile HVAC systems and, 1080
Thermosyphon solar water heater, 1332
Thin film photovoltaic (PV) systems, 1149–1150
    organic materials, 1150
    silicon, 1150
Third party financing (TPF), 567–568
3-E trilemma, 1147–1148
Three phase system, 699–700
Three phase transmission lines, 357–358
    terminals, 357
    types, 357–358
        high voltage alternating current, 357
        high voltage direct current lines, 357
Throttle, 1544
Throughput, 1652
Thyristors, 359
Tight buildings, ventilation problems and, 20–21
Time clocks, HVAC system at night, 518
Time domain, 117
Time intervals
    energy data structuring and, 619
    normalized, 619
    optimal, 619
Time of use (TOU) pricing, 1175–1177
    examples of, 1176–1177
    observations from, 1176
    rate development, 1177
Time of use rates, 1504
    off season, 1504
    seasonally differentiated, 1504
Time value of money, 930, 968
Timeline, energy benchmark preparation and, 692
Toilets, 1599–1601
Tools
    energy accounting and, 2–3
    software, 3
    energy project management and, 562–564
Top down approach, energy efficient buildings and, 525
    integrated design, 525–526
    sustainability, 526
Topping cycles, 1542
Total owning costs, 968
Total quality energy management, 465
TOU electric rate schedule, 1508
TOU large power electric rate schedule, 1508
TOU. *See* time of use.
Toyota, 467, 468
TPF. *See* third party financing.
TRACE 700 software, 1402
Tradable certificates for energy savings, 1433–1439
    characteristics of, 1435–1436
    implementation and results, 1436–1439
    policies, 1433–1435
Tradable emissions permits, 630
Trade-ally cooperation, energy demand-side management programs and, 290
Trading of electricity, difficulties in, 381–382
Traditional manufacturing, lean vs., 468–469
Traffic woes, U.S. transportation energy use and, 588
Training, 1029–1030
    lean manufacturing and, 470–471
Transducer sensor, 411
Transducer, 416
Transformer ownership, 1511
Transformer ratings, 705
Transformity, 420
    quality and, emergy and, 421
    specific and, emergy, 421
Transit service, land use impact and, 1450
Transmission absorption product, 1336
Transmission capacity, 362
Transmission components
    cables, 358
    lines, 357–358
    ratings, 358
    substations, 358–359
Transmission grid upgrading, energy engineering and, 136
Transmission lines, 357–358
    three phase, 357–358
Transmission rates, electric rate schedule, 1508
Transmittance of windows, 1623–1624
    damage-weighted transmittance, 1623–1624
    UV transmittance, 1623–1624
    visible transmittance, 1623–1624
Transmitter sensor, 411
Transparent surfaces, building system simulation process and, 120
Transport of air pollution, 718–721
Transportation energy use, U.S., 588–599
Transportation issues, coal to liquid fuels and, 168–169
Transportation systems, heat pipe applications and, 811
Transportation, coal and, 159, 161
Transportation, land use impact on, 1449–1451
Transportation, liquefied natural gas and, 1006–1008
Transportation, location efficiency, 1449–1453
Trash collection, 1530
Trends, interactive access to data and, 620–621
Tribal communities
    cultural awareness, 1458
    energy audits
        coordination, 1459
        funding, 1460
        geographic location, 1459
        information, 1460
        teamwork needed, 1461
        technical issues, 1459–1460
    governments, 1457–1458
    long term orientation, 1459
    sense of community, 1458–1459
Tribal lands, energy efficiency and, 1456–1462
TRNSYS software, 1402
True up, emissions trading and, 432
Tube-side heat transfer calculations, 1063
Turbine control, 1386
Turbine generators, 392

Index

Turbine inlet air cooling, 1546
  thermal efficiency, 1546
Turbines
  gas, 298
  micro type, 298
Turkey, 503–504
Two phase heat exchangers
  analysis of, 803–804
  condensers and evaporators, 803–804
    enthalpy calculations, 803
    maintenance of, 804
    materials used in, 804
    safe operation of, 804
Two phase run around heat recovery system, 1281–1282
Two phase working fluid, 1382
Two position controller, 415
TXV system, OT vs., mobile HVAC systems and, 1081

U

U Factor insulation, 1363
U.N. Framework Convention on Climate Change, 139
U.S. energy supplies, 1344, 1346–1347
U.S. energy use
  electricity generation, 598–599
    nuclear power, 598–599
  power grid concerns, 599
  solutions, 599
    energy usage reduction, 599
U.S. overview of energy use, 580–586
U.S. Partnership for The Decade of Education for Sustainable Development, 1408
U.S. transportation energy use, 588–599
  airport traffic woes, 588
  Artic Wildlife Reserve, 595
  cheap energy, effect of, 588, 589
  consumption, 595
  conversion to mass transportation, 599
  Federal funding of airports and roads, 591
  higher energy costs, 590
  housing growth, 590
  impact of China's economic growth, 592
  impact of India's economic growth, 592
  infrastructure strain, 590
  lack of coordination, 588–589
  lack of vision, 598
  land development policies, 588
  new technologies, 591
    Maglev system, 591
  oil company nationalization, 595
  oil consumed, 590
  oil in ground, 593
  oil reserve uncertainty, 592
  oil shale deposits, 594
  oil tankers, impact of, 591–592
  over reliance of oil, 596–597
  politics and, 594–595
  railroads, 590–591
  railway systems, ease of use, 591
  reacting to, 598
  rising prices and action taken, 595
  tar sand extraction, 594
  traffic woes, 588
  vulnerabilities of oil production, 592–593
UFAD. *See* underfloor air distribution.
Ultimate/elemental analysis, solid fuel properties and, 479
Ultrasonic leak detection, data collection and, 261–262
Ultraviolet light, 1559
Unbalanced system, 700–701
Underfloor air distribution (UFAD), 1463–1469
  benefits of, 1467
    air movement, 1467
    energy savings, 1467
    thermal comfort, 1467
    worker satisfaction, 1467
  cost effectiveness, 1468–1469
    churn factor, 1469
    operations and maintenance, 1469
    productivity and health, 1469
  history of, 1463
  limitations of, 1467–1468
    applicability of, 1468
    higher energy use, 1468
    minimal information, 1468
    plenum leakage, 1468
    thermal discomfort, 1468
    unfamiliar technology, 1468
  principles of, 1464–1467
    task/ambient conditioning, 1463–1464
  technology description, 1464
Underground coal mining, 144, 158
Underground thermal energy storage, 1420
Uninterruptible power supply (UPS), 214–215
  thermal compressed air energy storage, 214–215
Unit emergy values, 420–424
  common products, 426
  computing of, 422–423
  emergy per unit money, 420
  empower, 421
  non renewable energy sources, 425
  specific energy, 420
  transformity, 420
Unit of trade
  allowance, 432
  emissions and, 432
  emissions reduction credit, 432
  permit, 432
United Nations, sustainable development education and, 1408
United States coal production, 143–154
United States coal supply, 156–161
  coal to liquid fuels and, 168
United States Government Agencies, energy efficiency information sources and, 508–511
  Dept. of Energy, 508–510
  Environmental Protection Agency, 510
  Dept. of Commerce, 510–511
  Department of Defense, 511
United States
  alternate energy credit (AEC), 14
  electricity generation and, 581–582
  energy efficiency regulation state by state, 12–13
  ESCO and savings model used, 570
  renewable energy credits (REC), 14
Universal controller, 415
University programs, energy efficiency information sources and, 511–513
  Rensselaer Polytechnic Institute, 512
  Washington State University, 512–513
Uranium, 1112–1113
  availability of, 41
  conversion and enriching, 1113
  fabrication for nuclear fuel, 1113
  history of, 1112
  mining and milling of, 1112–1113
Urban sustainability policies, 1389–1390
URC. *See* utility report card.
Urinals, 1601–1603
Usable compressed air energy in storage, 230–233
  calculating, 231
User friendly energy audit report, 76–80
  assumptions used, 77–78
  customer feedback, 79
  definition of, 76
  short form type report, 79
  suggested format, 78
    appendix, 79
    assumptions and calculations, 78
    energy management plan, 78
    executive summary, 78
    report recommendations, 78–79
    technical supplement, 78
  writing of, 76–78
User-specific interactive access, 622
USGBC, LEED and, 945–946
Utility based energy service company, 307
Utilities
  electricity deregulation and
    challenges of, 385–386
    non utility generators, 385–386
    variation in rules, 385–386
  industry consolidation of, 392
  investor owned, 379–380
  long term planning, 1491–1495
    alternative resources, 1494–1495
    benefits of, 1492–1492
    demand forecasting, 1493–1494
    effective management of, 1493
    existing resources, 1493–1494
    optimal mix, 1495
    responsible parties for, 1492–1493
  profitability, 392
  role in society, 1491–1492
Utility bill
  analysis, 1471–1483
  calculating of, 1471–1472
  electric, calculating of, 1472–1473
  gas, calculating of, 1472
  weighted average cost of power, 1473–1478
    differing load profiles, 1478–1482
Utility billing factors
  adjustment charges, 1502
  basic service charge, 1501
  commodity charge, 1501
  demand charge, 1501
  energy adjustment charge, 1502
  minimum charge, 1501
  peak usage, 1501

ratchet adjustments, 1501–1502
taxes and fees, 1502
Utility bills, 691
Utility energy centers, energy efficiency information sources and, 514–515
California Statewide Emerging Technology Program, 514
Energy Design Resources, 514
Lighting Design Lab, 514–515
Sacramento Municipal Utility District, 515
Southern California Edison Energy Centers, 515
Southern California Gas Company Energy Resource Center, 515
Utility issues, hybrid electric vehicles, plug in type and, 852
Utility loads
baseloads, 1498
seasonal, 1498
types, 1497
off-peak, 1497–1498
peak, 1497–1498
Utility manager software, 3
Utility organizations, energy efficiency information sources and, 514–515
Alliance to Save Energy, 513
American Council for an Energy Efficient Economy, 513
Consortium for Energy Efficiency, 513
E Source Companies LLC, 513
Utility rate structures, 1497–1512
baseloads and, 1498
billing factors, 1501–1502
block rates, 1503
commodity function, distribution function vs., 1511–1512
competitive, 1506
cost allocation, 1500
demand/commodity, 1503–1504
deregulation, 1511–1512
electric rate, 1507–1510
end use, 1504
load factor, 1498–1500
measuring electric demand, 1510
metering point, 1511
natural gas, 1507
negotiated, 1505–1506
non firm service, 1504–1505
power factor, demand billing, 1510–1511
rate design strategies, 1502–1503
rate riders, 1507
real time pricing, 1506–1507
specialty, 1505–1506
time of use, 1504
transformer ownership, 1511
types of loads, 1497
usage profile, 1498–1500
wholesale market, 1506
Utility rates
blended, 967
commercial, 1413
Utility report cards (URC) program, 539–540
Utilization factor adjustments, energy balance and, 688
UV transmittance, of windows, 1623–1624

V

Vacuum systems, energy intensive, 463–464
Value added electricity services, IntelliGrid^{SM} and, 918
Value chain management, 1137–1138
Vapor compression heat pump, 817–818
Vapor retarders, 662
Vaporization, liquefied natural gas and, 1004–1006
Variable air volume (VAV), 51
Variable air volume control strategies, 58
airflow sensors and, 58
Variable air volume design types, 52
Variable air volume sensors, use savings, 59
Variable air volume supply, ventilation rate procedure and, 58
Variable air volume
airflow measurement devices, importance of, 52
steady-state CO_2 differentials, 52
Variable cost calculations, 958
Variable discount rates, 935
Variable flow
chiller capacity, 311–312
district cooling systems and, 310–311
Variable frequency, 705–706
Variable load applications, 867
Variable relationships, 1315–1316
Variable speed devices (VSD)
advantages of, 867
other applications of, 867
uses, 867
variable load applications, 867
Variable speed drives, 705–706, 862, 864
Variations, Six Sigma Methods and, 1313
VAST cycle, 1583
VASTIG cycle, 1583–1584
VAV. See variable air volume.
Vehicle air emission standards, 717
Vent selection, in attic insulation, 899–900
Ventilation problems, indoor air quality and
dilution, 19–20
increased outside air, 20
pollution sources, 20–21
relative humidity, 19
tight buildings, 20–21
Ventilation rate procedure (VRP),
ANSI/ASHRAE Standard 62.1–2004 compliance, 54, 55–60
advanced VAV control strategies, 58
breathing zone outdoor airflow, 57
calculations for, 56
Carbon dioxide levels, 56–57
demand control ventilation, 56
multiple zone recirculating system, 57
variable air volume supply, 58
Ventilation system requirements, ANSI/ASHRAE Standard 62.1–2004 and, 51–52
Ventilation, 839
advantages of, 21
indoor air quality and, 18–19
problems with, 19–20
standards of, 19
Verification of emissions, 431–432

Verification protocol, steam system optimization and, 1378–1379
Vertical stacking fan coil HVAC unit, 840
Visible transmittance, of windows, 1623–1624
Visual factory, lean manufacturing and, 473
VisualDOE software, 1402
VOC emission, printing inks and, 891
VOC. See volatile organic compounds.
Volatile organic compounds (VOC), 20
Volatile oxidation, 486–487
Volt amperes reactive, 1510
Voltage controls, lighting and, 989–990
Voltage levels, 702
Voltage, 255–256, 831
Voluntary energy efficiency programs, 442–443
ENERGY STAR, 442
government and private organizations, 442
home energy rating system, 442
VRP. See ventilation rate procedure.
VSD. See variable speed devices.

W

Walkthrough energy assessment, 64
Wall construction, 2 × 6, R values, 899
Wall insulation,
2 × 4, 896
problems and solutions, 897
2 × 6, 899
batts in, 896–897
concrete, 892
block cores, 892
strategies for use, 892
Wallbox dimmers, lighting control systems, 982
Walls and windows, building types, 1513
Walls, 1513–1518
exterior insulation finish systems, 1518
straw bale walls, 1518
structural insulated panels, 1518
thermal characteristics, payback of, 1521
thermal mass of, 1513
types, 1515–1518
masonry, 1516–1518
steel framed, 1516
wood framed, 1515–1516
Washington State University energy programs, 512–513
Waste disposal, nuclear energy and, 1103–1104
Waste energy emissions, exergy and, 659–660
Waste fuel suppliers, commercial, 1530
Waste fuel technologies, 1527–1530
new equipment, 1528–1530
anaerobic digesters, 1529–1530
bubbling fluid bed boiler, 1528
circulating fluid bed boiler, 1528
gasifiers, 1528–1529
retrofit applications, 1527
burner/engine, 1527
solid fuel boilers, 1527
waste heat recovery steam generators, 1527
Waste fuels project funding for, cogeneration, 1531
Waste fuels, 1523–1534
alternative types, 1523–1525
anaerobic digestion, 1527
combustion processes, 1525

disposal of, 1525
economics of, 1530–1533
  combined heat and power efficiencies, 1530
  disposal cost reduction, 1530
  opportunity fuels, 1530
  reduced cost alternates, 1530
    commercial waste fuel suppliers, 1530
    recycling centers, 1530
environmental considerations, 1525, 1527
equipment and maintenance average costs, 1532
handling of, 1525
non combustion processes, 1525, 1526
opportunity types, 1523–1525, 1526
project funding for, 1531–1534
  biomass programs, 1531
  greenhouse gas emission reduction, 1531, 1532
  renewable energy resources, 1531
Waste heat recovery steam generators, 1527
Waste heat recovery, 1536–1540
  absorption heat pumps, 1541–1547
  absorption refrigeration, 1544–1546
  building cooling applications, 1545–1546
  combined heat and power applications, 864–867, 1546–1547
  design sample calculations, 1538–1540
  engineering considerations, 1537–1538
  equipment, 1540
  feasibility of, 1536–1537
  industrial plant savings and, 864–867
  quality of, 1537–1537
  solar air conditioning, 1547
  turbine inlet air cooling, 1546
Waste heat steam recovery, 1540
Waste management of nuclear energy, 42–43
Waste management planning, 1593
Waste product disposal, nuclear energy and, 41
Waste products
  coal to liquid fuel plants and, 169–170
  energy form, 1091
  heating values, 1345
Waste summaries review, energy assessments and, 64
Waste to energy, chemical process industry and, 1305
Wastewater plants
  energy efficiency measures, 1551–1554
    adjustable speed drive pumps, 1553–1554
    dissolved oxygen mechanics, 1551–1553
    methane gas recovery, 1554
    renewable sources, 1554
    sludge processing, 1554
  energy savings in, 1548–1555
  energy usage, 1550–1551
    by amount, 1551
  goals of, 1548
  operations of, 1548–1549
  types, 1549–1550
    course bubble diffusion, 1549
    fine bubble diffusion, 1549
    mechanical aeration, 1549
    pure oxygen systems, 1549

Wastewater treatment plant (WWTP), case study of, 15
Wastewater treatment plant, carbon dioxide
  credits for, 15–17
  reduction in, 15–17
Wastewater, 1560–1562
  contemporary issues, 1562
  disposal of, 1561–1562
  history of, 1560
    cholera, 1560
  industrial waste, 1562
  sources and collection of, 1560–1561
  standards and monitoring, 1560
  treatment of, 1561
    carbon filtration, 1561
    gravity, 1561
    reactors, 1561
    screening, 1561
Water augmented gas turbine power cycles, 1574–1585
  air humidification, 1579, 1581–1582
  analyzing cycles, 1575
  Carnot efficiency, 1574–1575
  Combined, 1578–1579
    heat and power, 1584–1585
  combustor, 1582–1583
  compressor surge, 1576
  compressor train, 1579
  conventional Brayton cycles, 1576–1577
  design parameters, 1576
  modeling, 1575
  software for simulation and analyzing, 1575
Water conservation techniques, 1604–1605
Water consumption tracking, Energy Star Portfolio Manager software and, 578
Water cooled chiller efficiency, 309
Water electrolysis, 1442
Water element, chemical process industry and, 1305
Water injection, quasi isothermal compression, 1579
Water neutrality, coal to liquid fuel water quality and, 169
Water pollutants, geothermal energy and, 746
Water quality, coal to liquid fuels and, 169
Water source heat pump for modular building, case study, 1564–1573
Water treatment process, 1557–1559
  coagulation, 1557, 1558
  disinfection, 1557, 1559
  filtration, 1557, 1558–1559
  flocculation, 1557, 1558
  sedimentation, 1557, 1558
Water treatment
  history of, 1556–1557
    slow sand filter, 1557
  importance of, 1557
Water use efficiency
  LEED-C&S, rating system, 943
  LEED-CI, rating system, 939
  LEED-NC, rating system, 950
Water using equipment, 1587–1595
  boilers and steam generators, 1589
  clothes washers, 1588, 1603–1604
  conservation techniques, 1604–1605
  cooling towers, 1589–1590
  dishwashers, 1587–1588

dishwashers, 1603
faucets, 1596–1597
landscape irrigation, 1590–1592
leak detection, 1592–1593
showers, 1597–1599
single pass cooling equipment, 1588–1589
toilets, 1599–1601
urinals, 1601–1603
waste management planning, 1593
Water, 1556–1560
  environmental policy and, 627
  history of, cholera, 1557
  hydraulics, 1559–1560
  runoff, 1556
  sources, 1557
  standards and monitoring, 1557
    coliform test, 1557
    Safe Drinking Water Act of 1974, 1557
    Surface Water Treatment Rule, 1557
  transportation of, 1557
Wave energy, 1267
Wealth generation, energy's importance in, 444
Weather data, building system simulation process and, 119
Weather normalization, Energy Star Portfolio Manager software and, 576
Web publishing
  available tools, 538
    batch mail applications, 538
    charting tools, 538
    databases, 538
  energy information system processes and, 535–536
  purchase or self design, 538
  programming choices, 536–538
    client side programs, 537–538
    common gateway interface, 536
    server side programs, 536–537
Web-based compressor management, 212–213
  benefits of, 213
    air leakage measurement, 213
    cost metering, 213
    internet-based access, 213
    preventive service monitoring, 213
    real-time, data driven feedback, 213
    system efficiencies, 213
Weekend energy consumption, energy benchmark analysis and, 696
Weight reduction, aircraft fuel consumption and, 26–27
Weighted average cost of power, 1473–1478
  differing load profiles, 1478–1482
Weighted average unit cost, 920, 921–926
Well head pricing, 1083
Western United States coal production, 159
Westinghouse, 389
Wet bulb temperature, thermodynamic, 1184
Wet cooling towers, 246–247
  drawbacks to, 248–249
  heat exchanging, 248
Wet steam, 745
Wet-dry (hybrid) cooling towers, 250–251
Whole meter M&V procedures, 1051
Wholesale customers, regional transmission operator and, 380–381
Wholesale market energy rates, 1506
Wholesale power markets, 381–382

creation of trading rules, 381–382
liquid market structure, 381–382
maturing of, 382
results of, 382
Wick type, 805
Wicking, heat pipes and, 805–806
Wide area augmentation system, global positioning satellites and, 29
Wind farm towers, location of, 1610
Wind power plant, 954
Wind power, 1607–1615
   bird impact, 1612
   blades, importance of, 1614
   computer control systems, 1614
   control mechanisms, 1614
   cost of, 1609, 1611
   economics of, 1607–1608
   effect on electrical grid, 1611–1612
   electricity source, 1607
   energy source ranking, 1607
   environmental regulation of, 1615
   future of, 1610
   geographical use of, 1607
   government subsidies of, 1614–1615
   growth capacity, 1608
   history of, 1607
   impact on environment, 1611
   improving information availability, 1614
   installation costs, 1608
   maximizing production, 1610
   noise level of, 1612
   sensors in, 1614
   site selection, 1609–1610
   strengths as energy source, 1611
   technology of, 1612–1612
      blade diameter. 1612–1613
      tower height, 1612
   typical turbine, 1613
   variability, 1611
   visual impact of, 1612
   weaknesses as energy source, 1611–1612
Wind technology, 34–35, 1204
   development of, 35
   usage of, 35
Wind turbine design, 1612–1613
   blade diameter, 1612–1613
   tower height, 1612
Wind turbine
   example of, 1613
   site assembly of, 1613
Wind, coal vs., 957
Window economics, 1624
Window energy performance
   conductive heat flow, 1616
   irradiance in, 1616
   radiant flux, 1616
      spectral distribution, 1616
      spectral irradiance, 1616
Window energy, 1616–1625
Window film performance
   conventional applied, 1638
   spectrally selective, 1638
Window films
   CPFilms case study, 1626
      demonstration building, 1627
      DOE-2 simulation values, 1627
      energy savings from, 1627

   ideal qualities of, 1638
   IPMVP options C and D, 1626–1629
   savings before and after installation, 1628
   solar-control and insulating types, 1630–1636
   spectrally selective vs. conventional applied, 1637–1639. *See also* Spectrally selective window films.
Window glass, heat flow through, 1619
Window heat transfers, 1619–1620
Window orientation, path of sun as factor, 1618–1619
Window sash, glazing system and, 1620
Window shading options, 1619
Window size, 1616
Window technology, 1364–1365
Window thermal and solar properties, 1627
Windows, 1518–1521
Windows
   coatings and tints for energy efficiency, 1622
   condensation resistance, 1622–1623
   damage-weighted transmittance, 1623–1624
   daylight illumination from as energy saver, 1624
   direct heat entry of sunlight, 1619
   double-pane
      condensation in, 1620
      heat transfer, 1618
      insulating gas, 1619
   economics of, 1624
   energy basics, 1616–1618
      conduction heat transfer, 1617
      convection heat transfer, 1617–1618
      heat transfer, 1616–1617
      radiation heat transfer, 1617
   energy performance of, 1616, 1624
   Energy Star criteria, 1521
   energy transport, 1518–1520
   future improvements on, 1520–1521
   interior shades and, 1618
   low solar gain low-e coatings, 1621–1622
   low-e glass coatings, 1621–1622
   purpose of, 1616, 1641
   rating systems, 1520
   shading devices, 1641–1648
      acoustic performance, 1646
      control strategy for, 1644
      curtain wall systems, 1641
      decision-making framework, 1645, 1647–1648
      greenhouse effect, 1641
      performance parameters, 1646
      position of, 1643
      selection of, 1644–1645
      thermal performance, 1646
      variables, 1646
      visual performance, 1646
   solar radiant heat gain, 1623
   thermal characteristics, payback of, 1521
   UV transmittance, 1623–1624
   visible transmittance, 1623–1624
Wired external solutions
   existing LAN, 1654–1655
   phone line, 1655
Wireless applications, 1649–1658
   external communications, 1655–1656
   LAN vs., 1658

   mesh network, 1653–1654
   meter level, 1650, 1652
   phone line vs., 1658
   primer on, 1652
   serial communications networks, 1652–1653
   submetering primer, 1649–1650
   wired external solutions, 1654–1655
Wireless mobile thermostat climate control, 1660–1667
   annual savings, 1663–1666
   future of, 1667
   heating mode, 1663
   projected energy savings, 1663–1667
Wireless terms, 1652
   frequency, 1652
   interference, 1652
   mesh networks, 1652
   power rating, 1652
   repeaters, 1652
   throughput, 1652
Wiring harnesses, automobile technology evolution and, 675
Wisconsin, Energy Center of, 512
Wood as heating fuel, 1361
Wood framed walls, 1515–1516
Wood fuels, biomass energy and, 88
Work conversion, energy and, 604
Work heat energy principle, 1164–1168
   First Law of Thermodynamics, 1167
   Second Law of Thermodynamics, 1167
Work units, energy and, 605
Work, 1160–1161
   heat units and, 1162
Worker quality, automobile technology evolution and, 675
Working capital interest, 958–959
Working capital, rate of return regulation and, 1255
Working fluids, absorption, 1543–1544
Workmanship, lean manufacturing and, 470–471
Workplace organization, lean manufacturing and, 474
WWTP. *See* wastewater treatment plant.
Wyoming coal production, 159

X

XML, 108–110
XML. *See* extensible markup language.
XPS (extrusion polystyrene), as insulation material, 891, 894

Y

Yield ratio, emergy, 424–426

Z

Zone management, air distribution and, 211
   benefits of, 211–212
      air quality monitoring, 212
      end to end system management, 212
      measure zone leakage, 212
      reduced air usage, 211
      regulate air costs, 212
Zone switching, lighting based demand response and, 277

Volume 1

Accounting: Facility Energy Use / 1
Air Emissions Reductions from Energy Efficiency / 9
Air Quality: Indoor Environment and Energy Efficiency / 18
Aircraft Energy Use / 24
Alternative Energy Technologies: Price Effects / 31
ANSI/ASHRAE Standard 62.1-2004 / 50
Auditing: Facility Energy Use / 63
Auditing: Improved Accuracy / 69
Auditing: User-Friendly Reports / 76
Benefit Cost Analysis / 81
Biomass / 86
Boilers and Boiler Control Systems / 93
Building Automation Systems (BAS): Direct Digital Control / 104
Building Geometry: Energy Use Effect / 111
Building System Simulation / 116
Carbon Sequestration / 125
Career Advancement and Assessment in Energy Engineering / 131
Climate Policy: International / 137
Coal Production in the U.S. / 143
Coal Supply in the U.S. / 156
Coal-to-Liquid Fuels / 163
Cold Air Retrofit: Case Study / 172
Combined Heat and Power (CHP): Integration with Industrial Processes / 175
Commissioning: Existing Buildings / 179
Commissioning: New Buildings / 188
Commissioning: Retrocommissioning / 200
Compressed Air Control Systems / 207
Compressed Air Energy Storage (CAES) / 214
Compressed Air Leak Detection and Repair / 219
Compressed Air Storage and Distribution / 226
Compressed Air Systems / 236
Compressed Air Systems: Optimization / 241
Cooling Towers / 246
Data Collection: Preparing Energy Managers and Technicians / 255
Daylighting / 264
Demand Response: Commercial Building Strategies / 270
Demand Response: Load Response Resources and Programs / 279
Demand-Side Management Programs / 286
Desiccant Dehumidification: Case Study / 292
Distributed Generation / 296
Distributed Generation: Combined Heat and Power / 303
District Cooling Systems / 309
District Energy Systems / 316
Drying Operations: Agricultural and Forestry Products / 332
Drying Operations: Industrial / 338
Electric Motors / 349
Electric Power Transmission Systems / 356
Electric Power Transmission Systems: Asymmetric Operation / 364
Electric Supply System: Generation / 374
Electricity Deregulation for Customers / 379
Electricity Enterprise: U.S., Past and Present / 387
Electricity Enterprise: U.S., Prospects / 399
Electronic Control Systems: Basic / 407
Emergy Accounting / 420
Emissions Trading / 430
Energy Codes and Standards: Facilities / 438
Energy Conservation / 444
Energy Conservation: Industrial Processes / 458
Energy Conservation: Lean Manufacturing / 467
Energy Conversion: Principles for Coal, Animal Waste, and Biomass Fuels / 476
Energy Efficiency: Developing Countries / 498
Energy Efficiency: Information Sources for New and Emerging Technologies / 507
Energy Efficiency: Low Cost Improvements / 517
Energy Efficiency: Strategic Facility Guidelines / 524
Energy Information Systems / 535
Energy Management: Organizational Aptitude Self-Test / 543
Energy Master Planning / 549
Energy Project Management / 556
Energy Service Companies: Europe / 566

Volume 2

Energy Star® Portfolio Manager and Building Labeling Program / 573
Energy Use: U.S. Overview / 580
Energy Use: U.S. Transportation / 588
Energy: Global and Historical Background / 601
Enterprise Energy Management Systems / 616
Environmental Policy / 625
Evaporative Cooling / 633
Exergy: Analysis / 645
Exergy: Environmental Impact Assessment Applications / 655
Facility Air Leakage / 662
Facility Energy Efficiency and Controls: Automobile Technology Applications / 671
Facility Energy Use: Analysis / 680
Facility Energy Use: Benchmarking / 690
Facility Power Distribution Systems / 699
Federal Energy Management Program (FEMP): Operations and Maintenance Best Practices Guide (O&M BPG) / 707
Fossil Fuel Combustion: Air Pollution and Global Warming / 715
Fuel Cells: Intermediate and High Temperature / 726
Fuel Cells: Low Temperature / 733
Geothermal Energy Resources / 744
Geothermal Heat Pump Systems / 753
Global Climate Change / 760
Green Energy / 771
Greenhouse Gas Emissions: Gasoline, Hybrid-Electric, and Hydrogen-Fueled Vehicles / 787
Heat and Energy Wheels / 794
Heat Exchangers and Heat Pipes / 801
Heat Pipe Application / 807
Heat Pumps / 814
Heat Transfer / 822
High Intensity Discharge (HID) Electronic Lighting / 830
HVAC Systems: Humid Climates / 839
Hybrid-Electric Vehicles: Plug-In Configuration / 847
Independent Power Producers / 854
Industrial Classification and Energy Efficiency / 862
Industrial Energy Management: Global Trends / 872
Industrial Motor System Optimization Projects in the U.S. / 883
Insulation: Facilities / 890
Integrated Gasification Combined Cycle (IGCC): Coal- and Biomass-Based / 906
IntelliGrid^{SM} / 914
International Performance Measurement and Verification Protocol (IPMVP) / 919
Investment Analysis Techniques / 930
LEED-CI and LEED-CS: Leadership in Energy and Environmental Design for Commercial Interiors and Core and Shell / 937
LEED-NC: Leadership in Energy and Environmental Design for New Construction / 947
Life Cycle Costing: Electric Power Projects / 953
Life Cycle Costing: Energy Projects / 967
Lighting Controls / 977
Lighting Design and Retrofits / 993
Liquefied Natural Gas (LNG) / 1001
Living Standards and Culture: Energy Impact / 1009
Maglev (Magnetic Levitation) / 1018
Management Systems for Energy / 1027
Manufacturing Industry: Activity-Based Costing / 1042
Measurement and Verification / 1050
Measurements in Energy Management: Best Practices and Software Tools / 1056
Mobile HVAC Systems: Fundamentals, Design, and Innovations / 1061
Mobile HVAC Systems: Physics and Configuration / 1070
National Energy Act of 1978 / 1082
Natural Energy versus Additional Energy / 1088
Net Metering / 1096
Nuclear Energy: Economics / 1101
Nuclear Energy: Fuel Cycles / 1111
Nuclear Energy: Power Plants / 1115
Nuclear Energy: Technology / 1125
Performance Contracting / 1134
Performance Indicators: Industrial Energy / 1140